MW01076355

THE JOHN ZINK COMBUSTION HANDBOOK

INDUSTRIAL COMBUSTION SERIES

Edited by Charles E. Baukal, Jr.

PUBLISHED TITLES

Oxygen-Enhanced Combustion
Charles E. Baukal, Jr.

Heat Transfer in Industrial Combustion
Charles E. Baukal, Jr.

Computational Fluid Dynamics in Industrial Combustion
Charles E. Baukal, Jr., Vladimir Y. Gershtein, and Xianming Li

The John Zink Combustion Handbook
Charles E. Baukal, Jr.

THE JOHN ZINK COMBUSTION HANDBOOK

CHARLES E. BAUKAL, JR., PH.D., P.E.
Editor

ROBERT E. SCHWARTZ, P.E.
Associate Editor

John Zink Company, LLC
Tulsa, Oklahoma

CRC Press
Boca Raton London New York Washington, D.C.

Library of Congress Cataloging-in-Publication Data

The John Zink combustion handbook / Charles E. Baukal, editor ; Robert Schwartz, associate editor.
p. cm. — (Industrial combustion series)
Includes bibliographical references and index.
ISBN 0-8493-2337-1 (alk. paper)
1. Combustion engineering—Handbooks, manuals, etc. I. Baukal, Charles E.
II. Schwartz, Robert (Robert E.) III. John Zink Company. IV. Series.

TJ254.5 .J63 2000
621.502′3—dc21 00-049357
CIP

Direct all inquiries to CRC Press LLC, 2000 N.W. Corporate Blvd., Boca Raton, Florida 33431.

International Standard Book Number 0-8493-2337-1
Library of Congress Card Number 00-049357
Printed in the United States of America 6 7 8 9 0
Printed on acid-free paper

Dedication to David H. Koch

The staff at John Zink dedicates this book and gives special thanks to David H. Koch, president of John Zink Company from January 1998 to March 2000. David made significant contributions to this book from its inception to its completion. Without David's passion and excitement about science and technology, we could not have completed this book. David recognized the need for a comprehensive handbook regarding combustion that would address applications for the chemical, petrochemical, and power generation industries. He aided in developing a world-class research and development team, built a state-of-the-art testing center, and established alliances with the top combustion researchers in the world. David's technological leadership and strong support of the John Zink technical staff made this book possible. David, we thank you for your respect and dedication to the John Zink Company.

Foreword

As we enter the twenty-first century, the importance of energy for industry, transportation, and electricity generation in our daily lives is profound. Combustion of fossil fuels is by far the predominant source of energy today and will likely remain that way for many years to come.

Combustion has played major roles in human civilization, including both practical and mystical ones. Since man discovered how to create fire, we have relied on combustion to perform a variety of tasks. Fire was first used for heating and cooking, and later to manufacture tools and weapons. For all practical purposes, it was not until the onset of the Industrial Revolution in the nineteenth century that man started to harness power from combustion. We have made rapid progress in the application of combustion systems since then, and many industries have come into existence as a direct result of this achievement.

Demands placed on combustion systems change continuously with time and are becoming more stringent. The safety of combustion systems has always been essential, but emphasis on effective heat transfer, temperature uniformity, equipment scale-up, efficiency, controls, and — more recently — environmental emissions and combustion-generated noise has evolved over time. Such demands create tremendous challenges for combustion engineers. These challenges have been successfully met in most applications by combining experience and sound engineering practices with creative and innovative problem-solving.

Understanding combustion requires knowledge of the fundamentals: turbulent mixing, heat transfer, and chemical kinetics. The complex nature of practical combustion systems, combined with the lack of reliable analytical models in the past, encouraged researchers to rely heavily on empirical methods to predict performance and to develop new products. Fortunately, the combustion field has gained considerable scientific knowledge in the last few decades, which is now utilized in industry by engineers to evaluate and design combustion systems in a more rigorous manner. This progress is the result of efforts in academia, government laboratories, private labs, and companies like John Zink.

The advent of ever-faster and more powerful computers has had a profound impact on the manner in which engineers model combustion systems. Computational Fluid Dynamics (CFD) was born from these developments. Combined with validation by experimental techniques, CFD is an essential tool in combustion research, development, analysis, and equipment design.

Today's diagnostic tools and instrumentation — with capabilities unimaginable just a few years ago — allow engineers and scientists to gather detailed information in hostile combustion environments at both microscopic and macroscopic levels. Lasers, spectroscopy, advanced infrared, and ultraviolet camera systems are used to nonintrusively gather quantitative and qualitative information, including combustion temperature, velocity, species concentration, flow visualization, particle size, and loading. Advanced diagnostic systems and instrumentation are being transferred beyond the laboratory to implementation in practical field applications. The information obtained with these systems has considerably advanced our knowledge of combustion equipment and has been an indispensable source of CFD model validation.

Oil refining, chemical process, and power generation are energy-intensive industries with combustion applications in burners, process heaters, boilers, and cogeneration systems, as well as flares and thermal oxidizers. Combustion for these industries presents unique challenges related to the variety of fuel compositions encountered. Combustion equipment must be flexible to be able to operate in a safe, reliable, efficient, and environmentally responsible manner under a wide array of fuel compositions and conditions.

Combustion is an exciting and intellectually challenging field containing plenty of opportunities to enhance fundamental and practical knowledge that will ultimately lead to development of new products with improved performance.

This book represents the tireless efforts of many John Zink engineers willing to share their unique knowledge and experience with other combustion engineers, researchers, operators of combustion equipment, and college students. We have tried to include insightful and helpful information on combustion fundamentals, combustion noise, CFD design, experimental techniques, equipment, controls, maintenance, and troubleshooting. We hope our readers will agree that we have done so.

David H. Koch

Executive Vice President
Koch Industries

Preface

Combustion is described as "the rapid oxidation of a fuel resulting in the release of usable heat and production of a visible flame."[1] Combustion is used to generate 90% of the world's power.[2] Regarding the science of combustion, Liñán and Williams write,

> Although combustion has a long history and great economic and technical importance, its scientific investigation is of relatively recent origin. Combustion science can be defined as the science of exothermic chemical reactions in flows with heat and mass transfer. As such, it involves thermodynamics, chemical kinetics, fluid mechanics, and transport processes. Since the foundations of the second and last of these subjects were not laid until the middle of the nineteenth century, combustion did not emerge as a science until the beginning of the twentieth century.[3]

Chomiak writes, "In spite of their fundamental importance and practical applications, combustion processes are far from being fully understood."[4] Strahle writes, "combustion is a difficult subject, being truly interdisciplinary and requiring the merging of knowledge in several fields."[5] It involves the study of chemistry, kinetics, thermodynamics, electromagnetic radiation, aerodynamics and fluid mechanics including multiphase flow and turbulence, heat and mass transfer, and quantum mechanics to name a few. Regarding combustion research,

> The pioneering experiments in combustion research, some 600,000 years ago, were concerned with flame propagation rather than ignition. The initial ignition source was provided by Mother Nature in the form of the electrical discharge plasma of a thunderstorm or as volcanic lava, depending on location. ... Thus, in the beginning, Nature provided an arc-augmented diffusion flame and the first of man's combustion experiments established that the heat of combustion was very much greater than the activation energy — i.e., that quite a small flame on a stick would spontaneously propagate itself into a very large fire, given a sufficient supply of fuel.[6]

In one of the classic books on combustion, Lewis and von Elbe write,

> Substantial progress has been made in establishing a common understanding of combustion phenomena. However, this process of consolidation of the scientific approach to the subject is not yet complete. Much remains to be done to advance the phenomenological understanding of flame processes so that theoretical correlations and predictions can be made on the basis of secure and realistic models.[7]

Despite the length of time it has been around, despite its importance to man, and despite vast amounts of research, combustion is still far from being completely understood. One of the purposes of this book is to improve that understanding, particularly in industrial combustion applications in the process and power generation industries.

This book is generally organized in two parts. The first part deals with the basic theory of some of the important disciplines (combustion, heat transfer, fluid flow, etc.) important for the understanding of any combustion process and covers Chapters 1 through 13. While these topics have been satisfactorily covered in many combustion textbooks, this book treats them from the context of the process and power generation industries. The second part of the book deals with specific equipment design issues and applications in the process and power generation industries.

REFERENCES

1. Industrial Heating Equipment Association, *Combustion Technology Manual*, Fifth Edition Arlington, VA, Combustion Division of the Industrial Heating Equipment Association, 1994, 1.
2. N. Chigier, *Energy, Combustion, and Environment* McGraw-Hill, New York, 1981, ix.
3. A. Liñán and F.A. Williams, *Fundamental Aspects of Combustion* Oxford University Press, Oxford, 1993, 3.
4. Chomiak, *Combustion: A Study in Theory, Fact and Application*, 1.
5. W.C. Strahle, *An Introduction to Combustion* Gordon & Breach, Longhorne, PA, 1993, ix.
6. F.J. Weinberg, "The First Half-Million Years of Combustion Research and Today's Burning Problems," in the *Fifteenth Symposium (International) on Combustion*, The Combustion Institute, Pittsburgh, PA, 1974, 1.
7. B. Lewis and G. von Elbe, *Combustion, Flames and Explosions of Gases*, Third Edition, Academic Press, New York, 1987, xv.

Acknowledgments

The authors would like to collectively thank the John Zink Company, LLC for the consistent help and support provided during the preparation of this book. Many of our colleagues have helped in the ideas, the writing, and the preparation of figures and tables. We would like to especially thank the John Zink management including Steve Pirnat (President), Earl Schnell (Vice President of Burners), Roberto Ruiz (Vice President of Technology and Commercial Development), Jim Goodman (Vice President of the Systems Group), and Andy Barrieau (General Manager of Todd Combustion) for their interest and attention in this project and for providing the resources to complete it. The authors would like to thank Dr. David Fitzgerald who spent countless hours formatting, drawing figures, editing style, getting permissions, and collecting information for this book. The project would certainly have taken much longer without his help. The authors would also like to thank Kevin Hardison who drew some of the figures used in several chapters.

Chuck Baukal would like to thank his wife Beth and his daughters Christine, Caitlyn, and Courtney for their patience and help during the writing of this book. He would also like to thank the good Lord above, without whom this would not have been possible. Larry Berg would like to thank his wife, Betty, who has always encouraged him to be his very best at work. Without her support, much of his career would not have happened. He would also like to thank God who has blessed him with the talent and ability to work as a research engineer and participate in the creation of this book. Joe Colannino gives special thanks to his wife, Judy, for never complaining (not even once) regarding the myriad of evenings and weekends he devoted to this project. He knows that God is the giver of every good gift whenever he hears "Hi, honey, I'm happy you're home." Joe Gifford gives many thanks to his wife Barbara for typing, colleague Jim Heinlein, for suggestions and review, Kevin Hardison, for preparing figures, and to Charles Baukal and David Fitzgerald for guidance and editing. Bob Hayes would like to thank God, his family, and his wife, for all of their love and encouragement. Paul Melton thanks his wife, Toni, (and daughter, Angela, before she left for college) for her patience and understanding during the many years he has been involved with the art and science of burning wastes. Thanks also to all those who directly or indirectly helped him gain the experience and knowledge to contribute to his chapter. It is a continuing education process. Robert Schwartz thanks his wife, Stella, for her patience throughout the years. In addition, he would like to acknowledge the work of each of the authors and thank them for their contributions. A very special thank you to David Koch for the opportunity to work on this book and to Chuck Baukal and Dave Fitzgerald for their untiring efforts. Prem Singh thanks his wife and daughters for their constant encouragement and inquiry about this book that prompted him to prepare his chapters in time. Joseph Smith would like to thank the Lord for His continued grace in his life. He gives all credit to Him for anything noteworthy that he's done. Tim Webster would like to thank his parents, Lee and Marilyn Webster, for their continued support and encouragement, which has made all his personal and professional accomplishments possible. Jeff White thanks his wife and family who supported him even when he was very late at the office. He also thanks the John Zink Flare Group who continue to find new situations from which we all can learn something. Roger Witte would like to thank all his colleagues at John Zink Company for the wisdom and the knowledge they shared with him in writing this book. He would especially like to thank Dr. Robert R. Reed, Herschel Goodnight, Don Iverson, Harold Koons, and Bob Schwartz for the knowledge and wisdom they have shared through the years and their patience in letting him make the mistakes that one makes in learning the combustion business. He would also like to thank his wife, Nancy, for putting up with the long hours of working at home in writing his chapters.

The authors and especially the editor would like to give a special acknowledgement to Andrea Demby at CRC Press. This project would not have been nearly as successful without her tireless efforts working days, nights, and weekends on this project for many months. We salute her and thank her for her unending patience with our numerous revisions!

Tables and Figures[1]

CHAPTER 1

TABLE 1.1 Major Petroleum Refining Processes.
TABLE 1.2 Average Burner Configuration by Heater Type.
TABLE 1.3 Major Refinery Processes Requiring a Fired Heater.
TABLE 1.4 Major Fired Heater Applications in the Chemical Industry.

CHAPTER 2

TABLE 2.1 Alphabetical List of Atomic Weights for Common Elements. (From IUPAC Commission on Atomic Weights and Isotopic Abundances, Atomic Weights of Elements, 1995. *Pure Appl. Chem.*, 68, 2339, 1996. With permission.)
TABLE 2.2 Molar Ratios for Some Combustion Reactions and Products.
TABLE 2.3 Molecular Weights and Stoichiometric Coefficients for Common Gaseous Fuels.
TABLE 2.4 Combustion Data for Hydrocarbons.
TABLE 2.5 Adiabatic Flame Temperature.

CHAPTER 3

TABLE 3.1 Thermal Conductivity of Common Materials. (From S.C. Stultz and J.B. Kitto, Eds., *Steam: Its Generation and Use*, 40th ed., The Babcock & Wilcox Company, Barberton, OH, 1992. With permission.)
TABLE 3.2 Properties of Various Substances at Room Temperature.
TABLE 3.3 Properties of Selected Gases at 14.696 psi. (From F.P. Incropera and D.P. DeWitt, *Fundamentals of Heat and Mass Transfer*, 4th ed. Copyright© 1996. Reprinted by permission of John Wiley & Sons, Inc.)
TABLE 3.4 One-dimensional, Steady-State Solutions to the Heat Equation with No Generation[30].
TABLE 3.5 Typical Convective Heat Transfer Coefficients.
TABLE 3.6 Summary of Convection Correlations for Flow in a Circular Tube.[29] (From F.P. Incropera and D.P. DeWitt, *Fundamentals of Heat and Mass Transfer*, 4th ed. Copyright© 1996. Reprinted by permission of John Wiley & Sons, Inc.)
TABLE 3.7 Constants of Equation (3.80) for a Circular Cylinder in Cross Flow.
TABLE 3.8 Constants of Equation (3.85) for the Tube Bank in Cross Flow. (From F.P. Incropera and D.P. DeWitt, *Fundamentals of Heat and Mass Transfer*, 4th ed. Copyright© 1996. Reprinted by permission of John Wiley & Sons, Inc.)
TABLE 3.9 Spectrum of Electromagnetic Radiation.
TABLE 3.10 View Factors for Two-dimensional Geometries.
TABLE 3.11 View Factors for Three-dimensional Geometries.
TABLE 3.12 Normal Emissivities, Σ, for Various Surfaces.
TABLE 3.13 Mean Beam Lengths L_e for Various Gas Geometries.

CHAPTER 4

TABLE 4.1 Viscosity Conversion Table.
TABLE 4.2 Properties of U.S. Standard Atmosphere at Sea Level.
TABLE 4.3 Equivalent Roughness for New Pipes.
TABLE 4.4 Loss Coefficients for Various Fittings.

[1]Permission and source lines are in addition to or replacement of citations in the text.

CHAPTER 5

TABLE 5.1 Example Pipeline Quality Natural Gas. (From Gas Processors and Suppliers Association, *GPSA Engineering Data Book,* Vol. 1, 10th ed., Tulsa, OK, 1987, pp. 2–3. With permission.)

TABLE 5.2 Commercial Natural Gas Components and Typical Ranges of Composition.

TABLE 5.3 Composition of a Typical Refinery Gas.

TABLE 5.4 Typical Composition of Steam Reforming/PSA Tail Gas.

TABLE 5.5 Typical Composition of Flexicoking Waste Gas.

TABLE 5.6 Volumetric Analysis of Typical Gaseous Fuel Mixtures.

TABLE 5.7 Physical Constants of Typical Gaseous Fuel Mixtures.

TABLE 5.8 Physical Constants of Typical Gaseous Fuel Mixture Components.

TABLE 5.9 Quantitative Listing of Products Made by the U.S. Petroleum Industry. (From Gas Processors and Suppliers Association, *GPSA Engineering Data Book,* Vol. 1, 10th ed., Tulsa, OK, 1987, p. 6. With permission.)

TABLE 5.10 General Fraction Boiling Points.

TABLE 5.11 Requirements for Fuel Oils (per ASTEM D 396).

TABLE 5.12 Typical Analysis of Different Fuel Oils.

TABLE 5.13 Naphtha Elemental Analysis.

TABLE 5.14 Viscosity Conversion Chart.

CHAPTER 6

TABLE 6.1 Combustion Emission Factors (lb/10^6 Btu) by Fuel Type.

TABLE 6.2 Uncontrolled NOx Emission Factors for Typical Process Heaters.

TABLE 6.3 Reduction Efficiencies for NOx Control Techniques.

TABLE 6.4 NOx Control Technologies in Process Heaters.

TABLE 6.5 NOx Reductions for Different Low-NO Burner Types. (From A. Garg, *Chemical Engineering Progress,* 90, 1, 46–49. Reproduced with permission of the American Institute of Chemical Engineers. Copyright© 1994 AIChe. All rights reserved.)

CHAPTER 7

TABLE 7.1 The Ten Octave Bands.

TABLE 7.2 Octave and One-Third Octave Bands.

TABLE 7.3 Addition Rules.

TABLE 7.4 Sound Levels of Various Sources.

TABLE 7.5 OSHA Permissible Noise Exposures.

TABLE 7.6 Overall Sound Pressure Lever from Combustion.

TABLE 7.7 The Overall Sound Pressure Level (OASPL) Determined Experimentally and Using the Mathematical Model.

CHAPTER 8

TABLE 8.1 Exit Mach Number, Area Ratio, Driving Force Ratio, and Driving Force Percentage Increase for Various Gas Pressures ($\gamma = 1.33$ and $P_a = 14.3$ psia$^+$).

TABLE 8.2 Values for a 90° Mitered Elbow.

CHAPTER 9

TALBE 9.1 Current CFD Applications n the Chemical Process Industry.

TABLE 9.2 Universal "Empirical" Constants Used in k-Σ Turbulence Model.

TABLE 9.3 Cartesian Differential Equation Set.

TABLE 9.4 Cylindrical Differential Equation Set.

TABLE 9.5 Composition of Acid Gas Used in CFD Study.

TABLE 9.6 Limiting Cases Considering During RCl Combustion Study.

CHAPTER 10

TABLE 10.1 Flammability Limits for Common Fuels at Standard Temperature and Pressure. (Courtesy of R.J. Reed, *North American Combustion Handbook,* Vol. I, 3rd ed., North American Manufacturing Company, Cleveland, OH, 1986.)

TABLE 10.2 Minimum Ignition Temperatures for Common Fuels at Standard Temperature and Pressure. (Courtesy of R.J. Reed, *North American Combustion Handbook,* Vol. I, 3rd ed., North American Manufacturing Company, Cleveland, OH, 1986.)

TABLE 10.3 Flammability and Ignition Characteristics of Liquids and Gases. (Adapted from D.R. Lide, Ed., *CRC Handbook of Chemistry and Physics,* 80th ed., CRC Press, Boca Raton, FL, 1999. With permission.)

TABLE 10.4 Ignition Sources of Major Fires. (Adapted from the *Accident Prevention Manual for Industrial Operations,* National Safety Council, Itasca, IL, 1974.)

TABLE 10.5 Minimum Ignition Energies Required for Common Fuels.

TABLE 10.6 Benefits of a Successful Process Knowledge and Documentation Program. (Adapted from the *GPSA Engineering Data Book,* Vol. II, 10th ed., Gas Processors and Suppliers Association, Tulsa, OK, 1994.)

NO TABLES FOR CHAPTER 11

CHAPTER 12

TABLE 12.1 Gas Valve Data.

TABLE 12.2 Data for Characterizer.

CHAPTER 13

TABLE 13.1 Some Potential Factors Affecting NOx Response from a Burner.

TABLE 13.2 NOx as a Function of Burner Geometry and Operation.

TABLE 13.3 Transforms for Table 13.4.

TABLE 13.4 Transformed Data for Fuel-staged Burner.

TABLE 13.5 Generic ANOVA Table.

TABLE 13.6 F-Distribution, 99%, 95%, and 90% Confidence.

TABLE 13.7 ANOVA Table for Equation 13.2 Applied to Data of Table 13.4.

TABLE 13.8 ANOVA for Table 13.7 with Separate Effects.

TABLE 13.9 ANOVA for Table 13.7 with Pooled Effects.

TABLE 13.10 Factorial Design with Replicate Centerpoints.

TABLE 13.11 Generic ANOVA for Factorial Design with Replicates.

TABLE 13.12 ANOVA for Table 13.10 and Equation.

TABLE 13.13 ANOVA for Factorial Design with Centerpoint Replicates Case 2 of Table 13.11.

TABLE 13.14 ∫ Fractional Factorial [FF(3,-1,0)].

TABLE 13.15 FF(3,0,4) in Two Blocks.

TABLE 13.16 Experimental Design with Categorial Factors.

TABLE 13.17 ANOVA for Table 13.16.

TABLE 13.18 CC(3,0,6) Design.

TABLE 13.19 ~CC(3,0,3) Design.

TABLE 13.20 ANOVA for Table 13.19 and Equation (13.38).

TABLE 13.21 Example of an Orthogonal Subspace for a $q = 3$ Simplex.

TABLE 13.22 FF(7,-2,0) Design Generating a Combined Mixture-Factorial in Five Factors – Two at Four Levels and Three at Two Levels.

CHAPTER 14

TABLE 14.1 Tulsa Natural Gas (TNG) Composition and Properties.

TABLE 14.2 Example Refinery Gas.

TABLE 14.3 Comparison of Refinery Gas to Test Blend.

TABLE 14.4 Test Procedure Gas Specification Sheet.

TABLE 14.5 Example Test Procedure.

CHAPTER 15

TABLE 15.1 Burner Throat Area for Different Tile Dimensions.

CHAPTER 16

TABLE 16.1 Static Draft Effect per Foot of Height.
TABLE 16.2 Typical Flame Dimensions for Different Burner Types.
TABLE 16.3 Typical Excess Air Values for Gas Burners.
TABLE 16.4 Typical Excess Air Values for Liquid Fuel Firing.

CHAPTER 17

TABLE 17.1 Ratio of Upper and Lower Explosive Limits and Flashback Probability in Premix Burners for Various Fuels.
TABLE 17.2 Troubleshooting for Gas Burners.
TABLE 17.3 Troubleshooting for Oil Burners.

CHAPTER 18

TABLE 18.1 Typical NOx and CO Emissions from Duct Burners.

NO TABLES FOR CHAPTERS 19 AND 20

CHAPTER 21

TABLE 21.1 Typical Thermal Oxidizer Operating Conditions.
TABLE 21.2 Relative Characteristics of Centrifugal Blowers.

CHAPTER 1

FIGURE 1.1 Typical petroleum refinery.
FIGURE 1.2 Typical refinery flow diagram.
FIGURE 1.3 Offshore oil rig flare.
FIGURE 1.4 Duct burner flame.
FIGURE 1.5 Duct burner in large duct.
FIGURE 1.6 Front of boiler burner.
FIGURE 1.7 Thermal oxidizer.
FIGURE 1.8 Side- (a) and top-fired (b) reformers (elevation view).
FIGURE 1.9 Downfired burner commonly used in top fired reformers.
FIGURE 1.10 Elevation view of a terrace firing furnace.
FIGURE 1.11 Schematic of a process heater.
FIGURE 1.12 Typical process heater.
FIGURE 1.13 Fixed heater size distribution.
FIGURE 1.14 Sketch (elevation view) of center or target wall firing configuration.
FIGURE 1.15 Horizontal floor-fired burners.
FIGURE 1.16 Wall fired burner (side view).
FIGURE 1.17 Sketch (elevation view) of a horizontally mounted, vertically fired burner configuration.
FIGURE 1.18 Examples of process heaters.
FIGURE 1.19 Typical heater types.
FIGURE 1.20 Cabin heater.
FIGURE 1.21 Crude unit burners.
FIGURE 1.22 Typical burner arrangements (elevation view).
FIGURE 1.23 Process heater heat balance. (From Philip Conisbee, *Georges de La Tour and His World.* National Gallery of Art, Washington, D.C., 1996, 110. With permission.)

FIGURE 1.24 Burner (B) arrangement (plan view) in the floor of vertical cylindrical furnaces: (a) small-diameter furnace with a single centered burner and (b) larger diameter furnace with four burners symmetrically arranged at a radius from the center.
FIGURE 1.25 Burner (B) arrangement (plan view) in the floor of rectangular cabin heaters: (a) single row of burners in a narrower heater, (b) two rows of staggered burners in a slightly wider heater, and (c) two rows of aligned burners in an even wider heater.
FIGURE 1.26 Adiabatic equilibrium NO and CO as a function of the equivalence ratio for an air/CH_4 flame.
FIGURE 1.27 Typical combination oil and gas burner.
FIGURE 1.28 Schematic of flue gas recirculation.
FIGURE 1.29 Cartoon of a premixed burner.
FIGURE 1.30 Typical premixed gas burner. (From API Publication 535: *Burner for Fire Heaters in General Refinery Services,* 1st ed., American Petroleum Institute, Washington, D.C., July, 1995. With permission.)
FIGURE 1.31 Painting of a diffusion flame.
FIGURE 1.32 Cartoon of a diffusion burner.
FIGURE 1.33 Cartoon of a partially-premixed burner.
FIGURE 1.34 Cartoon of a staged-air burner.
FIGURE 1.35 Schematic of a typical staged air combination oil and gas burner.
FIGURE 1.36 Cartoon of a staged-fuel burner.
FIGURE 1.37 Schematic of a typical staged-fuel gas burner.
FIGURE 1.38 Typical natural-draft gas burner.
FIGURE 1.39 Natural draft burner.
FIGURE 1.40 Flames impinging on tubes in a cabin heater.
FIGURE 1.41 Flames pulled toward the wall.
FIGURE 1.42 Gas burners needing service.

CHAPTER 2

FIGURE 2.1 Typical cabin-style process heater.
FIGURE 2.2 Species concentration vs. excess air for the following fuels: (a) CH_4, (b) natural gas, (c) simulated refinery gas (25% H_2, 50% CH_4, 25% C_3H_8), (d) C_3H_8, (e) No. 2 oil, and (f) No. 6 oil.
FIGURE 2.2 (b) Natural gas
FIGURE 2.2 (c) Simulated refinery gas (25% H_2, 50% CH_4, 25% C_3H_8).
FIGURE 2.2 (d) Propane.
FIGURE 2.2 (e) Fuel oil #2.
FIGURE 2.2 (f) Fuel oil #6.
FIGURE 2.3 Species concentration vs. stoichiometric ratio for the following fuels: (a) CH_4, (b) natural gas, (c) simulated refinery gas (25% H_2, 50% CH_4, 25% C_3H_8), (d) C_3H_8, (e) No. 2 oil, and (f) No. 6 oil.
FIGURE 2.3 (b) Natural gas
FIGURE 2.3 (c) Simulated refinery gas (25% H_2, 50% CH_4, 25% C_3H_8).
FIGURE 2.3 (d) Propane.
FIGURE 2.3 (e) Fuel oil #2.
FIGURE 2.3 (f) Fuel oil #6.
FIGURE 2.4 Adiabatic equilibrium reaction process.
FIGURE 2.5 Adiabatic equilibrium calculations for the predicted gas composition as a function of the O_2:CH_4 stoichiometry for air/CH_4 flames, where the air and CH_4 are at ambient temperature and pressure.
FIGURE 2.6 Adiabatic equilibrium stoichiometric calculations for the predicted gas composition of the major species as a function of the air preheat temperature for air/CH_4 flames, where the CH_4 is at ambient temperature and pressure.
FIGURE 2.7 Adiabatic equilibrium stoichiometric calculations for the predicted gas composition of the minor species as a function of the air preheat temperature for air/CH_4 flames, where the CH_4 is at ambient temperature and pressure.
FIGURE 2.8 Adiabatic equilibrium stoichiometric calculations for the predicted gas composition of the major species as a function of the fuel preheat temperature for air/CH_4 flames, where the air is at ambient temperature and pressure.
FIGURE 2.9 Adiabatic equilibrium stoichiometric calculations for the predicted gas composition of the minor species as a function of the fuel preheat temperature for air/CH_4 flames, where the air is at ambient temperature and pressure.

FIGURE 2.10 Adiabatic equilibrium stoichiometric calculations for the predicted gas composition of the major species as a function of the fuel blend (H_2 + CH_4) composition for air/fuel flames, where the air and fuel are at ambient temperature and pressure.
FIGURE 2.11 Adiabatic equilibrium stoichiometric calculations for the predicted gas composition of the minor species as a function of the fuel blend (H_2 + CH_4) composition for air/fuel flames, where the air and fuel are at ambient temperature and pressure.
FIGURE 2.12 Adiabatic equilibrium stoichiometric calculations for the predicted gas composition of the major species as a function of the fuel blend (N_2 + CH_4) composition for air/fuel flames, where the air and fuel are at ambient temperature and pressure.
FIGURE 2.13 Adiabatic equilibrium stoichiometric calculations for the predicted gas composition of the minor species as a function of the fuel blend (N_2 + CH_4) composition for air/fuel flames, where the air and fuel are at ambient temperature and pressure.
FIGURE 2.14 Equilibrium calculations for the predicted gas composition of the major species as a function of the combustion product temperature for air/CH_4 flames, where the air and fuel are at ambient temperature and pressure.
FIGURE 2.15 Equilibrium calculations for the predicted gas composition of the minor species as a function of the combustion product temperature for air/CH_4 flames, where the air and fuel are at ambient temperature and pressure.
FIGURE 2.16 Adiabatic flame temperature vs. equivalence ratio for air/H_2, air/CH_4, and air/C_3H_8 flames, where the air and fuel are at ambient temperature and pressure.
FIGURE 2.17 Adiabatic flame temperature vs. air preheat temperature for stoichiometric air/H_2, air/CH_4, and air/C_3H_8 flames, where the air and fuel are at ambient temperature and pressure.
FIGURE 2.18 Adiabatic flame temperature vs. fuel preheat temperature for stoichiometric air/H_2, air/CH_4, and air/C_3H_8 flames, where the air is at ambient temperature and pressure.
FIGURE 2.19 Adiabatic flame temperature vs. fuel blend (CH_4/H_2 and CH_4/N_2) composition for stoichiometric air/fuel flames, where the air and fuel are at ambient temperature and pressure.
FIGURE 2.20 Adiabatic flame temperature vs. fuel blend (CH_4/H_2) composition and air preheat temperature for stoichiometric air/fuel flames, where the fuel is at ambient temperature and pressure.
FIGURE 2.21 Sample Sankey diagram showing distribution of energy in a combustion system.
FIGURE 2.22 Available heat vs. gas temperature for stoichiometric air/H_2, air/CH_4, and air/C_3H_8 flames, where the air and fuel are at ambient temperature and pressure.
FIGURE 2.23 Available heat vs. air preheat temperature for stoichiometric air/H_2, air/CH_4, and air/C_3H_8 flames at an exhaust gas temperature of 2000°F (1100°C), where the fuel is at ambient temperature and pressure.
FIGURE 2.24 Available heat vs. fuel preheat temperature for stoichiometric air/H_2, air/CH_4, and air/C_3H_8 flames at an exhaust gas temperature of 2000°F (1100°C), where the air is at ambient temperature and pressure.
FIGURE 2.25 Graphical representation of ignition and heat release.

CHAPTER 3

FIGURE 3.1 A typical fired heater.
FIGURE 3.2 Heat transfer through a plane wall: (a) temperature distribution, and (b) equivalent thermal circuit.
FIGURE 3.3 Equivalent thermal circuit for a series composite wall.
FIGURE 3.4 Temperature drop due to thermal contact resistance.
FIGURE 3.5 Temperature distribution for a composite cylindrical wall.
FIGURE 3.6 Transient conduction through a solid.
FIGURE 3.7 Thermal conductivity of (a) some commonly used steels and alloys and (b) some refractory materials.
FIGURE 3.8 Temperature-thickness relationships corresponding to different thermal conductivities.
FIGURE 3.9 Thermal boundary layer development in a heated circular tube.
FIGURE 3.10 Orthogonal oscillations of electric and magnetic waves in the oscillations in electromagnetic waves.
FIGURE 3.11 Spectrum of electromagnetic radiation.
FIGURE 3.12 Spectral blackbody emissive power.
FIGURE 3.13 Radiation transfer between two surfaces approximated as blackbodies.
FIGURE 3.14 Network representation of radiative exchange between surface *i* and the remaining surfaces of an enclosure.
FIGURE 3.15 View factor of radiation exchange between faces of area dA_i and dA_j.
FIGURE 3.16 View factor for aligned parallel rectangles.
FIGURE 3.17 View factor for coaxial parallel disks.

FIGURE 3.18 View factor for perpendicular rectangles with a common edge.
FIGURE 3.19 Infrared thermal image of a flame in a furnace.
FIGURE 3.20 Emission bands of (a) CO_2 and (b) H_2O.
FIGURE 3.21 Emissivity of water vapor in a mixture with nonradiating gases at 1-atm total pressure and of hemispherical shape.
FIGURE 3.22 Emissivity of carbon dioxide in a mixture with nonradiating gases at 1-atm total pressure and of hemispherical shape.
FIGURE 3.23 Radiation heat transfer correction factor for mixtures of water vapor and carbon dioxide.
FIGURE 3.24 Photographic view of a luminous flame.
FIGURE 3.25 Photographic view of a nonluminous flame.
FIGURE 3.26 Photographic view of a radiant wall burner.
FIGURE 3.27 Vertical heat flux distribution for oil and gas firing in a vertical tube furnace.
FIGURE 3.28 Distribution of dimensionless average radiant flux density at the tube surfaces for various flame lengths (L_f= flame length, L = heater height, Z = height).

CHAPTER 4

FIGURE 4.1 Variation in measured density with length scale.
FIGURE 4.2 Velocity profile of a fluid flowing along a solid surface.
FIGURE 4.3 Absolute viscosity vs. temperature for various fluids.
FIGURE 4.4 Temperature vs. viscosity for various hydrocarbons. (Courtesy of J.B. Maxwell, *Data Book on Hydrocarbons,* D. Van Nostrand, Princeton, NJ, 1950, 174.)
FIGURE 4.5 Viscosity of mid-continent oils. (Courtesy of J.B. Maxwell, *Data Book on Hydrocarbons,* D. Van Nostrand, Princeton, NJ, 1950, 164.)
FIGURE 4.6 Compressibility factor *Z* as a function of reduced pressure and reduced temperature for different gases.
FIGURE 4.7 U-tube manometer.
FIGURE 4.8 Inclined manometer .
FIGURE 4.9 Helium balloon attached to the ground.
FIGURE 4.10 A small packet of fluid from point A to B along an arbitrary path.
FIGURE 4.11 Pressure relief vessel venting to a flare.
FIGURE 4.12 An idealization of a small "differential" control volume.
FIGURE 4.13 Mass flow into and out of a volume in the *X*-direction.
FIGURE 4.14 Gravitational body force.
FIGURE 4.15 Normal or pressure forces.
FIGURE 4.16 Effect of shear stress on *X*-direction face.
FIGURE 4.17 Smoke from incense. (Courtesy of The Visualization Society of Japan, *Fantasy of Flow,* Tokyo, 1993, 93.)
FIGURE 4.18 Water exiting a faucet at low velocity. (Courtesy of The Visualization Society of Japan, *Fantasy of Flow,* Tokyo, 1993, 97.)
FIGURE 4.19 Leonardo daVinci's view of turbulence.
FIGURE 4.20 Osborn Reynolds' experimental apparatus used to study the transition from laminar to turbulent flow.
FIGURE 4.21 Water from faucet showing transition. (Courtesy of The Visualization Society of Japan, *Fantasy of Flow,* Tokyo, 1993, 97.)
FIGURE 4.22 Wake area showing mixing vortices. (Courtesy of The Visualization Society of Japan, *Fantasy of Flow,* Tokyo, 1993, 3.)
FIGURE 4.23 Laminar flow of smoke over a rectangular obstruction. (Courtesy of M. Van Dyke, *An Album of Fluid Motion,* The Parabolic Press, Stanford, CA, 1982, 10.)
FIGURE 4.24 Free jet structure. (Courtesy of M. Van Dyke, *An Album of Fluid Motion,* The Parabolic Press, Stanford, CA, 1982, 97.)
FIGURE 4.25 Free jet entrainment. (Courtesy of M. Van Dyke, *An Album of Fluid Motion,* The Parabolic Press, Stanford, CA, 1982, 99.)
FIGURE 4.26 Flow around the air intake of a jet engine in supersonic flow. (Courtesy of The Visualization Society of Japan, *Fantasy of Flow,* Tokyo, 1993, 23.)
FIGURE 4.27 Shock waves from a supersonic fighter. (Courtesy of The Visualization Society of Japan, *Fantasy of Flow,* Tokyo, 1993, 59.)
FIGURE 4.28 Choked flow test rig.
FIGURE 4.29 Types of flow of a fluid passing through a small orifice.
FIGURE 4.30 Photograph showing slow-moving streaks near a wall using the hydrogen bubble technique.
FIGURE 4.31 Moody diagram.
FIGURE 4.32 Factors that can influence the discharge coefficient.

CHAPTER 5

FIGURE 5.1 Simplified process flow diagram for hydrogen reforming/pressure swing absorption. (From R.A. Meyers, *Handbook of Petroleum Refining Processes,* 2nd ed., McGraw-Hill, New York, 1997, p. 6.27. With permission.)
FIGURE 5.2 Simplified process flow diagram for flexicoking. (From R.A. Meyers, *Handbook of Petroleum Refining Processes,* 2nd ed., McGraw-Hill, New York, 1997, p. 12.5. With permission.)
FIGURE 5.3 100% Tulsa natural gas flame.
FIGURE 5.4 90% Tulsa natural gas/10% nitrogen flame.
FIGURE 5.5 80% Tulsa natural gas/20% nitrogen flame. (From R.A. Meyers, *Handbook of Petroleum Refining Processes,* 2nd ed., McGraw-Hill, New York, 1997, p. 12.11. With permission.)
FIGURE 5.6 90% Tulsa natural gas/10% hydrogen flame.
FIGURE 5.7 75% Tulsa natural gas/25% hydrogen flame.
FIGURE 5.8 50% Tulsa natural gas/50% hydrogen flame.
FIGURE 5.9 25% Tulsa natural gas/75% hydrogen flame.
FIGURE 5.10 100% hydrogen flame.
FIGURE 5.11 50% Tulsa natural gas/25% hydrogen/25% C_3H_8 flame.
FIGURE 5.12 50% Tulsa natural gas/50% C_3H_8 flame.
FIGURE 5.13 100% C_3H_8 flame.
FIGURE 5.14 100% C_4H_{10} flame.
FIGURE 5.15 Simulated cracked gas flame.
FIGURE 5.16 Simulated coking gas flame.
FIGURE 5.17 Simulated FCC gas flame.
FIGURE 5.18 Simulated refoming gas flame.
FIGURE 5.19 100% Tulsa natural gas flame.
FIGURE 5.20 100% hydrogen.
FIGURE 5.21 100% propane.
FIGURE 5.22 50% hydrogen 50% Propane.
FIGURE 5.23 50% hydrogen 50% Tulsa natural gas.
FIGURE 5.24 50% propane 50% Tulsa natural gas.
FIGURE 5.25 25% hydrogen 75% Propane.
FIGURE 5.26 75% hydrogen 25% Propane.
FIGURE 5.27 25% hydrogen 75% Tulsa natural gas.
FIGURE 5.28 75% hydrogen 25% Tulsa natural gas.
FIGURE 5.29 25% propane 75% Tulsa natural gas.
FIGURE 5.30 75% propane 25% Tulsa natural gas.
FIGURE 5.31 25% hydrogen 25% propane 50% Tulsa natural gas.
FIGURE 5.32 25% hydrogen 50% propane 25% Tulsa natural gas.
FIGURE 5.33 50% hydrogen 25% propane 25% Tulsa natural gas.
FIGURE 5.34 Viewing oil flame through burner plenum.
FIGURE 5.35 Oil derrick, circa 1900.
FIGURE 5.36 Capping a burning oil well.
FIGURE 5.37 Refinery flow diagram.
FIGURE 5.38 Flow diagram of UOP fluid catalytic cracking complex.
FIGURE 5.39 Burner firing heavy oil (1).
FIGURE 5.40 Burner firing heavy oil (2).
FIGURE 5.41 Naphtha distillation curve.
FIGURE 5.42 Crude oil distillation curve.
FIGURE 5.43 Viscosity of fuel oils.

CHAPTER 6

FIGURE 6.1 Cartoon of NO exiting a stack and combining with O_2 to form NO_2.
FIGURE 6.2 Cartoon of acid rain.
FIGURE 6.3 Cartoon of photochemical smog formation.

FIGURE 6.4 Schematic of fuel NOx formation pathways.

FIGURE 6.5 Adiabatic equilibrium NO as a function of equivalence ratio for air/fuel flames.

FIGURE 6.6 Adiabatic equilibrium NO as a function of gas temperature for stoichiometric air/fuel flames.

FIGURE 6.7 Adiabatic equilibrium NO as a function of air preheat temperature for stoichiometric air/fuel flames.

FIGURE 6.8 Adiabatic equilibrium NO as a function of fuel preheat temperature for a stoichiometric air/CH_4 flames.

FIGURE 6.9 Adiabatic equilibrium NO as a function of fuel composition (CH_4/H_2) for a stoichiometric air/fuel flame.

FIGURE 6.10 Adiabatic equilibrium NO as a function of fuel composition (CH_4/N_2) for a stoichiometric air/fuel flame.

FIGURE 6.11 Sampling system schematic as recommended by the U.S. EPA.

FIGURE 6.12 Schematic of furnace gas recirculation.

FIGURE 6.13 Adiabatic equilibrium NO as a function of the fuel blend composition for H_2/CH_4 blends combusted with 15% excess air where both the fuel and the air are at ambient temperature and pressure.

FIGURE 6.14 Adiabatic equilibrium NO as a function of the fuel blend composition for C_3H_8/CH_4 blends combusted with 15% excess air where both the fuel and the air are at ambient temperature and pressure.

FIGURE 6.15 Adiabatic equilibrium NO as a function of the fuel blend composition for H_2/C_3H_8 blends combusted with 15% excess air where both the fuel and the air are at ambient temperature and pressure.

FIGURE 6.16 Ternary plot of adiabatic equilibrium NO (fraction of the maximum value) as a function of the fuel blend composition for H_2/CH_4/C_3H_8 blends combusted with 15% excess air where both the fuel and the air are at ambient temperature and pressure.

FIGURE 6.17 Raw gas (VYD) burner.

FIGURE 6.18 Test furnace.

FIGURE 6.19 Measured NOx (percent of the maximum ppmv value) as a function of the fuel blend composition for H_2/Tulsa natural gas blends combusted with 15% excess air where both the fuel and the air are at ambient temperature and pressure.

FIGURE 6.20 Measured NOx (percent of the maximum ppmv value) as a function of the fuel blend composition for C_3H_8/Tulsa natural gas blends combusted with 15% excess air where both the fuel and the air are at ambient temperature and pressure.

FIGURE 6.21 Measured NOx (percent of the maximum value in both ppmv and lb/MMBtu) as a function of the fuel blend composition for H_2/C_3H_8 blends combusted with 15% excess air where both the fuel and the air are at ambient temperature and pressure.

FIGURE 6.22 Measured NOx (fraction of the maximum value in both ppmv and lb/MMBtu) as a function of the fuel blend composition for Tulsa natural gas/H_2/C_3H_8 blends combusted with 15% excess air where both the fuel and the air are at ambient temperature and pressure for gas tip #2.

FIGURE 6.23 Measured NOx (fraction of the maximum value in both ppmv and lb/MMBtu) as a function of the fuel blend composition for Tulsa natural gas/H_2/C_3H_8 blends combusted with 15% excess air where both the fuel and the air are at ambient temperature and pressure for gas tip #4.

FIGURE 6.24 Measured NOx (fraction of the maximum value in both ppmv and lb/MMBtu) as a function of the fuel blend composition for Tulsa natural gas/H2/C3H8 blends combusted with 15% excess air where both the fuel and the air are at ambient temperature and pressure for gas tip #6.

FIGURE 6.25 Measured NOx (fraction of the maximum value in both ppmv and lb/MMBtu) as a function of the fuel blend composition for Tulsa natural gas/H_2/C_3H_8 blends combusted with 15% excess air where both the fuel and the air are at ambient temperature and pressure for a constant fuel gas pressure of 21 psig.

FIGURE 6.26 Measured NOx (fraction of the maximum value in ppmvd) as a function of the fuel pressure for all 15 different Tulsa natural gas/H_2/C_3H_8 blends (A through O) combusted with 15% excess air where both the fuel and the air are at ambient temperature and pressure.

FIGURE 6.27 Measured NOx (fraction of the maximum value in both ppmv and lb/MMBtu) as a function of the fuel blend composition, fuel gas pressure, and calculated adiabatic flame temperature for for Tulsa natural gas/H_2/C_3H_8 blends combusted with 15% excess air where both the fuel and the air are at ambient temperature and pressure.

FIGURE 6.28 Adiabatic equilibrium CO as a function of equivalence ratio for air/fuel flames.

FIGURE 6.29 Adiabatic equilibrium CO as a function of gas temperature for stoichiometric air/fuel flames.

FIGURE 6.30 Adiabatic equilibrium CO as a function of air preheat temperature for stoichiometric air/fuel flames.

FIGURE 6.31 Adiabatic equilibrium CO as a function of fuel preheat temperature for stoichiometric air/CH_4 flames.

FIGURE 6.32 Adiabatic equilibrium CO as a function of fuel composition (CH_4/H_2) for a stoichiometric air/fuel flame.

FIGURE 6.33 Adiabatic equilibrium CO as a function of fuel composition (CH_4/N_2) for a stoichiometric air/fuel flame.

CHAPTER 7

FIGURE 7.1 Pressure peaks and troughs.
FIGURE 7.2 Cross-section of the human ear.
FIGURE 7.3 Relationship of decibels to watts.
FIGURE 7.4 Sound pressure level at a distance *r*.
FIGURE 7.5 Threshold of hearing in humans.
FIGURE 7.6 Threshold of hearing and threshold of pain in humans.
FIGURE 7.7 A-weighted scale for human hearing threshold.
FIGURE 7.8 A-weighted burner noise curve.
FIGURE 7.9 Weighting curves A, B, C, and D.
FIGURE 7.10 Block diagram of a sound level meter.
FIGURE 7.11 Same sound spectrum on three different intervals.
FIGURE 7.12 Typical burner noise curve.
FIGURE 7.13 Same sound spectrum on 3 different intervals.
FIGURE 7.14 Typical noise signature emitted from a Flare.
FIGURE 7.15 Engineer measuring flare noise level.
FIGURE 7.16 Shadow photograph of a burning butane lighter.
FIGURE 7.17 A steam-assisted flare under normal and over-steamed conditions.
FIGURE 7.18 Sound pressure level burner with instability.
FIGURE 7.19 Development of orderly wave patterns.
FIGURE 7.20 Region of maximum jet mixing noise.
FIGURE 7.21 Shock waves downstream of an air jet.
FIGURE 7.22 Location of screech tone emissions.
FIGURE 7.23 Noise radiating from a valve.
FIGURE 7.24 Two enclosed flares.
FIGURE 7.25 A steam-assisted flare with a muffler.
FIGURE 7.26 Steam jet noise emitted with and without a muffler.
FIGURE 7.27 Noise spectrum from high pressure flare with and without water injection.
FIGURE 7.28 Sound pressure vs frequency with and without a muffler.
FIGURE 7.29 Burner noise example.
FIGURE 7.30 The sound-pressure-level spectrum of a high-pressure flare.
FIGURE 7.31 The noise contributions separately based on the mathematical model.
FIGURE 7.32 Effect of distance on flare noise.

CHAPTER 8

FIGURE 8.1 A typical eductor system.
FIGURE 8.2 Example eductor system.
FIGURE 8.3 Experimental apparatus that has been successfully used to determine the flow coefficients and eduction performance of various flare and burner eduction processes.
FIGURE 8.4 First and second generation steam flares.
FIGURE 8.5 Typical third generation steam flare tube layout.
FIGURE 8.6 Normalized plot showing the sonic-supersonic eduction performance in a single steam tube used in a typical steam-assisted flare.
FIGURE 8.7 Normalized eduction performance of a flare pilot operating on Tulsa natural gas with two different motive gas orifice diameters.
FIGURE 8.8 Experimental and theoretical results of the eduction performance of a particular radiant wall burner firing with two different orifice sizes and fuel gas compositions.
FIGURE 8.9 Simplified reactor modeling of a staged fuel burner.
FIGURE 8.10 Picture of a radiant wall burner.
FIGURE 8.11 Picture of a thermal oxidizer.
FIGURE 8.12 Sample results of simplified modeling for a premixed burner.
FIGURE 8.13 Sample results of simplified modeling for a thermal oxidizer.
FIGURE 8.14 Capacity curves that many burner manufactures use for sizing burners.

FIGURE 8.15 Burner designs typically consist of a muffler, damper, plenum, throat, and tile section.
FIGURE 8.16 Cold flow furnace of a test chamber (8 x 8 x 8) supported on legs approximately 7 feet above the ground.
FIGURE 8.17 Smoking and non-smoking flares.
FIGURE 8.18 Flame of a flare is divided into two parts: 1) the main body of the flame, and 2) the region near the tip.
FIGURE 8.19 Illustration showing the experimental setup utilized to obtain calibration and validation data.
FIGURE 8.20 Prediction of diluent flow for smokeless operation (a) Propylene (b) Propane.
FIGURE 8.21 Figure 8.20 with data points added (a) Propylene (b) Propane.
FIGURE 8.22 Steam-to-hydrocarbon ratios (per Leite).
FIGURE 8.23 Steam-to-hydrocarbon (large diameter flares).
FIGURE 8.24 One smokeless steam flare.
FIGURE 8.25 Typical air flare.
FIGURE 8.26 Effect of high velocity air: (a) blower off, (b) commence blower, and (c) blower on.
FIGURE 8.27 Aeration rate was determined and plotted against the scaling function.
FIGURE 8.28 Annular air-assisted flare (190,000 lb/hr propane).
FIGURE 8.29 Comparison of three different air flares to prediction.
FIGURE 8.30 Typical process heater oil flame.
FIGURE 8.31 Standard John Zink oil gun.
FIGURE 8.32 Schematic of a typical oil gun.
FIGURE 8.33 Comparison of predicted vs. actual oil and steam flow rates.
FIGURE 8.34 John Zink Co. LLC. (Tulsa, OK) Spray Research Laboratory.
FIGURE 8.35 Droplet size comparison between a standard and a newer oil gun.
FIGURE 8.36 High-efficiency new oil gun.
FIGURE 8.37 Heat transfer in a packed bed between the ceramic material and the air stream.
FIGURE 8.38 Installing small, type K thermocouple pairs into numerous ceramic saddles.
FIGURE 8.39 Summary of saddle data.
FIGURE 8.40 Rock temp distribution with time.
FIGURE 8.41 John Zink RTO test unit.

CHAPTER 9

FIGURE 9.1 Elements in CFD modeling.
FIGURE 9.2 Plot of the β-pdf for several values of $\langle f \rangle$ and $\langle f'^2 \rangle$.
FIGURE 9.3 Point measurement of scalar in a turbulent flow.
FIGURE 9.4 Rendered view of a CFD model of a John Zink Co. burner. This view illustrates the complex geometry that necessitates a variety of cell types. This mesh consists of hexahedral, pyramidal, and tetrahedral cell types.
FIGURE 9.5 Close-up view of primary tip. This view reveals the five fuel jets (indicated by the arrows on the image) issuing from the primary tip.
FIGURE 9.6 Rendered view inside an ethylene pyrolysis furnace showing flow patterns near the premixed radiant wall burners.
FIGURE 9.7 CFD model of an ethylene pyrolysis furnace. There are six burners shown in each row at the bottom of the furnace, and the tubes are approximately 35 feet long. The endwalls are not shown in this image.
FIGURE 9.8 Plot showing heat flux to the process tubes in the modeled ethylene furnace as a function of height above the furnace floor.
FIGURE 9.9 Geometry of a xylene reboiler. This view shows half (sliced vertically) of the furnace. Only three of the six burners are shown at the bottom of the image.
FIGURE 9.10 This view shows half of the furnace with unmodified burners. The "blob" in the furnace is the 50-ppm OH mole fraction iso-surface. This surface is colored according to its temperature ($^{\circ}$F)
FIGURE 9.11 This view shows half of the furnace with modified burners firing. The 50-ppm OH mole fraction iso-surface is shown as an indicator of the flame shape. This surface is colored according to its temperature ($^{\circ}$F)
FIGURE 9.12 Exterior geometry of the furnace is included in the model. The surface mesh is also shown.
FIGURE 9.13 Burner geometry. The acid swirl vanes are shown in red; the air swirl vanes are shown in green; and the start-up fuel tip is shown in purple.
FIGURE 9.14 Oxygen mass fractions viewed from above the furnace. The contour scale is logarithmic. The mass fractions are contoured on the mid-plane of the furnace.
FIGURE 9.15 H_2S mole fractions contoured on the midplane of the furnace.

FIGURE 9.16 Stoichiometric iso-surface colored by temperature (°C) for the initial burner design.
FIGURE 9.17 Stoichiometric iso-surface colored by temperature (°C) for the final burner design.
FIGURE 9.18 Midplane of geometry colored by temperature (°C). This view shows the burner quarl and the mixing regions of acid gas and air.
FIGURE 9.19 Temperature profiles (°C) exiting the reaction furnace for the initial (left) and final (right) burner geometries.
FIGURE 9.20 Geometric information describing the thermal oxidizer examined during this study.
FIGURE 9.21 Predicted centerline profiles for excess air case: (a) axial velocity (m/s), and (b) gas temperature (K) for the furnace section of the thermal oxidizer shown in Figure 9.20. Two distinct combustion zones are illustrated, with an exit temperature of about 1600 K (2400°F).
FIGURE 9.22 Predicted centerline profiles for excess air case: (a) methane concentration (ppmv), and (b) carbon monoxide concentration (ppmv) for the furnace section of the thermal oxidizer shown in Figure 9.20. These predictions depict the CO formation and oxidation zones common to most combustion processes.
FIGURE 9.23 Predicted centerline profiles for excess air case: (a) HCl concentration (ppmv), and (b) Cl_2 concentration (ppmv) for the furnace section of the thermal oxidizer shown in Figure 9.20. The predicted maximum Cl_2 concentration, nearly 3200 ppmv, occurs in the cooler reactor regions, while an exit Cl_2 concentration of about 100 ppmv is predicted.
FIGURE 9.24 Predicted centerline profiles for the stoichiometric case: (a) axial velocity (m/s), and (b) gas temperature (K) for the furnace section of the thermal oxidizer shown in Figure 9.20. A single combustion zone is indicated, with the local maximum temperature of 1450 K (2150°F) and an exit temperature of about 1350 K (1970°F)
FIGURE 9.25 Predicted centerline profiles for the stoichiometric case: (a) methane concentration (ppmv), and (b) carbon monoxide concentration (ppmv) for the furnace section of the thermal oxidizer shown in Figure 9.20. Here, the post-flame CO oxidation zone, shown in the first prediction, is not present; this results in a predicted exit CO concentration of 9000 ppmv.
FIGURE 9.26 Predicted centerline profiles for the stoichiometric case: (a) HCl concentration (ppmv), and (b) Cl_2 concentration (ppmv) for the furnace section of the thermal oxidizer shown in Figure 9.20. Dramatically less Cl_2 formation is predicted (local maximum of 7 ppmv and exit concentrations less than 1 ppmv) in this case due to excess H^+ radical present from the increased fuel gas.
FIGURE 9.27 Entrainment of flue gas.
FIGURE 9.28 Re-circulation region in the eductor throat.
FIGURE 9.29 Re-circulation zone starting to occur in eductor throat.
FIGURE 9.30 Re-circulation zone developing in eductor throat.
FIGURE 9.31 Contours of stream function with increasing backpressure.

CHAPTER 10

FIGURE 10.1 Fire tetrahedron.
FIGURE 10.2 Tube rupture in a fired heater. (Courtesy of R.E. Sanders, *Chemical Process Safety: Learning from Case Histories,* Butterworth–Heinemann, Woburn, MA, 1999.)
FIGURE 10.3 Trapped steam in a dead-end that can freeze and cause pipe failure.
FIGURE 10.4 CO detector: (a) permanent, (b) portable.
FIGURE 10.5 Flarestack explosion due to improper purging. (Courtesy of T. Kletz, *What Went Wrong: Case Histories of Process Plant Disasters,* 4th ed., Gulf Publishing, Houston, TX, 1998.)
FIGURE 10.6 Vapor pressures for light hydrocarbons. (Courtesy of M.G. Zabetakis, *AIChE-Inst. Chem. Engr. Symp., Ser. 2, Chem. Engr. Extreme Cond. Proc. Symp.* American Institute of Chemical Engineers, New York, 1965, 99–104.)
FIGURE 10.7 Ethylene oxide plant explosion caused by autoignition. (Courtesy of T. Kletz, *What Went Wrong: Case Histories of Process Plant Disasters,* 4th ed., Gulf Publishing, Houston, TX, 1998.)
FIGURE 10.8 Safety documentation feedback flow chart. (Courtesy of *GPSA Engineering Data Book,* Vol. II, 10th ed., Gas Processors and Suppliers Association, Tulsa, OK, 1994.)
FIGURE 10.9 Refinery damaged due to improper maintenance procedures. (Courtesy of R.E. Sanders, *Chemical Process Safety: Learning from Case Histories,* Butterworth–Heinemann, Woburn, MA, 1999.)

CHAPTER 11

FIGURE 11.1 Graph of sustainable combustion for methane.
FIGURE 11.2 Typical raw gas burner tips.
FIGURE 11.3 Typical premix metering orifice spud and air mixer assembly.
FIGURE 11.4 Typical gas fuel capacity curve.

FIGURE 11.5 Typical liquid fuel atomizer/spray tip configurations.
FIGURE 11.6 Typical liquid fuel capacity curve.
FIGURE 11.7 Typical throat of a raw gas burner.
FIGURE 11.8 Ledge in the burner tile.
FIGURE 11.9 Flame stabilizer or flame holder.
FIGURE 11.10 Swirler.
FIGURE 11.11 Round-shaped flame.
FIGURE 11.12 Flat-shaped flame.
FIGURE 11.13 Internal mix twin fluid atomizer.
FIGURE 11.14 Port mix twin fluid atomizer.
FIGURE 11.15 Regen tile and swirler.
FIGURE 11.16 Typical conventional raw gas burner.
FIGURE 11.17 Typical premix gas burner.
FIGURE 11.18 Typical round flame combination burner.
FIGURE 11.19 Typical round flame, high-intensity combination burner.
FIGURE 11.20 Typical staged-fuel flat flame burner.
FIGURE 11.21 Typical radiant wall burner.

CHAPTER 12

FIGURE 12.1 Programmable logic controller.
FIGURE 12.2 Touch screen.
FIGURE 12.3 Simplified flow diagram of a standard burner light-off sequence.
FIGURE 12.4 Simple analog loop.
FIGURE 12.5 Feedforward loop.
FIGURE 12.6 Double-block-and-bleed system.
FIGURE 12.7 Failsafe input to programmable logic controller.
FIGURE 12.8 Shutdown string.
FIGURE 12.9 Typical pipe rack.
FIGURE 12.10a Large control panel.
FIGURE 12.10b Small control panel.
FIGURE 12.11 Inside the control panel.
FIGURE 12.12a Pressure switch.
FIGURE 12.12b Pressure switch.
FIGURE 12.13 Pneumatic control valve.
FIGURE 12.14 Control valve characteristics.
FIGURE 12.15 Thermocouple.
FIGURE 12.16 Thermowell and thermocouple.
FIGURE 12.17 Velocity thermocouple.
FIGURE 12.18 Pressure transmitter (left) and pressure gauge (right).
FIGURE 12.19 Mechanically linked parallel positioning.
FIGURE 12.20 Electronically linked parallel positioning.
FIGURE 12.21 A variation of parallel positioning.
FIGURE 12.22 Fuel flow rate vs. control signal.
FIGURE 12.23 Typical butterfly-type valve calculation.
FIGURE 12.24 The required shape of the air valve characterizer.
FIGURE 12.25 Fully metered control scheme.
FIGURE 12.26 Fully metered control scheme with cross limiting.
FIGURE 12.27 O2 trim of air flow rate.
FIGURE 12.28 O2 trim of air setpoint.
FIGURE 12.29 Multiple fuels and O_2 sources.
FIGURE 12.30 Controller.
FIGURE 12.31 Analog controller with manual reset.
FIGURE 12.32 Analog controller with automatic reset.

CHAPTER 13

FIGURE 13.1 Contrast of classical experimentation and SED methods.
FIGURE 13.2 NOx contours for furnace temperature and oxygen concentration based on Equation 13.1.
FIGURE 13.3 A fuel-staged burner.
FIGURE 13.4 Municipal solid waste boiler using ammonia injection to control NOx.
FIGURE 13.5 Method of steepest ascent.
FIGURE 13.6 A combination burner capable of firing either oil or gas or both simultaneously.
FIGURE 13.7 Simplex design for $q = 3$.
FIGURE 13.8 Mixture factors, a transformation, and a combined mixture-factorial.
FIGURE 13.9 Flowchart showing a general sequential experimental strategy.
FIGURE 13.10 Some orthogonal designs for $f = 3$ arranged in a sequential strategy.

CHAPTER 14

FIGURE 14.1 John Zink Co., LLC, Research and Development Test Center, Tulsa, Oklahoma
FIGURE 14.2 Test furnace for simulation of ethylene furnace.
FIGURE 14.3 Test furnace for simulation of down-fired tests.
FIGURE 14.4 Test furnace for simulation of up-fired tests.
FIGURE 14.5 Test furnace for simulation of terrace wall reformers.
FIGURE 14.6 Test fuel storage tanks.
FIGURE 14.7 Forced-draft air preheater.
FIGURE 14.8 Heat flux probe schematic.
FIGURE 14.9 Ellipsoidal radiometer schematic.

CHAPTER 15

FIGURE 15.1 Heater cutout and burner bolt circle on a new heater.
FIGURE 15.2 Warped steel on the shell of a heater.
FIGURE 15.3 Burner improperly installed at an angle due to a warped shell.
FIGURE 15.4 Donut ring for leveling burner mounting onto the warped shell of a heater.
FIGURE 15.5 Typical burner drawing.
FIGURE 15.6 Burner mounted on the floor of a heater.
FIGURE 15.7 Burner mounted on the side of a heater.
FIGURE 15.8 Burner mounted on the top of heater.
FIGURE 15.9 Burner mounted in a common plenum.
FIGURE 15.10 Burner in a plenum box mounted to a heater.
FIGURE 15.11 Piping improperly loaded on the burner inlet.
FIGURE 15.12 Picture of a burner tile showing multiple tile pieces.
FIGURE 15.13 Sketch showing a round tile measured in 3 different diameters.
FIGURE 15.14 Sketch showing a square tile measured at different lengths and widths.
FIGURE 15.15 Oil tip in combination burner showing oil tip locations.
FIGURE 15.16 Welding rods in an oil tip.
FIGURE 15.17 VYD burner gas tip in a diffuser with a pilot.
FIGURE 15.18 VYD drawing showing the diffuser cone and the pilot tip.
FIGURE 15.19 Example of an air register.
FIGURE 15.20 Typical fuel gas piping system.
FIGURE 15.21 Typical heavy fuel oil piping system.
FIGURE 15.22 Typical light fuel oil piping system.
FIGURE 15.23 A pat-826 gas tip.
FIGURE 15.24 Example oil gun atomizer.
FIGURE 15.25 Catatlyst deposit within an oil burner tile.
FIGURE 15.26 Typical diffuser cone.
FIGURE 15.27 Typical spin diffuser.
FIGURE 15.28 Example of a damaged stabilizer.

FIGURE 15.29 Example of a damaged pilot tip.
FIGURE 15.30 New ST-1S pilot tip without an electronic ignitor.
FIGURE 15.31 New ST-1SE pilot with an electronic ignitor.

CHAPTER 16

FIGURE 16.1 Typical draft measurement points.
FIGURE 16.2 Inclined manometer.
FIGURE 16.3 Excess air indication by oxygen content.
FIGURE 16.4 Location for measuring excess oxygen.
FIGURE 16.5 Oxygen analyzer.
FIGURE 16.6 Cost of operating with higher excess oxygen levels (natural gas).
FIGURE 16.7 Cost of operating with higher excess oxygen levels (#6 oil).
FIGURE 16.8 Fuel gas pressure measurement.
FIGURE 16.9 Graph of fuel pressure vs. heat release.
FIGURE 16.10 Viscosity vs temperature for a range of hydrocarbons.
FIGURE 16.11 Velocity thermocouple.
FIGURE 16.12 Air control device schematic.
FIGURE 16.13 Picture of air control device.
FIGURE 16.14 Primary air door.
FIGURE 16.15 Burner ignition ledge.
FIGURE 16.16 Gas tips.
FIGURE 16.17 Oil tips.
FIGURE 16.18 Long narrow and short bushy flames.
FIGURE 16.19 Typical flame envelope with x-y-z axes.
FIGURE 16.20 Sodium ions in the flame.
FIGURE 16.21 Example of a good flame within the firebox.
FIGURE 16.22 Example of a very bad flame pattern in a fire box.
FIGURE 16.23 Typical draft profile in a natural draft heater.
FIGURE 16.24 Logic diagram for tuning a natural draft heater.
FIGURE 16.25 Logic diagram for tuning a balanced draft heater.
FIGURE 16.26 Unstable flame.
FIGURE 16.27 Broken burner tile.
FIGURE 16.28 Dark line or black streaks on hot refractory surface indicating air leaks.

CHAPTER 17

FIGURE 17.1 Coke deposit causes tube thinning.
FIGURE 17.2 Cracked gas tip causing an irregular flame pattern.
FIGURE 17.3 Damaged diffuser cone.
FIGURE 17.4 Effect of excess O_2 on NOx in raw gas burners.
FIGURE 17.5 Effect of combustion air temperature on NOx.
FIGURE 17.6 Effect of firebox temperature on NOx.
FIGURE 17.7 Effect of bound nitrogen in the liquid fuel on NOx.
FIGURE 17.8 Effect of burner model on NOx.
FIGURE 17.9 Staged air burncr.
FIGURE 17.10 Staged fuel burner.
FIGURE 17.11 Ultra low NOx burner.

CHAPTER 18

FIGURE 18.1 Typical plant schematic.
FIGURE 18.2 Cogeneration at Teesside, England. Courtesy of Nooter/Eriksen. St. Louis, MO. With permission.

FIGURE 18.3 Combination (oil and gas) fire duct burners at Dahbol, India. (Courtesy of Ms. Martha Butala, Dabhol Power Company, Bombay, India, as published in *Power Magazine*.)
FIGURE 18.4 Typical location of duct burners in an HRSG. (Courtesy of Deltak, Minneapolis, MN. With permission.)
FIGURE 18.5 Schematic of HRSG at Teesside, England. (Courtesy of Nooter/Eriksen, St. Louis, MO. With permission.)
FIGURE 18.6 Fluidized bed startup duct burner.
FIGURE 18.7 An inline burner.
FIGURE 18.8 Linear burner elements.
FIGURE 18.9 Gas flame from a grid burner.
FIGURE 18.10 Oil flame from a side-fired oil gun.
FIGURE 18.11 Approximate requirement for augmenting air.
FIGURE 18.12 Duct burner arrangement.
FIGURE 18.13 Comparison of flow variation with and without straightening device.
FIGURE 18.14 Physical model of burner.
FIGURE 18.15 Sample result of CFD modeling performed on an HRSG inlet duct.
FIGURE 18.16 Drilled pipe duct burner.
FIGURE 18.17 Low emission duct burner.
FIGURE 18.18 Flow patterns around flame stabilizer.
FIGURE 18.19 Effect of conditions on CO formation.
FIGURE 18.20 Typical main gas fuel train: single element or multiple elements firing simultaneously.
FIGURE 18.21 Typical main gas fuel train: multiple elements with individual firing capability.
FIGURE 18.22 Typical pilot gas train: single element or multiple elements firing simultaneously.
FIGURE 18.23 Typical pilot gas train: multiple elements with individual firing capability.
FIGURE 18.24 Typical main oil fuel train: single element.
FIGURE 18.25 Typical main oil fuel train: multiple elements.
FIGURE 18.26 Typical pilot oil train: single element.
FIGURE 18.27 Typical pilot oil train: multiple elements.

CHAPTER 19

FIGURE 19.1 Typical utility boilers. (Courtesy of Florida Power & Light.)
FIGURE 19.2 Typical single-burner industrial boiler. (Courtesy of North Carolina Baptist Hospital.)
FIGURE 19.3 Swirl burner.
FIGURE 19.4 Average flame length as a function of burner heat input.
FIGURE 19.5 A typical low-NOx burner, venturi-style.
FIGURE 19.6 A typical low-NOx burner, a venturi-style (second example).
FIGURE 19.7 A strong flame front established within a maximum of 0.5 diffuser diameters of the face of the diffuser.
FIGURE 19.8 The effects of boiler design on NOx.
FIGURE 19.9 The NOx of various boilers included in the database on oil and gas, respectively.
FIGURE 19.10 NOx generation for natural gas and No. 6 oil (0.5% N_f) vs. adiabatic flame temperature.
FIGURE 19.11 NOx vs. excess O_2 with FGR implementation.
FIGURE 19.12 NOx vs. excess O_2 (The TFM-94 boiler equipped with nine boilers, at ~94% load).
FIGURE 19.13 NOx vs. relative steam flow at the TGM-94 boilers (natural gas, O_2 = 1.2–1.6%).
FIGURE 19.14 NOx vs. relative steam flow at the TGM-94 boilers (No. 6 oil, O_2 = 1.2–1.6%).
FIGURE 19.15 NOx vs. relative steam flow with firing natural gas on the TGME-206 boiler equipped with Todd Combustion low-NOx Dynaswirl burners at O_2 = 0.8–1.0%.
FIGURE 19.16 NOx vs. load with firing natural gas on utility burners.
FIGURE 19.17 Degree of the power function NOx = f (load) vs. bounded nitrogen in No. 6 oil.
FIGURE 19.18 Relative NOx vs. relative load on industrial boilers firing natural gas and No. 6 oil with ambient air.
FIGURE 19.19 Relative NOx vs. relative load on industrial boilers firing natural gas and No. 6 oil with preheated air.
FIGURE 19.20 Effect of furnace cleanliness on NOx emissions.
FIGURE 19.21 Effect of HRA cleanliness on NOx emissions.
FIGURE 19.22 Effect of air in-leakage on the burner performance.
FIGURE 19.23 Improvement of mass flow distribution to burners (differences within ±2%).
FIGURE 19.24 Improvement of peripheral air flow distribution to burners (deviations ±10%)

FIGURE 19.25 Improvement of FGR flow distribution to burners.
FIGURE 19.26 A scaled, physical, aerodynamic simulation model.
FIGURE 19.27 Flame-to-flame similarity of appearance.
FIGURE 19.28 Premixing the FGR flow with the combustion air upstream of the windbox.
FIGURE 19.29 Relative NOx concentration vs. FGR flow rate with firing natural gas in the utility boilers at full load.
FIGURE 19.30 Relative NOx concentration vs. FGR flow rate with firing natural gas in the 800 MW_e boiler at full load.
FIGURE 19.31 NOx reduction vs. flue gas flow.
FIGURE 19.32 Full load NOx reduction data.
FIGURE 19.33 Overfire air flow rate influence on NOx reduction and CO emission with firing natural gas in the TGM-94 boiler at ~94% load and O_2 ~ 1.2%.
FIGURE 19.34 "Horizontal" imbalance intensifies the combustion process and especially burnout while firing oil.
FIGURE 19.35 Data comparing single and three-stage gas combustion.
FIGURE 19.36 NOx generation with firing natural gas and No. 6 oil (0.5% N_f) vs. the theoretical maximal flame temperature.
FIGURE 19.37 The ratio between NOx numbers obtained with gas and oil firing are considered as a function of the furnace space heat release.
FIGURE 19.38 Schematic for typical low NOx burner.
FIGURE 19.39 NOx vs. heat load with firing natural gas with ambient air in the packaged industrial boiler equipped with the Todd Combustion low NOx burner.
FIGURE 19.40 CO vs. heat load with firing natural gas with ambient air in the packaged industrial boiler equipped with the Todd Combustion low NOx burner.
FIGURE 19.41 NOx, O_2 and opacity vs. relative heat load with firing No. 6 oil with ambient air in the packaged industrial boiler equipped with the Todd Combustion low NOx burner.
FIGURE 19.42 Internal FGR impact on NOx with firing No. 6 oil with ambient air in the 30–100% load range.
FIGURE 19.43 NOx and CO emissions vs. injector gas flow rate/total gas flow rate ratio, at the boiler equipped with the Todd Combustion low-NOx burner firing natural gas with preheated air at full load.
FIGURE 19.44 An ultra-low emissions burner.
FIGURE 19.45 Relationship between the adiabatic flame temperature and thermal NOx formation.
FIGURE 19.46 HCN and NH_3 formation at three flame temperatures.
FIGURE 19.47 A nearly uniform fuel/air mixture at the ignition point.
FIGURE 19.48 Swirl vanes also serve as the gas injectors, and provide the burner's near-perfect fuel/air mixing.
FIGURE 19.49 Outer sleeve contains a second set of gas injector vanes attached to an outer gas reservoir.
FIGURE 19.50 NOx emission from the ultra-low emissions burner firing into the firetube boiler for ambient, 300°F preheat, and 500°F preheat as a function of FGR rate.
FIGURE 19.51 Data from the application of an ultra-low emissions burner on a new 230,000 lb/hr "A" type Nebraska boiler.
FIGURE 19.52 The ultra-low emissions burner can also be used on two-burner applications where NFPA 8501 guidelines are being followed.
FIGURE 19.53 Atomizers are fitted onto oil "guns".
FIGURE 19.54 Steam assist, air blast, or other "external mix" atomizers.

CHAPTER 20

FIGURE 20.1 Typical early 1950s flare performance.
FIGURE 20.2 First successful smokeless flare.
FIGURE 20.3 Major flaring event.
FIGURE 20.4 Typical elevated single point flare.
FIGURE 20.5 Typical pit flare installation.
FIGURE 20.6 A grade-mounted, multi-point LRGO flare system.
FIGURE 20.7 Elevated multi-point LRGO flare system.
FIGURE 20.8 Multiple ZTOF installation in an ethylene plant.
FIGURE 20.9 Combination ZTOF and elevated flare system.
FIGURE 20.10 Comparison of the flame produced by burning (a) 25 MW well head natural gas, (b) propane, and (c) propylene.
FIGURE 20.11 Combination elevated LRGO and utility flare system.
FIGURE 20.12 General arrangement of a staged flare system, including a ZTOF and an elevated flare.
FIGURE 20.13 John Zink Co. test facility in Tulsa, Oklahoma.

FIGURE 20.14 Liquid carryover from an elevated flare. (a) Start of flaring event. (b) Liquid fallout and flaming rain from flare flame. (c) Flaming liquid engulfs flare stack.
FIGURE 20.15 Thermogram of a flare flame.
FIGURE 20.16 API radiation geometry.
FIGURE 20.17 Comparison of stack height and relative cost for various radiation calculation methods.
FIGURE 20.18 Effectiveness of steam in smoke suppression.
FIGURE 20.19 Effectiveness of air in smoke suppression.
FIGURE 20.20 Steamizer™ steam-assisted smokeless flare.
FIGURE 20.21 Typical nonassisted flare.
FIGURE 20.22 Zink double refractory (ZDR) severe service flare tip.
FIGURE 20.23 Simple steam-assisted flare.
FIGURE 20.24 Perimeter:area ratio as a function of tip size.
FIGURE 20.25 Schematic of an advanced steam-assisted flare.
FIGURE 20.26 A comparison of the perimeter:area ratio for simple and advanced steam-assisted flares.
FIGURE 20.27 State-of-the-art Steamizer™ flare burner and muffler.
FIGURE 20.28 Air assisted smokeless flare with two blowers in a refinery.
FIGURE 20.29 Annular air flare. (Courtesy of Shell Canada Ltd.)
FIGURE 20.30 Hydra flare burner in an offshore location.
FIGURE 20.31 LRGO staging sequence during a flaring event from inception (a) to full load (g)
FIGURE 20.32 Multi-point LRGO system with a radiation fence.
FIGURE 20.33 A RIMFIRE® endothermic flare.
FIGURE 20.34 OWB liquid flare test firing 150 gpm.
FIGURE 20.35 Forced draft Dragon liquid flare.
FIGURE 20.36 Poseidon flare: water-assisted Hydra.
FIGURE 20.37 Fundamental pilot parts.
FIGURE 20.38 Conventional flame front generator.
FIGURE 20.39 Slip stream flame-front generator.
FIGURE 20.40 Self inspirating flame-front generator.
FIGURE 20.41 SoundProof acoustic pilot monitor.
FIGURE 20.42 Horizontal settling drum at the base of an air assisted flare.
FIGURE 20.43 Cyclone separator.
FIGURE 20.44 Schematic of a vertical liquid seal.
FIGURE 20.45 "Smoke signals" from a surging liquid seal.
FIGURE 20.46 Various seal head designs.
FIGURE 20.47 Airrestor™ velocity-type purge reduction seal.
FIGURE 20.48 Molecular Seal density-type purge reduction seal.
FIGURE 20.49 Schematic of a ZTOF.
FIGURE 20.50 Self-supported flare.
FIGURE 20.51 Guy wire-supported flare.
FIGURE 20.52 Derrick supported flare.
FIGURE 20.53 Demountable derrick.
FIGURE 20.54 Flare support structure selection guide.
FIGURE 20.55 Steam control valve station.
FIGURE 20.56 Staging control valve assembly.
FIGURE 20.57 Loop seal.
FIGURE 20.58 Purge control station.
FIGURE 20.59 Geometry for dispersion calculations.

CHAPTER 21

FIGURE 21.1 Typical natural-draft burner.
FIGURE 21.2 Typical medium pressure drop burner.
FIGURE 21.3 Typical high pressure drop burner.
FIGURE 21.4 Typical horizontal system with a preheat exchanger.

FIGURE 21.5 Watertube boiler.

FIGURE 21.6 Firetube boiler.

FIGURE 21.7 Typical all-welded shell-and-tube heat exchanger.

FIGURE 21.8 Regenerative preheat exchanger.

FIGURE 21.9 Organic fluid transfer system configuration.

FIGURE 21.10 Vertical, down-flow conditioning section.

FIGURE 21.11 Direct spray contact quench.

FIGURE 21.12 Submerged quench.

FIGURE 21.13 Adjustable-plug venturi quench.

FIGURE 21.14 Baghouse.

FIGURE 21.15 Dry electrostatic precipitator.

FIGURE 21.16 Horizontal venturi scrubber.

FIGURE 21.17 Wet electrostatic precipitator.

FIGURE 21.18 Simple packed column.

FIGURE 21.19 Two-stage acid gas removal system.

FIGURE 21.20 Combination quench/two-stage acid removal system.

FIGURE 21.21 Three-stage NOx reduction process.

FIGURE 21.22 Two-stage NOx reduction process.

FIGURE 21.23 Selective noncatalytic reduction system.

FIGURE 21.24 Common catalyst configuration.

FIGURE 21.25a Fan wheel designs.

FIGURE 21.25b Radial blade operating curve for 1780 RPM, 70°F, and 0.075 lb/ft^3 density.

FIGURE 21.25c Forward tip blade operating curve for 1780 RPM, 70°F, and 0.075 lb/ft^3 density.

FIGURE 21.25d Backward curved blade operating curve for 1780 RPM, 70° F, and 0.075 lb/ft^3 density.

FIGURE 21.25e Outlet damper flow control.

FIGURE 21.25f Radial inlet damper/inlet box damper flow control.

FIGURE 21.25g Blower speed control.

FIGURE 21.26 Simple thermal oxidizer.

FIGURE 21.27 Thermal oxidizer system generating steam.

FIGURE 21.28 Heat recovery thermal oxidation system.

FIGURE 21.29 Bypass recuperative system.

FIGURE 21.30 Horizontal thermal oxidizer with firetube boiler and HCl removal system.

FIGURE 21.31 Vertical thermal oxidizer with 180° turn quench section.

FIGURE 21.32 Chlorine reaction equilibrium vs. operating temperature.

FIGURE 21.33 Molten salt system.

FIGURE 21.34 Online cleaning with soot blowers.

FIGURE 21.35 Three-stage NOx system with packed column scrubber.

About the Editor

Charles E. Baukal, Jr., Ph.D., P.E., is the Director of the John Zink Company, LLC R&D Test Center in Tulsa, OK. He previously worked for 13 years at Air Products and Chemicals, Inc., Allentown, PA, in the area of oxygen-enhanced combustion. He has more than 20 years of experience in the fields of heat transfer and industrial combustion and has authored more than 50 publications in those fields. He is the editor of the books *Oxygen-Enhanced Combustion* and *Computational Fluid Dynamics in Industrial Combustion*, the author of the book *Heat Transfer in Industrial Combustion*, and the general editor of the *Industrial Combustion* series, all with CRC Press, Boca Raton, FL. He has a Ph.D. in mechanical engineering from the University of Pennsylvania, is a licensed Professional Engineer in the state of Pennsylvania, has been an adjunct instructor at several colleges, and has eight U.S. patents. He is a member of several *Who's Who* compilations. He is a member of the American Society of Mechanical Engineers and The Combustion Institute.

Contributors

John Ackland is a Test Engineer at the John Zink Company, Tulsa. He has worked in the field of combustion for two years and has a bachelor's degree in chemical engineering from the University of Tulsa.

Eugene Barrington is retired as a Senior Staff engineer for Shell Oil Company. He had primary responsibility for the design, selection, application, and performance improvement of fired equipment. He has taught public courses on this topic for many years, publishing extensive notes. He has also published in *Hydrocarbon Processing and Chemical Engineering Progress*. Mr. Barrington was heavily involved in writing API specifications for both fired and unfired heat transfer equipment. He has an M.S. degree in engineering science from the University of California at Berkeley and is a registered Professional Engineer in Texas.

Peter Barry is the Director of Duct Burners for the John Zink Company, LLC, Tulsa, OK. He has a B.S. in Mechanical Engineering from Lehigh University.

Lawrence D. (Larry) Berg is a Senior Development Engineer at the John Zink Company, LLC, Tulsa, OK. He has over ten years' experience as a research and product development engineer for the company, and has a Masters Degree in Mechanical Engineering from MIT. He holds four U.S. patents, has co-authored five publications, and authored numerous internal technical documents.

Wes Bussman is Research and Development Engineer at the John Zink Company, LLC, Tulsa, OK. He has worked in the field of combustion and fluid dynamics and has a Ph.D. in Mechanical Engineering from the University of Tulsa. He has authored five publications and has two patents. Honors achieved include Kappa Mu Epsilon Mathematical Society and Sigma Xi Research Society.

I-Ping Chung, Ph.D., is a Development Engineer in Technology and Commercial Development Group at the John Zink Company, LLC, Tulsa, OK. She has worked in the field of atomization and sprays, spray combustion, and laser diagnosis in combustion and has a Ph.D. degree in Mechanical and Aerospace Engineering. She has authored 14 publications and has two patents. She is a registered Professional Engineer of Mechanical Engineering in California and Iowa.

Michael G. Claxton is a Senior Principal Engineer in the Burner Process Engineering Group of the John Zink Company, LLC, Tulsa, OK. He has a B.S. in Mechanical Engineering from the University of Tulsa and has worked for the John Zink Company in the field of industrial burners and combustion equipment since 1974. He has co-authored a number of papers and presentations covering combustion, combustion equipment, and combustion-generated emissions, and is co-holder of several combustion-related patents.

Joseph Colannino, P.E., is the Director of Engineering and Design at the John Zink Company, LLC, Tulsa, OK, and is a registered professional chemical engineer with more than 15 years of experience regarding combustion and combustion-related emissions. He has over 20 publications to his credit and is listed in *Who's Who in Science and Engineering*, *Who's Who in California*, and *Who's Who in Finance and Industry*. He is a member of the American Institute for Chemical Engineers and the American Chemical Society.

Terry Dark is the Coordinator of Engineering and Technology Programs at Oklahoma State University's Tulsa campus, where he is currently pursuing a Masters of Science in Engineering and Technology Management. Terry worked previously for the John Zink Company as a Combustion Test Engineer, focusing on product testing, burner development, and combustion safety. He is a 1998 graduate of Oklahoma State University's School of Chemical Engineering.

Joe Gifford is a Senior Engineer, Instrumentation and Control Systems, at the John Zink Company, LLC, Tulsa, OK. He has worked in the field of control and facilities design for 40 years and has a B.S. in Physics. For many years, he has conducted company training classes for Control Engineers and Technicians. He has received numerous awards for innovative control system designs throughout his career, including the General Electric Nuclear Energy Division's Outstanding Engineering award for systems design over a 15-year period. Technical society memberships have included the Pacific Association of General Electric Scientists and Engineers (PAGESE), Instrument Society of America (ISA), American Society of Mechanical Engineers (ASME), and the National Fire Protection Association (NFPA).

Karl Graham, Ph.D., is currently the Process Engineering Director at UniField Engineering, Inc., Billings, MT. Karl was formerly the Manager of Flare Design and Development at the John Zink Company, LLC, Tulsa, OK. Karl's work has been in the design, testing, engineering, and specification of flare, incineration, and gas cleaning equipment. He has a Ph.D. in chemical engineering from MIT, is a member of the

American Institute of Chemical Engineers and is an active participant in the American Petroleum Institute.

John Guarco is a Combustion Specialist at the John Zink Company, LLC, Shelton, CT. He has worked with burners for utility and industrial boiler applications for seven years and has a Masters of Science in Mechanical Engineering. He has authored eight publications and holds three patents.

Robert Hayes is a Test Engineer at the John Zink Company, LLC, Tulsa, OK. He has worked in the fields of combustion, heat transfer, and experimentation for three years. He has an M.S. in mechanical engineering from Brigham Young University.

Jim Heinlein is a Senior Controls Engineer at John Zink Company, LLC, Tulsa, OK, where he has been employed for 10 years. He has also worked in Nuclear Engineering, Computer Systems Design, and Low Observables Engineering. He is a member of ISA, IEEE, and Tau Beta Pi. He is also qualified as a Naval Surface Warfare Expert.

Michael Henneke, Ph.D., is a CFD Engineer at John Zink Company, LLC, Tulsa, OK. His academic background is in the area of reacting flow modeling and radiative transport. He holds a Ph.D. in mechanical engineering from The University of Texas at Austin. He has published three refereed journal papers, as well as many non-refereed articles and has given a number of presentations on computational fluid dynamic modeling of industrial combustion systems.

Jaiwant D. Jayakaran (Jay Karan) is Director, Burner Technology at the John Zink Company, LLC, Tulsa, OK. He has worked in the fields of combustion, petrochemicals, and power, with responsibilities in R&D, plant operations, and engineering. Jay has an M.S. in mechanical engineering. He has authored several technical articles and papers over the years, and has several patents pending.

Jeff Lewallen is an Account Manager at John Zink Company, LLC, Tulsa, OK. He has worked in the field of combustion for eight years. He graduated from the University of Tulsa in 1992 and holds a B.S. in mechanical engineering.

Michael Lorra, Ph.D., has been a CFD engineer for the John Zink Company, LLC, Tulsa, OK, since 1999. Previous to that, he worked at Gaswaerme Institut, Essen, Germany, e.V for eight years, where he also finished his Ph.D. He gained experience in NO_x reduction techniques, especially in reburning technology. He developed his own software code for the computation of turbulent reacting flow problems using laminar flamelet libraries.

Paul Melton is a Senior Principal Engineer for the Thermal Oxidation Systems Group at the John Zink Company, LLC, Tulsa, OK. He received a BSME from Oklahoma State University and has worked in the field of combustion for more than 25 years. His specialty is combustion of all types of waste materials and he has several publications in that area. He has also presented information, by invitation, to the Oklahoma Senate Select Committee on Waste Incineration. Mr. Melton has been a Registered Professional Engineer for more than 20 years.

Robert E. Schwartz, P.E., is a Vice President at John Zink Company, LLC, Tulsa, OK. He has worked in the fields of combustion, heat transfer, and fluid flow for 40 years and has an M.S. in mechanical engineering. Mr. Schwartz has been granted 50 U.S. patents, has had a number of articles published and has spoken at many national society meetings. He is a registered professional engineer, a member of ASME and AIChE, and an associate member of Sigma Xi. He was elected to The University of Tulsa Engineering Hall of Fame in 1991 and was a 1988 recipient of the University of Missouri Honor Award for Distinguished Service to Engineering. He is a member of the Mechanical Engineering Industrial Advisory Board, The University of Tulsa. He has been a John Zink Burner School instructor for 30 years and served as Director of the School for 10 years.

Prem C. Singh, Ph.D., is a Test Engineer at the John Zink Company, LLC, Tulsa, OK. He has worked in the fields of combustion, energy engineering, and transport phenomena for over 20 years and has a Ph.D. in chemical engineering. He has authored more than 50 publications and is a contributor to a book on coal technology published by Delft University of Technology. He has worked as a reviewer for the *Canadian Journal of Chemical Engineering and Industrial and Engineering Chemistry Research*, is a member of AIChE, ASME, Sigma XI, and the New York Academy of Sciences, and has been cited in *Marquis's Who's Who in Science and Engineering*. He has taught courses in chemical and mechanical engineering to undergraduate and graduate classes during his long tenure as a faculty /visiting faculty member.

Joseph D. Smith, Ph.D., is Director of Flare Technology and Computational Fluid Dynamics at the John Zink Company, LLC, Tulsa, OK. He has worked in the field of CFD for nearly twenty years and has a Ph.D. in chemical engineering. He has authored 27 peer-reviewed publications, 16 invited lectures, 19 conference papers, two patents, and has organized and directed three special symposia. As a member of the American Institute of Chemical Engineering, he has served as National Chair of the Student Chapters Committee and as Local Chair of the Mid-Michigan AIChE section. Research topics include reaction engineering and turbulent reactive flow simulation. He has taught undergraduate and graduate courses in chemical engineering at the University of Michigan and at the University of Illinois/Urbana–Champaign.

Stephen L. Somers is a Senior Process Engineer at the John Zink Company, LLC, Tulsa, OK. He has 32 years of experience in combustion and process design with 15 years in sales and design of duct burners for supplementary firing. He has an M.S. in chemical engineering from the University of Oklahoma.

Lev Tsirulnikov, Ph.D., is a Senior Research Engineer for the John Zink Company, LLC, Shelton, CT. He has developed low-emission combustion technologies, burners, and other equipment for gas/oil-fired utility and industrial boilers. He has a Ph.D. in mechanical engineering. He holds 47 patents and has published more than 100 technical papers, including four books in the combustion/boiler field.

Richard T. Waibel, Ph.D., is a Senior Principal Engineer in the Burner Process Engineering Group at the John Zink Company, LLC, Tulsa, OK. He works in the field of burner design and development and has a doctorate in fuel science from The Pennsylvania State University. He has authored over 70 technical papers, publications, and presentations. Dr. Waibel has been the Chairman of the American Flame Research Committee since 1995.

Timothy Webster is the Sales Manager for New Technologies at the John Zink Company, LLC, Shelton, CT. He has worked in the field of industrial combustion for seven years and has a B.S. in mechanical engineering. He has authored three publications and is a licensed professional mechanical engineer in California.

Jeff White is the Senior Flare Design Consultant at the John Zink Company, LLC, Tulsa, OK. He has worked in the field of Flare System Design at John Zink Company for 19 years. He has an M.S. in mechanical engineering from The University of Texas at Austin. He has published two articles, one on flare radiation methods and the other on flow measurement by ASME nozzles.

Roger H. Witte is Director of Marketing and Sales-End Users at the John Zink Company, LLC, Tulsa, OK. He has worked at John Zink for 28 years in the development and application of Zink equipment and at Conoco for seven years in their refining and chemical operations. He is also Director of the John Zink Burner School which teaches the art of combustion. Roger has a B.S. in refining and chemical engineering from the Colorado School of Mines and has 35 years of experience in the fields of combustion and operations of combustion equipment. Roger is a member of the Tulsa Engineering Society.

Table of Contents

Prologue xxxvii

Chapter 1: Introduction 3
Charles E. Baukal, Jr.

Chapter 2: Fundamentals 33
Joseph Colannino and Charles E. Baukal, Jr.

Chapter 3: Heat Transfer 69
Prem Singh, Michael Henneke, Jaiwant D. Jayakaran, Robert Hayes, and Charles E. Baukal, Jr.

Chapter 4: Fundamentals of Fluid Dynamics 117
Lawrence D. Berg, Wes Bussman, and Michael Henneke

Chapter 5: Fuels 157
Terry Dark, John Ackland, and Jeff White

Chapter 6: Pollutant Emissions 189
Charles E. Baukal, Jr. and Joseph Colannino

Chapter 7: Noise 223
Wes Bussman and Jaiwant D. Jayakaran

Chapter 8: Mathematical Modeling of Combustion Systems 251
Lawrence D. Berg, Wes Bussman, Michael Henneke, and I-Ping Chung

Chapter 9: CFD Based Combustion Modeling 287
Michael Henneke, Joseph D. Smith, Michael Lorra, and Jaiwant D. Jayakaran

Chapter 10: Combustion Safety 327
Terry Dark and Charles E. Baukal, Jr.

Chapter 11: Burner Design 351
Richard T. Waibel and Michael Claxton

Chapter 12: Combustion Controls 372
Joe Gifford and Jim Heinlein

Chapter 13: Experimental Design for Combustion Equipment 401
Joseph Colannino

Chapter 14: Burner Testing 431
Jeffrey Lewallen, Robert Hayes, Prem Singh, and Richard T. Waibel

Chapter 15: Installation and Maintenance 449
Roger H. Witte and Eugene A. Barrington

Chapter 16: Burner/Heater Operations 469
Roger H. Witte and Eugene A. Barrington

Chapter 17 Troubleshooting 501
Roger H. Witte and Eugene A. Barrington

Chapter 18: Duct Burners 523
Peter F. Barry and Stephen L. Somers

Chapter 19: Boiler Burners 547
Lev Tsirulnikov, John Guarco and Timothy Webster

Chapter 20 Flares 589
Robert Schwartz, Jeff White, and Wes Bussman

Chapter 21: Thermal Oxidizers 637
Paul Melton and Karl Graham

Appendices
Appendix A: Physical Properties of Materials 695
Appendix B: Properties of Gases and Liquids 715
Appendix C: Common Conversions 725

Index 729

Prologue

Fred Koch and John Zink Pioneers in the Petroleum Industry

Fred Koch

John Zink

The early decades of the twentieth century saw the birth and growth of the petroleum industry in Oklahoma. Drilling derricks sprouted like wildflowers throughout the state, making it among the top oil producers in the nation and Tulsa the "Oil Capital of the World" by the 1920s.

Refining operations accompanied oil production. Many of the early refineries were so small that today they would be called pilot plants. They were often merely topping processes, skimming off natural gasoline and other light fuel products and sending the remainder to larger refineries with more complex processing facilities.

Along with oil, enough natural gas was found to make its gathering and sale a viable business as well. Refineries frequently purchased this natural gas to fuel their boilers and process heaters. At the same time, these refineries vented propane, butane, and other light gaseous hydrocarbons into the atmosphere because their burners could not burn them safely and efficiently. Early burner designs made even natural gas difficult to burn as traditional practice and safety concerns led to the use of large amounts of excess air and flames that nearly filled the fire box. Such poor burning qualities hurt plant profitability.

Among firms engaged in natural gas gathering and sales in the northeastern part of the state was Oklahoma Natural Gas Company (ONG). It was there that John Steele Zink, after completing his studies at the University of Oklahoma in 1917, went to work as a chemist. Zink's chemistry and engineering education enabled him to advance to the position of manager of industrial sales. But while the wasteful use of natural gas due to inefficient burners increased those sales, it troubled Zink, and awakened his talents first as an innovator and inventor, and then as an entrepreneur.

Seeing the problems with existing burners, Zink responded by creating one that needed less excess air and produced a

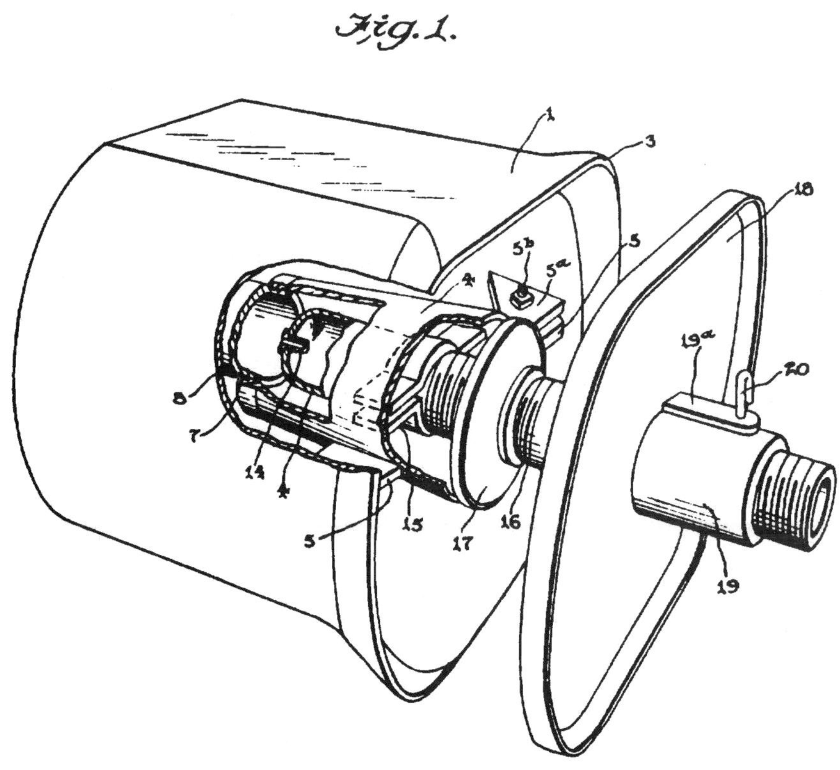

compact, well-defined flame shape. A superior burner for that era, it was technically a premix burner with partial primary air and partial draft-induced secondary air. The use of two airflows led to its trade name, BI-MIX®. The BI-MIX® burner is shown in a drawing from one of Zink's earliest patents.

ONG showed no interest in selling his improved burners to its customers, so in 1929 Zink resigned and founded Mid-Continent Gas Appliance Co., which he later renamed the John Zink Company.

Zink's BI-MIX® burner was the first of many advances in technology made by his company, which to date has seen almost 300 U.S. patents awarded to nearly 80 of its employees. He carried out early manufacturing of the burner in the garage of his Tulsa-area home and sold it from the back of his automobile as he traveled the Oklahoma oil fields, generating the money he needed to buy the components required to fabricate the new burners.

The novel burners attracted customers by reducing their fuel costs, producing a more compact flame for more efficient heater operation, burning a wide range of gases, and generally being safer to use. Word of mouth among operators helped spread their use throughout not only Oklahoma but, by the late 1930s, to foreign refineries as well.

Growth of the company required Zink to relocate his family and business to larger facilities on the outskirts of Tulsa. In 1935, he moved into a set of farm buildings on Peoria Avenue, a few miles to the south of the city downtown, a location Zink thought would allow for plenty of future expansion.

As time passed, Zink's company became engaged in making numerous other products, sparked by its founder's beliefs in customer service and solving customer problems. After World War II, Zink was the largest sole proprietorship west of the Mississippi. Zink's reputation for innovation attracted customers who wanted new burners and, eventually, whole new families of products. For example, customers began asking for reliable pilots and pilot igniters, when atmospheric venting of waste gases and emergency discharges was replaced by combustion in flares in the late 1940s. This in turn was followed by requests for flare burners and finally complete flare systems, marking the start of the flare equipment industry. Similar customer requests for help in dealing with gas and liquid waste streams and hydrocarbon vapor led the Zink Company to become a major supplier of gas and liquid waste incinerators and also of hydrocarbon vapor recovery and other vapor control products.

Mr. Zink's great interest in product development and innovation led to the construction of the company's first furnace for testing burners. This furnace was specially designed to simulate the heat absorption that takes place in a process heater. Zink had the furnace built in the middle of the employee parking lot, a seemingly odd placement. He had good reason for this because he wanted his engineers to pass the test furnace every day as they came and went from work as a reminder of the importance of product development to the Company's success.

Zink went beyond encouraging innovation and motivating his own employees. During the late 1940s, Zink and his technical team leader, Robert Reed (who together with Zink developed the first smokeless flare) sensed a need for an industry-wide meeting to discuss technologies and experiences associated with process heating. In 1950, they hosted the first of four annual Process Heating Seminars in Tulsa. Interest in the seminars was high, with the attendance level reaching 300. Attendees of the first Process Heating Seminar asked Zink and Reed to conduct training sessions for their operators and engineers. These training sessions, which combined lectures and practical hands-on burner operation in Zink's small research and development center, were the start of the John Zink Burner School®. The year 2000 marks the fiftieth anniversary of the original seminar and the fiftieth year in which the Burner School has been offered. Over the years, other schools were added to provide customer training in the technology and operation of hydrocarbon vapor recovery systems, vapor combustors, and flares.

Included among the 150 industry leaders attending the first seminar was Harry Litwin, former President and part owner of Koch Engineering Co., now part of Koch Industries of Wichita, Kansas. Litwin was a panelist at the closing session. Koch Engineering was established in 1943 to provide engineering services to the oil refining industry. In the early 1950s it developed an improved design for distillation trays and because of their commercial success the company chose to exit the engineering business. Litwin left Koch at that time and set up his own firm, the Litwin Engineering Co., which grew into a very sizeable business.

During the same period that John Zink founded his business, another talented young engineer and industry innovator, Fred C. Koch, was establishing his reputation as an expert in oil processing. The predecessor to Koch Engineering Co. was the Winkler–Koch Engineering Co., jointly owned by Fred Koch with Lewis Winkler, which designed processing units for oil refineries. Fred Koch had developed a unique and very successful thermal cracking process which was sold to many independent refineries throughout the United States, Europe, and the former Soviet Union. One of the first of these processing units was installed in a refinery in Duncan, Oklahoma, in 1928, one year before Zink started his own company.

While the two men were not personally acquainted, Koch and Zink's companies knew each other well in those early years. Winkler–Koch Engineering was an early customer for Zink burners. The burners were also used in the Wood River refinery in Hartford, Illinois. Winkler–Koch constructed this refinery in 1940 with Fred Koch as a significant part owner and the head of refining operations. Winkler–Koch Engineering, and later Koch Engineering, continued to buy Zink burners for many years.

Fred Koch and two of his sons, Charles and David, were even more successful in growing their family business than were Zink and his family. When the Zink family sold the John Zink Company to Sunbeam Corporation[2] in 1972, the company's annual revenues were US $15 million. By that time Koch Industries, Inc., the parent of Koch Engineering, had revenues of almost US $1 billion. Since then Koch has continued to grow, its revenues in the year 1999 were over US $30 billion.

When the John Zink Company was offered for sale in 1989, its long association with Koch made Koch Industries a very interested bidder. Acting through its Chemical Technology Group, Koch Industries quickly formed an acquisition team, headed by David Koch, which succeeded in purchasing the John Zink Company.

Koch's management philosophy and focus on innovation and customer service sparked a new era of revitalization and expansion for the John Zink Company. Koch recognized that the Peoria Avenue research, manufacturing, and office facilities were outdated. The growth of Tulsa after World War II had made Zink's facilities an industrial island in the middle of a residential area. The seven test furnaces on Peoria Avenue at the time of the acquisition, in particular, were cramped, with such inadequate infrastructure and obsolete instrumentation they could not handle the sophisticated research and development required for modern burners.

A fast-track design and construction effort by Koch resulted in a new office and manufacturing complex in the northeast sector of Tulsa and was completed at the end of 1991. In addition, a spacious R&D facility adjacent to the new office and manufacturing building replaced the Peoria test facility.

The initial multimillion dollar investment in R&D facilities included an office building housing the R&D staff and support personnel, a burner prototype fabrication shop, and an indoor laboratory building. Additional features included steam boil-

[2] Sunbeam Corporation was primarily known as an appliance maker. Less well known was Sunbeam's group of industrial specialty companies such as John Zink Company.

ers, fuel storage and handling, data gathering centers, and measurement instrumentation and data logging for performance parameters from fuel flow to flue gas analysis.

Koch has repeatedly expanded the R&D facility. When the new facility began testing activities in 1992, nine furnaces and a multipurpose flare testing area were in service. Today, there are 14 outdoor test furnaces and two indoor research furnaces. Control systems are frequently updated to keep them state of the art.

Zink is now able to monitor and control burner tests from an elevated Control Center that has a broad view of the entire test facility. The Control Center includes complete automation of burner testing with live data on control panels and flame shape viewing on color video monitors. Fuel mixtures and other test parameters can be varied remotely from the control panels inside the Control Center. Up to four separate tests in four different furnaces can be conducted and monitored simultaneously.

A new flare testing facility is under construction at the time of this writing to dramatically expand and improve Zink's capabilities. This project represents the company's largest single R&D investment since the original construction of the R&D facility in 1991. The new facilities will accommodate the firing of a wide variety of fuel blends (propane, propylene, butane, ethylene, natural gas, hydrogen, and diluents such as nitrogen and carbon dioxide) to reproduce or closely simulate a customer's fuel composition. Multiple cameras will provide video images along with the electronic monitoring and recording of a wide range of flare test data, including noise emissions. The facility will be able to test all varieties of flare systems with very large sustained gas flow rates at or near those levels which customers will encounter in the field. Indeed, flow capacity will match or exceed the smokeless rate of gas flow for virtually all customers' industrial plants, giving the new flare facility a capability unmatched in the world.

These world-class test facilities are staffed with engineers and technicians who combine theoretical training with practical experience. They use the latest design and analytical tools, such as Computational Fluid Dynamics, physical modeling, and a Phase Doppler Particle Analyzer. The team can act quickly to deliver innovative products that work successfully, based on designs which can be exactly verified before the equipment is installed in the field.

Koch's investment in facilities and highly trained technical staff carries on the tradition John Zink began more than 70 years ago: providing our customers today, as he did in his time, with solutions to their combustion needs through better products, applications, information, and service.

Robert E. Schwartz
October, 2000
Tulsa, Oklahoma

Chapter 1
Introduction

Charles E. Baukal, Jr.

TABLE OF CONTENTS

1.1 Process Industries 2
 1.1.1 Hydrocarbon and Petrochemical Industries 2
 1.1.2 Power Generation Industry 3
 1.1.3 Thermal Oxidation 5
1.2 Literature Review 5
 1.2.1 Combustion 5
 1.2.2 The Process Industries 6
 1.2.3 Combustion in the Process Industries 7
1.3 Fired Heaters 7
 1.3.1 Reformers 7
 1.3.2 Process Heaters 9
1.4 Burners 12
 1.4.1 Competing Priorities 13
 1.4.2 Design Factors 14
 1.4.3 General Burner Types 16
References 22

FIGURE 1.1 Typical petroleum refinery.

1.1 PROCESS INDUSTRIES

Process industries encompass the production of a wide range of products like fuels (e.g., oil and natural gas), glass, metals (e.g., steel and aluminum), minerals (e.g., refractories, bricks, and ceramics), and power, to name a few. The treatment and disposal of waste materials is another example of a process industry. In this book, only a few of these are considered and briefly discussed. The main focus of the book is on the hydrocarbon, petrochemical, power generation, and thermal oxidation industries.

1.1.1 Hydrocarbon and Petrochemical Industries

The hydrocarbon and petrochemical industries present unique challenges to the combustion engineer, compared to other industrial combustion processes. One of the more important challenges in these industries is the wide variety of fuels, which are usually off-gases from the petroleum refining processes that are used in a typical plant (see Figure 1.1). This differs significantly from most other industrial combustion systems that normally fire a single purchased fuel such as natural gas or fuel oil. Another important challenge is that many of the burners commonly used in the hydrocarbon and petrochemical industries are natural draft, where the buoyant combustion exhaust products create a draft that induces the combustion air to enter the burners. This is different from nearly all other industrial combustion processes, which utilize a combustion air blower to supply the air used for combustion in the burner. Natural draft burners are not as easy to control as forced draft burners, and are subject to things like the wind, which can disturb the conditions in a process heater.

According to the U.S. Dept. of Energy, petroleum refining is the most energy-intensive manufacturing industry in the U.S., accounting for about 7% of total U.S. energy consumption in 1994.[1] Table 1.1 shows the major processes in petroleum refining, most of which require combustion in one form or another. Figure 1.2 shows the process flow through a typical refinery.

The U.S. Dept. of Energy Office of Industrial Technologies has prepared a Technology Roadmap for industrial combustion.[2] For process heating systems, some key performance targets for the year 2020 have been identified for burners and for the overall system. For the burners, the targets include reducing criteria pollutant emissions by 90%, reducing CO_2 emissions to levels agreed upon by the international community, reducing specific fuel consumption by 20 to 50%, and maximizing the ability to use multiple fuels. For the heating system, the targets include reducing the total cost of combustion in manufacturing, enhancing system integration, reducing product loss rate by 50%, maximizing system robustness, and zero accidents. The following were identified as top-priority R&D needs in process heating: a burner capable of adjusting operating parameters in real time, advanced burner stabilization methods, robust design tools, and economical methods to premix fuel and air. The following were also identified as top-priority R&D needs in process heating: new furnace designs, advanced sensors, cost-effective heat recovery processes, and new methods to generate heat without environmental impact. Both the burners and the process heaters are considered in a number of chapters within this book.

Flares (see Figure 1.3) are used to dispose of unwanted gases or liquids. Usually, liquids are separated from the gas and burned in the liquid state or vaporized and burned as a gas. The unwanted material is generally composed of hydrocarbon gases, but may include hydrogen, carbon monoxide, hydrogen sulfide, certain other combustible gases, or some amount of an inert gas such as nitrogen or carbon dioxide. There are several conditions that may require flaring. The largest flaring events occur during emergency pressure relieving conditions associated with a sudden unavoidable failure, such as loss of electrical power, loss of cooling water, fire, or the like. Flaring also occurs when gases are vented in order to maintain control of a process or during start-up or shutdown of a plant. Yet another cause of flaring is the disposal of unwanted gases. Examples include unmarketable natural gas that is co-produced with oil and off-specification gases produced from a process. Whatever the reason, flares must reliably combust gases whenever they are called upon. One of the challenges for flares is initiating and maintaining ignition. Initiation of burning requires a pilot that can withstand high winds, rain, and inert surroundings. Another challenge is an extremely wide turndown ratio because of the variety of venting conditions. Environmental challenges include minimizing smoke, radiant head load and noise, and maximizing combustion efficiency. Flares are covered in Chapter 20.

1.1.2 Power Generation Industry

This book contains chapters on duct burners (Chapter 18) and boiler burners (Chapter 19) used in the power generation industries. Duct burners (see a typical flame in Figure 1.4) are burners that are inserted into large ducts (see Figure 1.5) to boost the temperature of the gases flowing through the ducts. These burners are frequently used in co-generation projects, electrical utility peaking stations, repowering programs, and in industrial mechanical driver systems employing gas turbines with site requirements for steam. They are also used in fluidized bed combustors and chemical process plants. The efficiency of a duct burner to supply additional heat approaches 95%, which is much higher than, for example, a backup boiler system in generating more steam. Duct burners are often easily retrofitted into existing ductwork. Several important factors in duct burner applications include low pollutant emissions, safe operation, uniform heat distribution from the duct burners to the gases flowing through the duct, getting uniform gas distribution through the duct burners, and having adequate turndown to meet fluctuating demands. Duct burners typically use gaseous fuels, but occasionally fire on oil.

Boiler burners (see Figure 1.6) are used to combust fuels, commonly natural gas or fuel oil, in the production of steam, which is often used to produce electrical energy for power generation. These burners produce radiation and convection used to heat water flowing through the boiler. The water is vaporized into steam. Sometimes, the steam is used in the plant in the case of smaller industrial boilers. Larger utility boilers produce steam to drive turbines for electrical energy production. While boiler burners have been around for many years, there have been many design changes in recent years

TABLE 1.1 Major Petroleum Refining Processes

Category	Major Process
Topping (separation of crude oil)	Atmospheric distillation
	Vacuum distillation
	Solvent deasphalting
Thermal and catalytic cracking	Delayed coking
	Fluid coking/flexicoking
	Visbreaking
	Catalytic cracking
	Catalytic hydrocracking
Combination/rearrangement of hydrocarbon	Alkylation
	Catalytic reforming
	Polymerization
	Isomerization
	Ethers manufacture
Treating	Catalytic hydrotreating/hydroprocessing
	Sweetening/sulfur removal
	Gas treatment
Specialty product manufacture	Lube oil
	Grease
	Asphalt

Source: From the U.S. Dept. of Energy.[1]

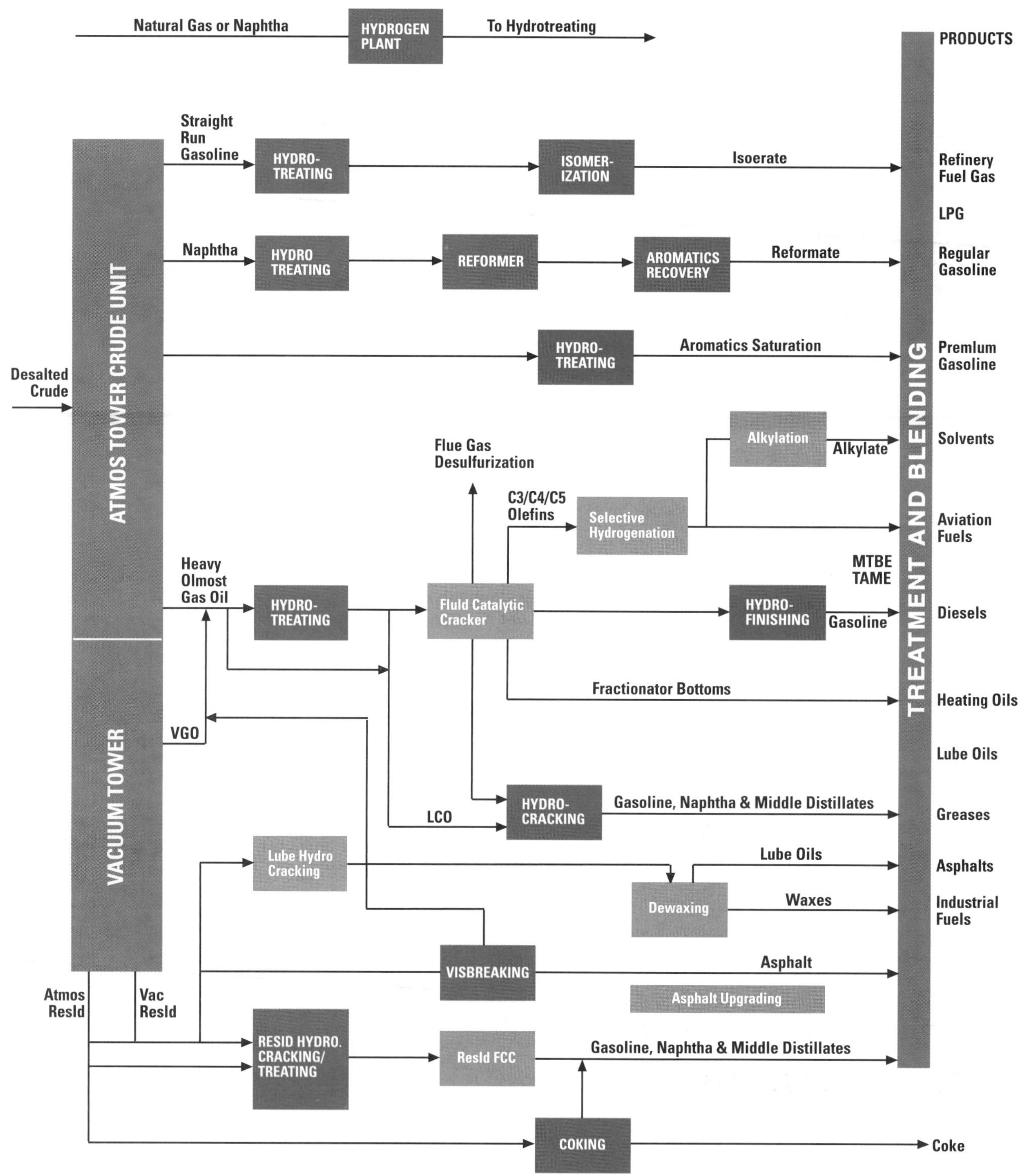

FIGURE 1.2 Typical refinery flow diagram. (From the U.S. Dept. of Energy.[1])

FIGURE 1.3 Offshore oil rig flare.

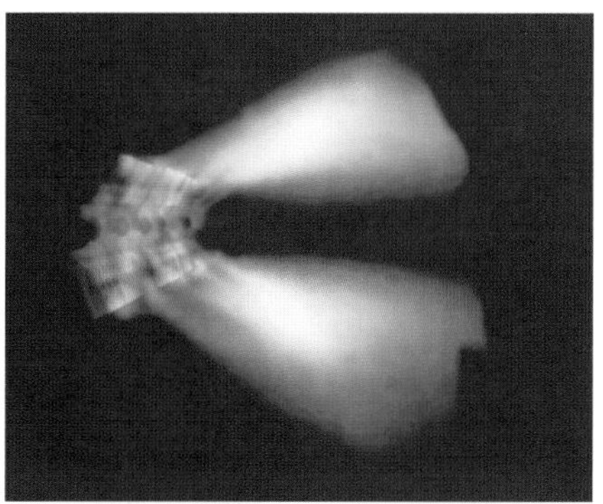

FIGURE 1.4 Duct burner flame.

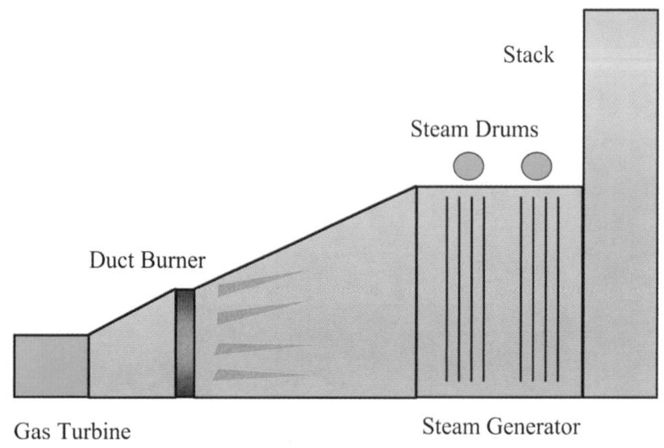

FIGURE 1.5 Duct burner in large duct.

FIGURE 1.6 Front of a boiler burner.

due to current emphasis on minimizing pollutant emissions. These burners are discussed in detail in Chapter 19.

1.1.3 Thermal Oxidation

Thermal oxidizers (see Figure 1.7) are used to treat unwanted by-product materials that may be solids, liquids, or gases. The composition of the by-products varies widely and may range from minute quantities of a contaminant up to 100%. These by-products come from a variety of industrial processes and often have some heating value, which aids in their thermal treatment.

There are often many options to choose from to eliminate the by-product materials. While the most preferable is recycling where the by-products are reused in the process, this is not always an option in some processes. Land-filling may be an option for some of the solid waste materials. However, it is often preferable to completely destroy the waste in an environmentally safe way. Many other methods are possible, but thermal treatment is often the most economical and effective. The waste products must be treated in a way that any emissions from the treatment process must be below regulatory limits. Thermal oxidation is discussed in Chapter 21.

1.2 LITERATURE REVIEW

Numerous books are available on the subjects of both combustion and the process industries considered here. However, few books have been written on the combination of the two. This section briefly surveys some of the relevant literature on the subjects of combustion, the process industries, and the combination of combustion in those industries. Most of these combustion books were written at a highly technical level for use in upper-level undergraduate or graduate-level courses. The books typically have broad coverage, with less emphasis on practical applications due to the nature of their target audience.

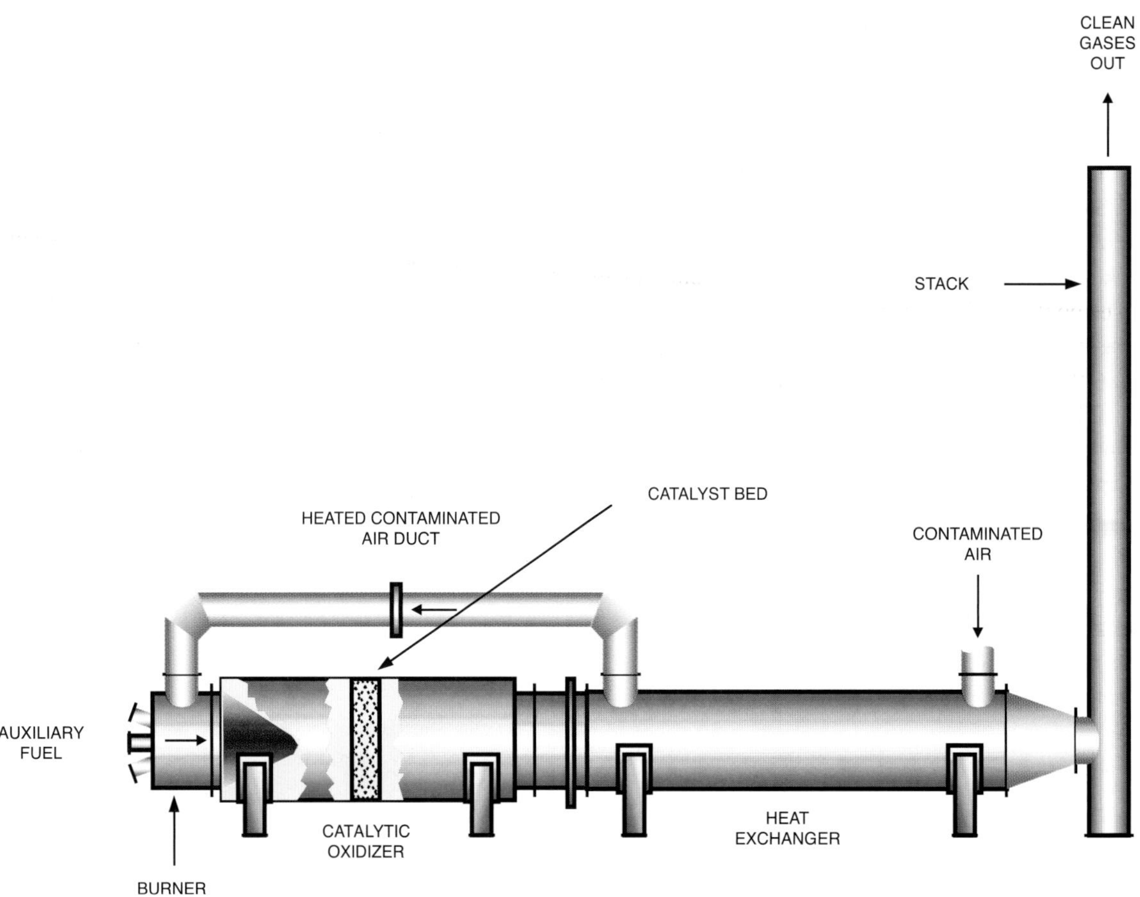

FIGURE 1.7 Thermal oxidizer.

1.2.1 Combustion

Many good textbooks are available on the fundamentals of combustion, but have little if anything on the hydrocarbon process and petrochemical industries.[3–8] A recent book by Turns (1996), designed for undergraduate and graduate combustion courses, contains more discussions of practical combustion equipment than most similar books.[9] Khavkin (1996) has written a book that combines theory and practice on gas turbines and industrial combustion chambers.[10] Of relevance here, the Khavkin book has a discussion of tube furnaces used in hydrogen production.

There have also been many books written on the more practical aspects of combustion. Griswold's (1946) book has a substantial treatment of the theory of combustion, but is also very practically oriented and includes chapters on gas burners, oil burners, stokers and pulverized-coal burners, heat transfer (although brief), furnace refractories, tube heaters, process furnaces, and kilns.[11] Stambuleanu's (1976) book on industrial combustion has information on actual furnaces and on aerospace applications, particularly rockets.[12] There is much data in the book on flame lengths, flame shapes, velocity profiles, species concentrations, and liquid and solid fuel combustion, with a limited amount of information on heat transfer. Perthuis' (1983) book has significant discussions of flame chemistry, and some discussion of heat transfer from flames.[13] Keating's (1993) book on applied combustion is aimed at engines and has no treatment of industrial combustion processes.[14] A recent book by Borman and Ragland (1998) attempts to bridge the gap between the theoretical and practical books on combustion.[15] However, the book has little discussion regarding the types of industrial applications considered here. Even handbooks on combustion applications have little if anything on industrial combustion systems.[16–20] The *Furnace Operations* book by Robert Reed is the only one that has any significant coverage of combustion in the hydrocarbon and petrochemical industries. However, this book was last updated in 1981 and is more of an introductory book with few equations, graphs, figures, pictures, charts, and references.

1.2.2 The Process Industries

Anderson (1984) has written a general introductory book on the petroleum industry, tracing its development from the beginning up to some projections for the future of oil.[21] There is no specific discussion of combustion in petroleum refining. Leffler (1985) has written an introductory book on the major processes in petroleum refining, including catalytic cracking, hydrocracking, and ethylene production, among many others.[22] The book is written from an overall process perspective and has no discussion of the heaters in a plant. Gary and Handwerk (1994) have written a good overview of petroleum refining.[23] The book discusses many of the processes involved in petroleum refining operations, including coking, catalytic cracking, and catalytic reforming, among others. However, it does not specifically discuss the combustion processes involved in heating the refinery fluids.

Meyers (1997) has edited a recently updated handbook on petroleum refining processes.[24] The book is divided into 14 parts, each on a different type of overall process, including catalytic cracking and reforming, gasification and hydrogen production, hydrocracking, and visbreaking and coking, among others. Each part is further divided into the individual subtypes and variations of the given overall process. Companies such as Exxon, Dow-Kellogg, UOP, Stone and Webster, and Foster-Wheeler have written about the processes they developed, which they license to other companies. Many aspects of the processes are discussed, including flow diagrams, chemistry, thermodynamics, economics, and environmental considerations, but there is very little discussion of the combustion systems.

1.2.3 Combustion in the Process Industries

The standard book on the subject of combustion in the hydrocarbon and petrochemical industries that has been used for decades is *Furnace Operations* by Robert Reed, formerly the chief technical officer of John Zink.[17] This book has been used in the John Zink Burner School for generations and gives a good introduction to many of the subjects important in burner and heater operation. However, it is somewhat outdated, especially with regard to pollution regulations and new trends in burner designs. The present book is designed to be a greatly expanded version of that book, with many more equations, figures, tables, references, and much wider coverage.

1.3 FIRED HEATERS

Fired or tubestill heaters are used in the petrochemical and hydrocarbon industries to heat fluids in tubes for further processing. In this type of process, fluids flow through an array of tubes located inside a furnace or heater. The tubes are heated by direct-fired burners that often use fuels that are by-products from processes in the plant and that vary widely in composition.

Using tubes to contain the load is somewhat unique compared to the other types of industrial combustion applications. It was found that heating the fluids in tubes has many advantages over heating them in the shell of a furnace.[25] Advantages include better suitability for continuous operation, better controllability, higher heating rates, more flexibility, less chance of fire, and more compact equipment.

One of the problems encountered in refinery-fired heaters is an imbalance in the heat flux in the individual heater passes.[26] This imbalance can cause high coke formation rates and high tube metal temperatures, which reduce a unit's capacity and can cause premature failures (see Chapter 10). Coke formation on the inside of the heater tubes reduces the heat transfer through the tubes, which leads to the reduced capacity. One cause of coking is flame impingement directly on a tube, which causes localized heating and increases coke formation there (see Chapter 17). This flame impingement can be caused by operating without all of the burners in service, insufficient primary or secondary air to the burner, operating the heater at excessive firing rates, fouled burner tips, eroded burner tip orifices, or insufficient draft. The problem of flame impingement shows the importance of proper design[27] to ensure even heat flux distribution inside the fired heater.

Recently, the major emphasis has been on increasing the capacity of existing heaters rather than installing new heaters. The limitations of overfiring a heater include:

- high tube metal temperatures
- flame impingement causing high coke formation rates
- positive pressure at the arch of the heater
- exceeding the capacity of induced-draft and forced-draft fans
- exceeding the capacity of the process fluid feed pump

Garg (1988) noted the importance of good heater specifications to ensure suitable performance for a given process.[28] Some of the basic process conditions needed for the specifications include heater type (cabin, vertical cylindrical, etc.), number of fluid passes, the tube coil size and material, fluid data (types, compositions, properties, and flow rates), heat duty required, fuel data (composition, pressure, and temperature), heat flux loading (heat flux split between the radiant and convection sections), burner data (number, type, arrangement, etc.), draft requirement, required instrumentation, as well as a number of other details such as the number of peepholes, access doors, and platforms.

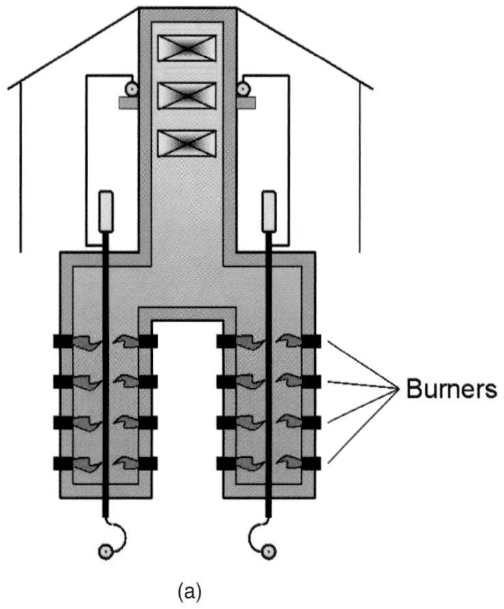

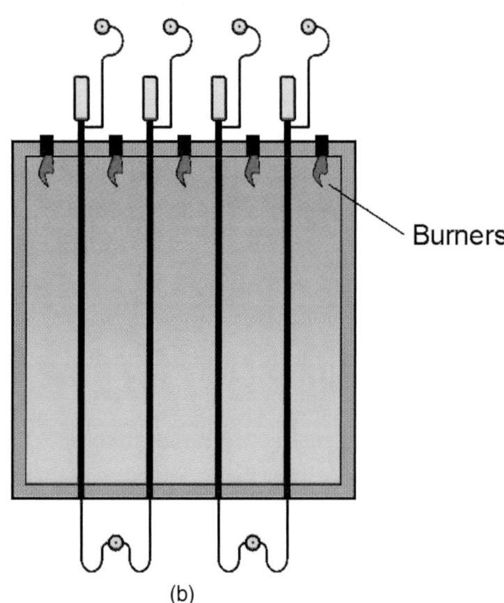

FIGURE 1.8 Side- (a) and top-fired (b) reformers (elevation view).[30]

1.3.1 Reformers

As the name indicates, reformers are used to reformulate a material into another product. For example, a hydrogen reformer takes natural gas and reformulates it into hydrogen in a catalytic chemical process that involves a significant amount of heat. A sample set of reactions is given below for converting propane to hydrogen[29]:

$$C_3H_8 \rightarrow C_2H_4 + CH_4$$
$$C_2H_4 + 2H_2O \rightarrow 2CO + 4H_2$$
$$CH_4 + H_2O \rightarrow CO + 3H_2$$
$$CO + H_2O \rightarrow CO_2 + H_2$$

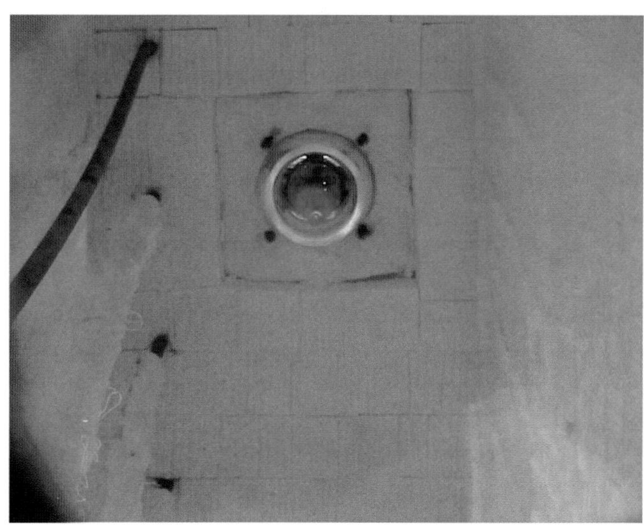

FIGURE 1.9 Down-fired burner commonly used in top-fired reformers.

The reformer is a direct-fired combustor containing numerous tubes, filled with catalyst, inside the combustor.[30] The reformer is heated with burners, firing either vertically downward or upward, with the exhaust on the opposite end, depending on the specific design of the unit. The raw feed material flows through the catalyst in the tubes which, under the proper conditions, converts that material to the desired end-product. The burners provide the heat needed for the highly endothermic chemical reactions. The fluid being reformulated typically flows through a reformer combustor containing many tubes (see Figure 1.8). The side-fired reformer has multiple burners on the side of the furnace with a single row of tubes centrally located. The heat is transferred primarily by radiation from the hot refractory walls to the tubes. Top-fired reformers have multiple rows of tubes in the firebox. In that design, the heat is transferred primarily from radiation from the flame to the tubes. Figure 1.9 shows a down-fired burner commonly used in top-fired reformers. In a design sometimes referred to as terrace firing, burners may be located in the side wall but be firing up the wall at a slight angle (see Figure 1.10). Foster Wheeler uses terrace wall reformers in the production of hydrogen by steam reformation of natural gas or light refinery gas.[31]

The reformer tubes are a critical element in the overall design of the reformer. Because they operate at pressures up to 350 psig (24 barg), they are typically made from a high-temperature and -pressure nickel alloy like inconel to ensure that they can withstand the operating conditions inside the reformer. Failure of the tubes can be very expensive because of the downtime of the unit, lost product, damaged catalyst, and, possibly, damaged reformer. New reactor technologies are being developed to improve the process for converting natural gas to precursor synthesis gas (syngas).[32]

1.3.2 Process Heaters

Process heaters are sometimes referred to as process furnaces or direct-fired heaters. They are heat transfer units designed to heat petroleum products, chemicals, and other liquids and gases flowing through tubes. Typical petroleum fluids include gasoline, naphtha, kerosene, distillate oil, lube oil, gas oil, and light ends.[33] The heating is done to raise the temperature of the fluid for further processing downstream or to promote chemical reactions in the tubes, often in the presence of a catalyst. Kern noted that refinery heaters can carry liquids at temperatures as high as 1500°F (810°C) and pressures up to 1600 psig (110 barg). The primary modes of heat transfer in process heaters are radiation and convection. The initial part of the fluid heating is done in the convection section of the furnace, while the latter heating is done in the radiant section (see Figure 1.11). Each section has a bank of tubes in it where the fluids flow through, as shown in Figure 1.12.[34] Early heater designs had only a single bank of tubes that failed prematurely because designers did not understand the importance of radiation on the process.[25] The tubes closest to the burners would overheat. Overheating caused the hydrocarbons to form coke inside the tube. The coke further aggravated the problem by reducing the thermal conductivity through the coke layer inside the tube. The reduced thermal conductivity prevented the process fluids from absorbing adequate heat to cool the tubes, resulting in overheating and failure of the tubes. One of the key challenges for the heater designer is to get even heat distribution inside the combustor to prevent coking inside the tubes. Bell and Lowy (1967) estimated that approximately 70% of the energy is transferred to the fluids in the radiant section of a typical heater and the balance to the convection section.[35] The tubes in the convection section often have fins to improve convective heat transfer efficiency. These fins are designed to withstand temperatures up to about 1200°F (650°C). If delayed combustion occurs in the convection section, the fins can be exposed to temperatures up to 2000°F (1100°C), which can damage the fins.[34]

Kern noted that process heaters are typically designed around the burners.[33] There may be anywhere from 1 to over 100 burners in a typical process heater, depending on the design and process requirements. In the refinery industry, the average number of burners in a heater varies by the heater type, as shown in Table 1.2.[36] On average, mechanical draft burners have higher firing rates than natural draft burners. For forced-draft systems, burners with air preheat typically have higher heat releases than burners without air preheat. According to one survey, 89.6% of the burners in oil refineries are natural draft, 8.0% are forced draft with no air preheat, and 2.4% are forced draft with air preheat.[37] The mean size of all process heaters is 72×10^6 Btu/hr (21 MW), which are mostly

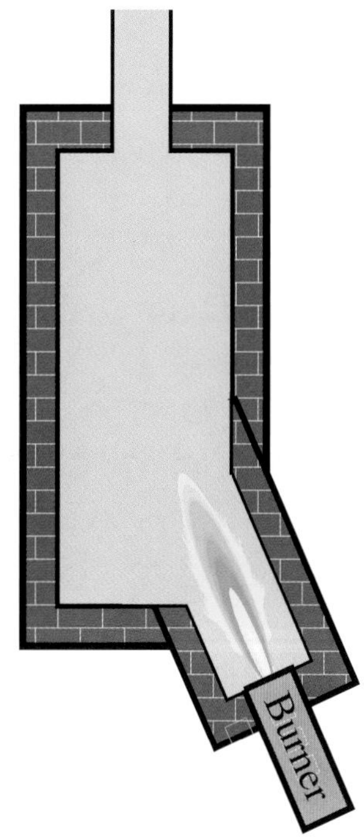

FIGURE 1.10 Elevation view of a terrace firing furnace.

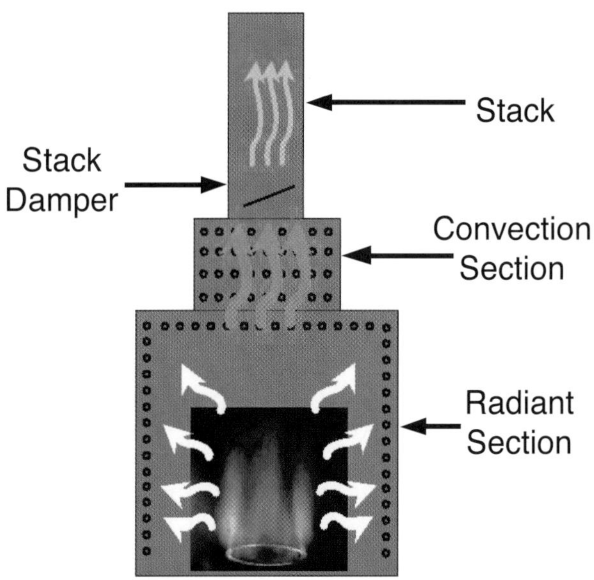

FIGURE 1.11 Schematic of a process heater.

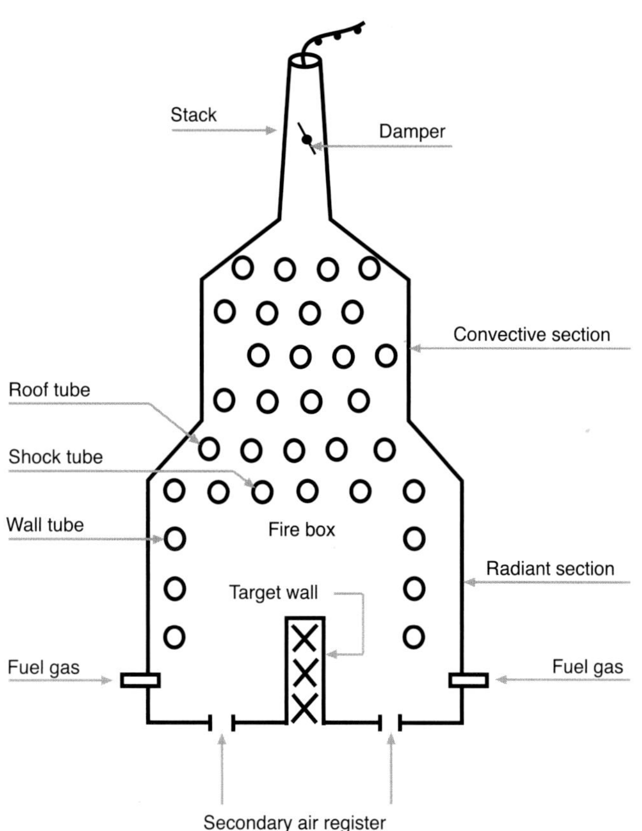

FIGURE 1.12 Typical process heater.[34]

TABLE 1.2 Average Burner Configuration by Heater Type

Heater Type	Ave. No of Burners	Ave. Design Total Heat Release (10^6 Btu hr^{-1})	Ave. Firing Rate per Burner (10^6 Btu hr^{-1})
Natural draft	24	69.4	2.89
Mechanical draft, no air preheat	20	103.6	5.18
Mechanical draft, with air preheat	14	135.4	9.67

From U.S. EPA.[36]

natural draft. The mean size of forced-draft heaters is 110×10^6 Btu/hr (32 MW). Figure 1.13 shows the distribution for the overall firing rate for fired heaters. Table 1.3 shows the variety of processes in a refinery that use fired heaters.

Table 1.4 shows the major applications for fired heaters in the chemical industry. These can be broadly classified into two categories: (1) low- and medium-firebox temperature applications such as feed preheaters, reboilers, and steam superheaters, and (2) high firebox temperature applications such as olefins, pyrolysis furnaces, and steam-hydrocarbon

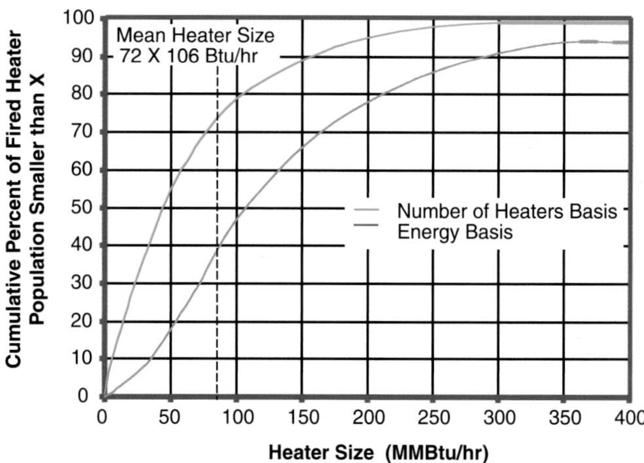

FIGURE 1.13 Fixed heater size distribution.[36]

reformers. The low and medium firebox temperature heaters represent about 20% of the chemical industry requirements and are similar to those in the petroleum refining industry.[38] The high firebox temperature heaters represent the remaining 80% of the chemical industry heater requirements and are unique to the chemical industry.

Berman (1979) discussed the different burner designs used in fired heaters.[39] Burners can be located in the floor, firing vertically upward. In vertical cylindrical (VC) furnaces, these burners are located in a circle in the floor of the furnace. The VC furnace itself serves as part of the exhaust stack to help create draft to increase the chimney effect.[40] In cabin heaters, which are rectangular, there are one or more rows of burners located in the floor. Burners can be at a low level, firing parallel to the floor. In that configuration, they may be firing from two opposite sides toward a partial wall in the middle of the furnace that acts as a radiator to distribute the heat (see Figures 1.14 and 1.15). Burners can be located on the wall, firing radially along the wall (see Figure 1.16), and are referred to as radiant wall burners. There are also combinations of the above in certain heater designs. For example, in ethylene production heaters, both floor-mounted vertically fired burners (see Figure 1.17) and radiant wall burners are used in the same heater.

Typical examples of process heaters are shown in Figures 1.18 and 1.19. A cabin heater is shown in Figure 1.20, burners firing in a crude unit are shown in Figure 1.21, and typical burner arrangements are shown in Figure 1.22. Berman (1979) noted the following categories of process heaters: column reboilers; fractionating-column feed preheaters, reactor-feed preheaters; including reformers, and heat supplied to heat transfer media (e.g., a circulating fluid or molten salt), heat supplied to viscous fluids; and fired reactors, including steam reformers

TABLE 1.3 Major Refinery Processes Requiring a Fired Heater

Process	Process Description	Heaters used	Process heat requirements KJ/liter	Process heat requirements 10³ Btu/bbl	Feedstock temperature outlet of heater, °F
	Distillation				
Atmospheric	Separates light hydrocarbons from crude in a distillation column under atmospheric conditions	Preheater, reboiler	590	89	700
Vacuum	Separates heavy gas oils from atmospheric distillation bottoms under vacuum	Preheater, reboiler	418	63	750–830
	Thermal Processes				
Thermal cracking	Thermal decomposition of large molecules into lighter, more valuable products	Fired reactor	4650	700	850–1000
Coking	Cracking reactions allowed to go to completion; lighter products and coke produced.	Preheater	1520	230	900–975
Visbreaking	Mild cracking of residuals to improve their viscosity and produce lighter gas oils	Fired reactor	961	145	850–950
	Catalytic Cracking				
Fluidized catalytic cracking	Cracking of heavy petroleum products; a catalyst is used to aid the reaction	Preheater	663	100	600–885
Catalytic hydrocracking	Cracking heavy feedstocks to produce lighter products in the presence of hydrogen and a catalyst	Preheater	1290	195	400–850
	Hydroprocessing				
Hydrodesulfurization	Remove contaminating metals, sulfur, and nitrogen from the feedstock; hydrogen is added and reacted over a catalyst	Preheater	431	65[a]	390–850
Hydrotreating	Less severe than hydrodesulfurization; removes metals, nitrogen, and sulfur from lighter feedstocks; hydrogen is added and reacted over a catalyst	Preheater	497	75[b]	600–800
	Hydroconversion				
Alkylation	Combination of two hydrocarbons to produce a higher molecular weight hydrocarbon; heater used on the fractionator	Reboiler	2500	377[c]	400
Catalytic reforming	Low-octane napthas are converted to high-octane, aromatic napthas; feedstock is contacted with hydrogen over a catalyst	Preheater	1790	270	850–1000

[a] Heavy gas oils and middle distillates.
[b] Light distillate.
[c] Btu bbl^{-1} of total alylate.
Source: From the U.S. EPA.[36]

and pyrolysis heaters.[41] Six types of vertical-cylindrical-fired heaters were given: all radiant, helical coil, crossflow with convection section, integral convection section, arbor or wicket type, and single-row/double-fired. Six basic designs were also given for horizontal tube-fired heaters: cabin, two-cell box, cabin with dividing bridgewall, end-fired box, end-fired box with side-mounted convection section, and horizontal-tube/single-row/double-fired.

Many commonly used process heaters typically have a radiant section and a convection section. Burners are fired in the radiant section to heat the tubes. Fluids flow through the tubes and are heated to the desired temperature for further processing. The fluids are preheated in the convection section and heated to the desired process temperature in the radiant section. Radiant heat transfer from the flames to the tubes is the most critical aspect of this heater because overheating of the tubes leads to tube failure and shutdown of the heater.[42] The tubes can be horizontally or vertically oriented, depending on the particular heater design.

A unique aspect of process heaters is that they are often natural-draft. This means that no combustion air blower is used. The air is inspirated into the furnace by the suction created by

TABLE 1.4 Major Fired Heater Applications in the Chemical Industry

Chemical	Process	Heater Type	Firebox Temperature (°F)	1985 Fired Heater Energy Requirement (10^{12} Btu yr^{-1})	% of Known Chemical Industry Heater Requirements
		Low- and Medium-Temperature Applications			
Benzene	Reformate extraction	Reboiler	700	64.8	9.9
Styrene	Ethylbenzine dehydrogenation	Steam superheater	1500–1600	32.1	4.9
Vinyl chloride monomer	Ethylene dichloride cracking	Cracking furnace	N/A	12.6	1.9
p-Xylene	Xylene isomerization	Reactor-fired preheater	N/A	13.0	2.0
Dimethylterephthalate	Reaction of *p*-xylene and methanol	Preheater, hot oil furnace	480–540	11.1	1.7
Butadiene	Butylene dehydrogenation	Preheater, reboiler	1100	2.6	0.4
Ethanol (synthetic)	Ethylene hydration	Preheater	750	1.3	0.2
Acetone	Various	Hot oil furnace	N/A	0.8	0.1
		High-Temperature Applications			
Ethylene/propylene	Thermal cracking	Pyrolysis furnace	1900–2300	337.9	51.8
Ammonia	Natural gas reforming	Steam hydrocarbon reformer	1500–1600	150.5	23.1
Methanol	Hydrocarbon reforming	Steam hydrocarbon	1000–2000	25.7	4.0
	Total Known Fired Heater Energy Requirement			652.4	100.0

Source: From U.S. EPA.[36]

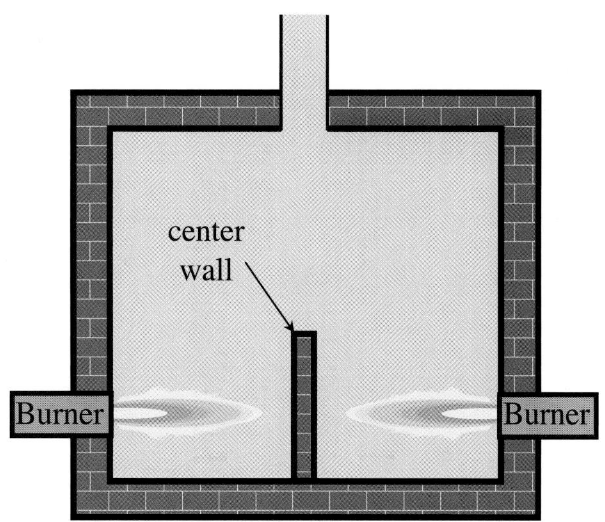

FIGURE 1.14 Sketch (elevation view) of center or target wall firing configuration.

FIGURE 1.15 Horizontal floor-fired burners.

the hot gases rising through the combustion chamber and exhausting to the atmosphere. Another unique aspect of these heaters is the wide range of fuels used, which are often byproducts of the petroleum refining process. These fuels can contain significant amounts of hydrogen, which has a large impact on the burner design. It is also fairly common for multiple fuel compositions to be used, depending on the operating conditions of the plant at any given time. In addition to hydrocarbons ranging up to C_5, the gaseous fuels can also contain hydrogen and inerts (like CO_2 or N_2). The compositions can range from gases containing high levels of inerts to fuels containing high levels of H_2. The flame characteristics for fuels with high levels of inerts are very different than for fuels with high levels of H_2 (see Chapters 2 and 5). Add to that the requirement for turndown conditions, and it becomes very challenging to design burners that will maintain stability, low emissions, and the desired heat flux distribution over the range of possible conditions. Some plants use liquid fuels, like No. 2 to No. 6 fuel oil, sometimes by themselves and sometimes in combination with gaseous fuels. So-called combination burners use both a liquid and a gaseous fuel, which are normally injected separately through each burner.

Shires gave a general heat balance for a process heater:[43]

$$\dot{Q}_f = \dot{Q}_g + \dot{Q}_l + \dot{Q}_p \tag{1.1}$$

where $\dot{Q}_f$ is the heat generated by combusting the fuel, $\dot{Q}_g$ is the heat going to the load, $\dot{Q}_l$ is the heat lost through the walls, and $\dot{Q}_p$ is the heat carried out by the exhaust products. This is shown schematically in Figure 1.23.

Talmor (1982) has written a book dealing with the prediction, control, and troubleshooting of hot spots in process heaters.[44] The book gives a method for estimating the magnitude and location of the maximum heat flux in the combustion zone. It takes into account the firing rate of each burner, the number of burners, the flame length, the flame emissivity, the spacing between the burner and the tubes, the spacing between the burners, and the geometry of the firebox. The book includes much empirical data specific to a variety of different process heaters and also gives many detailed examples that have been worked out.

1.4 BURNERS

The burner is the device that is used to combust the fuel with an oxidizer (usually air) to convert the chemical energy in the fuel into thermal energy. A given combustion system may have a single burner or many burners, depending on the size and type of the application. For example, in a vertical cylindrical furnace, one or more burners are located in the floor of a cylindrically shaped furnace (see Figure 1.24). The heat from the burner radiates in all directions and is efficiently absorbed by the tubes. Another type of heater geometry is rectangular (see Figure 1.25). This type of system is generally more difficult to analyze because of the multiplicity of heat sources and because of the interactions between the flames and their associated products of combustion.

There are many factors that go into the design of a burner. This section briefly considers some of the important factors to be taken into account for a particular type of burner, as well as how these factors impact things like heat transfer and pollutant emissions.

FIGURE 1.16 Wall-fired burner (side view). (Courtesy of John Zink Company LLC.)

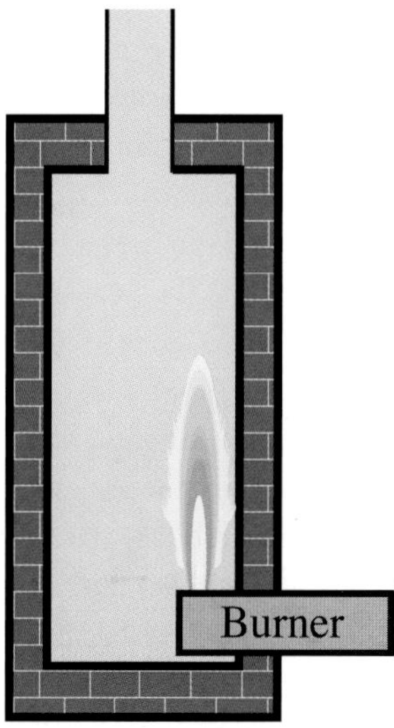

FIGURE 1.17 Sketch (elevation view) of a horizontally mounted, vertically fired burner configuration.

1.4.1 Competing Priorities

There have been many changes in the traditional designs that have been used in burners, primarily because of the recent interest in reducing pollutant emissions. In the past, the burner designer was primarily concerned with efficiently combusting the fuel and transferring the energy to a heat

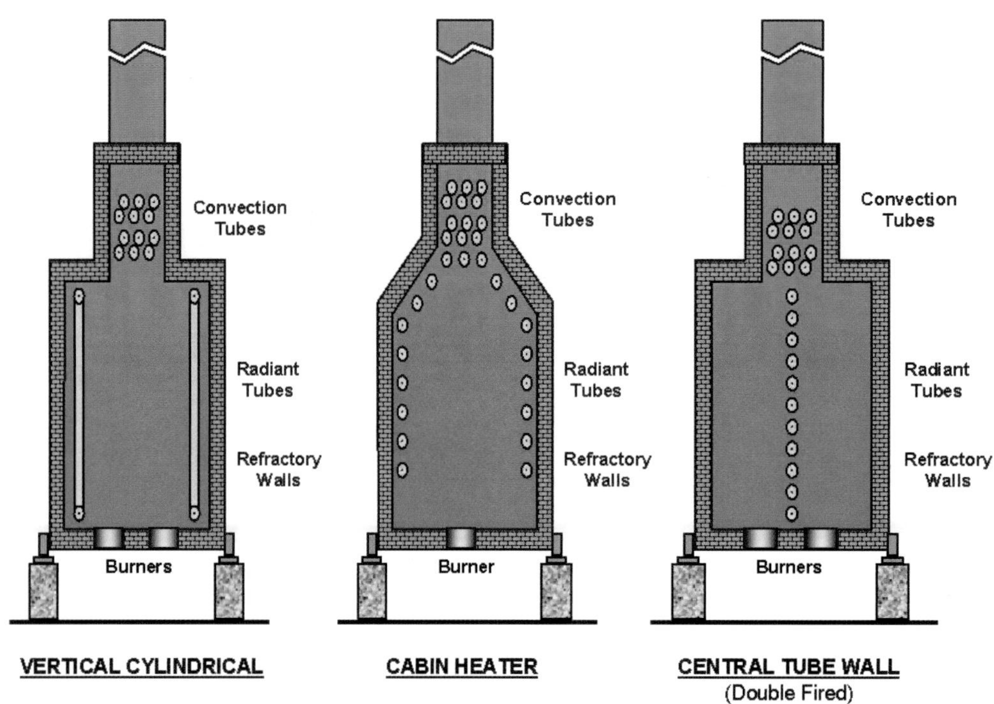

FIGURE 1.18 Examples of process heaters.[43]

load. New and increasingly more stringent environmental regulations have added the need to consider the pollutant emissions produced by the burner. In many cases, reducing pollutant emissions and maximizing combustion efficiency are at odds with each other. For example, a well-accepted technique for reducing NOx emissions is known as staging, where the primary flame zone is deficient in either fuel or oxidizer.[45] The balance of the fuel or oxidizer can be injected into the burner in a secondary flame zone or, in a more extreme case, can be injected somewhere else in the combustion chamber. Staging reduces the peak temperatures in the primary flame zone and also alters the chemistry in a way that reduces NOx emissions because fuel-rich or fuel-lean zones are less conducive to NOx formation than near-stoichiometric zones (see Chapter 6). NOx emissions increase rapidly with the exhaust product temperature (see Figure 6.5). Because thermal NOx is exponentially dependent on the gas temperature even small reductions in the peak flame temperature can dramatically reduce NOx emissions. However, lower flame temperatures often reduce the radiant heat transfer from the flame because radiation is dependent on the fourth power of the absolute temperature of the gases. Another potential problem with staging is that it may increase CO emissions, which is an indication of incomplete combustion and reduced combustion efficiency. However, it is also possible that staged combustion may produce soot in the flame, which can increase flame radiation. The actual impact of staging on the heat transfer from the flame is highly dependent on the actual burner design.

In the past, the challenge for the burner designer was to maximize the mixing between the fuel and the oxidizer to ensure complete combustion. If the fuel was difficult to burn, as in the case of low heating value fuels such as waste liquid fuels or process gases from chemical production, the task could be very challenging. Now, the burner designer must balance the mixing of the fuel and the oxidizer to maximize combustion efficiency while simultaneously minimizing all types of pollutant emissions. This is no easy task as, for example, NOx and CO emissions often go in opposite directions (see Figure 1.26). When CO is low, NOx may be high, and vice versa. Modern burners must be environmentally friendly, while simultaneously efficiently transferring heat to the load.

1.4.2 Design Factors

There are many types of burners designs that exist due to the wide variety of fuels, oxidizers, combustion chamber geometries, environmental regulations, thermal input sizes, and heat transfer requirements. Additionally, heat transfer requirements include things like flame temperature, flame momentum, and heat distribution. Garg (1989) lists the

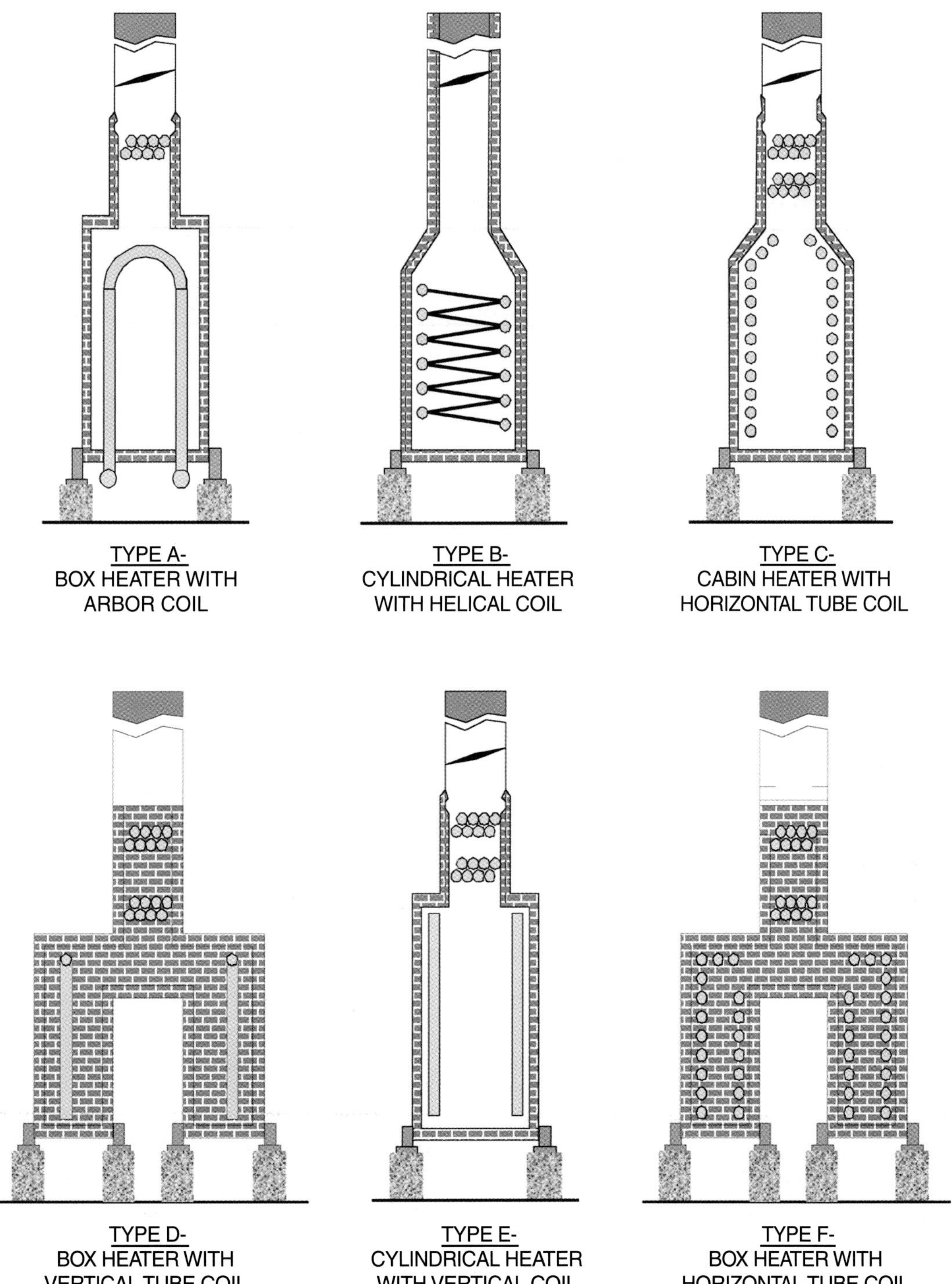

FIGURE 1.19 Typical heater types.

following burner specifications that are needed to properly choose a burner for a given application: burner type, heat release and turndown, air supply (natural draft, forced draft, or balanced draft), excess air level, fuel composition(s), firing position, flame dimensions, ignition type, atomization media for liquid fuel firing, noise, NOx emission rate, and whether waste gas firing will be used.[46] Some of these design factors are briefly considered next.

FIGURE 1.20 Cabin heater.

1.4.2.1 Fuel

Depending on many factors, certain types of fuels are preferred for certain geographic locations due to cost and availability considerations. Gaseous fuels, particularly natural gas, are commonly used in most industrial heating applications in the United States. In Europe, natural gas is also commonly used along with light fuel oil. In Asia and South America, heavy fuel oils are generally preferred, although the use of gaseous fuels is on the rise.

Fuels also vary depending on the application. For example, in incineration processes, waste fuels are commonly used either by themselves or with other fuels like natural gas. In the petrochemical industry, fuel gases often consist of a blend of several fuels, including gases like hydrogen, methane, propane, butane, propylene, nitrogen, and carbon dioxide.

The fuel choice has an important influence on the heat transfer from a flame. In general, solid fuels like coal and liquid fuels like oil produce very luminous flames which contain soot particles that radiate like blackbodies to the heat load. Gaseous fuels like natural gas often produce nonluminous flames because they burn so cleanly and completely without producing soot particles. A fuel like hydrogen is completely nonluminous as there is no carbon available to produce soot.

In cases where highly radiant flames are required, a luminous flame is preferred. In cases where convection heat transfer is preferred, a nonluminous flame may be preferred in order to minimize the possibility of contaminating the heat load with soot particles from a luminous flame. Where natural gas is the preferred fuel and highly radiant flames are desired, new technologies are being developed to produce

FIGURE 1.21 Crude unit burners.

more luminous flames. These include things like pyrolyzing the fuel in a partial oxidation process,[47] using a plasma to produce soot in the fuel,[48] and generally controlling the mixing of the fuel and oxidizer to produce fuel-rich flame zones that generate soot particles.[49]

Therefore, the fuel itself has a significant impact on the heat transfer mechanisms between the flame and the load. In most cases, the fuel choice is dictated by the customer as part of the specifications for the system and is not chosen by the burner designer. The designer must make the best of whatever fuel has been selected. In most cases, the burner design is optimized based on the choice of fuel.

In some cases, the burner may have more than one type of fuel. An example is shown in Figure 1.27.[50] Dual-fuel burners are designed to operate typically on either gaseous or liquid fuels. These burners are used, usually for economic reasons, where the customer may need to switch between a gaseous fuel like natural gas and a liquid fuel like oil. These burners normally operate on one fuel or the other, and sometimes on both fuels simultaneously. Another application where multiple fuels may be used is in waste incineration. One method of disposing of waste liquids contaminated with hydrocarbons is to combust them by direct injection through a burner. The waste liquids are fed through the burner, which is powered by a traditional fuel such as natural gas or oil. The waste liquids often have very low heating values and are difficult to combust without auxiliary fuel. This further complicates the burner design, wherein the waste liquid must be vaporized and combusted concurrently with the normal fuel used in the burner.

1.4.2.2 Oxidizer

The predominant oxidizer used in most industrial heating processes is atmospheric air. This can present challenges in some applications where highly accurate control is required due to the daily variations in the barometric pressure and humidity of ambient air. The combustion air is sometimes preheated to increase the overall thermal efficiency of a process. Combustion air is also sometimes blended with some of

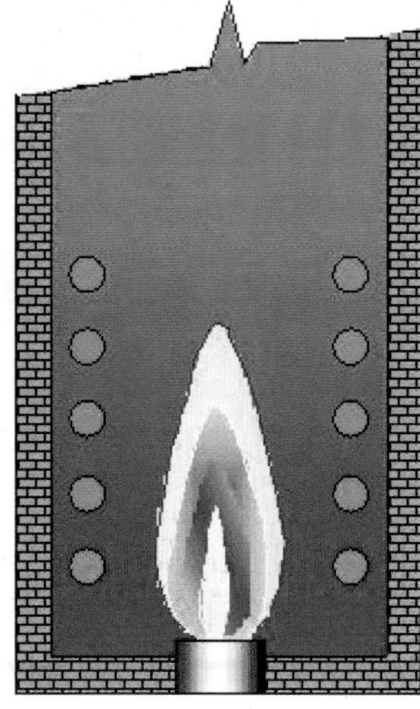

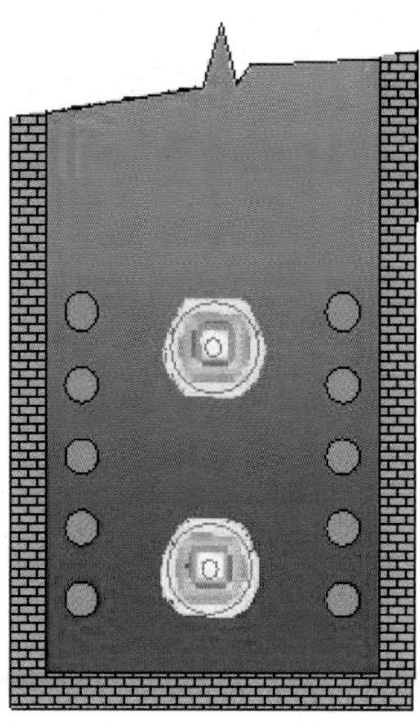

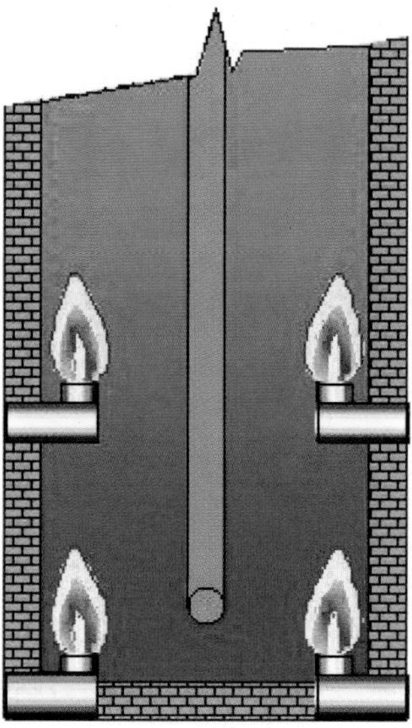

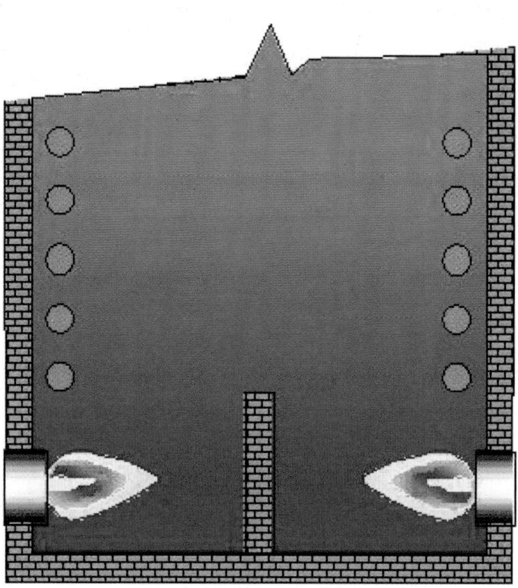

FIGURE 1.22 Typical burner arrangements (elevation view).

the products of combustion, a process usually referred to as flue gas recirculation (FGR).

FGR is used to both increase thermal efficiency and reduce NOx emissions. Capturing some of the energy in the exhaust gases and using it to preheat the incoming combustion oxidizer increases thermal efficiency. FGR also reduces peak flame temperatures resulting in reduced NOx emissions, because NOx emissions are highly temperature dependent.

1.4.2.3 Gas Recirculation

A common technique used in combustion systems is to design the burner to induce furnace gases to be drawn into

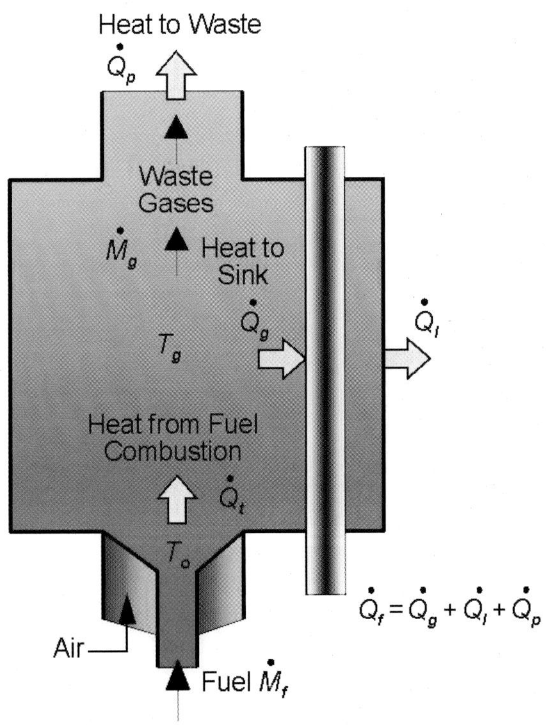

FIGURE 1.23 Process heater heat balance.[43]

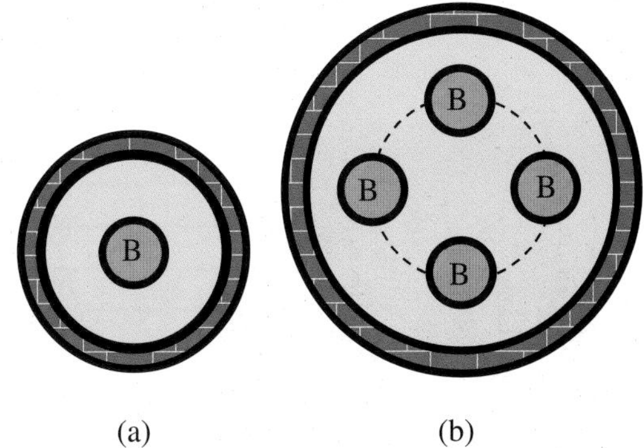

FIGURE 1.24 Burner (B) arrangement (plan view) in the floor of vertical cylindrical furnaces: (a) small-diameter furnace with a single centered burner and (b) larger diameter furnace with four burners symmetrically arranged at a radius from the center.

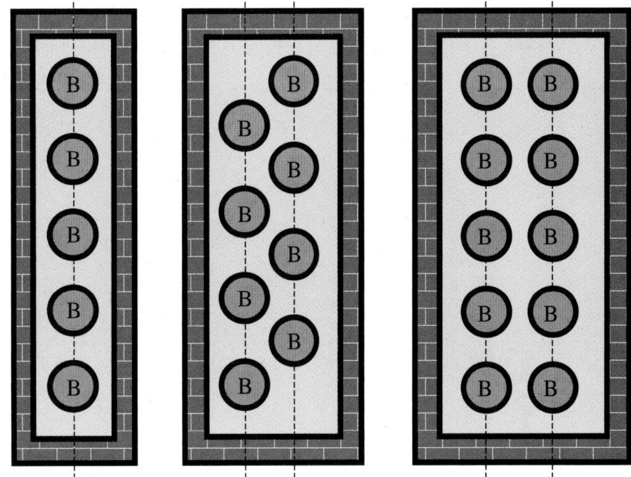

FIGURE 1.25 Burner (B) arrangement (plan view) in the floor of rectangular cabin heaters: (a) single row of burners in a narrower heater, (b) two rows of staggered burners in a slightly wider heater, and (c) two rows of aligned burners in an even wider heater.

the burner to dilute the flame, usually referred to as flue or furnace gas recirculation. Although the furnace gases are hot, they are still much cooler than the flame itself. This dilution can accomplish several purposes. One is to minimize NOx emissions by reducing the peak temperatures in the flame, as in flue gas recirculation. However, furnace gas recirculation may be preferred to flue gas recirculation (see Figure 1.28) because no external high-temperature ductwork or fans are needed to bring the product gases back into the flame zone. Another reason to use furnace gas recirculation may be to increase the convective heating from the flame because of the added gas volume and momentum. An example of furnace gas recirculation into the burner is shown in Figure 6.12.

1.4.3 General Burner Types

There are numerous ways that burners can be classified. Some of the common ones are discussed in this section, along with a brief description of implications for heat transfer.

1.4.3.1 Mixing Type

One common method for classifying burners is according to how the fuel and the oxidizer are mixed. In premixed burners, shown in a cartoon in Figure 1.29 and schematically in Figure 1.30, the fuel and the oxidizer are completely mixed before combustion begins. Radiant wall burners are usually of the premixed type. Premixed burners often produce shorter and more intense flames, compared to diffusion flames. This can produce high-temperature regions in the flame, leading to nonuniform heating of the load and higher NOx emissions.

In diffusion-mixed flames, the fuel and the oxidizer are separated and unmixed prior to combustion, which begins where the oxidizer/fuel mixture is within the flammability range. An example of a diffusion flame is a candle (see Figure 1.31). A diffusion-mixed gas burner, shown schematically in Figure 1.32, is sometimes referred to as a “raw gas” burner because the fuel gas exits the burner essentially as raw gas, having no

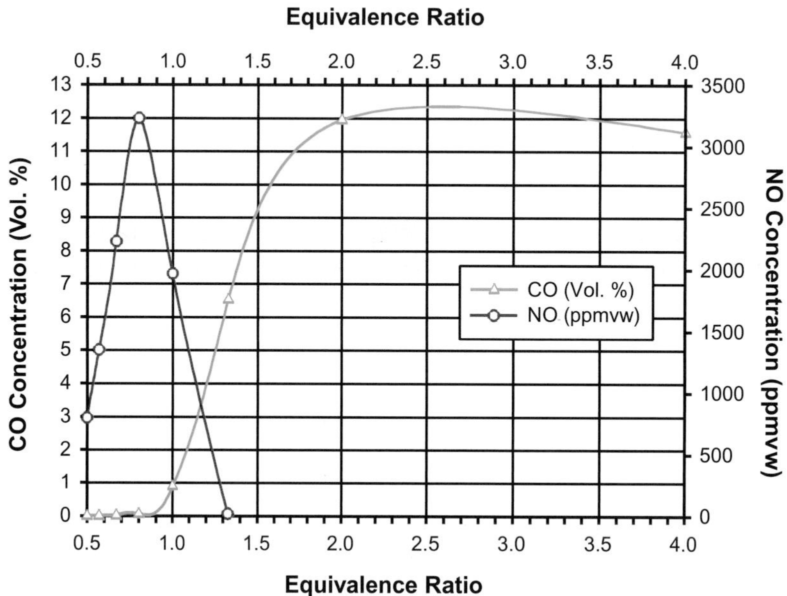

FIGURE 1.26 Adiabatic equilibrium NO and CO as a function of the equivalence ratio for an air/CH_4 flame.

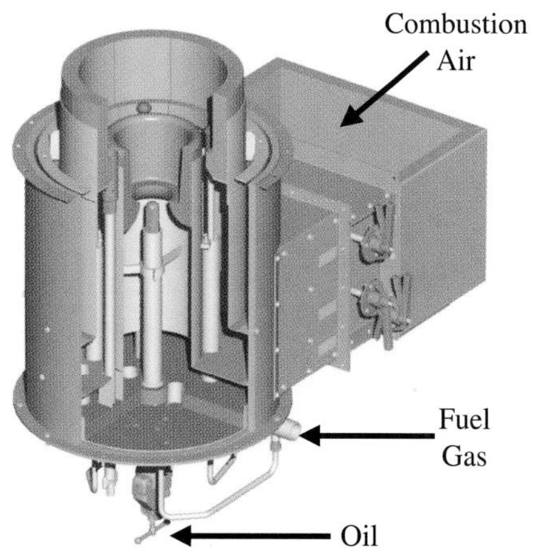

FIGURE 1.27 Typical combination oil and gas burner.

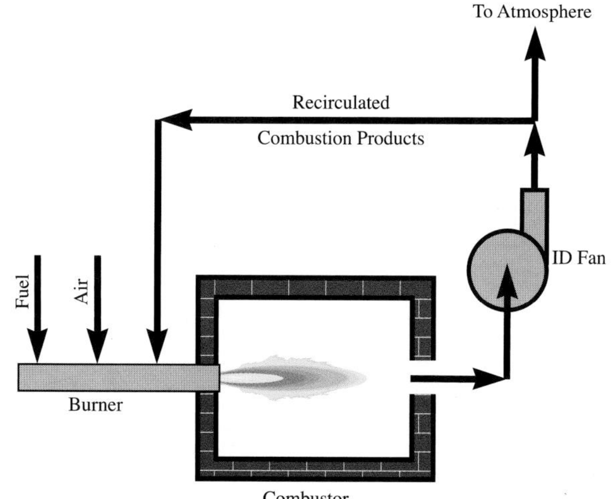

FIGURE 1.28 Schematic of flue gas recirculation.

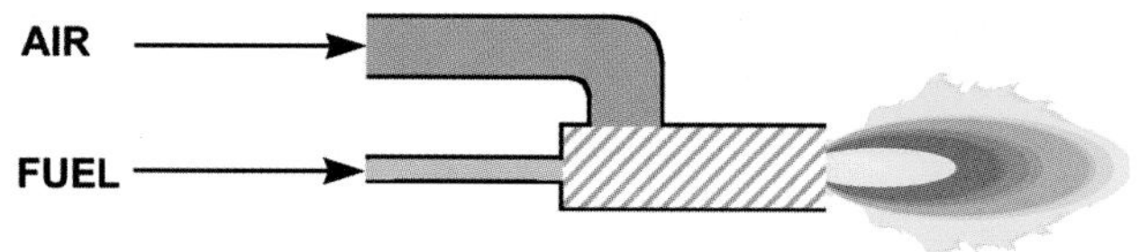

FIGURE 1.29 Cartoon of a premixed burner.

air mixed with it. Diffusion burners typically have longer flames than premixed burners, a lower temperature hot spot, and a more uniform temperature and heat flux distribution.

It is also possible to have partially premixed burners, shown schematically in Figure 1.33, where some fraction of the fuel is mixed with the oxidizer. Partial premixing is often done for stability and safety reasons because it helps anchor the flame, but also reduces the chance for flashback, which is sometimes a problem in fully premixed burners. This type of burner often has a flame length, temperature, and heat flux distribution that is somewhere between the fully premixed and diffusion flames.

Another burner classification based on mixing is known as staging — staged air and staged fuel. A staged air burner is shown in a cartoon in Figure 1.34 and schematically in Figure 1.35. A staged fuel burner is shown in a cartoon in

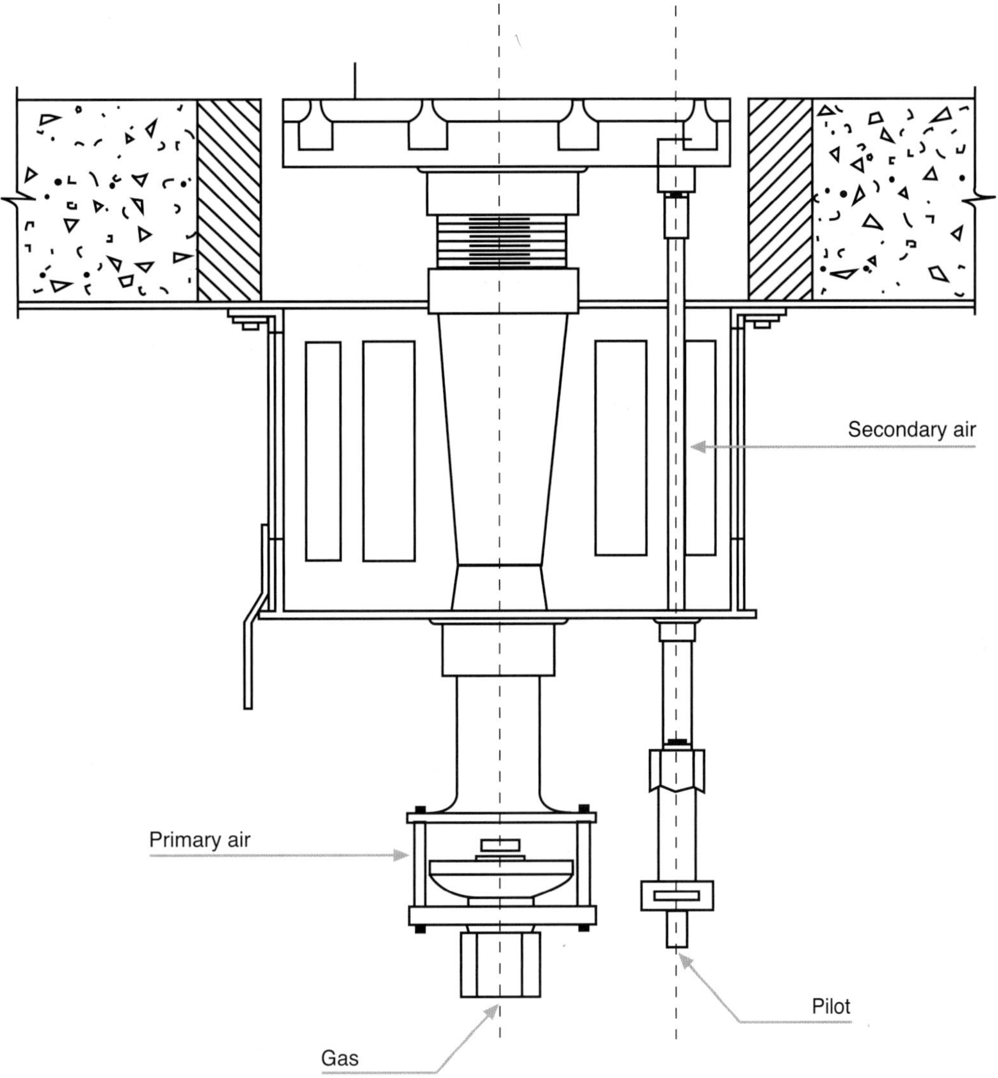

FIGURE 1.30 Typical premixed gas burner. (Courtesy of API.[50])

Figure 1.36 and schematically in Figure 1.37. Secondary and sometimes tertiary injectors in the burner are used to inject a portion of the fuel and/or the air into the flame, downstream of the root of the flame. Staging is often done to reduce NOx emissions and to produce longer flames. These longer flames typically have a lower peak flame temperature and more uniform heat flux distribution than non-staged flames.

1.4.3.2 Fuel Type

There are three common fuel classifications for burners used in the process industries and generally listed in order of increasing complexity as follows: gas, oil, or a combination of gas + oil. Gas burners are either diffusion (raw gas or no premixing), premixed, or partially premixed. The gas composition can vary widely, as it is often a by-product from the plant. Burners often need to be able to fire multiple fuels that may be produced by the plant, depending on the process conditions and on start-up vs. normal operation. These gaseous fuels often have significant amounts of methane, hydrogen, and higher hydrocarbons (e.g., propane and propylene). They may also contain inerts such as CO_2 and N_2. The heating value can range from 500 to 1500 Btu ft^{-3} (19 to 56 MJ m^{-3}). Burners firing oil require some type of liquid atomization, commonly mechanical (pressurizing the liquid high enough to force it through an atomizer), air, or steam. Steam is most commonly used because it is economical, readily available, and gives a wide turndown ratio and good flame control. Combination burners can usually fire 100% oil, 100% gas, or any combination in between.

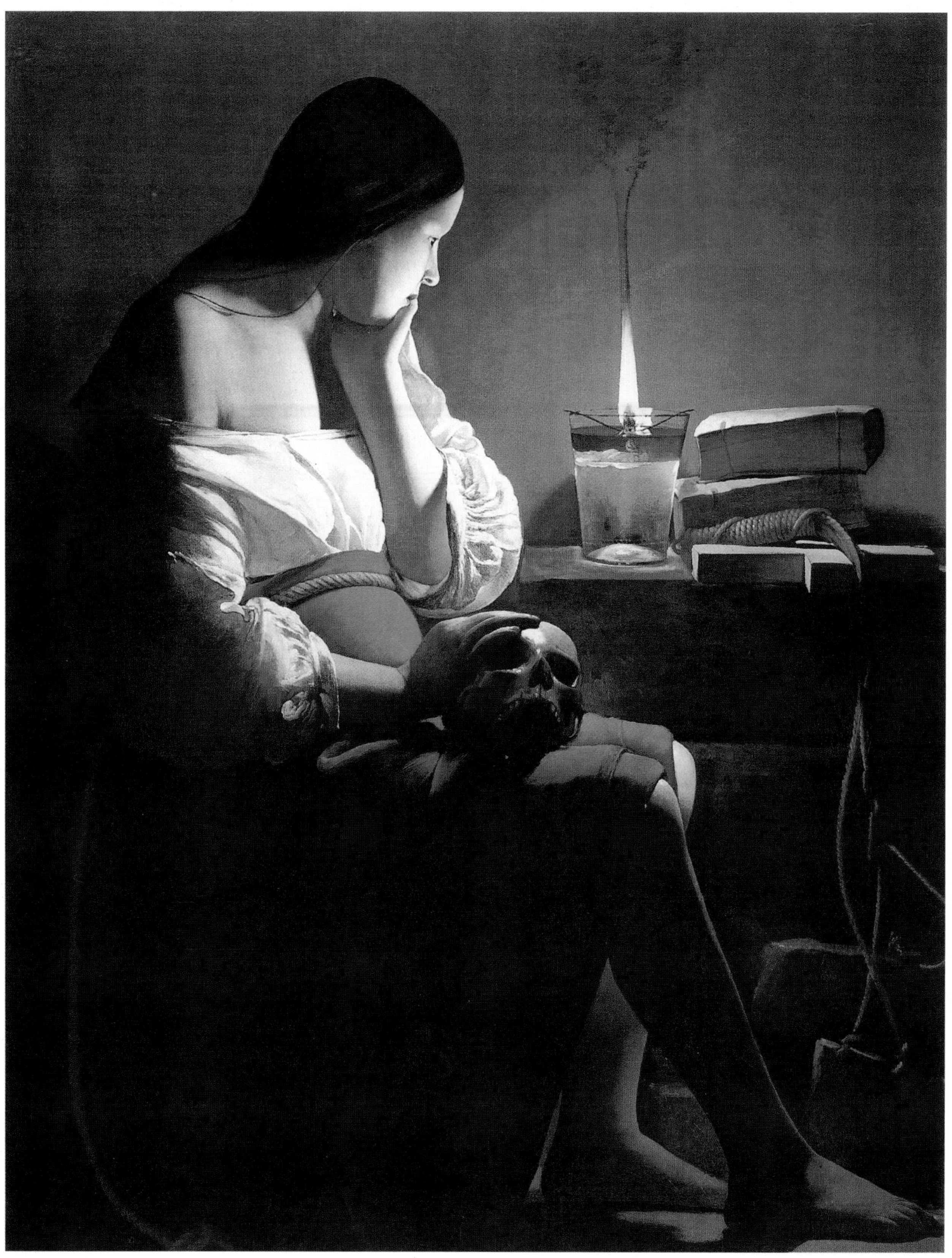

FIGURE 1.31 Painting of a diffusion flame by Georges de LaTour. (Courtesy of the Los Angeles County Museum of Art.)

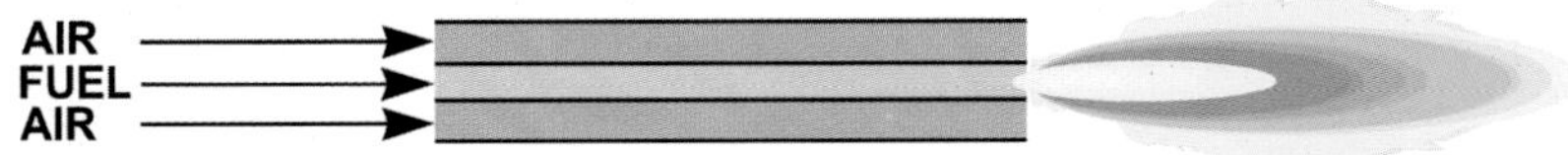

FIGURE 1.32 Cartoon of a diffusion burner.

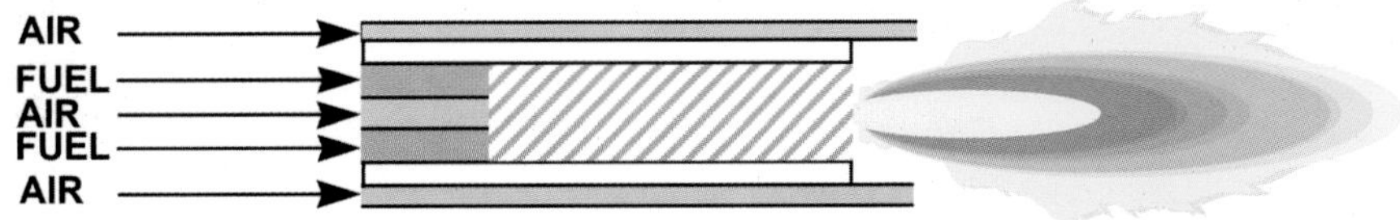

FIGURE 1.33 Cartoon of a partially premixed burner.

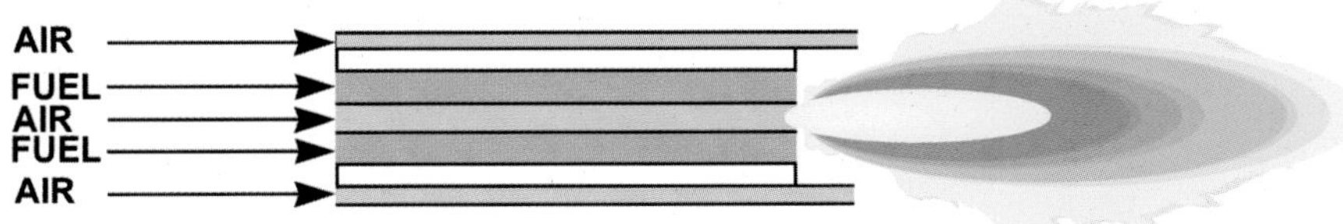

FIGURE 1.34 Cartoon of a staged air burner.

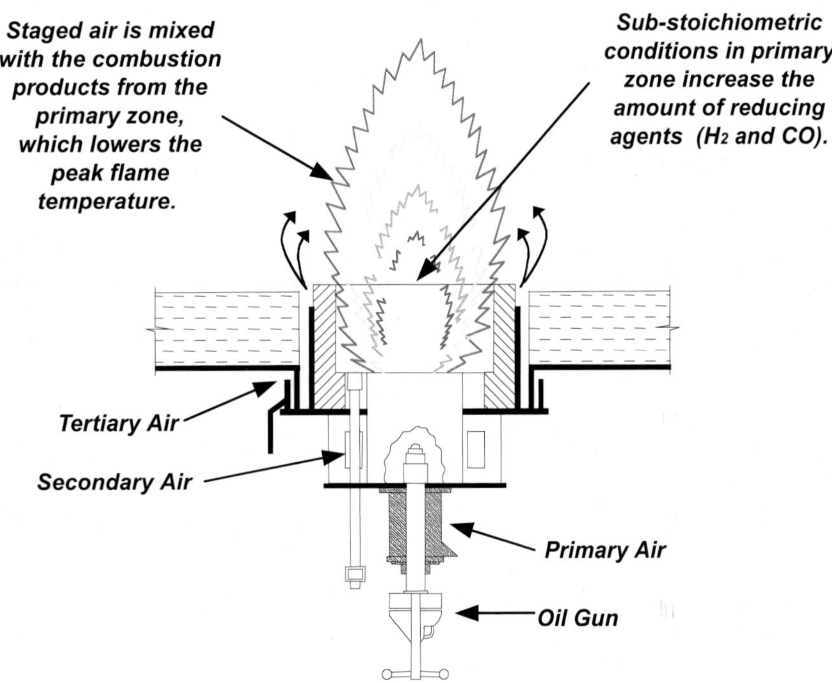

FIGURE 1.35 Schematic of a typical staged air combination oil and gas buner.

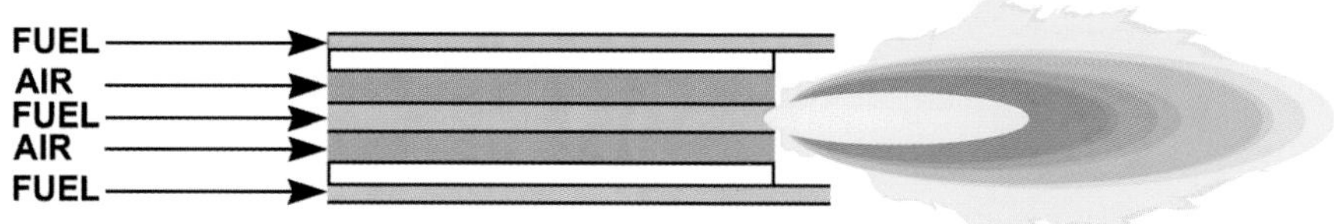

FIGURE 1.36 Cartoon of a staged fuel burner.

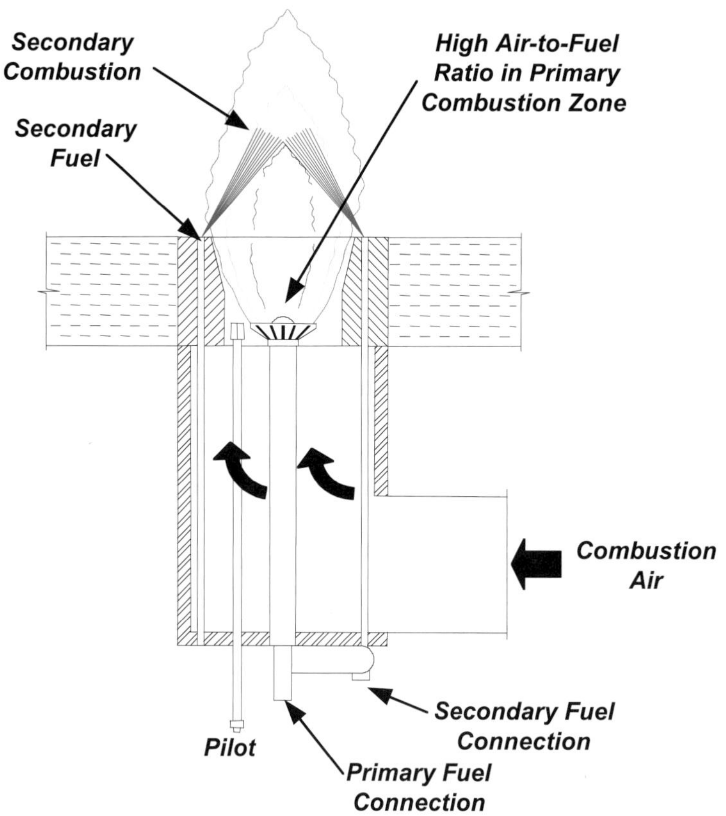

FIGURE 1.37 Schematic of a typical staged fuel gas burner.

1.4.3.3 Combustion Air Temperature

One common way of classifying the oxidizer is by its temperature. It is common in many industrial applications to recover heat from the exhaust gases by preheating the incoming combustion air, either with a recuperator or a regenerator. Such a burner is often referred to as a preheated air burner.

1.4.3.4 Draft Type

Most industrial burners are known as forced-draft or mechanical-draft burners. This means that the oxidizer is supplied to the burner under pressure. For example, in a forced-draft air burner, the air used for combustion is supplied to the burner by a blower. In natural-draft burners, the air used for combustion is induced into the burner by the negative draft produced in the combustor. A schematic is shown in Figure 1.38, and an example is shown in Figure 1.39. In this type of burner, the pressure drop and combustor stack height are critical in producing enough suction to induce enough combustion air into the burners. This type of burner is commonly used in the chemical and petrochemical industries in fluid heaters. The main consequence of the draft type on heat transfer is that the natural-draft flames are usually longer than the forced-draft flames so that the heat flux from the flame is distributed over a longer distance and the peak temperature in the flame is often lower.

1.4.3.5 Location

Process burners are often classified by their location in the furnace or heater. Floor or hearth burners are located in the

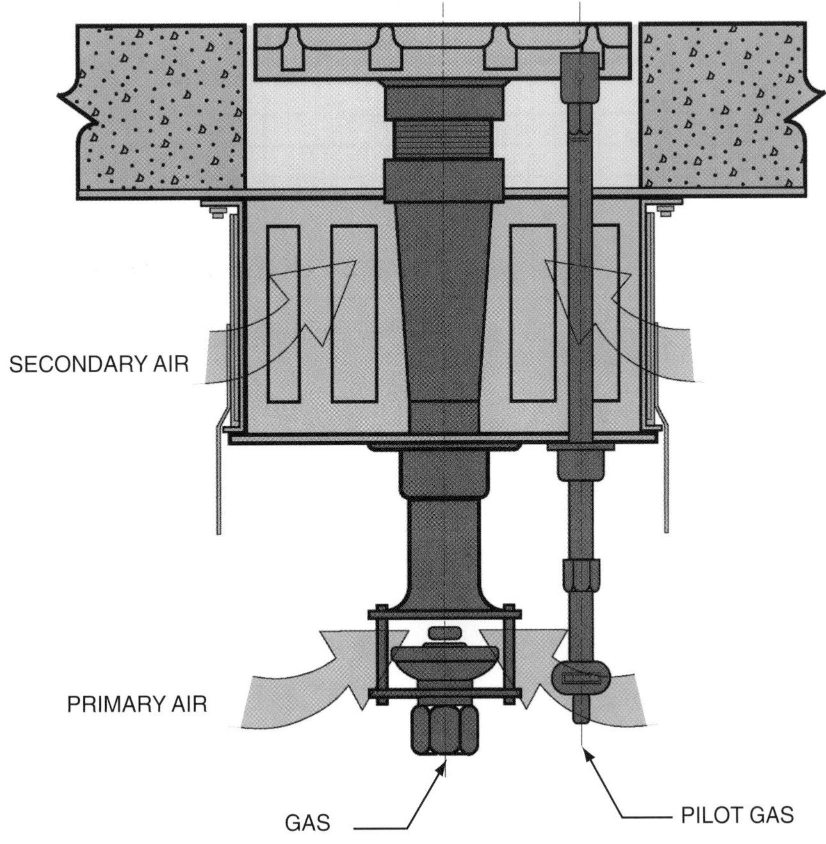

FIGURE 1.38 Typical natural-draft gas burner.

bottom of the combustor and fire vertically upward. Roof burners are located in the ceiling and fire vertically downward. Wall burners can be located in the wall or in the floor, firing along the wall. Their function is to heat a refractory wall to radiate heat to process tubes.

1.4.4 Potential Problems

There are many potential problems that could affect the performance of burners and therefore the performance of the heaters, boilers, and furnaces used to process the materials of interest to the end user. A few examples will illustrate some of the potential problems that may be encountered. Figure 1.40 shows flames impinging on the process tubes in a cabin heater. Flame impingement on the tubes can cause premature coking and significantly reduce the operational run time. Figure 1.41 shows flames pulled toward the wall of a heater. This may be caused by burner design problems or by gas flow currents in the furnace. While flames leaning away from the tubes may reduce coking, they also reduce performance because the heater is designed for vertical flames. Less heat is transferred to the tubes when the flames lean away from the

FIGURE 1.39 Natural-draft burner.

FIGURE 1.40 Flames impinging on tubes in a cabin heater.

FIGURE 1.41 Flames pulled toward the wall.

FIGURE 1.42 Gas flames needing service.

tubes. This reduces the throughput of the entire process. Figure 1.42 shows an oil burner that needs some type of service or adjustment. In all of the above examples, the performance of the combustion system is reduced. Chapter 17 discusses some of the common problems encountered and how to fix them.

1.2 CONCLUSIONS

This book considers all aspects of combustion, with particular emphasis on applications in the process industries including the petrochemical, hydrocarbon, power generation, and thermal oxidation industries. The fundamentals of combustion, heat transfer, and fluid flow are discussed from a more applied approach. Many other aspects of combustion, such as fuel composition, pollutant emissions, noise, safety, and control, are also discussed. Topics of specific interest to burners are also treated including design, testing, installation, maintenance, and troubleshooting. There are also very detailed considerations of process burners, flares, boiler burners, duct burners, and thermal oxidizers. Many of these topics have never been adequately covered in other combustion books. The extensive use of color illustrations further enhances the usefulness of this book as an essential tool for the combustion engineer.

REFERENCES

1. U.S. Dept. of Energy Office of Industrial Technology, Petroleum — Industry of the Future: Energy and Environmental Profile of the U.S. Petroleum Refining Industry, U.S. DOE, Washington, D.C., December 1998.
2. U.S. Dept. of Energy Office of Industrial Technology, Industrial Combustion Technology Roadmap, U.S. DOE, Washington, D.C., April 1999.
3. R.A. Strehlow, *Fundamentals of Combustion*, International Textbook Co., Scranton, PA, 1968.
4. F.A. Williams, *Combustion Theory*, Benjamin/ Cummings, Menlo Park, CA, 1985.
5. J.A. Barnard and J.N. Bradley, *Flame and Combustion*, 2nd edition, Chapman and Hall, London, 1985.
6. B. Lewis and G. von Elbe, *Combustion, Flames and Explosions of Gases*, 3rd edition, Academic Press, New York, 1987.
7. W. Bartok and A.F. Sarofim, Eds., *Fossil Fuel Combustion*, John Wiley & Sons, New York, 1991.
8. I. Glassman, *Combustion*, 3rd edition, Academic Press, New York, 1996.

9. S.R. Turns, *An Introduction to Combustion*, McGraw-Hill, New York, 1996.
10. Y.I. Khavkin, *Combustion System Design: A New Approach*, PennWell Books, Tulsa, OK, 1996.
11. J. Griswold, *Fuels, Combustion and Furnaces*, McGraw-Hill, New York, 1946.
12. A. Stambuleanu, *Flame Combustion Processes in Industry*, Abacus Press, Tunbridge Wells, U.K., 1976.
13. E. Perthuis, *La Combustion Industrielle*, Éditions Technip, Paris, 1983.
14. E.L. Keating, *Applied Combustion*, Marcel Dekker, New York, 1993.
15. G. Borman and K. Ragland, *Combustion Engineering*, McGraw-Hill, New York, 1998.
16. C.G. Segeler, Ed., *Gas Engineers Handbook*, Industrial Press, New York, 1965.
17. R.D. Reed, *Furnace Operations*, 3rd edition, Gulf Publishing, Houston, TX, 1981.
18. R. Pritchard, J.J. Guy, and N.E. Connor, *Handbook of Industrial Gas Utilization*, Van Nostrand Reinhold, New York, 1977.
19. R.J. Reed, *North American Combustion Handbook*, 3rd edition, Vol. I, North American Mfg. Co., Cleveland, OH, 1986.
20. IHEA, *Combustion Technology Manual*, 5th edition, Industrial Heating Equipment Assoc., Arlington, VA, 1994.
21. R.O. Anderson, *Fundamentals of the Petroleum Industry*, University of Oklahoma Press, Norman, OK, 1984.
22. W.L. Leffler, *Petroleum Refining for the Nontechnical Person*, PennWell Books, Tulsa, OK, 1985.
23. J.H. Gary and G.E. Handwerk, *Petroleum Refining: Technology and Economics*, 3rd edition, Marcel Dekker, New York, 1994.
24. R.A. Meyers, *Handbook of Petroleum Refining Processes*, 2nd edition, McGraw-Hill, New York, 1997.
25. W.L. Nelson, *Petroleum Refinery Engineering*, 2nd edition, McGraw-Hill, New York, 1941.
26. G.R. Martin, Heat-flux imbalances in fired heaters cause operating problems, *Hydrocarbon Processing*, 77(5), 103-109, 1998.
27. R. Nogay and A. Prasad, Better design method for fired heaters, *Hydrocarbon Processing*, 64(11), 91-95, 1985.
28. A. Garg and H. Ghosh, Good heater specifications pay off, *Chem. Eng.*, 95(10), 77-80, 1988.
29. H. Futami, R. Hashimoto, and H. Uchida, Development of new catalyst and heat-transfer design method for steam reformer, *J. Fuel Soc. Japan*, 68(743), 236-243, 1989.
30. H. Gunardson, *Industrial Gases in Petrochemical Processing*, Marcel Dekker, New York, 1998.
31. J.D. Fleshman, FW hydrogen production, in *Handbook of Petroleum Refining Processes*, 2nd edition, R.A. Myers, Ed., McGraw-Hill, New York, 1996, chap. 6.2.
32. J.S. Plotkin and A.B. Swanson, New technologies key to revamping petrochemicals, *Oil & Gas J.*, 97(50), 108-114, 1999.
33. D.Q. Kern, *Process Heat Transfer*, McGraw-Hill, New York, 1950.
34. N.P. Lieberman, *Troubleshooting Process Operations*, PennWell Books, Tulsa, OK, 1991.
35. H.S. Bell and L. Lowy, Equipment, in *Petroleum Processing Handbook*, W.F. Bland and R.L. Davidson, Eds., McGraw-Hill, New York, 1967, chap. 4.
36. E.B. Sanderford, Alternative Control Techniques Document — NOx Emissions from Process Heaters, U.S. Envir. Protection Agency Report EPA-453/R-93-015, February, 1993.
37. L.A. Thrash, Annual Refining Survey, *Oil & Gas J.*, 89(11), 86-105, 1991.
38. S.A. Shareef, C.L. Anderson, and L.E. Keller, Fired Heaters: Nitrogen Oxides Emissions and Controls, U.S. Environmental Protection Agency, Research Triangle Park, NC, EPA Contract No. 68-02-4286, June 1988.
39. H.L. Berman, Fired heaters. II: Construction materials, mechanical features, performance monitoring, in *Process Heat Exchange*, V. Cavaseno, Ed., McGraw-Hill, New York, 1979, 293-302.
40. A.J. Johnson and G.H. Auth, *Fuels and Combustion Handbook*, 1st edition, McGraw-Hill, New York, 1951.
41. H.L. Berman, Fired heaters. I: Finding the basic design for your application, in *Process Heat Exchange*, V. Cavaseno, Ed., McGraw-Hill, New York, 1979, 287-292.
42. V. Ganapathy, *Applied Heat Transfer*, PennWell Books, Tulsa, OK, 1982.
43. G.L. Shires, Furnaces, in *The International Encyclopedia of Heat & Mass Transfer*, G.F. Hewitt, G.L. Shires, and Y.V. Polezhaev, Eds., CRC Press, Boca Raton, FL, 1997, 493-497.
44. E. Talmor, *Combustion Hot Spot Analysis for Fired Process Heaters*, Gulf Publishing, Houston, 1982.

45. J.L. Reese, G.L. Moilanen, R. Borkowicz, C. Baukal, D. Czerniak, and R. Batten, State-of-the-art of NOx emission control technology, ASME Paper 94-JPGC-EC-15, *Proc. of Int. Joint Power Generation Conf.*, Phoenix, October 3-5, 1994.

46. A. Garg, Better burner specifications, *Hydrocarbon Processing*, 68(8), 71-72, 1989.

47. M.L. Joshi, M.E. Tester, G.C. Neff, and S.K. Panahi, Flame particle seeding with oxygen enrichment for NOx reduction and increased efficiency, *Glass*, 68(6), 212-213, 1990.

48. R. Ruiz and J.C. Hilliard, Luminosity enhancement of natural gas flames, *Proc. of 1989 Int. Gas Research Conf.*, T.L. Cramer, Ed., Government Institutes, Rockville, MD, 1990, 1345-1353.

49. A.G. Slavejkov, T.M. Gosling, and R.E. Knorr, Low-NOx Staged Combustion Device for Controlled Radiative Heating in High Temperature Furnaces, U.S. Patent 5,611,682, March 18, 1997.

50. API Publication 535: Burner for Fired Heaters in General Refinery Services, 1st edition, American Petroleum Institute, Washington, D.C., July 1995.

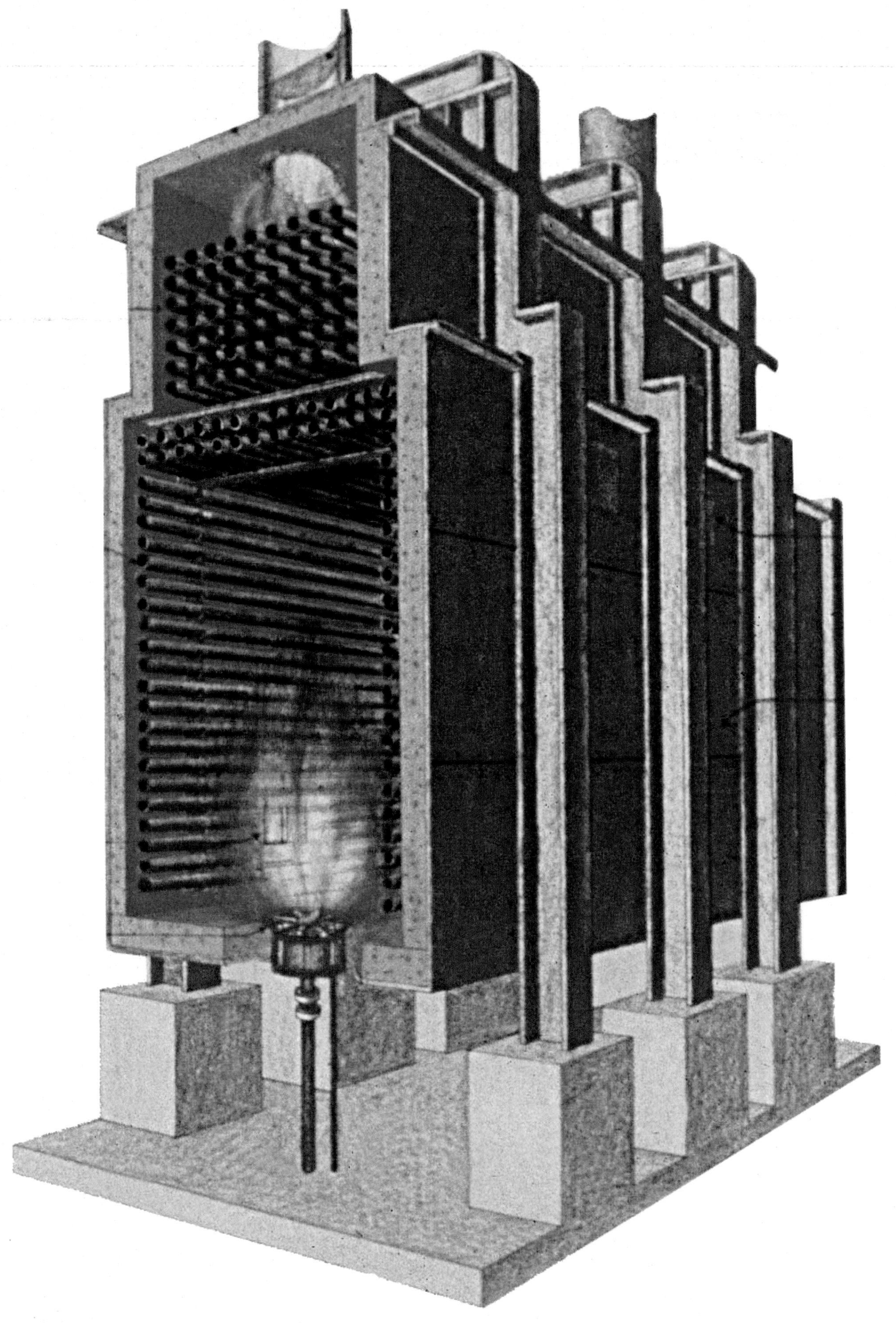

Chapter 2
Fundamentals

Joseph Colannino and Charles E. Baukal, Jr.

TABLE OF CONTENTS

2.1 Introduction ... 34
2.2 Uses for Combustion ... 34
2.3 Brief Overview of Combustion Equipment and Heat Transfer ... 34
2.4 Net Combustion Chemistry of Hydrocarbons ... 34
2.5 Conservation of Mass ... 35
2.6 The Ideal Gas Law ... 35
2.7 Stoichiometric Ratio and Excess Air ... 38
2.7.1 Heat of Combustion ... 38
2.7.2 Adiabatic Flame Temperature ... 46
2.8 Substoichiometric Combustion ... 46
2.9 Equilibrium and Thermodynamics ... 47
2.10 Substoichiometric Combustion Revisited ... 47
2.11 General Discussion ... 54
2.11.1 Air Preheat Effects ... 55
2.11.2 Fuel Blend Effects ... 57
2.12 Combustion Kinetics ... 60
2.12.1 Thermal NOx Formation: A Kinetic Example ... 60
2.12.2 Reaction Rate ... 60
2.12.3 Prompt-NOx Formation ... 61
2.12.4 The Fuel-Bound NOx Mechanism ... 61
2.13 Flame Properties ... 61
2.13.1 Flame Temperature ... 61
2.13.2 Available Heat ... 64
2.13.3 Minimum Ignition Energy ... 64
2.13.4 Flammability Limits ... 64
2.13.5 Flammability Limits for Gas Mixtures ... 66
2.13.6 Flame Speeds ... 67
References ... 67

2.1 INTRODUCTION

Combustion is the controlled release of heat from the chemical reaction between a fuel and an oxidizer. The fuels in the refining, petrochemical, and power generation industries are almost exclusively hydrocarbons. Hydrocarbons comprise only hydrogen (H) and carbon (C) in their molecular structure. Natural gas and fuel oil are examples of hydrocarbon fuels. Other fuels are described later in this chapter and in Chapter 5.

2.2 USES FOR COMBUSTION

Combustion is used either directly or indirectly to produce virtually every product in common use. To name a few, combustion processes produce and refine fuel, generate electricity, prepare foods and pharmaceuticals, and transport goods. Fire has transformed humankind and separated it from the beasts, illumined nations, and safeguarded generations. It has been used in war and peace, to tear down and build up; it is both feared and respected. It is a most powerful tool and worthy of study and understanding.

2.3 BRIEF OVERVIEW OF COMBUSTION EQUIPMENT AND HEAT TRANSFER

In the process industries, combustion powers gas turbines, process heaters, reactors, and boilers. The burner combusts fuel and generates products of combustion and heat. A firebox contains the flame envelope. The fire heats water in the tubes to boiling. The steam rises to a steam drum that separates the liquid and vapor phases, returning water to the tubes and passing steam. The steam may be further heated in a superheater. Superheaters raise the temperature of the steam above the boiling point, using either radiant and/or convective heat transfer mechanisms.

Radiant heat transfer requires a line-of-sight to the flame. Only this heat transfer mechanism can operate in a vacuum. For example, the Earth receives essentially all its heat from the sun through this mechanism. Convection requires the bulk movement of a hot fluid. In a boiler, hot combustion gases transfer heat to the outer tube wall via convection. Convection occurs naturally by means of buoyancy differences between hot and cool fluids — termed natural convection, or by motive devices such as fans or blowers — termed forced convection. Heat transfers from the outer to inner tube wall by conduction — the predominant heat transfer mode through metals. Inside the tube, convection is the predominant mode of heat transfer to the inside fluid. A more complete discussion of heat transfer is given in Chapter 3.

Most large boilers have water in the tubes and fire outside — called water-tube boilers. Fire-tube boilers put the fire and hot gases in tubes surrounded by water. This system is applicable to smaller, unattended boilers.

Process heaters are akin to water-tube boilers, but with some very important differences. First, process heaters contain a process fluid in the tubes, rather than water. The process fluid is usually a hydrocarbon, for example, crude oil. Process heaters come in two main varieties: vertical cylindrical (VC) and cabin style (see Chapter 1). VCs comprise a cylindrical flame zone surrounded by process tubes. Cabin-style heaters are rectangular with wall and roof tubes (see Figure 2.1). The radiant section comprises the space surrounded by tubes having a direct view of the flame. Most process heaters also have a convective section comprised of overhead tubes that cannot directly view the flame. Convective tubes receive their heat from the direct contact of the combustion gases. The transition from the radiant to convective sections is known as the bridgewall. Chapters 1 and 15 contain more detailed discussion of process heaters.

Reactors such as cracking furnaces and reforming furnaces are more extreme versions of process heaters. Here, the process fluid undergoes chemical transformations to a different substance. For example, in an ethylene cracking furnace, liquid or gas feedstock transforms to ethylene (C_2H_4), an intermediate in the production of polyethylene and other plastics. There are many specialized types of reactors using combustion as the heat source.

2.4 NET COMBUSTION CHEMISTRY OF HYDROCARBONS

Consider the combustion of methane (CH_4) and air. CH_4 is the major component of natural gas. The combustion of CH_4 produces carbon dioxide (CO_2) and water vapor (H_2O). Equation 2.1 summarizes the net reaction.

$$CH_4 + 2O_2 \rightarrow CO_2 + 2H_2O \tag{2.1}$$

Equation (2.1) is the stoichiometric equation. It gives the relative proportions of every element (e.g., C, H, and O) in each molecule, and the relative proportions of each molecule (e.g., CH_4, O_2, CO_2, and H_2O) in the reaction. A molecule is the smallest collection of chemically bound atoms that define a substance. An atom is the smallest building block of a molecule having a unique chemical identity. The arrow shows the direction of the reaction. Species to the left of the arrow are the reactants; those to the right are products.

In a stoichiometric equation, the subscripted numbers define the proportions of elements in a molecule. Equation (2.1) shows that the methane molecule comprises four hydrogen atoms for every carbon atom, and that an oxygen molecule comprises two oxygen atoms. Antecedent numbers (those that precede the molecular identity and are not subscripted) define the proportions of molecules in the reaction. The antecedent numbers in a stoichiometric equation are the stoichiometric coefficients. If a molecule appears without an antecedent number, the stoichiometric coefficient is assumed to be "1." Thus, Equation (2.1) shows two oxygen molecules reacting with one methane molecule to produce one molecule of carbon dioxide and two molecules of water.

Because stoichiometric equations deal only with proportions, they can be multiplied by any convenient basis to obtain more relevant or convenient units. One especially relevant unit is the mole. A mole comprises 6.02×10^{23} molecules and is abbreviated *mol*, where 6.02×10^{23} is known as Avogadro's number. The derivation of the mole follows a complicated history but results in some convenient properties. For example, 1 mole of hydrogen atoms has a mass of about 1 g (actually, 1.01 g). The mass of a mole of atoms is the atomic weight of the atom. Table 2.1 lists atomic weights for some common elements.[1] However, hydrogen does not exist as H under normal conditions, but rather as the molecule H_2. One easily derives the molecular weight as the sum of atomic weights of the molecule's elements. For example, from Table 2.1, calculate the molecular weight of H_2 as $2 \times 1.01 = 2.02$. The molecular weight of CH_4 is $12.01 + 4(1.01) = 16.05$.

2.5 CONSERVATION OF MASS

Stoichiometric equations are always balanced equations. That is, the number and kind of atoms in the reactants always equal the number and kind of atoms in the products. This is because chemical reactions, including combustion, have no ability to alter atomic identities, only molecular ones. The above discussion demands that combustion reactions conserve mass; that is, the mass of the reactants must equal the mass of the products.

The stoichiometric equation can be used to generate any convenient mass basis. According to Equation (2.1), 1 mole of CH_4 reacts with 2 moles of O_2 to produce 1 mole of CO_2 and 2 moles of H_2O. Using Table 2.1, it is possible to calculate that 16.05 lb methane react with 32.00 lb air to produce 44.01 lb CO_2 and 36.04 lb H_2O. According to the principle of conservation of mass, the mass of products equals the mass of reactants, or 48.05 lb.

Strictly speaking, the units must be in grams to refer to Avogadro's number of molecules. However, the use of grams, pounds, tons, or any convenient unit is justified as long as proportions are preserved. This has led to the very common and justified use of oxymorons like pound-mole (lbmol) and ton-mole that scale from, rather than refer to, Avogadro's number. One obvious basis for a combustion problem is to scale from the actual fuel flow.

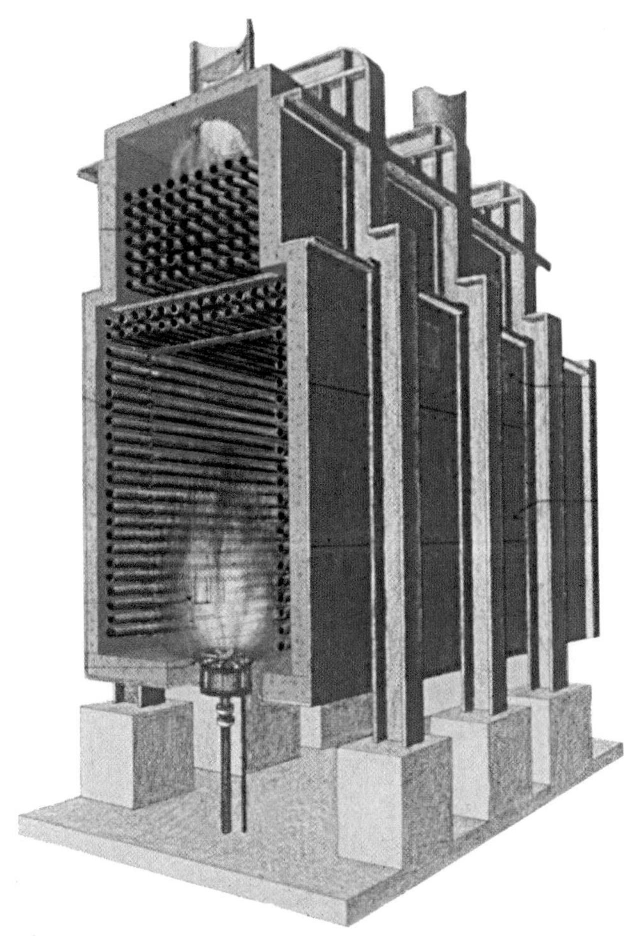

FIGURE 2.1 Typical cabin-style process heater.

In Equation (2.1), the moles of products equal the moles of reactants; but in general, moles are not conserved. For example, $2H_2 + O_2 \rightarrow 2H_2O$. Therefore, 3 moles of reactants yield only 2 moles of product. However, 36.04 lb reactants generate 36.04 lb products, confirming that mass is conserved, but not moles. Table 2.2 shows molar ratios for combustion reactants and products for some common fuels. Table 2.3 shows the moles of O_2 required to stoichiometrically combust 1 mole of some common fuels, and the number of moles of CO_2 and H_2O produced during that reaction.

2.6 THE IDEAL GAS LAW

The ideal gas law applies for typical combustion reactions and relates the pressure, volume, and number of moles:

TABLE 2.1 Alphabetical List of Atomic Weights for Common Elements[1]

Name	Symbol	At. no.	Atomic Weight	Footnotes
Actinium	Ac	89	[227]	
Aluminum	Al	13	26.981538(2)	
Americium	Am	95	[243]	
Antimony	Sb	51	121.760(1)	g
Argon	Ar	18	39.948(1)	g r
Arsenic	As	33	74.92160(2)	
Astatine	At	85	[210]	
Barium	Ba	56	137.32(7)	
Berkelium	Bk	97	[247]	
Beryllium	Be	4	9.012182(3)	
Bismuth	Bi	83	208.98038(2)	
Bohrium	Bh	107	[264]	
Boron	B	5	10.811(7)	g m r
Bromine	Br	35	79.904(1)	
Cadmium	Cd	48	112.411(8)	g
Calcium	Ca	20	40.078(4)	g
Californium	Cf	98	[251]	
Carbon	C	6	12.0107(8)	g r
Cerium	Ce	58	140.116(1)	g
Cesium	Cs	55	132.90545(2)	
Chlorine	Cl	17	35.4527(9)	m
Chromium	Cr	24	51.9961(6)	
Cobalt	Co	27	58.933200(9)	
Copper	Cu	29	63.546(3)	r
Curium	Cm	96	[247]	
Dubnium	Db	105	[262]	
Dysprosium	Dy	66	162.59(3)	g
Einsteinium	Es	99	[252]	
Erbium	Er	68	167.26(3)	g
Europium	Eu	63	151.964(1)	g
Fermium	Fm	100	[257]	
Fluorine	F	9	18.9984032(5)	
Francium	Fr	87	[223]	
Gadolinium	Gd	64	157.25(3)	g
Gallium	Ga	31	69.723(1)	
Germanium	Ge	32	72.61(2)	
Gold	Au	79	196.96655(2)	
Hafnium	Hf	72	178.49(2)	
Hassium	Hs	108	[269]	
Helium	He	2	4.002602(2)	g r
Holmium	Ho	67	164.93032(2)	
Hydrogen	H	1	1.00794(7)	g m r
Indium	In	49	114.818(3)	
Iodine	I	53	126.90447(3)	
Iridium	Ir	77	192.217(3)	
Iron	Fe	26	55.845(2)	
Krypton	Kr	36	83.80(1)	g m
Lanthanium	La	57	138.9055(2)	g
Lawrencium	Lr	103	[262]	
Lead	Pb	82	207.2(1)	g r
Lithium	Li	3	6.941(2)	g m r
Lutetium	Lu	71	174.967(1)	g
Magnesium	Mg	12	24.3050(6)	
Manganese	Mn	25	54.938049(9)	
Meitnerium	Mt	109	[268]	
Mendelevium	Md	101	[258]	
Mercury	Hg	80	200.59(2)	
Molybdenum	Mo	42	95.94(1)	g
Neodymium	Nd	60	144.24(3)	g
Neon	Ne	10	20.1797(6)	g m
Neptunium	Np	93	[237]	
Nickel	Ni	28	58.6934(2)	
Niobium	Nb	41	92.90638(2)	
Nitrogen	N	7	14.00674(7)	g r
Nobelium	No	102	[259]	
Osmium	Os	76	190.23(3)	g
Oxygen	O	8	15.9994(3)	g r
Palladium	Pd	46	106.42(1)	g
Phosphorus	P	15	30.973761(2)	
Platinum	Pt	78	195.078(2)	
Plutonium	Pu	94	[244]	
Polonium	Po	84	[209]	
Potassium	K	19	39.0983(1)	g
Praseodymium	Pr	59	140.90765(2)	
Promethium	Pm	61	[145]	
Protactinium	Pa	91	231.03588(2)	
Radium	Ra	88	[226]	
Radon	Rn	86	[222]	
Rhenium	Re	75	186.207(1)	
Rhodium	Rh	45	102.90550(2)	
Rubidium	Rb	37	85.4678(3)	g
Ruthenium	Ru	44	101.07(2)	g
Rutherfordium	Rf	104	[261]	
Samarium	Sm	62	150.36(3)	g
Scandium	Sc	21	44.955910(8)	
Seaborgium	Sg	106	[266]	
Selenium	Se	34	78.96(3)	
Silicon	Si	14	28.0855(3)	r
Silver	Ag	47	107.8682(2)	g
Sodium	Na	11	22.989770(2)	
Strontium	Sr	38	87.62(1)	g r
Sulfur	S	16	32.066(6)	g r
Tantalum	Ta	73	180.9479(1)	
Technetium	Tc	43	[98]	
Tellurium	Te	52	127.60(3)	g
Terbium	Tb	65	158.92534(2)	
Thallium	Tl	81	204.3833(2)	
Thorium	Th	90	232.0381(1)	g
Thulium	Tm	69	168.93421(2)	
Tin	Sn	50	118.710(7)	g
Titanium	Ti	22	47.867(1)	
Tungsten	W	74	183.84(1)	
Uranium	U	92	238.0289(1)	g m
Vanadium	V	23	50.9415(1)	
Xenon	Xe	54	131.29(2)	g m
Ytterbium	Yb	70	173.04(3)	g
Yttrium	Y	39	88.90585(2)	
Zinc	Zn	30	65.39(2)	
Zirconium	Zr	40	92.224(2)	g

g – geological specimens are known in which the element has an isotopic composition outside the limits for normal material. The difference between the atomic weight of the element in such specimens and that given in the table may exceed the stated uncertainty.

m – modified isotopic compositions may be found in commercially available material because it has been subjected to an undisclosed or inadvertent isotopic fractionation. Substantial deviations in atomic weight of the element from that given the table can occur.

r – range in isotopic composition of normal terrestrial material prevents a more precise atomic weight being given; the tabulated atomic weight value should be applicable to any normal material.

TABLE 2.1 (continued) Alphabetical List of Atomic Weights for Common Elements

This table of atomic weights is reprinted from the 1995 report of the IUPAC Commission on Atomic Weights and Isotopic Abundances. The Standard Atomic Weights apply to the elements as they exist naturally on Earth, and the uncertainties take into account the isotopic variation found in most laboratory samples. Further comments on the variability are given in the footnotes.

The number in parentheses following the atomic weight value gives the uncertainty in the last digit. An entry in brackets indicates that mass number of the longest-lived isotope of an element that has no stable isotopes and for which a Standard Atomic Weight cannot be defined because of wide variability in isotopic composition (or complete absence) in nature.

REFERENCE

IUPAC Commission on Atomic Weights and Isotopic Abundances, Atomic Weights of the Elements, 1995, *Pure Appl. Chem.*, 68, 2339, 1996.

Source: Courtesy of CRC Press.[1]

$$PV = nRT \tag{2.2}$$

where P = Pressure of the gas, psia
V = Volume of the gas, ft^3
n = Number of moles
R = Gas constant = 10.73 psia-ft^3/lbmol-°R
T = Absolute temperature, °R

Degrees Rankine (°R) are defined as the number of Fahrenheit degrees above absolute zero, the coldest possible theoretical temperature. Equation (2.2) shows that gas volume and moles are directly proportional.

Another useful form of the ideal gas law is:

$$PM = \rho RT \tag{2.3}$$

where ρ = Density of the gas, lb/ft^3
M = Molecular weight of the gas, lb/lbmol

Several units need further explanation. Molecular weights, whether given in g/mol, lb/lbmol, or tons/ton-mole, all have identical magnitude. That is, CH_4 has a molecular weight of 16.05 g/mol or 16.05 lb/lbmol. Because absolute zero is –459.67°F, to convert °F to °R, add 459.67. For example, 70°F ≈ 530°R. Finally, psia is defined as pounds force per square inch, absolute. Normal atmospheric pressure is 14.7 psia. However, pressure gauges read "0" at atmospheric pressure, denoted 0 psig — pounds force per square inch, gauge. Therefore, to convert from psig to psia, add local atmospheric pressure. For example, 35 psig ≈ 50 psia. Note that atmospheric pressure varies with elevation. For example, the normal atmospheric pressure in Denver (elevation ~5000 ft) is only 12.3 psia. Thus, a gauge reading of 35 psig in Denver equates only to about 47 psia. One must take into account elevation when performing combustion calculations.

An example best reinforces these points. A 1000 ft^3 vessel contains methane at 30 psig at 70°F. How many lbmol of methane does the vessel contain? What is the gas density? How much does the gas weigh? The solutions follow.

TABLE 2.2 Molar Ratios for Some Combustion Reactions and Products

Reaction	Moles Reactants	Moles Products
$H_2 + 0.5\ O_2 \rightarrow H_2O$	1.5	1.0
$CO + 0.5\ O_2 \rightarrow CO_2$	1.5	1.0
$CH_4 + 2\ O_2 \rightarrow CO_2 + 2\ H_2O$	3.0	3.0
$C_2H_2 + 3\ O_2 \rightarrow 2\ CO_2 + H_2O$	2.5	3.0
$C_3H_8 + 5\ O_2 \rightarrow 3\ CO_2 + 4\ H_2O$	6.0	7.0
$C_4H_{10} + 6.5\ O_2 \rightarrow 4\ CO_2 + 5\ H_2O$	8.5	9.0

Note: Most combustion reactions do not conserve moles.

TABLE 2.3 Molecular Weights and Stoichiometric Coefficients for Common Gaseous Fuels

Common Name	Formula	Molecular Weight	O_2 (moles)	CO_2 (moles)	H_2O (moles)
Hydrogen	H_2	2.02	0.5	0.0	1.0
Carbon monoxide	CO	28.01	0.5	1.0	0.0
Methane	CH_4	16.05	2.0	1.0	2.0
Ethane	C_2H_6	30.08	5.0	2.0	3.0
Ethene, ethylene	C_2H_4	28.06	4.0	2.0	2.0
Acetylene, ethyne	C_2H_2	26.04	3.0	2.0	1.0
Propane	C_3H_8	44.11	7.0	3.0	4.0
Propene, propylene	C_3H_6	42.09	6.0	3.0	3.0
Butane	C_4H_{10}	58.14	7.0	4.0	5.0
Butene, butylene	C_4H_8	56.12	8.0	4.0	4.0
Generic hydrocarbon	C_xH_y	12.01 x + 1.01 y	x + y/2	x	y/2

Example 2.1

From Equation (2.2),

$$n = \frac{PV}{RT} = \frac{(30+14.7)[\text{psia}] * 1000[\text{ft}^3]}{10.73\left[\dfrac{\text{psia ft}^3}{\text{lbmol °R}}\right] * (459.7+70)[\text{°R}]} = 7.86 \text{ lbmol}$$

From Equation (2.3) obtain

$$\rho = \frac{PM}{RT} = \frac{(30+14.7)[\text{psia}] * 16.05[\text{lb/lbmol}]}{10.73\left[\dfrac{\text{psia ft}^3}{\text{lbmol °R}}\right] * (459.7+70)[\text{°R}]}$$

$$= 0.126 \text{ lb/ft}^3$$

Finally, multiply the density by the volume to obtain the weight of gas, $m = \rho V = 0.126\ \text{lb/ft}^3 * 1000\ \text{ft}^3 = 126\ \text{lb}$.

2.7 STOICHIOMETRIC RATIO AND EXCESS AIR

The stoichiometric coefficient for oxygen identifies the theoretical oxygen required for combustion. To find the theoretical air comprising this amount of oxygen, it is necessary to define a mole of air as 0.21 lbmol O_2 + 0.79 lbmol N_2. Accordingly, Equation (2.1) is modified to account for air [in brackets]:

$$\begin{aligned} &CH_4 + 2[O_2 + 79/21\ N_2] \rightarrow \\ &CO_2 + 2\,H_2O + 2(79/21)\,N_2 \end{aligned} \tag{2.4}$$

Equation (2.4) is theoretical in that it presumes that all the oxygen and fuel react and that nitrogen does not. Actually, trace amounts of nitrogen will react with oxygen to form nitrogen oxides (NO_x). Although important in other contexts, the amount of reacting nitrogen is too small to consider here. In industrial practice, perfect mixing cannot be achieved. It is actually more cost-effective to ensure complete combustion with the addition of excess air. Excess air is that amount beyond theoretical added to ensure complete combustion of the fuel. To account for excess air, Equation (2.4) is modified with ε, the fraction of excess air.

$$\begin{aligned} &CH_4 + 2(1+\varepsilon)[O_2 + 79/21\ N_2] \rightarrow \\ &CO_2 + 2\ H_2O + 2\varepsilon\ O_2 + 2(1+\varepsilon)(79/21)\ N_2 \end{aligned} \tag{2.5}$$

Equation (2.5) shows two important chemical features of complete combustion: no carbon monoxide (CO) and some unreacted oxygen appear in the combustion products. To account for any hydrocarbon fuel, Equation (2.5) is modified by x, the H/C molar ratio. Equation (2.6) gives a generic equation for hydrocarbons with air.

$$\begin{aligned} &CH_x + (1+\varepsilon)(1+x/4)[O_2 + 79/21\ N_2] \rightarrow \\ &CO_2 + x/2\ H_2O + \varepsilon(1+x/4)\ O_2 \\ &+(1+x/4)(1+\varepsilon)(79/21)\ N_2 \end{aligned} \tag{2.6}$$

From Equation (2.6), it is possible to derive formulas relating volumes of flue gas species to excess air for a given fuel H/C:

oxygen-to-fuel ratio: $$O/F = (1+x/4)(1+\varepsilon) \tag{2.7}$$

air-to-fuel ratio: $$A/F = (100/21)(O/F) \tag{2.8}$$

total wet products: $$TWP = A/F + x/4 \tag{2.9}$$

total dry products: $$TDP = A/F - x/4 \tag{2.10}$$

In situ analyzers measure the flue gas species in the actual hot wet environment. In contrast, extractive analyzers remove the flue gas, condense the water, and measure the concentration of the flue gas species in the dry gas. Therefore, two sets of equations are needed for wet and dry measurements.

$$f_{O_2,wet} = \varepsilon(1+x/4)/TWP \qquad f_{O_2,dry} = \varepsilon(1+x/4)/TDP \tag{2.11}$$

$$f_{CO_2,wet} = 1/TWP \qquad f_{CO_2,dry} = 1/TDP \tag{2.12}$$

$$\begin{aligned} &f_{N_2,wet} = 79/21(1+\varepsilon)(1+x/4)/TWP \\ &f_{N_2,dry} = 79/21(1+\varepsilon)(1+x/4)/TDP \end{aligned} \tag{2.13}$$

$$f_{H_2O,wet} = x/(2\ TWP) \tag{2.14}$$

where f is the mole or volume fraction of the subscripted species, $0 < f < 1$, and the subscripts $_{wet}$ or $_{dry}$ refer to *in situ* or extractive measurements, respectively. Because of the strong relationship between oxygen and excess air, the excess oxygen can be used as a measure of excess air. For this purpose, it is usually easier to recast the equations for oxygen in the following forms:

$$f_{O_2,wet} = \frac{0.21\varepsilon}{K_{wet} + \varepsilon} \qquad f_{O_2,dry} = \frac{0.21\varepsilon}{K_{dry} + \varepsilon} \tag{2.15}$$

$$K_{wet} = \frac{4 + 1.21x}{4 + x} \qquad K_{dry} = \frac{4 + 0.79x}{4 + x} \tag{2.16}$$

$$\varepsilon = K_{wet}\left(\frac{f_{O_2,wet}}{0.21 - f_{O_2,wet}}\right) \qquad \varepsilon = K_{dry}\left(\frac{f_{O_2,dry}}{0.21 - f_{O_2,dry}}\right) \tag{2.17}$$

The equations are displayed graphically in Figure 2.2a–f for various fuels on a wet and dry basis.

2.7.1 *Heat of Combustion*

In addition to the conservation of mass, energy is also conserved in a combustion reaction. One measure of the chemical energy of a fuel is the *heat of combustion*. Table 2.4 gives heats of combustion for some typical fuels on a volume and mass basis.[2] Heat of combustion is reported as either net heating value (lower heating value, LHV) or gross heating value (higher heating value, HHV). To understand

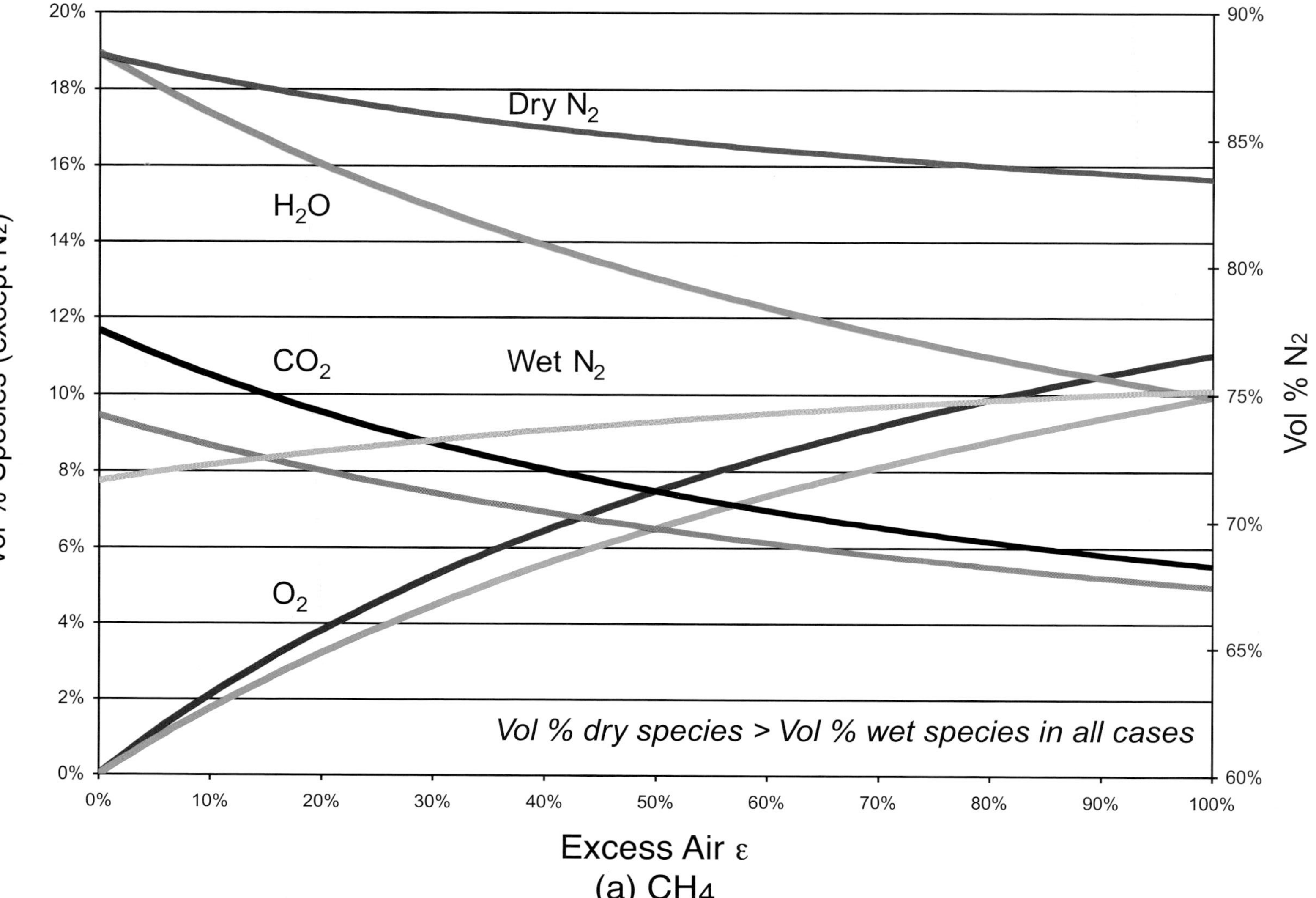

FIGURE 2.2 Species concentration vs. excess air for the following fuels: (a) CH_4, (b) natural gas, (c) simulated refinery gas (25% H_2, 50% CH_4, 25% C_3H_8), (d) C_3H_8, (e) No. 2 oil, and (f) No. 6 oil.

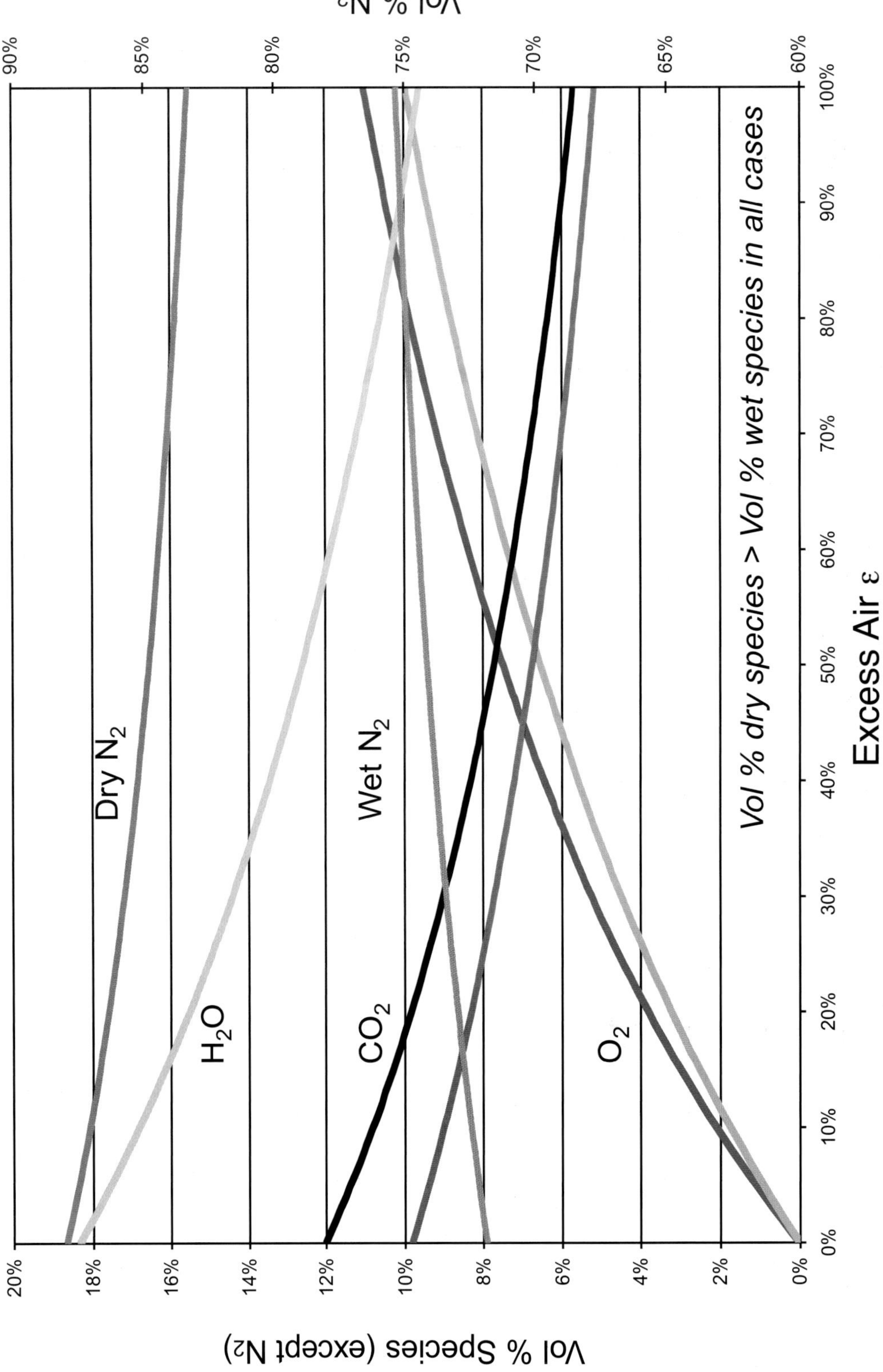

FIGURE 2.2 (continued)

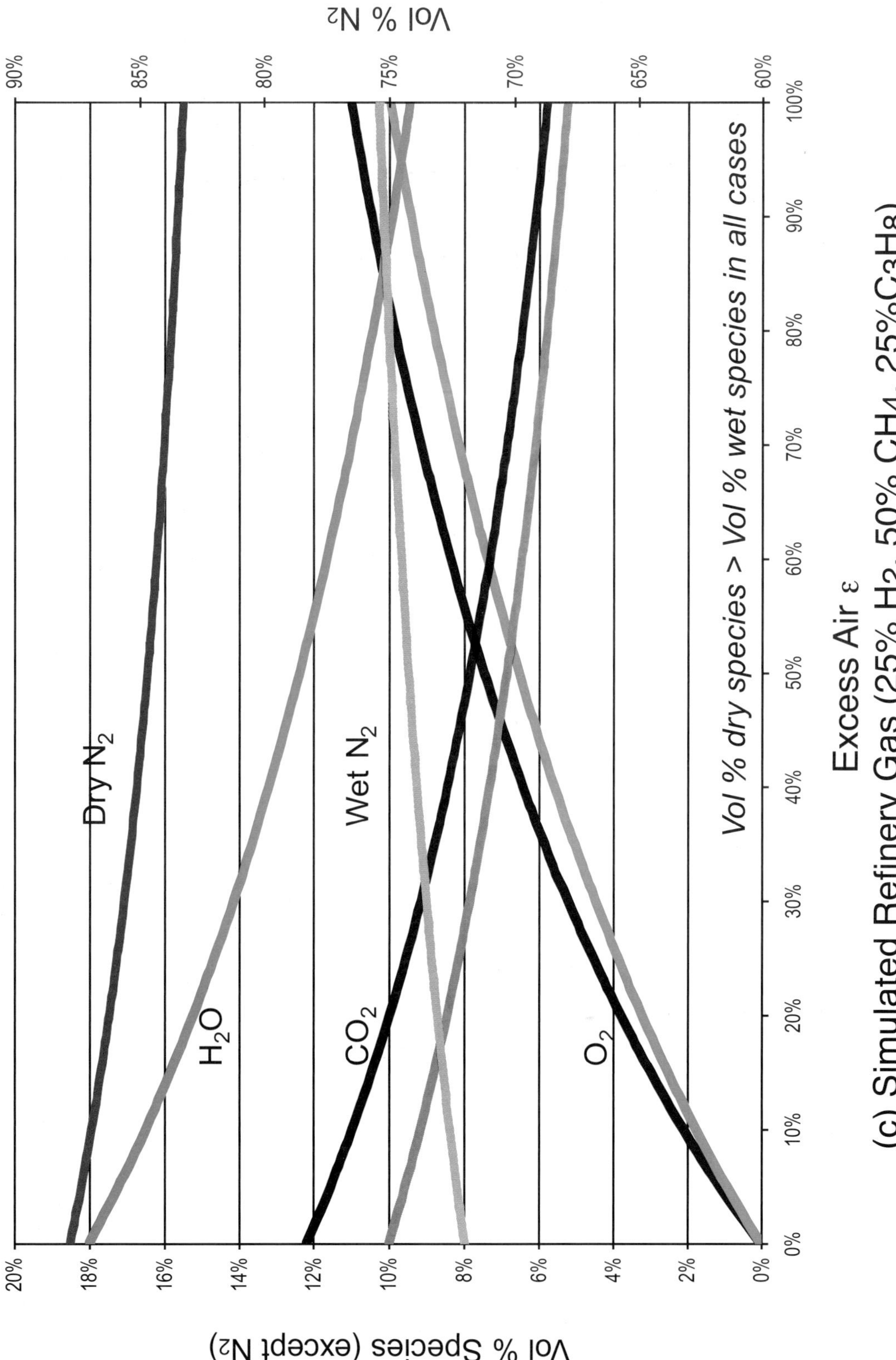

FIGURE 2.2 (continued)

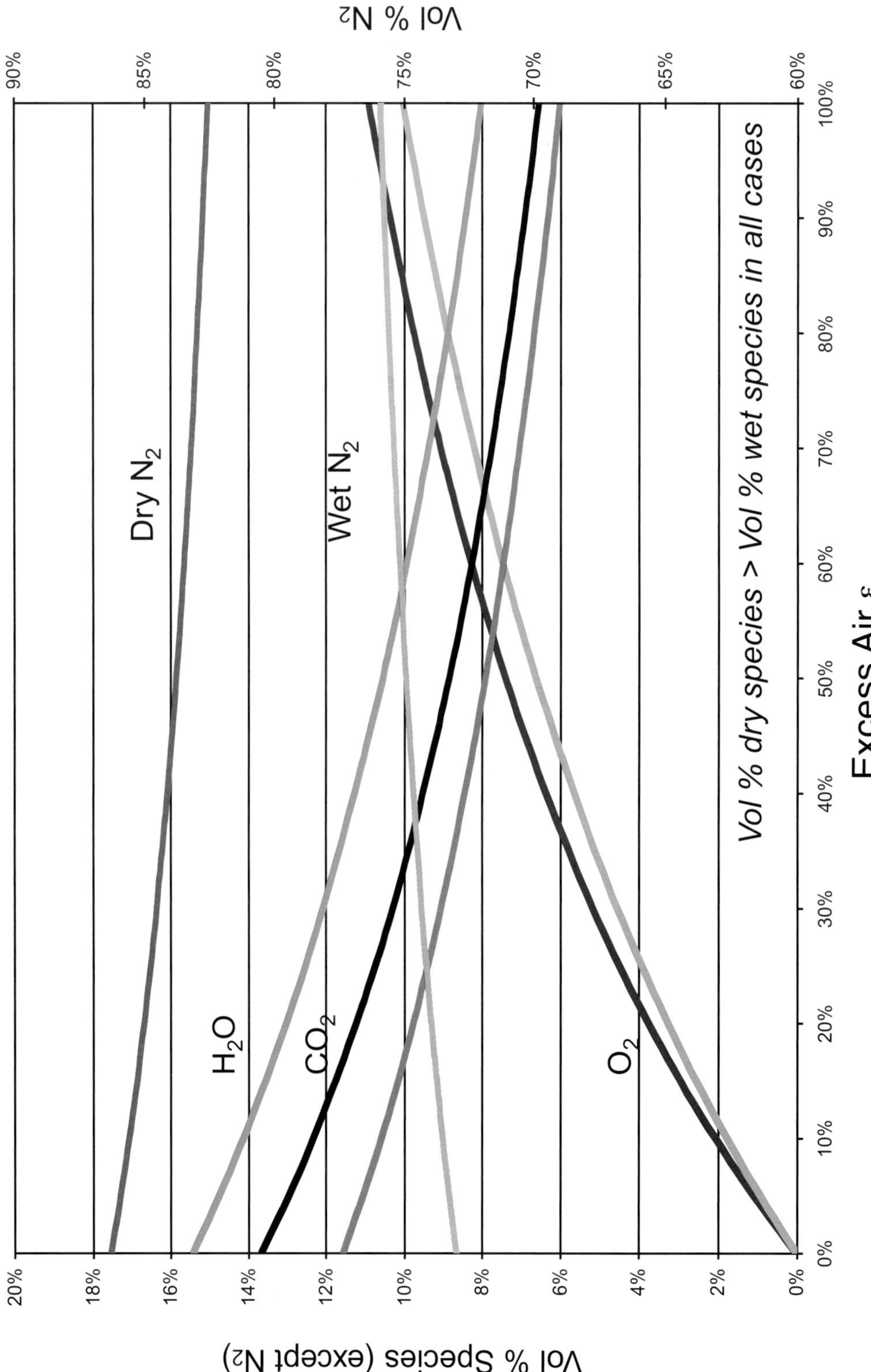

FIGURE 2.2 (continued)

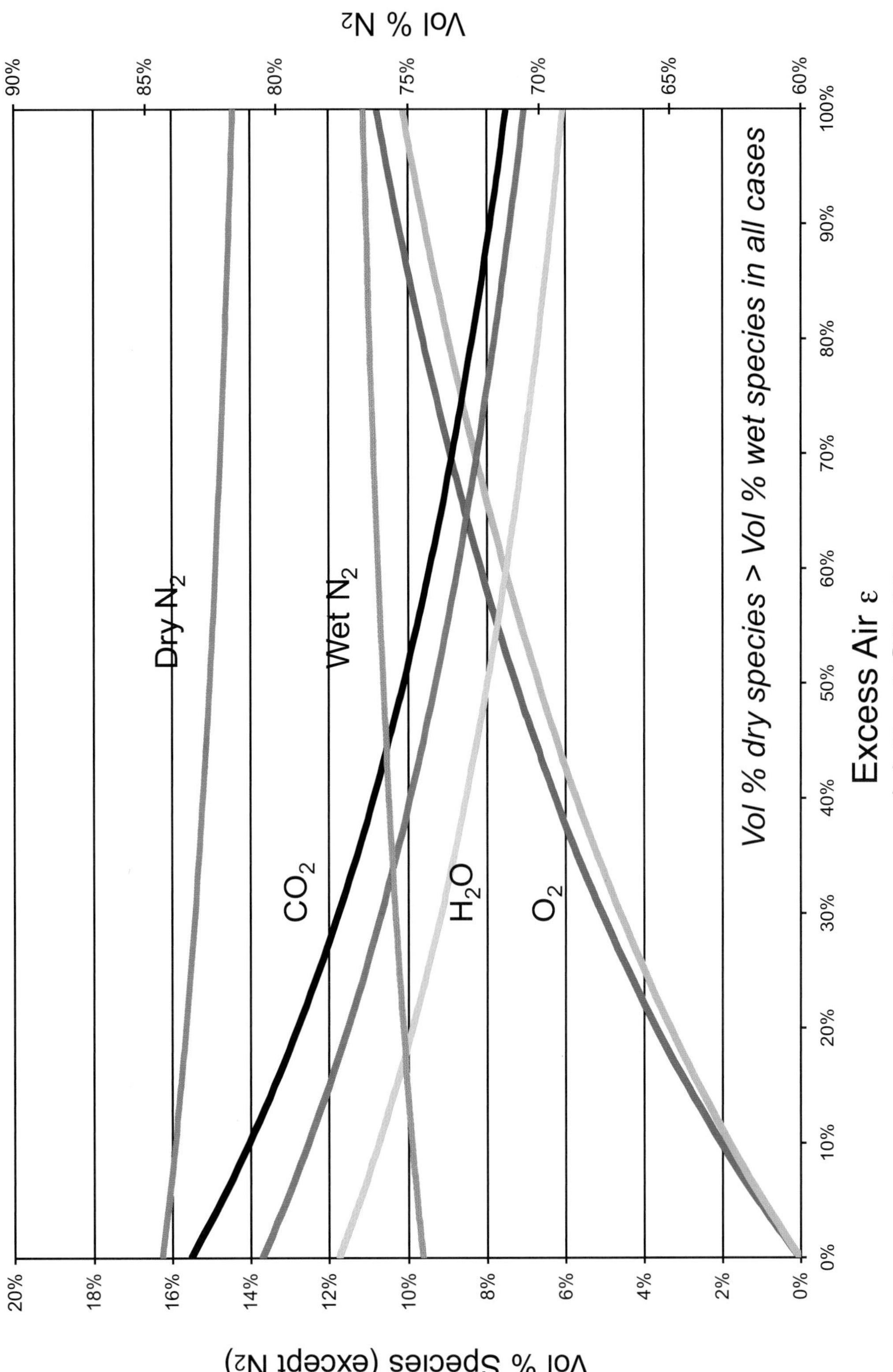

(e) Fuel Oil #2

FIGURE 2.2 (continued)

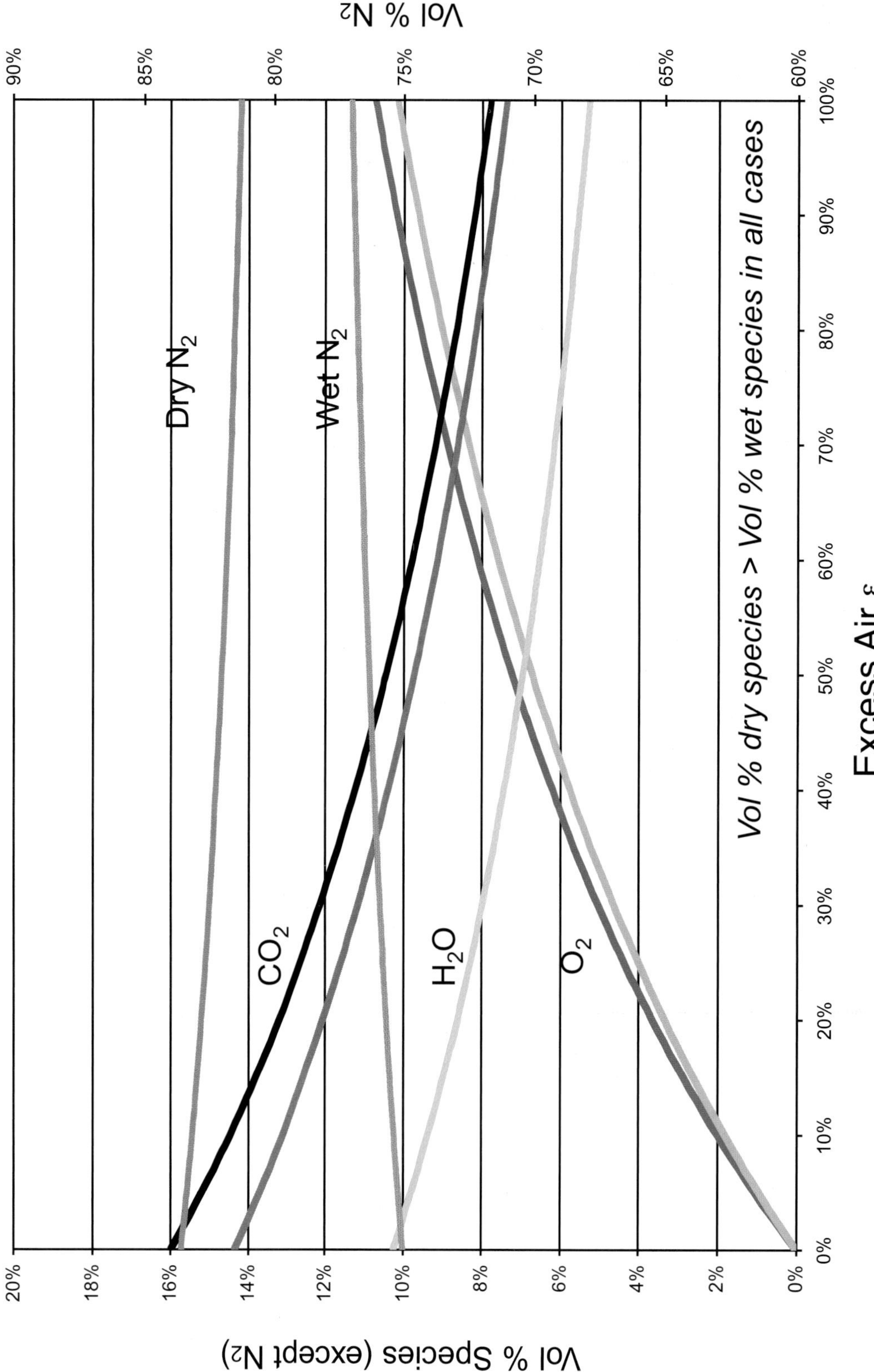

FIGURE 2.2 (continued)

TABLE 2.4 Combustion Data for Hydrocarbons

Hydrocarbon	Formula	Higher Heating Value (vapor), Btu lb_m^{-1}	Theor. Air/fuel Ratio, by mass	Max Flame Speed, (ft s^{-1})	Adiabatic Flame Temp (in air) (°F)	Ignition Temp (in air) (°F)	Flash Point (°F)	Flammability Limits (in air) (% by volume)	
Paraffins or Alkanes									
Methane	CH_4	23875	17.195	1.1	3484	1301	Gas	5.0	15.0
Ethane	C_2H_6	22323	15.899	1.3	3540	968–1166	Gas	3.0	12.5
Propane	C_3H_8	21669	15.246	1.3	3573	871	Gas	2.1	10.1
n-Butane	C_4H_{10}	21321	14.984	1.2	3583	761	–76	1.86	8.41
iso-Butane	C_4H_{10}	21271	14.984	1.2	3583	864	–117	1.80	8.44
n-Pentane	C_5H_{12}	21095	15.323	1.3	4050	588	< –40	1.40	7.80
iso-Pentane	C_5H_{12}	21047	15.323	1.2	4055	788	< –60	1.32	9.16
Neopentane	C_5H_{12}	20978	15.323	1.1	4060	842	Gas	1.38	7.22
n-Hexane	C_6H_{14}	20966	15.238	1.3	4030	478	–7	1.25	7.0
Neohexane	C_6H_{14}	20931	15.238	1.2	4055	797	–54	1.19	7.58
n-Heptane	C_7H_{16}	20854	15.141	1.3	3985	433	25	1.00	6.00
Triptane	C_7H_{16}	20824	15.151	1.2	4035	849	—	1.08	6.69
n-Octane	C_8H_{18}	20796	15.093	—	—	428	56	0.95	3.20
iso-Octane	C_8H_{18}	20770	15.093	1.1	—	837	10	0.79	5.94
Olefins or Alkenes									
Ethylene	C_2H_4	21636	14.807	2.2	4250	914	Gas	2.75	28.6
Propylene	C_3H_6	21048	14.807	1.4	4090	856	Gas	2.00	11.1
Butylene	C_4H_8	20854	14.807	1.4	4030	829	Gas	1.98	9.65
iso-Butene	C_4H_8	20737	14.807	1.2	—	869	Gas	1.8	9.0
n-Pentene	C_5H_{10}	20720	14.807	1.4	4165	569	—	1.65	7.70
Aromatics									
Benzene	C_6H_6	18184	13.297	1.3	4110	1044	12	1.35	6.65
Toluene	C_7H_8	18501	13.503	1.2	4050	997	40	1.27	6.75
p-Xylene	C_8H_{10}	18663	13.663	—	4010	867	63	1.00	6.00
Other Hydrocarbons									
Acetylene	C_2H_2	21502	13.297	4.6	4770	763–824	Gas	2.50	81
Naphthalene	$C_{10}H_8$	17303	12.932	—	4100	959	174	0.90	5.9

Note: Based largely on: "Gas Engineers' Handbook", American Gas Association, Inc., Industrial Press, 1967. For heating value in J kg^{-1}, multiply the value in Btu lb_m^{-1} by 2324. For flame speed in m s^{-1}, multiply the value in ft s^{-1} by 0.3048.

REFERENCES

American Institute of Physics Handbook, 2nd ed., D.E. Gray, Ed., McGraw-Hill Book Company, 1963.
Chemical Engineers' Handbook, 4th ed., R.H. Perry, C.H. Chilton, and S.D. Kirkpatrick, Eds., McGraw-Hill Book Company, 1963.
Handbook of Chemistry and Physics, 53rd ed., R.C. Weast, Ed., The Chemical Rubber Company, 1972; gives the heat of combustion of 500 organic compounds.
Handbook of Laboratory Safety, 2nd ed., N.V. Steere, Ed., The Chemical Rubber Company, 1971.
Physical Measurements in Gas Dynamics and Combustion, Princeton University Press, 1954.

the difference, reconsider Equation (2.1). From Equation (2.1), when methane burns, it produces two products: CO_2 and H_2O. The CO_2 will remain a gas under all conceivable industrial combustion conditions. However, H_2O can exist as either a liquid or a vapor, depending on how much heat is extracted from the process. If so much heat is extracted that the H_2O condenses, then the combustion yields its HHV. If water is released from the stack as a vapor, then combustion yields the LHV. A condensing turbine is an example of the former process, while a typical process heater is an example of the latter. Consequently, the process industry usually uses the LHV. Boiler and turbine calculations usually use the HHV. However, either measure can be used in combustion calculations as long as one is consistent. The inconsistent use of LHV and HHV is a major source of error in combustion calculations.

2.7.2 Adiabatic Flame Temperature

How hot can a flame be? First, there is a difference between heat (Q) and temperature (T). Heat is energy in transit. When a body absorbs heat, it stores it as another form of energy, increasing the body's temperature and expanding it. That is, the material uses some of the thermal energy to raise the temperature and some of the energy to expand the body against the atmosphere. The same amount of heat absorbed in different materials will yield different temperature increases and expansions.

For example, 100 Btu of heat will raise the temperature of 1 lb water by 100°F and expand the material from 62.4 ft^3 to 63.8 ft^3. The same 100 Btu of heat absorbed by 1 lb air will increase the temperature by 400°F and expand the material from 13.1 ft^3 to 15.6 ft^3. The total energy used to raise the temperature and increase the volume is called enthalpy (H). Enthalpy relates to temperature by a quantity known as the isobaric heat capacity, C_p. Table B-4 in the Appendix gives heat capacities for various gases. For a given mass of fuel, m, the quantities relate by Equation (2.18):

$$Q = m\Delta H = mC_p\Delta T \tag{2.18}$$

The symbol Δ is used to denote a difference between two states. Consider the following example.

Example 2.2

If 1 lb CH_4 combusts in 15% excess air, what is the maximum possible flame temperature.

Solution: From (2.6), the stoichiometric equation is

$$CH_4 + 1.15 * 2(O_2 + 79/21\ N_2) \rightarrow$$

$$CO_2 + 2\ H_2O + 0.30\ O_2 + 8.65\ N_2$$

Use the basis of 1 lb CH_4 and ratio all other components by the molecular weight of CH_4 to obtain the following:

IN		OUT	
CH_4:	1.00 lb	CO_2: 1*(44.01/16.05) =	2.74 lb
		H_2O: 2*(18.02/16.05) =	2.24
O_2: 1.15*2*(32.00/16.05) =	4.59	O_2: 0.30*(32.00/16.05) =	0.60
N_2: 1.15*2*79/21*(28.02/16.05) =	15.11	N_2: 8.65*(28.02/16.05) =	15.11
	20.7 lb		20.7

The highest possible flame temperature presumes no loss from the flame whatsoever. This is known as the adiabatic flame temperature. Equation (2.19) gives the total energy from combustion of 1 lb of fuel:

$$1\ \text{lb} * 22{,}000\ \text{Btu/lb} = 22{,}000\ \text{Btu} \tag{2.19}$$

Table B-4 in the Appendix gives the total heat capacity by the mass of each flue gas species. Rearranging Equation (2.17) for ΔT gives the following:

$$\Delta T = \Delta H / mC_p \tag{2.20}$$

Because there are several species in the flue gas, the contribution of each species must be used for mC_p. That is, $mC_p = m_{CO_2} C_{pCO_2} + m_{H_2O} C_{pH_2O} + m_{O_2} C_{pO_2} + m_{N_2} C_{pN_2}$. To illustrate the calculation, assume that an average heat capacity is ~0.30 lb/MMBTU. Then the adiabatic flame temperature becomes

$$\Delta T = 22{,}000 \frac{\text{Btu}}{\text{lb}}\left(\frac{\text{lb°F}}{0.30\ \text{Btu}}\right)\left(\frac{1}{20.7\ \text{lb}}\right) = 3543\text{°F}$$

For air and fuel and 60°F, the adiabatic fuel temperature becomes

$$\text{AFT} = 3543 + 60 = 3603\text{°F}$$

For more about adiabatic flame temperature, see Section 2.13 and Table 2.4.

Note that the actual flame temperature will be much cooler than this, because heat will transfer from the flame to the surroundings via convection and radiation. Also, at high temperatures, $CO_2 \rightarrow CO + 0.5\ O_2$ and $H_2O \rightarrow H_2 + 0.5\ O_2$, reducing the adiabatic flame temperature. Such dissociations have not been taken into account. Nonetheless, this calculation method is useful for calculating the approximate adiabatic flame temperature.

2.8 SUBSTOICHIOMETRIC COMBUSTION

The concept of excess air presumes air in addition to that required for combustion. However, if one does not provide enough air, combustion may still continue, generating large quantities of CO and combustibles. This is referred to as substoichiometric combustion. Process heaters and boilers should NEVER be operated in this mode. Suddenly adding air to such a hot mixture could result in explosion. Because substoichiometric combustion may have deadly consequences, it is useful to consider the process, observe its features, and learn to avoid it. The stoichiometric ratio, Φ, is a fuel:air ratio. It has the following relationship with ε.

$$\Phi = 1/(1+\varepsilon) \tag{2.21}$$

$$\varepsilon = (1-\Phi)/\Phi \tag{2.22}$$

Equation (2.23) shows Equation (2.6) modified for substoichiometric combustion.

$$\mathrm{CH}_x + a[\mathrm{O}_2 + 79/21\ \mathrm{N}_2] \rightarrow$$

$$b\,\mathrm{CO} + (1-b)\,\mathrm{CO}_2 + (x-c)/2\ \mathrm{H}_2$$

$$+ c/2\ \mathrm{H}_2\mathrm{O} + a\ 79/21\ \mathrm{N}_2 \qquad (2.23)$$

where a, x are specified and b, c are unknown having the relation $2a = 2 - b + c/2$. The reader should note that the formulation neglects soot. Turns[3] has pointed out that using an equilibrium calculation with the water gas shift reaction arrives at a good approximation for substoichiometric species. This is adequate for investigating the general features of substoichiometric combustion, which is done in Section 2.10.

2.9 EQUILIBRIUM AND THERMODYNAMICS

Equation (2.24) gives the water gas shift reaction:

$$\mathrm{CO} + \mathrm{H}_2\mathrm{O} \leftrightarrow \mathrm{CO}_2 + \mathrm{H}_2 \qquad (2.24)$$

The double-headed arrow indicates that the reaction proceeds in both directions simultaneously. When the rate of the forward reaction equals that of the reverse, the process is in dynamic equilibrium. Equilibrium is characterized by the following relation:

$$K = [\mathrm{CO}_2][\mathrm{H}_2]/[\mathrm{CO}][\mathrm{H}_2\mathrm{O}] \qquad (2.25)$$

where the brackets denote wet volume concentrations of the enclosed species. For substoichiometric combustion, it will be useful to define the following quantities: $\alpha = [\mathrm{H}_2]/[\mathrm{H}_2\mathrm{O}]$, $\beta = [\mathrm{CO}]/[\mathrm{CO}_2]$, then $K = \alpha/\beta$.

2.10 SUBSTOICHIOMETRIC COMBUSTION REVISITED

Now that equilibrium and the water gas shift reaction have been defined, one can revisit substoichiometric combustion. Solving the mass balance for C, H, and oxygen, in turn for α and β, and using the relation $K = \alpha/\beta$, one obtains the following equations:

$$\mathrm{CH}_x + \frac{1}{2}\left[\frac{2+\beta}{1+\beta} + \frac{x}{2(1+\beta K)}\right]\left(\mathrm{O}_2 + \frac{79}{21}\mathrm{N}_2\right) \rightarrow$$

$$\left(\frac{1}{1+\beta}\right)\mathrm{CO}_2 + \left(\frac{\beta}{1+\beta}\right)\mathrm{CO} + \frac{1}{2}\left(\frac{x}{1+\beta K}\right)\mathrm{H}_2\mathrm{O}$$

$$+\frac{1}{2}\left(\frac{\beta K x}{1+\beta K}\right)\mathrm{H}_2 + \left(\frac{79}{21}\right)\left(\frac{1}{2}\right)\left[\frac{2+\beta}{1+\beta} + \frac{x}{2(1+\beta K)}\right]\mathrm{N}_2 \qquad (2.26)$$

Now, by combining Equations (2.6), (2.10), (2.23), and (2.26) for the left side of the relation, one knows that a must have the following expression.

$$a = \frac{1}{\Phi}\left(1 + \frac{x}{4}\right) = \frac{1}{2}\left[\frac{2+\beta}{1+\beta} + \frac{x}{2(1+\beta K)}\right] \qquad (2.27)$$

One could solve for Φ and substitute into Equation (2.26). However, the equation is quadratic and complicated. An easier solution is to solve for both Φ and the desired species using a parametric relation in β. Equation (2.28) gives the relation for Φ:

$$\Phi = \frac{(x+4)(1+\beta)(1+\beta K)}{2(2+\beta)(1+\beta K) + x(1+\beta)} \qquad (2.28)$$

Solving for the species as a function of β gives the following:

$$\mathrm{TWP} = 1 + \frac{x}{2} + \frac{79}{21}\left[\frac{1}{1+\beta} + \frac{\beta}{2(1+\beta)} + \frac{x}{4(1+\beta K)}\right]$$

$$\mathrm{TDP} = 1 + \frac{1}{2}\left(\frac{\beta K x}{1+\beta K x}\right) + \frac{79}{21}\left[\frac{1}{1+\beta} + \frac{\beta}{2(1+\beta)} + \frac{x}{4(1+\beta K)}\right]$$

$$f_{\mathrm{O}_2,wet} = 0 \qquad f_{\mathrm{O}_2,dry} = 0 \qquad (2.29)$$

$$f_{\mathrm{CO}_2,wet} = \frac{1}{\mathrm{TWP}}\left(\frac{1}{1+\beta}\right) \qquad f_{\mathrm{CO}_2,dry} = \frac{1}{\mathrm{TDP}}\left(\frac{1}{1+\beta}\right) \qquad (2.30)$$

$$f_{\mathrm{CO},wet} = \frac{1}{\mathrm{TWP}}\left(\frac{\beta}{1+\beta}\right) \qquad f_{\mathrm{CO},dry} = \frac{1}{\mathrm{TDP}}\left(\frac{\beta}{1+\beta}\right) \qquad (2.31)$$

$$f_{\mathrm{H}_2\mathrm{O},wet} = \frac{1}{2\ \mathrm{TWP}}\left(\frac{x}{1+\beta K}\right) \qquad (2.32)$$

$$f_{\mathrm{H}_2,wet} = \frac{1}{2\ \mathrm{TWP}}\left(\frac{\beta K x}{1+\beta K}\right) \qquad f_{\mathrm{H}_2,dry} = \frac{1}{2\ \mathrm{TDP}}\left(\frac{\beta K x}{1+\beta K}\right) \qquad (2.33)$$

$$f_{\mathrm{N}_2,wet} = \frac{79}{21}\frac{1}{2\ \mathrm{TWP}}\left[\frac{2+\beta}{1+\beta} + \frac{x}{2(1+\beta K)}\right]$$

$$f_{\mathrm{N}_2,dry} = \frac{79}{21}\frac{1}{2\ \mathrm{TDP}}\left[\frac{2+\beta}{1+\beta} + \frac{x}{2(1+\beta K)}\right] \qquad (2.34)$$

Combining the excess air and substoichiometric equations, one can construct a graph of species concentrations vs. Φ, as shown in Figure 2.3(a–e) for various fuels on a wet and dry basis. In particular, the substoichiometric portion of the graphs use $K = 0.19$, which corresponds to a temperature of ~2200°F (1100°C). As Turns[3] has pointed out for propane, this gives excellent agreement with rigorous equilibrium calculations. Note that one can generate considerable CO and H_2 from substoichiometric combustion. If air is suddenly admitted to such a hot mixture, explosion is likely. The moral of the story? Do not run a process heater or boiler out of air!

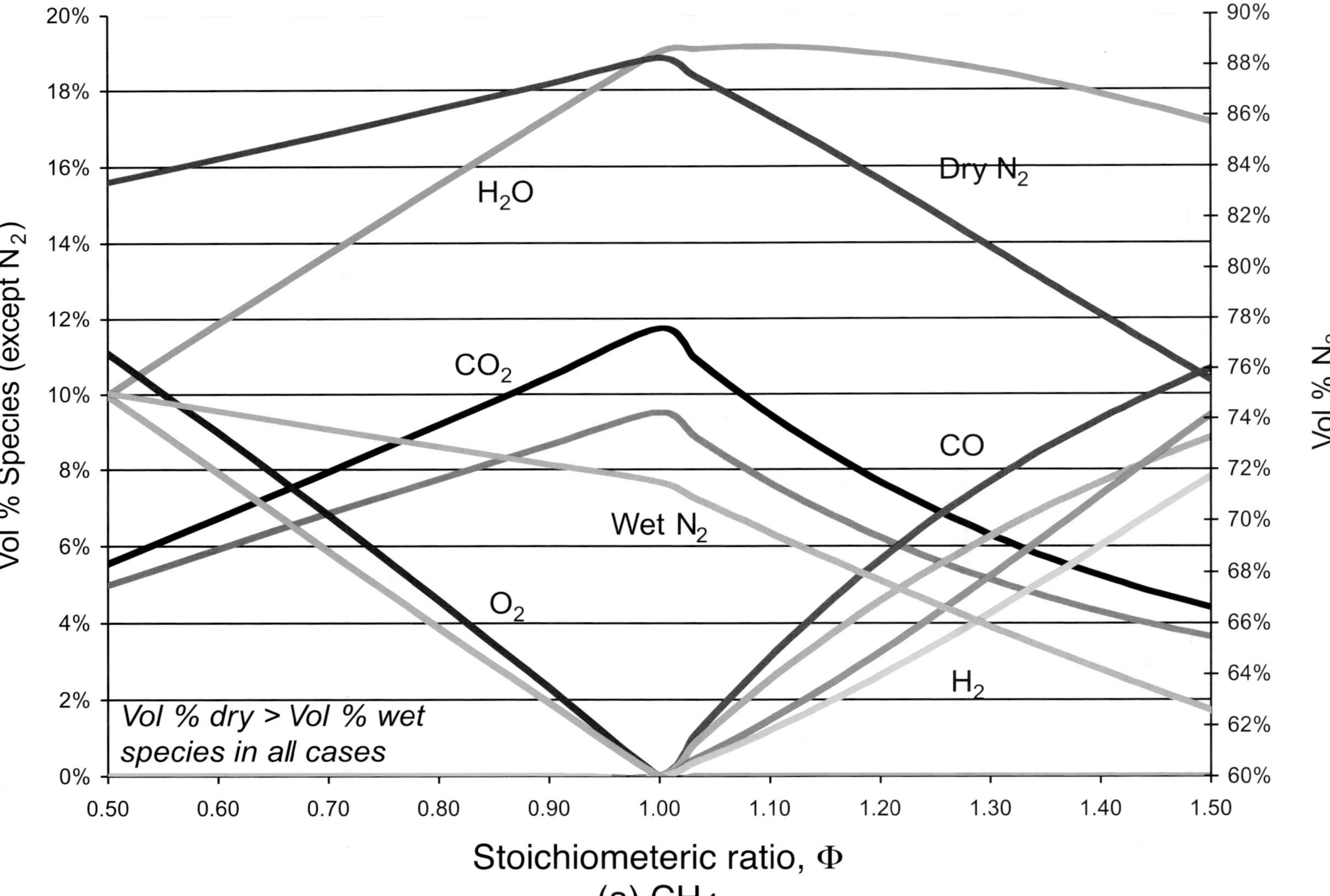

FIGURE 2.3 Species concentration vs. stoichiometric ratio for the following fuels: (a) CH_4, (b) natural gas, (c) simulated refinery gas (25% H_2, 50% CH_4, 25% C_3H_8), (d) C_3H_8, (e) No. 2 oil, and (f) No. 6 oil.

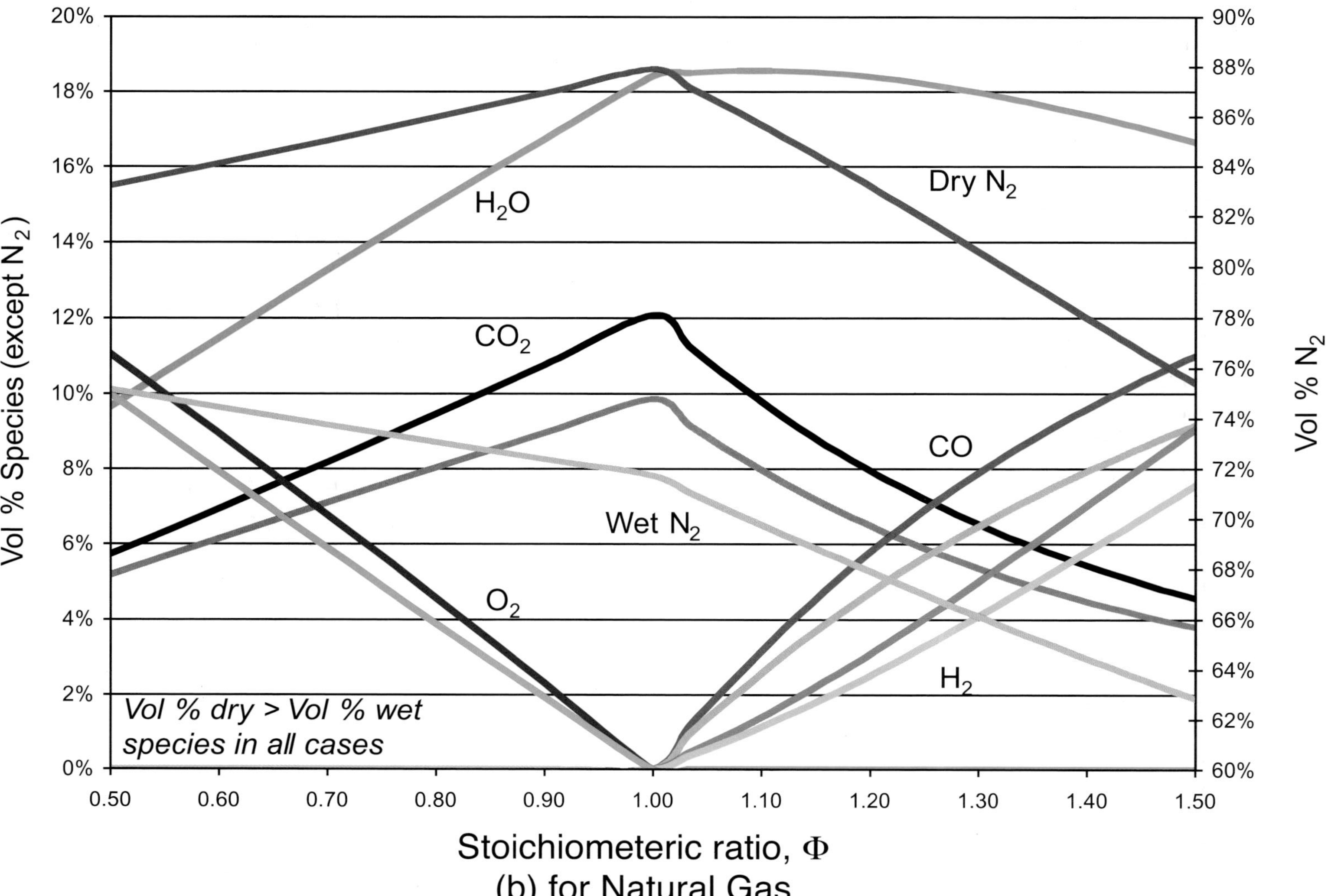

FIGURE 2.3 (continued)

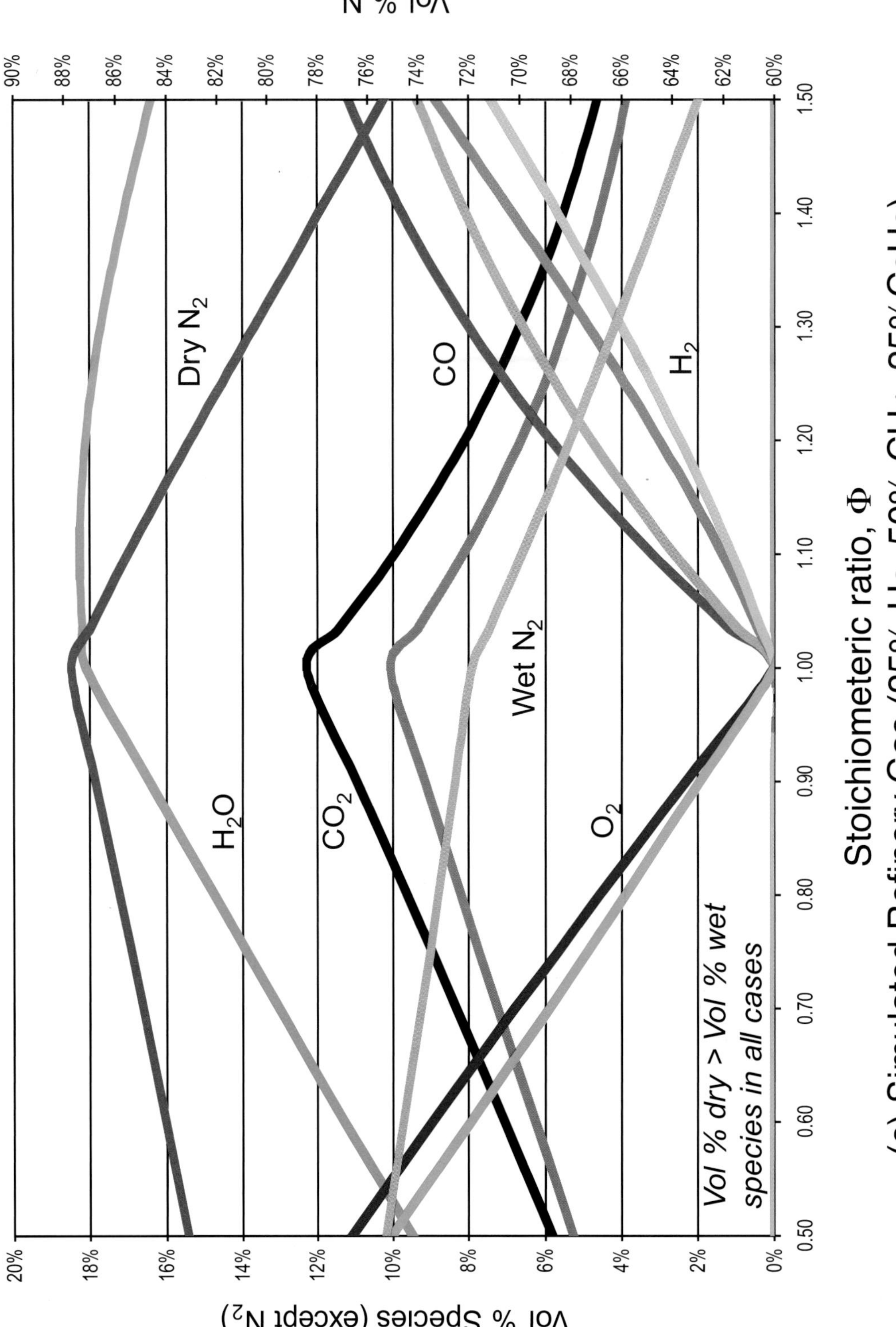

FIGURE 2.3 (continued)

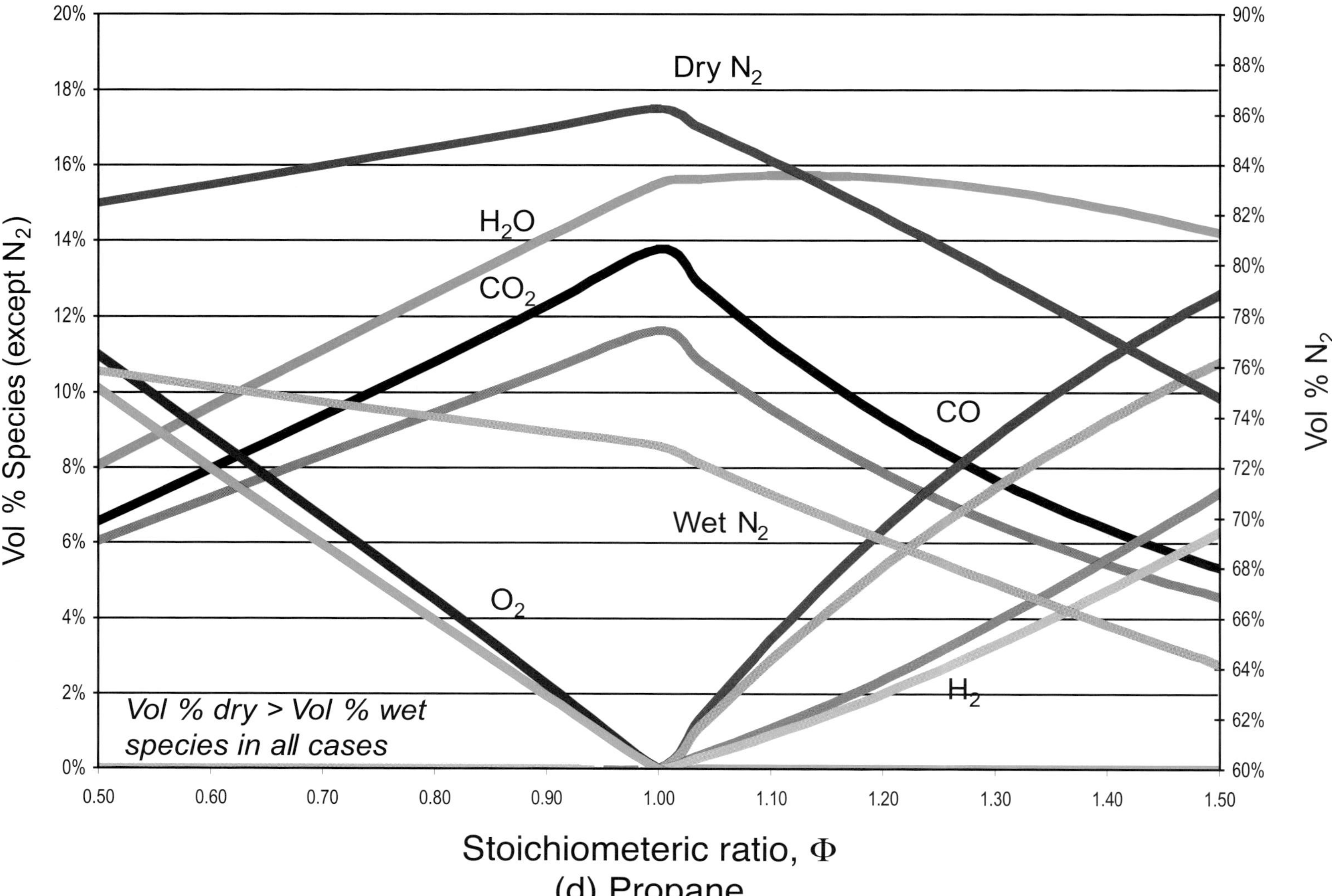

FIGURE 2.3 (continued)

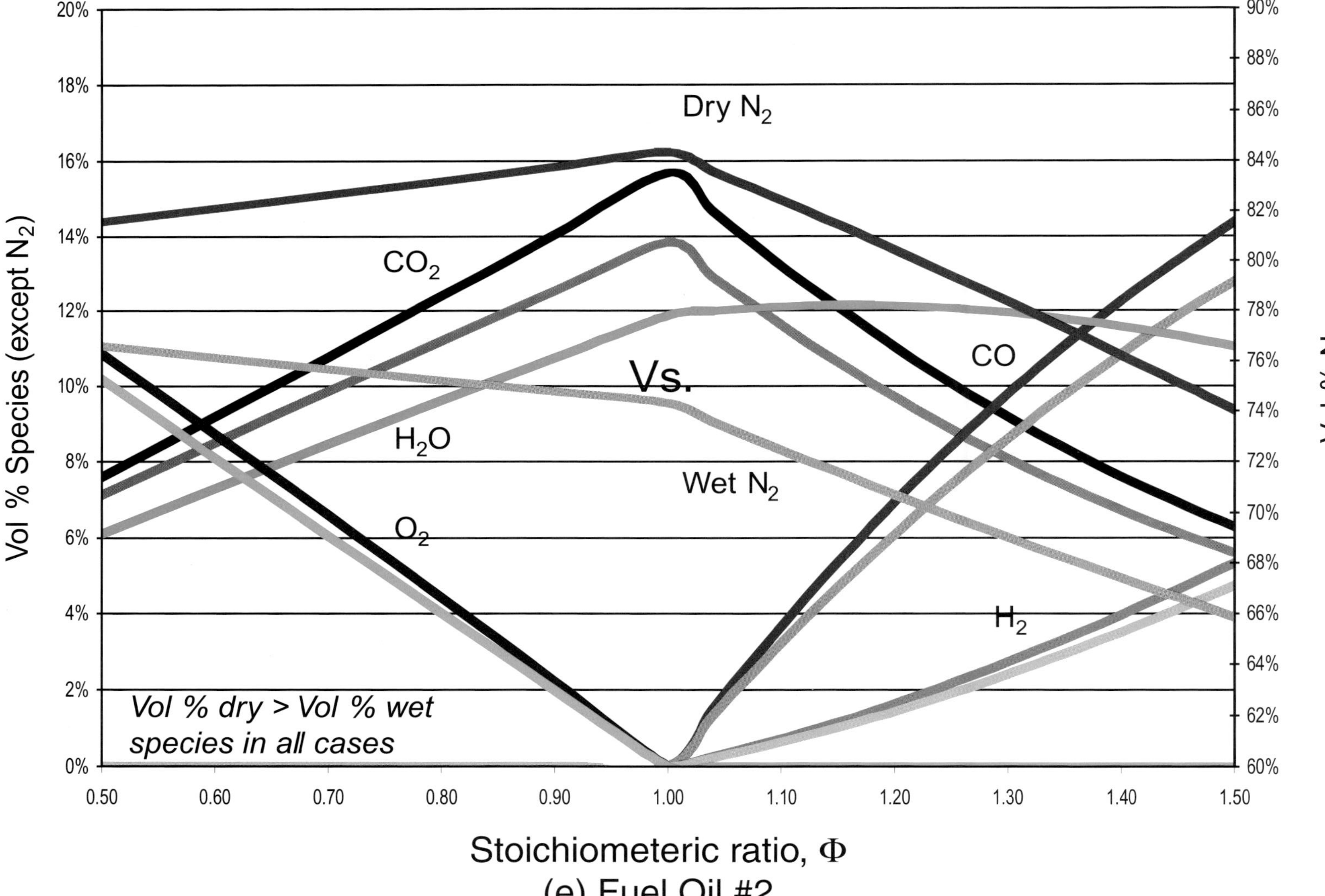

FIGURE 2.3 (continued)

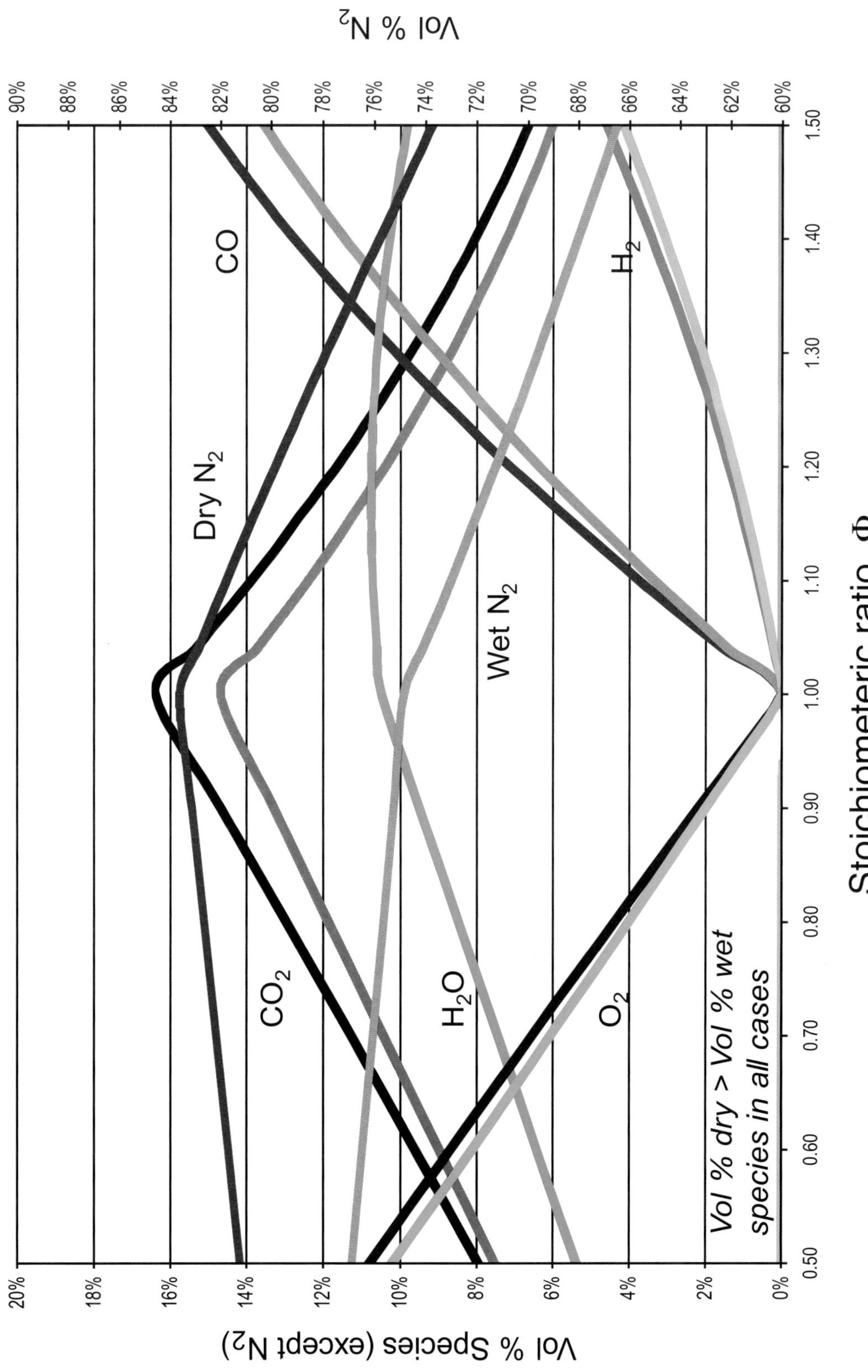

FIGURE 2.3 (continued)

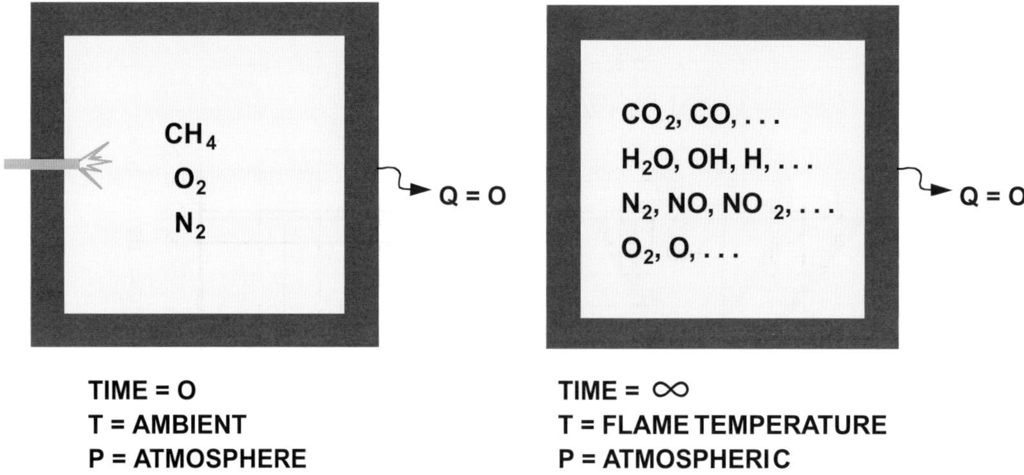

FIGURE 2.4 Adiabatic equilibrium reaction process.

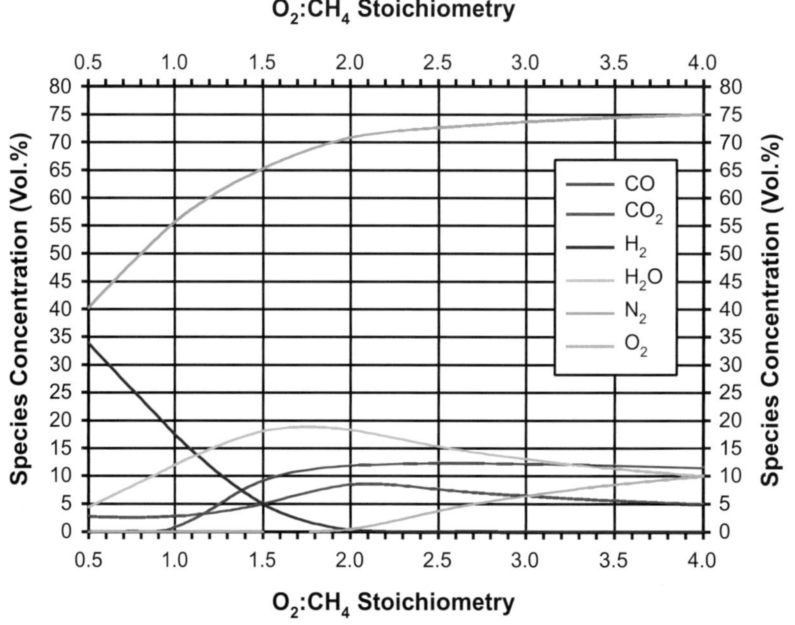

FIGURE 2.5 Adiabatic equilibrium calculations for the predicted gas composition as a function of the O_2:CH_4 stoichiometry for air/CH_4 flames, where the air and CH_4 are at ambient temperature and pressure.

2.11 GENERAL DISCUSSION

In this section, the concepts discussed so far are applied to combustion in general. Figure 2.4 shows a cartoon of an adiabatic equilibrium process. The boxes represent perfectly insulated enclosures, which do not exist in reality but are useful for illustrating the concept. The boxes are filled with a combustible mixture of a fuel and oxidizer, in this case methane and air, respectively. The left box represents the process at the time just before a spark is applied to ignite the mixture. The only species in the box are CH_4 and air (O_2 + 3.76N_2) in proportions to make the mixture flammable. A spark is then initiated to ignite the mixture. The right box represents the process an infinite time later to ensure all the reactions have gone to completion (i.e., reached equilibrium). In reality, most combustion reactions are completed in only a fraction of a second. Many species are then present after the reaction is completed. The exact composition depends on the ratio of the fuel to air. For example, if not enough air is present, then CO will be generated. If sufficient air is present, then little or no CO will be present. This is illustrated in Figure 2.5 which

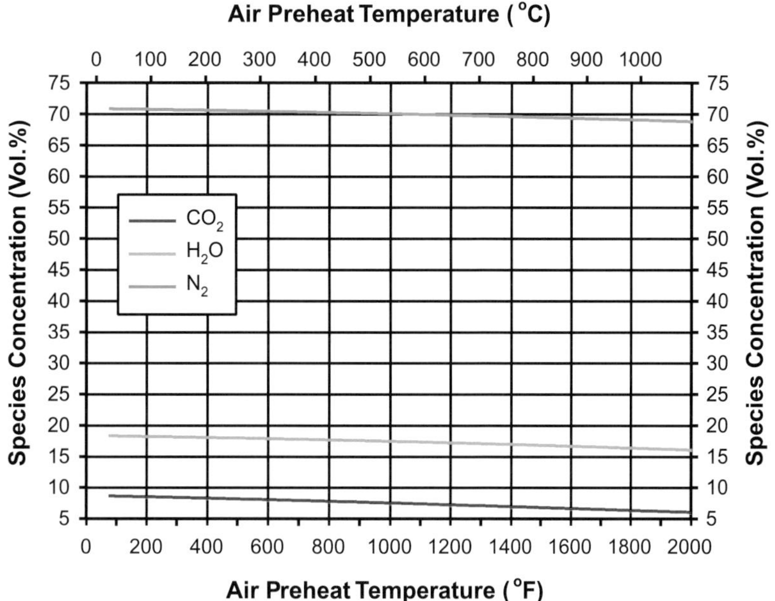

FIGURE 2.6 Adiabatic equilibrium stoichiometric calculations for the predicted gas composition of the major species as a function of the air preheat temperature for air/CH_4 flames, where the CH_4 is at ambient temperature and pressure.

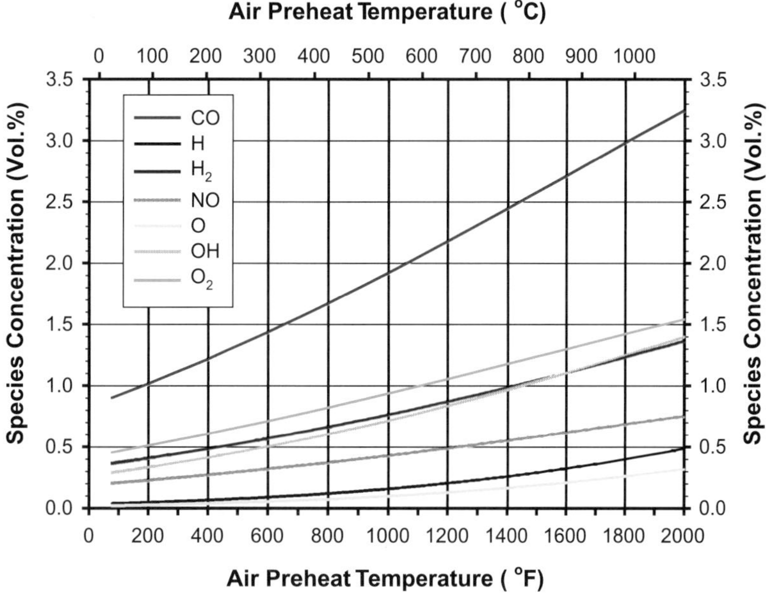

FIGURE 2.7 Adiabatic equilibrium stoichiometric calculations for the predicted gas composition of the minor species as a function of the air preheat temperature for air/CH_4 flames, where the CH_4 is at ambient temperature and pressure.

shows the predicted species for the adiabatic equilibrium combustion of methane and air as a function of the stoichiometry. For methane, the stoichiometric O_2:CH_4 ratio for theoretically perfect combustion is 2.0 as shown in Table 2.3. Stoichiometries less than 2.0 are fuel rich, as insufficient oxygen is present to fully combust the fuel. Stoichiometries greater than 2.0 are fuel lean, as excess oxygen is present. This figure shows that the exhaust product composition is highly dependent on the ratio of the fuel to the oxidizer.

2.11.1 Air Preheat Effects

Figure 2.6 shows the major species for the predicted exhaust gas composition for the stoichiometric combustion of

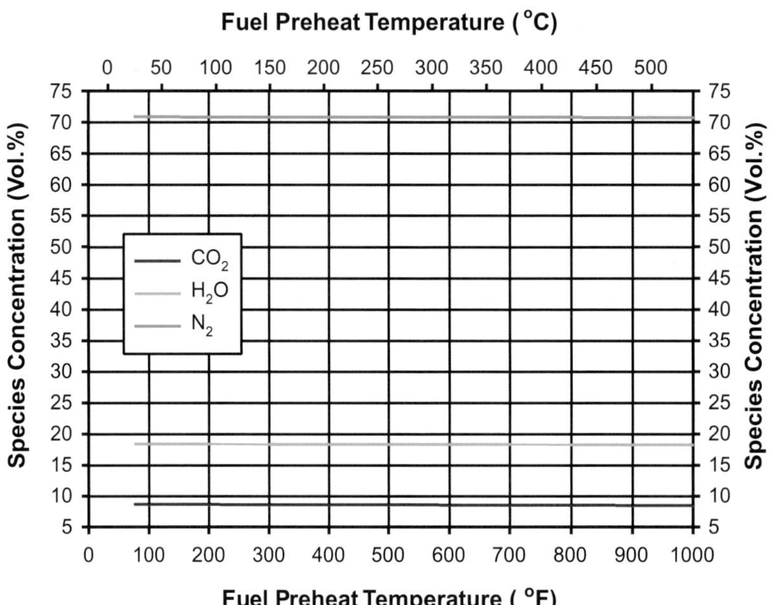

FIGURE 2.8 Adiabatic equilibrium stoichiometric calculations for the predicted gas composition of the major species as a function of the fuel preheat temperature for air/CH_4 flames, where the air is at ambient temperature and pressure.

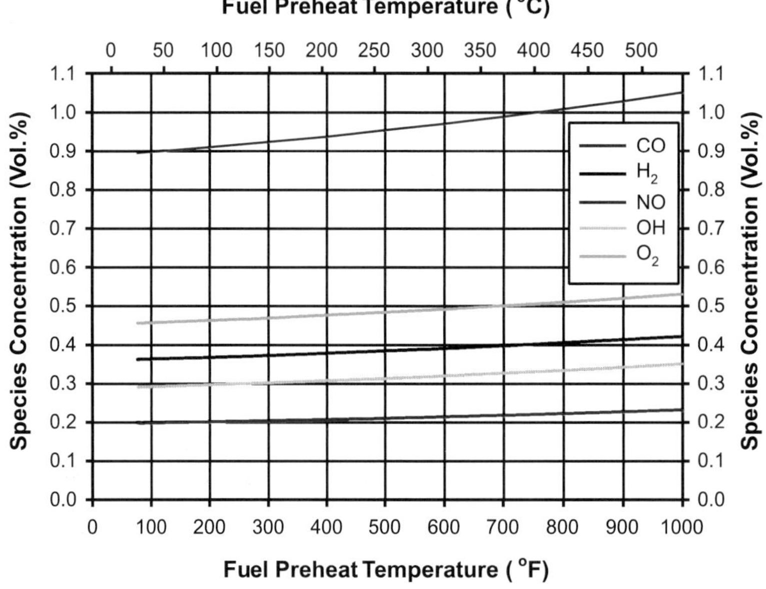

FIGURE 2.9 Adiabatic equilibrium stoichiometric calculations for the predicted gas composition of the minor species as a function of the fuel preheat temperature for air/CH_4 flames, where the air is at ambient temperature and pressure.

methane with preheated air. There is almost no change up to temperatures of about 1000°F (540°C), and only a relatively small change at higher temperatures. Figure 2.7 shows the predicted minor species in the exhaust gas for the same reaction of ambient temperature methane with preheated air. This graph shows that there is a dramatic increase in all the minor species as the air preheat temperature increases. This is due to chemical dissociation. Figure 2.8 shows the predicted major species in the exhaust products for the combustion of preheated methane with ambient air. There is very little change in the species concentration with fuel preheat. Note that higher fuel preheat temperatures present safety problems because of the auto-ignition temperature of methane, which is approximately 1200°F (650°C) in air.

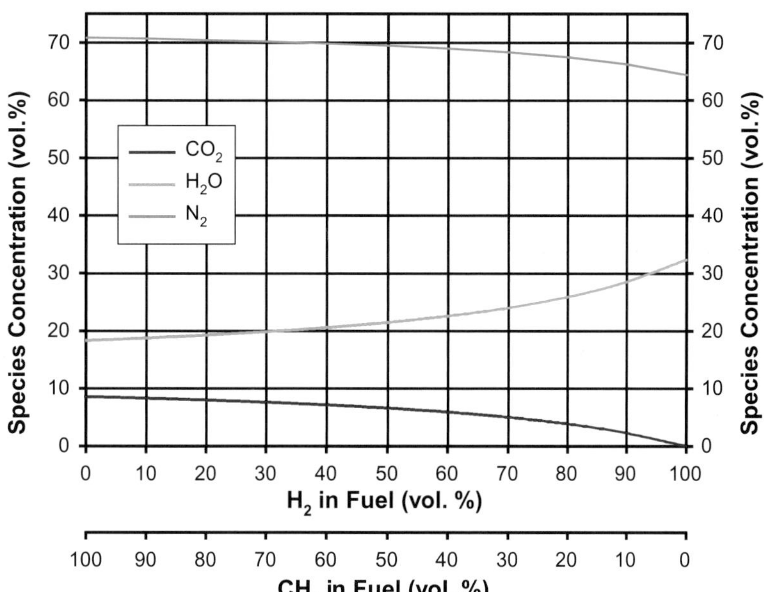

FIGURE 2.10 Adiabatic equilibrium stoichiometric calculations for the predicted gas composition of the major species as a function of the fuel blend (H_2 + CH_4) composition for air/fuel flames, where the air and fuel are at ambient temperature and pressure.

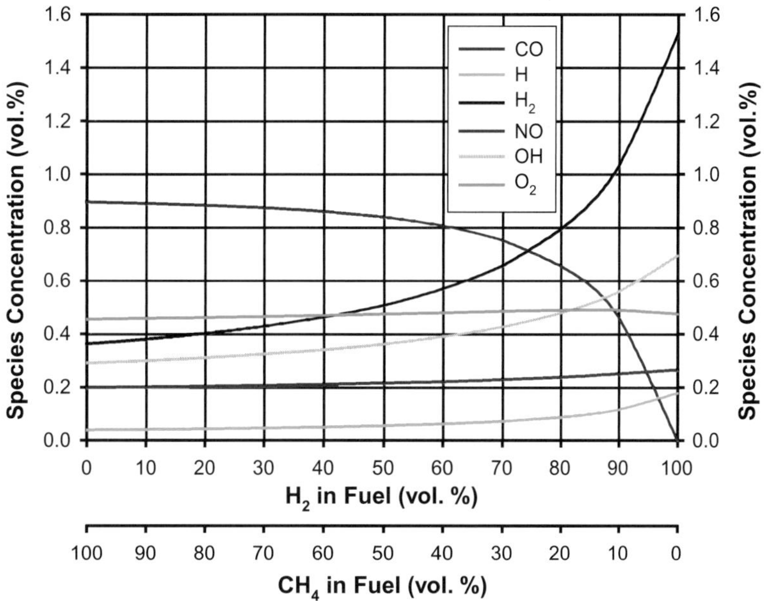

FIGURE 2.11 Adiabatic equilibrium stoichiometric calculations for the predicted gas composition of the minor species as a function of the fuel blend (H_2 + CH_4) composition for air/fuel flames, where the air and fuel are at ambient temperature and pressure.

Figure 2.9 shows that the predicted minor species concentrations increase with fuel preheat temperature.

2.11.2 *Fuel Blend Effects*

Fuel blends are particularly important in many of the hydrocarbon and petrochemical industries. Figure 2.10 shows the predicted major species for the combustion of air with fuel blends consisting of H_2 and CH_4. CO_2 and N_2 decline and H_2O increases as the H_2 content in the fuel increases. It is important to note that the species concentrations are not linear functions of the blend composition, where the change occurs more rapidly at higher H_2 compositions. Figure 2.11 is

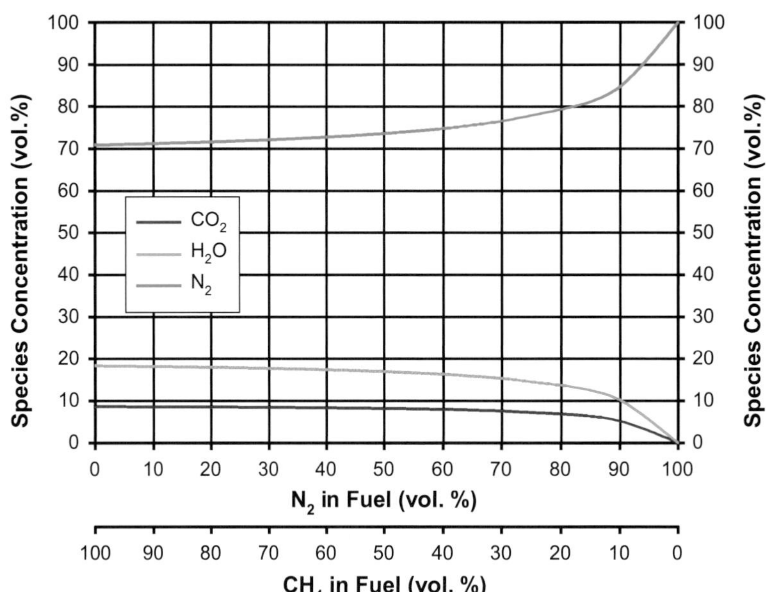

FIGURE 2.12 Adiabatic equilibrium stoichiometric calculations for the predicted gas composition of the major species as a function of the fuel blend (N_2 + CH_4) composition for air/fuel flames, where the air and fuel are at ambient temperature and pressure.

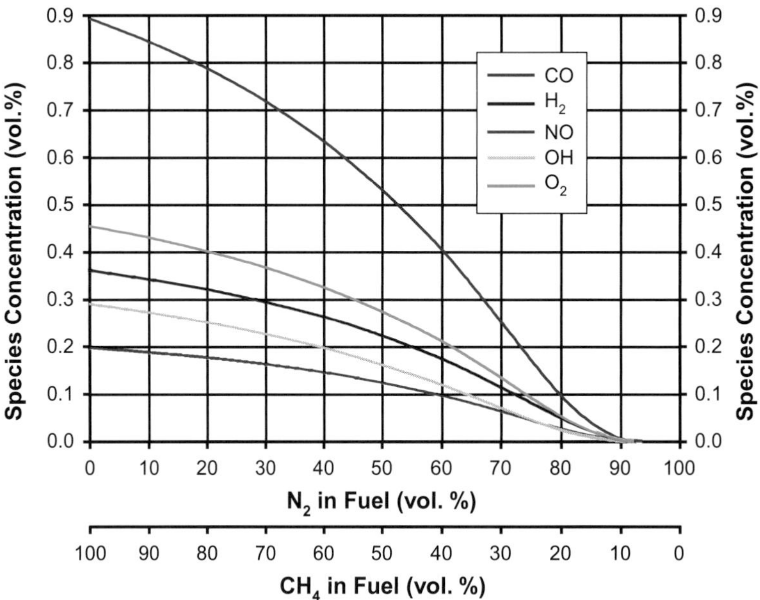

FIGURE 2.13 Adiabatic equilibrium stoichiometric calculations for the predicted gas composition of the minor species as a function of the fuel blend (N_2 + CH_4) composition for air/fuel flames, where the air and fuel are at ambient temperature and pressure.

a similar plot of the predicted minor species as functions of the H_2/CH_4 fuel blend. This graph also shows strong nonlinearities as the H_2 content increases. Figure 2.12 shows the predicted major species for the combustion of air with fuel blends consisting of an inert (N_2) and CH_4. At the extreme of 100% N_2, there is no fuel left in the "fuel blend" and no combustion takes place. There is a rapid change in the species concentrations as the N_2 content increases. Figure 2.13 shows the predicted minor species for the combustion of N_2/CH_4 fuel blends. This graph also shows a rapid decline in the species concentration, in this case for the minor species.

Real combustion processes are not adiabatic, as the whole intent is to transfer heat from the flame to some type of load. The amount of heat lost from the process determines the

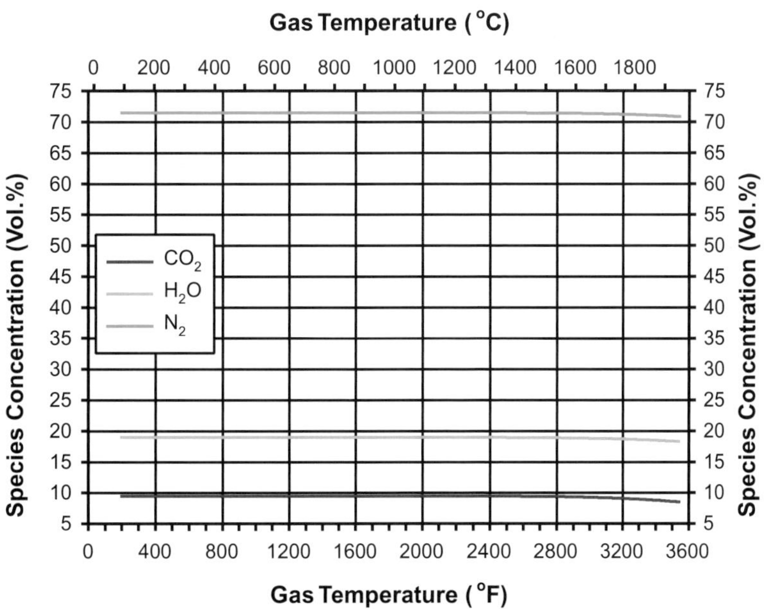

FIGURE 2.14 Equilibrium calculations for the predicted gas composition of the major species as a function of the combustion product temperature for air/CH_4 flames, where the air and fuel are at ambient temperature and pressure.

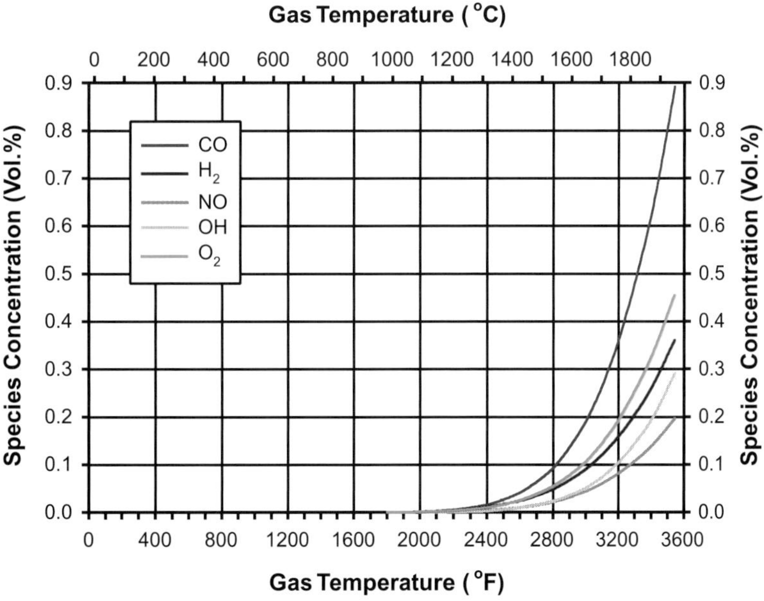

FIGURE 2.15 Equilibrium calculations for the predicted gas composition of the minor species as a function of the combustion product temperature for air/CH_4 flames, where the air and fuel are at ambient temperature and pressure.

temperature of the exhaust gases. The higher the heat losses from the flame, the lower the exhaust gas temperature. Figure 2.14 shows the predicted major species for the combustion of air and methane as a function of the exhaust gas temperature. The peak temperature is the adiabatic flame temperature. There is relatively little change in the major species concentration as a function of temperature. Figure 2.15 shows the predicted minor species for the combustion of air and methane as a function of the exhaust gas temperature. The concentrations are essentially zero up to temperatures of about 2000°F (1100°C) and rapidly increase up to the adiabatic flame temperature.

2.12 COMBUSTION KINETICS

The net results of stoichiometric equations have been considered thus far. However, the actual combustion mechanism is quite complex, involving very short-lived species that do not survive much beyond the flame. For example, the simplest system — hydrogen combustion — comprises about 20 elemental reactions. Elemental reactions denote the actual species involved in the reaction. Sometimes, liberties are taken with the antecedent numbers in order to balance the net equation. The equals (=) operator distinguishes the elemental reactions. The arrow operator (→) refers to the net results. Some important elemental reactions for the hydrogen-oxygen system are as follows:

$$2[\mathrm{H_2 + M = 2\,H + M}] \tag{2.35}$$

$$2[\mathrm{H + O_2 = OH + O}] \tag{2.36}$$

$$\mathrm{O + O + M = O_2 + M} \tag{2.37}$$

$$2[\mathrm{H + OH + M = H_2O + M}] \tag{2.38}$$

$$\text{Net: } \mathrm{2\,H_2 + O_2 \rightarrow 2\,H_2O} \tag{2.39}$$

In Equations (2.35 to 2.38), "M" refers to a third body such as the reactor wall or another nearby molecule that can absorb some of the reaction energy, but participates in no other way. The numbers preceding the brackets [ ] are used merely to balance the net reaction. Equations (2.35 to 2.38) show that the combustion of hydrogen is not as simple as 2 H_2 colliding with a single O_2 to form 2 H_2O, as the net reaction would suggest. About 20 elemental reactions exist for the hydrogen-oxygen system.

The situation worsens for hydrocarbons. It takes more than 100 elemental reactions to describe the combustion of refinery gases containing hydrogen, methane, and propane. Higher hydrocarbons and more complex mixtures require even more reactions. Fortunately, the net reaction is enough for most purposes. However, CO and NOx formation cannot be understood using only net reactions.

2.12.1 *Thermal NOx Formation: A Kinetic Example*

NOx is one pollutant that regulatory agencies are scrutinizing more and more (see Chapter 6). The "x" in NOx indicates a variable quantity. NOx from boilers and process heaters comprises mostly NO and very little NO_2.

Consider NO formed from the high-temperature reaction of N_2 and O_2, referred to as thermal NOx. There are more than 70 steps in the sequence. Often, a single elemental reaction paces an entire reaction sequence, just as the slowest car on a mountain road paces all the cars following it. This slowest step is the rate-limiting step. In the case of thermal NOx, the rate-limiting step is the rupture of the N≡N triple bond by an oxygen atom:

$$\mathrm{N_2 + O = NO + N} \tag{2.40}$$

In turn, the nitrogen radical reacts with available oxygen as follows:

$$\mathrm{N + O_2 = NO + O} \tag{2.41}$$

Adding these two equations together gives the overall reaction:

$$\mathrm{N_2 + O_2 \rightarrow 2\,NO} \tag{2.42}$$

2.12.2 *Reaction Rate*

Fundamental kinetic principles formulate the rate law. In general, the rate law for an elemental reaction $rR + sS = pP + qQ$ is given by the following differential equation:

$$\frac{d[P]}{dt} = k_f[R]^r[S]^s - k_r[P]^p[Q]^q \tag{2.43}$$

where P and Q are product species, R and S are reactant species, t is time, k_f is the forward reaction rate constant, k_r is the reverse reaction rate constant, the lowercase letters are the stoichiometric coefficients of the species (uppercase), and the brackets [ ] denote wet volume concentrations. For Equation (2.42), it is possible to neglect the reverse reaction and write:

$$\frac{d[\mathrm{NO}]}{dt} = k_f[\mathrm{N_2}][\mathrm{O}] \tag{2.44}$$

Unfortunately, [O] is not known, nor can it be conveniently measured. However, if it is presumed that a partial equilibrium exists between molecular and atomic oxygen, that is, 0.5 O_2 → O [see Equation (2.41)], then $[\mathrm{O}] = K\mathrm{O_2}^{0.5}$. The constants can be combined as $k = k_f * K$.

Rate constants follow the temperature relation $k = AT^n e^{-(b/T)}$, where A, b, and n are constants and T is the absolute temperature. For the case of thermal NOx, n = 0. Making these substitutions gives the final differential equation:

$$d[NO] = Ae^{-\frac{b}{T}}[N_2]\sqrt{[O_2]}dt \qquad (2.45)$$

This differential equation cannot be integrated because the actual temperature-oxygen-time path in a turbulent diffusion flame is not known. However, Equation (2.45) does show that increasing the temperature, oxygen concentration, or time increases NOx formation. Consequently, NOx reduction strategies usually attempt to reduce one or more of these factors.

2.12.3 Prompt-NOx Formation

Another NOx formation mechanism is prompt NOx. This occurs at the flame front and is responsible for no more than 20 ppm NOx in refinery or natural-gas fueled equipment. The mechanism can be summarized as:

$$CH_x + N_2 \rightarrow HCN \leftrightarrow CN \quad \text{(not balanced)} \quad (2.46)$$

$$HCN \leftrightarrow CN + O_2 \rightarrow NO + CO + H \quad \text{(not balanced)} \quad (2.47)$$

Both of these reactions are very fast and do not require high temperature. It would appear that the only way to reduce NOx from the prompt mechanism would be to dilute the HCN and CN species on the fuel side of the combustion zone, or reduce the available oxygen.

2.12.4 The Fuel-Bound NOx Mechanism

The fuel-bound NOx mechanism is similar to prompt NOx and proceeds through the same HCN-CN chemistry. However, the fuel-bound mechanism differs in the following ways.

1. The fuel-bound mechanism requires nitrogen as part of the fuel molecule.
2. At low fuel-nitrogen concentrations, all of the bound nitrogen converts to NOx.
3. The fuel-bound mechanism can be responsible for hundreds of ppm NOx, depending on the amount of nitrogen bound in the fuel.

The first steps in the chemistry differ in that the intermediates are formed directly from pyrolysis of the parent molecule. Ambient nitrogen is unimportant.

$$CH_xN_y \rightarrow HCN \leftrightarrow CN \quad \text{(not balanced)} \quad (2.48)$$

The subsequent chemistry (oxidation pathways for HCN and CN) is identical to prompt NOx.

Reducing the available oxygen, reducing the nitrogen content in the fuel, or diluting the fuel species with an inert gas reduces NOx.

TABLE 2.5 Adiabatic Flame Temperatures

	Air	
Fuel	**°F**	**°C**
H_2	3807	2097
CH_4	3542	1950
C_2H_2	4104	2262
C_2H_4	3790	2088
C_2H_6	3607	1986
C_3H_6	4725	2061
C_3H_8	3610	1988
C_4H_{10}	3583	1973
CO	3826	2108

2.13 FLAME PROPERITIES

The flame temperature is a critical variable in determining the heat transfer, as will be shown in Chapter 3. This section shows how the adiabatic flame temperature is affected by the fuel composition, the equivalence ratio, and the air and fuel preheat temperatures. As previously mentioned, real flame temperatures are not as high as the adiabatic flame temperature, but the trends are comparable and representative of actual conditions.

2.13.1 Flame Temperature

Table 2.5 shows the adiabatic flame temperature for common hydrocarbon fuels combusted with air. Figure 2.16 shows the adiabatic flame temperature as a function of the equivalence ratio for three fuels: H_2, CH_4, and C_3H_8. The peak temperature occurs at about stoichiometric conditions (ϕ = 1.0). In that case, there is just enough oxidizer to fully combust all the fuel. Any additional oxidizer absorbs sensible energy from the flame and reduces the flame temperature. In most real flames, the peak flame temperature often occurs at slightly fuel lean conditions (ϕ < 1.0). This is due to imperfect mixing where slightly more O_2 is needed to fully combust all the fuel. Nearly all industrial combustion applications are run at fuel-lean conditions to ensure that CO emissions are low. Therefore, depending on the actual burner design, the flame temperature may be close to its peak, a condition that is often desirable for maximizing heat transfer. One problem often encountered by maximizing the flame temperature is that high flame temperature maximizes NOx emissions. NOx increases approximately exponentially with gas temperature. This has led to many design concepts for reducing the peak flame temperature in the flame to minimize NOx emissions.[4]

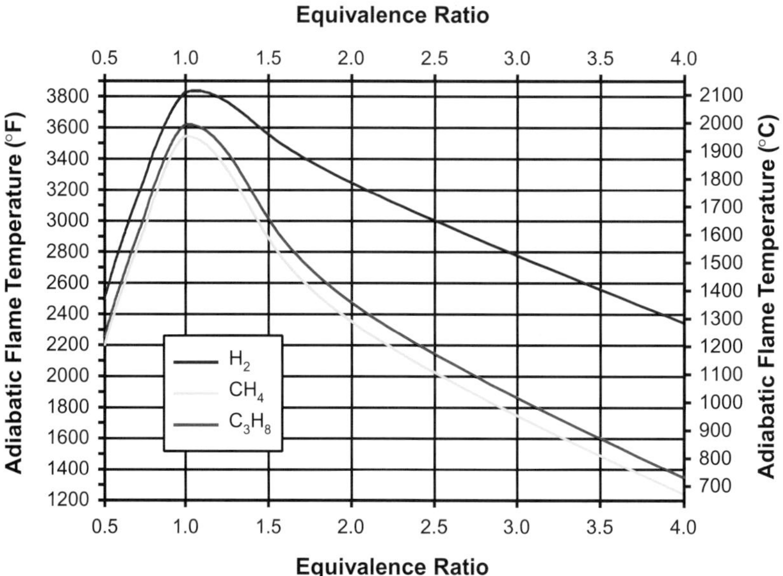

FIGURE 2.16 Adiabatic flame temperature vs. equivalence ratio for air/H_2, air/CH_4, and air/C_3H_8 flames, where the air and fuel are at ambient temperature and pressure.

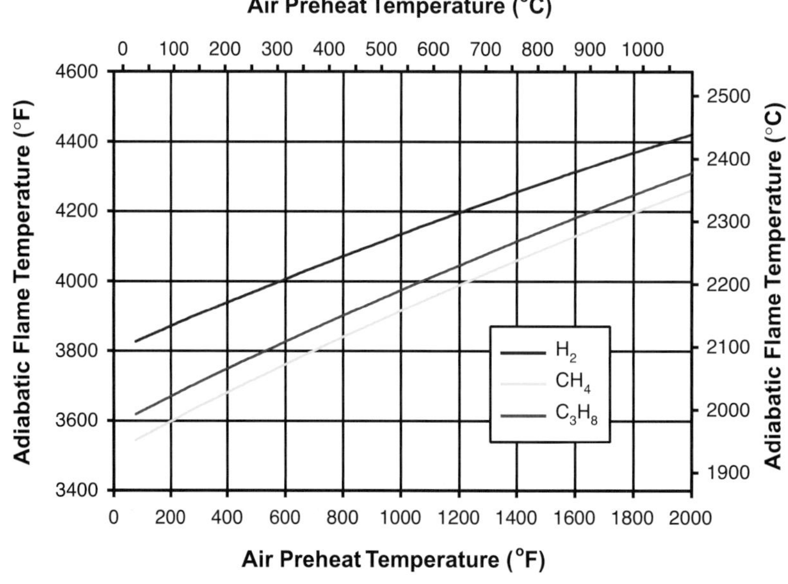

FIGURE 2.17 Adiabatic flame temperature vs. air preheat temperature for stoichiometric air/H_2, air/CH_4, and air/C_3H_8 flames, where the fuel is at ambient temperature and pressure.

Figure 2.17 shows how preheating the air in the combustion of the three fuels shown dramatically increases the adiabatic flame temperature. The increase is nearly linear for the air preheat temperature range shown. Air preheating is commonly done to both increase the overall system efficiency (which will be graphically shown later) and to increase the flame temperature, especially for higher temperature heating and melting processes like melting metal or glass. Figure 2.18 shows the effect of preheating the fuel on the adiabatic flame temperature. Again, there is a nearly linear rise in the flame temperature, but the magnitude of the increase is much less than for air preheating. This is due to the much larger mass of air compared to the mass of fuel in the combustion process. Preheating the air to a given temperature requires much more energy than preheating the fuel to that same temperature, because of the difference in mass.

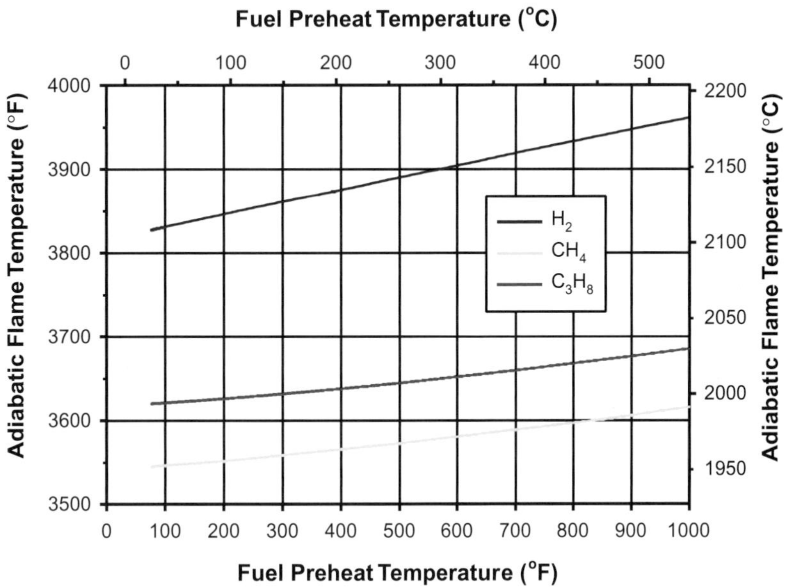

FIGURE 2.18 Adiabatic flame temperature vs. fuel preheat temperature for stoichiometric air/H_2, air/CH_4 and air/C_3H_8 flames, where the air is at ambient temperature and pressure.

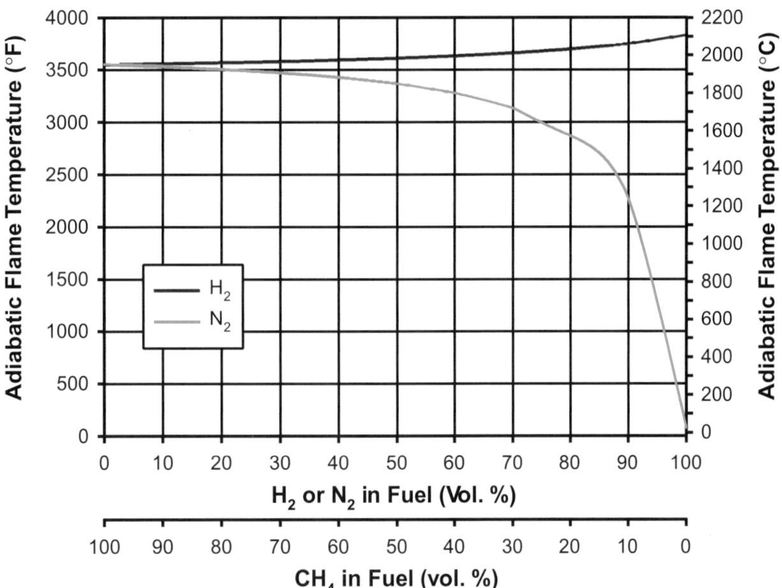

FIGURE 2.19 Adiabatic flame temperature vs. fuel blend (CH_4/H_2 and CH_4/N_2) composition for stoichiometric air/fuel flames, where the air and fuel are at ambient temperature and pressure.

Figure 2.19 shows how the flame temperature varies for fuel blends of H_2/CH_4 and N_2/CH_4. The flame temperature increases as the H_2 content in the blend increases. It is important to note that the increase is not linear; the increase is more rapid at higher levels of H_2. Because of the relatively high cost of H_2 compared to CH_4 and C_3H_8, it is not used in many industrial applications. However, high H_2 fuels are often used in many of the hydrocarbon and petrochemical applications for fluid heating. Because such fuels are by-products of the chemical manufacturing process, their use is much less expensive than purchasing H_2 from an industrial gas supplier as well as being more cost-effective than purchasing other fuels. The graph also shows that the adiabatic flame temperature decreases for N_2/CH_4 fuel blends as the N_2 content increases. Again, the decrease is not

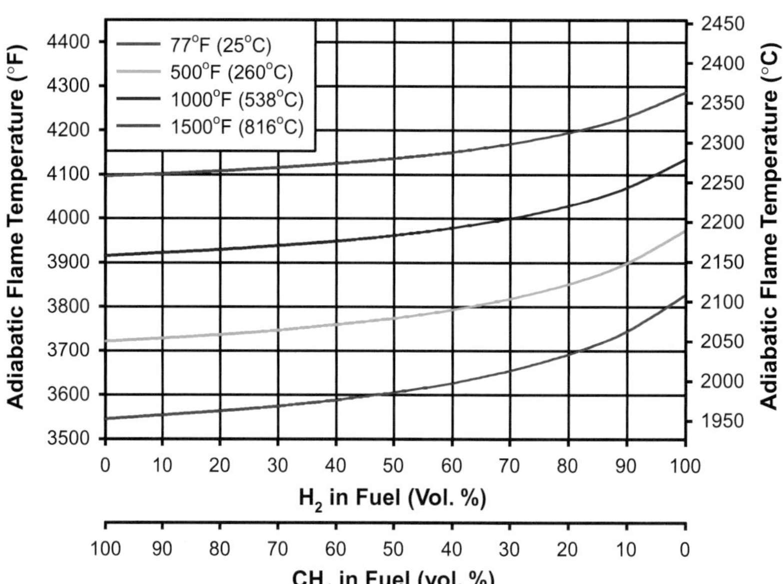

FIGURE 2.20 Adiabatic flame temperature vs. fuel blend (CH_4/H_2) composition and air preheat temperature for stoichiometric air/fuel flames, where the fuel is at ambient temperature and pressure.

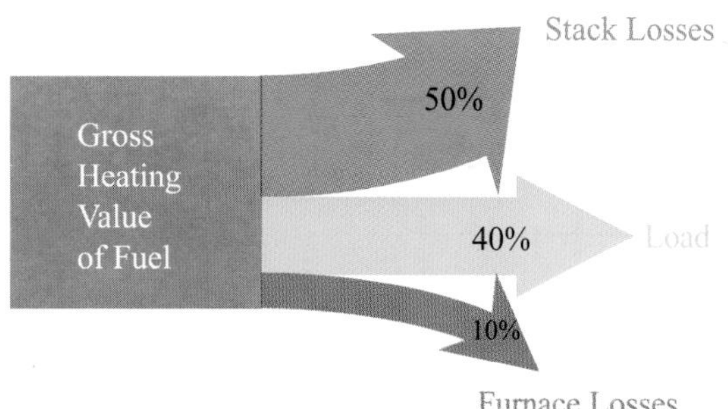

FIGURE 2.21 Sample Sankey diagram showing distribution of energy in a combustion system.

linear and rapidly decelerates at higher N_2 contents until no flame is present for a "fuel" having 100% N_2. Figure 2.20 shows how preheating the combustion air for fuel blends of H_2 and CH_4 increases the adiabatic flame temperature. However, the increase is not a dramatic rise from pure CH_4 to pure H_2. Again, the change in flame temperature with blend composition is nonlinear.

2.13.2 Available Heat

The available heat in a process is defined as the gross heating value of the fuel, minus the energy carried out of the exhaust stack by the flue gases. This difference is the energy that is available to do work. However, some of that energy will be lost by conduction through the heater walls, by radiation through openings, by air infiltration that will absorb sensible energy, as well as by other types of energy losses that are dependent on the burner and heater designs and by the process operations. The accounting of the distribution for where the energy goes in a process is sometimes graphically depicted using a Sankey diagram. Figure 2.21 presents a very simplified Sankey diagram showing that only 40% of the energy goes to the load in that example. The available heat for that example is 50%, which includes the 40% to the load and the 10% lost to various sources. Figure 2.22 shows the calculated available heat for three different fuels as a function of the exhaust or flue gas temperature. As expected, there is a rapid decrease in available heat as the exhaust gas temperature increases. This indicates that more and more energy is being carried out of the exhaust instead of being transferred to the load as the exhaust temperature increases. At the adiabatic flame temperature for each fuel, there is no available heat as all the energy was carried out in the exhaust. Figure 2.23 shows that the available heat increases with the air preheat temperature, which simply indicates that energy was recovered in the process and was used to preheat the combustion air. Figure 2.24 shows that preheating the fuel increases the efficiency, but to a much lesser extent than air preheating. The mass of air is much greater than the mass of fuel, so preheating the fuel is less effective than preheating the air if the preheat temperature is the same.

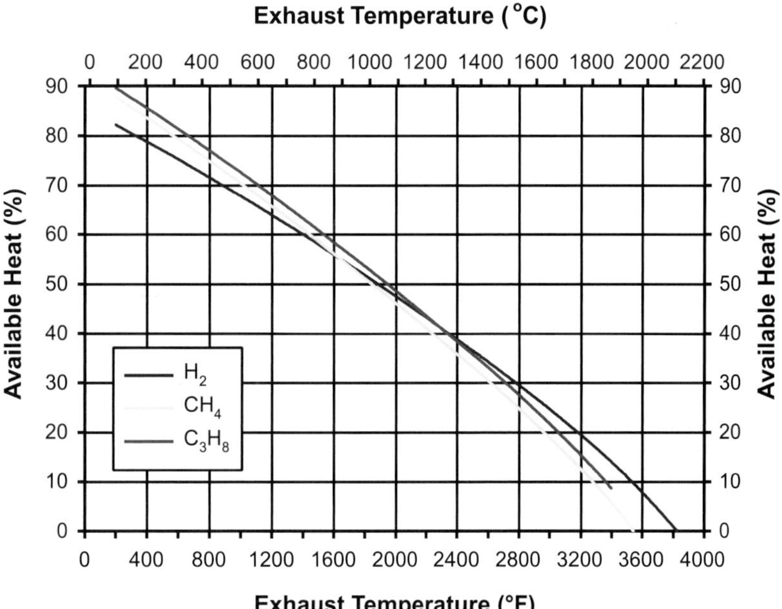

FIGURE 2.22 Available heat vs. gas temperature for stoichiometric air/H_2, air/CH_4, and air/C_3H_8, flames where the air and fuel are at ambient temperature and pressure.

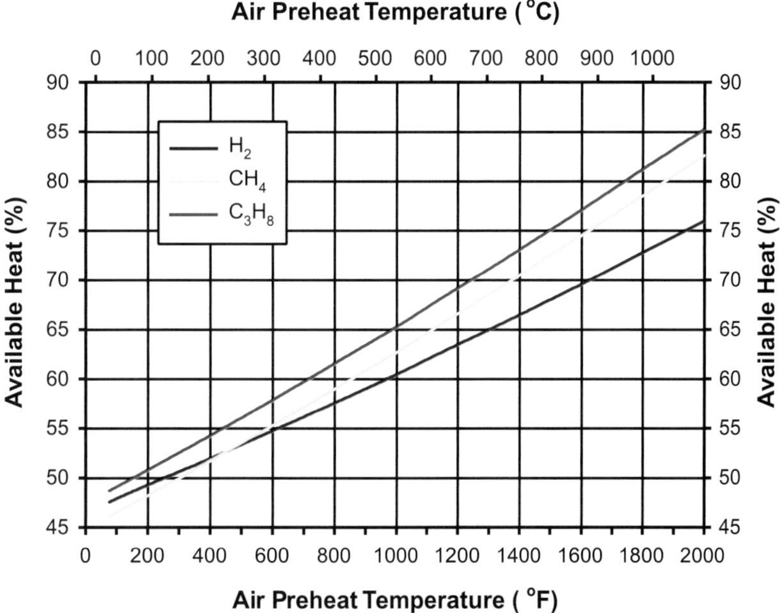

FIGURE 2.23 Available heat vs. air preheat temperature for stoichiometric air/H_2, air/CH_4, and air/C_3H_8 flames at an exhaust gas temperature of 2000°F (1100°C), where the fuel is at ambient temperature and pressure.

2.13.3 Minimum Ignition Energy

Ignition energy graphs usually have the vertical axis as the relative energy of the fuel mixture (see Figure 2.25). The reactants start from an initial state. If the minimum ignition energy is supplied, the reactant bonds will rupture, producing intermediate species such as CH_3, H, O, etc. Such species are extremely reactive and recombine to form the final products, CO_2 and H_2O. Since the net heat release is greater than the minimum ignition energy, the reaction, once started, will continue until virtually all of the reactants are consumed. The

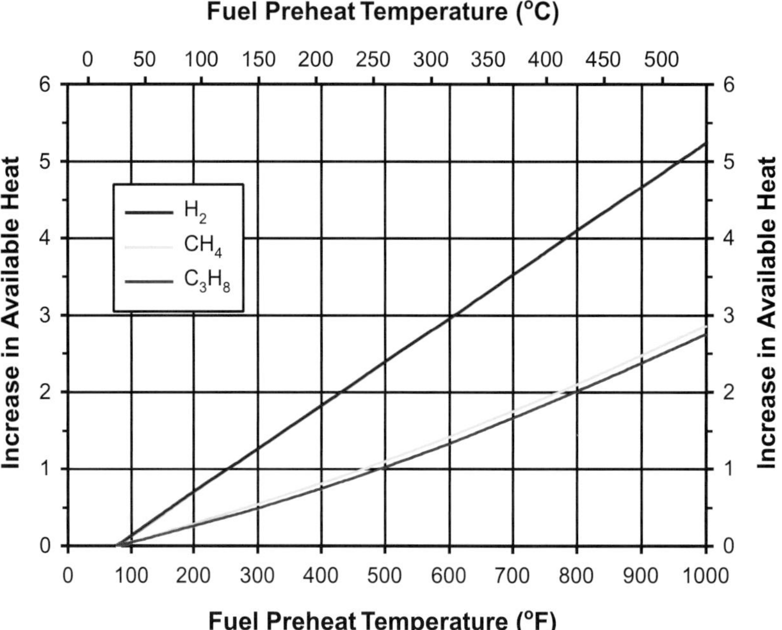

FIGURE 2.24 Available heat vs. fuel preheat temperature for stoichiometric air/H_2, air/CH_4, and air/C_3H_8 flames at an exhaust gas temperature of 2000°F (1100°C), where the air is at ambient temperature and pressure.

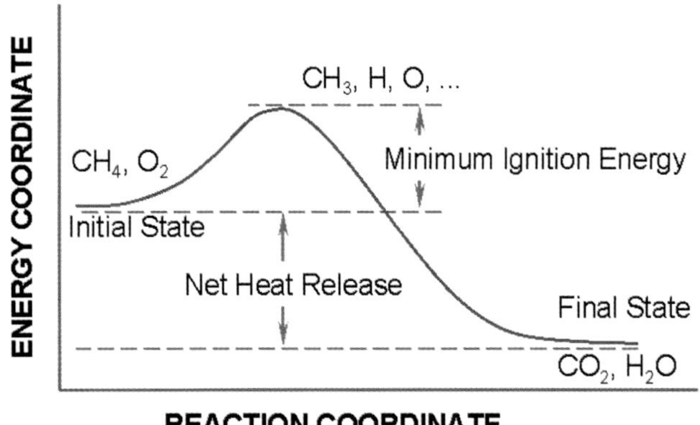

FIGURE 2.25 Graphical representation of ignition and heat release.

horizontal axis shows the progress of the reaction. At the upper left, the diagram shows that the fuel/air mixture has a high potential energy. At the lower right, it is noted that the products of combustion have relatively little remaining chemical energy. Because energy must be conserved, the difference between the upper and lower energy levels must be the amount of heat that the combustion reaction liberates. Note, however, that the energy diagram does not slope monotonically along the reaction coordinate, but contains a hump. This hump is the minimum ignition energy.

What the diagram says is that fuel and air comprising a very high chemical energy may exist in a metastable state, until one introduces a spark or flame of sufficient energy. Once the system reaches the minimum ignition energy, the reaction will be self-sustaining until the reaction consumes enough of the reactants. At that point, the reaction cannot liberate enough heat to supply the minimum ignition energy and the flame goes out.

2.13.4 *Flammability Limits*

Suppose that fuel and air are not provided in stoichiometric proportions, but have a great excess of fuel or air. Will the flame continue to propagate if the ignition source is removed? That depends on whether the fuel/air mixture has enough chemical energy to exceed the minimum ignition energy. If not, the flame will extinguish. This leads to a lower and upper flammability limit. The lower flammability limit (fuel lean) is where fuel is insufficient to achieve the minimum ignition energy. The upper flammability limit (fuel rich) is where there is insufficient air. Table 2.4 gives the flammability limits for many pure gases.

2.13.5 *Flammability Limits for Gas Mixtures*

For gas mixtures, one can use Le Chatelier's rule to estimate flammability limits for gas mixtures. Because this is only an estimate, one must confirm the flammability limit of the actual mixture. Such experiments are relatively inexpensive and many third parties exist that can perform this kind of analysis.

Le Chatelier's rule states that the flammability limit of a mixture is equal to the reciprocal of the sum of reciprocal flammability limits weighted by their mole fractions.

$$LFL = \frac{1}{y_1\left(\frac{1}{LFL_1}\right)+y_2\left(\frac{1}{LFL_2}\right)+y_3\left(\frac{1}{LFL_3}\right)+\cdots+y_n\left(\frac{1}{LFL_n}\right)} \tag{2.49}$$

$$UFL = \frac{1}{y_1\left(\frac{1}{UFL_1}\right)+y_2\left(\frac{1}{UFL_2}\right)+y_3\left(\frac{1}{UFL_3}\right)+\cdots+y_n\left(\frac{1}{UFL_n}\right)} \tag{2.50}$$

where LFL is the lower flammability limit, UFL is the upper flammability limit, LFL_i is the LFL for species i, and UFL_i is the UFL for species i.

2.13.6 *Flame Speeds*

The reaction between fuel and air can only occur at a finite speed. That finite speed depends on the speed of the reaction (chemical) and the amount of turbulence in the flame (physical). If the flame has a lot of turbulence, hot pockets of gas recirculate and the mixture burns faster. To first focus on the chemical part, suppose a long tube is filled with a flammable mixture. If one end of the tube is ignited, the flame front will move along the tube at a precise velocity. A flame that has no turbulence is a laminar flame. Accordingly, the flame speed of a laminar flame is known as the *laminar flame speed.* It is a function of the kinetics of the combustion reaction. Under standard conditions, this is a function of the fuel chemistry alone. The laminar flame speeds for various fuels are tabulated in Table 2.4. Now suppose that instead of a stationary fuel mixture with a moving flame front, the fuel is moved. If the fuel is metered exactly at its flame speed, the flame front will remain stationary. If the fuel is metered faster than the flame speed, the flame front will move forward (called liftoff). If the fuel is metered slower than its flame speed, the flame front will travel backward (called burnback or flashback).

Typical burners operate with fuel flows in excess of the laminar flame speed. To avoid liftoff, several devices are used. Consider premix burners first. Fuel flows across an orifice into the throat of a venturi. The venturi is designed to entrain air near the stoichiometric ratio. Gradual flow passages are used to avoid turbulence, and hot gases are recirculated back to the burner. The fuel/air mixture is supplied at velocities above the laminar flame speed. As the fuel jet issues from the burner, the velocity slows considerably. The flame front establishes where the flame and gas velocities are equal. The sudden expansion from the burner avoids liftoff as the velocity rapidly slows. The high fuel/air velocity avoids burnback. Sudden expansions of this type are used as flame holders because they stabilize the flame front and keep it from moving forward or backward.

Another concept used in premix burners is quench distance, the distance needed to remove sufficient heat from the flame to put it out. Here, burner slots or orifices have a finite thickness that exceeds the quench distance. Because the burner is cooler than the flame, if the flame does begin to burnback, the heavy metal will remove sufficient heat and cool the flame below its minimum ignition energy. Without this feature, a flame that finds its way into a premix burner could flashback. With flashback, the combustion occurs in the burner, rather than at the flame holder. Sustained burnback will destroy the burner in a short time.

Diffusion burners supply fuel with no premix chamber. The fuel meets the air outside the fuel nozzle. With diffusion burners, flashback is not an issue because the fuel alone cannot support combustion (i.e., the upper flammability limit is exceeded). However, liftoff is still a concern. If the flame lifts off the burner, it may travel to a place beyond the flammability limits and extinguish. Under certain conditions, the flame can repeatedly liftoff and reestablish. This behavior is dangerous because the fuel may burn incompletely during one part of the cycle and reignite later, causing an explosion. The cycle of liftoff and burnback can occur many times a second, causing rumble or vibration. Such rumble can be a sign of dangerous instabilities.

Modern burners are designed to give high heat release in short distances. This necessitates fuel velocities that greatly exceed the laminar flame speed. To stabilize such flames, various flame holders are used. For example, an ignition ledge on a burner is a type of flame holder known as a *bluff body.* Even if the air flows by the ledge at very high speed, the air speed very close to the ledge will be very slow. The flame will then establish very near the ignition ledge and be quite stable even over a wide range of firing rates. The burner tile itself is designed with a sudden expansion into the furnace, which also acts as a flame holder because the gas velocity decreases rapidly just after the expansion.

REFERENCES

1. D.R. Lide, Ed., *CRC Handbook of Chemistry and Physics*, 79th edition, CRC Press, Boca Raton, FL, 1998.
2. F. Kreith, Ed., *The CRC Handbook of Mechanical Engineering*, CRC Press, Boca Raton, FL, 1998.
3. S.R. Turns, *An Introduction to Combustion*, McGraw-Hill, New York, 1996.
4. J.L. Reese et al., State-of-the-Art of NOx Emission Control Technology, *Proc. Int. Joint Power Generation Conf.*, Phoenix, October 3-5, 1994.

Chapter 3
Heat Transfer

Prem Singh, Michael Henneke, Jaiwant D. Jayakaran,
Robert Hayes, and Charles E. Baukal, Jr.

TABLE OF CONTENTS

3.1 Introduction ... 70
3.2 Conduction ... 71
3.3 Thermal Conductivity ... 71
3.3.1 One-dimensional Steady-State Conduction ... 73
3.3.2 Transient Conduction: Lumped Capacitance ... 77
3.4 Convection ... 82
3.4.1 Newton's Law of Cooling ... 82
3.4.2 Laminar Flow Convection ... 83
3.4.3 Turbulent Internal Flow ... 84
3.4.4 Turbulent External Flow ... 85
3.5 Radiation ... 87
3.5.1 Blackbody Radiation/Planck Distribution ... 88
3.5.2 Radiant Exchange Between Black Surfaces ... 90
3.5.3 Radiant Exchange Between Gray/Diffuse Surfaces ... 91
3.5.4 View Factors for Diffuse Surfaces ... 92
3.5.5 Infrared Temperature Measurement ... 92
3.5.6 Radiation in Absorbing/Emitting/Scattering Media ... 93
3.5.7 Mean-Beam-Length Method ... 95
3.5.8 Equation of Radiative Transfer ... 97
3.5.9 Radiation Emitted by a Flame ... 101
3.6 Heat Transfer in Process Furnaces ... 102
3.6.1 Flame Radiation ... 105
3.6.2 Furnace Gas Radiation ... 105
3.6.3 Refractory Surface Radiation ... 106
3.6.4 Analysis of Radiation Heat Transfer ... 106
3.6.5 Heat Transfer Through the Wall of a Furnace ... 108

3.6.6 Heat Transfer in the Process Tube 109
3.6.7 Furnace Gas Flow Patterns 109
3.6.8 Role of the Burner in Heat Transfer 110
3.7 Conclusions 112
References 112
Nomenclature 114

3.1 INTRODUCTION

Heat transfer is one of the fundamental purposes of combustion in the process industries. The objective of many industrial combustion applications is to transfer energy, in the form of heat, to some type of load for thermal processing of that load.[1] An understanding of heat transfer is essential to the successful design and operation of fired equipment. The objective of this chapter is to review helpful concepts of heat transfer, focusing on those topics as applied to combustion, particularly in the process industries.

Numerous excellent books have been written on the subject of heat transfer. However, almost none of them have any significant discussion of combustion. This is not surprising as the field of heat transfer is very broad, making it very difficult for any work to be exhaustive. Many of the heat transfer textbooks have no specific discussion of heat transfer in industrial combustion but do treat gaseous radiation heat transfer.[1–7]

The heat transfer books written specifically about radiation often have sections covering heat transfer from luminous and nonluminous flames. Hottel and Sarofim's book has a good blend of theory and practice regarding radiation.[8] It also has a chapter specifically on applications in furnaces. Love's book on radiation has short theoretical discussions of radiative heat transfer in flames and measuring flame parameters, but no other significant discussions of flames and combustion.[9] Özisik's book focuses more on interactions between radiation, conduction, and convection, with no specific treatment of combustion or flames.[10] A short book by Gray and Müller is aimed toward more practical applications of radiation.[11] Sparrow and Cess have a brief chapter on nonluminous gaseous radiation, in which they discuss the various band models.[12]

Some of the older books on heat transfer are more practically oriented with less emphasis on theory. Kern's classic book *Process Heat Transfer* has a chapter devoted specifically to heat transfer in furnaces, primarily boilers and petroleum refinery furnaces.[13] Hutchinson gives many graphical solutions of conduction, radiation, and convection heat transfer problems, but nothing specifically for flames or combustion.[14] Hsu has helpful discussions on nonluminous gaseous radiation and luminous radiation from flames.[15] Welty discusses heat exchangers, but not combustors or flames.[16] Karlekar and Desmond give a brief presentation on nonluminous gaseous radiation, but no discussion of flames or combustion.[17] Ganapathy's book on applied heat transfer is one of the better ones concerning heat transfer in industrial combustion and includes a chapter on fired-heater design.[18] Blokh's book is also a good reference for heat transfer in industrial combustion and is aimed at power boilers.[19] It has much information on flame radiation from a wide range of fuels, including pulverized coal, oils, and gases. A few handbooks on heat transfer have been written, but these also tend to have little if anything on industrial combustion systems.[20–22]

Heat is a form of energy upon which the majority of all refinery processes are based. Heat transfer is that science which seeks to understand and predict the energy transfer between masses resulting from differences in temperature.[2,3] Heat transfer is commonly divided into three mechanisms or modes for classification: conduction, convection, and radiation.

These mechanisms all have importance as applied to combustion in the process industries. Consider a typical fired heater, as shown in Figure 3.1, that consists of tubes with flowing fluid (usually some type of oil) to be heated, and a burner (or group of burners) designed to provide the required energy for the desired process, a radiant section, and a convection section for heating. Major heat transfer processes in petrochemical or refinery furnaces include:

1. conduction through the furnace refractory and convection from the wall of the furnace to the surrounding air
2. radiation exchange between the flame, the surrounding walls, and process tubes
3. convection from the hot furnace gases to the process tubes and from process tube walls to the fluid flowing through the tubes

See Section 3.6 for a comprehensive discussion on the various heat transfer processes taking place in a furnace and how one can calculate various effects.

The consequences of the performance of these heat transfer mechanisms can significantly impact product throughput and quality, furnace efficiency, equipment lifetime, and safety. Other critical phenomena for consideration include the effect of heat transfer mechanisms on the fired equipment itself (e.g., heat transfer effects on burner fuel tips), or the effect of heat

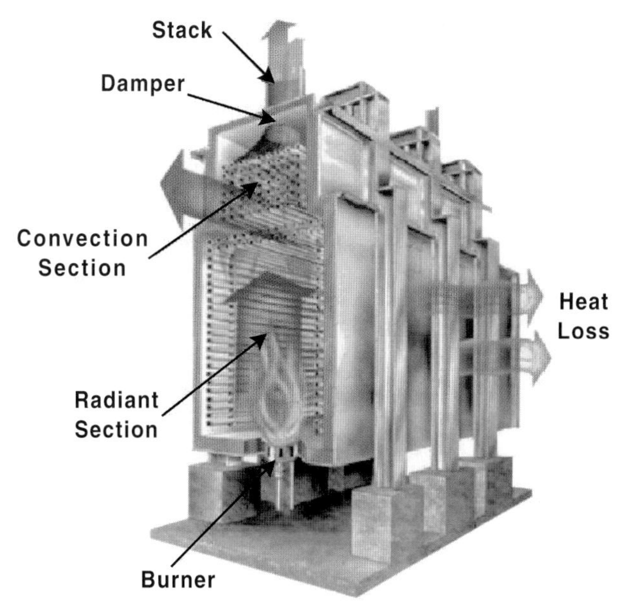

FIGURE 3.1 A typical fired heater.

transfer on the performance of the fired equipment with respect to NO_x emissions, flame stability, and flame shape.

3.2 CONDUCTION

Conduction heat transfer refers to the transfer of energy from the more energetic to the less energetic particles of a substance, resulting from interaction between the particles. Conduction is the net transfer of energy by random molecular motion — also called diffusion of energy. Conduction in gases and liquids is by such molecular motion, except that in liquids, the molecules are more closely spaced and the molecular interactions are stronger and more frequent. In the case of solids, conduction refers to the energy transfer by lattice waves induced by atomic motion. When the solid is a conductor, the translational motion of free electrons transfers energy. In nonconductors, the transfer of energy takes place only via lattice waves.

Heat conduction occurs in both stationary and moving solids, liquids, and gases. The primary postulate of classical heat conduction theory is that the rate of heat conduction in a material is proportional to the temperature gradient. This is consistent with the second law of thermodynamics, indicating that heat flows in the direction of decreasing temperature, or from hot bodies to cold bodies.

$$\vec{q} = -k\nabla T \tag{3.1}$$

Equation (3.1) states that heat flux is proportional to the temperature gradient, and the proportionality constant is called the thermal conductivity of the material transferring heat. More detailed information on thermal conduction heat transfer is available in books specifically written on that subject.[23–28]

TABLE 3.1 Thermal Conductivity of Common Materials

Material	Btu/h ft F	W/m C
Gases at atmospheric pressure	0.004 to 0.70	0.007 to 1.2
Insulating materials	0.01 to 0.12	0.02 to 0.21
Nonmetallic liquids	0.05 to 0.40	0.09 to 0.70
Nonmetallic solids (brick, stone, concrete)	0.02 to 1.5	0.04 to 2.6
Liquid metals	5.0 to 45	8.6 to 78
Alloys	8.0 to 70	14 to 121
Pure metals	30 to 240	52 to 415

TABLE 3.2 Properties of Various Substances at above 32°F (0°C) (except for steam as noted below)

	ρ Lb/ft^3	c_p Btu/lb °F	k Btu/h ft °F
	Metals		
Copper	559	0.09	223
Aluminum	169	0.21	132
Nickel	556	0.12	52
Iron	493	0.11	42
Carbon Steel	487	0.11	25
Alloy Steel 18Cr 8Ni	488	0.11	9.4
	Nonmetal Solids		
Limestone	105	~0.2	0.87
Glass Pyrex	170	~0.2	0.58
Brick K-28	27	~0.2	0.14
Plaster	140	~0.2	0.075
Kaowool	8	~0.2	0.016
	Gases		
Hydrogen	0.006	3.3	0.099
Oxygen	0.09	0.22	0.014
Air	0.08	0.24	0.014
Nitrogen	0.08	0.25	0.014
Steam[1]	0.04	0.45	0.015
	Liquids		
Water	62.4	1.0	0.32
Sulfur dioxide (liquid)	89.8	0.33	0.12

[1] Reference temperature for steam is 212°F (100°C). All other temperatures are 32°F (0°C).

3.3 THERMAL CONDUCTIVITY

Thermal conductivity is a material property that is expressed in Btu/(hr-ft-°F) or W/(m-K) and is dependent on the chemical composition of the substance. Typical values for some materials are shown in Tables 3.1 and 3.2.

The thermal conductivity of solids is generally higher than liquids, and liquids higher than gases. Among solids, the insulating materials have the lowest conductivities. The

TABLE 3.3 Properties of Selected Gases at 14.696 psi

Temperature °F	ρ lb/ft³	c_p Btu/lb °F	k Btu/h ft °F	ρ lb/ft³	c_p Btu/lb °F	k Btu/h ft °F
		Air			**CO_2**	
0	0.0855	0.240	0.0131	0.1320	0.184	0.0076
500	0.0408	0.248	0.0247	0.0630	0.247	0.0198
1000	0.0268	0.263	0.0334	0.0414	0.280	0.0318
1500	0.0200	0.276	0.0410	0.0308	0.298	0.042
2000	0.0159	0.287	0.0508	0.0247	0.309	0.050
2500	0.0132	0.300	0.0630	0.0122	0.311	0.055
3000	0.0113	0.314	0.0751	0.0175	0.322	0.061
		O_2			**N_2**	
0	0.0945	0.219	0.0133	0.0826	0.249	0.0131
500	0.0451	0.235	0.0249	0.0395	0.254	0.0236
1000	0.0297	0.252	0.0344	0.0260	0.269	0.0320
1500	0.0221	0.263	0.0435	0.0193	0.283	0.0401
2000	0.0178		0.0672	0.0156		0.0468
2500	0.0148		0.0792	0.0130		0.0528
3000	0.0127		0.0912	0.0111		
		H_2				
0	0.0059	3.421	0.1071			
500	0.0028	3.470	0.1610			
1000	0.0019	3.515	0.2206			
1500	0.0014	3.619	0.2794			
2000	0.0011	3.759	0.3444			
2500	0.0009	3.920	0.4143			
3000	0.0008	4.218	0.4880			
		Flue Gases				
	Natural Gas k	**Fuel Oil k**	**Coal k**			
0	—	—	—			
500	0.022	0.022	0.022			
1000	0.030	0.029	0.029			
1500	0.037	0.036	0.036			
2000	0.044	0.043	0.043			
2500	0.051	0.049	0.050			

thermal conductivities of pure metals typically decrease with an increase in temperature, while the conductivities of alloys can either increase or decrease (see Table 3.2). For many heat transfer calculations, it is sufficiently accurate to assume a constant thermal conductivity corresponding to the average temperature of the material.

The thermal conductivities of most nonmetallic liquids range from 0.05 to 0.15 Btu/hr-ft-°F (0.09 to 0.26 W/m-K), and the thermal conductivities of many liquids tend to decrease with increasing temperature.

The thermal conductivities of gases increase with temperature and are independent of pressure at the conditions at which most furnace cavities operate. Generally, gas thermal conductivities decrease with increasing molecular weight. Thus, a light gas such as hydrogen has a relatively high conductivity.

When calculating the thermal conductivity of nonhomogeneous materials, one must use the apparent thermal conductivity to account for the porous or layered construction of the material. In furnace refractory walls, the thermal conductivity can vary from site to site for the same material. This is because the thermal conductivity of these materials is strongly dependent on their apparent bulk density (mass per unit volume). For higher temperature insulations, the apparent thermal conductivity of fibrous insulations and insulating firebrick decreases as bulk density increases, because the denser material attenuates radiation. However, there is a limit at which any increase in density increases the thermal conductivity due to conduction in the solid material.

It is known that the specific heats of solids and liquids are generally independent of pressure. Table 3.2 also shows the specific heats of various metals, alloys, and nonhomogeneous materials at 68°F (20°C). These values can be used at other temperatures without significant error.

Gases, on the other hand, demonstrate more temperature dependence with regard to their specific heat. For all practical purposes, in furnace analyses one can neglect any pressure dependence. Table 3.3 gives the specific heat data for air and other gases at different temperatures. In the case of steam and water, the variation of both thermal conductivity and specific heat can be significant over the ranges of temperatures and pressures encountered in industrial steam systems. Refer to any standard steam tables for data on water and steam.

When using thermal insulators as a heat barrier, one must bear in mind that the effectiveness of an insulator depends significantly on the temperature of its cold face. Thus, it is not possible to protect a metal object in a furnace by insulating all around it, unless there is an adequate path for the heat to escape from the object to a cooler location, such as the atmosphere outside the furnace. Regardless of the thickness of insulation on an object that is in a furnace, if it is not attached to a cold sink, the object will eventually attain the furnace temperature. This heat-up time period is governed by the specific heat of the material and is merely the time (and heat input) required to heat the mass of insulation (and object) up to the furnace temperature. The quantity of heat required to reach furnace temperature is given by:

$$Q = mC_p \Delta T$$

where m is the mass of the material and C_p is the specific heat of material and ΔT is the temperature difference between ambient and furnace temperatures.

3.3.1 *One-dimensional Steady-State Conduction*

In a one-dimensional steady-state conduction situation, the temperature change occurs only in one direction. The system is described as steady state when the temperature at every point remains the same over time. This assumption is usually valid for analysis of a furnace wall under steady-state furnace operation, when the firing rate and temperature gradient through the furnace wall can be considered constant for all practical purposes. However, during startup and shutdown, the heat input and temperature gradients are changing over time and must be treated differently.

3.3.1.1 Plane Wall

It can be shown that the general heat equation in isotropic media is:

$$\frac{\partial}{\partial x}\left(k\frac{\partial T}{\partial x}\right)+\frac{\partial}{\partial y}\left(k\frac{\partial T}{\partial y}\right)+\frac{\partial}{\partial z}\left(k\frac{\partial T}{\partial z}\right)+\dot{q}=\rho c_p\frac{\partial T}{\partial t} \tag{3.2}$$

where k = thermal conductivity of the media
$\dot{q}$ = rate of energy generation within the system
$\rho c_p \frac{\partial T}{\partial t}$ = time rate of change of sensible energy of the system

Equation (3.2) is known as the heat diffusion equation. For steady-state conduction in a medium with constant thermal conductivity, the above equation, without heat generation, becomes:

$$\nabla^2 T = 0 \tag{3.3}$$

where ∇^2 is the Laplacian operator, defined in cartesian coordinates as:

$$\nabla^2 \equiv \partial^2/\partial x^2 + \partial^2/\partial y^2 + \partial^2/\partial z^2$$

For one-dimensional transfer of heat, Eq. (3.3) becomes:

$$\frac{d^2T}{dx^2} = 0 \tag{3.4}$$

Equation (3.4) for the plane wall shown in Figure 3.2 can be solved to obtain:

$$T = \left(T_{s,2} - T_{s,1}\right)\frac{x}{L} + T_{s,1} \tag{3.5}$$

The energy flux can be evaluated using Fourier's law of heat conduction:

$$q_x = \frac{kA}{L}\left(T_{s,1} - T_{s,2}\right) \tag{3.6}$$

Table 3.4 summarizes one-dimensional steady-state solutions to heat equations for different coordinate systems.

Equation (3.6) can also be written as:

$$q_x = \frac{\left(T_{s,1} - T_{s,2}\right)}{\frac{L}{kA}} \tag{3.7}$$

where the quantity L/kA has units of (K/W) and is called the thermal resistance. Figure 3.2 also includes the equivalent thermal circuit. One side of the plane wall ($x = 0$) is being heated by the surrounding fluid at $T_{\infty,1}$ and the other side ($x = L$) is being cooled by surrounding cold fluid at $T_{\infty,2}$. The thermal circuit in Figure 3.2 includes both convective heating and cooling, as well as conduction through the material of the wall.

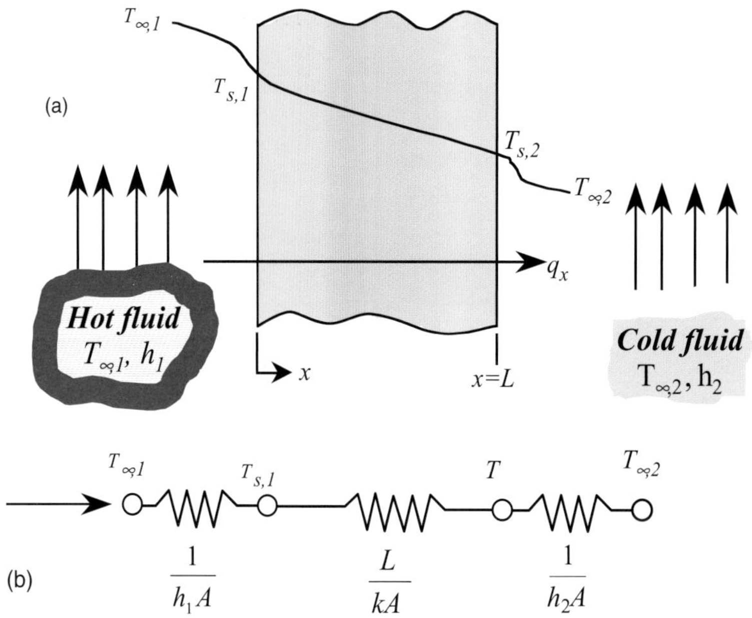

FIGURE 3.2 Heat transfer through a plane wall: (a) temperature distribution, and (b) equivalent thermal circuit.[29]

TABLE 3.4 One-dimensional, Steady-State Solutions to the Heat Equation with No Generation[29]

	Plane Wall	Cylindrical Wall[a]	Spherical Wall[a]
Heat equation	$\frac{d^2T}{dx^2}=0$	$\frac{1}{r}\frac{d}{dr}\left(r\frac{dT}{dr}\right)=0$	$\frac{1}{r^2}\frac{d}{dr}\left(r^2\frac{dT}{dr}\right)=0$
Temperature distribution	$T_{s,1}-\Delta T\frac{x}{L}$	$T_{s,2}+\Delta T\frac{\ln(r/r_2)}{\ln(r_1/r_2)}$	$T_{s,1}-\Delta T\left[\frac{1-(r_1/r)}{1-(r_1/r_2)}\right]$
Heat flux (q'')	$k\frac{\Delta T}{L}$	$\frac{k\Delta T}{r\ln(r_2/r_1)}$	$\frac{k\Delta T}{r^2\left[(1/r_1)-(1/r_2)\right]}$
Heat rate (q)	$kA\frac{\Delta T}{L}$	$\frac{2\pi Lk\Delta T}{\ln(r_2/r_1)}$	$\frac{4\pi k\Delta T}{(1/r_1)-(1/r_2)}$
Thermal resistance ($R_{t,\,cond}$)	$\frac{L}{kA}$	$\frac{\ln(r_2/r_1)}{2\pi Lk}$	$\frac{(1/r_1)-(1/r_2)}{4\pi k}$

[a] The critical radius of insulation is $r_{cr} = k/h$ for the cylinder and $r_{cr} = 2k/h$ for the sphere.
From F.P. Incropera and D.P. DeWitt, *Fundamentals of Heat and Mass Transfer*, 4th edition, John Wiley & Sons, New York, 1996. With permission.

3.3.1.2 Composite Wall

In industry, different furnace designs are used for different heat transfer operations. Design economics require that these furnaces often have several walls in series to increase strength, provide better insulation, or even to enhance appearance. The one-dimensional steady-state heat transfer analysis can also be applied to these cases. Multiple walls in series can be considered to be a composite wall, as shown in Figure 3.3. The heat flux in the x-direction is expressed as:

$$q_x = \frac{T_{\infty,1} - T_{\infty,2}}{\sum R_t} \tag{3.8}$$

where $T_{\infty,1}$ and $T_{\infty,2}$ are the surrounding temperatures and ΣR_t is the total thermal resistance of the system. The total thermal resistance is evaluated as:

$$\sum R_t = \frac{1}{h_1 A} + \frac{L_A}{k_A A} + \frac{L_B}{k_B A} + \frac{L_C}{k_C A} + \frac{1}{h_2 A} \tag{3.9}$$

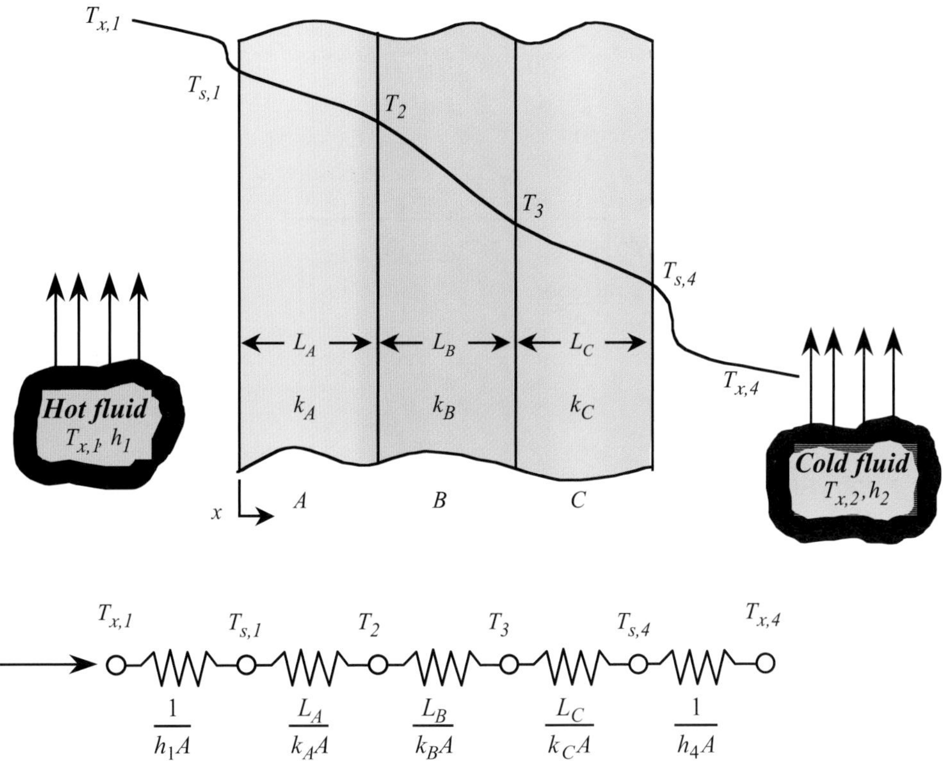

FIGURE 3.3 Equivalent thermal circuit for a series composite wall.[29]

where h_1 and h_2 are convection heat transfer coefficients on the two sides of the composite wall, respectively, and k_A, k_B, and k_C are the thermal conductivities of walls A, B, and C, respectively. Thus, the heat transfer rate can be expressed as:

$$q_x = \frac{T_{\infty,1} - T_{s,1}}{\frac{1}{h_1 A}} = \frac{T_{s,1} - T_2}{\frac{L_A}{k_A A}} = \frac{T_2 - T_3}{\frac{L_B}{k_B A}} = \ldots \quad (3.10)$$

Here A is the cross-sectional area for heat flow. An overall heat transfer coefficient U can then be defined:

$$U = \frac{1}{R_{tot} A} = \left[\left(\frac{1}{h_1}\right) + \left(\frac{L_A}{k_A}\right) + \left(\frac{L_B}{k_B}\right) + \left(\frac{L_C}{k_C}\right) + \left(\frac{1}{h_4}\right)\right]^{-1} \quad (3.11)$$

A circuit diagram of the thermal resistance of the composite walls is also shown in Figure 3.3.

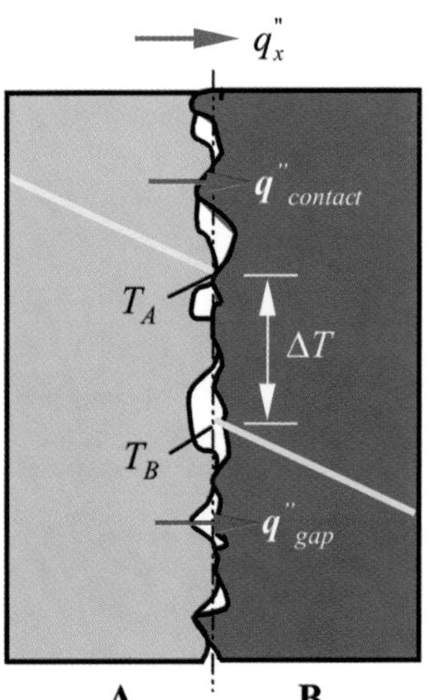

FIGURE 3.4 Temperature drop due to thermal contact resistance.

3.3.1.3 Contact Resistance

In composite systems, the temperature drop across the interface between the walls might be appreciable. This temperature drop, as caused by the contact resistance, $R_{t,c}$, between two solid materials A and B, is illustrated in Figure 3.4. If the heat flux for a unit area of interface is q_x'' ($q_x'' = q''_{\text{contact}} + q''_{\text{gap}}$), then the contact resistance can be defined as:

$$R''_{t,c} = \frac{T_A - T_B}{q''_x} \quad (3.12)$$

The thermal contact resistance for different combinations of solids is available in standard texts.[29]

3.3.1.4 Conduction in Radial Coordinate Systems

For steady-state heat conduction with no generation of energy during conduction, the transfer in a hollow cylinder of radius r is given as:

$$\frac{1}{r}\frac{d}{dr}\left(kr\frac{dT}{dr}\right) = 0 \quad (3.13)$$

Energy generation would refer to the exothermic or endothermic energy release within the conducting medium. This, of course, would not be the case in the wall of a pipe. However, if one were considering the transfer of heat through a gas that is reacting, as is often the case in combustion, then energy generation would have to be factored in.

Fourier's law of heat conduction for this case can be written as:

$$q_r = -kA\frac{dT}{dr} = -k(2\pi rL)\frac{dT}{dr} \quad (3.14)$$

where $2\pi rL$ is the area of the cylinder normal to the direction of heat flow. The temperature distribution in the cylinder can be determined by integrating Eq. (3.13) twice to give

$$T(r) = C_1 \ln r + C_2 \quad (3.15)$$

where C_1 and C_2 are constants of integration and can be determined by the application of boundary conditions $T(r_1) = T_{s,1}$ and $T(r_2) = T_{s,2}$.

Thus, the general equation for temperature distribution in a cylinder is:

$$T(r) = \frac{T_{s,1} - T_{s,2}}{\ln(r_1/r_2)} \ln\left(\frac{r}{r_2}\right) + T_{s,2} \quad (3.16)$$

The differentiation of Eq. (3.16) with respect to r and substitution in Eq. (3.14) yields the expression for heat transfer rate:

$$q_r = \frac{2\pi Lk(T_{s,1} - T_{s,2})}{\ln(r_2/r_1)} \quad (3.17)$$

From the above equation, the conduction thermal resistance can be given by:

$$R_{t,cond} = \frac{\ln(r_2/r_1)}{2\pi Lk} \quad (3.18)$$

The above concept can be extended to derive the equation for heat transfer in a system of multiple cylinders (see Figure 3.5) with radii r_1, r_2, r_3, and r_4; corresponding temperatures, $T_{s,1}$, $T_{s,2}$, $T_{s,3}$, and $T_{s,4}$; thermal conductivities of k_A, k_B, and k_C; and all having length, L:

$$q_r = \frac{T_{\infty,1} - T_{\infty,4}}{\frac{1}{2\pi r_1 LH_1} + \frac{\ln(r_2/r_1)}{2\pi k_A L} + \frac{\ln(r_3/r_2)}{2\pi k_B L} + \frac{\ln(r_4/r_3)}{2\pi k_C L} + \frac{1}{2\pi r_4 Lh_4}} \quad (3.19)$$

where $T_{\infty,1}$ and $T_{\infty,4}$ represent surrounding temperatures, and h_1 and h_4 represent convection heat transfer coefficients on surfaces 1 and 4, respectively. Figure 3.5 also shows the thermal circuit of the system.

In terms of the overall heat transfer coefficient U and total thermal resistance R_{tot},

$$q_r = \frac{T_{\infty,1} - T_{\infty,4}}{R_{tot}} = UA(T_{\infty,1} - T_{\infty,4}) \quad (3.20)$$

3.3.1.5 Cartesian Coordinate Systems

Generation of energy during the conduction process causes the general heat balance equation to become:

$$\frac{d^2T}{dx^2} + \frac{\dot{q}}{k} = 0 \quad (3.21)$$

where $\dot{q}$ is the rate of heat generation per unit volume. The boundary conditions can be formulated as $T(-L) = T_{s,1}$ and $T(L) = T_{s,2}$, which, when used with the solution of Eq. (3.21), gives the temperature distribution in the plane as:

$$T(x) = \frac{\dot{q}L^2}{2k}\left(1 - \frac{x^2}{L^2}\right) + \frac{T_{s,2} - T_{s,1}}{2}\frac{x}{L} + \frac{T_{s,1} - T_{s,2}}{2} \quad (3.22)$$

When both surfaces are maintained at the same temperature, T_s, the above equation simplifies to:

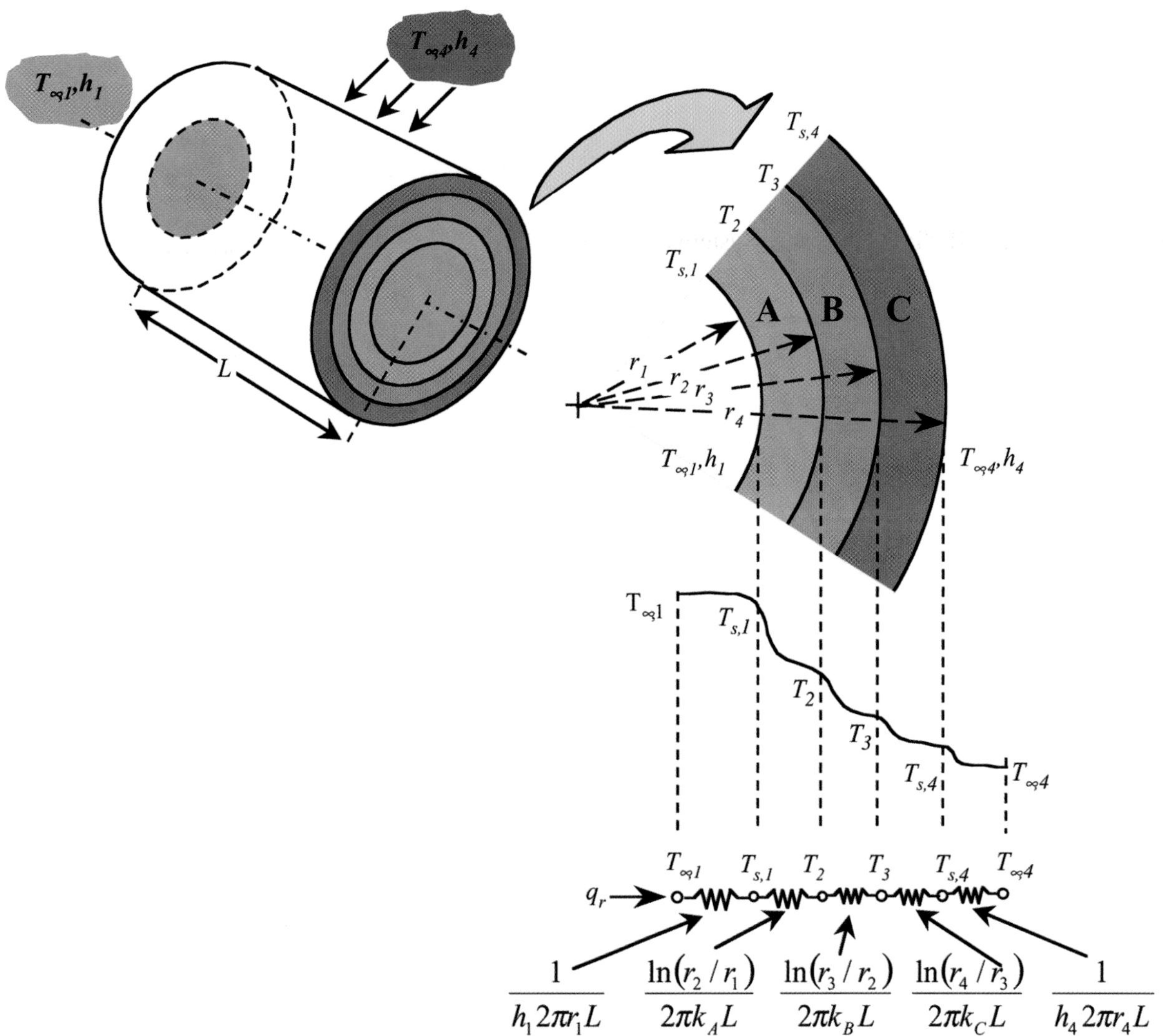

FIGURE 3.5 Temperature distribution for a composite cylindrical wall.[29]

$$T(x) = \frac{\dot{q}L^2}{2k}\left(1 - \frac{x^2}{L^2}\right) + T_s \tag{3.23}$$

3.3.1.6 Cylindrical Coordinate Systems

The steady-state conduction equation with heat generation in the cylindrical coordinate systems can be represented by:

$$\frac{1}{r}\frac{d}{dr}\left(r\frac{dT}{dr}\right) + \frac{\dot{q}}{k} = 0 \tag{3.24}$$

Here, again, the heat generated within the system per unit volume is given by $\dot{q}$. The solution to Eq. (3.24) with the boundary conditions $T(r_0) = T_s$, and $(dT/dr)_{r=0} = 0$ becomes:

$$T(r) = \frac{\dot{q}r_0^2}{4k}\left(1 - \frac{r^2}{r_0^2}\right) + T_s \tag{3.25}$$

Applying Fourier's law and differentiating the above equation with respect to r determines the heat transfer rate at any radius in the cylinder.

3.3.2 Transient Conduction: Lumped Capacitance

Unsteady-state conduction involves storage of heat. For example, in heating a furnace during startup, heat must be supplied to bring the walls to the operating temperature and to overcome the steady-state losses of normal operation. In a typical continuous furnace operation, the heat stored in the walls and the metal of the tubes is insignificant compared to the total heat input. In heaters that are heated and cooled periodically, such as in batch process work, the heat stored in the walls can be a significant cost.

Unsteady-state conduction occurs in heating or cooling processes where the temperatures change with time. Examples include operating regenerative heaters, raising boiler pressure, and turndown conditions on process furnaces. By introducing time as an additional variable, conduction analyses become more complicated.

Unsteady-state conduction problems can be easily solved if the temperature gradient within a solid material can be ignored. The conditions under which temperature gradients can be ignored are considered below. A common example would be dropping a hot metal sphere into a cold-water bath. The high thermal conductivity of the metal (compared with the convection coefficient as discussed below) usually allows analyzing the time-temperature history of the metal's temperature without regard to temperature variations within the sphere. The simplified approach, based on neglecting the temperature gradient in the metal, is called a *lumped capacitance* method.

In a furnace setting, it is usually possible to neglect the temperature gradient in metal pipe walls because the thermal conductivity is high compared to the rest of the heat transfer path. However, insulating refractory cannot be treated as lumped capacitance. Obviously, due to its simplicity, a lumped capacitance approximation is the first resort in the analysis of a transient problem. However, it should be noted that its simplicity also makes it the least accurate approach.

If a hot solid initially ($t < 0$) at a temperature T_i is cooled and attains any temperature $T(t)$ at any time ($t > 0$), a general heat balance equation can be written as:

$$-hA_s\left(T - T_\infty\right) = \rho Vc\left(\frac{dT}{dt}\right) \tag{3.26}$$

where h is the convection heat transfer coefficient, A_s is the surface area of the solid, ρ is the density of the solid, c is the specific heat of the solid, and T_∞ is the temperature of the surrounding medium. Solving the above differential equation gives:

$$\frac{\theta}{\theta_i} = \frac{T - T_\infty}{T_i - T_\infty} = \exp\left[-\left(\frac{hA_s}{\rho Vc}\right)t\right] \tag{3.27}$$

Here, the quantity ($\rho Vc/hA_s$) is called the thermal time constant, expressed as τ_t. It can also be written as:

$$\tau_t = \left(\frac{1}{hA_s}\right)(\rho Vc) = R_t C_t \tag{3.28}$$

where R_t is the resistance to convection heat transfer and C_t is the lumped thermal capacitance of the solid. The physical significance of Eq. (3.28) is the fact that any increase in the value of C_t or R_t will cause the system to respond more slowly for a given change in the temperature.

3.3.2.1 Applicability of Lumped Capacitance Method

Considering a plane wall with temperatures $T_{s,1}$ and $T_{s,2}$ ($T_{s,1} > T_{s,2}$) at sides 1 and 2, and a surrounding temperature of T_∞, the surface energy balance will give:

$$\frac{kA}{L}\left(T_{s,1} - T_{s,2}\right) = hA\left(T_{s,2} - T_\infty\right) \tag{3.29}$$

where A is the surface area and k is the thermal conductivity of the solid. Equation (3.29) can be rearranged to give:

$$\frac{\left(T_{s,1} - T_{s,2}\right)}{\left(T_{s,2} - T_\infty\right)} = \frac{\dfrac{L}{kA}}{\dfrac{1}{hA}} = \frac{hL}{k} \equiv \text{Bi} \tag{3.30}$$

The dimensionless quantity (hL/k) is called the Biot number (Bi). It is a measure of the temperature drop in a solid as compared to the temperature drop between the surface and the fluid. It can also be interpreted as the ratio of resistance due to conduction and resistance due to convection. When the conduction resistance is negligible as compared to the convection resistance (i.e., Bi << 1), the lumped capacitance assumption is valid. In the case of uneven surfaces or complicated shapes, the estimation of length, L, is difficult and, therefore, a characteristic length, L_c, is usually taken, which is the ratio of volume to surface area.

The exponent in Eq. (3.27) can now be rewritten as:

$$\frac{hA_s t}{\rho Vc} = \frac{hL_c}{k}\frac{k}{\rho c}\frac{t}{L_c^2} = \frac{hL_c}{k}\frac{\alpha t}{L_c^2} = \text{Bi} \cdot \text{Fo} \tag{3.31}$$

where α is the thermal diffusivity and Fo is the Fourier number $\equiv \dfrac{\alpha t}{L_c^2}$. Thus, the temperature distribution can be expressed as:

$$\frac{\theta}{\theta_i} = \frac{T - T_\infty}{T_i - T_\infty} = \exp(-\text{Bi} \cdot \text{Fo}) \tag{3.32}$$

If the condition of the solid is simultaneously affected by convection, radiation, applied surface heat flux, and internal energy generation, the situation could be complicated and difficult to solve. Thus, the more general form of Eq. (3.26) is:

$$-hA_s(T - T_\infty) + \varepsilon A_s \sigma(T^4 - T_{sur}^4) + E_g + q_s = \rho Vc \frac{dT}{dt} \quad (3.33)$$

where E_g is the heat generated within the system, q_s is the heat supplied to the system, ε is the emissivity of the solid, and σ is the Stefan-Boltzmann constant.

The above equation is a nonlinear, first-order, nonhomogeneous differential equation that cannot be integrated to obtain an exact solution. However, when there is no imposed heat flux and negligible convection compared to radiation, Eq (3.33) can be simplified and solved to finally give:

$$t = \frac{\rho Vc}{3\varepsilon A_s \sigma}\left(\frac{1}{T^3} - \frac{1}{T_i^3}\right) \quad (3.34)$$

Here it is assumed that the surrounding temperature (T_{sur}) is zero.

On the other hand, if radiation is negligible compared to convection, and the convection coefficient, h, is constant with respect to time, the differential equation can be solved to give:

$$\frac{T - T_\infty}{T_i - T_\infty} = \exp(-at) + \frac{b/a}{T_i - T_\infty}[1 - \exp(-at)] \quad (3.35)$$

where $a = \dfrac{hA_s}{\rho Vc}$

$$b = \frac{q_s + E_g}{\rho Vc}$$

The second term on the right-hand side of Eq. (3.35) is the outcome of the applied flux and the heat generated within the system.

3.3.2.2 General Solution for the Lumped Capacitance Method

The one-dimensional, unsteady-state heat balance equation, without any internal generation and with no energy input from the outside, is written as:

$$\frac{\partial^2 T}{\partial x^2} = \frac{1}{\alpha}\frac{\partial T}{\partial t} \quad (3.36)$$

This differential equation can be solved with the help of the following initial and boundary conditions:

$$\text{IC:} \quad T(x,0) = T_i \quad (3.37)$$

$$\text{BC 1:} \quad \left(\frac{\partial T}{\partial x}\right)_{x=0} = 0 \quad (3.38)$$

$$\text{BC 2:} \quad -k\left(\frac{\partial T}{\partial x}\right)_{x=L} = h[T(L,t) - T_\infty] \quad (3.39)$$

Using the following nondimensional quantities:

$$\theta^* = \frac{\theta}{\theta_i} = \frac{T - T_\infty}{T_i - T_\infty} \quad (3.40)$$

where $x^* = x/L$ and $t^* = (\alpha\, t/L^2)$. The final functional dependence of temperature can be shown as:

$$\theta^* = f(x^*, \text{Bi}, \text{Fo}) \quad (3.41)$$

Schneider[31] presented an exact solution for the case when a plane wall is subjected to cooling from both sides. Starting from the nondimensional form of Eq. (3.36), it was shown that:

$$\theta^* = \sum C_n \exp(-\zeta_n^2 \text{Fo})\cos(\zeta_n x^*) \quad (3.42)$$

where $\text{Fo} = \dfrac{\alpha t}{L^2}$

$$C_n = \frac{4\sin\zeta_n}{2\zeta_n + \sin 2\zeta_n}$$

The eigenvalues of ζ_n are positive roots of the transcendental equation:

$$\zeta_n \tan\zeta_n = \text{Bi} \quad (3.43)$$

which are given in mathematical tables.

3.3.2.3 Radial Coordinate Systems

Schneider[30] has given exact solutions for an infinite cylinder and sphere. For an infinite cylinder, the dimensionless temperature distribution is:

$$\theta^* = \sum C_n \exp(-\zeta_n^2 \text{Fo}) J_o(\zeta_n r^*) \quad (3.44)$$

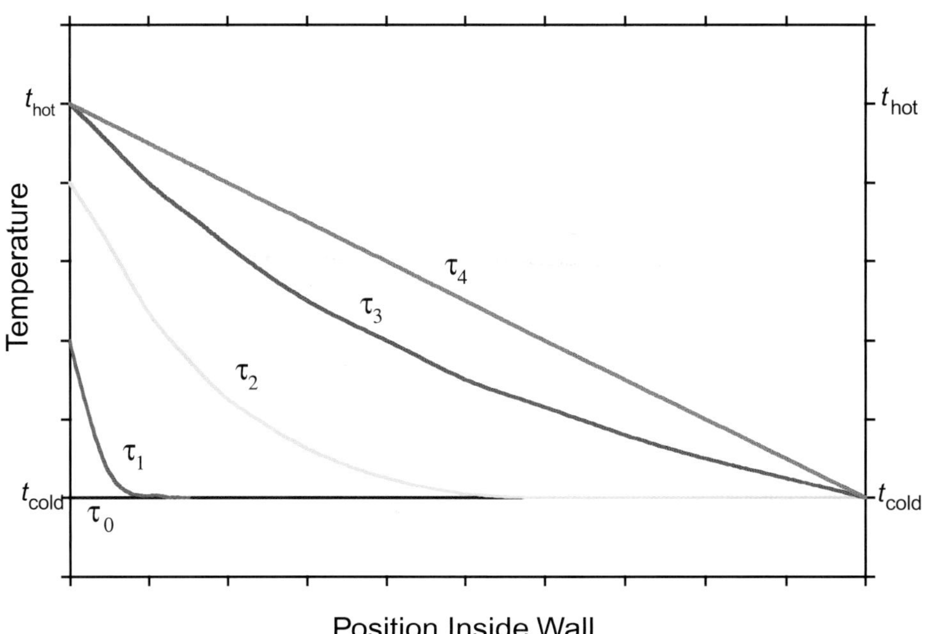

FIGURE 3.6 Transient conduction through a solid.

where $\mathrm{Fo} = \dfrac{\alpha t}{r_0^2}$

$$C_n = \frac{2}{\zeta_n} \cdot \frac{\mathrm{J}_1 \zeta_n}{\mathrm{J}_0^2 \zeta_n + \mathrm{J}_1^2 \zeta_n} \tag{3.45}$$

The eigenvalues of ζ_n are positive roots of the transcendental equation:

$$\frac{\zeta_n \mathrm{J}_1(\zeta_n)}{\mathrm{J}_0(\zeta_n)} = \mathrm{Bi} \tag{3.46}$$

J_1 and J_0 are Bessel functions of the first kind and their values are tabulated in standard mathematical tables.

For the spherical case, the dimensionless temperature distribution was given by Schneider[30] as follows:

$$\theta^* = \sum C_n \exp(-\zeta_n^2 \mathrm{Fo}) \frac{\sin(\zeta_n r^*)}{\zeta_n r^*} \tag{3.47}$$

where $\mathrm{Fo} = \dfrac{\alpha t}{r_0^2}$

$$C_n = \frac{4 \sin(\zeta_n) - \cos(\zeta_n)}{2\zeta_n - \sin 2\zeta_n} \tag{3.48}$$

The eigenvalues of ζ_n are positive roots of the transcendental equation:

$$1 - \zeta_n \cot \zeta_n = \mathrm{Bi} \tag{3.49}$$

given by Schneider.[30]

3.3.2.4 Transient Conduction in Semi-infinite Solids

The unsteady-state, one-dimensional conduction equation given by Eq. (3.36) and the initial condition defined by Eq. (3.37) can also be applied to the case of semi-infinite solids. Carslaw and Jaeger[31] and Schneider[30] have presented exact solutions for the following three different surface conditions:

Case (1): Constant Surface Temperature (T_s):

$$\frac{T(x,t) - T_s}{T_i - T_s} = \mathrm{erf}\left(\frac{x}{(4\alpha t)^{1/2}} \right) \tag{3.50}$$

Case (2): Constant Surface Heat Flux ($q_s'' = q_o''$):

$$T(x,t) - T_i = 2\left(q_0'' \frac{\left(\frac{\alpha t}{\pi}\right)^{1/2}}{k} \right) \exp\left(\frac{-x^2}{4\alpha t} \right) - \left(q_0'' \frac{x}{h} \right) \mathrm{erfc}\left(\frac{x}{(4\alpha t)^{1/2}} \right) \tag{3.51}$$

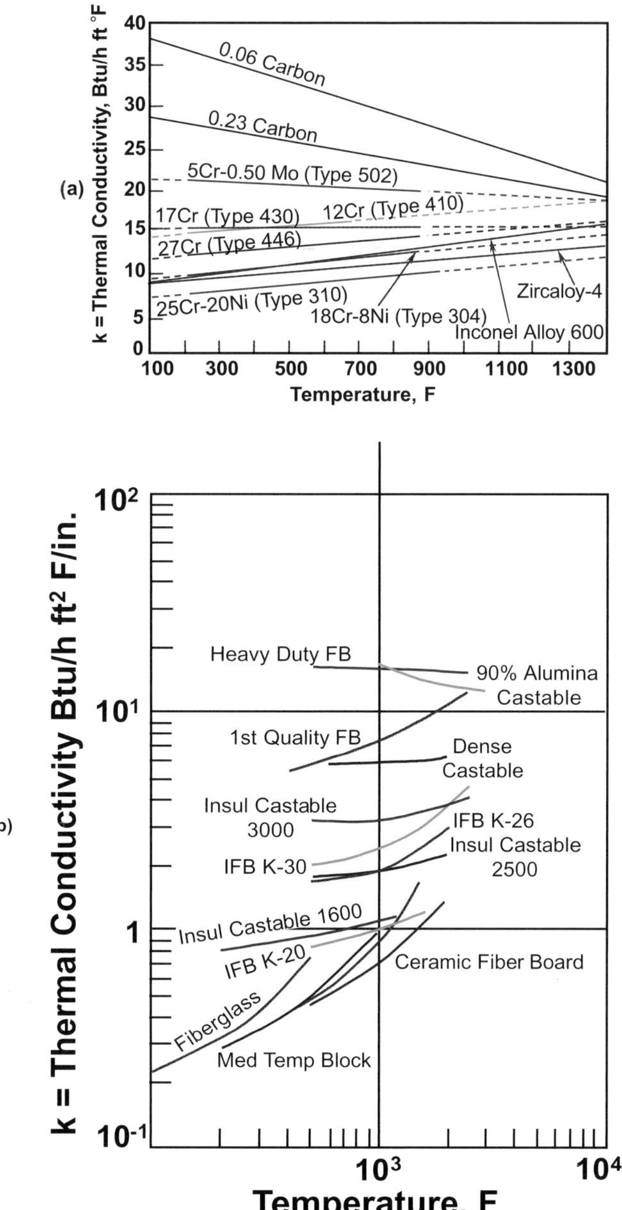

FIGURE 3.7 Thermal conductivity of (a) some commonly used steels and alloys and (b) some refractory materials.

Case (3): Surface Convection Condition:

$$-k\left(\frac{\partial T}{\partial x}\right)_{x=0} = \frac{h(T_\infty - T(0,t))(T(x,t)-T_i)}{T_\infty - T_i}$$

$$= \text{erfc}\left(\frac{x}{(4\alpha t)^{1/2}}\right) - \left[\exp\left(\frac{hx}{k} + \frac{h^2\alpha t}{k^2}\right)\right] \quad (3.52)$$

$$\left[\text{erfc}\left(\frac{x}{(4\alpha t)^{1/2}} + \frac{(h^2\alpha t)^{1/2}}{k}\right)\right]$$

where the complimentary error function (erfc m) has been defined as (1 – erf m).

As a typical example, case (2) has been shown graphically in Figure 3.6. In the above relationships, the thermal conductivity of the material has been assumed to be constant for most practical purposes if the temperature change is not appreciable. Figure 3.7(a) shows thermal conductivities of some commonly used steels and alloys as a function of temperature. Thermal conductivities of various refractory materials are shown in Figure 3.7(b). Figure 3.8 shows temperature variation in a slab, as a function of thickness, under three

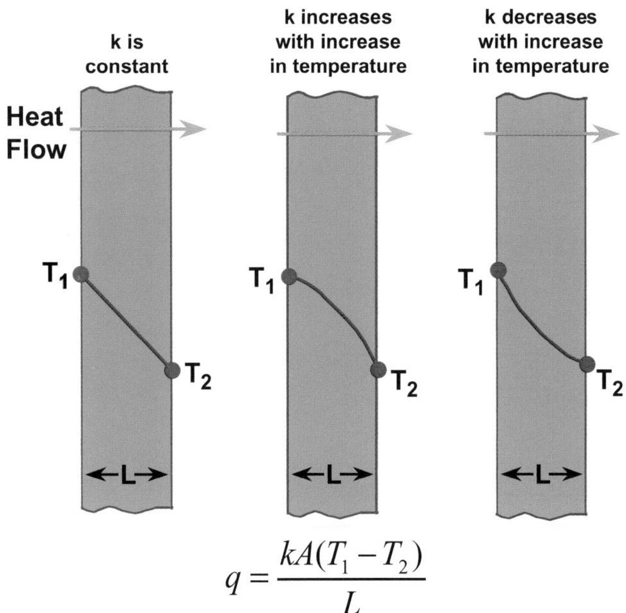

$$q = \frac{kA(T_1 - T_2)}{L}$$

FIGURE 3.8 Temperature-thickness relationships corresponding to different thermal conductivities.

TABLE 3.5 Typical Convective Heat Transfer Coefficients

Condition	Btu/h ft² F	W/m²C
Air, free convection	1 to 5	6 to 30
Air, forced convection	5 to 50	30 to 300
Steam, forced convection	300 to 800	1800 to 4800
Oil, forced convection	5 to 300	30 to 1800
Water, forced convection	50 to 2000	300 to 1200
Water, boiling	500 to 20,000	3000 to 120,000

different conditions: (a) thermal conductivity is constant, (b) thermal conductivity is increasing with temperature, and (c) thermal conductivity is decreasing with temperature.

3.4 CONVECTION

Convection heat transfer takes place in fluids. A combination of molecular conduction and macroscopic fluid motion contributes to convective heat transfer. Convection takes place adjacent to heated surfaces as a result of fluid motion past the surface. All convection processes fall into three categories: natural convection, forced convection, and mixed convection.

Natural convection occurs when fluid motion is created as a result of local density differences alone. Theoretical analyses of natural convection require the simultaneous solution of the coupled equations of motion and energy.

Forced convection results when mechanical forces from devices such as fans give motion to the fluid. Forced convection is the most commonly employed mechanism in the process industries. Schematically, hot and cold fluids, separated by a solid boundary, are pumped through the heat-transfer equipment, the rate of heat transfer being a function of the physical properties of the fluids, the flow rates, and the geometry of the system. Flow is generally turbulent and the flow duct varies in complexity from circular tubes to baffled and finned tubes. Theoretical analyses of forced convection heat transfer have been limited to relatively simple geometries and laminar flow. Usually, for complicated geometries, only empirical relationships are available. However, computational fluid dynamics CFD (see Chapter 9), the science of computer modeling flows and heat transfer, has advanced enough to provide good information based on the available semi-empirical CFD models. In forced convection, heat transfer coefficients are strongly influenced by the mechanics of flow occurring during forced-convection heat transfer. Intensity of turbulence, entrance conditions, and wall conditions are some of the factors that must be considered for greater accuracy.

Mixed convection refers to those situations when both natural and forced convection are at work. A good example would be the convective heat transfer process taking place on the outside surface of a furnace wall when there is some wind blowing. In the absence of wind, the wall would be cooled purely by natural convection; but with wind, both mechanisms are present simultaneously.

One example of the importance of forced convection in the process industries is the convection section in many process heaters. This is the downstream section of the heater that is heated by the combustion exhaust gases exiting the radiant or primary heating section. Not all heaters have a convection section, but Garg estimates that heater efficiency can be increased from 55–65% to 80% or more with the addition of a convection section.[32] A number of books are available that deal specifically with convection heat transfer.[33–41]

3.4.1 Newton's Law of Cooling

Any convective transfer of heat can be represented by a general heat balance equation called Newton's law of cooling:

$$q'' = h(T_s - T_\infty) \tag{3.53}$$

where q'' is the heat flux, h is the convective heat transfer coefficient, and $(T_s - T_\infty)$ is the temperature difference between the hot fluid and the cold fluid/surrounding. Table 3.5 gives typical values of convective heat transfer coefficients.

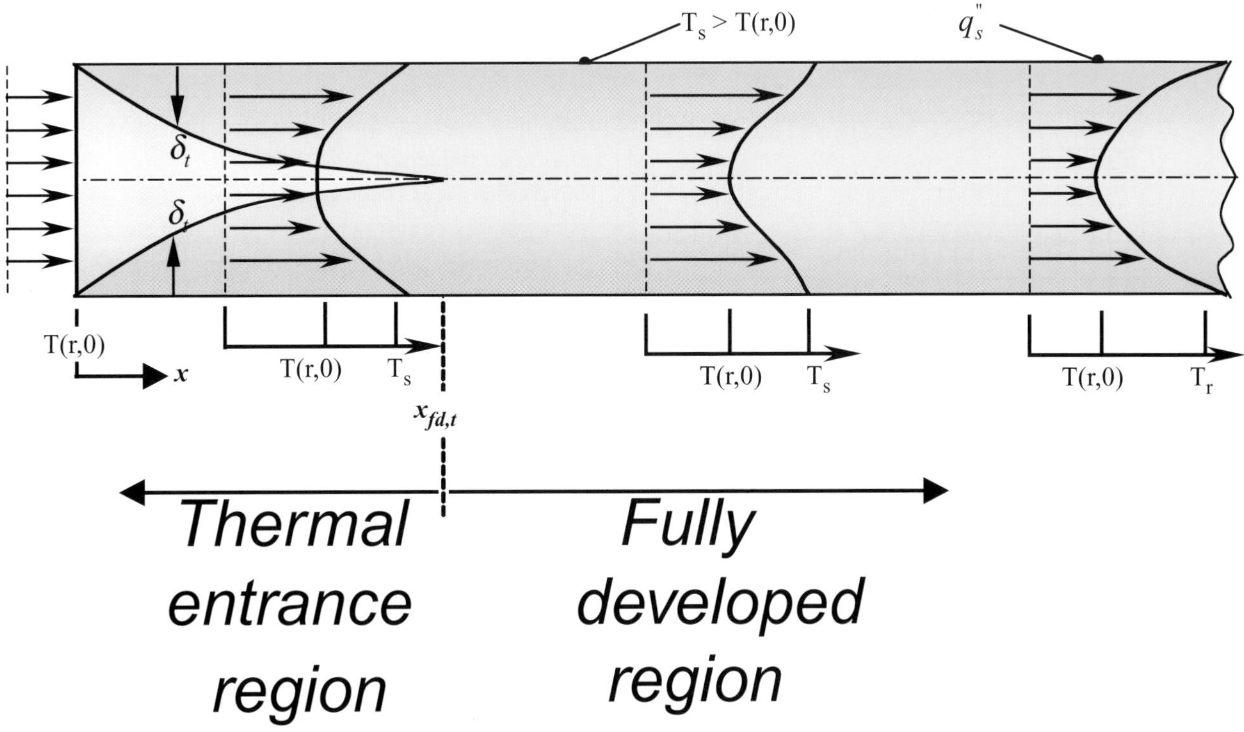

FIGURE 3.9 Thermal boundary layer development in a heated circular tube.[29]

3.4.2 Laminar Flow Convection

An important factor that influences convective heat transfer is the laminar sublayer. It is well-known that for a turbulent flow of a fluid past a solid, in the immediate neighborhood of the surface there exists a relatively quiet zone of fluid called the laminar sublayer. As one approaches the wall from the body of the fluid, the flow slows down, and this slowed-down region is known as the boundary layer. The boundary layer itself has fluid that is turbulent closer to the core of the flow, followed by a transition zone, and finally becomes laminar very close to the wall. The portion of the flow that is essentially laminar is called the laminar sublayer. In the laminar sublayer, the heat is transferred by molecular conduction. The resistance of the laminar layer to heat flow will vary according to its thickness and can range from 95% of the total resistance for some fluids to only 1% for others (e.g., liquid metals).

In highly turbulent flows, the sublayer is thinner and, thus, greater turbulence makes for better heat transfer in general. Similarly, surface roughness and other mechanisms, such as oscillating flow or phase change, will aid in heat transfer by disturbing the boundary layer. So, from a heat transfer point of view, it is of benefit in reformer and cracking tubes to have the rough surface finish that results from the spin-cast process used to make the tubes. Similarly, many boiler manufacturers and other heat exchanger manufacturers use heat transfer tubes with internal "rifling" to enhance the convective heat transfer.

In most cases, the boundary layer effect is dominant in gases. In a system transferring heat from a gas to a liquid, the resistance on the liquid side can usually be neglected because it is so much smaller than the resistance on the gas side.

Consider a fluid at a uniform temperature $T(\mathrm{r},0)$ entering a tube where the heat transfer takes place from the wall of the tube (maintained either at a constant temperature or with a constant wall heat flux); a thermal entrance region is formed as shown in Figure 3.9. For laminar flow conditions, the thermal entry length is given by Kays and Crawford[37] as:

$$\left(x_{\mathrm{fd,t}}/D_{\mathrm{lam}}\right) = 0.05\,\mathrm{Re}_D\,\mathrm{Pr} \tag{3.54}$$

where $x_{\mathrm{fd,t}}$ is the thermal entrance length, D is the diameter of the tube, Re_D is the Reynolds number based on tube diameter, and Pr is the Prandtl number. It is interesting that the equation for thermal entry length is very similar to the equation for hydrodynamic entry length, which is:

$$\left(x_{\mathrm{fd,h}}/D_{\mathrm{lam}}\right) = 0.05\,\mathrm{Re}_D \tag{3.55}$$

where $x_{\mathrm{fd,h}}$ is the hydrodynamic entry length.

3.4.2.1 Fully Developed Velocity and Temperature Profiles

For the particular situation when the velocity in a tube can be approximated as uniform and the temperature is given by a parabolic profile, that is:

$$u(r) = C_1 \quad \text{and} \quad T(r) - T_s = C_2\left[1 - \left(r/r_0\right)^2\right] \tag{3.56}$$

The convection coefficient, h, is given by:

$$h = \frac{q_s''}{T_s - T_m} \tag{3.57}$$

where T_s and T_m are tube surface temperature and the mean temperature of the fluid in the tube, respectively. Also, the mean temperature is given by:

$$T_m = \frac{2}{u_m r_0^2}\int_0^{r_0} uTr dr \tag{3.58}$$

Noting that the temperature profile is parabolic and the velocity is uniform throughout, Eq. (3.56) shows that:

$$q_s'' = k\left(\frac{\partial T}{\partial r}\right)_{r=r_0} = \frac{-2kC_2}{r_0} \tag{3.59}$$

Therefore,

$$h = \frac{q_s''}{T_s - T_m} = \frac{4k}{r_0} \tag{3.60}$$

or

$$\text{Nu}_D = \frac{hD}{k} = 8 \tag{3.61}$$

where Nu_D is the Nusselt number based on tube diameter.

Similarly, in a circular tube characterized by uniform surface heat flux and fully developed conditions, the Nusselt number is given as:

$$\text{Nu}_D = 4.36 \tag{3.62}$$

and for laminar, fully developed conditions, with a constant surface temperature, the Nusselt number is:

$$\text{Nu}_D = 3.66 \tag{3.63}$$

3.4.3 *Turbulent Internal Flow*

Internal or conduit flow is a flow field in which the fluid completely fills a closed stationary duct. On the other hand, external or immersed flow is where the fluid flows past a stationary immersed solid. With internal flow, the heat transfer coefficient is theoretically infinite at the location where heat transfer begins. The local heat transfer coefficient rapidly decreases and becomes constant, so that after a certain length, the average coefficient in the conduit is independent of the length. The local coefficient may follow an irregular pattern, however, if obstructions or turbulence promoters are present in the duct.

Because the analysis of turbulent flow heat transfer is quite complex, calculations must rely on empirical correlations. The Chilton-Colburn analogy provides an important correlation for the Nusselt number in turbulent flow heat transfer:

$$\text{ST Pr}^{2/3} = \frac{\text{Nu}_D}{\text{Re}_D \text{ Pr}}\text{Pr}^{2/3} = \frac{f}{8} \tag{3.64}$$

where St is Stanton number and f is the friction factor, given by

$$f = 0.184\left(\text{Re}_D\right)^{-1/5} \tag{3.65}$$

Thus,

$$\text{Nu}_D = 0.023\left(\text{Re}_D\right)^{4/5}(\text{Pr})^{1/3} \tag{3.66}$$

Dittus and Boelter[42] suggested a modification in Eq. (3.66) by replacing the exponent of the Prandtl number by n, where n is 0.4 for heating and 0.3 for cooling. It is to be noted that Eq. (3.66) or its modification, are good for cases where the temperature difference ($T_s - T_m$) is moderate. When the temperature difference is large, the equation suggested by Sieder and Tate[43] is recommended as follows:

$$\text{Nu}_D = 0.027\left(\text{Re}_D\right)^{4/5}(\text{Pr})^{1/3}\left(\mu/\mu_s\right)^{0.14} \tag{3.67}$$

where μ_s is the viscosity of the fluid determined at surface temperature, and all the other properties are measured at the mean temperature. The Dittus–Boelter and Sieder–Tate equations are applicable for cases of both uniform surface temperature and heat flux conditions. Petukhov[44] has given a correlation that gives more accurate results than the

Dittus–Boelter or Sieder–Tate equations, but is more complex to use. The correlation is:

$$Nu_D = \frac{(f/8)Re_D Pr}{1.07 + 12.7(f/8)^{1/2}\left(Pr^{2/3} - 1\right)} \quad (3.68)$$

where the friction factor, f, is obtained from the Moody diagram.[45]

For the special cases of liquid metals, where the Prandtl number is very small ($0.003 \leq Pr \leq 0.05$), Skupinski et al.[46] have given a correlation for heat transfer in fully developed turbulent flow. For constant surface heat flux:

$$Nu_D = 4.82 + 0.0185\ Pe_D^{0.827} \quad (3.69)$$

for $3.6 \times 10^3 < Re_D < 9.05 \times 10^5$ and $10^2 < Pe_D < 10^4$, where the Peclet number is defined as $Pe_D = Re_D \cdot Pr$. For constant surface temperature:[47]

$$Nu_D = 5.0 + 0.025\ Pe_D^{0.8} \quad (3.70)$$

Reed[48] has presented extensive literature on different correlations for heat transfer in laminar and turbulent flow conditions.

3.4.3.1 Noncircular Tubes/Sections

Although the correlations discussed thus far have been presented for circular tubes, these relationships can be extended to noncircular tubes and sections by simply replacing the tube diameter with the hydraulic diameter, defined as:

$$D_h = 4\frac{A_c}{P} \quad (3.71)$$

where A_c is the flow cross-sectional area and P is the wetted perimeter. Calculations of Reynolds number and Nusselt number are based on hydraulic diameter. In the case of a concentric annulus, the hydraulic diameter is given by:

$$D_h = \frac{4\left(\frac{\pi}{4}\right)\left(D_o^2 - D_i^2\right)}{\pi D_o + \pi D_i} = D_o - D_i \quad (3.72)$$

Table 3.6 summarizes the convection correlations in circular tubes.[29]

3.4.4 *Turbulent External Flow*

External flow or immersed flow occurs when a fluid flows past a stationary immersed solid. Similar to internal flow, the local coefficient immersed flow is again infinite at the point where heating begins. Subsequently, it decreases and may show various irregularities, depending on the configuration of the body. Usually, in this instance, the local coefficient never becomes constant as flow proceeds downstream over the body.

When heat transfer occurs during immersed flow, the rate depends on the configuration of the body, the position of the body, the proximity of other bodies, and the flow rate and turbulence of the stream. The heat transfer coefficient varies over the immersed body, because both the thermal and the momentum boundary layers change even with simple configurations immersed in an infinite flowing fluid. For complicated configurations and assemblages of bodies, such as found on the shell side of a heat exchanger, little is known about the local heat transfer coefficient; empirical relationships giving average coefficients are all that are usually available. Research conducted on local coefficients in complicated geometries has not been extensive enough to extrapolate into useful design relationships.

For turbulent flow with Reynolds numbers up to about 10^8, the local friction coefficient is given by:[49]

$$C_{f,x} = 0.0592\, Re_x^{-1/5} \quad (3.73)$$

The velocity boundary layer thickness is given by:

$$\delta = 0.37 \times Re_x^{-1/5} \quad (3.74)$$

where x is the distance in the direction of flow. Thus, the local Nusselt number for external turbulent flow is:

$$Nu_x = St\, Re_x\, Pr = 0.0296\, Re_x^{4/5}\, Pr^{1/3} \quad (3.75)$$

where St is the Stanton number.

Complications arise when the boundary layer formation on the external flow consists of both laminar and turbulent portions. Under these circumstances, neither laminar nor turbulent correlations are satisfactory. A reasonably good correlation for mixed boundary layer conditions is:

$$Nu_L = \left[0.037\, Re_L^{4/5} - 871\right] Pr^{1/3} \quad (3.76)$$

with the conditions that $0.6 < Pr < 60$; $5 \times 10^5 < Re_L < 10^8$; $Re_{x,c} = 5 \times 10^5$; and where Re_L is based on total length of the surface and $Re_{x,c}$ is the critical Reynolds number for transition from laminar to turbulent.

TABLE 3.6 Summary of Convection Correlations for Flow in a Circular Tube[29]

Correlation	Conditions
$f = 64/\mathrm{Re}_D$	Laminar, fully developed
$\mathrm{Nu}_D = 4.36$	Laminar, fully developed, uniform q''_s, Pr ≥ 0.6
$\mathrm{Nu}_D = 3.66$	Laminar, fully developed, uniform T_s, Pr ≥ 0.6
$\overline{\mathrm{Nu}}_D = 3.66 + \dfrac{0.0668(D/L)\mathrm{Re}_D \mathrm{Pr}}{1+0.04[(D/L)\mathrm{Re}_D \mathrm{Pr}]^{2/3}}$	Laminar, thermal entry length (Pr >> 1 or an unheated starting length), uniform T_s
or, $\overline{\mathrm{Nu}}_D = 1.86\left(\dfrac{\mathrm{Re}_D \mathrm{Pr}}{L/D}\right)^{2/3}\left(\dfrac{\mu}{\mu_s}\right)^{0.14}$	Laminar, combined entry length $\{[\mathrm{Re}_D\mathrm{Pr}/(L/D)]^{1/3}(\mu/\mu_s)^{0.14}\} \ge 2$; uniform T_s, 0.48 < Pr < 16,700; 0.0044 < (μ/μ_s) < 9.75
$f = 0.316\,\mathrm{Re}_D^{-1/4}$	Turbulent, fully developed; $\mathrm{Re}_D \le 2 \times 10^4$
$f = 0.184\,\mathrm{Re}_D^{-1/5}$	Turbulent, fully developed; $\mathrm{Re}_D \le 2 \times 10^4$
or	
$f = (0.790 \ln \mathrm{Re}_D - 1.64)^{-2}$	Turbulent, fully developed; $3000 \le \mathrm{Re}_D \le 5 \times 10^6$
$\mathrm{Nu}_D = 0.023\,\mathrm{Re}_D^{4/5}\,\mathrm{Pr}^n$	Turbulent, fully developed; 0.6 ≤ Pr ≤ 160; $\mathrm{Re}_D \ge 10{,}000$, $(L/D) \ge 10$; $n = 0.4$ for $T_s > T_m$ and $n = 0.3$ for $T_s < T_m$
or	
$\mathrm{Nu}_D = 0.027\,\mathrm{Re}_D^{4/5}\,\mathrm{Pr}^{1/3}\left(\dfrac{\mu}{\mu_s}\right)^{0.14}$	Turbulent, fully developed; 0.7 ≤ Pr ≤ 16,700, $\mathrm{Re}_D \ge 10{,}000$, $(L/D) \ge 10$
or	
$\mathrm{Nu}_D = \left(\dfrac{(f/8)(\mathrm{Re}_D - 1000)\mathrm{Pr}}{1+12.7(f/8)^{1/2}(\mathrm{Pr}^{2/3}+1)}\right)$	Turbulent, fully developed; 0.5 ≤ Pr ≤ 2000; $3000 \le \mathrm{Re}_D \le 5 \times 10^6$, $(L/D) \ge 10$
$\mathrm{Nu}_D = 4.82 + 0.0185(\mathrm{Re}_D \mathrm{Pr})^{0.827}$	Liquid metals, turbulent, fully developed, uniform q''_s, $3.6 \times 10^3 < \mathrm{Re}_D < 9.05 \times 10^5$, $10^2 < \mathrm{Pe}_D < 10^4$
$\mathrm{Nu}_D = 5.0 + 0.025(\mathrm{Re}_D \mathrm{Pr})^{0.8}$	Liquid metals, turbulent, fully developed; uniform T_s, $\mathrm{Pe}_D > 100$

Source: F.P. Incropera and D.P. DeWitt, *Fundamentals of Heat and Mass Transfer*, 4th edition, John Wiley & Sons, New York, 1996. With permission.

Similarly, the suitable correlation for friction coefficient in mixed boundary cases is given by:

$$C_{f,L} = [0.074/\mathrm{Re}_L^{1/5}] - [1742/\mathrm{Re}_L] \qquad (3.77)$$

with the conditions that $5 \times 10^5 < \mathrm{Re}_L < 10^8$ and $\mathrm{Re}_{x,c} = 5 \times 10^5$.

When L is very high compared to x_c (i.e., the entire surface is covered by turbulent layer) the correlation for heat transfer simplifies to:

$$\mathrm{Nu}_L = 0.037\,\mathrm{Re}_L^{4/5}\,\mathrm{Pr}^{1/3} \qquad (3.78)$$

and similarly, the friction coefficient becomes:

$$C_{f,L} = 0.074\,\mathrm{Re}_L^{-1/5} \qquad (3.79)$$

3.4.4.1 Convection Heat Transfer for Cylinders in Cross Flow

Hilpert[50] has presented a correlation for the average Nusselt number for convection heat transfer for cylinders in cross flow:

$$\mathrm{Nu}_D = \frac{hD}{k} = \mathrm{C}\,\mathrm{Re}_D^m\,\mathrm{Pr}^{1/3} \tag{3.80}$$

where the Nusselt and Reynolds numbers are based on the diameter of the cylinder, and constants C and m are as presented by Hilpert[50] and Knudsen and Katz (1958)[51] in Table 3.7.

In Eq. (3.81), Churchill and Bernstein[52] suggest a more comprehensive correlation covering a wider range of Reynolds and Prandtl numbers and suitable for the entire range of experimental data available:

$$\mathrm{Nu}_D = 0.3 + \frac{0.62\,\mathrm{Re}_D^{1/2}\,\mathrm{Pr}^{1/3}}{\left[1+(0.4/\mathrm{Pr})^{2/3}\right]^{1/4}}\left[1+\left(\frac{\mathrm{Re}_D}{282{,}000}\right)^{5/8}\right]^{4/5} \tag{3.81}$$

where all the physical properties are determined at film temperature.

3.4.4.2 Convection Heat Transfer in Banks of Tubes

Grimison[53] suggested a correlation for convection heat transfer in aligned or staggered banks of tubes for ten or more rows of tubes:

$$\mathrm{Nu}_D = 1.13\,\mathrm{C}_1\,\mathrm{Re}_{D,\max}^m\,\mathrm{Pr}^{1/3} \tag{3.82}$$

for the conditions $2000 < \mathrm{Re}_{D,\max} < 40{,}000$ and $\mathrm{Pr} \geq 0.7$

For the number of rows less than ten, a correction factor must be used.[29] In the above equation, the Reynolds number is based on the maximum fluid velocity occurring within the tube bank. The maximum velocity for the aligned arrangement is given by:

$$V_{\max} = \frac{S_T}{S_T - D}V \tag{3.83}$$

and the maximum velocity for the staggered arrangement is:

$$V_{\max} = \frac{S_T}{2(S_D - D)}V \tag{3.84}$$

where S_T = center-to-center distance in a transverse plane
S_D = distance between tube centers in a diagonal plane
D = diameter of tubes
V = freestream velocity

TABLE 3.7 Constants of Equation (3.80) for a Circular Cylinder in Cross Flow[29]

Re_D	C	m
0.4 – 4	0.989	0.330
4 – 40	0.911	0.385
40 – 4 × 10³	0.683	0.466
4 × 10³ – 4 × 10⁴	0.193	0.618
4 × 10⁴ – 4 × 10⁵	0.027	0.805

TABLE 3.8 Constants of Equation (3.85) for the Tube Bank in Cross Flow[29]

Configuration	$\mathrm{Re}_{D,\max}$	C	m
Aligned	10 – 10²	0.80	0.40
Staggered	10 – 10²	0.90	0.40
Aligned	10² – 10³	Approximate as a single (isolated) cylinder	
Staggered	10² – 10³		
Aligned ($S_T/S_L > 0.7$)[a]	10³ – 2 × 10⁵	0.27	0.63
Staggered ($S_T/S_L > 2$)	10³ – 2 × 10⁵	$0.35\,(S_T/S_L)^{1/5}$	0.60
Staggered ($S_T/S_L > 2$)	10³ – 2 × 10⁵	0.40	0.60
Aligned	2 × 10⁵ – 2 × 10⁶	0.021	0.84
Staggered	2 × 10⁵ – 2 × 10⁶	0.022	0.84

[a] For ($S_T/S_L < 0.7$), heat transfer is inefficient and aligned tubes should not be used.

Source: A. Zhukauskas, *Advances in Heat Transfer*, Vol. 8, J.P. Hartnett and T.F. Irvine, Jr., Eds., Academic Press, New York, 1972. With permission.

Zhukauskas[54] has presented a correlation that is more recent and widely used:

$$\mathrm{Nu}_D = C\cdot\mathrm{Re}_{D,\max}^m\,\mathrm{Pr}^{0.36}\left(\frac{\mathrm{Pr}}{\mathrm{Pr}_S}\right)^{1/4} \tag{3.85}$$

The equation is valid for the following conditions: number of rows, $N_L \geq 20$; $0.7 < \mathrm{Pr} < 500$; $1000 < \mathrm{Re}_{D,\max} < 2 \times 10^6$, where all the properties, except Pr_s, have been determined at the arithmetic mean of the inlet and outlet fluid temperatures, and constants C and m are listed in Table 3.8.

3.5 RADIATION

Thermal radiation heat transfer is the movement of energy by electromagnetic waves. Quantum theory describes electromagnetic energy as photons or quanta. Unlike conduction or convection, radiation does not require any intervening

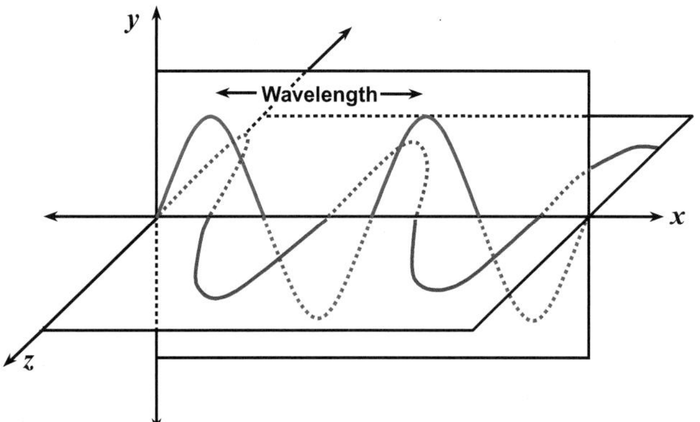

FIGURE 3.10 Orthogonal oscillations of electric and magnetic waves in the propagation of electromagnetic waves.

medium for transfer. Electromagnetic radiation, in the wavelength range of 0.1 to 100 micrometers, is produced solely by the temperature of a body. Energy at the body's surface is converted into electromagnetic waves that emanate from the surface and strike another body. Some of the thermal radiation is absorbed by the receiving body and reconverted into internal energy, while the remaining energy is reflected from or transmitted through the body. The fractions of radiation reflected, transmitted, and absorbed by a surface are known, respectively, as reflectivity ρ, transmissivity τ, and absorptivity α. The sum of these fractions equals one. Thermal radiation is emitted by all surfaces whose temperatures are above absolute zero.

Thermal radiation can pass through some gases, like N_2, without absorption taking place. Thus, these gases do not affect radiative transfer. On the other hand, gases like carbon dioxide, water vapor, and carbon monoxide affect radiation to some extent and are known as participating gases. These participating gases are, of course, common constituents of furnace flue gases and, as such, play a significant role in the transfer and distribution of heat to the heater tubes.

All surfaces emit radiation in amounts determined by the temperature and the nature of the surface. The perfect radiator, commonly known as a "blackbody," absorbs all the radiant energy reaching its surface and emits radiant energy at the maximum theoretical limit according to the Stefan-Boltzmann law:

$$Q_r = A\sigma T_s^4$$

where Q_r is the radiant energy, A is the area, σ is the Stefan-Boltzmann constant 0.1713×10^{-8} Btu/hr ft^2-R^4 (or 5.669×10^{-8} W/m^2-K^4), and T_s is the absolute temperature in Rankine (Kelvin). The product σT_s^4 is also known as the blackbody emissive power E_b.

Radiation plays an important role in many industrial processes that require heating, cooling, drying, combustion, and solar energy. Figure 3.10 shows a schematic of the propagation of electromagnetic waves. The electrical and magnetic oscillations can be seen to be orthogonal to each other. Figure 3.11 shows the spectrum of electromagnetic radiation and Table 3.9 gives approximate wavelengths, frequencies, and energies for selected regions of the spectrum. Several books dedicated to general radiation heat transfer are available.[8,55–60]

This section first treats thermal radiation heat transfer between surfaces in enclosures. Each surface is assumed to be isothermal, gray, and diffuse. The assumption that the surfaces are gray means that they emit and absorb thermal radiation without regard to the *wavelength* or *frequency* of the radiation. Because thermal radiation is a wave, it has all of the properties of a wave. It has a wavelength, a frequency, and a wavenumber. It can be reflected, refracted, and diffracted. The assumption of diffuse surfaces implies that the surfaces emit, absorb, and reflect radiation energy without regard to the direction (relative to the surface). All real surfaces are neither gray nor diffuse. Many real surfaces have surface imperfections or oxide layers that may be on the order of 1×10^{-6} m (1 micron) thick. This surface layer will interact with radiation of wavelengths near this value. Additionally, electromagnetic theory can be used to predict both the directional and wavelength dependence of pure metal surfaces. Analysis of radiation heat transfer in an industrial setting is complicated by these factors. Furthermore, the radiative properties of real surfaces can be strongly dependent on the surface preparation. For example, a polished metal will have an emissivity that may be an order of magnitude below the emissivity of the same surface with an oxide layer. Uncertainties in radiative properties are always a consideration when conducting radiation heat transfer analyses. However, when used appropriately and with care, values of properties obtained from literature surfaces can be used to make reasonably accurate heat transfer calculations.

3.5.1 Blackbody Radiation/Planck Distribution

A blackbody is an ideal body, useful as a reference in discussion of radiant heat transfer theory, that is both an ideal and diffuse emitter and absorber of radiant energy. It absorbs all incident radiation, regardless of wavelength and direction. No real surface can emit more energy than a blackbody at a given temperature and wavelength. Because the emission from a blackbody is diffuse, the intensity of a blackbody is given as:

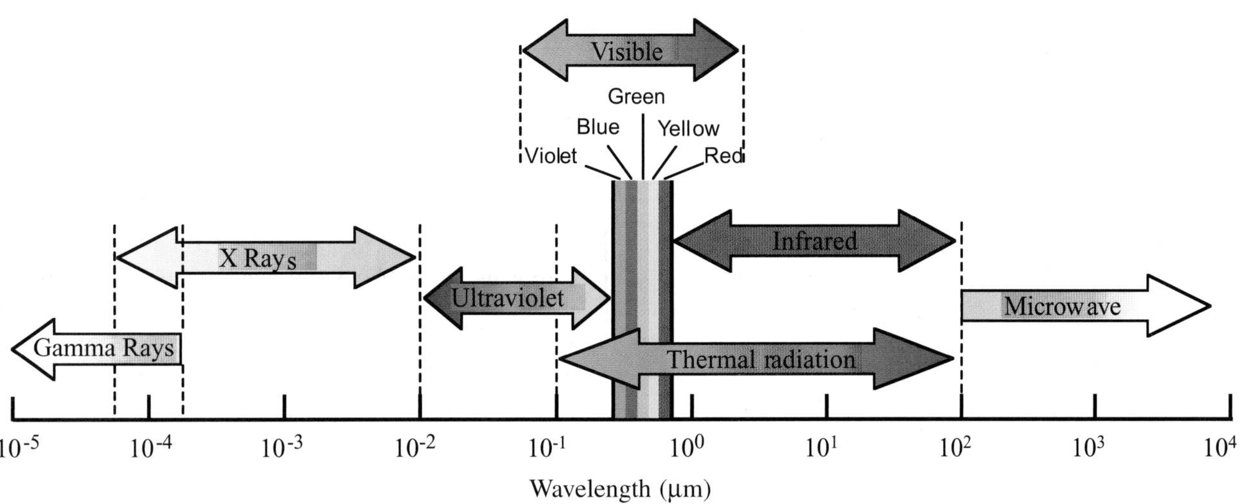

FIGURE 3.11 Spectrum of electromagnetic radiation.

TABLE 3.9 Spectrum of Electromagnetic Radiation

Region	Wavelength (Angstroms)	Wavelength (centimeters)	Frequency (Hz)	Energy (eV)
Radio	$>10^9$	>10	$<3 \times 10^9$	$<10^{-5}$
Microwave	10^9 to 10^6	10 to 0.01	3×10^9 to 3×10^{12}	10^{-5} to 0.01
Infrared	10^6 to 7000	0.01 to 7×10^{-5}	3×10^{12} to 4.3×10^{14}	0.01 to 2
Visible	7000 to 4000	7×10^{-5} to 4×10^{-5}	4.3×10^{14} to 7.5×10^{14}	2 to 3
Ultraviolet	4000 to 10	4×10^{-5} to 10^{-7}	7.5×10^{14} to 3×10^{17}	3 to 10^3
X-Rays	10 to 0.1	10^{-7} to 10^{-9}	3×10^{17} to 3×10^{19}	10^3 to 10^5
Gamma Rays	<0.1	$<10^{-9}$	$>3 \times 10^{19}$	$>10^5$

$$I_b = \frac{E_b}{\pi} \tag{3.86}$$

where E_b is the emissive power of the blackbody.

A relatively small opening to a cavity with a uniform interior surface temperature closely approximates the radiation characteristics of a blackbody. Radiation that enters the surface will be partially absorbed and partially reflected by the first internal surface of incidence. If the opening is small compared to the cavity dimension, then virtually all of the energy that enters the cavity will undergo multiple internal reflections and eventually be absorbed. Further, although the surfaces within the cavity are not black and they do not emit blackbody radiation, their *radiosity* (radiosity is the total radiation leaving a surface; in this case, it will be the emitted radiation plus the reflected radiation) will be that of a blackbody. Proof of this is given by Siegel and Howell.[59]

3.5.1.1 Planck Distribution

Spectral distribution of a blackbody emission was first determined by Planck[61] and is given by:

$$I_{\lambda,b}(\lambda, T) = \frac{2hc_0^2}{\lambda^5\left[\exp(hc_0/\lambda kT) - 1\right]} \tag{3.87}$$

where h = 6.6256×10^{-34} J·s (Planck's constant)
k = 1.3805×10^{-23} J/K (Boltzmann constant)
c_0 = 2.998×10^8 m/s (speed of light in vacuum)
T = absolute temperature of the blackbody, in K

On the assumption that the blackbody is a diffuse emitter, its spectral emissive power is given by:

$$E_{\lambda,b}(\lambda, T) = \pi I_{\lambda,b}(\lambda, T) = \frac{2h\pi c_0^2}{\lambda^5\left[\exp(hc_0/\lambda kT) - 1\right]} \tag{3.88}$$

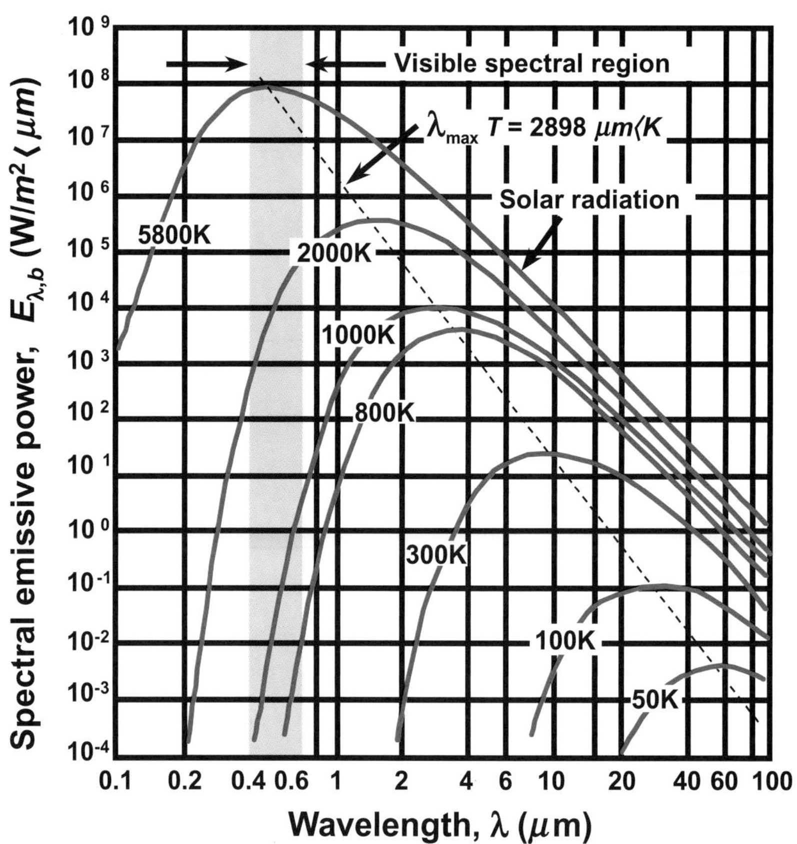

FIGURE 3.12 Spectral blackbody emissive power.

Equation (3.88) is known as the Planck distribution. Figure 3.12 shows the variation of spectral emissive power as a function of wavelength for selected temperatures. The figure indicates that as temperature increases, the blackbody emissive power at every wavelength increases and the wavelength of peak emission decreases. Radiation from the sun is approximated by radiation from a 5800 K blackbody source. The temperature at which radiant energy emissions from a surface become visible to the human eye is called the Draper point, occurring at approximately 800 K.

3.5.1.2 Wien's Displacement Law

From Figure 3.12 it is clear that the blackbody spectral distribution has a maximum and the corresponding λ_{max} depends on temperature. Differentiating Eq. (3.87) with respect to λ and setting the result equal to zero gives:

$$\lambda_{max} T = 2897.8\ \mu\text{m} \cdot \text{K} \tag{3.89}$$

Equation (3.89) is known as Wien's displacement law.

The dotted line in Figure 3.12 shows the locus of points of the maximum in the spectral distribution curves.

3.5.1.3 Stefan-Boltzmann Law

Integration of the Planck distribution equation shows that the emissive power of a blackbody is given as:

$$E_b = \sigma T^4 \tag{3.90}$$

where σ is the Stefan-Boltzmann constant and has the numerical value of 5.670×10^{-8} W/m²·K⁴. (0.1714×10^{-8} Btu/hr·ft²·R⁴) Equation (3.90) is known as the Stefan-Boltzmann law. The importance of this law is that the emissive power of a blackbody can be directly obtained for any temperature. Also, if the emissivity of any real surface is known, its emissive power can be calculated using the blackbody emissive power.

3.5.2 Radiant Exchange Between Black Surfaces

Because in a blackbody there is no reflection, energy leaves exclusively as emission and is absorbed completely by another blackbody. Figure 3.13 shows exchange between two black surfaces of arbitrary shape and size. If $q_{i \to j}$ is the rate at which radiation leaves surface *i* and is intercepted by surface *j*:

$$q_{i \to j} = \left(A_i J_i\right) F_{ij} \tag{3.91}$$

where F_{ij} is the shape factor. Because the radiosity of a black surface equals the emissive power,

$$q_{i \to j} = A_i F_{ij} E_{bi} \tag{3.92}$$

Similarly,

$$q_{j \to i} = A_i F_{ji} E_{bj} \tag{3.93}$$

Thus, the net exchange between the two black surfaces is:

$$q_{ij} = A_i F_{ij} \left(T_i^4 - T_j^4\right) \tag{3.94}$$

3.5.3 *Radiant Exchange Between Gray/Diffuse Surfaces*

The main problem in the radiation exchange between non-blackbodies is the surface reflection. Consider an exchange between surfaces in an enclosure. Assume that they are isothermal, opaque, and gray, with uniform radiosity and irradiation. The net rate of heat transfer from a surface is given by:

$$q_i = A_i \left(J_i - G_i\right) \tag{3.95}$$

where *J* is the radiosity and *G* is the irradiation. Also,

$$J_i = E_i + \rho_i G_i \tag{3.96}$$

where *E* is emissive power and ρ is the reflectivity of the surface. Thus,

$$q_i = A_i \left(E_i - \alpha_i G_i\right) \tag{3.97}$$

where $\rho_i = (1 - \alpha_i) = (1 - \varepsilon_i)$ for an opaque, diffuse, gray surface. Therefore, the radiosity is given as:

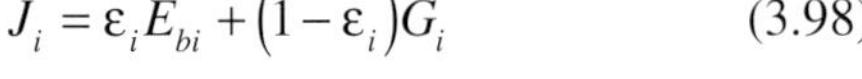

$$J_i = \varepsilon_i E_{bi} + \left(1 - \varepsilon_i\right) G_i \tag{3.98}$$

and

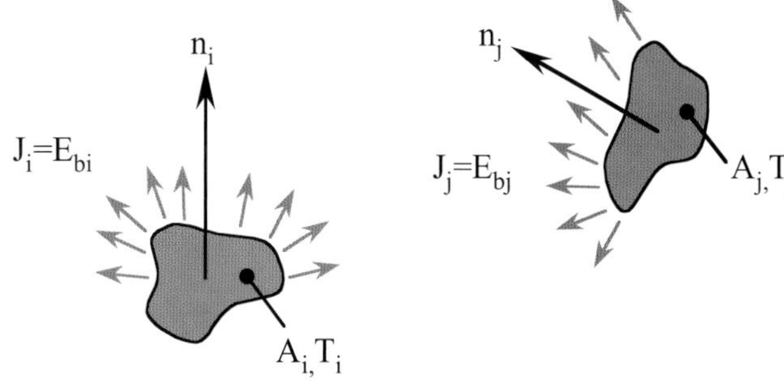

FIGURE 3.13 Radiation transfer between two surfaces approximated as blackbodies.

$$q_i = \frac{E_{bi} - J_i}{\left(1 - \varepsilon_i\right)\left(A_i \varepsilon_i\right)} \tag{3.99}$$

Thus, the total rate at which radiation reaches surface *i* from all surfaces is:

$$A_i G_i = \sum F_{ji} A_j J_j \tag{3.100}$$

Using the reciprocity and summation rule, the net rate of radiation transfer to surface *i* becomes:

$$q_i = \sum F_{ij} A_i \left(J_i - J_j\right) = \sum q_{ij} \tag{3.101}$$

Equation (3.101) is a relationship between the net rate of radiation transfer from surface *i* to the sum of radiant exchange with the other surfaces. The above exchange can also be represented as:

$$\frac{E_{bi} - j_i}{\left(1 - \varepsilon_i\right) / \left(A_i \varepsilon_i\right)} = \sum_{j=1}^{N} \frac{J_i - J_j}{\left(A_i F_{ij}\right)^{-1}} \tag{3.102}$$

A network representation of the above equation is shown in Figure 3.14.

In situations where the net radiation transfer rate is known and not the temperature, Eq. (3.101) is used in the alternate form:

$$q_i = \sum \frac{J_i - J_j}{\left(A_i F_{ij}\right)^{-1}} \tag{3.103}$$

The solutions of Eq. (3.101) or (3.103) are easily accomplished by matrix inversion or iteration methods.[29]

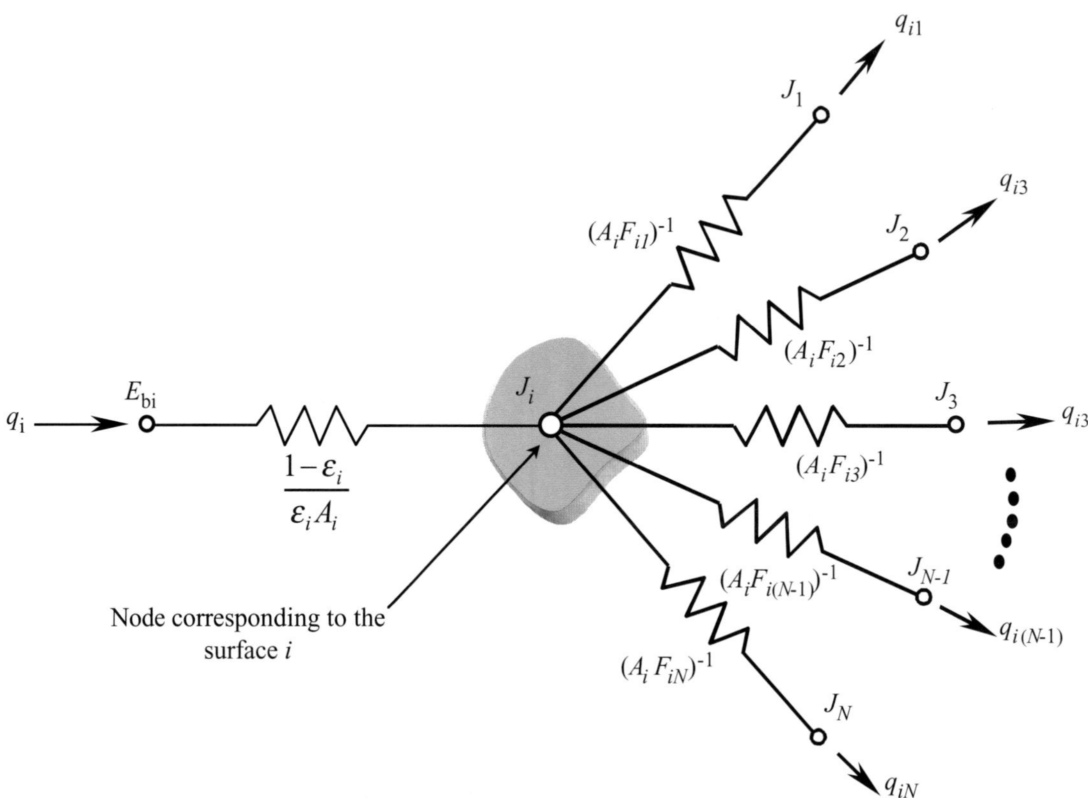

FIGURE 3.14 Network representation of radiative exchange between surface i and the remaining surfaces of an enclosure.[29]

3.5.4 *View Factors for Diffuse Surfaces*

View factor (also called shape factor or configuration factor) is defined as the fraction of the radiation leaving surface i that is intercepted by surface j. In Figure 3.15, two arbitrary surfaces are exchanging radiation. They have areas A_i and A_j, temperatures T_i and T_j, and are separated by a distance R. They are at angles θ_i and θ_j from normals n_i and n_j. It can be shown that the total rate at which radiation leaves surface i and is intercepted by j is:

$$q_{i \to j} = J_i \iint \frac{\cos\theta_i \cos\theta_j}{\pi R^2} dA_i dA_j \qquad (3.104)$$

Thus, the view factor, which is the fraction of radiation that leaves A_i and is intercepted by A_j, is given as:

$$F_{ij} = \frac{q_{i \to j}}{A_i J_i} \qquad (3.105)$$

or, assuming that the two surfaces are diffuse emitters and reflectors and have uniform radiosity, the shape factor is given as:

$$F_{ij} = \frac{1}{A_i} \iint \frac{\cos\theta_i \cos\theta_j}{\pi R^2} dA_i dA_j \qquad (3.106)$$

Similarly, view factor F_{ji} can be calculated for the radiation leaving surface j and intercepted by surface i.

View factors follow reciprocity and summation rules given as follows:

$$A_i F_{ij} = A_j F_{ji} \qquad \text{(reciprocity rule)}$$

$$\sum F_{ij} = 1 \qquad \text{(summation rule)}$$

Tables 3.10 and 3.11 show view factors for two- and three-dimensional geometries, respectively, and Figures 3.16, 3.17, and 3.18 show the view factors for three very common configurations.

3.5.5 *Infrared Temperature Measurement*

Planck's distribution relates the radiation emitted by a blackbody to its temperature. This relation is used in heat transfer analysis to determine how much energy is emitted by a surface. A further application of this relation is to mea-

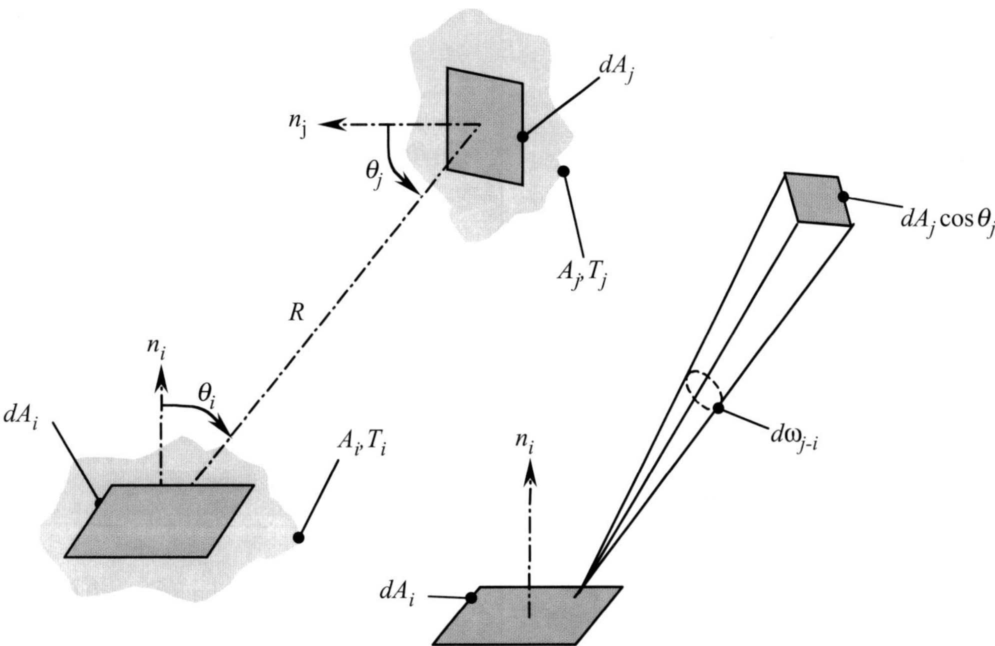

FIGURE 3.15 View factor of radiation exchange between faces of area dA_i and dA_j.[29]

sure the *intensity* of the emitted radiation and use this measurement to determine the surface temperature. In practice, a number of complicating factors make it impossible to use Planck's distribution to convert the measured intensity to the surface temperature. In reality, there are no perfectly black surfaces. Real surfaces are at best gray and not always diffuse. In addition, there will be radiation reflected from the surface that must be compensated for during measurement. While it is possible to make these corrections analytically, vendors of infrared temperature measuring devices invariably make extensive use of calibration. Calibration allows the measurement device to be corrected for spectral selectivity of the detector and for nonlinearities in the detector's response.

Figure 3.19 shows infrared temperature measurements made on a burner. By selecting an appropriate wavelength for the intensity measurement (in this case, the wavelength is 3.9 μm), the effects of CO_2 and H_2O between the emitting surface and the infrared camera are negated. Surface temperatures can then be readily measured "through" a flame. It is very difficult to make reliable gas temperature measurements by measuring the infrared emission because the emission from gases depends on the temperature of the gas volume, the composition of the gas volume, as well as the dimension of the gas volume. Because most real applications involve nonisothermal gas volumes (such as a flame in a furnace), IR measurements are not feasible.

3.5.6 Radiation in Absorbing/Emitting/Scattering Media

The foregoing discussion on radiation heat transfer was limited to *surface exchange*. Surface exchange is radiation heat transfer from one surface to another, assuming that the medium between the two surfaces is a vacuum or a transparent substance. The notion of surface exchange is actually an idealization. When radiant energy is incident on a surface, it actually penetrates that surface some distance when considered at the molecular level. For most metals, this distance is only several Angstroms (Å, 10^{-10} m), while for most nonmetals, it is several microns (μm, 10^{-6} m).

Radiation absorption and emission in gases is due to the quantum energy levels of the gas molecules. An in-depth analysis and discussion of the topic is beyond the scope of this section, but some understanding of gas spectra is necessary to understand gas radiation. Because air is primarily composed of symmetric diatomic molecules (which typically do not emit or absorb in the infrared) and inerts (N_2, O_2, and Ar), air is usually considered a transparent medium. Humid air, however, does absorb some radiation. Normally, this absorption is neglected as it is usually not significant. Other molecules that are commonly found in combustion applications, such as CH_4 and other hydrocarbons, CO_2, H_2O, CO, etc., do emit and absorb in the infrared. Unlike many solid surfaces, however, their emission and absorption do not smoothly vary with wavelength. Rather, their emission and absorption spectra oscillate

TABLE 3.10 View Factors for Two-dimensional Geometries

	Geometry	Relation
Parallel plates with midlines connected by a perpendicular	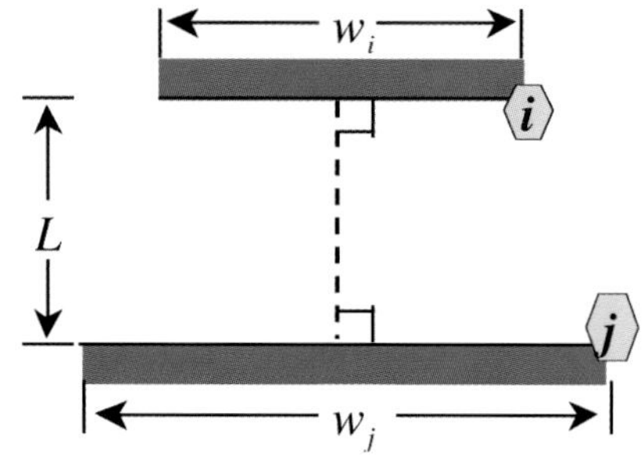	$F_{ij} = \dfrac{\left[(W_i + W_j)^2 + 4\right]^{1/2} - \left[(W_j + W_i)^2 + 4\right]^{1/2}}{2W_i}$ $W_i = w_i/L, \quad W_j = w_j/L$
Inclined parallel plates of equal width and a common edge	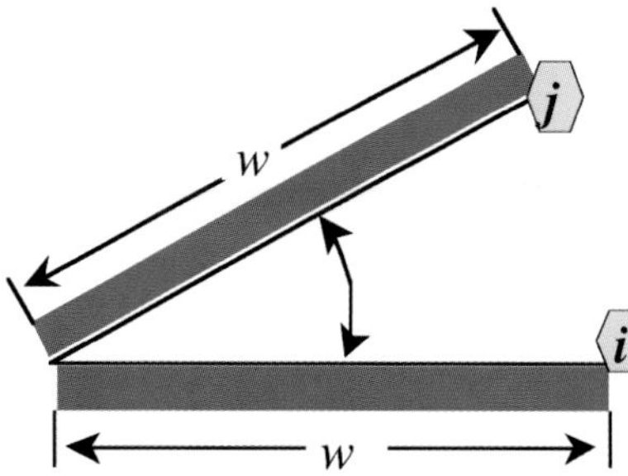	$F_{ij} = 1 - \sin\left(\dfrac{\alpha}{2}\right)$
Perpendicular plates with a common edge	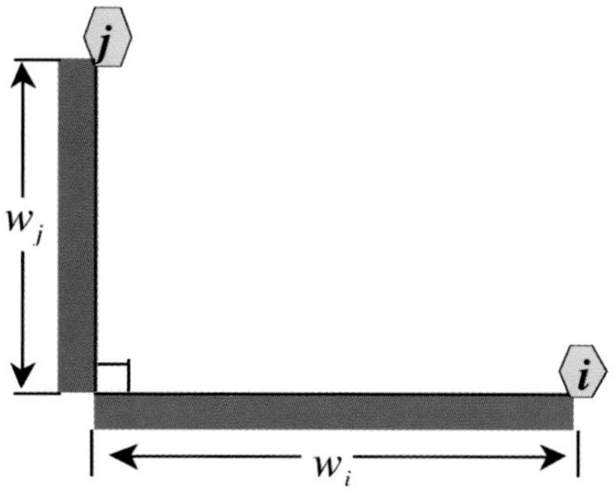	$F_{ij} = \dfrac{1 + (w_i/w_j) - \left[1 + (w_j/w_i)^2\right]^{1/2}}{2}$
Three-sided enclosure	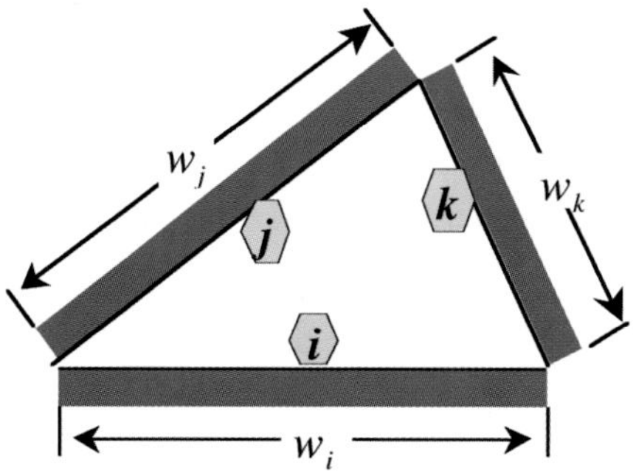	$F_{ij} = \dfrac{w_i + w_j - w_k}{2w_i}$

violently with wavelength, but only in narrow "bands" centered around wavelengths particular to the species under consideration.

If the medium between surfaces is not transparent to thermal radiation, it is called a participating medium. The notion of emissive power, so useful in analyzing surface exchange, is meaningless in a participating media. Instead, one must consider the *intensity* of radiation. From the study of surface exchange, one knows that the intensity of radiation emitted by a diffuse surface is independent of angle, while emissive power varies as the cosine of the normal angle. Section 3.5.8 briefly shows how radiant intensity is absorbed, emitted, and

TABLE 3.10 (continued) View Factors for Two-dimensional Geometries

	Geometry	Relation
Parallel cylinders of different radii	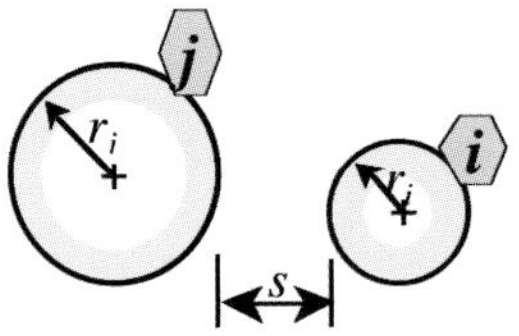	$F_{ij} = \frac{1}{2\pi}\left\{\pi + \left[C^2 - (R+1)^2\right]^{1/2} - \left[C^2 - (R-1)^2\right]^{1/2} \ldots + (R-1)\cos^{-1}\left[\frac{R}{C} - \frac{1}{C}\right] - (R+1)\cos^{-1}\left[\frac{R}{C} + \frac{1}{C}\right]\right\}$ $R = r_j/r_i, \quad S = s/r_i$ $R = r_j/r_i, \quad S = s/r_i$
Cylinder and parallel rectangle	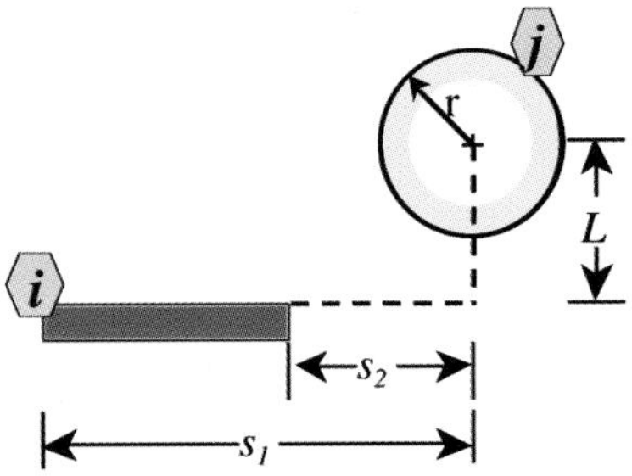	$F_{ij} = \frac{r}{s_1 - s_2}\left[\tan^{-1}\frac{s_1}{L} - \tan^{-1}\frac{s_2}{L}\right]$
Infinite plane and row of cylinders	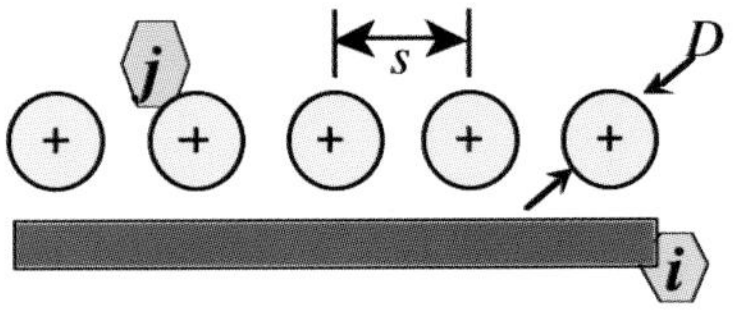	$F_{ij} = 1 - \left[1 - \left(\frac{D}{S}\right)^2\right]^{1/2} + \left(\frac{D}{S}\right)\tan^{-1}\left(\frac{s^2 - D^2}{D^2}\right)^{1/2}$

Source: Adapted from F.P. Incropera and D.P. DeWitt, *Fundamentals of Heat and Mass Transfer*, 4th edition, John Wiley & Sons, New York, 1996.

scattered by participating media. Radiant intensity within a participating medium is a function of location (typically three independent variables in a three-dimensional problem), direction (two independent angles are required to describe direction in a three-dimensional problem), and wavelength or wavenumber (one independent variable) if the problem is steady state. The fact that radiant intensity is a function of six independent variables immediately indicates that the analysis will be significantly more complicated than, for example, conduction heat transfer, where there are only three independent variables in a three-dimensional, steady-state problem. Further, if scattering is to be considered, the equation of radiative transfer will have an integro-differential form.

3.5.7 Mean-Beam-Length Method

Gases emit and absorb radiation in discrete energy bands dictated by the allowed energy states within the molecule. While the energy emitted by a solid shows a continuous spectrum, the radiation emitted and absorbed by a gas is restricted to bands. Figure 3.20 shows the emission bands of carbon dioxide and water vapor relative to blackbody radiation at 1090 K (1500°F). The emission of radiation for these gases occurs in the infrared region of the spectrum. The inert gases and diatomic gases of symmetrical composition such as O_2, N_2, and H_2 are transparent to thermal radiation. Important gases that absorb and emit radiation are polyatomic gases such as CO_2 and H_2O and asymmetric molecules such as CO. Determination of radiant flux from gases is highly complex, but it can be simplified by using Hottel's assumption[62] that involves determination of emission from a hemispherical mass of gas at temperature T_g to a surface element located at the center of the hemisphere's base as:

$$E_g = \varepsilon_g \sigma T_g^4 \tag{3.107}$$

where E_g is emissive power and ε_g is the gas emissivity. Figures 3.21 and 3.22 show the emissivity of water vapor and carbon dioxide, respectively, as a function of gas temperature. The figures are based on experimental data taken in hemispherical shape of the gas at 1 atm total pressure, in a mixture with nonradiating gases. For pressures other than 1 atm, corrections must be incorporated. When carbon dioxide and water vapor both appear together with nonradiating gases, the total emissivity of the gas is obtained by:

$$\varepsilon_g = \varepsilon_w + \varepsilon_c - \Delta\varepsilon \tag{3.108}$$

TABLE 3.11 View Factors for Three-dimensional Geometries

Geometry	Relation
Aligned parallel rectangles 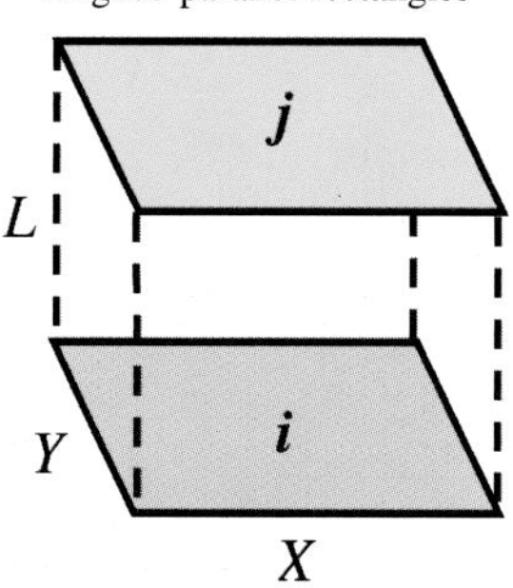	$\bar{X} = X/L \quad \bar{Y} = Y/L$ $F_{ij} = \frac{2}{\pi \bar{X}\bar{Y}} \left\{ \ln\left[\frac{(1+\bar{X}^2)(1+\bar{Y}^2)}{1+\bar{X}^2+\bar{Y}^2}\right]^{1/2} + \bar{X}(1+\bar{Y}^2)^{1/2} \tan^{-1}\frac{\bar{X}}{(1+\bar{Y}^2)^{1/2}} \ldots + \bar{Y}(1+\bar{X}^2)^{1/2} \tan^{-1}\frac{\bar{Y}}{(1+\bar{X})^{1/2}} - \bar{X}\tan^{-1}\bar{X} - \bar{Y}\tan^{-1}\bar{Y} \right\}$
Coaxial parallel disks 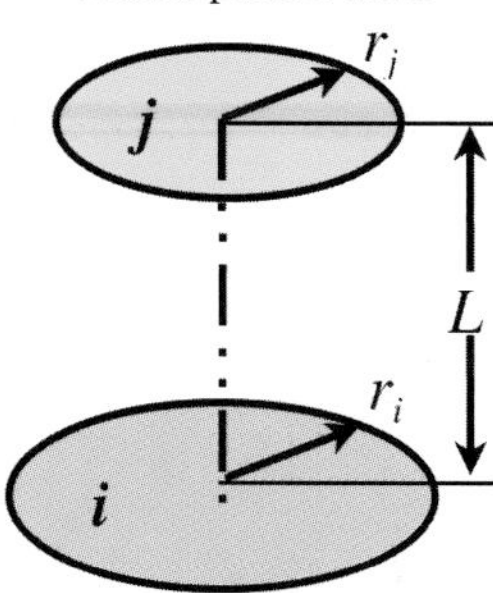	$R_i = r_i/L, \quad R_j = r_j/L$ $S = 1 + \frac{1+R_j^2}{R_i^2}$ $F_{ij} = \frac{1}{2}\left\{S - \left[S^2 - 4(r_j/r_i)^2\right]^{1/2}\right\}$
Perpendicular rectangles with a common edge 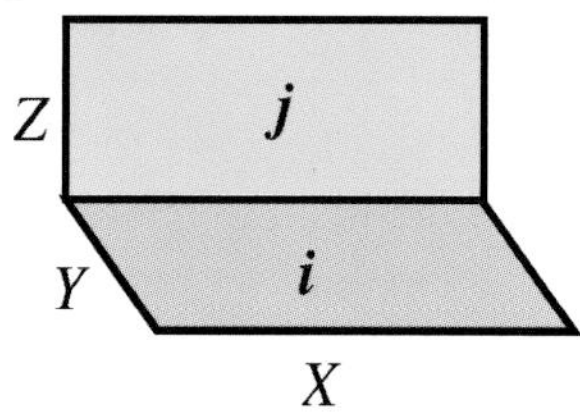	$H = Z/X, \quad W = Y/X$ $F_{ij} = \frac{1}{\pi W}\left(W\tan^{-1}\frac{1}{W} + H\tan^{-1}\frac{1}{H} - (H^2+W^2)^{1/2} \ldots \times \tan^{-1}\frac{1}{(H^2+W^2)^{1/2}} + \frac{1}{4}\ln\left\{\frac{(1+W^2)(1+H^2)}{(1+W^2+H^2)} \ldots \times \left[\frac{W^2(1+W^2+H^2)}{(1+W^2)(W^2+H^2)}\right]^{W^2} \times \left[\frac{H^2(1+H^2+W^2)}{(1+H^2)(H^2+W^2)}\right]^{H^2}\right\}\right)$

Source: Adapted from F.P. Incropera and D.P. DeWitt, *Fundamentals of Heat and Mass Transfer*, 4th edition, John Wiley & Sons, New York, 1996.

where the correction factor $\Delta\varepsilon$ can be obtained from Figure 3.23.

Normal emissivity of various surfaces is tabulated in Table 3.12.

The mean beam length, L_e, can be defined as the radius of a hemispherical gas mass whose emissivity is equivalent to that for the geometry of interest. Table 3.13 gives the mean beam length of numerous gas geometries and shapes from Hottel.[62] For geometries not covered in Table 3.13, the mean beam length can be approximated as

$$L_e = 3.4(\text{Volume})/(\text{Surface area}) \quad (3.109)$$

Using mean beam length L_e instead of L (the radius of hemisphere), gas emissivity is obtained, which in turn gives radiant heat transfer to a surface due to emission from an adjoining gas:

$$q = \varepsilon_g A_s \sigma T_g^4 \quad (3.110)$$

where A_s is the surface area. The net radiation exchange rate between the surface at temperature T_s and the gas at T_g is then given by

$$q_{net} = A_s\sigma\left(\varepsilon_g T_g^4 - \alpha_g T_s^4\right) \quad (3.111)$$

The relationship for determining absorptivity of carbon dioxide and water vapor from their respective emissivity data was given by Hottel[62] as follows:

$$\alpha_w = C_w\left(T_g/T_s\right)^{0.45}\varepsilon_w\left(T_s, p_w L_e T_s/T_g\right) \qquad (3.112)$$

$$\alpha_c = C_c\left(T_g/T_s\right)^{0.65}\varepsilon_c\left(T_s, p_c L_e T_s/T_g\right) \qquad (3.113)$$

When both carbon dioxide and water vapors are present, the total gas absorptivity is obtained by:

$$\alpha_g = \alpha_w + \alpha_c - \Delta\alpha \qquad (3.114)$$

where $\Delta\alpha = \Delta\varepsilon$.

3.5.8 Equation of Radiative Transfer

Consider the propagation of a "pencil" beam of radiant energy through a participating medium. The radiant energy is absorbed by the medium, decreasing the intensity of the radiant energy according to:

$$\left(\frac{\partial I_\lambda}{\partial s}\right)_{\text{absorption}} = -a_\lambda I_\lambda, \qquad (3.115)$$

where a_λ is the spectral absorption coefficient, and s is a coordinate along the path. Additionally, the intensity of the radiation is increased by emission from the medium. The increase in the radiant intensity is given by:

$$\left(\frac{\partial I_\lambda}{\partial s}\right)_{\text{emission}} = a_\lambda I_{b\lambda} \qquad (3.116)$$

A further effect to be considered when particulate media is present is scattering. When radiant energy strikes a solid particle within the medium, the radiation may be reflected or diffracted so that its direction changes. As radiation propagates, its intensity is decreased by *out-scattering* and increased by *in-scattering*. Attenuation by (out) scattering is described by:

$$\left(\frac{\partial I_\lambda}{\partial s}\right)_{\text{outscatter}} = -a_\lambda I_\lambda \qquad (3.117)$$

The increase in intensity due to in-scatter is given by:

$$\left(\frac{\partial I_\lambda}{\partial s}\right)_{\text{inscatter}} = \frac{\sigma_{s\lambda}}{4\pi}\int_{4\pi} I_\lambda(\mathbf{s}_i)\Phi(\mathbf{s}_i,\mathbf{s})\,d\Omega_i \qquad (3.118)$$

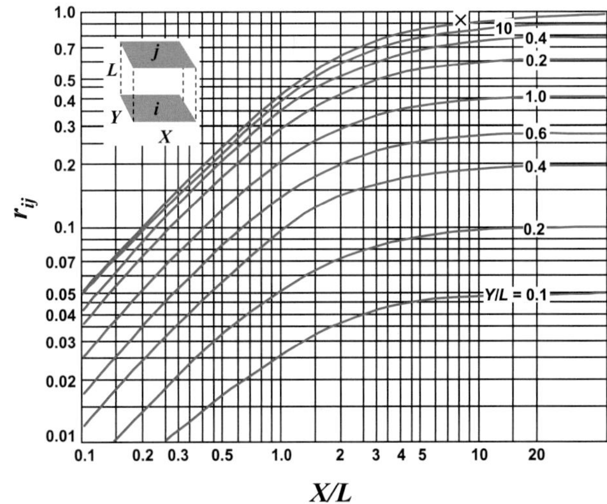

FIGURE 3.16 View factor for aligned parallel rectangles.[29]

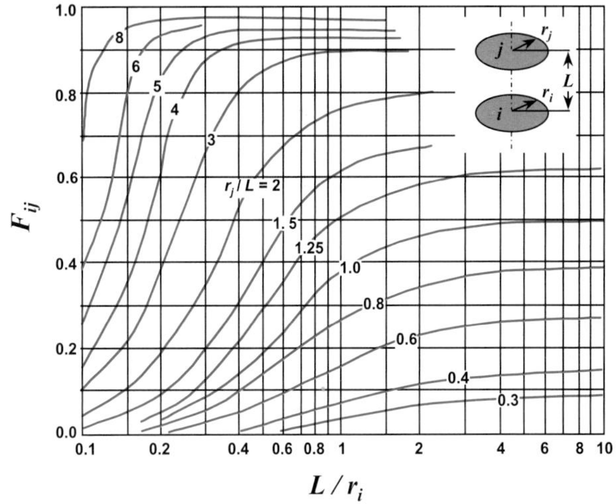

FIGURE 3.17 View factor for coaxial parallel disks.[29]

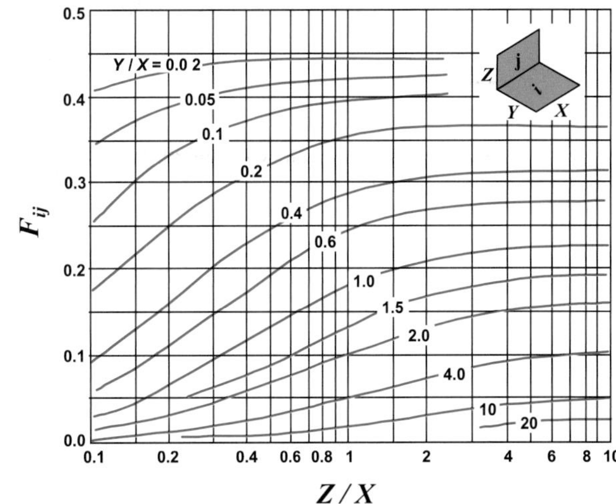

FIGURE 3.18 View factor for perpendicular rectangles with a common edge.[29]

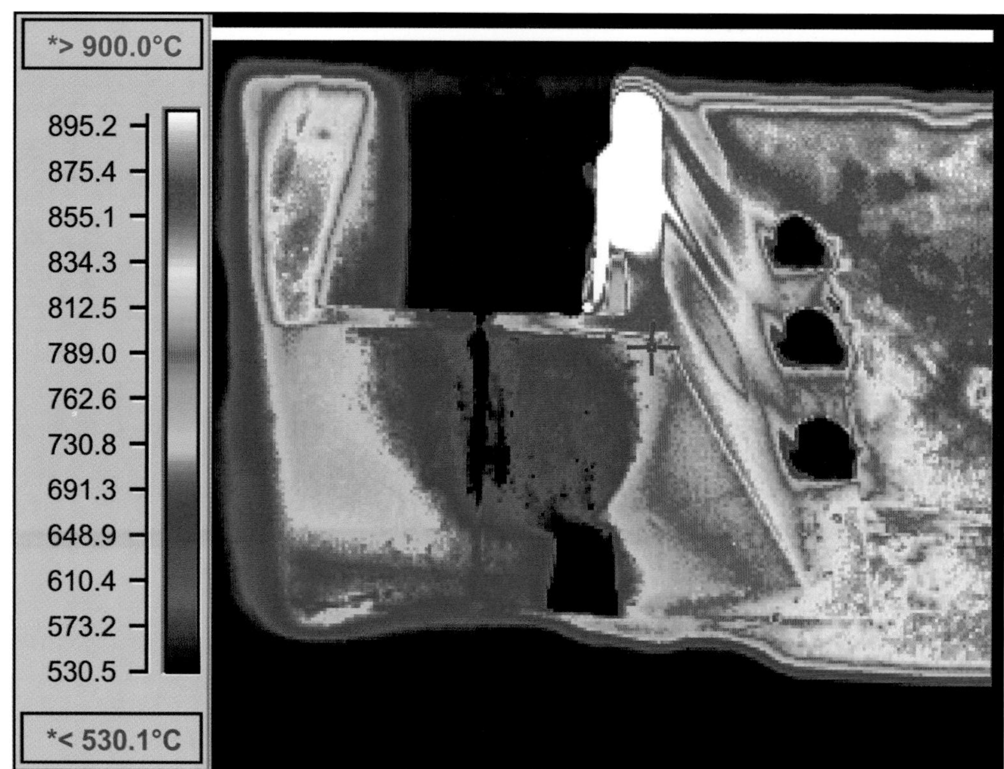

FIGURE 3.19 Infrared thermal image of a flame in a furnace. (Courtesy of John Zink Co.)

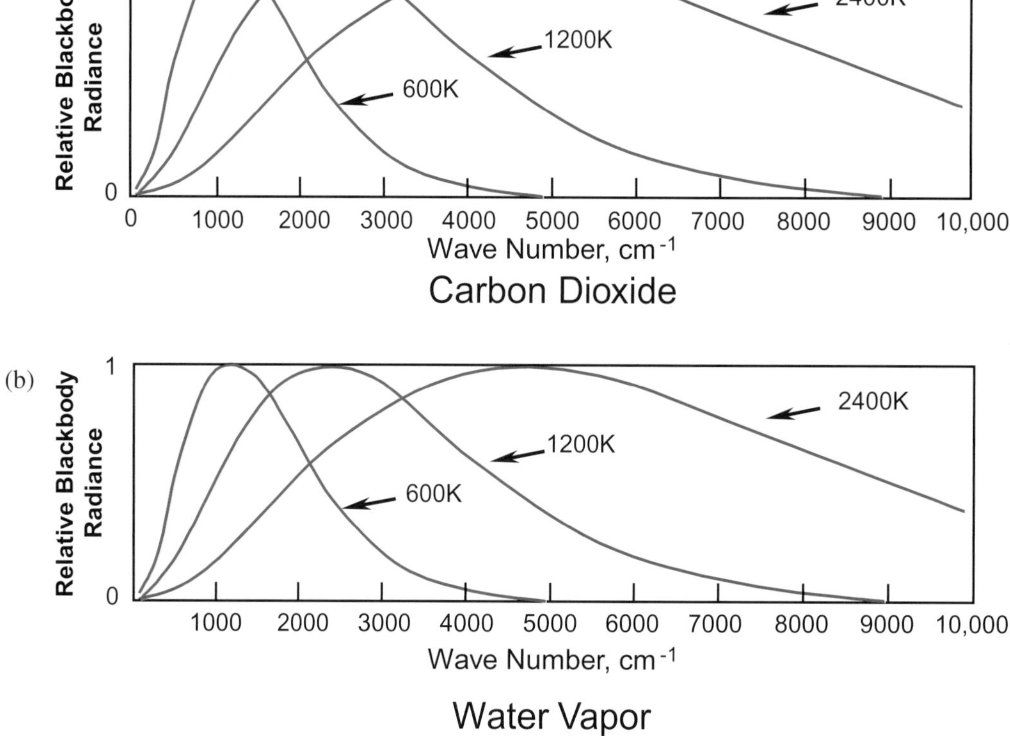

FIGURE 3.20 Emission bands of (a) CO_2 and (b) H_2O.[16]

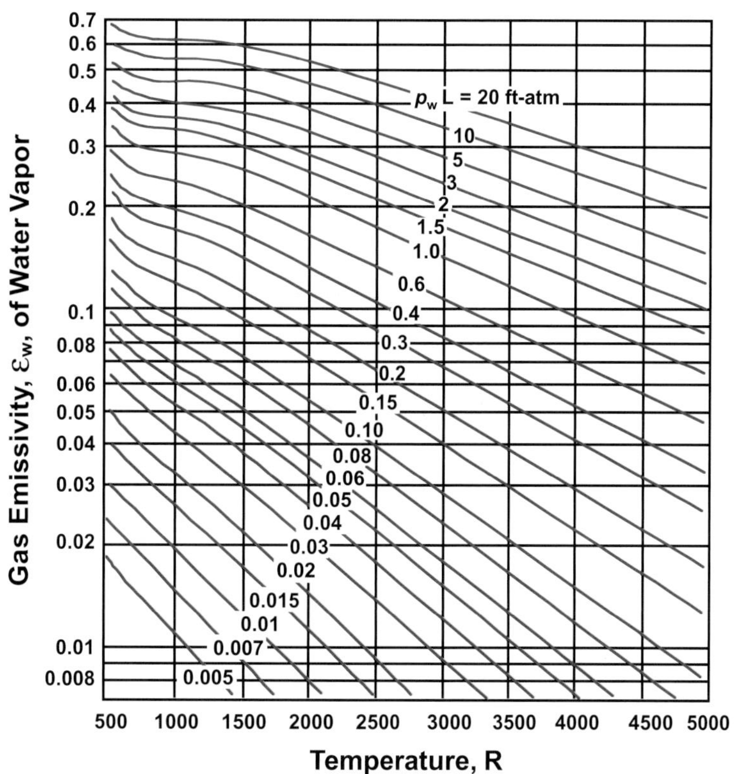

FIGURE 3.21 Emissivity of water vapor in a mixture with nonradiating gases at 1-atm total pressure and of hemispherical shape.[62]

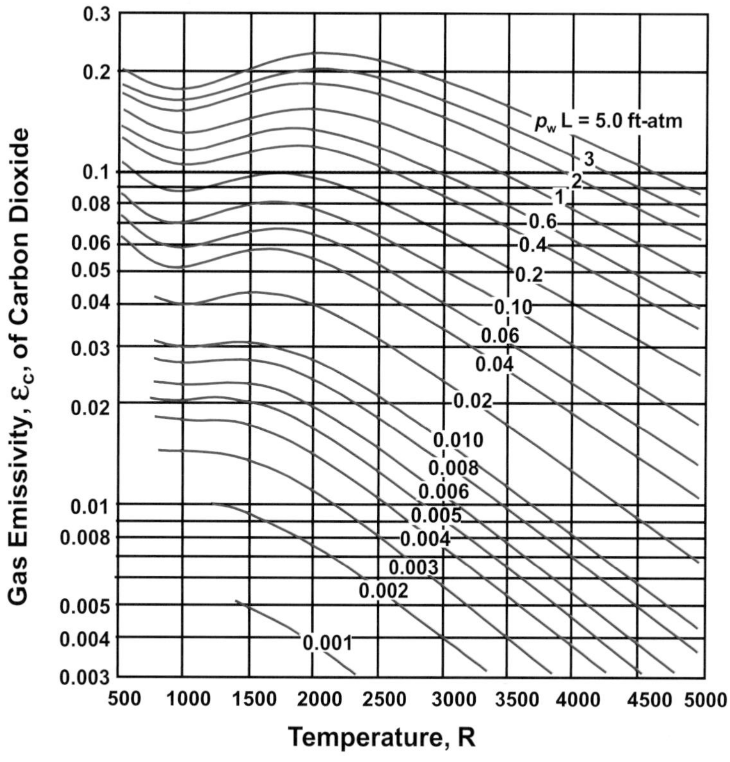

FIGURE 3.22 Emissivity of carbon dioxide in a mixture with nonradiating gases at 1-atm total pressure and of hemispherical shape.[62]

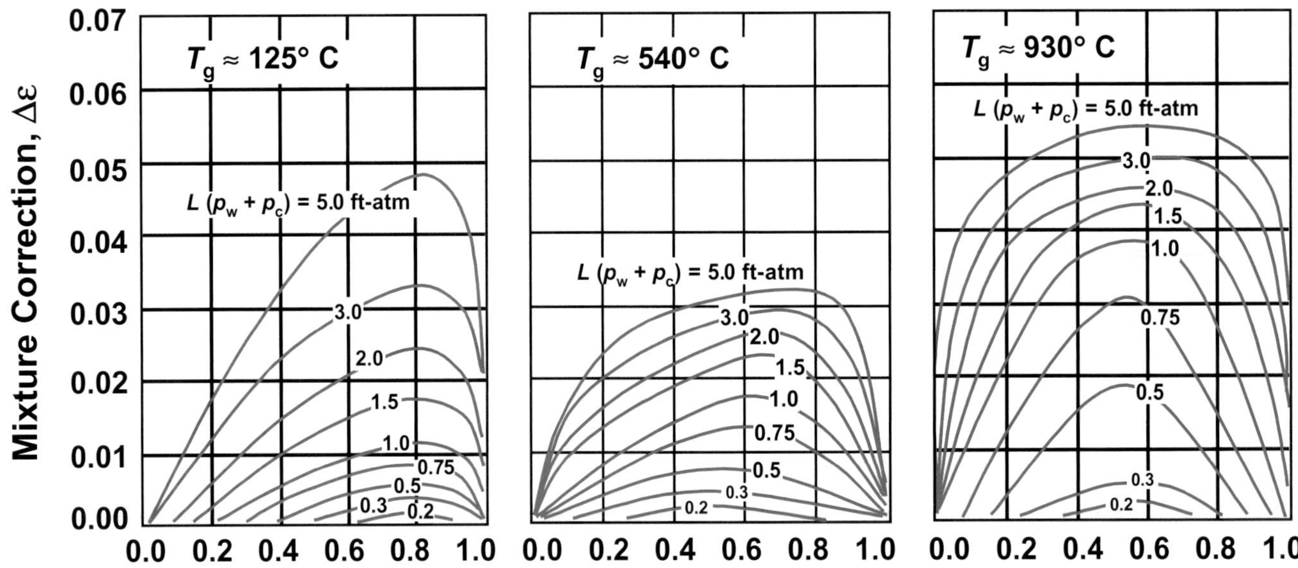

FIGURE 3.23 Radiation heat transfer correction factor for mixtures of water vapor and carbon dioxide.[63]

TABLE 3.12 Normal Emissivities, ε, for Various Surfaces

Material	Emissivity, ε	Temp. (°F)	Description
Aluminum	0.09	212	Commercial sheet
Aluminum oxide	0.63–0.42	530–930	
Aluminum paint	0.27–0.67	212	Varying age and Al content
Brass	0.22	120–660	Dull plate
Copper	0.16–0.13	1970–2330	Molten
Copper	0.023	242	Polished
Cuprous oxide	0.66–0.54	1470–2012	
Iron	0.21	392	Polished, cast
Iron	0.55–0.60	1650–1900	Smooth sheet
Iron	0.24	68	Fresh emeried
Iron oxide	0.85–0.89	930–2190	
Steel	0.79	390–1110	Oxidized at 1100F
Steel	0.66	70	Rolled Sheet
Steel	0.28	2910–3270	Molten
Steel (Cr-Ni)	0.44–0.36	420–914	18–8 rough, after heating
Steel (Cr-Ni)	0.90–0.97	420–980	25–20 oxidized in service
Brick, red	0.93	70	Rough
Brick, fireclay	0.75	1832	
Carbon, lampblack	0.945	100–700	0.003 in. or thicker
Water	0.95–0.963	32–212	

Note: 1. SI conversion: T, °C = 5/9 (°F – 32).

TABLE 3.13 Mean Beam Lengths L_e for Various Gas Geometries

Geometry	Characteristic Length	L_e
Sphere (radiation to surface)	Diameter (*D*)	0.65 *D*
Infinite circular cylinder (radiation to curved surface)	Diameter (*D*)	0.95*D*
Semi-infinite circular cylinder (radiation to base)	Diameter (*D*)	0.65*D*
Circular cylinder of equal height and diameter (radiation to entire surface)	Diameter (*D*)	0.60*D*
Infinite parallel planes (radiation to planes)	Spacing between planes (*L*)	1.80*L*
Cube (radiation to any surface)	Side (*L*)	0.66*L*
Arbitrary shape of volume V (radiation to surface of area A)	Volume to area ratio (*V*/*A*)	3.6(*V*/*A*)

Source: From C.J. Hoogendoorn, C.M. Ballintijn, and W.R. Dorrestijn, *J. Inst. Fuel*, 43, 511–516, 1970. With permission.

where the subscript *i* in the integrand denotes the incident direction. Physically, this integral can be described as a summation over all possible directions of the radiation entering a particular location, multiplied by the scattering phase function, which represents the fraction of radiation traveling in a particular direction $\mathbf{s}_i$ that is scattered into a new direction **s**.

Summing all of these effects results in the Equation of Transfer for radiation in a participating media:

$$\frac{\partial I_\lambda}{\partial s} = -\left(a_\lambda + \sigma_{s\lambda}\right)I_\lambda + a_\lambda I_{b\lambda} + \frac{\sigma_{s\lambda}}{4\pi}\int_{4\pi} I_\lambda(\mathbf{s}_i)\Phi(\mathbf{s}_i,\mathbf{s})d\Omega_i \tag{3.119}$$

This equation describes the propagation of radiation through absorbing/emitting/scattering media. It is an integro-differ-

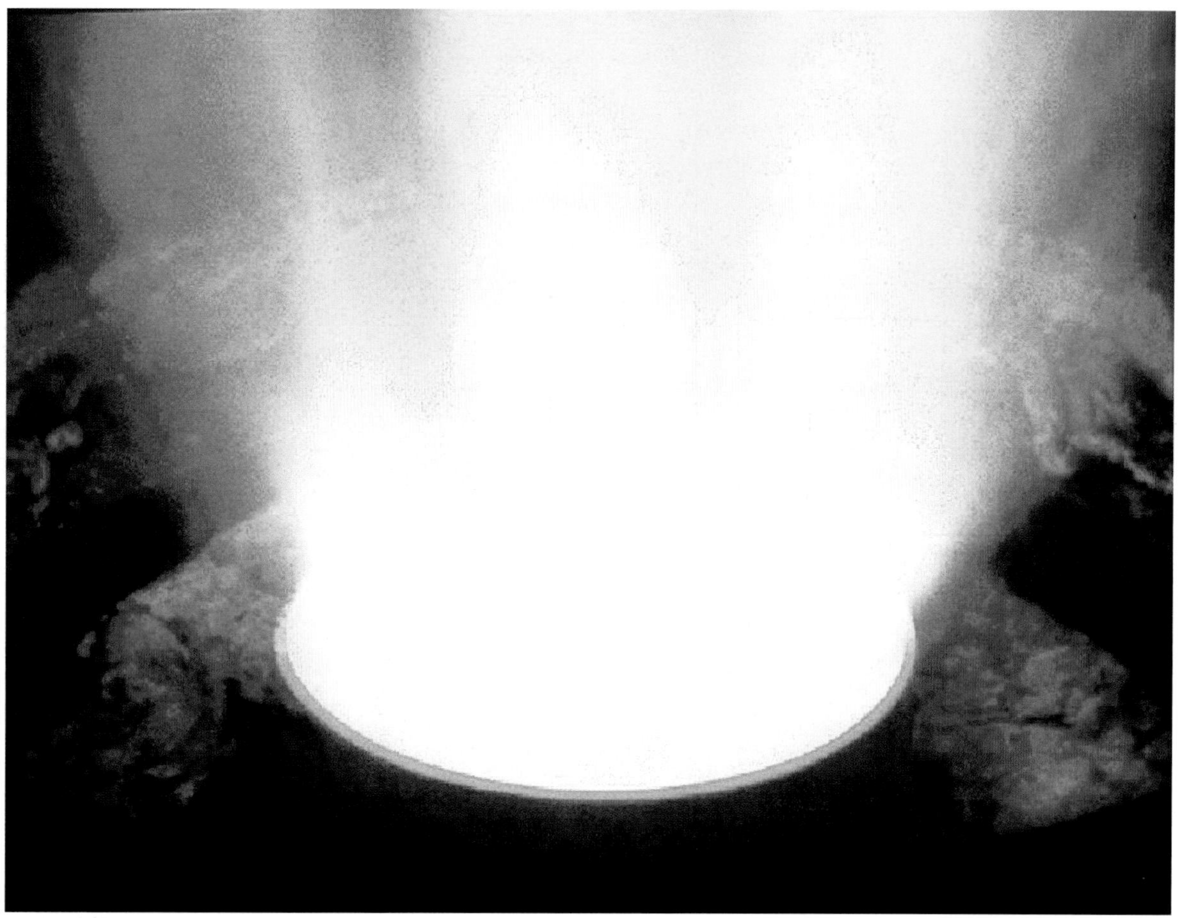

FIGURE 3.24 Photographic view of a luminous flame. (Courtesy of John Zink Co.)

ential equation when scattering is considered. Analytical solutions of the equation of transfer are possible only for very simple geometries and boundary conditions. For more complex geometries and boundary conditions, approximate solution techniques such as the spherical harmonics method and the discrete ordinates method can be used. These methods are discussed further in Chapter 9. These approximate solution techniques are more fully discussed in texts such as those by Modest[63] and Siegel and Howell.[59]

3.5.9 Radiation Emitted by a Flame

Accurate estimation of the heat emitted from a flame is very difficult, for reasons that include:

1. The flame temperature is not known. While one can readily calculate an adiabatic flame temperature for a given fuel, the actual flame temperature will be below this value because the flame emits radiant heat.
2. The often-used term "flame radiation" suggests that some special mechanism is at work within the reaction zone emitting radiant energy. This is not true. Radiant energy is emitted only by the gases and solids (particularly carbon) present in the flame. The gaseous combustion products H_2O and CO_2 are the gases that emit radiation in significant quantities, while any radical species (such as CO) in the flame are present at such small fractions and such thin pathlengths that their emission is typically negligible. However, determining the concentration and temperature of H_2O and CO_2 within the flame is nontrivial and, in fact, the gases are clearly nonisothermal.
3. The presence of solid carbon particles within the flame (which give flames a yellowish color) can dominate the radiant emission from the flame. Again, predicting the concentration (measured as a volume fraction) and temperature of these carbon particles is very difficult.

Figure 3.24 shows a yellow luminous flame. The yellowish color of the flame is due to broadband radiation by carbon particles. The flame shown was produced by combusting a fuel oil atomized by steam. Flames with significant soot

FIGURE 3.25 Photographic view of a nonluminous flame from a John Zink gas burner. (Courtesy of John Zink Co.)

fractions significantly radiate directly from the flame. In contrast, nonluminous or slightly luminous flames (as shown in Figure 3.25) emit only a small fraction of the energy liberated by the combustion process.

Figure 3.26 shows a radiant wall burner. In this burner, a mixture of fuel and air jets out radially from the burner. Very near the burner, the wall is dark because the flame "stands off" of the burner exit. Further away from the burner, the refractory surface is a bright yellow color. In this particular application, the temperature of this refractory is above 2000°F (1100°C). As illustrated in the photograph, the visible radiation from the hot refractory surface dominates any flame radiation in the visible region, rendering the flame invisible to the human eye. This burner is very common in ethylene pyrolysis furnaces, where the burner is used to heat a refractory wall. The primary heat transfer mode to the refractory wall is probably convective, although gas radiation plays a role. The hot wall then radiates energy to process tubes that run parallel to the wall at a distance of approximately 3 feet (1 m).

3.6 HEAT TRANSFER IN PROCESS FURNACES

A complete treatment of heat transfer in process furnaces is beyond the scope of this chapter, but the chapter would not be complete without describing, albeit briefly, the phenomena at play in process furnace heat transfer.

Process furnaces are a good example of systems that incorporate all the heat transfer mechanisms concurrently at work in gases, liquids, and solids. The challenge is to achieve good heat transfer in the radiation and convection mechanisms. Conduction plays only a minor role in getting the heat from the flame to the process fluid, but it is the primary mechanism at work in preventing heat loss to the surroundings.

For the following discussion, refer to Figure 3.1 at the beginning of this chapter. The main part of the furnace that contains the burner flames is the radiant section. The process heating tubes are located in the radiant section in various arrangements. In low temperature furnaces (gases exit the radiant section at less than 800°C (1500°F)), the process tubes

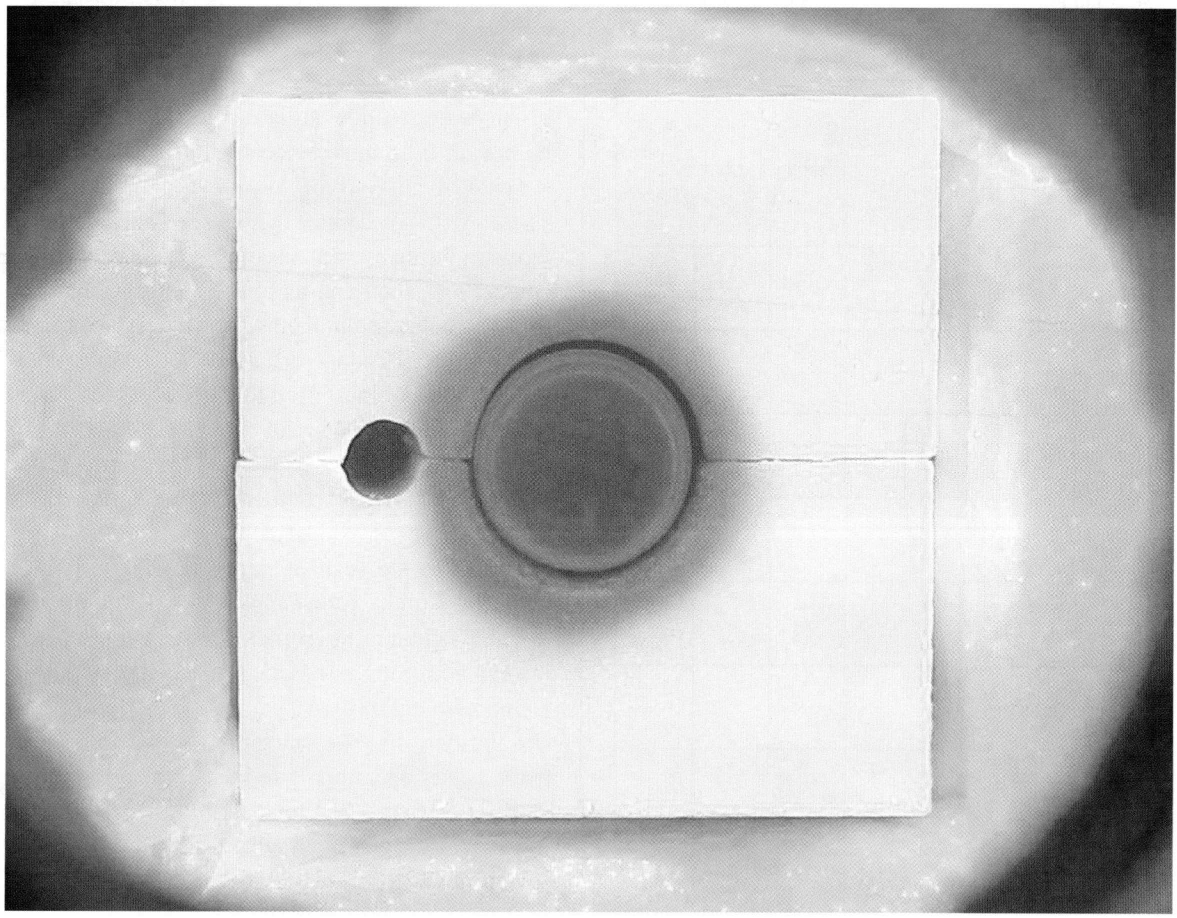

FIGURE 3.26 Photographic view of a radiant wall burner. (Courtesy of John Zink Co.)

are often located close to the walls of the furnaces, and in high temperature furnaces (gases exit the radiant section at more than 980°C (1800°F)), the tubes are usually suspended in the main furnace space, away from the walls.

In many furnace designs, a convection section is located downstream of the radiant section. The combustion gases that leave the radiant section flow through the convection section and then through the stack to be vented to atmosphere. Depending on the design, an air preheater (not shown in Figure 3.1) may be installed in the path to the stack to further extract heat from the flue gas. The convection section may be used to preheat the process fluid, generate steam, or heat another process fluid. Since the radiant section is the high temperature section of the furnace, the final passes of the heating process are located there.

Radiation is the dominant heat transfer mechanism in the radiant section. Both participating media (gaseous) radiation and surface exchange are significant. It should be noted that even though gas flames emit only a little in the visible spectrum, their emission into the infrared spectrum is quite large. In typical cracking furnaces and some boilers where temperatures are high (above about 1800°F or 1000°C), the furnace walls will glow bright orange or even yellow. In these furnaces, it is frequently difficult to see a gas flame visually. This is because the walls are radiating in the visual spectrum. However, even though the naked eye cannot see any gas radiation from the flame, radiation from the flame is still a significant contributor to heat transfer. Figure 3.12 (the blackbody emissive power graph) showed that blackbody curves do not cross. This means that at any given wavelength, the hotter the radiator is, the higher the blackbody emissive power. The governing factor in heat transfer in the radiation section is temperature because the radiant heat transfer coefficient is directly proportional to the fourth power of the temperature.

Convection also contributes to heat transfer in the radiant section. The furnace gases circulate vigorously inside the radiant section driven by the in-flow of combustion air, gas

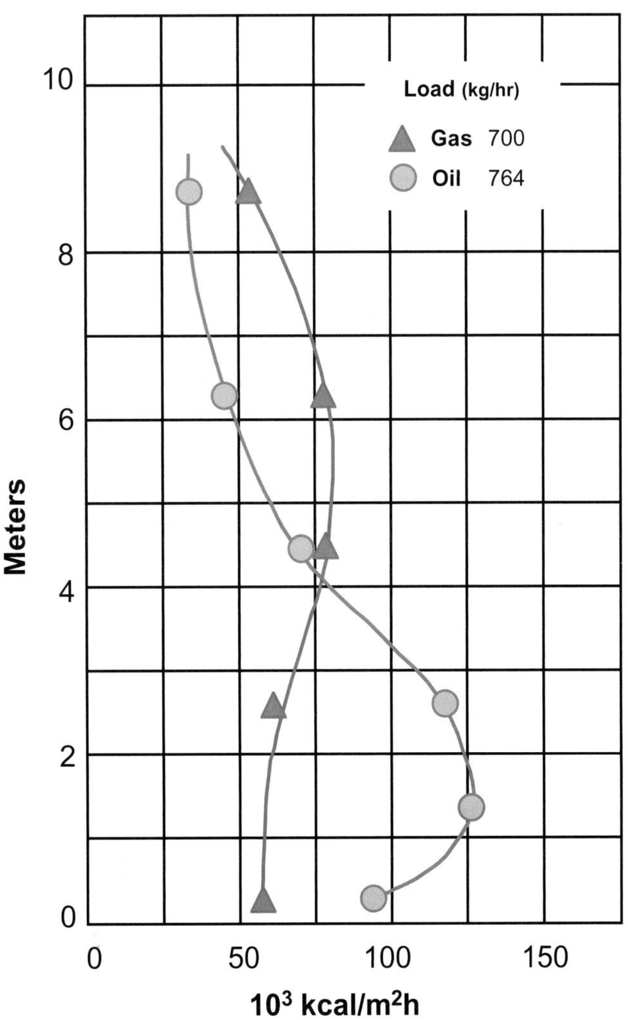

FIGURE 3.27 Vertical heat flux distribution for oil and gas firing in a vertical tube furnace.[66]

expansion due to combustion, and the temperature gradients through the furnace. These flow patterns are very difficult to predict *a priori*, but they are very important in assessing heat transfer to the process tubes. In the convection section of the furnace, most of the heat is transferred by convection. The first row or two of tubes in the convection section that have a "line of sight" view of the radiant section experience a lot of radiant transfer. Consequently, the first row of tubes is usually not equipped with fins, to avoid excessive localized heat flux. The subsequent rows of tubes in the convection section are often finned to maximize the convective heat transfer per unit tube length. The governing factor in heat transfer in the convection section is mass velocity, since the convective heat transfer coefficient is strongly dependent on the velocity.

In the average heater, about 80% of the total heat is transferred in the radiant section and approximately 20% comes from the convection section. With this ratio of heat transfer in the heater sections, it is obvious that the greatest benefit will result from improvements in the radiant section.

Detailed analysis of furnace heat transfer is complex. There exist well-defined methods to calculate the heat transferred in the many varieties of heat exchangers (e.g., parallel, counterflow, shell-and-tube, and compact), but furnace heat transfer calculation methods are less well-documented. Many furnace vendors do have proprietary methods for computing the heat transferred to the process load, but these are largely based on empirical data and are not documented in the open literature. Furthermore, furnace heat transfer seems to be the source of controversy and disagreement among engineering professionals in the field. Much disagreement exists as to the relative importance of gas radiation and surface radiation, for example.

The function of the burner is to deliver heat to the process load as uniformly as possible. Reaching this ideal condition is impossible. The burner equipment used must possess superior ability to disperse heat to the gaseous furnace atmosphere if the heater operation is to be satisfactory. It is the relative ability to disperse heat by a particular burner that may make it suited to firing in a particular heater. If the burner is applied to a different furnace design, it may not be as effective.[67]

Hoogendoorn et al. made heat flux measurements in a rectangular, vertical tube furnace with two round burners firing vertically upward.[65] Both oil and gas flames were tested. The objective of the study was to determine the validity of the assumption of a constant furnace temperature often used to calculate the heat flux to process tubes. A 1 in. (25 mm) diameter, water-cooled heat flux probe, with and without air screens, was used to measure radiation and total heat flux, respectively. Forced convection was calculated from the difference between the total and radiant heat flux measurements. The heat flux was found to be significantly nonuniform in the furnace. Gas flames were found to have a more uniform heat flux distribution than oil flames, as shown in Figure 3.27.

Selçuk et al. studied the effect of flame length on the radiative heat flux distribution in a process fluid heater.[66] The radiative heat flux distribution information is important in the design of the heater. It helps to prevent premature damage to the process tubes due to improper flame heights and helps optimize the heat transfer rate to the tubes, thereby maximizing thermal efficiency. A two-flux radiation model was used to predict the radiant heat transfer in the heater. The predictions were in good agreement with the experimental data. The results showed that the radiant flux at the tube surface was a strong function of the flame height, as shown in Figure 3.28.

3.6.1 Flame Radiation

The radiant sources in the furnace are the flame, the radiant furnace surfaces, and the radiant furnace gases (H_2O and CO_2). Gas flames produce some radiation directly from the flame. How much depends on how luminous the flame is. An oil flame can radiate three to four times as much as a gas flame due to the high quantities of soot formed in the flame that make it luminiscent. A gas flame also may produce soot under certain conditions of mixing and can radiate relatively more, but not as much as an oil flame. The distance from the flame to the heat transfer surface at various locations in the furnace varies widely. Since radiant transfer varies inversely as the square of the distance between the radiant and the absorptive bodies, flame radiation will not be uniformly delivered to all portions of the furnace. The heightened requirement of uniformity of heat flux in ethylene cracking furnaces has driven the design of furnaces with multiple small burners distributed uniformly over the furnace walls.

Flame radiation does not dominate the radiation process in the furnace. As an illustration, consider a combination burner that can provide both a gas and oil flame in the same location in a furnace. Even though the oil flame is three to four times as radiant as the gas flame, the furnace performance and tube surface temperatures are not significantly changed. Issues such as these introduce controversy in the various schools of thought regarding furnace heat transfer. Experimentally determining the true radiant behavior in the furnace is very difficult and would entail several compromises that would make the data questionable.

3.6.2 Furnace Gas Radiation

An analysis of the heat capacities of furnace gases indicates that gases that radiate in the infrared, carbon dioxide and water, only carry about 33% of the total heat released. Conversely, 63% of the heat is contained in other gases, namely, oxygen and nitrogen.

Consider the chemistry of burning methane at 10% excess air:

$$CH_4 + 2.20\ O_2 + 8.36\ N_2 = CO2 + 2H_2O + 0.20\ O_2 + 8.36\ N_2$$

Gaseous products are:

- 1 mol CO_2
- 2 mols H_2O
- 8.36 mols N_2
- 0.20 mols O_2

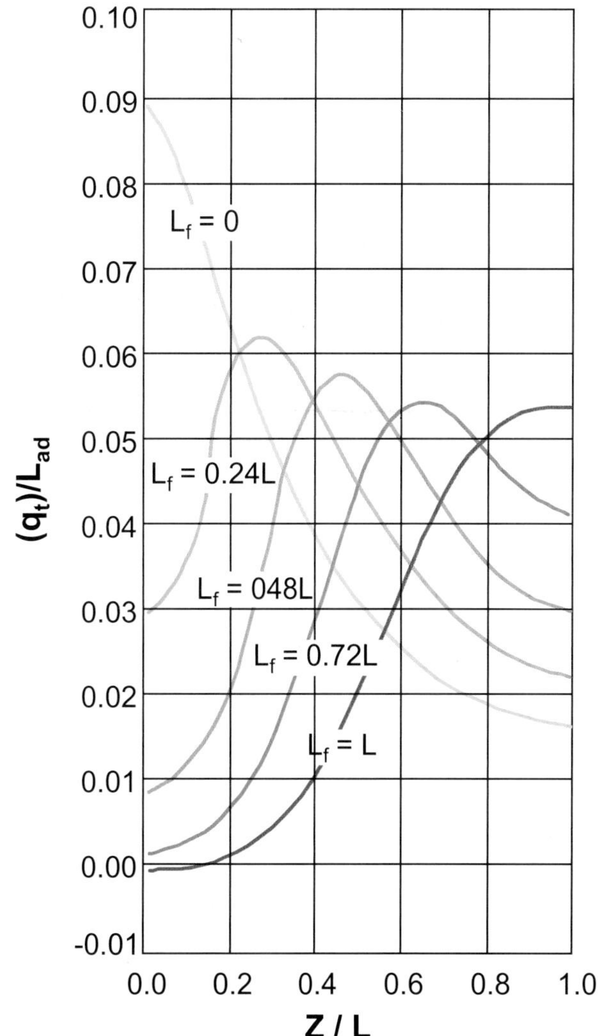

FIGURE 3.28 Distribution of dimensionless average radiant flux density at the tube surfaces for various flame lengths (L_f = flame length, L = heater height, Z = height).[67]

Consider a lb-mol of methane as 380 scf. At 910 Btu/scf, the lower heating value of a lb-mol of methane is 345,800 Btu. If it is presumed that 10% of the heat is radiated directly by the flame burst, the heat content of the gases is 311,220 Btu. The respective heat contents of the component gases based on their specific heats are as follows:

CO_2 = 37,700
H_2O = 67,200 = 104,900 Btu = 33.7%
O_2 = 4,800
N_2 = 201,520 = 206,320 Btu = 63.3%

The total heat content of the radiating gases, carbon dioxide and water, is 104,900 Btu. The heat content of the nitrogen and oxygen is 206,320 Btu immediately following the radiant part of the flame. These gases are in a homogeneous mixture in which a portion of the gases is radiant-capable and a portion is not. The heat energy of a portion of the gases is being dissipated by radiation to produce a steady decrease in heat content within these gases. The other portion that does not radiate then transfers its heat to the radiating gases as their temperature decreases.

However, the emissivity of the radiating gases is quite low. The quantity of heat radiated is a relatively small portion of the total heat content of the gases, so a significant amount of the heat transfer occurs when the gases come into close proximity of the heat transfer tubes. First, the contact of the gases with the surface of the tubes and refractory walls transfers heat by convection. Second, the close proximity of the gases to these surfaces makes the radiation transfer higher since distance is minimized. Thus, the heat transfer in this combination mode depends on the vigorous furnace hot gas currents.

3.6.3 *Refractory Surface Radiation*

The furnace interior refractory walls have a much greater surface area than the surface area of the heat transfer tubes. Therefore, proportionally greater energy is delivered to the furnace refractory surfaces. Refractory has a very high heat capacity. The refractory's ability to store heat exceeds that of the gases and tube materials. Thus, initially, a significant portion of the furnace heat up time is due to the heat capacity of the refractory.

Refer back to Table 3.3 for the specific heats of some common materials. One can see that an average refractory brick has a specific heat of approximately 0.2 Btu/lb-°F, almost double that of carbon steel. The largest heat storage occurs in the refractory. To illustrate the magnitude, let us consider a furnace that is 30 ft × 30 ft × 40 ft (9 m × 9 m × 12 m). With six in. (15 cm) of refractory thickness, the furnace now has approximately 200,000 lb (90,000 kg) of refractory. The quantity of heat stored in the refractory can be estimated using the formula:

$$Q = m \times C_p \times (T_1 - T_2)$$

Assume the refractory heats up from an ambient temperature of T_2 = 70°F to a steady-state average temperature of 500°F. The average is the temperature midway in the thickness of the refractory. The hot surface will be considerably hotter and the cold surface typically around 200°F. The amount of heat stored in the refractory is:

$$Q = 200{,}000 \text{ lb} \times 0.2 \text{ Btu/lb-°F} \times (500 - 70)\text{°F}$$
$$Q = 17.2 \times 10^6 \text{ Btu}$$

At steady state, the refractory reaches and maintains a thermal equilibrium. The amount of energy reaching the refractory is either re-radiated back or lost to the surroundings through conduction to the outside of the furnace. Analysis of the possible radiant heat transfer components that could transfer energy to the refractory do not account for all the heat that is in fact reaching the refractory. The difference, a significant amount, is therefore coming from convection. As the hot gases sweep down the walls of the furnace, they heat the walls by a combination of radiation and convection. To achieve these rates of convection, the gas velocity in the proximity of the walls has to be quite high. Reed stated that gas velocities could reach 50 ft/sec in the vicinity of the walls.

It is common to think of only the refractory areas that glow as being radiant. In reality, all the hot surfaces, whether they are visibly elevated in temperature or not, radiate. The visibly glowing surfaces are, of course, radiating more than the darker surfaces. Generally speaking, refractory surfaces possess high emissivity and thus readily deliver their heat by radiation.

3.6.4 *Analysis of Radiation Heat Transfer*

In this and following discussions, flame radiation is approximated by treating the flame as a isothermal cylinder of gases. These gases can be reasonably assumed to be 17% H_2O (by mole) and 8% CO_2 (by mole) at 1540°C (2800°F). For illustration purposes, the mean beam length of the flame is assumed to be 1 m (3.3 ft), however, this is only an approximation. Better accuracy requires more information to calculate the mean beam length. The pressure-pathlength for H_2O is then 0.56 atm-ft, and for CO_2, the pressure-pathlength is 0.264 atm-ft. From Figures 3.21 and 3.22, the emissivity of the water vapor is about 0.1, while the emissivity of the CO_2 is about 0.065. The total emissivity (uncorrected) is then 0.165. Figure 3.23 indicates a 5% correction to the combined emissivity, so the corrected flame emissivity is 0.157. At 1540°C (2800°F), the blackbody emissive power is $E_{b\text{-}flame} = \sigma T^4 = 5.67 \times 10^{-8} (1540 + 273)^4 = 610 \text{ kW/m}^2 = 194{,}000 \text{ Btu/ft}^2$.

Figure 3.29 is an illustration of a vertical/cylindrical furnace. In this style of furnace, the burners (shown here as a single flame) are surrounded by process tubes. The furnace shell is just outside the process tubes. The radiative circuit diagram in the figure shows how radiative heat flows from the flame to the tubes and refractory walls. To use this diagram, the emissivity values for the flame, refractory wall, and tube surfaces are required. For illustration purposes, the flame emissivity determined above (0.157 for a flame temperature of 2800°F [1540°C]), a typical refractory emissivity of 0.65, and a typical tube surface emissivity (oxidized metal) of 0.85

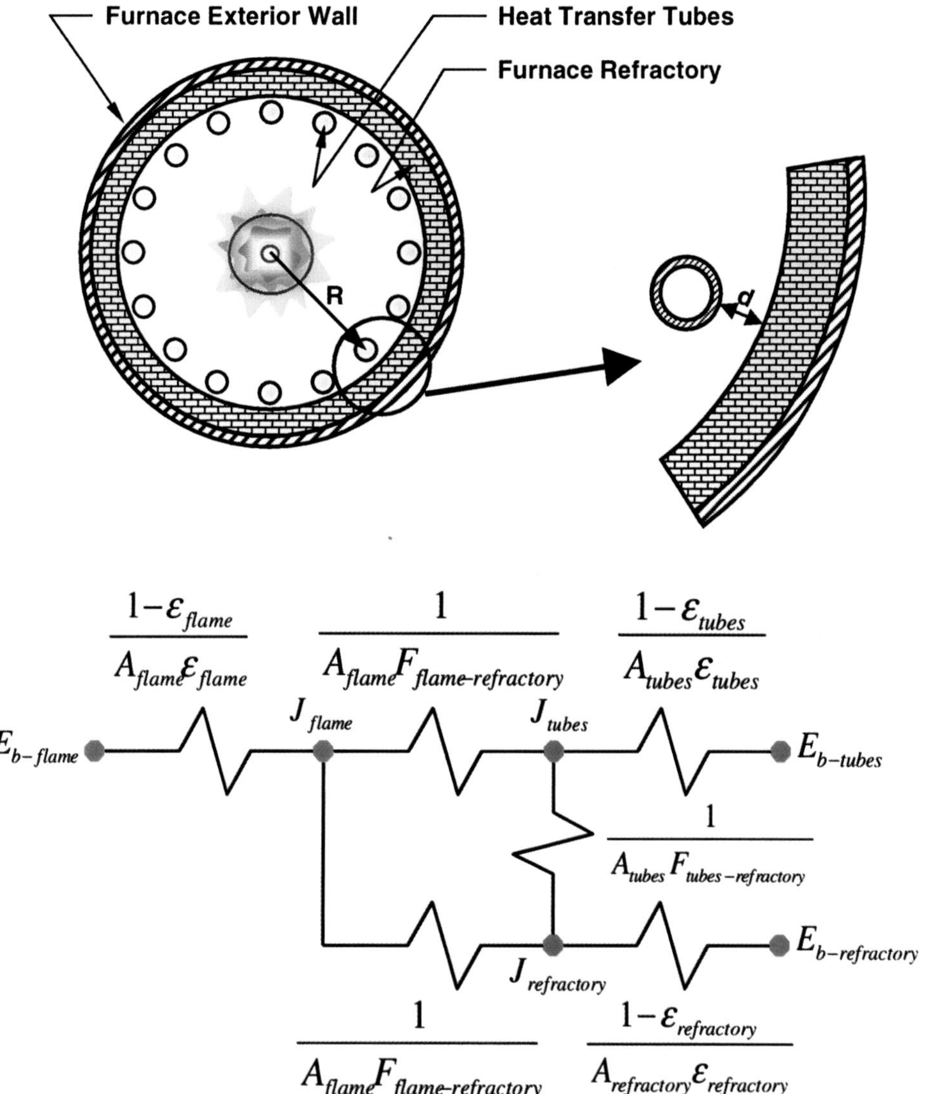

FIGURE 3.29 Radiation heat transfer in a cylindrical furnace.

will be used. To compute view factors, some dimensions need to be assumed, such as a furnace diameter of 10 m, a flame diameter of 1 m, and a tube diameter of 20 cm. Then, a calculation using the formula in Table 3.10 gives the view factor from the flame to a single tube as 0.022. If one assumes that there are 16 tubes in the furnace, then the view factor from the flame to the tubes is $F_{\text{flame-tubes}} = 16 \times 0.022 = 0.352$. Assume that the flame radiation that is not incident on the tubes is incident on the refractory, so that $F_{\text{flame-refractory}} = 1 - F_{\text{flame-tubes}} = 0.648$. Since there is no view factor catalog entry to help in computing the view factor from the tubes to the refractory, this value will be assumed to equal one. This neglects the view factor from the tubes to the flame (which is small) and the fact that the tubes "see each other" (which is not that small). For calculation purposes, an inside surface refractory temperature of 650°C (1200°F) and a tube surface temperature of 430°C (800°F) is assumed.

The circuit diagram shown in Figure 3.29 leads to a system of three linear equations. Since the height of the furnace has not been specified, all the results will be per unit height. The solution of these equations gives the radiative heat flux from the flame as 284 kW/m (294,000 Btu/hr-ft), the heat flux to the tubes is 251 kW/m (260,000 Btu/hr-ft), and the heat flux to the furnace refractory is 33 kW/m (34,000 Btu/hr-ft). These results mean, for instance, that the heat flux to the tubes is 260,000 Btu per foot of tube length. If the tubes were 50 ft long, the total heat flux into the tubes would be 13 x 10^6 Btu/hr.

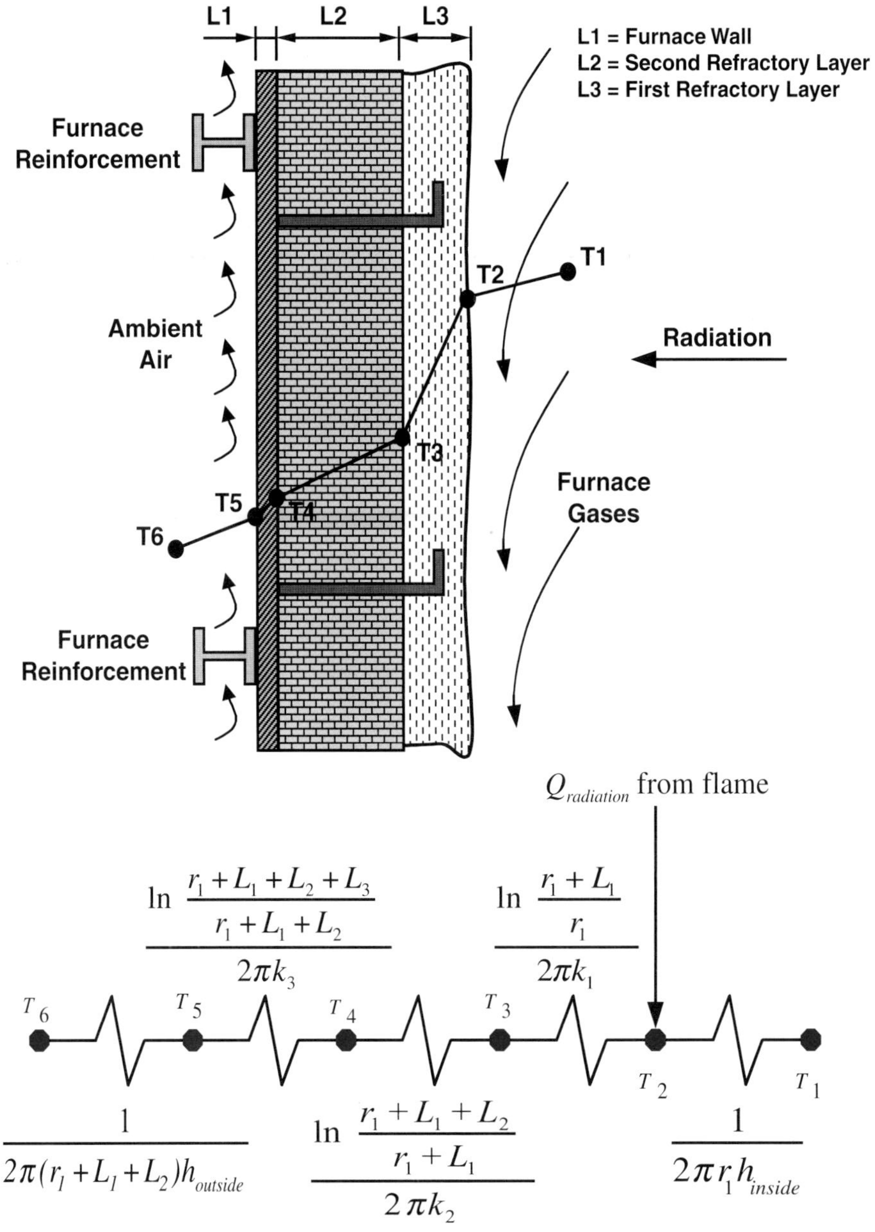

FIGURE 3.30 Cross section of furnace wall.

3.6.5 Heat Transfer Through the Wall of a Furnace

Figure 3.30 illustrates a typical furnace wall. The outer layer of the furnace wall is the steel furnace shell. The inner layers typically consist of refractory brick and, perhaps, soft refractory blanket. The circuit diagram on the figure indicates how the heat transfer through the wall can be analyzed. All three heat transfer mechanisms are indicated. The inside surface is subjected to both convective and radiative heat transfer from the flue gases and flame, respectively. This heat is conducted through the refractory and eventually convected away by natural and forced (wind) convection on the outside of the shell. A typical inside heat transfer coefficient is 30 W/m^2-K (5.3 Btu/hr-ft^2-°R). A typical outside heat transfer coefficient is 17 W/m^2-K (3 Btu/hr ft^2-°R). If the refractory (both blanket and brick) conductivity is assumed to be 0.5 W/m-K and the steel conductivity is 100 W/m-K, then the heat flux through the wall can be computed. Using the radiative heat flux from the previous analysis as 33 kW/m, assume a

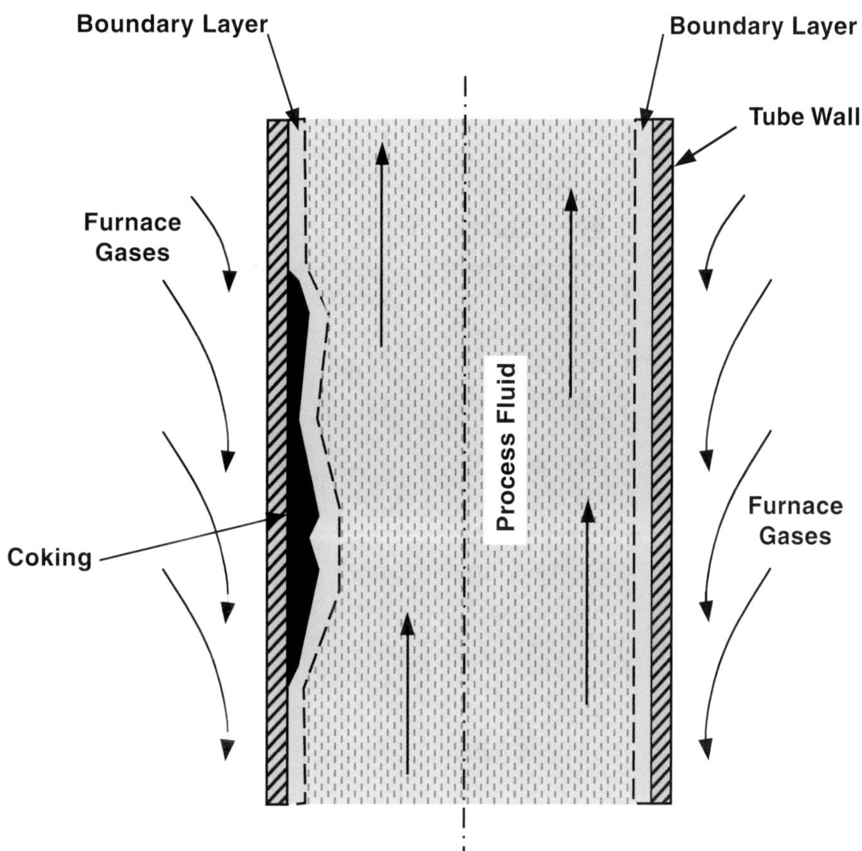

FIGURE 3.31 Cross section of process heat transfer tube.

refractory surface temperature of 650°C (1200°F) and a flue gas temperature of 675°C (1250°F). Assume that the blanket thickness (L_1) is 5 cm (2 in.),the brick thickness (L_2) is 5 cm (2 in.) and the steel thickness is 1.25 cm (0.5 in.); then the circuit analysis gives the total heat flux through the furnace wall as 56kW/m (58 Btu/hr ft), and the outer skin temperature is 92°C (200°F).

3.6.6 Heat Transfer in the Process Tube

Figure 3.31 shows a cross-sectional view of a process fluid flowing through a tube. Radiant heat is incident on the outer surface of the tube, along with convection heat transfer from the furnace gases. This heat is conducted through the wall of the tube. Any coking or scaling on the inside or outside surface of the tube will add to the heat transfer resistance, which will subsequently increase the outside surface temperature. Heat transfer into the process fluid can be analyzed using the formulae given earlier in this chapter. The circuit analysis shown in the previous two examples can also be applied to this example. Additionally, the effects of extra heat transfer resistance due to coke or scale buildup on the tube exterior surfaces can be readily added to get a more physically realistic calculation.

3.6.7 Furnace Gas Flow Patterns

As previously noted, in low temperature furnaces the tubes are usually located very close to the walls. Again, since the radiation is inversely proportional to the square of the distance, the radiation from the walls to the tube is significant when the tubes are close to the walls.

In the case of radiant wall furnaces used in ethylene cracking operations, the wall is directly heated by the flame in order to capitalize on the high heat capacity of the refractory. The quantity of ethylene produced is maximized when the heat is applied evenly to the entire length of the tube. The ideal way to accomplish this is to heat the wall and allow it to radiate to the tubes. The high heat capacity of the refractory acts as a huge capacitance that helps to smooth out peaks in the temperature profile.

In the low temperature furnaces, the heating of the refractory walls is caused by the hot gases sweeping down the walls. The flow of gases between the tubes and the walls is important, as the example below will illustrate. The flames heat the gases and buoyancy causes them to rise. As the gases close to the tubes deliver heat to the tubes and walls, they cool down, become denser, and flow down toward the bottom of the furnace. This establishes a circulation pattern within the furnace, such that the gases rise up from the flames and reverse direction higher up in the furnace and flow down the wall and tubes to the furnace floor. Upon reaching the floor, the gases are reheated to either make another circuit or to exit from the furnace en route to the stack. The benefit of recirculation is optimum when the tubes are on typical two-diameter centers, and decreases as the center-to-center distance is reduced to less than two diameters.

Gases in such recirculation flow pass over the entire tube areas as well as the wall behind the tubes, scrubbing the tube surfaces for heat transfer by convection. Far more importantly, they also scrub the refractory wall behind the tubes to continually deliver heat to the wall surface. Whether the tubes are horizontal or vertical does not seem to make much difference.

There are several methods to visualize furnace gas flow patterns. Today, CFD is the preferred engineering tool to study furnace flue gas patterns. Chapter 9 provides several examples of such studies. On the other hand, in the furnace one may use various powdery substances such as baking soda, particulate carbon, etc. to observe furnace flow patterns. The powder is usually introduced in the air stream to the burners and is seen to glow briefly in the furnace. The glowing particles trace the flow patterns. This is a useful but approximate technique because the persistence of the glowing is short, and if it is too short, it may adversely bias the conclusions being drawn.

Reed[67] mentions an incident to illustrate the importance of furnace gas flow between the tubes and walls. A heater had operated satisfactorily for years. As operation progressed, it was noted that the heater was rapidly losing heat transfer ability, despite the fact that there had been no change in operation which might account for the decrease. There was no change in pressure drop, so the possibility of coke laydown was rejected. The deterioration in performance came about in less than six months and was noticeable on a day-to-day basis as the heater operated.

The heater had been in service for many years and was due for repairs including replacing the refractory side walls which were sagging inwards. The walls, supported independently of the steel that supported the tubes, were gradually moving toward the tubes. The space between the walls and the back sides of the tubes was reduced to such a degree that most of the side walls were actually resting against the tubes.

The heater was shut down and the side walls were repaired to make the space between the tubes and the wall one full tube diameter, which in this case was four in. When the heater was put back into service, it had regained its original heat transfer capability. No change other than the refractory repair had been implemented. Because of this, it was surprisingly evident that the increased space between the tubes and the wall accounted for increased heat absorbing ability.

3.6.7.1 Tube-to-Wall Spacing

Reed conducted experiments at a reduced scale to evaluate the influence of tube to wall spacing.[68] A test heater was constructed using tubes with an outer diameter of 1/4 in. (6 mm). The test furnace dimensions were 18 × 18 × 27 in. (46 × 46 × 69 cm). There was a provision to accurately adjust the tube-to-wall spacing and the relationship of the burner to the tubes. Accurately metered, saturated air was passed through the 3/16-in. (4.8 mm) ID tubes as the source of heat absorption. A thermocouple was used in the air stream at the exit from the tubes to measure the temperature of the exiting air as well as the heat absorbed. The firing rate and excess air, as well as furnace temperature, were closely controlled to identical conditions for all tests.

Reed reported that for a tube spaced one-half diameter off the wall, the heat transfer to the tube is increased 13% over the condition where the tube is tangent to the wall. If the tube is spaced one diameter off the wall, the heat transfer to the tube is increased approximately 29% over the condition where the tube is tangent to the wall. Further increase of the tube-to-wall spacing to as much as three to four diameters provided no increase in heat transfer. Still greater spacing actually created a decrease in heat transfer.

To verify that the gas flow behind the tubes was the contributor to the enhanced heat transfer at a tube-to-wall distance of one tube diameter, an additional experiment was conducted. In this experiment, the tubes were spaced one diameter off the wall to produce the 29% increase in heat transfer. Strips of mica 0.003 in. (0.08 mm) thick were placed in the space between the tubes and the wall at the centerlines of the tubes to block the space between the tubes and the wall with material that is substantially transparent to infrared. The purpose was to avoid blocking radiant transfer while completely blocking the flow path for gases in the space between the tubes and the wall behind them. With the mica strips, the heat transfer to the tubes was exactly the same as was observed with the tubes tangent to the wall. In other words, the 29% performance gain was lost due to the blockage.

3.6.8 Role of the Burner in Heat Transfer

Simple release of an adequate amount of heat to the furnace atmosphere is not the only objective for the burner. Proper choice of burners is critical to the performance of the heater.

There are no hard and fast rules to govern the choice of burners. The design of the furnace and the burner must be matched carefully to achieve good overall performance. There is no single burner design that can be universally applied.

The function of the burner equipment is to deliver heat to the gas content of the furnace as uniformly as possible. Reaching this ideal condition would require an infinitely large number of small burners. Ethylene cracking and hydrogen reforming furnaces most closely approximate this ideal arrangement by using many small burners. Over the last 100 years, the quest for better heat transfer has resulted in a myriad of furnace designs. It follows that many burner designs were developed to fit the various furnace designs. In the past, when emissions were not regulated, the primary requirement for the burner was effective heat transfer. Typically, the old burner designs rapidly mixed the fuel and air, resulting in short flames. Emissions regulations have now driven the design of burners for the last two or three decades. The primary requirement is now to meet the emissions regulations without compromising furnace performance. This conflicting challenge has been met with considerable engineering ingenuity over the years. Low-NOx burners designed in the last two decades tend to have longer flames because the strategy for NOx reduction was to delay mixing and thereby reduce peak flame temperatures.

Over the years, furnace manufacturers as well as burner manufacturers have researched heat flux profiles in various burner–furnace combinations. Again, CFD is a great help in studying heat flux profiles, but even with today's sophisticated modeling capabilities and advanced instrumentation, exact measurements are not possible. Exact measurements are difficult to obtain because apart from the obvious problem of working in a high temperature zone, the geometry of the furnace and the re-radiation from various furnace surfaces make the analysis complicated. With wall fired burners, either floor mounted or wall mounted, it is somewhat easier to predict heat flux patterns. However, in furnaces that have free standing flames, the heat flux patterns tend to be specific to that burner–furnace combination.

If a desired heat flux pattern is identified, it is possible to engineer the flame shape to attempt to meet the requirement. Previous experience can help define the burner design required. On the other hand, the outcome can only be estimated if there is no previous experience with that particular furnace–burner combination. In such cases, some final testing and adjustment is usually required. Consequently, burners have been developed over the years with flames of every reasonably conceivable shape.

Burner flames must be shaped and directed to allow the required heat diffusion to the furnace gases without delivering excessive heat to any local heat transfer area. Local overheating and flame impingement must be avoided at all costs. Flame impingement does not occur solely due to burner performance. The burner has only limited control over the characteristic flow patterns of a furnace. It is possible to modify a burner to eliminate flame impingement by changing the fuel jet configurations, but there is only a narrow window of opportunity here, because radical modifications will require compromises in other areas of burner performance such as capacity or emissions.

Flame length is of utmost importance in burner design, although flame length and heat dispersion are not necessarily in a fixed relationship. Providing short flames exclusively for all applications is not the answer either, because some applications require long flames to reach further into large furnaces or to deliver heat to locations further away from the burner. For example, in the typical floor-fired steam reformer, it is necessary to drive the hot gases from the furnace floor to the top of the furnace to distribute heat to the tube areas where maximum heat density is demanded. Nowadays, most steam reforming furnaces are down-fired for this reason. The contrary is true in side-wall fired steam reformers.

In radiant wall firing, the burners are located in areas where maximum heat transfer is demanded. The flame is expected to remain close to the wall and not penetrate forward into the furnace at all. This is because the furnaces are narrow and the burners are located quite close to the tubes which may be either vertically or horizontally suspended at the center of the furnace. The "terrace wall" furnace design for the same application is a prime example of the differentiation between heat dispersion and flame length. The burners are mounted in terraces on the side wall in much the same way as a floor-mounted burner, and the flame is fired vertically up the wall. In this design, the flame is considerably longer than the small wall burners, yet the service performed is identical.

In a typical process heater, the demand for very precise control of heat density per linear foot of tube is not as great as in a steam-reformer furnace. It is possible to use a smaller number of larger burners to obtain satisfactory firing conditions and heat dispersion, but the burners must be suited to the service. Sometimes, burners capable of reasonably short flames have not had satisfactory heat dispersion characteristics and must be replaced to reduce tube damage.

Some designs have stand-alone flame burners that are mounted on the side walls of the furnace with the flame fired straight into the furnace space. This is not the same as the ethylene cracking or reforming radiant wall arrangements. Those are lower temperature furnaces where the design decision has been made to mount a stand-alone flame burner on the side wall instead of the floor to reduce the initial cost of the furnace. Side wall mounting costs less since the furnace does not have to be elevated to install burners below it, and, often, less burners are required. However, floor firing has some advantages over side wall firing. With floor-mounted burners the heater can typically be fired 25% harder. This is because floor mounting makes better use of the combustion volume and provides more uniform heat distribution.

Either way, there will be greater service from the heater when a relatively large number of small burners are used rather than a small number of large burners. If there is a relatively small number of large burners, there is a greater mass of gas issuing from each burner and a greater concentration of heat before the burner. This larger mass of gases and quantity of heat must then be dispersed evenly to the furnace atmosphere for good performance. It is far easier to disperse smaller amounts of gas and heat as issued by several smaller burners.

To conclude, in general, the heat transfer role of the burner in a furnace is to provide the required amount of heat with appropriate flame dimensions without localized hot spots or flame impingement. The sizing and selection of the burner must help make the temperature of the bulk of the furnace gases as uniform as possible in as short a distance as possible from the burner.

3.7 CONCLUSIONS

This chapter presented the fundamentals of heat transfer. Basic relations for conduction, convection, and radiation heat transfer were provided and discussed. This chapter, by necessity, is a very brief and dense presentation of the subject of heat transfer. The interested reader is strongly encouraged to consult with heat transfer texts (see references) for more detailed information or explanation. The subject of heat transfer is a vast and interesting one. The focus of this chapter is on heat transfer in combustion systems, but there are a multitude of other heat transfer applications to which these basic principles can be applied. The authors hope that this brief introduction has given the reader an interest in pursuing the subject further.

The difficulties that may arise when trying to apply heat transfer relations to furnace heat transfer problems were considered here. Some of the approximations that can be made to complete such an analysis were considered. Again, the reader interested in more information should consult the references given at the end of this chapter. In particular, the recent book by Baukal[64] provides a thorough overview of combustion heat transfer and cites many references.

REFERENCES

1. B. Gebhart, *Heat Transfer*, 2nd edition, McGraw-Hill, New York, 1971.
2. F. Kreith and M.S. Bohn, *Principles of Heat Transfer*, Harper & Row, New York, 1986.
3. J.P. Holman, *Heat Transfer*, 7th edition, McGraw-Hill, New York, 1990.
4. A. Bejan, *Heat Transfer*, John Wiley & Sons, New York, 1993.
5. F.P. Incropera and D.P. Dewitt, *Introduction to Heat Transfer*, 3rd edition, John Wiley & Sons, New York, 1996.
6. A.F. Mills, *Heat Transfer*, 2nd edition, Prentice-Hall, Englewood Cliffs, NJ, 1998.
7. W.X. Janna, *Engineering Heat Transfer*, 2nd edition, CRC Press, Boca Raton, FL, 2000.
8. H.C. Hottel and A.F. Sarofim, *Radiative Transfer*, McGraw-Hill, New York, 1967.
9. T.J. Love, *Radiative Heat Transfer*, Merrill Publishing, Columbus, OH, 1968.
10. M. Özisik, *Radiative Transfer and Interactions with Conduction and Convection*, John Wiley & Sons, New York, 1973.
11. W.A. Gray and R. Müller, *Engineering Calculations in Radiative Heat Transfer*, Pergamon, Oxford, U.K., 1974.
12. E.M. Sparrow and R.D. Cess, *Radiation Heat Transfer*, augmented edition, Hemisphere, Washington, D.C., 1978.
13. J.B. Dwyer, Furnace calculations, in *Process Heat Transfer*, D.Q. Kern, Ed., McGraw-Hill, New York, 1950.
14. F.W. Hutchinson, *Industrial Heat Transfer*, Industrial Press, New York, 1952.
15. S.T. Hsu, *Engineering Heat Transfer*, D. Van Nostrand Co., Princeton, NJ, 1963.
16. J.R. Welty, *Engineering Heat Transfer*, John Wiley & Sons, New York, 1974.
17. B.V. Karlekar and R.M. Desmond, *Engineering Heat Transfer*, West Pub. Co., St. Paul, MN, 1977.
18. V. Ganapathy, *Applied Heat Transfer*, PennWell Books, Tulsa, OK, 1982.
19. A.G. Blokh, *Heat Transfer in Steam Boiler Furnaces*, Hemisphere, Washington, D.C., 1988.
20. W.M. Rohsenow, J.P. Hartnett, and E.N. Ganic, *Handbook of Heat Transfer Applications*, McGraw-Hill, New York, 1985.

21. N.P. Cheremisinoff, Ed., *Handbook of Heat and Mass Transfer*, four volumes, Gulf Publishing, Houston, TX, 1986.

22. F. Kreith, Ed., *The CRC Handbook of Thermal Engineering*, CRC Press, Boca Raton, FL, 2000.

23. V.S. Arpaci, *Conduction Heat Transfer*, Addison-Wesley, Reading, MA, 1966.

24. M.N. Özisik, *Boundary Value Problems of Heat Conduction*, Dover, New York, 1968.

25. U. Grigull and H. Sandner, *Heat Conduction*, Hemisphere, Washington, D.C., 1984.

26. G.E. Myers, *Analytical Methods in Conduction Heat Transfer*, Genium Publishing, Schenectady, NY, 1987.

27. B. Gebhart, *Heat Transfer and Mass Diffusion*, McGraw-Hill, New York, 1993.

28. D. Poulikakos, *Conduction Heat Transfer*, Prentice-Hall, Englewood Cliffs, NJ, 1994.

29. F.P. Incropera and D.P. DeWitt, *Fundamentals of Heat and Mass Transfer*, 4th ed., John Wiley & Sons, New York, 1996.

30. P.J. Schneider, *Conduction Heat Transfer*, Addison-Wesley, Reading, MA, 1955.

31. H.S. Carslaw and J.C. Jaeger, *Conduction of Heat in Solids*, 2nd edition, Oxford University Press, London, 1959.

32. A. Garg, How to boost the performance of fired heaters, *Chem. Eng.*, 96(11), 239-244, 1989.

33. V.S. Arpaci, *Convection Heat Transfer*, Prentice-Hall, Englewood Cliffs, NJ, 1984.

34. C.S. Fang, *Convective Heat Transfer*, Gulf Publishing, Houston, TX, 1985.

35. S. Kakac, R.K. Shah, and W. Aung, Eds., *Handbook of Single-Phase Convective Heat Transfer*, John Wiley & Sons, New York, 1987.

36. L.C. Burmeister, *Convective Heat Transfer*, 2nd edition, John Wiley & Sons, New York, 1993.

37. W.M. Kays and M.E. Crawford, *Convective Heat and Mass Transfer*, 3rd edition, McGraw-Hill, New York, 1993.

38. A. Bejan, *Convection Heat Transfer*, 2nd edition, John Wiley & Sons, New York, 1994

39. M. Kaviany, *Principles of Convective Heat Transfer*, Springer-Verlag, New York, 1994.

40. S. Kakac and Y. Yener, *Convective Heat Transfer*, 2nd edition, CRC Press, Boca Raton, FL, 1995.

41. P.H. Oosthuizen, *An Introduction to Convective Heat Transfer*, McGraw-Hill, New York, 1999.

42. F.W. Dittus and L.M.K. Boelter, University of California, Berkeley, *Publications on Engineering*, 2, 443, 1930.

43. E.N. Sieder and G.E. Tate, *Ind. Eng. Chem.*, 28, 1429, 1936.

44. B.S. Petukhov, *Advances in Heat Transfer*, Vol. 6, T.F. Irvine and J.P. Hartnett, Eds., Academic Press, New York, 1970.

45. L.F. Moody, *Trans. ASME*, 66, 671, 1944.

46. E.S. Skupinski, J. Tortel, and L. Vautrey, *Int. J. Heat Mass Transfer*, 8, 937, 1965.

47. R.A. Seban and T.T. Shimazaki, *Trans. ASME*, 73, 803, 1951.

48. C.B. Reed, *Handbook of Single-Phase Convective Heat Transfer*, S. Kakac, R.K. Shah, and W. Aung, Eds., Wiley Interscience, New York, 1987, chap. 8.

49. H. Schlichting, *Boundary Layer Theory*, 6th edition, McGraw-Hill, New York, 1968.

50. R. Hilpert, *Forsch. Geb. Ingenieurwes.*, 4, 215, 1933.

51. J.D. Knudsen and D.L. Katz, *Fluid Dynamics and Heat Transfer*, McGraw-Hill, New York, 1958.

52. S.W. Churchill and M. Bernstein, *J. Heat Transfer*, 99, 300, 1977.

53. E.D. Grimison, *Trans. ASME*, 59, 583, 1937.

54. A. Zhukauskas, Heat transfer from tubes in cross flow, in *Advances in Heat Transfer*, Vol. 8, J.P. Hartnett and T.F. Irvine, Jr., Eds., Academic Press, New York, 1972.

55. W.A. Gray and R. Müller, *Engineering Calculations in Radiative Heat Transfer*, Pergamon, Oxford, U.K., 1974.

56. J.A. Wiebelt, *Engineering Radiation Heat Transfer*, Holt, Rinehart and Winston, New York, 1966.

57. E.M. Sparrow and R.D. Cess, *Radiation Heat Transfer*, Augmented Edition, Hemisphere, Washington, D.C., 1978.

58. D.K. Edwards, *Radiation Heat Transfer Notes*, Hemisphere, Washington, D.C., 1981.

59. R. Siegel and J.R. Howell, *Thermal Radiation Heat Transfer*, 2nd edition, McGraw-Hill, New York, 1981.

60. M.Q. Brewster, *Thermal Radiative Transfer and Properties*, John Wiley & Sons, New York, 1992.

61. M. Planck, *The Theory of Heat Radiation*, Dover, New York, 1959.

62. H.C. Hottel, Radiant heat transmission, in *Heat Transmission*, 3rd ed., W.H. McAdams, Ed., McGraw-Hill, New York, 1954.

63. M.F. Modest, *Radiative Heat Transfer*, McGraw-Hill, New York, 1993.

64. C.E. Baukal, *Heat Transfer in Industrial Combustion*, CRC Press, Boca Raton, FL, 2000.

65. C.J. Hoogendoorn, C.M. Ballintijn, and W.R. Dorresteijn, Heat-flux studies in vertical tube furnaces, *J. Inst. Fuel*, 43, 511-516, 1970.

66. N. Selçuk, R.G. Siddall, and J.M. Beér, Prediction of the effect of flame length on temperature and radiative heat flux distributions in a process fluid heater, *J. Inst. Fuel*, 43, 89-96, 1975.

67. R.D. Reed, *Furnace Operations*, 3rd edition, Gulf Publishing, Houston, TX, 1981.

NOMENCLATURE

Symbol	Description
A	Area, m^2
A_e	Cross sectional area, m^2
Bi	Biot number
C_t	Thermal capacitance, J/K
c	Speed of light, m/s
c_f	Friction coefficient
c_p	Specific heat at constant pressure, J/kg·K
D	Diameter, m
D_h	Hydraulic diameter, m
E	Thermal (internal) energy, J; emissive power, W/m^2
Eg	Rate of energy generation, W
F	Fraction of blackbody radiation in a wavelength band; view factor
F_o	Fourier number
f	Friction factor
G	Irradiation, W/m^2
g	Gravitational acceleration, m/s^2
h	Convection heat transfer coefficient, w/m^2·K; Planck's constant
J	Radiosity, W/m^2
k	Thermal conductivity, w/m·K
	Boltzmann's constant
L	Characteristic length, m
Nu	Nusselt number
P	Perimeter, m
Pe	Peclet number (Re·Pr)
Pr	Prandtl number
p	Pressure, N/m^2
Q	Energy transfer, J
q	Heat transfer rate, W
$\dot{q}$	Rate of energy generation per unit volume, W/m^3
q''	Heat flux, W/m^2
R	Cylinder radius, m
Re	Reynolds number
R_t	Thermal resistance, K/W
$R_{t,c}$	Thermal contact resistance, K/W
ro	Cylinder or sphere radium, m
r, φ, z	Cylindrical coordinates
r, θ, Φ	Spherical coordinates
St	Stanton number
S_D, S_T	Diagonal and transverse pitch of a Tube bank, m
T	Temperature, K
t	Time, s
U	Overall heat transfer coefficient, W/m^2·K
V	Volume, m^3
	Fluid velocity, m/s
x, y, z	Rectangular coordinates, m
x_c	Critical location for transition to turbulence, m
$x_{fd,\,h}$	Hydrodynamic entry length, m
$x_{fd,\,t}$	Thermal entry length, m

Greek Letters

α	Thermal diffusivity, m^2/s
δ	Hydrodynamic boundary layer thickness, m
δ_t	Thermal boundary layer thickness, m
ε	Emissivity
θ	Temperature difference, K
λ	Wavelength, µm
μ	Viscosity, kg/s·m
ρ	Mass density, kg/m^3
	Reflectivity
σ	Stefan–Boltzmann's constant
ω	Solid angle, sr

Chapter 4
Fundamentals of Fluid Dynamics

Lawrence D. Berg, Wes Bussman, and Michael Henneke

TABLE OF CONTENTS

4.1 Introduction ... 118
4.2 Fluid Properties ... 118
  4.2.1 Density ... 118
  4.2.2 Viscosity ... 119
  4.2.3 Specific Heat ... 125
  4.2.4 Equations of State ... 126
4.3 Fundamental Concepts ... 128
  4.3.1 Hydrostatics ... 128
  4.3.2 Bernoulli Equation ... 130
  4.3.3 Control Volumes ... 132
  4.3.4 Differential Formulation (Navier-Stokes Equations) ... 134
4.4 Different Types of Flow ... 138
  4.4.1 Turbulent and Laminar Flow ... 138
  4.4.2 Compressible Flow ... 144
4.5 Pressure Drop Fundamentals ... 148
  4.5.1 Basic Pressure Concepts ... 148
  4.5.2 Roughness ... 149
  4.5.3 Loss Coefficient ... 151
  4.5.4 Discharge Coefficient ... 152
References ... 153

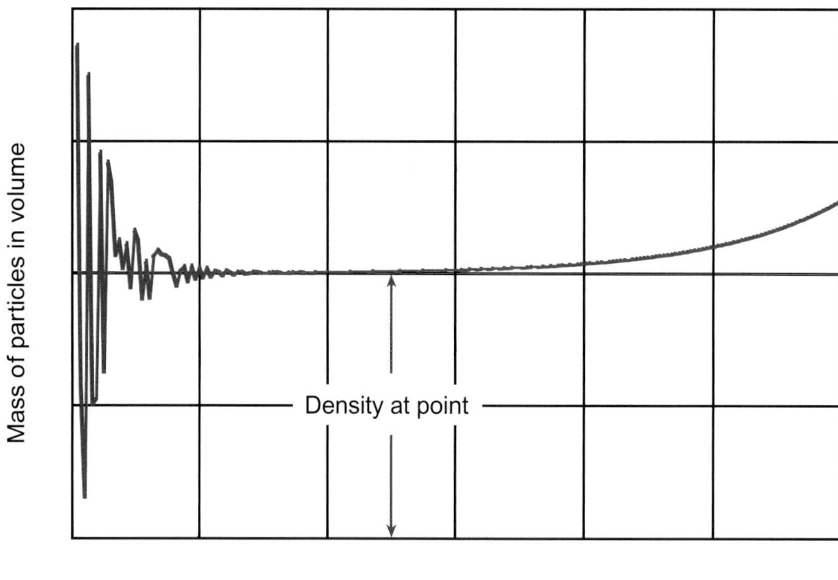

FIGURE 4.1 Variation in measured density with length scale.

4.1 INTRODUCTION

The study of fluid dynamics likely dates back to ancient times when early man hunted to keep himself alive. It is easy to imagine that he quickly discovered he could throw a streamlined spear much further than a blunt rock. Perhaps he also noticed that his horse ran a little faster if he ducked his head into the wake of the horse's neck. Through the course of scientific development, observations such as these, and countless others, have become a rigorous science.

The subject of this chapter is the science of fluid dynamics. There are many good textbooks on the subject; Panton,[1] White,[2] Fox and McDonald,[3] and Vennard and Street[4] are examples. The subject is so broad that it has a number of subfields, for example, turbulence, acoustics, and aerodynamics. The purpose of this chapter is to give the reader a fundamental understanding of some of the fluid mechanical concepts that are important in combustion systems, especially those in the petrochemical industry.

4.2 FLUID PROPERTIES

Several thermophysical properties are commonly used in fluid mechanics analysis. This section presents a brief description of these properties as they relate to combustion systems.

4.2.1 Density

Density is defined as the mass per unit volume of a fluid. Density is a point function, which means that it must be defined at every point within a continuum. Because density can vary from point to point within a flowing fluid, the question of how density is defined is not trivial. To measure the density of a fluid at a given point, a sample volume of the fluid is taken at the given point and the mass of the sample determined. For example, if a sample of one cubic centimeter (1 cm^3) of the fluid is taken and the mass of the sample is determined to be x grams, then the density of the fluid at that location would be x grams per cubic centimeter (x g/cm^3).

The problem with the previous example is the possibility that the measured value of density might depend on the chosen volume. If the chosen volume is too large, then density variations due to other fluid mechanical variables, such as temperature and pressure, could cause the density to vary within the chosen volume.

Figure 4.1 illustrates the problem. The x-axis is the length scale of the volume for measurement. As the length scale gets long, the measured density may increase or decrease because of variations within the sample. If the length scale gets too short, then the sample may contain only several molecules of the fluid, and the density will be inaccurate because of sampling. The proper definition of density is then:

$$\rho \equiv \lim_{L \to 0} \frac{m}{V} \tag{4.1}$$

where m is the measured mass of the sample and V is the volume sampled. The limit in Eq. (4.1) is understood to approach zero, but to remain much larger than molecular dimensions.

4.2.2 *Viscosity*

4.2.2.1 Definition

"Son, you are slower than molasses in January!" is an epithet that many a parent routinely applies to a child for his slow movement. While one hopes that age diminishes the truth of this description, time has not diminished the truth of the statement — molasses does pour very slowly when it is cold. This is true for a significant number of other liquids as well: oils, fluids with glycerin, polymeric, and others. At the same time, many other liquids seem to pour equally well at any temperature. Water and alcohol are good examples of this. These qualitative observations of the ease with which various liquids can be poured provide insight into the liquid property of viscosity. Indeed, at a conceptual level, the "pour-ability" of a liquid can be utilized to understand viscosity. For example, viscosity describes the rate at which a liquid pours out of a container.

In combustion processes, pouring rates are not of direct interest. However, due to requirements for maintaining certain heat releases, specific volumetric flows are required. In this case, viscosity could be descriptive of the power requirements to supply a specific volumetric flow through a pipe. From this description, it follows that as a property, viscosity is only useful in describing liquids or gases in motion. Therefore, viscosity is the difficulty that various fluids have maintaining motion.

While the above discussion is useful in understanding viscosity from a conceptual point of view, in order to be useful in engineering combustion solutions, a more exact definition that allows for quantitative analysis is required. For a fluid flowing in a pipe or along any solid surface, the velocity at the surface is zero. As distance from the surface increases, the velocity also increases, as shown schematically in Figure 4.2.

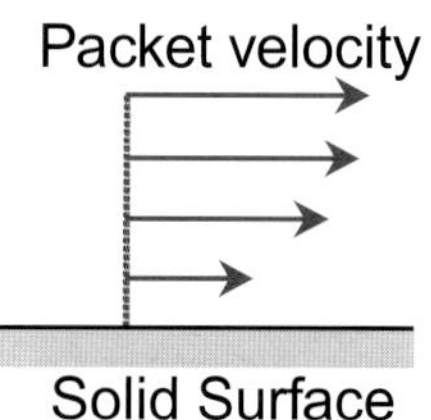

FIGURE 4.2 Velocity profile of a fluid flowing along a solid surface.

Assume that each of the arrows in the figure is a discrete "packet" of liquid. Because they are moving at different velocities, as time passes they will move relative to each other. This relative motion gives rise to a shear stress between the different layers. If there were no velocity difference between the layers, then no shear stress would exist. As the velocity difference increases, the shear stress increases. The implication is that shear stress is proportional to the velocity gradient, which is expressed mathematically as:

$$\tau \sim \frac{dV}{dY} \tag{4.2}$$

where τ is the shear stress, and the derivative is the velocity difference between different "layers" in the flow. The constant of proportionality between shear stress and the velocity is the viscosity. Conceptually, viscosity then is the difficulty that two fluid "layers" experience as they flow past each other. This is also expressed mathematically as:

$$\tau = \mu \frac{dV}{dY} \tag{4.3}$$

or

$$\mu = \frac{\tau}{\frac{dV}{dY}} \tag{4.4}$$

where τ = Shear stress
μ = Viscosity
$\frac{dV}{dY}$ = Velocity gradient

4.2.2.2 Units

4.2.2.2.1 Absolute Viscosity

The units of stress are (force/area) and the units for the gradient are (velocity/length). In English units, this is represented as follows:

$$\frac{\left(\frac{\text{lbf}}{\text{ft}^2}\right)}{\frac{\left(\frac{\text{ft}}{\text{sec}}\right)}{\text{ft}}} = \frac{\left(\frac{\text{lbf}}{\text{ft}^2}\right)}{\left(\frac{1}{\text{sec}}\right)} = \frac{(\text{lbf})(\text{sec})}{\text{ft}^2} \quad \text{or} \quad \frac{(\text{lbm})32.17}{(\text{ft})(\text{sec})} \tag{4.5}$$

In a similar manner, metric units for viscosity would, in general, be:

$$\frac{(\text{N})\cdot(\text{sec})}{\text{m}^2} \quad \text{or} \quad \frac{\text{kg}}{(\text{m})\cdot(\text{sec})} \tag{4.6}$$

where

sec = Seconds
N = Newtons
m = Meters
kg = Kilograms

The metric unit in either form, if multiplied by 10^{-1}, is called a poise. If the unit is multiplied by 10, it is called a centipoise. Most tabular data for viscosity are either in terms of poise, or the "lbm" formulation of English units. Occasionally, viscosity is tabulated in "lbf" form, or slugs utilized for the mass term. Because of the confusing nature of lbm vs. lbf, the reader is cautioned to carefully review the units utilized in tabular data prior to utilization. All information provided in the appendix is listed in centipoise.

4.2.2.2.2 *Kinematic Viscosity*

In addition to pure, absolute, or dynamic viscosity, viscosity information is often tabulated as viscosity divided by density. This is accomplished for convenience, as absolute viscosity often appears divided by density in real flow calculations. This fluid property is termed the "kinematic" viscosity, and is expressed mathematically as:

$$\nu = \frac{\mu}{\rho} \tag{4.7}$$

where

ν = Kinematic viscosity (m^2/sec)
μ = Absolute or dynamic viscosity
ρ = Fluid density

Meters squared per second are the dimensions for kinematic viscosity in metric units. When the m^2/sec value is divided by 10^{-4}, this property is termed "stokes." Most tabular data is available in terms of centistokes — the m^2/sec value divided by 10^{-6}. Kinematic viscosity is most useful when analyzing systems involving liquids in which the density is unlikely to change significantly with ambient conditions and temperature. In the case of gases, compression, atmospheric pressure change due to elevation, and fluid temperature changes will affect the density, and thus the kinematic viscosity. For these reasons, absolute viscosity should be utilized for any gas analysis.

4.2.2.3 Other Units

In addition to the units discussed above, there are a number of historical units employed in the hydrocarbon and petrochemical industries. These units are a consequence of various historic viscosity measurement techniques, especially for crude oil and various cuts. Reed's table[5] compares most of the various units and provides a method to convert from each of the units to a modern standard. This table, along with conversions, is provided in Table 4.1.

4.2.2.3.1 *Temperature Dependence and Multicomponent Liquids*

As a property, viscosity arises from molecular interactions at the atomic level. These interactions tend to be highly influenced by the fluid temperature and type. For liquids in which there are strong secondary intermolecular bonds, an increase in temperature would tend to weaken these bonds and viscosity would tend to decrease with higher temperature. For gases, intermolecular forces tend to be very weak and viscosity is due to an exchange of momentum between shear layers. In this case, as temperature increases, molecules will migrate to other areas of the flow at faster rates. The increased migration results in increased momentum transfer or greater viscosity. The graph in Figure 4.3 clearly shows these trends for several common liquids and gases.

In addition to a temperature dependence, the hydrocarbon and petrochemical industries have the additional challenge of determining the viscosity for multiple constituent liquids. Oils, fuel gases, and natural gases are rarely of a single molecular type, but rather a mixture with properties within certain boundaries. In this environment, empirical rules have been developed to provide reasonable viscosity estimates for given bulk properties and temperature. Figure 4.4 provides a graphical temperature dependence for hydrocarbon gases based on molecular weight. The following equation[6] provides a method for calculating the viscosity of gas mixtures if the individual components are known:

$$\text{Viscosity}_{Mixture} = \frac{\sum_i (\text{Viscosity}_i)(N_i)\sqrt{M_i}}{\sum_i (N_i)\sqrt{M_i}} \tag{4.8}$$

where

N = Mole fraction of component i
M = Molecular weight of component i

Liquid mixtures tend to be less predictable. Crude oils and heavy fuel oils, in particular, are mostly dependent on cut temperatures and origination of the crude. A typical chart (Figure 4.5) of viscosity vs. temperature for various oils is provided to illustrate the point. Normally, it is recommended that information provided by the supplier be utilized to determine viscosity for mixtures of liquids; however, Lederer[7] has given a correlation to determine the viscosity of a mixture of crudes and solvents, as follows:

$$\ln \mu_m = \chi_A \ln \mu_A + \chi_B \ln \mu_B \tag{4.9}$$

and

TABLE 4.1 Viscosity Conversion Table

Viscosity Centipoises	Seconds Saybolt Universal	Seconds Saybolt Fruol	Seconds Redwood	Seconds Redwood Admirality	Degrees Engler	Degrees Barbey
1.00	31		29		1.00	6200
2.56	35		32.1		1.16	2420
4.30	40		36.2	5.10	1.31	1440
5.90	45		40.3	5.52	1.46	1050
7.40	50		44.3	5.83	1.58	838
8.83	55		48.5	6.35	1.73	702
10.20	60		52.3	6.77	1.88	618
11.53	65		56.7	7.17	2.03	538
12.83	70	12.95	60.9	7.60	2.17	483
14.10	75	13.33	65.0	8.00	2.31	440
15.35	80	13.70	69.2	8.44	2.45	404
16.58	85	14.10	73.3	8.86	2.59	374
17.80	90	14.44	77.6	9.30	2.73	348
19.00	95	14.85	81.5	9.70	2.88	326
20.20	100	15.24	85.6	10.12	3.02	307
31.80	150	19.3	128	14.48	4.48	195
43.10	200	23.5	170	18.90	5.92	144
54.30	250	28.0	212	23.45	7.35	114
65.40	300	32.5	254	28.0	8.79	95
76.50	350	35.1	296	32.5	10.25	81
87.60	400	41.9	338	37.1	11.70	70.8
98.60	450	46.8	381	41.7	13.15	62.9
110	500	51.6	423	46.2	14.60	56.4
121	550	56.6	465	50.8	16.05	51.3
132	600	61.4	508	55.4	17.50	47.0
143	650	66.2	550	60.1	19.00	43.4
154	700	71.1	592	64.6	20.45	40.3
165	750	76.0	635	69.2	21.90	37.6
176	800	81.0	677	73.8	23.35	35.2
187	850	86.0	719	78.4	24.80	33.2
198	900	91.0	762	83.0	26.30	31.3
209	950	95.8	804	87.6	27.70	29.7
220	1000	100.7	846	92.2	29.20	28.2
330	1500	150	1270	138.2	43.80	18.7
440	2000	200	1690	184.2	58.40	14.1
550	2500	250	2120	230	73.00	11.3
660	3000	300	2540	276	87.60	9.4
770	3500	350	2960	322	100.20	8.05
880	4000	400	3380	368	117.00	7.05
990	4500	450	3810	414	131.50	6.26
1100	5000	500	4230	461	146.00	5.64
1210	5500	550	4650	507	160.50	5.13
1320	6000	600	5080	553	175.00	4.70
1430	6500	650	5500	559	190.00	4.34
1540	7000	700	5920	645	204.50	4.03
1650	7500	750	6350	691	219.00	3.76
1760	8000	800	6770	737	233.50	3.52
1870	8500	850	7190	783	248.00	3.32
1980	9000	900	7620	829	263.00	3.13
2090	9500	950	8040	875	277.00	2.97
2200	10000	1000	8460	921	292.00	2.82

Note: The viscosity is often expressed in terms other than centipoise. Formulas for the various viscosimeters are as follows:

Absolute viscosity in cp = $0.261 \times T1 - 188/T1$ (T1 = Redwood seconds)
Absolute viscosity in cp = $2.396 \times T2 - 40.3/T2$ (T2 = Redwood Admiralty seconds)
Absolute viscosity in cp = $0.22 \times T3 - 180/T3$ (T3 = Saybolt universal seconds)
Absolute viscosity in cp = $2.2 \times T4 - 203/T4$ (T4 = Saybolt Furol seconds)
Absolute viscosity in cp = $0.147 \times T5 - 374/T5$ (T5 = Degree Engler × 51.3)

Source: R.D. Reed, *Furnace Operations*, 3rd ed., Gulf Publishing, Houston, TX, 1981.

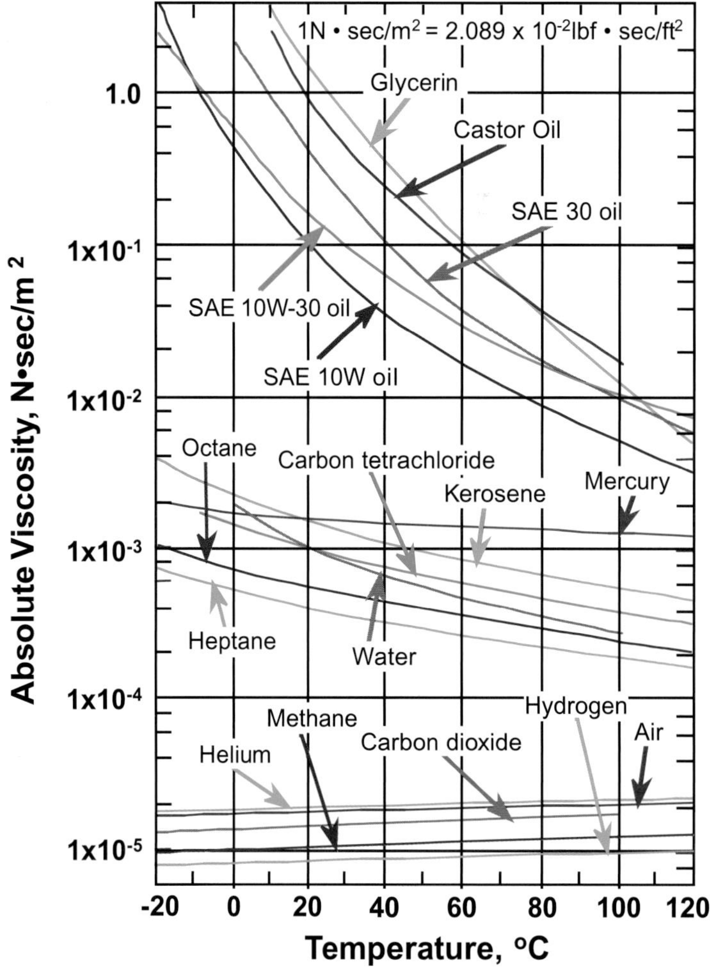

FIGURE 4.3 Absolute viscosity vs. temperature for various fluids.

$$\chi_A = (\alpha V_A)/(\alpha V_A + V_B) \tag{4.10}$$

$$\chi_B = 1 - \chi_A \tag{4.11}$$

where α is an empirical constant having a value between 0 and 1.0; A is the more viscous component and B is the less viscous component. V_A and V_B represent the volume composition for a given mixture. The above equation has been reported to give satisfactory results for other combinations and mixtures of petroleum fractions. Shu[8] developed the following correlation to obtain the empirical constant in the above equation:

$$\alpha = \frac{17.04\left[(\Delta\rho_m)^{0.5237}(\rho_A)^{3.2745}(\rho_B)^{1.6316}\right]}{\ln(\mu_A/\mu_B)} \tag{4.12}$$

where ρ_m, ρ_A, and ρ_B, are the densities of the mixture, component A, and component B, respectively. The value of μ_m obtained by the correlation compares very well with the experimental data, especially in the high viscosity region.

4.2.2.4 Kinetic Theory of Gases

Gases can be idealized as very small ping-pong balls bouncing around inside an enclosure. By systematically accounting for the distance that a ball must travel prior to hitting another ball (mean free path, or λ), the average molecular speed (c), and the number of balls per unit volume (n), various properties of a fluid can be derived. These properties include viscosity, thermal conductivity, diffusion, and others. Typically, these properties are called *transport* properties because they involve the movement of some quantity (momentum, heat, specie, etc.) throughout the fluid. For viscosity, these theories can be extended to a mixture of gases.

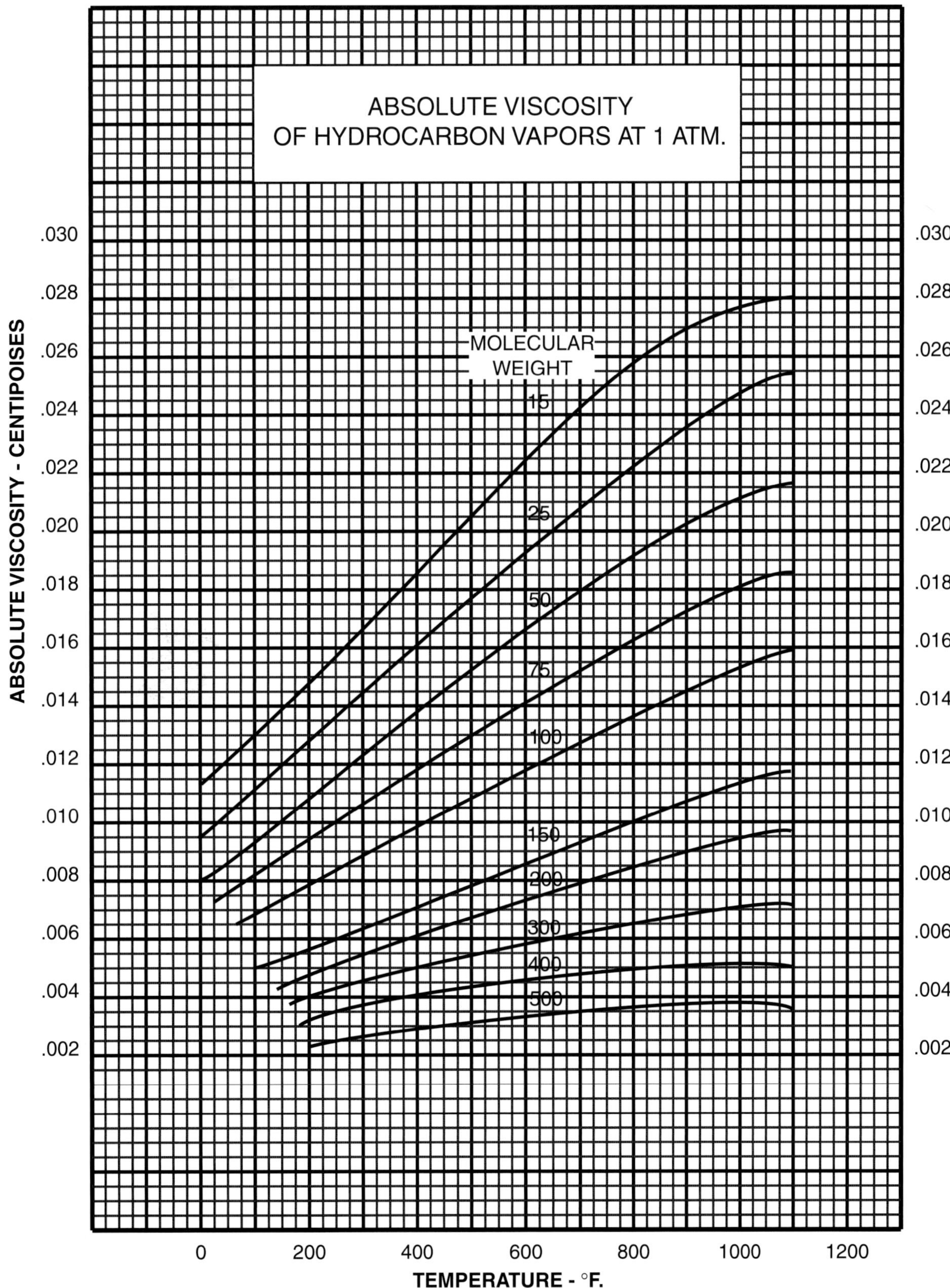

FIGURE 4.4 Temperature vs. viscosity for various hydrocarbons. (From J.B. Maxwell, *Data Book on Hydrocarbons*, D. Van Nostrand Company, Princeton, NJ, 1950.)

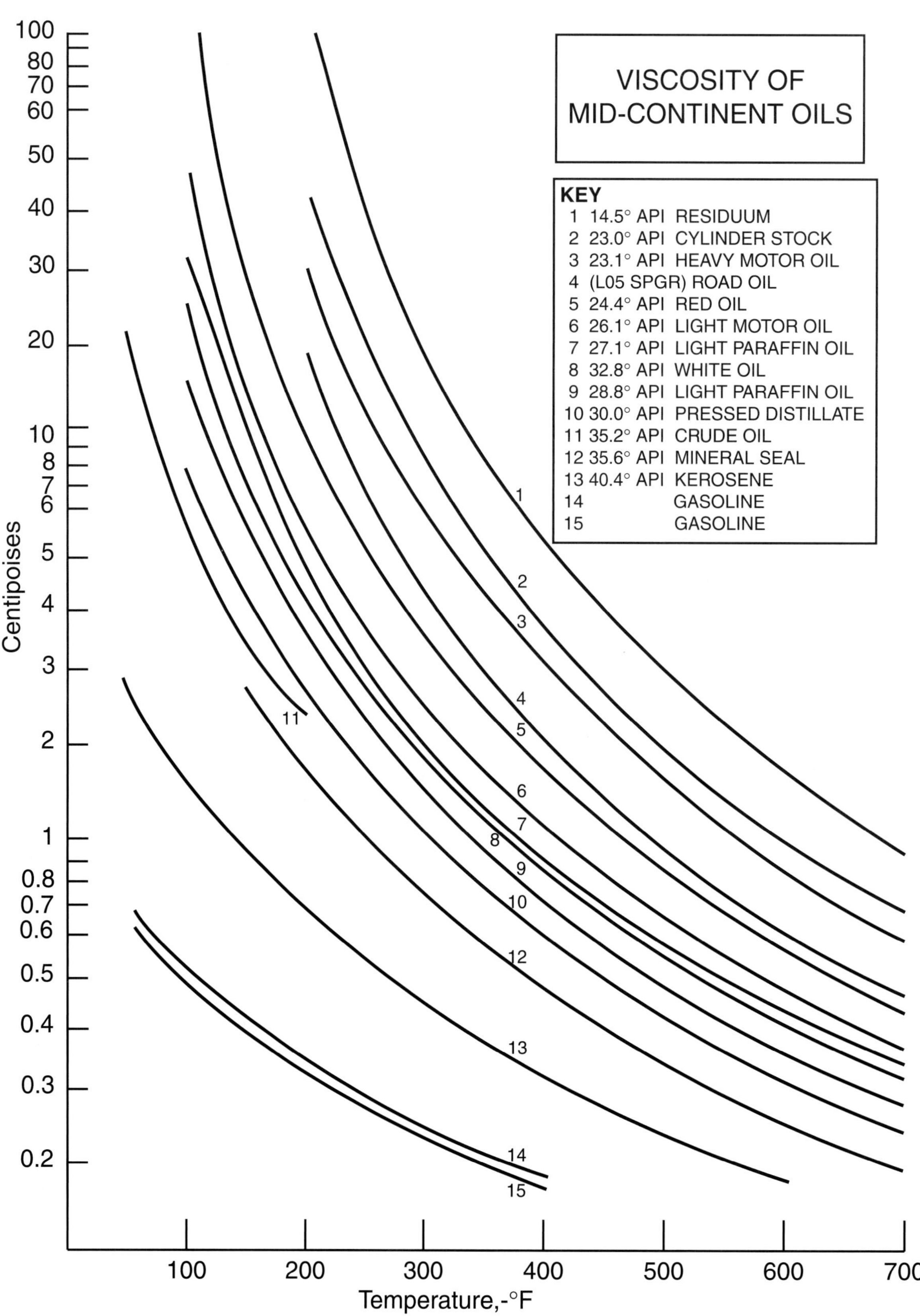

FIGURE 4.5 Viscosity of mid-continent oils. (From J.B. Maxwell, *Data Book on Hydrocarbons*, D. Van Nostrand Company, Princeton, NJ, 1950.)

Typical Data from CHEMKIN Database

```
N2                  121286N   2                 G   0300.00    5000.00    1000.00      1
   0.02926640E+02   0.14879768E-02  -0.05684760E-05  0.10097038E-09  -0.06753351E-13   2
  -0.09227977E+04   0.05980528E+02   0.03298677E+02  0.14082404E-02  -0.03963222E-04   3
   0.05641515E-07  -0.02444854E-10  -0.10208999E+04  0.03950372E+02                    4

H2O                  20387H   2O     1          G   0300.00    5000.00    1000.00      1
   0.02672145E+02   0.03056293E-01  -0.08730260E-05  0.12009964E-09  -0.06391618E-13   2
  -0.02989921E+06   0.06862817E+02   0.03386842E+02  0.03474982E-01  -0.06354696E-04   3
   0.06968581E-07  -0.02506588E-10  -0.03020811E+06  0.02590232E+02                    4

CO2                 121286C   1O     2          G   0300.00    5000.00    1000.00      1
   0.04453623E+02   0.03140168E-01  -0.12784105E-05  0.02393996E-08  -0.16690333E-13   2
  -0.04896696E+06  -0.09553959E+01   0.02275724E+02  0.09922072E-01  -0.10409113E-04   3
   0.06866686E-07  -0.02117280E-10  -0.04837314E+06  0.10188488E+02                    4
```

For most purposes, the following formula (Bird et al.[17]) can be applied to a mixture of gases:

$$\mu_{mix} = \sum (x_i \mu_i) / (x_j \phi_{ij}) \tag{4.13}$$

where

$$\phi_{ij} \sum \left[(1/8)^{0.5} \right] \left(1 + M_i / M_j \right)^{-1/2} \left[1 + \left(\mu_i / \mu_j \right)^{1/2} \left(M_j / M_i \right)^{1/4} \right]^2 \tag{4.14}$$

where i and j refer to two species.

4.2.3 *Specific Heat*

Heat capacity is defined as the heat input required to achieve a given temperature change. For gases, heat capacity is a function of the processes that the gas is undergoing. This occurs because as gases expand or compress they do pressure-volume work on the surroundings. In general, two processes are considered: constant pressure and constant volume, resulting in two different heat capacities for gases: C_v and C_p, which are defined below. In general, this is not the case for liquids and solids, which have a single heat capacity (or, for solids and liquids, $C_v = C_p$).

The specific heat for a constant volume process is defined as the change in internal energy with temperature at constant volume. Therefore, the specific heat for a constant volume process is given by:

$$C_v = (\partial u / \partial T)_v \tag{4.15}$$

where u is the internal energy and T is the gas temperature; for a constant pressure process, the specific heat is defined as the change in enthalpy with temperature at constant volume:

$$C_p = (\partial h / \partial T)_p \tag{4.16}$$

where h is the enthalpy of the gas and T is the gas temperature.

Values for the constants C_v and C_p are provided for a variety of substances in the Appendix.

4.2.3.1 Polynomial Expressions for Combustion Gases

For a calorically perfect gas, both C_v and C_p are constant with temperature. As noted, for a liquid, C_v and C_p have the same value and, in general, are fairly constant. For many applications, assuming a constant C_v and C_p is adequate. However, if large temperature variations are anticipated, or if the fluid is highly non-ideal, then temperature-dependent C_v and C_p values are desirable.

In the literature, many different expressions have been developed to allow for variation of specific heat with temperature. The NASA polynomials are recommended for any application. They are utilized in the CHEMKIN database and have a known standard form. There is an existing FORTRAN program to convert thermodynamic information into NASA polynomials, all of which are generally available.

To enhance the usefulness of this approach, a description of the coefficients (from Kee[10]) is provided. In addition, the coefficients for three of the most common gases (N_2, CO_2, and H_2O) are provided.

The NASA polynomial form was developed as part of the original NASA equilibrium program. As a result, the data supplied in the database is always in the form of four-line, representing the olden-day's four punch cards. The first line contains (1) the molecule, (2) the date of coding the information, (3) the number of different atoms, (4) the phase of the material, and then (5) the temperature range (in degrees Kelvin) of applicability of the polynomials. Most commonly, there are two ranges for the polynomials: a high-temperature range and a low-temperature range. If this is the case, the common temperature for high and low is provided. The common temperature is usually 1000 K (1300°F).

The second line contains coefficients a_1 to a_5 from the equations below. These coefficients are for the high-temperature range. The third line contains coefficients a_6 and a_7, for the high-temperature range (see below), and a_1, a_2, and a_3 for the low-temperature range. The fourth line contains coefficients a_4 to a_7 for the low-temperature range. Equations for using the polynomials are:

$$\frac{C_p}{R} = a_1 + a_2 T + a_3 T^2 + a_4 T^3 + a_5 T^4$$

$$\frac{h^0}{RT_{Units}} = a_1 + \frac{a_2}{2} T + \frac{a_3}{3} T^2 + \frac{a_4}{4} T^3 + \frac{a_5}{5} T^4 + \frac{a_6}{T}$$

$$\frac{s^0}{R} = a_1 \ln(T) + a_2 T + \frac{a_3}{2} T^2 + \frac{a_4}{3} T^3 + \frac{a_5}{4} T^4 + a_7$$

where

C_p = Specific heat at constant pressure
R = Gas constant
h^0 = Enthalpy
s^0 = Entropy
T = Temperature, in degrees K
T_{Units} = Temperature in preferred units (see below)

It should be noted that the left-hand side of the above equations is *dimensionless.* This means that the units of specific heat, enthalpy, or entropy obtained are dependent on the units of the gas constant and temperature. Temperature must be absolute temperature, either Rankine or Kelvin. Enthalpies and entropies are measured from a standard state, listed in Kee.[9] The above expression would be most useful when calculating changes in enthalpy or entropy. Finally, the expressions do not account for disassociation. Actual heat stored at elevated temperatures (above ~ 2000°F or 1100°C) must be evaluated by accounting for disassociation — either by including radical species (as in the NASA equilibrium code) or by modification of the polynomial expressions to account for the phenomena.

4.2.4 Equations of State

To accomplish many of the calculations required in the combustion field, a thorough knowledge of how density varies with pressure and temperature is required. In this section, some of the more common state equations are reviewed, along with comments on possible limitations. An exhaustive review of every type of equation of state is beyond the scope of this chapter. The interested reader is referred to Smith et al.,[10] Modell et al.,[11] or Van Wylen and Sonntag[15] for additional information on different equations of state, and additional information regarding those that have been summarized below.

4.2.4.1 Ideal Gas

The ideal gas law has the following form:

$$PV = nRT \tag{4.17}$$

where

P = Gas pressure
V = Volume under consideration
n = Number of moles
T = Temperature of gas (absolute, either Rankine or Kelvin)
R = Universal gas constant
= 8.314 kJ/(kmol-K)
= 1545 ft-lbf/(lbmole-°R)
= 1.986 Btu/(lbmole-°R)

In addition to the above form, if both sides of the equation are divided by the volume (V) term, and moles are converted to density by multiplying by molecular weight (MW), the equation becomes:

$$P = \rho\left(\frac{R}{\text{MW}}\right)T \tag{4.18}$$

The term (R/MW) is called the specific gas constant and, by inspection, is simply the universal gas constant divided by the molecular weight of the specific gas.

The ideal gas law can be derived from the kinetic theory of gases.[32] The two main assumptions that arise from the derivation are: (1) there exist no forces between molecules, and (2) all collisions are elastic. The fact that many gases frequently behave in this manner make this equation of state extremely useful. Unfortunately, gases do not always meet the two stated criteria, which gives rise to alternative equations of state discussed in the following sections.

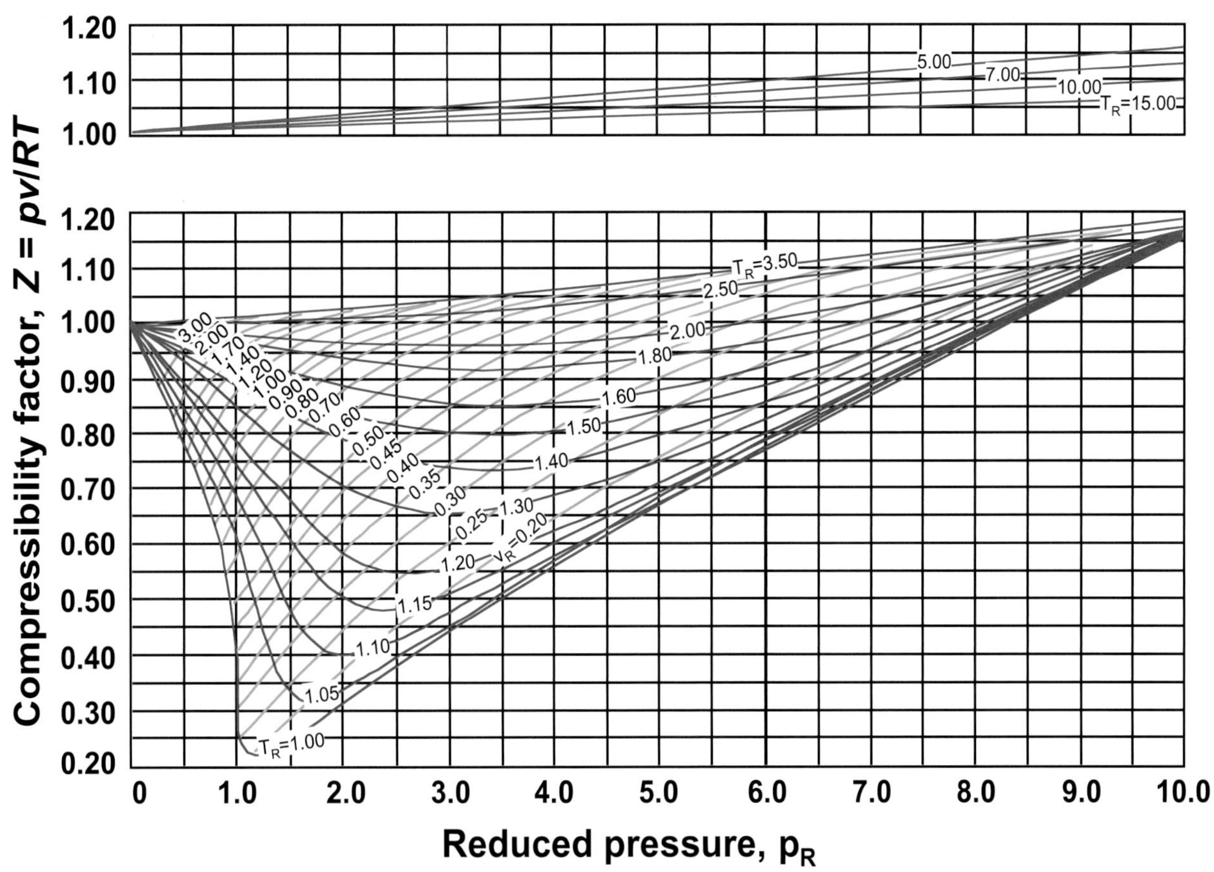

FIGURE 4.6 Compressibility factor, Z, as a function of reduced pressure and reduced temperature for different gases. (From F. Kreith, *The CRC Handbook of Mechanical Engineering*, CRC Press, Boca Raton, FL, 1998.)

4.2.4.2 Compressiblity

The ideal gas equation is very simple to use; however, it cannot be applied to many of the cases in refinery operations. The deviation of the gases from ideal behavior can accurately be accounted for by the introduction of a correction factor, called the compressibility factor, Z, as defined by:

$$Z = pv/RT \tag{4.19}$$

or

$$pv = ZRT \tag{4.20}$$

The compressibility factor can also be defined as:

$$Z = v_{\text{actual}}/v_{\text{ideal}} \tag{4.21}$$

Obviously, $Z = 1.0$ for the ideal gas; and the farther the value of Z deviates from 1.0, the more the gas deviates from ideal behavior.

All gases behave differently at a given temperature and pressure. However, their behavior has been shown to be similar when the temperature is divided, or normalized, by the critical temperature (T_c) and the pressure is normalized by the critical pressure (P_c). A critical property, in this case pressure or temperature, is the value of the stated property at the top of the vapor dome. This normalization of temperature and pressure is done by defining reduced pressure and temperature as follows:

$$P_R = P/P_{cr} \quad \text{and} \quad T_R = T/T_{cr} \tag{4.22}$$

where P_{cr} and T_{cr} are the critical pressure and temperature, respectively.

A plot of compressibility factor, Z, as a function of reduced pressure and reduced temperature for different gases is shown in Figure 4.6, which is a plot known as the generalized compressibility chart. It can be observed from this chart that at very low pressures ($P_R \ll 1$), all gases approximate ideal gas behavior. Also, at high temperatures ($T_R > 2$), ideal gas behavior can be assumed regardless of pressure, except when $P_R >> 1$. Finally, the deviation of a gas from ideal gas behavior is greatest near the critical point.

This method results in reasonably good results. Unfortunately, the functional form of the compressibility chart usually restricts this method to reading values from the chart.

4.2.4.3 Redlich-Kwong Equation

The development of cubic equations of state started from the Redlich-Kwong equation in the following form:

$$P = \frac{RT}{v-b} - \frac{a}{v(v+b)T^{0.5}} \tag{4.23}$$

where constants, a and b are determined from critical temperatures and pressures as shown:

$$a = 0.42748\frac{R^2 T_c^{2.5}}{P_c} \qquad b = 0.0866\frac{RT_c}{P_c} \tag{4.24}$$

In addition, mixing rules exist for the two constants for multiple constituent gases. From Modell and Reid,[11] they are as follows:

$$(a_m)^{1/2} = \sum_{i=1}^{n} y_i a_i^{1/2} \qquad b_m = \sum_{i=1}^{n} y_i b_i \tag{4.25}$$

Except near the critical point, the Redlich-Kwong equation will provide values that are reasonable within experimental uncertainties, making it useful for mixtures encountered in the hydrocarbon and petrochemical industries. The constants a and b can be determined from known critical temperatures, pressures, and mixture mole fractions.

4.3 FUNDAMENTAL CONCEPTS

4.3.1 *Hydrostatics*

4.3.1.1 Manometry

A manometer is an instrument that utilizes the displacement of a fluid column to evaluate pressure. Manometers can have different shapes and can be oriented at various angles, depending on the application. Two of the most common types of manometers used in the flare and burner industry are the U-tube and inclined-tube manometers.

A U-tube manometer consists of a tube shaped in the form of a U, as illustrated in Figure 4.7. In this illustration, water, in the U-tube manometer, is used to measure the pressure difference between points A and B within the pipe, as air passes through the restriction. Equations (4.26) and (4.27) relate the pressure at point a to the values of the other parameters on each side of the manometer:

$$P_a = h_1\gamma_{H_2O} + (h_3 - h_1)\gamma_{air} + P_A \tag{4.26}$$

$$P_a = h_2\gamma_{H_2O} + (h_3 - h_2)\gamma_{air} + P_B \tag{4.27}$$

where γ is the specific weight defined as $\gamma = \rho g$. Subtracting Eq. (4.27) from Eq. (4.26) gives

$$P_A - P_B = (h_2 - h_1)(\gamma_{H_2O} - \gamma_{air}) \tag{4.28}$$

Because the specific weight of air is small as compared to the specific weight of water, Equation (4.28) can be written as follows:

$$P_A - P_B = (h_2 - h_1)\gamma_{H_2O} \tag{4.29}$$

This equation states that the difference in pressure between points A and B is equal to the difference in the column height (h_2 – h_1) of water times the specific weight of water. For example, suppose that the difference in the column height of water is 1.0 foot (γ_{H_2O} = 62.4 lb/ft^3), the pressure difference between points A and B is

$$P_A - P_B = 1.0\,ft \times 62.4\frac{lb}{ft^3} = 62.4\frac{lb}{ft^2} = 0.433\frac{lb}{in.^2} \tag{4.30}$$

An inclined-tube manometer can be used to improve the accuracy of the pressure reading as compared to the vertical U-tube manometer. An inclined manometer consists of a tube oriented at a slope, as shown in the illustration in Figure 4.8. The pressure difference between points A and B can be written as Eq. (4.31):

$$P_A - P_B = (L\sin\phi)(\gamma_{H_2O}) \tag{4.31}$$

where the height of the column of air is neglected. Solving Eq. (4.31) for L gives:

$$L = \frac{P_A - P_B}{\gamma_{H_2O}\sin\phi} \tag{4.32}$$

Notice that as ϕ becomes small, the length L of the water column becomes larger for a given pressure difference between points A and B. Therefore, for relatively small angles of inclination, the differential pressure reading along the inclined tube can be made large, even for small pressure differences.

4.3.1.2 Buoyancy

When an object is placed in a fluid, it tends to float if the density of the object is less than the density of the fluid. The resultant force acting on the body is called the buoyant force. The buoyant force is equal to the weight of the fluid displaced by the object and is directed vertically upward. This phenomenon is referred to as Archimedes' principle, in honor of the Greek mathematician who first conceived of the idea. The buoyant force can be written mathematically as follows:

$$F_B = \gamma_{fluid} V \tag{4.33}$$

where F_B is the buoyant force, γ_{fluid} is the specific weight of the fluid, and V is the volume of the object. As an example, suppose a helium balloon with a volume of 62.8 ft^3 (1.78 m^3) is firmly attached to the ground as illustrated in Figure 4.9. Assume that the specific weights of the ambient air and helium are 0.0765 lb/ft^3 (1.22 kg/m^3) and 0.0106 lb/ft^3 (0.170 kg/m^3), respectively. Using Eq. (4.33), the buoyant force acting on the balloon will be $0.0765 \times 62.8 = 4.8$ lb (2.2 kg). The force acting in the upward direction at the ground, however, is not equal to 4.8 lb (2.2 kg) because the weight of the helium in the balloon creates a downward force (neglect the weight of the balloon skin). The force acting in the upward direction at the ground can be calculated by subtracting the weight of the helium from the buoyant force. For example, the weight of the helium is 0.0106 lb/ft^3 $\times$ 62.8 ft^3 = 0.666 lb. Therefore, the force acting in the upward direction at the ground will be 4.8 lb – 0.666 lb = 4.138 lb (1.9 kg). The force per unit area, or pressure force, at the location where the balloon is attached to the ground is 4.183 lb/($\pi \times 2^2/4$) ft^2 = 1.33 lb/ft^2 (6.5 kg/m^2). In summary, the force per unit area at the location where the balloon is attached to the ground can be written mathematically as:

$$P_{ground} = (\gamma_{air} - \gamma_{helium}) L_{balloon} \tag{4.34}$$

where γ_{air} and γ_{helium} are the specific weight of the atmospheric air and helium, respectively, and $L_{balloon}$ is the length of the balloon. Notice that the pressure force created at the ground is a function of the specific weight difference and the length of the balloon — not the diameter. Also notice that the pressure force at the ground increases linearly with the length

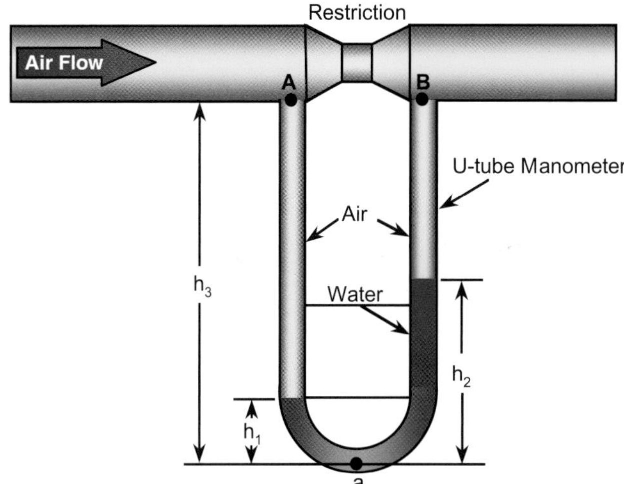

FIGURE 4.7 U-tube manometer.

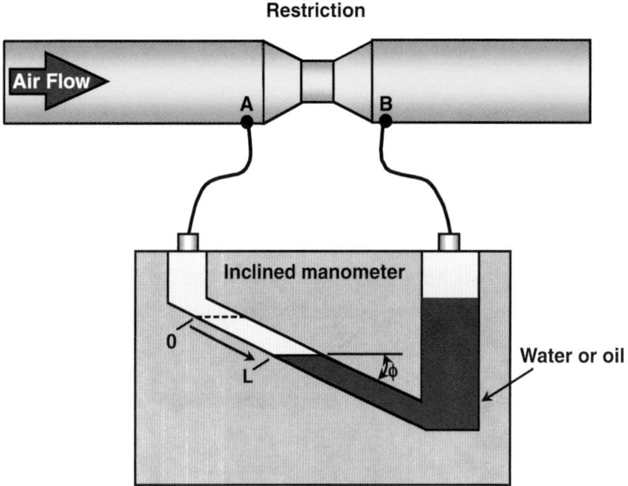

FIGURE 4.8 Inclined manometer.

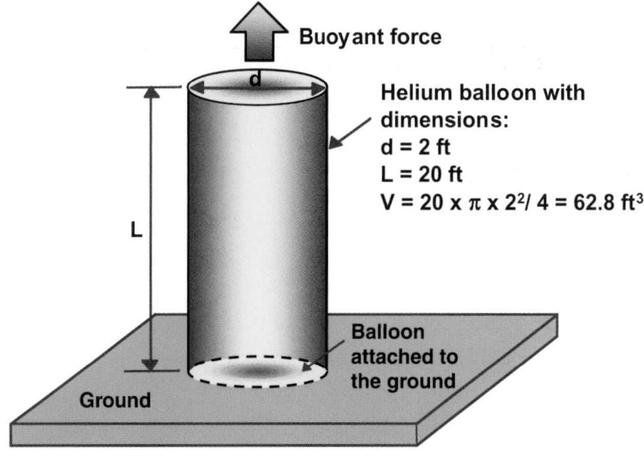

FIGURE 4.9 Helium balloon attached to the ground.

of the balloon. This example will be helpful when discussing draft (Section 4.3.1.3).

4.3.1.3 Draft

The term "draft" is commonly used to describe the pressure inside fired equipment. Draft is the pressure difference between the atmosphere and the interior of the fired equipment *at a particular elevation.* Since both the atmospheric pressure and the pressure inside the equipment vary with elevation as discussed previously, it is important to make these pressure measurements at the same elevation. For example, it is common to measure stack draft by connecting one leg of an inclined-tube manometer to a furnace and leaving the other leg open to the atmosphere. If the absolute pressure inside the furnace is less than the atmospheric pressure, then the liquid in the manometer will move. The difference in pressure is called the furnace draft. Typically, draft levels are measured in inches of water column pressure, generally written as "w.c." When recording draft, it is not necessary to place a minus sign in front of the numeric value. Customarily, it is understood that draft pressures are always negative values.

Draft is established in a furnace because the hot flue gases in the vertical stack have a lower specific weight than the outside air. The difference in specific weight of the flue gas and air creates a buoyant force, causing the flues gases to float vertically upward. The buoyant force is equal to the product of the difference in specific weights and the height of the stack, and can be written mathematically as:

$$P_{draft} = \left(\gamma_{\substack{atmospheric \\ air}} - \gamma_{\substack{hot \\ flue \\ gas}} \right) H_{stack} \tag{4.35}$$

where P_{draft} is the draft pressure, H_{stack} is the stack height, and $\gamma_{atmospheric\ air}$ and $\gamma_{hot\ flue\ gas}$ represent the specific weights of the ambient atmospheric air and the hot flue gas, respectively. Notice that this equation is very similar to the equation used to calculate the pressure at the ground in the helium balloon example discussed in the previous section.

As just shown, the draft created by a column of lighter-than-air gas is function of the specific weight difference between the atmospheric air and flue gas, as well as the height of the stack. The specific weight of the flue gas is a function of both the temperature of the gas in the stack and the composition. An increase in exhaust flue gas temperature decreases the specific weight, causing an increase in the draft. The specific weight of a hot flue gas can be calculated as follows:

$$\gamma_{\substack{hot \\ flue \\ gas}} = \gamma_{\substack{flue \\ gas \\ ST}} \left(\frac{520}{460 + T_{\substack{hot \\ flue \\ gas}} \ (°F)} \right) \tag{4.36}$$

where the subscript *flue gas ST* represents the specific weight of the flue gas at the standard temperature of 60°F (16°C), and the subscript *hot flue gas* represents the hot flue gas at the actual temperature.

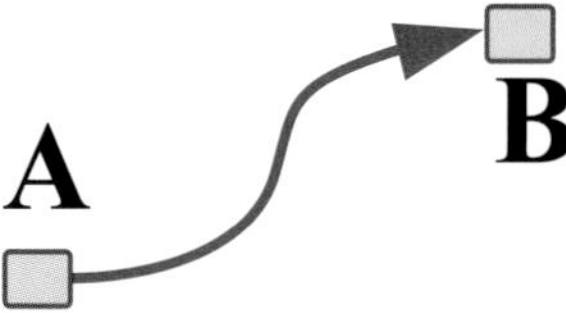

FIGURE 4.10 A small packet of fluid flows from point A to point B along an arbitrary path.

4.3.2 Bernoulli Equation

An equation that has proved useful in the calculation of flows for the combustion industry is the Bernoulli equation. In Figure 4.10, a small packet of fluid flows from point A to point B along an arbitrary path. Neglecting friction (i.e., there is negligible shear stress), the conservation of energy analysis yields the following:

(Change in pressure energy) + (Change in velocity energy) + (Change in potential energy) = 0

or

$$\Delta P + \Delta\left(\frac{1}{2g_c} \rho V^2 \right) + \Delta(\rho g h) = 0 \tag{4.37}$$

where

P = Pressure at fluid packet
ρ = Density of fluid packet
V = Velocity of fluid packet
g = Gravitational acceleration
h = Height of fluid packet
g_c = Gravitational constant

4.3.2.1 Total Pressure vs. Static Pressure

Often, the potential energy term can be neglected in the above equation, which results in just the pressure term and the velocity term. The combination of these two terms is often referred to as the total pressure. Pressure drop analysis in piping systems, discussed in Section 4.5, exclusively examines the total pressure loss of a system in order to prevent

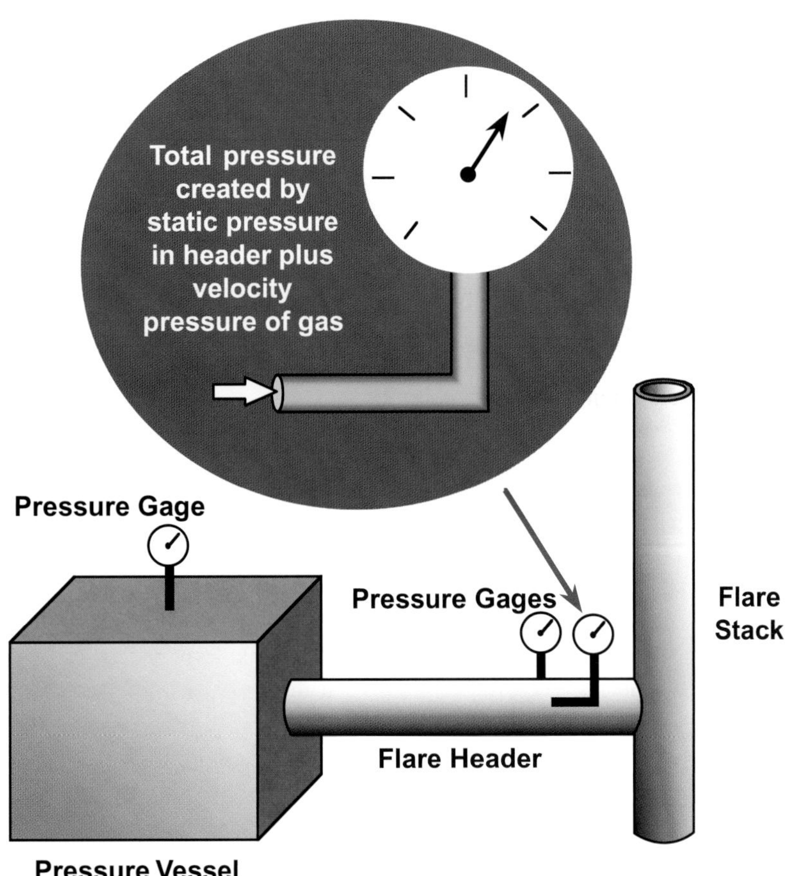

FIGURE 4.11 Pressure relief vessel venting to a flare.

errors from occurring. For example, it is known that there is an energy loss in a piping system if the pipe size increases, either suddenly or gradually. Despite the energy loss, the static pressure tends to increase if the pipe size is increased. According to the Bernoulli equation, as velocity is decreased, the pressure is expected to increase. This being the case, piping pressure drop analysis would be a meaningless term if only static pressure was included. Any analysis must include the total energy or total pressure of the system. Using total energy and total pressure, the following terms are defined:

Static pressure: the pressure term in the Bernoulli equation. This is normally understood as a pressure and is measured directly by a pressure gage.

Dynamic pressure: the velocity term ($^1/_2\rho V^2$) from the Bernoulli equation.

Total pressure: the sum of static and dynamic pressures. This term is utilized for most pressure loss analyses in a piping system. When liquids are discharged to the atmosphere, this term would also include potential energy changes.

The following is a simple example to illustrate the differences between these pressure concepts. Suppose there is a pressure relief vessel that is venting to a flare system as illustrated in Figure 4.11. Assume that the pressure relief vessel is maintained at 1 psig during the flaring event. As the gas exits the vessel, it enters the flare header. If a pressure gage were placed on the pressure vessel and on the flare header somewhere downstream of the vessel, these gages would not read the same pressure. At the vessel, the pressure would obviously read 1 psig (7 kPag); however, the pressure gage on the flare header would read less than 1 psig (< 7 kPag). The concepts of static, velocity, and total pressure are used to explain why the pressures are different.

The total pressure is defined as the static pressure plus the velocity pressure:

$$P_T = P_S + P_V \tag{4.38}$$

where P_T is the total pressure, P_S is the static pressure, and P_V is the velocity pressure. The static pressure is a measure of the pressure where the gas velocity is zero. Therefore, the static pressure of the vessel is obviously 1 psig (7 kPag). Because the velocity of the gas inside the vessel is zero, the velocity pressure must also be zero. From Eq. (4.38), the total pressure of the vessel must be equal to the static pressure of the vessel. The pressure gage in the flare header reads the pressure at the wall of the pipe where the gas velocity is zero; therefore, this pressure gage must be reading the static pressure in the flare header. Because the gas is moving in the pipe, the velocity pressure must be greater than zero. The velocity pressure in the flare header can be analyzed in two ways.

First, assume that the pressure energy losses from the vessel to the pressure gage on the flare header are zero. If no energy losses are assumed, then the total pressure on the flare header must be equal to the total pressure at the pressure vessel. It is also true that the pressure gage on the flare header must be reading the static pressure at that location. Therefore, from Eq. (4.38), the velocity pressure in the flare header must be equal to the total pressure minus the static pressure. For example, if the pressure gage on the flare header reads 0.8 psig (6 kPag) and the total pressure is 1.0 psig (7 kPag), then the velocity pressure equals 0.2 psig (1 kPag).

In reality, there will be pressure energy losses in the flare header due to friction between gas molecules and pipe wall. Therefore, the total pressure in the header will not equal the total pressure in the pressure vessel. One can measure the total pressure in the flare header by inserting a pitot tube into the header, as illustrated in the insert of Figure 4.11. The pitot tube will measure the static pressure plus the impact pressure, or velocity pressure, of the gas. This is a measure of the total pressure inside the flare header. The velocity pressure in the flare header can now be determined by subtracting the total pressure reading from the static pressure reading. For example, if the total pressure reads 0.9 psig (6 kPag) in the header and the static pressure reads 0.7 psig (5 kPag) in the header, then the velocity pressure in the header must be equal to 0.2 psig (1 kPag).

As just mentioned, the velocity pressure is the pressure created due to the impact of the gas molecules. The velocity pressure in a gas stream can be calculated as follows:

$$P_V = \frac{\rho V^2}{2g_c} \tag{4.39}$$

where ρ is the density of the flowing gas, V is the velocity, and g_c is the gravitational constant. Equation (4.39) is used extensively when determining the pressure drop through flare and burner systems.

4.3.2.1.1 "K" Factors

As implied in the previous section, the Bernoulli equation is logically utilized to analyze total pressure losses (or energy losses) for piping systems. To accomplish the analysis, the equation is modified as follows:

$$\Delta P + \Delta\left(\frac{1}{2g_c}\rho V^2\right) = \text{Total pressure loss} = K\left(\frac{1}{2g_c}\rho V^2\right)$$
$$= K(\text{Dynamic pressure}) = \text{KP}_v \tag{4.40}$$

The two assumptions utilized to derive the Bernoulli equation were: (1) flow along a stream line, and (2) no friction or shear stresses. Piping systems satisfy the first assumption, but do not satisfy the second. Shear stresses arise when a fluid flows over or past a solid object, as discussed in Section 4.2.2 of this book. Experience has shown (see Vennard and Street[4]) that the magnitude of the force will be proportional to the dynamic pressure term ($^1/_2\rho V^2$). The constant of proportionality commonly utilized is "*K*". Values of "*K*" for different piping configurations and different fittings have been extensively studied and reported in the Crane Piping manual[31] and in Idelchik.[12] In the literature, these constants are often referred to as "*K*" factors.

4.3.3 Control Volumes

A common analysis methodology is the utilization of control volumes — also termed the integral method. Conceptually, the integral method entails enclosing the region of interest with control surfaces, then writing the appropriate conservation equations for the enclosure. The method is discussed extensively in Fox and McDonald.[3] The interested reader is directed to that reference for additional information.

In general, there are two types of physical quantities of interest: extensive and intensive. For a given volume, extensive properties describe the entire volume. They include:

Total mass	=	M
Momentum	=	$M\vec{V}$
Angular momentum	=	$\vec{\Omega}$
Enthalpy	=	H
Energy	=	E
Entropy	=	S

Intensive properties are properties per unit mass. As a practical matter for systems of interest, an intensive property is determined by dividing the extensive property by the mass, or density times the volume, of the control volume. For mass, energy, and momentum conservation, the intensive variables associated with the appropriate extensive variables are:

Mass conservation : $M \Rightarrow 1$

Momentum conservation : $\Rightarrow \vec{V}$

Energy conservation : $E \Rightarrow e$ (enthalpy for flow applications)

If a control volume is drawn around an area of interest, at any given instant in time the volume is filled with a certain amount of mass, the control mass. At this instant in time, the control mass has the exact same shape as the control volume. The following conservation laws are valid for the control mass, which is coincident with the volume for the instant in time as indicated by:

$$\frac{d\text{M}}{dt} = 0 \tag{4.41}$$

4.3.3.1 Mass Conservation

Because mass is neither created nor destroyed, the total control mass will remain constant. This may seem confusing because there is a mass flux through the volume and the density of the volume may change with time. Remember that the mass of interest is the mass that happens to inhabit the control volume at a particular instant in time. During the next instant, while the mass just examined may not be in the volume, or even have the same shape as the volume, it still has the same mass. This makes the conservation equation simplistic. For a given piece of material (or mass), the total mass does not change with time. Without further discussion of this point, the other conservation equations are reproduced for the given piece of material (or mass):

4.3.3.2 Momentum Conservation

$$\frac{d\left(\text{M}\vec{V}\right)}{dt} = \sum_i \vec{F}_i \tag{4.42}$$

The time rate of change of momentum for the piece of mass is equal to the sum of the forces applied to the mass.

4.3.3.3 Energy Conservation

$$\frac{d(\text{E})}{dt} = \delta Q + \delta W \tag{4.43}$$

The time rate of change of energy for the piece of mass is equal to the sum of the heat transferred (Q) and work (W).

Conceptually, the above conservation laws are reasonably easy to understand, but very difficult to apply. To accomplish an analysis, it would be necessary to track and maintain information about all the possible pieces of mass that happen to flow through a control volume. The fundamental precept of utilizing control volumes for analysis is that it is possible to determine the time rate of change for a given extensive property by determining: (1) the time rate of change in the control volume of the corresponding *intensive* property, and (2) the flux in and out of the control volume of the same intensive property. If an arbitrary extensive property is denoted as N, and its corresponding intensive property is denoted as n, then, according to Fox and McDonald,[3] the time rate of change for **any** extensive property is given by:

$$\left(\frac{d\text{N}}{dt}\right)_{PieceofMass} = \frac{d}{dt}\int_{CV} \text{n}\rho dV + \int_{CS} \text{n}\rho\vec{V}\cdot d\vec{A} \tag{4.44}$$

An example of how this is applied is provided, but first some explanations of the meaning of the terms are in order:

$$\left(\frac{d\text{N}}{dt}\right)_{PieceofMass} \tag{4.45}$$

Equation (4.45), as discussed, is the total rate of change of any arbitrary extensive property of the piece of mass coincident with the control volume.

$$\frac{d}{dt}\int_{CV} \text{n}\rho dV \tag{4.46}$$

Equation (4.46) is the time rate of change of the arbitrary extensive property N within the control volume. This is expressed as a product of the associated intensive property n, the density, and the volume (dV). The CV on the integral represents the control volume.

$$\int_{CS} \text{n}\rho\vec{V}\cdot d\vec{A} \tag{4.47}$$

Equation (4.47) is the net flux of the extensive property N through the boundary of the control volume. The product of density, velocity, and area is the mass flow rate. Thus, the equation really represents the mass flow in and out of the control volume times the associated intensive property.

The boundary is called the control surface, or CS. The vector dot product of the velocity and the area signifies that only the velocity component normal or perpendicular to the surface can be used for calculation purposes. An example of how this is used practically can be seen from the conservation of mass as given by:

$$\frac{dM}{dt} = 0 \tag{4.48}$$

The associated intensive property for mass is simply one. Inserting the above information into the general control volume statement yields:

$$0 = \frac{d}{dt}\int_{CV} \rho dV + \int_{CS} \rho \vec{V} \cdot d\vec{A} \tag{4.49}$$

Notice the first term in Eq. (4.49). Practically speaking, this term takes into account the change in density of the control volume. For steady-state and incompressible analysis, this term becomes zero. The second term is just a statement of mass in and mass out. Thus, for steady-state and incompressible calculations, the equation reduces to:

$$(\text{Mass in})_{\text{Control Volume}} = (\text{Mass out})_{\text{Control Volume}}$$

For constant velocity across any opening, the equation becomes:

$$(\rho AV)_1 = (\rho AV)_2 \tag{4.50}$$

Equation (4.50) is the standard form for calculating mass balances. Additional examples of the method are found in Fox and McDonald[3] and Potter and Foss.[13] Application of the method to eductors is provided in Chapter 8 of this book. In addition to fixed control volumes, it is possible to construct control volumes such that all or part of the surfaces move. Required modifications are outlined in both Panton[1] and Potter and Foss.[14]

4.3.4 *Differential Formulation*

All of the previous fluid flow analysis required various simplifications and assumptions. The most fundamental formulation of fluid phenomena is the differential formulation. Numerous excellent references exist (e.g., Panton,[1] Potter and Foss,[13] and Kuo[14]) for the reader interested in greater detail on this subject. This treatment concentrates on the conceptual development of the formulation and the final form. It is the intent of this reference to provide insight into the meaning of the equations, how simplifications have been applied to them, and the limitations of the simplifications.

On a conceptual level, the equations of interest are partial differential equations that describe:

1. conservation of mass
2. conservation of momentum
3. conservation of energy
4. conservation of species

Similar to density, the main assumption regarding the differential derivation of these equations is that the volume reduces to a point, but a point much larger than the molecular length scale. Each of the conservation principles are discussed in turn.

Because it is fairly simple, a reasonably complete derivation of conservation of mass is provided. Insight gained from the conservation of mass derivation will be generalized as appropriate for the other conservation equations. Details will be referenced, as appropriate.

4.3.4.1 Conservation of Mass

Figure 4.12 is an idealization of a small "differential" control volume. For simplicity, Cartesian coordinates (X, Y, Z) are utilized; however, the results could be generalized to any other system. The little "cube" has dimensions of dx by dy by dz, and thus the area of any face is the same (i.e., $dx \cdot dy$), and the volume is $dx \cdot dy \cdot dz$. Reference will be made to these observations in all of the following sections.

If this small control volume is placed at a fixed location in a fluid flow field, then fluid will flow into and out of the volume continuously. In other words, conservation of mass states that the amount of mass going into the volume will equal the amount of mass leaving the volume plus any mass stored in the volume. The mass flow across any face into the volume is going to be the area of the face, times the velocity, times the density, or in equation form:

Mass flow through a face
= (Velocity component)·(Face area)·(Density)

From Section 4.2.1, it is observed that the differential volume is chosen so that the density is uniform throughout, but is not so small that molecular spacing is sparse. Choosing the X-direction for analysis, as shown in Figure 4.13, results in the following:

$$(\text{Mass flow})_X = U(dy \cdot dz)(\rho) \quad (4.51)$$

where

U = Velocity in the X-direction

ρ = Density

For conservation of mass analysis, however, the point of interest is ensuring that the mass entering, leaving, and being stored are the same. The real question that the equation must answer is: "What is the difference between mass entering and exiting in the X-direction?" or:

$$\begin{aligned}&(\text{Mass entering})_X - (\text{Mass leaving})_X \\ &= U\rho(dy\, dz) - (U\rho + d(U\rho)(dy\, dz)) \\ &= -d(U\rho)(dy\, dz) \quad (4.52)\end{aligned}$$

In Eq. (4.52), dU is simply the change in X-direction velocity that occurred over the distance dx, so the change in mass flow rate over distance dx is $d(U\rho)$.

The mass of a cube is the volume times the density. In Eq. (4.52), the rate at which mass is entering or leaving the cube is represented. This being the case, the conservation of mass for this system will be a rate equation, that is, how much has the mass changed with time. This is expressed as:

$$\begin{aligned}&\text{Mass stored/time} \\ &= \{(\text{Mass at time 1}) - (\text{Mass at time 2})\}/\text{time} \\ &= \{\rho_1(dx\, dy\, dz) - \rho_2(dx\, dy\, dz)\}/dt \\ &= (dx\, dy\, dz)\frac{d\rho}{dt} \quad (4.53)\end{aligned}$$

where

$d\rho$ = Change in density with time

dt = Differential change in time

$dx\, dy\, dz$ = Volume of cube

The above expression is very reasonable. The only way that mass can be stored in a fixed-volume container of any sort is for the density to change. Repeating the mass flow analysis for each of the other two directions, collecting the expressions, and setting them equal to the expression for the change in mass, results in the following general expression for conservation of mass for fluid flow:

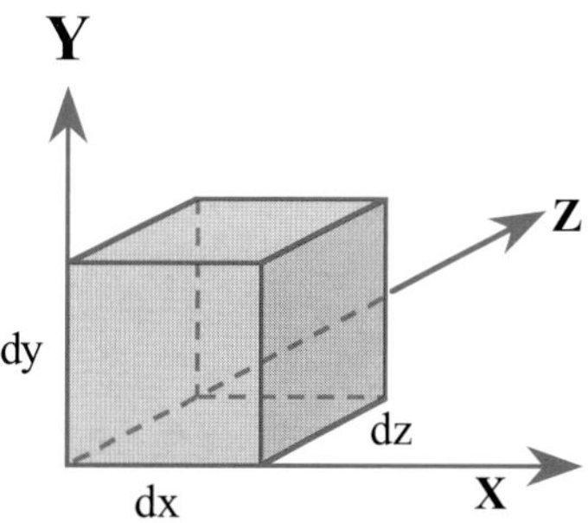

FIGURE 4.12 An idealization of a small "differential" control volume.

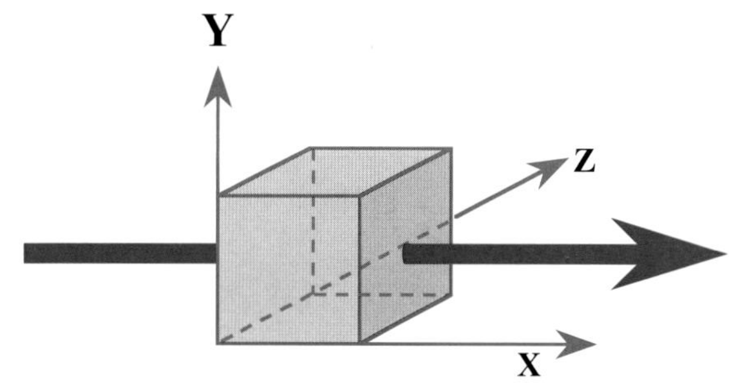

FIGURE 4.13 Mass flow into and out of volume in the X-direction.

$$\frac{d\rho}{dt}(dx\, dy\, dz) = -d(\rho U)(dy\, dz) - d(\rho V)(dx\, dz) - d(\rho W)(dx\, dy) \quad (4.54)$$

Dividing Eq. (4.54) by $(dx\, dy\, dz)$ and collecting terms on one side of the equation:

$$\frac{\partial(\rho)}{\partial t} + \frac{\partial(\rho U)}{\partial x} + \frac{\partial(\rho V)}{\partial y} + \frac{\partial(\rho W)}{\partial z} = 0 \quad (4.55)$$

where

U = X velocity component

V = Y velocity component

W = Z velocity component

The above expression has the following physical implications:

1. The time-dependent term $\frac{d\rho}{dt}$ represents the change of mass inside the volume.
2. The other terms represent the difference between what leaves and enters the volume.
3. This is summarized as:

$$\underbrace{\frac{\partial(\rho)}{\partial t}}_{\text{Internal changes}} + \underbrace{\frac{\partial(\rho U)}{\partial x} + \frac{\partial(\rho V)}{\partial y} + \frac{\partial(\rho W)}{\partial z}}_{\text{Mass difference between entering and leaving volume}} = 0 \tag{4.56}$$

For incompressible flow, this reduces to:

$$\frac{\partial(\rho)}{\partial t} + U\frac{\partial(\rho)}{\partial x} + V\frac{\partial(\rho)}{\partial y} + W\frac{\partial(\rho)}{\partial z} = 0$$

As it turns out, the general form of Eq. (4.56) can be used for any physical quantity of interest: momentum, energy, chemical species, etc. In general, this form of equation is said to represent the time variant and convection terms of the differential equations. In fact, this form appears so often that it is given the special name of "substantial derivative" and represented in shorthand as follows:

$$\frac{\mathrm{D}(\,)}{\mathrm{D}t} = \frac{\partial(\,)}{\partial t} + \frac{U\partial(\,)}{\partial x} + \frac{V\partial(\,)}{\partial y} + \frac{W\partial(\,)}{\partial z} \tag{4.57}$$

where different physical parameters can be substituted into the blank sets of parentheses. This would allow the conservation of mass for incompressible flow to be written as:

$$\frac{\mathrm{D}(\rho)}{\mathrm{D}t} = 0 \tag{4.58}$$

Use of the substantial derivative will be routinely referenced in the following sections.

4.3.4.2 Conservation of Momentum

Newton's second law of motion stated in words is: "the time rate of change of momentum is equal to the sum of all applied forces." In equation form, this is:

$$\frac{d(Momentum)}{dt} = \sum Applied\,forces \tag{4.59}$$

Instead of change of momentum, many physics textbooks refer to mass times acceleration. The two statements are equivalent for volumes with constant density; but for fluid considerations, changes in momentum are more appropriate. Momentum is a vector quantity with three components, so any equation derived for one direction can be generalized into three similar equations, one for each component (x, y, z directions). For simplicity, only the X-direction is considered here.

The rate of momentum change is represented as: (1) the difference between the momentum entering the volume and the momentum leaving the volume, and (2) the momentum stored. It can be shown that this is the same type of convection and storage term derived for conservation of mass, so the substantial derivative can be used to describe a change in momentum. For momentum, velocity times density is utilized; thus:

$$\frac{d(Momentum)}{dt} = \frac{\mathrm{D}(\rho U)}{\mathrm{D}t} \tag{4.60}$$

Forces that can be applied fall into two categories: (1) forces that are applied to the entire volume equally, and (2) forces that are applied to the surface of the volume. The first type of force is sometimes called a body force. Gravity is normally the only body force encountered, and will be the only one considered herein. Others, however, are possible, such as electromagnetic fields and acceleration. The second type of force, surface forces, can act either in a direction normal to the surface (pressure), or in a direction parallel to the surface (shear stress). The analytical form of each of these forces is summarized below.

Gravitational Body Force: The gravitational body force is illustrated in Figure 4.14 and is symbolized by F_x. If there is no body force in the X-direction, then this term is ignored. Common practice is to consider the Y-direction as up and down, so the gravitational body force term usually would only apply to the Y equation.

Normal or Pressure Forces: Normal or pressure forces are illustrated in Figure 4.15 and take the following form:

$$-\frac{P_x}{dx} \tag{4.61}$$

A net force will arise only if there is a difference in the direction of interest. The negative sign occurs because higher pressures in the direction of interest will induce a flow in the negative direction.

Parallel or Shear Stress: Parallel or shear stress is illustrated in Figure 4.16. For a Newtonian fluid, shear stress or force will be proportional to the velocity gradient (see Section 4.4.1.1). The force due to shear stress along the Y-direction on the X-plane is:

$$\tau = \mu\frac{dV}{dy} \tag{4.62}$$

However, what is of interest is the change in force (the force in this case being shear stress) along any one direction.

This change in force formulation gives rise to terms that look like:

$$\mu \frac{d^2V}{dy^2} \tag{4.63}$$

Collecting all of the above terms results in the following.

X-direction momentum equation:

$$\underbrace{\frac{D(\rho U)}{Dt}}_{\substack{\text{Momentum}\\ \text{changes}}} = \underbrace{\underset{\substack{\text{Body}\\ \text{force}}}{F_x} + \underset{\substack{\text{Pressure}\\ \text{force}}}{\frac{-\partial P}{\partial x}} + \underset{\substack{\text{Shear forces}\\ \text{(stresses)}}}{\mu\left(\frac{\partial^2 U}{\partial x^2} + \frac{\partial^2 V}{\partial y^2} + \frac{\partial^2 W}{\partial z^2}\right)}}_{\text{Summation of forces on volume}} \tag{4.64}$$

In a similar manner, equations for the *Y* and *Z* momentum can be derived. The analytical form of the above equation may at first be quite intimidating. However, it is nothing more than a logical extension of Newton's second law of motion from high school physics ($F = ma$), which has been applied to fluid flows. Strictly speaking, the above equation is only valid for incompressible flow. The interested reader is directed to previously listed references for additional details and an extension of the momentum equation to include compressibility effects.

4.3.4.3 Conservation of Energy

The conservation of energy equation derivation is similar to mass and momentum. First, start with a general conservation principle, in this case the first law of thermodynamics, as discussed in Van Wylen:[15]

$$\begin{aligned}&(\text{Heat transferred}) + (\text{Energy in}) + (\text{Energy generated})\\ &\quad = (\text{Energy stored}) + (\text{Energy out}) + (\text{Work})\end{aligned} \tag{4.65}$$

or, moving terms around:

$$\begin{aligned}&(\text{Energy stored}) + (\text{Energy out}) - (\text{Energy in})\\ &= (\text{Heat transferred}) + (\text{Energy generated}) - (\text{Work})\end{aligned} \tag{4.66}$$

The term on the left will again be the substantial derivative of energy into and out of the volume. Depending on the processes involved, terms on the right will vary, but may include all or none of the following:

1. convective heat transfer (heat transfer)
2. conductive heat transfer (heat transfer)
3. radiative heat transfer (heat transfer)
4. chemical heat release (energy generated)

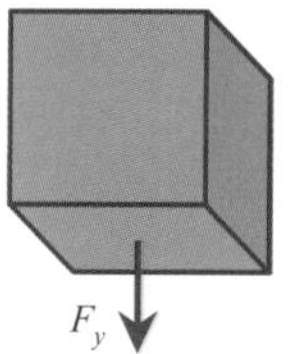

FIGURE 4.14 Gravitational body force.

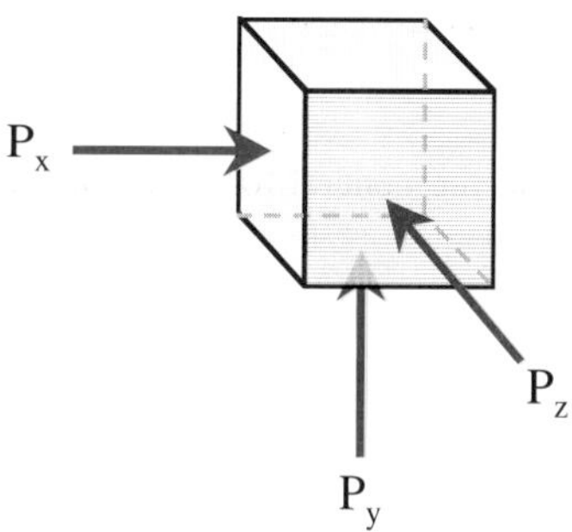

FIGURE 4.15 Normal or pressure forces.

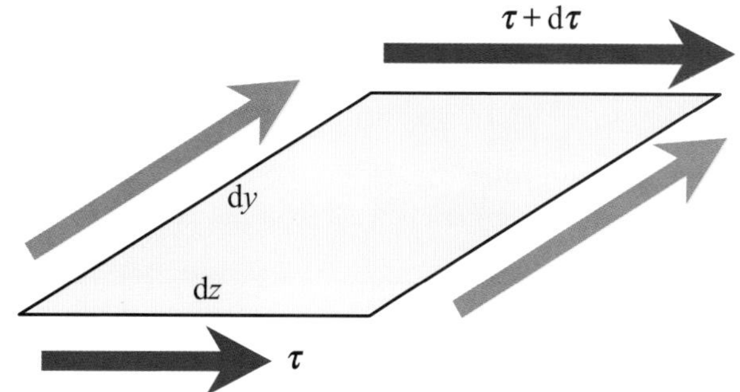

FIGURE 4.16 Effect of shear stress on *X*-direction face.

5. expansion (work)
6. wall friction (work)
7. viscous dissipation (work)

In general, for combustion in the petrochemical industries, the common terms utilized are convective and radiative heat transfer, chemical heat release, and wall friction. Inclusion of multiple chemical species complicates the formulation, as molecular diffusion of the various species provides an energy transport mechanism. Furthermore, the energy in a flowing fluid can be separated to consider thermal and kinetic energy separately. In low Mach-number flows (i.e., incompressible flows), conservation of kinetic energy is automatically satisfied whenever the momentum equation is satisfied (the conservation of kinetic energy equation can be derived by manipulating the momentum equation). The following (thermal) energy equation contains the terms that would normally be important in a combusting flow (numbers correspond to terms described above):

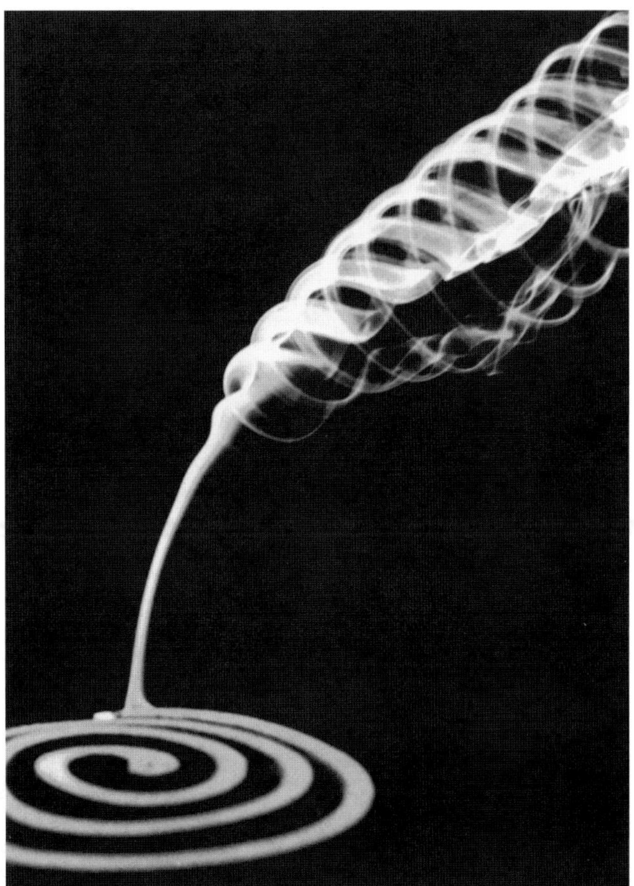

FIGURE 4.17 Smoke from incense. (From Visualization Society of Japan, *The Fantasy of Flow*, IOS Press, Amsterdam, 1993, 96.)

$$\underbrace{\rho C_P \frac{\mathrm{D}T}{\mathrm{D}t}}_{\text{Energy change}} = \underbrace{\nabla \cdot k\nabla T}_{\text{Molecular conduction}} - \underbrace{\sum_i C_{pi} Y_i \vec{\mathbf{V}}_i / C_p \cdot \nabla T}_{\text{Thermal diffusion}}$$
$$- \underbrace{\sum_i \omega_i h_{oi}}_{\text{Chemical reaction}} + \underbrace{\beta T \frac{\mathrm{D}P}{\mathrm{D}t}}_{\text{Pressure work}} + \underbrace{\Phi}_{\text{Viscous dissipation}} + \underbrace{\nabla \cdot \vec{\mathbf{q}}_r}_{\text{Thermal radiation}} \tag{4.67}$$

The reader should consult Bird et al.[16] for a much more complete discussion of energy conservation.

4.3.4.4 Conservation of Species

If there are no chemical reactions, then the implication is that the mass fraction of any particular species can only change due to mass flow into and out of the volume. Under this constraint, conservation of species becomes exactly like the conservation of mass equation, namely:

$$\frac{\mathrm{D(Chemical\ species)}}{\mathrm{D}t} = 0 \tag{4.68}$$

If reactions are occurring, then the change will be equal to the formation (or destruction) rate of the compound. Reaction rate equations are normally of the Arrhenius type (see Kuo[14]). This results in the conservation of species equation being:

$$\rho \frac{DY_i}{Dt} = \omega_i W_i - \mathbf{V}_i \cdot \nabla Y_i \tag{4.69}$$

4.4 DIFFERENT TYPES OF FLOW

4.4.1 Turbulent and Laminar Flow

For centuries, scientists have been fascinated with flows in nature. Visual observations of fluids in motion have resulted in two broad classifications of flows: laminar and turbulent. Laminar flows have no fluctuations and tend to flow for long distances with very little mixing occurring in the flow. Figures 4.17 and 4.18 illustrate two common examples of laminar flow. In Figure 4.17, smoke from burning incense slowly rises in still air for a distance of 6 to 12 in. (15 to 30 cm) before becoming unsteady. In Figure 4.18, water exits a water faucet at a very low velocity and falls in a straight line until it strikes an object or is disrupted by air currents. In both cases, the regions of smooth movement are examples of laminar flows.

Turbulent flows have fluctuations and eddies associated with them that increase mixing rates substantially. The legendary artist and scientist Leonardo da Vinci's view on turbulence is reproduced in Figure 4.19. As shown, the circulation, induced vortices and fluctuations allow, in this case, an introduced stream to mix rapidly with a still liquid. Most flows in nature will tend to become turbulent as flow rate or local velocity is increased.

4.4.1.1 Reynold's Number

During the 1880s, Osborn Reynolds quantified the previous qualitative observations of laminar and turbulent flows. His observations were first published in 1883.[17] A schematic of his experimental apparatus is shown in Figure 4.20. In this device, water flows from a tank through a bell-mouthed inlet into a glass pipe. Also, at the entrance to the glass pipe is the outlet of a small tube, which leads to a reservoir of dye.

Reynolds discovered that at low water velocities, the stream of dye issuing from the thin tube did not mix with the water. Instead, the dye became a distinct flow, parallel to the pipe centerline. As the valve was opened and the water velocity increased, it was observed that at some greater velocity, the dye would rapidly mix with the water, causing the entire flow to be colored with the dye. Reynolds deduced from these experiments that two flow regimes existed. In the first (laminar flow), the fluid streams flow past each other in parallel layers, or laminae, but do not mix with each other. The second regime, at higher velocities, is where the two streams mix rapidly. This mixing is caused by vortices, illustrated in Figures 4.21 and 4.22, that arise in the flow. Due to the presence of the vortices, this flow has been called turbulent. Figure 4.23 illustrates laminar flow.

Reynolds found that below a certain velocity, the flow always became laminar. Figure 4.23 shows laminar flow can even be achieved by flow over a cube. As the velocity was increased, turbulent flow could always be achieved, but the precise velocity depended on how still the water in the tank was prior to the experiment. In addition, Reynolds was able to generalize his results into a nondimensional parameter, which today is known as the Reynolds number, Re. It is defined as follows:

$$\mathrm{Re} = \frac{Vd\rho}{\mu} \tag{4.70}$$

where

V = Velocity in pipe
d = Pipe diameter
ρ = Density of fluid
μ = Fluid viscosity

As a practical matter, pipe flows having a Reynolds number less than 2300 are laminar, and flows having a Reynolds number greater than 4000 are turbulent. In addition to determining type of flow, Reynolds numbers have been proven to also scale the intensity of turbulence in a flow. That is, higher Reynolds numbers result in greater vortex generation and faster mixing rates. As a result, Reynolds number calculations are very common in the petrochemical industry. They are utilized to scale orifice coefficients,[19] friction factors (Vernard and Street,[4] and Section 4.5 of this book), heat transfer rates (Lienhard,[19] and Chapter 3 of this book), and mass transfer rates (Bird[16]).

4.4.1.2 Hydraulic Diameter

In addition to circular pipe flows, Reynolds numbers can be utilized, as described above, for geometries other than round. Square, triangular, and other enclosed flows scale with Reynolds numbers. For geometric variations with no obvious "diameter," a concept termed the hydraulic diameter is useful. Essentially, it is the ratio of the flow area to the flow perimeter, multiplied by four. In analytical form, this ratio is expressed as:

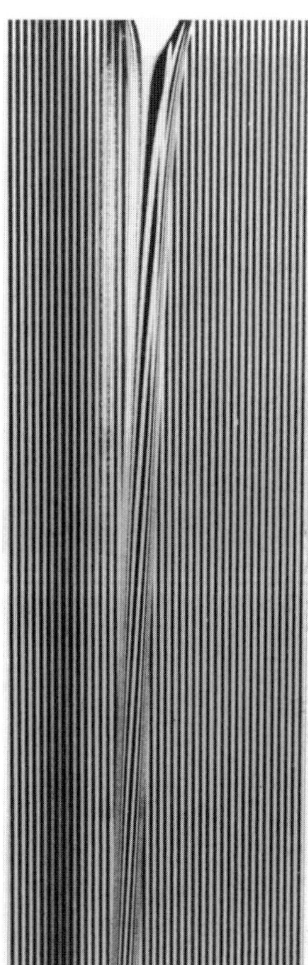

FIGURE 4.18 Water exiting a faucet at low velocity. (From Visualization Society of Japan, *The Fantasy of Flow*, IOS Press, Amsterdam, 1993, 96.)

$$\text{Hydraulic diameter} = \frac{\text{Area}}{\text{Perimeter}}(4) \tag{4.71}$$

It should be noted that the above definition of hydraulic diameter reduces to the definition of normal diameter for circular pipes. The above equation provides reasonable accuracy for calculations involving turbulent flow, but large errors will occur for laminar flow calculations. See Vernard and Street[4] for additional details.

4.4.1.3 Reynolds Averaging

Turbulent flows are characterized by a highly fluctuating instantaneous velocity. The flow fields generated are three-dimensional in nature and dependent on an unknown time function. To help understand the processes, Reynolds[34] suggested utilizing a velocity composed of two components: (1) a time averaged velocity, and (2) a fluctuating velocity. This is a logical substitution, because, for 99% of applications, only the

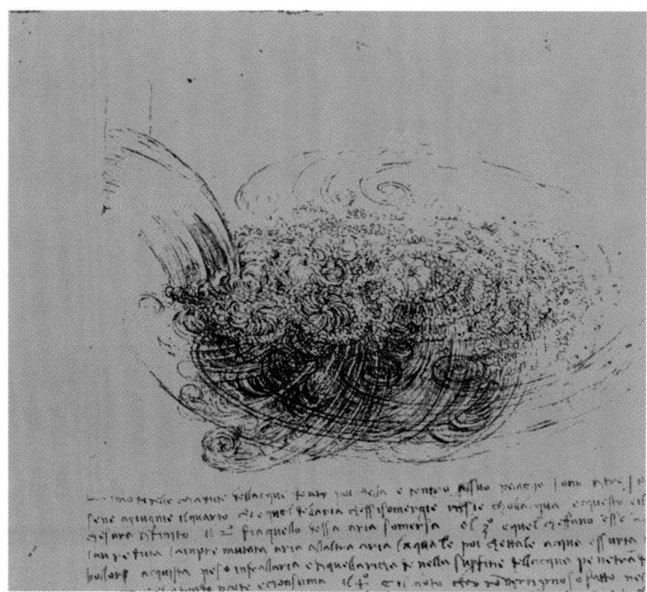

FIGURE 4.19 Leonardo da Vinci's view of turbulence.

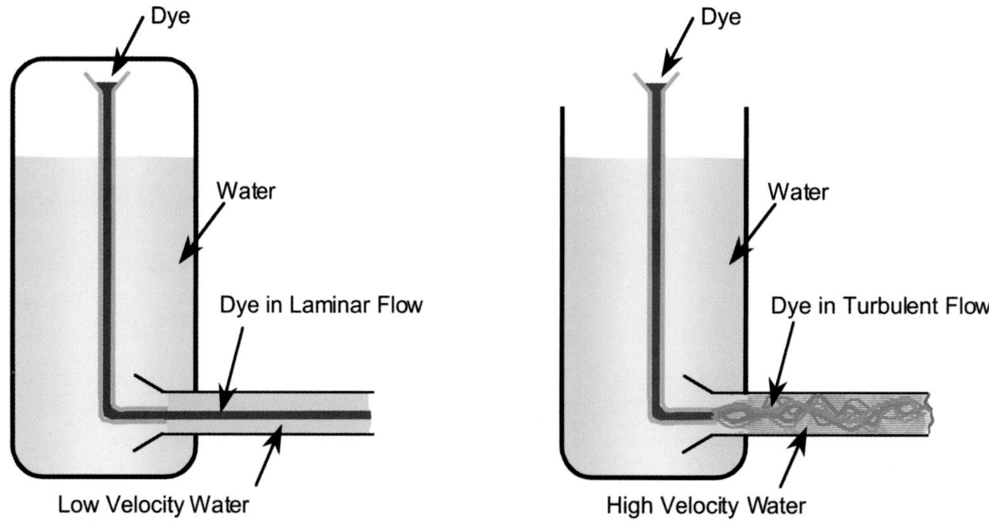

FIGURE 4.20 Osborn Reynolds' experimental apparatus used to study the transition from laminar to turbulent flow.

mean (or average) velocity is of interest, and the time average of the fluctuating velocity would be zero. This is expressed as:

$$\text{Velocity} = \overline{V} + V' \tag{4.72}$$

where

$\overline{V}$ = Average velocity (the bar represents "time averaged")

V' = Fluctuating velocity (where $\overline{V'} = 0$)

The two velocity terms can be substituted into the differential momentum equation (4.64), and the resulting equation averaged over time to obtain an expression for average velocity. It should be noted that while the time average of fluctuating velocity is zero, the time average of the product of two fluctuating velocities is not zero. Potter and Foss[13] present a good discussion of this observation.

Starting from Eq. (4.64) (the momentum equation), if body forces are neglected and density is considered constant, the "x" only equation can be rewritten as:

$$\frac{D(\rho U)}{Dt} = \frac{\partial \tau_{ix}}{\partial x_i} \tag{4.73}$$

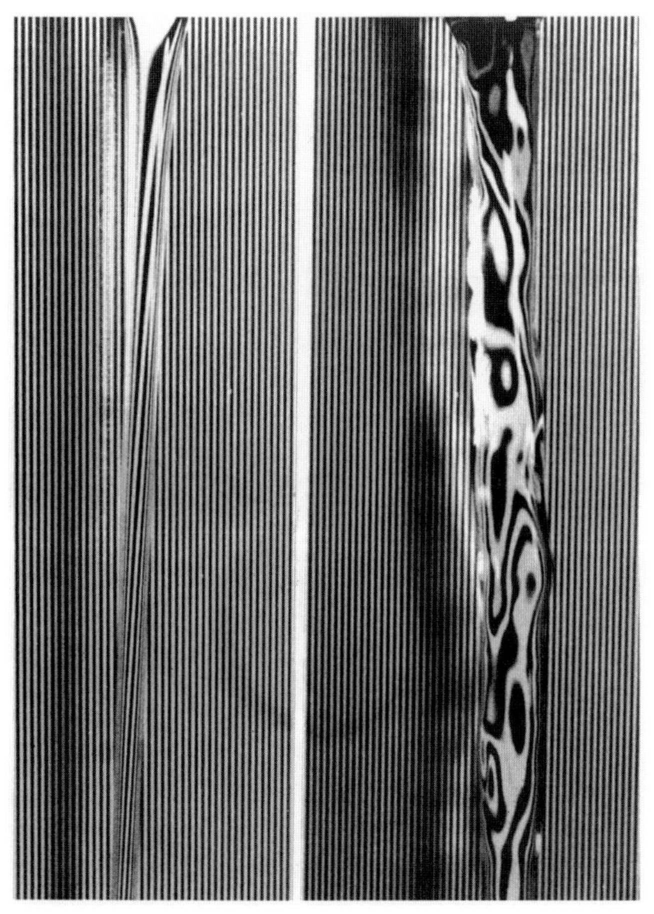

FIGURE 4.21 Water from faucet showing transition. (From Visualization Society of Japan, *The Fantasy of Flow*, IOS Press, Amsterdam, 1993, 96-97.)

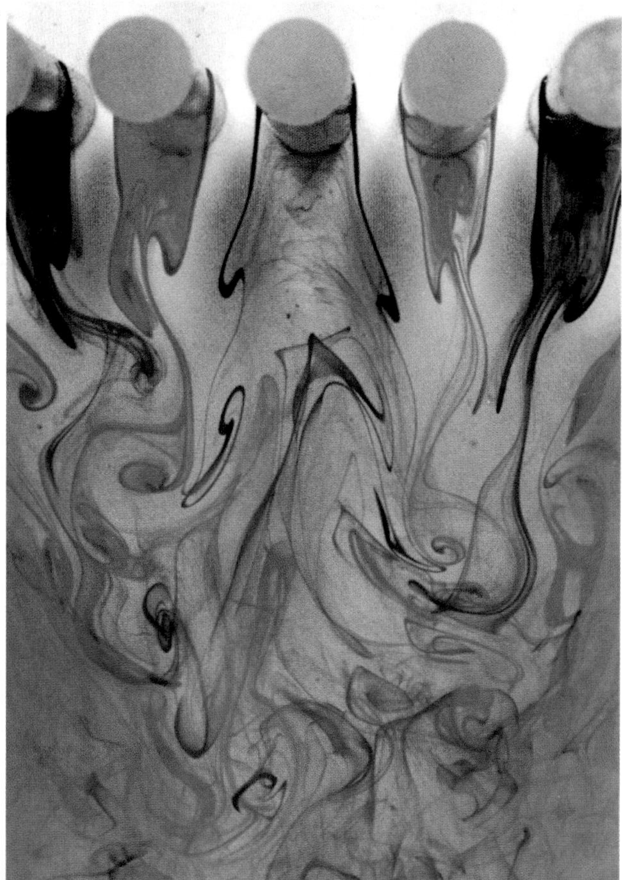

FIGURE 4.22 Wake area showing mixing vortices. (From Visualization Society of Japan, *The Fantasy of Flow*, IOS Press, Amsterdam, 1993, 3.)

FIGURE 4.23 Photograph showing the laminar flow of smoke over a rectangular obstruction. (From M. Van Dyke, *An Album of Fluid Motion*, The Parabolic Press, Stanford, CA, 1982, 10.)

where

$$\tau_{ix} = -P_x + \mu\left(\frac{\partial U}{\partial x} + \frac{\partial V}{\partial y} + \frac{\partial W}{\partial z}\right) \tag{4.74}$$

The τ part of the momentum equation contains no cross velocity terms. However, close examination of the substantial derivative (see Section 4.3.4 of this chapter) reveals that the left-hand side of Eq. (4.73) does contain the products of different velocities. Substitution of the decomposed velocity into the momentum equation, taking the time average (to develop a time-averaged momentum equation), and simplifying (see Potter and Foss[13] for details), results in the following x momentum equation:

$$\frac{\overline{D(\rho U)}}{Dt} = \frac{\partial \overline{\tau_{ix}}}{\partial x_i} - \rho\left[\frac{\partial \overline{U'^2}}{\partial x} + \frac{\partial \overline{U'V'}}{\partial y} + \frac{\partial \overline{U'W'}}{\partial z}\right] \tag{4.75}$$

where

$$\overline{\tau_{ix}} = -\overline{P_x} + \mu\left(\frac{\partial \overline{U}}{\partial x} + \frac{\partial \overline{V}}{\partial y} + \frac{\partial \overline{W}}{\partial z}\right) \tag{4.76}$$

Interestingly, the cross fluctuating velocity terms have the same derivative operator as the shear stress, so it is common to simply redefine the shear stress with those terms added to it. In that case, the momentum equation for time-averaged velocity will have the exact same form as before, namely:

$$\frac{\overline{D(\rho U)}}{Dt} = \frac{\partial \overline{\tau_{ix}}}{\partial x_i} \tag{4.77}$$

But the stress term for the x equation is now modified by the fluctuating velocity terms. It now has the following form:

$$\overline{\tau_{ix}} = -\overline{P_x} + \mu\left(\frac{\partial \overline{U}}{\partial x} + \frac{\partial \overline{V}}{\partial y} + \frac{\partial \overline{W}}{\partial z}\right) - \rho\left[\overline{U'^2} + \overline{U'V'} + \overline{U'W'}\right] \tag{4.78}$$

It should be noted that although the fluctuating terms are historically "lumped" with the stress and pressure terms (forces, from the derivation), they actually arise from the convective side of the momentum equation. If the same procedure is followed for the y and z momentum equations, a total of nine fluctuating velocities arise from the derivation. Noting, however, that velocity order can be interchanged, there are only six different terms. Collecting them in tensor form (terms from the three different equations are listed together), they are as follows:

$$\rho\begin{bmatrix} \overline{U'^2} & \overline{U'V'} & \overline{U'W'} \\ \overline{U'V'} & \overline{V'^2} & \overline{V'W'} \\ \overline{U'W'} & \overline{V'W'} & \overline{W'^2} \end{bmatrix} \tag{4.79}$$

Collectively, these terms are called "Reynolds stresses." The term "stress" is applied to them because they are associated with the stress term in the momentum equation, although they arise from the convective side of the momentum equation. Research over the last 100 years has not yet provided a completely satisfactory model for these terms. Computational fluid dynamics (CFD, see Chapter 9 of this book) makes extensive use of various assumption and modeling approximations. Occasionally, the ability to accurately model a flow field still depends on the modeler's understanding of the limitations of the chosen Reynolds stress model.

4.4.1.4 Jets

When a fluid emerges from a nozzle, it will interact with the surrounding fluids. This type of system, termed a "free jet," commonly occurs in combustion systems (Figure 4.24). High-pressure fuel from a nozzle, steam spargers, and liquid fuel sprays are all examples of free jets. Figure 4.25 illustrates the interaction of a free jet with the surrounding fluid. As the jet travels downstream from the nozzle, its diameter will increase as it captures ambient fluid into its stream. This phenomenon has been extensively studied, both experimentally and theoretically. For theoretical treatment, the interested reader is referred to Schlichting,[20] Hinze,[21] and Tennekes.[22] The historical treatment is to assume that fluctuating velocities have the same magnitude in each of the three directions (this is called *isotropic* turbulence), which reduces the Reynolds stresses from six terms to one term. The single stress term is then modeled as a function of the local velocity gradient and the nozzle diameter. This model has a single fitted parameter, which is deduced from experimental data.

Experimental characterization of nozzles has resulted in practically the same results as those obtained theoretically, except near-nozzle effects are quantified. Nevertheless, both treatments result in the following two equations describing jet velocities (from Beer and Chigier[23]):

$$\frac{\overline{U}_0}{\overline{U}_m} = 0.16\frac{x}{d_0} - 1.5 \tag{4.80}$$

where

U_0 = Initial velocity at nozzle (assuming plug flow)
U_m = Maximum (or centerline) velocity of the jet downstream

x = Distance from nozzle exit
d_0 = Diameter of nozzle

Equation (4.80) describes the velocity decay that occurs as a nozzle interacts with its surroundings. The above equations are dimensionless; the only requirement for accuracy is that consistent units for velocity and length be utilized. As the jet progresses downstream, it expands radially. As it expands, it develops a radial velocity profile. This profile can be described by a Gaussian[24] as follows:

$$\frac{\overline{U}_0}{\overline{U}_m} = \exp\left[-K_u\left(\frac{r}{x}\right)\right] \qquad (4.81)$$

where

U = Actual velocity
U_m = Maximum or centerline velocity at the particular x location
r = Radial distance from centerline
K_u = Gauss constant, which has a value of about 87[24]

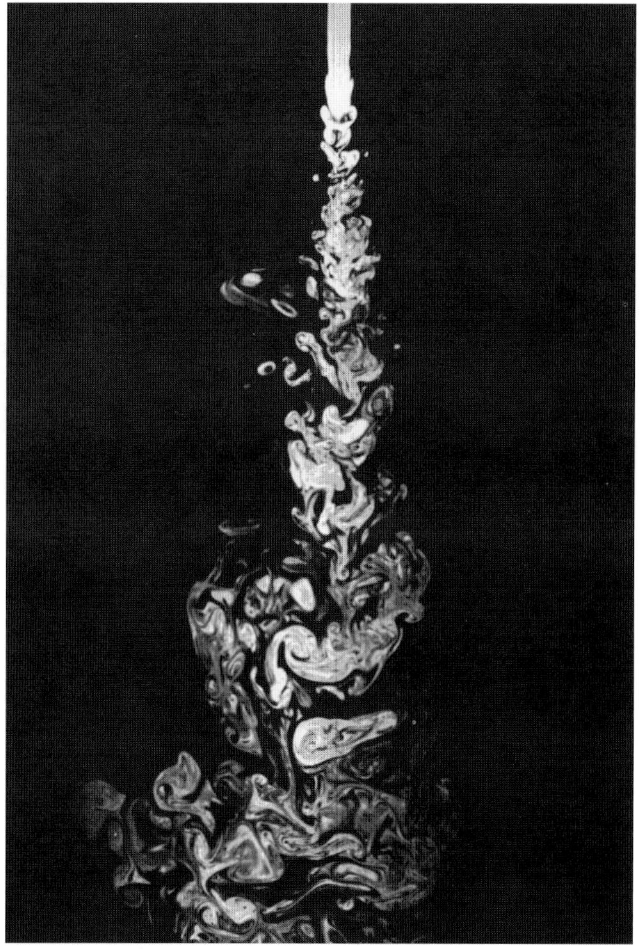

FIGURE 4.24 Free jet structure. (From M. Van Dyke, *An Album of Fluid Motion*, The Parabolic Press, Stanford, CA, 1982, 99.)

4.4.1.5 Entrainment

As a free jet interacts with surrounding fluids, it will "pick up" or entrain the ambient fluid and carry it downstream. It is this additional fluid that will cause the jet to expand and the velocity to decrease. Figures 4.25 illustrates the phenomenon of entrainment. Equations for concentration decay are similar to the velocity equations, and are given by:

$$\frac{\overline{C}_0}{\overline{C}_m} = 0.22\frac{x}{d_0} - 1.5 \qquad (4.82)$$

$$\frac{\overline{C}}{\overline{C}_m} = \exp\left[-K_u\left(\frac{r}{x}\right)\right] \qquad (4.83)$$

where

C = Actual concentration
C_0 = Initial concentration at nozzle (assuming flat profile)
C_m = Maximum (or centerline) concentration of the jet downstream
x = Distance from nozzle exit
d_0 = Diameter of nozzle
r = Radial distance from centerline
K_u = Gauss constant, which has a value of about 55.5[24]

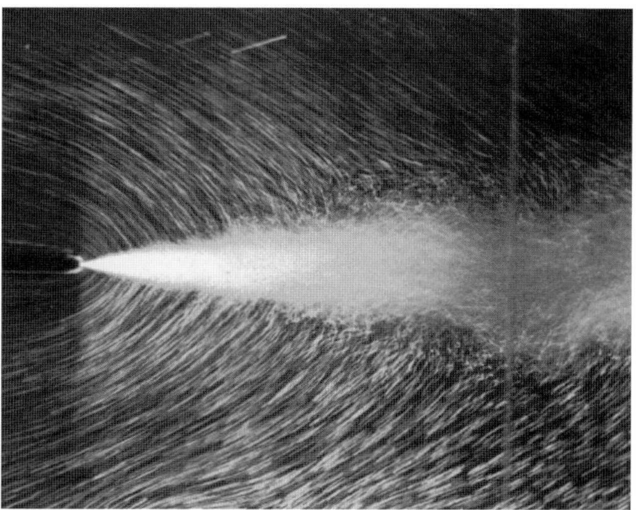

FIGURE 4.25 Free jet entrainment. (From M. Van Dyke, *An Album of Fluid Motion*, The Parabolic Press, Stanford, CA, 1982, 97.)

One unexpected result of the above jet laws is that concentration is independent of velocity. Intuition would lead one to believe that the faster a jet exits an orifice, the faster it mixes

FIGURE 4.26 Flow around the air intake of a jet engine in supersonic flow. (From Visualization Society of Japan, *The Fantasy of Flow*, IOS Press, Amsterdam, 1993, 23.)

with the surrounding fluid. While it is true that it is entraining more mass, it is not mixing any faster — downstream concentrations are not affected by velocity.

From these equations, the mass entrainment rate can also be deduced. The following equation[23] summarizes mass entrainment:

$$\frac{m_e}{m_0} = 0.32\frac{x}{d_0}\left(\frac{\rho_a}{\rho_0}\right) - 1 \tag{4.84}$$

where

m_e = Mass flow entrained by jet
m_0 = Initial mass flow from nozzle
x = Distance from nozzle exit
d_0 = Diameter of nozzle
ρ_a = Density of ambient fluid
ρ_0 = Density of initial jet fluid

4.4.2 Compressible Flow

All gases and liquids are compressible to some extent. The subject of this section, however, is the flow of fluids in which the density varies significantly due to pressure gradients within the flow. For engineering purposes, it is frequently assumed that a flow is incompressible as long as the density change is less than about 5%. Note here that density changes due to compression (or pressure forces) are being discussed. The density of a flowing fluid can also change because of temperature changes, such as in a reacting flow or a flow with heat transfer. In a typical reacting flow, the density of the reactants can be 500% of the products' density, yet the flow may still be analyzed as an incompressible flow. In this section, the Mach number is introduced and the Mach number will be used alone to determine whether or not a flow is compressible.

Compressible flow is a vast subject. Texts such as Saad[24] and Anderson[25] cover the subject in far more detail than space allows here. The purpose of this section is to describe enough of the subject that the reader understands the key assumptions behind the use of compressible flow relations in computing orifice flows. The choked orifice is an important flow measurement device because of its relative simplicity. In addition, many gas-fired burners and flares fire fuel through choked orifices. A further purpose is to give the reader some appreciation for the interesting flow phenomena, especially shock waves that occur in compressible flows with little or no analog in incompressible flows.

4.4.2.1 Basic Thermodynamics Relations

This section is limited to ideal gases, that is, gases that obey the relation $P = \rho RT$, where P is the absolute pressure, T is the absolute temperature, and R is the gas constant (the universal gas constant divided by the gas molecular weight) (see Section 4.2.4.1). For an ideal gas, one can readily show that the enthalpy, h, and internal energy, e, are functions of temperature alone.[15] One can also show that the specific heats

$$C_p = \left.\frac{\partial h}{\partial T}\right|_p = T\left(\frac{\partial s}{\partial T}\right)_p \tag{4.85}$$

and

$$C_v = \left.\frac{\partial e}{\partial T}\right|_v = T\left(\frac{\partial s}{\partial T}\right)_v \tag{4.86}$$

where v is the specific volume, are also functions of temperature only. The subscript notation indicates that differentiation is performed holding the specified variable constant. Since $h = e + P/\rho = e + RT$, $dh = de + RdT$. For an ideal gas, $dh = C_p dT$ and $de = C_v dT$. Combining these relations and dividing by dT reveals that $C_p = C_v + R$, a useful relationship that will be exploited later.

The Maxwell relation of classical thermodynamics is:

$$dh = Tds + dP/\rho \tag{4.87}$$

It applies to any "pure simple compressible substance." This means any substance composed of a single chemical species whose only mode of doing work is *PdV* work. Air is also considered a pure simple compressible substance because it is a nonreacting mixture of molecules that acts like a single species.[26] The Maxwell relation is important in the present context because it allows one to derive ideal gas relations for a constant entropy fluid. If entropy is constant, then Maxwell's relation reduces to $dh = dP/\rho$ or $C_p dT = dP/\rho$. Manipulation of this relation using the ideal gas law leads to the following relations for an ideal gas undergoing an isentropic change from state 1 to state 2:

$$\frac{P_2}{P_1} = \left(\frac{\rho_2}{\rho_1}\right)^{\gamma} = \left(\frac{T_2}{T_1}\right)^{\gamma/(\gamma-1)} \tag{4.88}$$

where γ is the ratio of specific heats, C_p/C_v.

4.4.2.2 Compressible Flow Concepts

The Mach number is the ratio of a fluid's velocity (measured relative to some obstacle or geometric feature), divided by the speed at which sound waves propagate through the fluid. A sound wave is a very weak pressure disturbance in a fluid. This small pressure wave in the fluid is accompanied by a density disturbance and a velocity disturbance. Analysis of the sound wave is straightforward and is given by Anderson.[25] The result is that the sound speed for an ideal gas is:

$$c = \sqrt{(\partial P/\partial \rho)_s} = \sqrt{\gamma R_u T/M}$$

For example, air at 77°F (25°C) has a sound speed of 1135 ft s^{-1} (346 m s^{-1} or 774 miles per hour).

Consider a fluid flowing around an obstacle, such as a circular cylinder placed in the flow. If the fluid is flowing at a very low speed compared to the sound speed, then the upstream fluid is able to "sense" (via pressure waves) the presence of the circular cylinder and adjust the flow well ahead of the cylinder. If the speed of the fluid is greater than its sound speed, no such adjustment can be made. Figure 4.26 shows shock waves at the sharply pointed air-intake of a jet engine, and Figure 4.27 shows shock patterns from a supersonic jet.

Compressible flows are typically very high Reynolds number flows. For example, because air has a sound speed of 1135 ft/s (346 m/s), air flowing at sonic conditions through a 1/8-in. (3.2 mm) orifice would have a jet Reynolds number of about 66,000. This is well into the turbulent regime. The subject of compressible flows has traditionally focused on aerodynamic problems, where length scales are considerably larger (typically tens of feet of modern supersonic aircraft). For this reason, compressible flows are almost always analyzed as inviscid flows. The term "inviscid" does not imply that the fluid itself has no viscosity, as air and other gases are viscous. Rather, the term implies that viscous effects are unimportant. Although the flow fields are mostly inviscid, it is known that in the region near solid surfaces (called the boundary layer), viscous forces are very important. Several clever analysis techniques have been developed to asymptotically patch the inviscid far-field solution to a boundary-layer solution. The interested reader should see Panton[1] for details.

FIGURE 4.27 Shock waves from supersonic fighter. (From Visualization Society of Japan, *The Fantasy of Flow*, IOS Press, Amsterdam, 1993, 59.)

4.4.2.3 Quasi-one-dimensional Isentropic Flow

A quasi-one-dimensional flow is a flow that is "nearly" one-dimensional. A truly one-dimensional flow must have a constant flow area. Here, the flow area is allowed to vary slowly with the *x*-coordinate. The *y* and *z* variations are neglected because the area varies gradually. Flow variables P, ρ, T, and u are treated as functions of x only. The isentropic assumption implies that the flow is adiabatic. The following summary is from Saad,[24] Howell,[26] and Ward-Smith.[27]

4.4.2.3.1 Basic Relations

The *x*-direction component of the Navier-Stokes equation can be used alone to describe a truly one-dimensional flow. Because the flow in this section is a quasi-one-dimensional flow, relations must be developed that correctly account for the fact that the flow area is changing. Anderson[25] has shown that momentum and energy conservation can be written as:

$$dP + \rho u du = 0 \tag{4.89}$$

and

$$dh + udu = 0 \tag{4.90}$$

Maxwell's relation, $Tds = dh + dP/\rho$, shows that these two relations are equivalent in an isentropic flow. The differential energy equation shown above can be integrated once to show that the total enthalpy, $h_o = h + u^2/2$, is a constant (h_o is the stagnation enthalpy, defined below).

4.4.2.3.2 Flow Through Converging-Diverging Nozzle
Conservation of mass in a variable-area duct requires that $d(\rho uA) = 0$. Expanding this differential and manipulating shows that $d\rho/\rho + du/u + dA/A = 0$. One can write $d\rho$ as $dP(d\rho/dP)$. Recall that the definition of the sound speed is $c^2 = (dP/d\rho)_s$. Using this relation and applying the differential momentum equation given above allows one to write energy conservation as $-u\ du/c^2 + du/u + dA/A = 0$. The Mach number is defined as $\mathrm{M} \equiv u/c$, and further manipulation allows mass conservation to be written as:

$$\frac{du}{u}\left(\mathrm{M}^2 - 1\right) = \frac{dA}{A}. \tag{4.91}$$

This is the area-velocity relation. This equation is significant in the study of compressible flow. Notice, for example, that when the Mach number is less than 1, area and velocity changes have opposite signs. This is the intuitive result that if the mass flow is constant, decreasing the available flow area must increase velocity. However, if the Mach number is greater than one, the relation gives a counterintuitive result: that to accelerate a flow, one must *increase* the available flow area. For this reason, a nozzle that is designed to accelerate a flow from rest (or low speed) to supersonic velocities must be a *converging-diverging* nozzle. Its area must decrease to accelerate the flow to a Mach number of 1, and then the area must begin increasing to accelerate the flow to supersonic ($\mathrm{M} > 1$) velocities. The minimum area of the duct is called the throat of the duct; at this location, the flow is sonic.

The analysis of flows is simplified by defining stagnation conditions (where the flow velocity is zero) and throat conditions (where the flow is sonic). Following Anderson,[26] the subscript *o* is used for stagnation conditions and the superscript * for throat conditions. Because the temperature varies along the length of the nozzle, the sound speed *c* varies with *x*. As long as the flow is isentropic, the stagnation properties and throat properties do not change.

Because the flow is isentropic, the thermodynamic relations shown above for an isentropic process between two states can be applied between the stagnation conditions and any downstream location. If one considers a calorically perfect gas (specific heats independent of temperature), then equating the stagnation enthalpy to the enthalpy at any location (denoted by subscript '1') within the duct gives:

$$h_o = h_1 + u_1^2/2 \tag{4.92}$$

because the stagnation velocity is zero. And because the specific heat is constant, one notes that $h_o - h_1 = C_p(T_o - T_1) = u_1^2/2$. Dividing by the sound speed squared gives $\frac{C_p(T_o - T_1)}{\gamma R T_1} = \frac{M_1^2}{2}$. From thermodynamics, one knows that $C_p - C_v = R$. Because $\gamma = C_p/C_v$, one can readily determine that $C_p/(\gamma R) = 1/(\gamma - 1)$. Now the temperature and Mach number can be related by

$$\frac{T_o}{T_1} = 1 + M_1^2\frac{\gamma - 1}{2} \tag{4.93}$$

Using the thermodynamic relations for ideal gases undergoing isentropic processes, one can determine additional relations that relate the pressure at any location (not subscripted now) to the Mach number:

$$\frac{P_o}{P} = \left(1 + \frac{\gamma - 1}{2}M^2\right)^{\gamma/(\gamma-1)} \tag{4.94}$$

and

$$\frac{\rho_o}{\rho} = \left(1 + \frac{\gamma - 1}{2}M^2\right)^{1/(\gamma-1)} \tag{4.95}$$

At the throat, the Mach number is 1; thus, the above relations [Eqs. (4.93), (4.94), and (4.95)] become, respectively:

$$\frac{T_o}{T^*} = 1 + \frac{\gamma - 1}{2} \tag{4.96}$$

$$\frac{P_o}{P^*} = \left(1 + \frac{\gamma - 1}{2}\right)^{\gamma/(\gamma-1)} \tag{4.97}$$

and

$$\frac{\rho_o}{\rho^*} = \left(1 + \frac{\gamma - 1}{2}\right)^{1/(\gamma-1)} \tag{4.98}$$

These relations allow the computation of fluid properties (u, ρ, P, and T) at any location within the duct. A common application of these equations is to ensure that a supersonic nozzle is properly expanded such that the fluid pressure at the

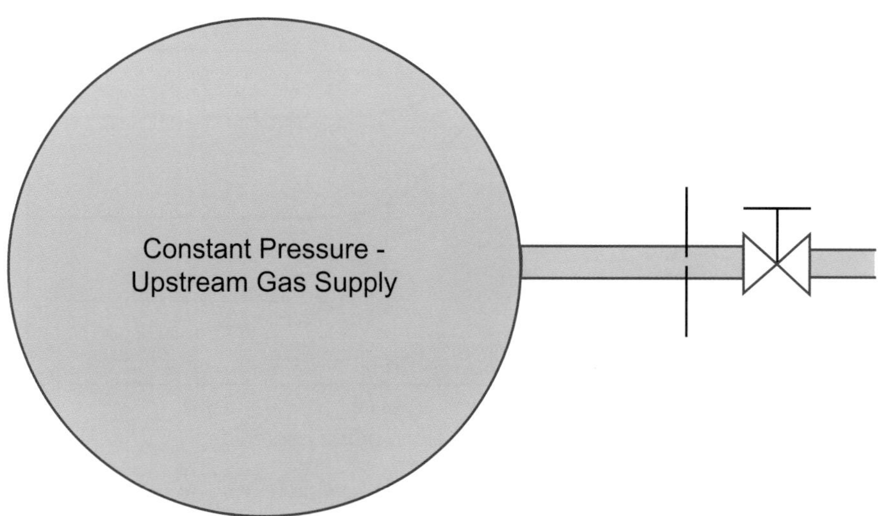

FIGURE 4.28 Choked flow test rig.

exit plane is the same as the surrounding pressure. A nozzle is termed "overexpanded" if the nozzle expands so that the pressure at the exit plane is lower than the surroundings. In this condition, either a normal shock wave will develop within the expanding portion of the nozzle, or a series of oblique shock waves will appear downstream of the exit plane. A nozzle is "underexpanded" if the pressure at the exit plane is higher than the surrounding pressure. Prandtl-Meyer expansion waves (discussed in the compressible flow texts cited in the introduction to this section and shown in Figure 4.26) are two-dimensional flow structures that will appear in this situation.

The mass flow at any location is $\dot{m} = \rho u A$. The mass flow rate is independent of the x-location, and thus one can write $\rho u A = \rho^* u^* A^*$. The throat velocity is $u^* = c^* = \sqrt{\gamma R T^*}$. Manipulation of these expressions using the above derived expressions for ρ_o/ρ^*, ρ_o/ρ, T_o/T^*, and T_o/T yields the area Mach number relation:

$$\left(\frac{A}{A^*}\right)^2 = \frac{1}{M^2}\left[\frac{2}{\gamma+1}\left(1+\frac{\gamma-1}{2}M^2\right)\right]^{(\gamma+1)/(\gamma-1)} \quad (4.99)$$

4.4.2.3.3 Choked Orifice Flows

This section examines the flow of an ideal gas through an orifice. Compressible orifice flows have numerous applications in petrochemical combustion. Ward-Smith[27] provides a thorough overview of the use of choked orifices as flowmeters.

Consider a fluid in a large reservoir at (stagnation) pressure P_0, as shown in Figure 4.28. If this fluid is allowed to pass through a small orifice where the pressure downstream of the orifice is the back pressure, P_b, there are four cases to consider, as shown in Figure 4.29.

1. If the back pressure and the stagnation pressure are equal, there is no flow through the orifice (i.e., the valve is closed).
2. If the back pressure is lower than the stagnation pressure, but above the choking pressure, then there will be flow through the orifice at a Mach number less than 1. The mass flow rate continues to increase as the back pressure is lowered.
3. If the back pressure is lowered to a value termed the choking pressure (P_c), then the velocity at the orifice will become sonic. Because the velocity is sonic, further decreases in the back pressure cannot be "communicated" upstream, and the mass flow rate through the orifice can no longer change. The mass flow at this pressure is the maximum (in reality, the mass flow will continue to increase somewhat because the assumption of quasi-one-dimensional flow is an idealization; see Ward-Smith[27] for details).
4. As the back pressure is lowered below the choking pressure, the mass flow rate does not continue to increase in the ideal analysis. In reality, a visually striking series of Prandtl-Meyer expansion waves, seen also in Figure 4.27, appear just downstream of the exit plane, producing a diamond pattern of high- and low-pressure zones. The density gradients in the flow cause the fluid's index of refraction to vary so that these diamond patterns are visible. It should be noted that as a practical matter, valves are not normally placed downstream of a choked orifice. Also, while it is true that for *a given upstream pressure*, mass flow cannot be increased, if the upstream pressure is increased, the mass flow will increase due to the increase in density.

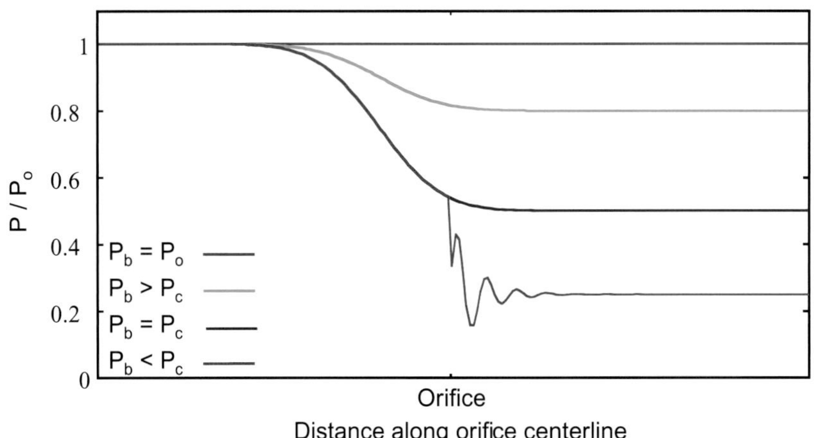

FIGURE 4.29 Types of flow of a fluid passing through a small orifice.

Given the stagnation pressure and the back pressure, the exit Mach number can be computed using the isentropic relation for P_0/P. However, if this computation reveals that the exit Mach number is greater than 1, it is realized that a simple orifice geometry will not accelerate the flow to supersonic velocities and the flow will be choked. The mass flow rate can be computed as $\dot{m} = \rho^* u^* A_{\text{orifice}}$. For example, consider air ($\gamma = 1.4$) supplied with stagnation pressure 30 psig (44.7 psia or 2 barg) and stagnation temperature 70°F (21°C) exhausted through a 0.25-in. (6.4-mm) orifice. Equation (4.87) gives the exit Mach number (assuming atmospheric pressure is 14.7 psia) as 1.36, indicating that the flow is choked, which means that the Mach number at the throat is 1. The throat conditions can then be computed using Eqs. (4.86) to (4.88) utilizing a Mach number of 1. The throat pressure is 23.6 psia (1.6 barg). The throat temperature is –18°F (–28°C), and the throat density is 0.144 lbm/ft^3. The sound speed at this temperature is 1029 ft/s (314 m/s). The (ideal) mass flow through this orifice is then 0.051 lbm/sec ($=\rho^* u^* A$). If the flow is not choked (the exit Mach number is less than 1), then the ideal mass flow can be computed using the fluid properties at the exit Mach number.

In computing real orifice flows, it is common to use a *discharge coefficient* (C_d). The discharge coefficient is defined as $C_d \equiv \dot{m}_{\text{actual}} / \dot{m}_{\text{ideal}}$. Ward-Smith[27] discusses discharge coefficients for various types of orifices. In general, the discharge coefficient for orifices is between 0.4 and 0.9. Fortunately, discharge coefficients do not vary significantly with Mach number. They are a function of orifice geometry, particularly the l/d of the orifice. For very thin-walled orifices (i.e., the l/d ratio is less than 1), the discharge coefficient will be very low. As the l/d ratio is increased to somewhere between 4 and 7, the discharge coefficient may increase to 0.85 or more. Further increases in the l/d ratio cause viscous effects to become significant. The analysis presented here is not appropriate for these high l/d orifices as viscous effects have been neglected. Viscous effects can readily be included in the analysis. Compressible flow with viscous effects in constant area ducts is called Fanno flow and is discussed at length in standard compressible flow texts such as Saad[24] and Anderson.[25]

4.5 PRESSURE DROP FUNDAMENTALS

4.5.1 *Basic Pressure Concepts*

4.5.1.1 Definition of Pressure

Pressure is created by the collision of gas molecules on a surface and is defined as the force exerted per unit area on that surface. The pressure of the air around us, called *atmospheric* pressure, is due to air molecules colliding with our bodies, objects, and the Earth's surface. Lower elevations at the Earth's surface will tend to have a higher atmospheric pressure because the height, and therefore the weight, of the column of air above us is greater. A higher column of air above the surface of the Earth will compress the air more, causing an increase in the number of collisions of air molecules at the Earth's surface. This increase in the number of collisions is the mechanism responsible for a higher atmospheric pressure. For example, at sea level, the atmospheric pressure is about 14.7 pounds per square inch (psi) or 1 bar. However, in Denver, Colorado, which is approximately 1 mile (1.6 km) above sea level, the atmospheric pressure is about 12.2 psi (0.83 bar).

4.5.1.2 Units of Pressure

Pressure is generally measured in units of Pascals (Pa = N/m^2). However, flare and burner system designers will use a variety of pressure units because the choice of units varies, depending on the customer. For example, customers in the United States will typically use units of psi or inches of water column, whereas in Europe and Asia the units are usually Pa and millimeters of water column. The conversion of these pressure units are as follows: 1 lb/in^2 = 27.68 in. water column = 6895 Pa = 703.072 mm water column.

4.5.1.3 Standard Atmospheric Pressure

Initially, engineers developed a *standard atmospheric pressure* so that the performance of aircraft and missiles could be evaluated at a standard condition. The idea of a standard atmospheric pressure was first introduced in the 1920s.[28] In 1976, a revised report was published that defined the U.S. standard atmosphere that is the currently accepted standard. This standard is an idealized representation of the mean conditions of the Earth's atmosphere in one year.

Table 4.2 lists several important properties of air for standard atmospheric conditions at sea level.

TABLE 4.2 Properties of U.S. Standard Atmosphere at Sea Level

Property	SI Units	English Units
Pressure	101.33 kPa (abs)	14.696 lb/in.^{2} (abs)
Temperature	15.0°C	59.0°F
Density	1.225 kg/m^3	0.07647 lb/ft^3

4.5.1.4 Gage and Absolute Pressure

If the Earth was in a perfect vacuum, there would be no column of air above its surface; hence, the atmospheric pressure would be zero. The *absolute pressure* is measured relative to a perfect vacuum. Therefore, when a pressure measurement is taken at the surface of the Earth, the absolute pressure is equal to the atmospheric pressure. When writing the units of pressure, it is customary to designate absolute pressure with the letter "a" or "abs" after the units. For example: psia, kPa, psi(abs), or kP(abs). The absolute pressure can never be less than zero; however, the *gage* pressure can.

The *gage pressure* is always measured relative to the atmospheric pressure. A gage pressure of less than zero can exist. For example, suppose there is a sealed container that holds a vacuum at 10 psia at sea level. The gage pressure, which is measured relative to the absolute pressure, would be (10 psia – 14.7 psia) = –4.7 psig. The letter "g" after the pressure units represents gage pressure. Now suppose the container is pressurized to 20 psia. The gage pressure will then be (20 psia – 14.7 psia) = 5.3 psig. Thus, the gage pressure can either be a positive or negative number, and is just the difference in pressure between the atmospheric pressure and the pressure of interest.

4.5.2 Roughness

The transport of a liquid or gas in a pipe system is very common in the flare and burner industry. Such applications include:

1. gas flow in a flare stack and header
2. steam flow in a pipe feeding a steam-assisted flare
3. air from a blower feeding an air-assisted flare
4. gas flow through a burner manifold
5. oil flow through a pipe feeding an atomizing gun

Although all of these systems are different, the governing equations used to describe the pressure drop are common. The purpose of this section is to discuss the basic concepts used to determine the pressure drop of a liquid or gas flowing through a straight pipe.

The roughness of the wall on the inside of a pipe influences the pressure drop of a fluid flowing through it if the flow is turbulent. When a fluid flows turbulent through a straight pipe, organized structures of fluid near the wall called slow-moving streaks, can suddenly move into the central region of the pipe by a phenomenon called a "turbulent burst." Figure 4.30 is a photograph showing the organized structure of a series of slow-moving streaks near a wall.[29]

When these slow-moving streaks burst from the wall, momentum interchanges between masses of fluid, extracting energy from the overall fluid in the form of heat. A pipe with a rough wall will experience more turbulent bursts than a pipe with a smooth wall operating under the same flowing conditions. Therefore, a pipe with a rough wall will have a larger pressure drop associated with it than a pipe with a smooth wall.

The pressure drop for a fully developed flow of fluid in a pipe can be calculated by relating the velocity pressure of the fluid to the pipe roughness and geometry as follows:

$$\Delta P = f \frac{L}{D} \frac{\rho V^2}{2g_c} \tag{4.100}$$

where the term ΔP is the pressure drop, f is the friction factor, L is the length of pipe, D is the inside pipe diameter, ρ is the density of the fluid in the pipe, and V is the mean velocity of the fluid in the pipe. Equation (4.100) is called the Darcy-Weisbach equation, named after two engineers of the nineteenth century. Weisbach first proposed the use of the friction factor term, f, and Darcy conducted numerous

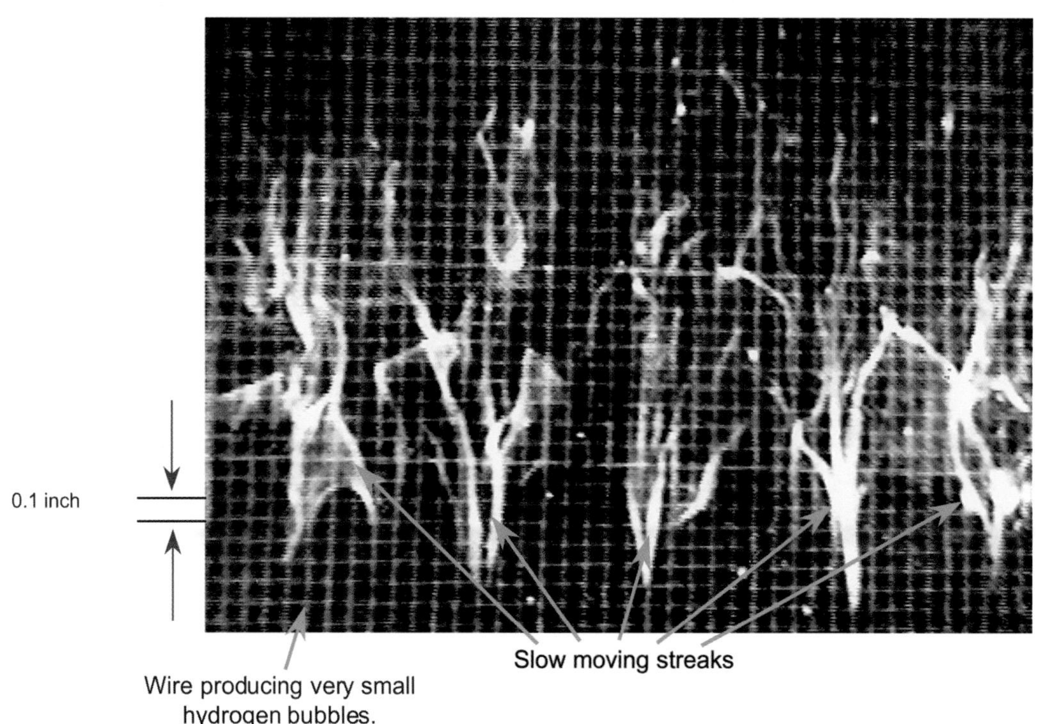

FIGURE 4.30 Photograph showing slow-moving streaks near a wall using the hydrogen bubble technique. (From W. Bussman, A theoretical and experimental investigation of near-wall turbulence in drag reducing flows, Ph.D. thesis, The University of Tulsa, Tulsa, OK, 1990.)

pressure drop experiments using water flowing through pipes with various wall roughness.

The friction factor is defined as the dimensionless fluid friction loss, in velocity heads, per diameter length of pipe. In 1933, Nikuradse conducted several experiments using artificially roughened pipes made by attaching grains of sand, of known size, to the inside of a pipe wall.[30] Nikuradse's experiments revealed two important characteristics of flow through rough pipes. First, if the flow is laminar, the pressure drop through the pipe is independent of the roughness of the pipe wall. Second, for highly turbulent flows, the friction factor is only dependent on the diameter of the pipe, D, and the height of the sand grains, ε. Nikuradse defined a nondimensional term, called the relative roughness, as ε/D.

For fully developed laminar flow in a pipe, the friction factor is independent of the relative roughness and can be simply written as:

$$f = \frac{64}{\text{Re}} \tag{4.101}$$

The term "Re" in Eq. (4.101) is the Reynolds number, defined as:

$$\text{Re} = \frac{VD}{\nu} \tag{4.102}$$

where ν is the kinematic viscosity of the fluid. Again, for fully developed laminar flow in a pipe, the friction factor — and hence pressure drop — are independent of the roughness of the pipe wall.

For fully developed turbulent flow in a pipe, the value of the friction factor can be determined using experimental data. Figure 4.31 is a plot, based on experimental data, showing f as a function of Re and ε/D. This plot is called the Moody chart or Moody diagram, in honor of L.F. Moody, who correlated Nikuradse's original data in terms of the relative roughness with commercial pipe material.

Notice that, for flow in the laminar regime, the friction factor is independent of the relative roughness. This straight line, in the laminar flow region, is a plot of Eq. (4.101). Also notice that, for very high Reynolds numbers, the friction factor no longer becomes a function of the Reynolds number as discovered by Nikuradse. However, for flows with moderate Reynolds numbers, the friction factor is a function of both the Reynolds number and the relative roughness. Also note that, for a smooth pipe ($\varepsilon = 0$), the friction factor is not zero. Therefore, regardless

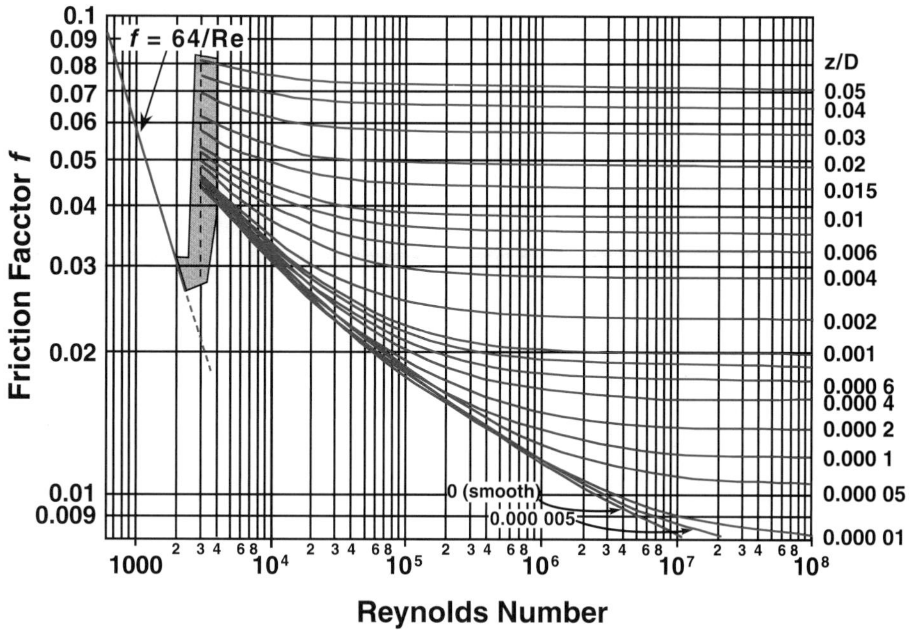

FIGURE 4.31 Moody diagram. (From J.A. Roberson and C.T. Crowe, *Engineering Fluid Mechanics*, Houghton Mifflin, Boston, MA, 1980.)

of the smoothness of a pipe wall, there will always be a pressure drop as the fluid flows through the pipe. There is no such thing as a perfectly smooth wall. On a microscopic level, a wall will always have a surface roughness.

Values of relative roughness are available for commercially manufactured pipe. Typical roughness values, for various pipe material and surfaces, are provided in Table 4.3.

It should be mentioned that the buildup of corrosion or scale on the inside of a pipe can significantly increase the relative roughness. Very old pipes can also be so badly eroded away on the inside that the effective diameter of the pipe is altered.

Several researchers have attempted to develop an analytical expression for the friction factor as a function of the Reynolds number and relative roughness. One well-known equation is the Colebrook formula:

$$\frac{1}{\sqrt{f}} = -2.0\log\left(\frac{\varepsilon/D}{3.7} + \frac{2.51}{\text{Re}\sqrt{f}}\right) \quad (4.103)$$

This formula is typically used to generate curves in the Moody diagram. The difficulty in using the Colebrook formula [Eq. (4.103)] is that, in order to solve for the friction factor f, an iterative scheme must be used. This is not too difficult, however, if a computer is used.

TABLE 4.3 Equivalent Roughness for New Pipes

Pipe	Equivalent Roughness, ε	
	(ft)	(mm)
Riveted steel	0.003–0.03	0.9–9.0
Concrete	0.001–0.01	0.3–3.0
Wood stave	0.0006–0.003	0.18–0.9
Cast iron	0.00085	0.26
Galvanized iron	0.0005	0.15
Commercial steel or wrought iron	0.00015	0.045
Drawn tubing	0.000005	0.0015
Plastic, glass	0.0 (smooth)	0.0 (smooth)

Source: J.A. Roberson and C.T. Crowe, *Engineering Fluid Mechanics*, Houghton Mifflin, Boston, MA, 1980.

4.5.4 *Loss Coefficient*

The previous section (Section 4.5.2) presented equations to calculate the pressure drop for a fully developed flow of a fluid through a straight pipe having a constant cross-sectional area. In the flare and burner industry, however, it is very common for pipe systems to include inlets, elbows, tees, and other fittings that can create additional pressure losses. The methods and procedures used for determining the pressure drop through fittings, however, are not as convenient as for straight pipe flow. Pressure losses through fittings are the result of additional turbulence and/or flow separation created by sudden changes in the fluid momentum. Therefore, pressure losses are significantly influenced by the geometry of the fitting. This

TABLE 4.4 Loss Coefficients for Various Fittings

Description	Sketch	Additional Data	K_L	
Contraction		D_2/D_1	$\phi = 60°$	$\phi = 180°$
		0.0	0.08	0.50
		0.20	0.08	0.49
		0.40	0.07	0.42
		0.60	0.06	0.32
		0.80	0.05	0.18
$\Delta P = K_L V^2/2 g_c$		0.09	0.04	0.10
Expansion		D_1/D_2	$\phi = 10°$	$\phi = 180°$
		0.0		1.00
		0.20	0.13	0.92
		0.40	0.11	0.72
		0.60	0.06	0.42
$\Delta P = K_L V^2/2 g_c$		0.80	0.03	0.16
90° Smooth Bend		r/d	K_L	
		1	0.35	
		2	0.19	
		4	0.16	
		6	0.21	
		8	0.28	
$\Delta P = K_L V^2/2 g_c$		10	0.32	

Source: J.A. Roberson and C.T. Crowe, *Engineering Fluid Mechanics*, Houghton Mifflin, Boston, MA, 1980.

section discusses the general procedure for estimating the pressure drop through fittings in piping systems.

A complete theoretical analysis for calculating the flow through fittings has yet to be developed. Thus, the pressure drop through fittings is based on equations that rely heavily on experimental data. The most common method used to determine the pressure loss is to specify the *loss coefficient*, K_L, defined as follows:

$$K_L = \frac{\Delta P g_c}{\frac{1}{2}\rho V^2} \tag{4.104}$$

Notice that the loss coefficient is dimensionless and is defined as the ratio of the pressure drop through a fitting to the approaching velocity pressure of the fluid stream. Solving Eq. (4.104) for ΔP relates the pressure drop through a fitting:

$$\Delta P = K_L \frac{\rho V^2}{2 g_c} \tag{4.105}$$

If the loss coefficient is equal to 1.0, then the pressure loss through that fitting will equal the velocity pressure of the approaching fluid stream, $\rho V^2/2$. The loss coefficient is strongly dependent on the geometry of the fitting and the Reynolds number in the pipe approaching the fitting. The loss coefficient for various fittings is given in Table 4.4. For additional information on loss coefficients through various fittings, refer to Idelchik[12] and Crane.[31]

4.5.3 Discharge Coefficient

Flare and burner engineers use equations based on the ideal gas law and assumptions of ideal flow to calculate the flow rate of a fluid through a burner nozzle or flare tip. To compensate for the results of these ideal equations and assumptions, a constant is introduced to account for the complexity of the flow that makes it non-ideal. This constant is called the *discharge coefficient.*

The discharge coefficient is defined as the ratio of the actual mass flow rate of a fluid through a nozzle to the ideal mass flow rate and is written as:

$$C_d = \frac{\dot{m}_{actual}}{\dot{m}_{ideal}} \tag{4.106}$$

The ideal mass flow rate is defined as the mass flow rate calculated using the ideal gas law and assumptions of ideal flow — no pressure losses due to the internals of the nozzle or tip. The value of the discharge coefficient for a burner nozzle or flare tip must be determined experimentally. Typically, the discharge coefficient varies from about 0.60 to 1.0, with 1.0 being ideal in most burner and flare applications. Factors that can affect the discharge coefficient include:

1. length-to-diameter ratio of the port
2. the Reynolds number of the fluid in the port
3. beta ratio
4. port angle
5. manufacturing tolerances

See Figure 4.32 for a description of these variables.

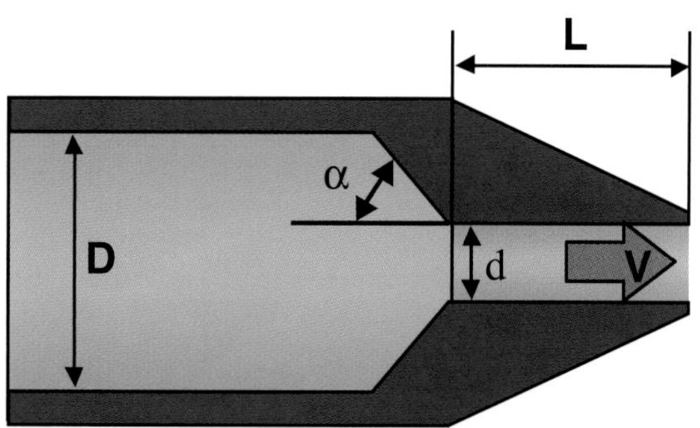

FIGURE 4.32 Factors that can affect the discharge coefficient.

REFERENCES

1. R.L. Panton, *Incompressible Flow*, John Wiley and Sons, New York, 1984.
2. F.M. White, *Viscous Fluid Flow*, McGraw-Hill, New York, 1991.
3. R.W. Fox and A.T. McDonald, *Introduction to Fluid Mechanics*, 2nd ed., John Wiley & Sons, New York, 1978.
4. J.K. Vennard and R.L. Street, *Elementary Fluid Mechanics*, 5th ed., John Wiley & Sons, New York, 1975.
5. R.D. Reed, *Furnace Operations*, 3rd ed., Gulf Publishing, Houston, 1981.
6. J.B. Maxwell, *Data Book on Hydrocarbons*, D. Van Nostrand Company, Princeton NJ, from the Standard Oil Development Company, 1950.
7. E.L. Lederer, Proceedings of the World Petrochemical Congress, London, 1933, 526.
8. W.R. Shu, A viscosity correlation for mixtures of heavy oil, bitumen, and petroleum fractions, *Soc. Petr. Engr. J.*, June 1984, 272.
9. R.J. Kee, F.M. Rupley, and J.A. Miller, The CHEMKIN Thermodynamic Data Base, Sandia Report SAND87-8215B, March 1990.
10. J.M. Smith, H.C. Van Ness, and M.M. Abbott, *Introduction to Chemical Engineering Thermodynamics*, 5th ed., McGraw-Hill, New York, 1996.
11. M. Modell and R. Reid, *Fundamentals of Classical Thermodynamics*, 2nd ed., John Wiley & Sons, New York, 1973.
12. I.E. Idelchik, *Handbook of Hydraulic Resistance*, Hemisphere Publishing, Washington, D.C., 1986.
13. M.C. Potter and J.F. Foss, *Fluid Mechanics*, Great Lakes Press, Okemos, MI, 1975.
14. K.K. Kuo, *Principles of Combustion*, John Wiley & Sons, New York, 1986.
15. G. Van Wylen and R. Sonntag, *Fundamentals of Classical Thermodynamics*, 2nd ed., John Wiley & Sons, New York, 1973.
16. R.B. Bird, W.E. Stewart, and E.N. Lightfoot, *Transport Phenomena*, John Wiley & Sons, New York, 1960.
17. O. Reynolds, An experimental investigation of the circumstances which determine whether the motion of water shall be direct or sinuous and of the law of resistance in parallel channels, *Phil. Trans. Roy. Soc.*, 174(III), 935, 1883.
18. *Mark's Standard handbook for Mechanical Engineers*, 9th ed., McGraw-Hill, New York, 1987.
19. J.H. Lienhard, *A Heat Transfer Textbook*, Prentice-Hall, Englewood Cliffs, NJ, 1987.
20. H. Schlichting, *Boundary Layer Theory*, McGraw-Hill, New York, 1979.
21. J.O. Hinze, *Turbulence*, Classic Textbook Re-issue series, McGraw-Hill, New York, 1987.
22. H. Tennekes, *A First Course in Turbulence*, MIT Press, Cambridge, MA, 1973.
23. J.M. Beer and N.A. Chigier, *Combustion Aerodynamics*, Krieger Publishing, Malabar, FL, 1983.

24. M.A. Saad, *Compressible Fluid Flow*, Prentice-Hall, Englewood Cliffs, NJ, 1985.
25. J.D. Anderson, *Modern Compressible Flow with Historical Perspective*, McGraw-Hill, New York, 1982.
26. J.R. Howell and R.O. Buckius, *Fundamentals of Engineering Thermodynamics*, 2nd ed., McGraw-Hill, New York, 1992.
27. A.J. Ward-Smith, Critical flowmetering: the characteristics of cylindrical nozzles with sharp upstream edges, *Int. J. Heat and Fluid Flow*, 1, 123–132, 1979.
28. R.R. Munson, D.F. Young, and T.H. Okiishi, *Fundamentals of Fluid Mechanics*, John Wiley & Sons, New York, 1990, 52.
29. W. Bussman, A Theoretical and Experimental Investigation of Near-Wall Turbulence in Drag Reducing Flows, Ph.D. thesis, The University of Tulsa, Tulsa, OK, 1990.
30. J. Nikuradse, *Stromungsgesetze in rauhen Rohren*, VDI-Forschungsheft, 1933, 361. Translation available in *N.A.C.A. Tech. Memorandum*, 1292.
31. Crane Engineering Division, Flow of Fluids through Valves, Fitting, and Pipe, Crane Co., New York, 1969.
32. B. Humiston, *General Chemistry — Principles and Structure*, John Wiley & Sons, New York, 1975, 169–177.
33. O. Reynolds, On the dynamical theory of imcompressible viscous fluids and the determination of the criterion, *Phil. Trans. Roy. Soc.*, 186, A123–164, 1895; and Sci Papers I, 355.
34. Visualization Society of Japan, *The Fantasy of Flow*, IOS Press, Amsterdam, 1993.
35. M. Van Dyke, *An Album of Fluid Motion*, The Parabolic Press, Stamford, CA, 1982.
36. F. Kreith, *The CRC Handbook of Mechanical Engineering*, CRC Press, Boca Raton, FL, 1998.
37. J.A. Roberson and C.T. Crowe, *Engineering Fluid Mechanics*, Houghton, Mifflin, Boston, MA, 1980.

Chapter 5
Fuels

Terry Dark, John Ackland, and Jeff White

TABLE OF CONTENTS

5.1 Gaseous Fuels 158
5.1.1 Introduction 158
5.1.2 Natural Gas 158
5.1.3 Liquified Petroleum Gas (LPG) 159
5.1.4 Refinery Gases 159
5.1.5 Combustible Waste Gas Streams 160
5.1.6 Physical Properties of Gaseous Fuels 165
5.1.7 Photographs of Gaseous Fuel Flames 165
5.2 Liquid Fuels 165
5.2.1 Introduction and History 165
5.2.2 Oil Recovery 173
5.2.3 Production, Refining, and Chemistry 175
5.2.4 Oils 178
5.2.5 Liquid Naphtha 179
5.2.6 Physical Properties of Liquid Fuels 179
5.3 Gas Property Calculations 183
5.3.1 Molecular Weight 183
5.3.2 Lower and Higher Heating Values 184
5.3.3 Specific Heat Capacity 184
5.3.4 Flammability Limits 184
5.3.5 Viscosity 185
5.3.6 Derived Quantities 185
5.4 Typical Flared Gas Compositions 185
5.4.1 Oil Field/Production Plant Gases 186
5.4.2 Refinery Gases 186
5.4.3 Ethylene/Polyethylene Gases 186
5.4.4 Other Special Cases 186
References 187

5.1 GASEOUS FUELS

5.1.1 Introduction

The term "gaseous fuel" refers to any combustible fuel that exists in the gaseous state under normal temperatures and pressures. Gaseous fuels are typically composed of a wide range of chemical compounds. Low boiling point hydrocarbons (both paraffins and olefins), hydrogen, carbon monoxide, and inert gases (nitrogen and carbon dioxide) are among the many chemical constituents of common gaseous fuels. The purpose of this section is to introduce many of the common fuel gas mixtures used as fuel in the hydrocarbon and petrochemical industries. Commonly occurring waste gas mixtures in flare systems are also described.

5.1.2 Natural Gas

Natural gas is a gaseous fossil fuel that is formed naturally beneath the Earth and is typically found with or near crude oil reservoirs. Proven natural gas reserves in the United States in 1993 totaled approximately 4.58×10^{12} m^3 (1.62×10^{14} ft^3).[1] In 1989, the U.S. Department of Energy estimated total natural gas consumption in the United States at 19.384 quadrillion Btu, 23.8% of the total U.S. energy consumption.[2]

Natural gas consists of a fluctuating range of low boiling point hydrocarbons. Methane is the primary chemical component, and can be present in amounts ranging from 70 to 99.6% by volume. Ethane can be present in amounts ranging from 2 to 16% by volume. Carbon dioxide, nitrogen, hydrogen, oxygen, propane, butane, and heavier hydrocarbons are also typically present in the fuel analysis.[3] The exact analysis usually varies somewhat depending on the source of the gas and on any heating value adjustments or supplementation.

Natural gas quality specifications have historically been negotiated in individual contracts between the natural gas producer and the purchaser or pipeline company. Specification parameters often include upper and lower limits for heating value, chemical composition, contaminants, water content, and hydrocarbon dew point. Table 5.1 outlines general specifications for pipeline-quality natural gas, as provided by the Gas Processors Suppliers Association.[4] Typical commercial natural gas compositions, listed by production region, are contained in Table 5.2.[3]

In addition to the primary combustible and inert chemical components discussed above, raw natural gas can also contain undesirable amounts of water, hydrogen sulfide, and/or carbon dioxide. Before the raw natural gas can be deposited into a pipeline transmission network, these undesirable components must be removed.

Failure to remove the water vapor from raw natural gas prior to introduction to the pipeline network will result in increased corrosion rates, formation of solid hydrate compounds that can restrict or interrupt gas flow, and freezing of valves and regulators during cold weather conditions.[5] Techniques for the dehydration of natural gases include:

1. Absorption with liquid desiccants: glycols (typically triethylene glycol) are used to absorb water vapor via countercurrent-flow, packed-bed absorption columns.[6]
2. Adsorption with solid desiccants: water vapor is adsorbed onto a bed of inorganic porous solid material (silica gel, alumina, molecular sieves, etc.).[6,7]
3. Dehydration with calcium chloride: solid anhydrous calcium chloride ($CaCl_2$) absorbs water from the wet natural gas and forms various calcium chloride hydrates ($CaCl_2 \cdot xH_2O$). These hydrates are removed from the natural gas stream as a calcium chloride brine solution.[6]
4. Refrigeration: a refrigeration coil is used to cool and condense water vapor from the wet natural gas stream. Separation of the liquid phase is accomplished via a two-phase, vapor/liquid separation drum.[5]

Hydrogen sulfide must be removed from the raw natural gas stream due to air pollution considerations and corrosion hazards. The hydrogen sulfide content of commercial natural gas rarely exceeds 1.0 grains per 100 std. ft^3 (0.023 g/m^3). A majority of pipeline companies responding to a 1994 poll limited hydrogen sulfide concentrations to less than 0.3 g per 100 std. ft^3 (0.007 g/m^3).[1] In addition, carbon dioxide is often removed from the raw gas because the inert component weakens the overall heating value of the gas stream.[5] There are numerous

TABLE 5.1 Example Pipeline Quality Natural Gas

	Minimum	Maximum
Major and Minor Components, vol%		
Methane	75%	—
Ethane	—	10.0%
Propane	—	5.0%
Butane	—	2.00%
Pentane and heavier	—	5.00%
Nitrogen and other inerts	—	3–4%
Carbon dioxide	—	3–4%
Trace Components	—	
Hydrogen sulfide	—	0.25–1.0 grains/100 scf
Mercaptan sulfur	—	0.25–1.0 grains/100 scf
Total sulfur	—	5–20 grains/100 scf
Water vapor	—	7.0 lb/mmcf
Oxygen	—	0.2–1.0 ppmv
Other characteristics		
Heating value, Btu/scf-gross saturated	950	1150
Liquids: Free of liquid water and hydrocarbons at delivery temperature and pressure.		
Solids: Free of particulates in amounts deleterious to transmission and utilization equipment.		

Adapted from Gas Processors and Suppliers Association, *GPSA Engineering Data Book*, Vol. 1, 10th ed., Tulsa, OK, 1987. With permission.

TABLE 5.2 Commercial Natural Gas Components and Typical Ranges of Composition

Fuel Gas Component	Sample Gas Compositions by Production Region (vol%)										
	Tulsa, OK	**Alaska**	**Algeria**	**Netherlands**	**Kuwait**	**Libya**	**North Sea**	**Alabama**	**Ohio**	**Missouri**	**Pennsylvania**
CH_4	93%	100%	87%	81%	87%	70%	94%	90%	94%	84%	83%
C_2H_6	3%	—	9%	3%	9%	15%	3%	5%	3%	7%	16%
C_3H_8	1%	—	3%	<1%	2%	10%	1%	—	<1%	—	—
C_4H_{10}	<1%	—	1%	<1%	1%	4%	<1%	—	<1%	—	—
C_5 & higher	—	—	—	—	—	—	—	—	—	—	—
CO_2	1%	—	—	1%	2%	—	<1%	—	1%	1%	—
N_2	2%	1%	<1%	14%	1%	1%	2%	5%	1%	8%	1%
O_2	—	—	—	—	—	—	—	—	<1%	—	—
H_2	—	—	—	—	—	—	—	—	<1%	—	—
Total	100%	100%	100%	100%	100%	100%	100%	100%	100%	100%	100%

Adapted from Reed, R.J., *North American Combustion Handbook*, Vol. 1, North American Mfg. Co., Cleveland, OH, 1986.

commercial processes (chemical reaction, absorption, and adsorption) for the removal of acidic components (H_2S and CO_2) from raw natural gas streams. The *Gas Processors Suppliers Association* (GPSA) *Engineering Data Book* discusses many of these hydrocarbon treatment processes in detail.[6] Hydrogen sulfide removed from the raw gas is generally converted to elemental sulfur via the Claus process.[6]

After the necessary purification processes have been completed, the commercial-grade natural gas is compressed to approximately 1000 psig (6.9 MPag) and is introduced to a natural gas pipeline distribution network.[5] The gas is recompressed along the path to the consumer as necessary. Operating pressure at an individual natural gas burner located at a process furnace inside a petrochemical or hydrocarbon processing facility is reduced to an operating pressure range that typically varies between 5 and 30 psig, depending on the furnace's heating requirements and the individual burner's design specifications.

5.1.3 Liquified Petroleum Gas (LPG)

Liquefied petroleum gas (LPG) is the general term used to describe a hydrocarbon that is stored as a liquid under moderate pressure but is a gas under normal atmospheric conditions. LPG is vaporized for use as a fuel. The primary chemical components of LPG are propane, propylene, normal butane, isobutane, and butylene.[1] The *Gas Processors Suppliers Association* (GPSA) *Engineering Data Book* contains industry standard product specifications for commercial propane (predominantly propane and/or propylene), commercial butane (predominantly butane and/or butylene), and commercial butane-propane mixtures.[4] LPG produced via the separation of heavier hydrocarbons from natural gas is mainly paraffinic, containing primarily propane, normal butane, and isobutane. LPG derived from oil-refinery gas may contain varying small amounts of olefinic hydrocarbons such as propylene and butylene.[1]

Most of the LPG used in the United States consists primarily of propane.[8] Due to their relatively high boiling point, LPG mixtures containing high concentrations of normal butane (boiling point = 31°F or –1°C at atmospheric pressure) or isobutane (boiling point = 11°F or –12°C at atmospheric pressure) are preferred for use in warm climates. Conversely, LPG mixtures containing high concentrations of propane (boiling point = –44°F or –49°C) are typically preferred for use in cold climates.[5]

LPG is often used in the hydrocarbon or petrochemical industry as a fuel gas supplement or as a standby/start-up fuel. However, due to its value as both a common petrochemical feedstock and a marketable commodity, LPG is not typically preferred as a primary processing fuel.[9]

5.1.4 Refinery Gases

Although commercial natural gas and LPG are often used as fuels in processing plants, internally generated refinery fuel gases serve as the primary fuel component for most refineries, petrochemical plants, and hydrocarbon facilities. It is not usual for a process unit to produce its own fuel supply. Often, fuel gas streams from various processing units are delivered to a common mixing point within the plant, before the new gas mixture is returned to the processing units as refinery gas. Refinery fuel gases contain an extremely wide variety of chemical constituents, including paraffins, olefins, diolefins, aromatics, mercaptans, organic sulfides, ammonia, hydrogen sulfide, carbon monoxide, carbon dioxide, etc. Because plants must operate in a manner best suited to maximize profit, the individual fuel gas streams originating at each process unit will vary in composition and quantity, depending on numerous economic and technical factors.[10] Table 5.3 contains

TABLE 5.3 Composition of a Typical Refinery Gas

Fuel Gas Component	Refinery Fuel Gas Source (Dry Gas) Cracked Gas	Coking Gas	Reforming Gas	FCC Gas	Combined Refinery Gas – Sample 1	Combined Refinery Gas – Sample 2
CH_4	65%	40%	28%	32%	36%	53%
C_2H_4	3%	3%	7%	7%	5%	2%
C_2H_6	16%	21%	28%	9%	18%	19%
C_3H_6	2%	1%	3%	15%	8%	6%
C_3H_8	7%	24%	22%	25%	20%	14%
C_4H_8	1%	—	—	—	—	—
C_4H_{10}	3%	7%	7%	0%	2%	1%
C_5 & Higher	1%	—	—	—	—	—
H_2	3%	4%	5%	6%	3%	3%
CO	—	—	—	—	—	—
CO_2	—	—	—	—	—	—
N_2	—	—	—	7%	8%	3%
H_2O	—	—	—	—	—	—
O_2	—	—	—	—	—	—
H_2S	—	—	—	—	—	—
Total	100%	100%	100%	100%	100%	100%

Adapted from Nelson, 1949.[11]

typical chemical compositions of fuel gas streams originating from various process units within a petroleum refinery.[11]

It is very important that the refinery fuel gas leaving the common mixing point is a homogenous mixture of the fuel gas streams supplied. If the individual fuel gas supply streams vary significantly in calorific value, and if the supply streams are not combined in a homogeneous manner, the calorific value of the nonhomogeneous refinery fuel gas mixture will also vary widely and often instantaneously. Unless the gas burners and control systems at each processing furnace have been designed to accommodate instantaneous changes in fuel gas calorific value, the process will likely be impossible to control. All of the combustion performance parameters — including burner stability, emissions control, heat transfer efficiency, and heat flux — will suffer as a result of the nonhomogeneous fuel mixture.[10] Static mixers are often used in various segments of industry to ensure a well-mixed, homogeneous fuel gas mixture. However, static mixers are often impractical in the petrochemical and hydrocarbon processing industries, typically due to pressure drop limitations of the refinery fuel gas system.

Another problem often associated with the combustion of refinery fuel gases is the presence of liquid hydrocarbons in the refinery fuel gas stream, which can accelerate the coking and plugging rates of downstream gas burner components. Sources of unwanted liquid hydrocarbons in refinery fuel gas streams include condensation of heavier fuel gas components (C_5 and higher) due to natural cooling of the fuel gas stream, liquid entrainment into absorber or fractionator overhead gas streams, and lubrication oil contamination of the fuel gas stream. Potential solutions for the problems associated with these liquid hydrocarbons include liquid extraction of the heavier chemical components (C_5 and heavier) and filtration/coalescence of liquid components from the gas stream. In addition, increasing the velocity of the flowing gas through burner components (tips, risers, etc.) has been proven to cool the hardware and inhibit the cracking reactions that eventually lead to plugging and coking.

Wet fuel gas can introduce problems in cooler climates associated with the condensation and subsequent freezing of water vapor inside the fuel gas system. If the water vapor reaches the dew point in a cold atmospheric environment, there is danger of frost stoppage, freezing, or bursting of lines — a considerable fire safety hazard that merits serious thought. Options to combat water present in the fuel gas system include dehydration systems (as discussed in Section 5.2.1) and steam/electric tracing of refinery fuel gas lines.[10]

5.1.5 Combustible Waste Gas Streams

The quantity and variety of combustible waste gas streams in the hydrocarbon and petrochemical industries are virtually unlimited. Many of these waste gas streams are relatively high in inert concentration, with large amounts of nitrogen and carbon dioxide often present. As a result, these waste fuels are often low in heat content, with lower heating values in the range of 400 to 800 Btu/scf (0.42 to 0.84 MJ/scm). For these reasons, waste fuels are not usually compressed into the main refinery fuel gas system. Two of the most widely used

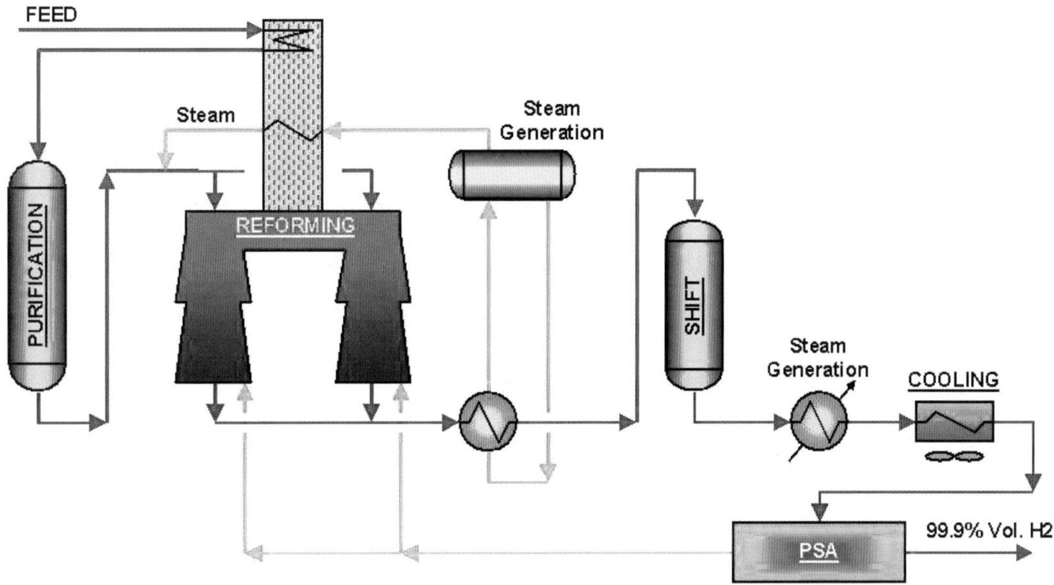

FIGURE 5.1 Simplified process flow diagram for hydrogen reforming/pressure swing adsorption. (Adapted from Meyers, 1997[12]).

combustible waste gas fuels, Pressure Swing Adsorption (PSA) tail gas and Flexicoking gas are discussed in detail in the sections below.

5.1.5.1 Pressure Swing Adsorption (PSA) Tail Gas

Pressure swing adsorption (PSA) tail gas is a low-pressure, low-Btu fuel gas produced as a by-product of a PSA process, a key purification component in the steam reforming hydrogen production process. Table 5.4 contains the approximate composition of a typical PSA tail gas fuel stream.

PSA is a cyclic process that uses beds of solid adsorbent to remove impurities such as carbon dioxide, carbon monoxide, methane, and nitrogen from the hydrogen production stream. A simplified process flow diagram of a typical steam reforming hydrogen production unit using PSA is shown in Figure 5.1.[12]

The steam reforming process is conducted in four stages:[8,9,12]

1. Feedstock preparation: feedstock (light hydrocarbons such as methane, propane, butane, and light liquid naphtha) at approximately 450 psig (31 bar) is preheated and purified to remove reformer catalyst poisons such as halogens and sulfur-containing compounds.
2. Reforming: the purified feedstock is reacted with steam to form carbon monoxide and hydrogen:

$$C_nH_m + nH_2O \xrightarrow{\text{1500°F \& Ni Catylst}} (n + m/2)H_2 + nCO \quad (5.1)$$

The reaction is endothermic and occurs within the process tubes of a reformer furnace in the presence of nickel catalyst at approximately 1500°F (815°C).

TABLE 5.4 Typical Composition of Steam Reforming/PSA Tail Gas

Fuel Gas Component	PSA Tail Gas Composition (vol%)
CH_4	17%
H_2O	<1%
H_2	28%
CO_2	44%
CO	10%
N_2	<1%
Total	100%

3. Shift conversion: the water-gas shift reaction is employed to convert the carbon monoxide produced in the reforming step into additional hydrogen and carbon dioxide:

$$CO + H_2O \rightarrow H_2 + CO_2 \quad (5.2)$$

The shift conversion step is exothermic and is conducted at approximately 650°F (343°C) in the presence of a chromium/iron oxide catalyst.

4. Hydrogen purification/PSA: following the shift conversion step, the hydrogen production stream enters the PSA portion of the process. Adsorbent beds remove the impurities (carbon dioxide, carbon monoxide, methane, and nitrogen) and a small portion of the product. Typical hydrogen recovery is 80% or greater, with product purity of approximately 99.9 vol%.

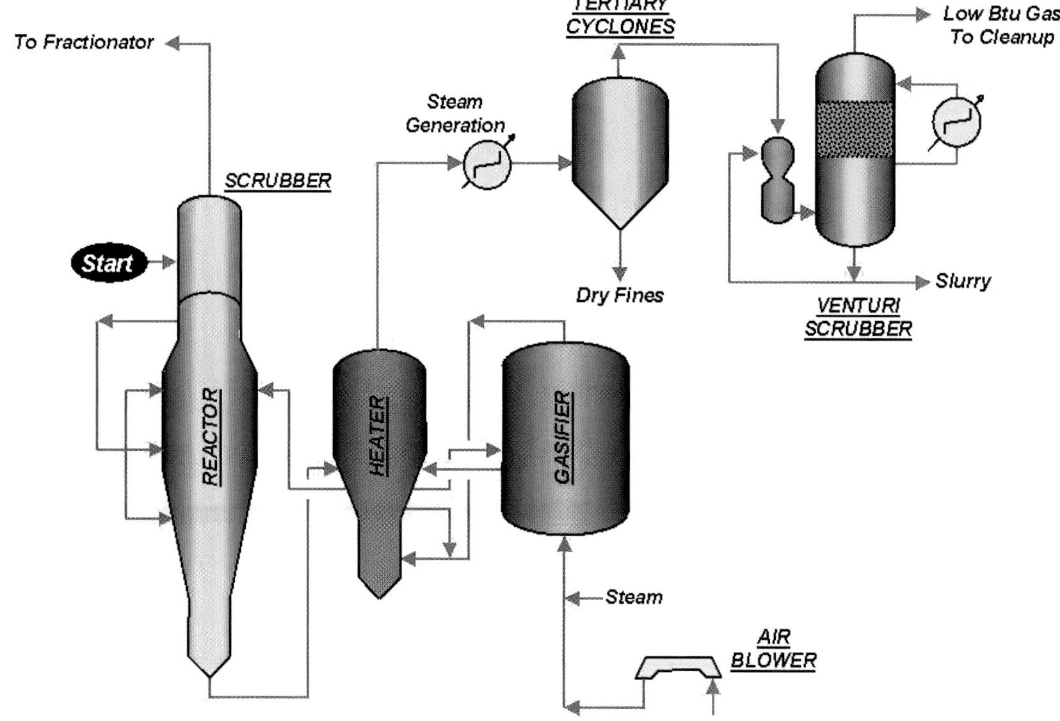

FIGURE 5.2 Simplified process flow diagram for Flexicoking.[13]

TABLE 5.5 Typical Composition of Flexicoking Waste Gas

Fuel Gas Component	Flexicoking Waste Gas Composition (by volume) Sample 1	Sample 2
CH_4	1.0%	0.8%
H_2	20.0%	21.0%
CO_2	10.0%	10.5%
CO	20.0%	18.6%
N_2	45.0%	45.6%
H_2O	4.0%	3.5%
H_2S	150 ppm	0
COS	120 ppm	120 ppm
Total	100%	100%

Adapted from Meyers, R.A., *Handbook of Petroleum Refining Processes*, 2nd ed., McGraw-Hill, New York, 1997.

The PSA unit must be frequently regenerated via depressurization of the adsorbent beds. When depressurization occurs, PSA tail gas (sometimes referred to as PSA waste gas) is produced at a pressure of about 5 psig (0.35 kg/cm^2) or less. The PSA tail gas consists of the impurities removed by the adsorbent beds, as well as the hydrogen that is not recovered in the product stream. The tail gas serves as the primary fuel for the reformer furnace burners. Due to flame stability problems associated with firing the low-pressure, high-inert-concentration (carbon dioxide and nitrogen) PSA tail gas alone, the PSA tail gas is typically supplemented by a light refinery fuel gas. The PSA and refinery fuel gases are fired in a dual-fuel burner specifically designed for the steam reforming/PSA process. In this arrangement, the PSA and refinery fuel gases enter the combustion zone through separate fuel connections and burner nozzles. The dual-fuel burners are capable of firing the two fuel mixtures separately or simultaneously, with PSA gas never providing more than 85 vol% of the total reformer fuel.

5.1.5.2 Flexicoking Waste Gas

Flexicoking waste gas is a low-pressure, low-Btu fuel gas produced by petroleum refiners as a by-product of the Exxon Flexicoking process. Flexicoking is a continuous fluidized-bed thermal cracking process used in the conversion of heavy hydrocarbon feedstocks (typically heavy gas oils from atmospheric and vacuum distillation) to various gaseous and liquid hydrocarbon products. Table 5.5 contains the approximate composition of two sample Flexicoking waste gas fuel streams.[13]

A simplified process flow diagram of the Flexicoking process is shown in Figure 5.2.[13] In the Flexicoking process,

TABLE 5.6 Volumetric Analysis of Typical Gaseous Fuel Mixtures

	Natural Gas				LPG		Refinery Gases (Dry)						Waste Gases	
Fuel Gas Component	Tulsa	Alaska	Netherlands	Algeria	Propane	Butane	Cracked Gas	Coking Gas	Reforming Gas	FCC Gas	Refinery Gas Sample 1	Refinery Gas Sample 2	PSA Gas	Flexicoking Gas
CH_4	93.4%	99%	81%	87%	—	—	65%	40%	28%	32%	36%	53%	17%	1%
C_2H_4	—	—	—	—	—	—	3%	3%	7%	7%	5%	2%	—	—
C_2H_6	2.7%	—	3%	9%	—	—	16%	21%	28%	9%	18%	19%	—	—
C_3H_6	—	—	—	—	—	—	2%	1%	3%	15%	8%	6%	—	—
C_3H_8	0.6%	—	0.4%	2.7%	100%	—	7%	24%	22%	25%	20%	14%	—	—
C_4H_8	—	—	—	—	—	100%	1%	—	—	—	—	—	—	—
C_4H_{10}	0.2%	—	0.1%	1.1%	—	—	3%	7%	7%	0%	2%	1%	—	—
C_5 & Higher	—	—	—	—	—	—	1%	—	—	—	—	—	—	—
H_2	—	—	—	—	—	—	3%	4%	5%	6%	3%	3%	28%	21%
CO	—	—	—	—	—	—	—	—	—	—	—	—	10%	20%
CO_2	0.7%	—	0.9%	—	—	—	—	—	—	—	—	—	44%	10%
N_2	2.4%	1%	14%	0%	—	—	—	—	—	7%	8%	3%	<1%	45%
H_2O	—	—	—	—	—	—	—	—	—	—	—	—	<1%	3%
O_2	—	—	—	—	—	—	—	—	—	—	—	—	—	—
H_2S	—	—	—	—	—	—	—	—	—	—	—	—	—	—
Total	100%	100%	100%	100%	100%	100%	100%	100%	100%	100%	100%	100%	100%	100%

Data compiled from a variety of sources.

TABLE 5.7 Physical Constants of Typical Gaseous Fuel Mixtures

	Natural Gas				LPG		Refinery Gases (Dry)						Waste Gases	
Fuel Gas Component	Tulsa	Alaska	Netherlands	Algeria	Propane	Butane	Cracked Gas	Coking Gas	Reforming Gas	FCC Gas	Refinery Gas Sample 1	Refinery Gas Sample 2	PSA Gas	Flexicoking Gas
Molecular weight	17.16	16.1	18.51	18.49	44.1	58.12	22.76	28.62	30.21	29.18	28.02	24.61	25.68	23.73
Lower heating value (LHV), Btu/SCF	913	905	799	1025	2316	3010	1247	1542	1622	1459	1389	1297	263	131
Higher heating value (HHV), Btu/SCF	1012	1005	886	1133	2517	3262	1369	1686	1769	1587	1515	1421	294	142
Specific gravity (14.696 psia/60°F, Air = 1.0)	0.59	0.56	0.64	0.64	1.53	1.1	0.79	0.99	1.05	1.01	0.97	0.85	0.89	0.82
Wobbe number, HHV/($SG^{1/2}$)	1318	1343	1108	1416	2035	3110	1540	1694	1726	1579	1538	1541	312	157
Isentropic coefficient (Cp/Cv)	1.30	1.31	1.31	1.28	1.13	1.10	1.24	1.19	1.19	1.20	1.21	1.23	1.33	1.38
Stoichiometric air required, SCF/MMBtu	10554	10567	10554	10525	10369	10371	10402	10379	10322	10234	10311	10375	9667	8265
Stoichiometric air required, lb_m/MMBtu	805	806	805	803	791	791	794	792	787	781	787	792	738	630
Air required for 15% excess air, SCF/MMBtu	12138	12152	12138	12104	11925	11926	11962	11936	11870	11769	11858	11931	11117	9505
Air required for 15% excess air, lb_m/MMBtu	923	924	923	920	907	907	910	908	903	895	902	907	845	723
Volume of dry combustion products, SCF/MMBtu	10983	10956	11141	10953	10962	10996	10890	10909	10871	10847	10911	10904	11722	13517
Weight of dry combustion products, lb_m/MMBtu	865	862	876	863	870	874	861	864	862	860	864	862	985	1103
Volume of wet combustion products, SCF/MMBtu	13257	13258	13415	13163	12788	12757	12935	12862	12771	12689	12821	12902	14198	15585
Weight of wet combustion products, lb_m/MMBtu	973	971	984	968	957	958	958	957	952	948	864	957	1102	1201
Adiabatic flame temperature, °F	3306	3308	3284	3317	3351	3351	3342	3348	3359	3371	3353	3345	3001	2856

Note: All values calculated using 60°F fuel gas and 60°F, 50% relative humidity combustion air.

TABLE 5.8 Physical Constants of Typical Gaseous Fuel Mixture Components

No.	Fuel Gas Component	Chemical Formula	Molecular Weight	Boiling Point 14.696 psia (°F)	Vapor Pressure 100°F (psia)	Specific Heat Capacity, C_p 60°F & 14.696 psia (Btu/lb$_m$/°F)	Latent Heat of Vaporization 14.696 psia & Boiling Point (Btu/lb$_m$)	Gas Density Ideal Gas, 14.696 psia, 60°F: Specific Gravity (Air = 1)	Gas Density (lb$_m$/ft^3)	Specific Volume (ft^3/lb$_m$)	Heating Value Btu/scf LHV (Net)	Btu/scf HHV (Gross)	Btu/lb$_m$ LHV (Net)	Btu/lb$_m$ HHV (Gross)	Unit Volume per Unit Volume of Combustible, Required for Combustion: O_2	N_2	Air	Flue Gas Products: CO_2	H_2O	N_2	SO_2	Unit Mass per Unit Mass of Combustible, Required for Combustion: O_2	N_2	Air	Flue Gas Products: CO_2	H_2O	N_2	SO_2	Theoretical Air Required (lb$_m$/10,000 Btu)	Flammability Limits (vol% in air mixture) Lower	Upper	No.
	Paraffin (alkane) Series (C_nH_{2n+2})																															
1	Methane	CH_4	16.04	−258.69	—	0.5266	219.22	0.554	0.042	23.651	912	1,013	21,495	23,845	2.0	7.547	9.547	1.0	2.0	7.547	—	3.989	13.246	17.235	2.743	2.246	13.246	—	7.219	5.0	15.0	1
2	Ethane	C_2H_6	30.07	−127.48	—	0.4097	210.41	1.038	0.079	12.618	1,639	1,792	20,418	22,323	3.5	13.206	16.706	2.0	3.0	13.206	—	3.724	12.367	16.092	2.927	1.797	12.367	—	7.209	2.9	13.0	2
3	Propane	C_3H_8	44.10	−43.67	190	0.3881	183.05	1.522	0.116	8.604	2,385	2,592	19,937	21,669	5.0	18.866	23.866	3.0	4.0	18.866	—	3.628	12.047	15.676	2.994	1.624	12.047	—	7.234	2.1	9.5	3
4	n-Butane	C_4H_{10}	58.12	31.10	51.6	0.3867	165.65	2.007	0.153	6.528	3,113	3,373	19,679	21,321	6.5	24.526	31.026	4.0	5.0	24.526	—	3.578	11.882	15.460	3.029	1.550	11.882	—	7.251	1.8	8.4	4
5	Isobutane	C_4H_{10}	58.12	10.90	72.2	0.3872	157.53	2.007	0.153	6.528	3,105	3,365	19,629	21,271	6.5	24.526	31.026	4.0	5.0	24.526	—	3.578	11.882	15.460	3.029	1.550	11.882	—	7.268	1.8	8.4	5
6	n-Pentane	C_5H_{12}	72.15	96.92	15.57	0.3883	153.59	2.491	0.190	5.259	3,714	4,017	19,507	21,095	8.0	30.186	38.186	5.0	6.0	30.186	—	3.548	11.781	15.329	3.050	1.498	11.781	—	7.267	1.4	8.3	6
7	isopentane	C_5H_{12}	72.15	82.12	20.44	0.3827	147.13	2.491	0.190	5.259	3,705	4,017	19,459	21,047	8.0	30.186	38.186	5.0	6.0	30.186	—	3.548	11.781	15.329	3.050	1.498	11.781	—	7.283	1.4	8.3	7
8	Neopentane	C_5H_{12}	72.15	49.10	35.9	0.3666	135.58	2.491	0.190	5.259	3,692	3,994	19,390	20,978	8.0	30.186	38.183	5.0	6.0	30.186	—	3.548	11.781	15.329	3.050	1.498	11.781	—	7.307	1.4	8.3	8
9	n-Hexane	C_6H_{14}	86.18	155.72	4.956	0.3664	143.95	2.975	0.227	4.403	4,415	4,767	19,415	20,966	9.5	35.846	45.346	6.0	7.0	35.846	—	3.527	11.713	15.240	3.064	1.463	11.713	—	7.269	1.2	7.7	9
	Napthene (cycloalkane) Series (C_nH_{2n})																															
10	Cyclopentane	C_5H_{10}	70.13	120.60	9.917	0.2712	137.35	2.420	0.180	5.556	3,512	3,764	19,005	20,368	7.5	27.939	35.180	5.0	5.0	28.939	—	3.850	11.155	14.793	3.146	1.283	11.155	—	7.262	—	—	10
11	Cyclohexane	C_6H_{12}	84.16	177.40	3.267	0.2901	153.25	2.910	0.220	5.545	4,180	4,482	18,849	20,211	9.0	33.528	42.970	6.0	6.0	33.528	—	4.620	13.386	17.750	3.146	1.283	11.155	—	7.848	1.3	8.4	11
	Olefin Series (C_nH_{2n})																															
12	Ethene (Ethylene)	C_2H_4	28.05	−154.62	—	0.3622	207.57	0.969	0.074	13.525	1,512	1,613	20,275	21,636	3.0	11.320	14.320	2.0	2.0	11.320	—	3.422	11.362	14.784	3.138	1.284	11.362	—	6.833	2.7	34.0	12
13	Propene (Propylene)	C_3H_6	42.08	−53.90	226.4	0.3541	188.18	1.453	0.111	9.017	2,185	2,336	19,687	21,048	4.5	16.980	21.480	3.0	3.0	16.980	—	3.422	11.362	14.784	3.138	1.284	11.362	—	7.024	2.0	10.0	13
14	1-Butene (Butylene)	C_4H_8	56.11	20.75	63.05	0.3548	167.94	1.937	0.148	6.762	2,885	3,086	19,493	20,854	6.0	22.640	28.640	4.0	4.0	22.640	—	3.422	11.362	14.784	3.138	1.284	11.362	—	7.089	1.6	9.3	14
15	Isobutene	C_4H_8	56.11	19.59	63.4	0.3701	169.48	1.937	0.148	6.762	2,868	3,069	19,376	20,737	6.0	22.640	28.640	4.0	4.0	22.640	—	3.422	11.362	14.784	3.138	1.284	11.362	—	7.129	1.6	—	15
16	1-Pentene	C_5H_{10}	70.13	85.93	19.115	0.3635	154.46	2.421	0.185	5.410	3,585	3,837	19,359	20,720	7.5	28.300	35.800	5.0	5.0	28.300	—	3.422	11.362	14.784	3.138	1.284	11.362	—	7.135	1.4	8.7	16
	Aromatic Series (C_0H_{2n-6})																															
17	Benzene	C_6H_8	78.11	176.17	3.224	0.2429	169.31	2.697	0.206	4.857	3,595	3,746	17,421	18,184	7.5	28.300	35.800	6.0	3.0	28.300	—	3.072	10.201	13.274	3.380	0.692	10.201	—	7.300	1.38	7.98	17
18	Toluene	C_7H_8	92.14	231.13	1.032	0.2598	154.84	3.181	0.243	4.118	4,296	4,497	17,672	18,501	9.0	33.959	42.959	7.0	4.0	33.959	—	3.125	10.378	13.504	3.343	0.782	10.378	—	7.299	1.28	7.18	18
19	o-Xylene	C_8H_{10}	106.17	291.97	0.264	0.2914	149.1	3.665	0.280	3.574	4,970	5,222	17,734	18,633	10.5	39.619	50.119	8.0	5.0	39.619	—	3.164	10.508	13.673	3.316	0.848	10.508	—	7.338	1.18	6.48	19
20	m-Xylene	C_8H_{10}	106.17	282.41	0.326	0.2782	147.2	3.665	0.280	3.574	4,970	5,222	17,734	18,633	10.5	39.619	50.119	8.0	5.0	39.619	—	3.164	10.508	13.673	3.316	0.848	10.508	—	7.338	1.18	6.48	20
21	p-Xylene	C_8H_{10}	106.17	281.05	0.342	0.2769	144.52	3.665	0.280	3.574	4,970	5,222	17,734	18,633	10.5	39.619	50.119	8.0	5.0	39.619	—	3.164	10.508	13.673	3.316	0.848	10.508	—	7.338	1.18	6.48	21
	Additional Fuel Gas Components																															
22	Acetylene	C_2H_2	26.04	−119	—	0.3966	—	0.899	0.069	14.572	1,448	1,499	20,769	21,502	2.5	9.433	11.933	2.0	1.0	9.433	—	3.072	10.201	13.274	3.380	0.692	10.201	—	7.300	2.5	80	22
23	Methyl alcohol	CH_3OH	32.04	148.1	4.63	0.3231	473	1.106	0.084	11.841	767	868	9,066	10,258	1.5	5.660	7.160	1.0	2.0	5.660	—	4.498	4.974	6.482	1.373	1.124	4.974	—	6.309	6.72	36.5	23
24	Ethyl alcohol	C_2H_5OH	46.07	172.92	2.3	0.3323	367	1.590	0.121	8.236	1,449	1,600	11,918	13,161	3.0	11.320	14.320	2.0	3.0	11.320	—	2.084	6.919	9.003	1.911	1.173	6.919	—	6.841	3.28	18.95	24
25	Ammonia	NH_3	17.03	−28.2	212	0.5002	587.2	0.588	0.045	22.279	364	441	7,966	9,567	0.75	2.830	3.582	—	1.5	3.330	—	1.409	4.679	6.008	—	1.587	5.502	—	6.298	15.50	27.00	25
26	Hydrogen	H_2	2.02	−423.0	—	3.4080	193.9	0.070	0.005	188.217	274.6	325.0	51,625	61,095	0.5	1.887	2.387	—	1.0	1.887	—	7.936	26.323	34.290	—	8.937	26.353	—	5.613	4.00	74.20	26
27	Oxygen	O_2	32.00	−297.4	—	0.2186	91.6	1.105	0.084	11.858	—	—	—	—	—	—	—	—	—	—	—	—	—	—	—	—	—	—	—	—	—	27
28	Nitrogen	N_2	29.16	−320.4	—	0.2482	87.8	0.972	0.074	13.472	—	—	—	—	—	—	—	—	—	—	—	—	—	—	—	—	—	—	—	—	—	28
29	Carbon monoxide	CO	28.01	−313.6	—	0.2484	92.7	0.967	0.074	13.546	321.9	321.9	4,347	4,347	0.5	1.877	2.387	1.0	—	1.887	—	—	1.897	2.468	1.571	—	1.870	—	5.677	12.50	74.20	29
30	Carbon dioxide	CO_2	44.01	−109.3	—	0.1991	238.2	1.519	0.116	8.621	—	—	—	—	—	—	—	—	—	—	—	—	—	—	—	—	—	—	—	—	—	30
31	Hydrogen sulfide	H_2S	34.08	−76.6	394.0	0.2380	235.6	1.177	0.090	11.133	595	646	6,537	7,097	1.5	5.660	7.160	—	1.0	5.660	1.0	1.410	4.682	6.093	—	0.529	4.682	1.880	8.585	4.30	45.50	31
32	Sulfur dioxide	SO_2	64.06	14.0	88	0.1450	166.7	2.212	0.169	5.923	—	—	—	—	—	—	—	—	—	—	—	—	—	—	—	—	—	—	—	—	—	32
33	Water vapor	H_2O	18.02	212.0	0.9492	0.4446	970.3	0.622	0.047	21.061	—	—	—	—	—	—	—	—	—	—	—	—	—	—	—	—	—	—	—	—	—	33
34	Air	—	29.97	−317.6	—	0.2400	92	1.000	0.076	13.099	—	—	—	—	—	—	—	—	—	—	—	—	—	—	—	—	—	—	—	—	—	34

hot (500 to 700°F or 260 to 370°C) gas oil is injected into the reactor vessel containing hot, fluidized coke particles. Thermal cracking reactions inside the reactor vessel produce fresh petroleum coke that is deposited as a thin film on the surface of existing coke particles inside the reactor bed. Cracked vapor products exit the Flexicoking process through the reactor vessel overhead stream for additional downstream processing. Coke from the reactor vessel is continuously injected into the top of a second fluidized vessel, the coke heater, where it is heated and recycled to maintain a reactor bed temperature of 950 to 1000°F (510 to 540°C). A portion of the coke fed into the top section of the coke heater is injected into the bottom of a third fluidized vessel, the gasifier. Inside the gasifier, the coke is reacted with air and steam at approximately 1500 to 1800°F (820 to 980°C), producing a low-Btu fuel gas, or Flexicoking gas, consisting primarily of nitrogen, hydrogen, carbon monoxide, and carbon dioxide. The Flexicoking gas flows from the top of the gasifier to the bottom of the heater, where it provides the heat necessary to maintain the reactor bed temperature and helps fluidize the coke heater bed. The high-temperature Flexicoking gas leaving the coke heater is used for high-pressure steam generation before entrained coke fines are removed in a cyclone/venturi scrubber system. Because the low-Btu gas stream leaving the Flexicoking process contains substantial concentrations of H_2S (~150 ppm by volume), the gas must first be sent through a hydrogen sulfide removal system before it can be burned as fuel.[9,13]

5.1.6 Physical Properties of Gaseous Fuels

Tables 5.6, 5.7, and 5.8 provide physical and combustion property data for a large variety of common fuel gas mixtures and their chemical components.

5.1.7 Photographs of Gaseous Fuel Flames

Figures 5.3 through 5.18 are photographs of a John Zink PSFG staged fuel gas burner, firing a wide variety of fuel gas mixtures into open air (i.e., not in a furnace). In each photograph, the burner is being operated at the same fuel flow rate (in terms of energy released per unit time) and under the same general ambient conditions. Fuel composition is the only parameter that is varied throughout the series. The images are provided to illustrate the differences in flame appearance (shape and color) produced by various fuel compositions.

Figures 5.19 to 5.33 show a similar series of photos of a John Zink VYD raw gas burner firing a wide variety of fuels inside a furnace all at the same firing rate.

There is a widely held misconception that yellow-flame burning is solely the direct result of combustion air deficiency. Inadequate or unsatisfactory fuel/air mixing will certainly result in the production of yellow flame. However, both yellow and blue flame burning can occur at virtually any condition of deficient or surplus combustion air.

Yellow flame burning is the direct result of the cracking of a hydrocarbon fuel into its hydrogen and carbon components, followed by separate burning of the two constituents. The hydrogen constituents are burned in a rapid process that produces a pale lavender-pink flame that is very difficult to see except against a dark background. When yellow flame burning occurs, the heavier carbon constituents burn in a relatively slower process that typically results in a luminescent yellow flame.

Blue flame burning is the direct result of progressive oxygenation of the fuel in a manner that does not allow uncombined carbon to be present in the reaction (flame) envelope. Inadequate fuel/air mixing can severely limit this reaction process, producing a greater tendency toward yellow flame. Both yellow and blue flame burnings are possible with any hydrocarbon fuel, and both kinds of flame produce equivalent quantities of heat.

The hydrogen-to-carbon weight ratio (H:C) is a good indicator of a fuel mixture's relative tendency to produce yellow flame burning, with low H:C ratios corresponding to an increased movement toward yellow flame burning. Pure hydrogen (H:C = ∞) typically burns as a pale lavender-pink flame that is very difficult to see except against a dark background. Pure methane (H:C = 0.33) typically burns as a light blue flame. Fuel mixtures containing propane (H:C = 0.22), butane (H:C = 0.21), and the olefins (H:C = 0.166) all have a greater tendency to exhibit yellow flame burning than pure methane fuels.[10]

5.2 LIQUID FUELS

5.2.1 Introduction and History

Liquid fuels are a key component in today's energy processes. An example of an oil flame is shown in Figure 5.34. During the Industrial Revolution, starting in the mid-18th century, the major energy source used in the world changed from charcoal (wood) to various forms of coal. As technology developed, the world began moving from the use of coal to crude oil (the most abundant liquid fuel used in industry today), and its derivatives, to provide the energy and heating requirements needed. The modern era of viable crude oil production and use began with commercial wells in the mid-1800s. An increasing need for oil products in technology (such as gasoline for the internal combustion engine and automobiles) spurred massive efforts in oil exploration and recovery in the early 1900s. Figure 5.35 depicts an oil derrick, circa 1900.

FIGURE 5.3 100% TNG flame.

FIGURE 5.4 90% TNG/10% N_2 flame.

FIGURE 5.5 80% TNG/20% N_2 flame.

FIGURE 5.6 90% TNG/10% H_2 flame.

Major oil deposits found in the United States prompted it to become a major world oil producer. The successes of American oil discovery and production inspired oil companies in other countries to start a worldwide exploration for oil reserves. In the mid-1950s, major U.S. oil companies provided approximately two thirds of the world's oil supply at prices near U.S. $1 per barrel (approximately 3000% lower than the current per barrel price).[14] In 1960, the Organization of Petroleum Exporting Countries (OPEC) was founded by the governments of major oil-exporting countries for the

FIGURE 5.7 75% TNG/25% H_2 flame.

FIGURE 5.8 50% TNG/50% H_2 flame.

FIGURE 5.9 25% TNG/75% H_2 flame.

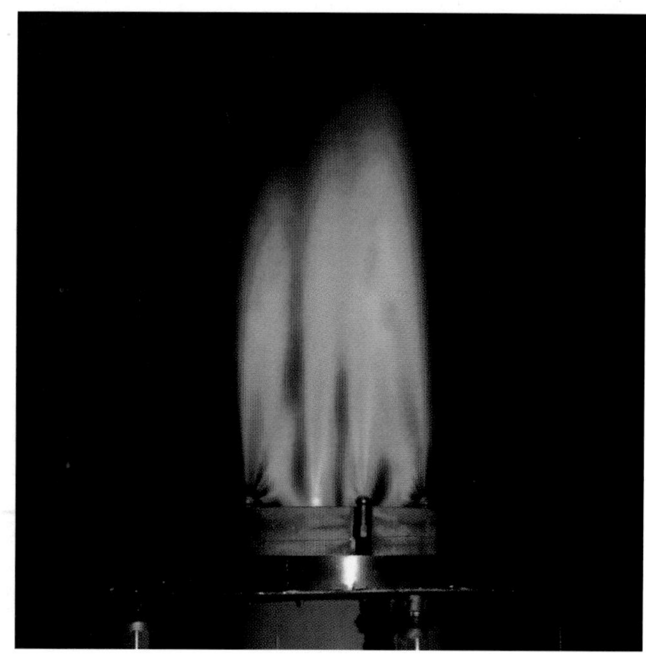

FIGURE 5.10 100% H_2 flame.

purpose of stabilizing oil production and prices. As the demand for oil increased, the production inevitably increased. The consumption of crude oil in 1998 was approximately 70 million barrels per day.[15] Technological developments in drilling and exploration techniques have identified and exploited oil reserves throughout the world. The vast majority of the known oil reserves in the world are located in the Middle East (approximately two thirds), while the United States ranks eighth on the known reserve list. The United States produces only about 17% of the world's oil, yet it consumes nearly

FIGURE 5.11 50% TNG/25% H_2/25% C_3H_8 flame.

FIGURE 5.12 50% TNG/50% C_3H_8 flame.

FIGURE 5.13 100% C_3H_8 flame.

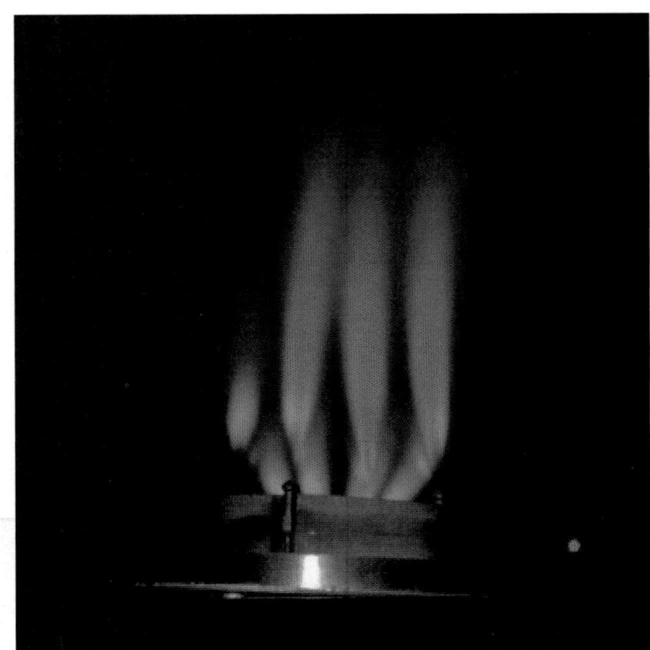

FIGURE 5.14 100% C_4H_{10} flame.

30%.[16] The Energy Information Administration (Office of Oil & Gas) estimates that the total U.S. crude oil stocks are 310 million barrels, excluding the strategic petroleum reserve, as of May 19, 2000.[14] Estimates of the United States' remaining oil reserves range from capacities of 25 to 100 billion barrels. Based on our current consumption patterns, these reserves could supply us with enough oil for only 10 to 30 more years.

FIGURE 5.15 Simulated cracked gas flame.

FIGURE 5.16 Simulated coking gas flame.

FIGURE 5.17 Simulated FCC gas flame.

FIGURE 5.18 Simulated reforming gas flame.

Oil exploration, which initially was confined to land, has led to recovery efforts on the bottom of the ocean floor. The most abundant forms of oil deposits found in the world today are oil shale, heavy oil deposits, and tar sands. However, difficulty with, and the high cost of extracting oil from, these complicated mediums keeps conventional crude oil recovery as the leading source of usable raw material for refining processes. Figure 5.36 shows the capping of a burning oil well.

FIGURE 5.19 100% Tulsa natural gas.

FIGURE 5.20 100% hydrogen.

FIGURE 5.21 100% propane.

FIGURE 5.22 50% hydrogen/50% propane.

FIGURE 5.23 50% hydrogen/50% Tulsa natural gas.

FIGURE 5.24 50% propane/50% Tulsa natural gas.

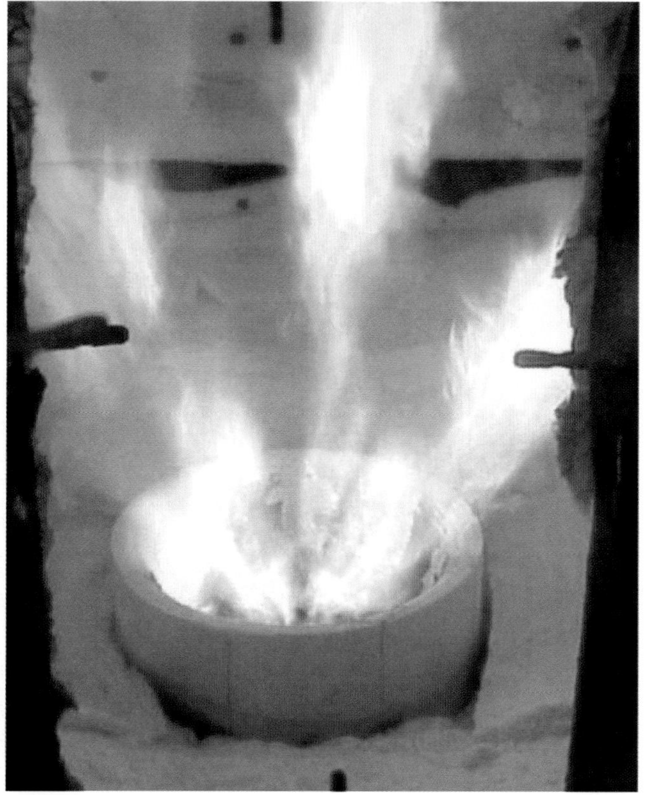

FIGURE 5.25 25% hydrogen/75% propane.

FIGURE 5.26 75% hydrogen/25% propane.

FIGURE 5.27 25% hydrogen/75% Tulsa natural gas.

FIGURE 5.28 75% hydrogen/25% Tulsa natural gas.

FIGURE 5.29 25% propane/75% Tulsa natural gas.

FIGURE 5.30 75% propane/25% Tulsa natural gas.

FIGURE 5.31 25% hydrogen/25% propane/50% Tulsa natural gas.

FIGURE 5.32 25% hydrogen/50% propane/25% Tulsa natural gas.

5.2.2 Oil Recovery

Crude oil is found in deep, high-pressure reservoirs, encased in rock, beneath the Earth's surface. Oil companies use complicated drilling techniques to tap into these pockets and bring the crude oil to the surface so that it can be collected. Oil drilling is an expensive process that can be complicated by the location of the oil in the earth. Therefore, oil companies spend millions of dollars annually in exploration and cost analysis of potential, new oil reserves. Incredibly hard rock and deep reserves (sometimes greater than 3000 ft below the surface) necessitate the use of specially designed drill bits that will stand up to the high pressures and constant mechanical trauma encountered in drilling. Once an oil reservoir is "hit," the oil, now having an avenue to expand, will rush out of the drilling channel that was cleared by the drilling rig. The oil will be continuously extracted until the reservoir becomes depleted to the extent that it is no longer economically viable for a company to spend time and money to retrieve it. When the oil pressure in the reservoir becomes too low for natural extraction, pumps can be used to help with the extraction. Other means of keeping reservoirs "active" include injecting water, steam, or chemicals into the reservoir to help make low-pressure, or viscous oil easier to extract.

Once the crude oil has been collected, and temporary storage facilities are nearing their capacity, it must be off-loaded

FIGURE 5.33 50% hydrogen/25% propane/25% Tulsa natural gas.

FIGURE 5.34 Viewing oil flame through a burner plenum.

FIGURE 5.35 Oil derrick, *circa* 1900.

FIGURE 5.36 Capping a burning oil well.

so that further collection is possible. The most common methods of off-loading and transporting crude oil is with pipelines (such as the Great Alaskan Pipeline), seafaring oil tankers, and barges. These transportation methods deliver the crude oil to locations around the world for refining into usable petroleum products.

5.2.3 Production, Refining, and Chemistry

The primary concern for a typical refinery is to convert a barrel of crude oil (42 U.S. gallons) into usable products. A barrel of crude oil can typically be refined to provide 11 gallons of gasoline, 5.3 gallons of kerosene, 20.4 gallons of gas-oil and distillates, and 5.3 gallons of heavier distillates.[9] The end products derived from crude oil number in the thousands. Table 5.9 provides a listing of many of these products. The processes that produce these different products are vast and complicated. Figure 5.37 provides a general refinery flow diagram.

The primary chemical components of crude oil are carbon, hydrogen, sulfur, oxygen, and nitrogen. The percentages of these elements found in a crude oil are most frequently used

TABLE 5.9 Quantitative Listing of Products Made by the U.S. Petroleum Industry

Product Classification	Number of Individual Products
Lubricating oils	1156
Chemicals, solvents, misc.	300
Greases	271
Asphalts	209
Waxes	113
White oils	100
Rust preventatives	65
Diesel and light fuel oils	27
Motor gasolines	19
Residual fuel oil	16
Liquified gases	13
Other gasolines	12
Transformer and cable oils	12
Kerosenes	10
Aviation gasolines	9
Jet fuels	5
Carbon blacks	5
Cokes	4
Fuel gas	1
Total	2347

From Gary, 1994.[9]

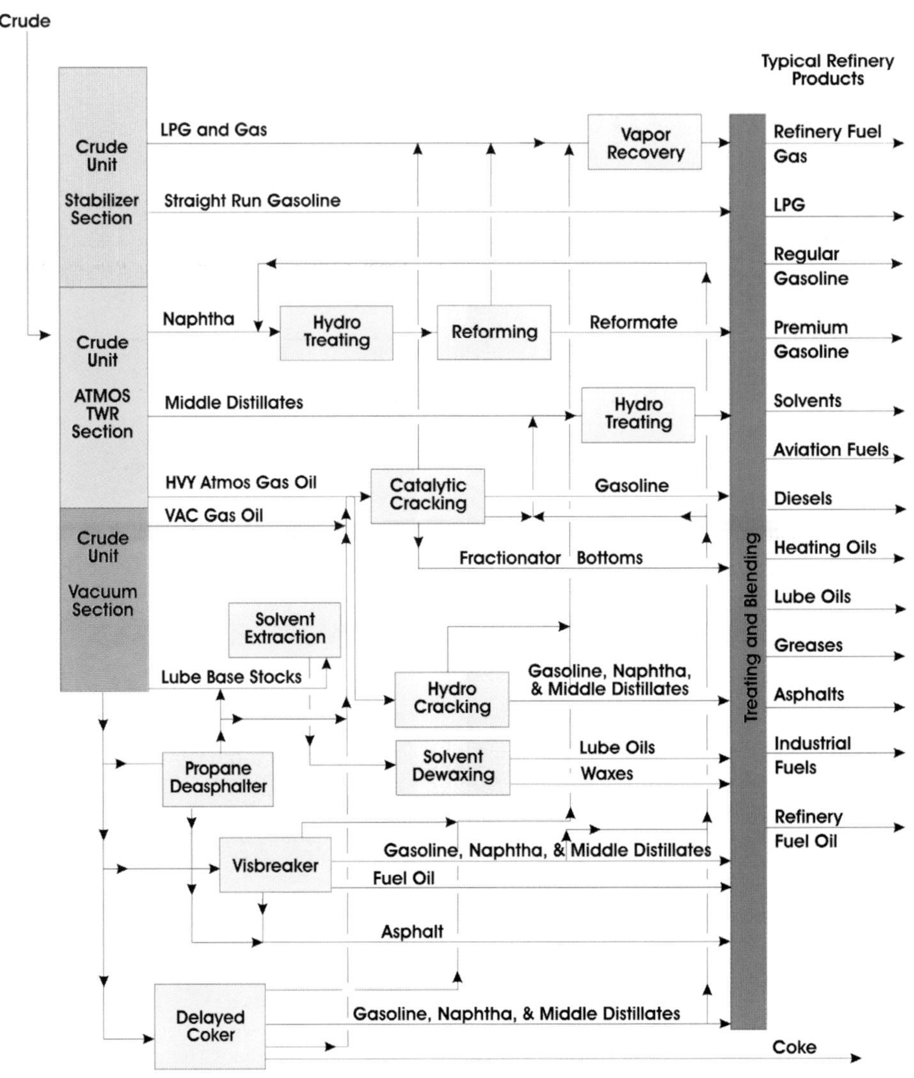

FIGURE 5.37 Refinery flow diagram.[9]

to characterize the oil. Two terms frequently used when referring to crude oil are "sweet" crude and "sour" crude. Sweet crude is oil that contains less than 0.5 wt.% sulfur, while sour crude contains greater than 0.5 wt.% sulfur. Sulfur content is of importance and concern, due to the sulfur oxides that are produced during combustion. SO_2, for example, is a gas that has been shown to contribute significantly to several different environmental problems — namely in acid rain formation and in its ready conversion to sulfuric acid, H_2SO_4. The nitrogen content of crude oil is of special interest to the combustion industry due to the high levels of nitrogen oxides or NOx (see Chapter 6) produced during combustion of these fuels (e.g., approximately 0.2 lb per MMBtu NOx or 142 ppm can be attributed to "Fuel NOx" for an oil that contains 0.47 wt.% nitrogen). Like SOx, NOx is an environmentally damaging group of gases. Any time a fuel is burned in air with a hot flame, NOx are produced. The greater the flame temperature of the combustion, the greater the amount of NO that will be produced. NO is then oxidized to form NO_2 (over a period of minutes or hours), which is a major contributor to photochemical smog. In general, the fate of SO_2 and NO are intertwined, as can be seen by the following reaction sequence[17]:

$$SO_2 + OH^{\bullet} \rightarrow HSO_3^{\bullet}$$

$$HSO_3^{\bullet} + O_2 \rightarrow SO_3 + HOO^{\bullet}$$

$$SO_3 + H_2O \rightarrow H_2SO_4(g)$$

$$H_2SO_4(g) \xrightarrow{H_2O} H_2SO_4(aq)$$

Below is a parallel reaction that takes place between nitrogen oxide and the hydroperoxy radical, thus producing more of the hydroxyl radical to feed the initial reaction above:

$$HOO^{\bullet} + NO^{\bullet} \rightarrow OH^{\bullet} + NO_2^{\bullet}$$

The overall reaction is then:

$$SO_2 + NO^{\bullet} + O_2 \xrightarrow{H_2O} H_2SO_4(aq)$$

Crude oil compositions are relatively constant. However, slight deviations in composition can result in vastly different refining methods. Crude oils also contain inorganic elements such as vanadium, nickel, and sodium, and usually contain some amount of water and ash (noncombustible material). The main hydrocarbon constituents of crude oils are alkanes (paraffins), cycloalkanes (naphthenes), and aromatics.

Alkanes (also called paraffins after the Latin *parum affinis*, "little affinity") are those chemical structures that are based on carbon atoms having only single bonds and that are completely saturated with hydrogen atoms. Some of the alkane hydrocarbons are listed in Table 5.8. The basic chemical formula for an alkane is C_xH_{2x+2}, where "x" is the number of carbon atoms present. Crude oils can contain structures with up to 70 carbon atoms.[9] However, the vast majority of the compounds contain 40 carbon atoms or less. When the number of different constitutional isomers (different chemical connectivity and different physical properties, yet identical chemical formulae) is considered (tetracontane [$C_{40}H_{82}$] has over 62 trillion possible isomers[18]), it is evident that the compositional diversity between differing crude oils is almost limitless.

Cycloalkanes (cylcoparaffins or naphthenes) are alkanes in which all or some of the carbon atoms are arranged in a ring. When a cycloalkane contains only one ring, the general formula is C_xH_{2x}. The most stable cycloalkane is cyclohexane, while cyclobutane and cyclopropane are the least stable. The properties of cycloalkanes are very similar to those of alkanes, as shown in Table 5.8.

Aromatic compounds are those compounds that contain at least one benzene-like ring. Benzene, discovered in 1825, has a chemical formula of C_6H_6, and is stable and nonreactive relative to alkanes and cycloalkanes. Aromatics, such as the heterocyclic compounds pyridine and furan, are composed of rings that contain elements other than carbon. For example, the benzene ring contains six carbon atoms, whereas the pyridine ring contains five carbon atoms and one nitrogen atom. Properties of some of the aromatic compounds are contained in Table 5.8.

TABLE 5.10 General Fraction Boiling Points

Distillation Fraction	Temperature Range
Butanes and lighter	<90°F
Gasoline	90–220°F
Naphtha	220–315°F
Kerosene	315–450°F
Fuel oils	450–800°F
Residue	>800°F

From Leffler, W.L., *Petroleum Refining for the Non-technical Person*, Penn Well Publishing, Tulsa, OK, 1985. With permission.

It is worth mentioning the group of compounds called alkenes (olefins). Alkene compounds do not occur naturally in crude oil, but are produced by reaction during the refining process. Therefore, it should be expected that a refined end product will have some percentage of ethylene, propylene, or butylene, for example. Alkenes have the general formula of C_xH_{2x} and contain a carbon-carbon double bond. Properties of some of the alkenes are contained in Table 5.8.

When a crude oil is refined, the first step is, invariably, distillation. The purpose of distillation is to separate lighter components from heavier ones, based on their respective volatility. The target of distillation is to separate the crude oil into different fractions. Each fraction consists of a boiling point range that will yield a mixture of hydrocarbons; see Table 5.10. Some of these mixtures can then be used as product (fuels, solvents, etc.) or further refined into gasoline or other desirable mixtures. Catalytic cracking is a typical process used to break down and rearrange alkane mixtures produced via distillation into smaller, highly branched alkanes by heating the mixtures to high temperatures in the presence of a variety of catalysts. Figure 5.38 shows a fluid catalytic cracking process. Due to the reactions that take place during catalytic cracking, the product streams are generally heavier than the feed streams. Alkanes that are more highly branched are desirable because they have a higher octane rating than their unbranched cousins.

Of particular interest are the liquid fuels produced during the various refining processes that are used by the hydrocarbon and petrochemical industries. Refineries frequently burn these liquid fuels in process heaters so that the heat liberated during combustion can be used to drive a more profitable process. Light fuel oils are relatively easy to burn and produce flames similar to gas flames, while heavier oils require a more complicated process and produce flames that are quite radiant and more highly dependent on atomization techniques than the light oils. Oils are fired in burners by themselves, or in combination with fuel gas, waste gas, or both. Naphtha is frequently fired in combination with a PSA or

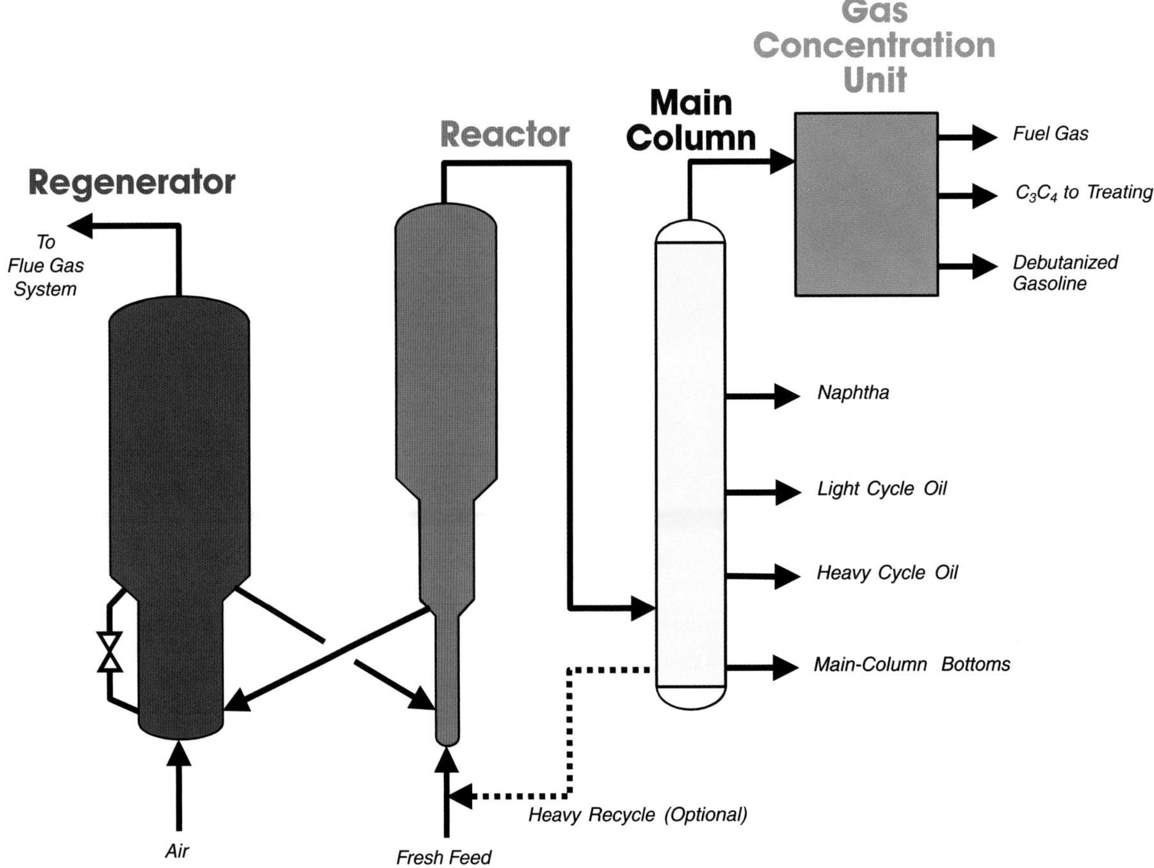

FIGURE 5.38 Flow diagram of UOP fluid catalytic cracking complex.[27]

other waste gas, and requires good vaporization to provide a quality flame.

5.2.4 Oils

According to the American Standard Testing Methods (ASTM) D-396, fuel oils are divided into grades, based on the types of burners for which they are suitable.[20] The grades are determined by those values determined to be most significant in figuring performance characteristics. The two classifications that separate these fuel oils are "distillates" and "residuals," where distillates indicate a distillation overhead product (lighter oils) and residuals indicate a distillation bottom product (heavier oils). Table 5.11 helps in differentiating between these various classifications; and Table 5.12 reveals typical analyses for these oils.

5.2.4.1 Light Oils

Grade 1 and 2 oils are light distillate (fuel) oils used primarily in applications that do not require atomization by air or steam in order to reduce droplet size for proper burning. No. 1 oil will typically vaporize when it comes into contact with a hot surface. No. 2 oil is quite frequently used as fuel for process burners because it will readily burn when injected through a nozzle into a combustion chamber. No. 2 oil is significantly easier to burn than residual oil due to the lack of atomization and preheating requirements. Atomization is the breaking apart of a liquid into tiny, more easily combustible, droplets using steam, air, fuel gas, or mechanical means. These light distillate oils will typically distill out between 450 and 800°F (230 and 430°C).

5.2.4.2 Heavy Oils

No. 4 oil is a heavy distillate oil typically blended from, and thus having characteristics of, both light distillates and residual oils. These oils do not readily combust and therefore require some type of atomization, but still fall into a viscosity range that does not require preheating prior to burning.

5.2.4.3 Residual Oils

No. 6 oil is a heavy residual oil sometimes referred to as Bunker C oil. This oil requires significant atomization for proper combustion. Due to its high viscosity, No. 6 oil

TABLE 5.11 Requirements for Fuel Oils (per ASTM D 396)

Classification	No. 1 Distillate	No. 2 Distillate	No. 4 Distillate (Heavy)	No. 6 Residual
Density (kg/m³) @ 60°F (15°C), max	850	876	—	—
Viscosity @ 104°F (40°C) mm/s²				
min	1.3	1.9	>5.5	—
max	2.1	3.4	24	—
Viscosity @ 212°F (100°C) mm/s²				
min	—	—	—	15
max	—	—	—	50
Flash point °F (°C), min	100 (38)	100 (38)	131 (55)	140 (60)
Pour point °F (°C), max	–0.4 (–18)	21 (–6)	21 (–6)	—
Ash, % mass, max	—	—	0.1	—
Sulfur, % mass, max	0.5	0.5	—	—
Water & sediment, % vol., max	0.05	0.05	0.5	2.0
Distillation temperature °F (°C)				
10% volume recovered, max	419 (215)	—	—	—
90% volume recovered, min	—	540 (282)	—	—
90% volume recovered, max	550 (288)	640 (338)	—	—

TABLE 5.12 Typical Analysis of Different Fuel Oils

	No. 1 Fuel Oil	No. 2 Fuel Oil	No. 4 Fuel Oil	No. 6 Fuel Oil (sour)
Ash (%)	<0.01	<0.01	0.02	0.05
Hydrogen (%)	13.6	13.6	11.7	11.2
Nitrogen (%)	0.003	0.007	0.24	0.37
Sulfur (%)	0.09	0.1	1.35	2.1
Carbon (%)	86.4	86.6	86.5	85.7
Heat of combustion (HHV), Btu/lb	20,187	19,639	19,382	18,343
Specific gravity 60/60°F	0.825	0.84	0.898	0.97
Density (lb/U.S. gal)	6.877	6.96	7.488	8.08

requires heating during handling and further heating prior to combustion chamber injection. No. 6 oil is usually preheated to 150 to 200°F (66 to 93°C), to decrease its viscosity, before being atomized and injected into the burner. John Zink recommends a maximum viscosity of 200 SSU (Seconds Saybolt Universal) for use in its standard oil guns. Figures 5.39 and 5.40 show burners firing heavy oil.

5.2.5 Liquid Naphtha

Liquid naphtha is similar in its characteristics to kerosene (Table 5.13). Figure 5.41 shows a typical naphtha distillation curve. In general, naphtha will boil out of a mixture between 220 and 315°F (100 and 157°C). Naphtha is categorized, based on its volatility, into light, intermediate, and heavy naphtha. Naphtha is a major constituent of gasoline; however, it generally requires further refining to make suitable quality gasoline. Prior to firing naphtha in a burner, care must be taken to vaporize it so that the combustion will be more complete and uniform.

5.2.6 Physical Properties of Liquid Fuels

When liquid fuels are encountered, there are certain properties that determine into which category they are divided, and for what processes they are suitable for.

5.2.6.1 Flash Point

The flash point of a liquid is the lowest temperature at which enough vapors are given off to form a mixture that will ignite when exposed to an ignition source. The standard method for determining flash point is ASTM D-93. Under certain conditions, ASTM D-56 can be used for light distillate oils. Some flash point values are provided in Table 5.11. The flash point is an important property for indication of volatility and for storage requirements.

5.2.6.2 Pour Point

The pour point of a liquid is determined by ASTM D-99 and indicates the lowest temperature at which an oil will flow at a controlled rate. If the fluid temperature goes below this point, flow will be inhibited.

FIGURE 5.39 Burner firing heavy oil (1).

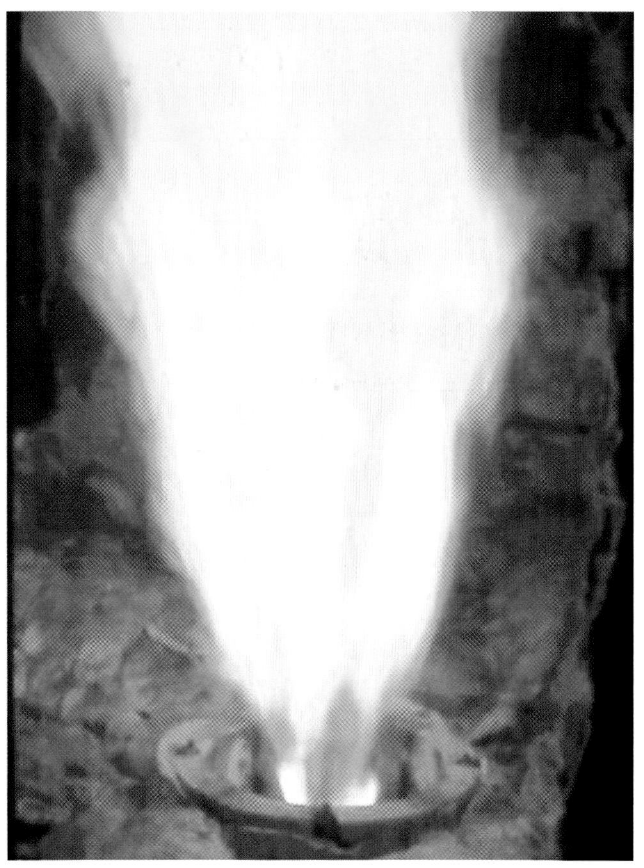

FIGURE 5.40 Burner firing heavy oil (2).

TABLE 5.13 Naphtha Elemental Analysis

Component	Vol. %
n-Heptane	1.610
Methylcyclohexane	2.433
2-methylheptane	5.618
4-methylheptane	1.824
3-methylheptane	4.841
1c,3-dimethylcyclohexane	3.252
1t,4-Dimethylcyclohexane	1.040
1t,2-Dimethylcyclohexane	1.169
n-Octane	16.334
1c,2-Dimethylcyclohexane	1.674
1,1,4-Trimethylcyclohexane	3.500
2,6-Dimethylheptane	2.094
1c,3c,5-Trimethylcyclohexane	2.638
m-xylene	2.426
p-xylene	0.797
2,3-Dimethylheptane	1.475
4-methyloctane	3.417
2-methyloctane	4.491
3-methyloctane	4.576
o-xylene	1.137
n-Nonane	10.120
Other	23.534
Total	100.000

5.2.6.3 Distillation

The distillation of a liquid gives an indication of its volatility, as well as the ease with which it can be vaporized. The test evaluates the vaporization range of a fuel between its end point (the point at which 100% of the volume has vaporized) and the initial boiling point (the point at which the liquid begins to vaporize). Figure 5.42 shows a typical crude oil distillation curve.

5.2.6.4 Viscosity

In layman's terms, the viscosity is a fluid's resistance to flow. Technically, the viscosity is the ratio of shear stress to shear rate of a fluid in motion. Most fluids under consideration in this chapter (gases, fuel oils) are Newtonian fluids because the ratio given above is constant with respect to time, at a given temperature and pressure. A very important factor in the determination of fluid flow is the dimensionless quantity called the Reynolds number. The Reynolds number is calculated as:

$$\mathrm{Re} = \frac{DV\rho}{\mu} \quad \text{or} \quad \mathrm{Re} = \frac{DV}{\nu} \tag{5.3}$$

where D = pipe diameter, V = fluid velocity, ρ = fluid density, μ = fluid absolute viscosity, and ν = fluid kinematic viscosity. When the Reynolds number is less than 2100, the flow is

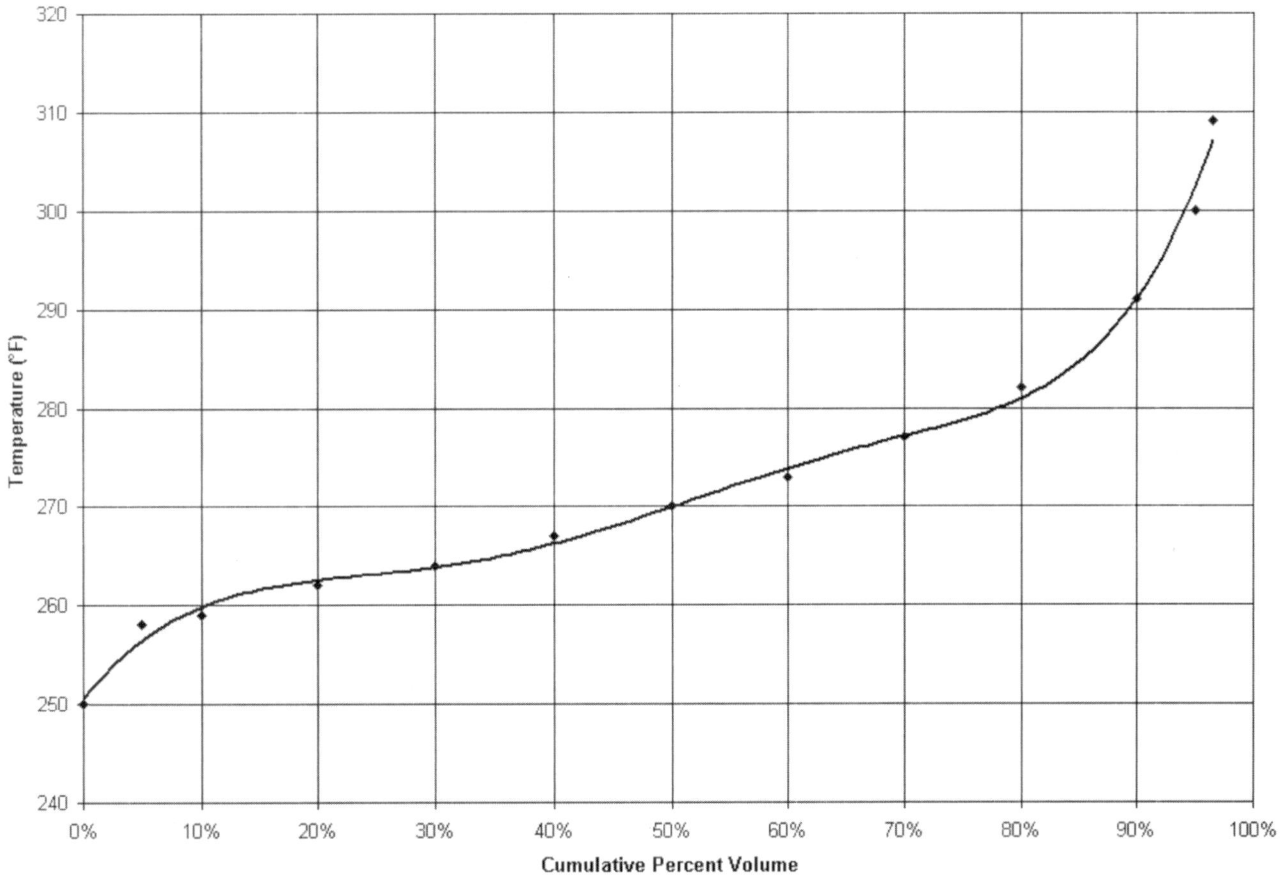

FIGURE 5.41 Naphtha distillation curve.

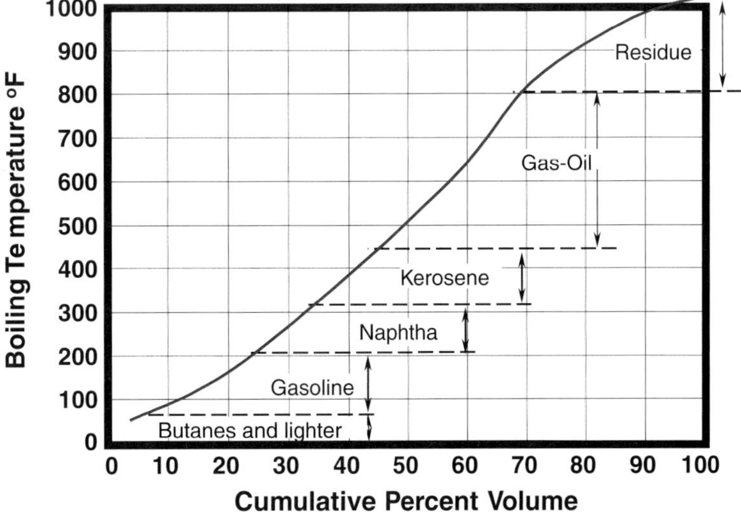

FIGURE 5.42 Crude oil distillation curve.

typically streamlined and smooth, and called laminar. However, when the Reynolds number increases above 2100, internal agitation takes place, and the flow is considered turbulent. As seen in the Eq. (5.3), as the viscosity increases, the flow becomes more laminar, assuming the other properties stay constant. Viscosity is divided into two different categories: kinematic viscosity and absolute viscosity.

Kinematic viscosity (ν) is dependent on fluid density, and has units of length^2 time^{-1}. Typical units for kinematic viscosity are stokes (0.001 m^2 s^{-1}), centistokes (stoke/100), Seconds Saybolt Universal (SSU), and Seconds Saybolt Furol (SSF). Because the density of a fluid is dependent on temperature, the viscosity of a fluid is likewise dependent on temperature. As the temperature increases, the viscosity of a fluid will decrease (become more fluid, or less viscous), and vice versa.

Absolute viscosity (μ) can be calculated by multiplying the kinematic viscosity by the density of the fluid. The most common units for absolute viscosity are the poise (1 Pa sec) and the centipoise (cp), which is poise/100.

The viscosity of oil is a very important consideration in proper burner design. As previously mentioned, the more viscous the fluid, the more preheating required prior to burning. Several useful conversions are listed below:

1 lb_m/ft hr =	0.00413 g/cm s
	0.000413 kg/m s
1 centipoise =	0.01 poise
	0.01 g/cm s
	0.001 kg/m s
	6.72×10^{-4} lb_m/ft s
1 stoke =	0.0001 m^2/s = 100 centistokes
centistokes =	$(0.266 \times SSU) - (195/SSU)$ for SSU 32 to 100
	$(0.220 \times SSU) - (135/SSU)$ for SSU > 100

See also Table 5.14 and Figure 5.43.

5.2.6.5 Density, Gravity, Specific Volume, and Specific Weight

Density is a fluid's mass per unit volume, and is important due to its effect on other properties, such as viscosity. Additionally, the density is used to calculate the heat capacity of an oil. The densities of liquids are frequently given as the °API or the specific gravity (SG). Density and specific gravity are related in that a liquid with a specific gravity of 1 has a density of 1 kg/m^3 (0.0624 lb/ft^3). The specific gravity of a liquid can be calculated by the formula:

$$SG = \rho/\rho_{ref} \tag{5.4}$$

where ρ is the density of the substance in question at specific conditions, and ρ_{ref} is the density of a reference substance at a specific condition. Water is frequently used as a reference substance and, at 60°F, has a specific gravity of 1.0 and a density of 1.94 slugs/ft^3 (999 kg/m^3), where 1 slug = 1 lb_f ft/s^2. Specific gravity for gases requires an additional assumption relating to pressure and temperature. Gas specific gravity is defined relative to air as the reference substance and is generally determined at a standard temperature and pressure. Under those conditions, gas-specific gravity can be calculated as the ratio of molecular weights.

°API runs opposite that of specific gravity; therefore, as °API increases, the density decreases. When a fluid and water are compared at 60°F, the °API can be calculated as:

$$°API = \frac{141.5}{SG} - 131.5 \tag{5.5}$$

The specific volume (volume per unit mass) is the reciprocal of the density, and is commonly used in thermodynamic calculations.

The specific weight of a fluid (γ) is defined as its weight per unit volume. The relationship that relates specific weight to the density is $\gamma = \rho \times g$, where ρ is the density, and g is the local acceleration (32.174 ft/s^2). The specific weight of water at 60°F is 62.4 lb_m/ft^3 (9.80 kN/m^3).

5.2.6.6 Heat Capacity (Specific Heat)

The heat capacity, or specific heat, of a fluid is defined as the amount of heat that is required per unit mass to raise the temperature by one degree. Typical units of heat capacity are Btu/(lb_m-°R) or kJ/(kg-K) in SI units. Heat capacity is temperature dependent, and is defined in terms of constant volume or constant pressure, as can be seen by the following equations:

$$C_p\left(\frac{\delta h}{\delta T}\right)_p \tag{5.6}$$

$$C_v\left(\frac{\delta h}{\delta T}\right)_v \tag{5.7}$$

where C_p = the heat capacity at constant pressure, C_v = the heat capacity at constant volume, δh = change in enthalpy, and δT = change in temperature.

To calculate the heat capacity of a petroleum liquid, to within 2 to 4% accuracy, the following equations can be employed:

$$C = \frac{0.388 + (0.00045 * °F)}{\sqrt{SG}} \quad \text{for units of Btu/(lb}_m - °\text{R)} \tag{5.8}$$

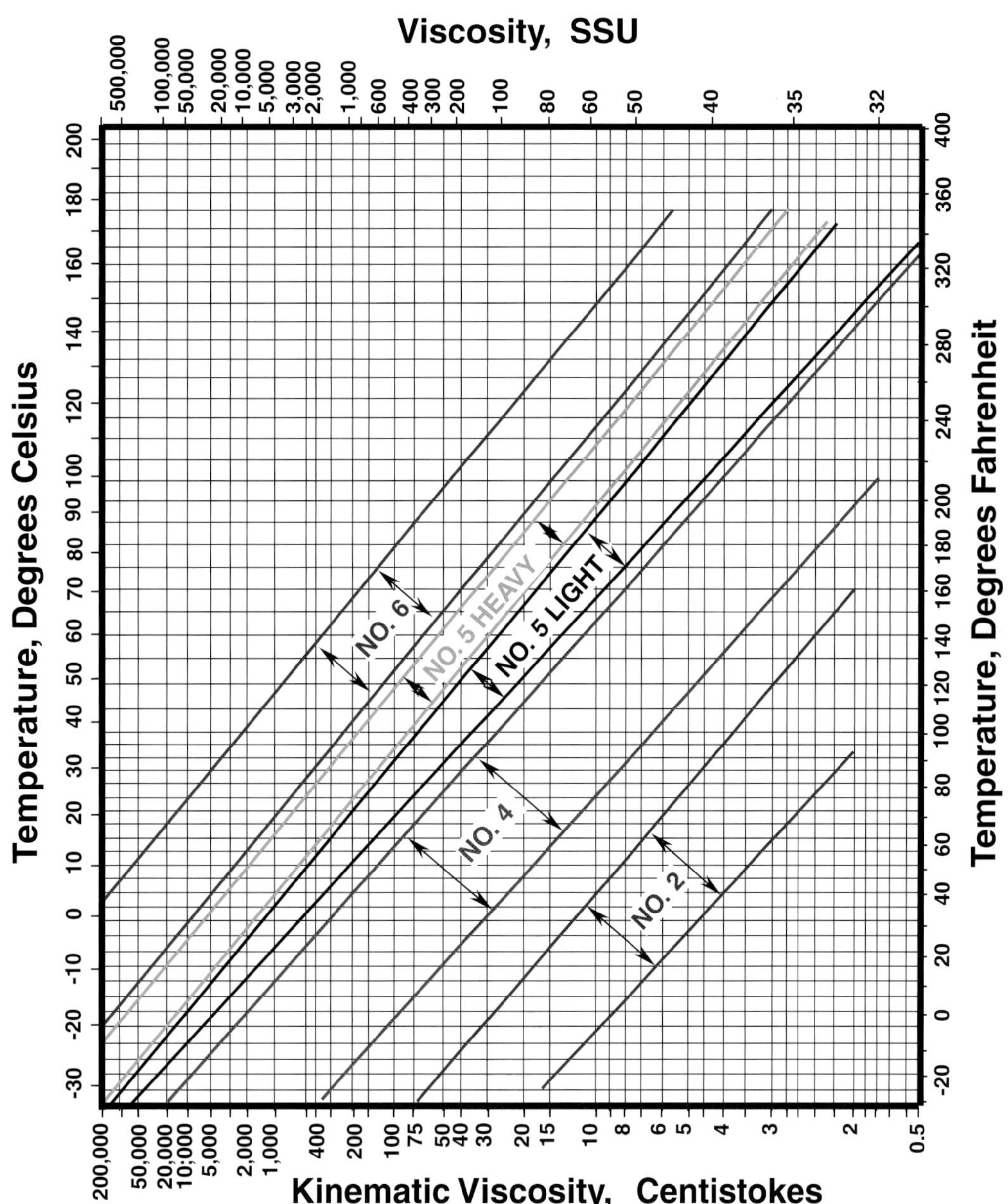

FIGURE 5.43 Viscosity of fuel oils.

$$C = \frac{1.685 + (0.039 * °C)}{\sqrt{SG}} \quad \text{for units of kJ/(kg – K)} \tag{5.9}$$

where C = heat capacity, and SG = specific gravity (relative density), so long as the liquid temperature is between 32 and 400°F (0 and 205°C) and the specific gravity is between 0.75 and 0.96 at 60°F (16°C).[1]

Further information about gaseous and liquid fuels and their properties can be obtained from the references listed at the end of this chapter.[21–24]

5.3 GAS PROPERTY CALCULATIONS

5.3.1 *Molecular Weight*

Molecular weight is the mass in grams of 1 gram-mole of a chemical compound. Avogadro's Number defines the number of molecules in a gram-mole to be 6.02252×10^{23}, a fundamental constant. To determine the molecular weight of a mixture of gases, it is necessary to know the molecular weight of each compound and the composition of the gases in terms of mole or mass fractions. Having assembled this information, the following formulae are used to calculate molecular weight:

$$\mathrm{MW} = \sum \mathrm{MW}_i \times y_i = \frac{1}{\sum \frac{x_i}{\mathrm{MW}_i}} \tag{5.10}$$

where MW = molecular weight of mixture, MW_i = molecular weight of component i, y_i = mole fraction of component i, and x_i = mass fraction of component i.

5.3.2 *Lower and Higher Heating Values*

The lower heating value (LHV) of a gas is the heat released by combustion of a specific quantity of that gas with the products of combustion remaining as vapor. The higher heating value (HHV) adds to the LHV the latent heat of any steam produced as a combustion product. It represents the total heat obtained by first burning a fuel and then cooling the products to standard temperature. Heating values may be provided on a volume basis, typically Btu/scf, or a mass basis such as $\mathrm{Btu/lb_m}$.

To determine the heating value of a mixture of gases, it is necessary to know the heating value of each compound and the composition of the gases in terms of mole or mass fractions. Having assembled this information, the following formulae are used to calculate heating values:

$$\mathrm{HV}_v = \sum \mathrm{HV}_{v,i} \times y_i \tag{5.11}$$

$$\mathrm{HV}_m = \sum \mathrm{HV}_{m,i} \times x_i \tag{5.12}$$

where HV_v = heating value of mixture, volume basis, $\mathrm{HV}_{v,i}$ = heating value of component i, volume basis; HV_m = heating value of mixture, mass basis; and $\mathrm{HV}_{m,i}$ = heating value of component i, mass basis.

5.3.3 *Specific Heat Capacity*

The specific heat capacity of a gas is the energy that must be added to a specific amount of the gas to raise its temperature by one (1) degree. If the gas is maintained at constant pressure during this heating process, the value is referred to as c_p. If the gas is maintained at constant volume, the value is referred to as c_v. Specific heat is not a constant for a given gas; it is a function of temperature. Specific heat can be defined on a volume basis, typically Btu/lbmole-°F; or on a mass basis such as Btu/lb-°F.

To determine the specific heat of a mixture of gases, it is necessary to know the specific heat of each compound at the mixture temperature and the composition of the gases in terms of mole or mass fractions. Having assembled this information, the following formulae are used to calculate specific heat (c_p and c_v formulae are analogous, only c_p formulae are shown):

$$c_p\left(\text{vol.}\right) = \sum c_{p,i}\left(\text{vol.}\right) \times y_i \tag{5.13}$$

$$c_p\left(\text{mass}\right) = \sum c_{p,i}\left(\text{mass}\right) \times x_i \tag{5.14}$$

where c_p (vol) = specific heat of mixture, volume basis; $c_{p,i}$ (vol) = specific heat of component i, volume basis; c_p (mass) = specific heat of mixture, mass basis; and $c_{p,i}$ (mass) = specific heat of component i, mass basis.

5.3.4 *Flammability Limits*

Flammability limits define the range of fuel concentrations in air that will sustain a flame without additional air or fuel. The upper flammability limit (UFL) is the maximum fuel concentration that can sustain a flame and the lower flammability limit (LFL) is the minimum. These limits are often tabulated for fuels at some standard temperature, typically 60°F. Flammability limits are not constants for a given gas; they are functions of the air/fuel mixture temperature. An extensive discussion of this subject can be found in Coward and Jones.[25]

Wierzba and Karim[26] present a method for estimating the flammability limits as a function of mixture temperature by calculating adiabatic flame temperature (AFT). First, the AFT for the standard temperature mixture is determined. Next, the mixture temperature is set to the desired level and the fuel concentration is varied until the calculated AFT for the non-standard temperature matches the AFT for the standard temperature. They provide an approximating method for calculating AFT for sub-stoichiometric mixtures.

Both the Coward and Jones manuscript and the Wierzba and Karim article indicate that a form of Le Chatelier's rule can be used calculate LFL and UFL for many combinations of fuels and inerts. Both references also mention that this rule fails to accurately predict for a few important situations. One notable example is a mixture of ethylene and carbon dioxide that differs substantially from normal calculated LFL and UFL. Another example is any mixture of chemicals that is prone to react with another at temperatures below the ignition point, such as ethylene and hydrogen. Mixtures involving significant amounts of inert compounds (for example, H_2O, N_2, and CO_2) require special treatment either by the AFT method described above or by grouping the inerts with fuel components in known proportions matching conditions for which LFL and UFL have been measured. This latter method is described in detail by Coward and Jones.[25]

With these exceptions in mind, the following mixing rules can be used to calculate LFL and UFL for most common gas mixtures:

$$LFL = \frac{100}{\sum \frac{y_i}{LFL_i}} \tag{5.15}$$

$$UFL = \frac{100}{\sum \frac{y_i}{UFL_i}} \tag{5.16}$$

5.3.5 *Viscosity*

Viscosity is discussed in detail in Chapter 4: "Fundamentals of Fluid Flow." A useful mixing rule Eq. (4.8) is also provided.

5.3.6 *Derived Quantities*

In addition to the specific properties described above, there are a number of useful derived parameters that may be of interest when studying combustion systems.

5.3.6.1 Partial Pressure

Partial pressure is the pressure exerted by a single component of a mixture when that component alone occupies the entire volume at the mixture temperature. Dalton's law states that the total pressure of a mixture is the sum of the partial pressures of the components. While this law has been demonstrated to be somewhat in error, especially at high pressures, it is often useful for estimating purposes to determine whether a more detailed analysis is justified. The basic relationship is:

$$p_i = y_i \leftrightarrow TP \tag{5.17}$$

Partial pressures are of interest when estimating the probability of forming condensate in a gas mixture. When the partial pressure of a component exceeds the vapor pressure of that component at the mixture temperature, condensation is likely.

5.3.6.2 Adiabatic Flame Temperature

The adiabatic flame temperature is the temperature at which the enthalpy of the products of combustion equals the sum of the enthalpy of the reactants plus the heat released by the combustion process. Heat loss due to radiation, convection, or conduction is not included; hence the reference to adiabatic. Accounting for dissociation of combustion products is important. Customarily, the adiabatic flame temperature is determined for a stoichiometric fuel/air mixture, although other mixtures such as LFL and UFL are sometimes studied for special purposes, as discussed in Section 5.3.4 above.

5.3.6.3 Heat Release

Heat release is the product of the flow rate and the heating value of the fuel using compatible units. This quantity is used throughout many areas of interest in combustion, including equipment sizing, radiation, and emissions. Unless the process involves the recovery of the heat of vaporization of the water vapor, the LHV is usually used when calculating heat release:

$$\text{HR} = w \leftrightarrow LHV_m = Q \leftrightarrow LHV_v \tag{5.18}$$

where HR = heat release (BTU/hr), w = mass flow (lb/hr), LHV_m = lower heating value (BTU/lb), Q = volumetric flow (SCFH), and LHV_v = lower heating value (BTU/scf).

5.3.6.4 Volume Equivalent of Flow

Volume equivalent of flow (V_{eq}) is the volumetric flow of air at standard temperature and pressure that produces the same velocity pressure in the same size line. This quantity is often used to provide generalized capacity curves for equipment that may need to handle several different gas streams. When designing equipment for this situation, the stream with the highest V_{eq} will often dominate the hydraulic design, unless the different streams have different allowable pressure drops. Caution should be used in cases where friction is expected to be a major factor in the system pressure drop because V_{eq} does not account for variations in viscosity.

$$V_{eq} = Q\sqrt{\frac{\text{MW}}{29}\frac{T_{gas}}{520}} = 13.1w\sqrt{\frac{29}{\text{MW}}\frac{T_{gas}}{520}} \tag{5.19}$$

where V_{eq} = volume equivalent (SCFH), and T_{gas} = gas temperature (°R).

5.4 TYPICAL FLARED GAS COMPOSITIONS

Gas compositions sent to flares include a large variety of individual compounds. The proportions of these compounds vary widely from one facility to another and even within a single facility from minute to minute. The following sections describe in general terms the kinds of gas streams commonly encountered in flare systems.

5.4.1 *Oil Field/Production Plant Gases*

Gases produced in oil fields generally consist of saturated hydrocarbon gases (paraffins), together with a certain amount of inerts. Oil field gases range in MW from 19 to 25. Such gases may contain significant amounts of H_2S (sour gas wells) or CO_2. In some cases, especially offshore, these associated gases are burned continuously in the immediate vicinity of the oil wells. In other cases, the gas is sent to a production plant where it is treated in preparation for pipeline use.

Production plants convert the raw associated gas into several, more valuable products. Undesirable components such as H_2S, CO_2, and water vapor are removed in treatment units. Depending on the composition of the feedstock, production plants may include a debutanizer, a depropanizer, and a deethanizer to separate the large majority of these valuable components. The remainder, mostly methane, becomes pipeline-quality natural gas after the addition of odorants such as mercaptans. Within the production plant, it may become necessary to flare the raw associated gas, the pipeline product, or the overhead streams from any of the separation units.

5.4.2 *Refinery Gases*

Refineries treat the liquids produced in the oil fields to generate many essential materials for public consumption as well as further chemical processing. As a result of various treatment processes, hydrogen and unsaturated hydrocarbon gases (olefins, diolefins, aromatics, etc.) are produced in abundance in a refinery. Due to the wide variety of treatment processes, the composition of flared gases in a refinery is almost entirely unpredictable. Refinery flaring generally involves hydrogen, paraffins up to decane, olefins up to hexene, diolefins up to butadiene, and aromatics up to ethylbenzene, as well as contaminants such as H_2S, CO_2, and water vapor.

5.4.3 *Ethylene/Polyethylene Gases*

Ethylene plants use cracking furnaces to convert feedstock into high-quality ethylene. Some plants use ethane as feedstock. The gas produced by such plants is often referred to as light cracked gas, and consists of approximately equal portions of hydrogen, ethane, and ethylene with relatively little else. Other plants use oil as feedstock and produce heavy cracked gas. Heavy cracked gas is also approximately equal portions of hydrogen, ethane, and ethylene, but a substantial fraction of the composition consists of heavy hydrocarbon gases.

Polyethylene plants take the ethylene from the ethylene plant and polymerize it in a variety of ways. In some cases, the ethylene is mixed with heavier hydrocarbons (pentane, hexane, hexene, etc.) to alter the properties of the polymer. Random mixtures of ethylene and other hydrocarbons may be sent to the flare from the main process area. In addition, reliefs from various special chemical storage areas may send relatively pure materials such as hexane or hexene to the flare.

5.4.4 *Other Special Cases*

Landfills and digester facilities produce an off-gas that must be disposed of to prevent odor problems in the community. The gas is generally a mixture of CO_2 and CH_4. Landfills are rarely above 30 to 40% methane, while digesters may be as high as 60 to 70% methane. In some landfills, perimeter wells are used to draw air into the edges of the landfill, which prevents the spread of anaerobic bacteria and methane. In these cases, the methane content is even lower and some air is also sent to the flare.

Marine and truck loading facilities burn the vapor displaced from the tankers or trucks during the loading operation. In many cases, the displaced vapor is mostly air with some amount of evaporated gasoline or diesel fuel. Depending on the ambient temperature, the resulting mixture could be very rich in hydrocarbon vapor, or very lean.

Medical equipment, such as bandages or hypodermic needles, is often sterilized by contact with ethylene oxide (ETO) vapors. ETO sterilizer flares are designed to receive the ETO vapor after the sterilization process is complete. The composition coming to these flares generally consists of a mixture of ETO and either air or nitrogen. It should be noted that ETO has a flammability range from 3 to 100% and a very low ignition temperature.

Flares are often used as backup equipment for incinerators during maintenance or malfunctions. In this type of service, the waste gas is usually enriched with a substantial amount of clean fuel gas to ensure reliable burning. Steel mills produce off-gases that consist mainly of H_2, H_2O, CO, CO_2, and air. These are generally low LHV mixtures that also require enrichment and supplemental fuel firing to maintain ignition. Fertilizer plants and other chemical plants produce ammonia, which may be sent to a flare in an emergency. Waste gases that are sent to flares in these facilities may be pure ammonia or diluted with nitrogen or water vapor.

The variety of gases and the hazards associated with each requires careful review of all aspects of system design to ensure that these fuels are safely handled, whether in a flare, a furnace, or an incinerator.

REFERENCES

1. Perry, R.H., Green, D.W., and Maloney, J.O., Eds., *Perry's Chemical Engineers' Handbook*, 7th ed., McGraw-Hill, New York, 1997, chap. 27.
2. U.S. Department of Energy, State Energy Data Report, Consumption Estimates 1960-1989, DOE/EIA-0214(89), May 1991.
3. Reed, R.J., *North American Combustion Handbook*, Vol. I, North American Mfg. Co., Cleveland, OH, 1986.
4. Gas Processors and Suppliers Association, *GPSA Engineering Data Book*, Vol. I, 10th ed., Tulsa, OK, 1987.
5. Austin, G.T., *Shreve's Chemical Process Industries*, 5th ed., McGraw-Hill, New York, 1984, chap. 6.
6. Gas Processors and Suppliers Association, *GPSA Engineering Data Book*, Volume II, tenth edition, Tulsa, OK, 1987.
7. McCabe, W.L., Smith, J.C., and Harriot, P., *Unit Operations of Chemical Engineering*, 5th ed., McGraw-Hill, New York, 1993.
8. Leffler, W.L., *Petroleum Refining for the Non-Technical Person*, PennWell Publishing, Tulsa, OK, 1985.
9. Gary, J.H. and Handwerk, G.E., *Petroleum Refining*, 3rd ed., Marcel Dekker, New York, 1994.
10. Reed, R.D., *Furnace Operations*, 3rd ed., Gulf Publishing, Houston, 1981.
11. Nelson, W.L., *Petroleum Refining Engineering*, 3rd ed., McGraw-Hill, New York, 1949.
12. Meyers, R.A., *Handbook of Petroleum Refining Processes*, 2nd ed., McGraw-Hill, New York, 1997, chap. 6.2.
13. Meyers, R.A., *Handbook of Petroleum Refining Processes*, 2nd ed., McGraw-Hill, New York, 1997, chap. 12.1.
14. Web site for Energy Information Administration, Office of Gas & Oil, www.eia.doe.gov, Crude Oil Watch, May 24, 2000.
15. *Microsoft Encarta Encyclopedia*, Microsoft Corporation, 1993–1999.
16. Press, F. and Siever, R., *Understanding Earth*, W.H. Freeman, New York, 1994.
17. Baird, C., *Environmental Chemistry*, W.H. Freeman, New York, 1995.
18. Solomons, T.W., *Organic Chemistry*, 5th ed., John Wiley & Sons, New York, 1992.
19. Peyton, K., *Fuel Field Manual*, McGraw-Hill, New York, 1998.
20. American Standard Testing Methods, ASTM D-396: Standard Specification for Fuel Oils, 1998.
21. Dean, J., *Lange's Handbook of Chemistry*, 14th ed., McGraw-Hill, New York, 1992.
22. Munson, B.R., Young, D.F., and Okiishi, T.H., *Fundamentals of Fluid Mechanics*, 2nd ed., John Wiley & Sons, New York, 1994.
23. Heald, C.C., *Cameron Hydraulic Data*, 18th ed., Ingersoll-Dresser Pumps, New Jersey, 1994.
24. Van Wylen, G.J., Sonntag, R.E., and Borgnakke, C., *Fundamentals of Classical Thermodynamics*, 4th ed., John Wiley & Sons, New York, 1994.
25. Coward, H.F. and Jones, G.W., Limits of Flammability of Gases and Vapors, U.S. Bureau of Mines, Dept. of Interior, Bulletin 503, Pittsburgh, PA, 1952.
26. Wierzba, I. and Karim, G.A., Prediction of the flammability limits of fuel mixtures, *AFRC/JFRC International Symposium*, October, Maui, Hawaii, 1998.
27. Meyers, R.A., *Handbook of Petroleum Refining Processes*, 2nd ed., McGraw-Hill, New York, 1997, chap. 3.3.

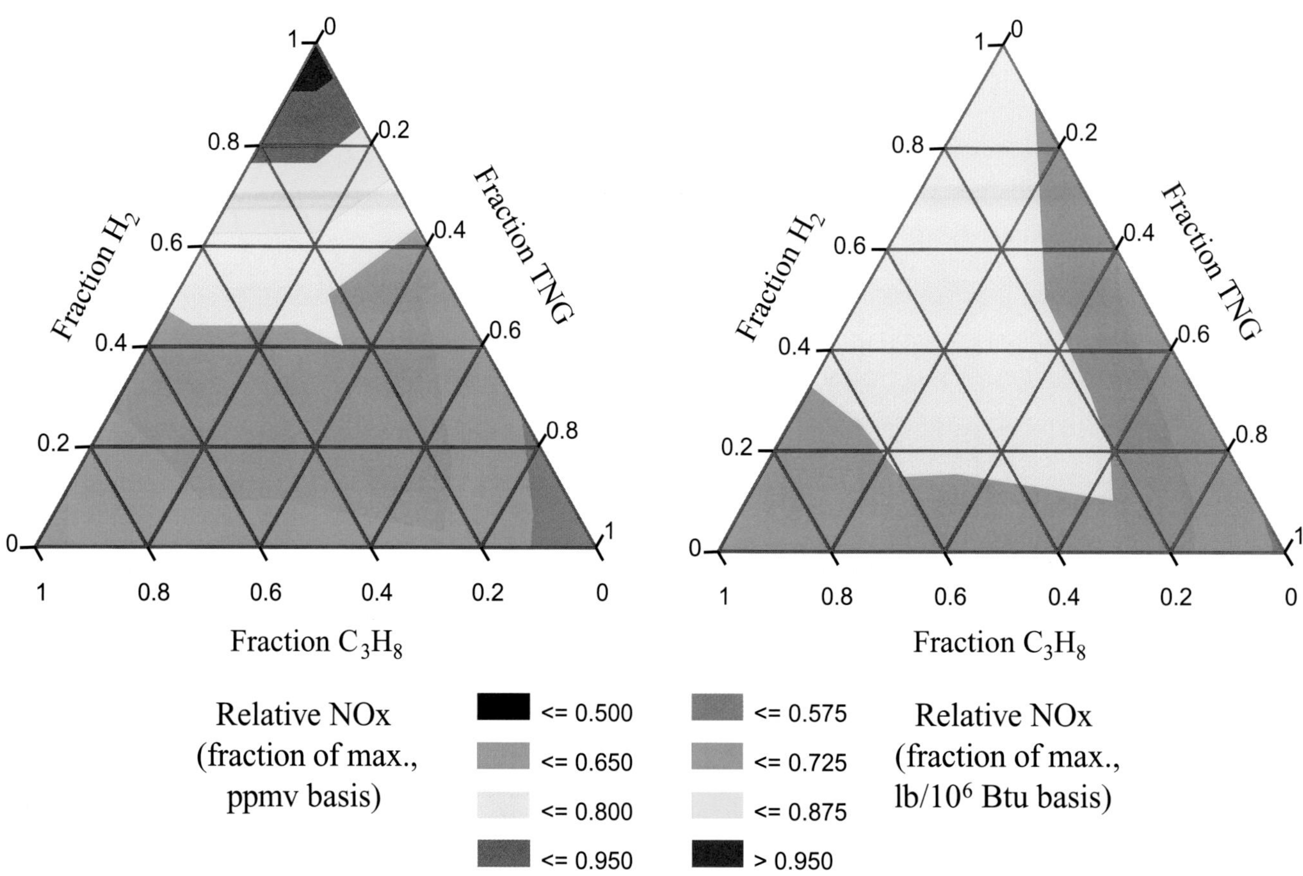

Fraction H$_2$
Fraction TNG
Fraction C$_3$H$_8$
1
0.8
0.6
0.4
0.2
0
Relative NOx (fraction of max., ppmv basis)
Relative NOx (fraction of max., lb/10^6 Btu basis)
<= 0.500
<= 0.575
<= 0.650
<= 0.725
<= 0.800
<= 0.875
<= 0.950
> 0.950

Chapter 6
Pollutant Emissions

Charles E. Baukal, Jr. and Joseph Colannino

TABLE OF CONTENTS

6.1 Introduction 190
6.1.1 Emissions in the Hydrocarbon and Petrochemical Industries 190
6.1.2 Conversions 190
6.2 Nitrogen Oxides (NOx) 191
6.2.1 Theory 192
6.2.2 Regulations 196
6.2.3 Measurement Techniques 197
6.2.4 Abatement Strategies 198
6.2.5 Field Results 204
6.3 Combustibles 214
6.3.1 CO and Unburned Fuel 214
6.3.2 Volatile Organic Compounds 215
6.4 Particulates 217
6.4.1 Sources 217
6.4.2 Treatment Techniques 218
6.5 Carbon Dioxide 218
6.6 SOx 219
6.7 Dioxins and Furans 219
References 219

TABLE 6.1 Typical Uncontrolled Combustion Emission Factors (lb/10^6 Btu) by Fuel Type

Fuel Type	SO_x	NO_x	CO	Particulates	VOCs
Distillate fuel	0.160	0.140	0.0361	0.010	0.002
Residual fuel	1.700	0.370	0.0334	0.080	0.009
Other oils	1.700	0.370	0.0334	0.080	0.009
Natural gas	0.000	0.140	0.0351	0.003	0.006
Refinery gas	0.000	0.140	0.0340	0.003	0.006
LPG	0.000	0.208	0.0351	0.007	0.006
Propane	0.000	0.208	0.0351	0.003	0.006
Steam coal	2.500	0.950	0.3044	0.720	0.005
Petroleum coke	2.500	0.950	0.3044	0.720	0.005
Electricity	1.450	0.550	0.1760	0.400	0.004

Source: From Table 1-11 on p.16 of U.S. Dept. of Energy, Energy & Environmental Profile of the U.S. Petroleum Refining Industry, 1998.

6.1 INTRODUCTION

The purpose of this chapter is to alert the interested reader to the potential effects on pollutant emissions of the combustion processes in the petrochemical and hydrocarbon industries. There continues to be increasing interest in reducing pollutant emissions of all types from all combustion processes. One prognosticator predicts this will continue well into the future.[1] These pollutants have deleterious effects on the environment and there is evidence they may have an impact on the health of humans and animals. Efforts are underway from a broad cross-section of organizations to improve existing techniques and to develop new techniques for minimizing pollution. While there are other pollutants potentially produced in the hydrocarbon and petrochemical industries, this chapter is only concerned with the air pollutants resulting from combustion processes.

There are numerous factors that affect the pollutant emissions generated from the combustion of fuels. The U.S. Dept. of Energy has classified emission factors by fuel type for petroleum refining, as shown in Table 6.1.[2] A U.S. Environmental Protection Agency (U.S. EPA) report identified the following heater design parameters that affect NOx emissions from process heaters: fuel type, burner type, combustion air preheat, firebox temperature, and draft type.[3] The important factors that influence pollution are considered here. A brief general discussion of emissions from heaters in refineries is given in API 560, section F.10.2.[40]

6.1.1 Emissions in the Hydrocarbon and Petrochemical Industries

The Western States Petroleum Association (WSPA) and the American Petroleum Institute (API) worked with the California Air Resources Board (CARB) to develop air toxics emission factors for the petroleum industry.[4] Source data was provided in 18 groups. Some of those groups of relevance here include both refinery gas-fired and fuel oil-fired boilers, and heaters fired on natural gas, refinery gas, oil, and a combination of natural gas and refinery oil. The U.S. EPA has compiled an extensive list of emission factors for a wide range of industrial processes.[5] Chapter 1 of U.S. EPA AP-42 concerns external combustion sources and focuses on the fuel type. Sections 1.3, 1.4, and 1.5 of AP-42 focus on fuel oil combustion, natural gas combustion, and liquefied petroleum gas combustion, respectively. Chapter 5 of AP-42 focuses on the petroleum industry, where the reader is referred to Sections 1.3 and 1.4 for boilers and process heaters using fuel oil and natural gas, respectively. Chapter 6 of AP-42 concerns the organic chemical process industry. Reis (1996) has written a general book on environmental issues in petroleum engineering, including drilling and production operations.[6]

6.1.2 Conversions

It is often necessary to convert pollutant measurements (e.g., NOx and CO) into a standard basis for both regulatory and comparison purposes. One conversion often necessary is from the measured O_2 level in the exhaust gases to a standard basis O_2 level. The method for converting measurements to a standard basis is given by[7]:

$$ppm_{corr} = ppm_{meas}\left(\frac{20.9 - O_{2_{ref}}}{20.9 - O_{2_{meas}}}\right) \tag{6.1}$$

where ppm_{meas} = Measured pollutant concentration in flue gases (ppmvd)
ppm_{corr} = Pollutant concentration corrected to a reference O_2 basis (ppmvd)
O_{2meas} = Measured O_2 concentration in flue gases (vol.%, dry basis)
O_{2ref} = Reference O_2 basis (vol.%, dry basis)

Example 6.1

Given: Measured NOx = 20 ppmvd; measured O_2 = 2% on a dry basis.

Find: NOx at 3% O_2 on a dry basis.

Solution: $ppm_{meas} = 20;\ O_{2_{meas}} = 2;\ O_{2_{ref}} = 3$

$$ppm_{corr} = 20\left(\frac{20.9 - 3}{20.9 - 2}\right) = 18.9 \text{ ppmvd}$$

This example shows that corrected NOx values will be lower when the basis O_2 is higher than the measured O_2 because higher O_2 levels mean more air dilution and therefore lower NOx concentrations. The reverse is true when the basis O_2 is lower than the measured O_2 level.

Another correction that may be required is to convert the measured pollutants from a measured furnace temperature to a different reference temperature. This may be required when a burner is tested at one furnace temperature and needs to be modified to find out the equivalent at another furnace temperature. The correction for temperature is:

$$ppm_{corr} = ppm_{meas}\left(\frac{T_{ref} - T_{basis}}{T_{meas} - T_{basis}}\right) \tag{6.2}$$

where ppm_{meas} = Measured pollutant concentration in flue gases (ppmvd)
ppm_{corr} = Pollutant concentration corrected to a reference temperature basis (ppmvd)
T_{ref} = Reference furnace temperature (°F)
T_{meas} = Measured furnace temperature (°F)
T_{basis} = Basis furnace temperature (°F)

Example 6.2
Given: Measured NOx = 20 ppmvd; measured furnace temperature = 1800°F.
Find: NOx at a reference temperature of 2000°F.
Solution: $ppm_{meas} = 20$; $T_{meas} = 1800°F$; $T_{ref} = 2000°F$ assume $T_{basis} = 400°F$

$$ppm_{corr} = 20\left(\frac{2000-400}{1800-400}\right) = 22.9 \text{ ppmvd}$$

There are two things to notice in the above example. The first is that the basis temperature was chosen as 400°F, which is an empirically determined value that applies to many burners commonly used in the hydrocarbon and petrochemical industries. However, this equation should be used with care for more unique burner designs and when there is a very large difference between the measured and the reference furnace temperatures. The second thing to notice is that the NOx increases when the reference temperature is higher than the measured temperature, and vice versa. As will be shown later, NOx generally increases with the furnace temperature.

These two corrections can also be combined into a single correction when both the measured O_2 level and furnace temperature are different from the reference O_2 level and furnace temperature:

$$ppm_{corr} = ppm_{meas}\left(\frac{20.9 - O_{2_{ref}}}{20.9 - O_{2_{meas}}}\right)\left(\frac{T_{ref} - T_{basis}}{T_{meas} - T_{basis}}\right) \tag{6.3}$$

where the variables are defined above.

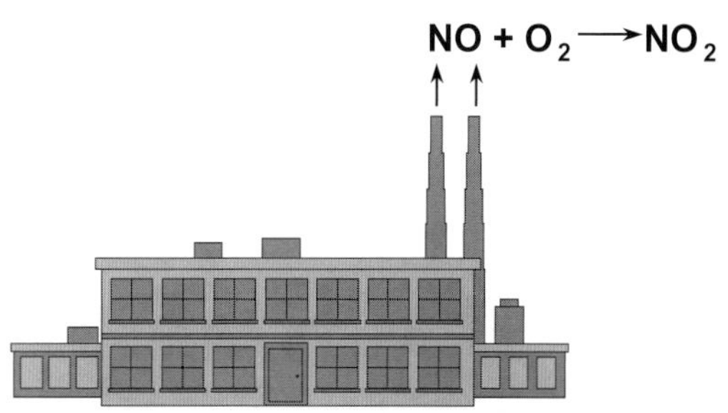

FIGURE 6.1 Cartoon of NO exiting a stack and combining with O_2 to form NO_2.

Example 6.3
Given: Measured NOx = 20 ppmvd; measured O_2 = 2% on a dry basis; measured furnace temperature = 1800°F.
Find: NOx at 3% O_2 on a dry basis at a reference temperature of 2000°F.
Solution: $ppm_{meas} = 20$; $O_{2_{meas}} = 2$; $O_{2_{ref}} = 3$; $T_{meas} = 1800°F$; $T_{ref} = 2000°F$ assume $T_{basis} = 400°F$

$$ppm_{corr} = 20\left(\frac{20.9-3}{20.9-2}\right)\left(\frac{2000-400}{1800-400}\right) = 21.6$$

In this case, the increase in NOx due to the temperature correction is greater than the reduction in NOx due to the higher O_2 reference.

6.2 NITROGEN OXIDES (NOx)

NOx refers to the oxides of nitrogen. These generally include nitrogen monoxide, also known as nitric oxide (NO), and nitrogen dioxide (NO_2). They may also include nitrous oxide (N_2O) (also known as laughing gas), as well as other less common combinations of nitrogen and oxygen such as nitrogen tetroxide (N_2O_4).

In most high-temperature heating applications, the majority of the NOx exiting the exhaust stack is in the form of nitric oxide (NO).[8] NO is a colorless gas that rapidly combines with O_2 in the atmosphere to form NO_2 (see Figure 6.1). In the lower atmosphere, NO reacts with oxygen to form ozone, in addition to NO_2. NO_2 is extremely reactive and is a strong oxidizing agent. NO_2 decomposes on contact with water to produce nitrous acid (HNO_2) and nitric acid (HNO_3), which are highly corrosive (see Figure 6.2). When NO_2 forms in the atmosphere and comes in contact with rain, acid rain is produced. Acid rain is destructive to anything it contacts, including plants, trees, and man-made structures like buildings, bridges, etc. In addition to acid rain, another problem with NO_2 is its contribution to smog. When sunlight contacts a

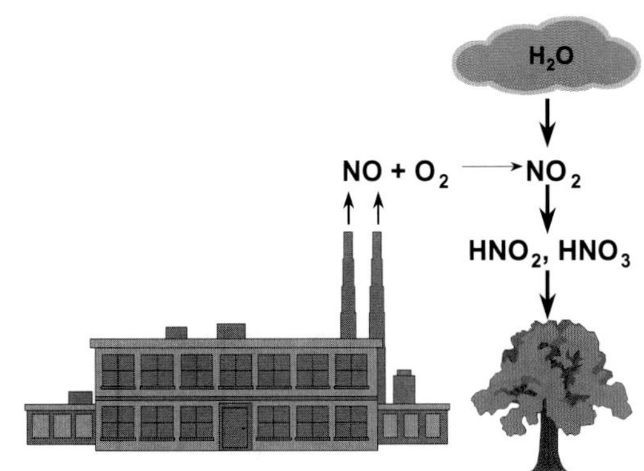

FIGURE 6.2 Cartoon of acid rain.

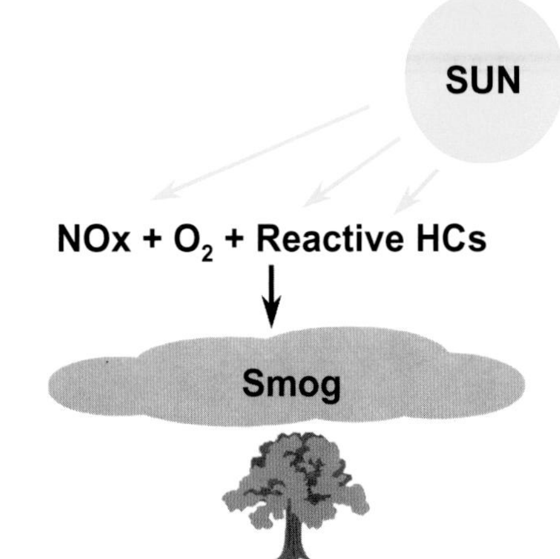

FIGURE 6.3 Cartoon of photochemical smog formation.

mixture of NO_2 and unburned hydrocarbons in the atmosphere, photochemical smog is produced (see Figure 6.3).

Many combustion processes are operated at elevated temperatures and high excess air levels. The combustion products may have long residence times in the combustion chamber. These conditions produce high thermal efficiencies and product throughput rates. Unfortunately, such conditions also favor the formation of NOx. NOx emissions are among the primary air pollutants because of their contribution to smog formation, acid rain, and ozone depletion in the upper atmosphere. It is interesting to note that only about 5% of typical NOx sources in an industrial region of the United States come from industrial sources, compared to 44% from highway and off-road vehicles.[10]

6.2.1 Theory

6.2.1.1 Formation Mechanisms

There are three generally accepted mechanisms for NOx production: thermal, prompt, and fuel. *Thermal NOx* is formed by the high-temperature reaction of nitrogen with oxygen, via the well-known Zeldovich mechanism,[11] as given by the simplified reaction:

$$N_2 + O_2 \rightarrow NO,\ NO_2 \tag{6.4}$$

Thermal NOx increases exponentially with temperature. Above about 2000°F (1100°C), it is generally the predominant mechanism in combustion processes, making it important in most high-temperature heating applications. This means that this mechanism becomes more important when air preheating or oxygen enrichment[22] of the combustion air are used, which normally increases the flame temperature.

Prompt NOx is formed by the relatively fast reaction between nitrogen, oxygen, and hydrocarbon radicals. It is given by the overall reaction:

$$CH_4 + O_2 + N_2 \rightarrow NO, NO_2, CO_2, H_2O, \text{trace species} \tag{6.5}$$

In reality, this very complicated process consists of hundreds of reactions. The hydrocarbon radicals are intermediate species formed during the combustion process. Prompt NOx is generally an important mechanism in lower temperature combustion processes.

Fuel NOx is formed by the direct oxidation of organonitrogen compounds contained in the fuel. It is given by the overall reaction:

$$R_xN + O_2 \rightarrow NO,\ NO_2,\ CO_2,\ H_2O,\ \text{trace species} \tag{6.6}$$

In reality, there are many intermediate reactions for this formation mechanism, as indicated in Figure 6.4. Fuel NOx is not a concern for high-quality gaseous fuels like natural gas or propane, which normally have no organically bound nitrogen. However, fuel NOx may be important when oil (e.g., residual fuel oil), coal, or waste fuels are used that may contain significant amounts of organically bound nitrogen. Table 6.2 shows typical thermal and fuel NOx emissions for process heaters.[12] The conversion of fuel-bound nitrogen to NOx ranges from 15 to 100%.[12] The conversion efficiency is generally higher the lower the nitrogen content in the fuel.

6.2.1.2 Important Factors Affecting NOx

There are many factors that have an impact on NOx formation. These include the oxidizer and fuel compositions and temperatures, the ratio of the fuel to the oxidizer, the burner and heater designs, the furnace and flue gas temperatures, and the operational parameters of the combustion system. Some of these are considered next.

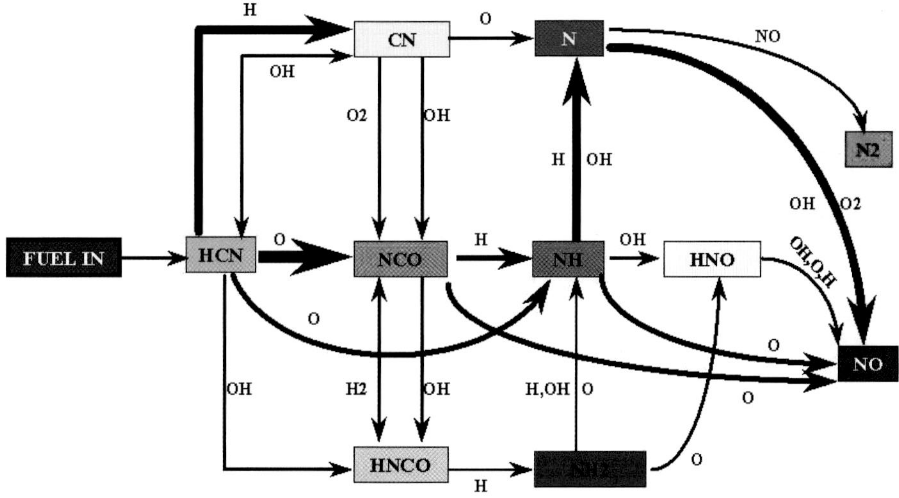

FIGURE 6.4 Schematic of fuel NOx formation pathways. (Adapted from W. Bartok and A.F. Sarofim, Eds., *Fossil Fuel Combustion: A Source Book*, John Wiley & Sons, New York, 1991.)

Figure 6.5 shows the predicted NO as a function of the flame stoichiometry for air/fuel flames. NO increases at fuel-lean conditions and decreases at fuel-rich conditions. Figure 2.16 shows a plot of the adiabatic equilibrium flame temperature for air fuel flames as a function of the flame equivalence ratio. There are several things to notice. The flame temperature for the air/CH_4 flame is very dependent on the stoichiometry. Figure 2.16 helps to explain why, for example, NOx is dramatically reduced under fuel-rich conditions. One reason is the dramatic reduction in the flame temperature; another reason concerns the chemistry. In a reducing atmosphere, CO is formed preferentially over NO. This is exploited in some of the NOx reduction techniques. An example is methane reburn.[13] The exhaust gases from the combustion process flow through a reduction zone that is at reducing conditions. NOx is reduced back to N_2. Any CO that may have formed in the reduction zone and other unburned fuels are then combusted downstream of the reduction zone. However, they are combusted at temperatures well below those found in the main combustion process. These lower temperatures are not favorable to NOx formation.

Figure 6.6 shows the importance of the gas temperature on thermal NOx formation. The NOx rises rapidly at temperatures

TABLE 6.2 Uncontrolled NOx Emission Factors for Typical Process Heaters

	Uncontrolled Emission Factor, lb/10^6 Btu		
Model Heater Type	Thermal NOx	Fuel NOx	Total NOx[a]
ND, natural gas-fired[b]	0.098	N/A	0.098
MD, natural gas-fired[b]	0.197	N/A	0.197
ND, distillate oil-fired	0.140	0.060	0.200
ND, residual oil-fired	0.140	0.280	0.420
MD, distillate oil-fired	0.260	0.060	0.320
ND, residual oil-fired	0.260	0.280	0.540
ND, pyrolysis, natural gas-fired	0.104	N/A	0.104
ND, pyrolysis, high-hydrogen fuel gas-fired[c]	0.140[d]	N/A	0.140

Note: N/A = Not applicable
ND = Natural draft
MD = Mechanical draft

[a] Total NOx = Thermal NOx + Fuel NOx
[b] Heaters firing refinery fuel gas with up to 50 mole percent hydrogen can have up to 20% higher NOx emissions than similar heaters firing natural gas.
[c] High-hydrogen fuel gas is fuel gas with 50 mole% or greater hydrogen content.
[d] Calculated assuming approximately 50 mole% hydrogen.

Table 2-1 on p. 2-3 of Sanderford, EPA document, 1993.

Source: From E.B. Sanderford, U.S. EPA Report EPA-453/R-93-015, February 1993.

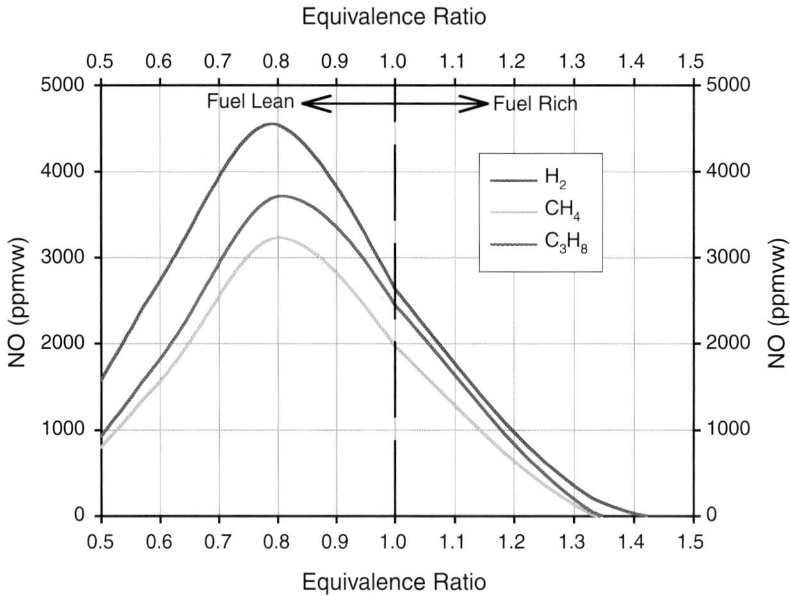

FIGURE 6.5 Adiabatic equilibrium NO as a function of equivalence ratio for air/fuel flames.

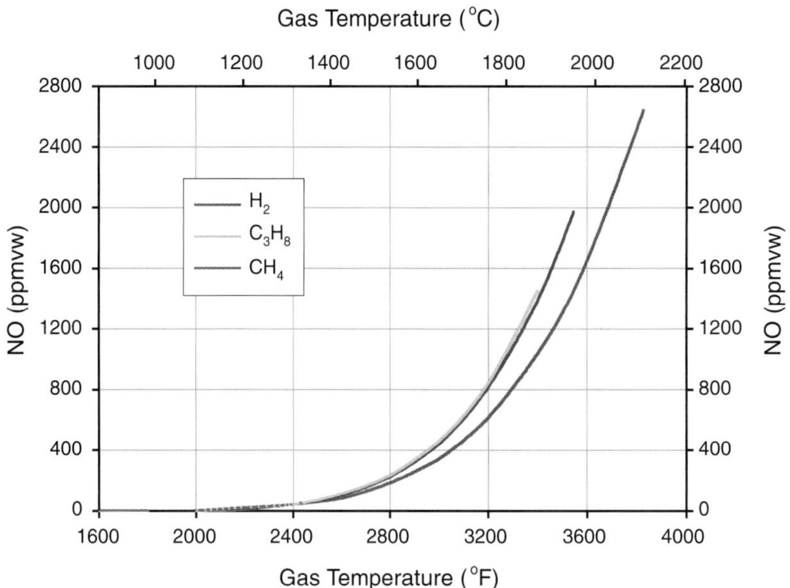

FIGURE 6.6 Adiabatic equilibrium NO as a function of gas temperature for stoichiometric air/fuel flames.

above 2000°F (1100°C) for all three fuels shown. This is a demonstration of the increase in thermal NOx as a function of temperature. Many combustion modification strategies for reducing NOx involve reducing the flame temperature because it has such a large impact on NOx. For example, one strategy is to inject water into the flame to reduce NOx by cooling down the flame to a lower temperature where NOx formation is less favorable.

Figure 6.7 shows how NOx increases when the combustion air is preheated. Air preheating is commonly performed to increase the overall thermal efficiency of the heating process. However, it can dramatically increase NOx emissions because of the strong temperature dependence of NO formation. Figure 2.17 shows how the adiabatic flame temperature increases with air preheating. The increase in NO emissions mimics the increase in flame temperature.

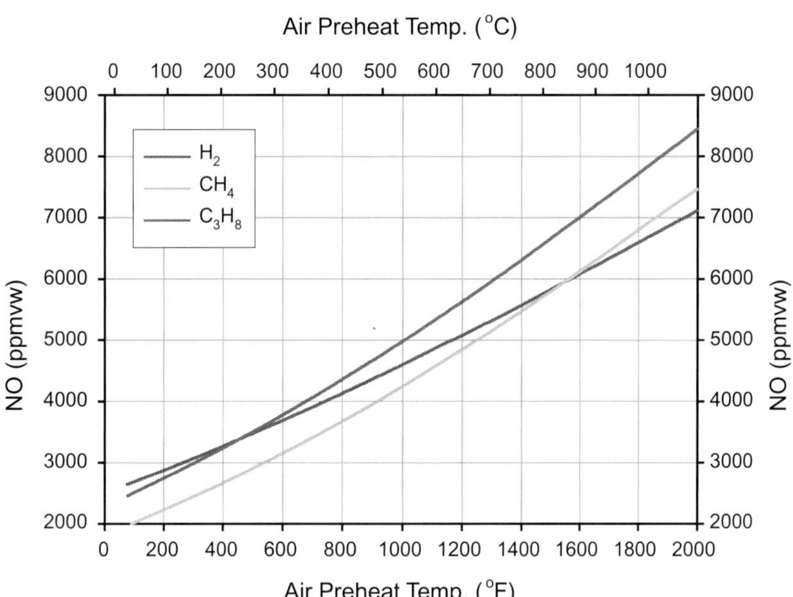

FIGURE 6.7 Adiabatic equilibrium NO as a function of air preheat temperature for stoichiometric air/fuel flames.

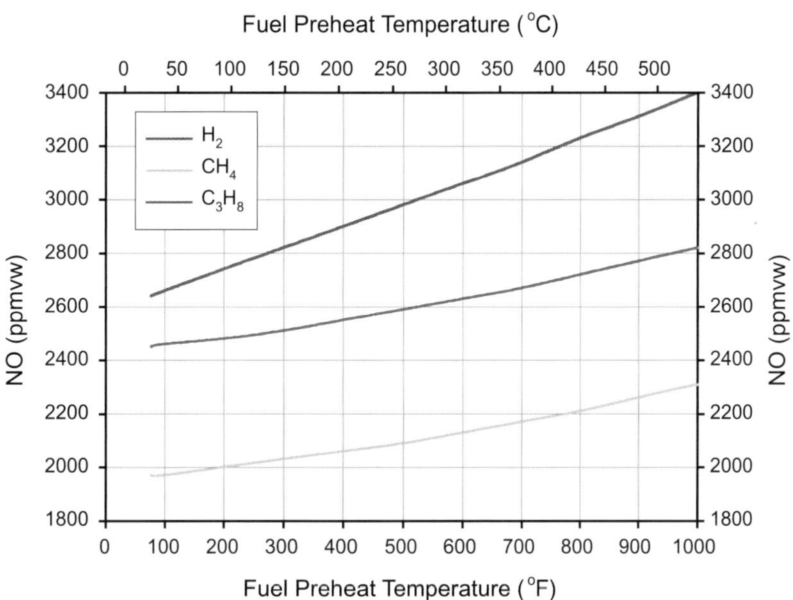

FIGURE 6.8 Adiabatic equilibrium NO as a function of fuel preheat temperature for a stoichiometric air/CH_4 flame.

Figure 6.8 shows how NOx increases with the fuel preheat temperature. Fuel preheating is another method used to improve the overall thermal efficiency of a heating process. Figure 2.18 shows how the adiabatic flame temperature increases due to fuel preheating. The increase in NOx emissions follows the same pattern as the increase in flame temperature.

Figure 6.9 shows how the fuel composition affects NO for a blend of CH_4 and H_2. First, it is important to note that NO increases as the H_2 content in the blend increases. This is similar to the effect on the adiabatic flame temperature as shown in Figure 2.19. The second thing to note is that the effect is not linear between pure CH_4 and pure H_2. NOx increases more rapidly as the H_2 content increases. The third thing to notice is that there is a significant difference between the two extremes as the NOx ranges from a little less than 2000 ppmvw to a little more than 2600 ppmvw.

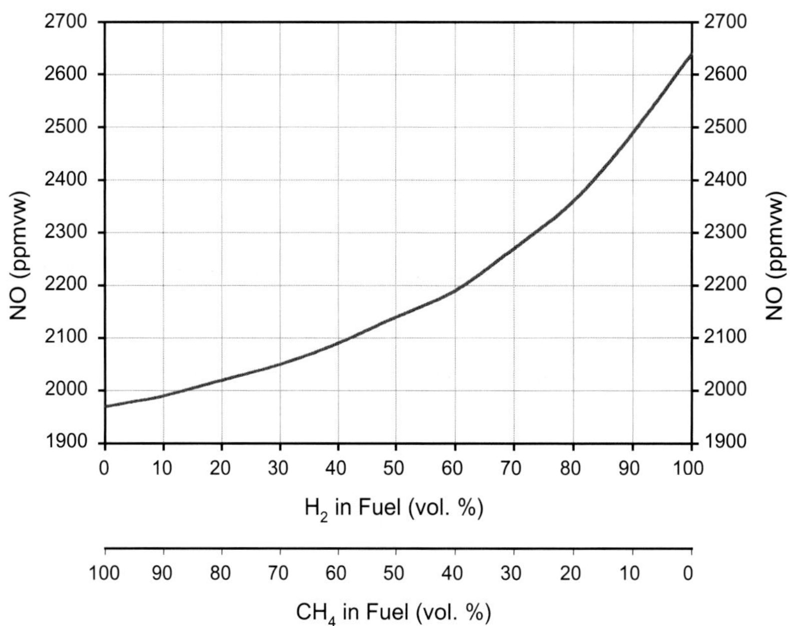

FIGURE 6.9 Adiabatic equilibrium NO as a function of fuel composition (CH_4/H_2) for a stoichiometric air/fuel flame.

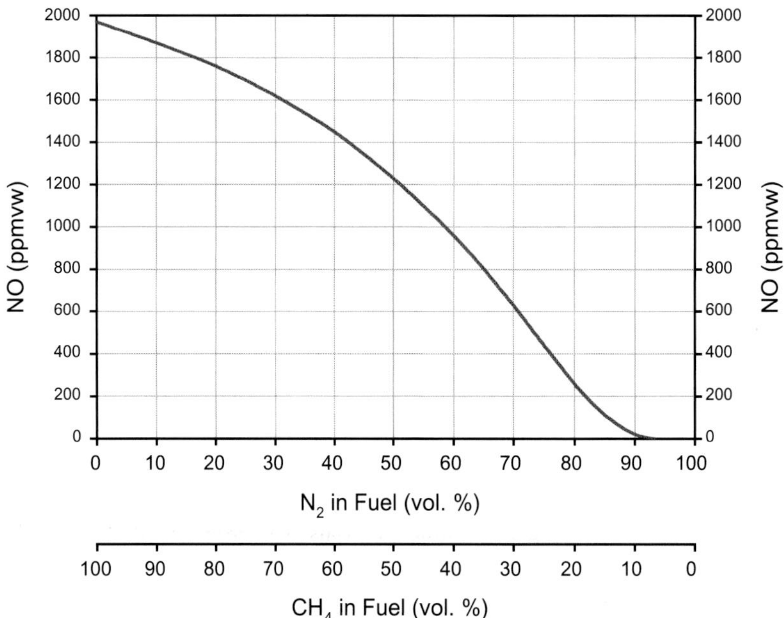

FIGURE 6.10 Adiabatic equilibrium NO as a function of fuel composition (CH_4/N_2) for a stoichiometric air/fuel flame.

Figure 6.10 shows how the fuel composition affects NO for a blend of CH_4 and N_2. NO (ppmvw) drops off rapidly as the N_2 in the fuel blend increases. At 100% N_2, the "fuel" produces no NO. The additional quantity of N_2 in the fuel does not increase NOx because of the increased availability of N_2 to make NOx since there is already plenty of N_2 available from the combustion air.

6.2.2 Regulations

Regulations for NOx vary by country and region. The United States, Japan, and Germany have some of the strictest regulations. Perhaps the most stringent standards in the world are those enforced by the South Coast Air Quality Management District (SCAQMD). SCAQMD governs the greater Los Angeles area and has proposed rules restricting NOx from burners to less than 5 ppmvd corrected to 3% O_2 for new sources. Currently, there are no burners that can meet these emissions without post-combustion controls.

6.2.2.1 Units

Baukal and Eleazer (1995)[14] have discussed potential sources of confusion in the existing NOx regulations. These sources of confusion can be classified as either general or specific. General sources of confusion include, for example, the wide variety of units that have been used, reporting on either a dry or wet sample basis, measuring NO but reporting NO_2, and reporting on a volume vs. a mass basis.

Historically, governing bodies have sprung up regionally for the purpose of regulating specific sources. The governing bodies have generally adopted units related to a traditional industry metric. This has led to a wide variety of NOx units. For example, internal combustion (IC) engines are generally regulated on a gram-per-brake-horsepower (g/bhp) basis — a mass-based unit normalized by the output power of the engine. Gas turbines, on the other hand, are generally regulated on a part-per-million (ppm) basis. Because this unit is volume based, it must be referenced to a standard condition. Gas turbines usually operate near 15% excess oxygen, and traditionally NOx measurement requires removal of water before analysis. Thus, gas turbines often use a ppm measurement referenced on a dry volume basis (ppmdv) to 15% oxygen.

In contrast, one typically operates industrial boilers and process heaters nearer to 3% excess oxygen. Thus, NOx emissions from those units are generally referenced as ppmvd corrected to 3% oxygen. However, these units can also be regulated on a mass basis normalized by the heat release of the burner, for example, pounds per million Btu (lb/MMBtu). Large electrical utilities operate their boilers under very tight oxygen limits. Therefore, some U.S. agencies regulate utility boilers on a pound-per-megawatt basis (lb/MW). A further complication is whether to normalize the unit by gross output power (Gross MW), or to subtract parasitic power losses (Net MW). Foreign regulatory agencies use SI units such as grams per normal cubic meter (g/Nm^3).

6.2.2.2 Regulations in the Hydrocarbon and Petrochemical Industries

The U.S. Environmental Protection Agency (U.S. EPA) regulates emissions in the hydrocarbon and chemical processing industries (HPI and CPI, respectively) nationwide. At the state level, additional agencies are free to adopt more stringent regulations. Examples are the California Air Resources Board (CARB) and the Texas National Resource Conservation Commission (TNRCC). The TNRCC is proposing a very strict sub-10-ppm regulation for burners. For reference, traditional burners generate ~100 ppm NOx when firing gaseous fuels.

Some states have even more local agencies such as SCAQMD regulating the greater Los Angeles area, or the Bay Area Air Quality Management District (BAAQMD) regulating the greater San Francisco area. Additionally, there are various voluntary standards recommended by various institutes. The general trend is toward more stringent regulation. The large number of governing bodies shows the general public support for stricter pollution control at all levels of government.

6.2.3 Measurement Techniques

Accurate measurements of pollutants, such as NO and CO, from industrial sources are of increasing importance in view of strict air-quality regulations. Based on such measurements, companies may have to pay significant fines, stop production, install expensive flue-gas treatment systems, buy NOx credits in certain non-attainment areas, or change the production process to a less polluting technology. If compliance is achieved, however, the company may continue their processes without interruption and, sometimes, sell their NOx credits. Mandel (1997) notes that the equipment cost for the gas analysis system is relatively small compared to the maintenance and repair costs.[15]

Numerous studies have been done and recommendations made on the best ways to sample hot gases from high-temperature furnaces. For example, U.S. EPA Method 7E[16] applies to gas samples extracted from an exhaust stack that are analyzed with a chemiluminescent analyzer. A typical sampling system is shown in Figure 6.11. The major components are: a heated sampling probe, heated filter, heated sample line, moisture removal system, pump, flow control valve, and then the analyzer. The U.S. EPA method states that the sample probe may be made of glass, stainless steel, or other equivalent materials. The probe should be heated to prevent water in the combustion products from condensing inside the probe.

The U.S. EPA method is appropriate for a lower temperature, nonreactive gas sample obtained, for example, from a utility boiler. However, this method should not be used to obtain samples from higher temperature industrial furnaces used in glass or metals production. Flue-gas temperatures from such furnaces, as well as from some incinerators, can be as high as 2400°F (1300°C). This would cause the probe to overheat and affect the measurements because of high-temperature surface reactions inside the probe.

The effects of probe materials, such as metal and fused quartz, as well as the probe cooling requirements, have been investigated for sampling gases in combustion systems.[17]

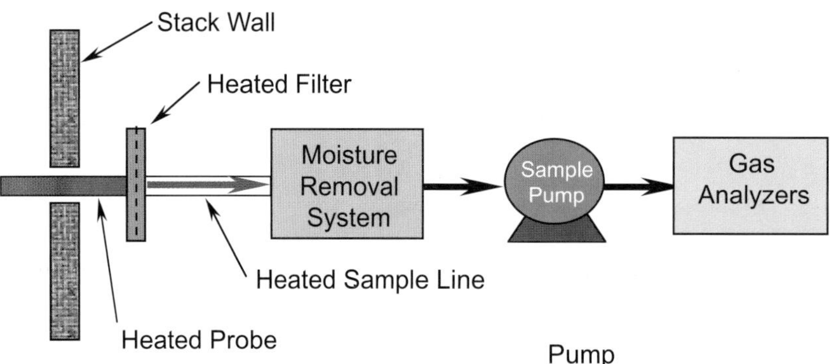

FIGURE 6.11 Sampling system schematic as recommended by the U.S. EPA.

TABLE 6.3 Reduction Efficiencies for NOx Control Techniques

Draft and Fuel Type	Control Technique	Total Effective NOx Reduction Percent
ND, distillate	(ND) LNB	40
	(MD) LNB	43
	(ND) ULNB	76
	(MD) ULNB	74
	SNCR[a]	60
	(MD) SCR	75
	(MD) LNB + FGR	43
	(ND) LNB + SNCR	76
	(MD) LNB + SNCR	77
	(MD) LNB + SCR	86
ND, residual	(ND) LNB	27
	(MD) LNB	33
	(ND) ULNB	77
	(MD)ULNB	73
	SNCR	60
	(MD) SCR	75
	(MD) LNB + FGR	28
	(ND) LNB + SNCR	71
	(MD) LNB + SNCR	73
	(MD) LNB + SCR	83
MD, distillate	(MD) LNB	45
	(MD) ULNB	74
	(MD) SNCR	60
	(MD) SCR	75
	(MD) LNB + FGR	48
	(MD) LNB + SNCR	78
	(MD) LNB + SCR	92
MD, residual	(MD) LNB	37
	(MD) ULNB	73
	(MD) SNCR	60
	(MD) SCR	75
	(MD) LNB + FGR	34
	(MD) LNB + SNCR	75
	(MD) LNB + SCR	91

Note: MD = mechanical draft, ND = natural draft, LNB = low-NOx burner, ULNB = ultra-low-NOx burner, SNCR = selective noncatalytic reduction, SCR = selective catalytic reduction, FGR = flue gas recirculation.

[a] Reduction efficiencies for ND or MD SNCR are equal.

Source: From E.B. Sanderford, U.S. EPA Report EPA-453/R-93-015, February 1993.

Several studies have found that both metal and quartz probe materials can significantly affect NO measurements in air/fuel combustion systems, especially under fuel-rich conditions with high CO concentrations.[18,19] However, the NO readings were not affected under fuel-lean conditions.

6.2.4 Abatement Strategies

Before air-quality regulations, the flue gases from combustion processes were vented directly to the atmosphere. As air-quality laws tightened and the public's awareness increased, industry began looking for new strategies to curb NOx emissions. The four strategies for reducing NOx are discussed next. Table 6.3 shows typical NOx reduction efficiencies as functions of the burner draft type (natural or forced), fuel (distillate or residual oil), and reduction technique.[12] The NOx emissions from gas-fired process heating equipment are highly variable (see Table 6.4).[29] Therefore, the technique or techniques chosen to reduce NOx emissions are very site and equipment dependent. This section is not intended to be exhaustive, but is comprehensive and includes many of the commonly used techniques for minimizing NOx emissions.

6.2.4.1 Pretreatment

The first NOx reduction strategy can be referred to as pretreatment. Pretreatment is a preventative technique to minimize NOx generation. In pretreatment, the incoming feed materials (fuel, oxidizer, and/or the material being heated) are treated in such a way as to reduce NOx. Some of these treatments include fuel switching, using additives, fuel treatment, and oxidizer switching.

6.2.4.1.1 Fuel Switching

Fuel switching is simply replacing a more polluting fuel with a less polluting fuel. For example, fuel oils generally contain some organically bound nitrogen that produces fuel NOx. Natural gas does not normally contain any organically bound

TABLE 6.4 NOx Control Technologies in Process Heaters

Control Technology	Controlled Emissions	Percent Reduction
Low-NOx burners	0.1–0.3 lb/10^6 Btu	25–65
Staged air lances	N/A	35–51
Fiber burner	10–20 ppm	
Ammonia injection	N/A	43–70
Urea injection + Low-NOx burner	N/A	55–70
Selective catalytic reduction	20–40 ppm	65–90
Selective catalytic reduction + Low-NOx burner	25–40 ppm	70–90

Note: Uncontrolled emissions are in the range of 0.1 to 0.53 lb/10^6 Btu. N/A = not available.

Source: From J. Bluestein, Gas Research Institute Report GRI-92/0374, Gas Research Institute, Chicago, IL, 1992.

nitrogen and usually has only low levels of molecular nitrogen (N_2). Partial or complete substitution of natural gas for fuel oil can significantly reduce NOx emissions by reducing the amount of nitrogen in the fuel. Figure 6.9 shows that CH_4 produces less NOx than H_2. Fuels composed entirely of hydrogen can produce twice as much NOx as fuels with no hydrogen.[20] Fuel switching may or may not be an option, depending on the availability of fuels and on the economics of switching to a different fuel.

6.2.4.1.2 Additives

Another type of pretreatment involves adding a chemical to the incoming feed materials (raw materials, fuel, or oxidizer) to reduce emissions by changing the chemistry of the combustion process. One example would be injecting ammonia into the combustion air stream as a type of *in situ* de-NOx process, but only under certain conditions (see Section 6.2.4.4.1). Several factors must be considered in determining the viability of this option. These include economics, the effects on the process, and the ease of blending chemicals into the process.

6.2.4.1.3 Fuel Pretreatment

A third type of pretreatment involves treating the incoming fuel prior to its use in the combustion process. An example would be removing fuel-bound nitrogen from fuel oil or removing molecular nitrogen from natural gas, which can reduce NOx in air- or O_2-fuel, and O_2-fuel combustion, respectively. This is normally an expensive process, depending on how much treatment must be done and how the fuel is treated. For example, it is generally more difficult to remove nitrogen from fuel oil than from natural gas. In Europe, some natural gas supplies have as much as 15% N_2 by volume. If only a few percent N_2 needs to be removed from that type of natural gas, this can usually be done relatively easily and inexpensively with adsorption or membrane separation techniques.

6.2.4.1.4 Oxidizer Switching

The fourth type of pretreatment is oxidizer switching, where a different oxidizer is used. Air is the most commonly used oxidizer. It can be shown that substantial NOx reduction can be achieved using pure oxygen, instead of air, for combustion.[21] For example, in the extreme case of combusting a fuel like CH_4 with pure O_2, instead of air that contains 79% N_2 by volume, it is possible to completely eliminate NOx as no N_2 is present to produce NOx. For example, if H_2 is combusted with pure O_2, the global reaction can be represented by:

$$2H_2 + O_2 \rightarrow 2H_2O \tag{6.7}$$

By drastically reducing the N_2 content in the system, NOx is minimized. However, there are significant challenges to using high-purity oxygen — instead of air — for combustion.[22] This technique has not been used widely in the hydrocarbon, petrochemical, and power generation industries, but could become more popular in the future as the cost of oxygen continues to decline as less expensive methods for separating oxygen from air are developed.

6.2.4.2 Combustion Modification

The second strategy for reducing NOx is known as combustion modification. Combustion modification prevents NOx from forming by changing the combustion process. There are numerous methods that have been used to modify the combustion process for low NOx. A popular method is a low NOx burner design in which specially designed burners generate less NOx than previous burner technologies. Low NOx burners may incorporate a number of techniques for minimizing NOx, including flue-gas recirculation, staging, pulse combustion, and advanced mixing. Common combustion modification techniques are discussed next.

6.2.4.2.1 Air Preheat Reduction

One combustion modification technique is reducing the combustion air preheat temperature. As shown in Figure 6.8, reducing the level of air preheat can significantly reduce NOx emissions. Air preheat greatly increases NOx for processes that use heat recuperation. However, reduction of air preheat also reduces the overall system efficiency, as shown in Figure 2.23. The loss of efficiency can be somewhat mitigated if the heater is equipped with a convection section. This is a fairly easy technique to implement and may be cost-effective if the lost efficiency is more than offset by alternative NOx reduction techniques.

6.2.4.2.2 Low Excess Air

As shown in Figure 6.5, excess air increases NOx emissions. The excess air generally comes from two sources: the combustion air supplied to the burner and air infiltration into the heater. Excess air produced by either source is detrimental to NOx emissions. Excess air increases NOx formation by providing additional N_2 and O_2 that can combine in a high-temperature reaction zone to form NO. In many cases, NOx can be reduced by simply reducing the excess air through the burners.

Air infiltration, sometimes referred to as tramp air, into a combustion system affects the excess air in the combustor and can affect NOx emissions. The quantity and location of the leaks are important. Small leaks far from the burners are not nearly as deleterious as large leaks near the flames. By reducing air infiltration (leakage) into the furnace, NOx can be reduced because excess O_2 generally increases NOx.

There is also an added benefit in reducing excess air. Reducing the excess O_2 in a combustion system is also useful for maximizing thermal efficiency because any unnecessary air absorbs heat that is then carried out of the stack with the exhaust products. However, there is a practical limit to how low the excess O_2 can be. Because the mixing of the fuel and air in a diffusion flame burner is not perfect, some excess air is necessary to ensure both complete combustion of the fuel and minimization of CO emissions. The limit on reducing the excess air is CO emissions. If the excess O_2 is reduced too much, then CO emissions will increase. CO is not only a pollutant, but also an indication that the fuel is not being fully combusted, resulting in lower system efficiencies.

There are some special techniques that control the O_2 in the flame to minimize NOx. One example is pulse combustion, which has been shown to reduce NOx because the alternating very fuel-rich and very fuel-lean combustion zones minimize NOx formation. The overall stoichiometry of the oxidizer and fuel is maintained by controlling the pulsations. Pulse combustion is not being used in many industrial combustion processes at this time due to some operational problems, especially the high-frequency cycling of the switching valves, that have not been satisfactorily resolved yet.

6.2.4.2.3 Staging

Staged combustion is an effective technique for lowering NOx. Staging means that some of the fuel or oxidizer, or both, is added downstream of the main combustion zone. The fuel, oxidizer, or both can be staged into the flame. For example, there may be primary and secondary fuel inlets where a portion of the fuel is injected into the main flame zone, and the balance of the fuel is injected downstream of that main flame zone. In fuel staging, some of the fuel is directed into the primary combustion zone, while the balance is directed into secondary and even tertiary zones in some cases (see Figures 1.36 and 1.37). This makes the primary zone fuel-lean, which is less conducive to NOx formation when compared to stoichiometric conditions. The excess O_2 from the primary zone is then used to combust the fuel added in the secondary and tertiary zones. While the overall stoichiometry can be the same as in a conventional burner, the peak flame temperature is much lower in the staged fuel case because the combustion process is staged over some distance while heat is simultaneously being released from the flame. The lower temperatures in the staged fuel flame help to reduce the NOx emissions. Thus, fuel staging is effective for two reasons: (1) the peak flame temperatures are reduced, which reduces NOx; and (2) the fuel-rich chemistry in the primary flame zone also reduces NOx. Waibel et al. (1986) have shown that fuel staging is one of the most cost-effective methods for reducing NOx in process heaters.[23]

In air staging, some of the combustion air is directed into the primary combustion zone, while the balance is directed into secondary and even tertiary zones in some cases (see Figures 1.34 and 1.35). This makes the primary zone fuel-rich, which is less conducive to NOx formation when compared to stoichiometric conditions. The unburned combustibles from the primary zone are then combusted in secondary and tertiary zones. While the overall stoichiometry may be the same as in a conventional burner, the peak flame temperature is much lower in the staged air case because the combustion process is staged over some distance while heat is simultaneously being released from the flame. The lower temperatures in the staged air flame help reduce the NOx emissions.

6.2.4.2.4 Gas Recirculation

Furnace gas recirculation is a process that causes the products of combustion inside the combustion chamber to be recirculated back into the flame (see Figure 6.12). This is sometimes referred to as internal flue gas recirculation. External flue gas recirculation (see Figure 1.28) is similar.

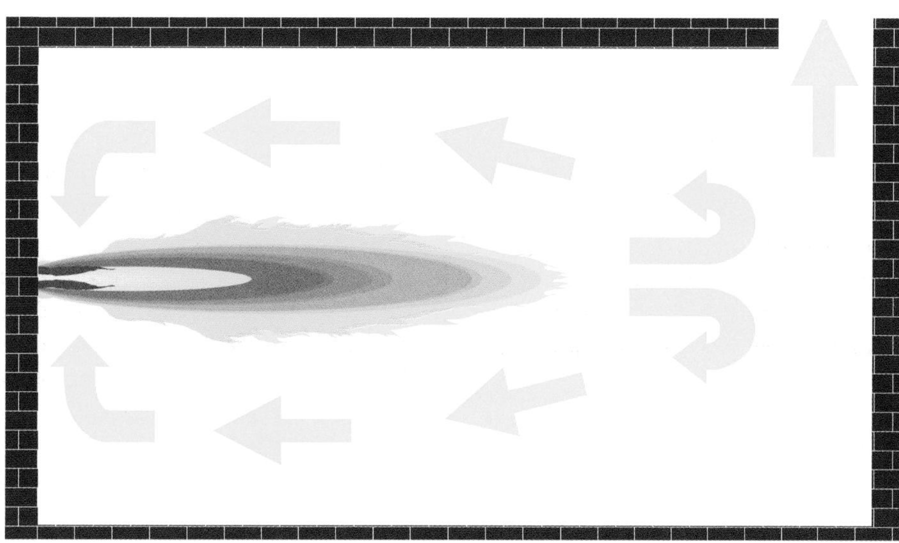

FIGURE 6.12 Schematic of furnace gas recirculation.

External flue gas recirculation causes the exhaust gases in the flue to be recirculated back through the burner into the flame via ductwork external to the furnace. Although the furnace or flue gases are hot, they are considerably cooler than the flame itself. The cooler furnace or flue gases act as a diluent, reducing the flame temperature, which in turn reduces NOx (see Figure 6.6). Advanced mixing techniques use carefully designed burner aerodynamics to control the mixing of the fuel and the oxidizer. The goals of this technique are to avoid hot spots and make the flame temperature uniform, to increase the heat release from the flame, which lowers the flame temperature, and to control the chemistry in the flame zone to minimize NOx formation.

External flue gas recirculation requires some type of fan to circulate the gases external to the furnace and back through the burner. The burner must be designed to handle both the added volume and different temperature of the recirculated gases that are often partially or fully blended with the combustion air. Garg (1992) estimates NOx reductions of up to 50% using flue gas recirculation.[24]

6.2.4.2.5 Water or Steam Injection

Many of the combustion modification methods attempt to reduce the temperature of the flame to lower NOx emissions. In many cases, this may result in a reduction of the combustion efficiency.[25] For example, if water is injected into the flame to lower NOx, the water absorbs heat from the flame and carries most of that energy out with the exhaust gases, thus preventing the transfer of much of that energy to the load. Combustion modification methods are usually less capital intensive than most post-treatment methods. In many cases, there is a limit to how much NOx reduction can be achieved using these methods.

Another form of water injection is to inject water in the form of steam. There are several reasons for this. One is that steam is much hotter than liquid water and already includes the latent heat of vaporization needed to change the liquid water to a vapor. When liquid water is injected into a combustion process, it can put a large heat load on the process because liquid water can absorb a large amount of energy before becoming a vapor due to its high latent heat of vaporization. Steam puts a much smaller load on the process because it absorbs less energy than liquid water. Another reason for using steam instead of liquid water is that steam is already in vapor form and mixes readily with the combustion gases. Liquid water must be injected through nozzles to form a fine mist to disperse it uniformly with the combustion gases. Therefore, it is often easier to blend steam into the combustion products compared to liquid water. At equivalent injection rates, water injection is usually more effective at reducing NOx than steam injection.

6.2.4.2.6 Reburning

Reburning is a technique similar to fuel staging but uses a different strategy. An example is methane reburn, where some methane is injected in the exhaust gases, usually well after the primary combustion zone, in which the gases are at a lower temperature. As previously shown in Figure 6.5, fuel-rich conditions are not favorable to NOx. As the exhaust gases from the combustion process flow through this fuel-rich reducing zone, NOx is reduced back to N_2. Any CO and

TABLE 6.5 NOx Reductions for Different Low-NOx Burner Types

Burner Type	Typical NOx Reductions (%)
Staged-air burner	25–35
Staged-fuel burner	40–50
Low-excess-air burner	20–25
Burner with external FGR[a]	50–60
Burner with internal FGR[a]	40–50
Air or fuel-gas staging with internal FGR[a]	55–75
Air or fuel-gas staging with external FGR[a]	60–80

[a] FGR = Flue gas recirculation.

Source: A. Garg, *Chem. Eng. Prog.*, 90(1), 46–49, 1994. With permission.

other unburned fuels in the exhaust gases are then combusted downstream of the reduction zone at temperatures well below those found in the main combustion process. These lower temperature reactions are not favorable to NOx formation, so the net effect is that NOx is reduced.

There are some challenges with this technique. One is to get proper injection of the reburning gas and the exhaust products. Another is that the reburn zone must be capable of sustaining combustion. It needs to be done in a lower temperature part of the process to minimize the subsequent formation of NOx. It may be done, for example, in a previously uninsulated portion of ductwork. This may require replacement with higher temperature materials and the new duct may need to be insulated. A third challenge is trying to take advantage of some of the energy produced during the reburning. A heat recovery system may need to be added.

6.2.4.2.7 Low NOx Burners

Garg (1994) discusses the use of various low NOx burners to achieve emissions reductions compared to standard gas burners.[26] Table 6.5 shows typical NOx reductions using various low NOx burner techniques. An EPA study found that ultra low-NOx burners were the most cost-effective means to reduce NOx.[12] Low NOx burners typically incorporate one or more of the techniques discussed in Section 6.2.4.2.

6.2.4.2.8 Burner Out-of-Service (BOOS)

This is a technique primarily used in boilers where the fuel is turned off to the upper burners, while maintaining the air flow to all.[27] The fuel removed from the upper burners is then redirected to the lower burners, while maintaining the same air flow to the lower burners. Therefore, the overall fuel and air flow to the boiler remains the same but is redistributed. This makes the lower burners fuel-rich, which is less conducive to NOx formation due to the lower flame temperatures and fuel-rich chemistry. The upper burners (air only) provide the rest of the air needed to fully combust the fuel. Rather than air staging in individual burners, the BOOS technique stages air over the entire boiler. This technique is relatively inexpensive to implement. Ensuring proper heat distribution is important to prevent overheating the tubes or derating the firing capacity.

6.2.4.3 Process Modification

There are a number of techniques that can be employed to change the existing process in such a way as to reduce NOx emissions. These methods are often more radical and expensive, and are not often employed except under somewhat unique circumstances. These must be analyzed on a case-by-case basis to see if they are viable.

6.2.4.3.1 Reduced Production

If the mass of NOx emitted from a plant is too high, an alternative is to reduce the firing rate, which means a corresponding reduction in production. The reduction in NOx is proportional to the reduction in firing rate as less fuel is burned and therefore less NOx is formed. However, this is generally not a preferred alternative, for obvious reasons, as less of the product being made is available to sell. Depending on the costs to reduce NOx, this may be the most economic alternative in some cases.

In boilers, reducing the firing rate reduces the overall temperature inside the boiler, which reduces thermal NOx formation.[28] This technique is known as derating and is not desirable if the boiler is capacity limited, but in certain limited applications it may be a viable alternative.

6.2.4.3.2 Electrical Heating

One process modification that is sometimes used to minimize or eliminate NOx emissions is to replace some or all of the fossil-fuel-fired energy with electrical energy. The electrical energy produces no NOx emissions at the point of use and moves the emissions to the power plant. In general, the resulting NOx emissions at the power plant are often lower than at an industrial site because of the strict limits imposed on the power plant and the various methods employed to minimize NOx, which are often more cost-effective on a unit mass basis because of the economies of scale.

There are a number of potential problems with this method. The first is that the economics are usually very unfavorable when replacing fossil fuels with electrical energy. In most hydrocarbon and petrochemical processes, the fuel used in the heaters is a by-product that is available at little or no cost. On the other hand, electrical energy is often much more expensive than even purchased fossil fuels like natural gas or oil. Besides the higher operating costs, there would be substantial capital costs involved in converting some or all of the existing fossil-energy heating to electricity. Besides the removal of the existing burners, there would be the cost of the new electrical heaters and often large costs of installing electrical substations that would be required for all the additional power. In many parts of the country, large additional sources of electricity are not readily available, so a new source of electricity may need to be built at the plant, such as a cogeneration facility. However, although the electrical costs can be reduced in that scenario because the transmission losses are much lower, the NOx emissions are now at different locations at the site and little may then be gained in reducing overall NOx emissions for the plant. It is likely in the future that regulations will consider the net NOx generated during the production of a product and would include the NOx formed in the generation of electricity. This will make replacement of fossil energy with electricity less attractive as most of the power generated in the United States is by fossil-fuel-fired power plants.

6.2.4.3.3 *Improved Thermal Efficiency*

By making a heating process more efficient, less fuel needs to be burned for a given unit of production. Because the firing rate is directly proportional to NOx emissions, less fuel used equals less NOx produced. There are many ways to improve the efficiency of a process. A few representative examples will be given. One is to repair the refractory and air infiltration leaks on an existing heater. This is often relatively inexpensive and saves fuel while reducing NOx. Another is to add heat recovery to a heating process that does not currently have it. The heat recovery could be in several forms. One method is to preheat the incoming combustion air. As previously discussed, this can increase NOx emissions due to the higher flame temperatures if it is not done properly. Another method is to add a convection section onto a heater that does not presently have it. This has other operational benefits as well and is often a good choice. A more drastic method of increasing the thermal efficiency of a heating process is to replace an old, existing heater with a new, more modern design. This may make sense if the existing heater is very old, is high maintenance, and is not easily repairable or upgradable. However, new sources often must meet more stringent NOx standards than existing sources.

6.2.4.3.4 *Product Switching*

Another radical process modification that can reduce NOx is to switch the product being produced to one that requires less energy to process. In a process heater, this would involve replacing the existing process fluid with one that requires less energy to heat. For example, the heavier crude oils require more energy to process than lighter, purer crudes, so less energy would be needed to process the purer crudes. Less energy consumption means less NOx generated. However, this is obviously not an option in most cases, and is only considered under extreme circumstances. In the above example, purer or "sweeter" crudes are much more expensive raw materials than less pure or more "sour" crudes. Therefore, the savings in energy may be more than offset by the higher raw material costs.

6.2.4.4 Post-treatment

The fourth strategy for minimizing NOx is known as post-treatment. Post-treatment removes NOx from the exhaust gases after the NOx has already been formed in the combustion chamber. Two of the most common methods of post-treatment are selective catalytic reduction (SCR) and selective noncatalytic reduction (SNCR).[29] Wet techniques for post-treatment include oxidation-absorption, oxidation-absorption-reduction, absorption-oxidation, and absorption-reduction. Dry techniques for post-treatment, in addition to SCR and SNCR, include activated carbon beds, electron beam radiation, and reaction with hydrocarbons. One of the advantages of post-treatment methods is that multiple exhaust streams can be treated simultaneously, thus achieving economies of scale. Most of the post-treatment methods are relatively simple to retrofit to existing processes.

Many of these techniques are fairly sophisticated and are not trivial to operate and maintain in industrial furnace environments. For example, catalytic reduction techniques require a catalyst that may become plugged or poisoned fairly quickly by dirty flue gases. Post-treatment methods are often capital intensive. They usually require halting production if there is a malfunction of the treatment equipment. Also, post-treatment does not normally benefit the combustion process in any way. For example, it does not increase production or energy efficiency. It is strictly an add-on cost. A good reference for post-treatment NOx control for heaters used in refineries is API 536.[39]

6.2.4.4.1 *Selective Catalytic Reduction (SCR)*

Selective catalytic reduction (SCR) involves injecting an NOx-reducing chemical into an exhaust stream in the presence of a catalyst within a specific temperature window. The chemical is typically ammonia and the temperature window is approximately 500 to 1100°F (230 to 600°C). The NOx and NH_3 react on the catalyst surface to form N_2 and H_2O. The important reactions are:[42]

$$6NO + 4NH_3 \rightarrow 5N_2 + 6H_2O$$

$$2NO + 4NH_3 + 2O_2 \rightarrow 3N_2 + 6H_2O$$

There are a number of potential problems and challenges with SCR techniques. The catalyst introduces a pressure drop into the system, which often increases the power requirements for the gas-handling equipment. The catalyst may become plugged or fouled in dirty exhaust streams, which is especially a challenge when firing liquid fuels like residual oil. The ammonia must be properly injected into the flue gases to get proper mixing, at the right location to be in the proper temperature window, and in the proper amount to get adequate NOx reduction without allowing ammonia to slip through unreacted. SCR systems are not very tolerant of constantly changing conditions, as a stable window of operation is required for optimum efficiency. Another problem is handling the spent catalyst. Regeneration is often most attractive but may be more expensive than buying new catalyst. Disposal of the spent catalyst may be expensive as it may be classified as a hazardous waste, especially if the catalyst contains vanadium, as is commonly the case. A U.S. EPA study found that SCR was the most expensive means to reduce NOx.[12]

6.2.4.4.2 Selective Noncatalytic Reduction (SNCR)

Selective noncatalytic reduction (SNCR) involves injecting NOx-reducing chemicals into the exhaust products from a combustion process within a specific temperature window.[39] No catalyst is involved in the process. The most commonly used chemicals are ammonia and urea. Other chemicals (e.g., hydrogen, hydrogen peroxide, and methanol) can be added to improve the performance and lower the minimum threshold temperature. The Exxon thermal de-NOx process is one common SNCR technique using ammonia, and is employed in a wide variety of industrial applications. A typical global reaction for this technique can be written as:

$$2NO + 4NH_3 + 2O_2 \rightarrow 3N_2 + 6H_2O$$

The optimum temperature window, without the addition of other chemicals to increase the temperature window, is 1600 to 2200°F (870 to 1200°C). The Nalco Fuel Tech NOxOUT® is a common SNCR technique employing urea:

$$CO(NH_2)_2 + 2NO + 1/2O_2 \rightarrow 2N_2 + CO_2 + 2H_2O$$

The optimum temperature window, without the addition of other chemicals to increase the temperature window, is 1600 to 2000°F (870 to 1100°C).

There are some problems with SNCR. The first is the cost, which is usually significantly more than non-post-treatment techniques like low-NOx burners. Although the use of SNCR decreases NOx, it may increase other undesirable emissions such as CO, N_2O and NH_3 (which can occur if the injected chemicals slip through the exhaust without reacting, referred to as ammonia slip when ammonia exits the stack).[12]

CO can be elevated with large quantities of NH_3 because the NH_3 competes for the OH radical, the main oxidation route for CO:[41]

$$CO + OH \rightarrow CO_2 + H$$

$$NH_3 + OH \rightarrow NH_2 + H_2O$$

There are also safety concerns with regard to the transport and storage of the ammonia (NH_3) used in SNCR. Other major challenges of this technology include finding the proper location in the process to inject the chemicals (the chemicals must be injected where the flue gases are within a relatively narrow temperature window for optimum efficiency); injecting the proper amount of chemicals (too much will cause some chemicals to slip through unreacted, and too little will not get sufficient NOx reductions); and getting proper mixing of the chemicals with the flue gas products (there must be both adequate mixing and residence time for the reactions to go to completion). Both physical and computer modeling are often used to determine the optimal place, amount, and method of injection.

6.2.5 Field Results

6.2.5.1 Conversions

It is important to be able to convert field measurements to specific units to determine whether the emissions from a specific burner or heater are below their allowable limits. In nearly all cases, NOx is measured on a ppmvd basis. The following examples will show how to convert these units to a specific basis.

Example 6.4

Given: Fuel = methane with a gross or higher heating value of 1012 Btu/ft^3; NO = 20 ppmvd; measured O_2 = 2% on a dry basis.

Find: NOx as NO_2 in lb/10^6 Btu (gross).

Solution: First calculate dry flue gas products.

Global chemical reaction:

$CH_4 + x(O_2 + 3.76N_2) = CO_2 + 2H_2O + yO_2 + 3.76xN_2$

where $O_2 + 3.76N_2$ is the composition of air (79% N_2, 21% O_2)

1. given 2% O_2 in dry flue gases:

$$\frac{y}{1 + y + 3.76x} = 0.02$$

2. O atom balance:

$$2x = 2 + 2 + 2y = 4 + 2y, \text{ or } x = 2 + y$$

Solving 1 and 2 simultaneously:

$$CH_4 + 2.188(O_2 + 3.76N_2) = CO_2 + 2H_2O + 0.188O_2 + 8.23N_2$$

This shows the moles of products for each mole of CH_4. Note that NO in the flue products has been ignored because it is only present in trace amounts. Assume that all NO is converted to NO_2 in the atmosphere.

$(988 \text{ ft}^3 \text{ CH}_4)(1012 \text{ Btu/ft}^3 \text{ CH}_4) = 1 \times 10^6$ Btu (gross)

$(988 \text{ ft}^3 \text{ CH}_4)(1 + 0.188 + 8.23) = 9305 \text{ ft}^3$ dry combustion products at STP per 10^6 Btu

Given 20 ppmvd NO_2 = $(20 \text{ ft}^3 \text{ NO}_2/10^6 \text{ ft}^3 \text{ dry products})(9305 \text{ ft}^3 \text{ dry products}/10^6 \text{ Btu}) = 0.186 \text{ ft}^3 \text{ NO}_2/10^6$ Btu

Density of $NO_2 = 0.111 \text{ lb/ft}^3$

Mass of NO_2 in exhaust products = $(0.186 \text{ ft}^3 \text{ NO}_2/10^6 \text{ Btu})(0.111 \text{ lb NO}_2/\text{ft}^3 \text{ NO}_2) = 0.021 \text{ lb NO}_2/10^6$ Btu (gross)

6.2.5.2 Fuel Composition Effects

The composition of the fuel supplied to a combustion system has a significant impact on the NOx emissions. In the petrochemical and chemical process industries, there is a very wide range of fuel blends used for process heating. These fuels are often by-products of a refining process. They typically contain hydrocarbons ranging from C1 to C4, hydrogen, and inert gases like N_2 and CO_2. In a given plant or refinery, burners used in process heaters may need to be capable of firing on multiple fuels that are present at different times (e.g., start-up, normal operation, upset conditions, etc.). In many cases, the NOx emissions from the heaters may not exceed a given value regardless of what fuel composition is being fired. Therefore, it is critical that the effects of the fuel composition on NOx emissions be understood and quantified to ensure that permitted values are not exceeded.

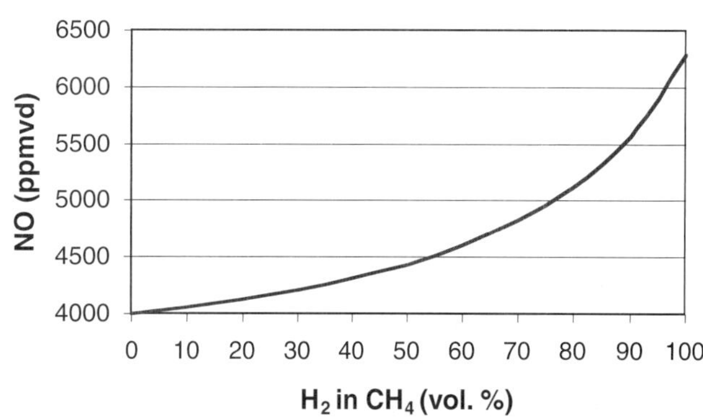

FIGURE 6.13 Adiabatic equilibrium NO as a function of the fuel blend composition for H_2/CH_4 blends combusted with 15% excess air where both the fuel and the air are at ambient temperature and pressure.

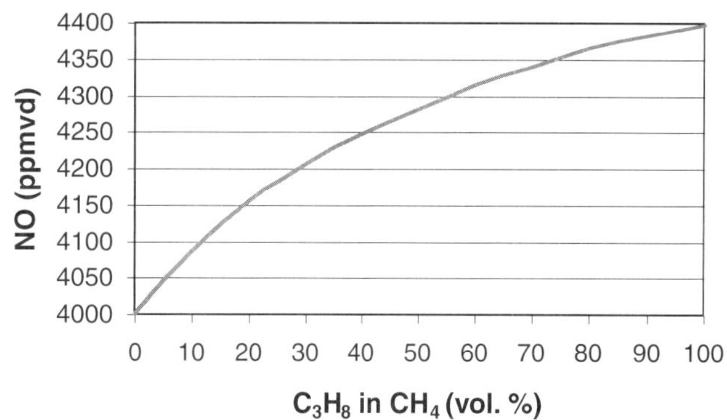

FIGURE 6.14 Adiabatic equilibrium NO as a function of the fuel blend composition for C_3H_8/CH_4 blends combusted with 15% excess air where both the fuel and the air are at ambient temperature and pressure.

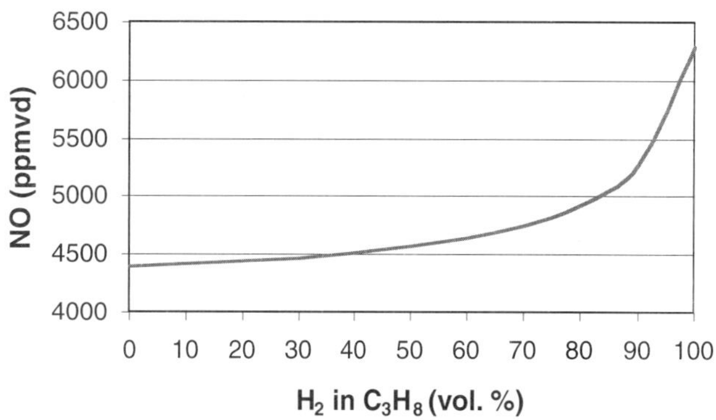

FIGURE 6.15 Adiabatic equilibrium NO as a function of the fuel blend composition for H_2/C_3H_8 blends combusted with 15% excess air where both the fuel and the air are at ambient temperature and pressure.

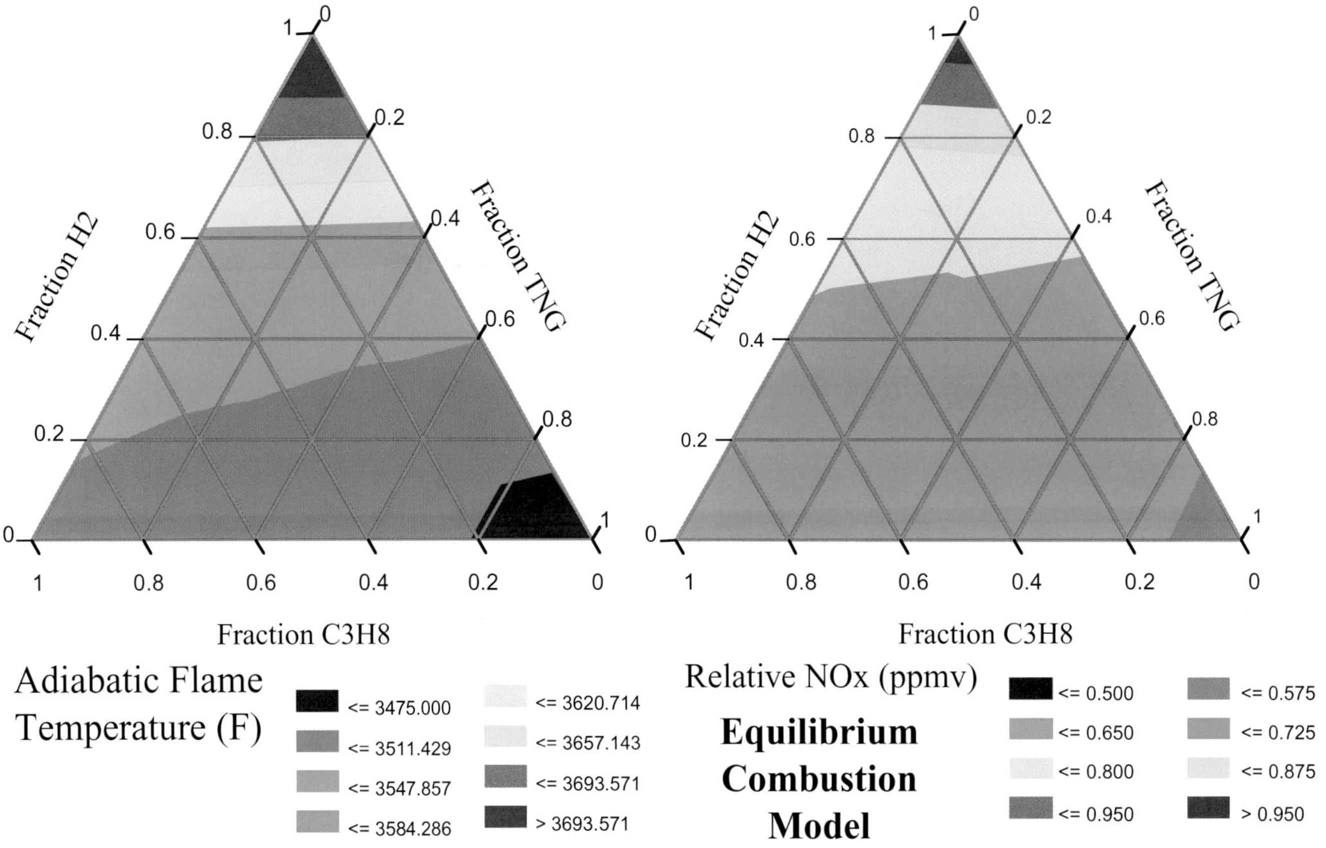

FIGURE 6.16 Ternary plot of adiabatic equilibrium NO (fraction of the maximum value) as a function of the fuel blend composition for $H_2/CH_4/C_3H_8$ blends combusted with 15% excess air where both the fuel and the air are at ambient temperature and pressure.

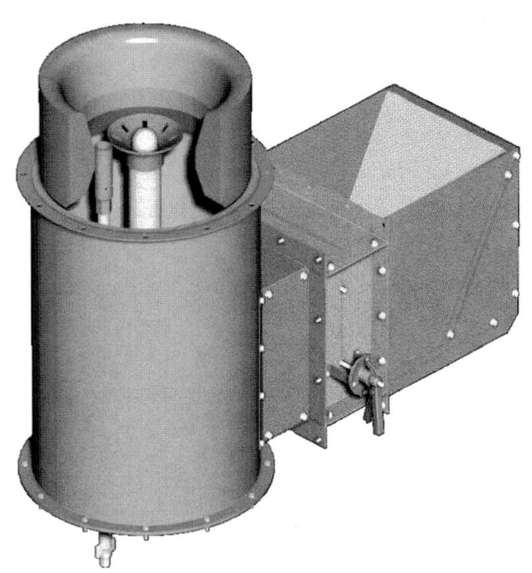

FIGURE 6.17 Raw gas (VYD) burner (Courtesy of John Zink Co.).

This section shows the results of an extensive series of tests to study the effects of fuel composition on NOx emissions from an industrial-scale burner.[43] The data provide additional insight into effects on NOx over the entire range of fuel compositions consisting of various fractions of three primary components: H_2, C_3H_8, and CH_4. Figures 6.13 – 15 show how NOx theoretically varies for two-component fuel mixtures of CH_4–C_3H_8, CH_4–H_2 and C_3H_8–H_2, respectively. These figures show the predicted adiabatic equilibrium NO concentrations for flames with 15% excess air. Figure 6.16 shows a ternary diagram of the calculated adiabatic flame temperatures (figure on the left) over the range of three-component fuel blends tested and another ternary diagram showing the predicted adiabatic equilibrium NO (figure on the right) for three-component fuel blends containing CH_4, C_3H_8 and H_2 combusted with 15% excess air.

Experiments were conducted using a conventional-type burner (see Figure 6.17) with a single fuel gas tip and flame-holder. The burner was fired vertically upward in a rectangular

FIGURE 6.18 Test furnace.

furnace (see Figure 6.18). The test furnace was a rectangular heater with internal dimensions of 8 ft (2.4 m) wide, 12 ft (3.7 m) long, and 15 ft (4.6 m) tall. The furnace was cooled by a water jacket on all four walls. The interior of the water-cooled walls was covered with varying layers of refractory lining to achieve the desired furnace temperature. The burner was tested at a nominal heat release rate of 7.5×10^6 Btu/hr (2.2 MW).

A velocity thermocouple (also known as a suction thermocouple or suction pyrometer — see Chapter 14) was used to measure the furnace and stack gas temperatures. The furnace draft was measured with an automatic, temperature-compensated, pressure transducer as well as an inclined manometer connected to a pressure tap in the furnace floor. Fuel flow rates were measured using calibrated orifice meters, fully corrected for temperature and pressure. Emission levels were measured using state-of-the-art continuous emissions monitors (CEMs) to measure emissions species concentrations of NOx, CO, and O_2.

The experimental matrix consisted of firing the burner at a constant heat release (7.5×10^6 Btu/hr or 2.2 MW) and excess air level (15%) with 15 different fuel blends comprised of varying amounts of H_2, C_3H_8, and Tulsa Natural Gas (TNG).* For testing and analysis purposes TNG was treated as a single fuel component for convenience. TNG, which is comprised of approximately 93% CH_4, is a more economical choice than pure CH_4 for experimental work, and the analysis is simplified by treating it as a single component. All 15 fuel compositions were tested on each of six different fuel gas tips, which differed in port diameter sizes, to enable the acquisition of additional information regarding effects resulting from differing fuel pressures.

Figure 6.19 shows the variation in relative measured NOx emissions resulting from different concentrations (volume basis) of H_2 in a fuel blend composed with a balance of TNG

*The nominal composition by volume of TNG is 93.4% CH_4, 2.7% C_2H_6, 0.60% C_3H_8, 0.20% C_4H_{10}, 0.70% CO_2, and 2.4% N_2.

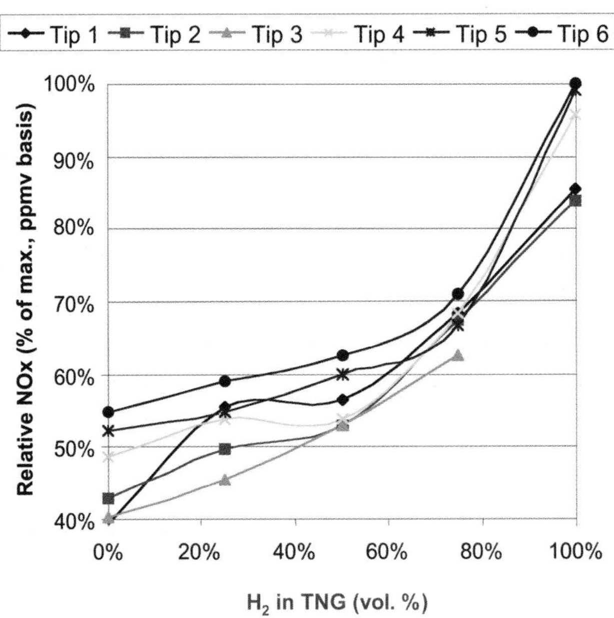

FIGURE 6.19 Measured NOx (percent of the maximum ppmv value) as a function of the fuel blend composition for H_2/TNG blends combusted with 15% excess air where both the fuel and the air were at ambient temperature and pressure.

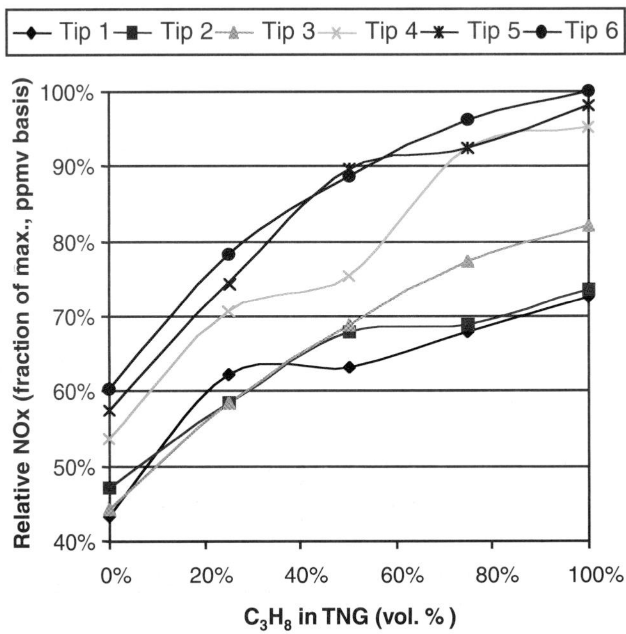

FIGURE 6.20 Measured NOx (percent of the maximum ppmv value) as a function of the fuel blend composition for C_3H_8/TNG blends combusted with 15% excess air where both the fuel and the air were at ambient temperature and pressure.

for each of the six different fuel gas tips tested. The plot, which illustrates NOx levels on a concentration basis, clearly shows the correlation between increased H_2 content and higher NOx emission levels. The slope of the profile is exponentially increasing, qualitatively similar to that predicted by the plotted theoretical calculations shown previously in Figure 6.13. The effect of H_2 is significant, with the sharpest increase in NOx levels taking place as concentration levels of H_2 in the fuel mixture rise from 75% to 100%.

The variation in relative measured NOx emissions resulting from different concentrations (volume basis) of C_3H_8 in a fuel blend composed with a balance of TNG is shown in

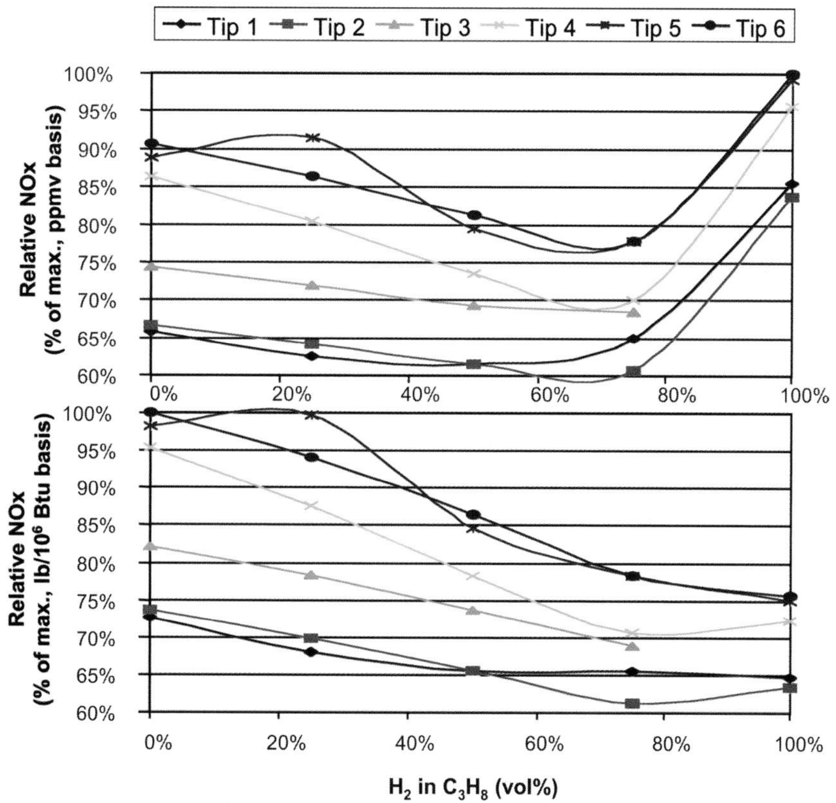

FIGURE 6.21 Measured NOx (percent of the maximum value in both ppmv and lb/MMBtu) as a function of the fuel blend composition for H_2/C_3H_8 blends combusted with 15% excess air where both the fuel and the air were at ambient temperature and pressure.

Figure 6.20. The slope of the increase in NOx levels corresponding to increased concentrations of C_3H_8 is shown to be relatively constant or slightly declining over the gradient in C_3H_8 concentration, in contrast with the exponentially increasing profile of the H_2–TNG plot in Figure 6.19. The profile showing the effect of C_3H_8 content is also seen to be similar to the corresponding calculated trends shown previously in Figure 6.14.

Figure 6.21 shows the final two-component fuel blend results being examined, which describe the variation in relative measured NOx emissions resulting from different concentrations (volume basis) of H_2 in a fuel blend composed with a balance of C_3H_8. The upper plot, which shows measured relative NOx on a volume concentration basis, illustrates that for a given tip geometry and port size, the measured NOx concentrations actually decrease slightly with increasing H_2 content up to 75% H_2 content, then sharply increase with H_2 concentration.

Due to the decrease in total dry products of combustion from the burning of H_2, expressing NOx in terms of concentration (ppmv) does not fully represent the actual mass rate of NOx emissions produced. The lower plot, which shows the variation in measured NOx levels on a mass per unit heat release basis, illustrates that the overall emissions of NOx on a mass basis decrease with increasing fuel hydrogen content and continue to decrease or remain relatively flat, even in the high-hydrogen content region which produced a sharp increase in NOx levels on a volume concentration basis.

6.2.5.3 Fuel Gas Tip Design

Three-component interaction results were also examined by considering results from several of the tested fuel gas tip designs. Figures 6.22 through 6.24 show contoured ternary plots of variation in relative measured NOx levels corresponding to different fractions of H_2, C_3H_8, and TNG in the fuel blend. Plots for three tip designs are shown, the tips differing only in fuel port area size, which results in different fuel pressures for a given heat release on each tip. The results are shown for tips in order of increasing port area size, or in other words, decreasing fuel pressure levels for the design heat release. Two plots are shown for each of three tips, with

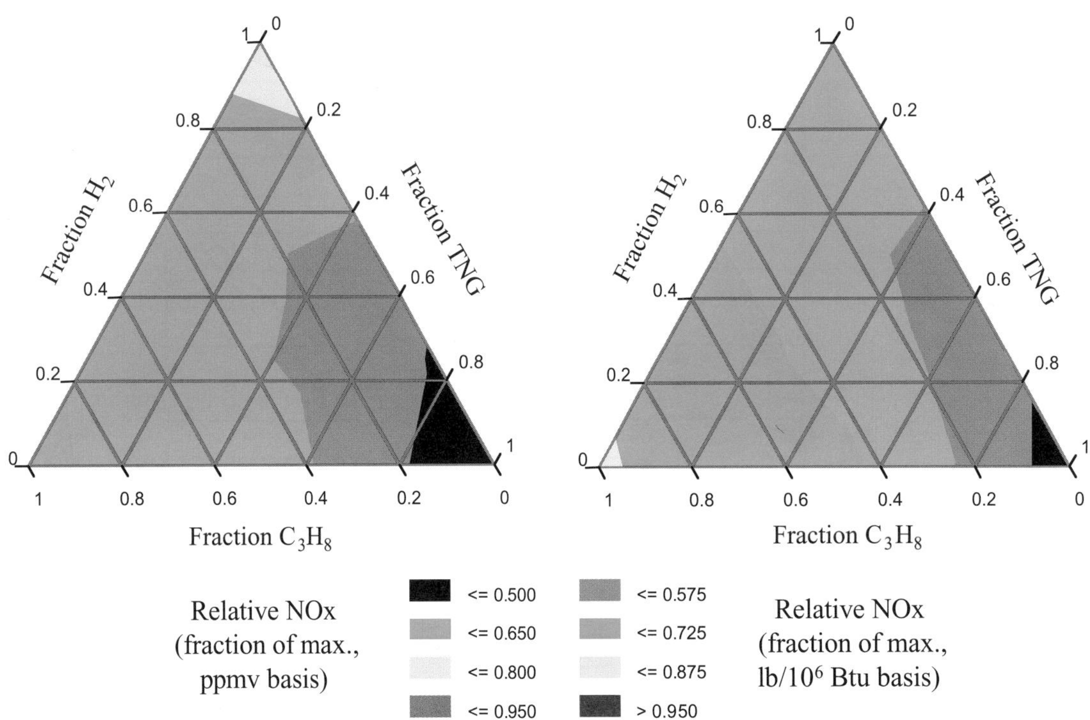

FIGURE 6.22 Measured NOx (fraction of the maximum value in both ppmv and lb/MMBtu) as a function of the fuel blend composition for TNG/H_2/C_3H_8 blends combusted with 15% excess air where both the fuel and the air were at ambient temperature and pressure, for gas tip #2.

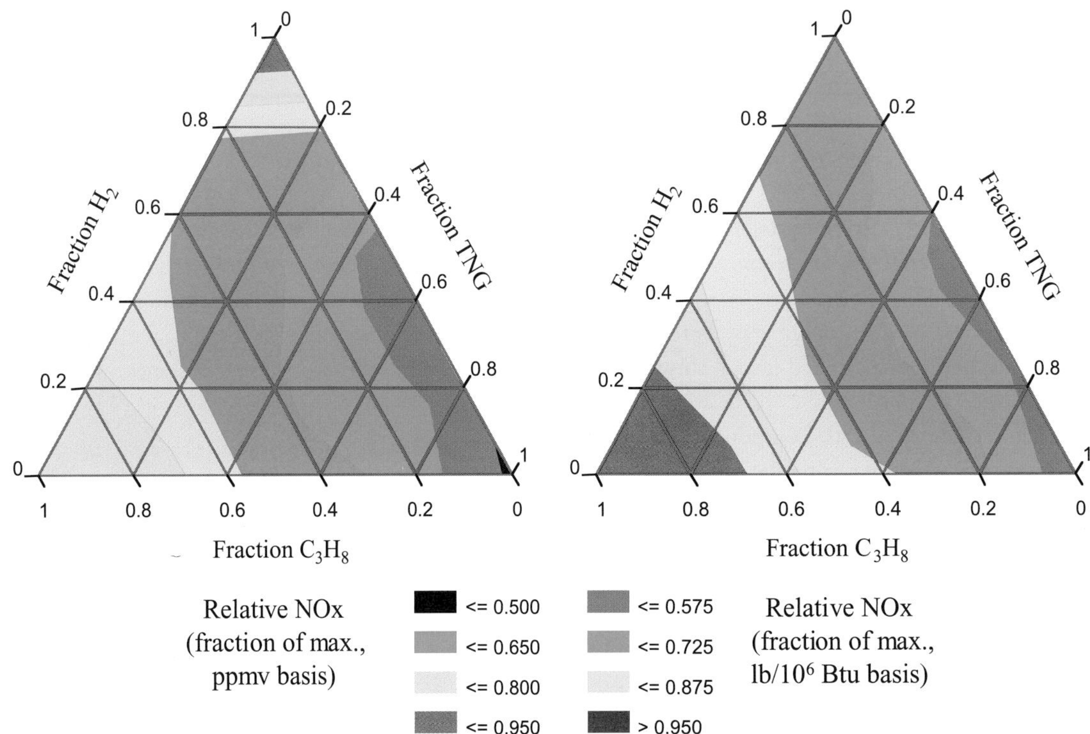

FIGURE 6.23 Measured NOx (fraction of the maximum value in both ppmv and lb/MMBtu) as a function of the fuel blend composition for TNG/H_2/C_3H_8 blends combusted with 15% excess air where both the fuel and the air were at ambient temperature and pressure, for gas tip #4.

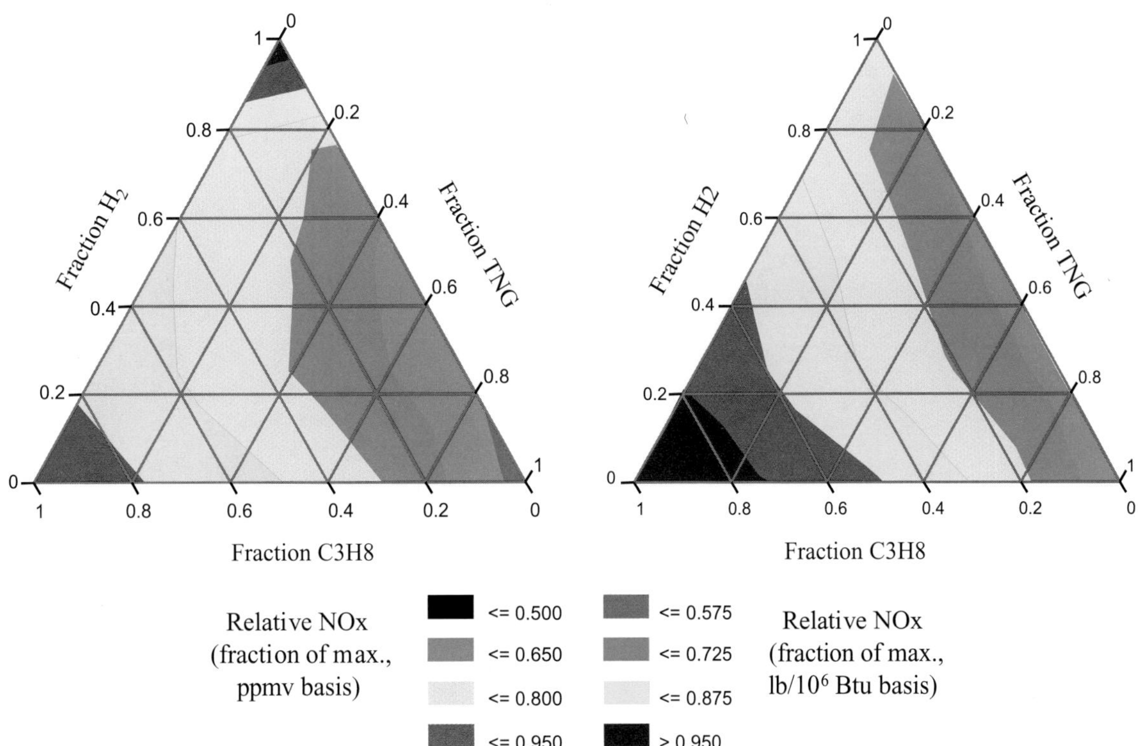

FIGURE 6.24 Measured NOx (fraction of the maximum value in both ppmv and lb/MMBtu) as a function of the fuel blend composition for TNG/H_2/C_3H_8 blends combusted with 15% excess air where both the fuel and the air were at ambient temperature and pressure, for gas tip #6.

one illustrating NOx levels on a volume concentration basis and the other illustrating NOx levels on a mass per unit heat release basis.

For each given tip, the highest NOx emissions on a concentration basis occur in the high-hydrogen content region, while the highest NOx emissions on a mass per unit heat release basis occur in the high-propane region. The contoured gradients illustrate the interaction of the three fuel components and how each of the components affects NOx emission in different regions of the fuel mixture, such as the steep NOx concentration gradients in the high-hydrogen content regions. The effect of C_3H_8 content can be seen to dominate the NOx level gradients on a mass per unit heat release basis with a relatively constant slope. It is also interesting to note that NOx levels overall appear to increase as fuel gas tips change from having less open fuel port area (higher fuel pressures for a given heat release) to having greater open fuel port area (lower fuel pressure for a given heat release).

Figure 6.25 shows ternary plots of fuel composition effects on NOx at a nominal constant fuel pressure of 21 psig (145 kPag). This analysis, made possible by testing a range of fuel gas tips, enables the examination of fuel composition effects on NOx emissions relatively independently from fuel pressure variations. A qualitative comparison of the plot on the left with the theoretical plots previously shown in Figure 6.16 reveals that, on a volume concentration basis, the change in NOx level as a function of fuel composition, for a relatively constant pressure and constant heat release, varies similarly to the trends predicted by the adiabatic flame temperature variation and predicted relative NOx concentrations from the equilibrium combustion model over the same regions. This result is expected due to the well-established correlation of the dependence of thermal NOx formation on flame temperature. The mass basis plot on the right in Figure 6.25, shows that variation in NOx levels with fuel composition, from a constant fuel pressure perspective, are less severe than seen in the analysis of a single fuel gas tip with fixed port sizes, for which fuel pressures may vary greatly to maintain a given heat release with fuel composition variation.

From both the two-component and three-component analyses, it is evident that fuel pressure has a significant effect on NOx emission levels. Figure 6.26 shows a plot of relative NOx levels vs. fuel pressure for each of the 15 different fuels

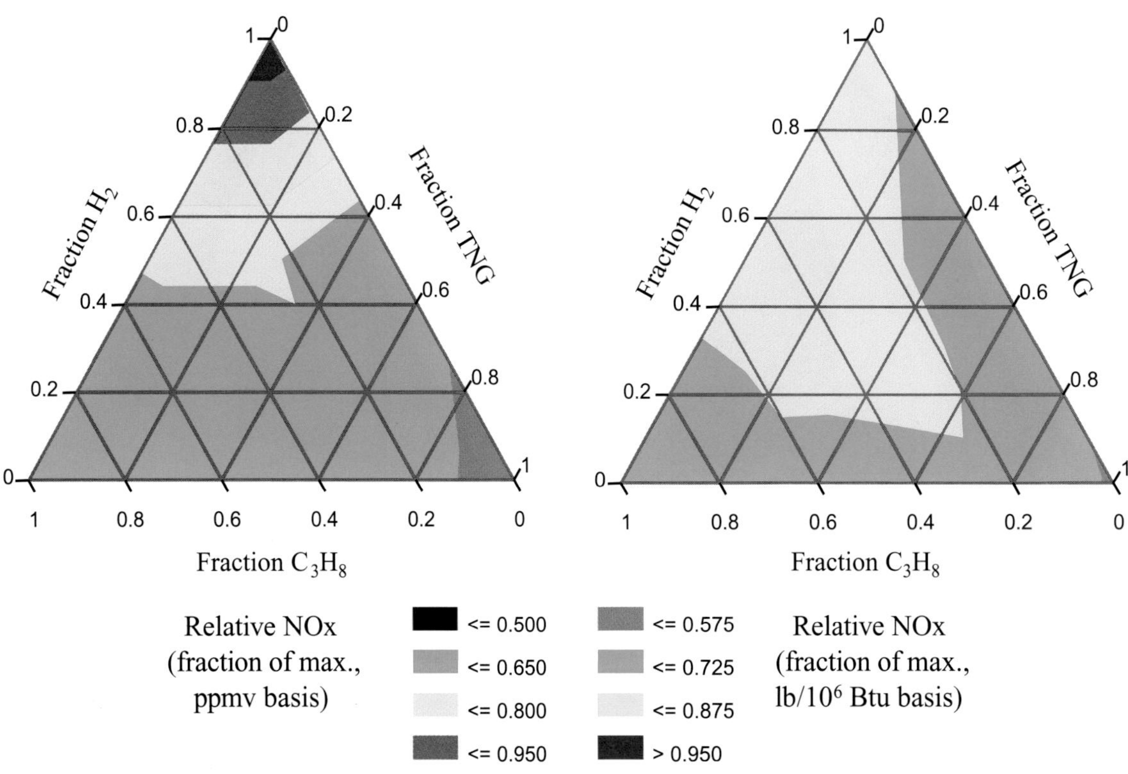

FIGURE 6.25 Measured NOx (fraction of the maximum value in both ppmv and lb/MMBtu) as a function of the fuel blend composition for TNG/H_2/C_3H_8 blends combusted with 15% excess air where both the fuel and the air were at ambient temperature and pressure, for a constant fuel gas pressure of 21 psig.

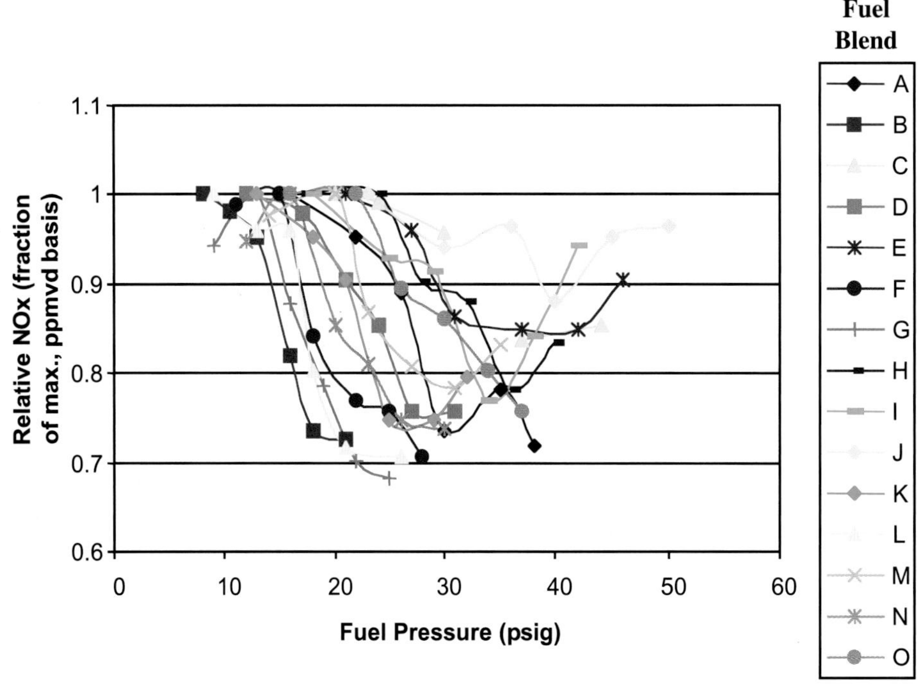

FIGURE 6.26 Measured NOx (fraction of the maximum value in ppmvd) as a function of the fuel pressure for all 15 different TNG/H_2/C_3H_8 blends (A through O) combusted with 15% excess air where both the fuel and the air were at ambient temperature and pressure.

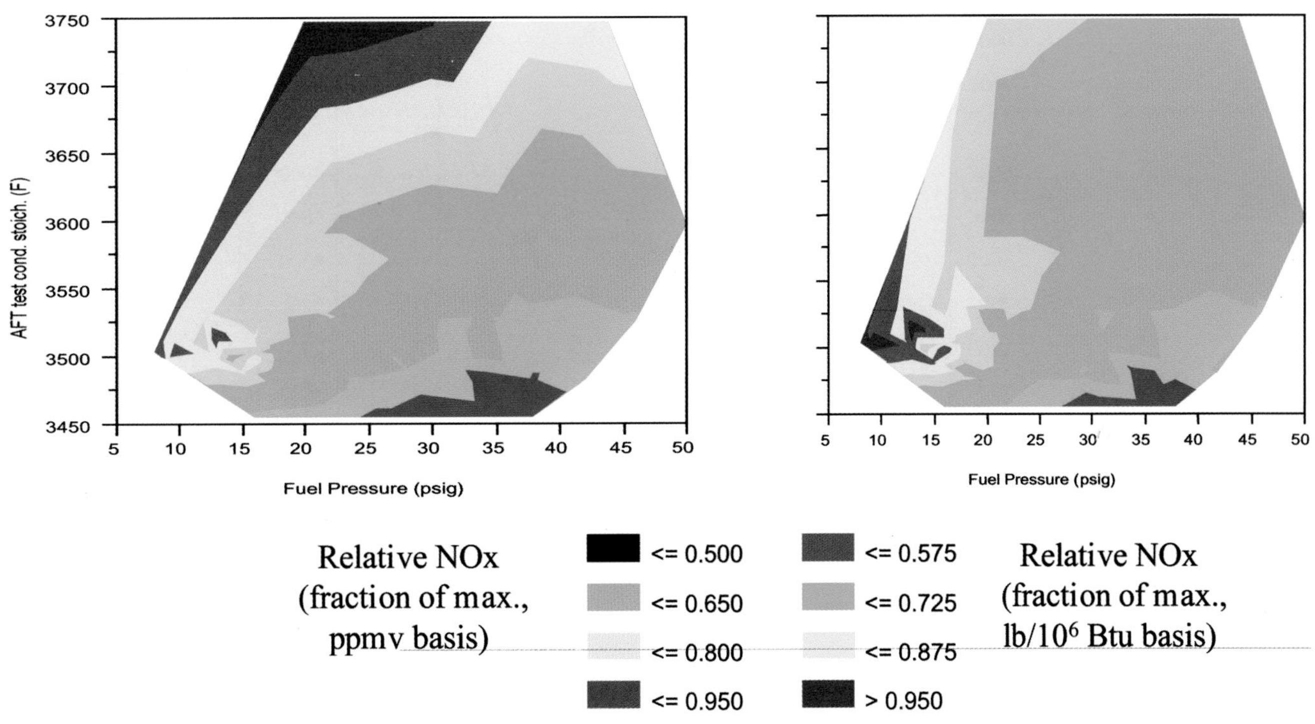

FIGURE 6.27 Measured NOx (fraction of the maximum value in both ppmv and lb/MMBtu) as a function of the fuel blend composition, fuel gas pressure, and calculated adiabatic flame temperature for TNG/H_2/C_3H_8 blends combusted with 15% excess air where both the fuel and the air were at ambient temperature and pressure.

tested. This plot shows a consistent decrease in NOx levels correlated with an increase in fuel pressure. This phenomena is explained by the burner configuration which allows significant amounts of inert flue gas to be entrained into the flame zone with increasing fuel jet momentum, thus decreasing thermal NOx formation.

6.2.5.4 Summary

Figure 6.27 shows an overall view of the data collected from all six tips with each of the 15 different fuel compositions (90 data points in total) from both a NOx volume concentration basis and mass per unit heat release viewpoint. The plots use fuel pressure and adiabatic flame temperatures as the primary axes to usefully illustrate some overall trends. The plot of relative NOx concentration levels shows the minimum NOx levels occur in the region with the lowest adiabatic flame temperature and highest fuel pressures. Inversely, the highest NOx concentration levels are found in the region of high adiabatic flame temperatures and low fuel pressures, when high concentrations of hydrogen are present. The mass per unit heat release NOx levels are also at a minimum in the same region as the concentration-based profiles, however the maximum NOx levels, when measured on a mass basis, are not found in the same region, but occur in areas of lowest fuel pressures with a mildly elevated adiabatic flame temperature, which correspond to high C_3H_8 concentration regions. These overall trends concur with the previously discussed results and agree with the correlations shown by the three-component and two-component interaction analyses.

Adiabatic flame temperature and fuel pressure are both identified as significant fundamental parameters affecting NOx emission levels when considering the effect of fuel composition on NOx levels. For a conventional burner, with NOx on a concentration basis, the adiabatic flame temperature

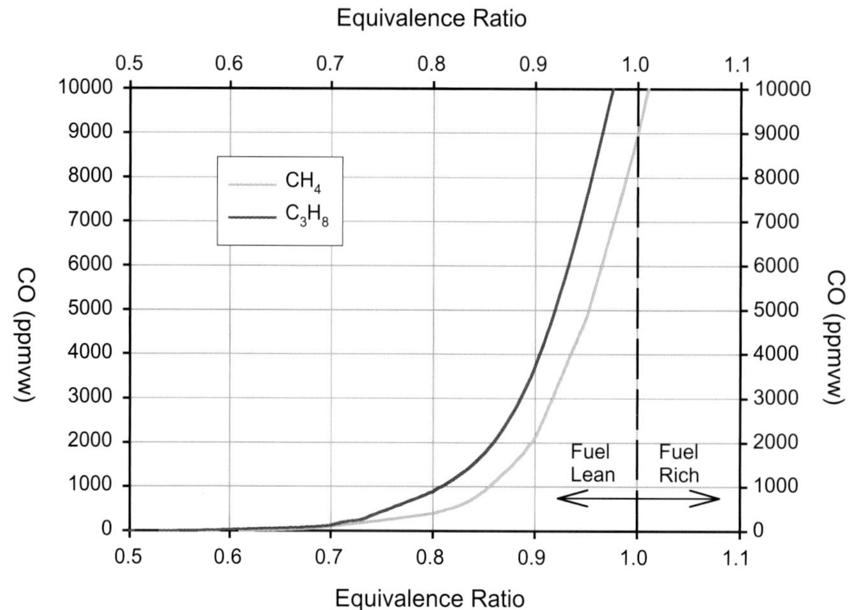

FIGURE 6.28 Adiabatic equilibrium CO as a function of equivalence ratio for air/fuel flames.

is dominant, with fuel pressure remaining significant in affecting NOx emission levels. The highest NOx levels on a volume concentration basis occurred at the highest hydrogen content fuel compositions at lower fuel pressures. On a mass per heat release basis, however, the highest relative NOx levels were achieved for fuel compositions containing large fractions of C_3H_8. This appears to result from some combined characteristics of a high-propane mixture including: very low fuel pressure for a given heat release in comparison with the other fuels; somewhat higher adiabatic flame temperature than CH_4; and a substantially larger amount of total dry products of combustion produced for a given heat release when compared with H_2. In summary, these results provide both quantitative and qualitative information to improve emission performance prediction and design of burners with application to a wide variation of fuel compositions.

6.3 COMBUSTIBLES

This section has been broken into two types of combustibles. The first involves the incomplete combustion of the fuel, which usually produces carbon monoxide, and in some limited cases, not all of the hydrocarbon fuel is consumed and passes through the combustor unreacted. The second type of combustible is volatile organic compounds (VOCs), which are generally only important in a limited number of processes, typically those involving contaminated or otherwise hazardous waste streams.

6.3.1 CO and Unburned Fuel

Carbon monoxide (CO) is generally produced in trace quantities in many combustion processes as a product of incomplete combustion (see Figure 6.14). CO is a flammable gas, which is nonirritating, colorless, odorless, tasteless, and normally noncorrosive. CO is highly toxic and acts as a chemical asphyxiant by combining with hemoglobin in the blood that normally transports oxygen inside the body. The affinity of carbon monoxide for hemoglobin is approximately 300 times more than the affinity of oxygen for hemoglobin.[30] CO preferentially combines with hemoglobin to the exclusion of oxygen so that the body becomes starved for oxygen, which can eventually lead to asphyxiation. Therefore CO is a regulated pollutant with specific emissions guidelines depending on the application and the geographical location.

CO is generally produced by the incomplete combustion of a carbon-containing fuel. Normally, a combustion system is operated slightly fuel lean (excess O_2) to ensure complete combustion and to minimize CO emissions. Figure 6.28 shows the calculated CO as a function of the equivalence ratio (ratio of 1 is stoichiometric, >1 is fuel rich, and <1 is fuel lean). Because these are adiabatic calculations with very high flame temperatures, the dissociation in the flame produces high quantities of CO even under fuel lean conditions. This is graphically shown in Figure 6.29 where much more CO is produced at higher gas temperatures while all other variables remain the same.

Figures 6.30 and 6.31 show the effects on CO production of air and fuel preheating, respectively. In both cases, the higher

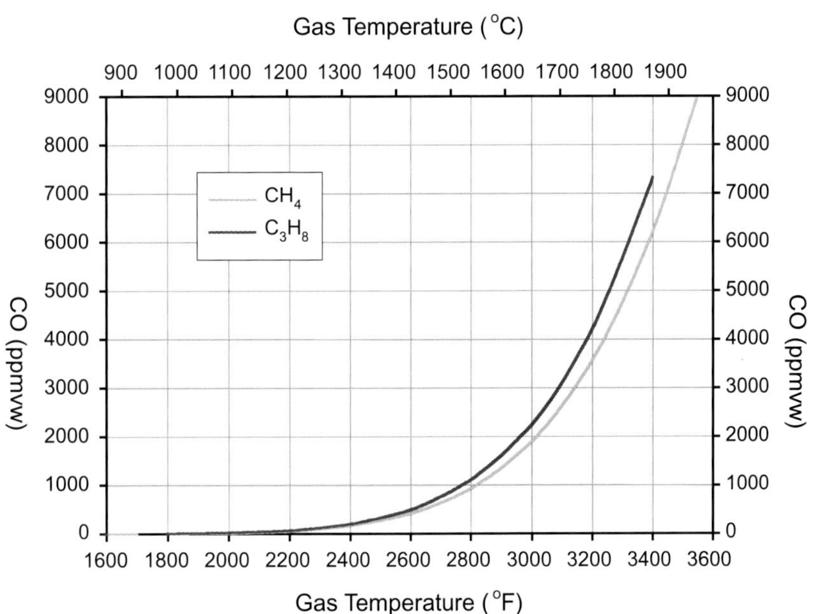

FIGURE 6.29 Adiabatic equilibrium CO as a function of gas temperature for stoichiometric air/fuel flames.

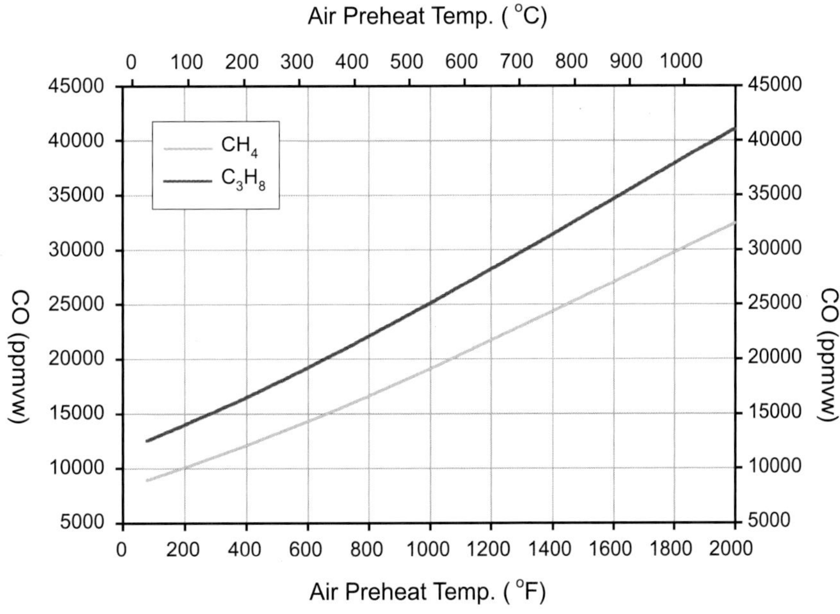

FIGURE 6.30 Adiabatic equilibrium CO as a function of air preheat temperature for stoichiometric air/fuel flames.

flame temperatures produced by preheating cause more CO formation as the preheat temperature increases. Figure 6.32 shows the effect of fuel composition for H_2/CH_4 blends. As expected, higher concentrations of H_2 produce less CO, and at pure H_2, no CO is generated. Similarly, Figure 6.33 shows the effect of fuel composition for CH_4/N_2 blends. Higher concentrations of N_2 reduce both the flame temperature and the concentration of carbon available to make CO, which both reduce CO generation.

6.3.2 *Volatile Organic Compounds*

Volatile organic compounds (VOCs) are generally low molecular weight aliphatic and aromatic hydrocarbons such

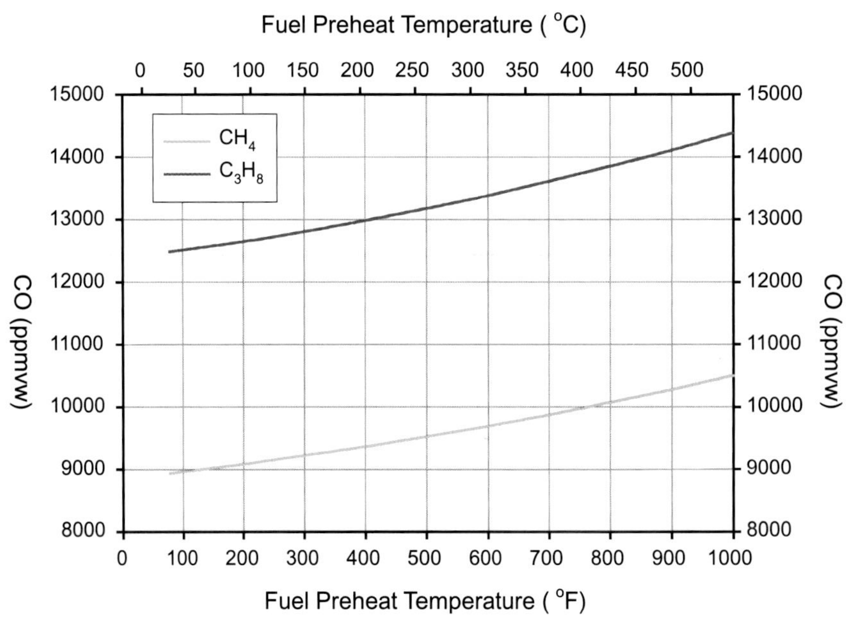

FIGURE 6.31 Adiabatic equilibrium CO as a function of fuel preheat temperature for a stoichiometric air/CH_4 flame.

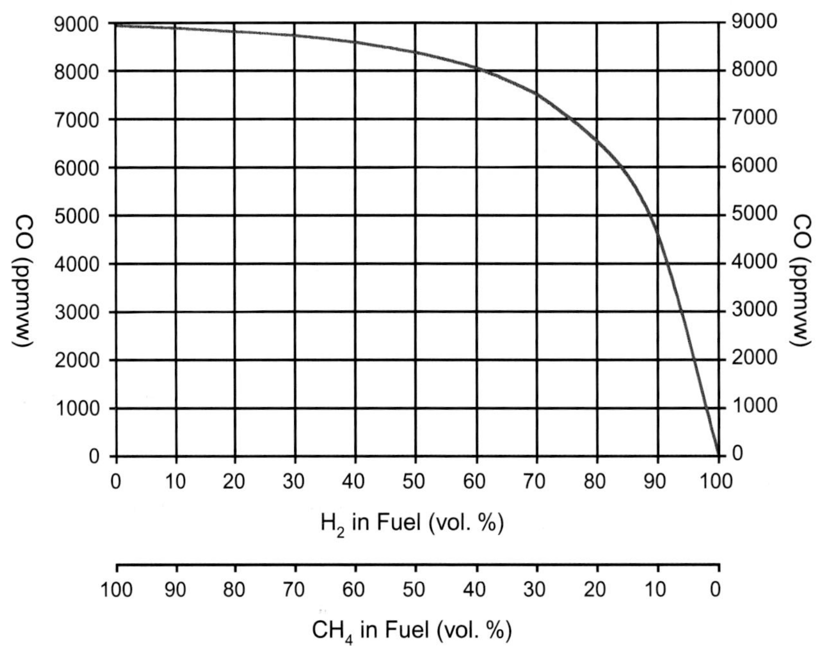

FIGURE 6.32 Adiabatic equilibrium CO as a function of fuel composition (CH_4/H_2) for a stoichiometric air/fuel flame.

as alcohols, ketones, esters, and aldehydes.[31] Typical VOCs include benzene, acetone, acetaldehyde, chloroform, toluene, methanol, and formaldehyde. These compounds are considered regulated pollutants because they can cause photochemical smog and depletion of the ozone layer if they are released into the atmosphere. They are not normally produced in the combustion process, but they may be contained in the material that is being heated, such as in the case of a contaminated hazardous waste in a waste incinerator. In that case, the objective of the heating process is usually to volatilize the VOCs out of the waste and combust them before they can be emitted to the atmosphere.

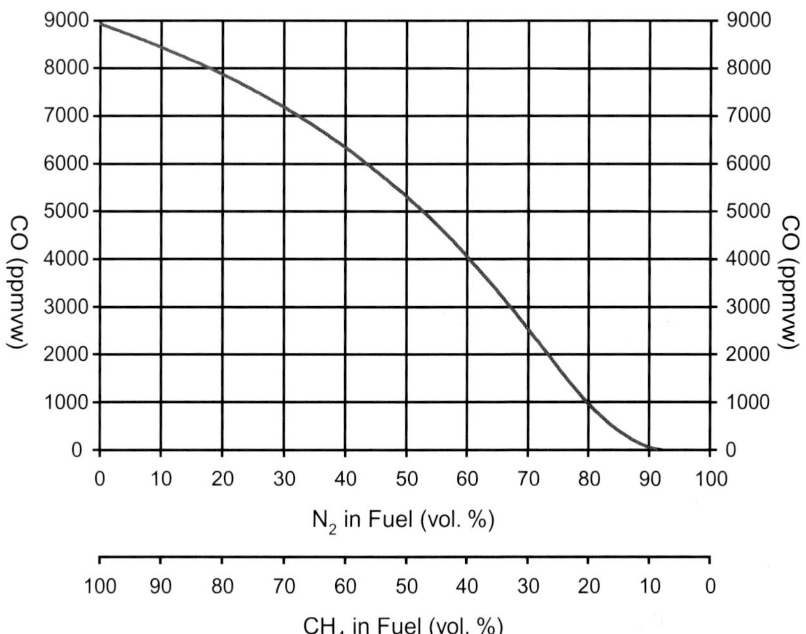

FIGURE 6.33 Adiabatic equilibrium CO as a function of fuel composition (CH_4/N_2) for a stoichiometric air/fuel flame.

There are two strategies for removing VOCs from the off-gases of a combustion process.[33] One is to separate and recover them using techniques like carbon adsorption or condensation. The other method involves oxidizing the VOCs to CO_2 and H_2O. This process includes techniques like thermal oxidation (see Chapter 21), catalytic oxidation, and bio-oxidation. One common way to ensure complete destruction of VOCs in waste incinerators is to add an afterburner or secondary combustion chamber, which may or may not have a catalyst, after the main or primary combustion chamber.[32]

6.4 PARTICULATES

There are two common sources of particulates that can be carried out of a combustion process with the exhaust gases. One is entrainment and carryover of incoming raw materials, and the other is the production of particles as a result of the combustion process. A particular health concern regarding particulate emissions is the hazardous materials that can condense on the particle surfaces and be carried into the atmosphere.[33] For example, heavy metals vaporized during a high-temperature combustion process can condense on solid particles and be carried out with the exhaust products.

6.4.1 Sources

Principal sources of particulates in most industrial combustion applications are:

1. dry fine particles being carried out of the process from the raw materials being processed
2. particles generated in the combustion process
3. fuel carryover, where some of the solid fuel passes through the combustor essentially unreacted
4. other particle carryover

The first and third mechanisms are not usually a major problem in most hydrocarbon and petrochemical applications.

6.4.1.1 Particle Entrainment

The gas flow through the combustor may entrain particles from the raw materials used in the process. This is often referred to as carryover. An example of this would be in the glass-making process where fine dust materials such as sand are used to make the glass and can be carried out of the glass furnace if the gas velocity in the combustion space is high enough. This type of particulate is expensive because not only must the particles be captured by some type of flue-gas scrubbing equipment, but some of the raw materials needed for the process are also lost.

6.4.1.2 Combustion-Generated Particles

The second method by which particles can be emitted from the combustion system is through the production of particles in the combustion process. For example, in the combustion of solid fuels, (e.g., coal), ash is normally produced. The air-borne portion of the ash, usually referred to as fly-ash, can be carried out of the combustor by the exhaust gases. Heavier oil flames also tend to generate particulates due to the high carbon contents and increased difficulty in fully oxidizing those particles prior to exiting the exhaust stack.

Another source of combustion-generated particles is soot that may be produced in a flame. Under certain conditions, even gaseous fuels can produce soot. To a certain extent, soot is desirable because it generally enhances the radiant heat transfer between the flame and the process. Fuels that have a higher carbon-to-hydrogen mass ratio tend to produce more soot than fuels with a lower ratio. For example, propane (C_3H_8), which has a C:H mass ratio of about 4.5, is more likely to produce soot than methane (CH_4), which has a C:H mass ratio of about 3.0. For clean-burning fuels like natural gas, it is much more difficult to produce sooty flames compared to other fuels (e.g., oil and coal) that have little or no hydrogen and a high concentration of carbon. Flames containing more soot are more luminous and tend to release their heat more efficiently than flames containing less soot, which tend to be nonluminous. Soot particles generally consist of high-molecular-weight polycyclic hydrocarbons and are sometimes referred to as "char."

Ideally, soot would be generated at the beginning of the flame so that it could radiate heat to the load, and then it would be destroyed before exiting the flame so that no particles would be emitted. Soot can be produced by operating a combustion system in a very fuel-rich mode or by incomplete combustion of the fuel due to poor mixing. If the soot particles are quenched or "frozen," they are more difficult to incinerate and more likely to be emitted with the exhaust products. The quenching could be caused by contact with much colder gases or possibly by impingement on a cool surface (e.g., a boiler tube). Soot particles tend to be sticky and can cling to the exhaust ductwork, clogging the ductwork and the pollution treatment equipment in the system. If the soot is emitted into the atmosphere, it can contribute to smog in addition to being dirty. The emitted soot particles become a pollutant because they produce a smoky exhaust that has high opacity. Most industrial heating processes have a regulated limit for opacity.

6.4.1.3 Solid Fuel Carryover

This mechanism for generating particulates involves some of the solid fuel passing through the combustor essentially unreacted. This is not usually a concern in the hydrocarbon and petrochemical industries where solid fuels are rarely used.

6.4.1.4 Other Particle Carryover

Particles are sometimes generated by scale formation in the piping. These scale particles (iron oxide) travel through the combustor and are emitted from the exhaust stack. Refractory particles also may be generated from refractory-lined combustors. This refractory dust can also be emitted into the atmosphere.

6.4.2 *Treatment Techniques*

There are a variety of techniques used to control particulate emissions from combustion processes. The specific method chosen will depend on many factors, including economics, particle size distribution and composition, volumetric flow rate, exhaust stream temperature, and particle moisture content. The preferred method in most cases is to minimize particulate formation in the first place by modifying the process. For example, substituting a gaseous fuel or lighter fuel oil for a heavy fuel oil can significantly reduce particulate emissions resulting from the fuel. Another strategy is to capture the particles for recycling back into the process. One example is a fluidized bed reactor, in which the majority of the particles are recirculated back into the process. Because of typically higher costs, the last choice is usually to remove the particles from the exhaust stream before they are emitted into the atmosphere. This can be done with electrostatic precipitators (wet or dry), filters (baghouses), or venturi scrubbers.[33]

6.5 CARBON DIOXIDE

Carbon dioxide (CO_2) is a colorless, odorless, inert gas that does not support life because it can displace oxygen and act as an asphyxiant. CO_2 is found naturally in the atmosphere at concentrations averaging 0.03%, or 300 ppmv. Concentrations of 3 to 6% can cause headaches, dyspnea, and perspiration. Concentrations of 6 to 10% can cause headaches, tremors, visual disturbances, and unconsciousness. Concentrations above 10% can cause unconsciousness, eventually leading to death.

Carbon dioxide emissions are produced when a fuel containing carbon is combusted near or above stoichiometric conditions. Some studies indicate that CO_2 is a greenhouse gas that may contribute to global warming. Many schemes have been suggested for "disposing" of CO_2, including injection deep into the ocean and deep-well injection for oil recovery. In some European countries, CO_2 emissions are considered a pollutant and as such are regulated. Any technique that improves the overall thermal efficiency of a process can significantly reduce CO_2 emissions because less fuel needs to be burned for a given unit of available heat output. Some predict that reductions in CO_2 emissions will become increasingly important for the petrochemical industry.[34]

6.6 SOx

Sulfur oxides, usually referred to as SOx, include SO, S_2O, S_nO, SO_2, SO_3, and SO_4, of which SO_2 and SO_3 are of particular importance in combustion processes.[35] SO_2 is preferably produced at higher temperatures, while SO_3 is favored at lower temperatures.[36] Because most combustion processes are at high temperatures, SO_2 is the more predominant form of SOx emitted from systems containing sulfur. Sulfur dioxide (SO_2) is a colorless gas with a pungent odor that is used in a variety of chemical processes. SO_2 can be very corrosive in the presence of water. It is considered a pollutant because of the choking effect it can cause on the human respiratory system. It is also damaging to green plants, which are more sensitive to SO_2 than people and animals. When SO_2 is released into the atmosphere, it can produce acid rain by combining with water to produce sulfuric acid (H_2SO_4). Sulfuric acid is very corrosive and can cause considerable damage to the environment.

It is often assumed that any sulfur in a combustor will be converted to SO_2, which will then be carried out with the exhaust gases.[37] The sulfur may come from the fuel or from the raw materials used in the production process. Fuels like heavy oil and coal generally contain significant amounts of sulfur, while gaseous fuels like natural gas tend to contain little or no sulfur. The two strategies for minimizing or eliminating SOx are: (1) removing the sulfur from the incoming fuel or raw materials, and (2) removing the SOx from the exhaust stream using a variety of dry and wet scrubbing techniques.[38] One dry scrubbing technique is limestone injection. After use, the combined limestone and sulfur can be used in gypsum board. New membrane separation technology is another reduction technique being developed.

6.7 DIOXINS AND FURANS

This class of pollutants includes the carbon-hydrogen-oxygen-halogen compounds and has received considerable attention from both the general public and from regulatory agencies because of the potential health hazards associated with them. Dioxins generally refer to polychlorinated dibenzo-*p*-dioxin (PCDD) compounds, while furans generally refer to polychlorinated dibenzofuran (PCDF) compounds. Some of the potential health risks include toxicity because of the poisoning effect on cell tissues, carcinogenicity because cancerous growth may be stimulated, mutagenicity because of possible mutations in cell structure or function, and teratogenicity because of the potential changes to fetal tissue.[32] The over 200 dioxin/furan compounds are regulated in certain industries, particularly in waste incineration, and also in certain geographical locations for a wide range of applications — especially in Europe.

In the vast majority of cases, dioxin/furan emissions result from some contaminant in the load materials being heated in the combustor. A quick scan of most of the textbooks on combustion shows that these emissions are essentially ignored because they are not generally produced in the flame, except in certain limited cases. This is primarily because usually there are not any halogens in either the fuel or the oxidizer to produce dioxins or furans. An exception is the case when waste materials are burned as a fuel by direct injection into a flame. One example is the destruction of waste solvents that may be injected into an incinerator through the burner.

REFERENCES

1. J.A. Stanislaw, Petroleum industry faces tectonic shifts changing global energy map, *Oil Gas J.*, 97(50), 8-14, 1999.
2. U.S. Dept. of Energy Office of Industrial Technology, Petroleum — Industry of the Future: Energy and Environmental Profile of the U.S. Petroleum Refining Industry, U.S. DOE, Washington, D.C., December 1998.
3. S.A. Shareef, C.L. Anderson, and L.E. Keller, Fired Heaters: Nitrogen Oxides Emissions and Controls, U.S. Environmental Protection Agency, Research Triangle Park, NC, EPA Contract No. 68-02-4286, June 1988.
4. D. Hansell and G. England, *Air Toxic Emission Factors for Combustion Sources Using Petroleum Based Fuels*, 3 volumes, Energy and Environmental Research Corp., Irvine, CA, 1998 (available at www.api.org/step/ piep.htm).
5. U.S. EPA, AP-42: Compilation of Air Pollutant Emission Factors, 5th ed., U.S. Environmental Protection Agency, January 1995.
6. J.C. Reis, *Environmental Control in Petroleum Engineering*, Gulf Publishing, Houston, TX, 1996.
7. American National Standards Institute/American Society Mechanical Engineering, Performance Test Code PTC 19.10, Part 10: Flue and Exhaust Gas Analyses, American Society of Mechanical Engineers, New York, 1981.
8. U.S. Environmental Protection Agency, Nitrogen Oxide Control for Stationary Combustion Sources, U.S. EPA Report EPA/625/5-86/020, 1986.
9. M. Sandell, Putting NOx in a box, *Pollution Engineering*, 30(3), 56-58, 1998.
10. M. Moreton and S. Beal, Controlling NOx emissions, *Pollution Engineering Int.*, Winter, 14-16, 1998.
11. Y.B. Zeldovich, *Acta Physecochem (USSR)*, 21, 557, 1946.

12. E.B. Sanderford, Alternative Control Techniques Document — NOx Emissions from Process Heaters, U.S. EPA Report EPA-453/R-93-015, February 1993.

13. U.S. Environmental Protection Agency, Alternative Control Techniques — NOx Emissions from Utility Boilers, U.S. EPA Report EPA-453/R-94-023, 1994.

14. C.E. Baukal and P.B. Eleazer, Quantifying NOx for Industrial Combustion Processes, *J. Air Waste Manage. Assoc.*, 48, 52-58, 1997.

15. S.B. Mandel, What is the total cost for emissions monitoring?, *Hydrocarbon Processing*, 76(1), 99-102, 1997.

16. U.S. Government, Code of Federal Regulations 40, Part 60, Revised July 1, 1994.

17. M.C. Drake, Kinetics of Nitric Oxide Formation in Laminar and Turbulent Methane Combustion, Gas Research Institute (Chicago, IL) Report No. GRI-85/0271, 1985.

18. M.F. Zabielski, L.G. Dodge, M.B. Colket, and D.J. Seery, The optical and probe measurement of NO: a comparative study, *Eighteenth Symp. (Int.) on Combustion*, The Combustion Institute, Pittsburgh, 1981, 1591.

19. A. Berger and G. Rotzoll, Kinetics of NO reduction by CO on quartz glass surfaces, *Fuel*, 74, 452, 1995.

20. H.M. Gomaa, L.G. Hackemesser, and D.T. Cindric, NOx/CO emissions and control in ethylene plants, *Environmental Progress*, 10(4), 267-272, 1991.

21. C.E. Baukal and A.I. Dalton, Nitrogen oxide measurements in oxygen enriched air-natural gas combustion systems, *Proc. 2nd Fossil Fuel Combustion Symp.*, ASME PD-Vol. 30, pp.75-79, New Orleans, LA, January 15, 1990.

22. C.E. Baukal, Ed., *Oxygen-Enhanced Combustion*, CRC Press, Boca Raton, FL, 1998.

23. R. Waibel, D. Nickeson, L. Radak, and W. Boyd, Fuel Staging for NOx Control, in *Industrial Combustion Technologies*, M.A. Lukasiewicz, Ed., American Society of Metals, Warren, PA, 1986, 345-350.

24. A. Garg, Trimming NOx, *Chem. Eng.*, 99(11), 122-124, 1992.

25. H.L. Shelton, Find the right low-NOx solution, *Environmental Engineering World*, Nov.–Dec., 24, 1996.

26. A. Garg, Specify better low-NOx burners for furnaces, *Chem. Eng. Prog.*, 90(1), 46-49, 1994.

27. J. Colannino, Low-cost techniques reduce boiler NOx, *Chem. Eng.*, 100(2), 100-106, 1993.

28. J. Colannino, NOx reduction for stationary sources, *AIPE Facilities*, 23(1), 63-66, 1996.

29. J. Bluestein, NOx Controls for Gas-Fired Industrial Boilers and Combustion Equipment: A Survey of Current Practices, Gas Research Institute (Chicago, IL) Report GRI-92/0374, 1992.

30. K. Ahlberg, Ed., *AGA Gas Handbook*, AGA AB, Lidingö, Sweden, 1985.

31. S. Setia, VOC emissions — Hazards and techniques for their control, *Chemical Engineering World*, XXXI(9), 43-47, 1996.

32. W.R. Niessen, *Combustion and Incineration Processes*, 2nd ed., Marcel Dekker, New York, 1995.

33. I. Ray, Particulate emissions: evaluating removal methods, *Chem. Eng.*, 104(6), 135-141, 1997.

34. M. Thorning, How climate change policy could shrink the federal budget surplus and stifle US economic growth, *Oil Gas J.*, 97(50), 22-26, 1999.

35. E.D. Weil, Sulfur compounds, in *Kirk-Othmer Encyclopedia of Chemical Technology*, 3rd ed., Vol. 22, John Wiley & Sons, New York, 1983.

36. C.T. Bowman, Chemistry of gaseous pollutant formation and destruction, in *Fossil Fuel Combustion*, W. Bartok and A. F. Sarofim, Eds., John Wiley & Sons, New York, 1991.

37. C.R. Bruner, *Handbook of Incineration Systems*, McGraw-Hill, New York, 1991.

38. S.R. Turns, *An Introduction to Combustion*, McGraw-Hill, New York, 1996.

39. API Recommended Practice 536: Post-Combustion NOx Control for Fired Equipment in General Refinery Services, 1 ed., American Petroleum Institute, Washington, D.C., March 1998.

40. API Recommended Practice 560: Fired Heaters for General Refinery Services, 2nd ed., American Petroleum Institute, Washington, D.C., September 1995.

41. J. Colannino, Results of a statistical test program to assess flue-gas recirculation at the Southeast Resource Recovery Facility (SERRF), Paper 92-22.01, presented at Air & Waste Management Association, 85th Annual Meeting and Exhibition, Kansas City, MO, June 21-26, 1992.

42. M. Takagi, T. Kawai, M. Soma, T. Onishi, and K. Tamaru, The Mechanism of the Reaction Between NOx and NH_3 and V2O5 in the Presence of Oxygen, *J. Catal.*, 50(3), 441-446, 1977.

43. R.R Hayes, C.E. Baukal, and D. Wright, Fuel composition effects on NOx, presented at the 2000 American Flame Research Committee International Symposium, Newport Beach, CA, September 2000.

Chapter 7
Noise

Wes Bussman and Jaiwant D. Jayakaran

TABLE OF CONTENTS

7.1 Fundamentals of Sound 224

7.1.1 Introduction 224

7.1.2 Basics of Sound 224

7.1.3 Measurements 228

7.2 Industrial Noise Pollution 231

7.2.1 OSHA Requirements 232

7.2.2 International Requirements 232

7.2.3 Noise Sources and Environment Interaction 234

7.3 Mechanisms of Industrial Combustion Equipment Noise 234

7.3.1 Combustion Roar and Combustion Instability Noise 234

7.3.2 Fan Noise 237

7.3.3 Gas Jet Noise 238

7.3.4 Valve and Piping Noise 239

7.4 Noise Abatement Techniques 239

7.4.1 Flare Noise Abatement Techniques 239

7.4.2 Burner Noise Abatement Techniques 242

7.4.3 Valve and Piping Noise Abatement Techniques 243

7.4.4 Fan Noise Abatement Techniques 243

7.5 Analysis of Combustion Equipment Noise 243

7.5.1 Multiple Burner Interaction 243

7.5.2 High-Pressure Flare 244

7.5.3 Atmospheric Attenuation 246

7.6 Glossary 246

References 248

Bibliography 248

7.1 FUNDAMENTALS OF SOUND

7.1.1 Introduction

Silence is golden

— Anonymous

Noise is a common by-product of our mechanized civilization and is an insidious danger in industrial environments. Noise pollution is usually a local problem and thus is not viewed on the same scale of importance as the more notorious industrial emissions like NOx, CO, and particulates. Nonetheless, it is an environmental pollutant that has significant impact.

Serious concern is merited when a pollutant can result in either environmental damage or human discomfort. Considering the impact on people, noise is most often a source of annoyance, but it can also have much more detrimental effects, such as causing actual physical injury. Noise-related injuries range from short-term discomfort to permanent hearing loss.

The sense of hearing is a fragile and vital function of the human body. It resembles vision, more than the other senses, because permanent and complete damage can be sustained quite commonly in an industrial environment. So it follows that noise pollution has been recognized as a safety concern for a long time and has been appropriately regulated.

Although personnel safety is the most important consideration, noise pollution has several other significant side effects. Combustion equipment designers are often asked why they would want to constrain the combustion process to reduce noise. Typically, these questions come from persons working in plants situated in remote areas who often do not realize that given the age and economic drivers of the petroleum refining and chemical industries, it is now common to find plants located in densely populated areas. With industry that is situated close to residential areas or busy commercial facilities, high levels of noise become objectionable to people in the neighborhood. These emissions eventually lead to government regulations to control noise. Within the industrial site itself, the immediate issue with noise is one of employee safety. Furthermore, it is not surprising to find that employee morale and performance improves when noise is reduced, since its presence increases stress level.

Equipment is also affected by noise. In most cases, these effects lie in the area of vibration control and are beyond the scope of this chapter. Suffice it to say, that noise is a form of vibration and eventually contributes to fatigue, which reduces equipment life. The effects of fatigue are frequently accepted as normal wear and tear if the equipment life cycle spans a reasonable duration. In extreme cases, the effects of vibration may be more rapidly manifested, such as in the case of cracking and falling of hard refractory linings in furnaces.

This chapter is written as a practical guide, as well as a reference on noise, for engineers involved in the design, operation, or maintenance of combustion equipment — be it burners, furnaces, flares (see Chapter 20), or thermal oxidizers (see Chapter 21). In addition, because this chapter provides comprehensive coverage of the fundamentals of sound, the creative engineer will also be able to extend his or her knowledge to analyze other noise-producing industrial equipment.

7.1.2 Basics of Sound

If a tree falls in the forest and nobody is around to hear it, does it still make a sound?

Webster's dictionary defines sound as "that which is heard." Obviously, an engineer will find this definition woefully inadequate for his or her purposes. The authors resort to the definition provided in many engineering handbooks: "Sound is the vibration of particles in a gas liquid or solid."[1]

Sound is propagated through any medium in waves that take the form of pressure peaks (compressions) and troughs (rarefactions), as illustrated in Figure 7.1. The pressure wave travels through a given medium at the speed of sound for that medium. The auditory system in humans and most animals senses the impingement of these pressure waves on a tissue membrane and converts them to electrical impulses that are then sent to the brain and interpreted there.

Figure 7.2 is a cross-section of the human ear. Sound is collected and funneled into the ear canal by the outer ear. At the end of the ear canal, the sound impinges on the eardrum. The bones of the middle ear convey the eardrum's vibration to the inner ear. The inner ear, or cochlea, consists of a fluid-filled membrane that has tiny hair cells on the inside. The hair cells sense the vibration conveyed to the cochlea and convert this vibration into electrical signals, which are then conveyed to the brain.

Any given sound can be uniquely identified by two of its properties: pressure level and frequency. Most naturally occurring sounds are composites of different pressure levels at various frequencies. A "pure tone" however, is a sound at only one frequency. A tuning fork is a good example of a pure tone generator. Such naturally occurring pure tone generators are rare. Even musical instruments create notes that have significant pressure levels at two or three multiples, or harmonics, of the fundamental frequency of the note.

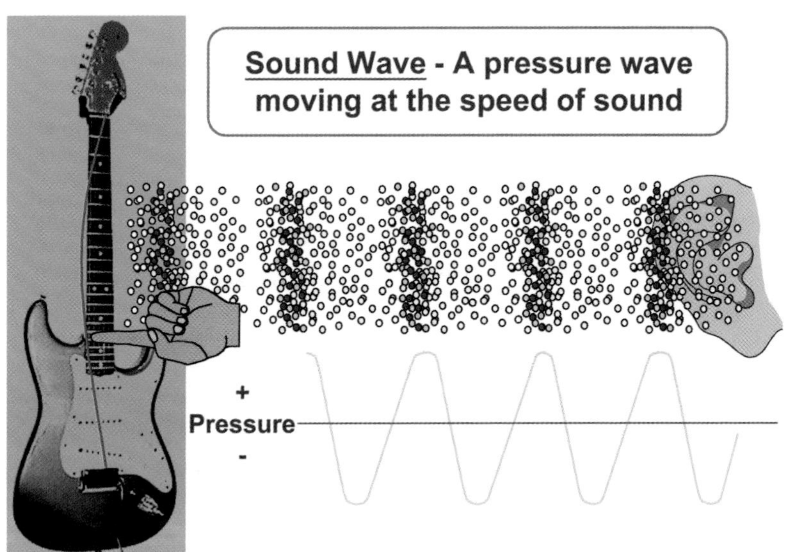

FIGURE 7.1 Pressure peaks and troughs.

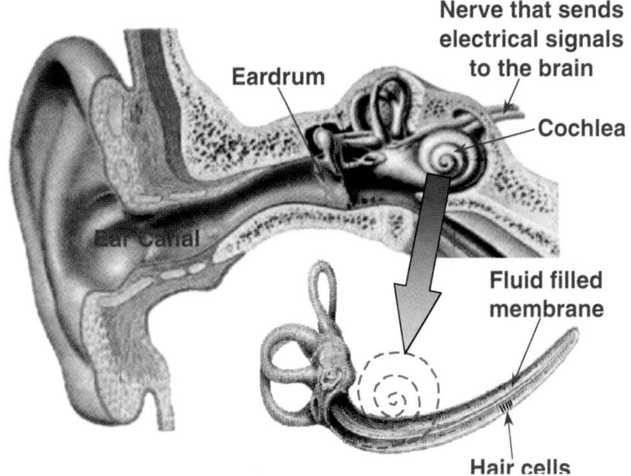

FIGURE 7.2 Cross-section of the human ear.

7.1.2.1 Sound Pressure Level and Frequency

Pressure level defines the loudness of the sound, while frequency defines the pitch or tone of the sound. Pressure level is the amplitude of the compression, or rarefaction, of the pressure wave. The common unit of pressure level is the decibel, abbreviated dB. Frequency is the number of pressure waves that pass by an arbitrary point of reference in a given unit of time. As such, the typical measure of sound frequency is cycles per second (cps), and as with electricity, the commonly used unit is the hertz (Hz); 1 Hz = 1 cps.

The typical range of human hearing extends from 20 Hz to 20 kHz. Young children can hear frequencies slightly higher than 20 kHz, but this ability diminishes with age. Loss of hearing in humans in the later stages of life typically manifests itself as diminished sensitivity to frequencies from 10 to 20 kHz. Mechanically, this is due to the deterioration of the fine hair cells in the cochlea.

It is important to note that the ear is not equally sensitive over the entire range from 20 Hz to 20 kHz. This is vital to understanding how noise affects humans and how noise control is implemented. The human ear is much less sensitive to sound at the extremes of low and high frequencies, as is discussed later in the chapter.

The wide range of frequencies in the human hearing range can be conveniently handled by breaking it up into octave bands. Each octave band represents a doubling in frequency.

TABLE 7.1 The Ten Octave Bands

Full Octave Band Standards	
Octave Band, Hz	**Center Frequency, Hz**
22–44	31.5
44–88	63
88–177	125
177–355	250
355–710	500
710–1420	1000
1420–2840	2000
2840–5680	4000
5680–11,360	8000
11,360–22,720	16,000

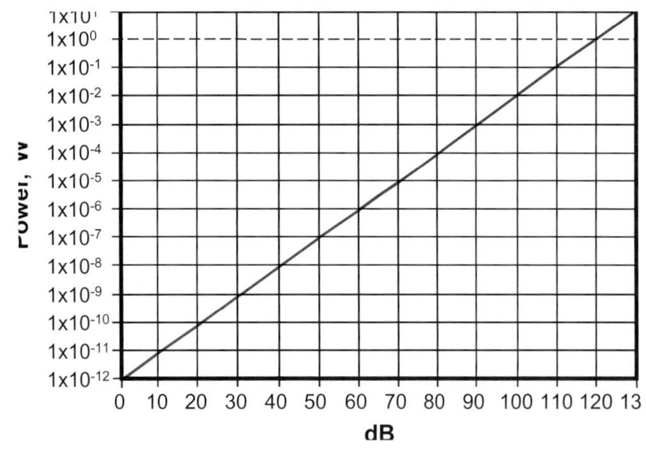

FIGURE 7.3 Relationship of decibels to watts.

Table 7.1 shows the ten octave bands that cover the human hearing range and the center frequencies that can be used to represent the octave band. Each octave band extends over seven fundamental musical notes.

7.1.2.2 The Decibel

The unit of pressure level, the decibel, can be difficult to conceptualize and merits some explanation. While it is possible to quantify the sound pressure level (SPL) in units of either power or pressure, neither unit is convenient to use because, in practice, one has to deal every day with sounds that extend over a very large range of power and/or pressure levels. For example, the sound power of a whisper is 10^{-8} watts (W), while the sound power of a jet plane is 10^5 W. The range of these two sound sources thus spans 10^{13} W. The term decibel that characterizes a dimensionless unit, was created to represent these large ranges conveniently. In the 1960s, Bell Laboratories coined the term "decibel." The "deci" stands for the base ten log scale on which the decibel is based. See Figure 7.3 for the relationship between decibel and watts. The "bel," of course, stands for Bell Labs.

In Figure 7.3, the *y*-axis represents power, in watts. The *y*-axis follows a base-10 scale. The *x*-axis gives dB values and the line provides the relationship; 120 dB is equal to 1 W. As an illustration of the $\log_{10}$ relationship, note that 110 dB is equal to one tenth of a W (0.1 W), and 100 dB is equal to a hundredth of a watt (0.01 W).

7.1.2.3 Sound Power Level

There is a subtle but important difference between the terms sound power level and sound pressure level. Sound power level is used to indicate the total energy-emitting ability of a sound source. In other words, sound power is an attribute of the source itself, while the sound pressure level (SPL) is used to indicate the intensity of sound received at any point of interest, from one or more sources. The illustration in Figure 7.4 shows the formula to calculate the sound pressure level that is expected at a distance *r* from a spherically radiating source of power level L_w:

$$L_p = L_w + 10\log_{10}\left(\frac{1}{4\pi r^2}\right) + 10.5 \qquad (7.1)$$

where r is in feet.

In practice, the sound intensity at the location of the listener is of interest, and this is easily achieved by making measurements at the point of interest. However, the sound pressure level can be analytically derived at different locations in complex industrial environments containing multiple sound sources if the power level of the sources is known. The equation can also be used to back-calculate the power level of a source from a measurement made at a known distance from the source.

The following are useful equations that can be used to calculate sound pressure and power levels, in dB, from the equivalent pressure and power units:

$$L_p(\text{dB}) = 20\log_{10} P/\left(2\times10^{-5}\right)$$

$$L_w(\text{dB}) = 10\log_{10} W/\left(1\times10^{-12}\right)$$

where

L_p = Sound pressure level, in dB
L_w = Sound power level, in dB
P = Pressure, in N/m²
W = Sound power level, in W

7.1.2.4 Threshold of Hearing

Figure 7.5 reveals a map of the threshold of hearing in humans. The *y*-axis represents SPL and the *x*-axis represents frequency.

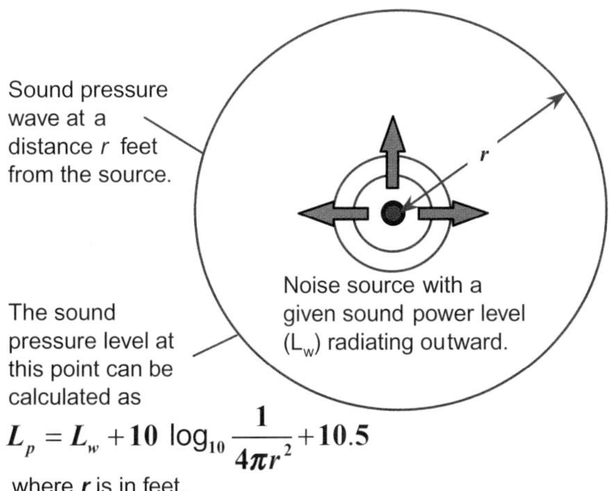

FIGURE 7.4 Sound pressure level at a distance *r*.

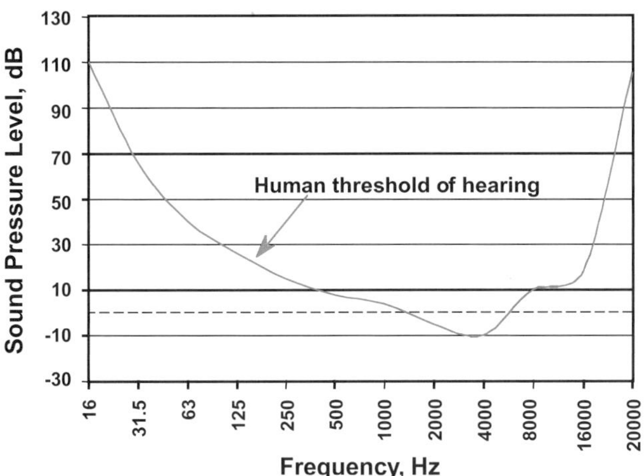

FIGURE 7.5 Threshold of hearing in humans.

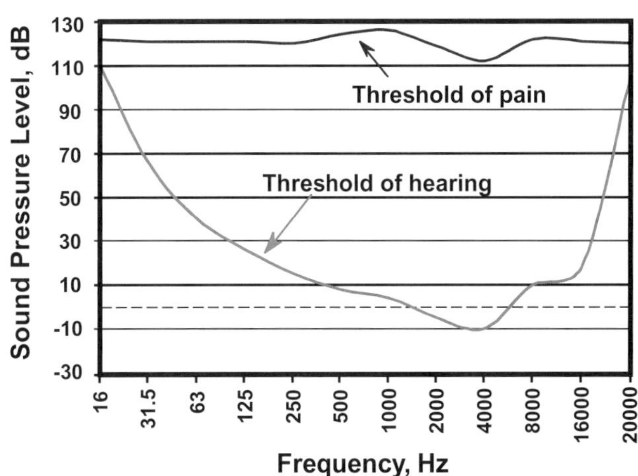

FIGURE 7.6 Threshold of hearing and threshold of pain in humans.

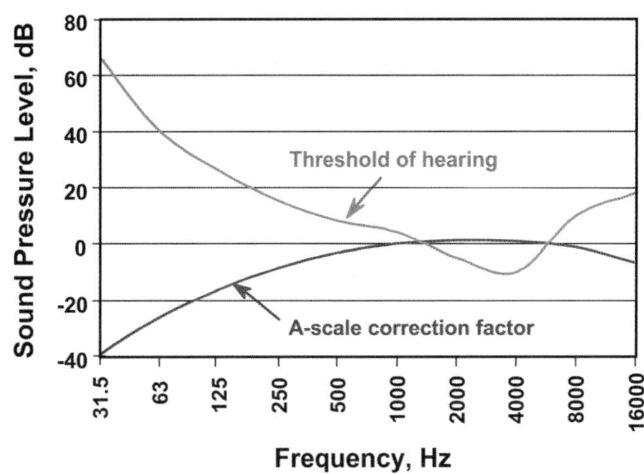

FIGURE 7.7 A-weighted scale for human hearing threshold.

Any SPL at a frequency that falls below the curve will be inaudible to humans. For example, a sound pressure level of 30 dB at 63 Hz will be inaudible; whereas an SPL of 70 dB at the same 63 Hz will be audible. Humans are most sensitive to sounds in the so-called "mid-frequencies" from 1 kHz to about 5 kHz. This is generally the region in which most of our everyday hearing activities take place. Additionally, at a constant level, sound with a very low or very high frequency will not have the same loudness sensation as that in the medium frequency range. For example, a 100-Hz tone at L = 50 dB gives the same loudness as a 1000-Hz tone at L = 40 dB.[2]

7.1.2.5 Threshold of Pain

Figure 7.6 shows the threshold of pain superimposed on the threshold of hearing. Fortunately, the threshold of pain is relatively flat. In general, a sound pressure level over 120 dB at any frequency will cause pain. An important observation that can be derived from the two curves is that if a sound is audible at very low or very high frequencies, persons subject to this sound are very close to experiencing pain.

7.1.2.6 Correction Scales

Sound meters are capable of measuring with equal sensitivity over the entire audible range. However, because humans do not hear with equal sensitivity at all frequencies, the sound meter's measurement needs to be modified to quantify what really affects humans. This can be done using a correction curve. The most common correction is the A-Scale. This is because, except for level, it resembles an idealized inverse of the threshold of hearing curve (refer to Figure 7.7). An

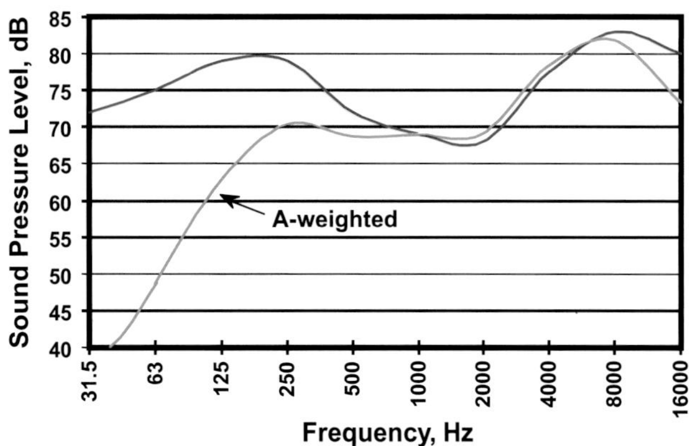

FIGURE 7.8 A-weighted burner noise curve.

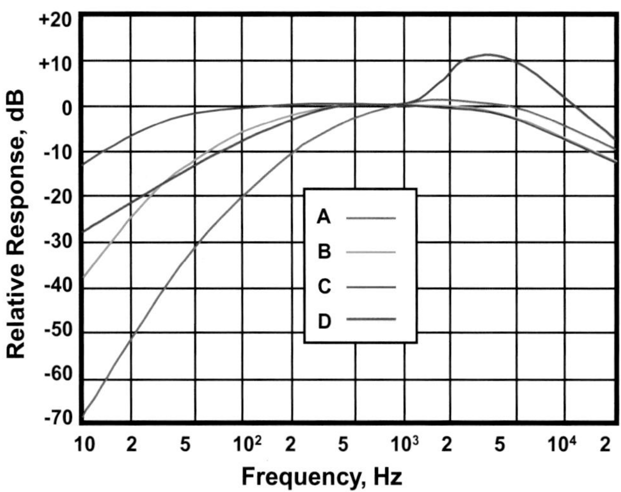

FIGURE 7.9 Weighting curves A, B, C, and D.

A-weighted sound level correlates reasonably well with hearing-damage risk in industry and with subjective annoyance for a wide category of industrial and community noises. After applying the A-scale correction, the unit of sound pressure level becomes dBA. Figure 7.8 shows a typical burner noise curve as measured by the noise meter (flat scale) and the result after applying the A-scale correction.

The other, less used correction scales are named, as might be expected, B, C, and D. Referring to Figure 7.9, one can see that the C-scale is essentially flat over the range of interest and the B-scale lies somewhere between the A- and C-scales. Given an understanding of the influence of low-frequency sounds, one finds that the B- and C-scales do not apply adequate correction in the lower frequencies. Finally, the D-scale is different from the others in that it has a pronounced correction in the range of 2 to 5 kHz. The D-scale was devised for the aircraft industry and is rarely used otherwise.

7.1.3 Measurements

A simple schematic of a noise meter is shown in Figure 7.10. The microphone is a transducer that transforms pressure variations in air to a corresponding electrical signal. Because the electrical signal generated by the microphone is relatively small in magnitude, a preamplifier is needed to boost the signal before it can be analyzed, measured, or displayed. Special weighting networks are used to shape the signal spectrum and apply the various correction scales discussed above. The weighted signal then passes through a second output amplifier into a meter. The meter and associated electronic circuits detect the approximate rms value of the signal and display it in dB.

Noise meters range from the simplest — microphone and needle gage — to sophisticated digital signal processing (DSP) equipped analyzers. The more sophisticated analyzers are equipped with fast Fourier transform (FFT) capabilities that aid in accurate narrow band analysis. In general, spectrum analyzers allow the user to map the sound pressure level at different frequencies, in other words, generate a curve of the sound over different frequencies. However, there is a significant difference between instruments that make one measurement per octave band and those that slice the octave band into several intervals and make a measurement at each interval. Typically, instruments are capable of:

1. octave band measurements
2. one-third octave band measurements
3. narrow band measurements

Table 7.2 shows the usual octave and one-third octave bands. As the name suggests, a one-third octave band instrument makes three measurements in each octave as opposed to the single measurement of an octave band instrument. A narrow band instrument, on the other hand, uses digital signal processing (DSP) to implement fast Fourier transform analysis (FFT), and in the current state-of-the-art, FFT analysis allows the octave band to be sliced into as many as 128 intervals.

Figure 7.11 provides a comparison of the same sound spectrum as analyzed using three different frequency band intervals: octave band, one-third octave band, and narrow band. This comparison shows that the additional resolution provided by narrower band methods is of vital importance. In this example, the level at 1 kHz, as recorded by the octave-band instrument, is 90 dB; on the one-third octave instrument, it is 85 dB; and on the narrow band instrument, it is 70 dB. The lower resolution measurements produce higher values due to the spill-over influence of the nearby peak at 1.8 kHz. In addition, in implementing noise control for this source, it is very valuable to know that it is the narrow peak at 1.8 kHz that is driving the maximum noise. This knowledge

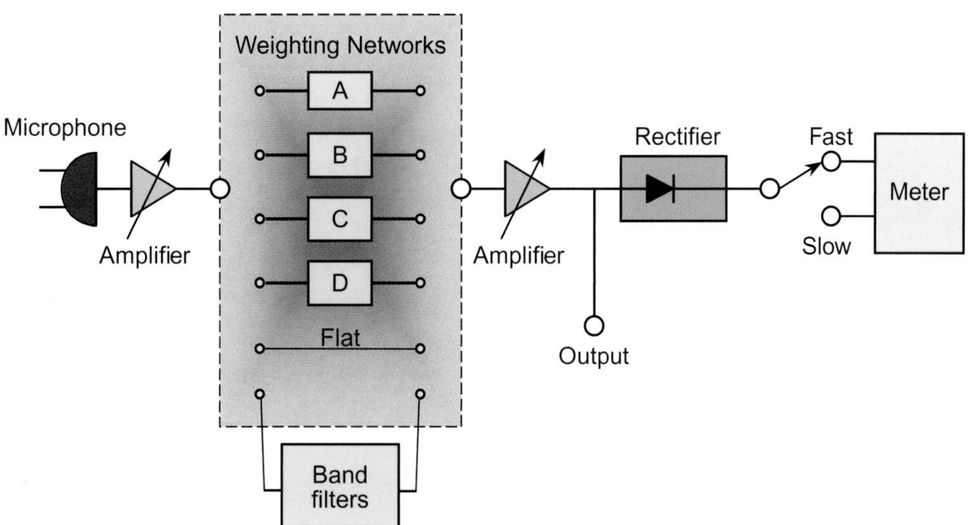

FIGURE 7.10 Block diagram of a sound level meter.

helps to zero-in on the source, which, for example may be an 1800-rpm motor or pump.

However, as with all things, there is a cost associated with high performance. For most purposes, a one-third octave analysis is usually quite adequate. The advantages of making broad band analyses using octave or one-third octave band filter sets are that less time is needed to obtain data and the instrumentation required to measure the data is less expensive.

In making sound measurements, several factors regarding the nature of the source should be considered. Whether the source is a true point source in space, radiating spherically, a hemispherical source close to one flat surface, or a quarter sphere between two flat surfaces, etc. will make a difference in the accuracy of the measurement. However, a detailed discussion of measurement issues is beyond the scope of this book, and the reader is encouraged to use some of the more comprehensive works in the list of references at the end of this chapter.

7.1.3.1 Overall Sound Level and How to Add dB Values

As mentioned, most sounds are composites of several different levels at different frequencies. This is especially true of industrial noise. A typical burner noise curve is shown in Figure 7.12. As can be seen, there are significant levels in two frequency zones, both of which will contribute to the apparent intensity experienced by a person working in the vicinity of the burner. It is difficult to describe this sound without using either a diagram like the one shown or a table listing various SPLs occurring in the different octave bands. The "overall sound level," a single number, has been devised to conveniently represent such composite sound curves. If a single number is to be used to represent the entire curve, then it should adequately represent the peaks in the curve, because the peaks have the most influence on the listener. Consequently, it is not practical to use the average of the various levels in the octave bands because this number would be less than the levels at the peaks. Therefore, one must not confuse the average level with the overall sound level.

The overall sound level is calculated by adding the individual levels in the various octave bands. In columns 1 and 2 of Table 7.3, the burner sound curve has been split up into its component levels in each octave band. In column 3, the A-weighted correction has similarly been split up and listed. Column 4 gives the A-corrected values for the sound curve by simply subtracting column 3 from column 2. Now, the values in column 4 must be added to obtain the overall sound level.

Because the decibel is based on a $\log_{10}$ scale, simple addition cannot be used. For example, if two values of equal magnitude are added, say 100 dB and 100 dB, the result is 103 dB. The formula to be used is:

$$L_{total} = 10\log_{10}\left(\Sigma_{i=1 \text{ to } n} 10^{0.1L_i}\right) \quad (7.2)$$

where

L_{total} = Total level
L_i = each individual level i
N = Number of levels to be added

Subtraction can be performed using

$$L_{diff} = 10\log_{10}\left(10^{0.1L_2} - 10^{0.1L_1}\right) \quad (7.3)$$

TABLE 7.2 Octave and One-Third Octave Bands

	Octave			One-Third Octave		
Band	**Lower Band Limit**	**Center**	**Upper Band Limit**	**Lower Band Limit**	**Center**	**Upper Band Limit**
12	11	16	22	14.1	16	17.8
13				17.8	20	22.4
14				22.4	25	28.2
15	22	31.5	44	28.2	31.5	35.5
16				35.5	40	44.7
17				44.7	50	56.2
18	44	63	88	56.2	63	70.8
19				70.8	80	89.1
20				89.1	100	112
21	88	125	177	112	125	141
22				141	160	178
23				178	200	224
24	177	250	355	224	250	282
25				282	315	355
26				355	400	447
27	355	500	710	447	500	562
28				562	630	708
29				708	800	891
30	710	1000	1420	891	1000	1122
31				1122	1250	1413
32				1413	1600	1778
33	1420	2000	2840	1778	2000	2239
34				2239	2500	2818
35				2818	3150	3548
36	2840	4000	5680	3548	4000	4467
37				4467	5000	5623
38				5623	6300	7079
39	5680	8000	11,360	7079	8000	8913
40				8913	10,000	11,220
41				11,220	12,500	14,130
42	11,360	16,000	22,720	14,130	16,000	17,780
43				17,780	20,000	22,390

The advantages of making broad band analyses of sound using octave or one-third octave band filter sets are that less time is needed to obtain data and the instrumentation required to measure the data is less expensive. The main disadvantage is the loss of detailed information about the sound which is available from narrow band (FFT) analyzers.

However, some simple rules of thumb can be used to perform quick estimates. They are as follows:

1. When adding dB values of equal magnitude, the sum is 3 dB added to one of the numbers.
2. When the two values are different by 3 dB or less, the sum is 2 dB added to the greater number.
3. When adding two values that differ by 7 dB or less, the sum is 1 dB added to the greater number.
4. For values that differ by 8 dB or more, the sum is just the larger number.
5. Always start with the smallest number in the list and add it to the next larger number.

To understand why these rules work, refer to the chart in Figure 7.3. From the chart it can be seen that 1 W is equal to 120 dB.

$$1\text{ W} = 120\text{ dB}$$
$$1\text{ W} = 120\text{ dB}$$
$$\overline{2\text{ W} = 123\text{ dB}}$$

On the chart, 2 W registers 123 dB on the line. Similarly, the reason that numbers 10 dB or more in difference are neglected is because:

$$1.0\text{ W} = 120\text{ dB}$$
$$0.1\text{ W} = 110\text{ dB}$$
$$\overline{1.1\text{ W} = 120\text{ dB}}$$

Because the 110 dB contributes only 0.1 W power, it is neglected in the approximation. The example becomes more vivid when adding two numbers that differ by 20 dB or more:

$$1.00 \text{ W} = 120 \text{ dB}$$
$$\underline{0.01 \text{ W} = 100 \text{ dB}}$$
$$1.01 \text{ W} = 120 \text{ dB}$$

Rule 5 is especially necessary when adding a list that contains several numbers that are almost equal in value and one or more that are 10 dB greater, such as in a list that contains six values of 90 dB and one value of 100 dB. If one begins to add from the 100-dB value, one will obtain a wrong result. Reader beware — the rules provided are approximations. For exact calculations, the formulae should be used.

Table 7.3 shows the effect of applying the addition rules to the values generated by breaking up the burner noise curve. At the end of the addition list, 1 dB has been added to compensate for any errors due to approximation.

7.1.3.2 Atmospheric Attenuation

When a sound wave travels through still air, it is absorbed or attenuated by the atmosphere. Over a couple of hundred feet, the atmosphere does not significantly attenuate the sound; however, over a few thousand feet, the sound level can be substantially reduced. The amount of sound attenuated in still air largely depends on the atmospheric temperature and relative humidity. Figure 7.13 depicts the atmospheric attenuation for aircraft-to-ground propagation in sound pressure level per 1000 ft (300 m) distance for center frequencies of 500, 1000, 2000, 4000, and 8000 Hz.[9] Notice that the atmospheric attenuation is more significant at higher frequencies than at lower frequencies. For example, suppose that we are 1000 ft (300 m) away from a noise source and that the atmospheric temperature and relative humidity are 80°F (27°C) and 10%, respectively. The plots in Figure 7.13 show that the atmospheric attenuation for 500 Hz is approximately 2 dB, whereas for 8000 Hz, the attenuation is 55 dB.

Atmospheric attenuation, outdoors, can also be affected by turbulence, fog, rain, and snow. Typically, the more turbulence present in the air, the more the sound is attenuated. There appears to be conflicting evidence as to whether or not fog attenuates sound. It is recommended that no excess attenuation be assigned to fog or light precipitation.

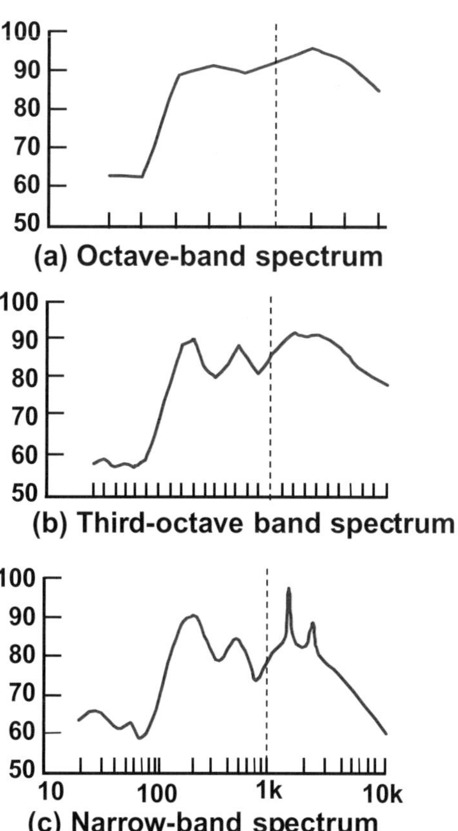

FIGURE 7.11 Same sound spectrum on three different intervals.

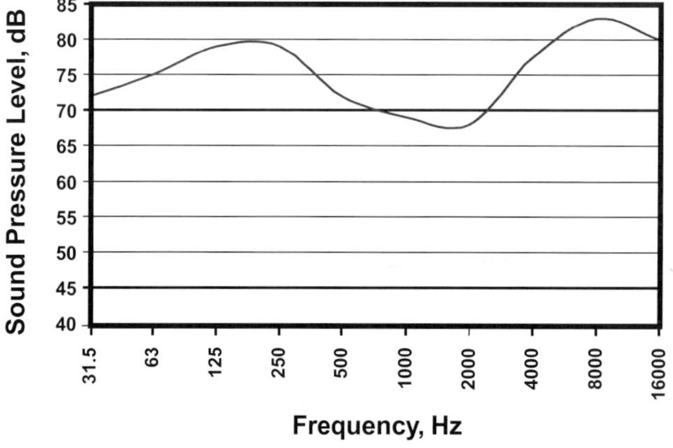

FIGURE 7.12 Typical burner noise curve.

7.2 INDUSTRIAL NOISE POLLUTION

Thus far, sound has been discussed. So what is noise? An all-encompassing definition would be that noise is any undesirable sound. By saying this, the concept is introduced that what is considered to be noise is somewhat relative and depends on several temporal and circumstantial factors.

For example, it is not unusual for a person to encounter sound pressure levels of 100 to 110 dB at a sporting event, in a stadium full of cheering fans, and yet not be perturbed by it. On the contrary, the barely 45-dB sound of a dripping faucet may cause considerable annoyance in the quiet of the night. Table 7.4 gives some typical noise levels for various scenarios.

TABLE 7.3 Addition Rules

What is the Overall dBA Level?

Frequency Hz	SPL dB	A-scale CF dB	SPL dBA	
31.5	72	–39	33	
63	75	–26	49	→49
125	79	–16	63	→63
250	79	–9	70	→71
500	72	–3	69	→73
1000	69	–0	69	→75
2000	68	–1	69	→76
4000	78	–1	79	→80
8000	83	–1	82	→84
16,000	80	–7	73	→85

Notes: Overall sound level = 85 dBA.

Caution: Overall SPL does not = average SPL.

Industrial noise pollution is a major concern for society as a whole. In a recent survey, the effects of exposure to noise in refinery workers was studied extensively. A cross-section of workers in different divisions/units was chosen. It was found that noise levels averaged 87 to 88 dBA in aromatic and paraffin facilities and 89 dBA in alkylation facilities. In comparison, workers in warehouses, health clinics, laboratories, and offices were not found to be exposed to the same levels.

Noise can damage hearing and cause physical or mental stress (increased pulse rate, blood pressure, nervousness, sleep disorders, lack of concentration, and irritability). Irreparable damage can be caused by single transient sound events with peak levels exceeding 140 dBA (e.g., shots or explosions). Long-duration exposure to noise exceeding 85 dBA can lead to short-term reversible hearing impairment, and long-term exposure to levels higher than 85 dBA can cause permanent hearing loss.

The following is a mathematical model based on empirical data (ISO 1999) used to calculate the maximum permissible continuous noise level at the workplace that will not lead to permanent hearing loss:

$$L_{A,m} < 85 + 10\log_{10}\left(24/T_n\right),\ \text{dBA} \tag{7.4}$$

where T_n is the daily noise exposure time in hours.

Wearing ear protection devices at continuous noise levels greater than 85 dBA can prevent or reduce the danger of permanent hearing damage.

7.2.1 OSHA Requirements

Title 29 CFR, section 1910.95 of the Occupational Safety and Health Act (OSHA) pertains to the protection of workers from potentially hazardous noise. Table 7.5 shows OSHA permissible noise exposure levels.

OSHA requires that the employer must provide protection against the effects of noise exposure when the sound levels exceed those shown in Table 7.6. When the daily noise exposure consists of two or more periods of noise exposure at different levels, their combined effect should be considered rather than the individual effects of each. According to OSHA, the exposure factor (EF) is defined as:[3]

$$EF = C_1/T_1 + C_2/T_2 + C_3/T_3 + \ldots + C_n/T_n \tag{7.5}$$

where C_n is the total time of exposure at a specific noise level and T_n is the total time of exposure permitted at that level (shown in Table 7.5). If the exposure factor exceeds 1.0, the employee's exposure is above OSHA limits. If OSHA identifies such a situation, a citation may be issued and a grace period defined in which the employer must correct the violation or face penalties as high as $10,000 per day.

7.2.2 International Requirements

Regulations aimed at protecting individuals from industrial noise pollution have been enforced in almost all industrialized countries. The noise caused in industry and the workplace is generally treated as a serious issue.

Most countries have adopted 85 dBA as the limit for permissible noise. At any work place with sound levels exceeding 85 dBA, ear protection devices must be worn, and workers exposed to this level should have their hearing level checked periodically.

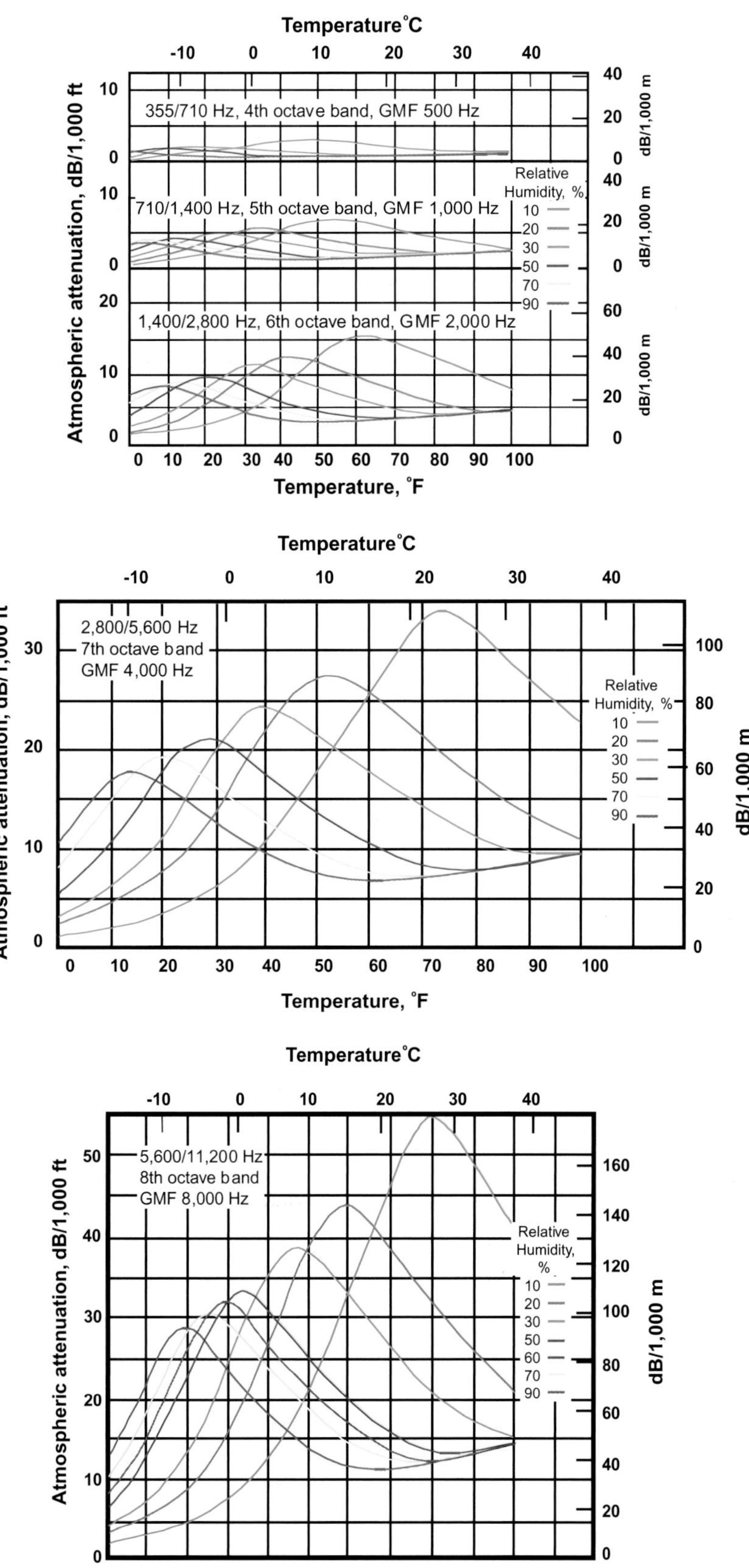

FIGURE 7.13 Same sound spectrum on three different intervals. (From Beranek.[9]).

TABLE 7.4 Sound Levels of Various Sources

Event	dBA Level
Threshold of hearing	0 dBA
Rustle of leaves	10 dBA
Normal conversation (at 1 m)	30 dBA
Minimum level in Chicago at night	40 dBA
City street, very busy traffic	70 dBA
Noisiest spot at Niagara Falls	85 dBA
Threshold of pain	120 dBA
Jet engine (at 50 m)	130 dBA
Rocket (at 50 m)	200 dBA

TABLE 7.5 OSHA Permissible Noise Exposures

Duration per Day (hours)	Sound Level, dBA (slow response)
8.0	90
6.0	92
4.0	95
3.0	97
2.0	100
1.5	102
1.0	105
0.5	110
≤ 0.25	115

Note: Exposure to impulsive or impact noise should not exceed 140 dBA.

7.2.3 Noise Sources and Environment Interaction

The predominant individual sources of noise in chemical and petrochemical plants are burners (process furnaces, steam boilers, and flares), fans, compressors, blowers, pumps, electric motors, steam turbines, gears, valves, exhausts to open air, conveyors and silos, airborne splash noise from cooling towers, coal mills, and loading and unloading of raw and finished materials.

Although noise pollution caused by the industrial sector is minor compared to that caused by road and rail traffic, industrial noise receives more attention due to public representation. ISO 1996 provides a set of international regulations for noise protection in residential neighborhoods located near industrial areas.

National or local authorities must enforce noise limits that should not be exceeded in the neighborhood. The magnitude of limiting values, additional fines for tonality and impulsive noise, and the legalities change, not only from country to country, but sometimes within different states and regions in the same country. In general, nighttime noise limits are 10 to 15 dB lower than that for the daytime.

7.3 MECHANISMS OF INDUSTRIAL COMBUSTION EQUIPMENT NOISE

There are four major mechanisms of noise production in combustion equipment. They can be categorized as either high-frequency or low-frequency sources. They are:

1. Low-frequency noise sources:
   a. Combustion roar and instability
   b. Fan noise
2. High-frequency noise sources:
   a. Gas jet noise
   b. Piping and valve noise

7.3.1 Combustion Roar and Combustion Instability Noise

To understand combustion roar, the mixing process taking place between the fuel and the oxidant on a very minute scale must be considered. It is known that a well-blended mixture of fuel and air will combust very rapidly if the mixture is within the flammability limits for that fuel. On the other hand, a raw fuel stream that depends on turbulence and momentum to mix in the ambient fluid, to create a flammable mixture, tends to create a slower combustion process. In either case, when regions in the mixing process achieve a flammable mixture and encounter a source of ignition, combustion takes place. The closer a mixture is to stoichiometry when it encounters the ignition source, the more rapid will be the combustion. Combustion occurring close to stoichiometry converts more of the energy release into noise. See the discussion on thermoacoustic efficiency in Section 7.3.1.1 for more details. The noise coming from each small region of rapidly combusting mixture adds up to create what is called combustion roar. Therefore, combustion roar is largely a function of how rapidly the fuel is being burned. In addition, in the context of combustion equipment like burners and flares, usually the larger the fuel release, the more the turbulence in the combustion process. Because turbulence directly influences the mixing rate, high turbulence processes also produce more combustion roar. Thus, it is more accurate to state that the level of combustion roar generated by a combustion process is a function of the amount of fuel being burned and how rapidly one arranges to burn it.

7.3.1.1 Flare Combustion Roar

It has been recognized for a long time that the noise emitted from a normal operating flare has two mechanisms at work: namely, combustion roar and gas jet noise. Combustion roar typically resides in the lower frequency region of the audible frequency spectrum, while gas jet noise occurs in the higher frequencies, as illustrated in Figure 7.14.

As mentioned, the amount of combustion roar emitted from a flare generally depends on how fast the waste gas stream mixes with the ambient air. A waste gas stream that exits a flare tip with low velocity and low levels of turbulence will mix slowly with the ambient air and burn relatively quietly. These types of flames are called buoyancy-dominated flames. A waste gas stream that exits a flare tip with high velocity and high levels of turbulence, however, will burn much faster and create substantially more combustion noise for the same heat release. These high-velocity flames are momentum-dominated flames. Increasing the rate at which the waste gas burns results in "bigger explosions" of the air/fuel mixture. These "bigger explosions" create larger disturbances in the atmosphere, resulting in higher levels of combustion roar.

High levels of turbulence in a flare flame are usually desirable because they help reduce radiation and increase the smokeless capacity of the flare. Unfortunately, such high turbulence levels also increase the combustion roar. Unlike the solution for flare radiation reduction, it is not practical to increase the height of a flare stack or boom to reduce combustion noise. This is because combustion roar is low-frequency sound and thus can travel a great distance without being substantially attenuated by the atmosphere. The signature of low-frequency combustion roar noise typically consists of a broadband spectrum with a single peak.

The combustion roar emitted from a stable burning flare typically peaks at a frequency of about 63 Hz. The combustion noise spectrum can be estimated by adjusting the sound pressure level emitted from combustion using the values in Table 7.6.[4] It is noted that, at frequencies above about 500 Hz, the noise contribution from flare combustion is relatively insignificant. A typical method for estimating the sound pressure level (SPL) emitted from a flare flame is to relate the energy released from the combustion of the waste gas stream (chemical energy) to the noise energy liberated by the combustion. The ratio of noise energy to chemical energy released from the combustion is called the thermoacoustic efficiency (TAE). For a stable-burning flare, the TAE typically varies between 1×10^{-9} to 3×10^{-6}. The value of the TAE largely depends on the turbulent mixing of the waste gas with ambient air and is usually determined experimentally.

In designing flares, the combustion noise emitted from flares operating under various conditions is usually measured to determine the TAE. This information can then be used in computer programs to model the level of combustion roar emitted from a flare. Figure 7.15 is a photograph of a John Zink engineer collecting noise levels from a flare using a real-time noise level meter.

A flare flame that is highly turbulent, such as the high-pressure flare shown in Figure 7.15, can have a TAE on the order of 1×10^{-6}. However, a flame with low levels of turbulence, such as the butane cigarette lighter shown in Figure 7.16, may have a TAE on the order of 1×10^{-9}. For every order of magnitude that the TAE changes, the sound pressure level will change by 10 dB.

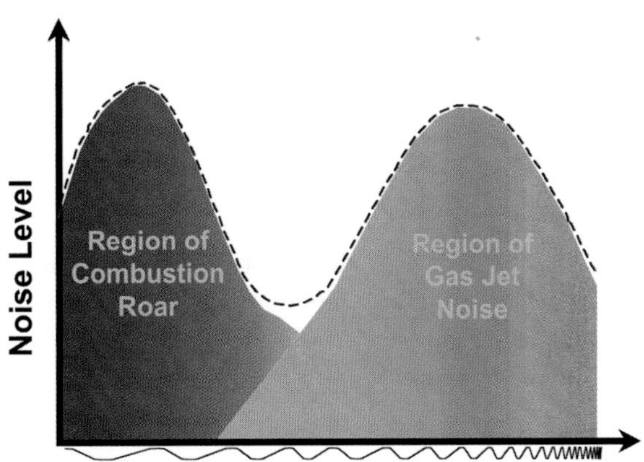

FIGURE 7.14 Typical noise signature emitted from a flare.

TABLE 7.6 Overall Sound Pressure Level from Combustion

Frequency (Hz)	Resultant Noise Spectrum (dB)[a]
31.5	OASPL - 5
63	OASPL - 4
125	OASPL - 9
250	OASPL - 15
500	OASPL - 20
1000	OASPL - 21
2000	OASPL - 24
4000	OASPL - 28
8000	OASPL - 34

[a] OASPL = Overall sound pressure level.

FIGURE 7.15 Engineer measuring flare noise level.

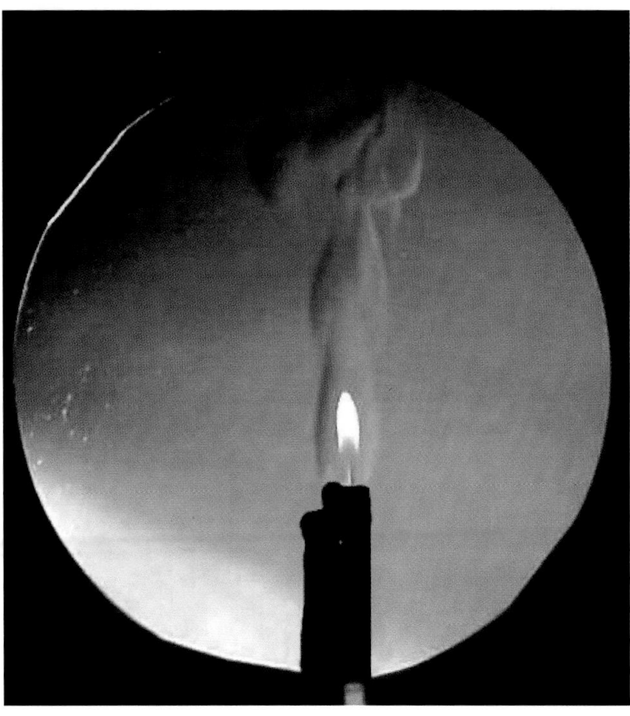

FIGURE 7.16 Shadow photograph of a burning butane lighter.

Example 7.1

Given: A flare burning a waste gas stream with a heat release of 5×10^9 BTU/h (1465 million W). Noise measurement shows that the sound pressure level 400 feet (120 m) from the flame is 100 dB.

Find: The TAE of the flare flame.

Solution: The sound power emitted from the flame, *W*, can be determined as follows:

$$W(\text{watts}) = 1\times10^{-12}\left[\text{anti}\log_{10}\left(\frac{L_p + 10\log_{10}(4\pi r^2) - 10.5}{10}\right)\right] \quad (7.6)$$

where L_p is the sound pressure level in dB (100 dB for this example), and *r* is the distance from the flame in feet (400 feet or 120 m for this example). Substituting these values into Eq. (7.6) gives W = 1792 watts. The TAE is then calculated to be:

$$\text{TAE} = \frac{\text{Acoustical power}}{\text{Thermal power}} = \frac{1792 \text{ watts}}{1465\times10^{6} \text{ watts}} = 1.2\times10^{-6} \quad (7.7)$$

Because the TAE is on the order of magnitude of 1×10^{-6}, one would expect that the flame would be highly turbulent and momentum-dominated.

The combustion roar emitted from a flare flame is not highly directional and is considered a monopole source. That is, it is analogous to a spherical balloon whose surface is expanding and shrinking at various frequencies and emitting uniform spherical waves.

7.3.1.2 Flare Combustion Instability Noise

If a flame lifts too far above a flare tip, it can become unstable. An unstable flame will periodically lift and then reattach to the flare tip and create a low-frequency rumbling noise. Typically, this low rumbling noise occurs in the frequency range of 5 to 10 Hz and is usually called combustion instability. Being as low in frequency as it is, combustion instability noise is usually inaudible and can travel over several miles without being substantially attenuated by the atmospheric air. When there are reports of shaking the walls and windows of buildings in the vicinity of a flare, it is usually due to combustion instability.

Combustion instability noise can occur if too much steam or air is added to the base of the flame. Thus, over-aerating the waste gas stream in a flare causes the flame to periodically lift from the flare tip. This periodic lifting and reattachment of the flame from the flare tip is the mechanism that drives the low-frequency rumbling noise. Combustion instability noise can usually be reduced by lowering the steam flow rate to a steam-assisted flare or by lowering the blower airflow rate to an air-assisted flare. Figure 7.17 graphically depicts a typical steam-assisted flare operating under both normal conditions as well as over-steamed conditions.[5] Note that the combustion noise frequency shifts substantially to a lower region and the level dramatically increases when the flare is over-steamed.

7.3.1.3 Burner Combustion Noise

Like flares, burner combustion noise is an unwanted sound associated with combustion roar and combustion instability. In many situations, the combustion noise can be the dominant source of noise emitted from a burner. Combustion roar and combustion instability are quite complex by nature. The literature contains a variety of combustion noise and combustion instability prediction techniques for burners operating in a furnace. Most of these prediction techniques are based on experimental studies that attempt to correlate the acoustic power radiated by the burner/furnace geometry, laminar burning velocity of the air/fuel mixture, and various turbulence parameters such as the turbulent length scale and intensity. This section does not attempt to discuss these prediction techniques in detail, but gives a broad and general discussion of combustion roar and combustion instability noise using some of the results from these studies.

Figure 7.12 is a plot showing a typical noise spectrum emitted from a burner operating under normal conditions in a furnace. Notice that the noise spectrum has two peak frequencies associated with it; the high-frequency noise contribution is from the fuel gas jets, while the low-frequency contribution is from the combustion roar. As with combustion roar emitted from flares, burner combustion roar is associated with a smooth broadband spectrum having relatively low conversion efficiency from chemical energy to noise — in the range of 1×10^{-9} to 1×10^{-6}. However, the combustion noise spectra associated with a burner and a flare are not similar. The reason is that a flame burning in the open atmosphere will behave differently compared to a flame that is burning in an enclosed chamber such as a furnace.

The combustion roar associated with flares typically peaks at a frequency of approximately 63 Hz, while the combustion roar associated with burners can vary in the 200 to 500 Hz range. Burner combustion roar can have a noise spectrum shape and amplitude that can vary with many factors. These factors include the internal shape of the furnace; the design of the burner muffler, plenum, and tile; the acoustic properties of the furnace lining; the transmission of the noise into the fuel supply piping; and the transmissive and reflective characteristics of the furnace walls and stack.

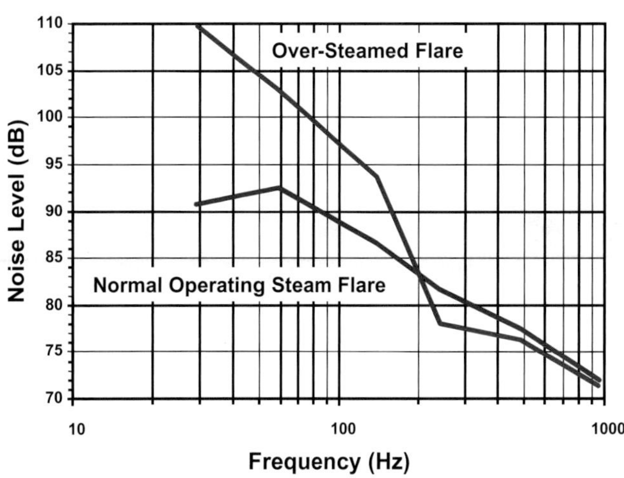

FIGURE 7.17 A steam-assisted flare under normal and over-steamed conditions.

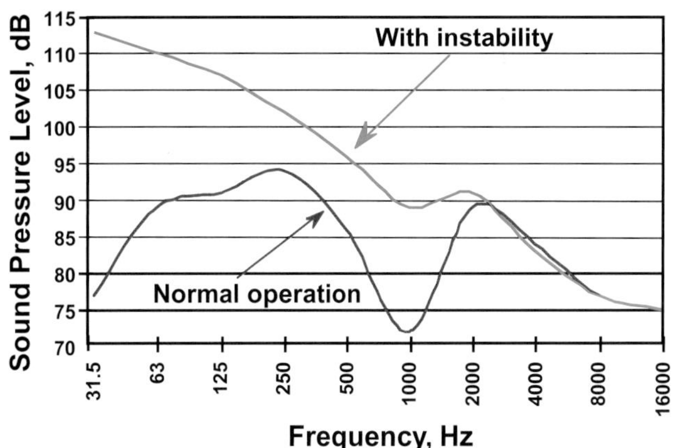

FIGURE 7.18 Sound pressure level burner with instability.

7.3.1.4 Burner Combustion Instability Noise

Combustion instability within a furnace is characterized by a high-amplitude, low-frequency noise resembling the puffing sound of a steam locomotive. This type of noise can create significant pressure fluctuations within a furnace that can cause damage to the structure and radiate high levels of noise to the surroundings.

Figure 7.18 is a plot showing the SPL for a gas burner operating under normal conditions and with instability. It is obvious that the sound pressure level increases substantially when the operation is accompanied by instability. Combustion instability noise has a high efficiency of conversion of chemical energy to noise. Typically, the TAE from burner combustion instability is in the range of 1×10^{-4}.[6]

The oscillations caused by combustion instability are naturally damped by pressure drop losses through the burner and furnace, and therefore cannot be sustained unless energy is provided. These steady oscillations are sustained by energy extracted from the rapid expansion of the air/fuel mixture upon reaction. Over the years, furnace operators have used several techniques in an attempt to eliminate combustion instability. Some of these techniques include modifying the (1) furnace stack height, (2) internal volume of the furnace, (3) acoustical properties of the furnace lining, (4) pressure drop through the burner by varying the damper position, (5) fuel port diameter, (6) location of the pilot, and (7) flame stabilization techniques.

7.3.2 Fan Noise

The noise emitted from industrial fans typically consists of two noise components: broadband and discrete tones. Vortex shedding of the moving blades and the interaction of the turbulence with the solid construction parts of the fan create the broadband noise. This broadband noise is of the dipole type, meaning that the noise is directional. On the other hand, the discrete tones are created by the periodic interactions of the rotating blades and nearby upstream and downstream surfaces.

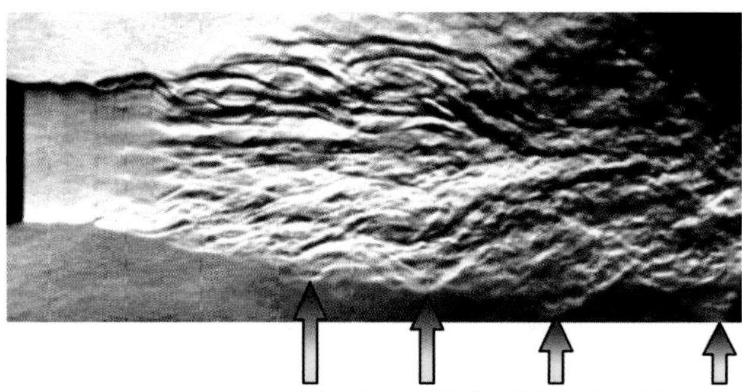

FIGURE 7.19 Development of orderly wave patterns.

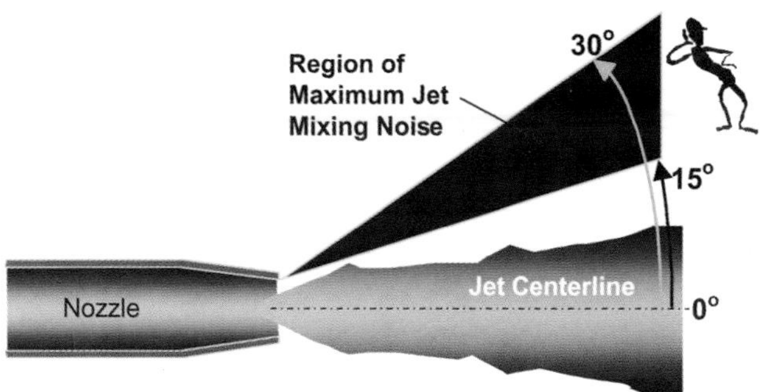

FIGURE 7.20 Region of maximum jet mixing noise.

Discrete tonal noise is usually the loudest at the frequency at which a blade passes a given point. The tonal frequency is easily calculated by multiplying the number of blades times the impeller rotation speed in revolutions per second.

The broadband and discrete tonal noise emitted from fans can radiate from both the suction and pressure side of a fan and through the fan casing. The noise can radiate downstream through the ducting and discharge into the environment at an outlet. Fan and duct systems should include provisions to control this noise if residential areas are located nearby. Installation of mufflers and silencers on the suction and the discharge sides of the fan, as well as wrapping of the casing and the ducts, are common methods for reducing fan noise.

7.3.3 Gas Jet Noise

Gas jet noise is very common in the combustion industry and in many instances it can be the dominant noise source within a combustion system. The noise created when a high-speed gas jet exits into an ambient gas usually consists of two principal components: gas jet mixing noise and shock-associated noise.[7]

7.3.3.1 Gas Jet Mixing Noise

Studies have shown that a high-speed gas jet exiting a nozzle will develop a large-scale orderly pattern, as shown in Figure 7.19. This orderly structure is known as the "global instability" or "preferred mode" of the jet. The presence of both the small-scale turbulent eddies within the jet and the large-scale structure is responsible for the gas jet mixing noise.

The source of gas jet mixing noise begins near the nozzle exit and extends several nozzle diameters downstream. Near the nozzle exit, the scale of the turbulent eddies is small and predominantly responsible for the high-frequency component of the jet mixing noise. The lower frequencies are generated further downstream of the nozzle exit where the large-scale orderly pattern of the gas jet exists.

Gas jet mixing noise consists of a broadband frequency spectrum. The frequency at which the spectrum peaks depends on several factors, including the diameter of the nozzle, the Mach number of the gas jet, the angle of the observer's position relative to the exit plane of the jet, and the temperature ratio of the fully expanded jet to the ambient gas. In the flare and burner industry, gas jet mixing noise typically peaks somewhere between 2000 and 25,000 Hz.

The overall sound pressure level created by gas jet mixing depends on several variables, including the distance from the gas jet, the angle of the observer relative to the gas jet centerline velocity vector, the Mach number, the fully expanded gas jet area, and the density ratio of the fully expanded jet to the ambient gas. The maximum overall SPL of gas jet mixing noise occurs at an angle between approximately 15° and 30° relative to the centerline of the gas jet velocity vector, as illustrated in Figure 7.20.[7]

As one moves in either direction from this angle, the noise level (in some cases) can drop off significantly. For example, the overall SPL created by gas jet mixing can be reduced as much as 25 dB when one moves from an angle of maximum noise level to an angle directly behind the nozzle (180°).

7.3.3.2 Shock-Associated Noise

When a flare or burner operates above a certain fuel pressure, a marked change occurs in the structure of the gas jet. Above a certain pressure called the critical pressure, the gas jet develops a structure of shock waves downstream of the nozzle, as shown in Figure 7.21. The critical pressure of a gas jet typically occurs at a pressure of 12 to 15 psig (0.8 to 1 barg), depending on the gas composition and temperature. These shock cells consist of compression and expansion waves that repeatedly compress and expand the gas as it moves downstream. Using Schlieren photography, several investigators have seen as many as seven shock cells downstream of a nozzle. These shock cells are responsible for creating two additional components of gas jet noise: screech tones and broadband shock-associated noise.

Screech tones are distinct narrow-band frequency sounds that can be described as "whistles" or "screeches." The literature reports that these tones are emitted from the fourth and fifth shock cells downstream of the nozzle exit, as shown in Figure 7.22.[8]

The sound waves from these shock cells propagate upstream, where they interact with the shear layer at the nozzle exit. This interaction then creates oscillating instability waves within the gas jet. When these instability waves propagate downstream, they interfere with the fourth and fifth shock cells, causing them to emit the screech tones.

Broadband shock-associated noise occurs when the turbulent eddies within the gas jet pass through shock waves. The shock waves appear to suddenly distort the turbulent eddies, which creates a noise that can range over several octave bands. The broadband, shock-associated peak frequency noise typically occurs at a higher frequency than the screech tone peak frequency.

7.3.4 Valve and Piping Noise

When a gas flowing steadily in a pipe encounters a valve, a change in the flow pattern and pressure will occur that can create turbulence and shock waves downstream of the valve. Typically, when valves are partially closed, creating a reduction in flow area, the small flow passage behaves much like an orifice and produces jet noise. As discussed above, turbulence and shock waves create mixing noise and shock-associated noise. This noise can radiate downstream through the pipe and exhaust into the environment at an outlet and/or radiate through the pipe wall into the space near the valve itself, as illustrated in Figure 7.23.

Usually, butterfly valves and ball valves are noisier than globe valves. Butterfly valves and ball valves typically have a smaller vena contracta than a globe valve operating at the same pressure drop, which results in higher levels of mixing and shock-associated noise. As a general guideline, when the pressure ratio across a valve is less than approximately 3, the mixing noise and shock-associated noise are within about the same order of magnitude. However, for pressure ratios greater than 3, shock noise usually dominates mixing noise.[9] There are several methods used for reducing the noise emitted from a valve. These include sound-absorptive wrapping of the pipes and the valve casings and the installation of silencers between the valve and connected pipes.

7.4 NOISE ABATEMENT TECHNIQUES

There are three places noise can be reduced: at the source, in the path between the source and personnel, and on the personnel.[10] The ideal place to stop noise is at the source. There are several techniques used in the flare and burner industry to reduce the noise at the source, but these techniques have limitations. Ear protection can reduce noise relative to the personnel using it; unfortunately, a plant operator cannot ask a surrounding community or workers within a nearby office building to wear ear protection when the noise levels become a problem. The most common method for reducing noise is in the path between the source and personnel, using silencers, plenums, and mufflers. The purpose of this section is to discuss the most common and effective noise abatement techniques utilized in the flare and burner industry.

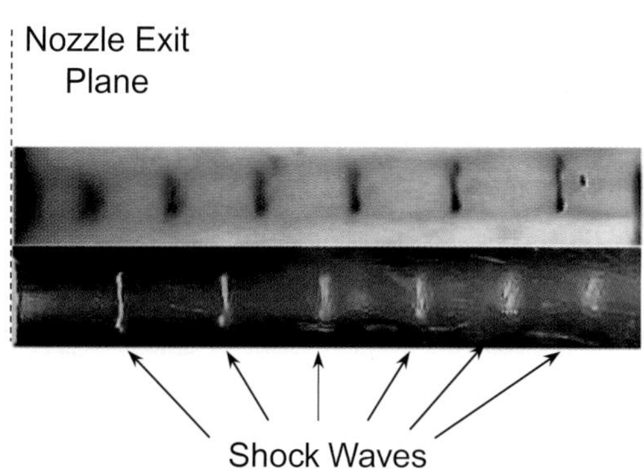

FIGURE 7.21 Shock waves downstream of an air jet.

7.4.1 Flare Noise Abatement Techniques

As previously discussed, the two principal sources of noise emitted from industrial flares are combustion roar and gas jet noise. Inhibiting the rate at which the air and fuel streams mix can reduce the level of combustion roar; however, this noise abatement technique generally tends to reduce the smokeless performance and increase thermal radiation and flame length. Reducing the mixing rate of the air and fuel stream in order to lower combustion roar levels usually does not justify the accompanying sacrifices in the performance of a flare.

In such cases, enclosed flares may provide one solution. Enclosed flares are designed to completely hide a flare flame in order to reduce noise and thermal radiation levels. The design of these flare systems typically consists of an insulated enclosure with a wind fence around the perimeter, as shown in the photograph in Figure 7.24. These types of flares can substantially reduce noise emissions as compared to open, elevated flares.

There are several abatement techniques commonly used to reduce the gas jet noise emitted from flares. Such techniques include mufflers, water injection, and modifications to the nozzle geometry. Mufflers are most commonly used on steam-

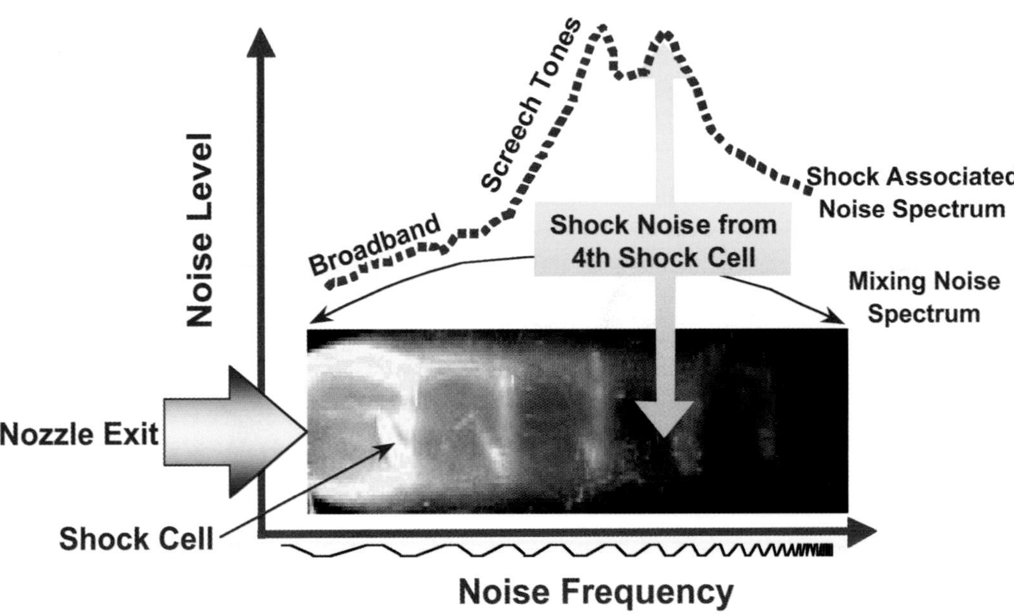

FIGURE 7.22 Location of screech tones emissions.

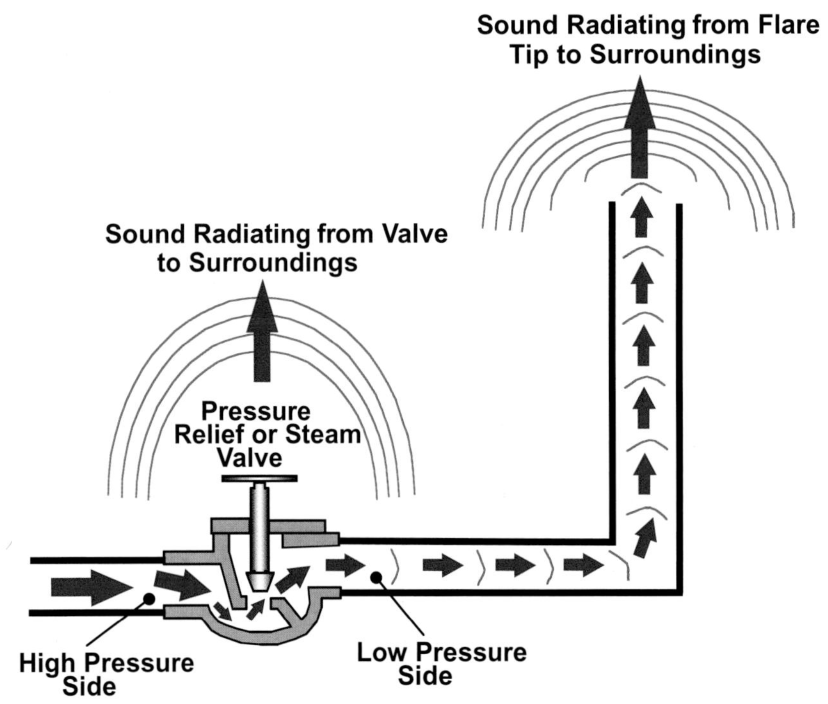

FIGURE 7.23 Noise radiating from a valve.

assisted flares to abate the high-pressure steam jet noise, as shown in Figure 7.25.

In most flare systems, steam is supplied to nozzles at a pressure of 100 to 150 psig (7 to 10 barg). These high-pressure steam jets produce high-frequency mixing and shock-associated noise. A number of flare muffler styles have been used in the industry with varying degrees of noise abatement performance. Many of these mufflers are designed with a fiber material several inches thick placed on the inside. Mufflers usually do a good job of absorbing the high-frequency

FIGURE 7.24 Two enclosed flares.

steam jet noise, as demonstrated by the data in Figure 7.26. This plot shows the noise spectrum emitted from a steam-assisted flare operating with and without a muffler on the lower steam jets. The data clearly show that mufflers are more efficient at absorbing higher noise frequencies than lower ones.

In high-pressure flaring applications, gas jet noise can be the major source of noise. Recently, the John Zink Company developed a unique method for reducing gas jet noise using water injection.[11] This method injects water into the waste gas stream near the nozzle exit. The water injection method substantially reduces the shock-associated noise, as shown in Figure 7.27. This plot depicts the noise spectrum emitted from a John Zink high-pressure flare operating with and without water injection. Schlieren photography shows that water injection does not eliminate the downstream shock cell structure, but does appear to alter its appearance. This suggests that water injection suppresses the feedback mechanism responsible for growth of the gas jet instability that leads to screech tones.

Gas jet noise reduction using water injection is more pronounced when flaring high-molecular-weight gases as compared to low-molecular-weight gases at the same operating pressure. Test data and computer modeling show that high-molecular-weight gases are more dominated by screech tone noise than low-molecular-weight gases operating at the

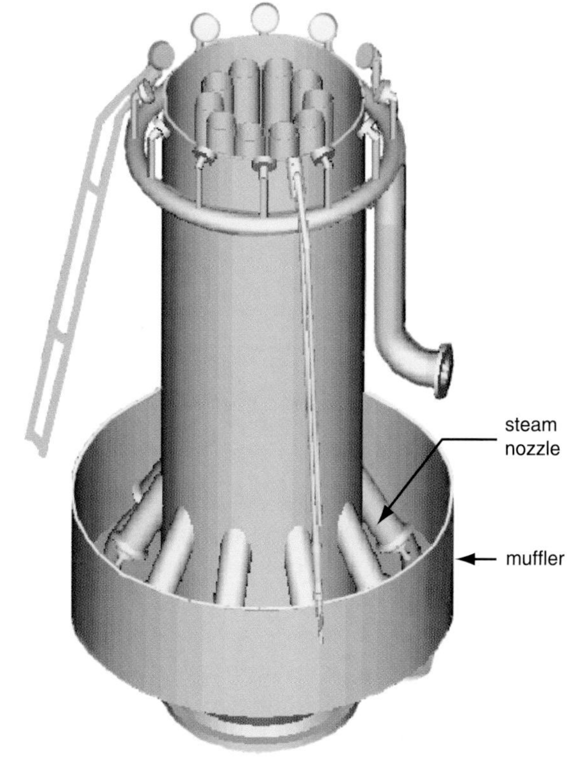

FIGURE 7.25 A steam-assisted flare with a muffler.

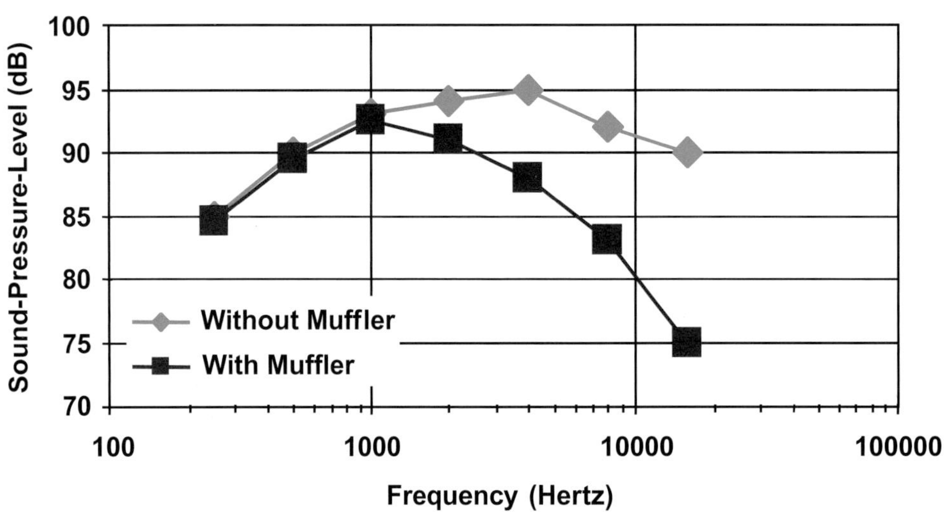

FIGURE 7.26 Steam jet noise emitted with and without a muffler.

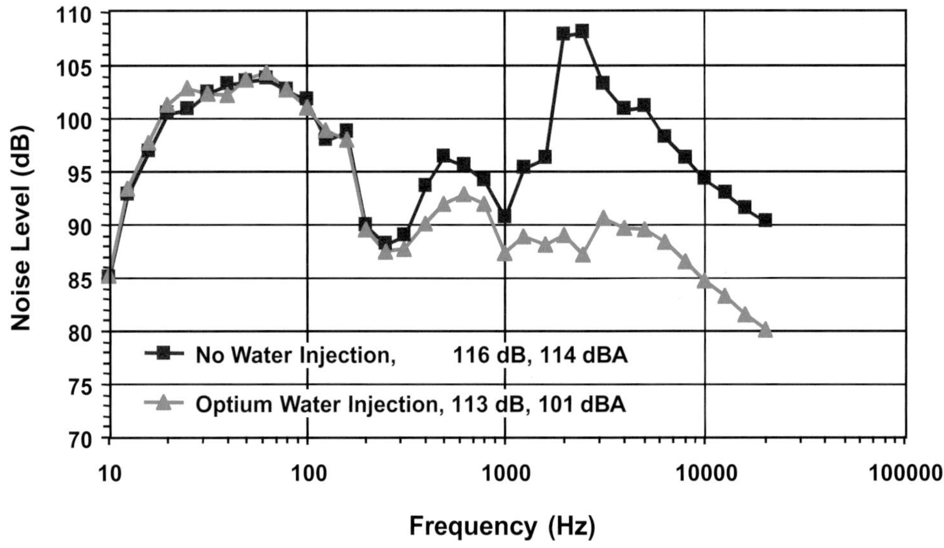

FIGURE 7.27 Noise spectrum from a high pressure flare with and without water injection.

same pressure, which explains why gas jet noise reduction using water injection is more pronounced when flaring high-molecular-weight gases.

It is very common in the flare industry to design a flare using several small-diameter nozzles to reduce the A-weighted gas jet noise level. Gas jet noise emitted from high-pressure flares usually peaks at a frequency between approximately 2000 and 16,000 Hz. The peak frequency is a function of several variables, but is most affected by the diameter of the nozzle. For example, a 4-in. gas jet nozzle will peak at a frequency between 2000 and 4000 Hz, whereas a 1-in. gas jet nozzle will peak between 8000 and 16,000 Hz. To the human ear, a group of several smaller-diameter gas jet nozzles will appear quieter than a single larger nozzle operating at the same pressure and mass flow rate, the primary reason being that the group of smaller nozzles will peak at a higher frequency, where the human ear is less sensitive. Designing a flare with many small-diameter nozzles is not always practical or economical to build. Some large-capacity flare designs require several thousand nozzles to substantially reduce the gas jet noise.

7.4.2 Burner Noise Abatement Techniques

Burners used in industrial heaters and furnaces emit a broadband spectrum of noise. The broadband noise spectrum consists of (1) combustion roar, which resides in the

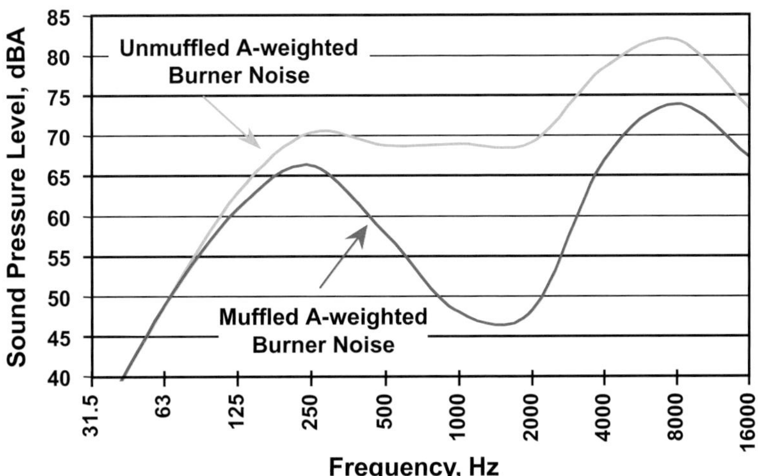

FIGURE 7.28 Sound pressure vs. frequency with and without a muffler.

frequency range of approximately 100 to 1000 Hz, and (2) gas jet noise, which typically ranges between 4000 and 16,000 Hz. The mid-to-high-frequency noise is the most annoying and damaging to the ear. Several techniques have been used to suppress the noise emitted within the mid-to-high frequencies. Four common techniques used to reduce noise in industrial burners are:

1. sound insulation in the burner plenum
2. mufflers at air inlets of natural-draft burners
3. acoustically optimized furnace wall construction
4. acoustical treatment of the air ducts in forced-draft burners

Figure 7.28 shows a plot of the sound pressure level vs. frequency for a burner operating with and without a muffler. Clearly, without the muffler, the noise level is higher — especially in the higher frequency region.

7.4.3 *Valve and Piping Noise Abatement Techniques*

Valve and piping noise abatements include sound-absorptive wrapping of the pipes and valve casings, and installation of silencers between the valves and the connecting pipes. Acoustical pipe lagging is similar to thermal pipe insulation. Acoustical pipe lagging also provides excellent thermal insulation, but many thermal insulations provide poor noise control. Rigid insulations for cold service (such as foam glass installed on smaller-diameter pipes) can actually aggravate the noise situation by easily conducting the noise to the outer surface. Although acoustical energy radiated per unit area of insulated and jacketed pipe is less than for the same noninsulated pipe, the surface area of an insulated and jacketed pipe is greater. The product of these two factors can cause larger-diameter jacketed pipes to radiate more noise than bare pipes.[12]

Piping requiring acoustical treatment in a typical petrochemical plant is often in cold service. These lagging systems have to be both thermal and acoustical insulators. For that reason, fibrous insulation followed by an outer leaded aluminum jacket is commonly used. Sometimes, very noisy pipes need a layer of impregnated vinyl sandwiched between layers of fibrous insulation, called a septum system.[12]

7.4.4 *Fan Noise Abatement Techniques*

Fan noise can usually be addressed similar to valve and piping noise:

1. Silencers can be installed at the suction and pressure sides of the fan, particularly for fans communicating with the atmosphere on either the suction or the pressure side, and thereby cut down on noise coming out of these portals.
2. Acoustically enclose the fan casing to address noise radiated from or transmitted through the casing surface.
3. Acoustically isolate the ductwork leading to and from a fan.

At the design stage, one can consider the use of low-noise motors (85 dBA or less) and the use of impellers with more blades and reduced tip speed, etc.

7.5 ANALYSIS OF COMBUSTION EQUIPMENT NOISE

7.5.1 *Multiple Burner Interaction*

A burner manufacturer will typically guarantee a burner noise level at a location 3 ft (1 m) directly in front of the

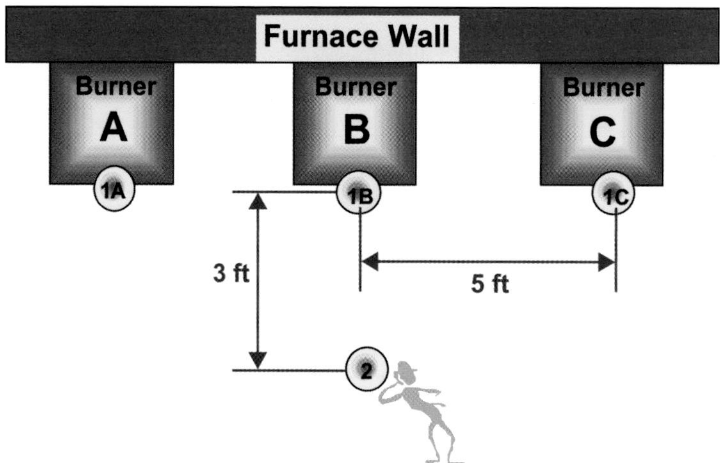

FIGURE 7.29 Burner noise example.

muffler. When several burners are installed in a furnace, however, the noise level 3 ft (1 m) from the burner may be higher than for a single burner, due to the noise contribution from surrounding burners. The purpose of this section is to give an example that illustrates the noise level increase due to noise emitted from surrounding burners.

Example 7.2

Given: Assume a furnace with a simple burner configuration, as illustrated in Figure 7.29, with burner B operating alone, and the noise level is 85 dB at location 2.

Find: How is the noise level determined at location 2 when all burners are operating?

Solution: First, find the sound power level, L_w, emitted from each burner, assuming that the source is emitted at the muffler exit at points 1A, 1B, and 1C. Assume that the noise spreads over a uniform sphere from each of these points. The sound power level can be calculated as:

$$L_w = L_p - 10\log_{10}\left(\frac{1}{4\pi r^2}\right) - 10.5 \tag{7.8}$$

where L_p is the sound pressure level and r is the distance from the source (in feet). The noise level 3 ft (1 m) from burner B (location 2) is 85 dB when it is operating alone. From Eq. (7.8), $L_{w_B} = 95.0$ dB. Assuming that all burners are operating at the same conditions, the sound power level must be 95.0 dB for each one. The sound pressure level contribution, L_p, can now be calculated at location 2 when burner A is operating alone by solving Eq. (7.9) for L_p:

$$L_p = L_w + 10\log_{10}\left(\frac{1}{4\pi r^2}\right) + 10.5 \tag{7.9}$$

For this case $L_{w_A} = 95.03$ and $r = (5^2 + 3^2)^{0.5} = 5.83$ ft. Substituting these values into Eq. (7.10) gives $L_{p_A} = 79.2$ dB. This is the sound pressure level contribution emitted from burner A measured at location 2. Because the distance from burner C to location 2 is the same, the sound pressure level contribution from burner C at location 2, L_{p_C}, is also 79.2 dB. The total sound pressure level at location 2 can be determined by adding the sound pressure level contribution from each burner (79.2 dB + 79.2 dB + 85 dB). The sound pressure levels can be added using the following equation:

$$L_{p_{\text{total}}} = 10\log_{10}\left(10^{L_{PA}/10} + 10^{L_{PB}/10} + 10^{L_{PC}/10}\right)$$
$$= 86.8 \text{ dB} \tag{7.10}$$

For this example, the noise level will be approximately 1.6 dB higher when all the burners are operating than if burner B is operating alone.

7.5.2 *High-Pressure Flare*

Figure 7.30 is a plot showing the sound-pressure-level spectrum of a high pressure flaring event burning Tulsa natural gas

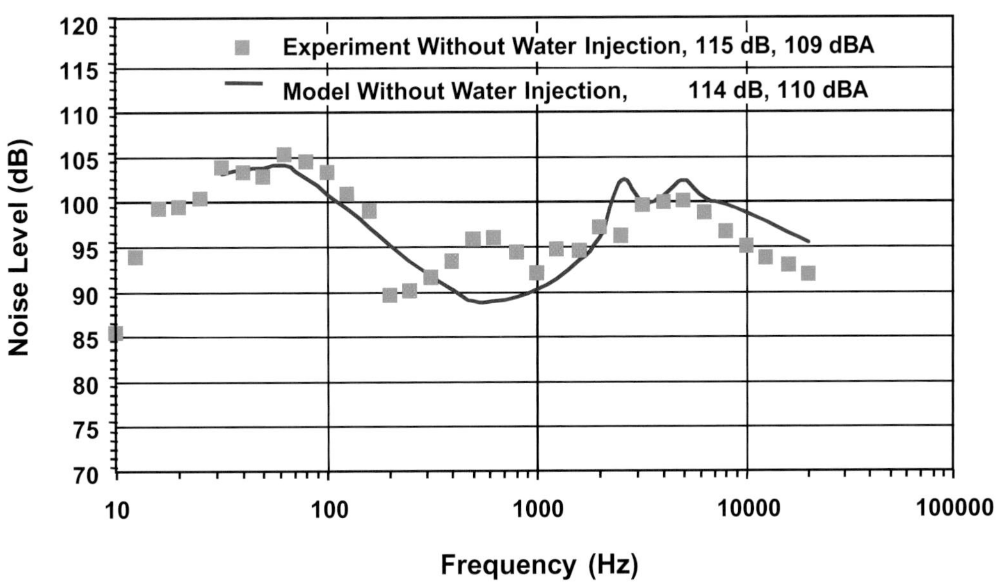

FIGURE 7.30 The sound-pressure-level spectrum of a high-pressure flare.

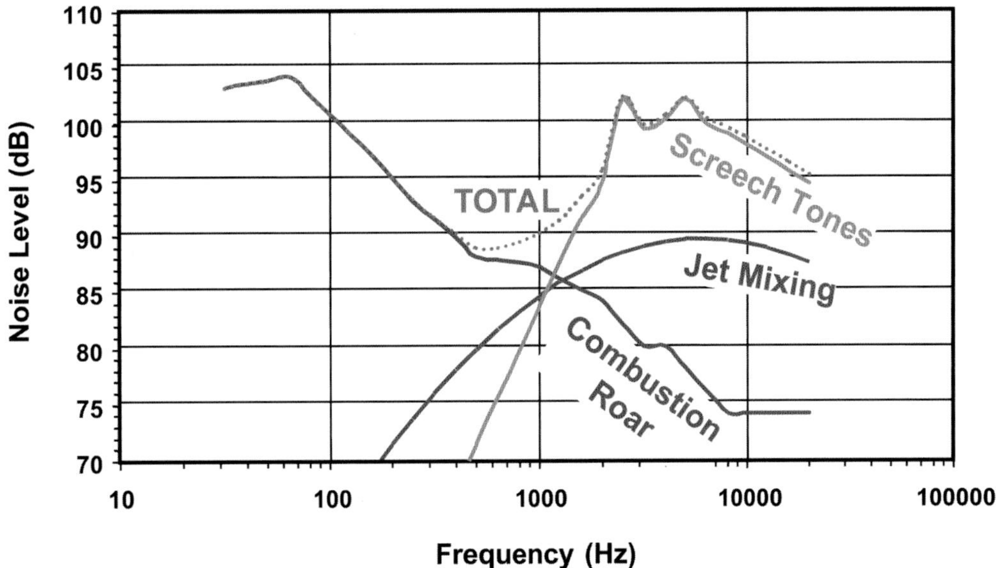

FIGURE 7.31 The noise contributions separately based on the mathematical model.

in a 3.5-in (8.9-cm) tip. The symbols and the lines represent the noise spectrum gathered using a real-time sound-level meter and mathematical modeling results, respectively. The sound pressure level spectrum consists of two major peaks: a low-frequency peak that corresponds to the combustion roar and a high-frequency peak that corresponds to the gas jet noise. The intermediate peak is a result of piping and valve noise. Notice that the combustion roar peaks at a frequency of approximately 63 Hz, which is typical for a stable burning open flare.

Figure 7.31 is a plot showing the noise contributions separately based on the mathematical model. Notice that the gas jet mixing noise is a broadband frequency spectrum, while the screech noise occurs over a fairly narrow bandwidth.

The screech noise would not exist if the flare operated below the critical gas pressure. Below the critical gas pressure, shock waves, which cause screech noise, do not form. The summation of the combustion roar, gas jet mixing noise, and screech noise provides the total sound pressure level prediction emitted from the flare.

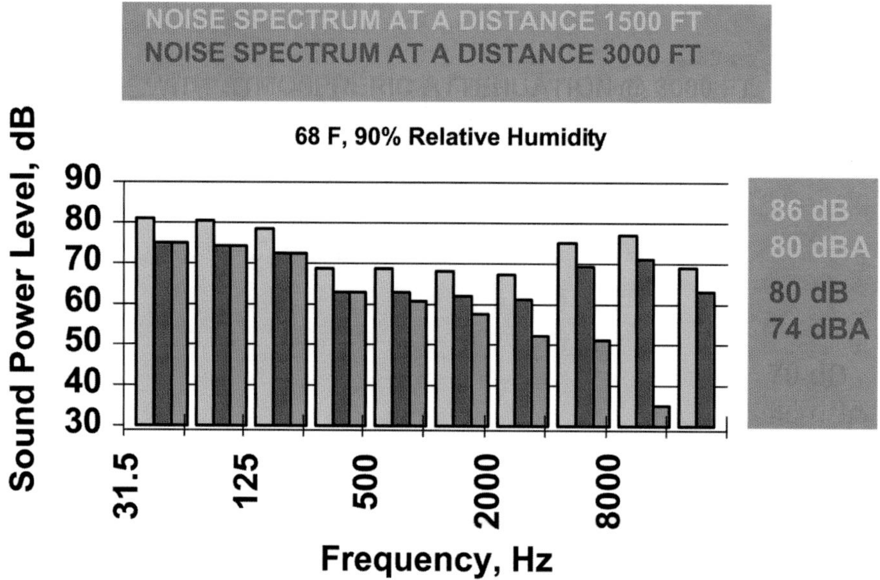

FIGURE 7.32 Effect of distance on flare noise.

TABLE 7.7 The Overall Sound Pressure Level (OASPL) Determined Experimentally and Using the Mathematical Model

	Jet Mixing Noise	Screech Noise	Combustion Roar	Total
		Model		
dB	105.7	105.1	113.0	114.3
dBA	105.2	105.3	97.4	108.6
		Experiment		
dB	—	—	—	113.7
dBA	—	—	—	109.2

The overall SPL (OASPL) determined experimentally and calculated using the mathematical model is summarized in Table 7.7. Notice that in this particular example, the OASPL, on a dBA scale, is dominated by the gas jet noise. If this 3.5-in. (8.9-cm) diameter flare were designed with several smaller-diameter ports having the same total exit area, then the gas jet noise would shift to higher frequencies. If the diameter of these ports were small enough to substantially shift the frequency of the gas jet noise, then the combustion noise would dominate on the dBA scale.

7.5.3 Atmospheric Attenuation

Figure 7.32 shows noise measurements emitted from a flare. Notice that at a distance of 1500 ft (460 m) from the flare, the noise level peaks at about 80 dBA, while at 3000 ft (910 m), the peak reduces to about 74 dBA. When the atmospheric attenuation is taken into account, depending on the ambient temperature and humidity level at the time of measurement, there is further reduction in noise levels. It is important to note that the contribution in each case is significant. Given the particular atmospheric conditions in this example, the attenuation has created a significant difference. The 10-dB attenuation (from 74 dBA to 64 dBA) amounts to the sound intensity reduction equal to one-tenth of its intensity at 3000 ft (910 m) without atmospheric attenuation. Hence, it should be noted that measurements may vary significantly on different days for the same equipment if the atmospheric conditions are significantly different.

7.7 GLOSSARY

Absorption: Conversion of sound energy into another form of energy, usually heat, when passing through an acoustical medium.

Absorption coefficient: Ratio of sound absorbing effectiveness, at a specific frequency, of a unit area of acoustical absorbent to a unit area of perfectly absorptive material.

Acoustics: Science of the production, control, transmission, reception, and effects of sound and of the phenomenon of hearing.

Ambient noise: All-pervasive noise associated with a given environment.

Anechoic room: Room whose boundaries effectively absorb all incident sound over the frequency range of interest, thereby creating essentially free field conditions.

Audibility threshold: Sound pressure level, for a specified frequency, at which humans with normal hearing begin to respond.

Background noise: Ambient noise level above which signals must be presented or noise sources measured.

Decibel scale: Linear numbering scale used to define a logarithmic amplitude scale, thereby compressing a wide range of amplitude values to a small set of numbers.

Diffraction: Scattering of radiation at an object smaller than one wavelength and the subsequent interference of the scattered wavefronts.

Diffuse field: Sound field in which the sound pressure level is the same everywhere, and the flow of energy is equally probable in all directions.

Diffuse sound: Sound that is completely random in phase; sound that appears to have no single source.

Directivity factor: Ratio of the mean-square pressure (or intensity) on the axis of a transducer at a certain distance to the mean-square pressure (or intensity) that a spherical source radiating the same power would produce at that point.

Far field: Distribution of acoustic energy at a much greater distance from a source than the linear dimensions of the source itself. *See also* diffraction.

Free field: An environment in which there are no reflective surfaces within the frequency region of interest.

Hearing loss: An increase in the threshold of audibility due to disease, injury, age, or exposure to intense noise.

Hertz (Hz): Unit of frequency measurement, representing cycles per second.

Infrasound: Sound at frequencies below the audible range, that is, below about 16 Hz.

Isolation: Resistance to the transmission of sound by materials and structures.

Loudness: Subjective impression of the intensity of a sound.

Masking: Process by which the threshold of audibility of one sound is raised by the presence of another (masking) sound.

Near field: That part of a sound field, usually within about two wavelengths of a noise source, where there is no simple relationship between sound level and distance.

Noise emission level: dB(A) level measured at a specified distance and direction from a noise source, in an open environment, above a specified type of surface; generally follows the recommendation of a national or industry standard.

Noise reduction coefficient (NRC)**:** Arithmetic average of the sound absorption coefficients of a material at 250, 500, 1000, and 2000 Hz.

Phon: Loudness level of a sound, numerically equal to the sound pressure level of a 1-kHz free progressive wave, which is judged by reliable listeners to be as loud as the unknown sound.

Pink noise: Broadband noise whose energy content is inversely proportional to frequency (–3dB per octave or –10 dB per decade).

Power spectrum level: Level of the power in a band 1 Hz wide referred to a given reference power.

Reverberation: Persistence of sound in an enclosure after a sound source has been stopped. Reverberation time is the time (in seconds) required for sound pressure at a specific frequency to decay 60 dB after a sound source is stopped.

Root mean square (RMS)**:** The square root of the arithmetic average of a set of squared instantaneous values.

Sabine: Measure of sound absorption of a surface. One metric sabine is equivalent to 1 m^2 of perfectly absorptive surface.

Sound: Energy transmitted by pressure waves in air or other materials which is the objective cause of the sensation of hearing. Commonly called noise if it is unwanted.

Sound intensity: Rate of sound energy transmission per unit area in a specified direction.

Sound level: Level of sound measured with a sound level meter and one of its weighting networks. When A-weighting is used, the sound level is given in dB(A).

Sound level meter: An electronic instrument for measuring the RMS of sound in accordance with an accepted national or international standard.

Sound power: Total sound energy radiated by a source per unit time.

Sound power level: Fundamental measure of sound power, defined as:

$$L_w = 10\log_{10}\frac{P}{P_0},\ \text{dB}$$

where P is the RMS value of sound power in watts, and P_0 is 1 pW.

Sound pressure: Dynamic variation in atmospheric pressure. The pressure at a point in space minus the static pressure at that point.

Sound pressure level: Fundamental measure of sound pressure defined as:

$$L_p = 20\log_{10}\frac{P}{P_0},\ \text{dB}$$

where P is the RMS value (unless otherwise stated) of sound pressure in pascals, and P_0 is 1 μPa.

Sound transmission loss: Ratio of the sound energy emitted by an acoustical material or structure to the energy incident on the opposite side.

Standing wave: A periodic wave having a fixed distribution in space that is the result of interference of progressive waves of the same frequency and kind. Characterized by the existence of maximum and minimum amplitudes that are fixed in space.

Thermoacoustic efficiency: A value used to characterize the amount of combustion noise emitted from a flame. Defined as the ratio of the acoustical power emitted from the flame to the total heat release of the flame.

Ultrasound: Sound at frequencies above the audible range, that is, above about 20 kHz.

Wavelength: Distance measured perpendicular to the wavefront in the direction of propagation between two successive points in the wave, which are separated by one period. Equal to the ratio of the speed of sound in the medium to the fundamental frequency.

Weighting network: An electronic filter in a sound level meter that approximates, under defined conditions, the frequency response of the human ear. The A-weighting network is most commonly used.

White noise: Broadband noise having constant energy per unit of frequency.

REFERENCES

1. A.P.G. Peterson, *Handbook of Noise Measurement,* 9th ed., GenRad, Concord, MA, 1980.
2. W. Daiminger, K.R. Fritz, E. Schorer, and B. Stüber, *Ullman's Encyclopedia of Industrial Chemistry,* Vol. B7, VCH, Weinheim, 1995, 384-401.
3. A. Thumann and R.K. Miller, *Secrets of Noise Control,* Fairmont Press, 1974.
4. O.C. Leite, Predict flare noise and spectrum, *Hydrocarbon Processing*, 68, 55, 1988.
5. W. Bussman and J. White, Steam-Assisted Flare Testing, John Zink Co. Internal Report, September 1996.
6. A.A. Putnam, Combustion Noise in the Handheld Industry, Battelle, Columbus Laboratories.
7. L.L. Beranek and I.L. Ve'r, *Noise and Vibration Control Engineering*, John Wiley & Sons, New York, 1992.
8. H. Shen and C.K.W. Tam, Numerical simulation of the generation of axisymmetric mode jet screech tones, *AIAA Journal*, 36(10), 1801, 1998.
9. L.L. Beranek, *Noise and Vibration Control*, McGraw-Hill, New York, 1971.
10. Allied Witan Co., Noise Facts and Control, 1976.
11. W.R. Bussman and D. Knott, Unique concept for noise and radiation reduction in high-pressure flaring, OTC Conference, Houston, TX, 2000.
12. L.D. Frank and D.R. Dembicki, Lower plant noise with lagging, *Hydrocarbon Processing*, 71(8), 83-85, 1992.

BIBLIOGRAPHY

Alberta Energy and Utilities Board, Calgary, Alberta, 1998.

American Petroleum Institute, 50, 125-146, 1972.

R.S. Brief and R.G. Confer, Interpreting noise dosimeter results based on different noise standards, *Am. Indust. Hygiene J.*, 36(9), 677-682, 1975.

S.C. Crow and F.H. Champagne, Orderly structure in jet turbulence, *J. Fluid Mech.*, 48(3), 547-591, 1971.

A.H. Diserens, Personal noise dosimetry in refinery and chemical plants, *J. Occupational Med.*, 16(4), 255-257, 1974.

A. Gharabegian and J.E. Peat, Saudi petrochemical plant noise control, *J. Environ. Eng.*, 112(6), 1026-1040, 1986.

HFP Acoustical Consultants, Effect of flow parameters on flare stack generator noise, *Proc. Spring Environmental Noise Conf.: Innovations in Noise Control for the Energy Industry*, Alberta, Canada, April 19-22, 1998.

International Electrochemical Commission, IEC Standard, Publication 651, Sound Level Meters, 1979.

ISO 1683 (E), Acoustics–Preferred Reference Quantities for Acoustic Levels, 1983.

ISO 532 (E), Acoustics Method for Calculating Loudness Level, 1975.

ISO 1996-1 (E), Acoustics: Description and Measurement of Environmental Noise.

ISO 3744 (E), Acoustics: Determination of Sound Power Levels of Noise Sources.

Engineering Methods for Free Field Conditions over a Reflecting Plane, 1081.

ISO/DIS 8297, Acoustics: Determination of Sound Power Levels of Multi-source Industrial Plants for the Evaluation of Sound Pressure Levels in the Environment-Engineering Method, 1988.

ISO 9614-1 (E), Determination of Sound Power Levels of Noise Sources Using Sound Intensity. Part I: Measurement at Discrete Points; Part II: Measurement at Planned Points, 1993.

W.W. Lang, Ed., A commentary on noise dosimetry and standards, *Proc. Noise Congress-75*, Gaithersburg, MD, Sept. 15-17, 1975.

J.C. Maling, Jr., Ed., Start-up silencers for a petrochemical complex, *Proc. Int. Conf. Noise Control Eng.*, December 3-5, 1984.

A. Powell, On the noise emanating from a two dimensional jet above the critical pressure, *The Aeronautical Quarterly*, 4, 103, 1953.

A.A. Putnam, Combustion noise in the hand glass industry, *Tenth Annu. Symp. on the Reduction Cost in the Hand Operated Glass Plants*, 1979.

R. Reed, *Furnace Operations*, Gulf Publishing, Houston, TX, 1981.

H.S. Ribner, Perspectives on Jet Noise, *AIAA Journal*, 19(12), 1513, 1981.

J.P. Roberts, Ph.D. thesis, London University, 1971.

G. Seebold and A.S. Hersh, Control flare steam noise, *Hydrocarbon Processing*, 51, 140, 1971.

G.K. Selle, Steam-assisted flare eliminates environmental concerns of smoke and noise, *Hydrocarbon Processing*, 73(12), 77-78, 1994.

B.N. Shivashankara, W.C. Strahle, and J.C. Henkley, Combustion noise radiation by open turbulent flames, Paper 73-1025, AIAA Aero-Acoustics Conference, Seattle, WA, 1973.

J.F. Straitz, Improved flare design, *Hydrocarbon Processing*, 73(10), 61-66, 1994.

A. Thomas and G.T. Williams, Flame noise: sound emission from spark-ignited bubbles of combustion gas, *Proc. Roy. Soc.*, A294, 449, 1966.

E. Zwicker and H. Fastl, *Psychoacoustics-Facts and Models*, Springer Verlag, Berlin, 1990.

Chapter 8

Mathematical Modeling of Combustion Systems

Lawrence D. Berg, Wes Bussman, Jianhui Hong,
Michael Henneke, I-Ping Chung, and Joseph D. Smith

TABLE OF CONTENTS

8.1 Overview ... 252

8.2 Eduction Processes ... 252

  8.2.1 Steam Flare Eduction Modeling ... 254

  8.2.2 Eduction Processes in Pilots ... 256

  8.2.3 Eduction Processes in Premixed Burners ... 257

8.3 Idealized Chemical Reactors and Combustion Modeling ... 258

  8.3.1 Plug Flow Reactor ... 258

  8.3.2 Perfectly Stirred Reactor ... 259

8.4 Burner Pressure Drop ... 260

8.5 Flare Smokeless Operation ... 266

  8.5.1 Predicting Flame Smoking Tendencies ... 266

  8.5.2 Application to Steam Flares ... 268

  8.5.3 Modeling Air-Assisted Flares ... 270

  8.5.4 Summary ... 273

8.6 Oil Gun Capacities ... 273

  8.6.1 Summary of Two-Phase Flow Analytical Development ... 274

  8.6.2 Results ... 277

8.7 Oil Gun Development ... 277

8.8 Regenerative Thermal Oxidizer (RTO) Performance ... 280

  8.8.1 Introduction ... 280

  8.8.2 RTO Model Development ... 280

8.9 Conclusion ... 282

References ... 283

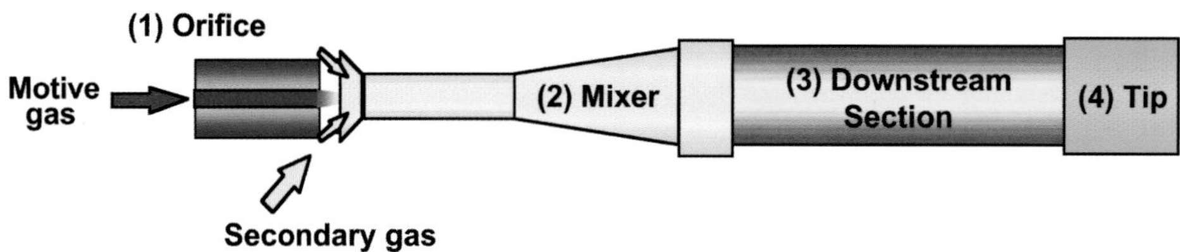

FIGURE 8.1 A typical eductor system.

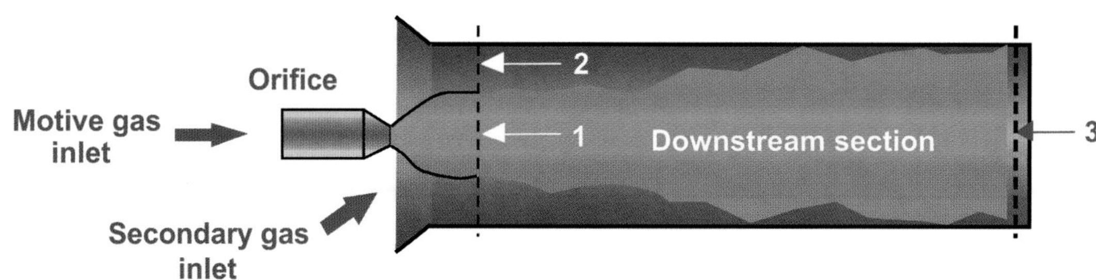

FIGURE 8.2 Example eductor system.

8.1 OVERVIEW

Over the course of time, the analytical modeling of combustion processes has become increasingly more important. This trend is a result of two simultaneous events: (1) the proliferation of powerful computers and (2) the development of a greater fundamental understanding of the underlying processes. The combination of these events has provided engineers with opportunities to model and understand physical processes in much greater detail than previously possible. This chapter highlights some of the applications developed and utilized at the John Zink Company. Some applications are straightforward extensions of previous techniques, while others are completely novel. The authors' purpose is to demonstrate how the practicing engineer can combine fundamental knowledge and computational methods to analyze combustion equipment and processes.

This chapter discusses the following:

- modeling of eductors
- modeling of chemical reactions
- modeling of burner pressure drop
- modeling of flare smokeless rates
- modeling of oil gun performance and improvement
- modeling of heat regenerator performance

As appropriate, technical literature will be referenced in order to provide the reader with additional details.

8.2 EDUCTION PROCESSES

Some of the first theoretical and experimental studies on eduction processes were performed in the early 1940s.[1] Since that time, a lot of work has been devoted to predicting and understanding the mechanisms governing the performance of various eductor systems. This information is important because these systems are widely used in industry for many applications.

A typical eductor system — sometimes called a venturi, inspirator, aspirator, ejector, or jet pump — is illustrated in Figure 8.1.

The objective of this device is to inspirate the surrounding (or secondary) gases into the mixer and then out through the tip. Mechanical energy is provided to the device by high-velocity gas (or motive gas) discharging from the orifice. An eductor system consists of four fundamental parts:

1. orifice
2. mixer
3. downstream section
4. tip

Each part plays a major role in the operating performance of the system, as follows. When the motive gas exits the orifice, it will entrain the surrounding air by creating a low-pressure zone in the area of the gas jet. The strength of this low-pressure zone depends on the energy of the gas jet and

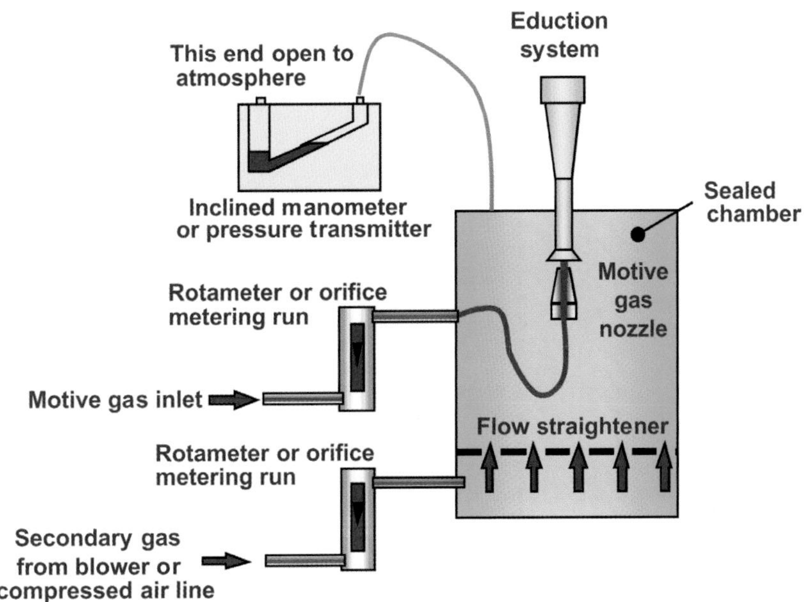

FIGURE 8.3 Experimental apparatus that has been successfully used to determine the flow coefficients and eduction performance of various flare and burner eduction processes.

the resistance of the overall system. The amount of resistance or energy lost through the eductor system governs its ability to entrain secondary gases. The four parameters that can influence the performance of an eductor system include the mass flow rate of the motive gas, pressure of the motive gas, pressure losses associated with the gas flowing through the system, and motive and secondary gas properties such as specific gravity and ratio of specific heat.

For illustrative purposes, consider the control volume (see Chapter 4.3.3 for a discussion of control volumes) analysis of an eductor system shown in Figure 8.2. The motive gas will expand fully somewhere downstream of the nozzle exit. The location of full expansion of the motive gas is designated as the inlet to the control volume. Location 1 represents the region where the motive gas and educted gas of location 2 have the same pressure. Location 2 represents the annular region of the secondary or educted gas within the eductor tube. Somewhere downstream of the orifice is the outlet from the control volume, represented as location 3. From control volume theory, one can write the conservation of mass for the process as:

$$\text{Mass}_1 + \text{Mass}_2 = \text{Mass}_3 \tag{8.1}$$

Similarly, the conservation of momentum can be written as:

$$\text{Momentum}_1 + \text{Momentum}_2 - \text{Momentum}_3 = \text{Losses} \tag{8.2}$$

and the conservation of energy can be written as:

$$\text{Energy}_1 + \text{Energy}_2 - \text{Energy}_3 = \text{Losses} \tag{8.3}$$

In addition to the conservation equations, the following constitutive equations and data are required:

- equation of state (see Chapter 4.2.4) to relate densities to the mass flow rates
- NASA nozzle performance data[2]
- saturated steam enthalpy and entropy

It is important that the losses (usually in the form of loss coefficients — see Chapters 4.3.2.2, 4.5.3, and 8.4 in this book) be accurately determined because they can have a significant influence on the eduction performance of the system. Typically, these losses must be measured experimentally. An experimental apparatus that has been successfully used to determine the loss coefficients and eduction performance of various flare and burner eduction processes is illustrated in Figure 8.3.

As shown in Figure 8.3, the primary gas nozzle is located inside a sealed chamber and the eduction system flows to the outside. A blower or compressed gas is used to deliver the secondary gas into the chamber. The mass flow rate of the secondary gas, entrained through the eduction system, is varied until the pressure inside the chamber is zero while the eduction system is in operation. The mass flow rate of the

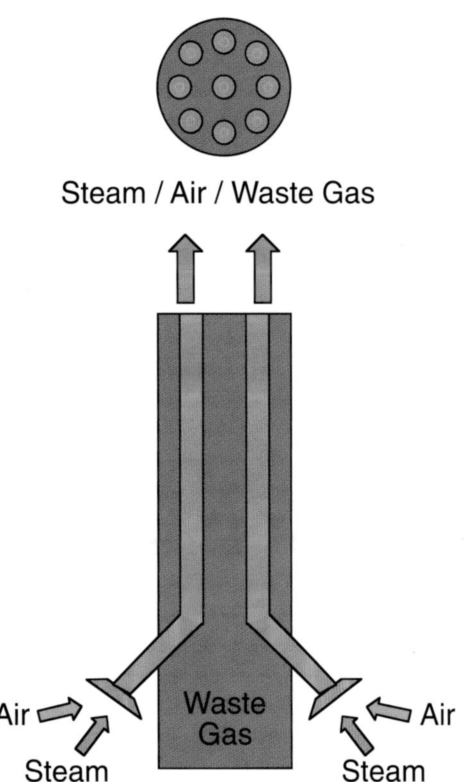

FIGURE 8.4 Steam flare with steam eductor.

FIGURE 8.5 Typical third-generation steam flare tube layout.

primary and secondary gas is measured using a flow measurement device such as a rotameter or orifice metering run. The pressure inside the chamber can be measured using an inclined manometer or pressure transducer.

Eduction processes are widely used in the flare and burner industry. Mathematical modeling of these processes, using the equations discussed above, has been used successfully to predict performance at the John Zink Company. Modeling examples of several flare and burner eduction processes are discussed and compared with experimental data.

8.2.1 Steam Flare Eduction Modeling

The ability to introduce air into the waste gas stream of a flare, prior to combustion, is important for reducing the thermal radiation and to improve the smokeless performance. Some steam-assisted flares rely on the motive energy from steam to inspirate air through several steam tubes. The steam-air mixture is then delivered to the flare tip in the central core of a waste gas stream. Figure 8.4 illustrates how this process is accomplished, and Figure 8.5 shows the tip outlet side of a large, steam-assisted flare tip equipped with eductor tubes. To estimate and optimize the radiation and smokeless performance of a steam-assisted flare, it is important to know the eduction performance of the steam tubes.

The eductor model previously discussed was adapted to help predict the air entrainment rate into a steam flare. A full-scale tube was also set up in a test rig similar to Figure 8.3 to obtain verification data. Figure 8.6 shows a sample of the data vs. model comparison. The vertical axis of the graph represents normalized air entrainment rates, and the horizontal axis represents normalized steam pressure. Experimental data vs. model predictions for sonic and supersonic nozzle air entrainment rates are shown for a typical steam-assisted flare eduction tube. As the steam pressure increases, the steam-to-air mass ratio decreases. The decrease in air eduction efficiency is due to the increase in pressure losses through the steam tube. In general, it can be shown that these losses increase with the gas velocity squared (see Eq. 8.17). The use of a supersonic nozzle partially mitigates this effect. As shown, at the maximum pressure, the increase in eduction performance is about 15%.

8.2.1.1 Analysis of Supersonic Eductor Performance

The experimental and modeling results just presented are somewhat surprising. Conventional wisdom would anticipate that a supersonic nozzle would inspirate more air than the incremental increase (~15%) observed. Steam at about 80 psig (a similar pressure to those reported) has an ideal expansion velocity of about Mach 2 (twice the speed of sound) and the velocity from a standard orifice exits at sonic velocity (see Chapter 4.4.2.2). For the approximate conditions at which many steam flares operate, it would seem that a supersonic nozzle would double the momentum (the supersonic nozzle has double the exit velocity of the sonic nozzle) and nearly double the entrainment rate would be anticipated.

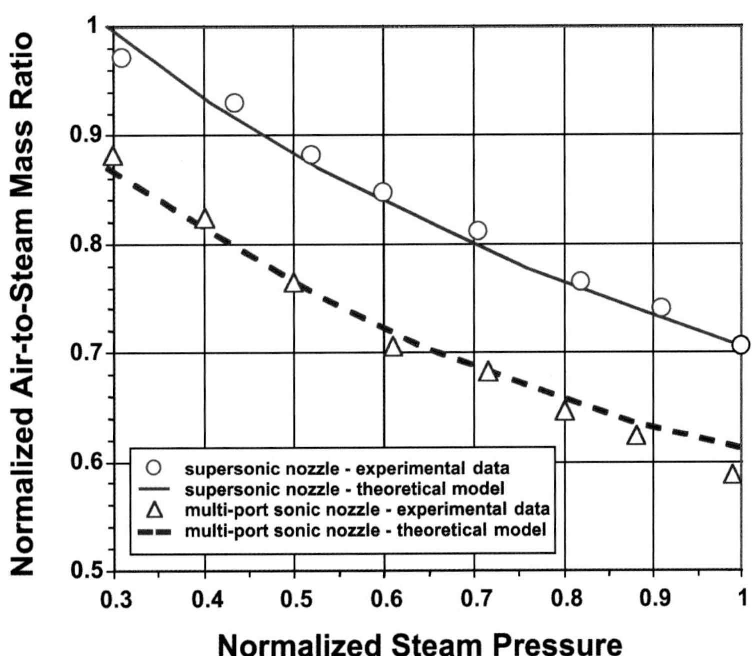

FIGURE 8.6 Normalized plot showing the sonic–supersonic eduction performance comparison in a single steam tube used in a typical steam-assisted flare.

The following analysis will show that for steam at 100 psig, an increase of only about 15% would be anticipated. It will also serve as a working example of how to apply control volumes and fundamental fluid mechanics (compressible flows) to analyze real equipment.

Consider Figure 8.2 again. Instead of the control volume inlet being set where the motive gas has the same pressure as the secondary gas, set the inlet at the *outlet of the nozzle* (see also McDermott and Henneke).[32] Applying the integral momentum balance yields the following:

$$G_j + G_e = G_m + \text{Losses} \tag{8.4}$$

where

G – Momentum flux normal to control surface = $\dot{m}v + PA$

Notice that the momentum equation now has to specifically include the pressure term, because the condition of equal pressure is no longer valid. Losses include viscous losses in the tube, entry losses, and exit losses due to non-plug flow velocity distributions. Since this section compares the driving force (i.e., the momentum flux of the motive gas G_j) from a converging–diverging (supersonic) nozzle to that from a converging (sonic) nozzle, losses will be neglected.

For a given inlet stagnation pressure P_o and an ambient pressure P_a, the maximum exit Mach number is achieved when the exit pressure P_e is equal to the ambient pressure P_a. Assuming isentropic flow in the supersonic nozzle, the maximum Mach number at the exit, M_e, is related to P_o and P_a by:[40]

$$\frac{P_o}{P_e} = \frac{P_o}{P_a} = \left(1 + \frac{\gamma - 1}{2} M_e^2\right)^{\gamma/(\gamma-1)} \tag{8.5}$$

where γ is the specific heat capacity ratio of the motive gas. Rearranging Eq. (8.5) gives the Mach number at the exit of a well-designed supersonic nozzle:

$$M_e = \sqrt{\frac{2}{\gamma - 1}\left[\left(\frac{P_o}{P_a}\right)^{(\gamma-1)/\gamma} - 1\right]} \tag{8.6}$$

The mass flow rate in a choked nozzle is:[40]

$$\dot{m} = A^* \frac{P_o}{\sqrt{T_o}} \sqrt{\frac{\gamma}{R}} \sqrt{\left(\frac{2}{\gamma + 1}\right)^{(\gamma+1)\,(\gamma-1)}} \tag{8.7}$$

The above equation is applicable to both sonic and supersonic nozzles. The velocity v at the exit of a nozzle is:

$$v = M_e c = M_e \sqrt{\gamma R T} \tag{8.8}$$

TABLE 8.1 Exit Mach Number, Area Ratio, Driving Force Ratio, and Driving Force Percentage Increase for Various Gas Pressures ($\gamma = 1.33$ and $P_a = 14.3$ psia[a])

P_o (psig)	M	A_e/A^*	Ratio	Increase
15	1.096	1.0076	1.003	0.3%
20	1.221	1.0390	1.014	1.4%
30	1.410	1.1287	1.040	4.0%
40	1.550	1.2288	1.063	6.3%
50	1.663	1.3306	1.083	8.3%
60	1.756	1.4315	1.100	9.2%
70	1.837	1.5308	1.114	11.4%
80	1.907	1.6281	1.126	12.6%
90	1.969	1.7234	1.137	13.7%
100	2.026	1.8167	1.147	14.7%

[a] Local ambient pressure of Tulsa, Oklahoma.

For a nozzle with throat diameter A^*, the driving force G_j is:

$$P_eA_e + \dot{m}v_e = P_o\left(1+\frac{\gamma-1}{2}M^2\right)^{-\gamma/(\gamma-1)}A_e + A^*\frac{P_o}{\sqrt{T_o}}\sqrt{\frac{\gamma}{R}}\sqrt{\left(\frac{2}{\gamma+1}\right)^{(\gamma+1)\ (\gamma-1)}}M_e\sqrt{\gamma RT}$$

$$= P_oA^*\left[\left(1+\frac{\gamma-1}{2}M_e^2\right)^{-\gamma/(\gamma-1)}\frac{A_e}{A^*} + \frac{\gamma\sqrt{T}}{\sqrt{T_o}}\sqrt{\left(\frac{2}{\gamma+1}\right)^{(\gamma+1)\ (\gamma-1)}}M_e\right]$$

$$= P_oA^*\left[\left(1+\frac{\gamma-1}{2}M_e^2\right)^{-\gamma/(\gamma-1)}\frac{A_e}{A^*} + \gamma\left(1+\frac{\gamma-1}{2}M_e^2\right)^{-1/2}\sqrt{\left(\frac{2}{\gamma+1}\right)^{(\gamma+1)/(\gamma-1)}}M_e\right] \quad (8.9)$$

For a choked sonic nozzle, the exit Mach number $M_e = 1$, and the exit area is the throat area; therefore, the above equation becomes

$$\left(P_eA_e + \dot{m}v_e\right)_s = P_oA^*\left[\left(1+\frac{\gamma-1}{2}\right)^{-\gamma/(\gamma-1)} + \gamma\left(1+\frac{\gamma-1}{2}\right)^{-1/2}\sqrt{\left(\frac{2}{\gamma+1}\right)^{(\gamma+1)/(\gamma-1)}}\right] \quad (8.10)$$

For an isentropic flow, the exit area of a supersonic nozzle needs to satisfy:

$$\frac{A_e}{A^*} = \frac{1}{M_e}\left[\left(\frac{2}{\gamma+1}\right)\left(1+\frac{\gamma-1}{2}M_e^2\right)\right]^{(\gamma+1)/[2(\gamma-1)]} \quad (8.11)$$

From Eqs. (8.9) and (8.10), the driving force ratio $(G_j)_{ss}/(G_j)_s$ is:

$$\frac{\left(P_eA_e + \dot{m}v_e\right)_{ss}}{\left(P_eA_e + \dot{m}v_e\right)_s} = \frac{\left[\left(1+\frac{\gamma-1}{2}M_e^2\right)^{-\gamma/(\gamma-1)}\frac{A_e}{A^*} + \gamma\left(1+\frac{\gamma-1}{2}M_e^2\right)^{-1/2}\sqrt{\left(\frac{2}{\gamma+1}\right)^{(\gamma+1)/(\gamma-1)}}M_e\right]}{\left[\left(1+\frac{\gamma-1}{2}\right)^{-\gamma/(\gamma-1)} + \gamma\left(1+\frac{\gamma-1}{2}\right)^{-1/2}\sqrt{\left(\frac{2}{\gamma+1}\right)^{(\gamma+1)/(\gamma-1)}}\right]} \quad (8.12)$$

The following procedure can be used to estimate the increase of driving force from the use of a supersonic nozzle compared to a sonic nozzle:

1. Calculate exit Mach number M_e from Eq. (8.6), assuming isentropic flow in the supersonic nozzle.
2. Calculate the ratio of the exit area A_e to the throat area from Eq. (8.11) (note: the nozzle needs to be designed to have the right exit area and proper converging–diverging profile).
3. Calculate the driving force ratio from Eq. (8.12).

The driving force ratios calculated from the above procedure are tabulated in Table 8.1 for a selected value of γ and ambient pressure P_a.

It can be seen from Table 8.1 that a Mach number of over 2 corresponds, at best, to a 14.7 percent increase of driving force. The increase in the $\dot{m}v$ term is largely offset by the decrease in the *PA* term. As presented previously, experiments conducted at the John Zink Company (Tulsa, OK) showed that the increase of air entrainment ratios roughly (within experimental errors) agree with the percentage increase of driving force listed in Table 8.1.

8.2.2 Eduction Processes in Pilots

The stability of the flame on a pilot is very critical. If the premixed air-to-fuel ratio is not within an appropriate range, then the flame could be easily extinguished.

Figure 8.7 shows the normalized eduction performance of a flare pilot operating on Tulsa Natural Gas (TNG) with two

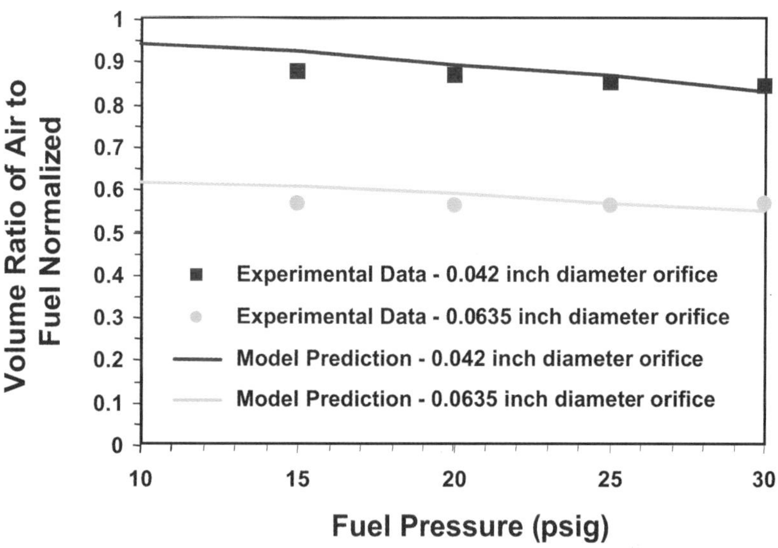

FIGURE 8.7 Normalized eduction performance of a flare pilot operating on Tulsa natural gas with two different motive gas orifice diameters.

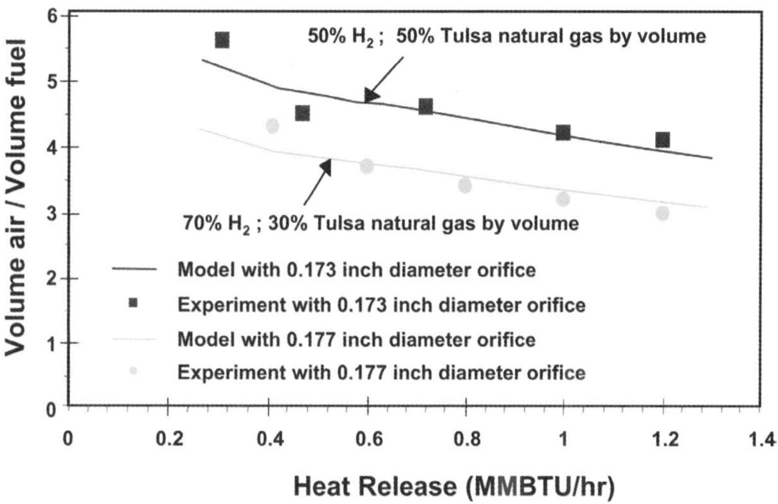

FIGURE 8.8 Experimental and theoretical results of the eduction performance of a particular radiant wall burner firing with two different orifice sizes and fuel gas compositions.

different motive gas orifice diameters. These results compare experimental data obtained from the test apparatus and theoretical modeling results. Here, the volumetric air-to-fuel ratio remains fairly constant throughout the pressure range tested. A pilot that operates with a constant air-to-fuel ratio is desirable because it offers better stability over a range of operating pressures. As the motive gas orifice diameter increases, the volumetric air-to-fuel ratio decreases for a given motive gas pressure. This demonstrates that slight modifications in the motive gas orifice diameter can significantly affect the eduction performance. In addition, it highlights the need for a thorough understanding of the fundamental fluid mechanics in the design of eductor-driven pilots.

8.2.3 Eduction Processes in Premixed Burners

Figure 8.8 is a plot showing experimental and theoretical results of the eduction performance of a particular radiant wall burner firing with two different orifice sizes and fuel gas compositions. For this particular burner, the volumetric air-to-fuel ratio remains fairly constant over the range of heat releases tested. The slope of the operating curve for the volumetric air-to-fuel ratio, however, can change significantly, depending on the fuel composition, mixer and tip design, and ambient conditions.

Eductor models have been developed and successfully applied to different types of combustion equipment at the John Zink Company. The experimental and theoretical methods described inevitably led to both greater insight into the operation of the equipment and better performance. The ability to measure and predict the eduction performance is crucial in the flare and burner industry. Without this ability, it is difficult — if not impossible — to consistently estimate burner and flare performance.

8.3 IDEALIZED CHEMICAL REACTORS AND COMBUSTION MODELING

Over the last 20 years, utilizing laminar flames and plug flow reactors, researchers have made substantial progress in detailing the chemical reaction pathways for combustion processes. Some of this progress has been summarized in Bartok (1991).[3] The comprehensive understanding of combustion chemistry has proven to be a daunting task. A simple methane–oxygen–nitrogen system requires on the order of 50 species and 300 chemical reactions. In general, a system of 50 species requires simultaneous solution of 50 ordinary differential equations. During the 1980s, a general-purpose FORTRAN package for calculation and standardized storage of chemical kinetic data, called CHEMKIN, was developed at Sandia National Laboratory (Kee et al.[4–6]).

For practical combustion equipment, reaction chemistry is closely coupled to the turbulent mixing occurring in the flame. Unfortunately, to effectively manipulate the large number of chemical reaction equations and constraints, it is very difficult to include accurate fluid mechanics. To circumvent this difficulty, CHEMKIN assumes extremely simplified fluid flows (perfectly stirred or plug flow), coupled to the detailed combustion reaction kinetics. This chapter summarizes some of these classic "reactor" assumptions, followed by subsequent utilization of these simplifications to assist in the performance analysis of real equipment. For additional details of these concepts, the interested reader is directed to Turns.[7]

8.3.1 Plug Flow Reactor

A true plug flow reactor is characterized by one-dimensional, diffusion-free flow. A fluid enters the reactor at a known temperature, pressure, and composition and flows axially through the reactor from inlet to outlet. Chemical reactions are allowed to take place, but it is assumed that fluid advection is the only transport process occurring. This also means that the mixing processes and recirculations are ignored. A plug flow model makes a number of assumptions. It is important that the user understand the limitations imposed by these assumptions.

8.3.1.1 Assumptions

- *The kinetic energy of the fluid flow is neglected.* This assumption is normally quite reasonable for burners and flares because in most gas flows, changes in the thermodynamic enthalpy due to chemical reactions and heat transfer processes are much larger than changes in the kinetic energy of the fluid flow.
- *Pressure is assumed constant.* Again, this is normally quite reasonable. In flows with moderate Mach numbers (e.g., 0.3 or above), this assumption becomes erroneous
- *No concentration gradients are present normal to the flow.*

8.3.1.2 Governing Equations

The mathematical description of the plug flow reactor requires solving the conservation of energy and the conservation of species equations. Because the flow is one-dimensional and steady-state, conservation of mass is trivial and simply requires that the mass flux is constant. Energy conservation requires:

$$\begin{aligned}&(\text{Energy change})\\&= (\text{Heat release due to chemical reactions})\\&\quad + (\text{Heat transferred into system})\end{aligned} \tag{8.13}$$

The rate of production or destruction of a species can be computed by summing its production or destruction rate in all the chemical reactions in which it participates. This overall rate is a function of all 300+ rate equations. The conservation statement for chemical species is:

$$\begin{aligned}&(\text{Change of species mass fraction})\\&= (\text{Overall reaction rate})\end{aligned} \tag{8.14}$$

If there are N chemical species, then N–1 ordinary differential equations (with independent variable x = *distance along reactor axis*) must be solved given an inlet condition at x = 0. These equations are frequently stiff. A system of ordinary differential equations is called stiff when the eigenvalues of the Jacobian matrix span several orders of magnitude. Press[8] gives a good introduction to the solution of these types of equations.

8.3.2 Perfectly Stirred Reactor

The perfectly stirred reactor (PSR), sometimes called a continuously stirred tank reactor, is a vessel of spatially uniform composition and temperature, and was first described by Longwell.[9] Experimentalists frequently approximate a PSR using high-velocity jets to introduce the unreacted mixture into a reaction vessel. In reality, since the entering mixture has a different composition than the mixture in the tank, the notion of a spatially uniform composition is somewhat idealistic. However, this setup closely approximates a PSR, which approximates the behavior of many real systems where fluid mixing rates are much larger than chemical reaction rates.

The PSR is characterized by either a volume or a residence time. These are related by:

$$\text{Residence time} = (\text{Volume})(\text{Density})/(\text{Mass flow rate})$$

The assumptions made in formulating the PSR model are numerous and impractical to list here, but, on a practical level, are the same as used for the plug flow reactor (PFR). The most important assumption is that the mixing rate is much faster than the reaction rate.

8.3.2.1 Governing Equations

The PSR model, like the PFR model, requires that conservation of energy and species be satisfied. Conservation of mass is satisfied by noting that the PSR has one inlet and one outlet, and these have the same mass flow rates.

The energy equation for the perfectly stirred reactor is:

$$(\text{Energy out}) - (\text{Energy in}) = (\text{Energy exchanged with surroundings}) \tag{8.15}$$

Similarly, the species conservation requirement is:

$$(\text{Species out}) - (\text{Species in}) = (\text{Species produced}) \tag{8.16}$$

The equations governing the PSR are nonlinear algebraic equations. When PSR calculations are made using detailed kinetics, the stiffness problem must again be considered. The PSR model supplied with the CHEMKIN[4,33] package uses a hybrid Newton/time-stepping algorithm to solve the set of nonlinear algebraic equations. While the Newton iteration is efficient, allowing the solution error to be reduced very quickly, it is not very robust and requires an initial solution estimate that lies within its domain of convergence. For readers who might consider using the CHEMKIN PSR model, the following numerical scheme is recommended.

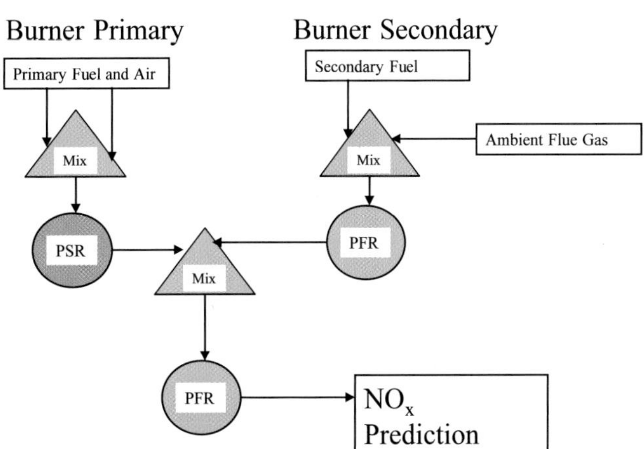

FIGURE 8.9 Simplified reactor modeling of a staged fuel burner.

Instead of Newton iteration, use only time-stepping to satisfy the governing equations. To optimize the solution algorithm, an adaptive time-step size to control discretization errors as outlined in Press[8] is recommended. The resulting algorithm, while not as efficient in some cases as the CHEMKIN PSR model, is more robust and rarely requires any tuning by the user.

8.3.3 Systems of Reactors

The above idealized reactors, either PSRs or PFRs, are rarely ever encountered in practical combustion systems. However, starting with Beer[10] and more recently Lutz et al.,[11,12] it has been shown that appropriately arranged systems of reactors can be utilized to simulate combustion processes. The basic concept is that a combustion system may be modeled as a series of parallel sequences of PSRs, PFRs and mixing regions. A mixing region provides for interaction with the environment by allowing additional material to be added to the composition and "freezing" the reactions during the process. Figure 8.9 schematically represents how this concept can be implemented for a simple staged fuel burner. Similar flow sequences have also been made for a radiant wall burner (shown in Figure 8.10) and NOx reduction thermal oxidizers (or NO_xIDIZERS©, shown in Figure 8.11). Sample results are shown in Figures 8.12 and 8.13. In Figure 8.12, trends from a PSR are compared to NOx from a premixed burner; and in Figure 8.13, predicted vs. actual

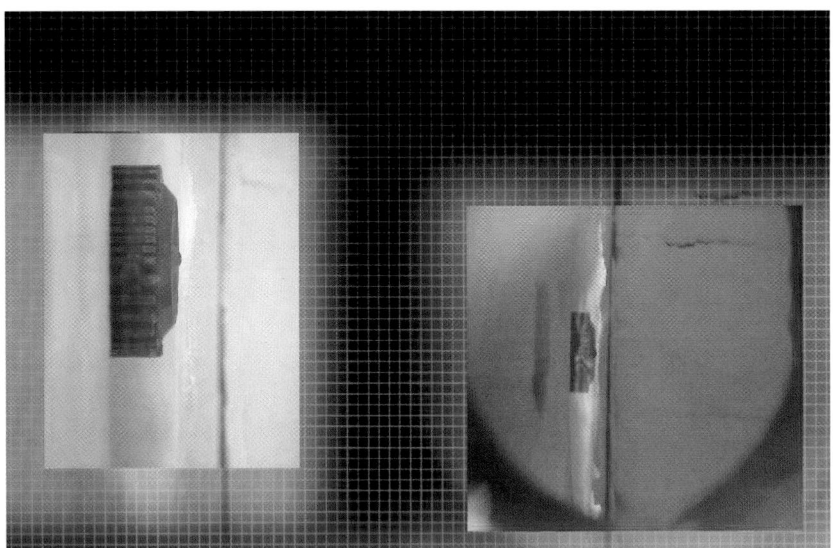

FIGURE 8.10 Picture of a radiant wall burner.

FIGURE 8.11 Picture of a thermal oxidizer.

NOx for NOx reduction thermal oxidizers is shown. Although the method is simplistic, trends are correctly predicted for the burner and excellent predictions for NOx reductions are possible. Because of the possibility of improved, fundamentally based performance predictions, improvement of this technique is an active research topic for many academic and industrial workers.

8.4 BURNER PRESSURE DROP

Figure 8.14 shows a typical capacity curve plot that many burner manufacturers use for sizing burners. Capacity curves describe the airside pressure drop through various burner sizes at different heat releases. The curves shown in this particular example are based on burners operating in the natural-draft mode with 15 percent excess air in the furnace at an

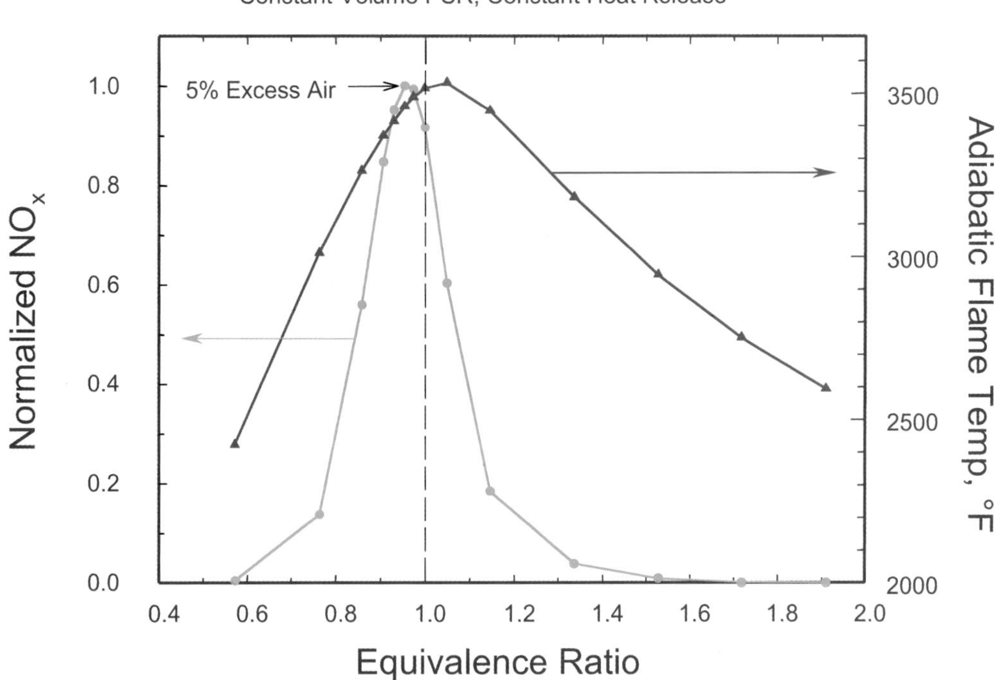

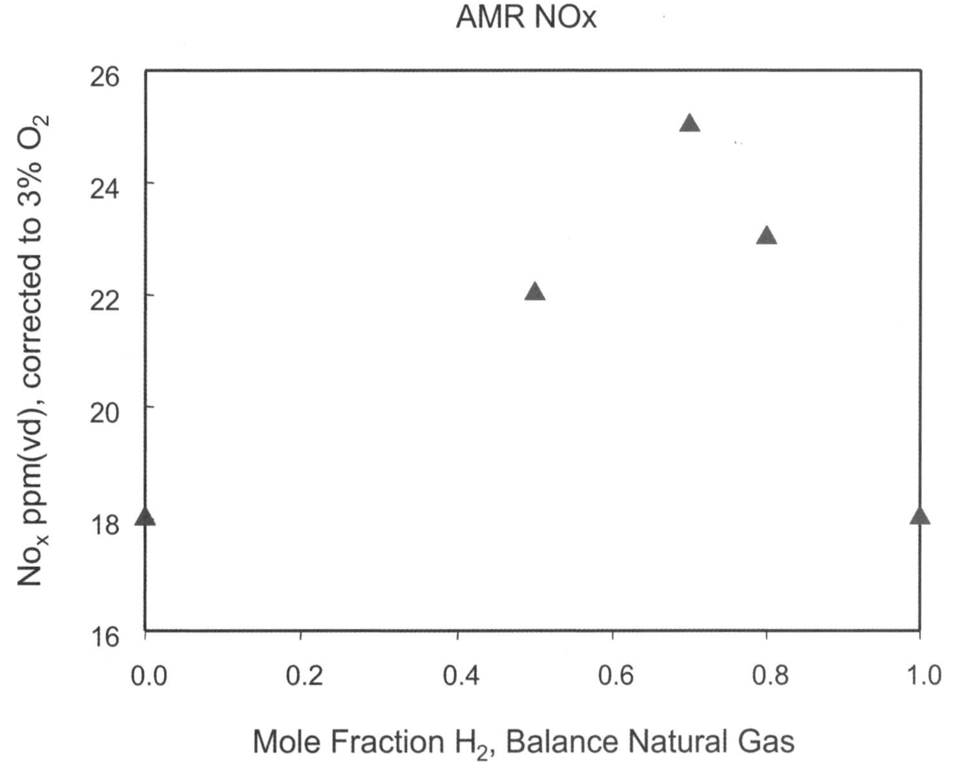

FIGURE 8.12 Sample results of simplified modeling.

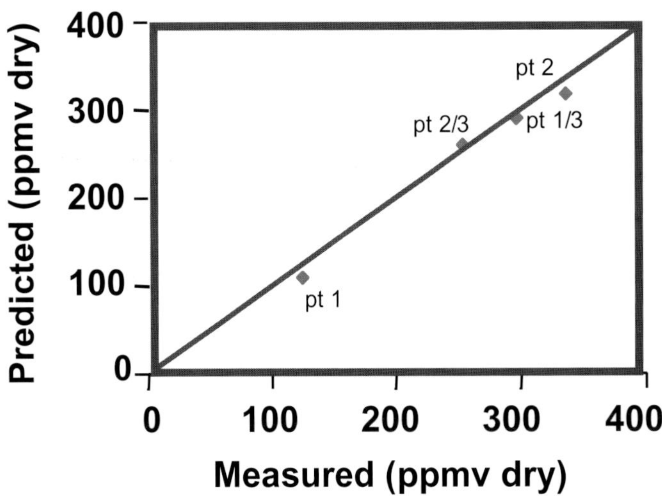

FIGURE 8.13 Sample results of simplified modeling for a thermal oxidizer.

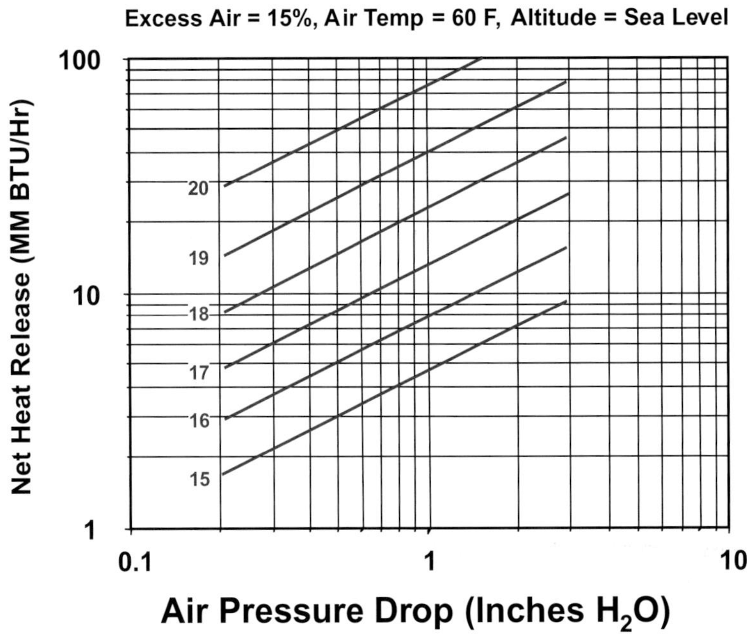

FIGURE 8.14 Capacity curves that many burner manufactures use for sizing burners.

atmospheric temperature and pressure of 60°F (16°C) and 14.7 psia (1 Bar), respectively.

When burners operate at different ambient conditions and excess air levels, the airside pressure drop obtained from the capacity curves must be corrected. The equation used to correct for the airside pressure drop can be derived as follows. The airside pressure drop through a burner is proportional to the velocity pressure of the air and can be written as (see Chapter 4.5.3):

$$\Delta P \propto \rho V^2 \tag{8.17}$$

where ΔP is the pressure drop through the burner, ρ is the density of the combustion air, and V is the mean velocity of the air at a particular location in the burner. The density of the combustion air can be related to the combustion air temperature, T, and atmospheric pressure, P, using the ideal gas law (Chapter 4.2.4):

$$\rho \propto \frac{P}{T} \tag{8.18}$$

The velocity, V, of the air through the burner is proportional to the mass flow of air going through the burner and the density. This can be written as:

$$V \propto \frac{\dot{m}}{\rho} \propto \frac{(100 + EA)}{\rho} \tag{8.19}$$

where $\dot{m}$ represents the mass flow and EA represents the percent excess air in the furnace. Substituting Eqs. (8.18) and (8.19) into Eq. (8.17) gives:

$$\Delta P \propto (100 + EA)^2 \times \frac{T}{P} \tag{8.20}$$

Equation (8.20) can be used to write the following equation to correct for the airside pressure drop at actual firing conditions:

$$\Delta P_{Actual} = \Delta P_{CC} \times \left(\frac{100 + EA_{Actual}}{100 + EA_{CC}} \right)^2 \times \left(\frac{T_{Actual}}{T_{CC}} \right) \times \left(\frac{P_{CC}}{P_{Actual}} \right) \tag{8.21}$$

where the subscript *Actual* represents the actual firing conditions and CC represents the variables from the capacity curves (i.e., T_{CC} = 460 + 60, P_{CC} = 14.7 psia, EA_{CC} = 15). Notice that as the temperature of the combustion air increases, the airside pressure drop through the burner also increases. This occurs because increasing the temperature reduces the density of the combustion air. A reduction in the density of the combustion air requires a higher volumetric air flow rate through the burner, which results in a pressure drop increase. Similarly, if the atmospheric pressure is reduced, the combustion air density is reduced and, hence, the airside pressure drop is increased.

As an example, consider the pressure drop through a size 15 burner with a 3 MMBtu/h heat release with 12 percent excess air, operating at a combustion air temperature of 100°F and an atmospheric pressure of 14.0 psia.

From the capacity curves (see Figure 8.14), for a heat release of 3 MMBtu/h at standard conditions, the pressure drop will be approximately 0.5 in. water column. This will be the pressure drop if the burner is operating at 15 percent excess air with the combustion air temperature at 60°F and an atmospheric pressure of 14.7 psia. To correct for the actual firing conditions we use Eq. (8.21) to obtain:

$$\Delta P_1 = 0.5 \times \left(\frac{100 + 12}{100 + 15} \right)^2 \times \left(\frac{460 + 100}{460 + 60} \right) \times \left(\frac{14.7}{14.0} \right) = 0.54$$

Although the percent excess air is reduced from 15 to 12 percent, the pressure drop through the burner has increased due to the reduced density of the combustion air.

Capacity curves are convenient for helping engineers size burners to be used for a specific burner design operating under certain conditions. However, capacity curves may not provide accurate estimates of airside pressure drop if a burner design or operation is modified. For example, several variables that can affect the airside pressure drop include: (1) fuel and atomizing gas tip drill angles, (2) fuel and atomizing gas tip position, (3) burner tolerances, (4) fuel and atomizing gas pressure and temperature, (5) flue gas recirculation, and (6) fuel splits in the primary and secondary combustion zones.

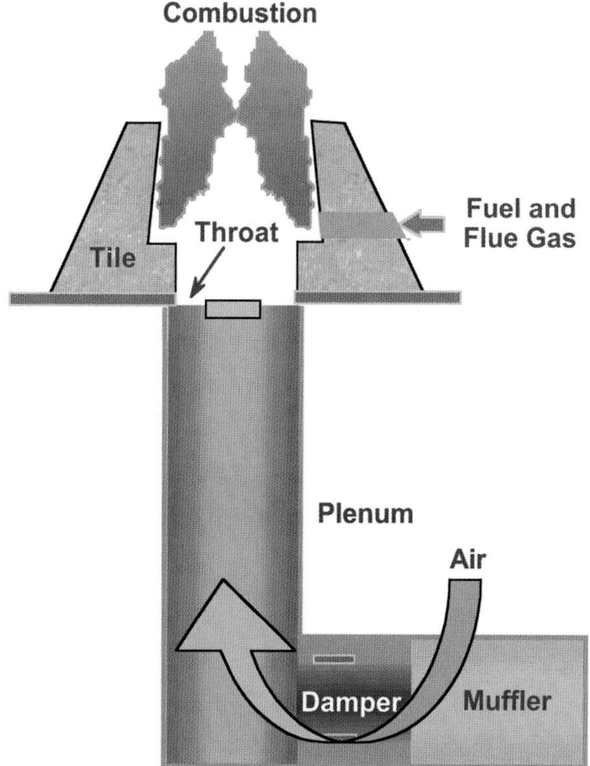

FIGURE 8.15 Burner designs typically consist of a muffler, damper, plenum, throat, and tile section.

FIGURE 8.16 Cold flow furnace consists of a test chamber (8 × 8 × 8 ft) supported on legs approximately 7 ft above the ground.

A method useful for capturing the possible variations of burner design is burner semi-empirical modeling, or just semi-empirical modeling. Burner designs typically consist of a muffler, damper, plenum, throat, and tile section, as illustrated in Figure 8.15. The strategy for determining the airside pressure drop, using the semi-empirical modeling technique, is to relate the airside pressure drop to the velocity pressure of the air going through each section of the burner. This strategy can usually be applied to all sections of a burner except the tile section. In the tile section, other factors can affect the airside pressure drop, depending on the design of the burner. For example, some burner designs use the pressure energy of the fuel to aspirate flue gas into the tile section where it mixes with the air and burns. If combustion occurs within the tile section, it can significantly increase the overall pressure drop through a burner. Other burner designs have fuel nozzles positioned inside the tile section that can help aspirate the air through the burner but may tend to either increase or reduce the pressure drop depending on the nozzle position and orientation. The advantage of using semi-empirical modeling is that it can take into account the effect that all of these variables have on the airside pressure drop.

John Zink engineers generally use a cold flow test rig specifically designed to study the airside pressure drop through various burner designs. The cold flow test rig consists of a test chamber on support legs located above the ground, as shown in the photograph in Figure 8.16. The test chamber is designed so that various sizes and types of burners can be mounted to the bottom. With this apparatus, burners can be tested in either the forced- or natural-draft mode. In the natural-draft mode, air flows through the burner and out of the test box. At the top, the test box transitions to a duct with flow straighteners, an orifice metering run, and dampers for controlling the air flow rate. The flow straighteners provide a uniform flow distribution of the air before it enters the orifice metering run. The data obtained from the cold flow test apparatus provide the necessary details to develop semi-empirical models used for estimating the airside pressure drop through burners.

A discussion of the methodology employed to develop the necessary information follows. For this discussion, Figure 8.15 will be utilized as the typical process burner. Details will vary from burner to burner, but the method will stay the same.

Starting with the muffler, as combustion air accelerates from a zero velocity to a given velocity at the muffler inlet, energy is lost. Energy is also lost due to the *vena contracta* created as the air enters the muffler. As the air flows through the *vena contracta*, it decelerates and loses additional energy. The acceleration and deceleration of the air as it moves through the muffler inlet has a pressure loss associated with it that can be related to the mean velocity pressure (see Chapter 4.5.3) as

$$\Delta P_{\substack{muffler \\ inlet}} = k_{\substack{muffler \\ inlet}} \frac{\rho_{air} V^2_{\substack{muffler \\ inlet}}}{2g_c} \tag{8.22}$$

where $\Delta P_{muffler\ inlet}$ = Pressure drop associated with the combustion air entering the muffler
ρ_{air} = Combustion air density
$V_{muffler\ inlet}$ = Mean velocity of the combustion air at the muffler inlet
$k_{muffler\ inlet}$ = Loss coefficient into the muffler
g_c = gravitational constant

The loss coefficient will vary depending on the muffler inlet design. For example, a square-edged inlet will have a loss coefficient equal to approximately 0.5. However, the loss coefficient can be as small as 0.02 if the inlet is well-rounded. For more information on loss coefficients through various fittings, see Chapter 4.4.5.3, the ASHRAE Handbook,[13] or Idelchik.[14] Independent of the reference, experimental validation of the loss coefficient will be accomplished in the cold flow test rig.

As the combustion air flows past the muffler inlet, it takes a sudden 90° turn through the muffler elbow. The pressure loss associated with the flow through an elbow can also be related to the mean velocity at the muffler inlet similar to the equation described in Eq. (8.17). The loss coefficient through the elbow is also available in the literature, with values shown as flow through mitered elbows provided in Table 8.2.[13] The values can also be determined/validated experimentally with the cold flow test rig, or determined using computational fluid dynamics (CFD, see Chapter 9).

Notice that the loss coefficient through a mitered elbow is a function of the Reynolds number. This suggests that the loss coefficient through the muffler can change with burner turndown conditions or draft levels. Capacity curves, as discussed earlier, assume a constant loss coefficient throughout the burner, regardless of turndown conditions or draft levels. That Reynolds dependence may be a hidden source of error for capacity curves is something to be aware of when sizing combustion equipment.

At this time, it is important to note a couple of points. First, as the combustion air flows through each burner component, the density will decrease due to the reduction in pressure. As discussed earlier, a reduction in the combustion air density will increase the pressure drop. The pressure drop through a burner system typically varies between 0.2 to 0.7 in. water column, which corresponds to an air density variation of about only 0.1 percent. This variation in density is not significant enough to be considered in the pressure drop calculations. However, if draft levels are high, the density variation through the burner should be considered. Second, as the combustion air approaches each section of the burner, the air velocity profile may not be fully developed. Most sources provide loss coefficients based on the assumption of a fully developed velocity profile of the gas upstream of the fitting. For practical combustion equipment, uniform velocity distributions may not exist, causing the loss coefficient reported in the literature to be lower than reality. When using literature loss coefficients, the reader should be aware of this possible difference and, to the maximum extent possible, utilize experimental validation for individual burner components.

In addition to the major components, dampers add possible pressure drop to a burner. Burners are typically sized with the damper blades positioned fully open because one would like to take as much pressure drop across the burner throat as possible to achieve good mixing between the air and fuel. With the damper blades positioned fully open, the pressure drop through the damper can usually be neglected. The pressure drop across the damper blades for various damper settings can be determined using the same technique as discussed above. The loss coefficient for each damper setting would need to be determined experimentally or by using CFD and based on the approach velocity of the combustion air in the damper section.

TABLE 8.2 Values for a 90° Mitered Elbow

	C'_0					
	W_1/W_0					
H_0/W_0	**0.6**	**0.8**	**1.2**	**1.4**	**1.6**	**2**
0.25	1.8	1.4	1.10	1.10	1.10	1.10
1	1.7	1.4	1.10	0.95	0.90	0.84
4	1.5	1.1	0.81	0.76	0.72	0.66
infinite	1.5	1.1	0.69	0.63	0.06	0.55

loss coefficient = $C'_0 Re_c$

where

C'_0 = values from table above
Re_c = Reynolds number correction factor (listed below)
Re = Reynolds number at inlet to elbow = $V \times D / \nu$
V = mean velocity into elbow
D = hydraulic diameter = $2 \times H_0 \times W_0/(H_0 + W_0)$
ν = kinematic viscosity

$Re \times 10^{-4}$	**1**	**2**	**3**	**4**	**6**	**8**	**10**	**>14**
Re_c	1.40	1.26	1.19	1.14	1.09	1.06	1.04	1.0

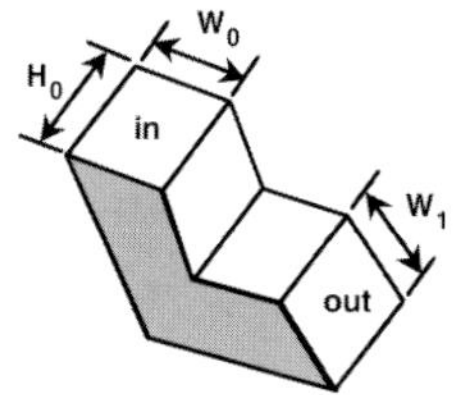

As the air enters into the plenum section, it turns 90°. The loss coefficient can be determined experimentally, using CFD modeling, or approximated using the loss coefficients given in the ASHRAE Data Handbook for a 90° mitered elbow. Just downstream of the elbow, the combustion air enters into the throat section of the burner.

The pressure loss through this section can be approximated as a sudden contraction. The loss coefficient for a sudden contraction depends on the design of the entrance and the throat-to-plenum area ratio. Table 4.4 shows typical loss coefficient information, generated from experimental data with a fully developed approach velocity profile. Again, due to the possibility of nonuniform velocity distributions, experimental validation is preferred.

When the air mixes with the fuel and reacts inside the tile section, the products of combustion expand to several times the volume of the original air/fuel mixture. The rapid increase in volume flow rate can have a significant effect on the total pressure drop through the burner. The John Zink engineers have correlated the pressure drop through the tile section based on the expansion and density of the products of combustion. These correlations are developed from both hot and cold flow furnace data and CFD analysis.

Fuel nozzles located around the tile section can either increase or decrease the overall pressure drop through the burner, depending of the location, orientation, and size of the fuel ports. Cold flow testing is required to provide the information necessary to correlate the effects of burner pressure drop for various nozzle configurations.

Semi-empirical modeling does require an attention to detail, but, as it turns out, it only requires about the same amount of time to accomplish as the capacity curve method. However, it has several advantages over the traditional method of using capacity curves. These models can (1) provide a consistent way of determining airside pressure drop for design engineers, (2) give better insight into what variables affect airside pressure drop, and (3) be used as a tool to help improve future burner designs. As a first approximation, capacity curves are an excellent engineering tool and are easily generated from semi-empirical models. Semi-empirical modeling, however, takes burner capacity and sizing to the next level and provides the burner design engineer unprecedented flexibility in meeting the needs of a diverse customer base.

8.5 FLARE SMOKELESS OPERATION

8.5.1 Predicting Flame Smoking Tendencies

Prior to 1947, venting of unburned hydrocarbons to the atmosphere was standard industry practice. After 1947, regulations required hydrocarbons to be burned (or "flared") due to serious health and safety hazards. Initially, flares burned the hydrocarbon waste gas stream directly at the vent exit. This method of flaring, however, often produced large clouds of black smoke that could be seen from miles away. In 1952, the John Zink Company patented and built the first smokeless flare (see Chapter 20). This flare eliminated smoke by injecting steam into the waste gas stream. Several years later, air flares and high-pressure flares were also developed and introduced into the industry.

The smokeless operating capacity of these early flares represented an improvement over a continuously smoking flare. Historically, the smokeless performance of a flare was estimated by vendor performance testing, field tests, or "rules of thumb" based on operating experience.

As smokeless flare capacity increased, vendor performance testing became impractical and performance predictions relied more heavily upon field data. Unfortunately, field experience was unreliable because flares were typically not monitored.

More recently, the environmental and regulatory agencies have required a minimum smokeless rate as part of plant operating permits. Customers have also begun installing flow meters and video cameras to monitor flare performance. This additional focus has resulted in better field data to assess actual flare performance. In many instances, the initial "rules-of-thumb" or other estimation methods have failed to accurately predict smokeless operation. Figure 8.17 highlights some of the difficulties of predicting flare performance. This shows that for the same flare, with the same gas, at nearly the same operating conditions, one flare smokes while the other does not.

To address previous shortcomings, a more fundamental technique for determining flare smokeless performance for commercial flare equipment has been developed. This technique is based on the hypothesis that some combination of nondimensional parameters can describe the smokeless burning of turbulent diffusion flames of known initial conditions (i.e., fuel type, diameter, exit velocity, etc.). This method divides the flame into two sections (see Figure 8.18) and focuses on the main body of the flame. Effects due to tip geometry are represented by other proprietary models.

8.5.1.1 Industrial Experience

Over time, John Zink engineers have identified several major factors that affect the smokeless operation of a flare: (1) fuel type, (2) tip diameter, (3) inerts, (4) flow velocity, (5) ambient conditions (e.g., wind speed, relative humidity, and temperature), and (6) total mass flow rate.

These factors, from a qualitative viewpoint, have a significant impact on the smokeless capacity, but the effect of each parameter is difficult to quantify. For example, the tendency to smoke was found[15] to roughly correlate with the fuel's hydrogen-to-carbon ratio (H:C) and lower heating value (LHV). For years, the H:C ratio and the LHV of the fuel were used to analyze the smoking tendency of a hydrocarbon fuel. This information, coupled with experience, were utilized to estimate the smokeless rate of a particular flare.

FIGURE 8.17 (a) Smoking and (b) non-smoking flares.

8.5.1.2 Combustion Literature

The orange and yellow color observed in a flame is produced by light emitted from glowing carbon particles, or soot, inside the flame. When these carbon particles cool, they turn black and are seen as soot or smoke. To eliminate soot, the particulate carbon must burn off at a faster rate than that at which it is produced. Currently, no directly applicable work has been documented that addresses the smokeless rates of industrial-scale flare flames. However, there is a substantial body of literature in related areas of flame lengths and soot formation rates.

To characterize the length of a flame, Hottel and Hawthorn[16] first noted that as the exit velocity of an external hydrocarbon jet increased, the flame length initially increased, while the flame color became more transparent (less yellow). The more translucent flame indicated that the carbon particles were burning off at a faster rate than they were being produced. They concluded that the aeration rate of the flame increased with exit velocity. This qualitative observation implies that flame length correlations might also be used for predicting carbon burn-off rates (Becker[17] and Blake[18]).

Numerous options have been identified for scaling in-flame soot formation rates, as summarized by Glassman[19] and Bartok.[3] A partial list includes: (1) sooting equivalence ratios, (2) number of C–C bonds, (3) critical oxygen partial pressures, (4) smoke heights, (5) maximum soot volume fraction, (6) H:C ratio (Reed[15]), and (7) maximum radiant fractions.

Considering the number of scaling parameters, several models are possible. All of the effects initially identified as important have been incorporated into John Zink's Flare Performance Model. In general, however, the model scales the smoke evolution from turbulent diffusion rates by taking the

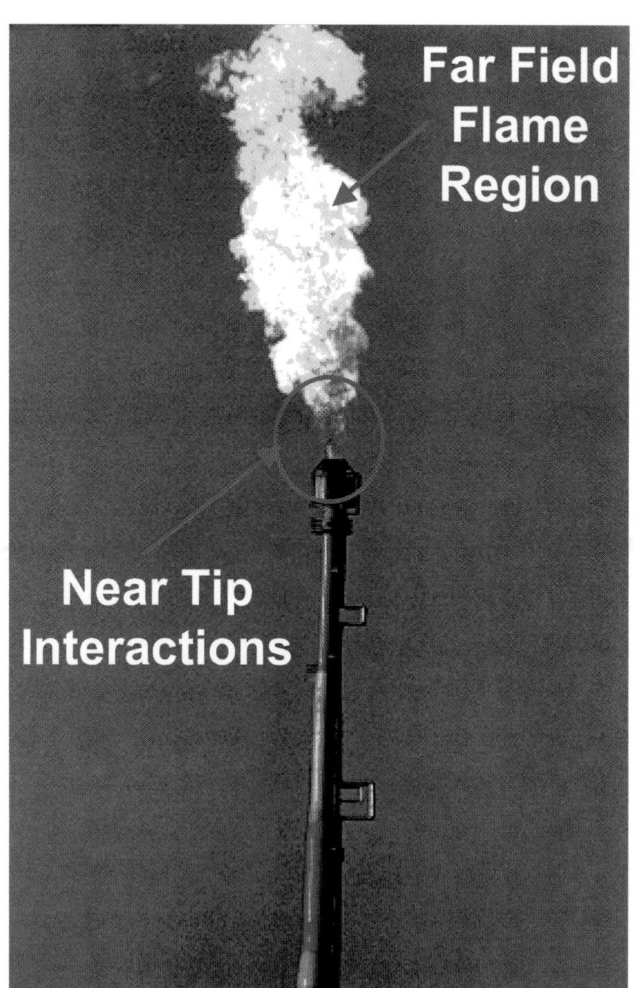

FIGURE 8.18 Flame of a flare is divided into two parts: near tip interactions and the far field flame region.

ratio of carbon burn-off, represented by flame length scaling, to the soot formation rate, which was represented by an appropriate set of scaling parameter requirements.

8.5.1.3 Calibration and Validation

To calibrate and validate the scaling of the performance model, the experimental setup shown in Figure 8.19 was used. Experiments consisting of five different nozzle sizes, three different diluents, and two different fuel types were conducted. The largest nozzle utilized had nearly 200 times the diameter of the smallest nozzle. Results are shown in Figures 8.20 and 8.21.

Figure 8.20 is a log-log plot of the predicted normalized hydrocarbon flow rate vs. the predicted normalized diluent flow rate. The lines on the graph represent the amount of diluent required to prevent the hydrocarbon stream from smoking. The "left" side of the graph represents the lowest mass flow rate of hydrocarbon and lowest nozzle exit velocity. As the mass flow of hydrocarbon is increased, the amount of diluent also increases. Flames on this part of the graph would be characterized by fairly low velocities, and smoke suppression is achieved by dilution. As the hydrocarbon flow continues to increase, a point is reached where a rapid decrease in the amount of diluent is required. Flames in this region are characterized by high velocities, and smoke suppression is achieved by increasing aeration rates. Thus, the model predicts that there are two different regimes that flare equipment could successfully operate in: (1) the dilution region and (2) the high-velocity or momentum region.

It has been historically observed that steam-assisted flares and air-assisted flare equipment (see Chapter 20) follow the general trends of the dilution region. It has also been observed that high-pressure flares (JZ LRGO flare and hydra flare) operate in the high-velocity region. Thus, the model as derived is capable of capturing historical knowledge and observations of flare operations. This matches previous qualitative observations for small flames as reported by Gollahalli and Parthasarathy.[20] In addition, the model provides insight into the picture sequence shown in Figure 8.17. This sequence is of a John Zink Company high-pressure flare. The model qualitatively predicts that for high-pressure flares near the no dilution required limit, small changes in operation might result in substantial changes in performance. This prediction is also verified by historical flare operations.

Figure 8.21 shows the same graphs as discussed above but with experimental data points superimposed. The comparison between the predicted and measured results is excellent. Thus, the model not only predicts historical flare operations accurately, but also quantitative data. The above documented validation has demonstrated that the new model can, with a high degree of confidence, predict whether a diffusion flame with known initial conditions will or will not smoke.

8.5.2 Application to Steam Flares

Steam-assisted flares were first introduced in 1952 to provide smokeless combustion for small to moderate hydrocarbon gas flow rates during a flaring event. These early steam flare designs were unable to provide good smokeless performance as relief capacity requirements increased and drove tip designs to larger diameters. During the 1960s and 1970s, a new steam flare design was introduced to improve smokeless performance. Equipped with steam eductor tubes, and shown in Figures 8.4 and 8.5, this design provided substantial benefits. Recently, the new prediction method for calculating smokeless rates has been applied to steam flares.

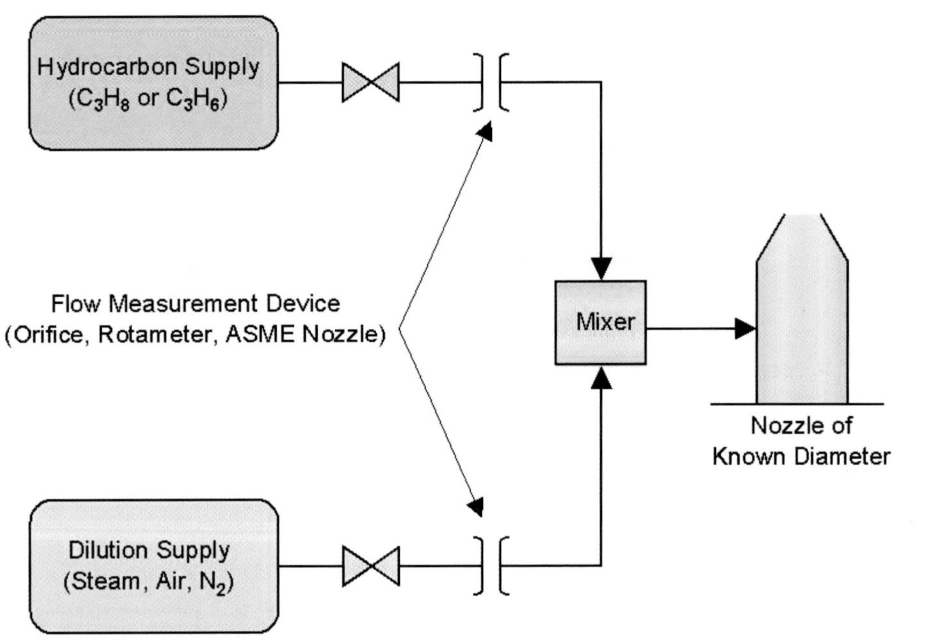

FIGURE 8.19 Illustration showing the experimental setup utilized to obtain calibration and validation data.

8.5.2.1 Predicting Steam Flare Smoking Tendencies

As previously discussed, John Zink engineers have developed a new predictive method to determine whether a flame will smoke or not. This method relates fundamental measures of the momentum of the air, steam, and flared gas, at the base of the flare tip, to smokeless capacity. This method can be used, in general, to predict the smokeless performance of any flare, as long as the mass flow rates and momentum of the various gas constituents can be accurately determined.

To apply the predictive tool to steam-assisted flares, one must first determine the initial conditions at the base of the flame. In this case, only the velocity and the air/steam mixed into the hydrocarbon stream are unknown.

For basic steam flares, the combined velocity of the hydrocarbon and steam mixture is determined from an overall momentum balance, and entrained air is estimated using free jet entrainment laws.[21]

For more complex steam flares, the same procedure is followed, except the air mixed into the hydrocarbon stream is enhanced by steam eduction tubes. The eductor model discussed in Chapter 8.2 is used to determine the additional air available at the base of the flame.

These predictive tools have been applied to estimating the smokeless performance of both basic and complex steam flares with good success. Such tools also indicate the possible error associated with using standard industry design guidelines for steam-assisted flares. Figure 8.22 is a plot of the steam-to-hydrocarbon ratio (on a mass basis) for various molecular weights of paraffinic hydrocarbons. The "data" points in the plot are from Leite.[22] These data points, originally published several years ago, are currently used in the industry as guidelines to estimate the steam-to-hydrocarbon ratio required for smokeless flaring of paraffinic hydrocarbons. The information likely represents a combination of small-scale testing and field observations. Assuming this data was gathered on flares with diameters between 16 and 24 in. (41 and 61 cm) and steam pressures of approximately 100 psig (6.8 barg), the data may be compared to predictions from the steam flare predictive tool. This comparison is also shown in Figure 8.23. The comparison between Leite's data and the predicted performance from the steam model, based on the most likely operating conditions, shows good agreement. However, experience suggests that for large-diameter flare tips, the steam-to-hydrocarbon ratios required for smokeless flaring can increase dramatically above the values provided in Figure 8.22. Figure 8.23 compares predictions from the new model and field data from flares with over three times the diameter. In this case, the graph clearly shows that there is excellent agreement between the model predictions and quality field data. The model, however, also dramatically demonstrates a limitation when using steam-to-hydrocarbon ratios for estimating steam flare smokeless performance — they (steam-to-hydrocarbon ratios) are a function of diameter.

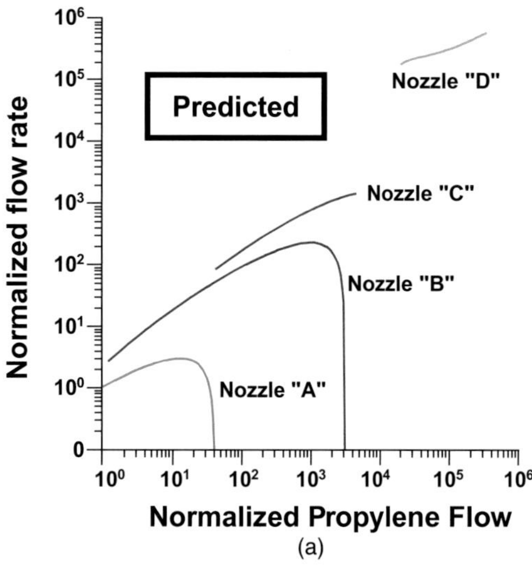

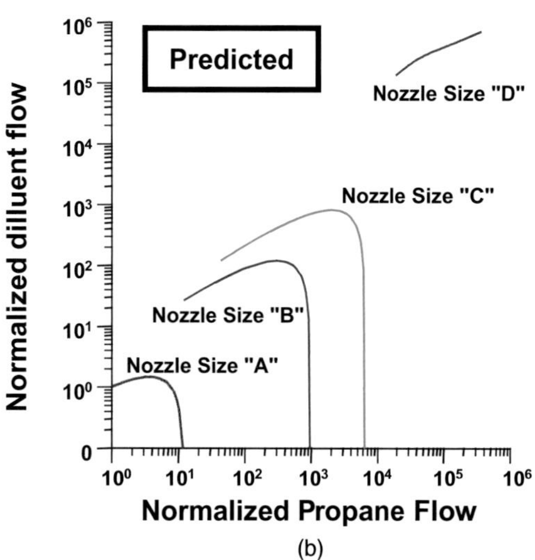

FIGURE 8.20 Prediction of diluent flow for smokeless operation.

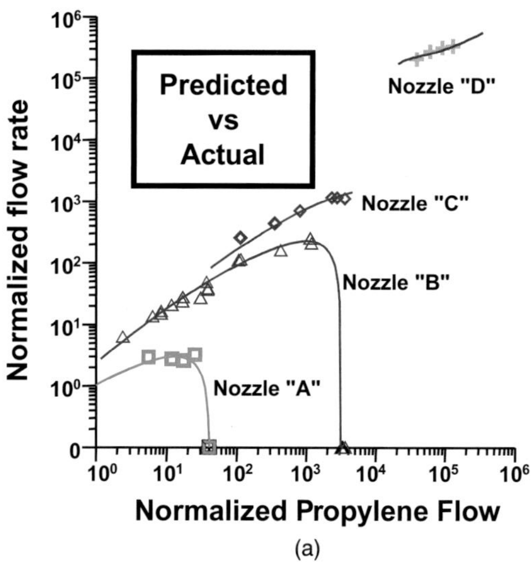

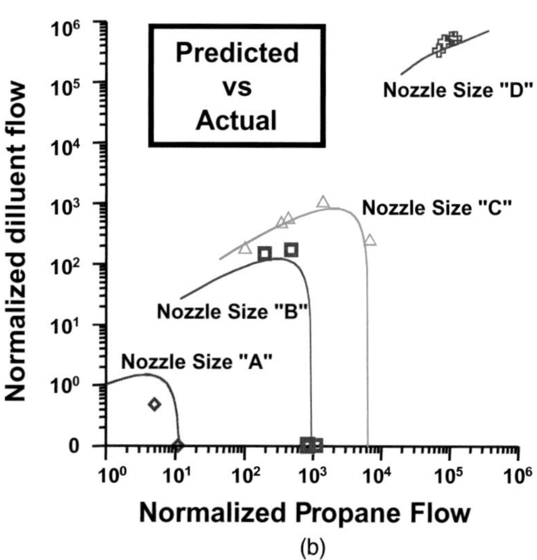

FIGURE 8.21 Figure 8.20 with data points added.

Figure 8.24 shows a steam flare in operation that was designed by this new method.

8.5.3 Modeling Air-Assisted Flares

An air-assisted flare uses air supplied from a high-pressure fan as a supplemental energy source to achieve smokeless combustion of a hydrocarbon stream. This is achieved by increasing the overall exit velocity of the stream and through increased aeration rates. A typical schematic is shown in Figure 8.25. Figure 8.26 shows the effect of the high-velocity air on an unsaturated hydrocarbon. The first photo is with the fan off; the second is the air starting to mix into the hydrocarbon as the fan comes up to speed; and the third is steady-state operation with a highly enhanced air mixing rate. Figure 8.26 also highlights the enhanced aeration effects on the flame; flame length decreases as smoking is reduced.

To apply the predictive tools to air-assisted flares, hydrocarbon and air velocity, along with a rate of aeration must be determined. The velocity increase of a hydrocarbon stream due to interactions with the supplied air from the blower is calculated from a momentum balance. The rate at which the supplied air is mixed into the hydrocarbon stream is more difficult, but may be assumed to be a turbulent diffusion

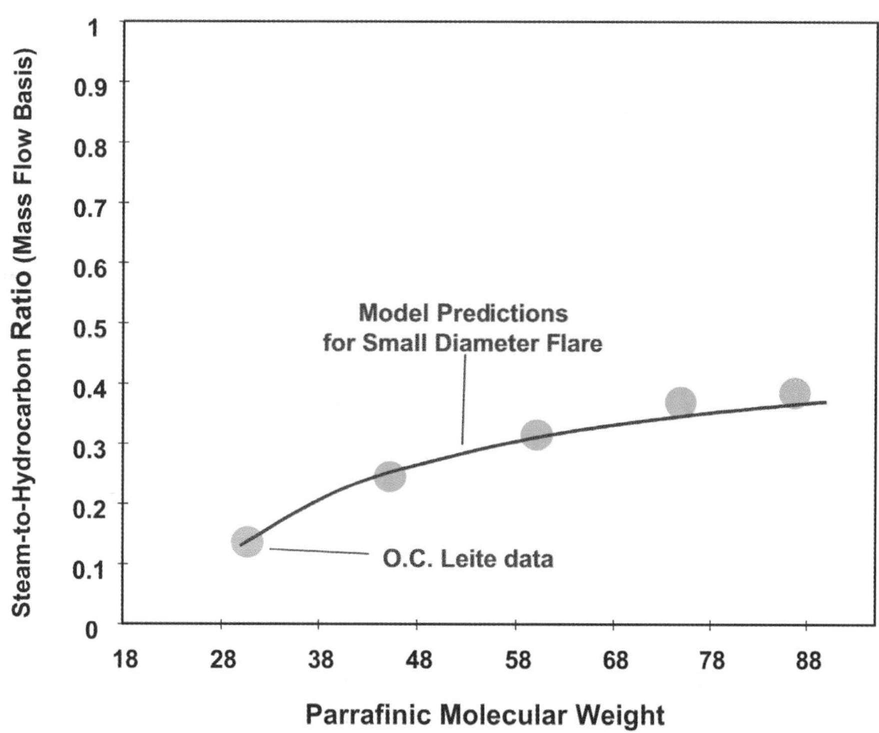

FIGURE 8.22 Steam-to-hydrocarbon ratios (per Leite).[22]

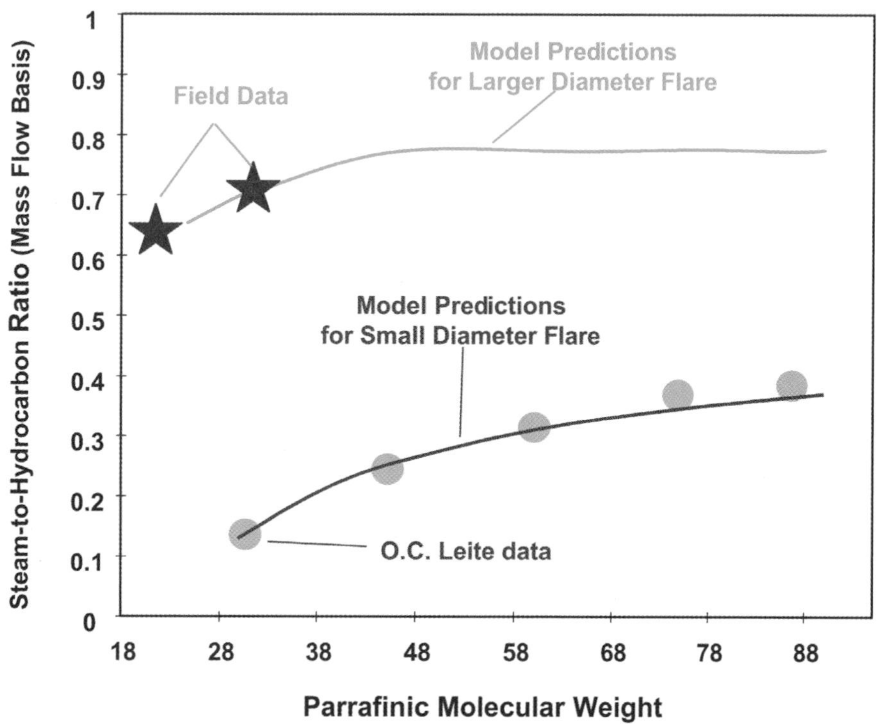

FIGURE 8.23 Steam-to-hydrocarbon ratios (large-diameter flares).

FIGURE 8.24 Smokeless steam flare designed with new method.

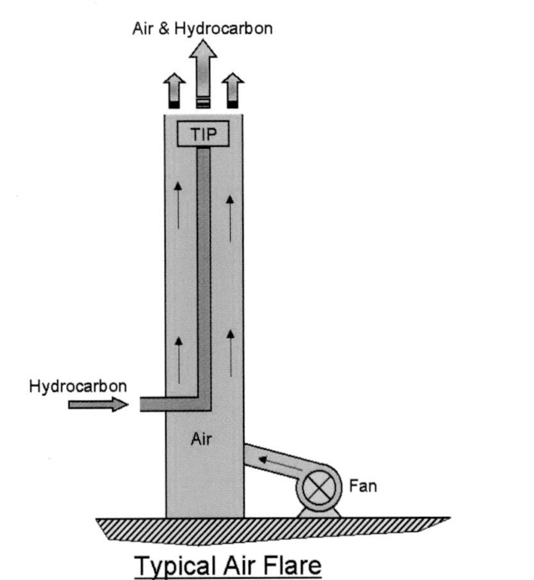

FIGURE 8.25 Typical air flare.

process. Turbulent diffusion for this system is a function of (1) velocity of fuel gas, (2) air velocity, (3) circumference perimeter, (4) characteristic dimension, and (5) turbulent mixing length.[21]

The turbulent mixing length, according to Rokke[23] and Schlichting,[21] is a function of (1) exit velocity, (2) reaction rate, (3) density, and (4) viscosity. These parameters can be combined into a "scaling functional," which allows the aeration rate of a hydrocarbon stream to be calculated.

A series of calibration tests were performed to evaluate this scaling functional. Tests were conducted on air flares with tip areas of 30 and 500 square in. For each test, the aeration rate was determined and plotted against the scaling functional (see Figure 8.27). The relationship shown for aeration rate was developed for a variety of hydrocarbon fuels (propane, propylene, natural gas, and mixtures) and for the various tip outlet areas. Combining the aeration parameter with the previously described turbulent flame model has improved our ability to calculate the smokeless performance for a wider variety of air flare designs, flow rates, and hydrocarbon types. This model has been successfully utilized to design numerous air-assisted flares worldwide (Figure 8.28 shows an air-assisted flare firing 190,000 lbs/hr of propane). Application of the model has also provided valuable insight into the operation and design of air flares as discussed below.

Leite[22] describes an alternative method for sizing air flares. This method predicts smokeless performance for an air flare if a specified fraction of the stoichiometric air requirement for a hydrocarbon fuel is supplied at the flare tip. No minimum velocity requirement is provided, but common practice is to provide an exit velocity of 120 ft/sec. This industry "rule of thumb" has been used for many years. Similar to the original steam flares, the vast majority of air flares (until recently) had relatively small exit areas (100 to 150 in.^{2}), enabling most vendor test facilities to evaluate air flare designs. However, as plant relief capacities and customer requirements have increased over the last several years, full vendor testing is not normally possible.

As demonstrated earlier, the standard rules of thumb were not applicable for large-diameter steam tips. The logical question that must be asked is whether the standard rule of thumb for air flare sizing applies to larger tip sizes as well as small tips.

Figure 8.29 shows the results generated using the air flare predictive tool with the aeration parameter included. Items in the box represent variables held constant. As the tip area is reduced, the exit velocity increases for a given hydrocarbon flow rate. The model clearly highlights the real effects of exit velocity on the smokeless rate of an air-assisted flare. The two cross symbols reflect actual experimental data gathered by John Zink R & D test engineers. The square symbols represent anticipated smokeless performance as predicted by Leite's method (assuming that the calibration of the Leite method was accomplished on a 125 in.^{2} air flare tip). This graph shows that the John Zink model accurately predicts the variation in performance trends for the various tip sizes tested. For a given hydrocarbon flow rate, the velocity decreases with an increase in tip diameter. As discussed earlier, the reduction in hydrocarbon velocity decreases the amount of air mixed with the hydrocarbon stream and decreases the overall stream velocity, resulting in decreased smokeless performance.

In Figure 8.29, all geometric variables were held constant except for the outlet area. Flare vendors typically fabricate larger tips to enhance air/hydrocarbon mixing. Although this helps mitigate the mixing limitation, it does not eliminate the problem, and serious performance problems can result. A variation of as much as a factor of three is observed when comparing the predicted smokeless performance using the Leite method vs. an advanced predictive tool.

8.5.4 Summary

An analytical model to predict smoke evolution from a turbulent diffusion flame has been developed and implemented by John Zink Company engineers. Recent applications of the method have resulted not only in optimized equipment but greater insights into equipment operation and possible failure modes as well. In every case, the most significant problems were observed when attempting to "scale-up" existing equipment to larger sizes. Even though the model predicted the possible performance problems with larger equipment, validation with large-scale data was still required. As described at the beginning of this section, reliance on field data may be problematic; ultimately, only larger test facilities can provide the data required to prevent serious under-sizing of equipment and verify the accuracy of predictive models. Having this experimental and analytical capability is a requirement for supplying reliable, high-capacity industrial flare equipment.

8.6 OIL GUN CAPACITIES

Oil firing of process heaters is common in most of the world. To efficiently combust oil, it must be "broken up" into very small droplets, or atomized (Figure 8.30 shows a typical oil flame). Typical applications utilize steam as an atomizing agent and require some type of equipment to effect mixing of the two streams. This oil/steam mixer or atomizer is normally referred to as an oil gun. Figure 8.31 shows the standard John Zink oil gun, and Figure 8.32 shows a typical oil gun schematic. Steam and oil are supplied separately, then mixed in a manner to enhance atomization. The equipment is operated

(a)

(b)

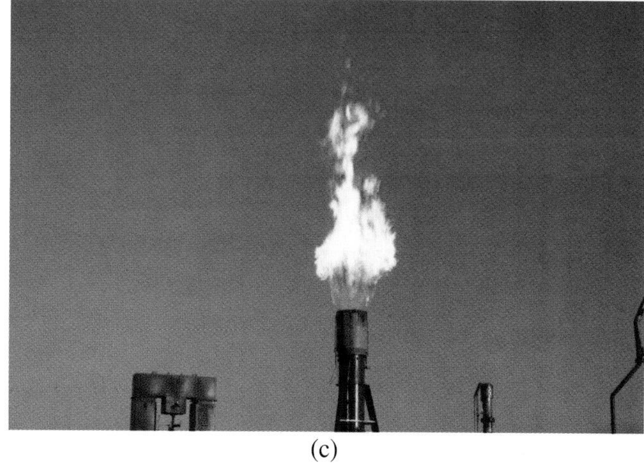

(c)

FIGURE 8.26 Effect of high velocity air: (a) blower off, (b) commence blower and (c) blower on (Courtesy of John Zink Co., LLC., Tulsa, OK).

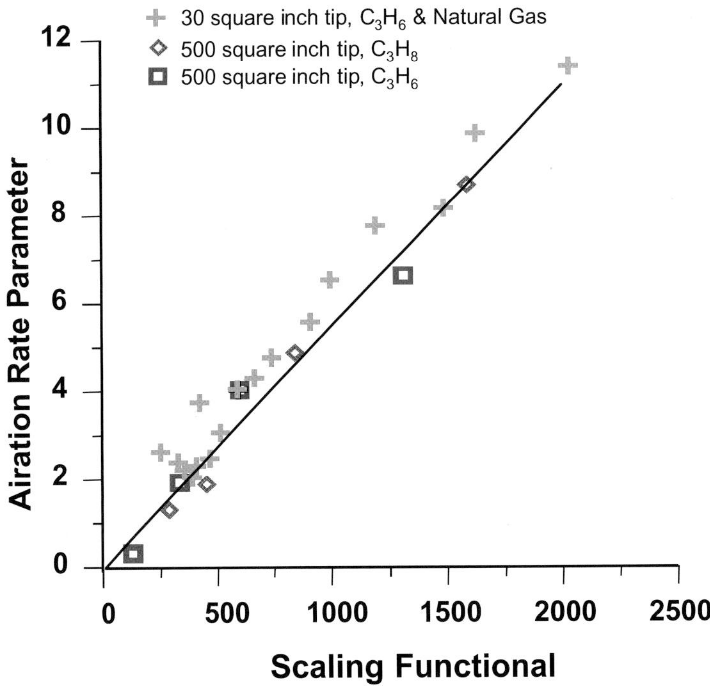

FIGURE 8.27 The aeration rate was determined and plotted against the scaling functional.

by setting the steam and oil pressures, with the steam having a higher pressure than the oil. Typically, these pressures will have a fixed differential; for example, setting the steam to 50 psig (3.4 barg) and the oil to 35 psig (2.4 barg) would correspond to a 15 psig (1 barg) differential.

For a given application, the customer is interested in knowing what heat release (or capacity) and steam usage to expect from given pressure settings. Typical oil gun capacity curves will show heat release as a function of available oil pressure only. These curves are based on limited experimental data from a specific oil, oil temperature, steam temperature, and pressure differential. For these types of curves, the only property that can be considered is the oil heating value. All other factors (e.g., oil temperature, specific gravity, steam temperature, or differential) cannot be considered because there is typically no basis to correct the capacity curves for these factors. In addition, it would be expected that steam usage (expressed as a fraction of the oil mass flow rate) would vary as a function of absolute oil pressure. But again, there is no basis for adjustment of anticipated steam flow rate based on a different operating pressure. Clearly, the need for a more thorough analytical model existed.

A recent characterization effort at the John Zink Company has focused on utilizing fundamental principles to develop a more general capacity prediction method for these devices. As can be seen from the schematic, there are several complicating issues in developing an analytical model of the system; two-phase flows, several choked flows, two-phase choked flow regions, atomization efficiency, and accurate determination of flow rates are just a few of the issues. This section summarizes the model formulation and results obtained.

The problem formulated and solved by this flow model is that of determining the required oil pressure to flow a specified amount of oil, atomized by steam, through a particular size proprietary gun. The required user inputs to this problem are the mass flow rate of oil (or, alternatively, the heat release), the steam differential (typical value ~30 psi), the grade of oil fired, and the steam and oil temperatures. This effort has resulted in a model that predicts the two-phase flow rates with greater accuracy than previously possible.

8.6.1 Summary of Two-Phase Flow Analytical Development

The description below follows the general development outlined in Kaviany,[24] Kaviany,[25] Chislom,[26] and Chislom.[27]

8.6.1.1 Conservation of Mass

In the model equations, the three components (oil, gaseous water, and liquid water) are considered to have different velocities. This complicates the description of the local composition because, as shown below, allowing each phase to have different velocities requires the definition not only of

FIGURE 8.28 Annular air-assisted flare; 190,000 lb/hr propane.

conventional mass and volume fractions, but also the definition of mass flux fractions. A mass flux fraction is defined as the mass flux of an individual component divided by the total mass flux. Mathematical description also requires the quality of the steam.

8.6.1.2 Choked Flow

From Chisolm,[28] the mass flux through a sonic orifice of a two-phase mixture is the square root of the pressure to specific volume derivative. Functionally, this is:

$$\text{Choked flow mass flux} = \sqrt{\frac{\partial(\text{Pressure})}{\partial(\text{Specific volume})}} \tag{8.23}$$

The model further assumes that the choked flow process is isentropic and that there is no condensation, boiling, or evaporation. For these assumptions, the only variable through the orifice that is a function of pressure is the specific volume of the steam. The above differential is then evaluated explicitly in terms of the steam specific volume.

8.6.1.3 Momentum Conservation

For an idealized, one-dimensional inviscid flow, the differential momentum equation can be written as in Anderson.[28]

$$dP + \rho u du = 0 \tag{8.24}$$

In the separated flow model,[27] the various phases are postulated to flow with separate velocities, but at any location, it is assumed that the pressure of each phase is the same. Following this line of reasoning, the above differential conservation equation is split into liquid and gas phases for the velocity and density terms. Under this model, this differential momentum equation can be written:

$$dP + (\rho u du)_{Steam} + (\rho u du)_{Liquid\ water} + (\rho u du)_{Oil} = 0 \tag{8.25}$$

8.6.1.4 Conservation of Energy

In an incompressible, constant-density flow, conservation of mass and momentum are sufficient to analyze the flow. The term incompressible does not imply constant density, but rather that the pressure forces arising within a fluid flow are

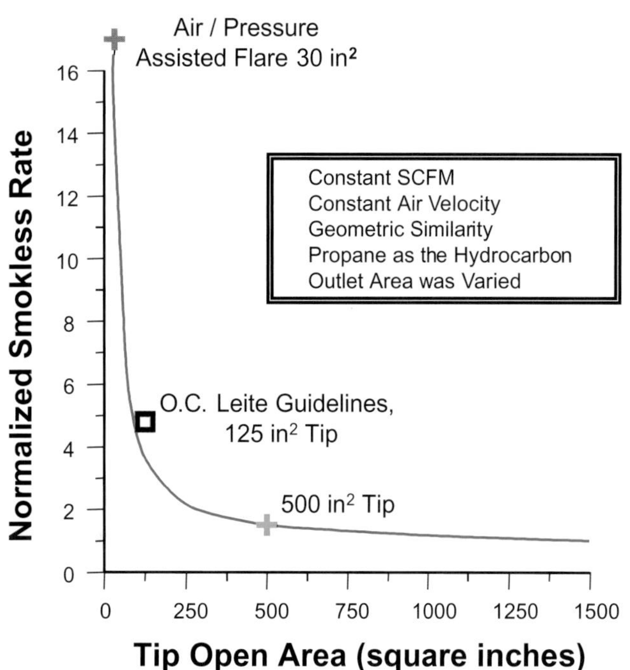

FIGURE 8.29 Comparison of three different air flares to prediction.

insufficient to change its density. Formally, an incompressible flow is one with Mach number $M^2 << 1$. When flow velocities are large enough that the arising pressures can change the density appreciably, it becomes necessary to consider energy conservation in analyzing a flow. Another way of thinking about this is that M^2 represents the ratio of the kinetic energy to the thermal energy. As the Mach number gets large, the conversion of thermal energy into kinetic energy becomes important.

For an idealized, one-dimensional compressible flow, energy conservation can be written as in Anderson:[28]

$$d(\mathit{Thermal\ energy} + \mathit{Kinetic\ energy}) = 0 \qquad (8.26)$$

In the three-component flows considered here, the kinetic energy flux (KE) of the mixture is computed by calculating the kinetic energy in the mass flow of each phase (oil, steam, and water). In a similar manner, the thermal energy (TE) is calculated by summing the thermal energy in each phase. This can be summarized as follows:

$$\mathrm{KE} = \mathrm{KE}_{\mathrm{Oil}} + \mathrm{KE}_{\mathrm{Steam}} + \mathrm{KE}_{\mathrm{Liquid\ water}}$$

$$\mathrm{TE} = \mathrm{TE}_{\mathrm{Oil}} + \mathrm{TE}_{\mathrm{Steam}} + \mathrm{TE}_{\mathrm{Liquid\ water}} \qquad (8.27)$$

FIGURE 8.30 Typical process heater oil flame.

The reference condition chosen is pure liquid at 373 K and 1 atm. Pressure changes and interfacial heat transfer will induce phase changes to the steam/water system. These effects are modeled by appropriate latent heat terms, thus accounting for condensation and evaporation effects. To conserve energy, the sum of the kinetic energy and the thermal energy must remain constant throughout the oil gun. Thus, a total energy (enthalpy) was defined as the sum of the thermal and kinetic energies

$$h_t \equiv TE + KE \tag{8.28}$$

This results in the following conservation equation:

$$d(h_t) = 0$$

The above system of coupled differential equations along with flow coefficient information for our proprietary equipment were then coded into a Visual Basic program utilizing a "shooting" Runge-Kutta solver.

8.6.2 Results

Typical results are shown in Figure 8.33. As can be seen, excellent agreement is achieved between experimental and theoretical flow rates for different oil pressures. Similar results were observed for different oils and pressure differentials. These efforts have resulted in a capacity prediction technique of unprecedented flexibility. John Zink application engineers now have the ability to optimize oil gun applications to meet a wider variety of customer conditions.

8.7 OIL GUN DEVELOPMENT

Following on the successful development of an oil gun model, it was decided to further optimize oil gun design by studying atomization. There are many ways to atomize a bulk liquid into small droplets. Normally, a high relative velocity between the liquid to be atomized and the surrounding air or gas is created. The high shear force between the liquid and the gas disrupts the liquid into droplets. Some atomizers accomplish this by discharging the high-velocity liquid into a relatively slow-moving stream of air or gas. Examples of this technique include pressure atomizers and rotary cup or disk atomizers. An alternative approach is to introduce a high-velocity gas stream into the liquid to assist atomization. This is generally known as a twin-fluid, air-assist, or airblast atomizer. In industrial furnaces, the steam-assist or air-assist atomizers are most common.

FIGURE 8.31 Standard John Zink oil gun.

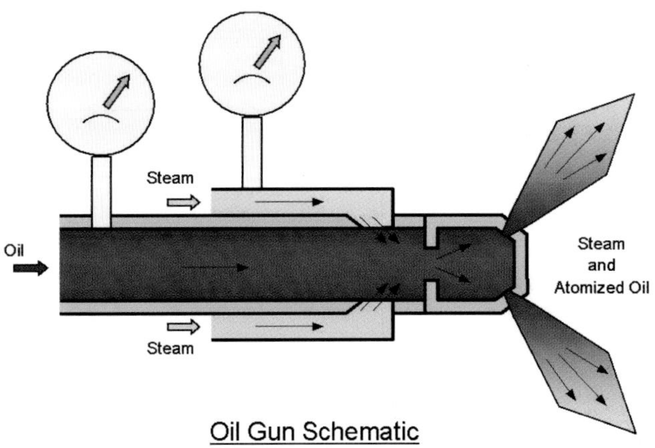

FIGURE 8.32 Schematic of a typical oil gun.

The performance of atomizers depends on their design, the physical properties of liquid fuel and atomizing medium, and the operating conditions (i.e., pressure and temperature). The liquid properties include viscosity, surface tension, and density. The effect of fluid properties on atomization can be found in Chung and Presser.[34] General atomizer design, atomizing medium properties, and operating conditions have been discussed in detail by Lefebvre.[35] Here, the focus will be on steam-assist oil gun characterization and improvement.

Oil gun design has a significant influence on the spray combustion performance and exhaust pollutant emissions. A good oil gun should generate a good flame shape, consume minimal atomization medium (steam or compressed air), and exhaust limited particulate and NOx emissions. The major factor that determines such performances is the atomization quality of the oil gun. The parameters that characterize the atomization quality include the mean droplet size, droplet size distribution, spray cone angle, and liquid distribution.

A fuel spray is usually composed of a wide range of droplet sizes. The biggest droplets in the spray may be 50 or 100 times the size of the smallest droplet. The actual sizes of the droplets represent the degree of spray fineness. When comparing the fineness of different sprays, it is useful to introduce some "mean droplet size." In spray combustion, the Sauter

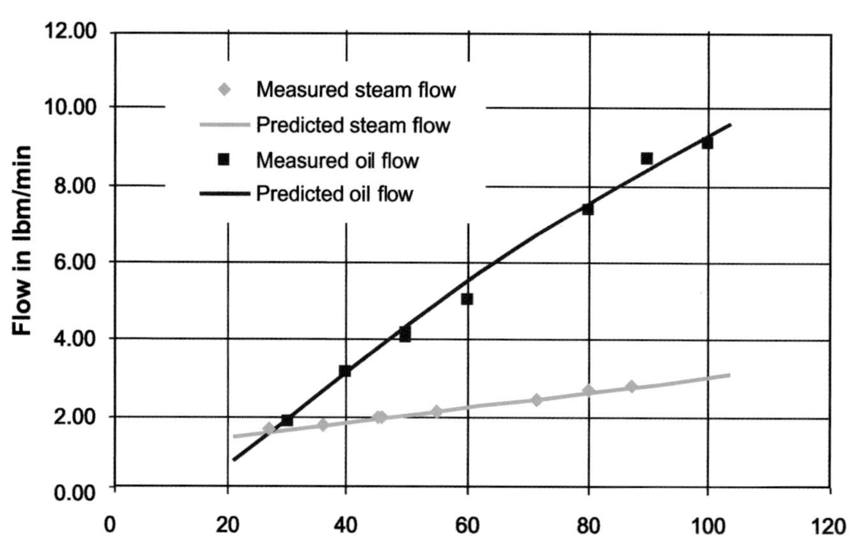

FIGURE 8.33 Comparison of predicted vs. actual oil and steam flow rates.

mean diameter (SMD) is usually used to represent the mean droplet size. The SMD is the diameter of the droplets whose ratio of volume to surface area is the same as that of the entire spray and defined as:

$$\mathrm{SMD} = \frac{\sum n_i d_i^3}{\sum n_i d_i^2} \tag{8.29}$$

where n_i is the number of droplets in size d_i. The SMD is well accepted because total droplet volume and surface are most significant in connection with the combustion process.

The range of droplet size variation (or size distribution) indicates the degree of spray uniformity. At present, the most widely used expression for the droplet size distribution is the Rosin–Rammler relationship.[36] It is expressed as:

$$1 - Q = \exp-\left(D/X\right)^q \tag{8.30}$$

where Q is the fraction of the total volume contained in droplets of diameter less than D, and X and q are constants. The constant X represents the droplet diameter, and the exponent q provides a measure of the spread of droplet size. The higher the value of q, the more uniform is the spray. If q is infinite, the droplets in the spray are all the same size. For most sprays, the value of q lies between 1.5 and 4.

For a given liquid fuel and a fixed environmental condition, the mean droplet size (SMD) and the size distribution (q) are two major parameters that affect the evaporation rate. It is well understood that the finer the droplets, the better the evaporation. As for droplet size distribution, a more complicated correlation exists. If one defines the initial size distribution before evaporation as q_o, then after sprays injecting into the hot furnace, q will vary with evaporation time. The general trend is for q to increase with evaporation time, and the effect is more significant for a spray having a low value of q_o. Usually, for the ignition of a fuel spray, the time required to vaporize 20% of the spray mass is important; whereas for combustion efficiency, the time required for vaporization of 90% of the spray mass is crucial. Chin et al.[37] reported that with a given mean droplet size, a spray of large q_o would have a low 90% evaporation time and a high 20% evaporation time. Thus, from a combustion efficiency viewpoint, it is desirable to have a fuel spray with a wide size distribution, but for good ignition performance, it is better to have a narrow size distribution. It is not possible to know *a priori* what is the optimal compromise between combustion efficiency and ignition; experimental performance will always be required.

The cone angle of a spray is usually defined as the angle between tangents to the spray envelope at the oil gun tip. The value to be selected for the cone angle of a spray will depend on the shape of the furnace and the conditions controlling the mixing of air and fuel. For furnaces with a high degree of air movement (i.e., swirl or forced-draft), sprays with a wide cone angle will give good results. On the other hand, furnaces with limited air movement (i.e., natural draft) will require sprays of narrow cone angle. If a short period of time between the beginning of injection and the beginning of combustion is desired, a compact and penetrating spray (i.e., narrow spray cone) should be used. However, for a short flame, a soft and well-dispersed spray (i.e., wide spray angle) is preferred.

The liquid fuel distribution within a spray has an important effect on pollutant generation. The technologies used to measure the liquid distribution can be found in Chung et al.[38] Nonuniformity in liquid fuel distribution can give rise to local pockets of fuel-lean or fuel-rich mixtures. The fuel-lean pockets have low burning rates and thereby produce high concentrations of carbon monoxide and unburned hydrocarbons. The fuel-rich mixtures are characterized by high soot formation and lead to high particulate emissions. The best oil guns exhibit excellent radial symmetry and circumferential maldistributions of less than 10%.

As can be readily understood from this discussion, steam-assist oil guns have numerous competing characteristics, all of which can affect performance. For instance, a well-atomized and well-dispersed spray generates a high-temperature flame. It is known that NOx formation strongly depends on the flame temperature. The high flame temperature has a tendency to produce high thermal NOx. On the other hand, large droplets have a poor evaporation rate and provide a favorable condition for soot formation. An inadequate atomization system can even reduce combustion efficiency and result in a low combustion turndown ratio. Therefore, actual spray combustion experiments are always required to determine the best compromise among the various parameters.

Prior to optimizing steam-assist oil gun characteristics, a method of characterizing the atomization was required. Several methods have been employed to measure the droplet sizes in sprays, for example, droplet collection on slides, molten-wax and frozen-drop techniques, cascade impactors, charged-wire and hot-wire techniques, high-speed photography, light scattering, etc. The most advanced and popular method in current use is the Phase Doppler Particle Analyzer (PDPA). The PDPA can simultaneously measure the droplet size and the velocity *in situ*. The combined information of droplet size and velocity is useful for calculating the spray cone angle and the liquid distribution. The John Zink PDPA instrument setup is shown in Figure 8.34. The PDPA uses the Doppler signal and its phase shift to simultaneously characterize the particle velocity and the droplet size. It counts every droplet that passes through the sample volume. The sample volume is the cross-section of two laser beams. Every measurement contains 10,000 sample points and statistically represents the mean droplet size, size distribution, standard deviation, and velocity distribution. The detailed theoretical description of PDPA can be found in Bachalo.[39]

An example of the use of a PDPA to compare two different designs of oil guns is provided in Figure 8.35. In the figure, the ordinate represents the droplet size measurements and the abscissa is the radial position of the spray.

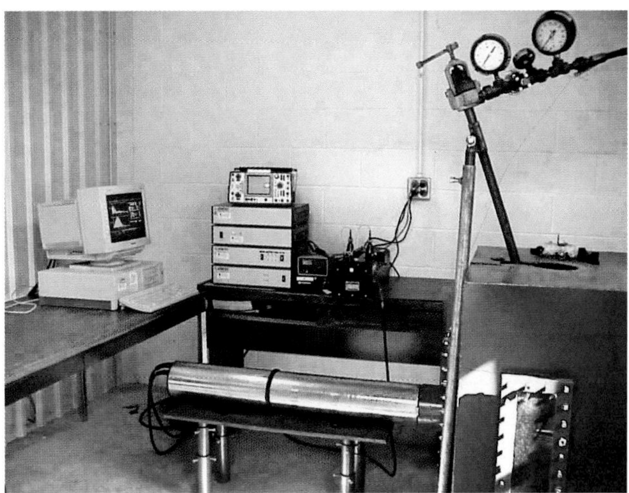

FIGURE 8.34 John Zink Co. LLC (Tulsa, OK) Spray Research Laboratory. (Courtesy of John Zink Co. LLC., Tulsa, OK.)

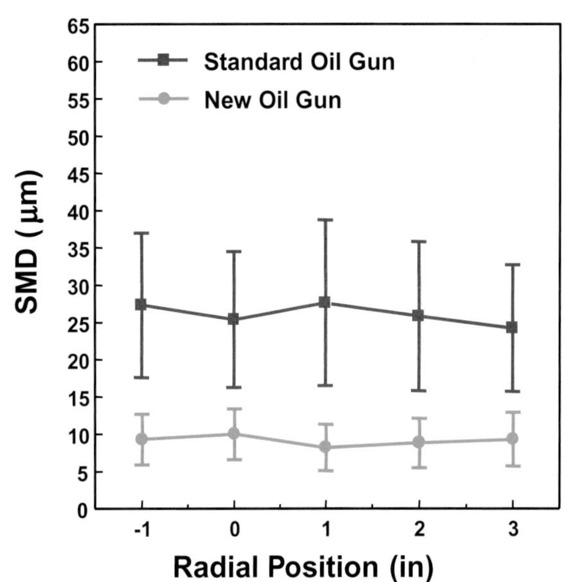

FIGURE 8.35 Droplet size comparison between a standard and a newer oil gun.

Utilizing the John Zink Company PDPA, the oil gun operating parameters were characterized. During the characterization efforts, it was found that the liquid jet diameter, or film thickness, was also an important parameter affecting atomization quality. To improve atomization, a pre-filming technique was developed and introduced into the design.

The new design was then optimized by combustion testing in the John Zink Company test furnaces. The optimized gun

FIGURE 8.36 High-efficiency new oil gun.

showed a high turndown ratio, low steam consumption, short flame length, and low NOx and particulate emissions. The pre-filming technique makes the best use of steam momentum, which reduces the steam consumption. The pre-filming technique improved the atomization control and generated optimal droplets. The droplets have a fast evaporation rate, resulting in high turndown ratios and low particulate emissions, and appropriate size distribution shortened the flame length and reduces the conversion of fuel-bound nitrogen to NOx, resulting in the reduction of total NOx emissions. The net result of this program was a major improvement in oil gun performance. The new gun is shown in Figure 8.36.

8.8 REGENERATIVE THERMAL OXIDIZER (RTO) PERFORMANCE

8.8.1 Introduction

For over 100 years, regenerators have been utilized in various process industries (steel and glass manufacturing are typical applications). Applications and performance solution methodologies are documented in Schmidt.[31] More recently, in response to air quality concerns, regenerators have been applied to incineration of high flow rate, low contamination (VOC) streams. Called regenerative thermal oxidizers (RTOs), they were first introduced in the early 1970s. Typical applications have VOC concentrations that vary from the ppm level to about 1 percent, or have a heating value from about 0 to 10 Btu/scf. These streams are typically produced by processes requiring ventilation, such as paint booths, printing, paper mills, and others. Many units are also equipped with a third bed to increase destruction and removal efficiency (DRE). When the inlet stream is switched to a different bed, the VOC-laden air is purged out of the bed with clean air for a period of time. Depending on design, the purge period may be shorter than the cycle time, allowing for numerous cycle variations. Thermal performance is measured in terms of heat recovery efficiency (HRE), and is defined as shown below:

$$\text{HRE} = \frac{T_{\textit{Combustion Chamber}} - T_{\textit{Outlet}}}{T_{\textit{Combustion Chamber}} - T_{\textit{Inlet}}} \tag{8.31}$$

The units typically can obtain an HRE of about 95 percent. For an industry standard 1500°F (820°C) combustion chamber temperature and 100°F (38°C) process inlet temperature, a 95 percent HRE would result in an exhaust temperature of about 170°F (77°C). One result of a high HRE is that the additional cost of regeneration equipment over the cost of a standard incinerator is typically recovered in about nine months due to lowered fuel consumption.

8.8.2 RTO Model Development

Figure 8.37 is a schematic of heat transfer in a packed bed between the ceramic material and the air stream. Neglecting heat release effects in the air stream, a first law of thermodynamics analysis for the air results in:

$$d\dot{Q}_{\text{a-w}} = \dot{m}c_{\text{p}}\left(dT_{\text{a}}\right) \tag{8.32}$$

where

$d\dot{Q}_{\text{a-w}}$ = Heat transferred from the ceramic to the air in distance "dx"
$\dot{m}$ = Mass flow of air
c_{p} = Specific heat of air
T_{a} = Air temperature
dT_{a} = Change of air temperature during distance "dx"

Similarly, a first law analysis of the ceramic results in the following:

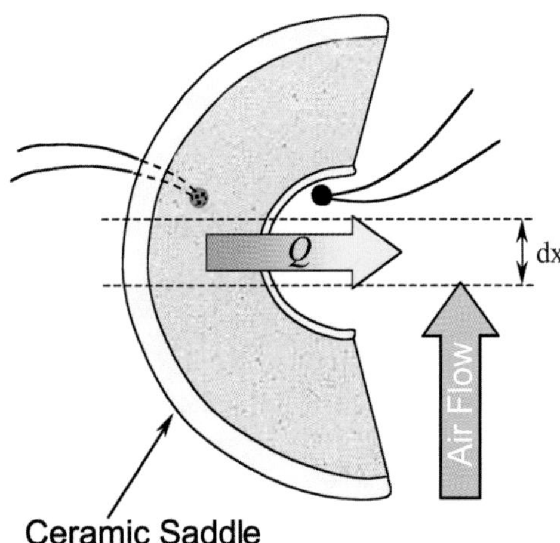

FIGURE 8.37 Heat transfer in a packed bed between the ceramic material and the air stream.

$$d\dot{Q}_{w-a} = mc(\partial T_w) \tag{8.33}$$

where

$d\dot{Q}_{w-a}$ = Heat transferred from the air to the ceramic in distance "dx"
m = Mass of ceramic
c = Specific heat of ceramic
T_w = Ceramic temperature
∂T_w = Change of ceramic temperature during time "dt"

Utilizing Newton's law of cooling for the heat transfer rate, and appropriate expressions for (1) mass of ceramic and (2) surface area of ceramic, the following coupled differential equations can be developed:

$$\frac{\mathrm{dT_a}}{\mathrm{d}x} = \frac{h\mathrm{SA}(T_w - T_a)}{\dot{m}c_p}$$

$$\frac{\mathrm{dT_w}}{\mathrm{d}t} = \frac{h\mathrm{SA}(T_a - T_w)}{mc} \tag{8.34}$$

where

T_w = Ceramic temperature
T_a = Air temperature
$\dot{m}$ = Mass flow of air
c_p = Specific heat of air
m = Mass of ceramic
c = Specific heat of ceramic
SA = Specific surface area of ceramic (area/volume)
h = Convective heat transfer coefficient

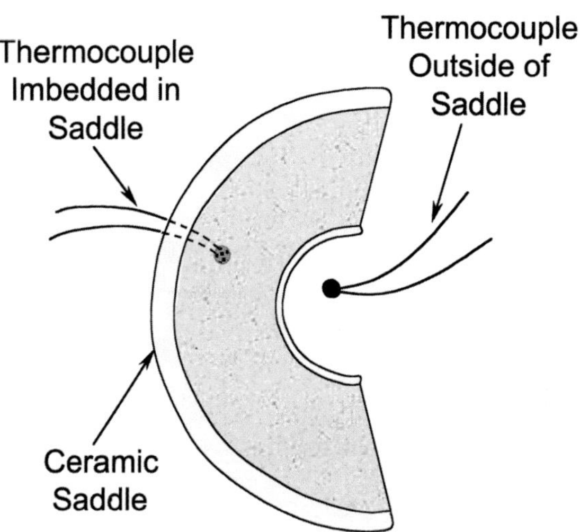

FIGURE 8.38 Installing small, type K thermocouple "pairs" into numerous ceramic saddles.

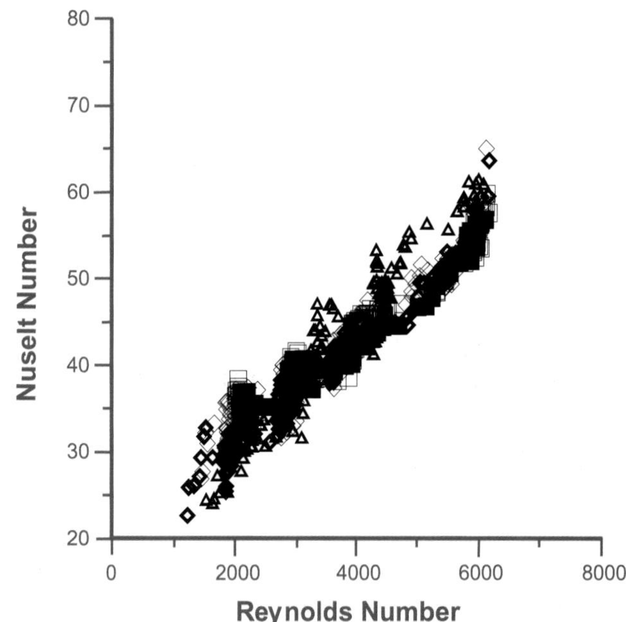

FIGURE 8.39 Summary of saddle data.

All of the above parameters are readily known, except for the convective heat transfer coefficient. Calibration information for this parameter was obtained by installing small, type K thermocouple "pairs" into numerous ceramic saddles as illustrated in Figure 8.38. One of the thermocouples is inside the ceramic, and the other is directly outside the saddle. Temperature variations with time in an actual RTO bed were measured and deconvoluted to obtain the value of the Nusselt number (nondimensional convective heat transfer coefficient) as a function of the local Reynolds number. Typical results are shown in the graph in Figure 8.39. The differential equations,

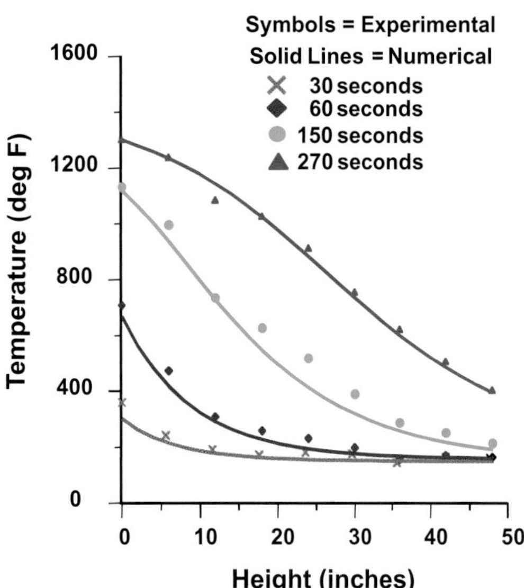

FIGURE 8.40 Rock temperature distribution with time.

along with a best fit for the Nusselt correlation, were then coded into a FORTRAN program, utilizing a Bulirsch-Stoer technique to solve the differential equations. This model was then compared to various scenarios; first, the bed with the thermocouple pairs was subjected to cooling and heating cycles. In-bed temperatures were measured as a function of time and position. The above model was then utilized to predict in-bed temperatures and compared to experimental data. Typical graphical results are shown in Figure 8.40. As can be seen, excellent agreement was obtained between the model and the experimental data. Figure 8.41 shows the John Zink Test RTO Test Unit.

Finally, the model was compared to heat recovery efficiencies (HREs) measured in the field, with results as follows.

Efficiency Comparison

1. 8.5 ft of 1.5-in. Berl saddles
   Inlet: 450°F, .25 Btu/scf,
   Combustion chamber: 1600°F

	Predicted	Actual
Outlet temperature	602°F	595°F
HRE	0.868	0.874

2. 36 in. Type 28H, 24 in. Type 48,
   42 in. (1.5 in.) Berl saddles
   Inlet: 450°F, .25 Btu/scf,
   Combustion chamber: 1600°F

	Predicted	Actual
Outlet temperature	538°F	549°F
HRE	0.923	0.914

FIGURE 8.41 John Zink RTO test unit.

This model is currently managed by Koch-Knight Division (a supplier of ceramic packing materials), and has been successfully utilized for numerous RTO bed retrofit applications. It is employed to estimate HRE changes for different flow rates and packing materials when customers want to change process conditions.

8.9 CONCLUSIONS

Numerous analytical techniques have been utilized to successfully model actual equipment performance by John Zink engineers. The following applications were reported in this chapter:

1. control volume analysis (eductor modeling — 8.2)
2. compressible flow analysis (eductor — 8.2, and oil gun modeling — 8.6)
3. detailed chemical kinetics (combinations of chemical reactors — 8.3)

4. extensions to classical fluid mechanics (burner pressure drops — 8.4)
5. novel scaling methodologies (flare smokeless operation — 8.5)
6. classic turbulence closure methods (flare smokeless operation — 8.5)
7. two-phase flow analysis (oil gun modeling — 8.6)
8. advanced experimental techniques (oil gun development — 8.7)
9. numerical ordinary differential equation solution techniques (oil gun modeling — 8.6, and RTO modeling — 8.8)

Combustion systems can be, and have been, modeled with good success. The broad range of applications and techniques reported in this chapter serve to illustrate the power of some of the various methods. Practicing engineers no longer have to be limited by narrow "rules of thumb." Rigorous application of fundamentals married to the power of modern computers can provide insight into equipment operation and yield performance predictions of greater precision than previously thought possible.

REFERENCES

1. Keenan, J.H. and Neumann, E.P., A simple air ejector, *J. Appl. Mech.*, A-75, 1942.
2. Ames Research Staff, Equation, Tables and Charts for Compressible Flow, NACA Report 1135, 1953.
3. Bartok, W. and Sarofim, A., *Fossil Fuel Combustion, A Source Book*, John Wiley & Sons, New York, 1991, 291-320.
4. Kee, R.J., Miller, J.A., and Jefferson, T.H., CHEMKIN: A General-Purpose, Problem-Independent, Transportable, FORTRAN Chemical Kinetics Code Package, Sandia Report SAND80-8003, 1980.
5. Kee, R.J., Rupley, F.M., and Miller, J.A., The CHEMKIN Thermodynamic Data Base, Sandia Report SAND87-8215B, 1990.
6. Kee, R.J. and Miller, J.A., A Structured Approach to the Computational Modeling of Chemical Kinetics and Molecular Transport in Flowing Systems, Sandia Report SAND86-8841, 1991.
7. Turns, S.R., *An Introduction to Combustion*, McGraw-Hill, New York, 1996.
8. Press, W.H., Teukolsky, S.A., Vetterling, W.T., and Flannery, B.P., *Numerical Recipes in FORTRAN — The Art of Scientific Computing*, Cambridge University Press, Cambridge, UK, 1992.
9. Longwell, J.P. and Weiss, M.A., High temperature reaction rates in hydrocarbon combustion, *Ind. Eng. Chem.*, 47, 1634–1643, 1955.
10. Beer, J.M. and Lee, K.B., The effect of the residence time distribution on the performance and efficiency of combustors, *Tenth Symposium on Combustion*, X, 1187–1202, 1965.
11. Lutz, A.E. and Broadwell, J.E., Simulation of Chemical Kinetics in Turbulent Natural Gas Combustion, GRI Report 92-0315, Gas Research Institute, 1992.
12. Lutz, A.E. and Broadwell, J.E., Simulation of Chemical Kinetics in Turbulent Natural Gas Combustion, GRI Report 94-0421.1, Gas Research Institute, 1994.
13. *ASHRAE Handbook*, 1985 Fundamentals, published by the American Society of Heating, Refrigeration and Air-Conditioning Engineers, Inc., Atlanta, GA.
14. Idelchik, I.E., *Handbook of Hydraulic Resistance*, Hemisphere, New York, 1986.
15. Reed, R.D., *Furnace Operations*, 3rd ed., Gulf Publishing, Houston, TX, 1981.
16. Hottel, H.C. and Hawthorne, W.R., Diffusion in laminar flame jets, *Third Symp. Combustion and Flame and Explosion Phenomena*, Williams and Wilkins Company, Baltimore, MD, 1949.
17. Becker, H.A. and Liang, D., Visible length of vertical free turbulent diffusion flames, *Comb. Flame*, 32, 115-137, 1978.
18. Blake, T.R. and McDonald, M., An Examination of Flame Length Data from Vertical Diffusion Flames, *Comb. Flame*, 94, 426-432, 1993.
19. Glassman, I., *22nd Int. Symp. Combustion*, Combustion Institute, Pittsburgh, PA, 1988, 295.
20. Gollahalli, S.R. and Parthasaranthy, R.P., Turbulent Smoke Points in a Cross-Wind, Research Testing Services Agreement No. RTSA 3-1-98, University of Oklahoma, Norman, OK, August 1999.
21. Schlichting, H., *Boundary-Layer Theory*, 7th ed., McGraw-Hill, New York, 1979.
22. Leite, O.C., Smokeless, efficient, nontoxic flaring, *Hydrocarbon Process.*, March 1991, 77-80.
23. Rokke, N.A., Hustad, J.E., and Sonju, O.K., A study of partially premixed unconfined propane flames, *Comb. Flame*, 97, 88-106, 1994.
24. Kaviany, M., *Principles of Heat Transfer in Porous Media*, Springer-Verlag, Berlin, 1991.
25. Kaviany, M., *Principles of Convective Heat Transfer*, Springer-Verlag, Berlin, 1994.

26. Chisolm, D., Flow of compressible two-phase mixtures through sharp-edged orifices, *J. Mech. Eng. Sci.*, 23, 45-48, 1981.

27. Chisolm, D., Gas–liquid flow in pipeline systems, in *Handbook of Fluids in Motion*, Cheremisinoff, N. P. and Gupta, R., Eds., Ann Arbor Science, Ann Arbor, MI, 1983.

28. Anderson, J.D., *Modern Compressible Flow: With Historical Perspective*, McGraw-Hill, New York, 1982.

29. Chung, I.P., Dunn-Rankin, D., and Ganji, A., Characterization of a spray from an ultrasonically modulated nozzle, *Atomization and Sprays*, 7, 295-315, 1997.

30. Bachalo, W.D., Method for measuring the size and velocity of spheres by dual-beam light scatter interferometry, *Appl. Opt.*, 19, 363-370, 1980.

31. Schmidt, F.W. and Willmott A.J., *Thermal Energy Storage and Regeneration*, Hemisphere/McGraw-Hill, Washington, D.C., 1981.

32. McDermott, R. and Henneke, M.R., "High Capacity, Ultra Low NOx Radiant Wall Burner Development," 12th Ethylene Forum, May 11–14, 1999, The Woodlands, TX.

33. Glarborg, P., Kee, R.J., Grear, J.F., Miller, J.A., PSR: A FORTRAN Program for Modeling Well-Stirred Reactors, Sandia Report SAND86-8209, 1986.

34. Chung, I.P. and Presser, C., Fluid properties effects on sheet disintegration of a simplex pressure-swirl atomizer, *AIAA J. Propulsion Power*, in press.

35. Lefebvre, A.H., *Atomization and Sprays*, Hemisphere, 1989.

36. Rosin, P. and Rammler, E., The law governing the fineness of powdered coal, *J. Inst. Fuel*, 7(31), 62-67, 1933.

37. Chin, J.S., Durrett, R., and Lefebvre, A.H., The interdependence of spray characteristics and evaporation history of fuel sprays, *ASME J. Eng. Gas Turbine Power*, 106, 639-644, 1984.

38. Chung, I.P., Dunn-Rankin, D., and Ganji, A., Characterization of a spray from an ultrasonically modulated nozzle, *Atomization and Sprays*, 7(3), 295-315, 1997.

39. Bachalo, W.D., Method for measuring the size and velocity of spheres by dual-beam light scatter interferometry, *Appl. Opt.*, 19(3), 363-370, 1980.

40. Saad, M.A., *Compressible Fluid Flow*, Prentice-Hall, Englewood Cliffs, NJ, 1993.

Chapter 9
Computational Fluid Dynamics (CFD) Based Combustion Modeling

Michael Henneke, Joseph D. Smith, Jaiwant D. Jayakaran, and Michael Lorra

TABLE OF CONTENTS

9.1 Overview 288
9.2 Introduction 288
 9.2.1 CFD Model Background 291
 9.2.2 The CFD Simulation Model 291
9.3 CFD-based Combustion Submodels 296
 9.3.1 Solution Algorithms 297
 9.3.2 Radiation Models 297
 9.3.3 Combustion Chemistry Models 299
 9.3.4 Pollutant Chemistry Models 301
 9.3.5 Turbulence Models 301
9.4 Solution Methodology 302
 9.4.1 Problem Setup: Preprocessing 302
 9.4.2 Solution Convergence 303
 9.4.3 Analysis of Results: Post-processing 303
9.5 Applications: Case Studies 305
 9.5.1 Case 1: Ethylene Pyrolysis Furnace 305
 9.5.2 Case 2: Xylene Reboiler 306
 9.5.3 Case 3: Sulfur Recovery Reaction Furnace 307
 9.5.4 Case 4: Incineration of Chlorinated Hydrocarbons 309
 9.5.5 Case 5: Venturi Eductor Optimization 319
9.6 Future Needs 319
9.7 Conclusion 319
9.8 Nomenclature 321
References 322

9.1 OVERVIEW

> If you want to win big, then you have to define the rules of the game.
>
> — M. Stahlman[1]

This concept has been used by companies and countries to control markets and economies. Changes to the rules are generally caused by significant technological advances. The invention of the telescope, the microscope, and the atom smasher are all examples of such paradigm shifting technological advances. Development of, and increased access to, supercomputers represents a significant advance that has opened new scientific frontiers and again changed the rules.

Several years ago, Nobel laureate and physicist Kenneth Wilson identified a collection of "grand challenges" to be addressed by researchers using supercomputers.[2] These challenges included designing efficient aircraft, simulating semiconductor materials, rationally designing drugs, understanding catalytic phenomena, studying air pollution, petroleum exploration, and analyzing fuel combustion. The chemical process industry (CPI) as a whole faces several of these "Grand Challenges." Future success of businesses in the process industry will depend on making advances in these "Grand Challenges."

Computational fluid dynamics (CFD) has been used in the CPI to solve various types of flow problems.[3,4] Krawczyk[5] has reviewed the applications and limitations of CFD modeling in the hydrocarbon processing industry (HPI). According to Krawczyk, the use of CFD in the HPI is relatively unknown because of a general lack of awareness of what CFD is and where it can be applied.

Past work by researchers in the chemical process industry[6] and the electric power industry[7] demonstrates the role that CFD can play in addressing important process questions (see Table 9.1). Specific work discussed in this chapter further illustrates how CFD has been used to investigate and solve important technical issues related to combustion processes. The cases described here are a small sampling of the many CFD simulations performed at the John Zink Company. The examples in this chapter deal with petrochemical process furnace optimization, venturi eductor design, sulfur recovery furnace simulation, and chlorinated hydrocarbon incineration.

As with any written work, it is a challenge to attempt to address a wide audience. Such is certainly the case for this chapter. In order to stay within the scope of this book, we have written this chapter to speak to that vital component of today's high-tech society, the engineer working in industry. The purpose of this chapter is to give the reader a concise introduction to both the science and application of CFD and to thereby

TABLE 9.1 Current CFD Applications in the Chemical Process Industry

* Rotary kiln incinerator	* Gas scrubber towers
* Drying ovens	* Thermal oxidation
* Packed catalysis beds	* Ventilation of foam scrap bin
* Crystallization	* Polymer extrusion
* Mixing tanks (liquid/liquid, etc.)	* Dust separation systems
* Impeller design	* Caustic evaporators
* Ceramic production	* Retention basin flow
* Membrane flow	* Liquid migration in diapers
* Caustic degasser inlet	* Nozzle design

communicate the power and potential of a cutting-edge 21st century technology. In writing about CFD it is difficult not to give the appearance of "wall to wall" equations. This is because, as the name implies, it is a computationally intensive science. In this chapter we limit the discussions to outlining the major topics, but even this necessitates providing several equations.

First, an overview of what CFD is presented, followed by a discussion of CFD software, with comments on the underlying foundations and assumptions. Next, the subprocesses involved in combustion simulation will be explained. In addition, the advantages and limitations of CFD in simulating fluid flows as well as combustion are also discussed. The latter part of the chapter is dedicated to case studies that will illustrate the problem-solving power of CFD. Finally, conclusions and a summary of issues that are important to the process industry regarding the future use of CFD as a process improvement tool will be presented.

9.2 INTRODUCTION

Knowledge in the field of fluid dynamics has evolved in three areas. First, *experimental* fluid dynamics emerged in France and England in the seventeenth century. Next, in the eighteenth and nineteenth centuries, *theoretical* fluid dynamics developed. For most of the twentieth century, fluid dynamics was practiced in these two realms. The advent of the high speed digital computer combined with the development of accurate numerical algorithms for solving physical problems has added a third approach to the study of fluid dynamics. This is *computational* fluid dynamics, or CFD. Today, computational fluid dynamics is an equal partner to the experimental and theoretical approaches. CFD synergistically complements the other two approaches nicely, but it will not replace them.[64]

CFD is based on the fundamental governing equations of fluid dynamics — the continuity, momentum, and energy equations. These equations *are* physics. So it is worthwhile to note at this point that much of CFD is based on fundamental physics and not on empirical functions. Therein lies CFD's power to extend to solving new flow problems. Simply stated, the fundamental physical principles underlying CFD, and all of fluid dynamics, are as follows:

1. Mass conservation
2. Newton's second law: F = ma
3. Energy conservation

These fundamental physical principles can be expressed in terms of mathematical equations, which in their most general form are either integral or partial differential equations. CFD is the art of replacing the integrals or the partial derivatives in these equations with discretized algebraic forms. These discrete equations are solved to obtain magnitudes for the different variables of interest (pressure, velocity, temperature, etc.) in the flow field at discrete points in time and space.

As the following sections will indicate, the equations are complex because they have to simultaneously deal with several interrelated transport mechanisms in great detail. Needless to say, they are also difficult to solve. Leading scientists have spent the last four decades developing models and approximations in order to quantify some of the terms in the equations that cannot be solved for directly. These models use a combination of fundamental physics, empirical functions and proven approximations to both reduce computational intensity and simplify the task at hand. One such model is the well-known k-ε turbulence model. The k-ε model is discussed in further detail later. Figure 9.1 is a graphical representation of the various elements that constitute CFD.

The CFD solution results in a collection of numbers that describe the flow field quantitatively. This matrix of values can then be queried for the values of interest, at the locations of interest, or, more commonly, represented graphically in contour or vector plots using color scales.

At this point the reader may have acquired the impression that because of its complexity, CFD is the realm of pure research and academics. It is true that this was once the case, but not any longer. The following analogy is given in an attempt to illustrate the current status of commercial implementation of CFD.

Let us consider the analogy of a piece of software, say, a spreadsheet software package. One may say that there are three levels of engineers involved with this spreadsheet software. First is the developer who develops the algorithms, software architecture, and basic code. Second is the intermediate user who uses the basic package to build sophisticated calculations for his or her specific needs, and in addition, can add modules of custom code or macros to the basic package to extend its abilities. Third would be the end-user who uses the spreadsheet tools developed by the intermediate user to perform calculations but does not add to the software.

These three levels of users may also be identified in CFD. The development engineer is highly specialized and must have several years of education in fluid dynamics and CFD. One may call this first level the CFD *specialist*. Next, the intermediate user in CFD not only uses the software, but has to have a sound understanding of CFD to extend its capabilities and critically evaluate the results it is providing. This second level may be termed the *CFD engineer*. The third level is the end-user, really the *customer* requiring the CFD. The customer knows enough about the capabilities of the technology to be able to identify when a problem may be a candidate for CFD and is able to consult with the CFD engineer about the technical details as well as the benefits and costs. In dealing with combustion simulation, the added complexity requires the CFD engineer (intermediate level in the example above) to have specialized knowledge to set up complex three-dimensional CFD simulations and interpret the results. As such, the skill set required for CFD combustion modeling leaves little or no distinction between a CFD specialist and a CFD engineer.

CFD first evolved, and still is strongest, in its ability to solve non-reacting flows. A wide variety of practical engineering problems can be solved by analyzing non-reacting flows. For example, the ability to predict the behavior of fluid flows through various flow passage configurations and conditions has promoted the rapid progress in the automotive and aircraft industries. However, in more recent decades, model development work in CFD has naturally extended into reacting flows because of the undeniable need.

Combustion is a complex reacting flow. Due to the inherent complexities of combustion and the relatively recent development of the combustion models, there are some limitations in the current state-of-the-art of CFD combustion modeling. These limitations are discussed later. Limitations and youth notwithstanding, currently available combustion modeling tools can provide tremendous, real-world problem-solving power.

The main intent of this chapter is to describe combustion modeling, and so, for brevity, the discussions of basic CFD flow modeling have been kept to a minimum. However, it must be kept in mind that reacting flow modeling is merely a superset of basic non-reacting flow modeling. In performing a CFD simulation of a reacting flow, the computations pertaining to reaction induced changes to fluid properties and composition are performed in conjunction with the same calculations that would be performed for any non-reacting flow simulation. Consequently, reacting flow modeling is considerably more resource intensive and challenging. In many cases it is possible

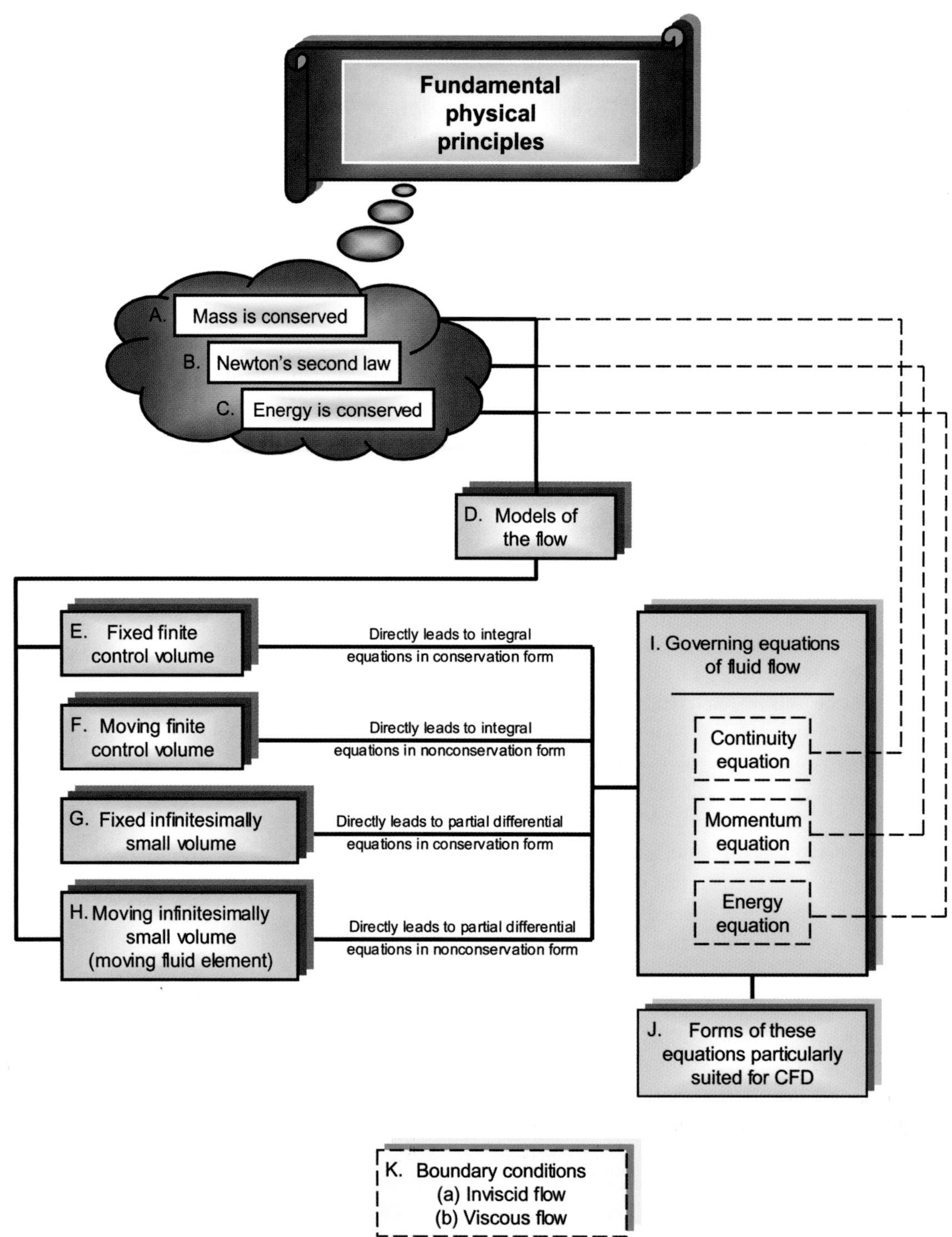

FIGURE 9.1 Elements in CFD modeling.

to use non-reacting flow simulation results to infer the reacting flow result. Therefore, frequently, in planning the solution approach, the CFD engineer must make a value vs. effort decision to choose between full combustion modeling and non-reacting flow modeling.

The reader will find that the emphasis of this chapter is gas-fired furnaces, with only brief discussion of oil-fired furnaces. This bias is reflective of the bias in the open literature where little discussion of modeling full-scale oil-fired furnaces can be found. Petrochemical furnaces are typically gas- or oil-fired. In the United States, most petrochemical furnaces are gas-fired, but in other parts of the world, various grades of oil are used. The examples discussed here rely on the ability to accurately simulate turbulent reacting flow with radiant heat transport in an enclosed space.

The reader will quickly observe that industrial-scale combustion modeling is an immature field (even after years of scientific effort), and may wonder why industries invest in such efforts. The simple answer is that there is no better alternative. In the past, experimentalists have tried to simulate the flow patterns inside industrial furnaces using scaled-down plexiglas representations of the geometry. Although the complex physics of the real furnace (with combustion, radiation, buoyancy, etc.) are completely neglected, these experimentalists have met with some success by being able to relate cold flow phenomena observed in the plexiglas model to the combustion phenomena in a furnace. However, this technique has its limitations because a good portion of the expected result is inferred and not based on detailed knowledge of the actual physics and chemistry. On the other hand, even with its numerous approximations about turbulence, radiation, and chemical reactions, a CFD model provides much more detailed information about the furnace process than a nonreacting, scaled-down plexiglas model.

Other possible options available to the engineer include either scaled reacting flow or full scale reacting flow experimentation. Full scale research is very expensive because industrial, multiple burner furnaces cost millions of dollars, and reacting flow experimentation on a scaled-down model has its share of disadvantages. Scaling combustion systems from the laboratory scale to the industrial scale is very difficult. No change scaling is well-understood theoretically. To have "similarity" as a combustion system is scaled, the Reynolds number and Damköhler numbers must remain unchanged. This is practically impossible, so some sort of incomplete scaling is used. Typically, combustion systems are scaled using a constant-velocity or constant-momentum flux method which does not provide predictable results in all cases.

Development of burners for petrochemical applications is usually done experimentally using a single burner at full scale. This method frequently produces a burner that performs well in the industrial setting; however, multiple burner firing typically produces higher NOx emissions than single burner firing.[8] Other flame interaction problems can arise as well. For example, flames from individual burners are occasionally observed to merge together, causing flame length to increase significantly when burner spacing is not sufficient. These problems are especially significant for ultra-low-NOx burners because these burners typically produce flame lengths significantly longer than a conventional burner. As the flame is "stretched out" by fuel staging strategies, it becomes more and more difficult to control the mixing between the fuel, air, and furnace gases (products of combustion). In addition, the stability of these ultra low NOx burner flames is problematic. With regard to full scale single burner development, it should be kept in mind that the dominant flow currents in an operating environment may be very different from those in a single burner furnace where the burner was developed. These operational problems, which otherwise could only be tackled by trial and error, can be systematically analyzed and solved using CFD.

Herein lies ample justification for the petrochemical industry to pursue CFD technology to better understand the performance of furnaces. Hopefully, this chapter will illustrate some of the potential value of CFD technology and stimulate further studies to better understand and extend the performance of CFD models in these applications.

9.2.1 CFD Model Background

In the work described below, two CFD software packages have been used, Fluent and PCGC3. Both codes have evolved from similar backgrounds, having been based on the SIMPLE (Semi-Implicit Method for Pressure Linked Equations) algorithm. SIMPLE solves a set of nonlinear coupled partial differential equations describing the conservation of mass, momentum, and energy, as described in more detail below. Originally, Fluent was designed to simulate basic fluid dynamics, while PCGC3 was designed to model pulverized coal combustion systems. More recently, both codes have been extended to address general combustion phenomena occurring in several diverse systems (e.g., pulverized coal combustor, hazardous waste incinerators, petrochemical process heaters, etc.). To illustrate the basis from which CFD codes are derived and to illustrate the capabilities and limitations, a brief review of the CFD code PCGC3 is given.

9.2.2 The CFD Simulation Model

Fluent is based on the original work done by Swithenbank and co-workers at the University of Sheffield.[65] For more than fifteen years, this code has been developed and extended by engineers at Fluent, Inc. PCGC3 was developed in the Advanced Combustion Engineering Research Center (ACERC) at Brigham Young University in the same time frame by various researchers.[7,9–18] PCGC3 describes a variety of reactive and nonreactive flow systems, including turbulent combustion and gasification of pulverized coal. To illustrate the model capabilities and limitations, a limited discussion of PCGC3 will be presented. A more detailed description is given by Smoot and Smith.[19]

9.2.2.1 Transport Equations

Nonreactive turbulent gas flow is modeled using the steady-state form of the Navier-Stokes equations by assuming a continuous flow field described locally by the general conservation of mass and momentum:[20]

$$\frac{D\rho}{Dt} = -\rho(\nabla \cdot \vec{u}) \tag{9.1}$$

$$\rho\frac{D\vec{u}}{Dt} = -\nabla p - \nabla \cdot \tau + \rho \underline{g} \tag{9.2}$$

where D/Dt is the total (or substantial) derivative. These equations can be simplified by assuming steady-state flow of a Newtonian fluid. If the fluid is assumed to be incompressible, the dilatation $(\nabla \cdot \vec{v})$ can also be used to further simplify the equations. An incompressible assumption implies that the code is applicable to low Mach number flows (i.e., Mach number $\lesssim 0.3$). Given these simplifying assumptions, the general conservation or transport equations for mass and momentum can be written in Cartesian tensor form as:

$$\frac{\partial(\rho u_j)}{\partial x_j} = 0 \tag{9.3}$$

$$\frac{\partial(\rho u_i u_j)}{\partial x_j} = -\frac{\partial P}{\partial x_i} + \frac{\partial \tau_{ij}}{\partial x_j} + \rho f_i \tag{9.4}$$

$$\tau_{ij} = \mu\left(\frac{\partial u_i}{\partial x_j} + \frac{\partial u_j}{\partial x_i}\right) + \left(\mu_B - \frac{2}{3}\mu\right)\frac{\partial u_k}{\partial x_k}\delta_{ij} \tag{9.5}$$

Similarly, a transport equation can be written for a conserved scalar, Φ, as:

$$\frac{\partial(\rho \Phi u_j)}{\partial x_j} = \frac{\partial}{\partial x_j}\left(\Gamma_\phi \frac{\partial \Phi}{\partial x_j}\right) + S_\Phi \tag{9.6}$$

Turbulent transport is characterized by both time and length scales, the smallest scales being too small to numerically resolve for practical problems. Therefore, Eqs. (9.3) through (9.6) are not solved directly, but are transformed by one of two averaging techniques: Reynolds averaging (most common operation applicable to incompressible flows) or Favre averaging (mass-weighted averaging applicable to compressible flows). Favre or mass-weighted averaging is accomplished by first decomposing the variable into a mean and a fluctuating term and then time-averaging the resulting equation set:

$$\Phi = \tilde{\Phi} + \phi'' \tag{9.7}$$

where $\tilde{\Phi} = \overline{\rho\Phi}/\bar{\rho}$ and $\overline{\rho\phi''} = 0$, but $\overline{\phi''} \neq 0$. Applying this decomposition to all variables except density and pressure, the conservation equations [Eqs. (9.3) through (9.6)] are transformed into the mass-averaged or Favre-averaged transport equations:

$$\frac{\partial(\bar{\rho}\tilde{u}_j)}{\partial x_j} = 0 \tag{9.8}$$

$$\frac{\partial(\bar{\rho}\tilde{u}_i\tilde{u}_j)}{\partial x_j} = -\frac{\partial \bar{P}}{\partial x_i} + \frac{\partial}{\partial x_j}\left(\bar{\tau}_{ij} - \overline{\rho u_i'' u_j''}\right) + \bar{\rho}\bar{f}_i \tag{9.9}$$

$$\frac{\partial(\bar{\rho}\tilde{\Phi}u_j)}{\partial x_j} = \frac{\partial}{\partial x_j}\left(\bar{\Gamma}_\phi \frac{\partial \tilde{\Phi}}{\partial x_j} - \overline{\rho\phi'' u_j''}\right) + \bar{S}_\Phi \tag{9.10}$$

These constitute the turbulent transport equation set for nonreacting flow. However, as a result of the averaging procedure, several additional variables called Favre stresses $(\overline{\rho u_i'' u_j''})$ and Favre fluxes $(\overline{\rho\phi'' u_j''})$ have been introduced. These stresses and fluxes represent the mean-momentum transport and the mean-scalar transport by turbulent diffusion. Additional ancillary equations are required to solve for these new turbulent transport variables. These extra equations make up the "turbulence" model.

9.2.2.2 Turbulence Equations

Although several turbulence models have been proposed (Nallasamy[21]), the k-ε turbulence model, originally proposed by Harlow and Nakayama,[22] remains the most widely used model to describe practical flow systems (Speziale[23]). The k-ε turbulence model employs a modified version of the Boussinesq hypothesis (Smoot and Smith[19]):

$$-\overline{u_i'' u_j''} = -\nu_t\left(\frac{\partial \tilde{u}_i}{\partial x_j} + \frac{\partial \tilde{u}_j}{\partial x_i}\right) - \frac{2}{3}\left(\nu_t \nabla \cdot \vec{v} + k\right)\delta_{ij} \tag{9.11}$$

where ν_t is known as the eddy diffusivity or turbulent viscosity. This approach allows the molecular viscosity to be replaced with the eddy diffusivity, which allows the instantaneous transport equations [Eqs. (9.3) through (9.6)] to be modeled using the mean-value equations [Eqs. (9.8) through

(9.10)]. A disadvantage to this approach is the need to assume isotropic eddy diffusivity. However, given this assumption and the specific velocity and length scales u' and l':

$$u' \approx \sqrt{k} \qquad l' \approx C_\mu \frac{k^{3/2}}{\varepsilon} \tag{9.12}$$

where the turbulent kinetic energy can be defined as:

$$k = \frac{1}{2}\overline{u_i u_i} = \frac{1}{2}\left(\overline{u_1 u_1} + \overline{u_2 u_2} + \overline{u_3 u_3}\right) \tag{9.13}$$

With these definitions, additional transport equations for the turbulent kinetic energy and the dissipation rate of turbulent kinetic energy, ε, can be written:

$$\frac{\partial k}{\partial t} + \bar{u} \bullet \nabla k = \frac{1}{\rho}\nabla \bullet \left(\frac{v_t}{\sigma_k}\nabla k\right) + G - \varepsilon$$

$$G = \frac{v_t}{\sigma_k}\left[2\nabla^2 \bar{u} + \sum_{i=1, i\neq j}^{3}\sum_{j=1}^{3}\left(\frac{\partial u_i}{\partial x_j} + \frac{\partial u_j}{\partial x_i}\right)\right] \tag{9.14}$$

$$\frac{\partial \varepsilon}{\partial t} + \bar{u} \bullet \nabla \varepsilon = \frac{1}{\rho}\nabla \bullet \left(\frac{v_t}{\sigma_\varepsilon}\nabla \varepsilon\right) + f_1 c_1 G\left(\frac{\varepsilon}{k}\right) - f_2 c_2\left(\frac{\varepsilon^2}{k}\right) \tag{9.15}$$

and the eddy diffusivity is defined as:

$$v_t = \frac{f_\mu c_\mu k^2}{\varepsilon} \tag{9.16}$$

Several key "empirical" constants are required by the k-ε turbulence model. The values used by PCGC3 are shown in Table 9.2.

The values shown in Table 9.2 were determined through comparison of turbulence and numerical optimization as discussed by Sloan et al.[12] These values are similar to those originally proposed by Launder and Spalding,[24] but differ slightly from those reported by other researchers (Nallasamy,[21] Lilleheie et al.,[25] Jones and Whitelaw,[26]). This may be due to the fact that these "empirical" constants are based on simple two-dimensional flows and adjustment may be required to simulate more complex flows. Regardless, this fact and the other simplifying assumptions suggest that the flow results be closely scrutinized when applying any CFD code using this turbulence model to simulate complex flow systems.

Application of the k-ε turbulence model requires boundary conditions for both k and ε. Boundary layer theory could be used to derive the equations for flow near the wall, but to reduce computer storage and runtimes the k-ε turbulence model uses wall functions instead. The Van Driest hypothesis on turbulent flow near walls is used to derive wall functions consistent with the logarithmic law of the wall.[24] These functions relate the dependent variables near the wall to those in the bulk flow field. Boundary conditions used in the turbulence model are discussed at length by Gillis and Smith.[18]

TABLE 9.2 Universal "Empirical" Constants Used in k-ε Turbulence Model

Constant:	C_μ	C_1	C_2	σ_k	σ_ε	κ
Value:	0.09	1.44	1.92	0.9	1.22	0.42

Given the turbulence model with the necessary boundary conditions, the full equation set can be written (see Tables 9.3 and 9.4). As shown, each equation is conveniently cast into a general convection-diffusion form with the off-terms collected on the right-hand side, and the specific terms that depend on the coordinate system selected. Examining the θ-momentum equation (see Table 9.4) helps illustrate the meaning of each term:

$$\begin{aligned}
& r\frac{\partial(\tilde{\rho}\tilde{u}\tilde{w})}{\partial x} + \frac{\partial(r\tilde{\rho}\tilde{v}\tilde{w})}{\partial r} + \frac{\partial(\tilde{\rho}\tilde{w}\tilde{w})}{\partial \theta} \\
& -r\frac{\partial}{\partial x}\left(\mu_e \frac{\partial \tilde{w}}{\partial x}\right) - \frac{\partial}{\partial r}\left(r\mu_e \frac{\partial \tilde{w}}{\partial r}\right) - \frac{\partial}{\partial \theta}\left(\frac{\mu_e}{r}\frac{\partial \tilde{w}}{\partial \theta}\right) = \\
& -\frac{\partial p}{\partial \theta} + r\frac{\partial}{\partial x}\left(\frac{\mu_e}{r}\frac{\partial \tilde{u}}{\partial \theta}\right) + \frac{\partial}{\partial r}\left(\mu_e \frac{\partial \tilde{v}}{\partial \theta} - \mu_e \tilde{w}\right) + \\
& \frac{\partial}{\partial \theta}\left[\left(\frac{\mu_e}{r}\right)\left(\frac{\partial \tilde{w}}{\partial \theta} + 2r\frac{\partial \tilde{v}}{\partial r}\right)\right] + \\
& \mu_e\left(\frac{\partial \tilde{w}}{\partial r} + \frac{1}{r}\frac{\partial \tilde{v}}{\partial \theta} - \frac{\tilde{w}}{r}\right) - \tilde{\rho}\tilde{v}\tilde{w} + r\tilde{\rho}g_\theta
\end{aligned} \tag{9.17}$$

The first three terms of Eq. (9.17) represent the net rate of momentum addition to a volume element by convection from the three direction components. The fourth, fifth, and sixth terms represent the corresponding diffusion terms. When the turbulence model solves for the individual Reynolds stresses, the diffusion terms do not strictly represent molecular diffusion, but also include momentum contributions due to the turbulent motion of the fluid. The first term on the right-hand side (RHS) of Eq. (9.17) represents the pressure force on the volume element. All other terms on the RHS of the equation represent either a source or sink term for momentum (e.g., gravity force, centripetal forces, etc.).

TABLE 9.3 Cartesian Differential Equation Set

$$\frac{\partial(\bar{\rho}\tilde{u}\phi)}{\partial x}+\frac{\partial(\bar{\rho}\tilde{v}\phi)}{\partial y}+\frac{\partial(\bar{\rho}\tilde{w}\phi)}{\partial z}-\frac{\partial}{\partial x}\left(\Gamma_\phi\frac{\partial(\phi)}{\partial x}\right)-\frac{\partial}{\partial y}\left(\Gamma_\phi\frac{\partial(\phi)}{\partial y}\right)-\frac{\partial}{\partial z}\left(\Gamma_\phi\frac{\partial(\phi)}{\partial z}\right)=S_\phi$$

Equation	ϕ	Γ_ϕ	S_ϕ
Continuity	1	0	0
X-momentum	$\tilde{u}$	μ_e	$-\frac{\partial p}{\partial x}+\frac{\partial}{\partial x}\left(\mu_e\frac{\partial\tilde{u}}{\partial x}\right)+\frac{\partial}{\partial y}\left(\mu_e\frac{\partial\tilde{v}}{\partial x}\right)+\frac{\partial}{\partial z}\left(\mu_e\frac{\partial\tilde{w}}{\partial x}\right)+\bar{\rho}g_x-\frac{2}{3}\bar{\rho}\tilde{k}$
Y-momentum	$\tilde{v}$	μ_e	$-\frac{\partial p}{\partial y}+\frac{\partial}{\partial x}\left(\mu_e\frac{\partial\tilde{u}}{\partial y}\right)+\frac{\partial}{\partial y}\left(\mu_e\frac{\partial\tilde{v}}{\partial y}\right)+\frac{\partial}{\partial z}\left(\mu_e\frac{\partial\tilde{w}}{\partial y}\right)+\bar{\rho}g_y-\frac{2}{3}\bar{\rho}\tilde{k}$
Z-momentum	$\tilde{w}$	μ_e	$-\frac{\partial p}{\partial z}+\frac{\partial}{\partial x}\left(\mu_e\frac{\partial\tilde{u}}{\partial z}\right)+\frac{\partial}{\partial y}\left(\mu_e\frac{\partial\tilde{v}}{\partial z}\right)+\frac{\partial}{\partial z}\left(\mu_e\frac{\partial\tilde{w}}{\partial z}\right)+\bar{\rho}g_z-\frac{2}{3}\bar{\rho}\tilde{k}$
Mixture fraction	$\tilde{f}$	$\frac{\mu_e}{\sigma_f}$	0
Mixture fraction variance	$\tilde{g}$	$\frac{\mu_e}{\sigma_g}$	$-\frac{C_{g1}\mu_e}{\sigma_g}+\left[\left(\frac{\partial\tilde{f}}{\partial x}\right)^2+\left(\frac{\partial\tilde{f}}{\partial y}\right)^2+\left(\frac{\partial\tilde{f}}{\partial z}\right)^2\right]-C_{g2}\bar{\rho}\tilde{g}\frac{\tilde{\varepsilon}}{\tilde{k}}$
Turbulent energy	$\tilde{k}$	$\frac{\mu_e}{\sigma_k}$	$G-\bar{\rho}\tilde{\varepsilon}$
Dissipation rate	$\tilde{\varepsilon}$	$\frac{\mu_e}{\sigma_\varepsilon}$	$\left(\frac{\tilde{\varepsilon}}{\tilde{k}}\right)(c_1G-c_2\bar{\rho}\tilde{\varepsilon})$

Note: $G=\mu_e\left\{2\left[\left(\frac{\partial\tilde{u}}{\partial x}\right)^2+\left(\frac{\partial\tilde{v}}{\partial y}\right)^2+\left(\frac{\partial\tilde{w}}{\partial z}\right)^2\right]+\left(\frac{\partial\tilde{u}}{\partial y}+\frac{\partial\tilde{v}}{\partial x}\right)^2+\left(\frac{\partial\tilde{u}}{\partial x}+\frac{\partial\tilde{w}}{\partial x}\right)^2+\left(\frac{\partial\tilde{v}}{\partial x}+\frac{\partial\tilde{w}}{\partial y}\right)^2\right\}$

TABLE 9.4 Cylindrical Differential Equation Set

$$r\frac{\partial(\bar{\rho}\tilde{u}\phi)}{\partial x}+\frac{\partial(r\bar{\rho}\tilde{v}\phi)}{\partial r}+\frac{\partial(\bar{\rho}\tilde{w}\phi)}{\partial\theta}-r\frac{\partial}{\partial x}\left(\Gamma_\phi\frac{\partial(\phi)}{\partial x}\right)-\frac{\partial}{\partial r}\left(r\Gamma_\phi\frac{\partial(\phi)}{\partial r}\right)-\frac{\partial}{\partial\theta}\left(\frac{\Gamma_\phi}{r}\frac{\partial(\phi)}{\partial\theta}\right)=S_\phi$$

Equation	ϕ	Γ_ϕ	S_ϕ
Continuity	1	0	0
X-momentum	$\tilde{u}$	μ_e	$-r\frac{\partial p}{\partial x}+r\frac{\partial}{\partial x}\left(\mu_e\frac{\partial\tilde{u}}{\partial x}\right)+\frac{\partial}{\partial r}\left(r\mu_e\frac{\partial\tilde{v}}{\partial x}\right)+\frac{\partial}{\partial\theta}\left(\mu_e\frac{\partial\tilde{w}}{\partial x}\right)+r\bar{\rho}g_x-\frac{2}{3}r\bar{\rho}\tilde{k}$
R-momentum	$\tilde{v}$	μ_e	$-r\frac{\partial p}{\partial r}+r\frac{\partial}{\partial x}\left(\mu_e\frac{\partial\tilde{u}}{\partial r}\right)+\frac{\partial}{\partial r}\left(r\mu_e\frac{\partial\tilde{v}}{\partial r}\right)+\frac{\partial}{\partial\theta}\left(\mu_e\frac{\partial\tilde{w}}{\partial r}-\frac{\tilde{w}}{r}\right)-\frac{2\mu_e}{r}\frac{\partial\tilde{w}}{\partial\theta}-\frac{2\tilde{v}\mu_e}{r}+\bar{\rho}\tilde{w}^2+r\bar{\rho}g_r-\frac{2}{3}\bar{\rho}\tilde{k}$
θ-momentum	$\tilde{w}$	μ_e	$-\frac{\partial\rho}{\partial\theta}+r\frac{\partial}{\partial x}\left(\frac{\mu_e}{r}\frac{\partial\tilde{u}}{2\theta}\right)+\frac{\partial}{\partial r}\left(\mu_e\frac{\partial\tilde{v}}{\partial\theta}-\mu_e\tilde{w}\right)$ $-\bar{\rho}\tilde{v}\tilde{w}+\frac{\partial}{\partial\theta}\left[\left(\frac{\mu_e}{r}\right)\left(\frac{\partial\tilde{w}}{\partial\theta}-2\tilde{v}\right)\right]+\mu_e\left(\frac{\partial\tilde{w}}{\partial r}+\frac{1}{r}\frac{\partial\tilde{w}}{\partial\theta}-\frac{\tilde{w}}{r}\right)+r\bar{\rho}g_\theta-\frac{2}{3}r\bar{\rho}\tilde{k}$
Mixture fraction	$\tilde{f}$	$\frac{\mu_e}{\sigma_f}$	0
Mixture fraction variance	$\tilde{g}$	$\frac{\mu_e}{\sigma_g}$	$-\frac{C_{g1}\mu_e r}{\sigma_g}+\left[\left(\frac{\partial\tilde{f}}{\partial x}\right)^2+\left(\frac{\partial\tilde{f}}{\partial r}\right)^2+\left(\frac{1}{r}\frac{\partial\tilde{f}}{\partial\theta}\right)^2\right]-C_{g2}r\bar{\rho}\tilde{g}\frac{\tilde{\varepsilon}}{\tilde{k}}$
Turbulent energy	$\tilde{k}$	$\frac{\mu_e}{\sigma_k}$	$r(G-\bar{\rho}\tilde{\varepsilon})$
Dissipation rate	$\tilde{\varepsilon}$	$\frac{\mu_e}{\sigma_\varepsilon}$	$\left(r\frac{\tilde{\varepsilon}}{\tilde{k}}\right)(c_1G-c_2\bar{\rho}\tilde{\varepsilon})$

Note: $G=\mu_e\left\{2\left[\left(\frac{\partial\tilde{u}}{\partial x}\right)^2+\left(\frac{\partial\tilde{v}}{\partial r}\right)^2+\left(\frac{1}{r}\frac{\partial\tilde{w}}{\partial\theta}+\frac{\tilde{v}}{r}\right)^2\right]+\left(\frac{\partial\tilde{u}}{\partial r}+\frac{\partial\tilde{v}}{\partial x}\right)^2+\left(\frac{1}{r}\frac{\partial\tilde{u}}{\partial\theta}+\frac{\partial\tilde{w}}{\partial x}\right)^2+\left(\frac{1}{r}\frac{\partial\tilde{v}}{\partial\theta}+\frac{\partial\tilde{w}}{\partial r}-\frac{\tilde{w}}{r^2}\right)^2\right\}$

9.2.2.3 Solution Technique

The equation set shown above is composed of several steady-state, second-order, nonlinear, elliptic partial differential equations (PDEs). Each of these continuous PDEs is transformed into a discrete finite difference equation (FDEs). Examining the equation set, seven equations (mass (1), momentum (3), turbulence (2), and conserved scalar (1)) with six unknowns (P, u, v, w, k, ε, and f) describe the turbulent flow system. Typically, the momentum equations are solved for each velocity component, the turbulence equations are solved for the respective turbulence variables, and the continuity equation is left for the pressure field. A key issue in CFD is solving for the pressure gradient source terms, found in the momentum equations, because the pressure fields for enclosed flows are usually unknown. PCGC3 uses variations of the SIMPLE (Semi-Implicit Method for Pressure Linked Equations) algorithm to solve the equations of motion and continuity in a decoupled fashion, by transforming the continuity equation into a pressure correction equation. A detailed description of the various forms of SIMPLE found in PCGC3 is given by MacArthur.[27]

In PCGC3, all equations are cast into one standard form so that a single solver is required. PCGC3 uses the tri-diagonal algorithm to solve the FDEs for each variable along a line on each plane of the computational space. The variables are solved for in succession, starting with a velocity component and ending with one of the turbulence variables. Because the FDEs are solved in a decoupled fashion, only four to five "micro-iterations" are required per variable. A complete cycle through the equation set, termed a macro-iteration, resolves the nonlinear coupling between equations to a prespecified convergence criterion. Overall convergence typically requires between 200 and 1000 macro-iterations.

9.2.2.4 Convergence Criteria

Although PCGC3 iterates on each equation individually, the equation coupling necessitates simultaneous convergence of the entire equation set. Various methods have been used to measure convergence, compare convergence rates of each equation, and determine when the required level of convergence is obtained. Typically, the error used to track convergence represents the residual for each FDE as shown:

$$R_\phi^0 = A_E\phi_E + A_W\phi_W + A_N\phi_N + A_S\phi_S + A_T\phi_T$$
$$+A_B\phi_B + S_U - A_P\phi_P \tag{9.18}$$

where $A_p\phi_p$ represents the computational node and the other $A_i\phi_i$ represent the neighboring nodes, S_u represents the source term (RHS of equations shown above), and R_ϕ represents the residual or relative error in the equation. As the solution converges, the residual is forced to zero and the convergence criteria is satisfied. Comparison of errors from each of the seven equations is difficult because of the relative magnitude of the coefficients (A_i) for each equation. Normalization is also difficult due to the range of variable and source term magnitudes within each equation. Without comparison of the convergence of each equation, it is impossible to determine when "overall" convergence is achieved or which equation is slowing the convergence process.

PCGC3 uses the largest term found in each variable's FDE to normalize the respective residual. This truncation term, defined as:

$$\psi_p = A_p\phi_p \tag{9.19}$$

is guaranteed to exceed the magnitude of the other terms in the FDE because of the requirement of diagonal dominance in the coefficient matrix (satisfied by the TDMA algorithm) and because of the manipulation of the source terms (RHS of equation). The final normalized equation error is thus calculated as:

$$R_\phi^\psi = \frac{\sum_{n=1}^{N} R_\phi^o}{\sum_{n=1}^{N} \psi_\phi} \tag{9.20}$$

This normalization allows for the comparison of equation error from the different equations and measures the closeness of computer round-off error to equation error. The total equation error, R_ϕ^ψ, ranges from approximately 1 to 10^{-nd}, where nd is the number of digits of computer accuracy. When the equation error reaches 10^{-nd}, computer round-off error prevents further reductions in equation error. Having determined the machine accuracy, PCGC3 determines and prints out the difference between this and the number of digits of computer accuracy, nd, for each equation and uses the term representing the greatest inaccuracy to judge when a calculation is converged.

9.2.2.5 Model Validation

The PCGC software has been extensively validated by direct comparison between model predictions and experimental data for several systems.[14] Fletcher[9] shows a comparison between mean velocity values obtained using laser Doppler velocimetry (LDV) techniques and predictions. This comparison shows reasonable agreement for a hydrogen/air diffusion flame for two different flow rates. Similarly, measured values of gas-phase concentrations of H_2, H_2O, and O_2 agree well with prediction. Measured RMS velocities show some disagreement with predicted values. This discrepancy illustrates the nonisotropic nature of the fluid mechanics, which are assumed isotropic in the turbulence model. However, the overall agreement between predictions and measurements demonstrates the applicability of the PCGC approach to simulating high-temperature reacting flow systems such as combustion.

9.2.2.6 Modeling Basis

The simulation results of reacting flow systems using PCGC software are based on two main assumptions:

1. uniform heat loss from the reactor (no heat loss indicates adiabatic operation)
2. local gas-phase chemistry is micro-mixing-limited and composition is determined using thermodynamic equilibrium.

The first assumption, based on the Crocco similarity,[7] is valid for cases where thermal diffusion and mass diffusion are equal. Some knowledge relative to the total reactor heat load and the related heat loss is required. The user must specify reactor heat loss as a fraction of the total energy in the system. This fraction is then extracted equally from each discrete computational cell in the overall simulation. Thus, this assumption neglects the effect of the temperature gradient that can produce artificially low local temperature regions while yielding appropriate exit temperatures.

The second assumption, generally valid for high-temperature combustion chemistry, suggests that the homogeneous kinetics are sufficiently fast so that gas mixing is controlling. This is commonly referred to as the "mixed-is-burnt" assumption. If this assumption applies, the local chemistry can be calculated from general thermodynamic calculations. The validity of this assumption is determined by the relative reaction and mixing (convective/diffusive) time scales, as expressed by the Damköhler number, Da:

$$\mathrm{Da} = \frac{t_{flow}}{t_{rxn}} = \frac{l_t/v'}{l_F/S_L} \tag{9.21}$$

where S_L is the burning velocity, l_F is the reaction zone thickness, v' is the turbulence intensity, and l_t is the turbulence length scale. Using this relationship, two physical limits have been identified.[28] The first, referred to as the "frozen" limit occurs when:

$$\mathrm{Da} \to 0 \tag{9.22}$$

In this case, the reaction time (t_{rxn}) is much larger than the flow time (t_{flow}), and kinetic effects are negligible compared to mixing effects. Conversely, the second limit, referred to as the "fast-chemistry" limit occurs when:

$$\mathrm{Da} \to \infty \tag{9.23}$$

In this case, the reaction time is very short (fast reactions) relative to the mixing time. Many diffusion flames are approximated well by the latter limit.

Given the large disparity between short reaction time scales and long mixing time scales, chemical activity may be confined to an infinitesimally thin layer, commonly referred to as a "flamelet" or "flamesheet."[29] This assumption allows flame chemistry to be approximated using local thermodynamic equilibrium without significant error. Well-known exceptions include NOx and CO chemistry where the reaction time scale and the mixing time scale are of similar magnitude:

$$\mathrm{Da} \approx O(1) \tag{9.24}$$

Here, finite-rate chemistry must be coupled with the turbulent fluid mechanics calculations. Because turbulent effects must be included in the kinetic scheme, global mechanisms are generally used to avoid solving individual transport equations for each specie in a detailed kinetics mechanism.[19]

9.3 CFD-BASED COMBUSTION SUBMODELS

The current state-of-the-art approaches in modeling petrochemical furnaces are described in this section. This is not a comprehensive discussion of all of the available models and algorithms available to the CFD analyst because not all of the models are commonly used in modeling furnaces. While a large number of algorithms and models are studied in academic circles, the majority of CFD analyses of large-scale furnaces are done with commercial CFD packages. There are currently several commercial CFD packages with very similar modeling capabilities. This section discusses the modeling approaches used by these codes, as well as several models currently being studied at the research level. These "under development" models will lead to improved combustion modeling capability in the future for industrial users.

There are a large number of approximations involved in modeling combustion processes in furnaces. Even in a gas-fired furnace, the multitude of important physics is daunting. The flow in the furnaces is turbulent flow with a very large integral length scale (the characteristic dimension of the furnace). The combustion chemistry in the furnace involves tens to hundreds of chemical species reacting with time scales from less than a microsecond to several seconds. Radiative transport from a non-gray gas (the products of combustion) to the furnace walls and tubes (with the process fluid flowing inside) whose emissivity is temperature dependent is the primary mode of heat transfer. The interaction between these physical processes is of considerable importance. The turbulence/chemistry interaction has been well-studied for many years, particularly for non-premixed systems. More recently, the interaction between turbulence and radiative emission from a non-gray gas has been studied.[30]

The geometries of the burners used in these furnaces are becoming increasingly complex. The dominant driver in most burner designs is NOx reduction, and this leads to burners that are more and more geometrically complex. Added to the geometric complexity is the chemical complexity. The preferred strategy for reducing NOx emissions from gas-fired burners is to use staged fuel systems and to use the fuel jets to entrain large amounts of the products of combustion into the flame zone. This means that to make a burner with lower NOx emissions, one has to make a turbulent flame that is comparatively less stable. For the CFD analyst, modeling the stability of a flame or its lift-off height is at present a very imposing problem. For this reason, CFD predictions of NOx emissions from ultra-low-NOx burners are typically poor. However, even if the quantitative NOx predictions are poor, the qualitative information from a CFD calculation can often be very useful. For example, the results of a CFD analysis can be used to study the entrainment of cooled furnace gases by fuel jets as well as the mixing of the fuel jets with combustion air. When used by an analyst who has a thorough knowledge of what real flames do, these quantitative measures, in conjunction with experimentation, can be used to solve equipment problems. As an example, McDermott and Henneke[31] used an axisymmetric CFD model to design turning vanes in a premixed burner. The problem being addressed was flashback, which is very difficult to model in a CFD study. However, by combining CFD analysis with knowledge of what conditions allow flashback, the authors were able to design a series of turning vanes that eliminated flashback for a wide range of operating conditions.

9.3.1 Solution Algorithms

This section discusses the algorithms used to solve the Reynolds- (or Favre-) averaged Navier-Stokes equations, the Reynolds- (or Favre-) averaged energy and species equations, and the radiative transfer equation. No discussion is given here of large eddy simulations (LES) or direct numerical simulations (DNS). LES may become a viable option for CFD modeling of furnaces in the near future, but at present the increase in computational cost of an LES calculation is usually not justifiable. Pope[32] notes that the appeal of LES in nonreacting flows is the expectation that the small scales of turbulence are universally related to the large scales. In a reacting flow, there is no similar expectation. LES does have the advantage of resolving the large-scale structures that challenge the Favre-averaged models, but many difficult problems remain to be addressed before LES will be a useful tool. Bray[33] notes that despite the fact that LES faces current difficulties, it will be successfully developed and will be a useful tool for the combustion modeler. DNS can be categorically neglected for this class of problems because the computational demands are far in excess of current computational resources.[33,34]

9.3.2 Radiation Models

Typical petrochemical furnaces consist of a radiant section and a convection section. These regions are so named because of the dominant mode of heat transfer. In the radiant section, refractory surface temperatures can be higher than 2200°F (1200°C). Radiant heat is incident on the process tubes, both from the high-temperature surfaces and directly from the flame. Accurate modeling of the heat delivered to the process fluid requires an accurate prediction of the radiant intensity inside the furnace. In addition, accurate prediction of radiation from the flame is necessary to accurately predict emissions. For example, Barlow[35] notes that the different radiation models can affect NOx predictions just as much as the different turbulence/chemistry interaction models that were evaluated by him.

Thermal radiation transport presents a difficult problem because of the number of independent variables. The radiation transport equation (RTE) describes radiation transport in absorbing, emitting, and scattering media. The equation is:[36]

$$\frac{\partial I_\lambda}{\partial s} = \kappa_\lambda I_{b\lambda} - \beta_\lambda I_\lambda + \frac{\sigma_{s\lambda}}{4\pi}\int_{4\pi} I_\eta(\vec{s}_i)\Phi(\vec{s}_i,\vec{s})d\Omega_i \quad (9.25)$$

Radiant intensity is a function of location (three coordinates in a three-dimensional problem), direction (two angular independent variables), and wavelength (one independent variable varying from 0 to infinity). This means that the problem of radiative transport is a six-dimensional problem. A common approach is to remove the wavelength dependence by making gray media approximations discussed below. The angular dependence can be treated by angular discretization, such as in the P-1 method or the discrete ordinates method. Methods such as the finite-difference and finite-volume methods discussed above treat the spatial dependence. Both the P-1 and discrete ordinates methods approximate the angular dependence of the equation of transfer. The Monte Carlo method takes a much different approach. In the Monte Carlo method, individual photons of radiant energy are emitted, reflected, and absorbed by both opaque surfaces and participating media using ray tracing algorithms. This method provides a very elegant approach to treating non-gray radiation as well as the directional dependence of radiation. Its use is limited by its computational cost.

Siegel and Howell[37] and Modest[36] provide extensive discussion of the solution methods for radiation in participating media. These texts discuss the accuracy, computational effort, and limitations of the various models. The reader should consult these books for further discussion of these solution methods.

9.3.2.1 Gas Radiation Properties

Molecular gas radiation is an important mode of heat transfer in gas-fired furnaces. Radiative emission from nonluminous hydrocarbon flames is mostly due to the H_2O and CO_2 species present in the products of combustion. Radiation from these gases is fairly well-understood, but a rigorous treatment of this radiation requires significant computational resources. For example, Mazumder and Modest[30] considered ten radiative bands in modeling emission from a hydrocarbon flame. This means that they solved the RTE for ten different intensities. In a large-scale furnace calculation, such a model would be extremely computationally demanding.

Quantum mechanics postulates that molecular gases emit and absorb gases only at distinct wavenumbers, called spectral lines. However, in reality, these distinct lines are broadened by several mechanisms, including collision broadening, natural line broadening, and Doppler broadening. These individual lines are characterized by a line strength and a line width. These lines are caused by quantum transitions in the vibrational or rotational state of a molecule. Frequently, vibrational and rotational transitions occur simultaneously, leading to a tightly clustered array of lines around a given vibrational transition. This subject is beyond the scope of the present chapter. The intent here is to illustrate the complexity of modeling a radiating gas.

9.3.2.2 Soot

The presence of soot in a flame can significantly increase the flame emissivity. Predicting soot formation within a flame is very difficult because soot is formed in fuel-rich regions of a flame when the temperature is high. Models such as those of Khan and Greeves[38] and Tesner et al.[39] allow the prediction of soot concentrations, but these models are very empirical and cannot be expected to provide quantitative results.

Soot within a flame is caused by the combustion of hydrocarbons under fuel-rich conditions. Soot is visually observed as a yellow-red brightness in the flame. C_2 hydrocarbons and higher have more tendency to soot, while methane does not normally produce a sooty flame. Soot has a strong impact on flame radiation. Emission from soot in flames is frequently much larger than the gas radiation emitted by the flame.[36] In some applications (oil-firing, in particular), soot emissions from the flame are regulated by environmental agencies. In flaring applications, smokeless (smoke results from unoxidized soot particles leaving the flame) operation is frequently guaranteed by the flare vendor for some range of conditions. In petrochemical applications, the gases flared are a wide range of hydrocarbons, typically ranging in molecular weight from 16 to 40. These gases have components such as ethylene and acetylene, which are known precursors to soot formation. Current CFD codes (limited by physical model availability) cannot predict smoking from these large, buoyant flare fires, but current LES work in this area appears promising.

9.3.2.3 Weighted Sum of Gray Gases

The weighted-sum-of-gray-gases (WSGG) model[37] provides formulae for computing the emissivity of a gas volume as a function of its temperature and partial pressures of CO_2 and H_2O. The model assumes the gas is a mixture of radiating gases that is transparent between the absorption bands. The WSGG model is probably the most widely used method to calculate radiation within combustion gases. Alternatives include band models discussed below. The computational cost of radiation transport can be very high compared to the flow solver portion of a simulation because of the large number of independent variables in the RTE. In practice, it is usually reasonable to lag the calculation of the RTE for a number of flow solver iterations, with the actual number dependent on the solver in use and stability requirements.

9.3.2.4 Turbulence/Radiation Interaction

The turbulence/radiation interaction plays an important role in predicting the radiative emission from a flame. Unfortunately, none of the available commercial products that we are aware of attempt to model this interaction. To appreciate the significance of this issue, consider the time-averaged radiative transport equation in an absorbing/emitting media:

$$\frac{\partial \bar{I}_\lambda}{\partial s} = \overline{\kappa_\lambda \left(I_{b\lambda} - I_\lambda \right)} \tag{9.26}$$

Frequently, the time-averaged emission is computed as $\overline{\kappa_\lambda I_{b\lambda}} \approx \overline{\kappa_\lambda}\, \overline{I_{b\lambda}}$, which neglects correlations between κ_λ and T as well as the effect of temperature fluctuations on the time-averaged emission (i.e., $\overline{T^4} \neq \overline{T}^4$). Mazumder and Modest[30] discuss the history of the turbulence/radiation interaction.

9.3.3 Combustion Chemistry Models

This section discusses the modeling of combustion chemistry in petrochemical applications. The focus of this section is on methods for modeling the interaction of turbulence with combustion chemistry. This is an area of intense current research, and some of this research is briefly discussed as it pertains to current CFD calculations as well as near-future CFD calculations. There are several relatively new turbulence/chemistry interaction models (such as CMC and joint-pdf transport models) that are not currently available for use in any of the commercial CFD packages. One can hope that this situation will change soon and these models will be available for more widespread use.

9.3.3.1 Regimes of Turbulent Combustion

Damköhler numbers are ratios of a fluid dynamical time scale to a chemical time scale.[40] In a turbulent flow, there are a variety of time scales, such as the integral scale (a convective scale) and the Kolmogorov scale (a viscous scale). There are also a variety of chemical time scales because of the many chemical reactions that accompany the combustion of even a simple molecule such as CH_4. Frequently, combustion problems are described as being in the high Damköhler or flamelet regime. The term "flamelet" is used because of the notion that within a turbulent non-premixed flame, the actual combustion reactions take place within small layers termed "flamelets." These flamelets are so thin that they are not affected by the turbulent motions within the fluid; instead, molecular diffusion effects dominate and the structure of the reaction zone is that of a laminar flame (albeit a strained laminar flame).

Following Bray,[33] the Damköhler number is defined as $\mathrm{Da} = \frac{t_T}{t_L^0} = \frac{k u_L^0}{\varepsilon l_L^0}$ where the subscript L and superscript 0 refer to an unstretched laminar flame, and the subscript T refers to the scale of the turbulence. In cases where non-premixed combustion is studied, it is common to use the velocity and length scales (the laminar premixed flame speed and thickness) as representative of the relevant chemical scales. The Karlovitz number is $\mathrm{Ka} = \frac{t_L^0}{t_k^0} = \frac{t_L}{\sqrt{\nu/\varepsilon}}$ where the subscript K refers to the Kolmogorov time scale. When the laminar flame time is less than the Kolmogorov scale (i.e., $\mathrm{Ka} < 1$), the flame is considered to be a laminar flame convected and stretched by a turbulent flow. Combustion in this regime is referred to as flamelet combustion. When the Damköhler number is less than 1, the time scale of larger turbulent eddies has become smaller than the chemical time scale. In these conditions, the combustion process is described as a well-stirred reaction zone. For intermediate values of Da and Ka, combustion is said to occur in distributed reaction zones. This term indicates that the turbulent flow can affect the structure of the reaction zone, in contrast to the flamelet regime, but the turbulent mixing is not so fast that the reaction can be considered to occur under well-stirred conditions.

9.3.3.2 Non-premixed Combustion

This section discusses modeling of non-premixed combustion systems. It is implicitly assumed that non-premixed combustion is being discussed, but the notion of non-premixed combustion is an idealization. In real combustion systems, mixing occurs simultaneously with combustion, and to call the combustion process non-premixed implies that the combustion takes place much faster than the mixing and that the flame is not lifted off or near extinction at any location. Although it is an idealization, the assumption that combustion is non-premixed provides very useful insight into the combustion processes occurring in real systems.

There are a multitude of computational models for non-premixed (also called diffusion) flames. One of the earliest models to appear is the eddy-breakup model of Spalding.[41] The model of Magnussen and Hjertager[42] limits the reaction rate according to the local mass fractions of the reactant concentrations or product concentrations. The ratio of the turbulent kinetic energy k to the dissipation rate ε is used as the time scale of the turbulent eddies controlling mixing. These models give physically reasonable predictions of species concentrations in non-premixed systems, but do not consider the important effect of turbulent fluctuations on reaction rates. The model can be extended to consider finite-rate chemistry, but the model is a moment model, using the time-averaged temperature in the Arrhenius rate expression. This limitation is severe in light of the large temperature fluctuations observed in flames.

The mixture fraction concept plays a central role in reducing a turbulent non-premixed flame to a mixing problem. The mixture fraction is a conserved scalar, meaning that it is convected and diffused by fluid motions and gradients, but it is neither created nor destroyed. The mixture fraction, f, represents the mass fraction of fluid at a particular location that originated with the fuel stream. The pure fuel stream then will have $f = 1$, while the oxidant stream will have $f = 0$.

In a turbulent flow, the mixture fraction f fluctuates at a given point with time. A probability density function (pdf) for these fluctuations can be defined so that the probability

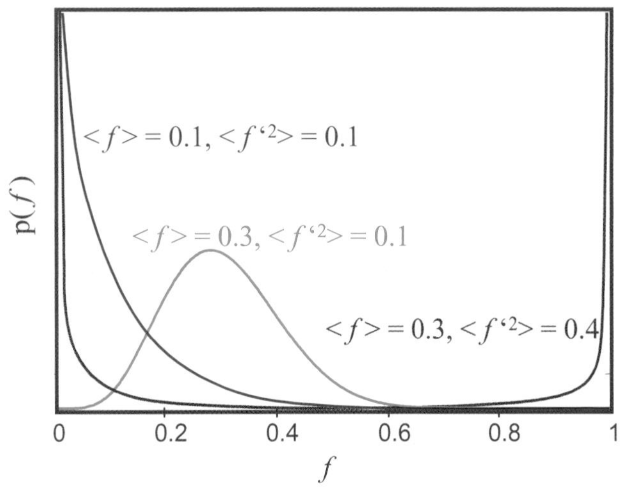

FIGURE 9.2 Plot of the β-pdf for several values of $\langle f \rangle$ and $\langle f'^2 \rangle$.

of f lying between x and $(x + dx)$ is $P(x)dx$. The pdf has some additional properties:[43]

$$\int_0^1 P(f)df = 1 \tag{9.27}$$

$$\langle f \rangle = \int_0^1 fP(f)df \tag{9.28}$$

$$\langle f'^2 \rangle = \int_0^1 (f - \langle f \rangle)^2 P(f)df \tag{9.29}$$

where the $\langle f \rangle$ notation indicates the expectation value (or ensemble average, equivalent to the time average in a statistically stationary flow) of f, and f' is the turbulent fluctuation of f. $\langle f'^2 \rangle$ is the variance of f.

In CFD calculations of large-scale furnaces with non-premixed burners, the most common combustion model used is the assumed-pdf model with equilibrium chemistry. In this model, the shape of the pdf of f is assumed. The β-pdf is a commonly used function that describes the probability of finding the instantaneous fluid to have a specific mixture fraction. The β-function is given by:

$$P(f) = \frac{\langle f \rangle^{\alpha-1}(1 - \langle f \rangle)^{\beta-1}}{\int_0^1 \langle f \rangle^{\alpha-1}(1 - \langle f \rangle)^{\beta-1} d\langle f \rangle} \tag{9.30}$$

where

$$\alpha = \langle f \rangle \left[\frac{\langle f \rangle (1 - \langle f \rangle)}{\langle f'^2 \rangle} - 1 \right] \tag{9.31}$$

$$\beta = (1 - \langle f \rangle) \left[\frac{\langle f \rangle (1 - \langle f \rangle)}{\langle f'^2 \rangle} - 1 \right] \tag{9.32}$$

Other pdf shapes, such as a clipped Gaussian function and a double delta-function, are discussed in Jones and Whitelaw.[44] The equilibrium chemistry assumption is poor in flames that are lifted or flames near extinction. Figure 9.2 shows the shape of the β-pdf for several values of $\langle f \rangle$ and $\langle f'^2 \rangle$.

An alternative to using the equilibrium chemistry assumption is to use the laminar flamelet model. In this model, the relationship between the state of the mixture and the mixture fraction f is determined by a laminar diffusion flame calculation. Peters[45] introduced this idea, which assumes that the reaction length scale, L_R, is much smaller than the Kolmogorov length scale, L_K. Bilger[46] has criticized the classical flamelet method, claiming that for most non-premixed flames of interest, the flamelet criterion, $L_R < L_K$, is violated. Bish and Dahm (1995)[47] discuss the concept further and attempt to eliminate what they view as a key limitation of the method: its assumption that the reaction layers are bounded by pure fuel on one side and pure oxidizer on the other. Their SDRL model is based on the one-dimensionality of the reaction layer, but does not assume the reaction layer to be thin relative to the dissipative scales.

The classical flamelet model's assumption that the reaction zones are bounded by pure fuel on one side and pure oxidizer on the other is severe in light of the NOx control strategies used in practical combustion systems. NOx control is predicated on entraining cooled combustion products into the reaction zone, and the proportion of these gases entrained varies along the length of the flame. The effects of this flue gas entrainment are to reduce flame temperatures and dilute the reactants. Both of these effects are effective in reducing NOx formation.

Research of models of non-premixed combustion continues at a fervent pace. Pope's[32] joint pdf methods appear promising because they have the ability to treat finite-rate kinetics and eliminate the closure problems. Bilger's[48] conditional moment closure (CMC) method is also a promising model for non-premixed combustion modeling. Both of these models are applicable to premixed combustion as well.[33] These models are still subjects of active research and academic debate and are not implemented in any of the commercial packages of which these authors are aware.

9.3.3.3 Premixed Combustion

Most petrochemical applications use non-premixed combustion because of safety issues in premixed combustion. There are some important exceptions to this statement, however. One important class of premixed burners in the petrochemical industry are venturi-based radiant wall burners. These burners use high-pressure fuel to educt combustion air from the ambient environment. The fuel and combustion air are then mixed in a tube prior to the combustion zone.

Turbulent premixed flames have proven to be much more difficult to model than their non-premixed counterparts.[32] In a turbulent, mixing-limited, non-premixed flame, the flame structure is governed by turbulent mixing, a reasonably well-understood phenomenon. The ideal turbulent premixed flame consists of a flame sheet propagating at some flame speed with respect to the fluid around it, which is itself undergoing turbulent motions. The consequence of superposing flame propagation and turbulent fluid motions is that premixed flame modeling is much more challenging than modeling non-premixed flames.[34] For this reason, most commercial CFD codes only include limited support for premixed flame modeling.

The model of Magnussen and Hjertager[42] can be used to simulate a premixed flame. The model is unsatisfactory in many ways, however. It has no means of modeling the effect of temperature fluctuations on the reaction rate and no description of the turbulent flame as an ensemble of premixed flamelets.

A significant limitation of many of the flamelet models for premixed combustion (see, e.g., Warnatz et al.[34]) is that they assume that the combustion process is adiabatic. In operating furnaces, heat losses from the flame to the load are an integral part of the process. The inability to adequately model premixed flames is a significant limitation. Pdf methods such as those discussed by Pope[32] may allow improved simulations in the future. In addition, it may be possible to include heat losses in the flamelet models of premixed combustion.

9.3.4 *Pollutant Chemistry Models*

Pollutant emissions are among the most important drivers in the petrochemical industry, especially in the United States (see Chapter 6). The U.S. EPA allowed levels of NOx and SOx emissions from petrochemical plants and refineries continue to decrease. To respond to this challenge, burner manufacturers strive to develop burners that produce less and less emissions. In addition, furnace manufacturers and other vendors develop post-combustion technologies such as SCR (selective catalytic reduction) and SNCR (selective noncatalytic reduction) to deal with NOx in the stack. Sulfur scrubbers are used to reduce SOx levels after the combustion process.

A recent paper by Barlow et al.[35] evaluated NOx predictions using two different models for the turbulence/ chemistry interaction: the probability density function (pdf) model of Pope (see Pope,[32] for example) and the conditional moment closure (CMC) model of Bilger.[48] The pdf model here is not the assumed-pdf discussed above. Instead, the method used solves for the transport and production of the scalar joint pdf and is extremely computationally expensive because a Monte Carlo solution algorithm must be used. The particularly interesting thing about this article is the comment in the introduction that "a realistic target for agreement between experiment and prediction might be ±20 to ±30%." The article goes on to discuss how sensitive NOx predictions are to the radiation model used. The flame studied in this paper is a simple diluted hydrogen jet flame. If the most sophisticated turbulence/chemistry models currently under research applied to a very simple flame in a very simple geometry can only be expected to yield an accuracy of ±30%, then how accurately can one realistically expect to predict NOx emissions in more complex flames?

9.3.5 *Turbulence Models*

A turbulent flow is a flow with a wide range of temporal and length scales. Figure 9.3 is an example of a typical point measurement (e.g., pressure or velocity) within a turbulent flow. Within a turbulent flow, the quantities of interest such as pressure and velocity fluctuate in an apparently random fashion. Analysis reveals that these quantities are not truly random.[43] Information revealed by spectral analysis of point measurements reveals that there are ranges of temporal and length scales that contain significant energy (the large or integral scales) and smaller scales where this turbulent energy is dissipated by viscous processes. The energy cascade is the mechanism by which energy is moved from the large scales to the small scales. For more information on the physics of turbulent flows, the reader should refer to Libby.[43]

Prediction of turbulent flow from the Reynolds- or Favre-averaged conservation equations requires closure approximations. This is because the time-averaged conservation equations contain terms that are not known. In the case of the momentum equations, the time-averaging of the convection terms leads to the following Reynolds stresses:

$$\overline{\rho u'^2},\ \overline{\rho v'^2},\ \overline{\rho w'^2},\ \overline{\rho u'v'},\ \overline{\rho u'w'},\ \overline{\rho v'w'}$$

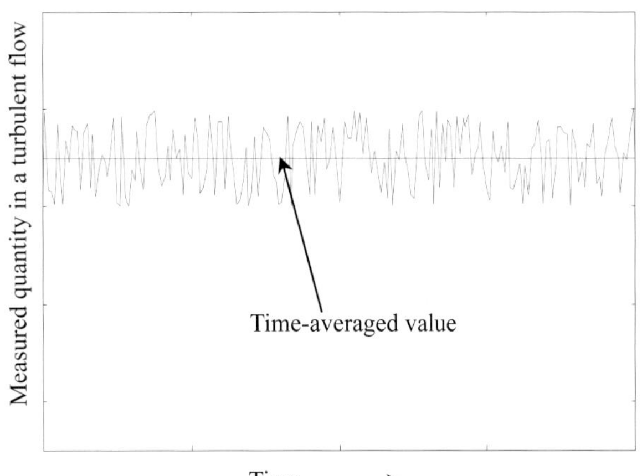

FIGURE 9.3 Point measurement of scalar in a turbulent flow.

Closure approximations are required to solve the Reynolds-averaged conservation equations.

The workhorse turbulence model used in furnace simulations is the k-ε model.[24] The popularity of this model can be ascribed to its relative simplicity (compared to a Reynolds stress model, for example) and its good performance in a variety of engineering flows. Its weaknesses include its performance in unconfined flows, in rotating and swirling flows, and in flows with large strains, such as curved boundary layers. The Reynolds stress model (RSM) addresses some of these performance issues. RSM is much more computationally demanding because it involves seven extra partial differential equations rather than the two of the k-ε model. However, in a typical combustion calculation, the number of PDEs solved is typically quite large, so adding five more can be easily justified if the quality of the predictions improve.

A number of variants of the classical k-ε model exist. The classical k-ε model uses a single eddy viscosity in all directions. The nonlinear k-ε model by Speziale[49] addresses this assumption, which is known to be poor even in relatively simple flows. Another development in k-ε modeling is the Renormalization Group (RNG) k-ε model of Yakhot et al.[50] Its performance in complex flows has been promising, so much so that several of the commercial CFD code vendors have implemented the RNG k-ε model. The Realizable k-ε model[51] represents yet another variant recently introduced. The advantages and limitations of these turbulence models are discussed in more detail in Veersteeg and Malalasekera.[52]

9.4 SOLUTION METHODOLGY

There are many schemes used to discretize the partial differential equations of fluid flows onto different types of meshes. Because the primary focus of this chapter is applied CFD where mesh types, by necessity, include tetrahedra, the two important discretization schemes are the finite-volume method and the finite-element method. The finite-volume method is clearly the method of choice in the industry today for large-scale computations of turbulent flows. The dominant software products commercially available for these problems almost exclusively use the finite-volume method. There are occasions when other methods, such as the finite-difference method, are used. Most advanced combustion models are first implemented in academic CFD codes, which, as mentioned in the introduction, are rarely intended to model complex geometry. For this reason, industrial CFD analysts rarely have access to advanced combustion models.

9.4.1 Problem Setup: Preprocessing

The preprocessing phase of a problem includes all the steps from the initial problem definition through the beginning of computations. In typical problems, this includes geometry creation, mesh generation, model selection, fluid property specification, and enabling and setting up the models (such as k-ε for turbulence).

Problem definition is critical. At this phase, the scope of the geometry to be studied should be considered. In many cases, it is difficult to determine where to place the outside boundaries of a CFD model. Flow condition must be known at inlets. Thus, for example, putting a flow inlet just downstream of an elbow would probably be a poor choice because it would be difficult to know the velocity and pressure profiles at such a location. This issue is particularly important if heat transfer is to be considered. Thermal boundary conditions are typically difficult to specify, requiring considerable physical insight into a problem.

It is also important to consider the capabilities of the software and computer hardware to be used when specifying a problem. For example, if the software's only turbulence model is the k-ε model, then studying a highly swirling flow (where k-ε is known to perform poorly) may generate useless results. On the other hand, if one is aware of this limitation and recognizes that the turbulence model will not accurately predict the decay of the swirl, a conscious decision can be made to neglect the portions of the solution that are expected to be poor and only use the results that are expected to be meaningful.

9.4.1.1 Complex Geometry

The capability of a CFD package to treat complex geometries is an important consideration for industrial applications. The geometries encountered in low emissions burners frequently employ complicated shapes and jet angles. The purpose of these geometries is to precisely control when and where the fuel is oxidized. These combustion control strategies are critical to the performance of the equipment. CFD models must be able to accurately capture the effect of these complex geometries in order to be useful. This means, as a practical matter, that cell types other than hexahedra are required. All current-generation commercial codes are compatible with a variety of cell types, including hexahedra, prisms, pyramids, and tetrahedra. Figure 9.4 shows a rendered view of a CFD model of a hearth burner. The burner shown in Figure 9.4 is vertically fired, and the view shown is looking down at the burner from above. There are four gas tips, two in the primary combustion zone (inside the tile) and two firing secondary fuel (these tips are sitting on the tile). Each primary tip has five small orifices firing fuel in different directions, as shown in Figure 9.5. The orifice diameters range from 0.0625 in. (1.59 mm) to about 0.25 in. (6.4 mm). Each secondary tip has three fuel orifices of similar size. The figure illustrates the necessity of a CFD package to treat complex geometry to accurately model the performance of an industrial burner.

Generating a computational mesh for these geometries is a well-known bottleneck in CFD analysis. Mesh generation can consume well over half of the time budgeted for a CFD project. Improvements in mesh generation technology greatly benefit industrial CFD users as they allow more and more of the actual geometry to be included in the CFD model. In addition, mesh generation improvements frequently simplify the process of modifying an existing geometry. Making geometric modifications to a geometry after the initial meshing can be nearly as time-consuming as generating the initial mesh.

New combustion and turbulence models are frequently developed only for simple Cartesian grids. Only much later, after considerable proof of concept in the simple geometries, are these models adapted for use on mesh types used in industrial analysis. In the present authors' opinions, the ability to treat complex geometry is the main reason that industries turn to commercial CFD vendors instead of using codes developed at national laboratories and universities.

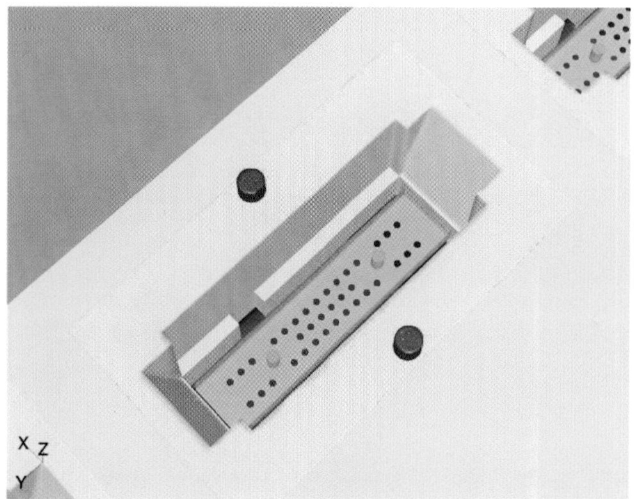

FIGURE 9.4 Rendered view of a CFD model of a John Zink Co. burner. This view illustrates the complex geometry that necessitates a variety of cell types. This mesh consists of hexahedral, pyramidal, and tetrahedral cell types. (Courtesy of John Zink Co.)

9.4.2 Solution Convergence

After generating the mesh and setting up the problem for solution, the calculations begin. This phase of the CFD analysis does not usually require much effort unless severe convergence problems are encountered. Normally, all that is required of the analyst is to observe the progress of the CFD code toward convergence and perhaps adjust under-relaxation factors and adapt the grid. It is always the stated goal to obtain a solution that is grid-independent, but in practice it usually is too time-consuming to refine the mesh such that the obtained solution can be proven grid-independent.

9.4.3 Analysis of Results: Post-processing

A typical CFD simulation provides on the order of 10^6 to 10^8 discrete numerical outputs. For example, a simulation with 500,000 nodes and 11 variables per node (pressure, density, three velocity components, k, ε, temperature, mixture fraction, variance of mixture fraction, and irradiation) would generate 5,500,000 numbers. If the various chemical species are considered as well as the detailed results of a discrete ordinates model, the number of variables per node could easily exceed 50, leading to 25,000,000 numerical results. The generation of x-y plots (e.g., temperature vs. position along the burner centerline), contour plots, velocity vector plots, streamline plots, and combinations and animations of these outputs are necessary for the analyst to understand the results of a simulation. The production of these different sorts of outputs becomes very important in communicating the results of a simulation. This is especially true when the intended audience is not composed of CFD specialists. Current post-processing packages have the ability to add lighting to a model, which makes the

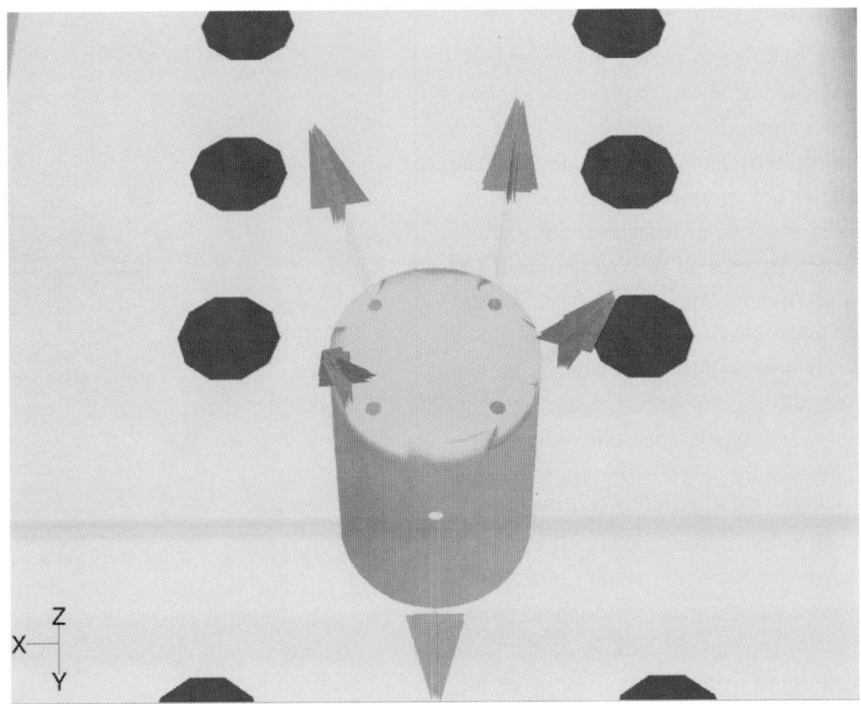

FIGURE 9.5 Close-up view of primary tip. This view reveals the five fuel jets (indicated by the arrows on the image) issuing from the primary tip.

FIGURE 9.6 Rendered view inside an ethylene pyrolysis furnace showing flow patterns near the premixed radiant wall burners. (Courtesy of John Zink Co.)

images more realistic to the viewer. Figure 9.6 shows an example of using the rendering capabilities of a CFD package to generate an image with photo-realistic qualities. Images such as Figure 9.6 can take anywhere from several seconds to several minutes for current generation scientific workstations to render, depending on the number of lights applied, the number of surfaces in the scene, and the complexity of these surfaces. High-performance virtual reality environments must be able to regenerate these scenes many times per second.

In addition to still images, animations can be effectively used to illustrate CFD results. Animated velocity vectors and streamlines accurately illustrate the path of fluid flow in internal and external flow problems. Sweeping planes showing either velocity vectors or filled contour maps of a scalar result can quickly present information about an entire three-dimensional simulation.

Generating effective presentations, including still images and animations, is a time-consuming task. Creating a suitable image to make a specific argument frequently requires the analyst to look at and reject a large number of candidate images. It also requires significant expertise from the CFD analyst. It is certainly true that CFD results can be misinterpreted or misapplied to lead to an incorrect conclusion. In addition, in an industrial setting, the audience will frequently not have the expertise required to assess the quality of a simulation.

FIGURE 9.7 CFD model of an ethylene pyrolysis furnace. There are six burners shown in each row at the bottom of the furnace, and the tubes are approximately 35 feet long. The endwalls are not shown in this image.

9.5 APPLICATIONS: CASE STUDIES

This section describes several applications of CFD in the petrochemical industry. CFD can address a wide variety of problems in this industry. The applications discussed here relate to fired heaters and incinerators. Many other applications (e.g., flare systems) exist in petrochemical plants where CFD analysis is valuable.

9.5.1 Case 1: Ethylene Pyrolysis Furnace

Ethylene pyrolysis furnaces produce ethylene and propylene from feedstock containing ethane, propane, butane, and hydrocarbons including naphtha. The process entails rapidly heating the feedstock for a short time (less than 1 second is typical) to a temperature of about 1600°F (870°C). The feed gases are then rapidly cooled and subjected to a number of separation processes.

This section focuses on the modeling of the pyrolysis furnace. Typical pyrolysis furnaces are approximately 10 ft (3 m) wide, 30 ft (9 m) long, and 40 ft (12 m) tall. There are two rows of "flat flame" burners that directly fire onto the walls of the furnace. These fired walls then radiate heat to the process tubes in the center of the furnace. Figure 9.6 illustrates this geometry. In the figure, the process fluid tubes extend from the floor of the furnace to the roof of the radiant section. In the image, only the radiant section is shown because the radiant section is where the combustion occurs. In a production furnace, the products of combustion would leave the radiant section and enter a convection section where heat is recovered from the products of combustion.

Figure 9.4 shows a view of the burner geometry. The burner is a staged-fuel gas burner. This example illustrates the disparity in scales in a furnace analysis. The furnace has a height of approximately 30 ft (9 m), while the fuel orifices can be as small as 0.0625 in. (1.59 mm) in diameter. The ratio from the largest dimension to the smallest is then greater than 5000. In this example, a nonconformal mesh interface was used to reduce the cell requirements.

The CFD model of the ethylene pyrolysis furnace includes detailed information about all of the fuel jets in the burner. In this particular burner, there are five fuel jets on each of the two primary tips (Figure 9.5) and four fuel jets on each secondary tip.

Figure 9.8 shows the predictions of heat flux to the process tubes as a function of height above the furnace floor. These

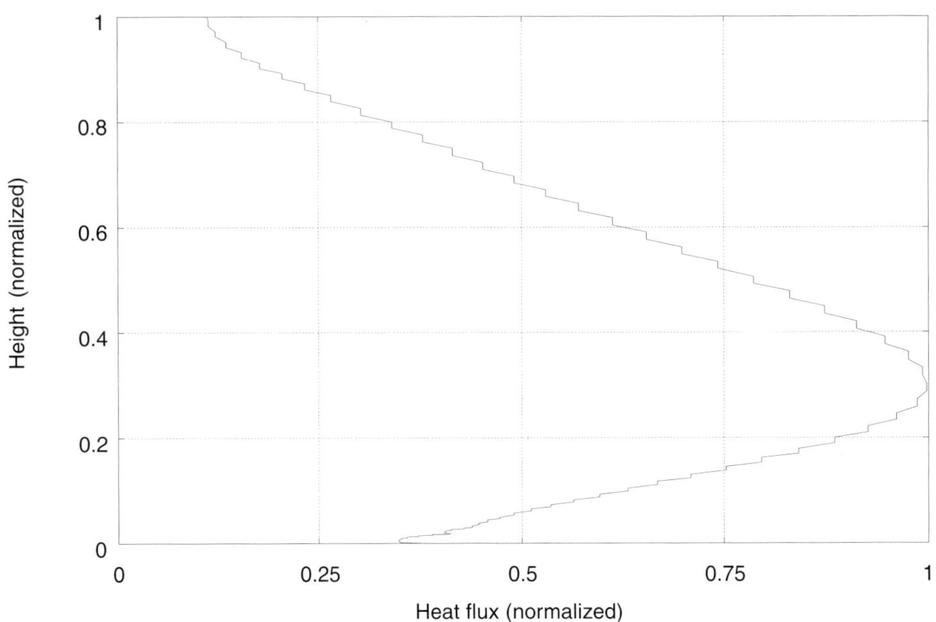

FIGURE 9.8 Plot showing heat flux to the process tubes in the modeled ethylene furnace as a function of height above the furnace floor.

heat flux profiles drive the design of modern ethylene pyrolysis furnaces. CFD is being used in these designs more and more as the results of the model become better validated. These results have not been validated, and it seems unlikely that such data will become available, given the difficulty of acquiring data in operating furnaces. Availability of data is a significant limitation in further use of CFD in petrochemical applications, as discussed in the introduction.

9.5.2 Case 2: Xylene Reboiler

This study involves an operating furnace in a refinery. The problem observed in the furnace was that the flames from the ultra-low-NOx burners were very long and had the potential to damage process tubes in the top of the furnace. The authors have observed this phenomena in several vertical/cylindrical furnaces with ultra-low-NOx burners. The problem is related to the flow pattern within the furnace as it does not allow complete mixing of the combustion air with the fuel, but rather distorts the flame prior to burnout.

The geometry of the vertical cylindrical furnace is shown in Figure 9.9. The small wall around the burners is a reed wall and is used to heat the cold flue gases coming from the tubes. The periodicity of the furnace was used to simplify the model. The computational model of the vertical/cylindrical heater consisted of only one burner with periodic boundary conditions applied. This model has the shape of a very tall slice of pie. The images shown here are created by rotating the results about the vertical axis of the furnace.

The combustion model used in these calculations is an assumed pdf of mixture fraction. Because heat transfer to the tubes and furnace temperature are known to be important, a nonadiabatic mixture fraction table was constructed. The independent variables in the lookup table are mixture fraction, variance of mixture fraction, and enthalpy. Radiation was modeled using the discrete ordinates model with 32 ordinates. All solid surfaces were assumed to be radiatively black. Gas radiation properties were computed using the weighted-sum-of-gray-gases method.

Figure 9.10 shows a CFD simulation of the burners as they were originally installed. The figure shows an iso-surface of OH, which is a good indicator of flame shape in this case. The results reveal that the flames from adjacent burners merge together to produce a single long flame, which is confirmed by observations of the operating furnace. This burner has two primary fuel tips that fire fuel inside the tile and four secondary fuel tips. The solution to this flame interaction problem was to change the burner so that only three of the secondary tips actually fired. The CFD results for that configuration are shown in Figure 9.11. This solution was implemented and tested in the operating furnace and found to yield qualitatively the same result: the flames became distinct and burned out at the appropriate height.

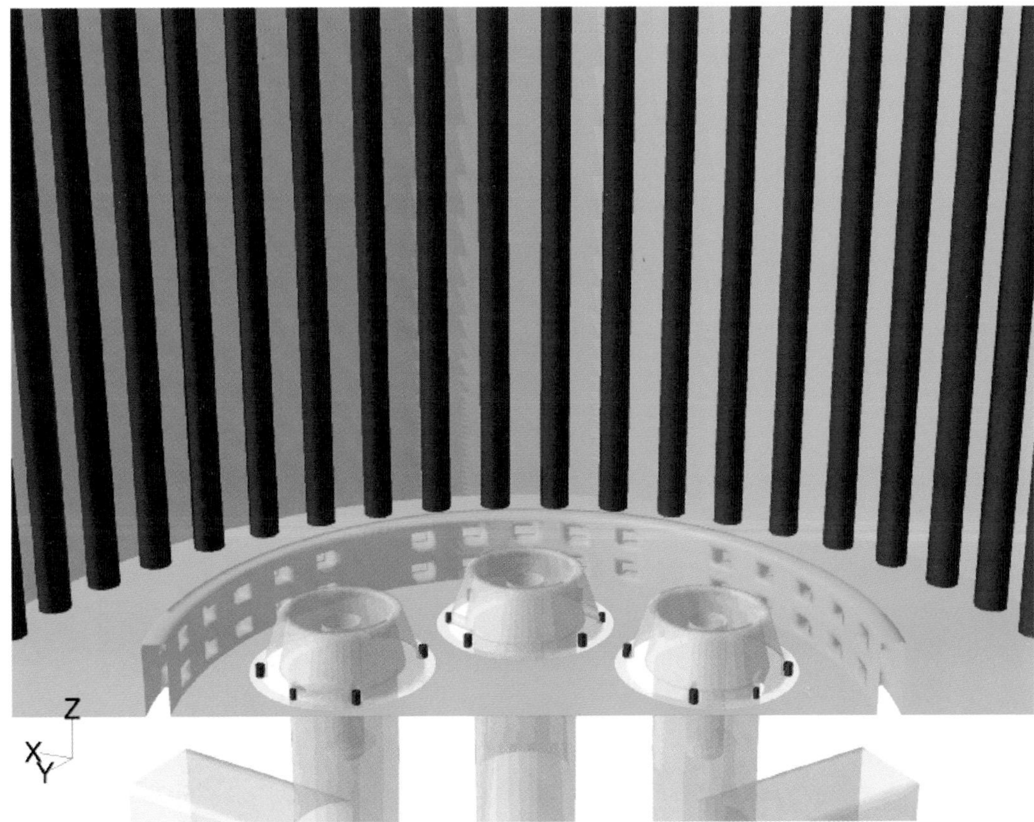

FIGURE 9.9 Geometry of a xylene reboiler. This view shows half (sliced vertically) of the furnace. This view shows only three of the six burners at the bottom of the image.

9.5.3 Case 3: Sulfur Recovery Reaction Furnace

This case study considers incineration of an acid gas to produce atomic sulfur. The reaction furnace is part of a Klaus process for sulfur recovery. The composition of the acid gas is given in Table 9.5. During the course of the study, the burner geometry was modified in order to shorten the flame and create better mixing within the reaction furnace.

In the present study, the "realizable *k*-ε" turbulence model is used to simulate the effects of turbulence on transport within the domain. The combustion process in this study is modeled using the assumed pdf (probability density function) of the mixture fraction model. In this model, a transport equation for a conserved scalar, called the mixture fraction, is solved. The effect of turbulence on the chemistry is simulated by solving a transport equation for the variance of mixture fraction. An assumed form for the pdf (a beta-pdf, which is a common choice in combustion simulations) allows for the creation of a lookup table by assuming that chemical equilibrium exists. The lookup table gives the composition and density at any location in the domain as a function of mean mixture fraction, mixture fraction variance, and enthalpy. Radiation heat transfer was not considered.

Figure 9.12 shows the geometry of the reaction furnace. The exterior of the burner is shown in the lower left corner of the figure. The furnace is approximately 5 m (16 ft) in diameter and 16 m (52 ft) in length. It is operated at about 60 kPa of positive pressure.

Figure 9.13 shows the geometry of the burner from the furnace looking at the burner throat. The green swirl vanes swirl the combustion air, while the acid gas is carried in the light blue tube and swirled in the opposite direction by the red swirl vanes. The initial burner geometry did not have the red acid gas swirl vanes or the yellow bluff body in the acid gas passageway.

The simulations predict all the species concentrations throughout the burner and furnace. Figure 9.14 shows the predicted O_2 mass fractions in the furnace. The plot shows a top-down view of the furnace with the O_2 mass fractions contoured at the mid-plane. This figure shows that the oxygen does not penetrate through the combustion zone, but is consumed near

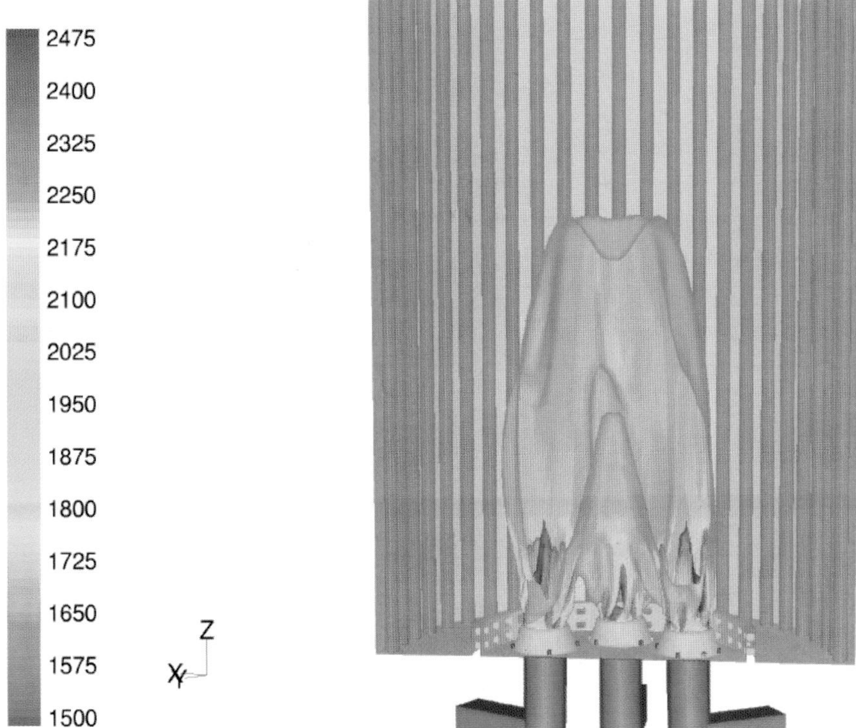

FIGURE 9.10 This view shows half of the furnace with unmodified burners. The "blob" in the furnace is the 50-ppm OH mole fraction iso-surface. This surface is colored according to its temperature (°F).

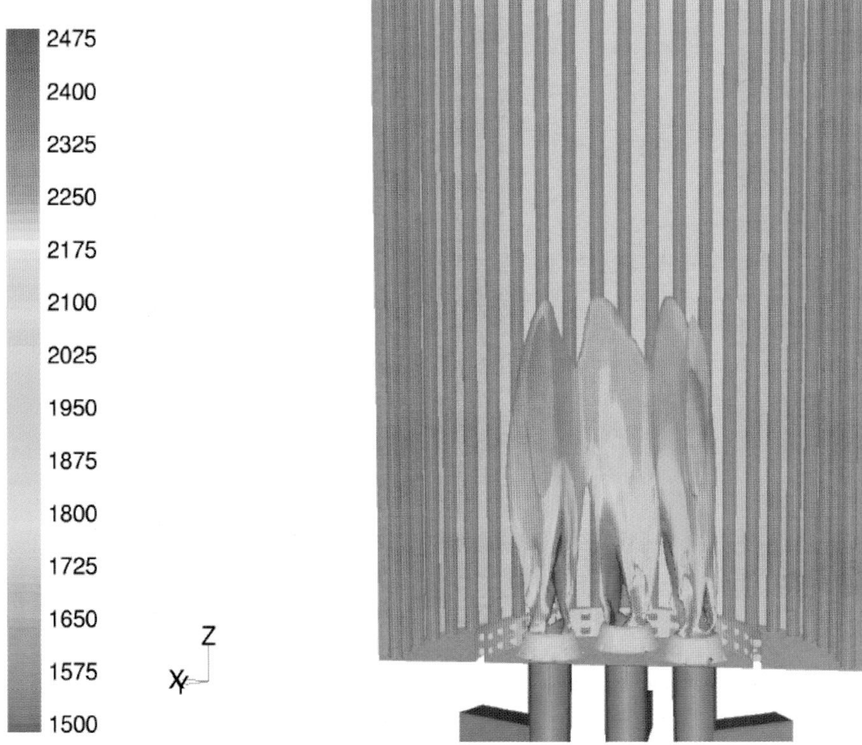

FIGURE 9.11 This view shows half of the furnace with the modified burners firing. The 50-ppm OH mole fraction iso-surface is shown as an indicator of the flame shape. This surface is colored according to its temperature (°F).

the burner. The asymmetry observed is due to the swirling flow. The swirling flow creates a swirling flame in the furnace.

Figure 9.15 shows the same view of the furnace, only the mid-plane is colored according to the mole fraction of H_2S. Figures 9.16 and 9.17 show the stoichiometric iso-surfaces for the initial and final burner design, respectively. The initial design did a poor job of mixing the acid gas and the combustion air. This resulted in a long corkscrewing flame that did not completely burn out, even by the end of the reaction furnace.

Figure 9.18 shows a close-up view of the temperature results near the burner quarl. Temperature contours in the mid-plane of the burner are also shown, clearly showing the reaction and mixing regions. The figure also shows velocity vectors, which reveal the swirl both in the combustion air and acid gas.

Figure 9.19 shows temperature profiles exiting the reaction furnace for the initial and final design. The figure shows that the final design produces much better temperature uniformity exiting the furnace. In this case study, uniformity of the exiting temperature profile was critical because this furnace had a waste heat recovery boiler directly downstream of the reaction furnace. Nonuniformity in the exiting temperature profile would have produced significant deterioration of the boiler tube metal.

9.5.4 *Case 4: Incineration of Chlorinated Hydrocarbons*

Research performed by VanDell and co-workers[53–55] and by workers at other research laboratories (Choudhry et al.;[56] Shaub and Tsang;[57] Graham et al.;[58] Senkan;[59] Taylor and Dellinger;[60] Altwicker et al.;[61] Young and Voorhees[62]) have investigated the chemical processes leading to products of incomplete combustion (PICs) formation. Although earlier work has identified specific mechanisms leading to PIC formation, the quantity of PICs formed depends greatly on local process conditions.[63] Thus, developing and implementing a methodology to predict local conditions inside a thermal oxidizer may significantly enhance the ability to predict PIC formation during incineration. This example describes a study where the PCGC code was applied to a thermal oxidizer to study the formation of HCl/Cl_2 resulting from the combustion of several RCl feed streams.

Given the necessary inputs to the CFD model (e.g., reactor geometry (see Figure 9.20), inlet flowrates and compositions, boundary conditions, etc.), a base case was established. Next, several separate simulations were performed for various combinations of the feed stream flow rates. Each case represented a combination of the nominal and/or maximum flow rate of fuel gas, of combustion air, and of organic feed streams. The predicted HCl, Cl_2, O_2, Cl, H outlet concentrations and average exit temperature for each case were recorded and evaluated.

Although performing actual test burns for the cases identified during this study would be prohibitively expensive and time-consuming, the CFD-based incineration model was able to assess the impact that various operating conditions might have on HCl/Cl_2 products in the effluent gas. From these simulations, the estimated range of expected HCl and Cl_2 production was:

TABLE 9.5 Composition of Acid Gas Used in CFD Study

Component	Mole (%)
N_2	0.005
CH_4	0.194
CO_2	17.981
C_2H_6	0.035
H_2S	75.223
COS	0.001
H_2O	6.500
C_3H_8	0.009
C_4H_{10}	0.003

Expected Cl_2 concentration range: 0.01–221 ppmv

Expected HCl concentration range: 3.9–6.4 mol%

Having performed the matrix of incineration simulations, two cases were identified as representative of the limiting scenarios for HCl/Cl_2 production. To better understand the causes for the predicted dramatic impact on Cl_2 production rates and on exit gas temperature, these cases were further investigated. One advantage of the CFD-based incineration model is its ability to generate local profiles of velocity, temperature, and species (e.g., CH_4, CO, HCl, and Cl_2). With these profiles, results from the limiting cases were closely examined to better understand the differences between each.

9.5.4.1 Excess Oxygen Condition

The first case considered represented an excess oxygen scenario. Important input data for this case are shown in Table 9.6.

Both the axial velocity and local gas temperature are shown first (Figure 9.21(a) and (b)). A maximum velocity of nearly 60 m/s is predicted near the reactor entrance. This is likely caused by the temperature driven gas expansion associated with combustion (Figure 9.21). A weak recirculation zone is predicted near the reactor wall in the quarl section (lower left-

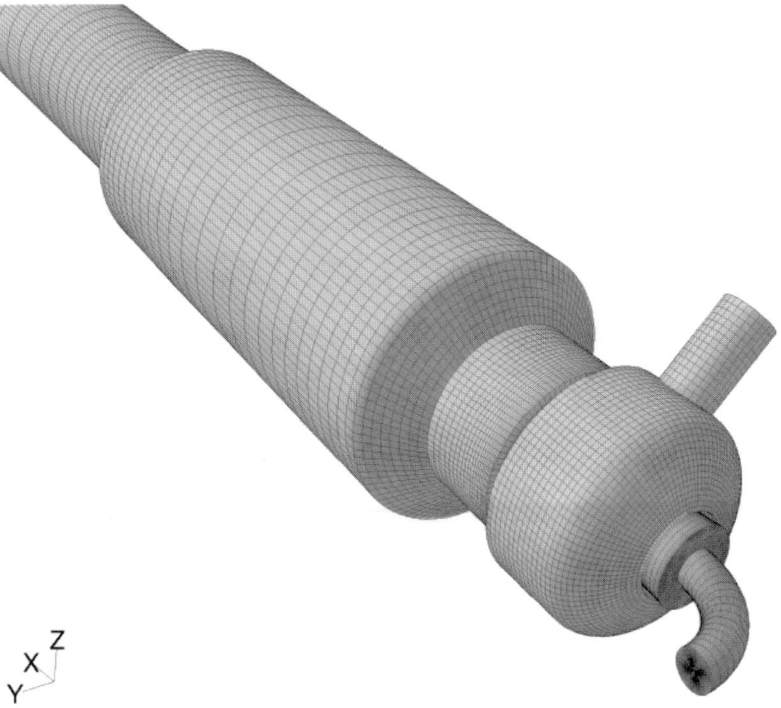

FIGURE 9.12 Exterior geometry of furnace included in model. The surface mesh is also shown.

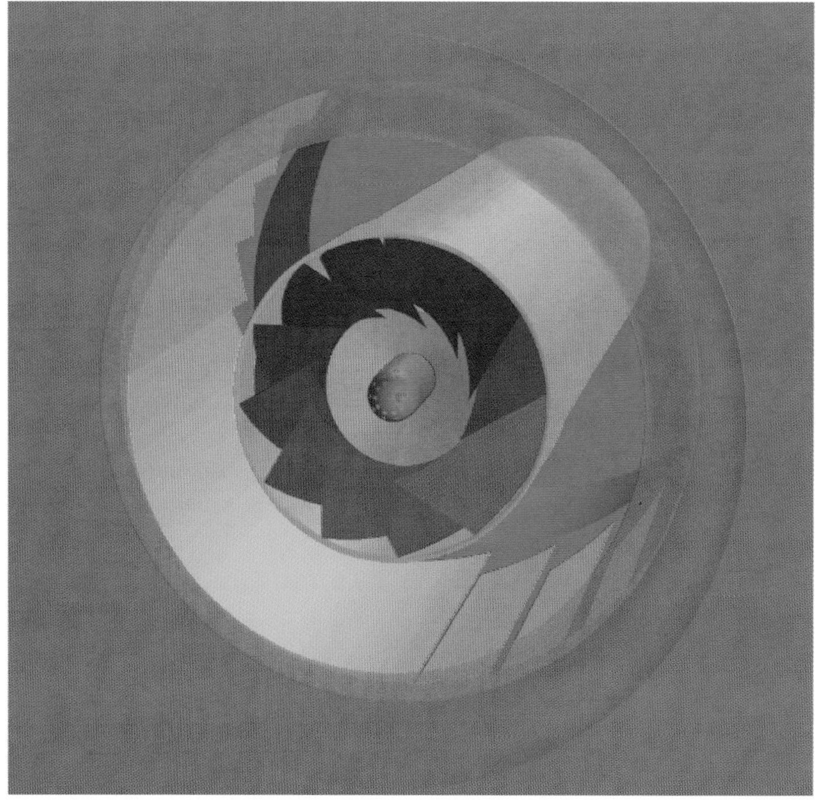

FIGURE 9.13 Burner geometry. The acid swirl vanes are shown in red; the air swirl vanes are shown in green; and the start-up fuel tip is shown in purple.

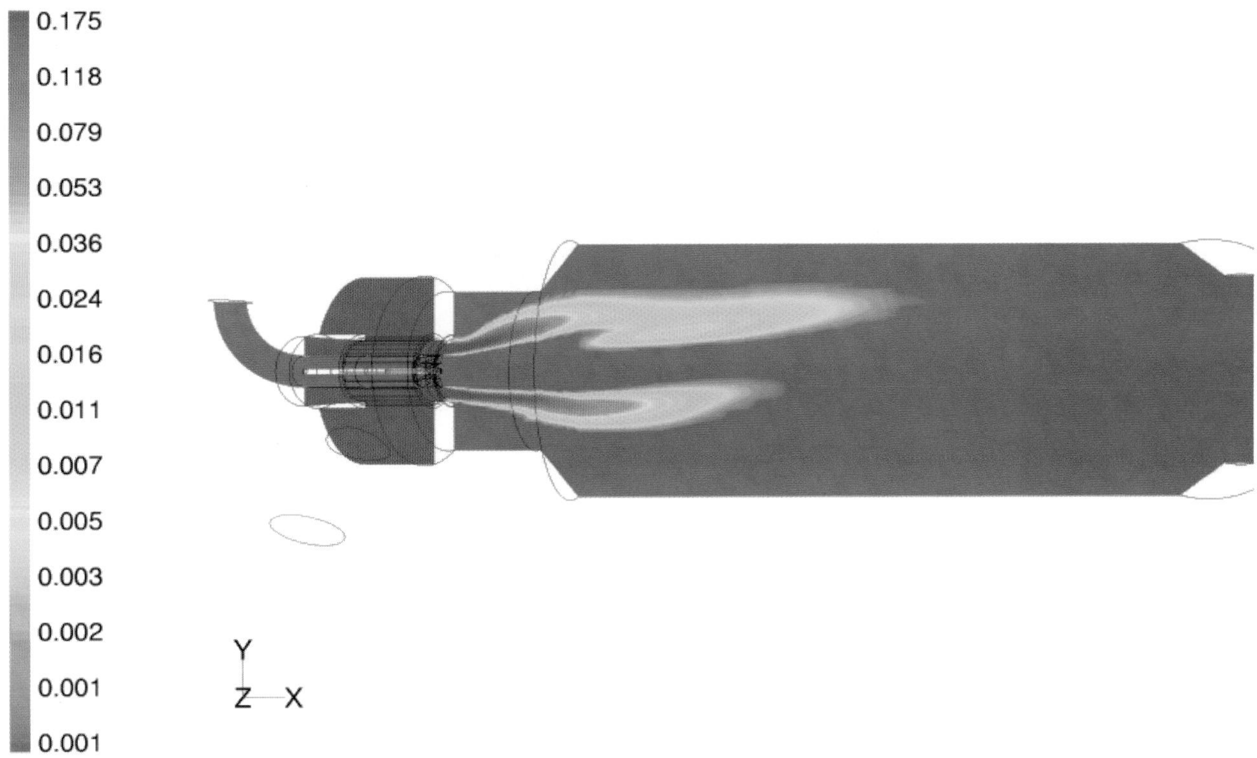

FIGURE 9.14 Oxygen mass fractions viewed from the above the furnace. The contour scale is logarithmic. The mass fractions are contoured on the mid-plane of the furnace.

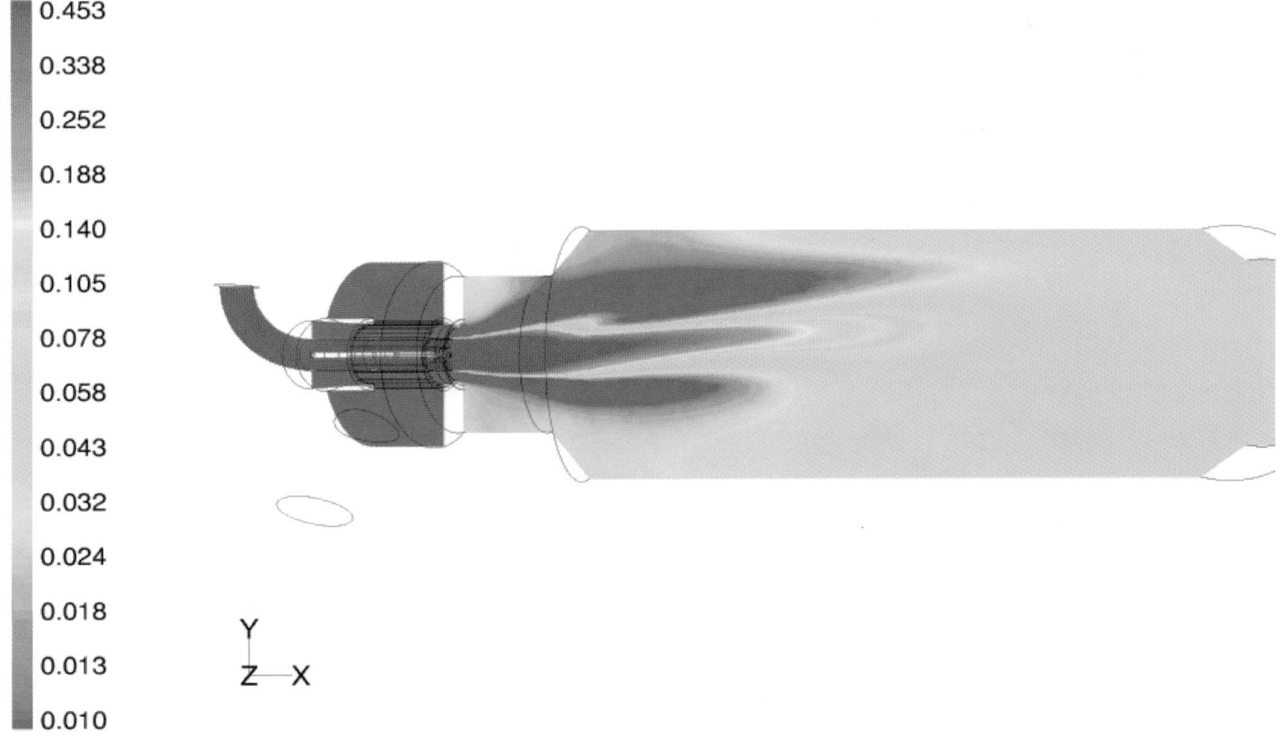

FIGURE 9.15 H_2S mole fractions contoured on the mid-plane of the furnace.

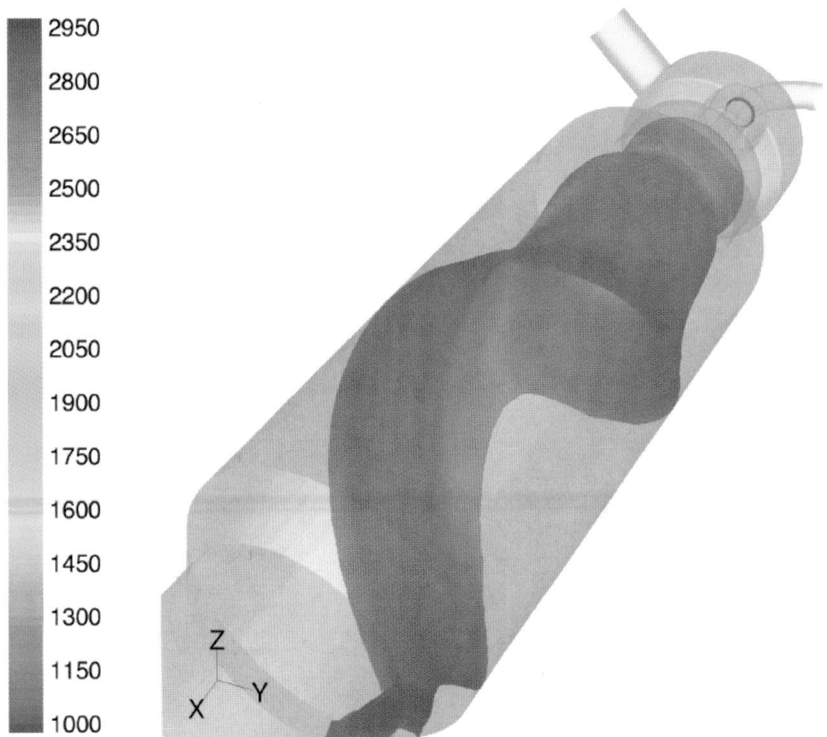

FIGURE 9.16 Stoichiometric iso-surface colored by temperature (°C) for the initial burner design.

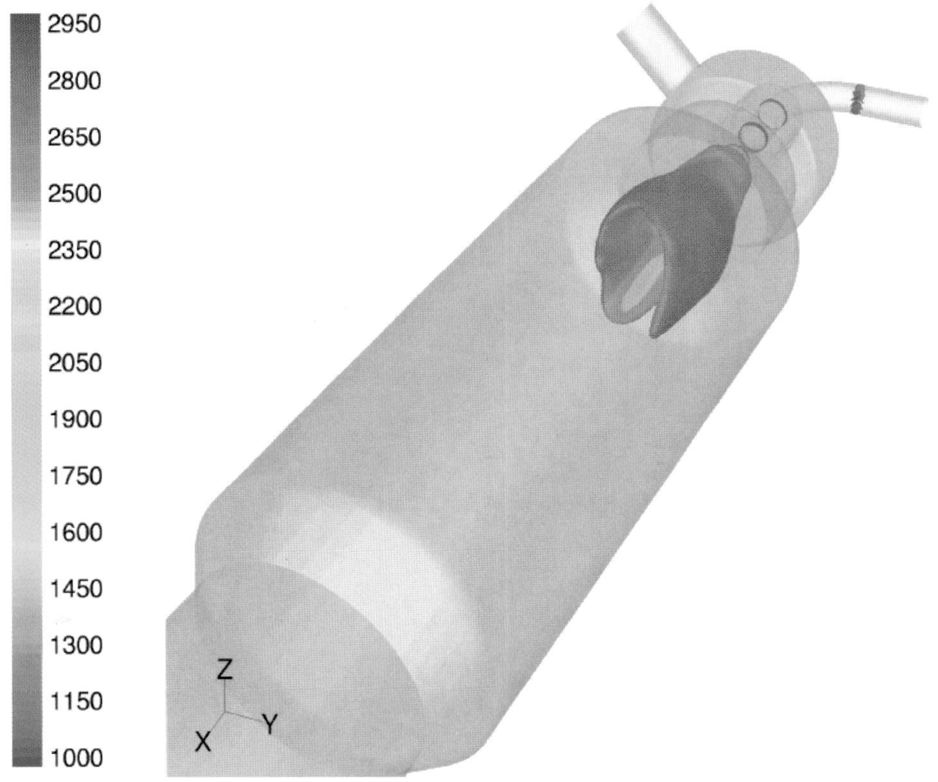

FIGURE 9.17 Stoichiometric iso-surface colored by temperature (°C) for the final burner design.

FIGURE 9.18 Mid-plane of geometry colored by temperature (°C). This view shows the burner quarl and the mixing regions of acid gas and air.

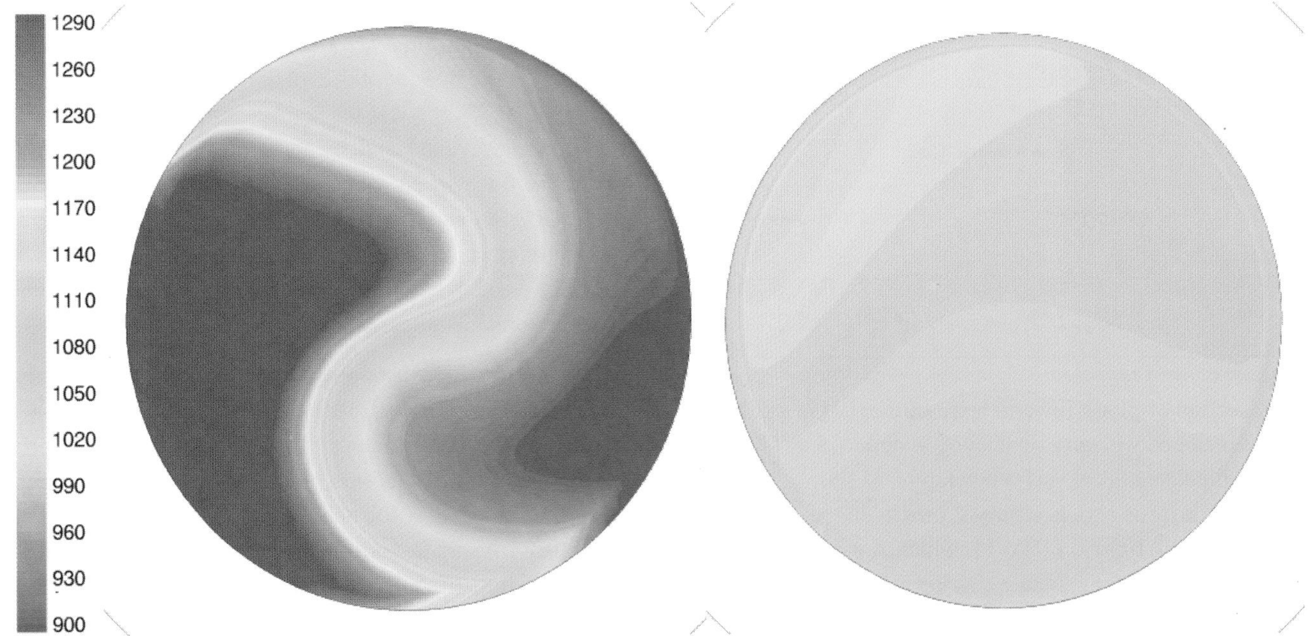

FIGURE 9.19 Temperature profiles (°C) exiting the reaction furnace for the initial (left) and final (right) burner geometries.

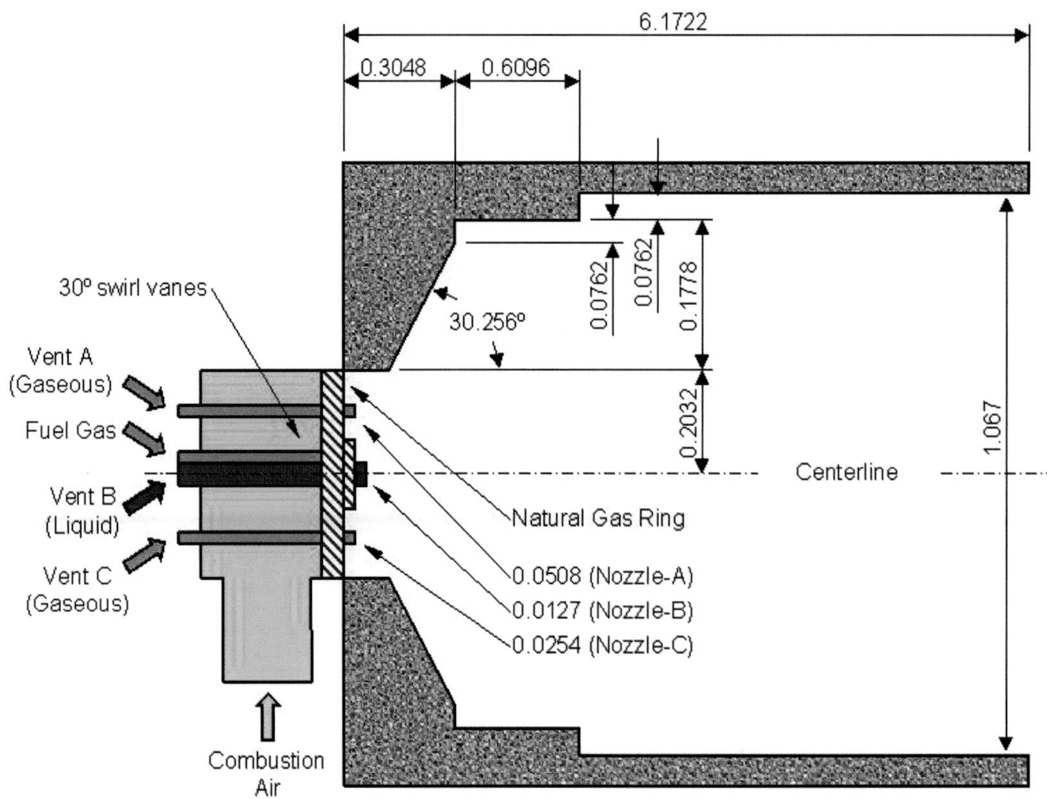

FIGURE 9.20 Geometric information describing the thermal oxidizer examined during this study.

TABLE 9.6 Limiting Cases Considered During RCl Combustion Study

Excess Oxygen Case	
Organic feed (g/s): 390	Inlet temperature (K): 298
Combustion air feed (g/s): 1955	Swirl number (–)[a]: 0.313
Fuel gas feed rate (g/s): 1.4	Heat loss (%): 35%
Stoichimetric Case	
Organic feed rate (g/s): 390	Inlet temperature (K): 298
Fuel gas feed rate (g/s): 1987	Swirl number (–)[a]: 0.313
Fuel gas feed rate (g/s): 69.4	Heat loss (%): 35%

[a] The swirl number is a measure of tangential velocity in the secondary inlet stream. (–), time-averaged value.

hand section of plot). Toward the reactor exit, the gas flow is fully developed with an exit velocity of about 15 m/s (49 ft/s).

The predicted local gas temperature is shown in Figure 9.21(b). The quarl region, clearly shown in this plot, has a wall temperature of 1000°C (1800°F), which is a preset boundary condition. A high-temperature envelope (>1000°C or 1800°F) is predicted near the reactor entrance, associated with the initial combustion zone. Methane is consumed in this region (see Figure 9.22(a)). A cooler region exists near the reactor wall in the quarl region. This may be caused by the recirculation of the cooler gas fed with the secondary feed stream. Also, the abnormally low prediction for gas temperature may be a result of using the uniform reactor heat loss. Finally, a second high-temperature region (>1600°C or 2900°F) is predicted near the reactor exit. This is caused by the combustion of remaining fuel (organic vents) and by the CO oxidation reaction that occurs in this portion of the reactor. CO formation and oxidation (CO_2 formation) are shown in Figure 9.20(b). The spatial nature of these predictions is illustrated by the high CO levels near the reactor centerline (>4000 ppmv) that decreases toward the reactor walls to essentially zero.

Finally, predicted values of HCl and Cl_2 concentrations are shown in Figure 9.23(a) and (b). Figure 9.23(a) clearly shows the localized nature of the HCl/Cl_2 chemistry in the combustion zone. The maximum predicted HCl concentration (20.6 mol%) is located early in the reactor (high-temperature region), compared to a predicted exit HCl concentration of 5.3 mol%. Also, the maximum Cl_2 level (3600 ppmv) occurs just beyond the reactor quarl wall (low-temperature region), compared to an exit Cl_2 concentration of 75 ppmv. This indicates how dramatically the local con-

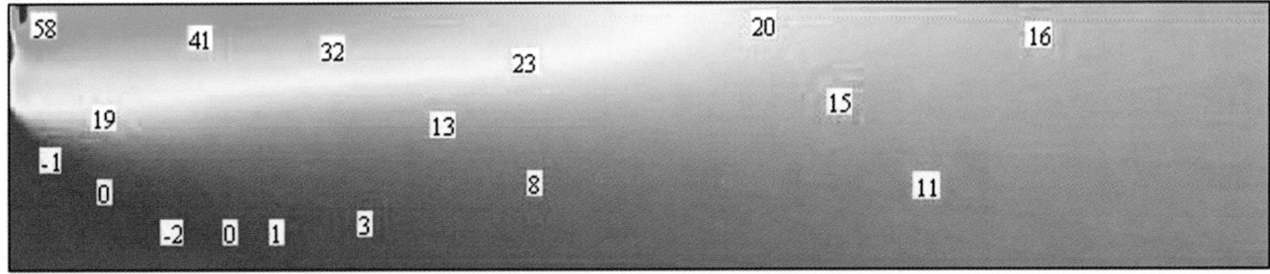

(a) Axial velocity (m/sec)

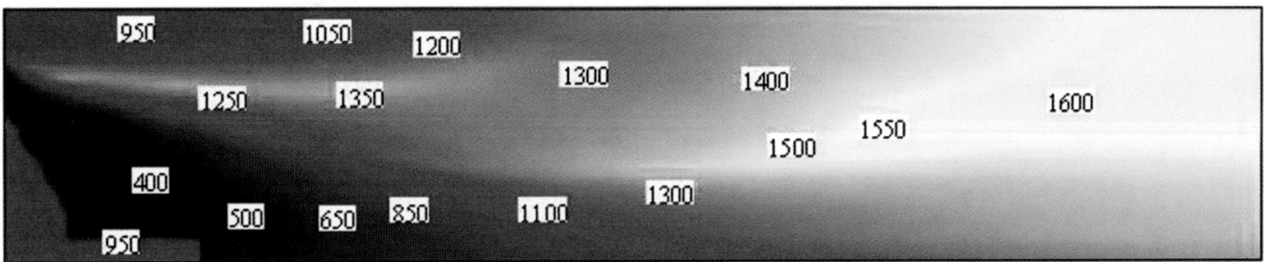

(b) Gas temperature profile (K)

FIGURE 9.21 Predicted centerline profiles for excess air case: (a) axial velocity (m/s), and (b) gas temperature (K) for the furnace section of the thermal oxidizer shown in Figure 9.20. Two distinct combustion zones are illustrated, with an exit temperature of about 1600 K (2400°F).

ditions affect the HCl/Cl_2 levels. This indicates that those process variables affecting local regions of the reactor can be used to optimize emissions levels.

9.5.4.2 Stoichiometric Oxygen Condition

The second case considered here represents an overall stoichiometric condition inside the reactor. Predictions from this case indicated a much lower Cl_2 emission and a lower exit temperature. Therefore, this case was carefully analyzed to understand the reasons for the apparent difference. Important input data for this case are also shown in Table 9.6.

Although the axial velocity profiles for this case are similar to those of the first case, there are some interesting differences. First, the velocity region near the reactor inlet is higher than before, and the recirculation zone near the wall in the quarl region is larger than before (see Figure 9.24(a)). Both differences are caused by dissimilar temperature profiles throughout the reactor (see Figure 9.24(b)). First, the early centerline gas temperature appears to be about the same (~950°C or 1740°F) as before. Instead of two distinct high-temperature zones (near the entrance and near the exit), in this case a single flame zone extends from the centerline reactor entrance to just beyond the quarl section. The flame also expands to the reactor wall and has a maximum temperature of approximately 1450°C (2640°F). The low-temperature zone (~400°C or 750°F) in the quarl section of the reactor and the weak recirculating flow zone (~–1 m/s or –3 ft/s) from the earlier case have also changed. The gas temperature has increased (>1200°C or 2200°F), as has the recirculating velocity (~–5 m/s or –16 ft/s). The expansion of the high-velocity region near the reactor entrance, the stronger recirculation zone in the quarl region, and the increased gas temperature in the quarl region appear to correlate with the relative amounts of natural gas and organic feeds considered in the respective cases. Two factors can help explain the different results from these cases.

First, while this case represents a near-stoichiometric reactant mixture, the previous case represents a reactant mixture with excess air. Examining Figure 9.25(b), the CO oxidation that occurs midway down the reactor (due to the excess oxygen and turbulent mixing) is not observed here (see Figure 9.25(b)). Thus, higher exit CO levels (>8000 ppmv) are predicted, which results in less energy release from the exothermic oxidation reaction of CO (ΔH_r = –282 KJ/mol), which leads to a lower exit gas temperature (1398°C or 2548°F).

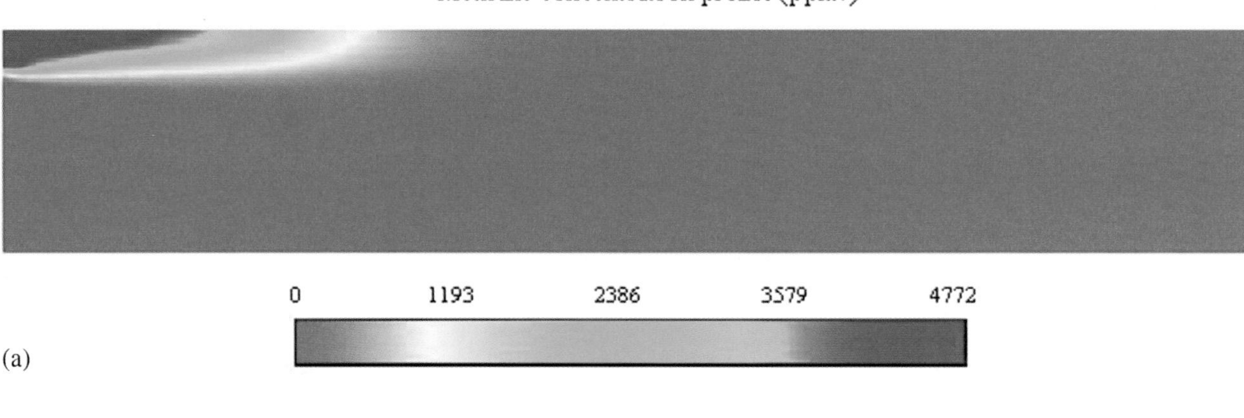

(a)

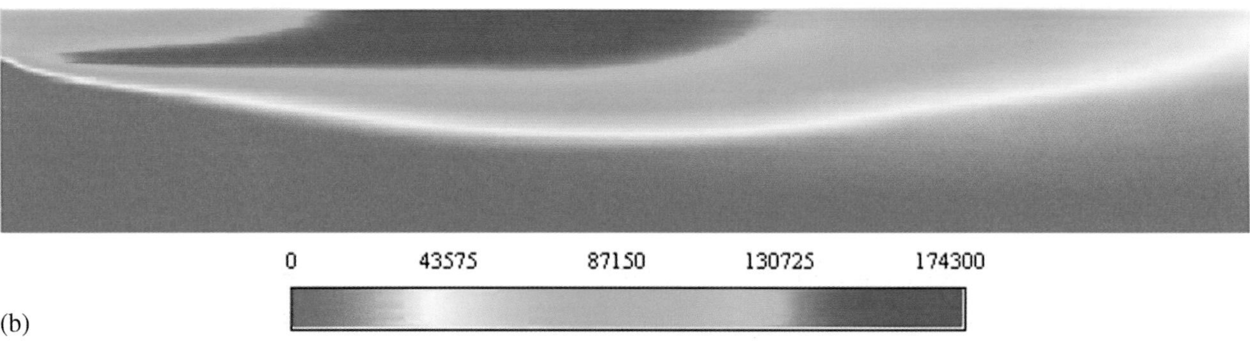

(b)

FIGURE 9.22 Predicted centerline profiles for excess air case: (a) methane concentration (ppmv) and (b) carbon monoxide concentration (ppmv) for the furnace section of the thermal oxidizer shown in Figure 9.20. These predictions depict the CO formation and oxidation zones common to most combustion processes.

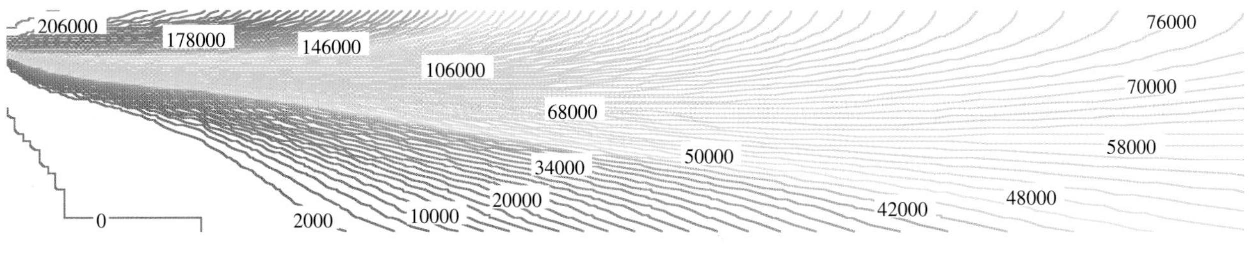

HCl concentration profile (ppmv)

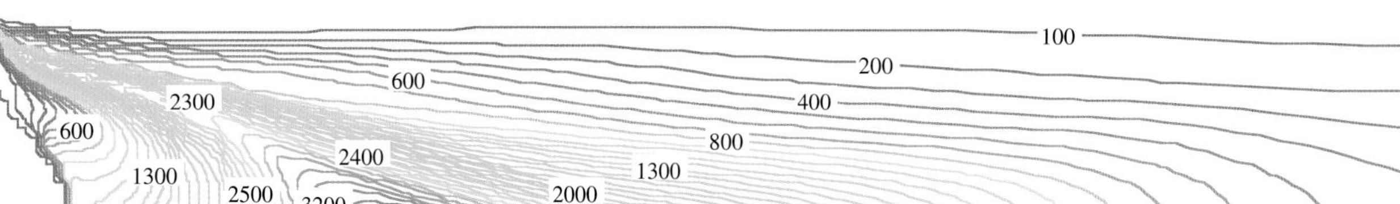

Cl2 concentration profile (ppmv)

FIGURE 9.23 Predicted centerline profiles for excess air case: (a) HCl concentration (ppmv) and (b) Cl_2 concentration (ppmv) for the furnace section of the thermal oxidizer shown in Figure 9.20. The predicted maximum Cl_2 concentration, nearly 3200 ppmv, occurs in the cooler reactor regions, while an exit Cl_2 concentration of about 100 ppmv is predicted.

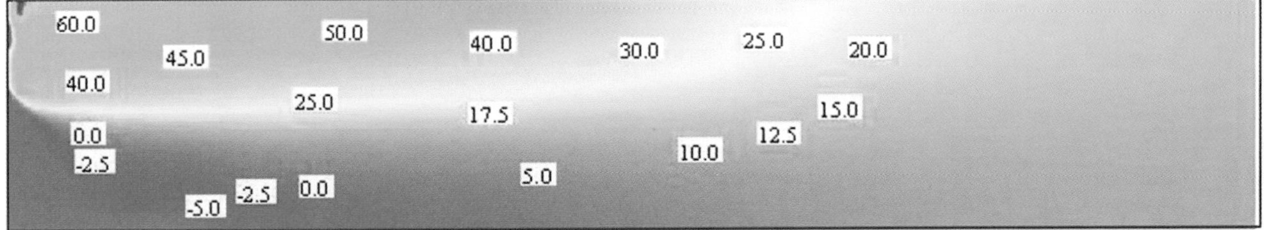

(a)

(b)

FIGURE 9.24 Predicted centerline profiles for stoichiometric case: (a) axial velocity (m/s), and (b) gas temperature (K) for the furnace section of the thermal oxidizer shown in Figure 9.20. A single combustion zone with the local maximum temperature of 1450 K (2150°F) and an exit temperature of about 1350 K (1970°F).

Another possible cause for the different temperature profiles might be related to how and where the organic vents are burned. In the first case, the organic vents appear to be burned after the methane is consumed. This would lead to two reaction zones: one where fuel gas is burned (early flame zone), and one where organic vents ignite, resulting in a secondary flame. These two high-temperature regions, observed in Figure 9.21(b), would lead to a higher exit temperature. In either case, the relative amounts of fuel gas and oxidizer are critical to the predicted combustion characteristics inside the reactor. More importantly, they dramatically affect the HCl/Cl_2 chemistry in the reactor, as seen in Figure 9.25(a) and (b).

In both cases, the predicted exit HCl concentration is greater than 5 mol%. However, the predicted exit Cl_2 concentration in Figure 9.26(b) is nearly 2 orders of magnitude less than that for the initial case. The same general trends show up in both cases: maximum Cl_2 concentrations near the outer wall just past the quarl section that is reduced to uniform exit concentrations toward the reactor exit. Two factors explain the significant differences in the HCl/Cl_2 concentrations for the different cases:

1. HCl is favored over Cl_2 at higher temperatures
2. Lower O_2 concentrations favor Cl_2

For the initial case, the early maximum of Cl_2 can also be due to poor mixing of O_2 with the fuel and the accompanying low gas temperatures (see Figure 9.21(b)). Similarly, for the present case, the global maximum is in the same region, but now the effect of higher gas temperature (favors HCl formation) results in significantly less Cl_2 formation. Thus, both the temperature effect and the oxygen effect are important.

This prediction is most interesting when considering the slight increase in HCl production accompanied by the dramatic decrease in Cl_2 production. In an attempt to validate this predicted behavior, the two conditions were reproduced in the field by adjusting the fuel gas and the organic feed rates accordingly. The flame inside the thermal oxidizer was visually monitored along with the exit gas temperature during the two tests. For the high RCl feed rate/low fuel-gas feed rate scenario, the flame appeared to nearly fill the entire combustion zone of the thermal oxidizer. As the fuel gas was increased while holding the organic feed rate constant, the visible flame front appeared to retreat toward the front of the burner. Also,

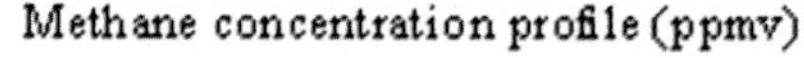

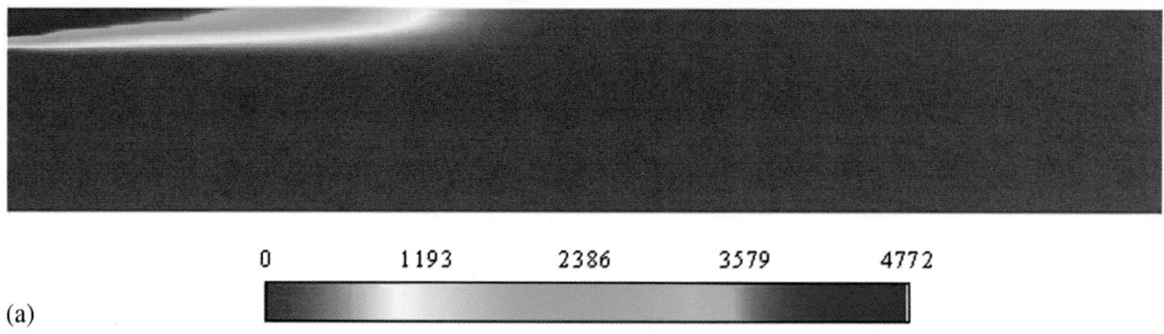

(a)

Carbon monoxide concentration profile (ppmv)

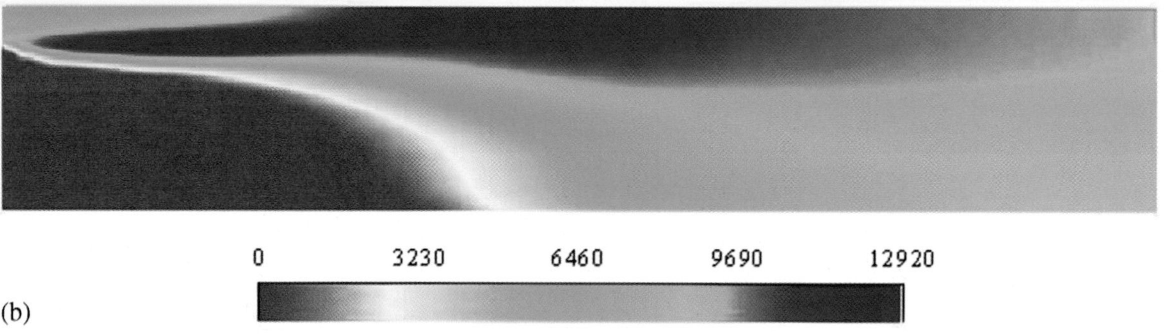

(b)

FIGURE 9.25 Predicted centerline profiles for stoichiometric case: (a) methane concentration (ppmv), and (b) carbon monoxide concentration (ppmv) for the furnace section of the thermal oxidizer shown in Figure 9.19. Here, the post-flame CO oxidation zone, shown in the first prediction, is not present; this results in a predicted exit CO concentration of 9000 ppmv.

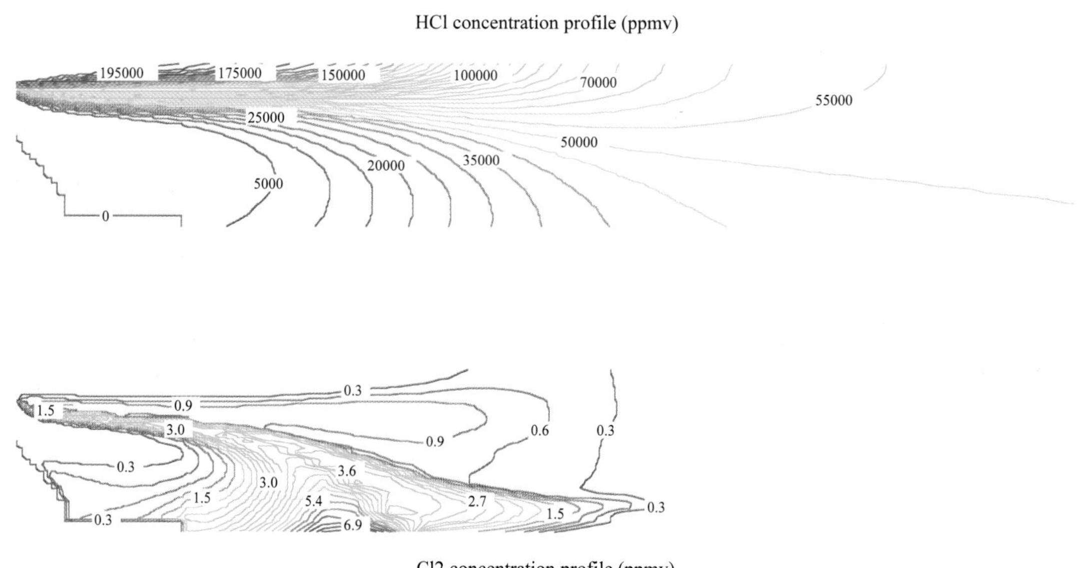

FIGURE 9.26 Predicted centerline profiles for stoichiometric case: (a) HCl concentration (ppmv), and (b) Cl_2 concentration (ppmv) for the furnace section of the thermal oxidizer shown in Figure 9.20. Dramatically less Cl_2 formation is predicted (local maximum of 7 ppmv and exit concentrations less than 1 ppmv) in this case due to excess H^+ radical present from the increased fuel gas.

the exit temperature decreased by nearly 200°C (400°F) during this test. Although this is only a limited amount of data, it provides some degree of validation of the predictions.

9.5.5 Case 5: Venturi Eductor Optimization

Natural draft, premixed, radiant wall burners use a venturi to educt air and mix it with the fuel, leading to a combustible mixture which will be burned in a furnace. This example illustrates the use of CFD in optimizing the performance of venturis.

New burner design technologies use the motive energy of fuel gas injection to entrain flue gas for the purpose of diluting the combustible mixture to reduce NOx emissions. Using an eductor, these designs entrain flue gas from above the convection section of a furnace, as shown in Figure 9.27. A large pressure drop through the upstream and downstream piping of the eductor system can create a recirculation pattern in the eductor throat. This re-circulation results in a reduction of the flue gas entrainment rate. Semi-empirical models, based on the conservation of mass, momentum and energy, have been used in the past to estimate the entrainment performance of an eductor system. These models, however, cannot reliably estimate the entrainment performance of an eductor system when a re-circulation zone occurs in the throat. The eduction performance is dependent on the size of the re-circulation zone in the eductor throat as illustrated in Figure 9.28.

The dimensions of the recirculation zone are a function of the ratios of fuel orifice-to-eductor throat diameter, motive and educted gas densities, and flow rates. Numerical modeling has been used to study the effects of entrainment performance based on these parameters. Figures 9.29 and 9.30 show the results of a numerical model with boundary conditions leading to a re-circulation pattern in the eductor throat. Several calculations performed for a variety of different boundary conditions such as fuel pressures, eductor and orifice diameters, and backpressures. These results are shown as a series in Figure 9.31.

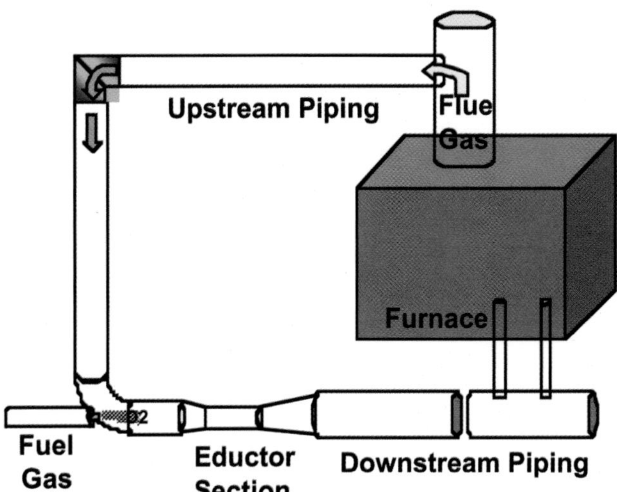

FIGURE 9.27 Illustration showing entrainment of flue gas using the motive energy of the fuel gas.

9.6 FUTURE NEEDS

It seems fitting to summarize the needs for future development. Throughout this chapter there have been discussions of model limitations and the need for better basic physical models. These needs are relatively clear and easily understood. The focus of this section is on issues that hinder CFD from becoming a valued design and troubleshooting tool in the petrochemical industry.

Given the importance of NOx emissions in the installation of new combustion equipment, NOx predictions may be the single most important improvement in the CFD analysis of industrial systems. Accurate NOx predictions will continue to be a challenge due to the nature of the physical problem (i.e., the strong coupling between the relatively slow chemical reactions and the turbulence in the flame).

The ability of CFD models to treat stability problems in non-premixed combustion is another area needing improvement. As discussed above, the presumed PDF model with equilibrium chemistry is probably the most accessible model to the industrial analyst, yet significant departures from equilibrium are observed in flames near extinction.[34] Given the chemical complexity of the problem, finding a model that can be valid for a range of fuels will be a significant challenge, but such a model would benefit the industry. It seems likely that a model that can capture these effects is a prerequisite to improving NOx predictions due to the strategies employed in industrial combustion systems for reducing NOx.

The inability of current-generation commercial products to model premixed flames with heat losses is a significant limitation. Premixed combustion is significantly more challenging than non-premixed combustion because of the coupling of flame propagation and turbulent motions in the fluid. Improvement in models of premixed combustion that are applicable to petrochemical burners would benefit the industrial CFD user.

9.7 CONCLUSION

CFD modeling of industrial furnaces is a valuable tool that can be used profitably. CFD modeling can help identify the cause of problems and it can be used to test solutions. In addition, CFD

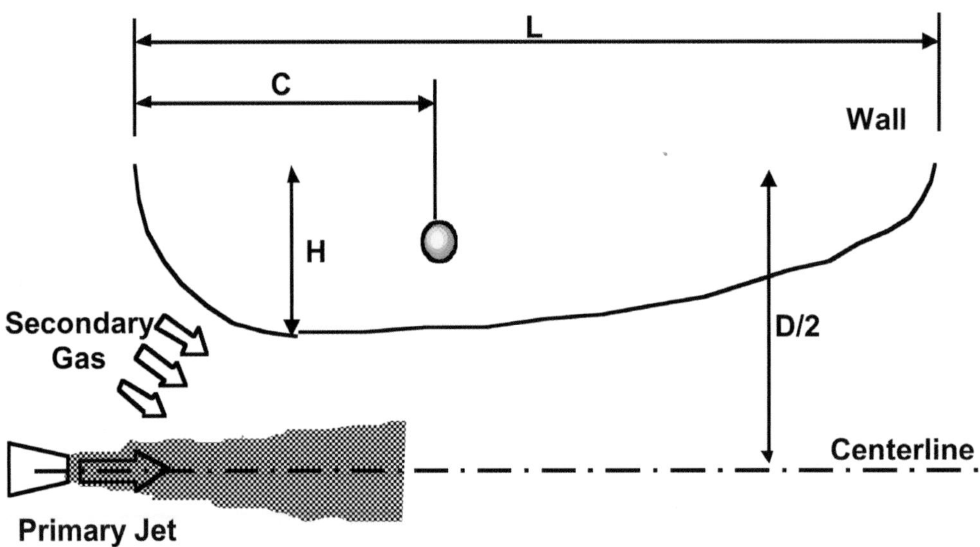

FIGURE 9.28 Illustration showing re-circulation region in the eductor throat.

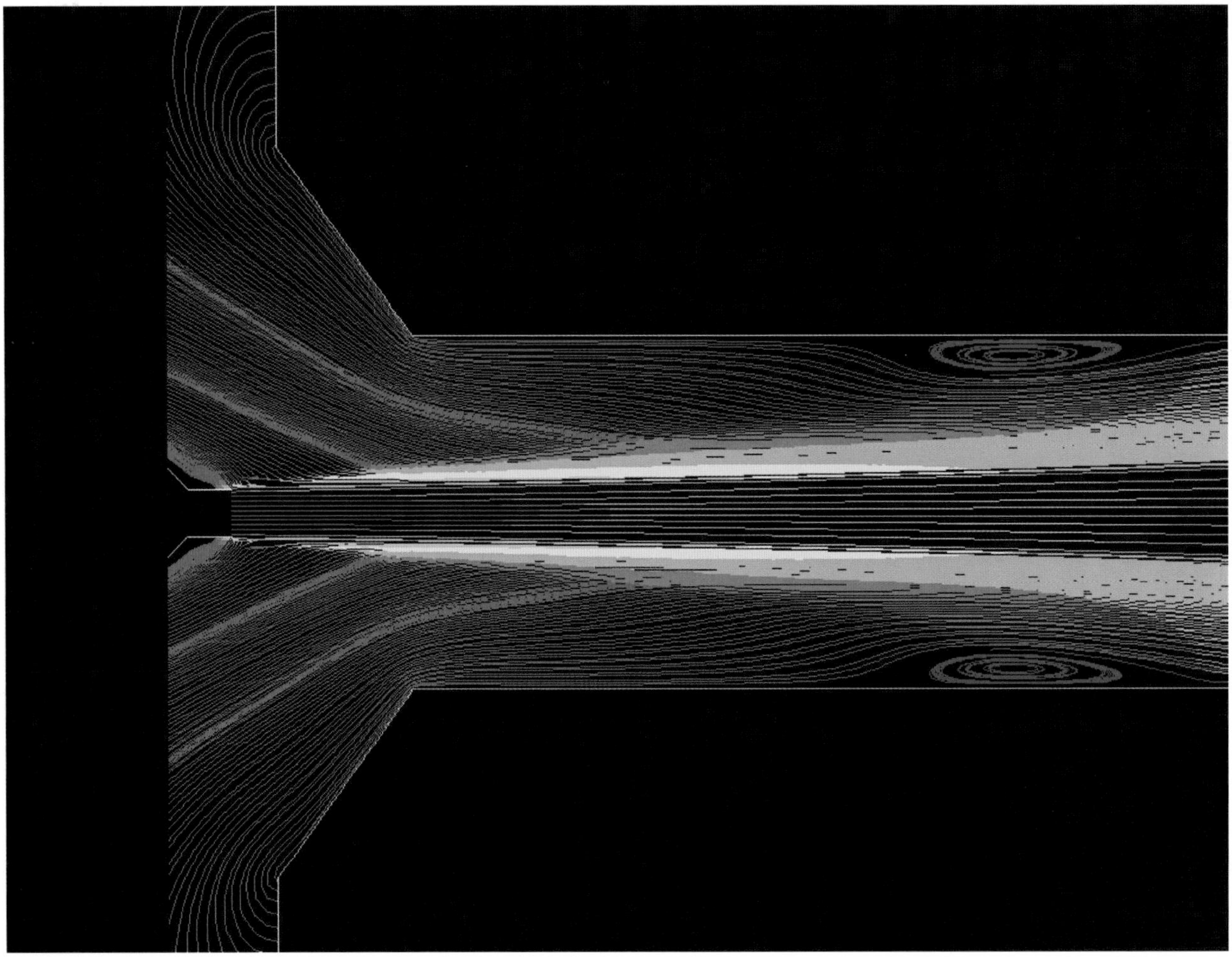

FIGURE 9.29 Re-circulation zone starting to occur in eductor throat.

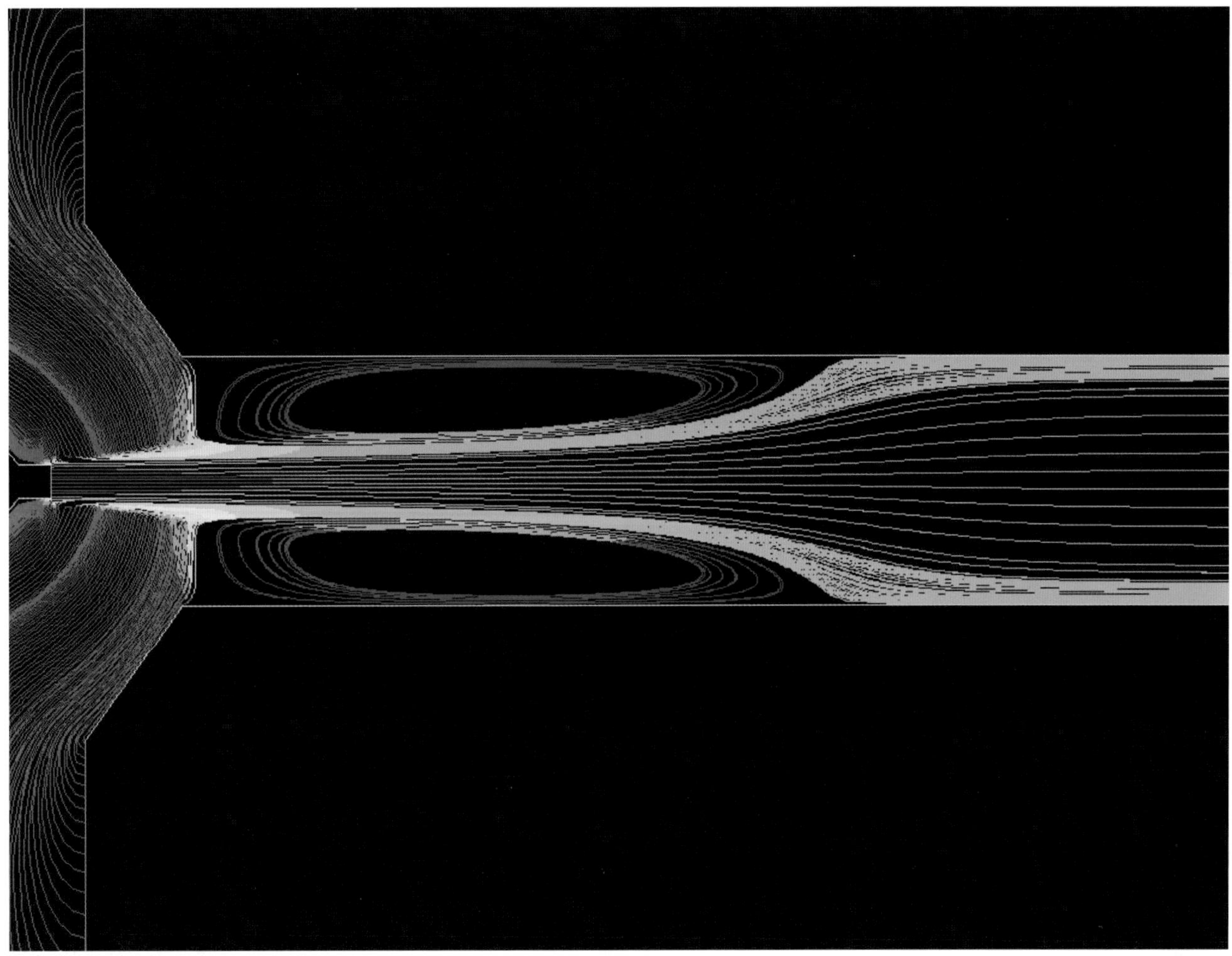

FIGURE 9.30 Re-circulation zone developing in eductor throat.

modeling can be a valuable design tool for combustion equipment in the petrochemical industry. It is also clear that CFD has not achieved the status of stress analysis in terms of ease of use. In many cases, engineers without advanced understanding of the physics do stress analysis of mechanical designs and obtain reasonable results. With CFD analysis, especially the study of combustion systems, this is not the case. In a typical furnace model, the science involved is multi-disciplinary, involving heat transfer, fluid flow, and combustion kinetics. Understanding and interpreting the results of a CFD model require a thorough understanding of the underlying physics.

9.8 NOMENCLATURE

Symbol (Units)	Description
A (-)	Difference coefficient composed of convection/diffusion terms
Da (-)	Damköhler number: ratio of reaction time to flow time
f (-)	General body force in momentum equation
g (m/s^2)	Gravity
I (W/m^2 sr)	Radiation intensity
k (m^2/s^2)	Turbulent kinetic energy
l_F (m)	Reaction zone thickness used to define Damköhler number
l_t (m)	Turbulent length scale used to define the Damköhler number
p (N/m^2)	Pressure
r (m)	Cylindrical coordinate position variable
R (-)	Residual or relative equation error
S_ϕ	Source term in conservation equations for general property
S_L(m/s)	Laminar flame speed
t_{flow} (s)	Characteristic time for flow to adjust to imposed shear

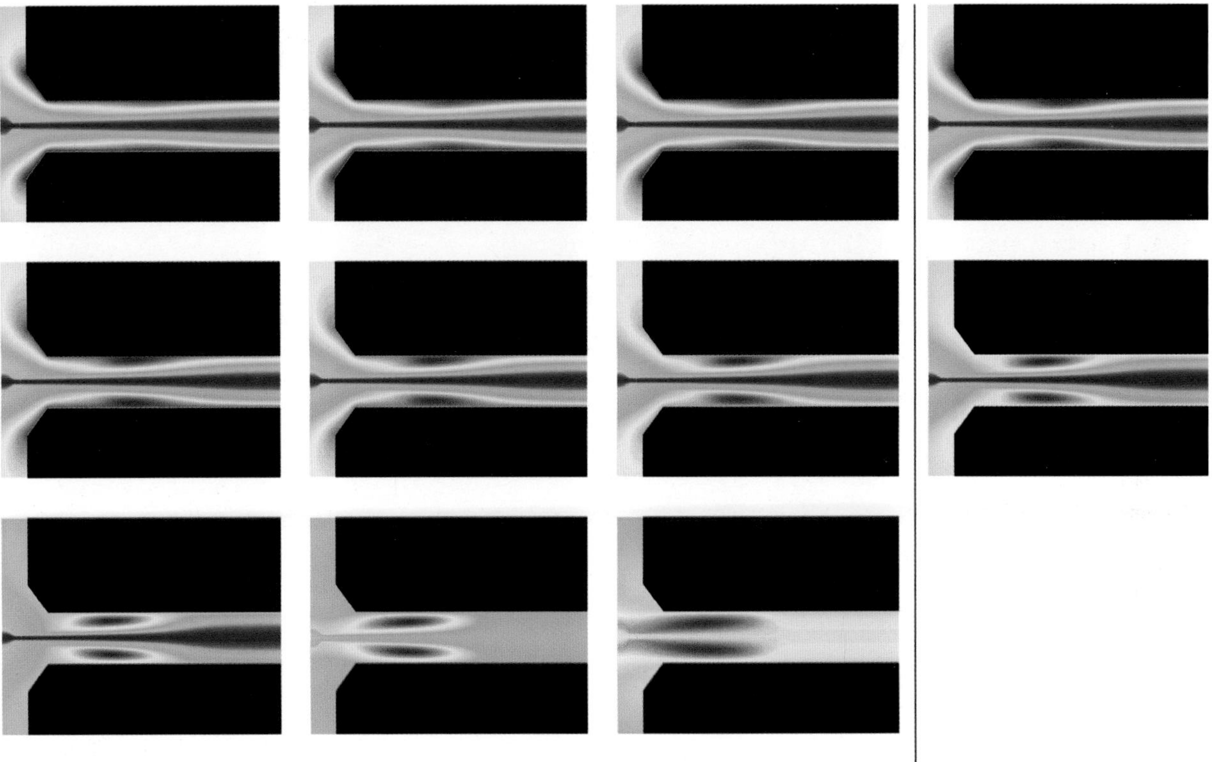

FIGURE 9.31 Plots of contours of streamfunction with increasing backpressure at the burner tip (left to right).

t_{rxn} (s)	Characteristic time for chemical species to react with each other
u (m/s)	Axial gas velocity
v (m/s)	Radial gas velocity
v' (-)	Turbulence intensity used to define Damköhler number
w (m/s)	Tangential gas velocity
x (m)	Cartesian coordinate position variable
y (m)	Cartesian coordinate position variable
z (m)	Cartesian coordinate position variable
δ_{ij} (-)	Kronecker delta
ε (m^2/s^2 s)	Kinetic energy dissipation rate
μ (kg/m s)	Viscosity
μ_b (kg/m s)	Bulk viscosity
μ_e (kg/m s)	Eddy viscosity
ν (m^2/s)	Eddy diffusivity used in the k-ε turbulence model
ρ (kg/m^3)	Density
τ (kg/m s^2)	Viscous stress tensor
Γ_ϕ(-)	General transport coefficient for transport property ϕ
Φ (-)	Conserved scalar
ϕ (-)	General transport property
Ω (-)	Represents a solid angle in radiation transport equation
ψ (-)	Dimensionless position variable, difference equation truncation ψ (-); Truncation error from difference equation

Overlines

-	Time-averaged value
~	Favre- or mass weighted-averaged value
$\rightarrow$	Vector quantity

Superscripts

″	Fluctuating portion of instantaneous value
O	Initial value
p	Center point in Difference scheme

Subscripts

i,j,k	Indices representing coordinate directions in 3-space
ϕ	General transport property
E,W,N,S,T,B	East, West, North,South, Top, Bottom — relative directions in grid

REFERENCES

1. M. Stahlman, Timeline: ten years of successes, *SunWorld, Special Commemorative Issue*, Fall 1992.

2. M. Grossman, Supercomputers/emerging application: modeling reality, *IEEE Spectrum*, 56-60, Sept. 1992.

3. E.A. Foumeny, and F. Benyahlya, Application of computational fluid dynamics for process equipment design, *Heat Exchanger Engineering*, Vol. 2, Foumeny and Heggs, Eds., Ellis Horwood Ltd., 1991, 155-164.

4. E.A. Foumeny, and F. Benyahlya, Can CFD improve the handling of air, gas, and gas-liquid mixtures?, *Chem. Eng. Progress*, 89(2), 21-26, 1993.

5. J.R. Krawczyk, Applications and limitations of CFD in the HPI, *Hydrocarbon Processing*, 76(7), 53-56, 1997.

6. J.D. Smith, Application of Computational Fluid Dynamics to Industrial "Opportunities" at The Dow Chemical Company, invited lecture at the Basel World CFD User Days: Second World Conference in Applied Computational Fluid Dynamics, Basel, Switzerland, May 1–5, 1994.

7. P.J. Smith, T.H. Fletcher, and L.D. Smoot, Model for pulverized coal fired reactors, *18th Symp. (Int.) Combustion*, The Combustion Institute, Pittsburgh, PA, 1981, 1285.

8. K.A. Davis, M.J. Bockelie, P.J. Smith, M.P. Heap, R.H. Hurt, and J.P. Klewicki, Optimized fuel injector design for maximum in-furnace NOx reduction and minimum unburned carbon, presented at the First Joint Power and Fuel Systems Contractor Conference, Pittsburgh, July 1996.

9. T.H. Fletcher, A two-dimensional model for coal gasification and combustion, Ph.D. dissertation, Department of Chemical Engineering, Brigham Young University, Provo, UT, 1983.

10. S.C. Hill, L.D. Smoot, and P.J. Smith, Prediction of nitrogen oxide formation in turbulent coal flames, *20th Symp. (Int.) Combustion*, The Combustion Institute, Pittsburgh, PA, 1984, 1391.

11. L.L. Baxter, P.J. Smith, P.J., and L.D. Smoot, Comparison of model predictions and experimental data for coal-water mixture combustion, *Western States Section, The Combustion Institute*, San Antonio, TX, April 1985.

12. D.G. Sloan, P.J. Smith, and L.D. Smoot, Modeling of swirl in turbulent flow systems, *Prog. Energy Combust. Sci.*, 12, 163, 1986.

13. A.S. Jamaluddin, and P.J. Smith, Prediction of radiative heat transfer in cylindrical furnaces, *Western States Section, The Combustion Institute*, Banff, Alberta, Canada, April 1986.

14. M.W. Rasband, PCGC-2 and the data book: a concurrent analysis of data reliability and code performance, M.S. thesis, Department of Chemical Engineering, Brigham Young University, Provo, UT, 1988.

15. L.L. Baxter, and P.J. Smith, Turbulent dispersion of particles, *Western States Section, The Combustion Institute*, Salt Lake City, UT, 1988.

16. R.D. Boardman, Development and evaluation of a combined thermal and fuel nitric oxide predictive model, Ph.D. dissertation, Department of Chemical Engineering, Brigham Young University, Provo, UT, 1990.

17. J.D. Smith, A detailed evaluation of comprehensive simulation software describing pulverized-coal combustion and gasification using advanced sensitivity analyses techniques, Ph.D. dissertation, Department of Chemical Engineering, Brigham Young University, Provo, UT, 1990.

18. P.A. Gillis, and P.J. Smith, An evaluation of 3-D computational combustion and fluid-dynamics for industrial furnace geometries, *Twenty-Third Symp. (Int.) Combustion*, The Combustion Institute, Pittsburgh, PA, July 1990.

19. L.D. Smoot, and P.J. Smith, *Coal Combustion and Gasification*, Plenum Press, New York, 1985.

20. R.B. Bird, W.E. Stewart, and E.N. Lightfoot, *Transport Phenomena*, John Wiley & Sons, New York, 1960.

21. M. Nallasamy, A Critical Evaluation of Various Turbulence Models as Applied to Internal Fluid Flows, NASA TP - 2474, Marshall Space Flight Center, Alabama, May 1985.

22. F.H. Harlow, and P.I. Nakayama, Transport of turbulence energy decay rate, Rpt. # LA-3854, Los Alamos Scientific Laboratory, Los Alamos, NM, January 1968.

23. C.G. Speziale, On Nonlinear k-l and k-ε models of turbulence, *J. Fluid Mech.*, 178, 459, 1987.

24. B.E. Launder, and D.B. Spalding, *Mathematical Models of Turbulence*, Academic Press, London, 1972.

25. N.I. Lilleheie, I. Ertesvag, T. Bjorge, S. Byggstoyl, and B.F. Magnussen, Modeling and Chemical Reactions: Review of Turbulence and Combustion Models, NEI-DK--286, DE90 760249, SINTEF/The Norwegian Institute of Technology, Division of Thermodynamics, July 11, 1989.

26. W.P. Jones, and J.H. Whitelaw, Calculation methods for reacting turbulent flows: a review, *Combustion and Flame*, 48, 1, 1982.

27. J.W. MacArthur, Development and implementation of robust direct finite-difference methods for the solution of strongly coupled elliptic transport equations, Ph.D. dissertation, University of Minnesota, May 1986.
28. C.K. Law, Heat and mass transfer in combustion: fundamental concepts and analytical techniques, *Progress in Energy and Combustion Science*, 10, 295-318, 1984.
29. N. Peters, Length scales in laminar and turbulent flames, *Prog. Astro. Aero.*, 35, 155-183, 1991.
30. S. Mazumder and M.F. Modest, Turbulence radiation interactions in nonreactive flow of combustion gases, *ASME J. Heat Transfer*, 121, 726-729, 1999.
31. R. McDermott and M.R. Henneke, High capacity, ultra low NOx radiant wall burner development, *12th Ethylene Forum*, May 11-14, 1999, The Woodlands, TX.
32. S.B. Pope, Computations of turbulent combustion: progress and challenges, *23rd Symp. (Int.) on Combustion*, The Combustion Institute, 1990.
33. K.N.C. Bray, The challenge of turbulent combustion, *26th Symp. (Int.) on Combustion*, The Combustion Institute, 1996.
34. J. Warnatz, U. Mass, and R.W. Dibble, *Combustion: Physical and Chemical Fundamentals, Modeling and Simulation, Experiments, Pollutant Formation*, Springer-Verlag, Berlin, 1995.
35. R.S. Barlow, Nitric oxide formation in dilute hydrogen jet flames: isolation of the effects of radiation and turbulence-chemistry submodels, *Combustion and Flame*, 117, 4, 1999.
36. M.F. Modest, *Radiative Heat Transfer*, McGraw-Hill, New York, 1993.
37. R. Siegel and J.R. Howell, *Thermal Radiation Heat Transfer*, Hemisphere, Washington, D.C., 1992.
38. I.M. Khan, and G. Greeves, A method for calculating the formation and combustion of soot in diesel engines, in *Heat Transfer in Flames*, N.H. Afgan and J.M. Beer, Eds., Scripta, Washington D.C., 1974, chap. 25.
39. P.A. Tesner, T.D. Snegiriova, and V.G. Knorre, Kinetics of dispersed carbon formation, *Combustion and Flame*, 17, 253-260, 1971.
40. F.A. Williams, *Combustion Theory*, Addison-Wesley, 1985.
41. D.B. Spalding, Mixing and chemical reaction in steady confined turbulent flames, *13th Symp. (Int.) on Combustion*, The Combustion Institute, 1970.
42. B.F. Magnussen and N.H. Hjertager, On mathematical modeling of turbulent combustion with special emphasis on soot formation and combustion, *16th Symp. (Int.) on Combustion*, The Combustion Institute, 1976.
43. P.A. Libby, *Introduction to Turbulence*, Taylor and Francis, New York, 1996.
44. W.P. Jones and J.H. Whitelaw, Calculation methods for reacting turbulent flows: a review, *Combustion and Flame*, 48, 1-26, 1982.
45. N. Peters, *Progress in Energy and Combustion Science*, 10, 319, 1984.
46. R.W. Bilger, The structure of turbulent nonpremixed flames, *22nd Symp. (Int.) on Combustion*, The Combustion Institute, 1988.
47. E.S. Bish and W.J.A. Dahm, Strained dissipation and reaction layer analyses of nonequilibrium chemistry in turbulent reaction flows, *Combustion and Flame*, 100, 3, 1995.
48. R.W. Bilger, Conditional moment closure for turbulent reacting flow, *Physics of Fluids A*, 5, 436, 1993.
49. C.G. Speziale, On non-linear k-l and k-ε models of turbulence, *J. Fluid Mechanics*, 178, 459-478, 1987.
50. V. Yakhot, et al., Development of turbulence models for shear flows by a double expansion technique, *Physics of Fluids A*, 4(7), 1510-1520, 1992.
51. T.H. Shih, W.W. Liou, A. Shabbir, and J. Zhu, A new k-ε eddy-viscosity model for high Reynolds number turbulent flows — model development and validation, *Computers Fluids*, 24(3), 227-238, 1995.
52. H.K. Versteeg and W. Malalasekera, *An Introduction to Computational Fluid Dynamics, The Finite Volume Method*, Addison-Wesley Longman Limited, England. 1995.
53. R.D. Van Dell and L.A. Shadoff, Relative rates and partial combustion products from the burning of chlorobenzenes and chlorobenzene mixtures, *Chemosphere*, 13, 177, 1984.
54. R.D. Van Dell and N.H. Mahle, The role of carbon particle surface area on the products of incomplete combustion (PICs) emissions, in *Emissions from Combustion Processes: Origin, Measurement, Control*, Clement, R.E. and Kagel, R.O., Eds., Lewis, Boca Raton, FL, 1990, 93-107.

55. R.D. Van Dell and N.H. Mahle, A study of the products of incomplete combustion and precursors produced in the flame and their post flame modification from the combustion of *o*-dichlorobenzene in a high resolution laboratory thermal oxidizer, *Comb. Sci. Tech.*, 85, 327, 1992.

56. C.G. Choudhry, K. Olie, and O. Hutzinger, Mechanisms in the thermal formation of chlorinated compounds including polychorinated dibenzo-*p*-dioxins, in *Chlorinated Dioxins and Related Compounds: Impact on the Environment*, Hutzinger, O., Frei, R.W., Merian, E., and Pocchiari, F., Eds., Pergamon Press, Oxford, 1982, 275-301.

57. W.M. Shaub and W. Tsang, Dioxin formation in incinerators, *Environ. Sci. Technol.*, 17, 721, 1983.

58. J.L. Graham, D.L. Hall, and B. Dellinger, Laboratory investigation of thermal degradation of a mixture of hazardous organic compounds, *Environ. Sci. Technol.*, 20, 703, 1986.

59. S.M. Senkan, Thermal destruction of hazardous wastes: the need for fundamental chemical kinetic research, *Environ. Sci. Technol.*, 22, 368, 1988.

60. P.H. Taylor and B. Dellinger, Thermal degradation characteristics of chloromethane mixtures, *Environ. Sci. Technol.*, 22, 438.

61. E.R. Altwicker, R. Kumar, N.V. Konduri, and M.S. Milligan, The role of precursors in formation of polychloro-dibenzo-*p*-dioxins and polychloro-dibenzofurans during heterogeneous combustion, *Chemosphere*, 20(10-12), 1935, 1990.

62. C.M. Young and K.J. Voorhees, Thermal decomposition of 1,2-dichlorobenzene. II. Effect of feed mixtures, *Chemosphere*, 24(6), 681, 1992.

63. D.P.Y. Chang, W.S. Nelson, C.K. Law, R.R. Steeper, M.K. Richards, and G.L. Huffman, Relationships between laboratory and pilot-scale combustion of some chlorinated hydrocarbons, *Env. Prgs.*, 8(3), 152, 1989.

64. J.D. Anderson, *Computational Fluid Dynamics: The Basics with Applications*, McGraw-Hill, New York, 1995.

65. J. Swithenbank, A. Turan, and P.J. Felton, Three-dimensional, two-phase mathematical modeling of gas turbine combustors, paper presented at Project SQUID, Purdue University, 1978, 1-89.

Chapter 10
Combustion Safety

Terry Dark and Charles E. Baukal, Jr.

TABLE OF CONTENTS

10.1 Introduction 328
10.2 Overview 329
10.2.1 Definitions 329
10.2.2 Combustion Tetrahedron 330
10.2.3 Fire Hazards 331
10.2.4 Explosion Hazards 334
10.2.5 Process Hazard Analysis (PHA) 337
10.2.6 Codes and Standards 338
10.3 Design Engineering 339
10.3.1 Flammability Characteristics 339
10.3.2 Ignition Control 342
10.3.3 Fire Extinguishment 344
10.3.4 Safety Documentation and Operator Training 344
10.4 Sources of Further Information 346
References 347

10.1 INTRODUCTION

Fires and explosions are a major concern in hydrocarbon and petrochemical plants as the consequences can be very severe and very public because of the high volume of flammable liquids and gases handled in those plants.[1,2] The process industries have invested much money in equipment, instrumentation, training, and procedures to enhance safety. Considerable progress has been made in the recent past in improving the safety of their operations. Unfortunately, the industry is not immune to accidents, as evidenced by explosions that have been documented.[3,4] *Loss Prevention Bulletin* has listed all the major incidents worldwide that occurred from 1960 to 1989 in the hydrocarbon chemical process industries, including refineries, petrochemical plants, gas processing plants, and terminals.[5] Some of these involved large property losses and deaths. These types of events have heightened the safety consciousness of these industries to both prevent such incidents and to effectively handle them if they should occur.[6] The moral, social, economic, environmental, and legal ramifications of an accident make combustion safety a critical element in plant design and operation. Preventing an incident is definitely preferred to protecting people and equipment from the consequences of an incident if it occurs.[7] While fires and explosions can occur at many different processes in a plant, this chapter deals specifically with the fired heaters section.

There are many factors that can contribute to an accident:[8]

- human error,[9,10]
- equipment malfunction
- upset plant conditions
- fire or explosion near the apparatus
- improper procedures
- severe weather conditions

In a report prepared by the American Petroleum Institute,[11] the following causes were noted for 88 incidents that occurred in refining and chemical unit operations from 1959 to 1978:

- 28% equipment failures
- 28% human error
- 13% faulty design
- 11% inadequate procedures
- 5% insufficient inspection
- 2% process upsets
- 13% education

Uehara (1991) analyzed the risks to Japan's petrochemical plants in the event of a large earthquake, which has a stronger likelihood in Japan due to the high frequency of seismic activity.[12] There are also many potential dangers caused by fires and explosions: flying shrapnel, pressure waves from a blast, high heat loads from flame radiation,[13–15] and high temperatures. All of these can have severe consequences for both people and equipment and may need to be considered in minimizing the potential impact of an incident. Fry[16] showed how computer models can be used to simulate fires and explosions in the chemical process industry to help design appropriate measures to prevent these incidents and how to respond if they should occur. Ogle[17] presented a method for analyzing the explosion hazard in an enclosure that is only partially filled with flammable gas. Ogle showed that an explosion pressure at the stoichiometric condition is approximately 50 times greater than the failure pressure of most industrial structures. This obviously can have catastrophic results.

A number of good books are available on safety in combustion systems and in the chemical and petrochemical industries.[18–28] Crowl and Louvar have written a textbook designed to teach and apply the fundamentals of chemical process safety.[29] King has written a large book on safety for the process industries, including the chemical and petrochemical industries, with specific emphasis on U.K. and European standards and regulations.[30] Kletz has written an encyclopedia-format book on safety and loss prevention, containing small articles on about 400 different topics.[31] Nolan has written an extensive guide to understanding and mitigating hydrocarbon fires and explosions.[32] Nolan characterized accidents or failures into the following basic areas: ignorance, economic considerations, oversight and negligence, and unusual occurrences. However, it is noted that nearly all incidents are preventable. Nolan listed the following principles as the general philosophy for fire and explosion protection for oil, gas, and related facilities:

1. Prevent the immediate exposure of individuals to fire and explosion hazards.
2. Provide inherently safe facilities.
3. Meet the prescriptive and objective requirements of governmental laws and regulations.
4. Achieve a level of fire and explosion risk that is acceptable to the employees, the general public, the petroleum and related industries, the local and national governments, and the company itself.
5. Protect the economic interest of the company for both short- and long-range impacts.
6. Comply with a corporation's policies, standards, and guidelines.
7. Consider the interest of business partners.
8. Achieve a cost-effective and practical approach.
9. Minimize space (and weight if offshore) implications.

10. Respond to operational needs and desires.
11. Protect the reputation of the company.
12. Eliminate or prevent the deliberate opportunities for employee- or public-induced damages.

10.2 OVERVIEW

It is the intent of this section to provide a general education regarding the many hazards associated with the unsafe operation of combustion equipment, as well as to discuss the evaluation tools and regulations used to eliminate hazards and unsafe practices from combustion system operation.

10.2.1 Definitions

Crowl and Louvar,[29] Nolan,[32] and NFPA 86[33] all provide extensive definitions of common combustion safety vocabulary. Some of the most commonly used definitions related to fire and explosion phenomena are provided below.

Autoignition: The process through which a flammable liquid's vapors are capable of extracting enough energy from the environment to self-ignite, without the presence of a spark or flame.

Autoignition temperature: The minimum temperature at which a flammable liquid is capable of autoignition.

Autooxidation: The process of slow oxidation, resulting in the production of heat energy, sometimes leading to autoignition if the heat energy is not removed from the system.

Burner: A device or group of devices used for the introduction of fuel and oxidizer into a furnace at the required velocities, turbulence, and mixing proportion to support ignition and continuous combustion of the fuel.

Combustion: A chemical process that is the result of the rapid reaction of an oxidizing agent and a combustible material. The combustion reaction releases energy (in the form of heat and light), part of which is used to sustain the combustion reaction.

Combustible: In general, a material capable of undergoing the combustion process in the presence of an oxidation agent and a suitable ignition source.

Combustible liquid: A liquid having a flash point at or above 140°F (60°C) and below 200°F (93°C). A combustible liquid basically becomes a flammable liquid when the ambient temperature is raised above the combustible liquid's flash point.

Confined explosion: An explosion occurring within a confined space, such as a building, vessel, or furnace.

Detonation: An explosion that results in a shock wave that moves at a speed greater than the speed of sound in the unreacted medium.

Deflagration: An explosion that results in a shock wave that moves at a speed less than the speed of sound in the unreacted medium.

Explosion: A rapid expansion of gases that results in a rapidly moving shock wave.

Fire: The generic term given to the combustion process.

Flame: A controlled fire produced by a burner.

Flammable: In general, a material that is capable of being easily ignited and burning rapidly.

Flammable liquid: A liquid having a flash point below 140°F (60°C) and having a vapor pressure below 40 psia (2000 mmHg) at 100°F (38°C).

Flare: A device incorporating a large burner, typically on top of a large exhaust stack, used for the burning of combustible exhaust gases vented from an industrial process.

Flash point (FP): The lowest temperature of a liquid at which it gives off enough vapor to form an ignitable mixture with air immediately over the surface of the liquid.

Ignition: The process of initiating the combustion process through the introduction of energy to a flammable mixture.

Lower flammability limit (LFL): The minimum concentration of a combustible gas or vapor in air, below which combustion will not occur upon contact with an ignition source; sometimes referred to as the lower explosive limit (LEL).

Minimum ignition energy (MIE): The minimum energy required to initiate the combustion process.

Overpressure: The pressure generated by an explosive blast, relative to ambient pressure.

Shock wave: A pressure wave moving through a gas as the result of an explosive blast. The generation of the shock wave occurs so rapidly that the process is primarily adiabatic.

Spontaneous combustion: The combustion process resulting from autooxidation and subsequent autoignition of a flammable liquid.

Upper flammability limit (UFL): The maximum concentration of a combustible gas or vapor in air, above which combustion will not occur upon contact with an ignition source; sometimes referred to as the upper explosive limit (UEL).

Vapor pressure: The pressure exerted by a volatile liquid as determined by the Reid Method (ASTM D-323-58); measured in terms of pounds per square inch (absolute).

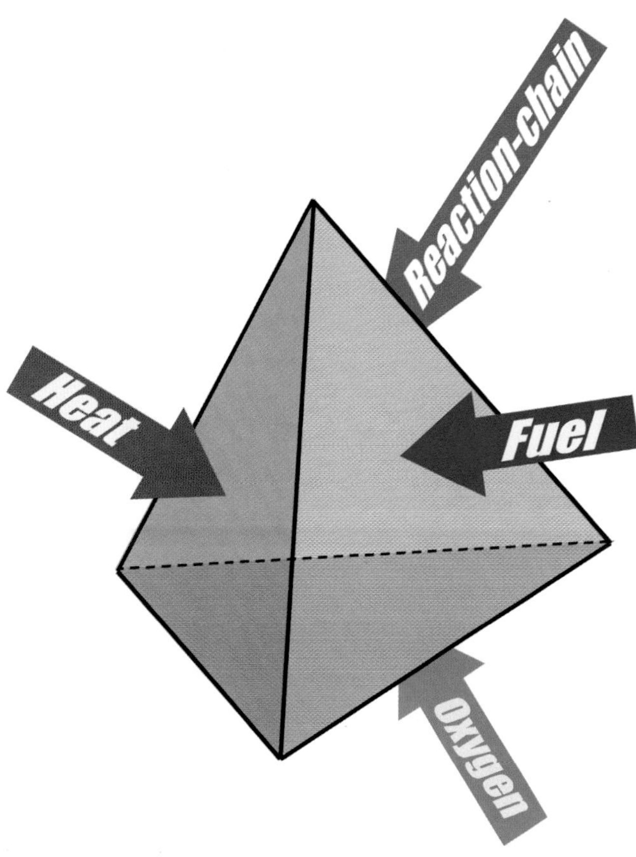

FIGURE 10.1 Fire tetrahedron.

10.2.2 Combustion Tetrahedron

Four basic elements must be present for a combustion process:

1. fuel
2. oxidizer
3. heat
4. reaction chain

This is usually referred to as the "fire tetrahedron" as shown in Figure 10.1. Fire can be defined as a rapid chemical reaction between a fuel and an oxidant, where there is sufficient heat to both initiate and sustain the reaction. The typical distinction between a flame and a fire is that a flame is controlled and desirable, while a fire is uncontrolled and undesirable. To prevent or extinguish a fire, one or more of the four legs of the combustion tetrahedron must be removed. Fuel may be in the form of a solid, liquid, or gas. Typical solid fuels include coal, wood dust, fibers, metal particles, and plastics. Typical liquid fuels include gasoline, acetone, ether, fuel oil, and pentane. Typical gaseous fuels include natural gas, propane, hydrogen, acetylene, and butane. If the fuel source is removed, then the fire goes out. For example, in a fired heater, the flame goes out when the fuel supply to the burner is shut off.

The second face of the fire tetrahedron is the oxidizer. Oxidizers are also present in solid, liquid, and gaseous forms. Solid oxidizers include metal peroxides and ammonium nitrate. Liquid oxidizers include hydrogen peroxide, nitric acid, and perchloric acid. Gaseous oxidizers include oxygen, fluorine, and chlorine. Oxygen (contained in air) is the oxidizer used in industrial combustion. A fire will be extinguished if the oxidizer is removed. This can be done, for example, by smothering the fire with a blanket or by injecting an inert gas like N_2 or CO_2 in and around the fire to displace the oxidizer. In nearly all cases, the oxidizer is oxygen that is present in normal air at about 21% by volume. Higher concentrations of oxygen can cause a flame to burn more rapidly and violently.[34] For example, pure oxygen is used to enhance many high-temperature industrial combustion processes.[35] Even metals can burn in pure oxygen. In reality, a fuel and oxidizer can be in the presence of an ignition source without combusting, if the mixture is not within the flammability limits. Table 10.1 provides the flammability limits for a few common fuels.[36] For example, the lower flammability limit for methane (CH_4) in air is 5.0% CH_4 by volume, with the balance being air. The upper flammability limit for CH_4 in air is 15.0% CH_4 by volume. If the mixture contains less than 5.0% or more than 15.0% CH_4 by volume, then the mixture is outside the flammability limits and will not combust at standard temperature and pressure.

The third face of the tetrahedron involves an energy source to both initiate and sustain the combustion reactions. Table 10.2 provides the minimum ignition temperature required to initiate the combustion reaction of various gaseous fuels and oxidizers at stoichiometric conditions and standard temperature and pressure. For example, the minimum ignition temperature for an air/CH_4 mixture is 1170°F (632°C). In normal burner operation, the flame is ignited with either a pilot or a spark igniter. A fire or explosion can be initiated if the fuel/oxidizer mixture contacts a hot surface, an unintended spark, or static electricity.[37] A common fire prevention step is to eliminate all ignition sources from an area containing known fuel sources. After the fire has been initiated, energy is still required to sustain the flame. That source is normally the exothermic heat release from the reaction itself that makes most flames self-sustaining. A common method of extinguishing fires is to deluge the area with water, which eventually cools all surfaces below the ignition point. Water is inert, inexpensive, and has a very high heat capacity relative to other liquids, which makes it a good extinguishing agent.

TABLE 10.1 Flammability Limits for Common Fuels at Standard Temperature and Pressure

Fuel	Flammability Limits (% fuel gas by volume)			
	Lower, in air	Upper, in air	Lower, in O_2	Upper, in O_2
Butane, C_4H_{10}	1.86	8.41	1.8	49
Carbon monoxide, CO	12.5	74.2	19	94
Ethane, C_2H_6	3.0	12.5	3	66
Hydrogen, H_2	4.0	74.2	4	94
Methane, CH_4	5.0	15.0	5.1	61
Propane, C_3H_8	2.1	10.1	2.3	55
Propylene, C_3H_6	2.4	10.3	2.1	53

Source: Reed, R.J., *North American Combustion Handbook*, Vol. I, 3rd ed., North American Manufacturing Company, Cleveland, OH, 1986. With permission.

TABLE 10.2 Minimum Ignition Temperatures for Common Fuels at Standard Temperature and Pressure

Fuel	Minimum Ignition Temperature			
	In air (°F)	In air (°C)	In O_2 (°F)	In O_2 (°C)
Butane, C_4H_{10}	761	405	541	283
Carbon monoxide, CO	1128	609	1090	588
Ethane, C_2H_6	882	472		
Hydrogen, H_2	1062	572	1040	560
Methane, CH_4	1170	632	1033	556
Propane, C_3H_8	919	493	874	468

Source: Reed, R.J., *North American Combustion Handbook*, Vol. I, 3rd ed., North American Manufacturing Company, Cleveland, OH, 1986. With permission.

The fourth face of the tetrahedron is a reaction chain to sustain combustion. If any of the steps in the chemical chain reaction are broken, the flame can be extinguished. This is the principle behind certain types of fire extinguishers, such as dry chemical or halogenated hydrocarbon. These extinguishing agents inactivate the intermediate products of the flame reactions. This reduces the combustion rate of heat evolution, which eventually extinguishes the flame by removing the heat source that makes the flame self-sustaining.

10.2.3 Fire Hazards

10.2.3.1 Heat Damage

One of the most devastating causes of fires in process plants is process tube rupture (see Figure 10.2). A furnace tube rupture feeds the furnace firebox with an uncontrolled amount of fuel, usually resulting in enormous damage and, sometimes, loss of life. As with all safety issues, prevention is preferred to remediation. Although there are numerous safety features of process equipment that can minimize damage when a tube failure occurs, the most desirable, safest, and least costly mode of operation is to prevent the occurrence in the first place.

Furnace operators should be carefully trained to monitor and assess the state of tubes as a regular part of the operational routine. Every occasion of tube overheating, whether it is caused by excessive firing or inadequate cooling action by the process fluid, should be recorded and assessed to determine the likelihood of tube failure. Tubes are normally designed to last 10 or more years under normal operation, but excessive temperatures can shorten the life of the tubes to a few days or less. Coking in the tubes, if not identified and removed expediently, can cause hot spots on the tubes that will result in premature failure. Another mechanism for tube failure is trapped liquid that freezes and expands (see Figure 10.3).

Prevention of tube failure requires adequate instrumentation and proper furnace design to allow the tube condition to be monitored continuously and accurately. Accurate temperature measurement of tubes can be a difficult problem. Adequate instrumentation is expensive, but inadequate monitoring of tube condition can be dangerous and much more expensive. Temperature measurements of the process

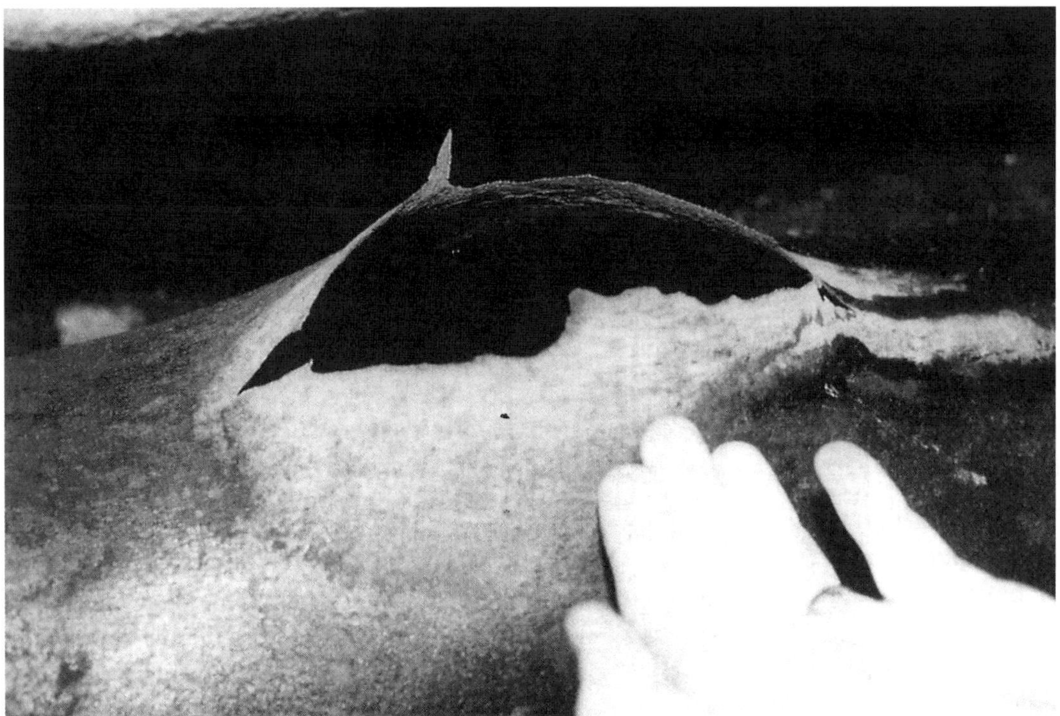

FIGURE 10.2 Tube rupture in a fired heater. (Courtesy of Butterworth Heinemann.[2])

fluid should be possible at several points in the tube layout. Furnace wall temperatures should be measurable at several locations in the firebox, as well as the convection section. Interlocks should be capable of shutting down the fuel supply if the process fluid temperature is too high or if the process fluid pressure rises above safe levels.

Furnaces should be provided with adequate viewing ports and operators should be trained to visually assess tube condition during furnace operation. Flame impingement on tubes is the most common cause of tube coking and failure, and is usually discovered only by visual observation. When impingement occurs, adjustments in furnace operations should be made immediately. Tubes intended to last 10 years can fail in a matter of hours if they are heated more than 100°F (38°C) above their designed operating temperature. It is important to be able to assess flame impingement visually, because even well-instrumented systems cannot be certain of measuring the tubes at their point of highest temperature. If the furnace is shut down for maintenance, the opportunity should be used to make a careful, close-up inspection of the tubes. Alterations in the furnace structure, layout, or operating procedure should only be undertaken with the advice and consent of an engineer qualified to assess the proposed changes.

If, despite careful precautions, tube failure occurs, adequate safety features should be in place to minimize the hazard and the damage. Process liquid flow should be controllable from a location that is an adequate distance from the furnace. Remote isolation valves should be included in the furnace design and should be inspected and tested regularly. Remote stop buttons should be located at a sufficient distance from the furnace to allow circulation pumps and any other equipment that feeds any form of fuel or oxidizer to the furnace area to be shut down, even in a worst-case scenario.

Sanders[38] reported two cases of tube failures in process heaters that illustrate the necessity of the above guidelines. In one case, a fire occurred during the start-up procedure of a plant during a cold winter morning. A natural gas-fired heater was used to provide the heat energy to a heat-transfer fluid. The combustible heat-transfer fluid was to be circulated through piping in a gaseous phase reactor until a start-up temperature of 500°F (260°C) for the process was reached. Once the start-up temperature was reached, the exothermic nature of the reaction was used to sustain the reaction and the furnace was taken offline.

After start-up, the process fluid was switched to circulate through the reactor and a cooler to remove excess heat from the reactor. On this occasion, the weather had been unseasonably cold. All of the piping in the system, except for the piping that was inside the furnace, was steam traced and insulated. Apparently, while the system was not in use, the heat-transfer

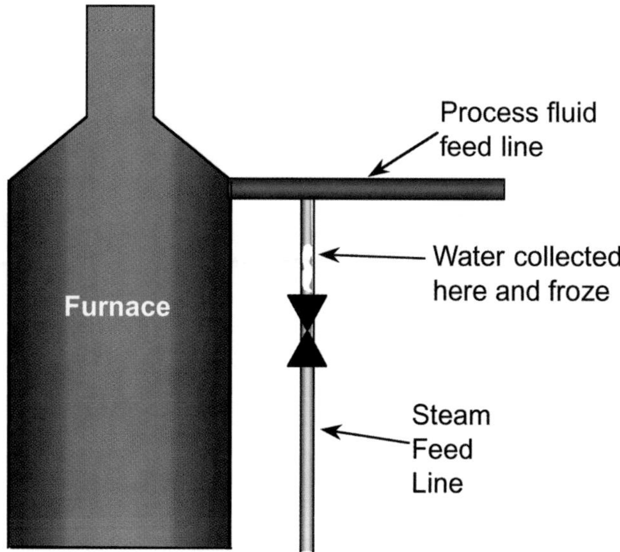

FIGURE 10.3 Trapped steam in a dead-end that can freeze and cause pipe failure. (Courtesy of Gulf Publishing.[46])

fluid had frozen in the piping inside the furnace. The start-up procedure called for the operator to establish a flow of the heat-transfer fluid through the piping in the furnace, light the furnace to heat the transfer fluid, then switch the flow of the heated fluid through the piping in the reactor.

When the circulation pump was started, a flow could not be established through the furnace. Correctly assuming that the fluid in the furnace had frozen, the operator started a small fire in the furnace in the hopes of thawing the fluid and establishing a flow. This procedure had been used successfully in the past. About 1 hour after the operator had ignited the furnace and established a low fuel rate to thaw the pipes, the operations foreman entered the control room and noticed that the natural gas flow to the furnace was much higher than it should be for the thawing operation. The rate was immediately cut to about one fourth of the existing flow.

Approximately 15 minutes later, black smoke and fire were reported coming from the heater stack. The fire was quickly extinguished, but inspection of the heater revealed that two tubes had ruptured. Fire investigators could find no one who would admit to increasing the fuel flow to the furnace. The flow recorder for the natural gas supply to the heater was not working properly, so neither the exact flow rate nor the times of flow rate change could be ascertained.

An investigation revealed that there were no written instructions for thawing frozen pipes in the heater. The operations crew was not fully aware of the hazards of lighting the furnace before a flow was established in the piping. The heater tubes had carbon buildup, which restricted the flow of the heat-transfer fluid. There was evidence of thinning of the tubes in the higher heat flux zones. Because the same technique for thawing the tubes had been used in the past, damage to the tubes may have accumulated over time.

In this case, several common safety precautions were not observed, including:

- Adequate instrumentation for measuring the temperature of the process fluid in the heater tubes was not available.
- The flow recorder for the fuel to the heater was not in working order.
- An alternative startup procedure, which had apparently been used on more than one occasion, was not properly documented.
- The operators and foreman had failed to adequately oversee the furnace operation during a time in which the furnace was being used for a purpose for which it was not designed.

In the second incident reported by Sanders, a similar tube failure in a furnace caused about $1.5 million in property damage and over $4 million in business interruption. The furnace in this incident performed a function similar to the furnace in the aforementioned incident. A combustible heat-transfer fluid was heated in the furnace, then circulated through gas-phase reactors where solvents were produced. Because the gas-phase reaction is self-sustaining once the operating temperature is reached, a single furnace was used to provide a start-up supply of heated fluid to each of five reactors, one at a time. The piping and operational procedures for the system were complex. When a reactor was brought online, the operator was to align the valves to allow circulation of the heat-transfer fluid from the heater to the reactor being started, start the circulation pump, and ignite the heater. On this occasion, the operator erred by starting the heater while the heater tubes were isolated from the circulation pump by closed block valves. About 30 minutes after the heater was fired, a water sprinkler system tripped, followed shortly thereafter by a heater-flame failure alarm and the rupture of a heat-transfer fluid pipe in the heater. Heat from the resulting 50-foot flames was spread throughout the unit by 10 to 12 mph winds.

The 15-year-old heater had been relatively well-designed to prevent an accident of the type that occurred. Automatic shutdown equipment and alarms were provided to respond to flame failure, high tube-wall temperature, low fuel supply pressure, and high heat-transfer fluid pressure. Although records indicated that some maintenance of the shutdown equipment had been performed, there was no systematic program of inspection and testing that would have ensured that the equipment was properly adjusted and would operate dependably in an emergency. The specific system that provided for automatic shutdown due to high tube-wall temperature was equipped with a setpoint adjustment that allowed the operator to set the temperature at which alarms would sound and automatic shutdown would occur. For some reason, the setpoint temperature had been set to 1600°F (870°C). The alarm temperature should have been set at 830°F (440°C), and the shutdown temperature should have been set at 850°F (450°C). No reason was ever discovered for the unreasonable and clearly dangerous temperature setting.

Accidents occur due to both human error and equipment failure. Safety instrumentation is designed to prevent human error from creating a dangerous situation. In this case, the safety equipment that would have prevented the human error from producing a destructive incident had been defeated by improper management of the equipment. An appropriate safety program that included routine inspection and testing of the safety equipment would have caught the improper setting and prevented a costly accident.

10.2.3.2 Smoke Generation

Smoke is produced in most uncontrolled fires. Smoke is generated by incomplete oxidation of the fuel, caused by insufficient mixing of air and the combustible materials. Smoke contains fine particles made primarily of solid carbon. In many fires, more people die from smoke inhalation than from the heat produced in the fire. There are several potential problems with smoke. One problem is elevated temperatures, which can damage the lungs upon inhalation. Another problem is the deposition of smoke particles on the lungs, which can hinder breathing. Smoke can also block or impair vision, which can hinder escape from the fire. Only trained personnel with adequate breathing and eye protection should ever deliberately enter smoky conditions produced by a fire.

Another product of incomplete combustion is carbon monoxide (CO), which is an extremely toxic gas that can quickly kill humans via respiratory failure. Smoke generation is an indicator of the probability of the presence of CO. CO kills by blocking the ability of hemoglobin in the blood to carry oxygen to the cells in the body. CO has a 300 times greater affinity for hemoglobin than does oxygen. Unfortunately, CO is colorless and odorless, so one must be careful to avoid possible situations where it may be present, such as in smoky fires. Fortunately, inexpensive CO detectors are available to warn of its presence (see Figure 10.4).

10.2.4 Explosion Hazards

Danger of explosion may come from many sources, but explosions most often occur when the equipment involved is in a state of change such as start-up, shutdown, or maintenance. Because

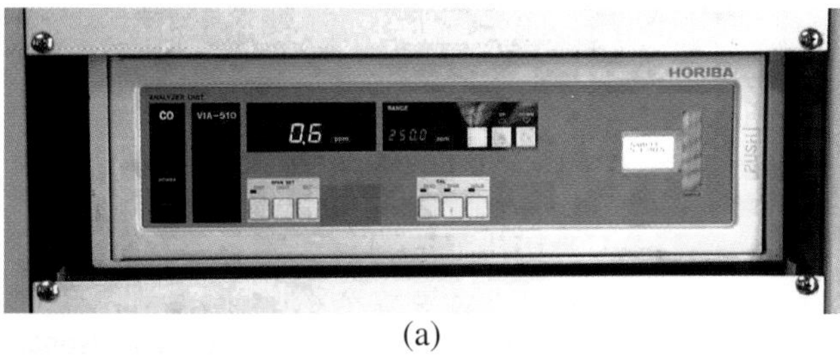

(a)

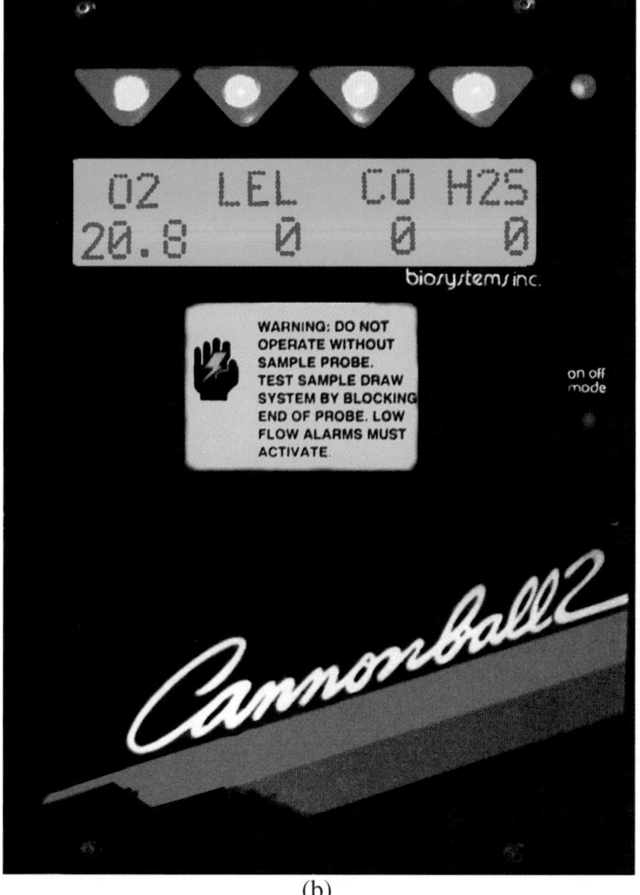

(b)

FIGURE 10.4 CO detector: (a) permanent, (b) portable, (c) area monitor.

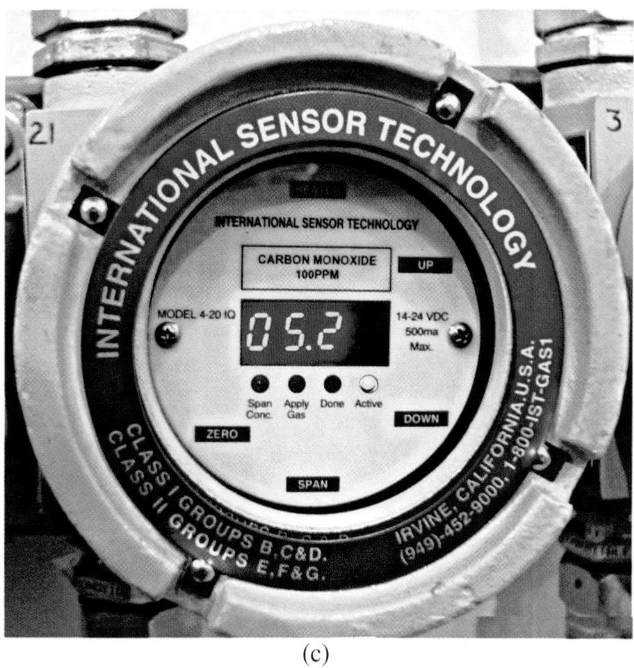

(c)

furnaces are made primarily of metals, maintenance often involves welding. The welding process provides an ignition source for any combustible gases or liquids that might remain in the work area. Typically, tests for the presence of flammable gases or liquids are conducted before any maintenance is allowed to commence. However, thorough testing to ensure the absence of flammable materials can be complex and difficult.

Two principles should be followed when testing for combustibles in a planned, enclosed maintenance area. First, testing should be done not only in the immediate work area, but also in any connecting plumbing or equipment. Second, testing should be done immediately before the work begins and periodically during the time the ignition source is in use.

10.2.4.1 Explosions in Tanks and Piping

For example, liquid storage tanks should be thoroughly flushed, cleaned, and tested before any maintenance begins. Yet, even storage tanks that have been thoroughly flushed can have vapor buildup over time from the evaporation of liquid residue in the cracks, seams, and structural members of the tank. It is almost impossible to completely clean tanks that have contained heavy oils or polymers. Such tanks can test completely clean with a combustible gas detector, yet fill with explosive vapors when maintenance activities heat the residues to the vapor point.

Pipes containing a number of bends, low points, or attached equipment may test clear of flammable materials in a proposed work area, yet contain significant amounts of explosive materials that can be vaporized by welding or migrate to the work area from elsewhere in the piping. Low points or dips in piping are particularly dangerous because they may store liquids that can vaporize and be ignited by maintenance work. Storage tanks, furnaces, pipes, or other metallic containers that have been in contact with acid can have hydrogen buildup due to the action of the acid on the metal.

10.2.4.2 Explosions in Stacks

Stacks are designed to vent exhaust gases from industrial processes. If the gases are combustible, flare stacks are used to

FIGURE 10.5 Flarestack explosion due to improper purging. (Courtesy of Gulf Publishing.[46])

burn the exhaust gases as they are vented into the atmosphere. In the case of combustible gases, operational procedures should be used that prevent the infiltration of oxygen into the stack where it could mix with the exhaust gases and cause an explosion. Figure 10.5 shows the results of an explosion caused by improper purging of a flare stack, allowing air to leak in the stack, and leading to the explosion when fuel gas was introduced back into the stack. Stacks should be built with welded seams rather than bolted joints. Bolted joints, especially joints having surfaces that are not machined, can leak air into the stack in areas of negative pressure. Operational procedures should be used that ensure that a continuous flow of gas is moving up the stack. A continuous flow of gas will sweep away small leaks of air and prevent air from moving down into the stack from the top. If the industrial process does not provide a continuous flow of gas, purge gases should be used to ensure a flow velocity of 1 to 3 in./s (8 cm/s). Oxygen sensors should be used to ensure that the oxygen content of the exhaust gases does not rise above 5%. A lower percentage is safer if hydrogen is present in the stack gases.

Flare stacks always include some form of pilot burner or other ignition source. However, other ignition sources exist even in vent stacks where ignition of the stack gases is not intentional. For example, in furnaces that are improperly operated with too little oxygen for combustion, the stack gases can contain unburned fuel. Spectacular incidents have occurred in which lightning has ignited stack gases over a distance of 100 to 200 ft (30 to 60 m) above the top of a furnace stack. As long as there is no oxygen in the stack itself and the gases are ignited above the top of the stack, such incidents, although disquieting, are not particularly dangerous. Other incidents have occurred in flare stacks in which fuel gases have reached combustion temperature and exploded inside the stack. The only way to ensure that stack explosions do not occur is to prevent oxygen from infiltrating and mixing with the stack gases.

10.2.4.3 Explosions in Furnaces

The most common source of furnace explosions is the use of an improper lighting procedure. Even procedures that have been used for years without incident can cause explosions if conditions change. For example, if the fuel source is not completely isolated from the furnace before the ignition source is inserted, a leaky valve can allow enough gas into the furnace to cause an explosion. A safe furnace lighting procedure for a furnace containing piloted burners should include the following generalized steps:

1. Confirm that the burner fuel lines are completely isolated from the furnace (either by disconnection, blind flange, or a double block and bleed valve assembly).
2. Confirm that all auxiliary furnace equipment is functioning properly, including all instrumentation and measurement devices. Open both the burner air inlet register/damper and the furnace stack damper to the fully open position.
3. Purge the furnace of any combustible or flammable substances. Follow the NFPA recommendation of purging the furnace with four furnace volumes of fresh air or inert gas (N_2 or CO_2).[34]
4. Test the atmosphere inside the furnace to ensure that there are no combustibles present.
5. Connect the pilot fuel line(s) to the burner. Activate the permanent pilot igniter or insert a portable pilot igniter or premixed ignition torch.
6. Slowly open the pilot fuel control valve to the manufacturer-specified pilot fuel pressure; visually confirm stable ignition of the pilot flame(s).
7. Reestablish the main burner fuel supply (either by connecting the main fuel line(s), removing a blind flange, or reversing the double block and bleed valve assembly) to provide a fuel source to the furnace.
8. Slowly open the main burner fuel control valve to supply the burner with the manufacturer-specified ignition fuel pressure. Visually confirm stable ignition of the main burner flame.

If the burner ignition attempt is unsuccessful, or if the procedure is aborted prior to successful burner ignition, the burner fuel supply should be immediately disconnected.

Subsequent attempts to successfully light the burner *must* begin at step 1.

In multi-burner furnaces, the operator must be certain to follow the designed lighting procedures. In some furnaces, one burner may be cross-ignited by another burner until all burners are lit. In others, however, the burners are too far apart for one to be safely lit from another. Although the eight steps given are a general procedure for a typical furnace, the designed lighting procedure provided by the manufacturer of the furnace should always be followed. The manufacturer of the furnace should preapprove any deviation from the designed procedure.

Another source of explosions in furnaces is improper air management. If the furnace is starved for air, a pulsating huffing sound may result. The flame will be unsteady, changing from long to short or wide to narrow. The variations are the result of the available air being completely consumed without burning all of the fuel. The air-starved flame will then be reduced in volume until more air is available. It will then increase in size until the available air is again consumed. This alternating cycle causes the huffing sound.

The correct action, when a furnace is huffing, is to reduce the flow of fuel until there is enough air for full combustion. If the operator incorrectly increases the air without first reducing the fuel flow, the increased air supply can mix with the large volume of unburned fuel already in the furnace, causing an explosion.

If low fuel pressure causes the flame to be extinguished in a furnace that burns fuel oil, careful tests for combustible vapors should be made immediately before attempting to relight the furnace. Although most furnaces are equipped with automatic valves that shut off the fuel flow when the flame is lost, the piping between the valve and the furnace can still contain fuel that can be vaporized into an explosive source. The vaporization process can take time. Fuel that is still in its liquid state at the time of a test can cause a furnace to test clear of combustibles. Yet, the vaporization process can cause the furnace to be filled with explosive vapors 10 minutes later.

Obtaining an accurate test with a gas detector can be problematic when oil is being used as fuel. Vapors in the furnace can condense back into their liquid state in the tube of the gas detector, preventing the vapors from reaching the detector head and being recorded. If the vaporization temperature of the fuel oil being used is near the ambient temperature, an accurate reading may be difficult to obtain. Where inaccurate readings are suspected, the furnace should be purged until the operator is certain that all unburned fuel oil has evaporated and has exited the furnace.

10.2.5 Process Hazard Analysis (PHA)

The increasing quantities of hazardous materials used at a given plant and the increasing complexity of the plants have made safety analyses both more difficult and more important because of the possibility of catastrophic accidents.[39] There are various types of analyses used for a process hazard analysis (PHA) of the design of equipment and processes, including the effects of human error. Qualitative methods include checklists, what-if reviews, and HAZOP. Quantitative methods include event trees, fault trees, and failure modes and effect analysis (FMEA). All of these methods require rigorous documentation and implementation to ensure that all potential safety problems and the associated recommendations are addressed.

HAZOP (hazard and operability) and what-if reviews are two common qualitative methods used to conduct process hazard analyses in the chemical and petrochemical industries.[29,40–43] These methods are used to review equipment and process designs to identify potential hazards[44] and to minimize the risk of accidents.[45]

In a what-if review, many questions are asked to determine what the consequences might be in the event of a particular incident. Some of the parameters commonly assessed include pressure, temperature, and flow. Questions are then asked as to what happens if that parameter is too high or too low, for example, to see what the consequence would be. If the consequence could cause an accident, then preventative actions can be taken either to prevent that parameter from ever reaching that state, or to add a safety system to respond in the event that the parameter does reach that state. For example, if a furnace temperature exceeds a predetermined level indicating a problem, then the burners can automatically be shut off to prevent damaging the furnace. The system is designed to prevent the parameter, temperature in this case, from ever reaching a critical level. Another example is causing a relief valve to vent to reduce the pressure, if the pressure becomes too high in a system.

Kletz[46] lists four circumstances that are frequent causes of accidents or dangerous conditions:

1. performing or preparing for maintenance
2. making modifications to furnace design
3. human error
4. labeling errors or labeling omissions

When preparing for maintenance, it is important to remove hazards from the maintenance area, isolate the area and/or equipment from operational equipment, and carefully follow maintenance procedures. When modifying the furnace design, even when the modification seems minor, the proposed modification should go through design procedures similar to those used for the original installation of the equipment. Without careful, detailed analysis, it is often difficult to determine how a seemingly small change will affect the entire process.

Human error is sometimes caused by inattention or poor training, but is frequently caused by a deliberate attempt to shortcut a cumbersome procedure or to make an inconvenient piece of equipment more convenient to use. Accidents caused by labeling are frequently the result of out-of-date labeling, incorrect labeling, or no labeling at all, thus resulting in the incorrect operation of equipment.

10.2.6 Codes and Standards

There is often confusion regarding the differences between codes and standards. The National Fire Protection Association (NFPA)[47] defines codes and standards in the following manner:

Code: A standard that is an extensive compilation of provisions covering broad subject matter or that is suitable for adoption into law independently of other codes and standards.

Standard: A document, the main text of which contains only mandatory provisions using the word "shall" to indicate requirements and which is in a form generally suitable for mandatory reference by another standard or code or for adoption into law.

10.2.6.1 NFPA Codes and Standards

The National Fire Protection Association (NFPA) publishes a variety of codes and standards that address key safety issues related to fire protection. The NFPA Web site[48] contains a complete listing and description of all available codes and standards. However, the following NFPA codes and standards are essential to the safe operation of combustion equipment.

1. *NFPA 86: Standard for Ovens and Furnaces, 1999 Edition.*[33] NFPA 86 is the primary standard that addresses fire and explosion hazards related to the operation and design of fired equipment used for heat utilization. Many of the components of Chapters 3 to 5 can be directly applied to process furnaces in the hydrocarbon and petrochemical industries:
   - Chapter 3: Location and Construction — the NFPA's recommendations regarding fired equipment and its proximity to personnel, buildings, and external combustible materials. Also addressed are design considerations such as structural integrity, explosion relief, observation port locations, skin temperature restrictions, etc.
   - Chapter 4: Furnace Heating System — the furnace heating system refers to both the heating source as well as all associated piping and electrical wiring. The standard provides the NFPA's rules on the design and selection of burners, fuel piping, fittings, valves, and flue ventilation devices.
   - Chapter 5: Safety Equipment and Application — the NFPA's detailed guidance concerning the design of automated safety systems (burner management systems) and the process conditions (safety interlocks) that should trigger an automated emergency shutdown (ESD). Requirements regarding the design and placement of automated fuel gas safety shutoff valves, high and low fuel pressure switches, flame supervision, excess temperature limit controllers, burner pilots, and flame proving devices are included. Guidelines for burner preignition procedures and ignition trials are also well-documented.
2. *NFPA 70: National Electric Code (NEC), updated annually.*[49] NFPA 70 provides "practical safeguarding of persons and property from hazards arising from the use of electricity." The NEC covers the installation of electric conductors and associated equipment in both the public and private sectors, including all electrical wiring associated with fired equipment. The NEC is accepted as law throughout the United States.
3. *NFPA 497: Classification of Flammable Liquids, Gases, or Vapors and of Hazardous (Classified) Locations for Electrical Installations in Chemical Process Areas, 1997 Edition.*[50] NFPA 497 recommends steps to determine the location, type, and scope of hazards presented by electrical installations in operations where flammable or combustible liquids, gases, or vapors are processed or handled. NFPA 497 can be considered a companion standard to NFPA 70: National Electrical Code (NEC).
4. *NFPA 54: National Fuel Gas Code, 1999 Edition.*[51] NFPA 54 sets minimum safety requirements for fuel gas piping systems, fired equipment, flue-gas ventilation systems, and related equipment. The NFPA considers fuel gas to include natural gas fuel, manufactured gas, and liquefied petroleum gas (propane/butane). NFPA 54 is an American National Standard, appearing as designation Z223.1.
5. *NFPA 58: Liquefied Petroleum Gas Code, 1998 Edition.*[52] NFPA 58 provides minimum safety requirements for the design and operation of liquefied petroleum gas (LPG) facilities, including fuel tank locations, piping, etc. NFPA 58 is the basis of LPG law for the United States.
6. *NFPA 30: Flammable & Combustible Liquids Code, 1996 Edition.*[53] NFPA 30 provides the minimum safety requirements for liquid fuel installations, including requirements for bulk storage tanks, spill control, emergency relief ventilation, etc. NFPA 30 is accepted as law in 35 U.S. states.
7. *NFPA 921: Guide for Fire and Explosion Investigations, 1998 Edition.*[54] NFPA 921 provides guidance for the safe and systematic investigation and analysis of fires and explosions. Topics include fire science, fire patterns, plan-

ning and execution of investigations, origin and root cause determination, etc.

10.2.6.2 Additional Standards and Guidelines

In addition to the aforementioned NFPA codes and standards, several voluntary standards and guidelines address the design and operation of combustion devices. These standards and guidelines include:

1. *European Committee for Standardization (CEN).* The multi-national European organization develops standards addressing industrial safety concerns (including fuel handling and combustion) for its 19 national member countries.
2. *CSA International.* The independent, not-for-profit organization is the largest standards development organization in Canada. The CSA has published many standards addressing combustion and the petroleum refining industry.
3. *American Petroleum Institute (API).* The API publishes a wide variety of standards applicable to combustion processes. *API Publication 535: Burners for Fired Heaters in General Refinery Service* provides guidelines for the selection and/or evaluation of burners installed in fired process heaters.

10.2.6.3 Industrial Insurance Carriers

1. *Industrial Risk Insurers (IRI).* The IRI provides comprehensive insurance protection for industrial losses due to fire, explosion, hail, lightening, windstorm, smoke, etc. The IRI generally requires adherence to NFPA codes and standards. However, the IRI often supplements the NFPA with its own requirements.[55]
2. *Factory Mutual (FM).* The Factory Mutual consists of three insurance firms, as well as the Factory Mutual Research Corporation. The FM Research Corporation conducts reliability and efficiency testing on a variety of equipment. The "FM Approval" label provides consumers with the confidence that equipment bearing that label has been rigorously tested and found worthy of use in a fire protection system. The *Factory Mutual Approval Guide* contains a listing of FM-approved items, as well as the details regarding the application and installation criteria for which the equipment is approved. Similar to the IRI, Factory Mutual will often supplement the NFPA codes and standards with its own requirements.[55]

10.3 DESIGN ENGINEERING

Several resources are available with general discussions of safety equipment used for industrial combustion applications.[36,56–58] The intent of this section is not to detail how equipment should be designed, but to point out the factors that should be considered in the design.

10.3.1 Flammability Characteristics

There are three primary parameters used to measure the relative flammability of a substance: the flash point (FP), the upper flammability limit (UFL), and the lower flammability limit (LFL). The flash point is the lowest temperature of a liquid at which it evaporates enough vapor to form an ignitable mixture with the air immediately over the surface of the liquid. The upper and lower flammability limits bracket the ignitable concentration range of a gas or vapor mixed with air. Table 10.3 contains experimentally determined flash point temperatures, as well as upper and lower flammability limits for a wide range of pure component substances in air.[59]

10.3.1.1 Liquids

The flash point of pure component liquids is usually determined experimentally. However, flash point estimates can be obtained for multi-component mixtures containing a single combustible species if both the flash point and the molar concentration of the combustible component are known. Raoult's law is used to determine the vapor pressure of the pure component in the diluted mixture (P^{SAT}), based upon the vapor pressure of the combustible species at its flash-point temperature (p):

$$p = xP^{SAT} \tag{10.1}$$

where

P^{SAT} = Vapor pressure of the combustible component present within the mixture
x = Mole fraction of the combustible component present within the mixture
p = Vapor pressure of the pure combustible component at its flash point

Once the vapor pressure of the combustible component present within the mixture (P^{SAT}) has been calculated, the resulting flash-point temperature of the mixture can be determined using a vapor pressure vs. temperature diagram.[29] Figure 10.6 is a vapor pressure vs. temperature diagram for light hydrocarbon fuels.[60]

Experimental methods are recommended for flash-point determination of multi-component mixtures involving two or more combustible components.[29]

10.3.1.2 Vapors

Similar to the flash point for pure component liquids, the upper and lower flammability limits are also determined experimentally. For multi-component gas mixtures, the Le Chatelier equation[61] is used to estimate the upper and lower flammability limits of gaseous fuels:

TABLE 10.3 Flammability and Ignition Characteristics of Liquids and Gases

Compound	Normal Boiling Point at 14.695 psia (°F)	Flash Point (°F)	LFL (vol% in air)	AFL (vol% in air)	Autoignition Temperature (°F)
Acetone	133	–4	2.5	12.8	869
Acetylene	–120	—	2.5	100	581
Acrolein	127	–15	2.8	31	428
Aniline	363	158	1.3	11	1139
Benzene	176	12	1.2	7.8	928
Butane	31	–76	1.9	8.5	549
Carbon monoxide	–313	—	12.5	74	1128
Chlorobenzeze	269	82	1.3	9.6	1099
Cyclohexane	177	–4	1.3	8.0	473
Ethane	–127	—	3.0	12.5	882
Ethyl alcohol	173	55	3.3	19	685
Ethylene	–155	—	2.7	36	842
Ethyl oxide	51	–4	3.0	100	804
Ethyl ether	–13	–42	3.4	27	662
Formaldehyde	–2	185	7.0	73	795
Heptane	209	25	1.1	6.7	399
Hexane	156	–8	1.1	7.5	437
Hydrogen	–423	—	4.0	74	—
Isopropyl alcohol	180	54	2.0	12.7	750
Isopropyl ether	45	–35	2.0	10.1	374
Methane	–259	—	5.0	15	999
Methyl acetate	134	14	3.1	16	849
Methyl alcohol	148	52	6.0	36	867
Methyl chloride	–11	—	8.1	17.4	1170
Methyl ethyl ketone	175	16	1.4	11.4	759
Methyl isobutyl ketone	242	54	1.2	8.0	838
Methyl propyl ketone	216	45	1.5	8.2	846
Napthalene	424	174	0.9	5.9	979
Octane	258	55	1.0	6.5	403
Pentane	97	–40	1.4	8.0	500
Phenol	359	174	1.8	8.6	1319
Propane	–44	–155	2.1	9.5	842
Propylene	–54	—	2.0	11.1	851
Propylene dichloride	206	70	3.4	14.5	1035
Styrene	293	88	0.9	6.8	914
Toluene	231	39	1.1	7.1	896
o-Xylene	292	90	0.9	6.7	865
m-Xylene	282	81	1.1	7.0	981
p-Xylene	281	81	1.1	7.0	982

Source: Lide, D.R., Ed., *CRC Handbook of Chemistry and Physics*, 80th ed., CRC Press, Boca Raton, FL, 1999.

$$LFL_{MIX} = \frac{1}{\sum_{i=1}^{n} \frac{y_i}{LFL_i}} \tag{10.2}$$

$$UFL_{MIX} = \frac{1}{\sum_{i=1}^{n} \frac{y_i}{UFL_i}} \tag{10.3}$$

where

LFL = Lower flammability limit for component *i* (vol%)

UFL = Upper flammability limit for component *i* (vol%)

y_i = Mole fraction of component *i* on a combustible basis

n = Number of combustible species present within the fuel mixture

Flammability limit data is often provided at process conditions of 77°F (25°C) and 14.695 psia (1 atm). However, the flammability limit ranges increase dramatically with temperature. The following empirical equations describe the temperature dependency of the flammability limits in air[62]:

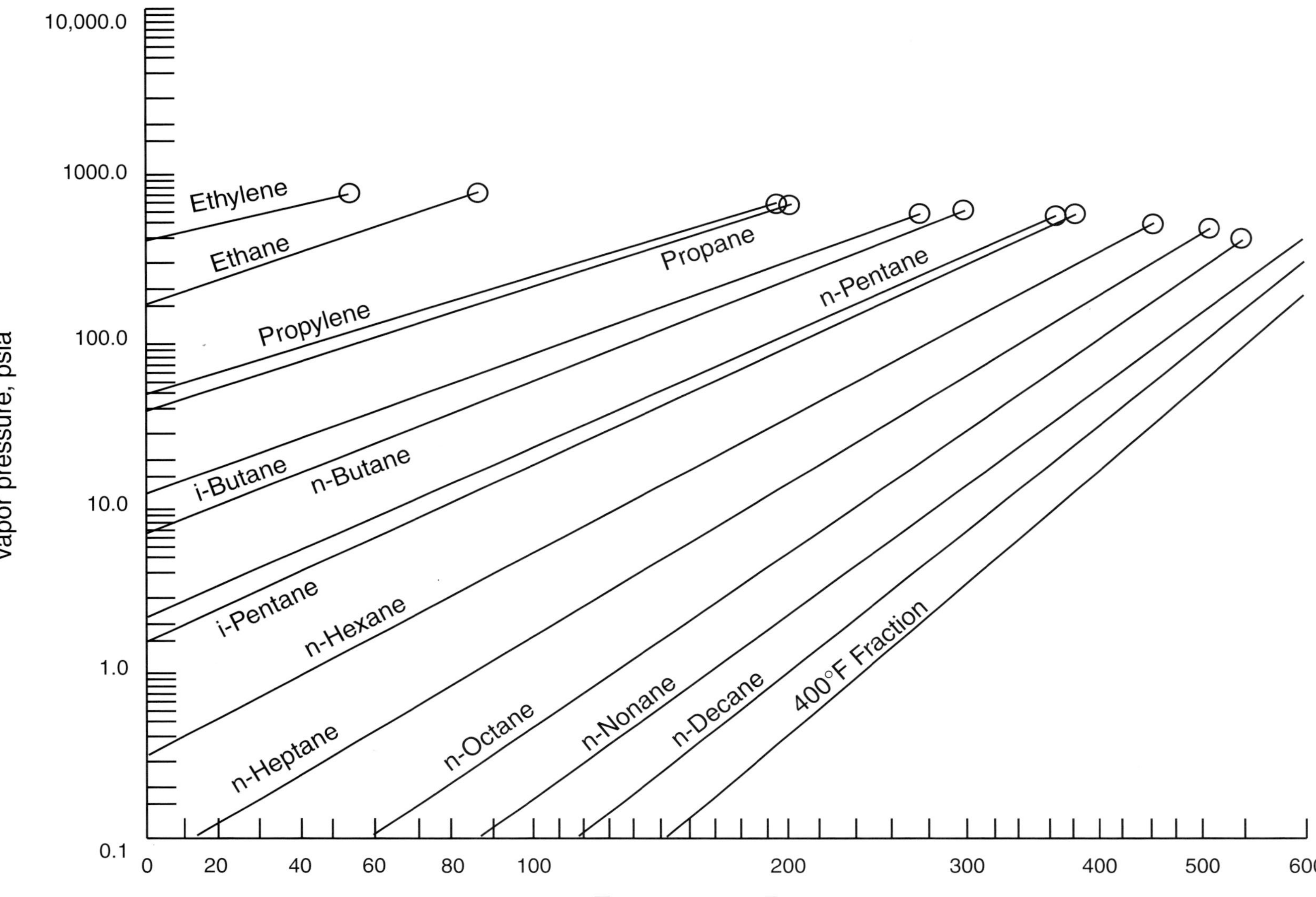

FIGURE 10.6 Vapor pressures for light hydrocarbons. (Adapted from *GPSA Engineering Data Book*, Vol. II, 10th ed., Gas Processors and Suppliers Association, Tulsa, OK, 1994.)

TABLE 10.4 Ignition Sources of Major Fires

Electrical (wiring of motors)	23%
Smoking	18%
Friction (bearings or broken parts)	10%
Overheated materials (abnormally high temperatures)	8%
Hot surfaces (heat from boilers, lamps, etc.)	7%
Burner flames (improper use of torches, etc.)	7%
Combustion sparks (sparks and embers)	5%
Spontaneous ignition (rubbish, etc.)	4%
Cutting and welding (sparks, arcs, heat, etc.)	4%
Exposure (fires jumping into new areas)	3%
Incendiarism (fires maliciously set)	3%
Mechanical sparks (grinders, crushers, etc.)	2%
Molten substances (hot spills)	2%
Chemical action (processes not in control)	1%
Static sparks (release of accumulated energy)	1%
Lightning (where lightning rods are not used)	1%
Miscellaneous	1%

Source: National Safety Council, *Accident Prevention Manual for Industrial Operations*, National Safety Council, Itasca, IL, 1974.

TABLE 10.5 Minimum Ignition Energies Required for Common Fuels

Compound	Pressure (atm)	Minimum Ignition Energy (mJ)
Methane	1	0.29
Propane	1	0.26
Heptane	1	0.25
Hydrogen	1	0.03
Propane (mol%)		
$[O_2/(O_2 + N_2)] \times 100\%$		
1.0	1	0.004
0.5	1	0.012
0.21	1	0.15
1.0	0.5	0.01

Source: Zabetakis, M.G., *AIChE-Inst. Chem. Engr. Symp., Ser. 2, Chem Engr. Extreme Cond. Proc. Symp.*, American Institute of Chemical Engineers, New York, 1965, 99–104.

$$LFL_T = LFL_{25}\left[\frac{1-0.75(T-25)}{\Delta H_C}\right] \tag{10.4}$$

$$UFL_T = UFL_{25}\left[\frac{1+0.75(T-25)}{\Delta H_C}\right] \tag{10.5}$$

where

ΔH_C = the heat of combustion (kcal/mole)
LFL_T = lower flammability limit at temperature, T (vol%)
LFL_{25} = lower flammability limit at 25°C (vol%)
UFL_T = upper flammability limit at temperature, T (vol%)
UFL_{25} = upper flammability limit at 25°C (vol%)
T = actual process temperature (°C)

Variation in pressure does not significantly affect the LFL except at very low pressures (below 27 in. w.c. absolute [50 mmHg absolute]). However, increases in pressure can significantly raise the UFL. The following empirical expression describes the pressure dependence of the UFL in air[63]:

$$UFL_P = UFL_{1\text{ atm}} + 20.6\left(\log_{10} P + 1\right) \tag{10.6}$$

where

UFL_P = Upper flammability limit at pressure P (vol%)
$UFL_{1\text{ atm}}$ = Upper flammability limit at 1 atm (vol%)
P = Actual process pressure (MPa, absolute)

The presence of pure oxygen as the oxidizing agent (as opposed to air) has very little effect on the LFL, as the oxygen concentration of air is in excess of that required for combustion at the LFL. However, the UFL of most hydrocarbon fuels in pure oxygen is increased by approximately 45 to 55% when compared to the equivalent UFL in air.[64] Table 10.1 compares flammability limits of several common fuels using both air and pure oxygen as the oxidizing agent.[36]

10.3.2 Ignition Control

Ignition is the process through which combustion is initiated, and occurs when a flammable mixture of fuel and oxidizer comes in contact with a suitable ignition source. The minimum ignition energy is the minimum energy required to initiate combustion, and can be obtained through a variety of sources: direct contact with a spark or flame, static electricity, autoignition, autooxidation, and adiabatic compression. Table 10.4 lists the ignition sources tabulated from over 25,000 fires by the Factory Mutual Engineering Corporation.[65] Table 10.5 contains the minimum ignition energy (MIE) required for several common fuels.[66]

In general, the MIE decreases with pressure and increases with inert gas concentration. MIEs for hydrocarbon fuels are relatively low when compared to ignition sources. Walking across a rug can produce an electrostatic discharge of 22 mJ. An internal combustion engine's spark plug can generate an electrical discharge energy of 25 mJ.[29]

Direct contact with a spark or flame is a very common energy source often used for the intentional ignition of industrial combustion equipment. For process burners in fired heaters, the ignition source may be in the form of a small premixed pilot burner, a portable electrostatic ignitor, or a portable premixed gas torch. Flare burners typically use a continuous flare pilot that is ignited by a flame front generator (FFG)

FIGURE 10.7 Ethylene oxide plant explosion caused by autoignition. (Courtesy of Gulf Publishing.[46])

(see Chapter 20). An FFG is a custom-built device that sends a small burst of flame to the top of a tall flare stack, allowing the operator to ignite the pilot burner from ground level. FFG devices have been proven to safely ignite flare pilot burners a distance of 1 mile away from the flame front generator. Regardless of the source, all ignition devices should be designed for the particular equipment and the specific set of process conditions for which they will be used.

Static electricity is a common ignition source of fires and explosions in chemical processing plants. An electrostatic charge is formed whenever two dissimilar surfaces move relative to each other. A relevant example is liquid flowing through a pipeline, moving past the walls of the pipe. In this example, one charge is formed on the pipe surface, while another equal but opposite charge is formed on the surface of the moving liquid. When the voltage becomes strong enough, the static electricity will discharge in the form of an electrical spark. The spark can ignite combustible and flammable materials present. Crowl and Louvar[29] and the NFPA[67] present detailed explanations of design fundamentals for the prevention of fires and explosions due to electrostatic discharge. Kletz[46] discusses several case histories of fires and explosions ignited by electrostatic discharge.

Autoignition is the process through which a flammable liquid's vapors are capable of extracting enough energy from the environment to self-ignite, without the presence of a spark or flame. The ability of a flammable liquid to autoignite is characterized by the liquid's autoignition temperature. Table 10.3 contains common autoignition temperatures for a variety of flammable liquids. Autoignition temperatures depend on a number of factors, including fuel vapor concentration, fuel volume, system pressure, presence of catalytic material, and flow conditions. Because the autoignition temperature is a function of so many process variables, it is important that the autoignition temperature is determined experimentally under the conditions that most closely simulate actual process conditions.[29] Figure 10.7 shows the wreckage of an ethylene oxide plant explosion caused by autoignition leading to fire and explosion.[46]

Autooxidation is the process of slow oxidation, resulting in the production of heat energy, sometimes leading to autoignition if the heat energy is not removed from the system. The most common example of this potential ignition process

TABLE 10.6 Benefits of a Successful Process Knowledge and Documentation Program

- Preserves a record of design conditions and materials of construction for existing equipment, which helps ensure that operations and maintenance remain faithful to the original intent
- Allows recall of the rationale for key design decisions during inception, design, and construction of major capital projects, which is useful for a variety of reasons (i.e., an aid in future projects and modifications)
- Provides a basis for understanding how the process should be operated and why it should be run in a given way
- Offers a "baseline" for use in evaluating a process change
- Records accident/incident causes and corrective actions and other operating experience for future guidance
- Protects the company against unjustified claims of irresponsibility and negligence
- Retains basic research and development information on process chemistry and hazards to guide future research effort

Source: Center for Chemical Process Safety, *Plant Guidelines for Technical Management of Chemical Process Safety*, rev. ed., American Institute of Chemical Engineers, New York, 1995. With permission.

is when rags saturated with oils are discarded or stored in a warm area. If allowed to autooxidize, the increased temperatures can result in autoignition of the rags, and a damaging fire or explosion can result. Relatively high-flash-point materials are the most susceptible to the autooxidation process, while low-flash-point materials can often evaporate without ignition. Fuel leaks that saturate thermal insulation or other absorbent materials should be isolated immediately, and the contaminated absorbent should be removed promptly and discarded in a suitable manner.[64]

Adiabatic compression of combustible or flammable materials can result in high temperatures, which in turn may result in autoignition of the compressed fuel. Examples of adiabatic compression include internal combustion engines and gas compressors. The temperature rise associated with the adiabatic compression of an ideal gas can be determined using thermodynamic relationships:

$$\frac{T_2}{T_1} = \left(\frac{P_2}{P_1}\right)^{\frac{(\gamma-1)}{\gamma}} \tag{10.7}$$

where

T_2 = Final absolute temperature
T_1 = Initial absolute temperature
P_2 = Final absolute pressure
P_1 = Initial absolute pressure
γ = Specific heat ratio, C_p/C_v

10.3.3 Fire Extinguishment

In premixed burners, there is a chance that the fire can flash back into the burner. Flashback can occur when the velocity of the fuel mixture leaving the combustor is exceeded by the flame speed (speed with which the flame front burns back towards the fuel source). To prevent flashback in premixed burners, flame arrestors are commonly used.[68] The primary applications for flame arrestors are to protect people and equipment from flashbacks, fires, and catastrophic explosions. A flame arrestor is designed to stop (extinguish) a flame or explosion from further propagation past the arrestor. It is a special type of heat exchanger that cools the flame, thus removing one of the legs of the fire tetrahedron. Time is required to dissipate the heat, so the design and construction of the quenching media are important.[69] In fuel piping systems, the arrestor must be perforated or porous to allow gas flow through it. The technique often used in flame arrestors is to cool the propagating flame or explosion enough to extinguish the fire. Thermal mass, usually in the form of metal, is used to extract enough energy from the reacting gases that the flame can no longer be supported and is extinguished. Many different arrestor designs are available, including gauzes, perforated plates, expanded metal, sintered metal, metal foam, compressed wire wool, loose filling, hydraulic arrestors, stacked plate, and crimped ribbon.[70]

10.3.4 Safety Documentation and Operator Training

Safety documentation and operator training provide the backbone of a strong safety program, and are absolutely essential in order to maintain a safe combustion working environment. Table 10.6 illustrates some of the benefits of a successful process documentation program.[71]

The AIChE Center for Chemical Process Safety publishes several titles that address implementation of process safety documentation.[71–74] Safety documentation for combustion-related processes includes design information, process hazard analysis (PHA) reports, standard operating procedures, and training documentation. Feedback from each of theses documentation elements are linked together as part of a plant's overall process safety program. Figure 10.8 visually describes the documentation feedback linkage suggested by the AIChE.[72]

10.3.4.1 Design Information

Design information, or process knowledge, refers to all of the documents that pertain to the safe design of the combustion system. This set of information can include, but is not limited to:[71,72]

- process information:
  - detailed information regarding the design criteria of the combustion system (i.e., heat load, process flow rates, temperatures and pressures, fuel composition, etc.)
  - process chemistry

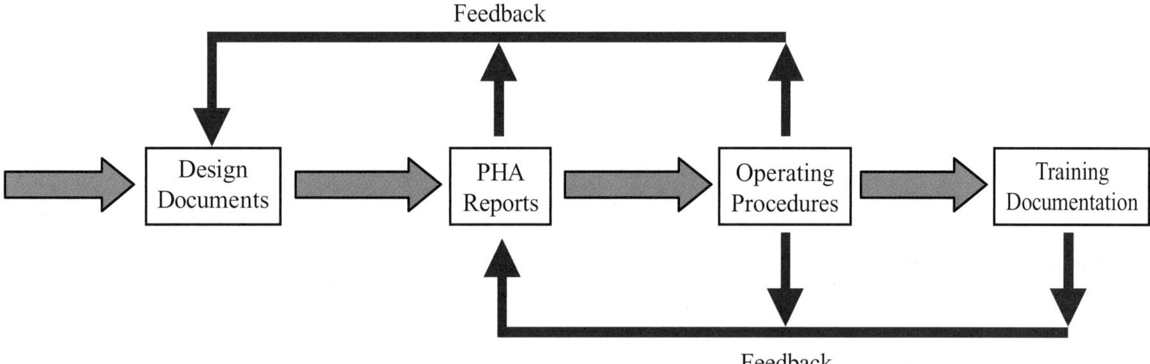

FIGURE 10.8 Safety documentation feedback flow chart. (Adapted from Center for Chemical Process Safety, *Guidelines for Process Safety Documentation*, American Institute of Chemical Engineers, New York, 1995.)

  - safe operating ranges for process conditions (flow, temperature, pressure, compositions, etc.)
  - the known hazardous effects of deviation from the stated safe operating conditions
- equipment information:
  - process flow diagrams (PFDs)
  - piping and instrumentation diagrams (P&IDs)
  - detailed equipment drawings (i.e., heater, flare, or incinerator drawings illustrating details such as tube locations, burner orientation, etc.)
  - electrical wiring schematics and electrical classification data
  - manufacturer's equipment manuals (including design criteria, safe operation recommendations, etc.)
  - equipment, valve, and instrumentation specification sheets
  - maintenance manuals
  - automated safety interlock system details

10.3.4.2 Process Hazard Analysis (PHA) Reports

Regardless of the evaluation technique chosen, the PHA process (see Section 10.2.5) puts the combustion system (as well as the entire processing unit) under the microscope, systematically analyzing the entire system piece by piece. To ensure that all process safety recommendations are executed, thorough documentation must be conducted in an organized manner. The PHA process is conducted both prior to initial start-up of the system, and periodically (usually every 5 years) during the system's lifetime. This cyclic approach guarantees continuous safety feedback for the operators and engineers entrusted with the safe operation of the combustion system. The AIChE provides detailed guidance on the selection and execution of various PHA evaluation techniques.[74]

10.3.4.3 Standard Operating Procedures (SOPs)

Day in and day out, standard operating procedures (SOPs) provide operators with clear, detailed, sequenced instructions regarding the safe operation and maintenance of combustion equipment. In addition to providing detailed instructions for the operation of a combustion system, SOPs also assist in the training of both new and existing plant personnel. The AIChE provides guidance regarding the preparation, revision, content, and distribution of SOPs.[72]

As with any set of SOPs, written procedures for combustion systems should include pre-startup, start-up, normal operation, shutdown, and emergency shutdown procedures, as well as additional procedures for preventative maintenance operations. The equipment manufacturer should always be consulted regarding the proper ignition and operation of individual combustion devices. Figure 10.9 shows the charred wreckage of a refinery where an accident was caused due to improper maintenance procedures.

10.3.4.4 Operator Training and Documentation

Training is essentially the successful communication and transfer of knowledge and skills. Applied to combustion systems, training is the successful communication and transfer of knowledge regarding basic combustion science, safety hazards associated with fires and explosions, and skillful execution of standard operating procedures. In general, training exercises should emphasize the need to complete all tasks in a safe manner. Training is not a substitute for poor combustion system design. Rather, the design, documentation, and training should complement each other to provide a safe working environment.

Employees should be selected for a level of training that is commensurate with the level of exposure they will encounter on the job. New employees should successfully complete

FIGURE 10.9 Refinery damaged due to improper maintenance procedures. (Courtesy of Butterworth-Heinemann.[2])

their training regimen before being allowed to perform training-required tasks unsupervised. Training may involve formal (attendance and examination required), informal (safety discussions, demonstrations, seminars), self-study, and on-the-job training methods.[72] Training sessions can include but are not limited to the following topics:

- start-up operations (including combustion device ignition)
- normal operations
- shutdown operations (including combustion device extinguishment)
- maintenance
- OSHA HAZWOPER (Hazardous Waste Operations and Emergency Response)
- fuels handling
- emergency procedures and evacuation
- confined space entry
- lock-out and tag-out (hazardous energy sources)
- hazard communication
- blood-borne pathogens
- fire extinguishment

The safety training program must be thoroughly documented to ensure that it is conducted in an acceptable and timely manner. Management and trainers should solicit continuous feedback in order to evaluate and improve program effectiveness. The AIChE provides guidance on the successful management and implementation of a comprehensive safety training program.[71,72]

10.4 SOURCES OF FURTHER INFORMATION

In addition to the references cited in this chapter, there are many organizations that have good information on safety. Many of these have general safety information, but some also have specific information related to combustion. Typical organizations are listed below:

American Society of Safety Engineers (ASSE)
1800 E. Oakton Street
Des Plaines, IL 60018
(847) 699-2929
www.asse.org

American National Standards Institute (ANSI)
11 West 42nd Street, 13th Floor
New York, NY 10036
(212) 642-4900
www.ansi.org

Board of Certified Safety Professionals (BCSP)
208 Burwash Avenue
Savoy, IL 61874
(217) 359-9263
www.bcsp.com

Center for Chemical Process Safety (CCPS)
American Institute of Chemical Engineers
Three Park Avenue
New York, NY 10016-5991
(212) 591-7319
www.aiche.org/cps

DuPont Safety Resources
Christiana Executive Campus
131 Continental Drive, Suite 307
Newark, DE 19713
(800) 532-SAFE
www.dupont.com/safety

Occupational Safety & Health Administration (OSHA)
U.S. Department of Labor
Office of Public Affairs - Room N3647
200 Constitution Avenue
Washington, D.C. 20210
(202) 693-1999
www.osha.gov

National Fire Protection Association (NFPA)
1 Batterymarch Park
P.O. Box 9101
Quincy, MA 02269-9101
(800) 344-3555
www.nfpa.com

Society of Fire Protection Engineers (SPFE)
7315 Wisconsin Avenue, Suite 1225 W
Bethesda, MD 20814
(301) 718-2910
www.sfpe.org

REFERENCES

1. Center for Chemical Process Safety, *Explosions in the Process Industries*, American Institute of Chemical Engineers, New York, 1994.
2. R.E. Sanders, *Chemical Process Safety: Learning from Case Histories*, Butterworth-Heinemann, Boston, 1999.
3. T. Richardson, Learn from the Phillips explosion, *Hydrocarbon Proc.*, 70(3), 83-84, 1991.
4. Workplace Health, Safety and Compensation Commission of New Brunswick, Explosion and Fire at the Irving Oil Refinery in Saint John, New Brunswick: Interim Report - EDB 98-22, Fredericton, New Brunswick (Canada), published by Government of New Brunswick, 1998.
5. Anonymous, A thirty year review of large property damage losses in the hydrocarbon chemical process industries, *Loss Prevention Bulletin*, Part 99, 3-25, 1991.
6. Fire Protection Association, The hydrocarbon processing industry: fire hazards and precautions, *Fire Protection Assoc. J.*, No. 82, 252-264, 1969.
7. F.K. Crawley and G.A. Dalzell, Fire and explosion hazard management in the chemical and hydrocarbon processing industry, *IMechE Conf. Trans.: Management of Fire & Explosions*, 5, 61-72, 1997.
8. T. Kletz, *Learning from Accidents*, 2nd ed., Butterworth-Heinemann, Oxford, 1994.
9. T. Kletz, *An Engineer's View of Human Error*, 2nd ed., Institution of Chemical Engineers, Rugby, U.K., 1991.
10. Center for Chemical Process Safety, *Guidelines for Preventing Human Error in Process Safety*, American Institute of Chemical Engineers, New York, 1994.
11. American Petroleum Institute, *Safety Digest of Lessons Learned: Section 2, Safety in Unit Operations*, Publication 758, Washington, D.C., 1979.
12. Y. Uehara, Fire safety assessments in petrochemical plants, *Fire Safety Sci., Proc. 3rd Int. Symp., Int'l Assoc. Fire Safety Sci.*, Edinburgh, U.K., 1991, 83-96.
13. W.P. Crocker and D.H. Napier, Thermal radiation hazards of liquid pool fires and tank fires, *IChemE Symp. Series No. 97*, Institute of Chemical Engineers, Pergamon Press, Oxford, U.K., 1986, 159-84.
14. Center for Chemical Process Safety, *Thermal Radiation 1: Sources and Transmission*, American Institute of Chemical Engineers, New York, 1989.
15. Center for Chemical Process Safety, *Thermal Radiation 2: The Physiological and Pathological Effects*, American Institute of Chemical Engineers, New York, 1996.
16. M.A. Fry, Benefits of fire and explosion computer modeling in chemical process safety, *Proc. of American Chemical Industries Week '94*, Oct. 18-20, Philadelphia, PA, 1994.
17. R.A. Ogle, Explosion hazard analysis for an enclosure partially filled with a flammable gas, *Process Safety Progress*, 18(3), 170-177, 1999.
18. D.R. Stull, *Fundamentals of Fire and Explosion*, AIChE Monograph Series, No. 10, Vol. 73, American Institute of Chemical Engineers, New York, 1977.
19. W. Bartknecht, *Explosions*, Springer-Verlag, New York, 1980.
20. F.T. Bodurtha, *Industrial Explosion Prevention and Protection*, McGraw-Hill, New York, 1980.
21. F.P. Lees, *Loss Prevention in the Process Industries*, Vol. 1, Butterworths, London, 1980.

22. G.L. Wells, *Safety in Process Design*, George Goodwin, London, 1980.
23. H.H. Fawcett and W.S. Wood, *Safety and Accident Prevention in Chemical Operations*, 2nd ed., John Wiley & Sons, New York, 1982.
24. Center for Chemical Process Safety, *Guidelines for Engineering Design for Process Safety*, American Institute of Chemical Engineers, New York, 1993.
25. Center for Chemical Process Safety, *Guidelines for Safe Process Operations and Maintenance*, American Institute of Chemical Engineers, New York, 1995.
26. Center for Chemical Process Safety, *Guidelines for Process Safety Fundamentals in General Plant Operations*, American Institute of Chemical Engineers, New York, 1995.
27. American Petroleum Institute (API), RP 750, Management of Process Hazards, First Edition, API, Washington, D.C., 1990, reaffirmed 1995.
28. R. Skelton, *Process Safety Analysis*, Gulf Publishing, Houston, TX, 1997.
29. D.A. Crowl and J.F. Louvar, *Chemical Process Safety: Fundamentals with Applications*, Prentice-Hall, Englewood Cliffs, NJ, 1990.
30. R. King, *Safety in the Process Industries*, Butterworth-Heinemann, London, 1990.
31. T.A. Kletz, *Critical Aspects of Safety and Loss Prevention*, Butterworths, London, 1990.
32. D.P. Nolan, *Handbook of Fire and Explosion Protection Engineering Principles for Oil, Gas, Chemical, and Related Facilities*, Noyes Publications, Westwood, NJ, 1996.
33. National Fire Protection Association, *NFPA 86 Standard for Ovens and Furnaces*, 1999 edition, NFPA, Quincy, MA, 1999.
34. M.A. Niemkiewicz and J.S. Becker, Safety overview, in *Oxygen-Enhanced Combustion*, C.E. Baukal, Ed., CRC Press, Boca Raton, FL, 1998, 261-278.
35. C.E. Baukal, Ed., *Oxygen-Enhanced Combustion*, CRC Press, Boca Raton, FL, 1998.
36. R.J. Reed, *North American Combustion Handbook*, Vol. I: Combustion, Fuels, Stoichiometry, Heat Transfer, Fluid Flow, 3rd ed., North Amer. Mfg. Co., Cleveland, OH, 1986.
37. T.H. Pratt, *Electrostatic Ignitions of Fires and Explosions*, Burgoyne Inc., Marietta, GA, 1997.
38. R.E. Sanders, *Chemical Process Safety: Learning from Case Histories*, Butterworth-Heineman Publishers, Boston, 1999.
39. J.H. Burgoyne, Reflections on process safety, *IChemE Symp. Series No. 97*, Inst. of Chem. Engineers, Pergamon Press, Oxford, U.K., 1986, 1-6.
40. Center for Chemical Process Safety (CCPS), *Guidelines for Chemical Process Quantitative Risk Analysis*, 1st ed., AIChE, New York, 1989.
41. Center for Chemical Process Safety (CCPS), *Guidelines for Hazard Evaluation Procedures*, 2nd ed., AIChE, New York, 1992.
42. D.P. Nolan, *Application of the HAZOP and What-If Safety Reviews to the Petrochemical and Chemical Industries*, Noyes Publications, Park Ridge, NJ, 1994.
43. T. Kletz, *Hazop and Hazan*, 4th ed., Taylor & Francis, Philadelphia, PA, 1999.
44. G. Wells, *Hazard Identification and Risk Assessment*, Gulf Publishing, Houston, TX, 1996.
45. G. Wells, *Major Hazards and Their Management*, Gulf Publishing, Houston, TX, 1997.
46. T. Kletz, *What Went Wrong: Case Histories of Process Plant Disasters*, 4th ed., Gulf Publishing, Houston TX, 1998.
47. From the NFPA website: http://www.nfpa.org/Codes/Background/background.html.
48. From the NFPA website: http://www.nfpa.org/.
49. National Fire Protection Association, *NFPA 70: National Electric Code (NEC)*, 2000 edition, NFPA, Quincy, MA, 2000.
50. National Fire Protection Association, *NFPA 497: Classification of Flammable Liquids, Gases, or Vapors and of Hazardous (Classified) Locations for Electrical Installations in Chemical Process Areas*, 1997 Edition, NFPA, Quincy, MA, 1997.
51. National Fire Protection Association, *NFPA 54: National Fuel Gas Code*, 1999 edition, NFPA, Quincy, MA, 1997.
52. National Fire Protection Association, *NFPA 58: Liquefied Petroleum Gas Code*, 1998 edition, NFPA, Quincy, MA, 1998.
53. National Fire Protection Association, *NFPA 30: Flammable & Combustible Liquids Code*, 1996 edition, NFPA, Quincy, MA, 1996.

54. National Fire Protection Association, *NFPA 921: Guide for Fire and Explosion Investigations*, 1998 edition, NFPA, Quincy, MA, 1998.

55. J.W. Coons, *Fire Protection Design Criteria, Options, Selection*, R. S. Means Company, Kingston, MA, 1991.

56. J.R. Cornforth, Ed., *Combustion Engineering and Gas Utilisation*, E&FN, London, 1992.

57. IHEA, *Combustion Technology Manual*, 5th ed., Industrial Heating Equipment Assoc., Arlington, VA, 1994.

58. M.A. Niemkiewicz and J.S. Becker, Equipment design, in *Oxygen-Enhanced Combustion*, C.E. Baukal, Ed., CRC Press, Boca Raton, FL, 1998, 279-313.

59. D.R. Lide, Ed., *CRC Handbook of Chemistry and Physics*, 80th ed., CRC Press, Boca Raton, FL, 1999.

60. *GPSA Engineering Data Book*, Vol. II, 10th ed., Gas Processors and Suppliers Association, Tulsa, OK, 1994.

61. H. Le Chatelier, Estimation of firedamp by flammability limits, *Ann. Mines*, 19(8), 388-395, 1891.

62. M.G. Zabetakis, S. Lambiris, and G.S. Scott, Flame temperatures of limit mixtures, *Seventh Symp. Combustion*, Butterworths, London, 1959, 484.

63. M.G. Zabetakis, Fire and explosion hazards at temperature and pressure extremes, *AIChE-Inst. Chem. Engr. Symp., Ser. 2, Chem. Engr. Extreme Cond. Proc. Symp.*, 1965, 99-104.

64. R.H. Perry, D.W. Green, and J.O. Maloney, Eds., *Perry's Chemical Engineers' Handbook*, 7th ed., McGraw-Hill, New York, 1997.

65. National Safety Council, *Accident Prevention Manual for Industrial Operations*, National Safety Council, Itasca, IL, 1974.

66. M.G. Zabetakis, Flammability Characteristics of Combustible Gases and Vapors, U.S. Bureau of Mines Bulletin 627, USNT AD 701, 576, 1975.

67. National Fire Protection Association, *NFPA 77: Recommended Practice on Static Electricity*, 1993 edition, NFPA, Quincy, MA, 1993.

68. V.A. Mendoza, V.G. Smolensky, and J.F. Straitz, Don't detonate — arrest that flame, *Chem. Eng.*, 103(5), 139-142, 1996.

69. V.A. Mendoza, V.G. Smolensky, and J.F. Straitz, Understand flame and explosion quenching speeds, *Chem. Eng. Prog.*, 89, 38-41, 1993.

70. H. Phillips and D.K. Pritchard, Performance requirements of flame arresters in practical applications, *IChemE Symp. Series No. 97*, Inst. of Chem. Engineers, Pergamon Press, Oxford, U.K., 1986, 47-61.

71. Center for Chemical Process Safety, *Plant Guidelines for Technical Management of Chemical Process Safety*, rev. ed., American Institute of Chemical Engineers, New York, 1995.

72. Center for Chemical Process Safety, *Guidelines for Process Safety Documentation*, American Institute of Chemical Engineers, New York, 1995.

73. Center for Chemical Process Safety, *Guidelines for Technical Management of Chemical Process Safety*, American Institute of Chemical Engineers, New York, 1989.

74. Center for Chemical Process Safety, *Guidelines for Hazard Evaluation Procedures*, 2nd ed., American Institute of Chemical Engineers, New York, 1992.

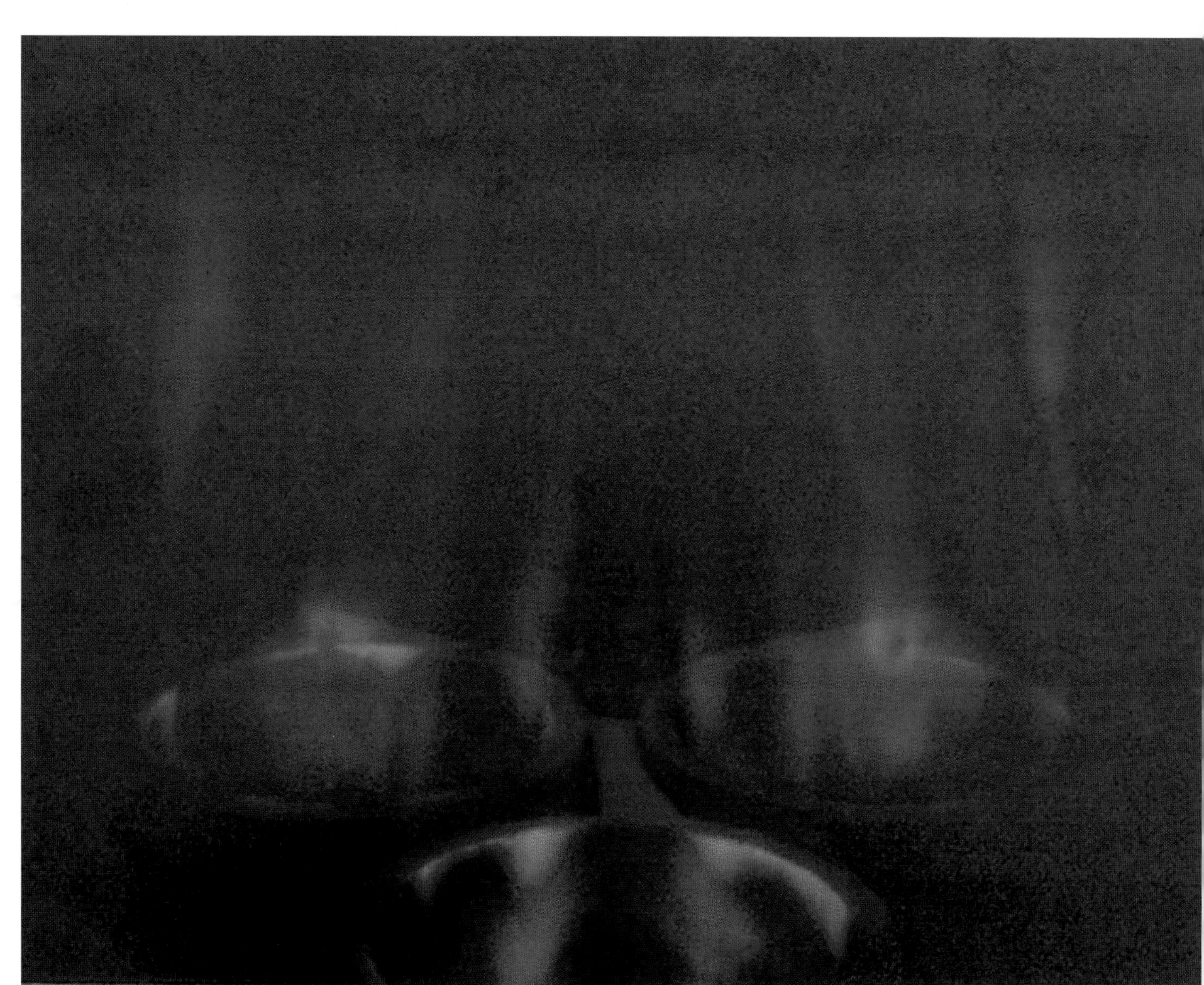

Chapter 11
Burner Design

Richard T. Waibel and Michael Claxton

TABLE OF CONTENTS

11.1 Introduction 352
11.2 Combustion 352
11.3 Burner Design 353
  11.3.1 Metering: Fuel 353
  11.3.2 Metering: Air (Combustion O_2) 356
  11.3.3 Air Control 358
  11.3.4 Mixing Fuel/Air 359
  11.3.5 Maintain (Ignition) 360
  11.3.6 Mold (Patterned and Controlled Flame Shape) 361
  11.3.7 Minimize (Pollutants) 362
11.4 Burner Types 362
  11.4.1 Premix and Partial Premix Gas 363
  11.4.2 Raw Gas or Nozzle Mix 364
  11.4.3 Oil or Liquid Firing 364
11.5 Configuration (Mounting and Direction of Firing) 367
  11.5.1 Conventional Burner, Round Flame 367
  11.5.2 Flat Flame Burner 368
  11.5.3 Radiant Wall 369
  11.5.4 Downfired 369
11.6 Materials Selection 369
References 370

11.1 INTRODUCTION

What is a burner? In its simplest form, a burner is a device used to provide heat. Specifically, it is a device used to provide a controlled exothermic oxidation reaction.

Using this definition, one could argue that a wooden torch is a burner — even if there is some question as to the use of the exothermic reaction condition being met. However, it is also beneficial to assume that the device is not itself consumed by the reaction. As a more reasonable approach, a burner is a device that provides three basic design functions:

1. A burner must provide for controlled mixing of the reactant, fuel, and the oxidizing agent, in most cases air.
2. The burner must provide a stable and self-renewing ignition source.
3. The burner should provide for a controlled region of reaction or controllable flame shape.

The modern concept of patterning the reaction flame is a direct outgrowth of continuous hydrocarbon and petrochemical processes. Processing of naturally occurring crude oils and organic by-products was initially accomplished through "batching." In this method, the necessity of controlling the actual flame to precise dimensions was not critical. If, at the end of a single batch process, there were carbon residues or undesirable deposits, the fire was extinguished and the vat cleaned. With the advent of tubed, continuous throughput process furnaces, it was no longer economically desirable, and in some cases physically possible, to perform mechanical cleaning. This made it necessary to reduce or eliminate the carbon residues and deposits generated by localized overheating of the fluid being processed. Direct conductive heat transfer from flame impingement is a major source of this localized overheating.

11.2 COMBUSTION

Combustion as a controlled process is often considered a "black art." While there are substantial amounts of experience-based information and design "rules of thumb" involved, there are very real chemical and physical laws involved. The layman's view of flames and combustion stability are actually soundly based in the principles of chemistry, chemical reactions, and fluid flow.

Combustion has been defined as a relatively fast exothermic gas-phase chemical reaction. It can occur in either flame or non-flame mode. For the purposes of discussion in burner design, only flame mode will be considered and is defined as a dual reactant flame. This flame is an exothermic reaction propagating subsonically through the mixture of two reactants. The two reactants are, of course, the fuel and the oxidant. The oxidant is usually atmospheric air. However, the combustion O_2 can be from a number of alternative sources, including:

- *Ambient atmospheric air* is the predominant source.
- *Preheated atmospheric air* is commonly used to improve the thermal efficiency and typically requires a forced-draft application.
- *Turbine exhaust gas* (TEG) is the high-temperature, reduced O_2 content exhaust gas from a gas turbine. In general, this is a good source for combustion O_2; however, low oxygen content with low-temperature exhaust gas can result in an oxidant that cannot support combustion.
- *Diesel engine exhaust* is used periodically as an alternative to TEG. The major problem with this as a source of combustion O_2 is in the low temperatures associated with lowered oxygen levels and the pulsing flow associated with internal combustion exhaust.
- *Kiln and drier off-gas* is seldom seen, but can be a usable source for combustion O_2. Again, attention must be paid to the O_2/temperature relationship.
- *Oxygen-enriched streams* are not currently common in typical industrial combustion applications; however, they do exist in specialized processes.[1]

Each of these sources for combustion O_2 has been designed by burner designers and utilized by industry. For the purposes of this chapter, the majority of the discussion is limited to the use of ambient atmospheric air. "Special" considerations within a burner design that are particular to one or more of the alternative sources are periodically noted.

The temperature vs. $\%O_2$ (gross) relationship with respect to sustainable combustion is particularly unforgiving. Figure 11.1 shows the theoretical limitations of methane combustion with a combustion O_2 stream varying in oxygen and temperature. Practical application does not allow for operation near the theoretical limit.

Practical writings on combustion often include discussions of the 3-Ts of combustion. The 3-Ts are simply a condensation of fluid flow and chemical reaction principles:

1. *Turbulence* is the interaction between two fluid streams required to achieve intermixing of the two.
2. *Temperature* is the required energy for the initiation of a chemical reaction — oxidation.
3. *Time* is the period for the reaction to reach completion.

In other words, mix the fuel with the air and ignite the mixture by heating it to sufficient temperature, and the result is a flame with a specific volume dictated by the time for the reaction to complete. The principles applied to burner design are consistent.

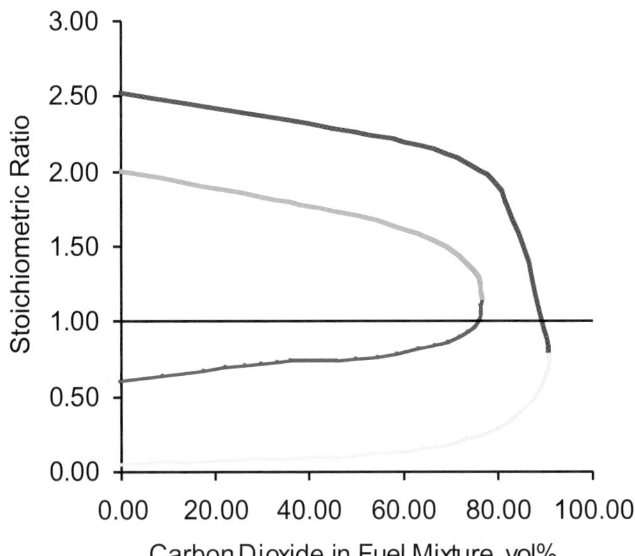

FIGURE 11.1 Graph of sustainable combustion for methane.

The major differences in burner design come from the specific requirements addressed by the design. Reaction rates, or flame lengths and diameters, are directly related to the 3-Ts. They are also dependent on the fuels themselves. The composition, distribution, and port velocity of the fuel can and will produce different results. The flame generated by a combustion system on one fuel will not duplicate the flame generated by that same system on another fuel. Their general flame shape, length, and diameter or width, may be within acceptable tolerances; however, they will not be completely identical. The wider the variance in the physical properties of the fuels, the greater the deviation in flame properties.

11.3 BURNER DESIGN

Specialization of processes, and furnace designs to meet the demands of those processes, has resulted in the necessity of specialized burners. A burner is designed to provide stable operation and an acceptable flame pattern over a specific set of operating conditions. In addition, there may be a specified maximum level of pollutant emissions that can be generated through the combustion process. The American Petroleum Institute gives some guidelines for burners used in fired heaters.[2] Specifications of operating conditions include:

- specific types of fuels
- specific range of fuel compositions
- maximum, normal, and minimum heat release rates
- maximum fuel pressures available
- maximum atomizing medium pressures available for liquid firing
- fuel temperature
- oxidant source, either ambient air, exhaust gases, or enriched
- available combustion air pressure, whether forced (positive) or induced (negative)
- combustion air temperature (ambient or preheated)
- furnace firebox temperature
- furnace dimensions for flame size restrictions
- type of flame (configuration or shape)

To provide acceptable operation, the burner must be designed to perform the 5-Ms:

1. *Meter* the fuel and air into the flame zone.
2. *Mix* the fuel and air to efficiently utilize the fuel.
3. *Maintain* a continuous ignition zone for stable operation over the range.
4. *Mold* the flame to provide the proper flame shape.
5. *Minimize* pollutant emissions.

11.3.1 Metering: Fuel

Typically, the furnace operating system is able to monitor only the total flow of fuel to a furnace. A typical process heater has multiple burners installed to provide the proper heat distribution. The fuel system must then be designed to ensure that the fuel is properly distributed to all burners. Uniform fuel pressures to each burner are critical to the proper operation of the burners. The burner designer then ensures that each burner takes the correct amount of fuel from the system. Controlling the proper amount of fuel flow is accomplished through a system of metering orifices designed for each burner. These ports are specifically designed to act as metering and limiting orifices, passing a specified and known amount of fuel at a given fuel pressure.

11.3.1.1 Gas Fuel

The system of ports provided by the burner designer allows him to provide the operator with a capacity curve that specifies heat release vs. pressure for a given fuel composition and temperature. For gaseous fuels, in compressible or incompressible flow, the calculations for the mass flow through a given orifice are dependent on:

- P_o, the fuel pressure immediately upstream of the orifice
- P_a, the downstream pressure (generally atmospheric pressure)
- T_o, the fuel temperature upstream of the orifice

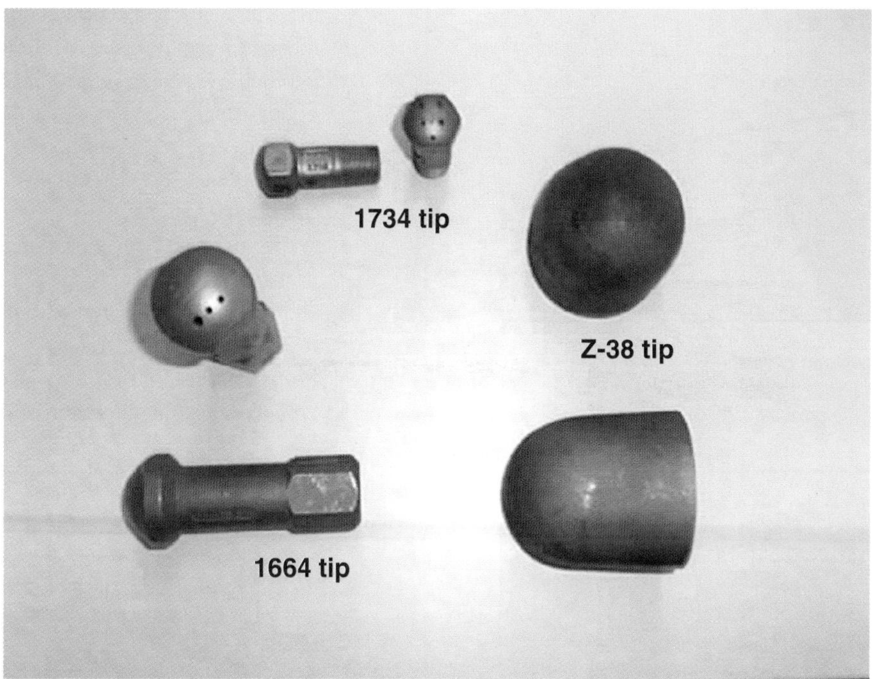

FIGURE 11.2 Typical raw gas burner tips.

- K, the fuel's ratio of specific heats, which is dependent on the composition of the fuel (this is a factor used in calculating the compressibility of the fuel gas)
- A, the area of the port
- C_d, the discharge coefficient, which depends on the design of the orifice port

The fuel metering orifices for gaseous fuels in raw gas burners are typically installed at the point where the flame is formed. This is generally located in a region of the burner tile often termed the "burner throat." The fuel injector tips with the fuel orifices can be centered in the burner throat or located on the periphery. Figure 11.2 shows typical raw gas burner tips.

For premix or partial premix burners, the metering orifice serves dual functions. It is generally a single port located at the entrance of a venturi eductor. The fuel gas discharging from the orifice is used to entrain air for combustion. A typical premix metering orifice spud and air mixer assembly is shown in Figure 11.3.

Capacity curves are generally presented in terms of heat release vs. fuel pressure. The heat release is calculated from the mass (or volume) flow of the fuel, multiplied by the heating value per unit mass (or volume) of the fuel. Figure 11.4 presents a typical gas fuel capacity curve.

11.3.1.2 Liquid Fuel

Liquid fuels must be vaporized in order to burn. Burners designed for firing liquid fuels include an atomizer designed to produce a spray of small droplets, which enhances the vaporization of the fuel. The design of the atomization system will have a significant impact on the liquid fuel flow metering of the burner. With liquid fuels, the metering design is more complicated because of the need to "mix" the oil with an atomizing medium. This results in fuel metering orifices and atomizing media (typically steam or air) metering orifices in combination with orifices designed to flow the mixture. Figure 11.5 depicts typical liquid fuel atomizer/spray tip configurations for various liquid fuel systems. Based on the burner designer's knowledge of the oil metering and atomization system, the designer is capable of providing the operator with a capacity curve that specifies heat release vs. fuel pressure for a given liquid fuel.

Many factors are important in the operation of a liquid fuel atomization system. Variations in any of a number of factors must be considered when designing the system of ports for the metering of flow for a liquid fuel, including the:

- temperature/viscosity relationship of the fuel
- atomization medium temperature
- fuel/atomizing medium pressure ratio
- temperature/vaporization relationship of the fuel

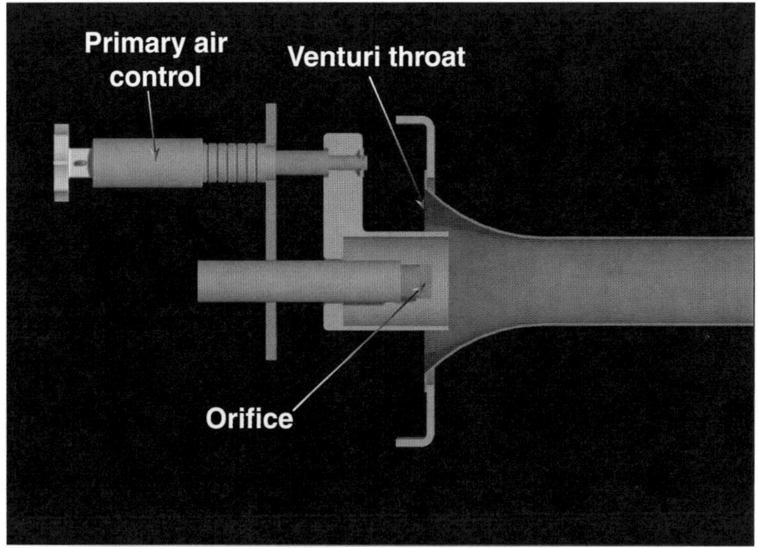

FIGURE 11.3 Typical premix metering orifice spud and air mixer assembly.

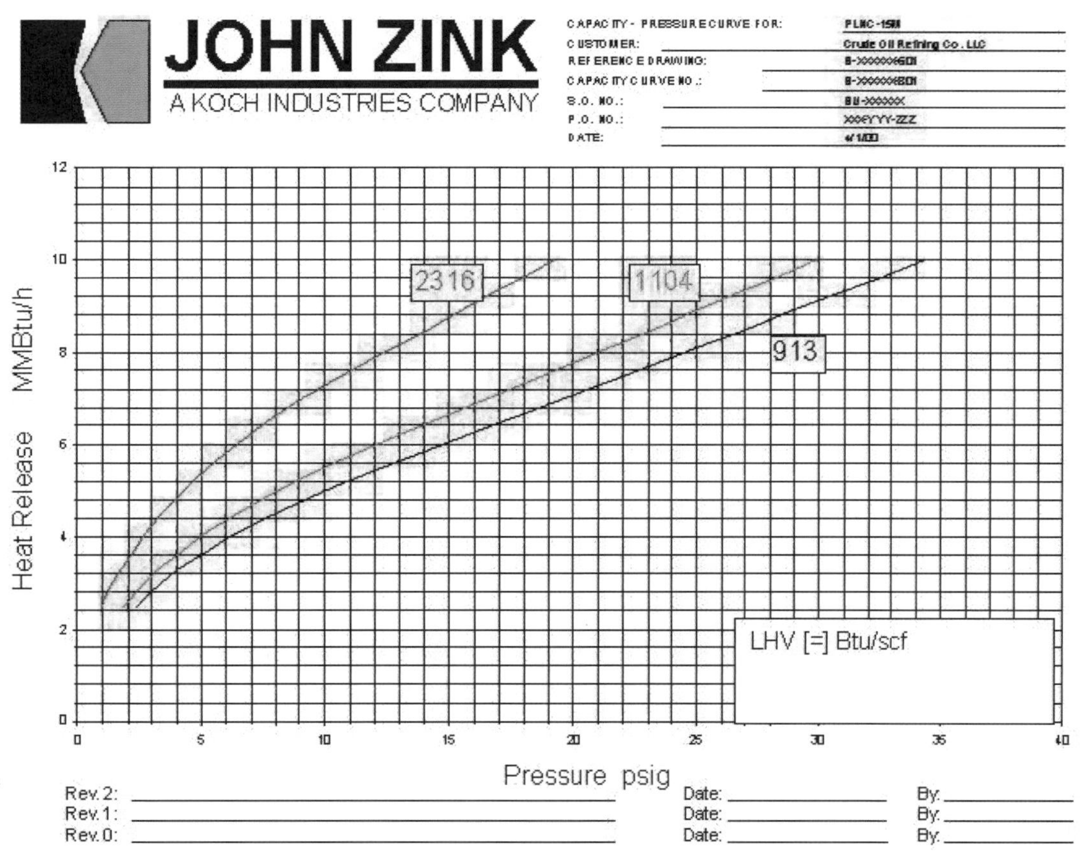

FIGURE 11.4 Typical gas fuel capacity curve.

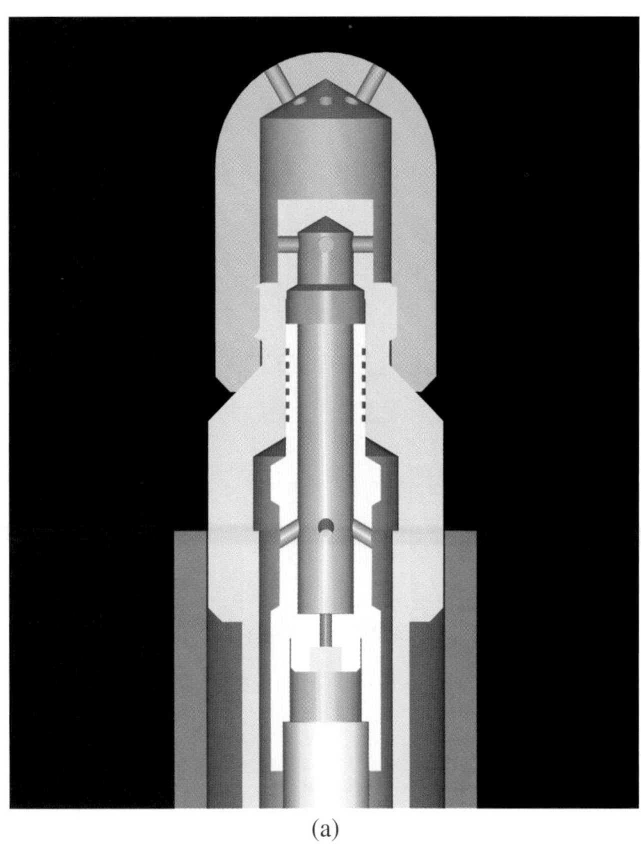

(a)

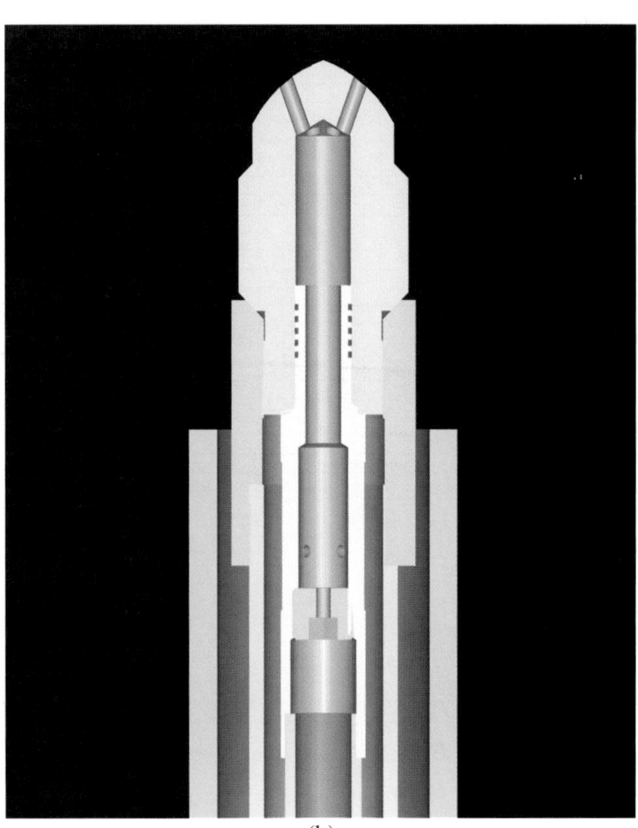

(b)

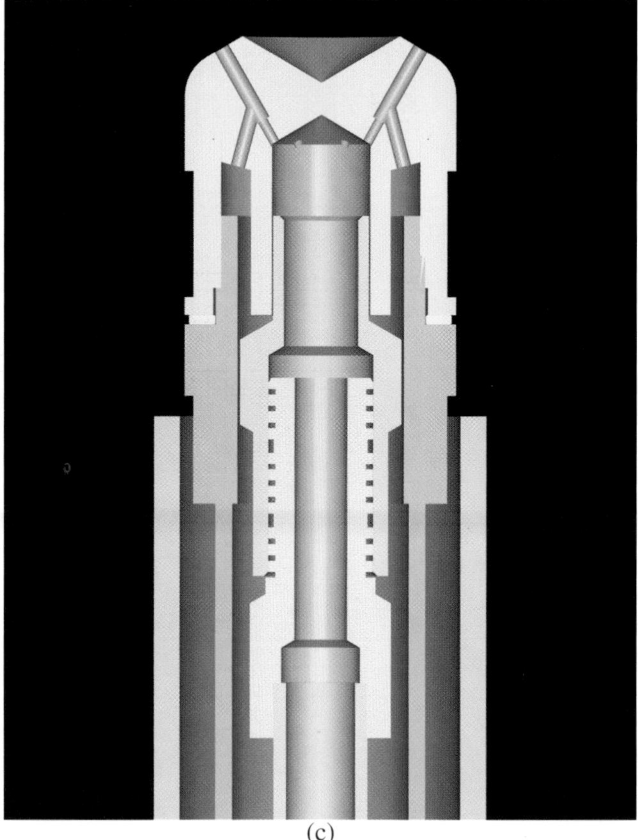

(c)

FIGURE 11.5 Typical liquid fuel atomizer/spray tip configurations.

Because capacity curves are generally presented in terms of heat release vs. fuel pressure, the heat release is calculated from the mass/volume capacity of the system, multiplied by the heating value per unit mass of the fuel. Figure 11.6 presents a typical liquid fuel capacity curve.

11.3.2 Metering: Air (Combustion O_2)

The burner designer must ensure that the required air flow for the design operating conditions can be achieved with the air pressure that is available. The design of the burner throat is critical for two independent reasons. First, there is the requirement of achieving the proper flow of combustion O_2 to meet the demands of the fuel. Second, the burner throat is important in controlling the pattern of the flame.

The primary air flow metering point and the point at which the bulk of the air pressure loss is utilized is designated as the "throat." In most burners, especially raw gas or nozzle mix and liquid fuel burners, it is the point at which the fuel is first injected into the air stream. In a premix or partial premix burner, it is the point at which the fuel/air mixture is

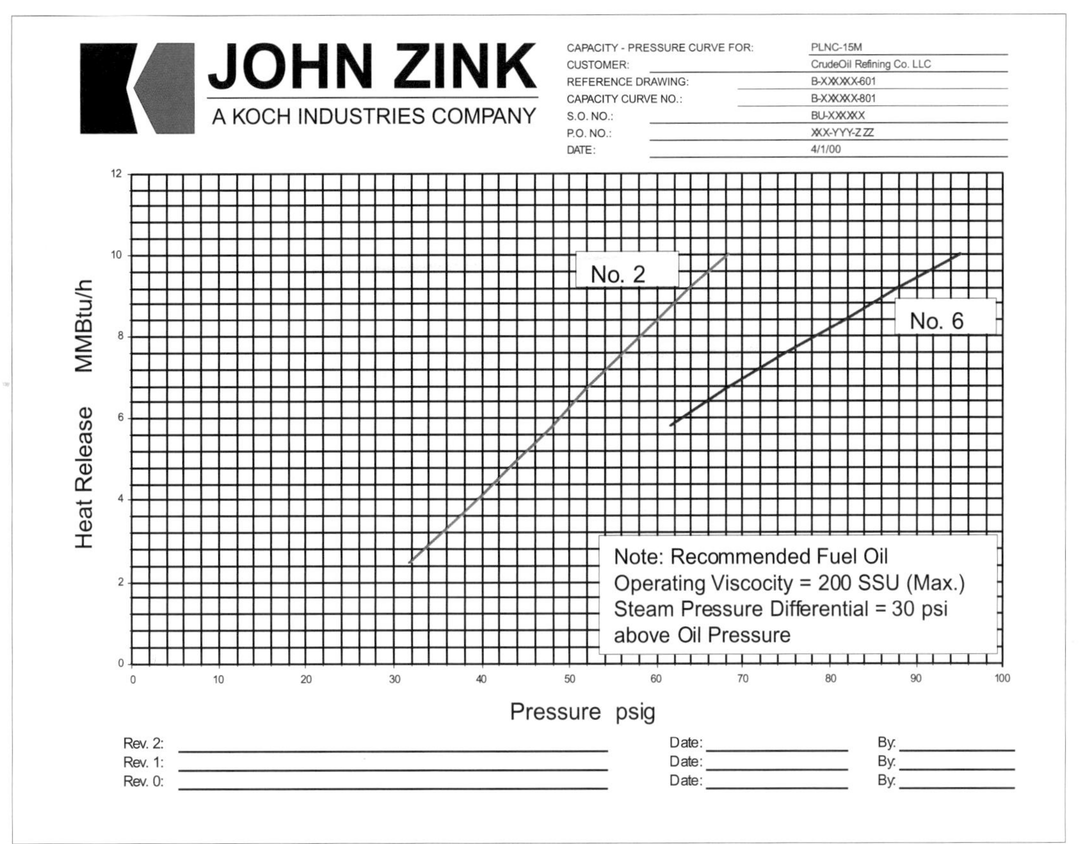

FIGURE 11.6 Typical liquid fuel capacity curve.

injected and, if necessary, secondary or completion air is introduced. In all cases, it is the location at which the initiation of combustion — or ignition — occurs. Figure 11.7 illustrates the throat of a typical raw gas burner.

11.3.2.1 Natural Draft

Many burners installed in refinery or petrochemical process heater service utilize the natural draft available in the furnace as the motive force for drawing combustion air through the burner. This negative pressure is typically less than 1 in. water column (<1.0 in. w.c.) at the burner location. This pressure is dependent on the height of the radiant firebox, the temperature of the radiant firebox, and the level of draft (or negative pressure) maintained at the top of the radiant firebox. For burners required to operate under a natural-draft application, the burner throat is designed to pass the correct amount of combustion air required at the design heat release, utilizing only the available negative pressure provided by the furnace. In some cases, the natural draft can be enhanced by an induced draft fan; however, the available pressure still rarely exceeds 1 in. (1 in. w.c.) of draft.

FIGURE 11.7 Typical throat of a raw gas burner.

11.3.2.2 Forced Draft

Some burner applications provide the combustion O_2 stream at a positive pressure. These are generally applications either with preheated air, turbine exhaust gas, or another alternative oxidant source. Applications requiring smaller flame volumes can also utilize increased air pressure drop and enhanced mixing. If the application operates on forced draft at all times, the available pressure loss is often in the range of 4 to 10 in. water column (4 to 10 in. w.c.). With this level of available pressure loss, the flame dimensions are considerably smaller due to the increased turbulent mixing in the flame. In addition, it is often possible to control the combustion O_2 stream flow over a wider heat release range. This enhanced mixing and improved control allows operation with low excess O_2 over a wider range of fuel input, resulting in improved combustion efficiencies across the heat release range.

In some forced-draft applications, it is specified that the burner must also operate under an ambient air, natural-draft operating mode. For such a dual operational mode specification, it is most common that the natural-draft case is the limiting design. Burners designed for natural-draft and forced-draft operation seldom require forced-draft pressures greater than 2 in. water column (<2 in. w.c.). The exception to this is when the forced-draft operation uses the lowered oxygen concentrations of some of the alternative oxidant sources. For this design condition, the increased mass flow of the combustion O_2 source, with oxygen content lower than air, along with the increased temperature of that stream, generally increase the required burner pressure to levels greater than 2 in. water column (>2 in. w.c.).

11.3.3 Air Control

As a primary metering restriction, a majority of the pressure loss is expected to be utilized within the burner throat. Engineering groups such as the American Petroleum Institute (API),[2,3] process design companies, and some furnace manufacturers provide burner design guidelines that define the use of the available airside pressure drop at that point.

Depending on the level of accuracy desired, there are limitations to the practical range of control, especially when applied to burners designed for natural-draft applications. The percentage of available draft utilized for throat metering can leave little room for the design of an adequate control mechanism.

Example 11.1

Natural-draft burner

Refinery fuel gas: 1200 Btu/scf, 0.70 sp.gr. (47.27 MJ/Nm^3, 11288.8 kcal/Nm^3)

Maximum design duty: 10.0 MMBtu/h (2.93 MW, 2.52 MMkcal/h)

Normal duty: 8.0 MMBtu/h (2.344 MW, 2.016 MMkcal/h)

Minimum design duty (turndown): 2.0 MMBtu/h (0.586 MW, 0.504 MMkcal/h)

Draft: 0.5 in. w.c. (124.5 Pa, 1.25 mbar, 12.7 mm H_2O)

20% Excess air design, 10% excess air operation

Design to meet API 560 and 535

Reference: API 560 *Section 10.1.12*

"The burner shall be selected to use no less than 90% of the maximum draft available for the maximum specified heat release."

Reference: API 560 *Appendix A — Equipment Data Sheets Section b. Burner data sheets, Sheet 3 of 3 Note 1*

"At design condition, minimum of 90% of the available draft with the air register fully open shall be utilized across the burner. In addition, a minimum of 75% of the airside pressure drop with the air registers full open shall be utilized across burner throat."

Reference: API 535 *Section 8.2 — Dampers and Registers*

"Dampers and burner registers shall be sized such that the air rate can be controlled over a range of at least 40 to 100% of burner capacity."

Assume that, through the efforts of the burner designer, the burner actually requires between the specified 90 to 100% of the available draft, and that the specified 75% is measurable as static loss across the throat. The following tables indicate the:

- operating heat release (column 1, Heat Release)
- desired operational excess air (column 3, Percent X-Air)
- required pressure loss across the throat (column 4, Throat Drop)
- required pressure loss across the register or air control device (column 5, Control Drop)
- ratio percentage of register or damper opening required (column 7, Percent Control Open)

Table 11.1 shows the theoretical percentage of air control opening based on no change in control flow coefficient and 3% leakage at the full closed position. Table 11.2 is based on 6% leakage. These tables assume that the variable orifice, the air control mechanism, has a constant discharge coefficient throughout its full range of operation. This is a simplification and is not totally correct in practical application.

These charts (Tables 11.1 and 11.2) provide insight into another common fallacy in burner specification — the error in excess design capacity. Many who specify combustion equipment seek to provide excess capacity as a cushion against reduced operation with burners out of service, upset

TABLE 11.1 Theoretical Air Control Opening, Based on No Change in Control Flow Coefficient and 3% Leakage at the Full Closed Position

Heat Release (MMBtu/hr)	Percent Design (%)	Percent X-Air (%)	Throat Drop (in. w.c.)		Control Drop (in. w.c.)		Total dP (in. w.c.)	Percent Control Open % (w/3% leakage)	
			90% Total	100% Total	90% Total	100% Total		90% Total	100% Total
10.0	100	15	0.338	0.375	0.162	0.125	0.50	86.3	Full open
10.0	100	10	0.309	0.343	0.191	0.157	0.50	75.9	83.8
9.0	90	10	0.250	0.278	0.250	0.222	0.50	59.4	63.1
8.0	80	10	0.198	0.220	0.302	0.280	0.50	47.7	49.6
7.0	70	10	0.151	0.168	0.349	0.332	0.50	38.6	39.6
6.0	60	10	0.111	0.124	0.389	0.376	0.50	31.0	31.6
5.0	50	10	0.077	0.086	0.423	0.414	0.50	24.5	24.8
4.0	40	20	0.059	0.065	0.441	0.435	0.50	20.7	20.9
3.0	30	40	0.045	0.050	0.455	0.450	0.50	17.6	17.7
2.0	20	80	0.033	0.037	0.467	0.463	0.50	14.7	14.8
0	0	3	0.00031	0.00034	0.49969	0.49966	0.50	Full closed	Full closed

TABLE 11.2 Theoretical Air Control Opening, Based on No Change in Control Flow Coefficient and 6% Leakage at the Full Closed Position

Heat Release (MMBtu/hr)	Percent Design (%)	Percent X-Air (%)	Throat Drop (in. w.c.)		Control Drop (in. w.c.)		Total dP (in. w.c.)	Percent Control Open % (w/6% leakage)	
			90% Total	100% Total	90% Total	100% Total		90% Total	100% Total
10.0	100	15	0.338	0.375	0.162	0.125	0.50	84.8	Full open
10.0	100	10	0.309	0.343	0.191	0.157	0.50	74.4	82.3
9.0	90	10	0.250	0.278	0.250	0.222	0.50	57.9	61.6
8.0	80	10	0.198	0.220	0.302	0.280	0.50	46.2	48.1
7.0	70	10	0.151	0.168	0.349	0.332	0.50	37.1	38.1
6.0	60	10	0.111	0.124	0.389	0.376	0.50	29.5	30.1
5.0	50	10	0.077	0.086	0.423	0.414	0.50	23.0	23.3
4.0	40	20	0.059	0.065	0.441	0.435	0.50	19.2	19.4
3.0	30	40	0.045	0.050	0.455	0.450	0.50	16.1	16.2
2.0	20	80	0.033	0.037	0.467	0.463	0.50	13.2	13.3
0	0	6	0.0012	0.0014	0.4988	0.4986	0.50	Full closed	Full closed

operation, process design contingencies, and future desired capacity. The fallacy is in the ability to control excess air and to utilize the available pressure drop for the normal operation. The air and fuel pressures are both reduced in the mixing zone and the resulting flame quality will suffer.

11.3.4 Mixing Fuel/Air

Mixing is a general term used to describe the function of bringing the fuel (reactant) into close molecular proximity with the air (oxidant). The higher the level of turbulence and shear between the streams, the more uniform the fuel and air mixture and the more rapid and complete the combustion reaction. It is generally accepted that the higher the level of mixing, the more intense and more complete and better the combustion. However, there are a number of conditions and factors that should be considered, including flame shaping and control of pollutant emissions that lead to some exceptions to this rule. Intimate mixing does not always produce the most desirable results.

Designing a combustion system for special applications (e.g., high inert composition fuels) can also set limitations on the desired level of mixing. All combustion reactions have a rate at which they will proceed and a minimum temperature that is required to initiate or sustain that reaction. The introduction of inert components to the fuel can generate two conditions that will affect burner stability. First, the inert components slow the reaction. This slowing of the flame speed can result in the stabilization point being translated out of the desired position. Second, the inerts introduce a heat absorption component that narrows the flammability limits and reduces the flame temperature. Quenching of the flame, to extinction, is an important consideration when inert components are present in the fuel or the combustion O_2 stream.

Many combustion O_2 streams with greater inert compositions than ambient air are the result of a combustion process.

As such, they are typically available at an elevated temperature. This elevated thermal energy will often partially offset the potential quenching effect of the elevated inert levels. However, there are limits, as discussed in Section 11.1.

An effective form of emissions control is through the delaying of the combustion process. Highly mixed fuel and combustion O_2 in the proper proportions will generate the maximum flame temperature. This will result in the formation of a large quantity of nitrogen oxides (NOx), which is a highly regulated pollutant (see Chapter 6). Certain burner designs will utilize a reduced level of fuel and air mixing, or a delayed mixing, extending the reaction zone to achieve reductions in these emissions.

In the case of a dual reactant flame, there are only two sources from which to obtain the energy required for mixing. First is the relatively high-mass, low-velocity combustion O_2 stream, and second is the low-mass, relatively high-velocity fuel stream. Each stream will contribute energy to the work required to mix in proportion to its mass and its velocity.

The intimate mixing of two or more dissimilar fluid streams under flowing conditions occurs in a turbulent shear zone defined by the intersection of the two streams. This region can be described as the dynamic interaction of each stream's mass and velocity, or momentum. This surface of intersection will vary in its turbulence in proportion to the magnitude of the shear forces developed.

Four basic mechanisms are available for the development of fuel and air mixing:

1. entrainment
2. co-flow mixing
3. cross flow mixing
4. flow streamline disruption or eddy formation

11.3.4.1 Entrainment

Entrainment is an effective demonstration of the law of conservation of momentum. One stream, usually the fuel, is utilized to inspirate the other. This motive fluid's energy and momentum are conserved. As the velocity of the fuel jet is dissipated, the mass of the fuel stream must increase by the entrainment of ambient air to satisfy the conservation of momentum.

A well-designed premix burner primarily utilizes entrainment as its fuel/air mixing mechanism. Intimately mixed fuel and oxidant prior to the distribution tip results in the highest possible rate of reaction.

11.3.4.2 Co-flow

When the streams are effectively parallel in directional flow, the amount of shear force developed is proportional to only the differential mass velocity of the streams. At high-velocity differentials, the interface of these streams becomes a turbulent region in which the components of the streams become intermixed. However, in low-velocity differentials, as in all transitional or laminar boundary layers, the relative thickness of this interface is small in comparison with the total cross-section. Co-flow mixing with low-velocity differential conditions is — by its dynamic configuration — slow mixing.

11.3.4.3 Cross-flow

The shear energy generated between two flowing streams is greater anytime the streams are intersecting. The work required to redirect the combined flow of the streams results in turbulent intermixing of those streams. The included angle of intersection of the streams, the relative differential velocity of the streams, and the mass densities of the streams are all factors in the resulting direction and the turbulence generated for the mixing of the streams.

Included angles of intersection closer to perpendicular result in higher levels of mixing. However, quick mixing is not always the most important function being sought. Flame shape, stability, and emissions are all primary functions that can be affected by the rate and direction of combustion.

11.3.4.4 Flow Stream Disruption

Additional turbulence can be developed through the disruption of the "path" of a flowing fluid. Strategically located obstructions (e.g., bluff bodies) and sudden expansions (e.g., tile ledges) provide forced changes in flow streams, generating turbulence. If these disruptions are located in a region where both fuel and air streams are present, mixing will occur.

11.3.5 Maintain (Ignition)

The most important function that a burner performs is to provide for the continuous and reliable ignition of the fuel and air passing through the burner over a specified range of operating conditions. Each burner is designed to provide a specific location in which a portion of the fuel and air is continuously introduced in near-stoichiometric proportions at velocities at or below the mixture flame speed, thereby allowing for a continuous "ignition zone." At this location, a continuous flame is maintained and ignites the fresh fuel/air mixture as it is introduced. The ignition zone is designed to operate over the specified range of operation of the burner. The flame from this ignition zone is then used to ignite the remainder of the fuel and air mixture.

In natural-draft burners, the ignition zone is often situated in the eddy formed downstream of a step or ledge in the burner tile (see Figure 11.8), or in the wake of a bluff body "flame stabilizer or flame holder" (see Figure 11.9) located in the center of the combustion air stream. Forced-draft burners can utilize similar techniques or other aerodynamic meth-

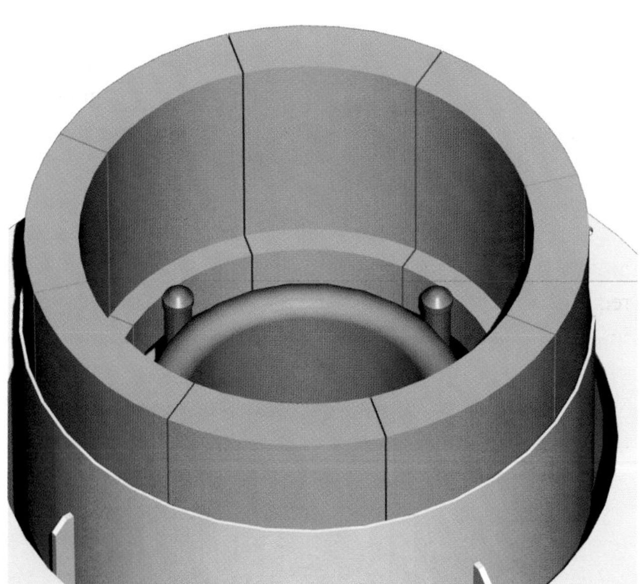

FIGURE 11.8 Ledge in the burner tile.

ods, such as a swirler (see Figure 11.10), to generate reverse flow zones that recirculate hot combustion products that assist in providing continuous ignition zones.

Safe ignition of the fuel/air mixture depends on the fuel and combustion O_2 source composition. These factors are primary in affecting the flammability limits, ignition temperatures, and stoichiometry of the mixture. The rate of reaction (flame speed) is controlled by the stoichiometry of the mixture and combustion characteristics of the fuel. The ability to sustain a single point of ignition for a fuel/air mixture is controlled by both the stoichiometry and the velocity of the mixed stream. Wide ranges in capacity, excess air requirements, and fuel compositions make strict dependence on the mixed stream for stability unreliable. The development of low-velocity zones and recirculating eddies ensures a single point for the kindling of the flame over a broad range of operating conditions. Each burner is designed for a limited range of fuel (and "air") compositions and rates, and one should not attempt to operate the burner outside this range.

11.3.6 Mold (Patterned and Controlled Flame Shape)

The design of combustion equipment, in the form of burners, for the hydrocarbon processing industry (HPI), the chemical processing industry (CPI), and to a major extent the power generation industry (PGI) has critical restrictions on the shape, size, and consistency of the flame generated. In fact, all burners providing energy to a process will be restricted in flame qualities by the design of the chamber, or furnace, into which it is fired.

FIGURE 11.9 Flame stabilizer or flame holder.

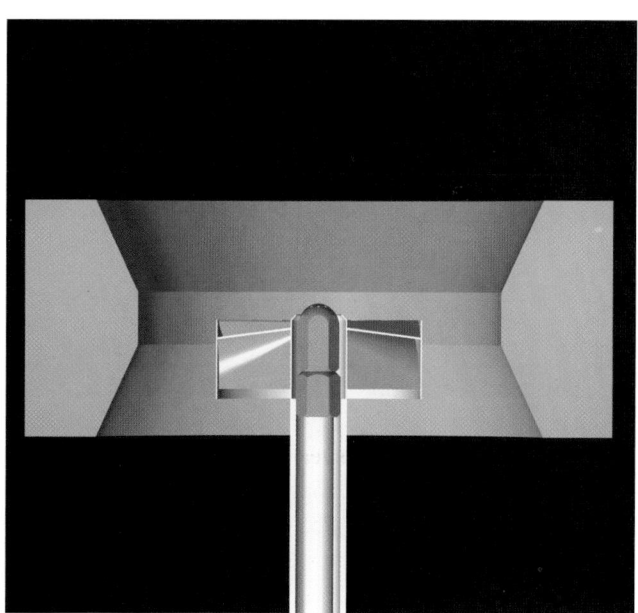

FIGURE 11.10 Swirler.

The furnace chamber and burner flame in every process furnace are each designed to provide for the efficient transfer of heat to the process load. Flame size and shape for the

HPI is especially critical due to the sensitivity of the hydrocarbons being processed to overheating. The rate of heat transfer to the process tubes must be limited to prevent overheating of the process tubes leading to the formation of carbon or coke inside the process tubes. As a result, there are generally strict guidelines for the flame dimensions. Typical specifications for flames include maximum flame lengths and widths. The number, heat release, and layout of the burners in the furnace are designed to provide the proper heat transfer pattern.

Patterning of the air flow through approach distribution, tile throat sizing and shape, and the tile exit configuration provides the most reliable method for flame pattern control. Introduction of the fuel into the established air flow streams provides the primary function in a raw gas burner. The proper flame pattern is generated by the combination of fuel injection pattern provided by the fuel injectors and the burner tile and flame holder which controls the air flow. The fuel injectors are also often called *spuds* or tips. The injectors have fuel injection ports that introduce the main portion of the fuel into the air stream in a manner that generates the desired flame pattern or shape. In conjunction, the air stream must be shaped in an appropriate manner by the air flow passages provided by the shape of the tile and flame holder. In many cases, the flame shape is round or brush-shaped and acceptable in length and diameter (see Figure 11.11). In this case, the burner tile is typically round and the fuel is injected symmetrically. Some furnaces require a fan-shaped flame, often termed a flat flame. In that case, the burner tile is generally rectangular and the fuel is injected in a manner producing a flame that is essentially rectangular or "flat" rather than round (see Figure 11.12).

11.3.7 Minimize (Pollutants)

Most societies have come to the point at which the environment is of foremost concern. Therefore, the governments of most nations and localities are very critical of any source that puts certain undesirable materials into the air, soil, or water. Environmental regulations limiting air pollution have direct impact on the design of combustion equipment.

The challenge presented to burner design by these restrictions comes from the thermo-chemical reactions that form the regulated emissions. Emission control issues are discussed in Chapter 6.

11.4 BURNER TYPES

Burners are typically classified based on the type of fuel being burned. A subdivision in burner type often includes the method of combustion O_2 supply. Therefore, there can be as many as eight basic design criteria.

1. Gas — premix and partial premix, natural draft and/or low pressure drop air
2. Gas — raw gas or nozzle mix, natural draft and/or low pressure drop air

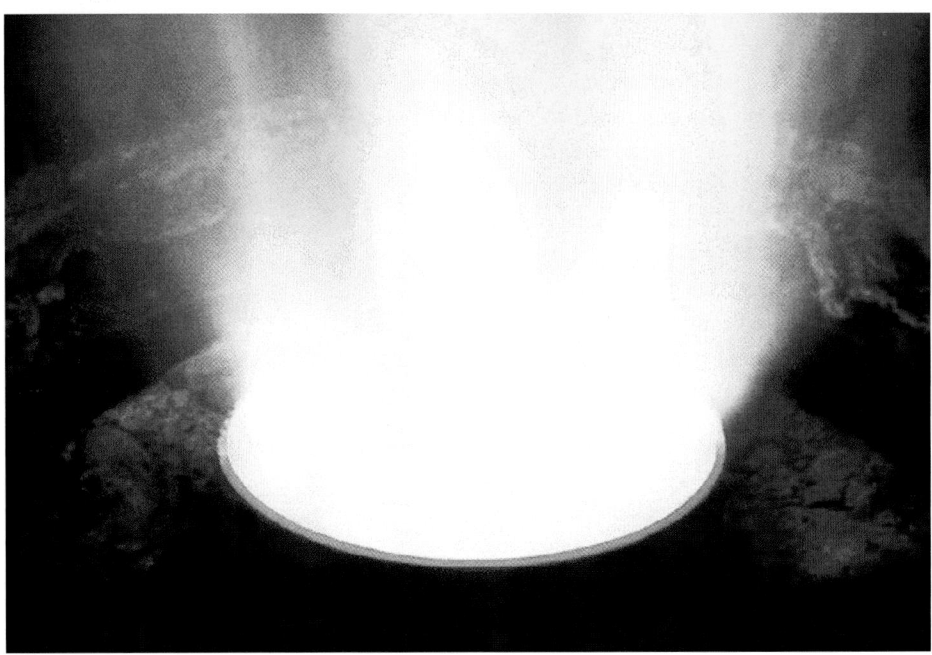

FIGURE 11.11 Round-shaped flame.

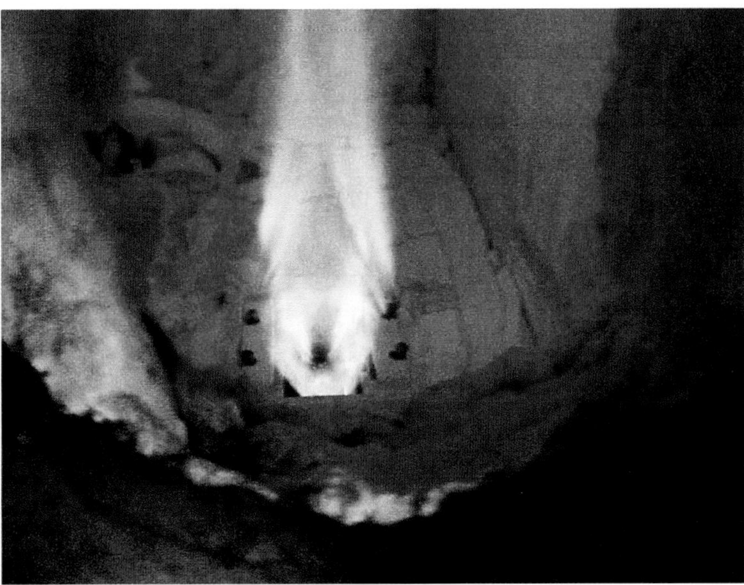

FIGURE 11.12 Flat-shaped flame.

3. Gas — raw gas or forced nozzle mix, forced draft high pressure drop air
4. Liquid — natural draft and/or low pressure drop air
5. Liquid — forced draft high pressure drop air
6. Solid fuel — forced draft high pressure drop air (typically)
7. Combination gas and liquid — raw gas and oil (typically), natural draft and/or low pressure drop air
8. Combination gas and liquid — raw gas and oil (typically), forced draft high pressure drop air

The final two designs listed are simply extensions of other designs. The combination of any two or more burners is simply a matter of basing the design on the most difficult of the fuel types and adapting the other fuel distribution systems to the base design.

11.4.1 Premix and Partial Premix Gas

Premix is a term applied to burners that inspirate part or all of the total air required for combustion. This type of burner provides for intimate mixing of the fuel and combustion O_2 prior to the ignition zone.

In inspiration, the motive energy is supplied by the low-mass, high-potential-energy fuel. The fuel gas is metered through one or more orifices at the entrance to a venturi or mixer. The entrained air stream is made available at zero, or virtually zero, velocity at the same location. The conversion of the potential energy, pressure of the fuel stream, to kinetic energy, jet velocity, is achieved within the zone of air supply. The "free jet" of the fuel immediately begins to expand and decelerate. Conservation of momentum requires that the reduction in velocity be balanced by an increase in mass of the moving stream. This additional mass is entrained air.

Burners designed with an inspirating arrangement typically require fewer adjustments to the air control. The utilization of the mass and velocity of the fuel results in that portion of the air inspirated being proportional to the gas flow. Therefore, with the reduction in fuel mass flow, there is a resulting reduction in the entrained air. The efficiency of the venturi and the restriction imposed by the fuel/air distribution nozzle are the limits to the capacity (volume) of air that can be inspirated.

Another benefit to this design is in the volume of the flame generated. Since the majority of the air is initially intimately mixed with the fuel, the resulting flame volume of this premixed burner will be much smaller than that of any other low air pressure drop burner. Conversely, if the efficiency of this premixing is sufficiently low, the flame will actually be larger than a raw gas burner. This is a result of the reduced secondary air mixing energies. By utilizing the majority of the available energy of the fuel stream to achieve the primary premix, the remaining energies for any required secondary air mixing are reduced to only that available from the air due to the draft loss.

Further benefits provided by this style of burner lie in the fuel metering configuration. Because all of the fuel is metered through a single or minimal number of orifices, the size of those orifices will be maximized for the conditions. Larger-diameter orifices, as long as they do not jeopardize the function of the burner, are a benefit because they minimize the chances of plugging from dirty fuels.

One of the basic limitations to this type of burner is founded in the burning characteristics of fuels. Each fuel chemical compound has its "rate of reaction" with oxygen. This rate is dependent on concentration levels of the fuel, the oxygen, and any inert components. This rate is also dependent on the temperature of the mixture. Another way to describe this rate is in terms of "flame speed," or the velocity at which the flame will propagate. The design of the distribution tip/system into the ignition and combustion zones is dependent on this same burning characteristic of the design fuel. Changes in the firing rate or the fuel will result in a change in the volume, velocity, and burning characteristics at the distribution tip. If the fuel/air mixture is within combustible limits, and the velocity is not maintained above the flame propagation speed, the result is a translation of the flame front back along the fuel/air flow streams. In the worst case, this flash-back condition may result in flames being translated completely outside of the designed ignition and flame zone, causing mechanical and structural thermal damage.

11.4.2 Raw Gas or Nozzle Mix

Raw gas (nozzle-mix) burners are designed to introduce the fuel and air separately into the combustion zone. They provide all mixing of fuel and combustion O_2 at or after the ignition point of the burner. Typically, these burners provide for the major, or metering, pressure drop for both the fuel and air immediately prior to the ignition zone.

With all of the fuel maintained as segregated from the combustion O_2 until mixed in the ignition zone, there is no possibility of flashback. Therefore, the raw gas style of design can effectively handle a wide range of fuels without concern for equipment damage or personnel safety. Turndown on a raw gas burner is limited only to the method of stabilization, the flammability limits of the fuel, and the controllable and safety lower pressure limits of the fuel system.

With all of the air being supplied through and metered by the throat of the burner, it becomes necessary to adjust the air flow for efficient operation. Independent of whether the burner is designed for natural draft, low forced draft, high induced draft, or high forced draft, the ability to efficiently combust the fuel and transfer the heat is directly related to the amount of excess air. Therefore, it is important to control the air with every change in fuel flow or fuel composition or there will be a related change in efficiency.

Another problem intrinsic to the raw gas design condition comes from the fuel distribution system, especially in burners designed for natural draft and/or low combustion O_2 pressure drop. Patterning of the flame, distribution of the fuel into the combustion air stream, and mixing of the fuel with the oxygen will typically require multiple points of fuel injection. This breaking up of the fuel flow into multiple metering orifices while maintaining high potential energy, pressure, can require small orifices. These small orifices are highly subject to fouling problems due to fuel quality and foreign material in the piping.

Natural draft on the low pressure drop, combustion O_2 side often requires a burner design utilizing quiescent zones for stabilization. These low flow, low pressure zones can be generated through the use of flow stream disruption. Cones or bluff bodies located in the throat of the burner are a common form of flow stream disruption. Flow disrupters or shields around fuel tips and ledges or sharp changes in tile profile are also common. In all cases, these mechanisms are basically designed to prevent a portion of the combustion O_2 from leaving the designed ignition zone prior to the introduction of fuel. They provide "pockets" of a continuously renewing combustible mixture at a location in which the velocity is lower than the flame propagation speed.

High combustion O_2 pressure drop designs have another available form of flame stabilization — vortex recirculation. Through the utilization of swirl mechanisms and tile profiles, the combustion O_2 stream can be forced into a vortex. The physical properties of a vortex provide a low pressure zone on the interior of the rotation that actually generates a backflow to the point of origin. The shape and strength of the vortex is determined by the flow and pressure loss through the swirler, the flow around the perimeter of the swirler, and the profile of the surrounding air throat. By injecting fuel into the swirling O_2 stream, a portion of the fuel and oxygen is continuously recirculated to the centralized ignition zone.

11.4.3 Oil or Liquid Firing

Oil firing is more complicated than fuel gas firing because the oil must be "atomized" and vaporized before it can be properly burned. Atomization involves producing a relatively fine spray of droplets that will vaporize quickly. Industrial oil burners typically utilize twin fluid atomizers and employ steam as the atomizing medium. Compressed air, or even high-pressure fuel gas, can be utilized in some applications rather than steam.

Some *internal mix twin fluid atomizers* form a gas/liquid emulsion which then issues from the fuel tip. This type of atomizer is shown in Figure 11.13. The oil issues from a single port at the entry to the atomizer section. Steam is injected into the oil stream through multiple ports, forming the emulsion. This oil steam emulsion then travels to the chamber between the atomizer section and the exit ports. Multiple exit ports are placed on the tip, which allows for

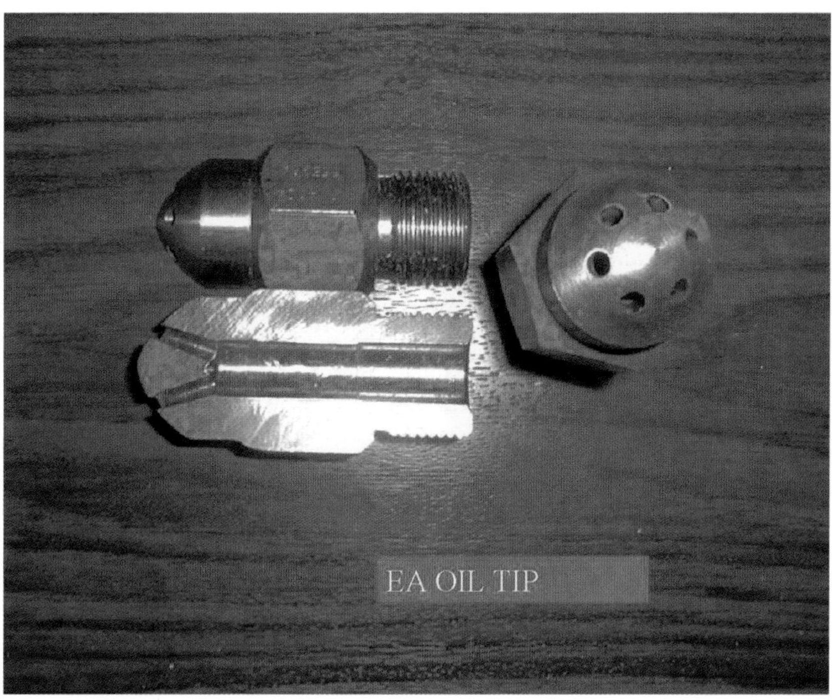

FIGURE 11.13 Internal mix twin fluid atomizer.

enhanced mixing of the spray and the combustion air, flame shaping, and stabilization. Internal mix atomizers often utilize a steam pressure maintained at 10 to 30 psig (70 to 210 kPa) above the oil pressure over the operating range of the burner. The oil pressure may be 100 to 120 psig (700 to 800 kPa) at maximum firing rate.

Port mix twin fluid atomizers (see Figure 11.14) use the atomizing media to form a thin liquid annulus on the inner surface of the port. The high velocity of the atomizing media pushes the liquid along the length of the port shearing and stretches the liquid into a thin film. These atomizers generally operate with a constant steam pressure, typically 120 to 150 psig (800 to 1000 kPa) over the entire operating range of the burner. The oil pressure may be 100 to 150 psig (700 to 1000 kPa) at maximum firing rate.

In each case, internal mix or port mix, droplets are formed as the liquid forms a sheet as it exits the port, and this sheet then breaks up as it expands. The size of the droplets formed depends on the relative velocity of the liquid sheet and the surrounding media. It also depends on the viscosity, surface tension and density of the liquid, size of the port, momentum of the fuel, and ratio of fuel to atomizing media.

11.4.3.1 High Viscosity Liquid Fuels

The bulk of the industrial liquid fuel (see Chapter 5) fired is high-viscosity or heavy fuel oil. Heavy oils include No. 6 fuel oil, bunker C oil, residual fuel oil, pitch, tar, and vacuum tower bottoms. Most heavy fuel oils are the residue from the oil refining process and are considered a low value by-product. This viscous oil must be heated to be efficiently atomized. Typically, the best atomization requires the viscosity of the oil to be in the range of 100 to 250 SSU. This

FIGURE 11.14 Port mix twin fluid atomizer.

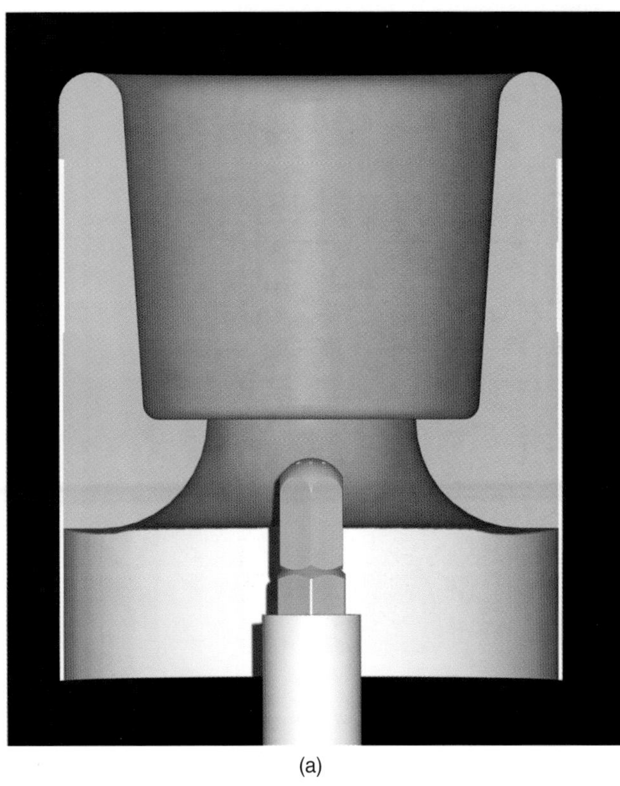

(a)

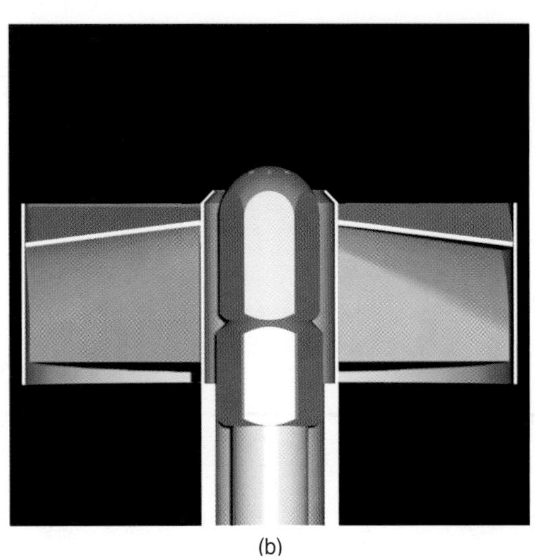

(b)

FIGURE 11.15 (a) Regen tile and (b) swirler.

often requires heavy oil to be heated to the range of 200 to 250°F (90 to 120°C). However, a heavy pitch may require a temperature of 600°F (320°C)to achieve the proper viscosity for atomization.

Heavy fuel oils have high boiling points and are therefore difficult to vaporize. Many heavy oil burners have special refractory tiles that redirect the intense flame radiation back to the root of the spray to enhance vaporization rates and help stabilize the oil flame. Some other heavy oil burners utilize swirlers to promote internal recirculation of the hot combustion products back into the flame root to enhance heating of the spray (see Figure 11.15).

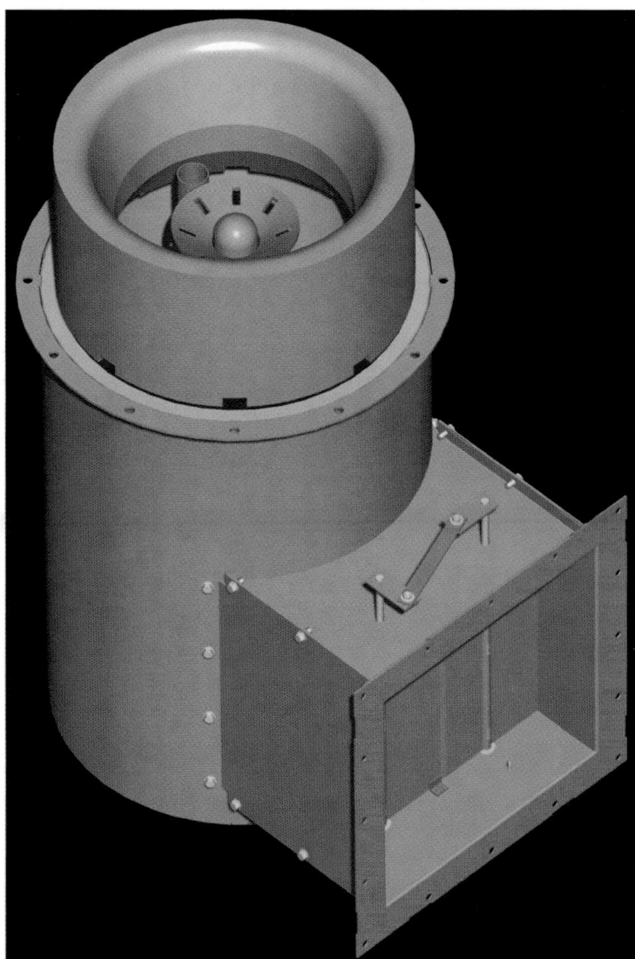

FIGURE 11.16 Typical conventional raw gas burner.

11.4.3.2 Low-Viscosity Liquids

Low-viscosity liquid fuels include light oils such as No. 2 fuel oil, naphtha, and by-product waste liquids containing a variety of hydrocarbon by-products such as alcohols. These liquids are also typically fired using twin fluid atomizers. However, care must be taken to avoid overheating low-boiling-point fuels. Vaporization of the liquid within the oil gun can cause disruption in flow within the atomizer and unsteady operation. Many light liquids require compressed air rather than steam atomization. Because light oils generally have lower

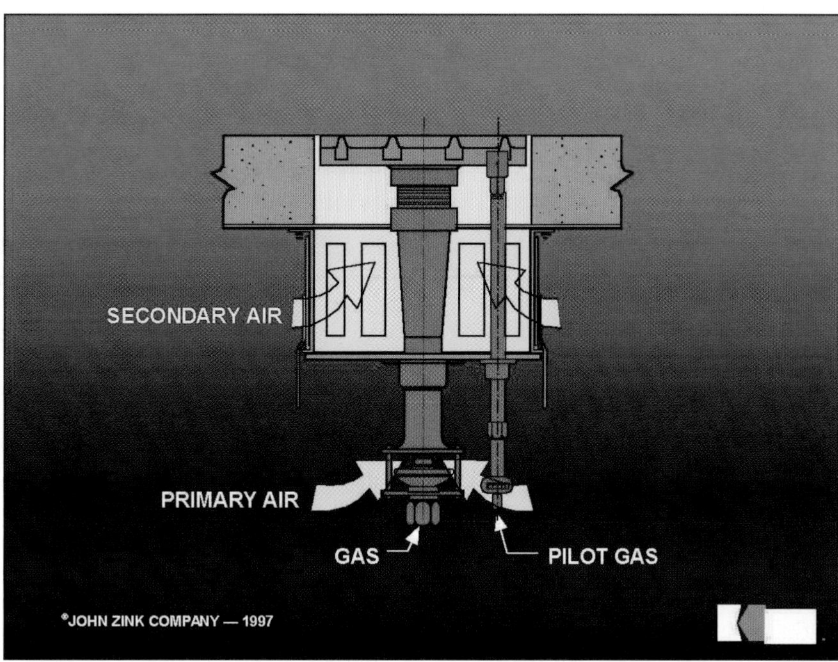

FIGURE 11.17 Typical premix gas burner.

boiling points, they are easier to vaporize and their flames can often be stabilized using simple bluff-body stabilizers rather than swirlers or special tiles.

11.5 CONFIGURATION (MOUNTING AND DIRECTION OF FIRING)

Burners can be mounted in the furnace or heater floor to fire vertically upward; in the heater wall to fire horizontally; or in the roof to fire vertically downward. The major consideration required for burner design is to ensure proper support for the burner tile. Burner blocks for floor-mounted service can typically be simply placed on the furnace steel or burner mounting plate. Burner blocks for horizontal mounting must be supported by tile case assemblies and must be held in place so that they do not move if subjected to vibration. Roof-mounted tiles must have support surfaces cast into them so that they can hang from steel supports in the furnace roof.

11.5.1 Conventional Burner, Round Flame

The following figures illustrate the different types of burners that are typically used in refinery and petrochemical furnaces. The round flame burner is the most universal design and used in many applications. Figure 11.16 illustrates a typical raw gas conventional burner. This burner is used where NOx emissions are not a primary concern and a short flame is desired.

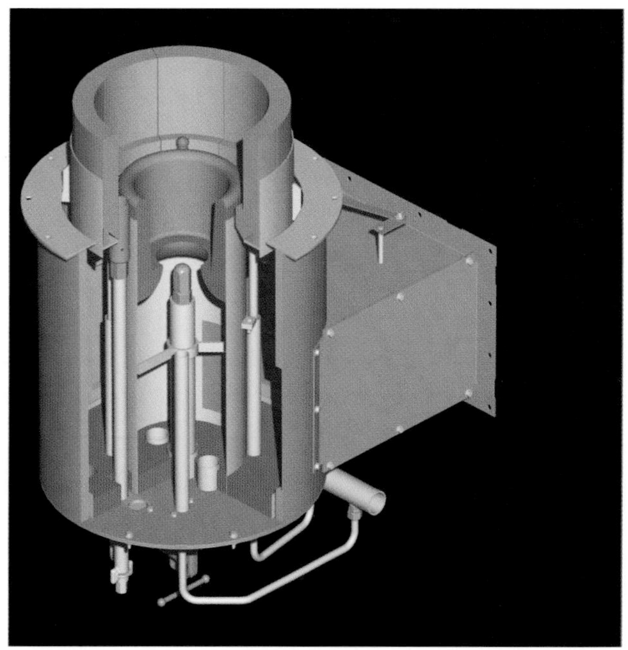

FIGURE 11.18 Typical round flame combination burner.

Figure 11.17 shows a typical premix gas burner. A premix round flame burner is useful when a short heater does not

FIGURE 11.19 Typical round flame, high-intensity combination burner.

have enough draft to supply the required combustion air. The premix burner uses the fuel jet as a motive force to allow the burner to pull in part or all of its combustion air.

Figure 11.18 shows a typical round/conical flame combination oil and gas burner. This burner can be used to burn gas or liquid fuels. This versatility is desirable to applications with liquid fuels or where gas fuels may be in short supply at various times throughout the year and an alternative fuel must be fired to maintain a process.

Figure 11.19 shows a typical round/conical flame high intensity combination oil and gas burner. This burner is used in applications where a high heat release per burner is required but a short flame length is required.

11.5.2 Flat Flame Burner

Some applications require a flat or fan-shaped flamed due to the close proximity of process tubes or due to the fact that the burner is fired along a wall. Figure 11.20 shows a typical staged-fuel flat flame gas burner. This burner produces a freestanding flame and is used in applications where the process tubes are close to the centerline of the burner.

The advantage in using this particular burner is that because of fuel-staging, the concentration of NOx reduces extensively. Based on fired capacity, one can utilize single or multiple primary and staged injectors to effect the proper fuel distribution to produce the desired flame pattern. A flat flame burner can be fired in two basic configurations — wall fired or freestanding.

11.5.2.1 Wall Fired

A wall-fired flat flame burner is similar to a freestanding flat flame burner except that it is installed and fired against a refractory wall. By directing the fuel jets toward the wall, the flame heats the refractory wall, which in turn radiates heat to the process tubes facing the wall. Wall-fired burners are typically used in ethylene furnaces.

11.5.2.2 Freestanding

A free-standing flat flame burner is used in applications where it is necessary to fire a burner between two sets of process tubes. The tube spacing often requires the use of a flat flame burner. A staged-fuel, flat flame is shaped by firing stage fuel jets opposite one another. This makes the flame shape into a fan. The flame thickness is typically less than or equal to the burner tile width.

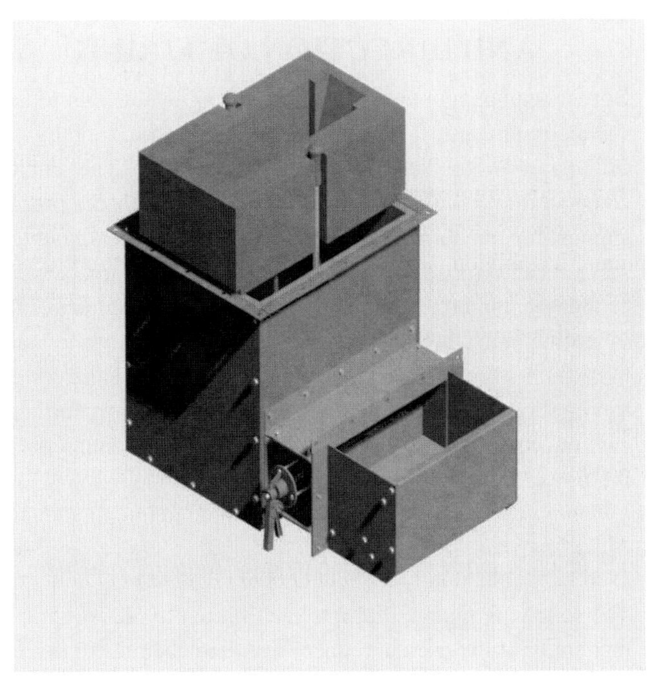

FIGURE 11.20 Typical staged-fuel flat flame burner.

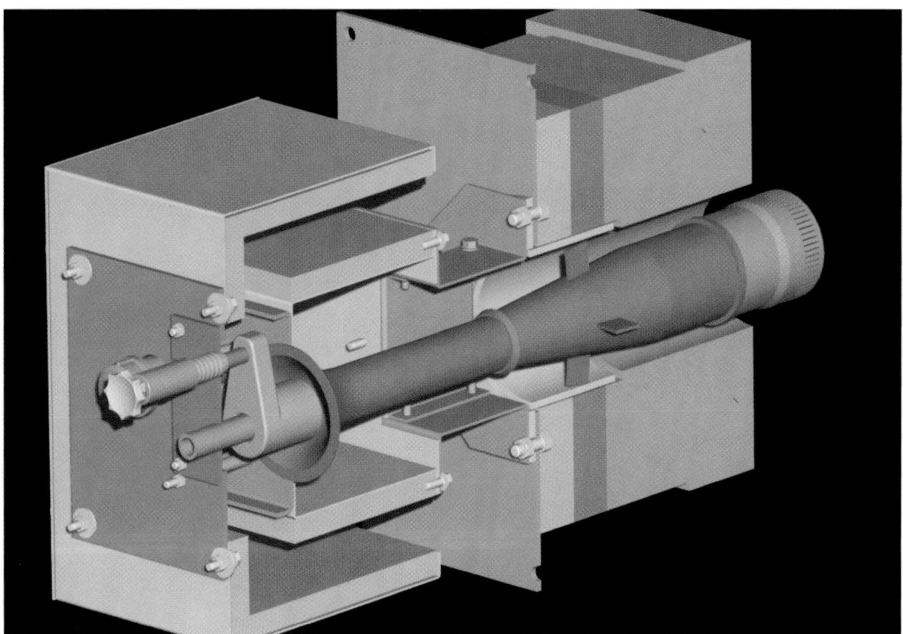

FIGURE 11.21 Typical radiant wall burner.

11.5.3 Radiant Wall

Radiant wall burners are used in cracking furnaces. These burners are mounted through the furnace wall and produce a thin, flat circular disk of flame adjacent to the wall. There are typically several rows of burners, with burners equally spaced in a grid pattern on walls. The burners uniformly heat the walls, which radiate heat to the process tubes on the centerline of the furnace. Figure 11.21 shows a typical radiant wall burner.

Many applications, such as reformers and cracking furnaces, require high furnace temperatures in the radiant section of the heater to produce the required high process temperatures. In addition, very uniform heat transfer to the process tubes is needed. In such operations, a flat flame burner that fires adjacent to a refractory wall is utilized.

11.5.4 Downfired

A downfired burner is a burner that is fired vertically into a furnace from the ceiling. Typically, these burners produce a flame that is round/conical in shape. These burners are utilized in applications such as hydrogen reformers or ammonia reformers. Downfired burners are often dual-fuel burners firing a makeup fuel and a waste gas. The makeup fuel can be natural gas, naphtha, No. 2 fuel oil, diesel oil, or a propane/butane gas fuel.

11.6 MATERIALS SELECTION

Burner materials are selected for strength, corrosion resistance, and in many cases temperature resistance. Carbon steel is used for most metal parts unless temperature or corrosion considerations require more resistant alloys.

Cast iron or carbon steel can be utilized for fuel gas manifolds. The fuel gas riser material between the manifold and the fuel injector tip is generally carbon steel for ambient air service and 304 SS for parts in contact with air temperatures higher than 700°F (370°C). If H_2S is present in the fuel gas and preheated air over 300°F (150°C) is used, 316L stainless steel may be required for the gas piping passing through the windbox.

Fuel injector tips for premix burners may be cast iron. Cast steel or cast stainless steel may be required for fuel gases containing appreciable levels of hydrogen (typically >50 vol%).

Raw gas burners typically use 300 series stainless steel gas tips. Tips for oil burners are normally 416 SS. For oils containing erosive particles, T-1 tool steel is generally used. Atomizers for oil service are normally brass, or 303 SS for oils containing sulfur. Other oil injector parts are carbon steel, although nitride-hardened parts can be used for erosive liquids.

Flame stabilizer cones and swirlers are normally 300 series stainless steel.

Mineral wool is commonly used for noise reduction in plenums and as insulation in preheated air service up to 1000°F (540°C). Ceramic fiber insulation is used for higher temperature applications.

High-strength, low-alloy structural steel (ASTM A 242/A 242M) or 304 SS is used for other metal parts subject to >700°F (370°C) air preheat.

Burner block material is typically 55 to 60% alumina refractory with a 3000°F (1650°C) service temperature. Primary oil tiles may require 90% alumina refractory if the combined vanadium and sodium in the oil is greater than 50 ppm by weight.

REFERENCES

1. C.E. Baukal, Ed., *Oxygen-Enhanced Combustion*, CRC Press, Boca Raton, FL, 1998.
2. American Petroleum Institute, Burners for Fired Heaters in General Refinery Services, API 535, 1st ed., Washington, D.C., July 1995.
3. American Petroleum Institute, Fired Heaters for General Refinery Services, API 560, Washington, D.C., November 1996.

Chapter 12
Combustion Controls

Joe Gifford and Jim Heinlein

TABLE OF CONTENTS

12.1 Fundamentals 374
12.1.1 Control Platforms 374
12.1.2 Discrete Control Systems 376
12.1.3 Analog Control Systems 376
12.1.4 Failure Modes 378
12.1.5 Agency Approvals and Safety 379
12.1.6 Pipe Racks and Control Panels 381
12.2 Primary Measurement 383
12.2.1 Discrete Devices 383
12.2.2 Analog Devices 385
12.3 Control Schemes 389
12.3.1 Parallel Positioning 389
12.3.2 Fully Metered Cross Limiting 392
12.4 Controllers 394
12.5 Tuning 398
References 399

12.1 FUNDAMENTALS

This chapter discusses the various control system components, concepts, and philosophies necessary for understanding how control systems work, what the systems are designed to accomplish, and what criteria the controls engineer uses to design and implement a system. The interested reader can find further information on controls in numerous references.[1–11]

The purpose of the control system is to start, operate, and shut down the combustion process and any related auxiliary processes safely, reliably, and efficiently. The control system consists of various physical and logical components chosen and assembled according to a control philosophy and arranged to provide the user with an informative, consistent, and easy-to-use interface.

A combustion system typically includes a fuel supply, a combustion air supply, and an ignition system that all come together at one or more burners. During system startup and at various times during normal operation, the control system needs to verify or change the status of these systems. During system operation, the control system needs various items of process information to optimize system efficiency. Additionally, the control system monitors all safety parameters at all times and will shut down the combustion system if any of the safety limits are not satisfied.

12.1.1 Control Platforms

The control platform is the set of devices that monitors and optimizes the process conditions, executes the control logic, and controls the status of the combustion system. There are several different types of platforms and several different ways that the tasks mentioned above are divided among the types of platforms. Following is a list and a brief description of the most commonly used platforms.

12.1.1.1 Relay System

A relay consists of an electromagnetic coil and several attached switch contacts that open or close when the coil is energized or de-energized. A relay system consists of a number of relays wired together in such a way that they execute a logical sequence. For example, a relay system can define a series of steps to start up the combustion process. Relays can tell only if something is on or off and have no analog capability. They are generally located in a local control panel.

Relays have several advantages. Relays are simple, easily tested, reliable, and well-understood devices that can be wired together to make surprisingly complex systems. They are modular, easily replaced, and inexpensive. They can be configured in fail-safe mode so that if the relay itself fails, combustion system safety is not compromised.

There are also a few disadvantages. Once a certain complexity level is reached, relay systems can quickly become massive. Although individual relays are very reliable, a large control system with hundreds of relays can be very unreliable. Also, relays take up a lot of expensive control panel space. Because relays must be physically rewired to change the operating sequence, system flexibility is poor.

12.1.1.2 Burner Controller

A variety of burner controllers is available from several different vendors. They are prepackaged, hardwired devices in different configurations to operate different types of systems. Burner controllers will execute a defined sequence and monitor defined safety parameters. They are generally located in a local control panel. Like relays, they generally have no analog capability.

Advantages of burner controllers include the fact that they are generally inexpensive, compact, simple to hook up, require no programming, and are fail-safe and very reliable. They are often approved for combustion service by various safety agencies and insurance companies.

There are also some disadvantages. Burner controllers cannot control combustion systems of much complexity. System flexibility is nonexistent. If it becomes necessary to change the operating sequence, the controller must be rewired or replaced with a different unit. Controllers also require the use of attached peripherals from the same vendor, so some design flexibility is lost.

12.1.1.3 Programmable Logic Controller (PLC)

A programmable logic controller (PLC) is a small modular computer system that consists of a processing unit and a number of input and output modules that provide the interface to the combustion components. PLCs are usually rack-mounted and modules can be added or changed (see Figure 12.1). There are many types of modules available. Unlike the relays and burner controllers above, PLCs have analog control capability. They are generally located in a local control panel.

PLCs have the advantage of being a mature technology. They have been available for over 20 years. Simple PLCs are inexpensive and PLC prices are generally very competitive. They are compact, relatively easy to hook up, and because they are programmable, they are supremely flexible. They can operate systems of almost any level of complexity. PLC reliability has improved over the years and is now very good.

Disadvantages of PLCs include having to write software for the controller. Coding can be complex and creates the possibility of making a programming mistake, which can compromise system safety. The PLC can also freeze up, much like a desktop computer freezes up, where all inputs and

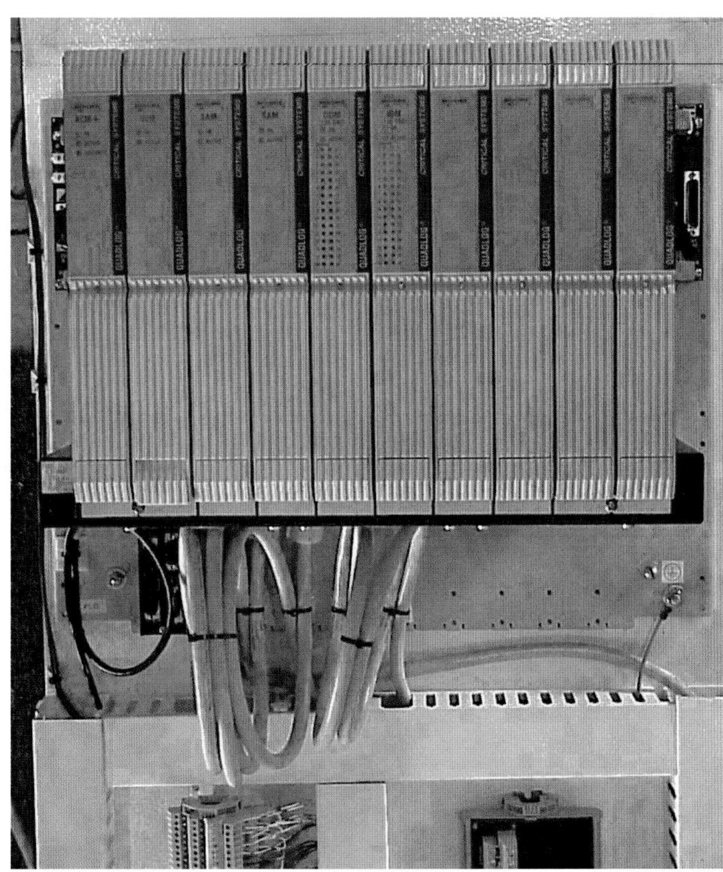

FIGURE 12.1 Programmable logic controller.

outputs are ignored and the system must be reset in order to execute logic again. Because of this possibility, standard PLCs should never be used as a primary safety device. Special types of redundant or fault-tolerant PLCs are available that are more robust and generally accepted for this service, but they are very expensive and generally difficult to implement.

12.1.1.4 Distributed Control System (DCS)

A distributed control system (DCS) is a larger computer system that can consist of a number of processing units and a wide variety of input and output interface devices. Unlike the systems described above, when properly sized, a DCS can also control multiple systems and even entire plants. The DCS is generally located in a remote control room but peripheral elements can be located almost anywhere.

DCSs have been around long enough to be a mature technology and are generally well-understood. They are highly flexible and are used for both analog and discrete (on/off) control. They can operate systems of almost any level of complexity and their reliability is excellent.

However, DCSs are often difficult to program. Each DCS vendor has a proprietary system architecture, so the hardware is expensive and the software is often different from any other vendor's software. Once a commitment is made to a particular DCS vendor, it is extremely difficult to change to a different one.

12.1.1.5 Hybrid Systems

If one could combine several of the systems listed above and build a hybrid control system, the advantages of each system could be exploited. In practice, that is what is usually done. A typical system uses relays to perform the safety monitoring, a PLC to do the sequencing, and either dedicated controllers or an existing DCS for the analog systems control. Sometimes, the DCS does both the sequencing and the analog systems control, and the safety monitoring is done by a fault-tolerant logic system. Most approval agencies and insurers require the safety monitoring function to be separate from either of the other functions.

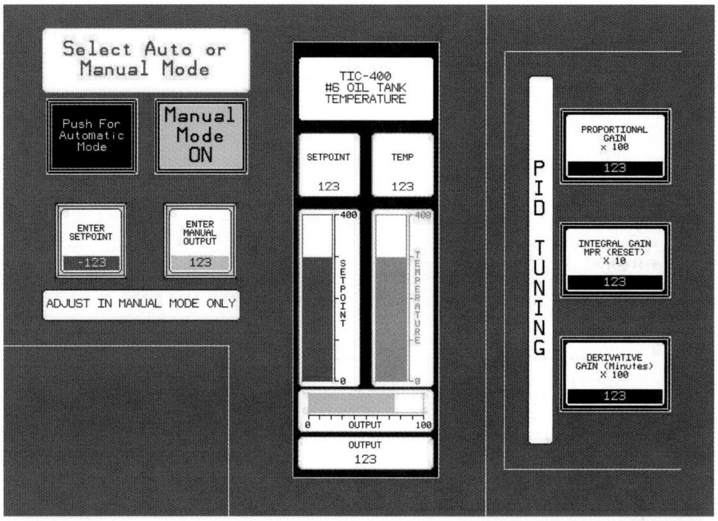

FIGURE 12.2 Touchscreen.

12.1.1.6 Future Systems

Over the next decade or so, it is expected that embedded industrial microprocessors using touchscreen video interfaces (see Figure 12.2) will start to appear in combustion control. These interfaces will communicate with field devices such as valves and switches via a single communications cable. They will use a digital bus protocol such as Profibus or Fieldbus. These systems are becoming common on factory floors around the world. Because establishment of a single standard has not yet happened and combustion standards are slow to change, these systems have not yet achieved widespread acceptance in the combustion world.

12.1.2 Discrete Control Systems

The world of discrete controls is black and white. Is the valve open or shut? Is the switch on or off? Is the button pressed or not? Is the blower running or not? There are two basic types of discrete devices: (1) input devices (sensors) that have electrical contacts that open or close depending on the status of what is being monitored; and (2) output devices, or final elements, that are turned on or off by the control system.

In a typical control system, sensors such as pressure switches, valve position switches, flame scanners, and temperature switches do all the safety and sequence monitoring. These devices tell the control system what is happening out in the real world. They are described in more detail in Section 12.1.3.

The final elements carry out the on/off instructions that come from the control system. These are devices such as solenoid valves, relays, indicating lights, and motor starters. These devices allow the control system to make things happen in the real world, and are described in more detail in Section 12.1.3.

The discrete control system does safety monitoring and sequencing. Typically, the system monitors all of the discrete inputs, and if they are all satisfactory, allows combustion system startup. If a monitored parameter is on when it should be off, or vice versa, the startup process is aborted and the system must be reset before another startup is permitted. The system also controls such things as which valves are opened in what order, if and when the pilot is ignited, and if and when main burner operation is allowed. Once the system starts, the discrete system has little to do other than monitor safety parameters. If any of the defined safety parameters are not satisfactory, the system immediately shuts down. Figure 12.3 is a simplified flow diagram showing a standard burner light-off sequence.

12.1.3 Analog Control Systems

The world of analog controls is not black and white — it is all gray. How far open is that valve? What is the system temperature? How much fuel gas is flowing?

There are two categories of analog devices with familiar names: (1) sensors, which measure some process variable (e.g., flow or temperature) and generate a signal proportional to the measured value; and (2) final elements (e.g., pumps and valves), which change their status (speed or position, for example) in response to a proportional signal from the control system.

In contrast to the discrete control system, the analog control system usually has few tasks to perform until the system completes the startup sequence and is ready to maintain normal operation. Most analog devices are part of a control loop. A simple loop consists of a sensor, a final element, and a controller. The controller reads the sensor, compares the

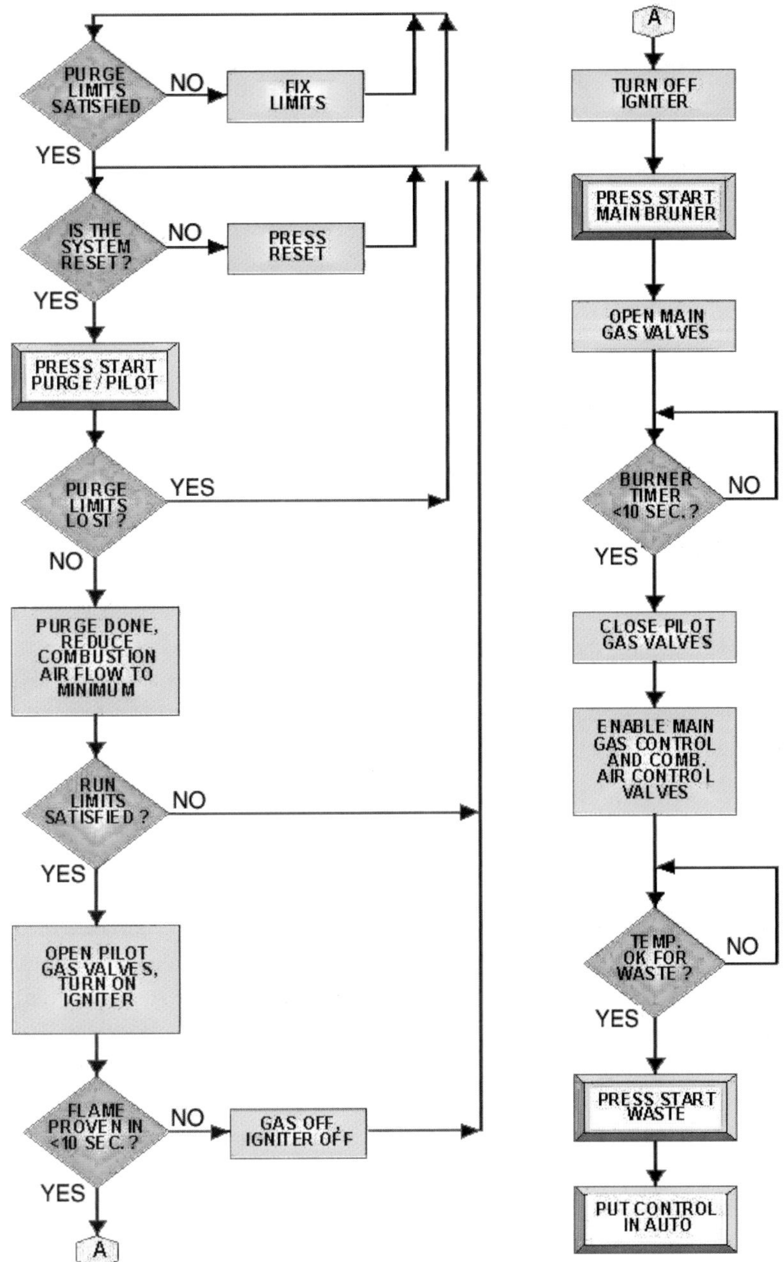

FIGURE 12.3 Simplified flow diagram of a standard burner light-off sequence.

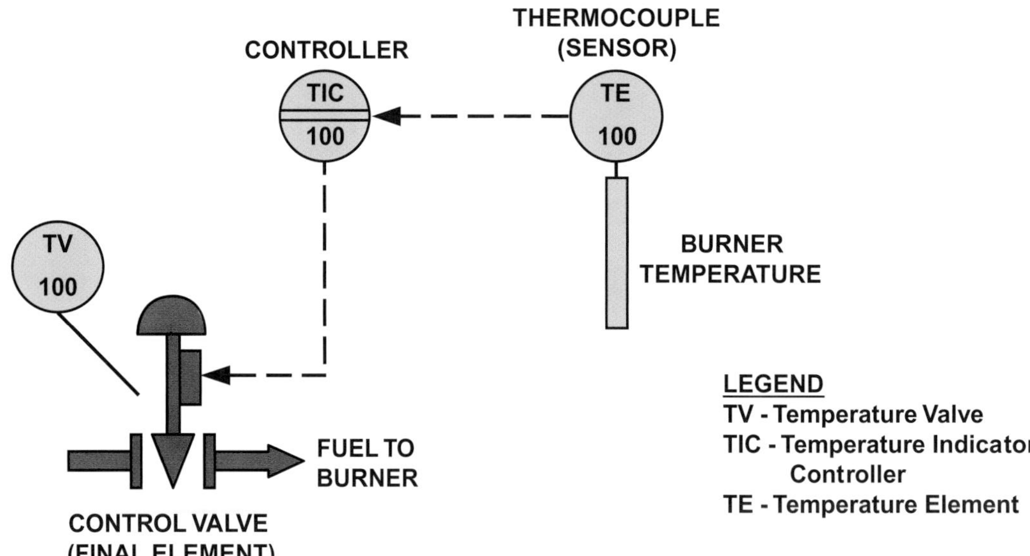

FIGURE 12.4 Simple analog loop.

measured value to its setpoint set by the operator, and then positions the final element to make the measured value equal the setpoint. Figure 12.4 illustrates a simple analog loop.

In this case, the thermocouple transmits the temperature to the controller. If the temperature is higher than the setpoint, the controller will decrease its signal to the control valve. This will decrease the fuel flow to the burner, thus lowering the temperature. In this way, the loop works to maintain the desired temperature — also known as the setpoint.

The previous illustration is a good example of a simple feedback system. After the controller adjusts the control valve, the resulting change in temperature is fed back to the controller. This way, the controller "knows" the result of the adjustment and can make a further adjustment if it is required. Another good example of feedback takes place whenever one drives a car. If one gets on the expressway and decides to drive at 60 mph, one presses the accelerator and watches the speedometer. Near 60 mph, one begins to ease off the accelerator so as not to overshoot. From then on, one glances at the speedometer every now and then and adjusts one's foot position as necessary.

Feedback alone is not enough, however. What if there is traffic congestion? One begins to slow down in anticipation. This is called feedforward, which occurs when one changes the operating point because some future event is about to happen and one needs to prepare for it. Feedforward is commonly used in combustion control systems. A good example from the world of combustion is waste flow. In a waste destruction system, what happens if the waste flow suddenly doubles? There will no longer be enough combustion air in the system to allow destruction of all the waste. Unburned waste will burn at the tip of the smokestack and clouds of smoke will billow from the stack. Phone calls from irate neighbors will soon begin to accumulate. Using feedforward, as shown in Figure 12.5, the waste flow is measured. When it doubles, the combustion airflow setpoint immediately increases by a similar amount, avoiding all of the unpleasant consequences listed above.

12.1.4 Failure Modes

Almost everything fails eventually. No matter how well the components of a control system are designed and built, some of them will fail from time to time. One of the primary tasks of the controls engineer is to design the control system so that failure of one or even several components will not cause a safety problem with the combustion system.

All components used in a control system have one or more defined failure modes. For example, if a discrete sensor fails, it will most likely cause the built-in switch contacts to fail open. To design a safe system, the controls engineer must choose and install the sensor so that when an alarm condition is present, the switch contacts will open. Thus, the alarm condition coincides with the failure mode. If it did not, the sensor could fail and the control system would still think everything was normal and attempt to keep operating as before — a condition that could be catastrophic.

In addition to sensors, final control elements also have failure modes. The controls engineer can usually select the

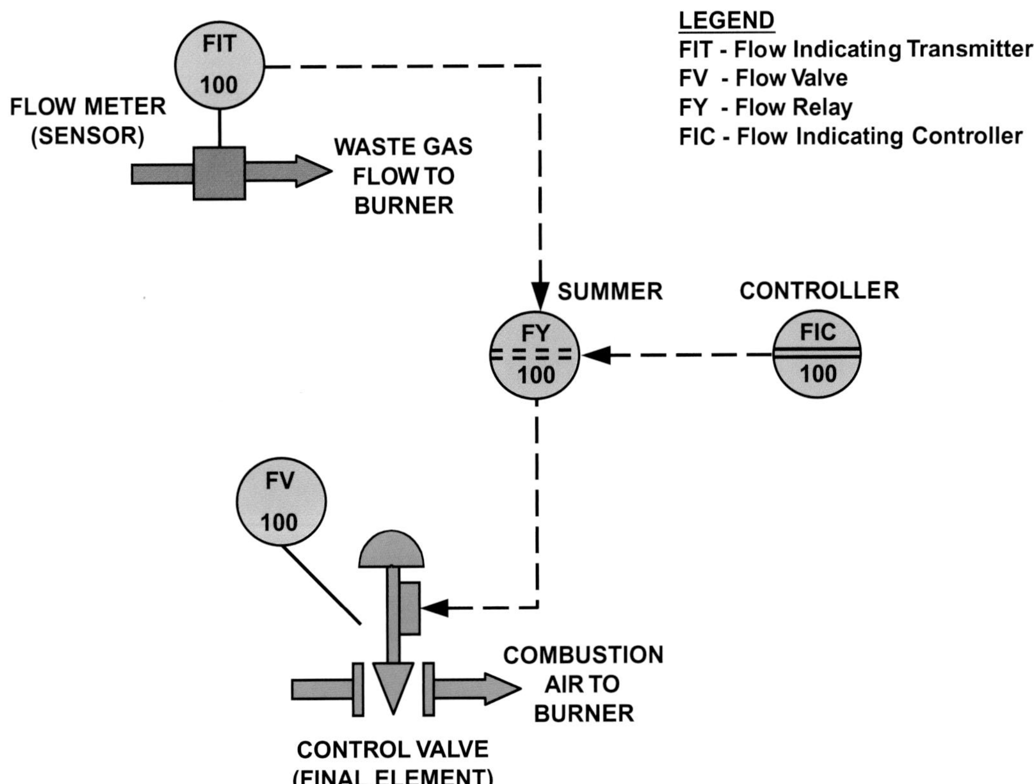

FIGURE 12.5 Feedforward loop.

desired failure mode. If there is an actuated valve that turns the fuel gas supply on and off, the actuator is installed so that the valve will spring closed (fail shut) upon loss of air. In addition, assume there is a solenoid valve that turns air to the actuator on and off. The solenoid valve should be selected and installed to rapidly dump air from the actuator upon loss of electrical power, thus closing the valve. These designs ensure that the two most likely circumstances of component failure enhance system safety.

Construction of a well-designed system ensures that every component that can fail is installed so that component failures do not compromise system safety. At its core, that is what controls engineering is all about.

12.1.5 Agency Approvals and Safety

Worldwide, there are hundreds of private, governmental, and semi-governmental safety organizations. Each ostensibly has the proper implementation of safety at the top of its agenda. Some agencies are concerned with the electrical safety and reliability of the components used in a control system; others are concerned about preventing explosions caused by sparking equipment in a gaseous atmosphere; and still others are concerned with the proper design of control systems to ensure safe operation of various combustion processes. No single organization does all of the things listed.

The design of combustion systems in the United States should include specifications that meet National Fire Protection Association (NFPA) and National Electrical Code (NEC) standards. In accordance with the applicable standards and years of experience in the field, systems should be designed with some or all of the following safety features.

12.1.5.1 Double-Block-and-Bleed for Fuel Supply

This means that there are two fail-shut safety shutoff block valves with a fail-open safety shutoff vent valve located between them, as shown in Figure 12.6. Each of the three safety shutoff valves (SSOVs) in the double-block-and-bleed system has a position switch not shown in the figure. For a system purge to be valid, the block valves must be shut and verified. For burner light-off, the vent valve is shut. After the vent valve position switch confirms that the valve is shut, the two block valves are opened. If there is a system failure, all three of the valves de-energize and return to their failure positions. Note that if the upstream block valve ever leaks, the leakage will preferentially go through the open vent valve and vent to a safe location rather than into the burner.

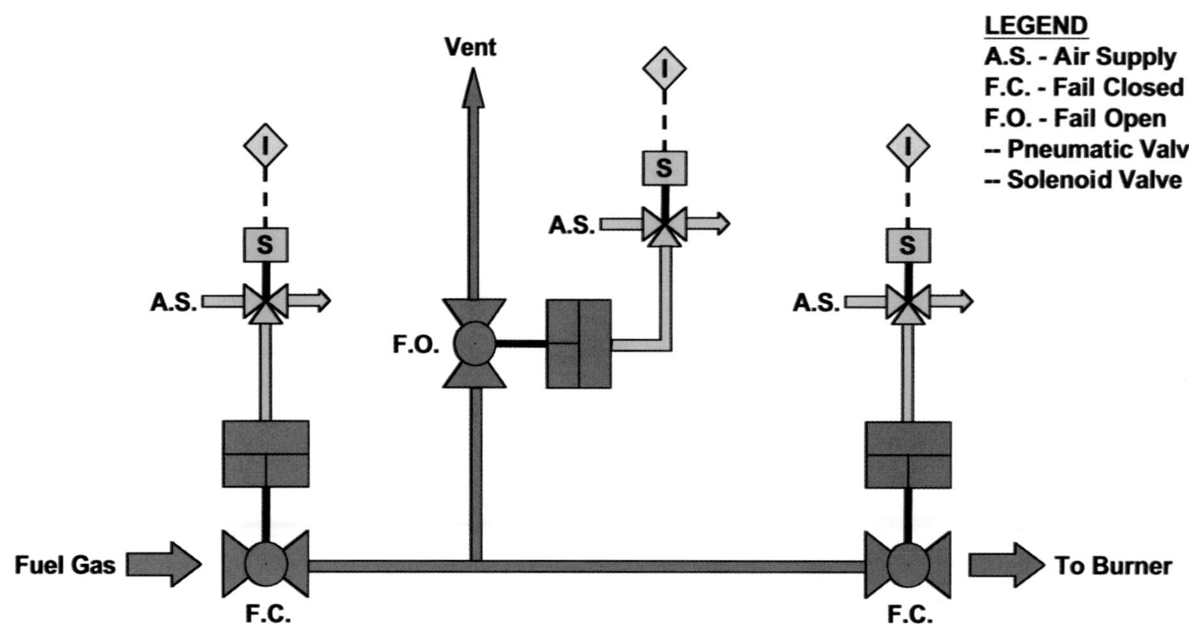

FIGURE 12.6 Double-block-and-bleed system.

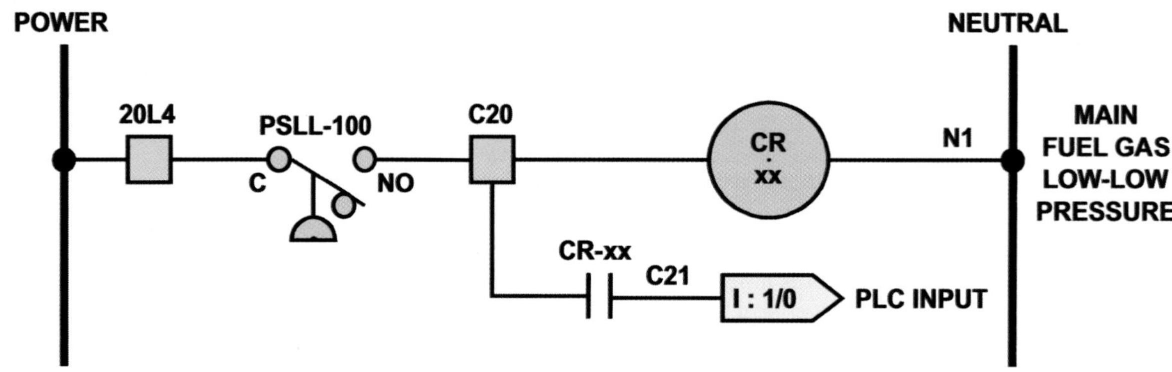

FIGURE 12.7 Failsafe input to programmable logic controller (PLC).

12.1.5.2 Unsatisfactory Parameter System Shutdown

An unsatisfactory parameter for any critical input immediately shuts down the system. The control system typically receives critical input information as shown in Figure 12.7. The pressure switch PSLL-03073 is wired so that if it fails, the voltage is interrupted to the relay (CR-xx) and the programmable logic controller (PLC). The PLC will then shut down the system. If either the switch or the relay fails, the system shuts down.

In addition, the relay has another contact in series with all the other critical contacts. If any of these contacts open, the power cuts off to all ignition sources (all fuel valves, igniters, etc.), immediately shutting down the combustion system.

In Figure 12.8, if there is a failure anywhere in the circuitry, the system shuts down. Even if damage to the PLC occurs, the relays will shut down the system. This is an excellent example of redundancy, fail-safe design principles, and effective design philosophy.

12.1.5.3 Local Reset Required after System Shutdown

After a system shutdown caused by an alarm condition, the system allows a remote restart only after an operator has pressed a reset button located at the combustion system. The operator should perform a visual inspection of the system and verify the correction of the condition that caused the shutdown.

12.1.5.4 Watchdog Timer to Verify PLC Operation

If the PLC logic freezes, a separate timer fails to receive an expected reset pulse from the PLC and shuts down the system.

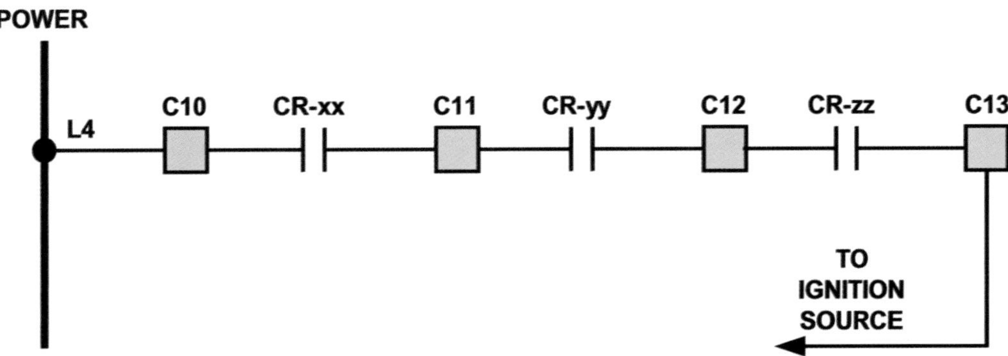

FIGURE 12.8 Shutdown string.

FIGURE 12.9 Typical pipe rack.

12.1.6 Pipe Racks and Control Panels

For most combustion control systems, two major assemblies comprise the bulk of the system: the pipe rack and the control panel. A pipe rack is shown in Figure 12.9. Sometimes called a skid, the pipe rack is a steel framework that has a number of pipes and associated components attached to it. Usually, most of the combustion process feeds such as air, fuel, and waste have their shutoff and control elements located on the pipe rack. This makes maintenance and troubleshooting more convenient and reduces the amount and complexity of wiring systems required to connect all of the components.

Typical control panels are shown in Figures 12.10a and b. Figure 12.11 shows the inside of the large control panel. The control panel is usually attached to the pipe rack. All of the devices on the pipe rack, as well as the field devices, are electrically connected to the control panel. The control system usually resides inside the control panel. In addition to the wiring, maintenance, and troubleshooting benefits mentioned above, another benefit to packaging the control system is far more important — the people who designed and built the system can test and adjust it at the factory. When the control system arrives at the job site, installation consists mostly of

(a)

FIGURE 12.10a Large control panel.

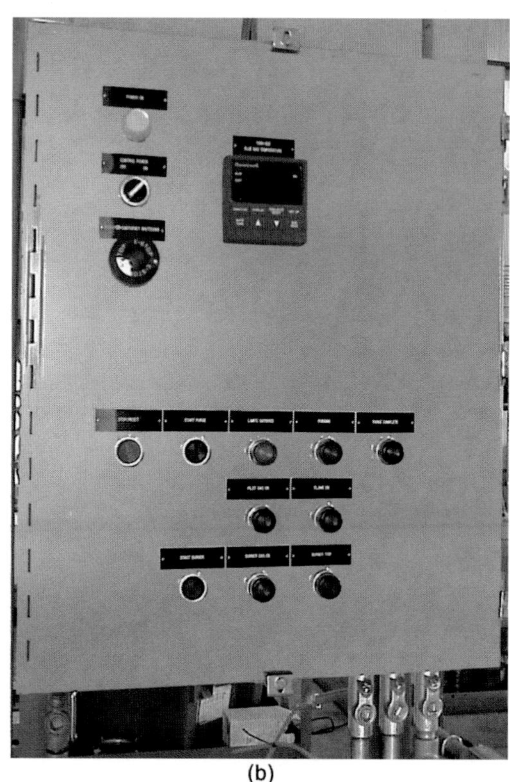

(b)

FIGURE 12.10b Small control panel.

FIGURE 12.11 Inside the control panel.

hooking up utilities and the interconnecting piping, minimizing the amount of expensive on-site troubleshooting, and tuning.

12.2 PRIMARY MEASUREMENT

This section describes a number of different analog and discrete devices used to provide the interface between combustion systems and control systems. All of these devices are available from numerous suppliers around the world. There are vast differences in price, quality, and functionality among the different devices and suppliers.

12.2.1 Discrete Devices

12.2.1.1 Annunciators

An annunciator is a centralized alarm broadcasting and memory device. Typically, all of the alarms in a control system are routed to the annunciator. When any alarm is triggered, the light associated with that alarm will flash and the annunciator horn will sound. If the annunciator is a "first-out" type, any subsequent alarms that occur will trigger their associated lights to come on solidly — rather than flashing, as the first alarm does. This is very useful for diagnosing system problems. When a safety shutdown occurs, other alarms are usually triggered while the system is shutting down. With a first-out capability, the original cause of the shutdown can easily be determined.

12.2.1.2 Pressure Switches

Pressure switches (see Figure 12.12a and b) are sensors that attach directly to a process being measured. They can be used to detect absolute, gage, or differential pressure. The switches generally have a pressure element such as a diaphragm, tube, or bellows that expands or contracts against an adjustable spring as pressure changes. The element attaches to one or more sets of contacts that open or close upon reaching the setpoint. The devices are used in a number of ways, but in combustion systems, they are usually used to test for high and low fuel gas pressure. They are normally set so the contacts are open when in the alarm condition.

12.2.1.3 Position Switches

Also called limit switches, these sensors attach to or are built into valves, insertable igniters, and other devices. Position switches usually employ a mechanical linkage, but proximity sensors are also quite common. They are adjustable, and can tell the control system if a valve is open, closed, or in some defined intermediate position. Position switches are not usually used for alarms. In combustion systems, they are generally used to check valve positions during purges and burner light-off sequences. Position switches are often used with integrated beacons and other visual devices. They are normally installed so that their contacts are closed only when the valve is in the desired safety position.

(a)

(b)

FIGURE 12.12 Pressure switches.

12.2.1.4 Temperature Switches

Temperature switches are usually attached to auxiliary equipment such as tanks or flame arresters. These sensors generally do not have the range to test for combustion system temperatures, so those applications use other devices. The switches usually use a bimetallic element, where the differential expansion of two different metals generates physical movement. The movement opens or closes one or more sets of contacts. The failure mode of temperature switches is not always predictable. Generally, installation requires open contacts when the switch is in the alarm condition.

12.2.1.5 Flow Switches

Flow switches are sensors that generally insert into the pipe or duct in which flow is measured. Because of the lack of a quantitative readout and the improved reliability of analog transmitters in this service, these devices are becoming less common. Their failure mode is not always predictable. Usual installation requires open contacts when the switch is in the alarm condition.

12.2.1.6 Run Indicators

A run indication sensor shows whether or not a pump or fan is running. It is usually possible to order a motor starter with a built-in set of signal contacts that close when the starter motor contacts are closed. However, that does not always ensure that the pump is running and pumping fluid. A magnetic shaft encoder rotates a magnetic slug past a pickup sensor every revolution and provides positive indication of shaft revolution, but that too does not always ensure that the pump is pumping fluid. It is usually preferable to have a pressure or flow indicator that shows that the system is functioning normally and moving fluid.

12.2.1.7 Flame Scanners

Flame scanners are crucial to the safe operation of a combustion system. If the flame is out, the fuel flow into the combustion enclosure is stopped and the area is purged before a re-light can be attempted. Flame scanners come in two main varieties: infrared and ultraviolet. The name tells which section of the electromagnetic spectrum it is designed to see. Generally, ultraviolet scanners are preferred because they are more sensitive and quicker to respond. The detector is a gas-filled tube that scintillates in the presence of flame ultraviolet radiation and emits bursts of current, called an avalanche, several hundred times per second as long as the flame continues. When the flame stops, the current stops. There is a 2- to 4-second delay, to minimize spurious shutdowns, and then the contacts open to designate the alarm condition. Most systems have two flame scanners and both scanners must fail to achieve system shutdown. Use of infrared scanners is desirable if there is a waste stream, such as sulfur, that absorbs ultraviolet light and makes operation of ultraviolet scanners unreliable. Self-checking scanners are usually used. They have output contacts that open on either loss of flame or failure of the self-check. Usually, the contacts are part of an amplifier/relay unit located in the control panel, but some newer systems have everything located in the scanner housing, which mounts on the burner end plate. One limitation with flame scanners is the possibility of the power wire to the scanner inducing false flame indications in the signal wire from the scanner. If there are separate wires for scanner power and scanner signal, they must run in separate conduits or shielding to prevent false signals caused by induction.

12.2.1.8 Solenoid Valves

Solenoid valves are turned on or off by the presence or absence of voltage from the control system. A solenoid valve has a relay coil that links mechanically to a valve disc mechanism. Energizing the solenoid causes the linkage to push against a spring to reposition the valve disc. De-energizing the solenoid allows the spring to force the valve to the failure position. The most common types of solenoid valves are two-way and three-way valves. Two-way valves have two positions. They either allow flow or they do not. They are often used to turn pilot gas on and off. Three-way solenoid valves have three ports but still only two positions. If ports are labeled A, B, and C, energizing the valve may allow flow between ports A and B, while de-energizing the valve may allow flow between ports A and C. It is very important to carefully select, install, and test three-way solenoid valves. Three-way solenoid valves typically attach to control valves and safety shutoff valves (SSOVs). In the case of control valves, when the solenoid valve is energized, the control valve is enabled for normal use. When the solenoid valve is de-energized, the instrument air is dumped from the control valve actuator diaphragm, causing the control valve to go to its spring-loaded failure position. For safety shutoff valve (SSOV) service, the solenoid valve is hooked up so that when energized, instrument air is allowed to reposition the SSOV actuator away from its spring-loaded failure position. When the solenoid valve is de-energized, the air is dumped from the SSOV actuator, causing the control valve to go to its spring-loaded failure position. The failure modes of the solenoid valve, control valve, and SSOV are coordinated to maximize system safety no matter which component fails.

12.2.1.9 Ignition Transformers

Ignition transformers supply the high voltage necessary to generate the spark used to ignite the pilot flame during system light-off. The type of transformer usually used converts standard AC power to a continuous 6000 V DC. This voltage then continuously jumps the spark gap at the igniter, which is located at the head of the pilot burner. High-energy igniters provide a more intense spark. A high-energy igniter is similar to the transformer mentioned above except that a capacitor is included to store energy and release it in spurts, resulting in a more intense spark. Both types of transformers are usually located close to the burner in a separate enclosure and hooked to the igniter using coaxial cable similar to the spark plug wire used in cars.

12.2.2 Analog Devices

12.2.2.1 Control Valves

Control valves are among the most complex and expensive components in any combustion control system. Numerous books document the nearly infinite variety of valves. Misapplication or misuse of valves compromise system efficiency and safety. Controls engineers cannot simply pick control valves from a catalog because they are the right size for the line where they will be used. Control valves must be engineered for their specific application. A typical pneumatic control valve is shown in Figure 12.13.

As shown in Figure 12.14, the type of service and control desired determines the selection of different flow characteristics and valve sizes. Controls engineers use a series of calculations to help with this selection process. A typical control valve consists of several components that are mated together before installation in the piping system.

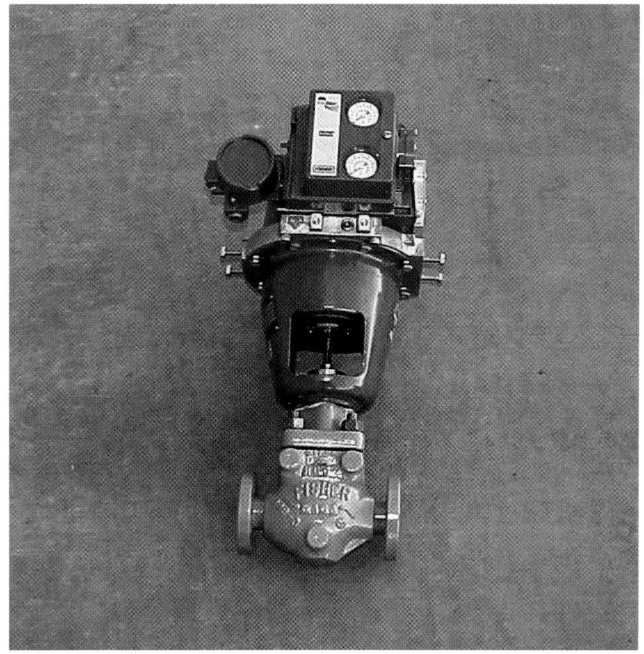

FIGURE 12.13 Pneumatic control valve.

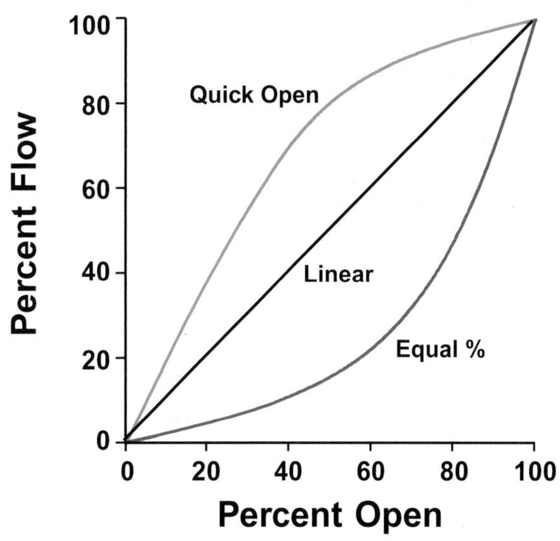

FIGURE 12.14 Control valve characteristics.

12.2.2.1.1 Control Valve Body

The control valve body can be a globe valve, a butterfly valve, or any other type of adjustable control valve. Usually, special globe valves of the equal percent type are used for fuel gas control service or liquid service. Control of combustion air and waste gas flows generally require the use of butterfly valves — often the quick-opening type. Because the combustion air or waste line usually has a large diameter, and because the cost of globe valves quickly becomes astronomical after line size exceeds 3 or 4 in., butterfly valves are usually the most economical choice. In Section 12.3, a discussion of parallel positioning describes how controls engineers use a globe valve and a butterfly valve together to work smoothly for system control.

12.2.2.1.2 Actuator

The actuator supplies the mechanical force to position the valve for the desired flow rate. For control applications, a diaphragm actuator is preferred because compared to a piston-type actuator, it has a relatively large pressure-sensitive area and a relatively small frictional area where the stem touches the packing. This ensures smooth operation, precision, and good repeatability. Proper selection of the actuator must take into account valve size, air pressure, desired failure mode, process pressure, and other factors. Actuators are usually spring-loaded and single-acting, with control air used on one side of the diaphragm and the spring on the other. The air pressure forces the actuator to move against the spring. If air pressure is lost, the valve fails to the spring position, so the actuator is chosen carefully to fail to a safe position (i.e., closed for fuel valves, open for combustion air valves).

12.2.2.1.3 Current-to-Pressure Transducer

The current-to-pressure transducer, usually called the I/P converter, takes the 24 V DC (4 to 20 mA) signal from the

FIGURE 12.15 Thermocouple.

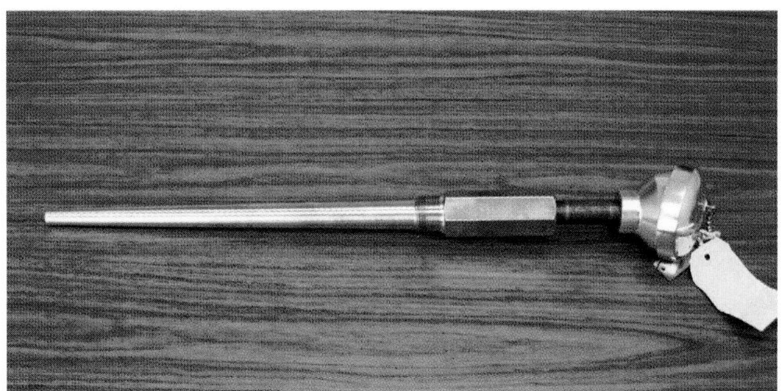

FIGURE 12.16 Thermowell and thermocouple.

controller and converts it into a pneumatic signal. The signal causes the diaphragm of the actuator to move to properly position the control valve.

12.2.2.1.4 Positioner

The positioner is a mechanical feedback device that senses the actual position of the valve as well as the desired position of the valve. It makes small adjustments to the pneumatic output to the actuator to ensure that the desired and the actual positions are the same. Current conventional wisdom states that positioners should be used only on "slow" systems and not on "fast" systems, where they can actually degrade performance. There is no defined border between "fast" and "slow," but virtually all combustion control applications are considered to be "slow," so positioners are almost always used in these systems.

12.2.2.1.5 Three-way Solenoid Valve

When energized, the three-way (3-way) solenoid valve admits air to the actuator. When de-energized, it dumps the air from the actuator. Because single-acting actuators are generally used, the spring in the actuator forces the valve either fully open or fully closed, depending on the engineer's choice of failure modes when specifying the valve. Obviously, a control valve that supplies fuel gas to a combustion system should fail closed, while the control valve that supplies combustion air to the same system should fail open. In an application in which the failure mode of the valve is irrelevant — and there are some — solenoid valves are not used.

12.2.2.1.6 Mechanical Stops

Mechanical stops are used to limit how far open or shut a control valve can travel. If it is vital that no more than a certain amount of fluid ever enters a downstream system, an "up" stop is set. If it is necessary to ensure a certain minimum flow (e.g., for cooling purposes), a "down" stop is set. In the case of a fuel supply control valve, the "down" stop is set so that during system light-off, an amount of fuel ideal for smooth and reliable burner lighting is supplied. After a defined settling interval, usually 10 seconds, the 3-way solenoid valve is energized and normal control valve operation is enabled.

12.2.2.2 Thermocouples

Whenever two dissimilar metals come into contact, current flows between the metals, and the magnitude of that current flow, and the voltage driving it, vary with temperature. This phenomenon is called the Seebeck effect. If both metals are carefully chosen and are of certain known alloy compositions, the voltage will vary in a nearly linear manner with temperature over some known temperature range. Because the temperature and voltage ranges vary depending on the materials employed, engineers use different types of thermocouples for different situations. In combustion applications, the K type thermocouple (0 to 2400°F, or –18 to 1300°C) is usually used. When connecting a thermocouple (see Figure 12.15) to a transmitter, the transmitter should be set up for the type of thermocouple employed. Installing thermocouples in a protective sheath know as a thermowell (see Figure 12.16), prevents the sensing element from suffering the corrosive or erosive effects of the process being measured. However, a thermowell also slows the response of the instrument to changing temperature and should be used with care.

12.2.2.3 Velocity Thermocouples

Also known as suction pyrometers, the design of velocity thermocouples attempts to minimize the inaccuracies in temperature measurement caused by radiant heat. Inside a combustor, the thermocouple measures the gas temperature. However, the large amount of heat radiated from the hot surroundings significantly affects this measurement. If a thermocouple is shielded from its surroundings by putting it in a

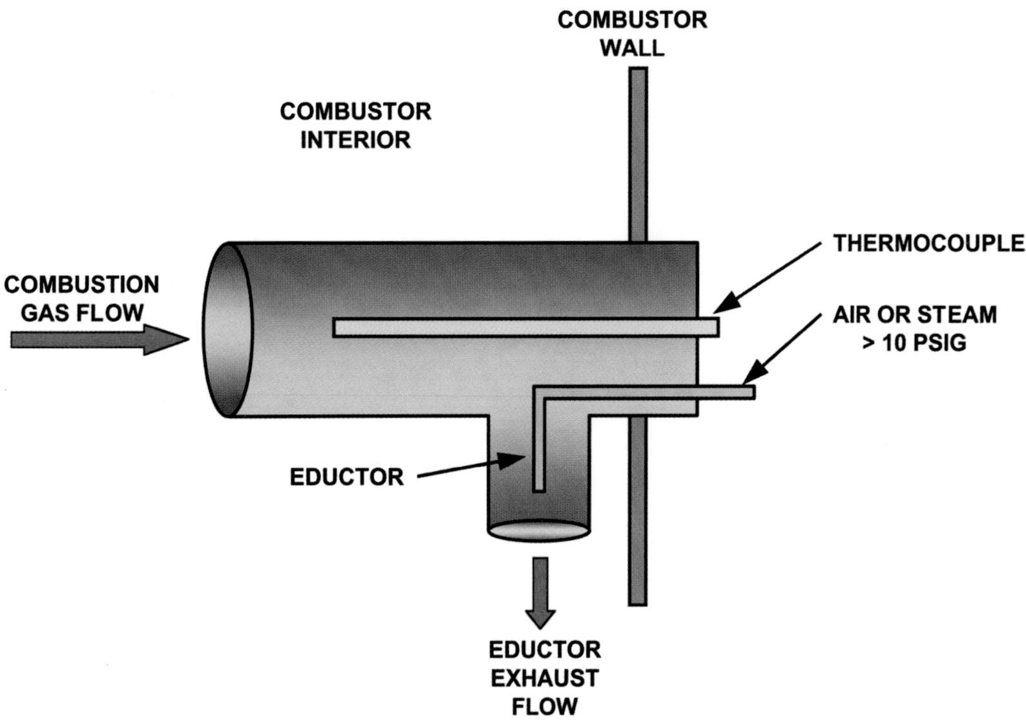

FIGURE 12.17 Velocity thermocouple.

hollow pipe as shown in Figure 12.16, the response time is slowed because the thermocouple is now located in a shield-created low-flow zone. Drawing suction on the shield quickly pulls gas in from the combustor and the response time improves. Using velocity thermocouples (see Figure 12.17) provides a high degree of precision in combustion temperature measurement.

12.2.2.4 Resistance Temperature Detectors (RTDs)

The resistance of any conductor increases with temperature. If a specific material and its resistance are known, it is possible to infer the temperature. Similar to the thermocouples described above, the linearity of the result depends on the materials chosen for the detector and their alloy composition. Engineers sometimes use RTDs in place of thermocouples when higher precision is desired. Platinum is a popular material for RTDs because it has good linearity over a wide temperature range. Like thermocouples, installation of RTDs in thermowells is common.

12.2.2.5 Pressure Transmitters

A pressure transmitter (see Figure 12.18) is usually used to provide an analog pressure signal. These devices use a diaphragm coupled to a variable resistance, which modifies the 24 V DC loop current (4 to 20 mA) in proportion to the range in which it is calibrated. In recent years, these devices have become more accurate and sophisticated, with onboard intelligence and self-calibration capabilities. They are available in a wide variety of configurations and materials, and can be used in almost any service. It is possible to remotely check and reconfigure these "smart" pressure transmitters using a handheld communicator.

12.2.2.6 Flow Meters

There are many different types of flow meters and many reasons to use one or another for a given application. The following is a list of several of the more common types of flow meters, how they work, and where they are used.

12.2.2.6.1 Vortex Shedder Flow Meter

A vortex shedder places a bar in the path of the fluid. As the fluid goes by, vortices (whirlpools) form and break off constantly. An observation of the water swirling on the downstream side of bridge pilings in a moving stream reveals this effect. Each time a vortex breaks away from the bar, it causes a small vibration in the bar. The frequency of the vibration is proportional to the flow. Vortex shedders have a wide range, are highly accurate, reasonably priced, highly reliable, and useful in liquid, steam, or gas service.

FIGURE 12.18 Pressure transmitter (left) and pressure gage (right).

12.2.2.6.2 Magnetic Flow Meter

A magnetic field, a current-carrying conductor, and relative motion between the two create an electrical generator. In the case of a magnetic flow meter, the meter generates the magnetic field and the flowing liquid supplies the motion and the conductor. The voltage produced is proportional to the flow. These meters are highly accurate, very reliable, and have a wide range, but are somewhat expensive. They are useful with highly corrosive or even gummy fluids, as long as the fluids are conductive. Only liquid flow is measured.

12.2.2.6.3 Orifice Flow Meter

Historically, almost all flows were measured using this method and it is still quite popular. Placing the orifice in the fluid flow causes a pressure drop across the orifice. A pressure transmitter mounted across the orifice calculates the flow from the amount of the pressure drop. Orifice meters are very accurate but have a narrow range. They are reasonably priced, highly reliable, and are useful in liquid, steam, or gas service.

12.2.2.6.4 Coriolis Flow Meter

The Coriolis flow meter is easily the most complex type of meter to understand. The fluid runs through a U-shaped tube that is being vibrated by an attached transducer. The flow of the fluid will cause the tube to try to twist because of the Coriolis force. The magnitude of the twisting force is proportional to flow. These meters are highly accurate and have a wide range. They are generally more expensive and their reliability is not as good as some other types.

12.2.2.6.5 Ultrasonic Flow Meter

When waves travel in a medium (fluid), their frequency shifts if the medium is in motion relative to the wave source. The magnitude of the shift, called the Doppler effect, is proportional to the relative velocity of the source and the medium. The ultrasonic meter generates ultrasonic sound waves, sends them diagonally across the pipe, and computes the amount of frequency shift. These meters are reasonably accurate, have a fairly wide range, are reasonably priced, and are highly reliable. Ultrasonic meters work best when there are bubbles or particulates in the fluid.

12.2.2.6.6 Turbine Flow Meters

A turbine meter is a wheel that is spun by the flow of fluid past the blades. A magnetic pickup senses the speed of the rotation, which is proportional to the flow. These meters can be very accurate but have a fairly narrow range. They must be very carefully selected and sized for specific applications. They are reasonably priced and fairly reliable. They are used in liquid, steam, or gas service.

12.2.2.6.7 Positive Displacement Flow Meters

Positive displacement flow meters generally consist of a set of meshed gears or lobes that are closely machined and matched to each other. When fluid is forced through the gears, a fixed amount of the fluid is allowed past for each revolution. Counting the revolutions reveals the exact amount of flow. These meters are extremely accurate and have a wide range. Because there are moving parts, the meters must be maintained or they can break down or jam. They also cause a large pressure drop, which can sometimes be important for certain applications.

12.2.2.7 Analytical Instruments

There are many different types of analytical instruments used for very specific applications. Unlike the sensors described previously, these devices are usually systems. They are a combination of several different sensors linked together by a processor of some sort that calculates the quantity in question. Unlike a pressure transmitter, most analytical instruments sample and chemically test the process in question. Because the process takes time, the engineer, when designing the system, must plan for a delayed response from the analytical instrument. A detailed discussion of the design and operation of analytical instruments is beyond the scope of this book; however a list of several of the more common types and their uses is given below.

12.2.2.7.1 pH Analyzer

Almost any combustion system occasionally requires the scrubbing of effluent or other similar processes. pH monitoring is needed to ensure that the water going into the scrubber is the correct pH to neutralize the acidity or alkalinity of the effluent. The analyzer sends information to a controller that is responsible for opening or closing valves that add alkaline chemicals to the water to raise pH.

12.2.2.7.2 Conductivity Analyzer

Conductivity analyzers are often used in conjunction with pH analyzers. Where the pH analyzer system functions to raise pH, the output from the conductivity analyzer is usually sent to a controller responsible for opening or closing valves that dilute the water to lower pH.

12.2.2.7.3 Oxygen (O_2) Analyzer

Oxygen (O_2) or combustibles analyzers monitor the amount of oxygen or combustibles in the exhaust of a combustion system. The analyzer sends data back to the control system, which uses it to tightly control the amount of combustion air coming into the system. This has the dual result of making the system more efficient and reducing the amount of pollutants that result from the combustion process. Different models have varying methodologies, accuracies, and sample times, but there are two major types: (1) *in situ* analyzers carry out the analysis at the probe; and (2) extractive analyzers remove the sample from the process and cool it before analysis.

12.2.2.7.4 Nitrogen Oxides (NOx) Analyzer

Nitrogen oxides (NO, NO_2, etc.; see Chapter 6) are one of the main components of smog and are the result of high-temperature combustion. Noxidizers are combustion systems that use an extended low-temperature combustion process designed to minimize the formation of nitrous oxide compounds. Noxidizers use NOx analyzers. To properly control the process, the NOx analyzer output goes to a controller that controls airflow into the system, minimizing NOx formation.

12.2.2.7.5 Carbon Monoxide (CO) Analyzer

CO is also an undesirable pollutant and is a product of incomplete combustion. The output of the CO analyzer (see Figure 10.4a) is often used in the analysis of system efficiency or to control airflow to the combustion system.

12.3 CONTROL SCHEMES

Other chapters of this book present the combustion process and the definition of the terms used to describe it. This section describes methods used to control the process. Generally, controlling the process means controlling the flow of fuels and combustion air.

12.3.1 Parallel Positioning

Designers use analog control schemes to modulate valve position and control fan and pump speeds to achieve the required mix of fuel and oxygen in a combustion system. Simple systems often use parallel positioning of fuel and air valves from a single analog signal.

12.3.1.1 Mechanical Linkage

A common method of parallel positioning is to mechanically link the fuel and air valves to a single actuator. Adjustment of a cam located on the fuel valve supplies the proper amount of fuel throughout the air valve operating range. Figure 12.19 shows the arrangement.

In the figure, the temperature indicating controller (TIC) operates an actuator attached to the air valve. A mechanical linkage and an adjustable cam operate the fuel valve in parallel with the air valve. Springs or weights attached to the air valve shaft force a full open position of the air valve if the mechanical connection to the air valve fails. The system uses a fail-closed actuator to ensure that a low fire failure mode results from loss of signal or loss of actuator power.

Mechanical linkage is simple in operation but requires considerable adjustment at startup to obtain the correct fuel:air ratio over the entire operating range. Predictable flow rates of fuel and air throughout valve position require a fixed supply pressure to the valves and constant load geometry downstream

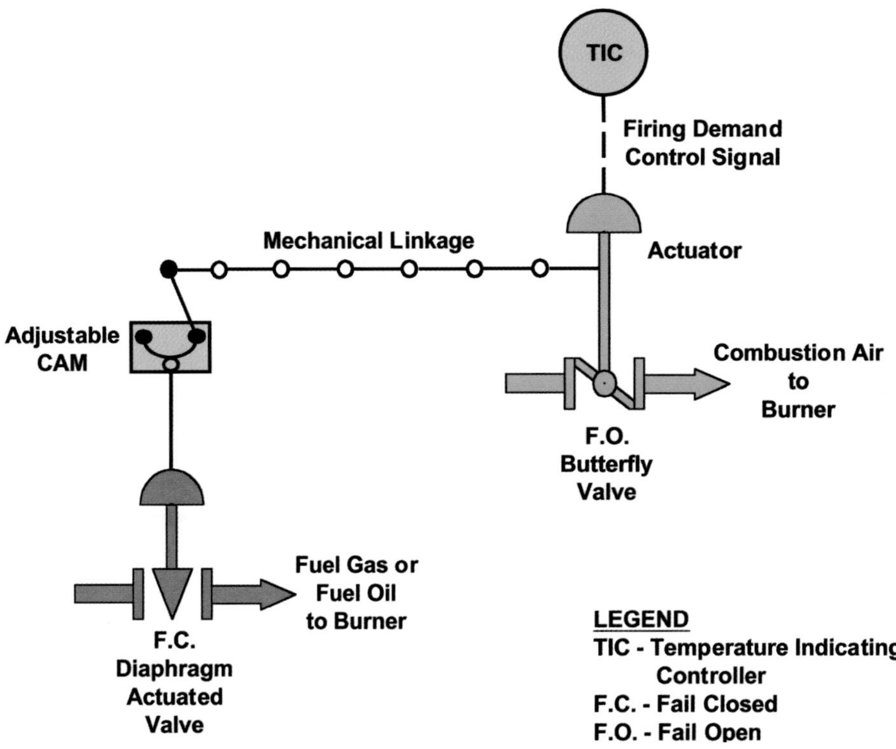

FIGURE 12.19 Mechanically linked parallel positioning.

of the valves. Analytical feedback to control fuel gas or combustion air supply pressure can make dynamic corrections for fuel variations, temperature changes, and system errors. Dynamic adjustments should be small, trimming adjustments, rather than primary control parameters.

12.3.1.2 Electronic Linkage

Electronically linked fuel and combustion air valves for parallel positioning have many advantages over mechanically linked valves. Figure 12.20 illustrates the scheme.

In the example, a TIC (temperature indicating controller) generates a firing rate demand. The controller applies an output of 4 to 20 mA to the fuel valve and to a characterizer in the air valve circuit. Electronic shaping of the characterizer output positions the air valve for correct airflow. Predictable and repeatable valve positions require the use of positioners at each valve. Without positioners, valve hysteresis causes large errors in flow rate.

Signal inversion (1 minus the value being measured) is sometimes integral to the characterizer. Signal inversion is necessary because the air valve fails open and the fuel valve fails closed. Safety concerns dictate failure modes. Fuel should always fail to minimum and air should fail to maximum.

Electronically linked parallel positioning works well if properly designed. Good design requires valves with known coefficients throughout valve position and the use of high-performance positioners. Supply pressure of fuel and air to each valve must be constant or repeatable. System load downstream of the valves must be of fixed geometry. Section 12.3.1.3 shows an example of how to calculate and configure a characterizer for the air valve.

Figure 12.21 shows a variation of parallel positioning that permits use of the combustion air valve for the multiple purposes of:

1. supplying combustion air during normal operation
2. supplying quench air when burning exothermic waste
3. using another heat source requiring quench air

When showing a range of milliampere signals, the first value is the minimum valve position and the second value is the maximum valve position. This convention aids system analysis and is especially useful for complex systems. The TIC output is split-ranged. The top half (12 to 20 mA) is used for firing fuel gas. When burning exothermic waste requiring quench air, the temperature controller output decreases, providing low fire fuel at 12 mA, then quench air below 12 mA.

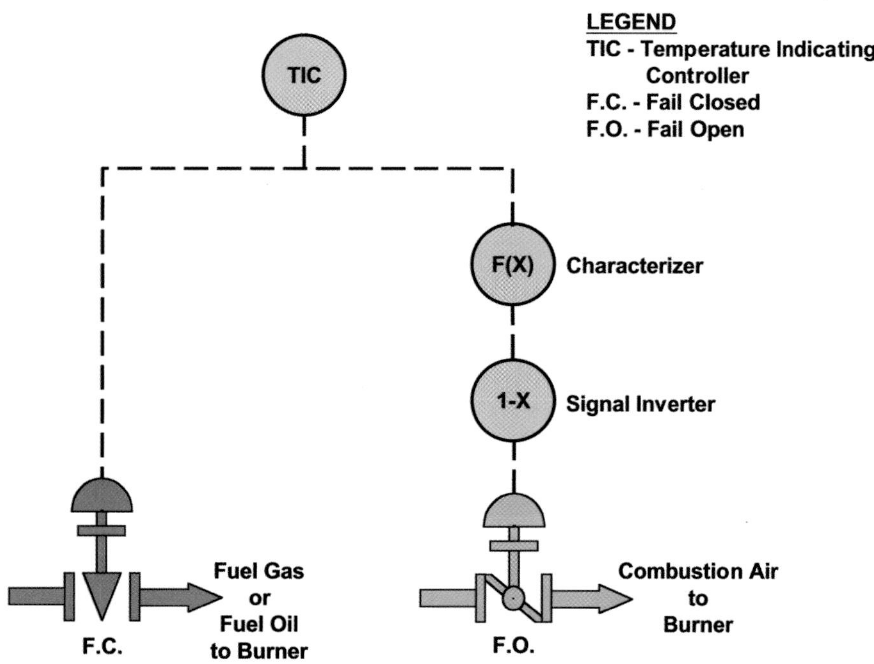

FIGURE 12.20 Electronically linked parallel positioning.

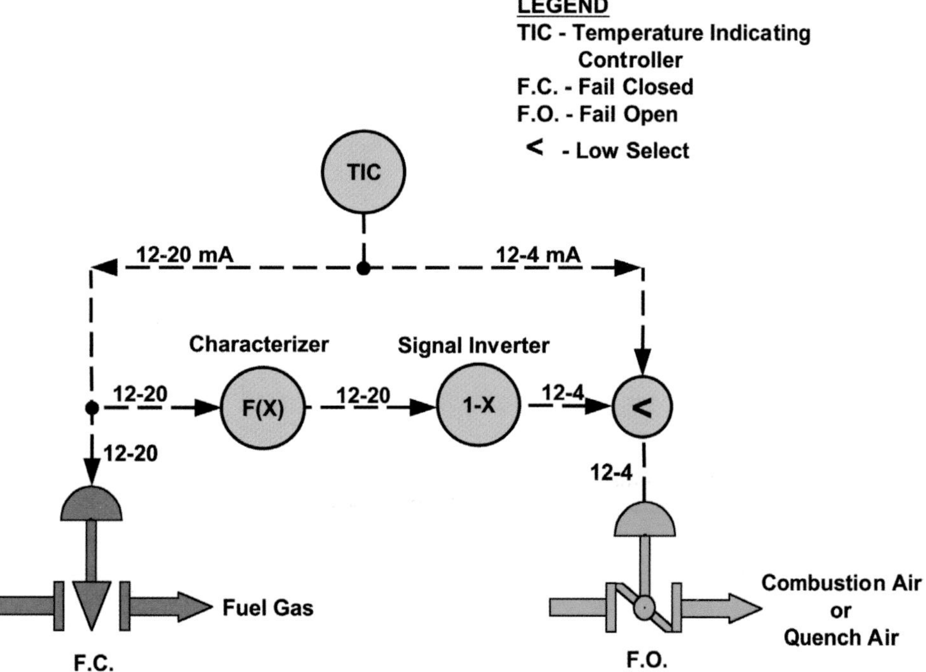

FIGURE 12.21 A variation of parallel positioning.

The TIC output is actually 4 to 20 mA. The description of the action of the receivers uses the term "split-ranged." For example, the TIC applies the entire 4 to 20 mA range to the fuel gas valve, but the valve is configured to respond only to the partial range of 12 to 20 mA.

12.3.1.3 Characterizer Calculations

Parallel positioning of a globe-type fuel gas valve and a butterfly-type combustion air valve requires characterizer calculations as described below. Figure 12.20 shows the control scheme.

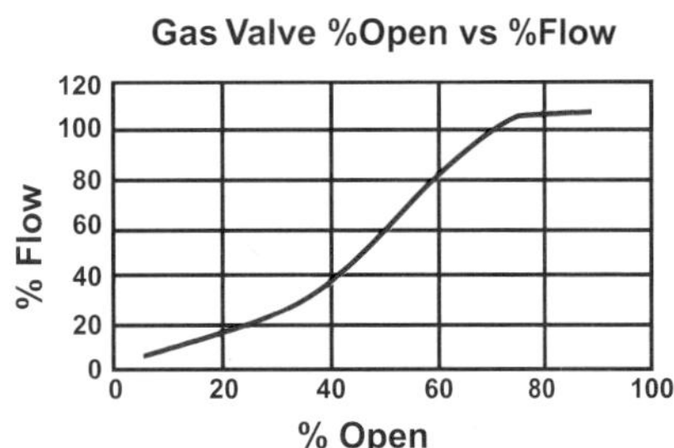

FIGURE 12.22 Fuel flow rate versus control signal.

TABLE 12.1 Gas Valve Data

Control Signal TIC Output %	Gas Valve % Open	Fuel Gas Flow Rate %
10	10	10
20	20	16
30	30	25
40	40	39
50	50	58
60	60	83
70	70	100
80	80	107
90	90	109
100	100	110

Three general steps are required to define the characterizer:

1. Calculate and graph fuel flow rate vs. control signal.
2. Calculate and graph combustion air flow rate vs. control signal.
3. Tabulate and graph air valve characterizer.

12.3.1.3.1 Step 1: Fuel Flow Rate vs. Control Signal
Predictable and repeatable calculations of fuel gas flow rate vs. control valve position require:

1. pressure regulator upstream of control valve to provide constant inlet pressure (varying inlet pressure can be used only if it is repeatable with flow rate.)
2. constant temperature and composition of fuel gas
3. high-quality positioner on the control valve to eliminate hysteresis and to ensure that valve percent open equals percent control signal
4. knowledge of valve coefficient vs. valve position throughout the control valve range, including the pressure recovery factor
5. fixed and known pressure drop geometry downstream of the control valve
6. subsonic regime throughout the flow range

Results of Step 1 are shown in Figure 12.22 for a typical fuel gas valve with equal percent trim. Low fire position of the valve is approximately 25% open for many applications. Maximum firing rate occurs between 70 and 80% open, resulting in a near-linear function of flow rate vs. valve position throughout the firing range. The linear function is not necessary for configuring a combustion air characterizer, but is useful for the application of a dynamic fuel:air ratio correction to the control circuit.

Use of a positioner on the fuel gas valve establishes equality between percent control signal and percent valve opening. Columns 1 and 2 of Table 12.1 show gas valve data.

12.3.1.3.2 Step 2: Air Flow Rate vs. Air Valve Position
Calculation of air flow rate versus vs. position that is predictable and repeatable requires:

1. known and repeatable valve inlet pressure vs. flow rate
2. near-constant temperature
3. high-quality positioner on the valve
4. knowledge of valve coefficient and pressure recovery factor of the air valve at all valve positions
5. fixed and known flow (pressure drop) geometry downstream of the control valve

Figure 12.23 shows the results of a typical butterfly-type valve calculation for Step 2. The low fire mechanical stop is normally set at approximately 20% open.

12.3.1.3.3 Step 3: Air Valve Characterizer
Table 12.2 combines data from Figure 12.23 with data from Table 12.1. Air valve graph data are tabulated in columns 3 and 4. Figure 12.24 is a plot of the data from columns 1 and 4 and represents the required shape of the air valve characterizer. The TIC output signal is the characterizer input and is plotted on the x-axis. The characterizer output is the percent open of the air valve and is shown on the y-axis. Many characterizer instruments are available that will model a curved response using straight-line segments. This characterizer is sufficiently defined using three segments.

12.3.2 Fully Metered Cross Limiting

Development of a fully metered control scheme for modulating fuel and air to a burner begins with the electronically linked parallel positioning scheme as previously shown in Figure 12.20. Figure 12.25 adds flow meters and flow controllers.

Flow meters are linear with flow rate. Meter output signal scaling provides the firing rate and air:fuel ratio required for the application. The combustion air characterizer used for parallel positioning is not required because the transmitters are linear with flow rate.

In the illustration, the temperature controller TIC output sets the firing rate by serving as setpoint to each flow controller. Signal inversion, shown as (1 minus parameter value) in the parallel positioning scheme, is not required. Instead, controller output mode is configured to match the valve failure mode.

Controller output mode, reverse or direct acting, defines the change in output signal direction with respect to process variable changes. For example, if the controller output increases as the process variable increases, the controller mode is direct acting. In combustion control schemes, fail-closed fuel valves require a reverse-acting flow controller, while fail-open combustion air valves require direct acting flow controllers. From controller mode definitions, it is clear that the temperature controller (TIC) should be reverse acting. That is, the TIC output should decrease, reducing the firing rate, in response to an increase in temperature, the process variable.

Addition of high and low signal selectors provides cross limiting of the fully metered control scheme, as shown in Figure 12.26. The low signal selector (<) compares demanded firing rate from the TIC to the actual combustion air flow rate and applies the lower of the two signals as the setpoint to the fuel flow controller. The low signal selector ensures that the fuel setpoint cannot exceed the amount of air available for combustion.

The high signal selector (>) compares demanded firing rate from the TIC to actual fuel flow rate and applies the higher of the two signals as the setpoint to the airflow controller. This ensures that the air setpoint is never lower than required for combustion of actual fuel flow rate.

Together, the high and low signal selectors ensure that unburned fuel does not occur in the combustion system. Unburned fuel accumulations can cause explosions. Cross limiting by the signal selectors causes air flow to lead fuel flow during load increases and for air flow to lag fuel flow during fuel decreases. This lead/lag action explains why the fully metered cross-limiting control system is often called "lead-lag" control. Whatever the name, the system performs the function of maintaining the desired air:fuel mixture during load changes. The system also provides fuel flow rate reduction in the event air flow is lost or decreased.

It is possible to trim the control scheme using measurement of flue gas oxygen content, as illustrated in Figure 12.27. For most systems, the oxygen signal should be used to "trim," and not be a primary control. Many oxygen analyzers are high maintenance and/or too slow in response to be used as a primary control in the combustion process. As shown, the oxygen controller is utilized for setpoint injection and provides tuning parameters to help process customization. High and low signal limiters restrict the oxygen controller output to a trimming function, normally 5 to 10% of the normal combustion air flow rate.

A multiplication function (X) in the combustion airflow transmitter signal makes the oxygen trim adjustment. The multiplier gives a fixed trim gain. Substituting a summing function for the multiplier would result in high trim gain at low flow rates and could produce a combustion air deficiency.

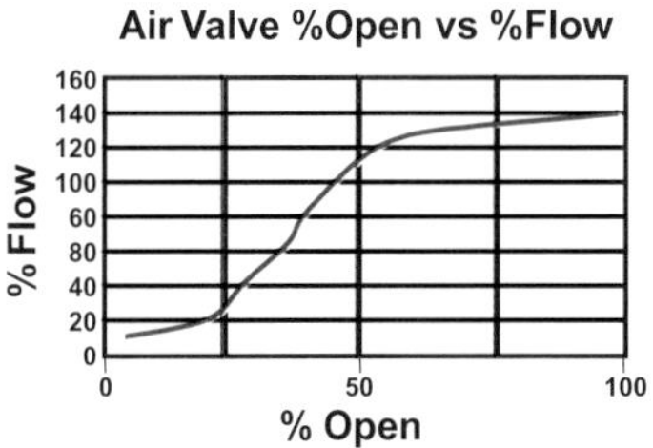

FIGURE 12.23 Typical butterfly-type valve calculation.

TABLE 12.2 Data for Characterizer

Control Signal TIC Output %	Fuel Gas Flow Rate %	Combustion Air Flow Rate %	Air Valve % Open
10	10	10	5
20	16	16	13
30	25	25	22
40	39	39	29
50	58	58	33
60	83	83	41
70	100	100	46
80	107	107	48
90	109	109	49
100	110	110	50

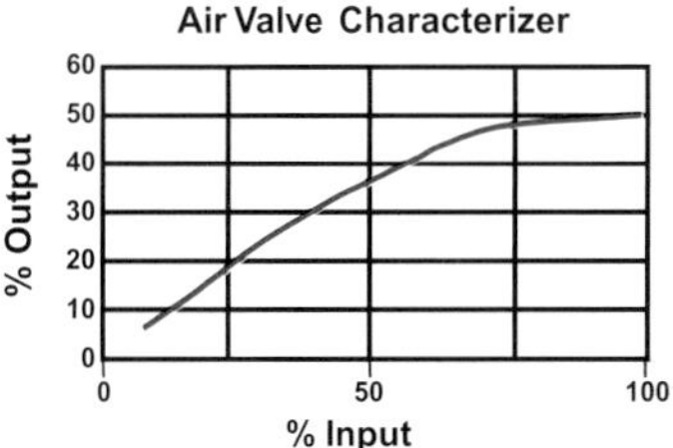

FIGURE 12.24 The required shape of the air valve characterizer.

Oxygen trim may be applied to the combustion airflow controller setpoint rather than the flow transmitter signal. If this technique is used, the airflow signal to the low signal selector must retain trim modification (see Figure 12.28 for the scheme).

Multiple fuels and oxygen sources are accommodated by the cross-limiting scheme, as shown in Figure 12.29. When multiple fuels are used, heating values must be normalized by adjusting flow transmitter spans or by addition of heating value multipliers. Similar methods are used to normalize oxygen content for multiple air sources.

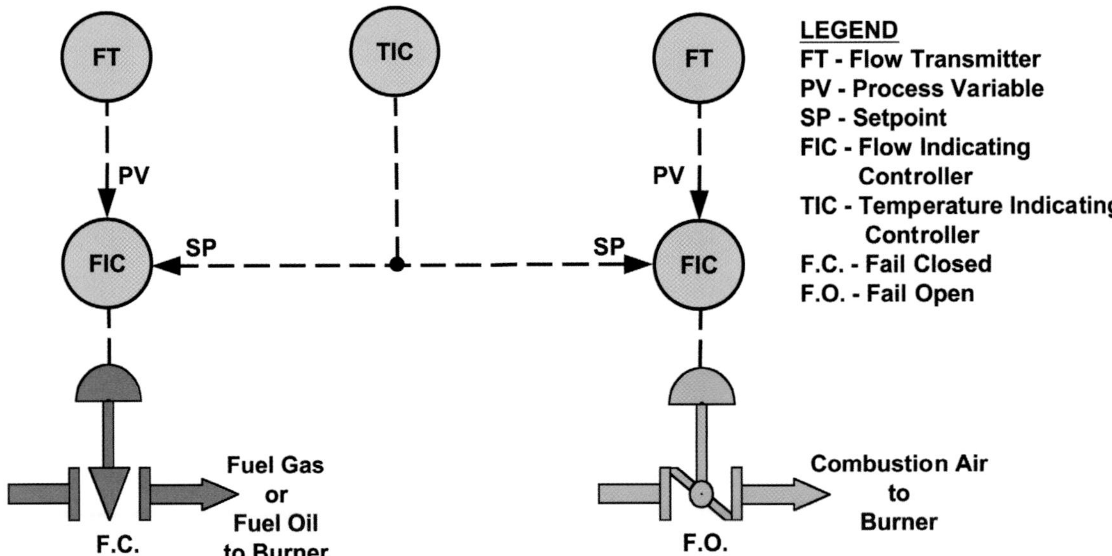

FIGURE 12.25 Fully metered control scheme.

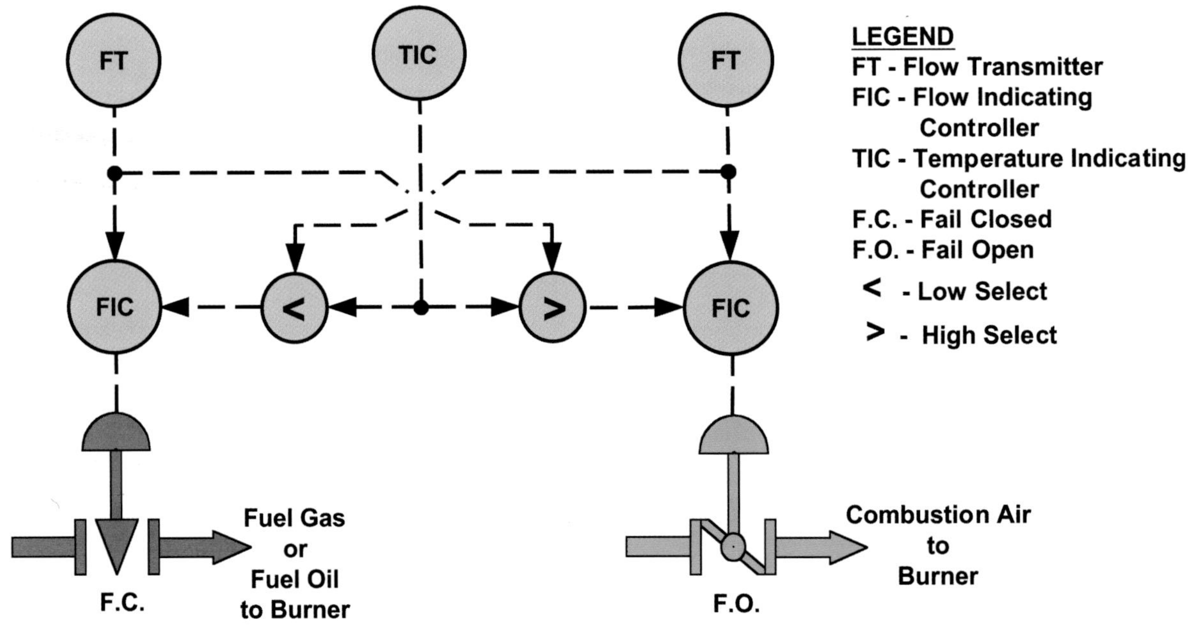

FIGURE 12.26 Fully metered control scheme with cross limiting.

12.4 CONTROLLERS

Controllers have historically been called analog controllers because the process and I/O signals are usually analog. Controller internal functions performed within a computer or microprocessor by algorithm are sometimes called digital controllers, although the I/O largely remains analog. Some digital controllers communicate with other devices via digital communication, but for the most part, controllers connect other devices by analog signals. The analog signal is usually 4 to 20 mA, DC.

In Figure 12.30, the setpoint is a signal representing the desired value of a process. If the process is flow rate, the setpoint is the desired flow rate. Setpoint signals can be generated internally within the controller, called the local setpoint, or may be an external signal, called the remote setpoint.

Controller output, called the controlled variable (CV) or the manipulated variable (MV), connects to a final element

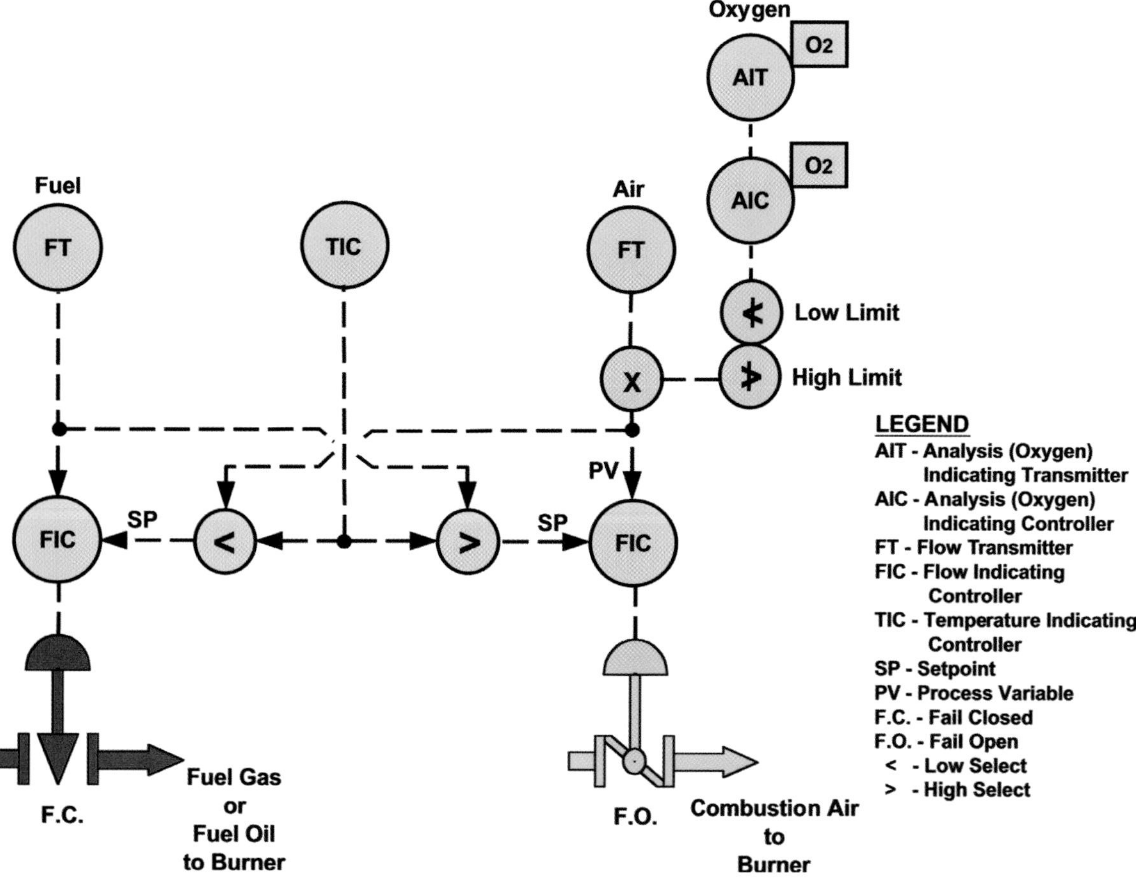

FIGURE 12.27 O_2 trim of air flow rate.

in the process. In this example of flow control, the final element is probably a control valve. Feedback from the process, called the process variable (PV) in this example, could be the signal from a flow meter.

The controlled variable (CV) is generated within the controller by subtraction of feedback (PV) from the setpoint (SP), generating an error signal *e*, which is multiplied by a gain *K*. The product *eK* is the controller output (CV).

$$\text{Output(CV)} = (\text{SP} - \text{PV})(K) = eK \qquad (12.1)$$

This simple controller is an example of the first controller built in the early 20th century and is called a proportional controller. The output is proportional to the error signal. The proportional factor is the gain *K*. Proportional controllers require an error (*e*) to produce an output. If the error is zero, the controller output is zero. To obtain an output that produces the correct value of the process variable, the operator is required to adjust the setpoint higher than the desired process variable in order to create the requisite error signal. Reduction of error by increasing controller gain is limited by controller instability at high gain.

Offset is the term given to the difference between the setpoint and process variable. Correction of offset was the first improvement made to the original controller. Offset correction was accomplished by adding bias to the controller output:

$$\text{Output(CV)} = eK + \text{Bias} \qquad (12.2)$$

Bias adjustment required operator manipulation of a knob or lever on the controller, which added bias until setpoint and process variables were equal. The operator considered the controller "reset" when equality occurred. Each setpoint change or process gain change required a manual reset of the controller. Figure 12.31 illustrates the proportional controller with manual reset.

Many operators prefer the term "proportional band" when describing controller gain. Proportional band is defined as:

$$\text{Proportional band (PB)} = \frac{100}{\text{Gain}} = \frac{100\%}{K} \qquad (12.3)$$

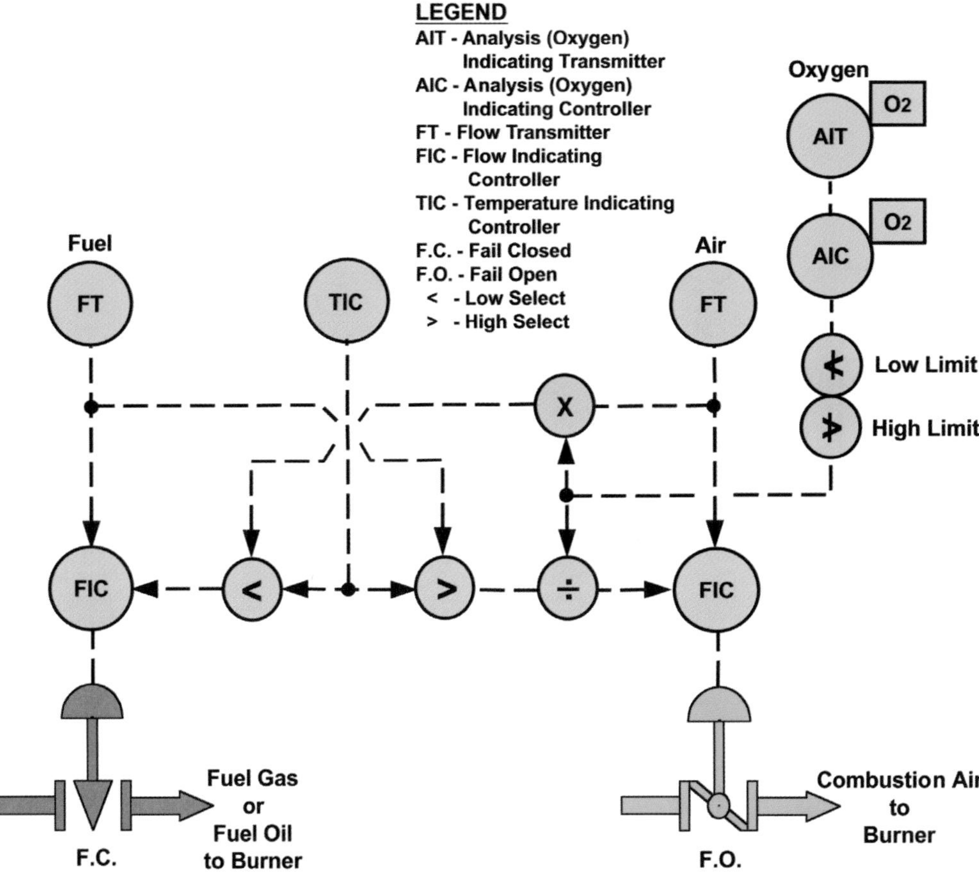

FIGURE 12.28 O_2 trim of air setpoint.

Proportional band represents the percent change in the process variable (PV) required to change the controller output 100%. For example, a controller gain of 1 ($K = 1.0$) requires a PV change of 100% to obtain a 100% change in controller output. Proportional bands for combustion process variables are generally in the range of 1000% to 20% (Gain = 0.1 to 5.0). Flow controller gains are always less than unity. Temperature controller gains vary from 0.1 to 3.0, depending on the process gain. Pressure controllers generally have gains higher than those of flow or temperature controllers. Controller output becomes unstable (oscillatory) when the gain is too high. When instability occurs, the controller operating mode must be changed from automatic to manual to stabilize the process and prevent equipment damage. A reduction of controller gain must occur before a return to automatic mode.

Automatic reset was the next improvement to the process controller. This was a most welcome addition that eliminated the need for manual reset. Automatic reset is a time integral of the error signal, summed with the proportional gain signal to produce the controller output. Integral gain (controller reset) is highest with large errors and continues until the error is reduced to zero. Figure 12.32 illustrates automatic reset.

Automatic reset is the "I" component of a PID (three-mode) controller. P is proportional gain and D is derivative (rate) gain. I is expressed as repeats per minute (RPM), or as minutes per repeat (MPR), depending on the controller manufacturer's choice of terms. Some controllers permit user selection of the term. Controller output is the same regardless of terminology, but the operator must know and apply proper tuning constants. For example, if tuning requires an I of 2 RPM, the operator must enter 0.5 MPR into the controller if MPR is the terminology in use. Integral gain of 0.5 MPR means that automatic reset equal to the proportional gain will be applied at the controller output each 30 seconds. Integral gain is a smooth continuous process that contributes phase lag to the system. Additional phase lag contributes to system instability (oscillation), which prohibits high values of integral gain.

Derivative gain D is a function of how fast the process variable is changing. For slow changing processes, derivative gain is of little use. Derivative gain is not used on flow control loops with head meters or on other loops with noisy process

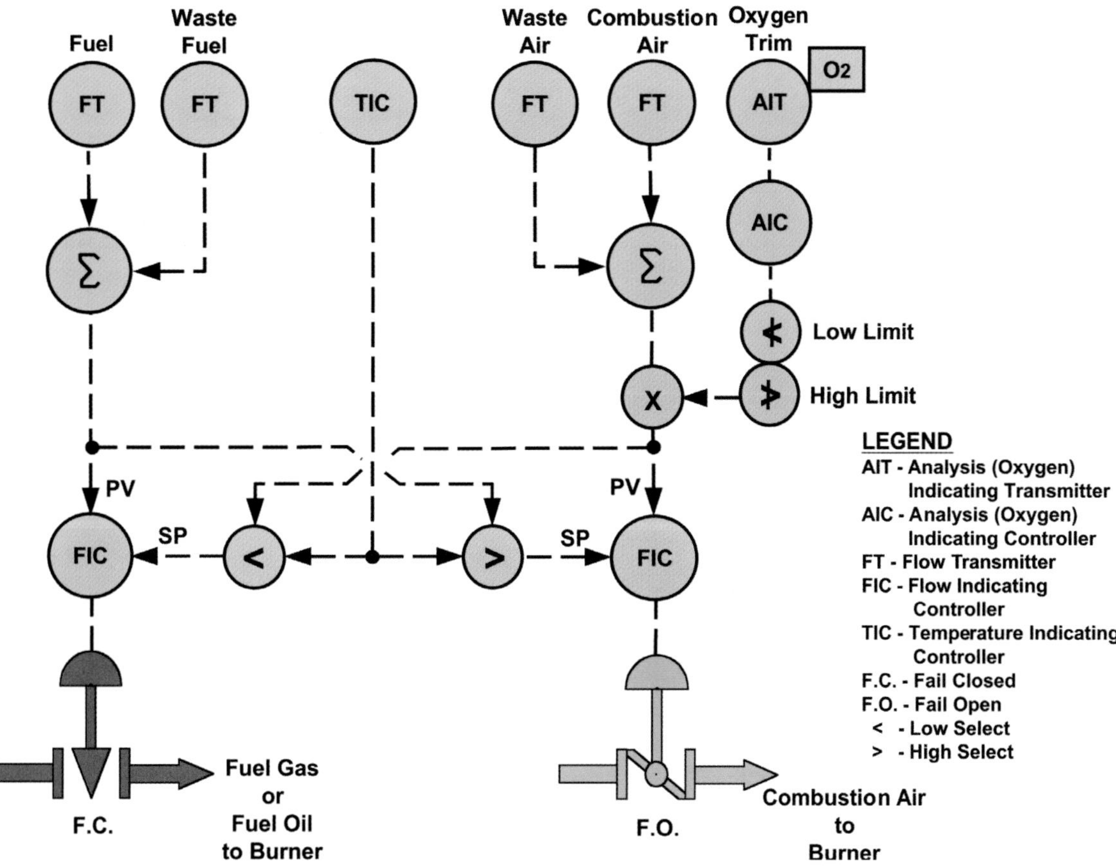

FIGURE 12.29 Multiple fuels and O_2 sources.

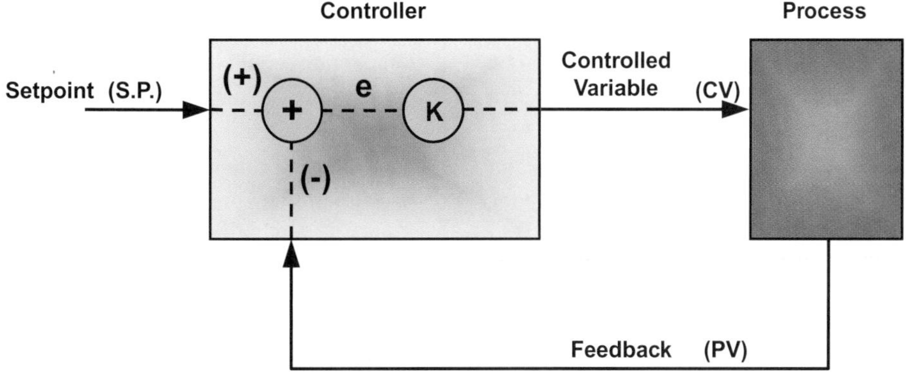

FIGURE 12.30 Controller.

variable signals. High noise levels will drive derivative gain to instability. Derivative gain contributes phase lead that can sometimes be beneficial.

Controllers have many modes of operation. P, I, D, automatic, and manual modes have been discussed. Reverse or direct mode is another choice that must be configured to match the process. Reverse or direct describes the change in direction of controller output when the process variable changes. Reverse acting means the controller output decreases if the process variable increases. An example illustrates how to select reverse or direct.

In this example of flow control, the process variable (flow rate) increases when the final element (control valve) opens. In addition, the flow meter output or process variable

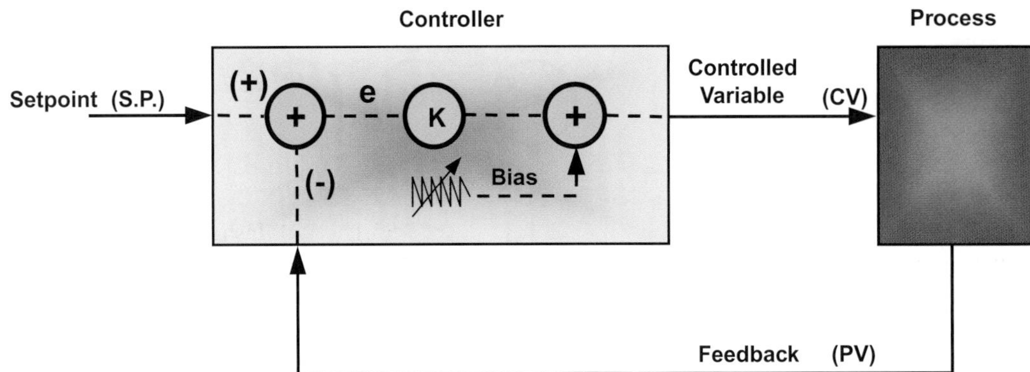

FIGURE 12.31 Analog controller with manual reset.

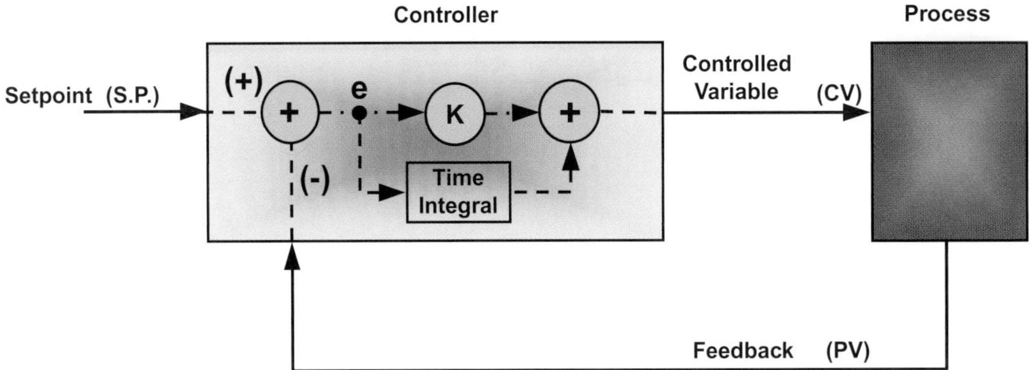

FIGURE 12.32 Analog controller with automatic reset.

(PV) increases with increased flow rate. If the control valve fails closed, and opens on increasing signal (increasing controller output), the controller mode must be reverse acting. That is, if PV increases, the controller output must decrease to close the valve and restore flow rate to the correct value. If the control valve fails open (closes on increasing signal), the controller mode must be direct acting. This example illustrates the need to know if each element in a control loop is reverse or direct acting, including transmitters, isolators, transducers (such as I/Ps), positioners, and actuators. Proper selection and configuration of loop elements provide not only proper operation, but also proper failure mode. Reversal of any two elements within a loop will not affect loop response, but failure modes will change.

12.5 TUNING

Many modern controllers have built-in automatic tuning routines. Tuning parameters calculated automatically require a loop upset to enable calculation. Parameters are normally detuned considerably from optimum because process gains are often nonlinear. Variable loop gain can also be a problem for manual tuning. A controller tuned at low flow rates or low temperature could become unstable at high flow rates or high temperatures. Control of most process loops benefits from addition of feedforward components that relieve the feedback controller of primary control. Operation improves if the feedback controller functions as setpoint injection and error trimming of the feedforward system.

Many processes controlled by a current proportional controller successfully use the tuning procedure below. The process must be upset to produce oscillations of the process variable. A graphic recorder should be used to determine when the oscillations are constant and to ascertain the time for one cycle (oscillation).

1. In manual mode, adjust the output to bring the process variable (PV) near the desired value.
2. Set the Rate Time to 0 minutes and set Reset Time to the maximum value (50.00 min), or set repeats per minute (RPM) to the minimum value to minimize reset action.
3. Increase gain (decrease proportional band PB) significantly. Try a factor of 10.

4. Adjust local setpoint to equal PV and switch to automatic mode.
5. Increase the setpoint by 5 or 10% and observe PV response.
6. If the process variable oscillates, determine the time for one oscillation. If it does not oscillate, return to the original setpoint, increase the gain again by a factor of 2, and repeat Step 5.
7. If the oscillation of Step 6 dampens before cycle time is measured, increase the gain slightly and try again. If the oscillation amplitude becomes excessive, decrease gain slightly and try again.
8. Record the current value of gain, and record the value for one completed oscillation of PV.
9. Calculate gain, reset, and rate:
   a. For PI (two-mode controller):
      Gain = Measured gain × 0.5
      Reset time = Measured time/1.2 (MPR)
   b. For PID (three-mode controller):
      Gain = Measured gain × 0.6
      Reset time = Measured time/2.0 (MPR)
      Rate = Measured time/8.0 (min)
10. Enter the values of gain, reset, and rate into the controller.
11. Make additional trimming adjustments, if necessary, to fine-tune the controller.
12. To reduce overshoot: less gain, perhaps a longer rate time.
13. To increase overshoot or increase speed of response: more gain, perhaps shorter rate time.

REFERENCES

1. J.O. Hougen, *Measurement and Control Applications for Practicing Engineers*, CAHNERS Books, Barnes & Noble Series for Professional Development, 1972.
2. *Combustion Control*, 9ATM1, Fisher Controls, Marshalltown, IA, 1976.
3. Boiler Control, Application Data Sheet 3028, Rosemount, Inc., Minneapolis, MN, 1980.
4. Instrumentation Symbols and Identification, ANSI/ISA – S5.1 – 1984, Instrument Society of America, Research Triangle Park, NC, 1984.
5. Temperature Measurement Thermocouples, ANSI – MC96.1 – 1984, Instrument Society of America, Research Triangle Park, NC, 1984.
6. F.G. Shinskey, *Process Control Systems, Application, Design, and Tuning*, 3rd ed., McGraw-Hill, New York, 1988.
7. M.J.G. Polonyi, PID controller tuning using standard form optimization, *Control Engineering*, March, 102-106, 1989.
8. D.W. St. Clair, Improving control loop performance, *Control Engineering*, Oct., 141-143, 1991.
9. Controller Tuning, Section 11, UDC 3300 Digital Controller Product Manual, Honeywell Industrial Automation, Fort Washington, PA, 1992.
10. F.Y. Thomasson, Five steps to better PID control, *CONTROL*, April, 65-67, 1995.
11. API Recommended Practice 556: Instrumentation and Control Systems for Fired Heaters and Steam Generators, 1st ed., American Petroleum Institute, Washington, D.C., May 1997.

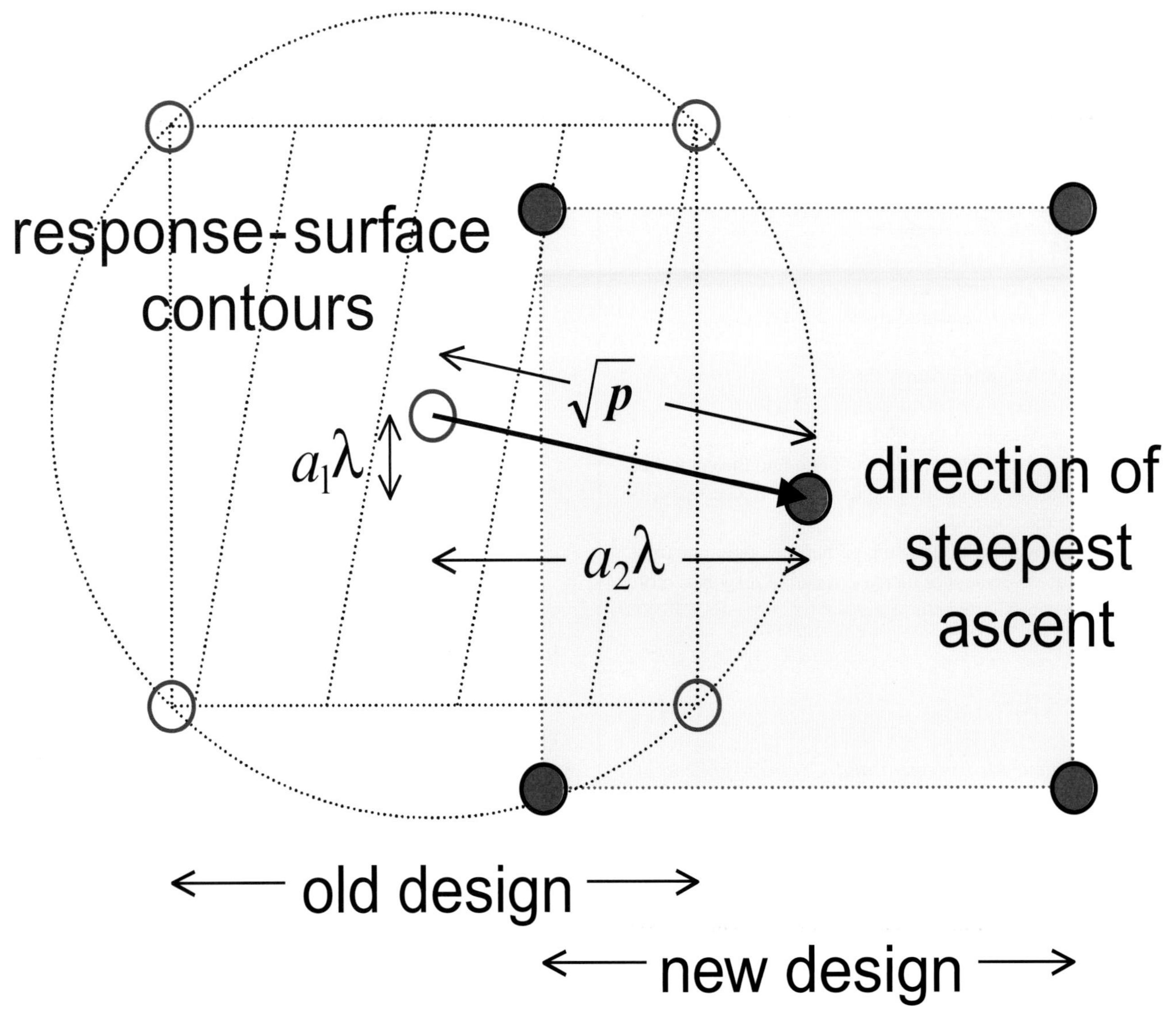
response-surface
contours
$\sqrt{p}$
$a_1\lambda$
direction of
steepest
ascent
$a_2\lambda$
old design
new design

Chapter 13

Experimental Design for Combustion Equipment

Joseph Colannino

TABLE OF CONTENTS

13.1 Introduction to Experimental Design 402

13.1.1 The Power of SED: A Burner NOx Example 402

13.1.2 Statistical Experimental Design Principles 404

13.1.3 The Method of Least Squares 405

13.1.4 Matrix Solution 406

13.1.5 Linear Transformations 407

13.2 Important Statistics 408

13.2.1 The Analysis of Variance (ANOVA) 409

13.2.2 The F-distribution 409

13.2.3 ANOVA with Separate Model Effects 409

13.2.4 Pooling Insignificant Effects 411

13.2.5 Standard Errors of Effects 412

13.3 Two-level Factorial Designs 412

13.3.1 Interactions 412

13.3.2 Pure Error and Bias 412

13.3.3 Two-level Fractional Factorials 414

13.3.4 Screening Designs 415

13.3.5 Method of Steepest Ascent 416

13.3.6 Serial Correlation and Lurking Factors 416

13.3.7 Foldover 416

13.3.8 Orthogonal Blocking 417

13.3.9 Including Categorical Factors 418

13.4 Second-order Designs 419

13.4.1 Central-composite Designs 419

13.4.2 Practical Considerations 420

13.5 Accounting for Fuel Mixtures 420
13.5.1 Experimental Designs for Mixtures 421
13.5.2 Orthogonal Mixture Designs 423
13.5.3 Combining Mixture and Factorial Designs 423
13.5.4 Building 2^n-level Factorials from Two-level Factorials 424
13.6 Combining Domain Knowledge with SED 424
13.6.1 Practical Considerations 425
13.6.2 Semi-empirical Models 425
13.6.3 Sequential Experimental Strategies 426
13.7 Linear Algebra Primer 427
13.7.1 Taylor and MacLaurin Series Approximations 427
13.7.2 Matrix Multiplication 428
13.7.3 Identity, Inverse, and Transpose 428
13.7.4 Matrix Addition 429
References 429

13.1 INTRODUCTION TO EXPERIMENTAL DESIGN

Statistical experimental design (SED) is a method for constructing experiments that will mute the muddling effect of experimental error and increase experimental efficiency. SED leads to better and less-expensive data analysis. It is far superior to the classical one-factor-at-a-time experimentation often taught in school. Classical experimentation cannot account for variable interactions. Classical experimentation often leads to false conclusions, for example, that one has arrived at an optimal place, when in fact one has not. Figure 13.1 contrasts classical and SED methods. SED is a powerful tool used by too few engineers. This disuse is due to several factors. First, engineers and scientists can successfully (though not as efficiently) experiment without SED. SED is a power tool. In the hands of a skilled practitioner, it reduces the time for experimentation and squeezes the most information from the data.

Second, SED vocabulary contains alien terms because statisticians first applied the methods to agricultural problems. Terms like "treatment" and "block" have obvious meanings in agriculture and obscure meanings to the practicing engineer. Engineers prefer terms like variable and experimental series. This chapter is written from an engineering perspective, rather than a statistical perspective. However, the bulk of SED knowledge is in the statistical literature. Therefore, this chapter contains a judicious choice of vocabulary to allow the interested reader to consult statistical treatises for further reference.

Third, statisticians rather than engineers write most SED texts. Therefore, SED texts usually do not incorporate the domain knowledge that engineers find so indispensable. Domain knowledge is a specialized understanding in a nonstatistical branch of science or engineering — for example, NOx formation or combustion fundamentals. This chapter provides a cursory overview of SED related to the performance characterization of burners and combustion equipment.

13.1.1 The Power of SED: A Burner NOx Example

Consider a manufacturer that makes many different burners. To compete, the company must make certain performance guarantees. These guarantees could concern heat release, turndown, flame length, heat flux profile, or combustion-related pollutants like NOx and CO. Uncertainty in any of these areas may force the company to decline to bid or to increase the bid price to cover the risk of redesign. Neither of these alternatives is attractive. Declining to bid surrenders the job to a competitor. Inflating the price to cover risk will make the burner less competitive. Even if the company wins the job, it may spend more than it anticipated achieving the performance it guaranteed. Thus, performance uncertainty translates directly to lost profit and lost opportunity.

Suppose a burner generates too much NOx. Many factors influence NOx response; Table 13.1 provides a partial list. The terms "response" and "factor" are used in a very specific

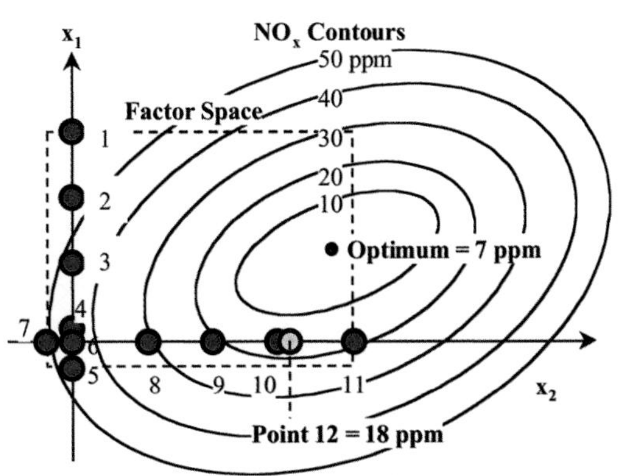

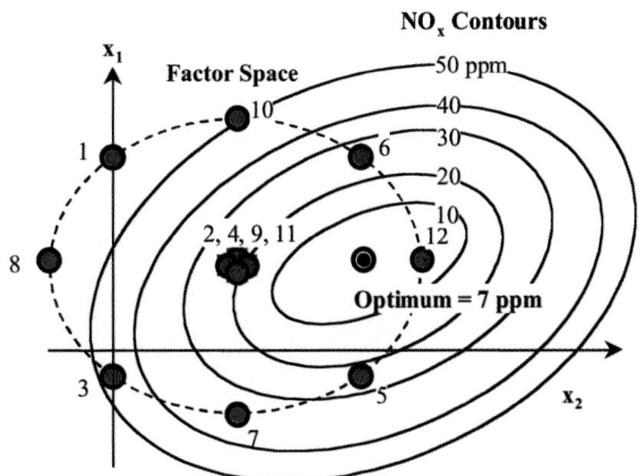

Classical (one-factor-at-a-time)	SED
General Procedure 1. Run a series of experimental points in sequential increments holding all other factors constant. 2. Search for the "optimum" along that factor. 3. Hold the factor constant at its "optimum" and repeat Steps 1 and 2 for the next factor. 4. Continue in this way for all factors, until arriving at the "optimal" point. **Specific Example** NO_X (contours shown above) is a function of two factors (x_1 and x_2) in the above hypothetical example (perhaps firing rate and excess oxygen, or whatever). The investigator seeks to minimize NO_X, stepping along x_1. The NO_X continues to decrease along x_1 until reaching point 5. So, the investigator backtracks and arrives at an "optimum" (~45ppm, point 6). The investigator holds x_1 at its "optimum" level and proceeds to investigate along x_2. In like manner, the investigator arrives at point 12 and declares the point "optimal." **General Considerations** 1. The investigator is unaware that the "optimum" is specious – more improvement is available. 2. The design does not account for interaction between factors. 3. The unbalanced design poorly maps the region: classical designs require more points than SED designs to generate the same information. This is especially so as the number of factors increase.	**General Procedure** 1. Design a balanced experimental series using SED principles. 2. Run the experimental points in random order. 3. Construct an interpolating function and perform various statistical tests. 4. To arrive at an optimum, use validated SED tools (e.g., interpolation, method of steepest ascent, sequential design, etc.). **Specific Example** The investigator begins with a factorial design (points 1-6). SED uses the centerpoints (2, 4) to test (and reveal) curvature. So, the investigator adds complementary points (7-12) forming a central composite design. This design gives a full second-order interpolating polynomial for the region and properly estimates the optimum. **General Considerations** 1. The investigator better estimates the true optimum in the experimental space. 2. The design gives a second-order polynomial that accounts for curvature and interactions between the factors. 3. The balanced design gives good estimates over the entire experimental region, not just near the factor axes. 4. In general, SED gives better information and more statistically valid models than classical designs.

FIGURE 13.1 Contrast of classical experimentation and SED methods.

sense. The response is any dependent variable or output of interest. A factor is any input or independent variable that affects the response. A given system of interest can have many factors and associated responses.

Specifying a response and its factors is a necessary first step, but it is not sufficient to solve the problem. The interaction of the factors must be known. What is the form of the model? What explicit equation should be used? What are the coefficient values? Proceeding from a purely theoretical basis, it would not be possible to arrive at a reliable explicit formulation for NOx performance. However, using statistical methods, the solution is quite tractable. Equation (13.1) gives the

TABLE 13.1 Some Potential Factors Affecting NOx Response from a Burner

Operating Factors (Constrained by Process)	Burner Factors (Controlled by Burner Manufacturer)	Furnace Factors (Controlled by Heater Design)	Ambient Factors (Uncontrollable)
Degree of air preheat	Burner throat diameter	Available air-side pressure drop	Ambient humidity
Firing rate	Degree of air staging	Burner-to-burner spacing	Ambient air temperature
Fuel composition	Degree of fuel staging	Burner-to-furnace wall spacing	Barometric pressure
Furnace temperature	Fuel port arrangement	Heat release/furnace vol. ratio	
Fuel pressure	Fuel port diameter		
Oxygen concentration	Multiple combustion zones		
Process fluid flow rate	Number of fuel ports		

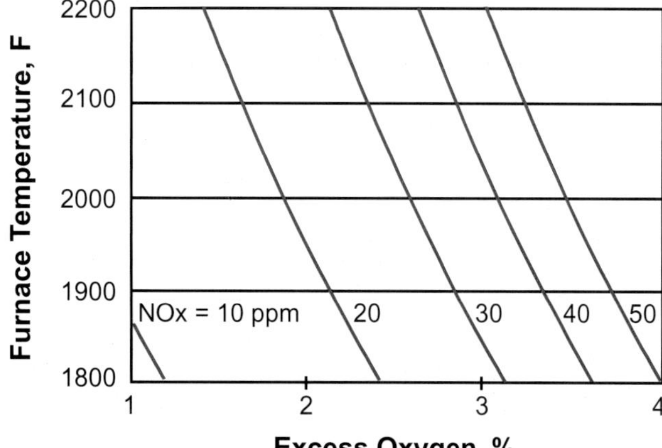

FIGURE 13.2 NOx contours for furnace temperature and oxygen concentration based on Eq. (13.1).

relation for a family of burners manufactured by John Zink, and Figure 13.2 shows a graphical representation for two of the factors, furnace temperature and excess oxygen.

$$\begin{aligned} y = 3.062[\pm 0.10] + 0.107[\pm 0.011]x_1 \\ -0.096[\pm 0.011]x_2 - 0.142[\pm 0.020]x_3 \\ +0.576[\pm 0.018]x_4 + 0.083[\pm 0.014]x_5 \\ +0.067[\pm 0.011]x_6 + 0.057[\pm 0.013]x_7 \\ +0.027[\pm 0.013]x_8 \end{aligned} \qquad (13.1)$$

where y is the natural log of the NOx mole fraction (ppm). The values in brackets [ ] are the standard errors associated with each coefficient, which are described in Section 13.2. The factors have the following associations.

x_1: the reciprocal of the burner spacing
x_2: the reciprocal preheat temperature
x_3: the reciprocal furnace temperature
x_4: the oxygen concentration
x_5: the heat release density
x_6: the hydrogen mole fraction in the fuel
x_7: the mole fraction of alkenes (olefins)
x_8: the mole fraction of non-methane alkanes (saturates).

Equation (13.1) reveals:

1. an explicit functional relationship of the response to the given factors
2. the basis for derivative figures and graphs such as Figure 13.2
3. the statistical significance of each coefficient (using the standard errors)
4. safety margin for NOx guarantees
5. an equation for feedforward NOx control[1]

One can use this knowledge to derive a competitive advantage or control process units with simultaneous constraints for product and performance.

13.1.2 Statistical Experimental Design Principles

A team of engineers proposes to correlate NOx data collected from the operation of a fuel-staged burner. Fuel staging segregates the combustion into two or more distinct zones and lowers NOx. Figure 13.3 shows a fuel-staged burner. Based on experience, the team proposes the model of Eq. (13.2).

$$y = a_0 + a_1\xi_1 + a_2\xi_2 + a_3\xi_3 + a_{12}\xi_1\xi_2 + \varepsilon \qquad (13.2)$$

where y is the NOx response (ppm)
$a_0 - a_{12}$ are the coefficients
ξ_1 is the heat release (10^6 Btu/hr)

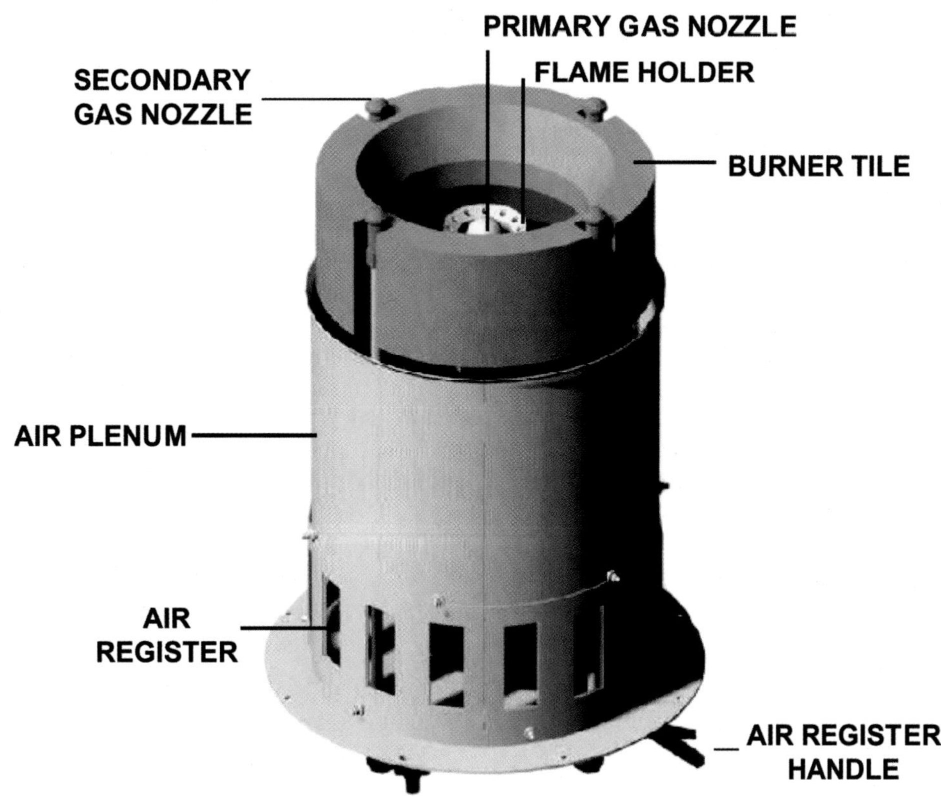
PRIMARY GAS NOZZLE
FLAME HOLDER
SECONDARY GAS NOZZLE
BURNER TILE
AIR PLENUM
AIR REGISTER
AIR REGISTER HANDLE
Model PSFG Staged Fuel burner, manufactured by John Zink, Tulsa, Oklahoma. The fuel first combusts in the primary zone. Secondary nozzles add fuel at a higher elevation. Fuel staging lowers NO_x by diluting fuel concentrations and staging fuel over a longer pathway to reduce peak flame temperatures.

FIGURE 13.3 Fuel-staged burner.

ξ_2 is the oxygen concentration (%)

ξ_3 is the percent of fuel to the primary combustion zone (%)

$\xi_1\xi_2$ is the interaction between the heat release and the oxygen (% × 10^6 Btu/hr)

ε is the experimental error (ppm)

Interaction accounts for synergy between factors — something classical experimentation cannot do. In the present case, the investigators proposed Eq. (13.2) based on their experience. However, with SED, one arrives at the form directly, as Section 13.2.3 will demonstrate.

13.1.3 The Method of Least Squares

Table 13.2 gives two replicates of the NOx response for nominal values of each factor. The replicates differ presumably by experimental error. The value of the coefficients is found using the technique of least squares. The technique of least squares calculates coefficients such that they minimize the sum of the squared deviations from the presumed model. The model need not be linear and may contain quadratics, transcendental functions, etc. Table 13.2 comprises eight experimental conditions and 16 NOx values. Equation (13.3) explicitly indexes each response.

TABLE 13.2 NOx as a Function of Burner Geometry and Operation

Run	NOx (ppm)	Firing Rate, [10^6 Btu/hr (ξ_1)]	Oxygen, Conc., [% (ξ_2)]	Fuel to Primary, [% (ξ_3)]
1	13, 14	7	1	20
2	18, 19	13	1	20
3	27, 24	7	5	20
4	26, 24	13	5	20
5	18, 19	7	1	50
6	21, 22	13	1	50
7	28, 29	7	5	50
8	29, 27	13	5	50

$$\begin{pmatrix} y_1 \\ y_2 \\ \vdots \\ y_{16} \end{pmatrix} = \begin{pmatrix} a_0 + a_1\xi_{1,1} + a_2\xi_{2,1} + a_3\xi_{3,1} + a_{12}\xi_{1,1}\xi_{2,1} \\ a_0 + a_1\xi_{1,2} + a_2\xi_{2,2} + a_3\xi_{3,2} + a_{12}\xi_{1,1}\xi_{2,2} \\ \vdots \\ a_0 + a_1\xi_{1,16} + a_2\xi_{2,16} + a_3\xi_{3,16} + a_{12}\xi_{1,16}\xi_{2,16} \end{pmatrix} + \begin{pmatrix} \varepsilon_1 \\ \varepsilon_2 \\ \vdots \\ \varepsilon_{16} \end{pmatrix} \quad (13.3)$$

Although there are 16 observations of the response, only five coefficients — a_0, a_1, a_2, a_3, and a_{12} — are calculated. Equation (13.4) is expressed in matrix form. For a brief review of matrices, see Section 13.7.

$$\begin{pmatrix} y_1 \\ y_2 \\ \vdots \\ y_{16} \end{pmatrix} = \begin{pmatrix} 1 & \xi_{1,1} & \xi_{2,1} & \xi_{3,1} & \xi_{1,1}\xi_{2,1} \\ 1 & \xi_{1,2} & \xi_{2,2} & \xi_{3,2} & \xi_{1,2}\xi_{2,2} \\ \vdots & \vdots & \vdots & \vdots & \vdots \\ 1 & \xi_{1,16} & \xi_{2,16} & \xi_{3,16} & \xi_{1,16}\xi_{2,16} \end{pmatrix} \begin{pmatrix} a_0 \\ a_1 \\ a_2 \\ a_3 \\ a_{12} \end{pmatrix} + \begin{pmatrix} \varepsilon_1 \\ \varepsilon_2 \\ \vdots \\ \varepsilon_{16} \end{pmatrix} \quad (13.4)$$

To obtain a compact notation, denote the response vector by $\underline{y}$, the test matrix by $\underline{\underline{\xi}}$, the coefficient vector by $\underline{a}$, and the error vector by $\underline{\varepsilon}$. The single underline denotes a vector (single column), and a double underline denotes a matrix comprising any number of rows and columns.

$$\underline{y} = \underline{\underline{\xi}}\,\underline{a} + \underline{\varepsilon} \quad (13.5)$$

Equations (13.2 through 13.5) represent identical models. The various forms are used as needed. With the method of least squares, the intent is to drive the sum of the squared errors as close to zero as possible:

$$\sum_{i=1}^{n=16} \varepsilon_i^2 = \sum_{i=1}^{n=16} \left[y_i - \left(a_0 + a_1\xi_{1,i} + a_2\xi_{2,i} + a_3\xi_{3,i} + a_{12}\xi_{1,i}\xi_{2,i} \right) \right]^2 \to 0 \quad (13.6)$$

Equating first derivatives to zero gives the minimum. For simplicity, drop the indices in Eqs. (13.7) through (13.11):

$$\frac{\partial}{\partial a_0} \sum \left[y - \left(a_0 + a_1\xi_1 + a_2\xi_2 + a_3\xi_3 + a_{12}\xi_1\xi_2 \right) \right]^2 = 0 \quad (13.7)$$

$$\frac{\partial}{\partial a_1} \sum \left[y - \left(a_0 + a_1\xi_1 + a_2\xi_2 + a_3\xi_3 + a_{12}\xi_1\xi_2 \right) \right]^2 = 0 \quad (13.8)$$

$$\frac{\partial}{\partial a_2} \sum \left[y - \left(a_0 + a_1\xi_1 + a_2\xi_2 + a_3\xi_3 + a_{12}\xi_1\xi_2 \right) \right]^2 = 0 \quad (13.9)$$

$$\frac{\partial}{\partial a_3} \sum \left[y - \left(a_0 + a_1\xi_1 + a_2\xi_2 + a_3\xi_3 + a_{12}\xi_1\xi_2 \right) \right]^2 = 0 \quad (13.10)$$

$$\frac{\partial}{\partial a_{12}} \sum \left[y - \left(a_0 + a_1\xi_1 + a_2\xi_2 + a_3\xi_3 + a_{12}\xi_1\xi_2 \right) \right]^2 = 0 \quad (13.11)$$

Equations (13.7) through (13.11) reduce to five simultaneous equations, one for each coefficient. Equation (13.12) expresses them in a single (symmetrical) matrix form:

$$\begin{pmatrix} \sum y \\ \sum \xi_1 y \\ \sum \xi_2 y \\ \sum \xi_3 y \\ \sum \xi_1\xi_2 y \end{pmatrix} = \begin{pmatrix} n & \sum \xi_1 & \sum \xi_2 & \sum \xi_3 & \sum \xi_1\xi_2 \\ \sum \xi_1 & \sum \xi_1^2 & \sum \xi_1\xi_2 & \sum \xi_1\xi_3 & \sum \xi_1^2\xi_2 \\ \sum \xi_2 & \sum \xi_1\xi_2 & \sum \xi_2^2 & \sum \xi_2\xi_3 & \sum \xi_1\xi_2^2 \\ \sum \xi_3 & \sum \xi_1\xi_3 & \sum \xi_2\xi_3 & \sum \xi_3^2 & \sum \xi_1\xi_2\xi_3 \\ \sum \xi_1\xi_2 & \sum \xi_1^2\xi_2 & \sum \xi_1\xi_2^2 & \sum \xi_1\xi_2\xi_3 & \sum \xi_1^2\xi_2^2 \end{pmatrix} \begin{pmatrix} a_0 \\ a_1 \\ a_2 \\ a_3 \\ a_4 \end{pmatrix} \quad (13.12)$$

Applying the data of Table 13.2 to Eq. (13.12) gives Eq. (13.13):

$$\begin{pmatrix} 358 \\ 3622 \\ 1214 \\ 12950 \\ 12158 \end{pmatrix} = \begin{pmatrix} 16 & 160 & 48 & 560 & 480 \\ 160 & 1744 & 480 & 5600 & 5232 \\ 48 & 480 & 208 & 1680 & 2080 \\ 560 & 5600 & 1680 & 23200 & 16800 \\ 480 & 5232 & 2080 & 16800 & 22672 \end{pmatrix} \begin{pmatrix} a_0 \\ a_1 \\ a_2 \\ a_3 \\ a_{12} \end{pmatrix} \quad (13.13)$$

Equation (13.13) yields its solution by the usual rules of matrix arithmetic (see next section):

$$\begin{pmatrix} a_0 \\ a_1 \\ a_2 \\ a_3 \\ a_{12} \end{pmatrix} = \begin{pmatrix} 3.19 \\ 0.85 \\ 4.06 \\ 0.12 \\ -0.19 \end{pmatrix} \quad (13.14)$$

13.1.4 Matrix Solution

Consider once again Eq. (13.5). Multiply both sides of the equation by the transpose of the test matrix $\left(\underline{\underline{\xi}}^T\right)$ (see Section 13.7.3) to obtain Eq. (13.15):

$$\underline{\underline{\xi}}^T \underline{y} = \underline{\underline{\xi}}^T \underline{\underline{\xi}}\,\underline{a} + \left(\underline{\underline{\xi}}^T \underline{\varepsilon} = 0 \right) \quad (13.15)$$

The reader can verify that Eq. (13.15) and Eq. (13.12) are the same least-squares matrix. Thus, Eq. (13.15) provides a shortcut for obtaining the least-squares equations without the need for differential calculus. Premultiplying by the inverse matrix $\left(\underline{\underline{\xi}}^T\underline{\underline{\xi}}\right)^{-1}$ yields the solution for the coefficient vector:

$$\left(\underline{\underline{\xi}}^T\underline{\underline{\xi}}\right)^{-1}\underline{\underline{\xi}}^T\underline{y}=\underline{a} \qquad (13.16)$$

Equations (13.15) and (13.16) are general solutions. This result is quite important because most computer spreadsheets can perform the transpose, matrix multiply, and matrix invert operations.

But what do the coefficients mean? Because they come from simultaneous equations, it is impossible to gauge the effect of a single coefficient viewed alone. Their magnitudes cannot be compared because the associated factors have different ranges and units. Moreover, if the model is modified, all coefficients must be recalculated because Eq. (13.12) is a system of simultaneous equations. Some method of scaling the factors and transforming the equations to an independent system is desirable.

13.1.5 Linear Transformations

Transforming the factors to a common dimensionless unit scale allows a direct comparison of the coefficients and simplifies the analysis. Linear transforms are preferred because they do not bend or nonuniformly stretch the data and they are easy to invert. The following transformation satisfies the conditions:

$$x_i=\frac{\xi_i-\overline{\xi}_i}{\frac{1}{2}\left(\xi_{i+}-\xi_{i-}\right)} \qquad (13.17)$$

where x_i is the ith variable, transformed to a dimensionless range of ±1

ξ_i is the ith untransformed variable in the conventional metric (e.g., Btu/hr, %, etc.)

$\overline{\xi}_i$ is the mean value of ξ_i in the conventional metric

ξ_{i+} is the maximum value of ξ_i in the conventional metric

ξ_{i-} is the minimum value for ξ_i in the conventional metric

This transform normalizes ξ_i to x_i, which is dimensionless and spans the range –1 to +1. Table 13.3 gives the transforms, and Table 13.4 shows the transformed factors. Since we understand they are unit values, only their sign need be shown. With these transforms, the off-diagonal values in Eq. (13.12) sum to zero. For the data of Table 13.2, the transformation leads to Eq. (13.18), a diagonal matrix. (For convenience, the zero elements are omitted.)

TABLE 13.3 Transforms for Table 13.4

i	Factor Description	Raw Value (ξ_i)	Transform (ξ_i to x_i)	Transformed (x_i)
1	Firing rate, 10^6 Btu/h	7 13	$x_1=\frac{\xi_1-10\times10^6\text{ Btu/h}}{3\times10^6\text{ Btu/h}}$	–1 +1
2	Oxygen concentration, %	1 5	$x_2=\frac{\xi_2-3\%}{2\%}$	–1 +1
3	Fuel to primary, %	20 50	$x_3=\frac{\xi_3-35\%}{15\%}$	–1 +1

TABLE 13.4 Transformed Data for Fuel-staged Burner

Run	x_1	x_2	x_3	y
1	–	–	–	13, 14
2	+	–	–	18, 19
3	–	+	–	27, 24
4	+	+	–	26, 24
5	–	–	+	18, 19
6	+	–	+	21, 22
7	–	+	+	28, 29
8	+	+	+	29, 27

$$\begin{pmatrix}358\\14\\70\\28\\-18\end{pmatrix}=\begin{pmatrix}16&&&&\\&16&&&\\&&16&&\\&&&16&\\&&&&16\end{pmatrix}\begin{pmatrix}a_0\\a_1\\a_2\\a_3\\a_{12}\end{pmatrix} \qquad (13.18)$$

If x^Tx generates a diagonal matrix, then x is orthogonal. Orthogonal matrices generate independent least-squares equations. The solution is as easy as dividing each element of $\underline{y}$ by each diagonal element of $\underline{\underline{x}}$:

$$\begin{pmatrix}a_0\\a_1\\a_2\\a_3\\a_{12}\end{pmatrix}=\begin{pmatrix}22.375\\0.875\\4.375\\1.750\\-1.125\end{pmatrix} \qquad (13.19)$$

Note also that removing any terms (e.g., eliminating the term a_1x_1) does not affect the value of the remaining coefficients. This is not so for the simultaneous system, Eqs. (13.12) to

(13.14). One can directly compare all the coefficients because they are dimensionless and have uniform range. The coefficients provide the following:

1. the average NOx is 22.375 ppm for this experimental series
2. the firing rate influences the NOx by ±0.875 ppm over its range
3. the oxygen influences NOx by ±4.375 ppm over its range
4. the percentage of fuel to the primary zone influences NOx by ±1.75 ppm over its range
5. firing rate and oxygen interact to moderate their effect on NOx by ∓1.125 ppm over their joint range

From an examination of the coefficients, oxygen has the greatest influence on NOx, followed by the percentage of fuel to the primary combustion zone, the interaction between firing rate and oxygen, and the firing rate. Actually, the coefficient for the firing rate seems small. Is it significant? How can one be sure that the effect is real? How does the coefficient compare to the experimental noise? The next section addresses these questions.

13.2 IMPORTANT STATISTICS

It is not enough to know the form of a model, or even calculate its coefficients explicitly. Some indication of the influence of background noise and its relationship to the coefficients is also necessary. The terms background noise, noise, error, experimental error, pure error, and random error are used interchangeably in this chapter. $\underline{\hat{y}}$ is the best approximation of the true but unknown model

$$\underline{\hat{y}} = \underline{\underline{\xi}}\,\underline{a} \tag{13.20}$$

The problem is to recover Eq. (13.20) from Eq. (13.5). The data contain n total observations and the model contains a coefficient vector, $\underline{a}$, comprising p coefficients. Substituting Eq. (13.20) into Eq. (13.5) and solving for $\underline{\varepsilon}$ yields Eq. (13.22):

$$\underline{\varepsilon} = \underline{y} - \underline{\hat{y}} \tag{13.21}$$

Equation (13.21) represents a vector of n values implicitly comprising p coefficients. A single number (statistic) to quantify the error vector is desired. Summing the vector $\left(\sum \underline{y} - \underline{\hat{y}}\right)$ is a logical place to start. However, if $\underline{\varepsilon}$ is truly random and unbiased, then $\sum \underline{y} - \underline{\hat{y}} \sim 0$. (For the present case, $\sum \underline{y} - \underline{\hat{y}} = \sum_{i=1}^{n}\left(y_i - \hat{y}_i\right)$. The simpler notation is used for convenience.)

The sum of the squares, $\sum\left(\underline{y} - \underline{\hat{y}}\right)^2$, has the advantage of a non-zero sum that increases as the error grows larger. However, $\sum\left(\underline{y} - \underline{\hat{y}}\right)^2$ should grow larger with n and smaller with p. If possible, the noise statistic should be independent of the number of observations or the number of parameters in the model. This reasoning leads to Eq. (13.22), called the mean square residual (MSR), as a logical measure of random error:

$$\text{MSR} = \frac{\sum\left(\underline{y} - \underline{\hat{y}}\right)^2}{n - p} \tag{13.22}$$

The divisor $(n - p)$ is called the *degrees of freedom*.

In the worst case, $\hat{y}$ is worthless and none of its factors actually influence the response. In this case, the data merely represent n replicates differing only by random error. The model is no better than the *mean* of all observations, $\bar{y}$, defined by Eq. (13.23).

$$\bar{y} = \frac{\sum \underline{y}}{n} \tag{13.23}$$

Equation (13.23) is the simplest possible model, having only a single parameter ($p = 1$); the parameter is $\bar{y}$, or a_0 for the orthogonal designs considered. In such a case, Eq. (13.22) reduces to Eq. (13.24), called the mean square total or MST:

$$\text{MST} = \frac{\sum\left(\underline{y} - \bar{y}\right)^2}{n - 1} \tag{13.24}$$

The squared deviation normalized by its degrees of freedom is the *variance*. The estimated mean and variance ($\bar{y}$ and s^2, respectively) are unbiased estimators for the actual mean and variance (μ and σ^2, respectively). In an experimental context, the actual mean and variance are usually unknown. One problem with s^2 is that it does not have the same units as y. The problem is remedied by taking the square root of Eq. (13.22) or Eq. (13.24).

TABLE 13.5 Generic ANOVA Table

Component	SS	DF	MS	F
Model (*M*)	$SSM = \sum(\hat{y}-\bar{y})^2$	DFM = p – 1	MSM = SSM/DFM	MSM/MSR
Residual (*R*)	$SSR = \sum(y_i-\hat{y})^2$	DFR = n – p	MSR = SSR/DFR	
Total (*T*)	$SST = \sum(y_i-\bar{y})^2$	DFT = n – 1	MST = SST/DFT	

$$s = \begin{cases} \pm\sqrt{MST} = \pm\sqrt{\dfrac{\sum(y-\bar{y})^2}{n-1}}, & \text{if } \hat{y} = \bar{y} \\ \pm\sqrt{MSR} = \pm\sqrt{\dfrac{\sum(y-\hat{y})^2}{n-p}}, & \text{if } \hat{y} \neq \bar{y} \end{cases} \tag{13.25}$$

In Eq. (13.25), s is the *standard deviation*, having the same units as the response. The hypothesis that $\hat{y} = \bar{y}$ is called the *null hypothesis* (H_0).

13.2.1 The Analysis of Variance (ANOVA)

Table 13.5 arranges the variances in a very convenient form, referred to as the ANOVA table. The table has the following column headings: SS is the sum of squares, DF is the degrees of freedom, MS is the mean square, and F is a special ratio described below. The row headings are M for model, R for residual, and T for total. The column and row headings are combined to obtain the cell heading. For example, the cell located at column SS and row M is SSM, standing for sum-of-squares, model.

The mean squares (MS) are calculated by dividing the sum of squares (SS) by the appropriate degrees of freedom (DF). One should divide SST by $(n - 1)$, the total number of data points, less one for the mean. SSM should be divided by DFM $= (p - 1)$, the number of coefficients in the model, less one for the mean. Finally, SSR should be divided by the remaining (or residual) degrees of freedom, DFR $= (n - p)$. Now the statistics needed to explore tests for significance are in place.

MSR estimates the noise. The ratio MSM/MSR is called F. If F ~ 1, then $\hat{y} = \bar{y}$ and the null hypothesis is accepted. If F >> 1, then H_0 is rejected and the model is significant. To find how large F must be to reject H_0, the properties of the F distribution must be known.

13.2.2 The F-distribution

Ratios of variance have an *F-distribution*. The distribution depends on three quantities:

1. the desired confidence in rejecting H_0
2. the degrees of freedom for the numerator (DFM in the present case)
3. the degrees of freedom for the denominator (DFR in the present case)

These quantities have the notation F_{conf}(DFM, DFR), where *conf* is the percent confidence. For example, $F_{95}(4,11) = 3.36$ means an F-distribution with DFM = 4 and DFR = 11 will reject the null hypothesis with at least 95% confidence if it is greater than or equal to 3.36. The source for the critical value of 3.36 is Table 13.6.

With this background, it is possible to test Model 13.2 (Eq. 13.2) for significance. Table 13.7 gives the ANOVA for the data of Table 13.4 and the transformed Eq. (13.2). It appears that the model is significant, accounting for ~50 times the variance of the residual error. The calculated F-ratio of 48.7 with 6 and 9 degrees of freedom is compared to $F_{95}(6, 9) = 3.37$. Because 48.7 is much greater than 3.37, the null hypothesis is rejected. There is greater than 95% confidence that the model is significant. In fact, the confidence level is greater than 99%, because $48.7 > F_{99}(6, 9) = 5.80$. It is also possible to estimate $\sigma^2 \approx MSR = 1.33 \left(\sigma \approx \pm\sqrt{MSR} = \pm 1.2 \text{ ppm } NO_x\right)$. However, although the model as a whole is significant, some of the individual terms of the model may not be significant. To see consider each term separately as presently described.

13.2.3 ANOVA with Separate Model Effects

A single ANOVA table comprising separate model effects is only possible with orthogonal data. This is another reason to use SED. To test nonorthogonal data, look at ANOVA tables for every possible model combination comprising one to four effects. Test each model in a separate ANOVA table. This requires 2^p (16) different ANOVA tables.

For orthogonal data, one can construct a single table, Table 13.8, comprising every possible factor and two-factor interaction. The ANOVA shows that only x_1, x_2, x_3, and x_1x_2 are significant. The significant entries are shown in bold type. In the previous section, $s = 1.2$ ppm was obtained. But if some

TABLE 13.6 F-Distribution, 99%, 95%, and 90% Confidence

		DF ν_1																
DF ν_2	**Conf**	**1**	**2**	**3**	**4**	**5**	**6**	**7**	**8**	**9**	**10**	**15**	**20**	**25**	**30**	**50**	**100**	**Infinity**
	99%	4052	4999	5404	5624	5764	5859	5928	5981	6022	6056	6157	6209	6240	6260	6302	6334	6366
1	95%	161	199	216	225	230	234	237	239	241	242	246	248	249	250	252	253	254
	90%	39.86	49.50	53.59	55.83	57.24	58.20	58.91	59.44	59.86	60.19	61.22	61.74	62.05	62.26	62.69	63.01	63.33
	99%	98.50	99.00	99.16	99.25	99.30	99.33	99.36	99.38	99.39	99.40	99.43	99.45	99.46	99.47	99.48	99.49	99.50
2	95%	18.51	19.00	19.16	19.25	19.30	19.33	19.35	19.37	19.38	19.40	19.43	19.45	19.46	19.46	19.48	19.49	19.50
	90%	8.53	9.00	9.16	9.24	9.29	9.33	9.35	9.37	9.38	9.39	9.42	9.44	9.45	9.46	9.47	9.48	9.49
	99%	34.12	30.82	29.46	28.71	28.24	27.91	27.67	27.49	27.34	27.23	26.87	26.69	26.58	26.50	26.35	26.24	26.13
3	95%	10.13	9.55	9.28	9.12	9.01	8.94	8.89	8.85	8.81	8.79	8.70	8.66	8.63	8.62	8.58	8.55	8.53
	90%	5.54	5.46	5.39	5.34	5.31	5.28	5.27	5.25	5.24	5.23	5.20	5.18	5.17	5.17	5.15	5.14	5.13
	99%	21.20	18.00	16.69	15.98	15.52	15.21	14.98	14.80	14.66	14.55	14.20	14.02	13.91	13.84	13.69	13.58	13.46
4	95%	7.71	6.94	6.59	6.39	6.26	6.16	6.09	6.04	6.00	5.96	5.86	5.80	5.77	5.75	5.70	5.66	5.63
	90%	4.54	4.32	4.19	4.11	4.05	4.01	3.98	3.95	3.94	3.92	3.87	3.84	3.83	3.82	3.80	3.78	3.76
	99%	16.26	13.27	12.06	11.39	10.97	10.67	10.46	10.29	10.16	10.05	9.72	9.55	9.45	9.38	9.24	9.13	9.02
5	95%	6.61	5.79	5.41	5.19	5.05	4.95	4.88	4.82	4.77	4.74	4.62	4.56	4.52	4.50	4.44	4.41	4.36
	90%	4.06	3.78	3.62	3.52	3.45	3.40	3.37	3.34	3.32	3.30	3.24	3.21	3.19	3.17	3.15	3.13	3.10
	99%	13.75	10.92	9.78	9.15	8.75	8.47	8.26	8.10	7.98	7.87	7.56	7.40	7.30	7.23	7.09	6.99	6.88
6	95%	5.99	5.14	4.76	4.53	4.39	4.28	4.21	4.15	4.10	4.06	3.94	3.87	3.83	3.81	3.75	3.71	3.67
	90%	3.78	3.46	3.29	3.18	3.11	3.05	3.01	2.98	2.96	2.94	2.87	2.84	2.81	2.80	2.77	2.75	2.72
	99%	12.25	9.55	8.45	7.85	7.46	7.19	6.99	6.84	6.72	6.62	6.31	6.16	6.06	5.99	5.86	5.75	5.65
7	95%	5.59	4.74	4.35	4.12	3.97	3.87	3.79	3.73	3.68	3.64	3.51	3.44	3.40	3.38	3.32	3.27	3.23
	90%	3.59	3.26	3.07	2.96	2.88	2.83	2.78	2.75	2.72	2.70	2.63	2.59	2.57	2.56	2.52	2.50	2.47
	99%	11.26	8.65	7.59	7.01	6.33	6.37	6.18	6.03	5.91	5.81	5.52	5.36	5.26	5.20	5.07	4.96	4.86
8	95%	5.32	4.46	4.07	3.84	3.69	3.58	3.50	3.44	3.39	3.35	3.22	3.15	3.11	3.08	3.02	2.97	2.93
	90%	3.46	3.11	2.92	2.81	2.73	2.67	2.62	2.59	2.56	2.54	2.46	2.42	2.40	2.38	2.35	2.32	2.29
	99%	10.56	8.02	6.99	6.42	6.06	5.80	5.61	5.47	5.35	5.26	4.96	4.81	4.71	4.65	4.52	4.41	4.31
9	95%	5.12	4.26	3.86	3.63	3.48	3.37	3.29	3.23	3.18	3.14	3.01	2.94	2.89	2.86	2.80	2.76	2.71
	90%	3.36	3.01	2.81	2.69	2.61	2.55	2.51	2.47	2.44	2.42	2.34	2.30	2.27	2.25	2.22	2.19	2.16
	99%	10.04	7.56	6.55	5.99	5.64	5.39	5.20	5.06	4.94	4.85	4.56	4.41	4.31	4.25	4.12	4.01	3.91
10	95%	4.96	4.10	3.71	3.48	3.33	3.22	3.14	3.07	3.02	2.98	2.85	2.77	2.73	2.70	2.64	2.59	2.54
	90%	3.29	2.92	2.73	2.61	2.52	2.46	2.41	2.38	2.35	2.32	2.24	2.20	2.17	2.16	2.12	2.09	2.06
	99%	9.65	7.21	6.22	5.67	5.32	5.07	4.89	4.74	4.63	4.54	4.25	4.10	4.01	3.94	3.81	3.71	3.60
11	95%	4.84	3.98	3.59	3.36	3.20	3.09	3.01	2.95	2.90	2.85	2.72	2.65	2.60	2.57	2.51	2.46	2.40
	90%	3.23	2.86	2.66	2.54	2.45	2.39	2.34	2.30	2.27	2.25	2.17	2.12	2.10	2.08	2.04	2.01	1.97
	99%	9.33	6.93	5.95	5.41	5.06	4.82	4.64	4.50	4.39	4.30	4.01	3.86	3.76	3.70	3.57	3.47	3.36
12	95%	4.75	3.89	3.49	3.26	3.11	3.00	2.91	2.85	2.80	2.75	2.62	2.54	2.50	2.47	2.40	2.35	2.30
	90%	3.18	2.81	2.61	2.48	2.39	2.33	2.28	2.24	2.21	2.19	2.10	2.06	2.03	2.01	1.97	1.94	1.90
	99%	9.07	6.70	5.74	5.21	4.86	4.62	4.44	4.30	4.19	4.10	3.82	3.66	3.57	3.51	3.38	3.27	3.17
13	95%	4.67	3.81	3.41	3.18	3.03	2.92	2.83	2.77	2.71	2.67	2.53	2.46	2.41	2.38	2.31	2.26	2.21
	90%	3.14	2.76	2.56	2.43	2.35	2.28	2.23	2.20	2.16	2.14	2.05	2.01	1.98	1.96	1.92	1.88	1.85
	99%	8.86	6.51	5.56	5.04	4.69	4.46	4.28	4.14	4.03	3.94	3.66	3.51	3.41	3.35	3.22	3.11	3.00
14	95%	4.60	3.74	3.34	3.11	2.96	2.85	2.76	2.70	2.65	2.60	2.46	2.39	2.34	2.31	2.24	2.19	2.13
	90%	3.10	2.73	2.52	2.39	2.31	2.24	2.19	2.15	2.12	2.10	2.01	1.96	1.93	1.91	1.87	1.83	1.80
	99%	8.68	6.36	5.42	4.89	4.56	4.32	4.14	4.00	3.89	3.80	3.52	3.37	3.28	3.21	3.08	2.98	2.87
15	95%	4.54	3.68	3.29	3.06	2.90	2.79	2.71	2.64	2.59	2.54	2.40	2.33	2.28	2.25	2.18	2.12	2.07
	90%	3.07	2.70	2.49	2.36	2.27	2.21	2.16	2.12	2.09	2.06	1.97	1.92	1.89	1.87	1.83	1.79	1.76
	99%	8.53	6.23	5.29	4.77	4.44	4.20	4.03	3.89	3.78	3.69	3.41	3.26	3.16	3.10	2.97	2.86	2.75
16	95%	4.49	3.63	3.24	3.01	2.85	2.74	2.66	2.59	2.54	2.49	2.35	2.28	2.23	2.19	2.12	2.07	2.01
	90%	3.05	2.67	2.46	2.33	2.24	2.18	2.13	2.09	2.06	2.03	1.94	1.89	1.86	1.84	1.79	1.76	1.72

TABLE 13.6 (continued) F Distribution, 99%, 95%, and 90% Confidence

DF v_2	Conf	DF v_1: 1	2	3	4	5	6	7	8	9	10	15	20	25	30	50	100	Infinity
17	99%	8.40	6.11	5.19	4.67	4.34	4.10	3.93	3.79	3.68	3.59	3.31	3.16	3.07	3.00	2.87	2.76	2.65
	95%	4.45	3.59	3.20	2.96	2.81	2.70	2.61	2.55	2.49	2.45	2.31	2.23	2.18	2.15	2.08	2.02	1.96
	90%	3.03	2.64	2.44	2.31	2.22	2.15	2.10	2.06	2.03	2.00	1.91	1.86	1.83	1.81	1.76	1.73	1.69
18	99%	8.29	6.01	5.09	4.58	4.25	4.01	3.84	3.71	3.60	3.51	3.23	3.08	2.98	2.92	2.78	2.68	2.57
	95%	4.41	3.55	3.16	2.93	2.77	2.66	2.58	2.51	2.46	2.41	2.27	2.19	2.14	2.11	2.04	1.98	1.92
	90%	3.01	2.62	2.42	2.29	2.20	2.13	2.08	2.04	2.00	1.98	1.89	1.84	1.80	1.78	1.74	1.70	1.66
19	99%	8.18	5.93	5.01	4.50	4.17	3.94	3.77	3.63	3.52	3.43	3.15	3.00	2.91	2.84	2.71	2.60	2.49
	95%	4.38	3.52	3.13	2.90	2.74	2.63	2.54	2.48	2.42	2.38	2.23	2.16	2.11	2.07	2.00	1.94	1.88
	90%	2.99	2.61	2.40	2.27	2.18	2.11	2.06	2.02	1.98	1.96	1.86	1.81	1.78	1.76	1.71	1.67	1.63
20	99%	8.10	5.85	4.94	4.43	4.10	3.87	3.70	3.56	3.46	3.37	3.09	2.94	2.84	2.78	2.64	2.54	2.42
	95%	4.35	3.49	3.10	2.87	2.71	2.60	2.51	2.45	2.39	2.35	2.20	2.12	2.07	2.04	1.97	1.91	1.84
	90%	2.97	2.59	2.38	2.25	2.16	2.09	2.04	2.00	1.96	1.94	1.84	1.79	1.76	1.74	1.69	1.65	1.61
25	99%	7.77	5.57	4.68	4.18	3.85	3.63	3.46	3.32	3.22	3.13	2.85	2.70	2.60	2.54	2.40	2.29	2.17
	95%	4.24	3.39	2.99	2.76	2.60	2.49	2.40	2.34	2.28	2.24	2.09	2.01	1.96	1.92	1.84	1.78	1.71
	90%	2.92	2.53	2.32	2.18	2.09	2.02	1.97	1.93	1.89	1.87	1.77	1.72	1.68	1.66	1.61	1.56	1.52
30	99%	7.56	5.39	4.51	4.02	3.70	3.47	3.30	3.17	3.07	2.98	2.70	2.55	2.45	2.39	2.25	2.13	2.01
	95%	4.17	3.32	2.92	2.69	2.53	2.42	2.33	2.27	2.21	2.16	2.01	1.93	1.88	1.84	1.76	1.70	1.62
	90%	2.88	2.49	2.28	2.14	2.05	1.98	1.93	1.88	1.85	1.82	1.72	1.67	1.63	1.61	1.55	1.51	1.46
40	99%	7.31	5.18	4.31	3.83	3.51	3.29	3.12	2.99	2.89	2.80	2.52	2.37	2.27	2.20	2.06	1.94	1.80
	95%	4.08	3.23	2.84	2.61	2.45	2.34	2.25	2.18	2.12	2.08	1.92	1.84	1.78	1.74	1.66	1.59	1.51
	90%	2.84	2.44	2.23	2.09	2.00	1.93	1.87	1.83	1.79	1.76	1.66	1.61	1.57	1.54	1.48	1.43	1.38
50	99%	7.17	5.06	4.20	3.72	3.41	3.19	3.02	2.89	2.78	2.70	2.42	2.27	2.17	2.10	1.95	1.82	1.68
	95%	4.03	3.18	2.79	2.56	2.40	2.29	2.20	2.13	2.07	2.03	1.87	1.78	1.73	1.69	1.60	1.52	1.44
	90%	2.81	2.41	2.20	2.06	1.97	1.90	1.84	1.80	1.76	1.73	1.63	1.57	1.53	1.50	1.44	1.39	1.33
100	99%	6.90	4.82	3.98	3.51	3.21	2.99	2.82	2.69	2.59	2.50	2.22	2.07	1.97	1.89	1.74	1.60	1.43
	95%	3.94	3.09	2.70	2.46	2.31	2.19	2.10	2.03	1.97	1.93	1.77	1.68	1.62	1.57	1.48	1.39	1.28
	90%	2.76	2.36	2.14	2.00	1.91	1.83	1.78	1.73	1.69	1.66	1.56	1.49	1.45	1.42	1.35	1.29	1.21
Infinity	99%	6.63	4.61	3.78	3.32	3.02	2.80	2.64	2.51	2.41	2.32	2.04	1.88	1.77	1.70	1.52	1.36	1.00
	95%	3.84	3.00	2.60	2.37	2.21	2.10	2.01	1.94	1.88	1.83	1.67	1.57	1.51	1.46	1.35	1.24	1.00
	90%	2.71	2.30	2.08	1.94	1.85	1.77	1.72	1.67	1.63	1.60	1.49	1.42	1.38	1.34	1.26	1.18	1.00

TABLE 13.7 ANOVA Table for Equation 13.2 Applied to Data of Table 13.4

Component	SS	DF	MS	F	$F_{95}(6,9)$
Model (**M**)	389.75	6	64.95	48.7	3.37
Residual (**R**)	12.00	9	1.33		
Total (**T**)	401.75	15			

model effects are insignificant, it might be best to move them from the model to the residual and re-estimate σ with a greater DFR.

13.2.4 Pooling Insignificant Effects

To pool insignificant effects, their SS and DF are moved to the residual, generating Table 13.9. If the null hypothesis is true for these factors, then the MSR should not change significantly. In fact, as the table shows, s^2 remains nearly unchanged, decreasing from 1.33 to 1.27. This changes the estimate of σ from 1.2 to 1.1 ppm. In the absence of replicate

TABLE 13.8 ANOVA for Table 13.7 with Separate Effects

Component		SS	DF	MS	F	$F_{95}(1,9)$
Model (**M**):	a_1	**12.25**	**1**	**12.25**	**9.19**	**5.12**
	a_2	**306.25**	**1**	**306.25**	**229.69**	**5.12**
	a_3	**49.00**	**1**	**49.00**	**36.75**	**5.12**
	a_{12}	**20.25**	**1**	**20.25**	**15.18**	**5.12**
	a_{13}	1.00	1	1.00	0.75	5.12
	a_{23}	1.00	1	1.00	0.75	5.12
Residual (**R**)		12.00	9	1.33		
Total (**T**)		401.75	15			

TABLE 13.9 ANOVA for Table 13.7 with Pooled Effects

Component		SS	DF	MS	F	F_{95} (1,11)
Model (*M*):	a_1	**12.25**	**1**	**12.25**	**9.63**	**4.84**
	a_2	**306.25**	**1**	**306.25**	**240.63**	**4.84**
	a_3	**49.00**	**1**	**49.00**	**38.50**	**4.84**
	a_{12}	**20.25**	**1**	**20.25**	**15.91**	**4.84**
Residual (*R*)		14.00	11	1.27		
Total (*T*)		401.75	15			

observations, insignificant effects are the only available estimates for the error.

13.2.5 Standard Errors of Effects

An estimate for σ provides the standard errors for each factor's coefficient. The standard error of each effect (s_i) is estimated by the product of s and the respective diagonal element of the inverse least-squares matrix: $s_i = s(x^T x)^{-1}_{i,i}$. For orthogonal data, the diagonal elements of the inverse matrix are simply the reciprocals of the diagonal elements of $x^T x$. The reader may verify that the final model becomes Eq. (13.26) with the standard errors enclosed in brackets [ ]:

$$y = 22.375[\pm 0.28] + 0.875[\pm 0.28]x_1 + 4.375[\pm 0.28]x_2 + 1.750[\pm 0.28]x_3 - 1.125[\pm 0.28]x_1 x_2 \tag{13.26}$$

The ratio of an effect to its standard error is known as the *t*-ratio. It distributes as the square root of the F(1,*n*)-distribution. For example, for x_1, the 95% confidence interval is $x_1 \pm t_{95}(11)s_1$, or 0.875 ± 0.62. This follows from $t_{95}(11)s_1 = \pm\sqrt{F_{95}(1,11)}s_1 = \pm 2.20(0.28) = \pm 0.62$. Because 0.875 is greater than 0.62, the confidence interval does not include zero, and one rejects the null hypothesis. The *t*- and F-ratios provide equivalent tests with identical confidence.

13.3 TWO-LEVEL FACTORIAL DESIGNS

The test matrix given in Table 13.2 is a special one known as a two-level factorial design. This type of test matrix offers special benefits. Two-level factorial designs are powerful methods that allow for very efficient experimentation. That is, they maximize the information available for a given set of factors. To construct them requires a minimum of 2^f runs (experimental tests), where f is the number of factors in the investigation. For $f > 5$, these designs can involve many runs. Later, Section 13.3.3 introduces *fractional factorials* as a way to reduce the required testing. As their name suggests, two-level designs comprise factors at two levels: high (+) and low (–).

Consider the design used tacitly in Table 13.2. Table 13.4 gives the transformed matrix in a standard order comprising 2^3 runs. A general factorial design in f factors will require 2^f runs. The sign of the factors will alternate in blocks of 2^{a-1}, where a is the factor subscript. For example, the sign of the first factor alternates every run (2^{1-1}); the sign of the second factor alternates in blocks of two (2^{2-1}); the sign of the third factor alternates in blocks of four (2^{3-1}); etc. This continues for all f factors. After construction of the design, domain knowledge is used to assign values to the high and low levels.

13.3.1 Interactions

Factorial designs comprising 2^f unique points can specify models with 2^f coefficients, as in Eq. (13.28):

$$y = a_0 + \sum a_i x_i + \sum_{i<j} \sum_{j} a_{ij} x_i x_j + \sum_{i<j} \sum_{j<k} \sum_{k} a_{ijk} x_i x_j x_k + \cdots \tag{13.27}$$

The summations continue until all 2^f terms are specified. The additive sequence reaches 2^f coefficients when it reaches the f-factor interaction. For example, if $f = 6$, the last term in the series for Eq. (13.27) will be a six-factor interaction, $a_{123456}\, x_1 x_2 x_3 x_4 x_5 x_6$. At that time, the series will have grown to comprise (2^6) 64 terms. Third- or higher-order interactions are rarely a concern. If the researcher does not require higher interactions he can reduce the number of required runs. Section 13.3.3 shows how.

13.3.2 Pure Error and Bias

One problem not yet discussed is that in addition to noise, the residual may contain bias (also called lack of fit). Statisticians sometimes use the term "pure error" to distinguish random error from the entire residual, which may contain bias. For example, omission of model terms that should have been included will bias the residual. A biased residual inflates the error estimate. Larger error estimates make the F-test less sensitive, and increase the probability of falsely accepting the null hypothesis. Replicating some of the runs quantifies pure error. Subtracting pure error from the residual gives an estimate of the bias. It is not necessary to replicate every design point as was done in Table 13.4. Adding replicate centerpoints allows testing for both bias and pure error. Factorial

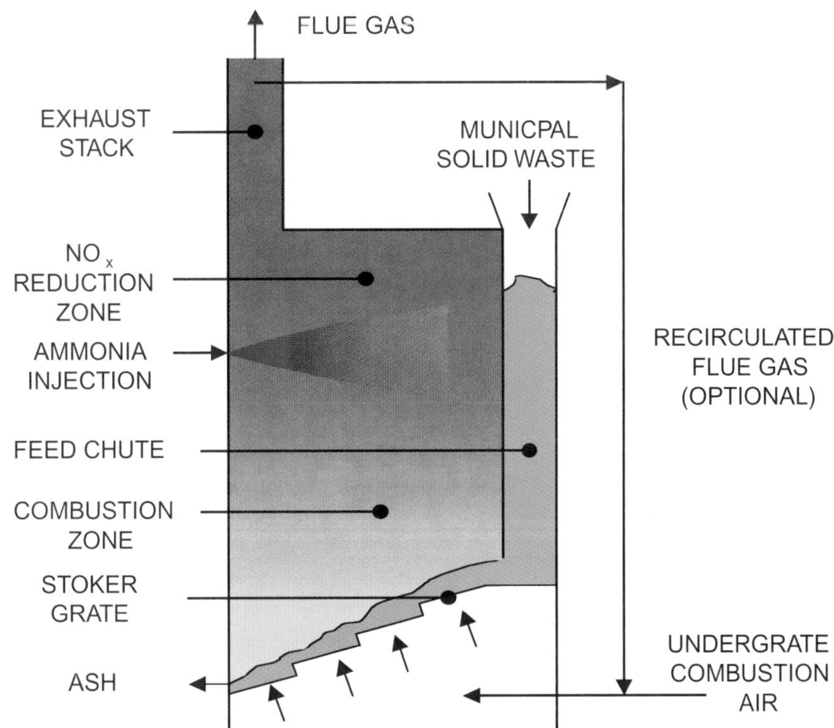
FLUE GAS
EXHAUST STACK
MUNICPAL SOLID WASTE
NO x REDUCTION ZONE
AMMONIA INJECTION
RECIRCULATED FLUE GAS (OPTIONAL)
FEED CHUTE
COMBUSTION ZONE
STOKER GRATE
UNDERGRATE COMBUSTION AIR
ASH
Municipal solid waste (MSW) enters the feed chute onto a perforated stoker grate. Combustion air enters under the grate. The burning the MSW exits the stoker as ash, but creates a hot combustion zone that generates NOX. A series of nozzles inject ammonia, which chemically reduces NOX. Varying the injector pressure changes the penetration of ammonia into the hot furnace and affects NOX reduction. Optionally, flue gas may be added to the combustion air as an additional NOX reduction strategy.

FIGURE 13.4 Municipal solid waste boiler using ammonia injection to control NOx.

designs remain orthogonal with any number of replicate centerpoints.

Consider a municipal solid waste boiler[2] that injects NH_3 to reduce NOx as in Figure 13.4. It is necessary to correlate the injection pressure (x_1) and the NH_3/NOx ratio (x_2) with the NOx concentration (y):

$$y = a_0 + a_1 x_1 + a_2 x_2 + a_{12} x_1 x_2 \tag{13.28}$$

Table 13.10 gives the data, comprising $m = 5$ unique points with one centerpoint replicated three times. The "0" values represent the centerpoints.

To test separately for bias in the ANOVA table, partition SSR into two parts: one for bias (SSB) and the other for pure error (SSE). The degrees of freedom (DF) for each are DFE $= n - m$, and DFB $= m - p$, where m is the number of unique points. Table 13.11 gives the generic ANOVA for replicated data, and Table 13.12 gives ANOVA for the case at hand. Use the following two sums to separate the residual into bias and pure error components.

TABLE 13.10 Factorial Design with Replicate Centerpoints

Run	ξ_1 Pressure, psig	ξ_2 NH_3/NOx	x_1	x_2	y NOx, lb/h
1	20	1.4	–	–	31.0
2	40	1.4	+	–	25.0
3	20	1.8	–	+	16.4
4	40	1.8	+	+	18.1
5	30	1.6	0	0	23.0
6	30	1.6	0	0	25.9
7	30	1.6	0	0	25.2

- Calculate SSE as the squared difference between the actual response (y_i) and the replicate means ($\bar{y}_i$). When there are no replicates, then $\bar{y}_i = y_i$ for the purposes of the summation, adding 0 to SSE. For the case at hand, only three centerpoints are replicated. They generate a single replicate mean [(23.0 + 25.9 + 25.2)/3 = 24.7]. Therefore, SSE = $(23.0 - 24.7)^2 + (25.9 - 24.7)^2 + (25.2 - 24.7)^2 = 4.58$.

TABLE 13.11 Generic ANOVA for Factorial Design with Replicates

Component		SS	DF	MS	F, Case 1	F, Case 2
Model (*M*)		$\sum_{i=1}^{n}(\hat{y}_i - \bar{y})^2$	$p - 1$	SSM/DFM	MSM/MSE	MSM/MSR
Residual	Bias (*B*)	$\sum_{i=1}^{n}\sum_{k}(\hat{y} - \bar{y}_k)^2$	$m - p$	SSB/DFB	MSB/MSE	
	Pure Error (*E*)	$\sum_{i=1}^{n}\sum_{k}(y_i - \bar{y}_k)^2$	$n - m$	SSE/DFE		
Total (*T*)		$\sum_{i=1}^{n}(y_i - \bar{y})^2$	$n - 1$			

Case 1: F = MSM/MSE if MSB/MSE ≥ F_{95}(DFB, DFE)
Case 2: F = MSM/MSR if MSB/MSE < F_{95}(DFB, DFE)
Note: for Case 2, MSR = (SSB + SSE)/(DFB + DFE) = SSR/DFR

TABLE 13.12 ANOVA for Table 13.10 and Equation

Component		SS	DF	MS	F	$F_{95}(1,2)$
Model (*M*)	a_1	4.62	1	4.62	2.02	18.51
	$\boldsymbol{a_2}$	**115.56**	**1**	**115.56**	**50.46**	**18.51**
	a_{12}	14.82	1	14.82	6.47	18.51
Residual (*R*)	Bias (*B*)	7.38	1	7.38	3.22	18.51
	Pure error (*E*)	4.58	2	2.29		
Total (*T*)		146.97	6			

Note: MSB/MSE < $F_{95}(1,2)$.

TABLE 13.13 ANOVA for Factorial Design with Centerpoint Replicates Case 2 of Table 13.11

Component		SS	DF	MS	F	$F_{95}(1,3)$
Model (*M*)	a_1	4.62	1	4.62	1.16	10.13
	$\boldsymbol{a_2}$	**115.56**	**1**	**115.56**	**28.96**	**10.13**
	a_{12}	14.82	1	14.82	3.71	10.13
Residual (*R*)		11.96	3	3.99		
Total (*T*)		146.97	6			

- Calculate SSB by finding SSR in the usual way and subtracting SSE. Alternatively, calculate SSB directly as the squared difference between the predicted values ($\hat{y}_i$) and the replicate means ($\bar{y}_i$). Substitute the value $\bar{y}_i = y_i$ whenever a point comprises only a single value. However, $\hat{y}_i - \bar{y}_i$ will not necessarily be zero in these cases.

Table 13.12 gives the MSB/MSE ratio (F = 3.22) and shows that bias is not significant at the 95% confidence level. The conclusion of no significant bias allows pooling of the bias and pure error. Then one performs the F-test on the entire residual, as in Table 13.13. The larger number of degrees of freedom should better estimate σ^2. Table 13.13 shows that besides the mean, only a_2 is significant. In a like manner, partition the residual of Table 13.9 as MSE = 11.00/8 = 1.375, and MSB = 3.00/3 = 1.000. An F-test proves the model contains no significant bias (F = 1.000/1.375 = 0.72, $F_{95}(3,8)$ = 4.07, 0.72 < 4.07). This justifies the use of the entire residual for the earlier F-tests.

13.3.3 Two-Level Fractional Factorials

Fractional factorial designs always result in fewer runs than full factorials. Specifically, fractional factorials comprise n_{FF} runs where:

$$n_{FF} = 2^{f-d} + c \geq f + 1 + c \tag{13.29}$$

The "fraction" is 2^{-d}. For example, the quarter fraction comprises 2^{f-2} runs ($d = 2$, $2^{-2} = 1/4$). The nomenclature FF(f, $-d$, c) is used to specify a full or fractional factorial comprising $2^{f-d} + c$ runs, where c is the number of centerpoints. Sometimes it is necessary to study a large group of factors, without knowing in advance which are likely to be influential. Other times, constraints limit the total number of possible experiments. The maximum fractionation ($d_{max,FF}$) is determined by solving Eq. (13.30) for d, which must be an integer:

$$d_{max,FF} = f - \text{Ceil}\left[\log_2(f+1)\right] \tag{13.30}$$

where *Ceil* is the ceiling operator. It specifies the smallest integer greater than or equal to [$\log_2(f + 1)$]. To calculate $\log_2(x)$ observe that $\log_2(x) = \ln(x)/\ln(2)$. For example, if $f = 12$, $\log_2(12 + 1) = 3.7$, then Ceil[$\log_2(12 + 1)$] = 4, and f-Ceil[$\log_2(f)$] = 8; thus, the maximum fractionation is $d = d_{max,FF} = 8$.

Consider a full factorial for $f = 4$. It requires 2^4 runs. Suppose, however, only eight runs are possible — half the number needed. The ½ fraction given in Table 13.14 comprises the required number of runs.

How is it generated? First, generate the full factorial for f-d factors (columns x_1 to x_3 in Table 13.14). Then, *alias* the fourth factor with the highest order interaction ($x_1x_2x_3$). The fourth factor is *aliased* by multiplying the values of x_1, x_2, and x_3 together and assigning the result to x_4. For the first row, this gives (–1)(–1)(–1) = (–1). Continue for each of the eight rows to generate the full matrix shown.

Because x_4 is just another name for $x_1x_2x_3$, $x_1x_2x_3$ is *aliased* with x_4. That is, the factor levels for the $x_1x_2x_3$ interaction are identical to x_4. Thus, the response to $x_1x_2x_3$ is *confounded* with that of x_4. So, the one effect is impossible to distinguish from the other. The price for reducing the number of runs is to give up the ability to separately assess the three-factor interaction from x_4. With careful inspection, it is also possible to determine the following aliases. Simply reporting the subscript of the factor sufficiently identifies the alias structure.

$$0 \leftrightarrow 1234,\quad 1 \leftrightarrow 234,\quad 2 \leftrightarrow 134,\quad 3 \leftrightarrow 124,\quad 4 \leftrightarrow 123,$$
$$12 \leftrightarrow 34 \quad 13 \leftrightarrow 24 \quad 14 \leftrightarrow 23,$$

The symbol "↔" means *is aliased with*. Thus, if the 1234 interaction is significant it will be completely *aliased* with the overall mean.

Likewise, if the 234 interaction is significant, its effect will be completely buried in the a_1 coefficient. At most, it is only possible to estimate seven effects plus the mean. This makes sense because the design comprises only eight runs.

The reader may notice a pattern to this confounding. When multiplying the alias groups together, the *word* 1234 is always obtained. This word is referred to as the *defining contrast* (I). The *word length* is the number of factors the word contains. In this case, the word length of 1234 is four.

The alias structure can be computed with a simple algorithm: multiply each effect by I — a squared effect cancels. For example, $1 \times I = 1 \times 1234 = 1^2234 = 234$. So, x_1 is aliased with $x_2x_3x_4$. To figure out the alias structure for x_1x_2, use $12 \times I = 34$. Thus, it is understood that $x_1x_2 \leftrightarrow x_3x_4$. Because the defining contrast is 1234, generate the alias structure using the relation 123 = 4. Therefore, $I = 123 \times 4 = 1234$; and conversely, $123 \times I = 4$.

The shortest word length among the defining contrasts is known as the *resolution*. For the example at hand, the shortest (and only) defining contrast is $I = 1234$. Therefore, the design is Resolution IV (typically expressed in Roman numerals). In a Resolution IV design, the *main effects* (factors to the first order) alias with three-factor interactions. In a Resolution III design, the *main effects* alias with two-factor interactions. As long as the resolution is at least III, no main effect will be aliased with any other main effect. Resolution IV designs or higher have main effects and two-factor interactions clear of each other.

TABLE 13.14 ½ Fractional Factorial [FF(3,–1,0)]

Run	x_1	x_2	x_3	x_4
1	–	–	–	–
2	+	–	–	+
3	–	+	–	+
4	+	+	–	–
5	–	–	+	+
6	+	–	+	–
7	–	+	+	–
8	+	+	+	+

A fractional factorial requires specification of d defining contrasts. For the quarter fraction, $d = 2$, and two defining contrasts, I_1 and I_2, are needed. Suppose there are five factors to study in eight runs. It can be done with a ¼ fraction. The contrasts must be chosen, and the choice has consequences. If I_1 and I_2 are chosen, $I_3 = I_1 \times I_2$ is also tacitly picked. In general, choose defining contrasts to be as large, as equal, and with as little overlap as possible. This suggests something like $I_1 = 123$ and $I_2 = 345$. Then $I_3 = I_1 \times I_2 = 123^245 = 1245$. Suppose $I_1 = 1234$ and $I_2 = 2345$ were chosen. They would generate $I_3 = 15$, a Resolution II design. This is a very poor defining contrast, aliasing x_1 with x_5! With $I_1 = 123$ and $I_2 = 345$, then $I_3 = 1245$. The design is Resolution III. Therefore, no main effect will be aliased with any other. Typically, studies require estimates for main effects and two-factor interactions. Three-factor and higher interactions are usually negligible. Reporting only main effects and two-factor interactions gives the alias structure: 1↔23, 2↔13, 3↔12↔45, 4↔35, 5↔34, 14↔25, and 15↔24 when $I_1 = 123$ and $I_2 = 345$.

13.3.4 Screening Designs

Screening designs comprise a minimal number of runs to find statistically significant factors among a large body of possible factors. One kind of screening design is the *saturated* factorial. When $f - d = \log_2(f + 1)$, the design becomes saturated. Saturated designs have exactly $f + 1$ runs to specify $f + 1$ coefficients. Such designs are also called *simplex* designs. Actually, one can construct a simplex design[3] for any integer value of f. Saturated designs, on the other hand, are a subset of simplex designs restricted to $f = 2^J - 1$, where J is an integer greater than 1 (e.g., $f = 3, 7, 15, \ldots$). Nonetheless, it is recommended that for $f \neq 2^J - 1$, one should not use the simplex, but the next larger fractional factorial, also known as a *highly fractionated* design. For highly fractionated designs, Eq. (13.31) gives the degree, d_{HF}.

$$d_{HF} = f - \text{Ceil}\left[\log_2(f+1)\right] \tag{13.31}$$

Thus, for $f = 12$, the FF(12, –8,0) design comprises 16 runs.

For highly-fractionated designs, construct the ANOVA as shown in Section 13.3.3. For saturated and other simplex designs, there are no degrees of freedom to calculate MSR. However, with so many factors, it is likely that several will be insignificant. Therefore, to construct the residual, look for factors with approximately the same low magnitude, presuming that they represent the noise.

13.3.5 Method of Steepest Ascent

Sometimes, the initial screening finds several factors, but the response does not meet expectations. It may be that the range for the initial factors was too conservative. The *factor space* is the region delimited by the joint range of the factors. In this case, it is necessary to expand the range of factors, or at least move to a new location in factor space. The method of steepest ascent is an efficient way to do this. After screening the important variables, the result is a first-order model:

$$y = a_0 + \sum a_i x_i \tag{13.32}$$

To find the steepest path, take the first derivatives of Eq. (13.32). This generates the coefficient vector of Eq. (13.33).

$$\frac{\partial y}{\partial x_i} = \begin{pmatrix} a_1 \\ a_2 \\ \vdots \\ a_n \end{pmatrix} \tag{13.33}$$

So then, the path of steepest ascent is along the coefficient vector of Eq. (13.33). Consider Eq. (13.34) as an example:

$$y = 5.1 - 0.9x_1 + 4.4x_2 \tag{13.34}$$

The direction of steepest ascent is –0.9 units in the x_1 direction for every 4.4 units in the x_2 direction. To minimize the response, step in the opposite direction, for example, +0.9 units in the x_1 direction for every –4.4 units in the x_2 direction.

A good rule for step size, in the absence of other criteria, is to move the new design center $\sqrt{p}$ units from the old one along the path of steepest ascent. For the current case, $p = 2$. To scale the coefficient vector to $\sqrt{2}$ length, multiply by λ, defined by Eq. (13.35):

$$\lambda = \sqrt{\frac{p}{\sum a_i^2}} \tag{13.35}$$

This results in $\lambda = \sqrt{\frac{2}{-0.9^2 + 4.4^2}} = 0.31$. Accordingly, move the new design center from (0, 0) to $0.31 \cdot (-0.9, 4.4) \approx (-0.3, 1.4)$. Figure 13.5 shows the procedure graphically.

13.3.6 Serial Correlation and Lurking Factors

Experimental designs are generated in standard order. However, running the designs in standard order can introduce error caused by *serial correlation* and *lurking variables*. Lurking variables are influential but unknown factors. Serial correlation refers to responses that correlate with their run order. The subject system may contain hysteresis — a previous condition that influences the response. Alternatively, it may be that influential factors correlate with time; for example, ambient temperature or humidity. Running experiments in random order virtually eliminates the effects of serial correlation. With randomization, lurking factors may inflate the background noise, but will not bias the results. *Foldover* and *blocking* techniques, described in the next two sections, are used to reduce the effect of background noise.

13.3.7 Foldover

Sometimes, a complementary experimental series is run after learning something from the initial one. Perhaps it is later suspected that a particular interaction is significant, but the current design aliases it to another important effect. Conducting a second complement of runs with the signs reversed from the first comprises a procedure called *foldover*. This increases the resolution of the design and de-aliases the interactions. However, because the two groups of designs are run at different times, the researcher must consider the possibility that some important changes may have occurred. For example, combustion tests with two different batches of fuel oil, or tests run during different seasons of the year may contain systematic differences. To provide for differences between one block of experiments and another, use the technique of *orthogonal blocking*. It generates an experimental series whose effects are orthogonal to the factors. Therefore, batch

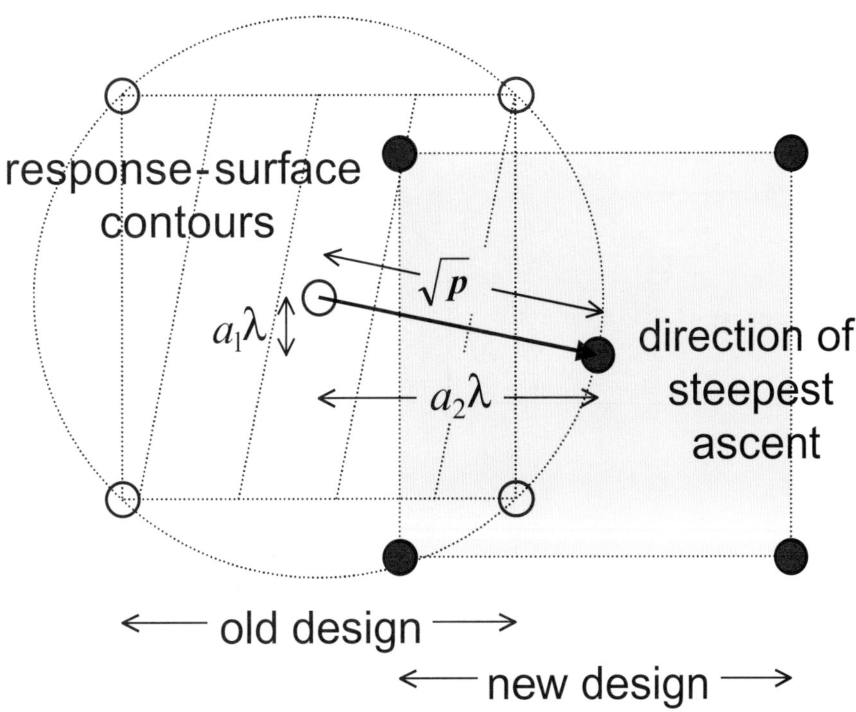

The method of steepest ascent uses the derived coefficients (a_1, a_2) to find the direction of greatest increase in the response. λ scales the direction vector to move the design $\sqrt{p}$ units along this path. One may reapply the method as many times as desired. However, near the optimum, the first order design is no longer sufficient and one must augment the design to determine second-order behavior.

FIGURE 13.5 Method of steepest ascent.

effects are prevented from confounding the results. Foldover automatically creates an orthogonal block. The next section shows how.

13.3.8 Orthogonal Blocking

Consider a test conducted with a heavy oil burner using steam to atomize the oil as shown in Figure 13.6. The purpose of the test is to correlate flame length with oxygen concentration, steam:oil ratio, and firing rate. If it is feasible to run 12 tests, simply run a two-level three-factor factorial design of eight runs and add four centerpoint replicates to give measures of bias and pure error. However, what if the available oil tank will only accommodate six or seven tests?

One alternative is to run some tests with the current batch of oil, then run more tests when a new batch is available. Analytical testing could even ensure that the new batch is similar to the current one. However, it is not possible to know for sure that the new batch will not change in some important but unknown way. Heavy oil is notorious for batch-to-batch variation, and subtle changes in the batch could affect the response. Running the low steam:oil ratios on Heavy Oil Batch I, and the higher steam:oil ratios on Heavy Oil Batch II, will completely confound the batch and steam:oil effects. The preferred test series would ensure that the batch effects are orthogonal to the steam:oil, oxygen, and firing-rate effects. How is this accomplished?

First, construct the experimental design in the usual way to evaluate two levels (low and high) of steam:oil ratio (x_1), oxygen (x_2), and firing rate (x_3) (see Table 13.15). Because the fuel oil batch may have some effect, study it as an additional factor. The study of the oil is not interesting in itself. The oil is more of a nuisance factor, changing unpredictably from batch to batch, but the block factor accounts for the systematic difference between batches. The burner and its performance at different oxygen levels, steam:oil ratios, and firing rates is the real point of interest. Factors of real interest to the study are called *fixed* effects. Factors that represent an uncontrolled variation are called *random* effects.

Table 13.15 gives the experimental structure. All runs within Block = I (1, 4, 6, 7, 9, and 10) are randomized and performed. The remaining runs, Block = II (2, 3, 5, 8, 11, and 12), are randomized and run for the new batch of oil. To derive the

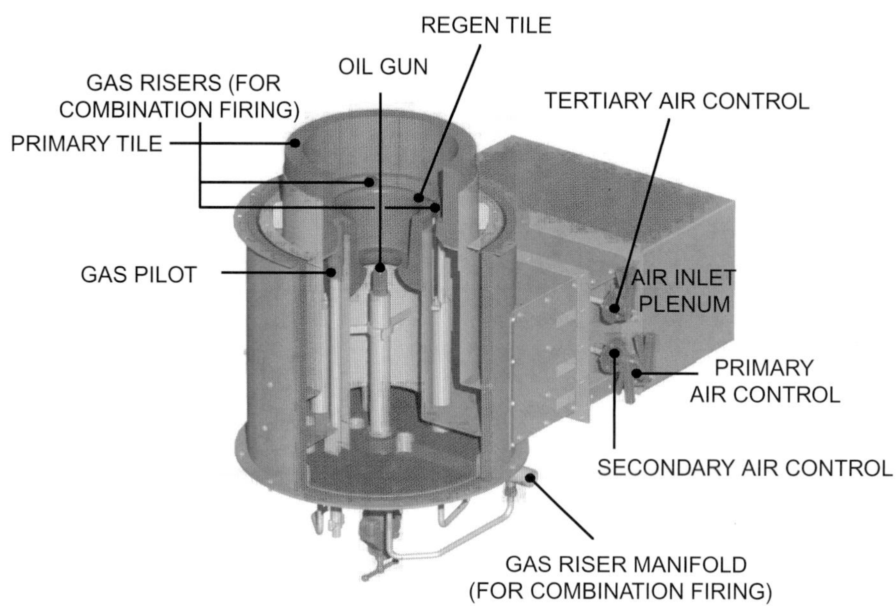

FIGURE 13.6 John Zink PLNC combination burner capable of firing either oil or gas, or both simultaneously. Oil flows through a central oil gun and is stabilized by a regen tile. A surrounding primary tile stabilizes the gas fire. All air flows through the inlet plenum. However, this version of the PLNC has air control to three zones — primary, secondary, and tertiary — via respective dampers whose handles are shown. A gas pilot is used for startup but not required for operation.

TABLE 13.15 FF(3,0,4) in Two Blocks

Run	x_1	x_2	x_3	Block
1	–	–	–	I
2	+	–	–	II
3	–	+	–	II
4	+	+	–	I
5	–	–	+	II
6	+	–	+	I
7	–	+	+	I
8	+	+	+	II
9	0	0	0	I
10	0	0	0	I
11	0	0	0	II
12	0	0	0	II

matrix, put the first three factors in standard order and add the centerpoint replicates at the end — half of which are assigned to Block I and the other half to Block II. Generate the blocking variable with the *blocking generator* $B = 123$, using the algorithm for defining contrasts in Section 13.3.3. Assign $x_1x_2x_3 = -1$ to Block I, and $x_1x_2x_3 = +1$ to Block II. Now, the block effect is aliased with the $x_1x_2x_3$ interaction but with no main effects. Next construct the ANOVA with entries for a_1, a_2, a_3, a_{12}, a_{13}, a_{23}, Block, Bias, and Pure Error. The blocking variable reduces the residual variance, so our F-tests will be more sensitive. Hunter et al.[4] summarize the general philosophy: "Block what you can. Randomize what you cannot." If needed, construct more orthogonal blocks using additional blocking generators while taking care to account for their mutual interactions.

13.3.9 Including Categorical Factors

The previous discussion presumed continuously distributed factors. For example, in principle, oxygen can take on any value over its range. This contrasts with discrete factors that can only take specific values. Even some discrete factors can be treated as continuous. For example, factors like pipe size or burner throat diameter may only exist in certain discrete sizes. Nonetheless, for experimental design purposes, they are treated as continuous because intermediate values are theoretically conceivable if not practical. However, some factors such as burner type are clearly discrete and cannot be considered continuous.

Consider again two different burners. The task is to characterize flame length (y) as a function of the burner type (ξ_I), oil pressure (ξ_2), and the differential steam pressure (ξ_3). To signify categorical factors, use uppercase Roman numeral subscripts to signify a categorical value. A *categorical factor,* ξ_I, distinguishes one category or type of burner from another. Transform ξ_l to x_l by arbitrarily assigning the values $x_l = -1$ for Burner A and $x_l = +1$ for Burner B. Use capital letters to distinguish among levels of ξ_l. The lack of a continuous scale for ξ_l excludes many of the experimental designs considered thus far. However, there are some options:

1. Substitute a continuous factor for the categorical one. This is possible if the burners differ by some influential characteristic, measure, or scale such as throat diameter or length:diameter ratio. Instead of the categorical factor, use the continuous characteristic dimension and apply any of the previous experimental designs.
2. Design the matrix as a full- or fractional-factorial design in *f-k* factors, where *k* is the number of categorical factors. Randomize the $k\ 2^{f-k-d}$ runs, and conduct the tests.

Table 13.16 contains the data for a FF(3,0,0) design with a categorical factor. It generates the following equation and the ANOVA of Table 13.17:

$$y = \begin{cases} 9.08[\pm 0.33] & \text{if } x_l \equiv \text{Burner A} \\ 14.30[\pm 0.33] & \text{if } x_l \equiv \text{Burner B} \end{cases}$$

$$+1.16[\pm 0.33]x_2 - 1.04[\pm 0.33]x_3 \quad (13.36)$$

Table 13.17 shows that burner type is significant, along with steam:oil ratio and oxygen, and Eq. (13.36) quantifies the effects. The analysis can be extended to more than two categorical factors.

13.4 SECOND-ORDER DESIGNS

The discussion so far includes first order designs and first-order designs augmented with second- and higher-order interactions. However, these designs cannot account for *curvature* — pure quadratic factor effects. To account for curvature, a full second-order model of the form given in Eq. 13.37 is needed:

$$y = a_0 + \sum_i a_i x_i + \sum_{i<j} \sum_j a_{ij} x_i x_j + \sum_i a_{ii} x_i^2 \quad (13.37)$$

Unfortunately, none of the designs studied thus far can determine the pure quadratic coefficients, a_{ii}, because none of the designs have any factor at three or more levels. A design of Resolution V or greater comprising at least three levels per factor is needed. A full second-order model requires n_{SO} runs, as in Eq. (13.38):

$$n_{SO} = \frac{(f+1)(f+2)}{2} \quad \text{(full second-order model)} \quad (13.38)$$

TABLE 13.16 Experimental Design with Categorical Factors

Run	ξ_l = Burner Type	ξ_2 = Oil Pressure (psig)	ξ_3 = Differential Pressure (psid)	y = Flame Length (ft)
1	A	60	20	8.7
2	A	80	20	10.6
3	A	60	30	8.0
4	A	80	30	9.0
5	B	60	20	13.3
6	B	80	20	16.3
7	B	60	30	10.1
8	B	80	30	13.5

TABLE 13.17 ANOVA for Table 13.16

Component		SS	DF	MS	F	$F_{95}(1,4)$
Model (***M***)	x_l	**35.70**	**1**	**35.70**	**40.98**	**7.71**
	x_2	**10.81**	**1**	**10.81**	**12.41**	**7.71**
	x_3	**8.61**	**1**	**8.61**	**9.88**	**7.71**
Residual (***R***)		3.49	4	0.87		
Total (***T***)		58.61	7			

13.4.1 Central-Composite Designs

Preferably, the second-order design should build upon the two-level designs studied thus far. If the two-level design detects significant bias, additional runs can be added to create a full second-order model. *Central-composite* designs are denoted with a nomenclature similar to the factorials — CC(*f*, –*d*, *c*).

Central-composite designs require n_{CC} runs as shown in Eq. (13.39):

$$n_{CC} = 2^{f-d} + 2f + c \geq \frac{(f+1)(f+2)}{2} + c \quad (13.39)$$

(central-composite)

Equation (13.40) gives the maximum fractionation, $d_{max,CC}$:

$$d_{max,CC} = \text{Ceil}\left[f + 1 - \log_2\left(f^2 - f + 2\right)\right] \quad (13.40)$$

(central-composite)

A central-composite design uses a two-level design (fractional or full) and augments the design with 2*f* axial runs. For example, if *f* = 3, at least ten runs are required for a full second-order model according to Eq. (13.38). Equation (13.40) gives $d_{max,CC}$ = 1, and Eq. (13.39) shows that a CC(3,–1,0) is the minimum design needed.

Table 13.18 represents a CC(3,0,6) design for the same purpose, comprising 20 runs. Except for the cost of the runs,

TABLE 13.18 CC(3,0,6) Design

Run	x_1	x_2	x_3
1	–	–	–
2	+	–	–
3	–	+	–
4	+	+	–
5	–	–	+
6	+	–	+
7	–	+	+
8	+	+	+
9	$-\alpha$	0	0
10	$+\alpha$	0	0
11	0	$-\alpha$	0
12	0	$+\alpha$	0
13	0	0	$-\alpha$
14	0	0	$+\alpha$
15	0	0	0
16	0	0	0
17	0	0	0
18	0	0	0
19	0	0	0
20	0	0	0

TABLE 13.19 CC(3,0,3) Design

Run	x_1	x_2	x_3	y
1	–	–	–	31.0
2	+	–	–	25.1
3	–	+	–	16.4
4	+	+	–	18.1
5	–	–	+	31.1
6	+	–	+	29.6
7	–	+	+	18.0
8	+	+	+	17.1
9	–2	0	0	28.2
10	+2	0	0	25.9
11	0	–2	0	33.1
12	0	+2	0	11.6
13	0	0	–	25.9
14	0	0	+	24.9
15	0	0	0	23.0
16	0	0	0	25.9
17	0	0	0	25.2

adding them loses nothing. However, extra degrees of freedom to estimate the residual, pure error, and bias.

A central composite design augments the ±1 structure of the factorials with axial points at $\pm\alpha$. So long as $\alpha \neq 1$, each factor can be tested at five levels with only $2f$ additional runs. The choice of α affects the error structure. Consider four criteria.

1. Choose α so that $\pm\alpha$ lies on an n-dimensional sphere $\left(\alpha = \pm\sqrt{f}\right)$ (e.g., 1.732 for $f = 3$).
2. Choose α so that the error structure is spherical (called *rotatability*, $\alpha = \sqrt[4]{2^{f-d}}$); for example, 1.682 for CC(3,0,6).
3. Choose α to make the pure quadratics mutually orthogonal. This requires a numerical evaluation. (If the experiment is run in a single randomized block, $\alpha = 1.525$; if a sequential strategy is used in three randomized blocks, then $\alpha = 1.633$.)
4. Choose α so that the values are convenient, for example, ±2.

Because f, d, and c are restricted to integer values, it is often not possible to satisfy all criteria simultaneously.

13.4.2 Practical Considerations

The general recommendation is to choose values of α that are orthogonal and approximately rotatable. In practical situations, even this can be difficult. Table 13.19 revisits the stoker-fired boiler using ammonia injection to reduce NOx (Figure 13.4). The mass flow of NOx(y) is a function of the nozzle pressure (x_1), the NH_3:NOx ratio (x_2), and fraction of flue-gas recirculated, FGR (x_3). One should choose a single value for α; say, $\alpha = \sqrt[4]{8}$ for rotatability, or perhaps $\alpha = 2$ for convenience. However, FGR = on and FGR = off were the most important states. A central composite gives only one test at each of these extreme points. To account for this, the investigator chose $\alpha_1 = \alpha_2 = 2$ and $\alpha_3 = 1$. Consequently, the design is not rotatable and without mutually orthogonal quadratic coefficients. The pure quadratics are not orthogonal to one another or the mean, and they cannot be considered separately. However, they are orthogonal to the rest of the model. Therefore, it is still possible to use a single ANOVA for the model of Eq. (13.37), providing the quadratics are considered as a group. From the ANOVA of Table 13.20, the only significant term besides a_0 is x_2. The omitted quadratics are not orthogonal to a_0, therefore, a_0 must be recalculated for the truncated model. The reader may verify that the final model becomes NOx (ppm) = 24.13 – 5.61[±0.43] x_2. The term x_2 ranges from –2 to +2. Therefore, in the extreme, NH_3 injection reduces NOx from 35 ppm to 13 ppm.

13.5 ACCOUNTING FOR FUEL MIXTURES

Until now, independent or unconstrained factors have been considered. However, the performance of combustion equipment depends not only on the burner and the furnace, but also on the fuel composition. A *mixture* is a collection of q species whose properties depend only on their relative proportions.

$$z_i = \frac{\xi_i}{\sum_{i=1}^{q} \xi_i} \tag{13.41}$$

TABLE 13.20 ANOVA for Table 13.19 and Equation (13.38)

Component		Coef	Std err	SS	DF	MS	F	F_{95}
a_0		**25.16** (**24.13** without pure quadratics) Std err = **0.93**						
Model (***M***) SSM = 550.52	x_1	−0.71	0.43	7.98	1	7.98	2.72	5.99
	$\boldsymbol{x_2}$	**−5.61**	**0.43**	**502.88**	**1**	**502.88**	**171.62**	**5.99**
	x_3	0.43	0.54	1.85	1	1.85	0.63	5.99
	x_1x_2	1.04	0.61	8.61	1	8.61	2.94	5.99
	x_1x_3	0.24	0.61	0.45	1	0.45	0.15	5.99
	x_2x_3	−0.51	0.61	2.10	1	2.10	0.72	5.99
	x_{11}	0.385	0.36	26.65	3	10.76	3.67	4.35
	x_{22}	−0.765	0.36					
	x_{33}	−1.149	0.87					
Residual (***R***) SSR = 20.51	Bias (Lack of Fit)			15.93	5	3.19	1.39	19.30
	Pure Error			4.58	2	2.29		
	Upon noting no significant bias, the total residual estimates the error: (15.93 + 4.58)/(5 + 2) = 2.93. This is the denominator in all the F-tests except for bias. Bias uses the pure error as the denominator in the F-ratio. Significance at 95% confidence: no effect exceeds F_{95} but x_2.							
Total (***T***)				571.03	16			

In Eq. (13.41), z_i is the fraction of the ith component, and ξ_i is the ith component in the original metric (e.g., %, SCFH, etc). The fractions must sum to unity, by definition:

$$\sum_{i=1}^{q} z_i \equiv 1 \tag{13.42}$$

Of the q dependent factors, the constraining relation Eq. (13.42) reduces the degrees of freedom by one to give f independent factors:

$$f = q - 1 \tag{13.43}$$

The fuel components may be pure (e.g., hydrogen) or blends (e.g., natural gas, heavy fuel oil, refinery fuel gas, or others).

13.5.1 Experimental Designs for Mixtures

Mixture designs are constructed from SED principles but can differ in important ways from factorial designs. Simplex designs occupy a central role for mixtures, and third- and higher-order interactions are often quite important. Suppose it is desirable to characterize a burner for the three fuel blends below:

1. a simulated refinery gas comprising 25% H_2, 25% C_3H_8, and 50% Tulsa natural gas
2. a simulated gas low-BTU waste (LBG) comprising 25% H_2, 25% C_3H_8, 50% CO_2
3. natural gas comprising mostly CH_4 with some higher hydrocarbons

As defined, the system is a three-component system. The relation is illustrated graphically in Figure 13.7. The relation of Eq. (13.42) removes a degree of freedom and constrains $q = 3$ components to $f = 2$ independent factors. A response will be mapped to the $f = 2$ system by testing with a sprinkling of points (p_1, p_2 …).

The factor space representing q components is known as the *simplex*. *Simplex designs* are experimental designs with at least q data points at the extreme vertices of the region.[5] See p_1, p_2, p_3 in Figure 13.7. If the centerpoint, p_{123}, is included, the design becomes a *simplex-centroid*. If the edge points or binary blends, p_{12}, p_{13}, p_{23}, are added, the simplex-centroid design increases in order. If more components are included, then additional points for higher-order interactions such as ternary, quaternary, quinary, etc. can be added. The design becomes a *simplex-axial* design if one adds axial points, such as p_{1123}, p_{1223}, p_{1233}. In general, response functions are mapped as a subset of Eq. (13.44):

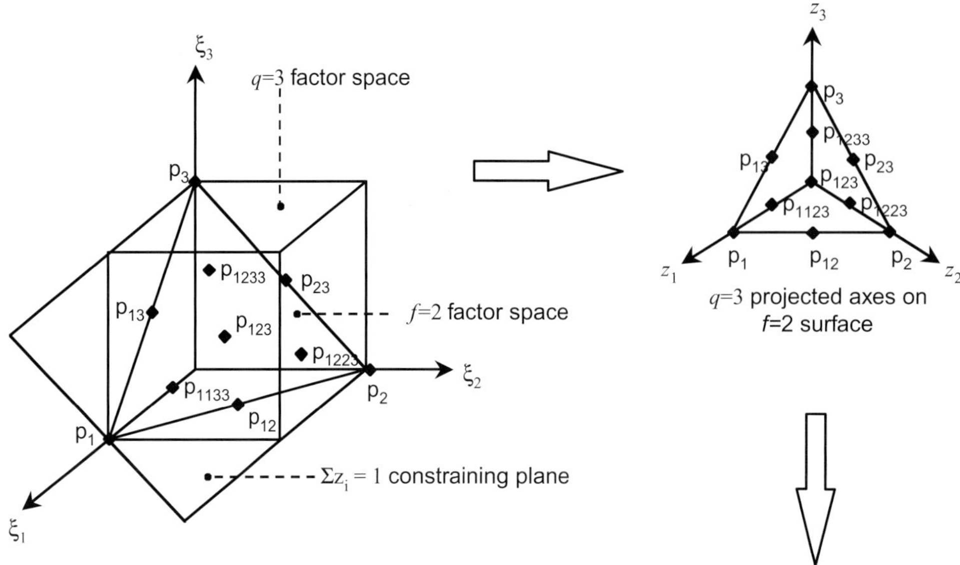

D. COORDINATES FOR q=3, f=2

pt	z_1	z_2	z_3	x_1	x_2
p_1	1	0	0	2	-1
p_2	0	1	0	-1	2
p_3	0	0	1	-1	-1
p_{123}	1/3	1/3	1/3	0	0
p_{12}	1/2	1/2	0	1/2	1/2
p_{13}	1/2	0	1/2	1/2	-1
p_{23}	0	1/2	1/2	-1	1/2
p_{1123}	2/3	1/6	1/6	1	-1/2
p_{1223}	1/6	2/3	1/6	-1/2	1
p_{1233}	1/6	1/6	2/3	-1/2	-1/2

C. RIGHT-ANGLE SIMPLEX, q=3

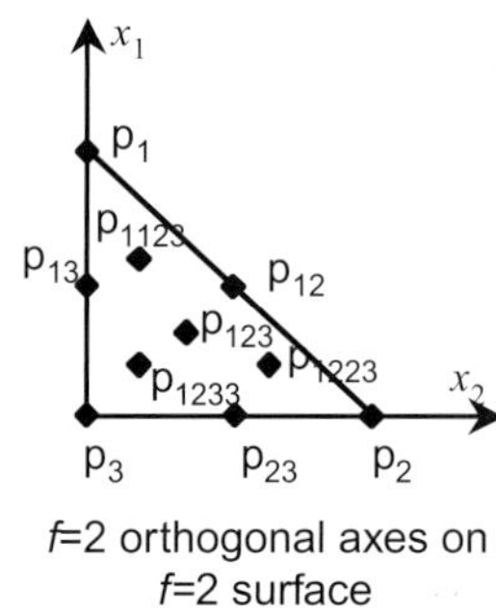

1. A three-component system is constrained by the mixture relation which requires all components to sum to unity. The intersection of the constraining plane and three-component space produces an equilateral triangular region (the *simplex*).
2. The simplex represents all possible mixture fractions: $z_1 + z_2 + z_3 = 1$.
3. We may also transform the region into a right angle triangle by considering q-1 independent factors. The linear transform $x_i = 3z_i - 1$ gives convenient values. Any two factors may form this region; we have arbitrarily chosen x_1 and x_2.
4. A table gives the coordinates for the points (solid diamonds) indicated on the diagrams for comparison.

FIGURE 13.7 Simplex design for $q = 3$.

$$y = \sum_{i=1}^{q} b_i z_i + \sum_{i<j}^{q-1} \sum_{j=1}^{q} b_{ij} z_i z_j + \sum_{i<j<k}^{q-2} \sum_{j<k}^{q-1} \sum_{k=1}^{q} b_{ijk} z_i z_j z_k + \cdots \quad (13.44)$$

For clarity, b and z, as opposed to a and x, are used to distinguish the coefficient and component, respectively, for mixture variables. The same subscript notation adopted earlier for factorials is used. Because Eq. (13.44) has $q - 1$ components, the constant coefficient (b_0) is eliminated to generate a solvable matrix.

In general, a q-component design will have n interactions of order k according to Eq. (13.45).

$$n = \binom{q}{k} = \frac{q!}{k!(q-k)!} \quad (13.45)$$

Equation (13.46) gives the total number of all mixture terms:

$$\sum_{k=1}^{q} \binom{q}{k} = 2^q - 1 \quad (13.46)$$

The simplex-axial design of Figure 13.7 generates the special quartic Eq. 13.47.

$$\begin{aligned} y &= b_1 z_1 + b_2 z_2 + b_3 z_3 + b_{12} z_1 z_2 + b_{13} z_1 z_3 + b_{23} z_2 z_3 \\ &+ b_{123} z_1 z_2 z_3 + b_{1123} z_1^2 z_2 z_3 + b_{1223} z_1 z_2^2 z_3 + b_{1233} z_1 z_2 z_3^2 \end{aligned} \quad (13.47)$$

This equation represents the response as mapped by the regular simplex of Figure 13.7B. If preferred, the equation can be recast in terms of two of the three independent factors, by choosing z_1 and z_2, and obtaining z_3 by the relation $z_3 = 1 - z_1 - z_2$. The region can also be transformed into a quasi-orthogonal one using any convenient linear transform, such as $x_i = 3z_i - 1$. With these transforms, Eq. (13.48) depicted in Figure 13.7C is obtained.

$$\begin{aligned} y &= a_0 + a_1 x_1 + a_2 x_2 + a_{11} x_1^2 + a_{12} x_1 x_2 + a_{22} x_2^2 \\ &+ a_{112} x_1^2 x_2 + a_{122} x_1 x_2^2 + b_{1112} x_1^3 x_2 + a_{1222} x_1 x_2^3 \end{aligned} \quad (13.48)$$

In general, it is not possible to have an orthogonal matrix over the entire simplex for designs higher than first-order, even with linear transforms. Unlike factorial designs, mixtures very often involve ternary interactions or higher. However, for a region of interest within the simplex, one can have orthogonal data sets.

13.5.2 Orthogonal Mixture Designs

There are two convenient ways to generate orthogonal factor space. If the region of interest is small, $0 < z_i < 1/(q-1)$, then use it directly with the transform $x_i = 2fz_i - 1$ to yield an orthogonal factor space in $f = q - 1$ factors. There are q such overlapping spaces in any simplex. However, any one region of interest comprises the $2(q-1)^{1-q}$ fraction of the factor space. For example, if $q = 3$, then $0 < z_i < 1/2$, $x_i = 4z_i - 1$, comprising $1/2$ the factor space.

Another method uses selected ratios. The disadvantage of the method is that ratio transforms are nonlinear so they distort the factor space, especially near a zero denominator. However, in many cases, the component ratio is the factor of interest rather than the mixture fraction. Moreover, with judicious selection of the ratios one can engulf virtually all of the simplex. The next section gives an example.

TABLE 13.21 Example of an Orthogonal Subspace for a $q = 3$ Simplex

Run	z_1	z_2	z_3	$\rho_1 = z_1/(z_1+z_2)$	$\rho_2 = z_3$	$x_1 = 6\rho_1 - 3$	$x_2 = 8\rho_2 - 3$
1	0	1	0	0	0	−3	−3
2	1/3	2/3	0	1/3	0	−1	−3
3	2/3	1/3	0	2/3	0	1	−3
4	1	0	0	1	0	3	−3
5	0	3/4	1/4	0	1/4	−3	−1
6	1/4	1/2	1/4	1/3	1/4	−1	−1
7	1/2	1/4	1/4	2/3	1/4	1	−1
8	3/4	0	1/4	1	1/4	3	−1
9	0	1/2	1/2	0	1/2	−3	1
10	1/6	1/3	1/2	1/3	1/2	−1	1
11	1/3	1/6	1/2	2/3	1/2	1	1
12	1/2	0	1/2	1	1/2	3	1
13	0	1/4	3/4	0	3/4	−3	3
14	1/12	1/6	3/4	1/3	3/4	−1	3
15	1/6	1/12	3/4	2/3	3/4	1	3
16	1/4	0	3/4	1	3/4	3	3

Note: See Figure 13.8 for a graphical representation.

13.5.3 Combining Mixture and Factorial Designs

Orthogonal simplex designs may combine mixture and factorial points in several ways to form a combined factorial. For example:

1. replicate the full simplex at each factorial point
2. replicate the full simplex at each fractional factorial point
3. combine an orthogonal simplex and factorial design and fractionate the entire design as in Section 13.3.3

To illustrate the last option, the first three columns of Table 13.21 specify a grid of points for $q = 3$. Figure 13.8A depicts the design. The next two columns of Table 13.21 give a ratio transformation that can be orthogonalized to give x_1 and x_2. The orthogonal coordinates are given in the last two columns of Table 13.21 and depicted in Figure 13.8B.

Now suppose that in addition to x_1 and x_2, there are three non-mixture factors — x_3, x_4, and x_5. The system in Figure 13.8C can be depicted as a mixture design replicated at each factorial point. The first option would be to run the complete design comprising $16 \times 2^3 = 128$ experiments. A second option would be to use a $1/2$-fractional factorial with the full mixture design comprising $16 \times (1/2)2^3 = 64$ points. If even this is too many, the third option is to combine the design and use the quarter fraction. This results in $(1/4) \cdot 4^2 \cdot 2^3 = 32$ tests

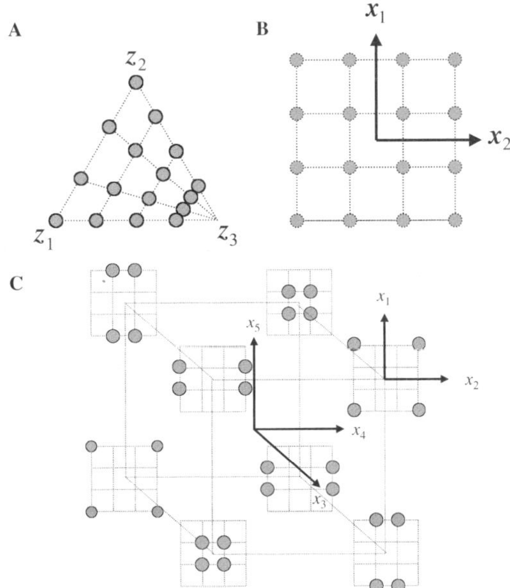

The figure shows a ternary mixture design (A) transformed into an orthogonal one (B). The orthogonal mixture design combines with a factorial design in three factors and fractionates to give the combined fractional mixture-factorial design (C).

FIGURE 13.8 Mixture factors, a transformation, and a combined mixture-factorial.

— a more manageable number. However, the reader will note that the mixture design is a four-level factorial. Thus far, only two-level factorials have been studied. The next section describes how to use the principles of two-level factorials to generate any 2^n-level factorial.

13.5.4 Building 2^n-Level Factorials From Two-level Factorials

It is possible construct 2^n-level factorials from two-level factorials having n factors. To illustrate, consider the four-level mixture design of Section 13.5.3. First, split the design into four factors labeled x_{1a}, x_{1b}, x_{2a}, and x_{2b}. Then build a 2^4 design in the usual way, comprising the first 16 runs of Table 13.22. The four possible combinations of x_{1a} and x_{1b} generate the four values for x_1. The four combinations are – –, – +, – –, + +. Use Eq. (13.49) to assign these combinations to the values –3, –1, 1, and 3 for x_1.

$$x_i = \sum_{j=a,b,\ldots n}^{n} 2^{n-i} x_{ia} \tag{13.49}$$

In the present case, this equates to $x_1 = 2x_{1a} + x_{1b}$. This keeps the matrix orthogonal and preserves the two-unit difference between adjacent levels.

TABLE 13.22 FF(7,–2,0) Design Generating a Combined Mixture-Factorial in Five Factors — Two at Four Levels and Three at Two Levels

Run	x_{1a}	x_{1b}	x_{2a}	x_{2b}	x_1	x_2	x_3	$x_4 = x_{1a}x_{1b}x_3$	$x_5 = x_{2a}x_{2b}x_3$
1	–	–	–	–	–3	–3	–	–	–
2	+	–	–	–	–1	–3	–	+	–
3	–	+	–	–	1	–3	–	+	–
4	+	+	–	–	3	–3	–	–	–
5	–	–	+	–	–3	–1	–	–	+
6	+	–	+	–	–1	–1	–	+	+
7	–	+	+	–	1	–1	–	+	+
8	+	+	+	–	3	–1	–	–	+
9	–	–	–	+	–3	1	–	–	+
10	+	–	–	+	–1	1	–	+	+
11	–	+	–	+	1	1	–	+	+
12	+	+	–	+	3	1	–	–	+
13	–	–	+	+	–3	3	–	–	–
14	+	–	+	+	–1	3	–	+	–
15	–	+	+	+	1	3	–	+	–
16	+	+	+	+	3	3	–	–	–
17	–	–	–	–	–3	–3	+	+	+
18	+	–	–	–	–1	–3	+	–	+
19	–	+	–	–	1	–3	+	–	+
20	+	+	–	–	3	–3	+	+	+
21	–	–	+	–	–3	–1	+	+	–
22	+	–	+	–	–1	–1	+	–	–
23	–	+	+	–	1	–1	+	–	–
24	+	+	+	–	3	–1	+	+	–
25	–	–	–	+	–3	1	+	+	–
26	+	–	–	+	–1	1	+	–	–
27	–	+	–	+	1	1	+	–	–
28	+	+	–	+	3	1	+	+	–
29	–	–	+	+	–3	3	+	+	+
30	+	–	+	+	–1	3	+	–	+
31	–	+	+	+	1	3	+	–	+
32	+	+	+	+	3	3	+	+	+

To add an additional factor, x_3, double the number of runs to 32. The generators $I_1 = 1_a 2_a 34$ and $I_2 = 1_b 2_b 35$ specify the final two columns. Overall, the FF(7,–2,0) design generates a combined mixture-factorial design in five factors. Two of the factors (x_1, x_2) have four levels and three of the factors (x_3, x_4, x_5) have two levels. Analyze the design in the usual way for a 2^{7-2} design. The colored points in Figure 13.8C represent the FF(7,–2,0) design given in Table 13.22.

13.6 COMBINING DOMAIN KNOWLEDGE WITH SED

Although SED is a powerful tool, there is no substitute for intelligence. It is important to think carefully about the problem and the desired outcomes. Combining domain knowledge with SED is a very powerful way to tackle experimental problems. Even in complex or poorly understood cases, domain knowledge often supplies the form of the model, leaving SED to fit the adjustable parameters. The resulting *semi-empirical* model is often more parsimonious and has better properties than a purely empirical one.

As an example, consider NOx formation having the following rate law:

$$[NO_x] = A\int_0^t e^{-\frac{b}{T}}[N_2]\sqrt{[O_2]}d\theta \qquad (13.50)$$

where A and b are constants, T is absolute temperature, θ is time, and the brackets denote volume concentrations of the enclosed species. The rate law shows that NOx is a function of temperature, oxygen, and time. Unfortunately, it is not possible to integrate this function over the tortured path of an industrial burner flame. However, presuming that combustion takes place at near stoichiometric conditions[6] and using a mixture fraction approach results in the following model:[7]

$$y = a_0 + a_1x_1 + a_2x_2 \qquad (13.51)$$

where y = Natural log of the NOx fraction, ln[ppm]
a_0–a_2 = Regressed constants
x_1 = Turndown of the burner
x_2 = Oxygen concentration in the flue gas

The turndown is the reciprocal of the heat release of the burner normalized by the full firing rate. Turndown usually varies from 1 to no more than 10 for boilers, and no more than 3 or 4 for process burners.

The techniques of this chapter are used to fit the coefficients of Eq. (13.51). Because the reciprocal and log functions require an infinite polynomial series, the same model cast in terms of heat release and NOx concentration would look like Eq. (13.52).

$$\begin{aligned}&\beta_1\psi + \beta_2\psi^2 + \beta_3\psi^3 + \ldots \\ &= \alpha_0 + \alpha_1\xi_1 + \alpha_2x_2 + \alpha_{11}\xi_1^2 + \ldots\end{aligned} \qquad (13.52)$$

Obviously, Eq. (13.52) would require many more coefficients ($\alpha_0 - \alpha_{11}$ ...) to adequately represent NOx (ψ) as a function of heat release (ξ_1). Clearly, Eq. (13.51) is a more parsimonious model than Eq. (13.52).

13.6.1 Practical Considerations

The contrast between Eq. (13.51) and Eq. (13.52) underscores the value of applying SED in concert with domain knowledge. However, it would be misleading to imply that the form of the model must be exactly correct. All models are wrong, but some are useful. Power transformations, to which the reciprocal and log functions belong,[8] have little effect unless the ratio of maximum to minimum values is at least 3 or greater. Most burners do not have this extreme NOx response, although some do. Centerpoint replicates and lack-of-fit estimates usually give good guidance as to the appropriateness of the model. Nonetheless, one should spend quality time thinking about the factors, and particularly about factor relationships. Formulation of theoretical models will reduce experimental time, keep the investigator focused on the problem, and partner with SED to generate parsimonious semi-empirical models. As an aid to the reader, some semi-empirical forms relating to NOx formation and reduction strategies[9] are itemized in the next section.

13.6.2 Semi-empirical Models

Equation (13.53) represents a family of semi-empirical models that well-represent NOx for a variety of combustion and NOx abatement scenarios:[7]

$$y = a_0 + a_1x_1 + a_2x_2 \left(+ a_3x_3 + a_4x_4 + \sum_{i\geq 5} a_{1i}x_1x_i + \sum_{j\neq i} a_{2j}x_2x_j \right) \qquad (13.53)$$

where y is the log of the NOx concentration, ln[NOx]

a_0, a_1, a_2, a_3, a_4, a_{1i}, and a_{2j} are the constant coefficients

x_1 is the unit turndown

x_2 is the oxygen concentration in the flue gas

x_3 is the log of the windbox oxygen concentration due to FGR

x_4 is the NH_3:NOx ratio for ammonia injection (with or without a catalyst)

x_{1i} is any of the following: the degree of water injection, steam injection, other diluent injection

x_{2j} is any of the following: the weight fraction of nitrogen in the fuel, degree of air staging, degree of fuel staging, or fraction of burners out of service

To construct the model, include the first three terms of Eq. (13.53) and any other pertinent effects. For example, NOx from combustion of high-nitrogen fuel oil in a burner using air staging and steam injection generates Eq. 13.54:

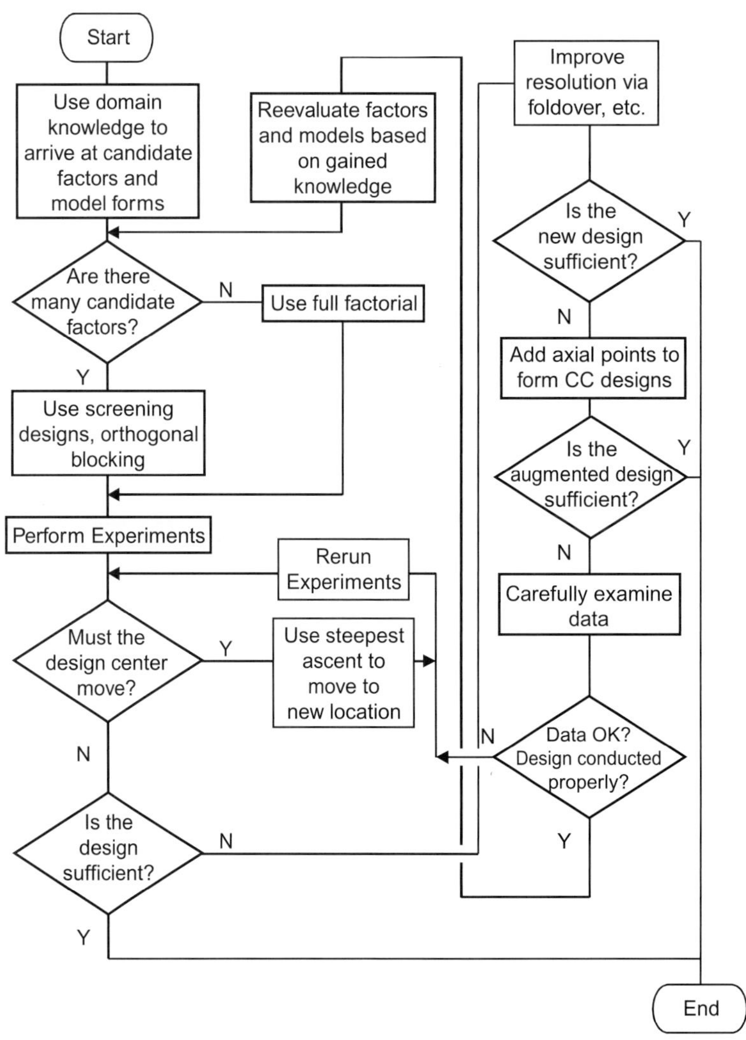

FIGURE 13.9 Flowchart showing a general sequential experimental strategy.

$$y = a_0 + a_1x_1 + a_2x_2 + a_{15}x_1x_5 + a_{26}x_2x_6 + a_{27}x_2x_7 \quad (13.54)$$

where x_5 is the degree of steam injection
x_6 is the fraction of nitrogen in the fuel
x_7 is the degree of air staging
a_{15}, a_{26}, and a_{27} are the constant coefficients

13.6.3 Sequential Experimental Strategies

Paradoxically, the best time to plan experiments is after one has performed them. Only then does one understand which are the right experiments. Unfortunately, investigators never have that luxury. Despite the experimenter's best efforts, even the best science contains trial, error, and serendipity. As a rule, it is best not to spend more than 25% of the budget in the first experimental series. That way, subsequent experiments are planned in light of knowledge gained. These reflections point to developing some strategy for learning during the experimental process.

Figure 13.9 outlines a general experimental strategy. It starts by using domain knowledge to identify important candidate factors and conjecture about their relationship. Experience shows that while investigators identify significant factors, they often overlook important candidates. In general,

A. Fractional Factorial Design, FF(3,-1,2)

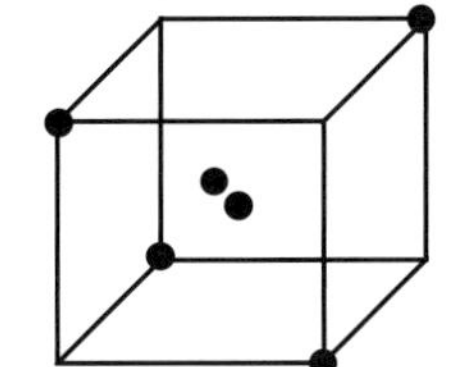

Can generate:

$$y = a_o + \sum_{i=1}^{3} a_i x_i$$

B. Full Factorial Design, FF(3,0,4)

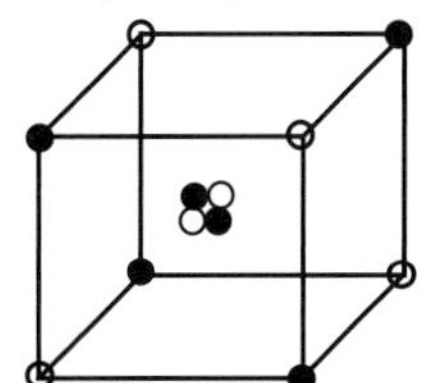

Can generate:

$$y = a_o + \sum_{i=1}^{3} a_i x_i + \sum_{i<j}^{2} \sum_{j=1}^{3} a_{ij} x_i x_j$$

C. Central Composite Design, CC(3,0,6)

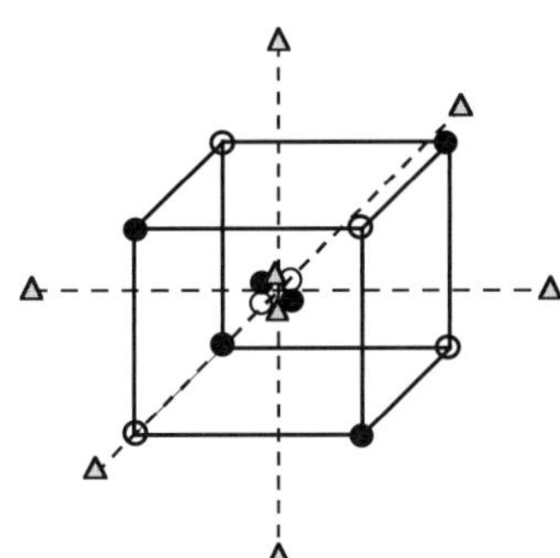

Can generate:

$$y = a_o + \sum_{i=1}^{3} a_i x_i + \sum_{i<j}^{2} \sum_{j=1}^{3} a_{ij} x_i x_j + \sum_{i=1}^{3} a_{ij} x_i^2$$

A sequential design in three orthogonal blocks, starting with a firs order design and ending with a second-order one.

FIGURE 13.10 Some orthogonal designs for $f = 3$ arranged in a sequential strategy.

it is best to err on the side of including too many factors rather than too few. Screening designs such as those described in Section 13.3.4 can reduce the number of tests. Foldover and blocking in, covered in Section 13.3.7, increase the resolution of the factorial design. If diagnostics show a lack of fit as in Section 13.3.2, one may augment the design with axial points to form a second-order design as in Section 13.4. As an optimum is approached, second-order designs often become necessary. If the resulting models are still insufficient, the investigators should carefully examine the way they performed the experiments, review candidate factor lists, and enlist the aid of colleagues. Be especially alert for aberrant responses from quality data. Often, the data point the way to influential but previously unknown factors and relations.

Figure 13.10 shows one sequential experimental strategy for $f = 3$. The investigator starts with a fractional-factorial design and two centerpoint replicates in Figure 13.10A, generating a linear model. If the model is insufficient, the investigator can move in the direction of improved response and try again. In this figure, the investigator has found a region of interest that requires better resolution. He augments the design via foldover in Figure 13.10B and generates several interaction terms. Two additional centerpoint replicates confirm significant lack of fit. Therefore, in Figure 13.10C, the investigator augments the design with axial points to form a central-composite design that accounts for curvature.

13.7 LINEAR ALGEBRA PRIMER

This section reviews Taylor and MacLaurin series approximations and their relevance to SED. It also reviews elementary matrix operations, for example, matrix multiplication, identity and inverse, and addition. These linear algebra topics are implicit and fundamental to SED calculations.

13.7.1 Taylor and MacLaurin Series Approximations

Why do SED methods work? One can use first-, second-, and higher-order curves to approximate engineering functions because a Taylor series allows it. Consider an arbitrary analytic function $y(\xi)$. An interval centered at a can be approximated using a Taylor series:

$$y(\xi) = y(a) + \frac{dy}{d\xi}(\xi - a) + \frac{d^2y}{d\xi^2}\frac{(\xi - a)^2}{2!} + \cdots \quad (13.55)$$

If one lets $x = \xi - a$, then the series becomes a MacLaurin series with no loss of generality:

$$y(x) = y(0) + \frac{dy}{dx}(x) + \frac{d^2y}{dx^2}\frac{(x)^2}{2!} + \cdots \quad (13.56)$$

If one uses two or more variables, then one must account for mixed derivatives. Usually, a second-order function (at most) is sufficient to correlate any response over the interval of interest. Equation (13.57) gives the MacLaurin series for n factors:

$$\begin{aligned} y(x_1, x_2, \ldots, x_n) &= y(0, 0, \ldots, 0) \\ &+ \sum_i \frac{\partial y}{\partial x_i} x_i + \sum_{i<j}\sum_j \frac{\partial^2 y}{\partial x_i \partial x_j} x_i x_j \quad (13.57) \\ &+ \frac{1}{2!}\sum_i \frac{\partial^2 y}{\partial x_i^2} x_i^2 + \cdots \end{aligned}$$

Often, the function y is unknown (or unknowable). This makes it impossible to derive the partial derivatives analytically. However, Eq. (13.37), developed in Section 13.4, gives the general second-order regression model:

$$y = a_0 + \sum_i a_i x_i + \sum_{i<j}\sum_j a_{ij} x_i x_j + \sum_i a_{ii} x_i^2 \quad (13.37)$$

Its coefficients correspond one-for-one with the partial derivatives in Eq. (13.57) up to the second order. Therefore, the coefficients (a_i, a_{ij}, and $2!a_{ii}$) estimate the partial derivatives in the MacLaurin series. One can estimate the coefficients for Eq. (13.37) using the least-squares technique.

Consider $\underline{\underline{a}}$, a general $m \times n$ matrix (read m by n):

$$\underline{\underline{a}} = \begin{pmatrix} a_{11} & a_{12} & \cdots & a_{1n} \\ a_{21} & a_{22} & \cdots & a_{2n} \\ \vdots & \vdots & \ddots & \vdots \\ a_{m1} & a_{m2} & \cdots & a_{mn} \end{pmatrix} \quad (13.58)$$

Each matrix element is indexed with a double subscript. The first subscript refers to the row, the second to the column. Thus, a_{2n} refers to the second row, nth column in matrix $\underline{\underline{a}}$. A double underline indicates a matrix. A matrix having a single column or row (also known as a *vector*) uses a single underline. In general, a matrix can contain any number of rows and columns, and the number of rows and columns need not be equal.

13.7.2 Matrix Multiplication

Matrix multiplication proceeds from left to right with the result being the sum of the product of rows and columns taken in order per Eq. (13.59). Therefore, to multiply two matrices, the number of columns, n, in the first must equal the number of rows in the second. The result of multiplying an $m \times n$ and $n \times p$ matrix is an $m \times p$ matrix:

$$\begin{pmatrix} a_{11} & a_{12} & \cdots & a_{1n} \\ a_{21} & a_{22} & \cdots & a_{2n} \\ \vdots & \vdots & \ddots & \vdots \\ a_{m1} & a_{m2} & \cdots & a_{mn} \end{pmatrix} \begin{pmatrix} b_{11} & b_{12} & \cdots & b_{1p} \\ b_{21} & b_{22} & \cdots & b_{2p} \\ \vdots & \vdots & \ddots & \vdots \\ b_{n1} & b_{n2} & \cdots & b_{np} \end{pmatrix} = \begin{pmatrix} \sum_{i=1}^{n} a_{1i}b_{i1} & \sum_{i=1}^{n} a_{1i}b_{i2} & \cdots & \sum_{i=1}^{n} a_{1i}b_{ip} \\ \sum_{i=1}^{n} a_{2i}b_{i1} & \sum_{i=1}^{n} a_{2i}b_{i2} & \cdots & \sum_{i=1}^{n} a_{2i}b_{ip} \\ \vdots & \vdots & \ddots & \vdots \\ \sum_{i=1}^{n} a_{mi}b_{i1} & \sum_{i=1}^{n} a_{mi}b_{i2} & \cdots & \sum_{i=1}^{n} a_{mi}b_{ip} \end{pmatrix} \quad (13.59)$$

In general, matrix multiplication is not commutative ($\underline{\underline{ab}} \neq \underline{\underline{ba}}$, in general). It is associative, $\underline{\underline{a}}(\underline{\underline{ba}}) = (\underline{\underline{ab}})\underline{\underline{c}}$. One can multiply a constant and a matrix, that operation being commutative $\underline{\underline{a}}k = k\underline{\underline{a}}$:

$$\begin{pmatrix} a_{11} & a_{12} & \cdots & a_{1n} \\ a_{21} & a_{22} & \cdots & a_{2n} \\ \vdots & \vdots & \ddots & \vdots \\ a_{m1} & a_{m2} & \cdots & a_{mn} \end{pmatrix} k = \begin{pmatrix} ka_{11} & ka_{12} & \cdots & ka_{1n} \\ ka_{21} & ka_{22} & \cdots & ka_{2n} \\ \vdots & \vdots & \ddots & \vdots \\ ka_{m1} & ka_{m2} & \cdots & ka_{mn} \end{pmatrix} \quad (13.60)$$

13.7.3 Identity, Inverse, and Transpose

The *identity matrix* (I) has the property that $\underline{\underline{I}}\underline{\underline{a}} = \underline{\underline{a}}\underline{\underline{I}} = \underline{\underline{a}}$. The identity matrix comprises a diagonal matrix of unit values (and zero values elsewhere). For convenience, the zero elements are omitted, but their existence is understood implicitly:

$$\begin{pmatrix} 1 & 0 & \cdots & 0 \\ 0 & 1 & \cdots & 0 \\ \vdots & \vdots & \ddots & \vdots \\ 0 & 0 & \cdots & 1 \end{pmatrix} = \begin{pmatrix} 1 & & & \\ & 1 & & \\ & & \ddots & \\ & & & 1 \end{pmatrix} \tag{13.61}$$

If $m = n$ then $\underline{\underline{a}}$ is square. For a square matrix the *inverse* ($\underline{\underline{a}}^{-1}$), if it exists, has the property $\underline{\underline{a}}^{-1}\underline{\underline{a}} = \underline{\underline{a}}\underline{\underline{a}}^{-1} = I$. The transpose operation, denoted $\underline{\underline{a}}^{\mathrm{T}}$, switches rows and columns:

$$\text{if } \underline{\underline{a}} = \begin{pmatrix} a_{11} & a_{12} & \cdots & a_{1n} \\ a_{21} & a_{22} & \cdots & a_{2n} \\ \vdots & \vdots & \ddots & \vdots \\ a_{m1} & a_{m2} & \cdots & a_{mn} \end{pmatrix},$$

$$\text{then } \underline{\underline{a}}^T = \begin{pmatrix} a_{11} & a_{21} & \cdots & a_{m1} \\ a_{12} & a_{22} & \cdots & a_{m2} \\ \vdots & \vdots & \ddots & \vdots \\ a_{1n} & a_{2n} & \cdots & a_{mn} \end{pmatrix} \tag{13.62}$$

$$\text{if } \underline{b} = \begin{pmatrix} b_1 \\ b_2 \\ \vdots \\ b_n \end{pmatrix}, \quad \text{then } \underline{b}^T = \begin{pmatrix} b_1 & b_2 & \cdots & b_n \end{pmatrix} \tag{13.63}$$

In general, the inverse is soluble both analytically[10] and numerically.[11] However, for diagonal matrices, the inverse is simply the reciprocal of the diagonal elements.

$$\text{if } \underline{\underline{a}} = \begin{pmatrix} a_{11} & & & \\ & a_{22} & & \\ & & \ddots & \\ & & & a_{nn} \end{pmatrix},$$

$$\text{then } \underline{\underline{a}}^{-1} = \begin{pmatrix} \frac{1}{a_{11}} & & & \\ & \frac{1}{a_{22}} & & \\ & & \ddots & \\ & & & \frac{1}{a_{nn}} \end{pmatrix} \tag{13.64}$$

13.7.4 Matrix Addition

One can add two matrices if both their rows and columns are equal. The resulting $m \times n$ matrix comprises the sum of the individual matrix elements:

$$\begin{pmatrix} a_{11} & a_{12} & \cdots & a_{1n} \\ a_{21} & a_{22} & \cdots & a_{2n} \\ \vdots & \vdots & \ddots & \vdots \\ a_{m1} & a_{m2} & \cdots & a_{mn} \end{pmatrix} + \begin{pmatrix} b_{11} & b_{12} & \cdots & b_{1n} \\ b_{21} & b_{22} & \cdots & b_{2n} \\ \vdots & \vdots & \ddots & \vdots \\ b_{m1} & b_{m2} & \cdots & b_{mn} \end{pmatrix}$$
$$= \begin{pmatrix} a_{11}+b_{11} & a_{12}+b_{12} & \cdots & a_{1n}+b_{1n} \\ a_{21}+b_{21} & a_{22}+b_{22} & \cdots & a_{2n}+b_{2n} \\ \vdots & \vdots & \ddots & \vdots \\ a_{m1}+b_{m1} & a_{m2}+b_{m2} & \cdots & a_{mn}+b_{mn} \end{pmatrix} \tag{13.65}$$

Matrix addition is both commutative, $\underline{\underline{a}} + \underline{\underline{b}} = \underline{\underline{b}} + \underline{\underline{a}}$, and associative, $\underline{\underline{a}} + (\underline{\underline{b}} + \underline{\underline{c}}) = (\underline{\underline{a}} + \underline{\underline{b}}) + \underline{\underline{c}}$.

REFERENCES

1. J. Colannino, Control of NOx using MRSM, *1998 American-Japanese Flame Research Committees International Symposium*, Maui, October 11-15, 1998.
2. J. Colannino, Results of a Statistical Test Program to Assess Flue-Gas Recirculation at the Southeast Resource Recovery Facility (SERRF), Paper 92-22.01, presented before the *Air & Waste Management Association, 85th Annual Meeting & Exhibition*, Kansas City, MO, June 21-26, 1992.
3. G.E.P. Box and N.R. Draper, *Empirical Model Building and Response Surfaces*, John Wiley & Sons, New York, 1987.
4. G.E.P. Box, W.G. Hunter, and J.S. Hunter, *Statistics for Experimenters*, John Wiley & Sons, New York, 1978.
5. J.A. Cornell, *Experiments with Mixtures, Designs Models, and the Analysis of Mixture Data*, 2nd ed., ISBN 0-471-52221-X, John Wiley & Sons, New York, 1990.
6. I. Glassman, *Combustion*, 3rd ed., Academic Press, San Diego, 1996, 273.
7. J. Colannino, NOx and CO: Semi-empirical models for boilers, Session 49.T28.d, presented at the *American Power Conference, 59th Annual Meeting*, Chicago, April 1-3, 1997.
8. G.E.P. Box and D.R. Cox, An analysis of transformations, *J. Roy. Statistical Assoc.*, B26, 211-252, April 8, 1964.
9. J. Colannino, Using modified response surface methodology (MRSM) to control NOx, *Institute of Clean Air Companies Seminar: Cutting NOx Emissions*, Durham, North Carolina, March 81-20, 1998.
10. E. Kreyszig, *Advanced Engineering Mathematics*, John Wiley & Sons, New York, 1979, chap. 19.
11. A.J. Pettofrezzo, *Matrices and Transformations*, Dover, New York, 1978.

Chapter 14
Burner Testing

Jeffrey Lewallen, Robert Hayes, Prem Singh, and Richard T. Waibel

TABLE OF CONTENTS

14.1 Introduction 432

14.2 Burner Test Setup 433

- 14.2.1 Application 434
- 14.2.2 Test Furnace Selection Criteria 434
- 14.2.3 Selection of Test Fuels 434

14.3 Instrumentation and Measurements 437

- 14.3.1 Measuring Air-side Pressure and Temperature 437
- 14.3.2 Furnace Gas Temperature Measurement 438
- 14.3.3 Emissions Analysis 438
- 14.3.4 Fuel Flow Rate Metering 441
- 14.3.5 Flame Dimensions 441
- 14.3.6 Heat Flux 442

14.4 Test Matrix (Test Procedure) 442

- 14.4.1 Heater Operation Specifications 442
- 14.4.2 Performance Guarantee Specifications 442
- 14.4.3 Definition of Data to be Collected 444

References 444

FIGURE 14.1 John Zink Co., LLC, R&D Test Center, Tulsa, Oklahoma. (With permission.)

14.1 INTRODUCTION

Burner testing provides an opportunity to gather and verify valuable information such as operating parameters, pollutant emissions, flame dimensions, the heat flux data, safety limitations, and noise data. Information from test data is often essential for performance verification of customer applications as well as being vital to research and development efforts. Empirical data collected from burner testing is a valuable source of information that can be used to improve the predictive capabilities of CFD models, which are becoming more prevalent tools used in the research, development, and design of combustion equipment at the forefront of technology in the industry. At state-of-the-art test facilities as shown in Figure 14.1, testing is done year-round to provide furnace designers with the data they need to improve heater designs and operate their heaters and furnaces more efficiently as well as to develop new technology to meet the ever-increasing demands of customer processes and environmental regulation.

While designing a burner appears to involve relatively simple calculations, it is difficult to predict how a burner will operate over a broad range of operating conditions. Considering the multitude of heater applications, the wide range of fuels available to be burned, the required pollutant levels to be met, and the different ways to supply air, the variations between burner designs are nearly infinite. Through full-scale testing, specific conditions can be simulated and the actual operational performance of a burner can be measured accurately. Testing allows a burner manufacturer to optimize a burner design to closely meet the requirements of a specific application.

The operating parameters that can be obtained through testing include the heat release envelope of the burner. Burners are sized for maximum heat release with a specified turndown, or a minimum rate at which a burner can be safely operated. Turndown is defined as the ratio of maximum heat release to minimum heat release. For example, if the maximum heat release of a burner is 5 MMBtu/hr and the minimum heat release of that burner is 1 MMBtu/hr, then the turndown is 5:1. Another variable operators and engineers may need to know is what happens to a burner if it is fired beyond its maximum designed heat release. With this performance information, a customer can set a target oxygen level in the flue gas to stay above or set an upper pressure limit for a given fuel to stay below to ensure that the burner does not exceed the designed parameters. More importantly, test data can determine the upper heat release value at which a burner can be safely operated for short durations until a process upset can be corrected.

An operator also needs to know the point at which a burner will become unstable if fired below the minimum heat release. The rate at which a burner can be fired below the designed minimum heat release is defined as the absolute minimum.

A lower pressure limit can be set for the fuel gas to ensure safe heater operation. This information is especially useful in determining how many burners should be fired and at what heat release for special operations. Decoking and furnace startup to cure refractory are examples of special operations in which a heater is fired at a level different from its usual designed rate.

Along with defining the firing envelope of a burner, the proper air door or air damper settings can be determined through testing to ensure the efficient operation of a heater by controlling the excess oxygen in the flue gas. By running at lower excess oxygen concentrations, fuel savings can be realized, leading to higher heater efficiency. In complex furnaces such as ethylene heaters, which may have hundreds of burners in operation at once, advanced knowledge of air door settings for various operating conditions can save time in trimming out the excess air during actual operation.

Other information that can be collected during a burner test or demonstration includes emissions of pollutants such as NOx, CO, and unburned hydrocarbons (UHCs). Based on theory and field experience, although it is easy to predict emissions for a single fuel, modern burners are often expected to burn a wide range of fuels. As a result, some fuels may not be fired at their optimum pressures, and variables such as fuel pressure can significantly affect the emission performance of a burner. By testing a burner in an operating furnace prior to final installation, the expected emissions for different operating conditions can be predicted and anticipated with a greater degree of accuracy. It is likely that the emission tests in a furnace might differ in the field to some extent due to factors such as interaction with other burners, furnace conditions, and changes in fuel compositions.

When firing burners with a wide variety of fuels, flame dimensions can change, depending on the fuel fired and the operating fuel pressure, as well as the heat release, because the mixing energy available can significantly affect the volume or shape of a flame. By conducting a burner test over the normal operating envelope of a burner, the dimensions of the flame can be determined for all conditions. The flame dimensions are important for ensuring there is no flame impingement on the process tubes in the furnace.

Another valuable piece of data that can be collected is noise data. New plants built today, as well as existing plants, see stricter requirements for noise levels. Depending on the severity of the requirement, mufflers can be designed to attenuate the burner noise to acceptable levels.

Some burners are designed to heat a furnace wall. For these burners, heat flux profiles can be determined through testing to provide heater manufacturers with information about the transfer of heat radiated from the wall to the process tubes. By optimizing the heat flux profile, cycle times between decoking procedures can be increased, improving the heater's runtime and thus its efficiency.

Testing can provide a variety of data concerning a burner's performance. But without a proper setup, the correct instrumentation and measurement methods, and a well-defined test procedure (test matrix), the data collected during a test may be meaningless. This chapter discusses the proper elements required for conducting a test. Items to be covered include identifying the application of the burner, selecting the correct test furnace, and determining the test fuels to be utilized during testing.

This chapter also discusses the instrumentation necessary to record consistent and accurate data — in particular, the concentrations of NOx, CO, O_2 (wet and dry), unburned hydrocarbons, and particulates in the flue gas, heat flux, and noise emissions from the burner. Fuel flow metering and flame measurement are covered as well.

With input from the heater manufacturer and end user, a meaningful test procedure can be put together that will yield valuable data in determining a burner's performance under different operating conditions and fuels. The test procedure is designed to answer a specific set of questions regarding the performance of the burner. By closely matching the conditions of operation expected in the field, data can be collected that will aid operators in running their furnaces.

Finally, this chapter discusses data analysis. Once a test is run, it must be determined if the burner has met the criteria outlined in the test procedure. The criteria include performance guarantees and operating parameters. With the data collected, the test engineer can optimize the burner to improve emissions, flame dimensions, stability, and air flow distribution.

Armed with the knowledge described above, customers will have a greater understanding of what to expect from a burner test, as well as what goes into setting up and conducting a test that will provide meaningful data. API 535 gives some good guidelines for specifications and data required for burners used in fired heaters.[1]

14.2 BURNER TEST SETUP

One of the most important aspects of a burner test is the setup. This includes the selection of a test furnace, which is determined by the type of burner to be tested and its installation configuration. Typically, test furnaces are built with one of two methods of cooling: a water-cooled jacket or a series of water-cooled tubes. A water jacket is simply a furnace surrounded by two shells (inner and outer) of carbon steel that contains circulating water between the shells. This keeps cooling water on the four vertical surfaces to transfer heat. The

FIGURE 14.2 Test furnace for simulation of ethylene furnaces.

other method utilizes cooling tubes that run either horizontally or vertically along one or two of the furnace walls.

Burners are designed to cover a wide range of applications. They are vertically up-fired, vertically down-fired, or horizontally fired. They may have round or rectangular flames and may be free-standing or fired along a refractory wall. The criteria for selecting a burner normally include the fuel to be fired, air supply method, emission requirements, and heater configuration. Fuels can be gas, oil, waste gas, or some combination of the three. The air supply can be either natural (induced) draft, forced draft, turbine exhaust gas, or other sources of oxygen. Emission requirements are primarily based on NOx, but can include unburned hydrocarbons, carbon monoxide, SOx, and particulates.

The test fuel blend is a critical component of a successful test. Without the proper blend, a simulated fuel may not provide data that will aid engineers and operators when starting up new units or evaluating the performance of new burners installed in an existing unit. Test fuels are typically blended to closely simulate the heating value, flame temperature (adiabatic), specific gravity, and major components of the fuels to be used in the actual application.

14.2.1 Application

Although any fuel (solid, liquid, or gas) can be used in a burner designed for a specified fuel type, this chapter is limited to gas and liquid fuel firing, as they are by far the most common found in the hydrocarbon and petrochemical industries. When firing a fuel, the normal products of combustion are CO_2, H_2O, N_2, O_2, and the energy or heat released during a combustion process. Unfortunately, there are also other less desirable products that may be released as well. These commonly include unburned hydrocarbons (UHCs), particulates, NOx, SOx, and CO.

14.2.2 Test Furnace Selection Criteria

The selection of a test furnace is important. The furnace should be large enough to contain the flame without impingement on the walls or ceiling of the furnace. Also, it is important to select the proper furnace to keep the furnace gas temperature close to the customer's expected furnace temperature. The following figures show some typical furnaces necessary for burner testing.

The furnace shown in Figure 14.2 can be tested in a variety of configurations. Wall-fired burners can be tested at the floor level, and radiant wall burners can be tested at higher elevations in the furnace to match the customer's configurations. The number of burners can be changed to simulate the setup in the field as well as to achieve a certain furnace temperature.

The furnaces shown in Figure 14.3 can be tested in the down-fired configuration to simulate certain types of reformers where the burners are installed on the roof of the furnace and fired down in between the process tubes. Another type of furnace is a vertical cylindrical water-jacketed furnace as shown in Figure 14.4.

Because test facilities are not built for the purpose of heating an oil to a desired temperature or creating products, the heat released from the burners must be absorbed by some method. The furnace shown in Figure 14.4 is surrounded by a shell filled with water. Figure 14.5 shows a test furnace used to demonstrate burners for terrace wall reformers.

14.2.3 Selection of Test Fuels

The main criteria for fuel selection include:

- similarity in combustion characteristics with the actual fuel specified for the application
- economics
- availability
- compatibility with the systems, operations, and equipment

FIGURE 14.3 Test furnace for simulation of down-fired tests.

Figure 14.6 shows both permanent and portable fuel storage tanks. Portable tanks can be used when testing is required on specialty fuels.

Probably the most critical component of successfully testing a burner is the selection of the test fuel. Without matching key components in the customer's fuel, the emissions, stability, and flame shape shown during a burner test can vary significantly when compared to the field results. Fuel can be blended to match the heating value and molecular weight as specified or as mutually agreed upon with the customer. Hydrogen and the diluent content of the gas should be similar in volumetric proportion to the specified actual service fuel gas if these proportions significantly impact burner performance. The Wobbe index is often used as a criteria for specifying a test blend that will be used to simulate a fuel. The Wobbe index is the higher heating value (HHV) of a fuel, divided by the square root of its specific gravity (SG).

$$\text{Wobbe index} = \frac{\text{HHV}}{\sqrt{\text{SG}}} \tag{14.1}$$

The specific gravity (SG) for a gas is the ratio of the molecular weight of a gas to the molecular weight of air. The specific gravity for a liquid is the ratio of the density of a liquid to the density of water. It is important to note that the two fluids should be compared at the same temperature. Two fuels will

FIGURE 14.4 Test furnace for simulation of up-fired tests.

FIGURE 14.5 Test furnace for simulation of terrace wall reformers.

FIGURE 14.6 Test fuel storage tanks.

provide the same heat release from a gas tip at a given supply pressure if the Wobbe index is the same.

While the Wobbe index is a good indicator to see if a fuel is similar, it is important to try and match the lower heating value (LHV), molecular weight, and adiabatic flame temperature to ensure a good simulation. Commonly available gaseous fuels available for blending at well-equipped test facilities include natural gas, propane, propylene, butane, hydrogen, nitrogen, and carbon dioxide. The composition of natural gas varies by geographic location. As an example, Tulsa natural gas has a typical composition as shown in Table 14.1.

TABLE 14.1 Tulsa Natural Gas (TNG) Composition and Properties

CH_4 (volume%)	93.4	C_4H_{10} (volume%)	0.20
C_2H_4 (volume%)	2.70	CO_2 (volume%)	0.70
C_3H_8 (volume%)	0.60	N_2 (volume%)	2.40
LHV (Btu/scf)	913	HHV (Btu/scf)	1012
Molecular weight	17.16	Specific heat ratio @ 60°F	1.3
Adiabatic flame temperature (°F)			3452

TABLE 14.2 Example Refinery Gas

Fuel Component	Formula	Volume%
Methane	CH_4	8.13[a]
Ethane	C_2H_6	19.9[a]
Propane	C_3H_8	0.30
Butane	C_4H_{10}	0.06
Ethylene	C_2H_4	32.0[b]
Propylene	C_3H_6	0.78
Butylene	C_4H_8	0.66
1-Pentene	C_5H_{10}	0.07
Benzene	C_6H_6	0.12
Carbon monoxide	CO	0.22
Hydrogen	H_2	37.8[c]

[a] Balance of fuel is primarily methane and ethane.
[b] Level of olefins in the fuel.
[c] Hydrogen content.

TABLE 14.3 Comparison of Refinery Gas to Test Blend

Property	Refinery Fuel	Test Fuel
LHV (Btu/scf)	1031	1026
HHV (Btu/scf)	1124	1121
Molecular weight	18.09	18.38
Specific heat ratio @ 60°F	1.27	1.26
Adiabatic flame temperature (°F)	3481	3452
Wobbe index	1422	1407

Table 14.2 illustrates an example of a refinery gas and the points of interest in determining a test blend that will effectively simulate the fuel-handling properties, burning characteristics (tendency of a fuel to coke, etc.), and emission levels of the customer's fuel composition.

Based on the available fuels for blending, the hydrogen content is matched, propylene is used to substitute the ethylene content, and Tulsa natural gas (TNG) is used to simulate the methane content. By holding the hydrogen content fixed at 38%, TNG and propylene are balanced to obtain a match of the lower heat value (LHV) and molecular weight. By attempting to balance the LHV, molecular weight, and adiabatic flame temperature, a test fuel blend of 34% TNG, 28% C_3H_6, and 38% H_2 would be acceptable to simulate the refinery fuel gas illustrated in Table 14.2. Table 14.3 gives a side-by-side comparison of the fuel properties.

With the test fuel(s) established, it is time to determine what measurements will need to be taken and what instruments will be required.

14.3 INSTRUMENTATION AND MEASUREMENTS

Measurements generally required during testing (but not limited to) are fuel pressure and temperature, air pressure drop and temperature, fuel flow measurement, flame dimensions, and emissions measurements.

14.3.1 Measuring Air-side Pressure and Temperature

Most applications for the hydrocarbon and petrochemical industries being tested today are natural-draft applications where practical measurement of the air flow cannot be done without impacting the quality of the data recorded.

14.3.1.1 Natural Draft

The combustion air for natural-draft burners is induced through the burner, either by the negative pressure inside the furnace or by fuel gas pressure that educts the air through a venturi. Natural-draft burners are the simplest and least expensive burners, and are most commonly found in the hydrocarbon and petrochemical industries. Because the energy available to draw air into the burner is relatively low, there is no practical way to measure the air flow through the burner. As a result, the temperature of the air, the ambient air pressure, the fuel flow, and the excess air measurements are critical for accurately calculating the air flow through a natural-draft burner.

14.3.1.2 Forced Draft

Forced-draft burners are supplied with combustion air at a positive pressure. The air is supplied by mechanical means (air fans/blowers). These burners normally operate at an air-side delivery pressure that can be in excess of 2 inches of water column (0.5 kPa). They utilize the air pressure to provide a superior degree of mixing between fuel and air. Also, with forced-draft systems, air control can be better maintained, thus allowing furnaces to operate at lower excess air rates over a wide firing range and allowing the operator to realize economic savings. Figure 14.7 shows an example of a mobile air preheater used during forced-draft testing.

With the use of an air delivery system, the air flow can be measured to provide a direct method of measuring the air flow to validate the air flow through a burner. Fuel flow metering is still used to also determine the air flow. By knowing the

FIGURE 14.7 Forced-draft air preheater.

amount of fuel burned and the excess air exiting the furnace, the amount of air consumed by combustion at the burner can be calculated.

14.3.1.3 Turbine Exhaust Gas (TEG)

Some applications use turbine exhaust gas, often mixed with air, as the source of oxygen for the burners. The turbine exhaust stream or mixture normally contains between 13 and 17 mole % oxygen. These burners are also forced-draft type burners. When test firing a TEG simulation, it is important to match the customer's TEG stream pressure, temperature, and oxygen content approaching the burner. During this type of test, a second set of probes must be arranged to measure the TEG stream composition.

14.3.2 Furnace Gas Temperature Measurement

A suction pyrometer (also known as a suction thermocouple or velocity thermocouple) is widely considered the preferred method for obtaining accurate gas temperature measurements in the harsh environment of a furnace. If a bare thermocouple is introduced into a hot furnace environment for the measurement of gas temperature, measurement errors can arise due to the radiative exchange between the thermocouple and its surroundings. A suction pyrometer is typically comprised of a thermocouple recessed inside a radiation shield. An eductor rapidly aspirates the hot gas across the thermocouple. This configuration maximizes the convective heat transfer to the thermocouple while minimizing radiation exchange between the thermocouple and its surroundings, ensuring that the equilibrium temperature is nearly that of the true gas temperature.

14.3.3 Emissions Analysis

Emission analysis is an important criterion for burner testing. The main pollutants in the combustion products are NOx, CO, unburned hydrocarbons, and particulates.

14.3.3.1 NOx

The chemiluminescent method is most widely used for NOx analysis.[2] This method is capable of measuring oxides of nitrogen from sub-parts per million to 5000 ppm. Newer detector models are free from the disadvantages inherent in analog systems and provide for increased stability, accuracy, and flexibility. The principle of operation of these analyzers is based on the reaction of nitric oxide (NO) with ozone:

$$NO + O_3 \rightarrow NO_2 + O_2 + h\nu \qquad (14.2)$$

The sample, after it is drawn into the reaction chamber, reacts with the ozone generated by the internal ozonator. The above reaction produces a characteristic luminescence with an intensity proportional to the concentration of NO. Specifically, light emission results when electronically excited NO_2 molecules decay to lower energy states. The light emission is detected by a photomultiplier tube, which in turn generates a proportional electronic signal. The electronic signal is processed by the microcomputer into an NO concentration reading. To measure the NOx (NO + NO_2) concentration, NO_2 is transformed to NO before reaching the reaction chamber. This transformation takes place in a converter heated to about 625°C (1160°F). Upon reaching the reaction chamber, the converted molecules along with the original NO molecules react with ozone. The resulting signal represents the NOx. Further details of the workings of a chemiluminescent gas analyzer can be found in any standard text.

14.3.3.2 Carbon Monoxide

The carbon monoxide (CO) exiting a burner will initially increase slowly as the excess air rate decreases. The increase will accelerate as excess air levels continue to decline to near-zero. Typical control points range between 150 and 200 ppm CO. This range usually results in the best overall heater efficiency. Certain localities may require lower emission levels. The presence of unsaturated hydrocarbons can lead to pyrolysis and polymerization reactions, resulting in a greater possibility that CO will be produced. Burners with greater swirl and/or higher combustion air pressure drop (such as forced-draft burners) typically have lower CO emissions at equivalent excess air levels. The reason is that these burners provide a superior degree of mixing to allow improved combustion at lower excess air levels.

Although CO can be continuously monitored by chromatographic analysis, using thermal conductivity detectors, or by FTIR spectroscopy methods, individual analysis is best accomplished using a nondispersive infrared technique. The main advantages of this technique are that it is highly specific to CO and has lower ranges with a wider dynamic range, increased sensitivity and stability, and easy operation because of microcomputer control diagnostics. An added advantage of the technique is that the changes in temperature and pressure of the sample gas are immediately compensated for by the microcomputer, and the results are thus not affected by fluctuations in the operating conditions. The basic principle of these analyzers is based on the radiation from an infrared source passing through a gas filter alternating between CO and N_2 due to rotation of the filter wheel. The CO gas filter acts to produce a reference beam that cannot be further attenuated by CO in the sample cell. The N_2 side of the filter wheel is transparent to the IR radiation and, therefore, produces a measured beam that can be absorbed by CO in the cell. These analyzers can measure 0.1 to 1000 ppm CO under well-controlled conditions. The detailed workings of an IR analyzer can be obtained from a standard text on the subject.

14.3.3.3 O_2 (Wet and Dry)

The oxygen concentration is also conveniently measured by chromatographic techniques using thermal conductivity detectors and also by low resolution FTIR spectroscopy. Individual measurements of oxygen concentration are most widely done by analyzers based on standard polarographic techniques. The detailed working of such analyzers can be found in related texts.

14.3.3.4 Unburned Hydrocarbons (UHCs)

Unburned hydrocarbons (UHCs) increase as the excess air rate decreases. The combustion of hydrogen and paraffin-rich fuel will produce a minimum of combustibles. The presence of unsaturated hydrocarbons leads to pyrolysis and polymerization reactions, resulting in more combustibles. Unsaturated hydrocarbons, chlorides, amines, and the like can plug or damage burner tips, disrupting the desired fuel-air mixing. This can cause a further increase in the combustibles level. Heavy oils are more likely to produce greater levels of combustibles than lighter oils. Heavier components are not as easily atomized and ignited, and therefore polymerization and pyrolysis reactions are more likely to occur. Forced-draft burners provide a better mixing of the fuel-air mixture and therefore produce reduced combustibles at equivalent excess air rates.

Chromatographic techniques are the most widely used for VOC determination in refinery off-gases. Their use as a multicomponent, completely automated, and continuous emissions monitor is not documented in the literature.[3] Coleman et al. have discussed the use of a gas chromatography-based continuous emission monitoring system for the measurement of VOCs using a dual-column (with DB-5 and PoraPlot U, respectively) gas chromatograph equipped with thermal conductivity detectors, in which separation was optimized for fast chromatography. In this system, nine different VOCs plus methane and CO_2 were separated and analyzed every 2 minutes. Because permits are issued to report emission in pounds or tons of pollutants emitted and not on the basis of parts per million (ppm), the setup was equipped with a continuous mass flow measurement device. The data thus collected can be converted to pounds or mass of VOCs emitted. The DB-5 column separates ethanol, isopropanol, *n*-propanol, methyl ethyl ketone, isopropyl acetate, heptane, *n*-propyl acetate, and toluene. The PoraPlot U column separates methane and carbon dioxide. A chromatographic technique using two fused silica columns — one with Dura-Bond and the other with Gas Solid-Q-PLOT — equipped with a flame ionization detector was used by Viswanath to measure VOCs in air.[4]

A technique reported by Pleil et al.[5] uses the fact that the compounds once identified by retention time in the chromatographic analysis can be confirmed by determining a second dimension, such as its mass fragmentation pattern or its infrared absorption spectrum from a highly specific detector such as a mass selective detector (MSD) or a Fourier transform infrared system (FTIR). Even with this combination, care should be taken to avoid occasional confusion among isomeric, co-eluting compounds with similar, strongly absorbing functional groups. Using this technique, Pleil et al.[5] were able to identify and successfully determine more than 40 compounds in the VOCs. They used a cross-linked methyl silicon megabore capillary column with both flame ionization detector and electron capture detector simultaneously. A similar study of VOCs was reported by Siegel et al.[6] They used a DB-1 column with flame ionization detector and a mass selective detector (GC/MSD).

The U.S. EPA guidelines, as presented in "Compendium of Methods for the Determination of Toxic Organic Compounds in Ambient Air, Method TO-14," is slowly becoming the criterion for VOCs.[7] The recommended method uses cryogenic preconcentration of analytes with subsequent gas chromatographic separation and mass spectrometric detection. The methodology requires detecting nanogram and subnanogram quantities. To obtain this high sensitivity, Method TO-14 recommends the use of a selective ion monitoring (SIM) spectrometric technique. The details of the method are discussed by Pleil et al.[7] Evans et al.[8] have also discussed the use of a cryogenic GC/MSD system to measure the VOCs in air in different parts of the country. The sample first passes through a fused silica column to resolve the target compounds. The column exit flow splits such that one-third of the flow is directed to the chromatographic column (with the flame ionization detector) and two-thirds of the flow goes to the mass selective detection system (MSD). The method was found to effectively detect 0.1 ppb by volume of about 25 VOCs.

Larjava et al.[9] have recently reported a comprehensive technique for the determination of nitric oxide (NO), sulfur dioxide (SO_2), carbon monoxide (CO), carbon dioxide (CO_2), and total hydrocarbons (C_xH_y) in the air. The technique used single-component gas analyzers in parallel with a low-resolution Fourier transform infrared (FTIR) gas analyzer. This technique successfully demonstrated that the results obtained by single-component analyzers and FTIR were very close. Online analysis of stack gases with FTIR spectrometry has recently received considerable attention because of the multi-component analysis capability and sensitivity of the method.[9–11] A typical low-resolution FTIR spectrometer uses spectral resolution, BaF_2 optics, a Peltier-cooled semi-conductor detector, and a temperature-controlled multi-reflection gas cell. The advantages of low-resolution FTIR over conventional high-resolution FTIR include its rugged design, high signal-to-noise ratio without liquid nitrogen-cooled detectors, reduced data storage requirements, and increased dynamic range for quantitave analysis.[9]

Jayanti and Jay[12] have summarized studies on estimating VOCs by different techniques employed by various workers.

14.3.3.5 Particulates

Proper combustion of gaseous fuels does not generate significant quantities of combustion-generated particulates. Particulate emissions generally occur with the burner of heavy fuel oils. Burners with greater swirl and/or higher combustion air pressure (such as forced-draft burners) are less likely to produce particulates. They provide a superior degree of mixing to reduce the formation of particulates. Greater atomization of fuel oil into finer particles will reduce particulate emissions. High-intensity burners can considerably reduce particulates. The high degree of swirl, coupled with the high-temperature reaction zone, induces superior combustion of the particulates. However, these burners also emit an increased amount of NOx.

The particulates from hydrocarbon industries are the pollutants emitted by the effluent gases. The most important criterion for the evaluation of particulates is the particle size. It has been observed that different results are obtained using different techniques of collection and analysis.

The U.S. EPA[13] recommended procedures suggest that sampling ports be located at least 8 × duct diameters downstream and 2 × diameters upstream from any flow disturbance. Flue gas should be drawn through a U.S. EPA sampling train.[13] It is important to maintain isokinetic sampling conditions.

Particles can be collected by filtration, impaction, and impingement. Glass fiber and membrane filters are efficient for 0.3-μm particles. These filters can be used in an inline filter holder. The filter holder can be kept inside the sampling port such that the filter attains the temperature of the gas stream.

For particulate collection by impaction, an Anderson type in-stack sampler is used. In cases where the sampler cannot be accommodated inside the sampling port flange, it can be put outside, with an arrangement to heat it to prevent condensation within the sampler. In this type of sampler, the collecting plates are coated by a thin film of silicone grease formed by immersing the plates in a 1% solution of silicone grease dissolved in benzene and drying them overnight at 100°C (212°F).

The collection of particulates by impingement consists of using a series of three or four liquid impingers. These impingers each contain 250 ml distilled water. A common practice is to use impaction and impingement techniques, followed by glass fiber backup filters.

Sampling times also vary according to the technique employed; 1 to 5 minute samples are common for filters and Anderson-type units, while 20 minutes are needed for impingers.

Filter samples are analyzed by light microscopy and scanning electron microscopy (SEM). The liquid samples from the wet impingement device are filtered onto 0.2-μm membrane filters and examined by SEM. The samples from an Anderson sampler can be analyzed by the recommended procedure or by calculations based on Ranz and Wong equations. The Anderson plates can also be examined by SEM to determine the range of particles trapped on the plates. Particle counts can be done by light microscopy using an oil-immersion lens system.[14] Individual particles are compared on the basis of equal area to previously calibrated circle sizes contained on a size comparator. In an exhaustive study, Byers[14] took data on particulates from a refinery effluent and suggested that membrane filters should be preferred when the gas being sampled is at a temperature less than 300°F (150°C). For higher temperatures, the Anderson sampler is suitable provided the plates are suitably coated. Sampling techniques causing agglomeration, such as glass fiber filters, wet collection, bulk grab samples, and scrapping of deposits from collecting surfaces, should be avoided.

14.3.4 Fuel Flow Rate Metering

One of the most important aspects of burner testing is fuel metering. When firing a natural-draft burner, it is difficult to measure the flow rate of combustion air. Therefore, accurately metering the flow rate of each individual component used to make up a fuel blend is necessary to measure the heat release of the burner and its performance. There are many ways to measure flow: differential pressure, magnetic, mass, oscillatory, turbine, and insertion flow meters, just to name a few. For purposes of burner testing, the differential flow meter is discussed. Even limiting this discussion to differential flow meters, there are still several different methods of measurement available. Measuring the differential pressure across a known orifice plate is a common method of measuring the gaseous fuel and steam flow. For liquid fuel firing, a coriolis meter is often used.

The orifice plate is the most commonly applied method.[15] The advantage of using orifice plates is that they are versatile and can be changed to match the flow rate and fuel to be metered. Also, there is a significant amount of data concerning measuring fuel flow via an orifice plate.[15] Finally, there are no moving parts to wear out. The drawbacks to orifice plates are that they are precision instruments and the following must be considered: the flatness of the plate, the smoothness of the plate surface, the cleanliness of the plate surface, the sharpness of the upstream orifice edge, the diameter of the orifice bore, and the thickness of the orifice edge.[15] The critical inaccuracies due to these items can be alleviated by the purchase of ASTM-approved plates, rather than machining the plates. Another drawback is loss of accuracy when measuring flow rates of dirty fuels. While dirty fuels are a way of life for the refining industry, test fuels are clean (no liquid or solid particles in the gaseous fuels), and this concern is minimized. Two items that should be verified when testing with orifice plates are that they are installed in the right direction (the paddle usually indicates the inlet side) and that the correct orifice bore is in the correct flow run. While orifice plates can be used to meter liquids, coriolis meters are often preferred for measuring liquid fuel flows such as No. 6 oil or diesel oil.

The coriolis meter operates on the basic principle of motion mechanics.[15] The coriolis meter is able to measure the mass by measuring the amount of vibration the tube carrying the fluid is undergoing. The coriolis meter is a more expensive means of measurement, but this is often offset by its degree of accuracy and its low maintenance requirements.

14.3.5 Flame Dimensions

The flame shape and dimensions are determined by the burner tile, the drilling of the gas tip, the fuel, and the aerodynamics of the burner. Round burner tiles are used to produce a conical or cylindrical shape. Flat flame burners are designed with rectangular burner tiles and produce fishtail-shaped flames. Many of the liquid fuel burners are designed with round burner tiles and produce a conical flame. The drilling of the oil tip determines the shape and length of the flame. The normal included angle of a burner tip is 40 to 70°. With a 50° included angle, the flame length will be approximately 2 ft per MMBtu/hr for natural-draft burners. Reducing the angle to 40° produces a longer, narrower flame. Increasing the angle to 70° produces a shorter, bushier flame. Forced-draft burners produce a shorter flame because of the better mixing between the fuel and the air. Firing in combination with both liquid and gas fuels will increase the length and volume of the flame and can cause coking of the oil and gas tips.

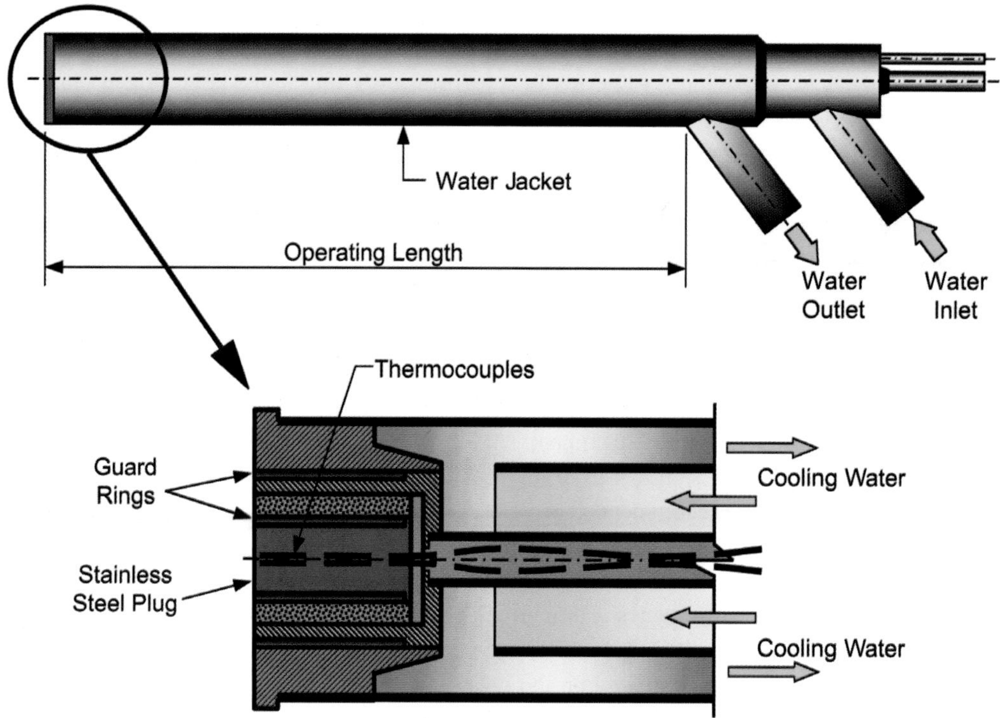

FIGURE 14.8 Heat flux probe schematic.

14.3.6 Heat Flux

Several techniques have been developed to measure heat flux levels at different locations within a furnace. The instruments designed to successfully obtain heat flux data in the hostile environment of a full-scale furnace are typically water-cooled probes, which are inserted through a furnace port at the location of interest. The probes may utilize pyrometers that measure radiant or total (radiant + convective) heat flux levels. The sensing element is typically composed of a thermopile-type sensor that outputs a voltage proportional to the temperature difference between the area of the element exposed to heat transfer from the furnace and the area that is cooled and kept at a relatively constant temperature per the element design. Sensor element designs differ chiefly between the geometry and configuration of the thermopile-type sensing element. Figures 14.8 and 14.9 show schematics of typical heat flux and radiometer designs, respectively. Common designs utilize a plug-shaped thermopile element with the exposed face at one end and the opposite end cooled by contact with a heat sink. Others use a disk-shaped sensor with the temperature gradient existing between the center of the disk receiving radiant energy and the radial edge, which is cooled by contact with a heat sink. A sensor designed to measure only the radiant component of heat flux (radiometer) can utilize a crystal window, gas screen, or a mirrored ellipsoidal cavity to negate convective heat transfer to the sensor. A radiometer is also often equipped with a gas purge in an effort to keep the crystal window or mirrored ellipsoidal cavity clean and free from fouling. Critical parameters to consider when using a heat flux meter include the ruggedness, sensitivity, calibration method, and view angle of the instrument.

14.4 TEST MATRIX (TEST PROCEDURE)

14.4.1 Heater Operation Specifications

Some of the parameters normally measured in burner testing are fuel pressure, airside pressure drop, noise emissions, NOx emissions, CO emissions, UHC emissions, particulate emissions, heat flux profiles, and flame dimensions.

14.4.2 Performance Guarantee Specifications

The primary reasons for conducting a burner test is to determine the operating envelope of the burner as well as the emissions performance. With this data collected, the burner's performance in the field will be more predictable and easier to operate.

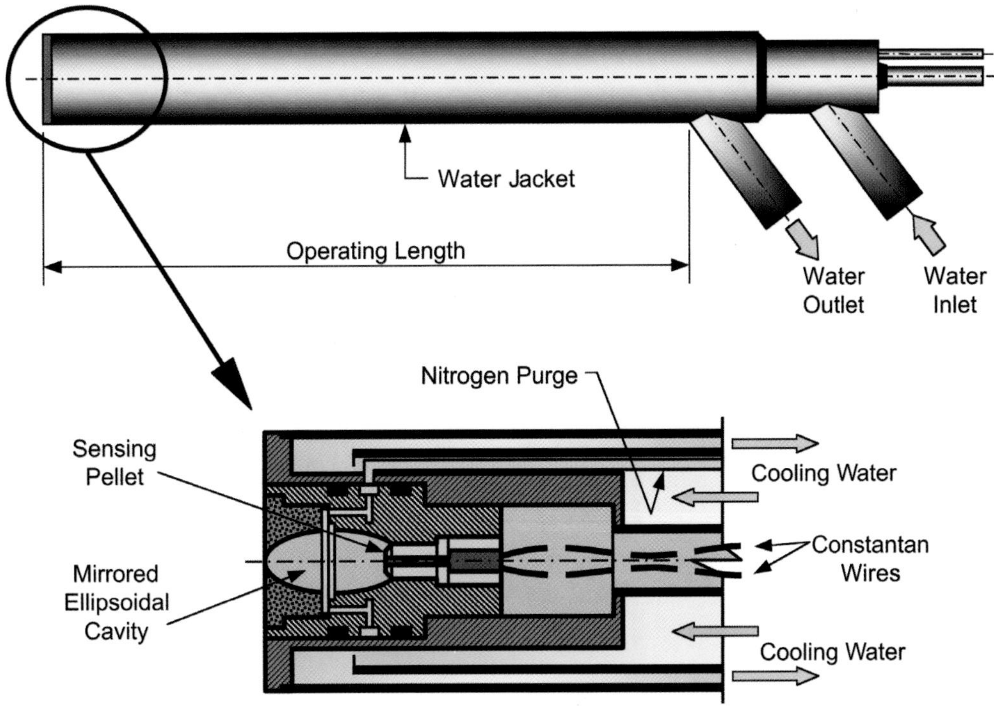

FIGURE 14.9 Ellipsoidal radiometer schematic.

14.4.2.1 Emissions Guarantees

It is important to identify which fuels are the operation or online fuels and which fuels are for start-up or emergency use only. By identifying which fuels the emissions guarantees apply to, the burner can be better optimized to run on the operation fuels.

14.4.2.2 Noise

Noise emissions are becoming as important as stack emissions. With some refineries located near populated areas, it is important to keep noise to a minimum. Burner testing is usually conducted on a single burner and noise emissions are usually measured 1 m (3 ft) from the burner. Data collected during the test include an overall dBA measurement and an octave band measurement ranging from 31.5 to 8000 Hz. When collecting noise data, it is important to measure it with the burner operating and without it operating, to obtain the background noise, which may or may not be required to determine the noise contribution from the burner.

14.4.2.3 Fuel and Air-side Pressure Drop

The fuel and air-side pressure drop also need to be verified during the test. The test confirms that the burner will have the correct capacity for proper operation.

The fuel-side pressure drop is displayed during the test. It is important that the test engineer ensure that the customer's fuel will meet the design pressure requirement based on the data collected on the test fuels.

When verifying the air-side pressure drop, the test engineer must determine the elevation and the range of the ambient air temperatures that the burners will be subjected to once installed in the field.

14.4.2.4 Flame Dimension Guarantees

Flame dimensions in a full-scale furnace are typically made by subjective measurement. The flame envelope is most often determined by visual observation. This operating parameter is important to ensure that the flame will not impinge on the furnace process tubes or interact with another burner's flame. Flame impingement on the tubes can damage the process tubes and cause the furnace to prematurely shut down for repairs — at great expense to the operator. Flame interaction between two or more burners can result in longer, more uncontrollable flames and higher emissions. It is important to identify the burner spacing, the furnace dimensions, and the customer's desired flame dimensions. With this information, the test engineer can fine-tune the flame envelope to improve the burner's performance in the customer's heater.

14.4.3 Definition of Data to be Collected

Prior to installation and testing a burner, a test matrix (test procedure) must be developed. A sample fuel gas specification is shown in Table 14.4. A typical test procedure might resemble that shown in Table 14.5. With a well-developed test procedure, the data collected from a test will be meaningful and will assist the operator in running the furnace and predicting the performance from the furnace.

REFERENCES

1. *Burners for Fired Heaters in General Refinery Services*, API Publication 535, 1st ed., American Petroleum Institute, Washington, D.C., July 1995.
2. B.K. Gullett, M.L. Lin, P.W. Groff, and J.M. Chen, NOx removal with combined selective catalytic reduction and selective non catalytic reduction: pilot scale test results, *J. Air Waste Manag. Assoc.*, 44, 1188, 1994.
3. W.M. Coleman, L.M. Dominguez, and B.M. Gordon, A gas chromatographic continuous emission monitoring system for the determination of VOCs and HAPs, *J. Air Waste Manag. Assoc.*, 46, 30, 1996.
4. R.S. Viswanath, Characteristics of oil field emissions in the vicinity of Tulsa, Oklahoma, *J. Air Waste Manag. Assoc.*, 44, 989, 1994.
5. J.D. Pleil, K.D. Oliver, and W.A. McClenny, Ambient air analyses using nonspecific flame ionization and electron capture detection compared to specific detection by mass spectroscopy, *J. Air Waste Manag. Assoc.*, 38, 1006, 1988.
6. W.O. Siegel, R.W. McCabe, W. Chun, E.W. Kaiser, J. Perry, Y.I. Henig, F.H. Trinker, and R.W. Anderson, Speciated hydrocarbon emission from the combustion of single component fuels. I. Effect of fuel structure, *J. Air Waste Manag. Assoc.*, 42, 912, 1992.
7. J.D. Pleil, T.L. Vossler, W.A. McClenny, and K.D. Oliver, Optimizing sensitivity of SIM mode of GC/MS analysis for EPA's TO-14 air toxics method, *J. Air Waste Manag. Assoc.*, 41, 287, 1991.
8. G.F. Evans, T.A. Lumpkin, D.L. Smith, and M.C. Somerville, Measurement of VOCs from the TAMS network, *J. Air Waste Manag. Assoc.*, 42, 1319, 1992.
9. K.T. Larjava, K.E. Tormonen, P.T. Jaakkola, and A.A. Roos, Field measurements of flue gases from combustion of miscellaneous fuels using a low resolution FTIR gas analyzer, *J. Air Waste Manag. Assoc.*, 47, 1284, 1997.
10. K. Wülbern, On line messung von rauchgasen mit einen FTIR spektrometer, *VGB Kraftwerkstechnik*, 72, 985, 1992.
11. J.C. Demirgian and M.D. Erickson, The potential of continuous emission monitoring of hazardous waste incinerators using FTIR spectroscopy, *Waste Management*, 10, 227, 1990.
12. R.K.M. Jayanti and B.W. Jay (Jr.), Measurement of toxic and related air pollutants, *J. Air Waste Manag. Assoc.*, 40, 1631, 1990.
13. Standard Performance for New Stationary Sources, Environmental Protection Agency, Federal Register, 36, No. 247, December 23, 1971.
14. R.L. Byers, Evaluation of effluent gas particulate collection and sizing methods, *API Proc. Division Refining*, 53, 60, 1973.
15. D.W. Spitzer, Practical Guides for Measurement and Control, Instrument Society of America, 1991.

TABLE 14.4 Test Procedure Gas Specification Sheet

John Zink Company	**Burner Performance Demonstration**
Tulsa, OK (English)	**Burner Specification (Gas)**

Date:	2/4/00	Rev. No:	0	J.Z. Quote No:
Customer:				Customer P.O No:
Burner:		PSFFG-45M		J.Z. Sales Order No:
Drawing:				Capacity Curve No:
User:				Project Engineer
Jobsite:				Test Engineer

Customer Heater Data

Spec Reference:		Direction of Firing:	
Item No:		Setting Thickness:	
Quantity of Burners:	60	Burner Spacing:	39 inches (1 meter)
Type of Heater:	Ethylene	Pilot:	ST-1S manual pilot
Firebox Dimensions:		Elevation:	<1000 feet ASL
Draft Type:	induced		

Specifications

Fuel Composition	LHV	M.W.	**Fuel A**	**Fuel B**	**TNG**	
Component:	Btu/scf	#/# mol	vol%	vol%	vol%	vol%
Tulsa natural gas	913.0	17.160	34.00	80.00	100.00	
Hydrogen	273.8	2.022	38.00	20.00		
Propane	2314.9	44.100				
Propylene	2181.8	42.080	28.00			
Carbon Dioxide		44.010				
Butane	3010.8	58.120				
Lower Heating Value:		Btu/scf				
Molecular Weight:		(#/# mol)	18.385	14.132	17.160	
Isentropic Coefficient:		(Cp/Cv)	1.2600			
Temperature:		(degF)	60	60	60	
Pressure Available:		psig	20.0	20.0	20.0	
Heat Release per Burner:						
Design Maximum:		(MMBtu/hr)	6.800	6.800	5.913	
Normal:		(MMBtu/hr)	5.913	5.913	1.360	
Minimum:		(MMBtu/hr)	1.360	1.360	6.800	
Flame Dimensions:	@	ft				
Cross Section (Dia.):			4.000	4.000	4.000	
Length:			16.000	16.000	16.000	
Turndown:			5:1	5:1	5:1	
Excess Air @ Design:			10	10	10	
Conditions @ Burner:						
Heater Draft Available:		("w.c.)	0.80	0.80	0.80	
Burner dP @ Design:		("w.c.)	0.80	0.80	0.80	
Combustion Air Temp.:		(degF)	100	100	100	
Guarantees:						
NOx:		ppm(vd)	100	100	100	
CO:		ppm(vd)	50	50	50	
Note 1 Particulate:		#/MMBtu (lhv)				
Note 1 UHC:		#/MMBtu (lhv)				
DSCF/MMBtu @ 3% O2(d)		(lhv)				
Noise:		dBA (spl) @ 3 ft.	85	85	85	
Conditions of Guarantees:						
% Oxygen Corrected to:			3	3	3	
Combustion Air Temp.:		(degF)	100	100	100	
Heat Release:		(MMBtu/hr)	4.500	4.500	4.500	
Furnace Temperature:		(degF)	2100	2100	2100	

Comments: General — Heat Releases are shown as Net or Lower Heating Value
Note 1: Particulate and UHC are NOT measured during Test Demonstration

Courtesy of John Zink Co., Tulsa, OK.

TABLE 14.4 (continued) Test Procedure Gas Specification Sheet

Typical Refinery Fuel Composition

Fuel Component		
Name	**Formula**	**volume %**
Methane	CH_4	8.13*
Ethane	C_2H_6	19.9*
Propane	C_3H_8	0.30
Butane	C_4H_{10}	0.06
Ethylene	C_2H_4	32.0**
Propylene	C_3H_6	0.78
Butylene	C_4H_8	0.66
1-Pentene	C_5H_{10}	0.07
Benzene	C_6H_6	0.12
Carbon Monoxide	CO	0.22
Hydrogen	H_2	37.8***

Tulsa Natural Gas Composition

CH_4 (Volume %)	93.4	C_4H_{10} (Volume %)	0.20
C_2H_4 (Volume %)	2.70	CO_2 (Volume %)	0.70
C_3H_8 (Volume %)	0.60	N_2 (Volume %)	2.40
LHV (Btu/scf)	913	HHV (Btu/scf)	1012
Molecular weight	17.16	Specific heat rate	1.3
Adiabatic flame temperature (F)			3452

	Refinery fuel:	**Test fuel:**
LHV (Btu/scf)	1031	1026
HHV (Btu/scf)	1124	1121
Molecular weight	18.09	18.38
Specific heat ratio @ 60F	1.27	1.26
Adiabatic flame temperature (F)	3481	3452
Wobbe Index	1422	1407

* Balance of fuel is primarily methane and ethane.

** Note: level of olefins in the fuel.

*** Note: hydrogen content

TABLE 14.5 Example Test Procedure

John Zink Company		**Burner Performance Demonstration**			**Date:**	**2/4/00**
Tulsa, OK (English)		**Burner Test Procedure**			**Rev. No:**	
J.Z. SO No:		**Customer:**	**User:**	**Burner Drawing No:**		**Furn.**
Burner:	PSFFG-45M	**P.O. No:**	**Jobsite:**	**Capacity Curve:**		

Data Point	Fuel	Liberation (MMBtu/hr)	Air Temp. (°F)	Excess O2 (%O2 (dry))	Draft (in. w.c.)	deltaP (in. w.c.)	Description/Comments
1	TNG		60		0.80		Burner light off on Tulsa natural gas (TNG).
2	A	6.800	60	2.0	0.80		Maximum heat release, set air to 10% excess air, record noise and heat flux profile.
3	A		60		0.80		Above maximum heat release, increase fuel flow until CO > 250 ppm.
4	A	5.910	60	2.0	0.80		Reduce heat release to normal heat release, set air damper to maintain 10% excess air.
5	A	1.360	60		0.80		With damper set for normal heat release, reduce fuel flow to minimum heat release.
6	A		60		0.80		Determine absolute minimum heat release.
7	B	6.800	60	2.0	0.80		Maximum heat release, set air to 10% excess air, record noise and heat flux profile.
8	B		60		0.80		Above maximum heat release, increase fuel flow until CO > 250 ppm.
9	B	5.910	60	2.0	0.80		Reduce heat release to normal heat release, set air damper to maintain 10% excess air.
10	B	1.360	60		0.80		With damper set for normal heat release, reduce fuel flow to minimum heat release.
11	B		60		0.80		Determine absolute minimum heat release.

General Comments:

1) Refer to the schematic of the equipment set-up for approximate instrument and sight port locations.
2) The firebox temperature will be within the range of 1900°F and 2100°F for Normal and Maximum Heat Release. It will be lower at reduced rates.
3) Once the O2 has stabilized, at the target value for a given test point data will be recorded and adjustments for the next point will begin.
4) All data points will be run with the furnace draft as specified, controlling excess O_2 with the blower (forced draft), or with the damper (register(s)) (natural draft).
5) The proposed time frame for the duration of the test is approximate and could be longer or shorter depending on equipment operation and/or weather.
6) Standard tolerances on measurements will be as follows: Air Temp. ±20°F, Fuel Temp. ±20°F, O_2 ±0.2%, Draft or dP ±6% of specified.

Data to be Recorded:

Fuel Flow	×	Burner dP	×
Fuel Press.	×	NOx	×
Fuel Temp.	×	CO	×
Air Temp.	×	O_2	×
Draft	×	Noise	×
Box Temp.	×		

Test Procedure Acceptance:			
_ Approved _ Approved as Noted	**Company:**	**Signature:**	**Date:**
Test Acceptance:			
Name:	**Company:**	**Signature:**	**Date:**

Courtesy of John Zink Co., Tulsa, OK.

Chapter 15
Installation and Maintenance

Roger H. Witte and Eugene A. Barrington

TABLE OF CONTENTS

15.1 Introduction 450

15.2 Installation 450

15.2.1 Prepare the Heater 450

15.2.2 Burner Pre-installation Work 450

15.2.3 Burner Mounting 451

15.2.4 Tile Installation 453

15.2.5 Connecting the Burner to the Heater 455

15.2.6 Burner Installation Inspection 456

15.2.7 Air Control 456

15.2.8 Fuel Piping Design 458

15.3 Maintenance 459

15.3.1 Gas Tip and Orifice Cleaning 462

15.3.2 Oil Tip and Atomizer Cleaning 463

15.3.3 Tile 463

15.3.4 Flame Stabilizer 463

15.3.5 Air Registers and Dampers 464

15.3.6 Pilot Burners 465

References 465

15.1 INTRODUCTION

The correct installation of burners into a heater is essential for good performance. During installation — whether in the initial heater construction or at turnaround — the critical dimensions and orientations of the burners should be observed. Adequate maintenance will allow the burners to deliver the design performance for many years.

15.2 INSTALLATION

Improperly installed burners will not operate efficiently and may damage the heater during service. The burners must be installed in accordance with the burner manufacturer's recommended procedure.

15.2.1 Prepare the Heater

15.2.1.1 Safety

A safety checklist for the work to be performed should be prepared and reviewed by all personnel working on the heater. This will vary, depending on the type of project and the requirements of the installing and purchasing companies. A common requirement is unimpeded access for the final burner placement. A permit detailing the immediate safety requirements, will normally be issued by the user company.

15.2.1.2 New Heaters

Installing new burners on new heaters is usually easier than retrofitting burners to an existing heater. The installer compares the heater manufacturer's and burner manufacturer's drawings for the required cutout in the heater steel and the burner mounting bolt pattern, including bolt circle and size (see Figure 15.1). These are compared to the field measurements of the same dimensions. If differences are discovered, they must be resolved before proceeding with the installation work.

The heater steel is checked for interferences with the burner steel and for flatness. Any necessary work should be completed before attempting burner installation.

15.2.1.3 Existing Heaters

Installing burners on existing heaters is usually more difficult because some refractory must often be removed, the new burners may be of different sizes than the old burners, and/or the heater steel will often be warped after years of service (see Figure 15.2). The warping may make it difficult to maintain the installation tolerances required for proper burner performance. Figure 15.3 depicts a burner improperly installed at an angle. Such an installation can cause undesirable flame characteristics and make it difficult to operate burner components such as air registers.

The burner steel should be level so that tolerances can be maintained when the burner is installed. If the heater steel is more than 0.25 in. (6 mm) out of level, a "donut ring" or adapter plate should be installed to provide a level burner mount. The adapter plate is a steel ring installed in the opening of the burner and welded to the heater steel to provide a level mounting surface for the burner (see Figure 15.4).

15.2.2 Burner Pre-installation Work

15.2.2.1 Bill of Materials

The burner should be unpacked and inspected to ensure that all parts are in accordance with the bill of materials included with the burner. Missing parts or parts that appear to be incorrect or damaged should be immediately reported to the burner manufacturer for correction. The bill of materials will list the main burner parts, such as the burner assembly and burner tile, and other miscellaneous parts. Confirm that any preassembled parts are as shown on the burner drawing.

15.2.2.2 Burner Drawing

The burner manufacturer provides a drawing of each type of burner supplied to each customer. Do not try to install a burner without the appropriate drawings and instructions. The burner drawing will show the outside tile dimensions; burner orientation; and positions of the gas tips, the pilot tip, the oil tip (if provided), and the burner tile(s). The burner mounting bolt pattern will also be shown. All of this information is required to ensure proper installation.

15.2.2.3 Installation and Operating Instructions

With each burner project, the manufacturer typically sends a copy of installation and operating instructions for the burner and any auxiliary equipment provided. These instructions will provide the following basic information:

1. safety summary
2. design specifications
3. reference drawings
4. receiving and handling of the equipment
5. installation of the burner and auxiliary equipment
6. operation of the burner and auxiliary equipment
7. troubleshooting recommendations
8. maintenance instructions
9. recommended spare parts list
10. service available

The burner installer should be familiar with the installation and operating manual as it provides the information necessary for a satisfactory installation. If old burners are being rein-

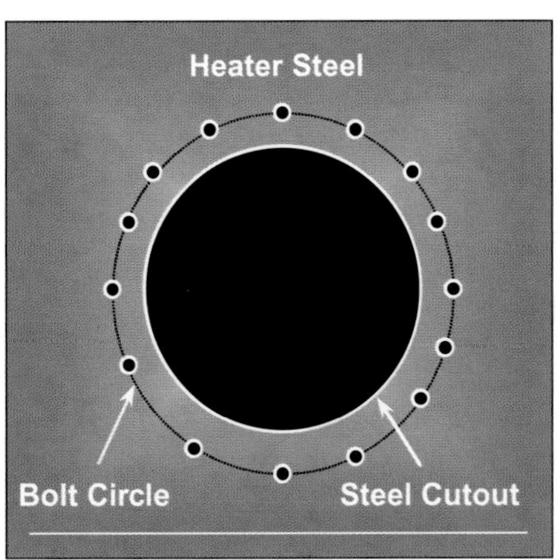

FIGURE 15.1 Heater cutout and burner bolt circle on a new heater.

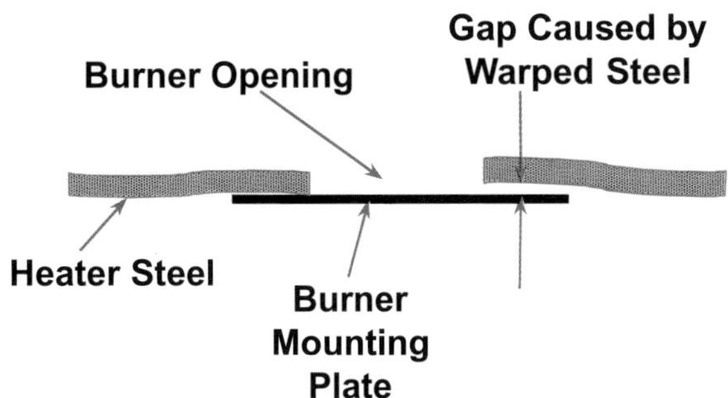

FIGURE 15.2 Warped steel on the shell of a heater.

stalled on a heater, a new set of installation instructions should be requested from the burner manufacturer.

15.2.3 Burner Mounting

15.2.3.1 Heater

Burners are mounted on the heater steel in a variety of ways. The design of the heater, the heater manufacturer's practices, and the process to which the heater is applied all impact how the burners are mounted. Some mounting options include:

- on the floor steel, firing vertically upward (see Figure 15.6)
- on the wall steel, firing horizontally (see Figure 15.7)
- on the radiant-zone roof steel, firing vertically downward (see Figure 15.8)
- on a common air or noise reduction plenum, which is in turn attached to the heater steel (see Figure 15.9)
- on an individual noise reduction plenum, which is attached to the heater steel (see Figure 15.5)

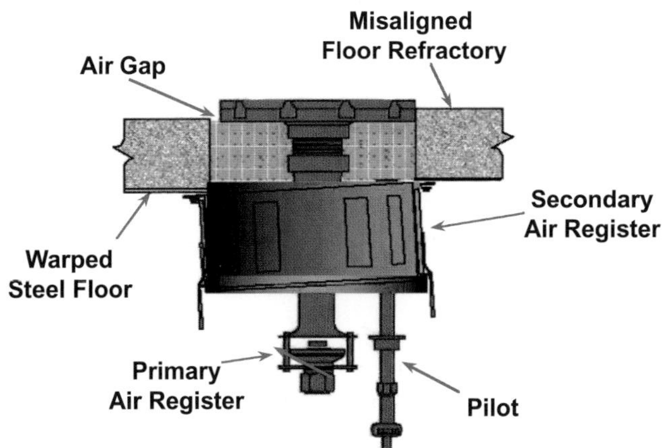

FIGURE 15.3 Burner improperly installed at an angle due to a warped shell.

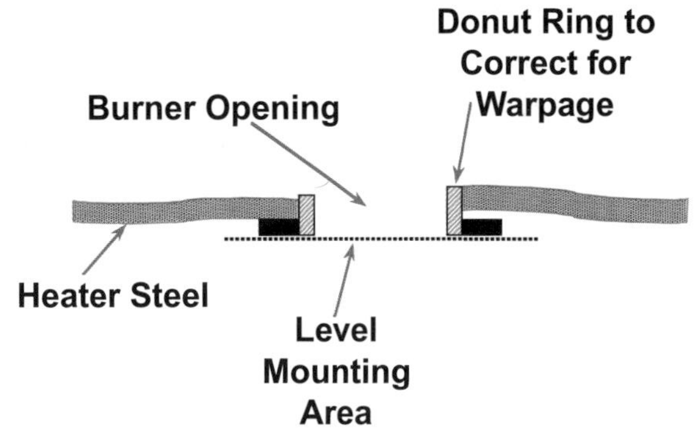

FIGURE 15.4 Donut ring for leveling burner mounting onto the warped shell of a heater.

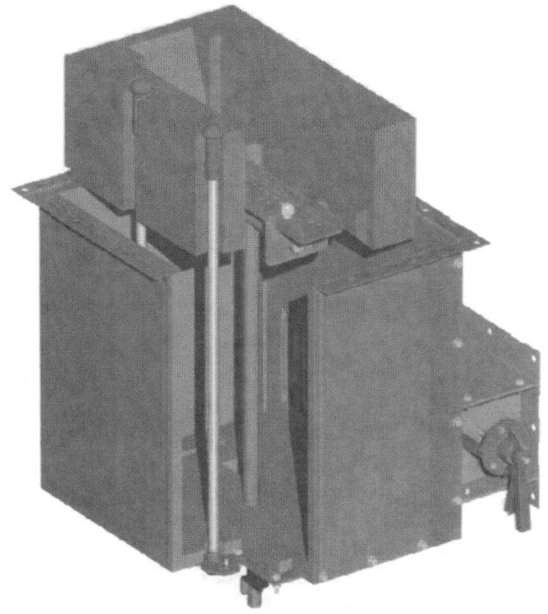

FIGURE 15.5 Individual noise reduction plenum.

FIGURE 15.6 Burner mounted on the floor of a heater.

FIGURE 15.7 Burners mounted on the side of a heater.

FIGURE 15.8 Burner mounted on the top of a heater.

FIGURE 15.9 Burner mounted in a common plenum.

15.2.3.2 Burner

The burner is commonly attached to the heater steel in one of three ways:

1. The air register mounting flange can be bolted directly to the steel casing. The burner air register supports the burner parts in the throat of the burner (see Figure 15.6).
2. The burner front plate is bolted to the air plenum, which is attached to the steel casing. Figure 15.9 shows a burner mounted in a plenum in common with other burners. The plenum may be provided for noise abatement or for distribution of pressurized and/or preheated air from a forced draft system.
3. The burner with integral individual plenum box for noise reduction is bolted to the heater steel (see Figure 15.10).

The type of mounting is a function of the burner model selected jointly by the heater and burner manufacturers, considering the process and heater design characteristics, and local regulations. Complete a dimensional check of the burner mounting plate and the heater steel mounting location to identify any problems. If the burner is plenum mounted, the air register and plenum depths should be checked prior to attempting assembly to identify any possible problems. This is particularly important if burners are provided for more than one heater. Ensure that the proper burners are allocated to each heater.

15.2.3.3 Burner Piping

The fuel and atomizing medium piping to the burners should have zero loading on the burner connection; the burners are not designed to carry piping loads. Piping loads can cause burner misalignment (see Figure 15.11) and failure to maintain tolerances or position on the gas and oil tips within the burner. Unanticipated loads can also cause air registers to bind and be difficult or impossible to operate. All piping should be supported independently of the burner.

The piping should be installed with consideration for the need to remove the complete burner or individual parts with minimum piping rework. Flanges or unions should be located so that the piping can be easily removed when necessary. If the piping is to be insulated, it must be installed with the necessary clearances to allow the covering to be installed. The piping should not interfere in any way with the operator's access to the burner viewports or operating functions (e.g., air register adjustment). Inspection should confirm all of these features.

The burner piping connections may be either threaded or welded, as the purchasing company's standards require. There should be enough flexibility in the piping to easily move the burner gas and oil tips into and out of the burner (for proper positioning) when the piping is connected. Sometimes, flexible metal hoses are used for the burner fuel connections to make the positioning easier. These flexible connectors should be made of steel to meet the required temperature and pressure ratings and provide reasonable durability. The end connections of flexible steel hoses are points of weakness unless designed and fabricated by reliable suppliers. The hoses should have a braided armor covering to resist impact and rough handling. Care should be taken in handling and installation to avoid sharp bends or "kinks" that can be the cause of catastrophic hose failure.

15.2.4 Tile Installation

The burner tile is key to forming the correct flame pattern within the firebox. It also forms the orifice that controls the flow of air to the combustion reaction. If not installed in the heater as designed by the burner manufacturer and shown on the drawings, the flame patterns and heat flux distribution within the heater are likely to be adversely affected. Depending on the model of the burner, the burner tile can be designated as primary, secondary, and in some cases, tertiary tile.

FIGURE 15.10 Burner in an individual plenum box mounted to a heater.

FIGURE 15.11 Piping improperly loaded on the burner inlet.

15.2.4.1 Heater Refractory

The refractory on the inside surface of the heater casing is provided to protect the steel casing and structure from the internal temperatures and to reduce heat loss from the heater. The refractory materials and thickness are commonly selected to limit the casing steel outside temperature to a specified value. API Standard 560 provides guidance as to the selection of the casing design temperature and refractory materials.[1] The heater designer determines the

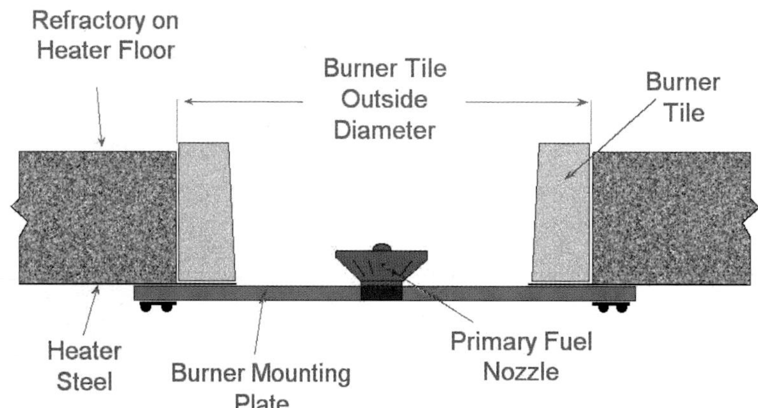

FIGURE 15.12 Cross section of a burner tile.

heater refractory thickness. The burner secondary tile projection from the casing is at least as great as the heater refractory thickness. Some burner tiles may extend an additional 1.5 in. beyond the heater refractory.

15.2.4.2 Secondary Tile

The secondary tile is normally the main burner tile attached to the heater floor, walls, or roof, depending on burner orientation. If the burner is floor-mounted, the secondary tile rests on the steel floor around the burner opening. If wall-mounted, the burner is supported by both the heater steel and the refractory wall of the heater. The secondary tile of a roof-mounted burner is specifically designed to be hung from the roof steel of the heater firebox.

The burner tile is commonly designed for a maximum service temperature of 2400 to 3000°F (1425 to 1650°C) and fabricated in one or more pieces, depending on the size and design of the burner. Figure 15.12 reveals a cross-sectional view of a round burner tile with a number of tile sections. The tile is normally supplied by the burner manufacturer.

Because the secondary tile forms the air orifice to control the airflow to the combustion reaction, its dimensions are critical. The installation must not alter the outside dimensions shown on the burner manufacturer's drawing.

15.2.4.2.1 Tile Installation

Most commonly, the burner manufacturer will specify the outside diameter of the burner tile to the heater manufacturer and installer. The manufacturer or heater installer will form the refractory covering the surface to which the burner mounts, leaving an opening for the burner that typically includes an expansion gap of 0.5 in. (1.3 cm) around the periphery. The burner installer will then install the secondary tile on the centerlines of the burner opening and place ceramic fiber refractory in the expansion gap between the tile and the refractory. The heater refractory should be installed prior to burner tile placement.

An alternative secondary tile installation method involves the burner manufacturer placing the tile into a (usually metallic) tile case. In that case, the burner manufacturer is responsible for maintaining the burner tile dimensions and tolerances. The installation contractor attaches the tile case to the heater and installs the ceramic fiber insulation in the expansion gap around the tile case.

15.2.4.2.2 Tile Tolerances

Consult the manufacturer's burner drawing for the burner tile dimensional tolerances; the tolerances for the secondary tile are typically ± 0.5 in. (± 1.3 cm). Each piece of a tile is normally sealed to its neighbors with relatively thin mortar completely covering and sealing the joined sides. If too much mortar is applied, the tile tolerances cannot be maintained to specification. After all tile sections are installed and before the mortar sets, the burner dimensions are checked. On a round flame burner, the tile outside diameter is measured in four directions to ensure that the dimensions are within the tolerances (see Figure 15.13). If the tile is rectangular, check the dimensions at two locations on each of the long and short sides (see Figure 15.14). After all other interior work on the heater is complete, recheck the tile condition and dimensions and confirm that no loose material is blocking the throat.

All dimensions of the tile are on the burner drawing, and the tile installation must be in accordance and within the range of tolerances. Table 15.1 shows the difference in area of a burner that has different tile dimensions. The flow rate and distribution of air through the burner are a function of the tile open area and shape. Larger open areas allow a higher flow rate of air than smaller areas for the same pressure drop or draft loss. Improper tile areas result in maldistribution of air flow between multiple burners.

15.2.4.2.3 Tile Height

The height of the tile installed in the burner opening is shown on the manufacturer's drawing. The tile height is related to the thickness of the adjacent heater refractory. Some burner models have the tile height the same as the refractory thickness. For other models, the tile height is 1 to 1.5 in. (3 to 4 cm) greater than the adjacent refractory thickness. The installer and the inspector must confirm that the tile height and the adjacent heater refractory thickness are in accordance with the heater drawings. If the burner tile height is not correct, the burner may exhibit flame instability, resulting in potentially unsafe operating conditions.

It is not good practice to install a burner without a tile, using only the heater refractory to form the burner opening. Burner tile is much more durable than common heater refractories and less likely to be damaged in service. If the opening is damaged or worn, the air distribution and flame quality suffer.

15.2.4.2.4 Expansion Joints

The burner tile is surrounded by the heater refractory. When the burner is fired, the flame burns on or near the tile, and the tile temperature becomes greater than that of the adjacent refractory. The burner tile expands into the ceramic fiber-packed expansion gap mentioned previously. If the expansion gap is not provided, the burner tile may be crushed by the adjacent refractory. The tile may crumble in spots and the dimensions of the air flow passage may change, thus adversely affecting the burner performance and potentially causing a poor flame pattern and unsafe conditions. A burner tile surrounded by ceramic fiber heater refractory does not require an expansion gap.

15.2.4.2.5 Oil Burner Tile (or Primary Tile)

Fuel oil tiles are an integral part of some combination burners, so named because they can fire a combination of fuels, liquid or gas. This tile is located in the center of the burner secondary tile and is normally some distance below the top of the secondary tile. These tiles are often set in or poured in a steel tile case that is mounted to the burner front plate.

15.2.5 Connecting the Burner to the Heater

The burner is commonly bolted to the heater casing steel. The number and size of bolts and the bolt pattern are established by the burner manufacturer. The bolts support and properly locate the burner on the heater steel and are instrumental in holding the proper tolerances of burner installation in the center of the tile opening.

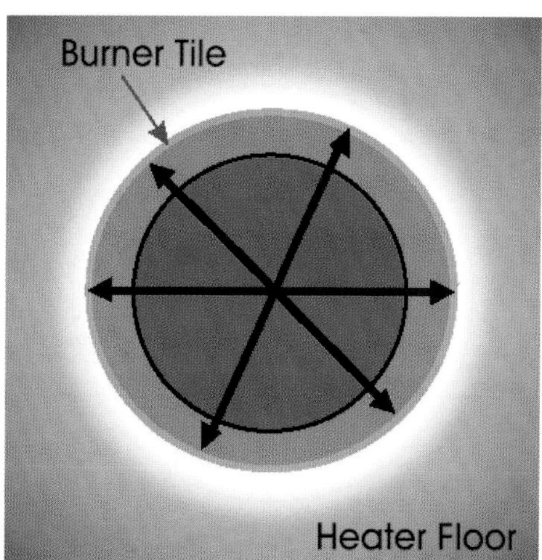

FIGURE 15.13 Sketch showing a round tile measured in three different diameters.

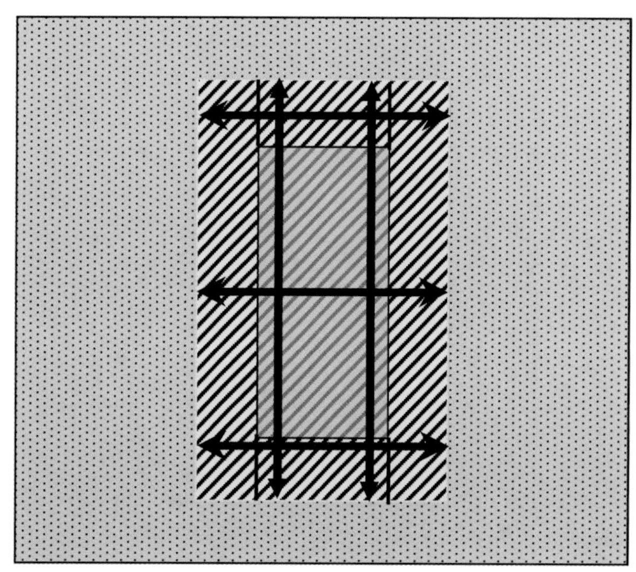

FIGURE 15.14 Sketch showing a rectangular tile measured at different lengths and widths.

On new heaters, bolts are mounted on flat steel surfaces and the connection of the burner to the heater is easily within tolerances. On heaters that have seen service, the heater steel is often not flat and the bolts and bolt pattern may not fit well to the new burner. The diameter of the selected bolt holes may have to be enlarged in the field. The modifications must not compromise the dimensions and tolerance of the finished installation or poor burner performance may result.

TABLE 15.1 Burner Throat Area for Different Tile Dimensions

Diameter in Inches	Area in Square Inches	Diameter in Centimeters	Area in Square Centimeters
10	78.54	25.40	506.71
12	113.10	30.48	729.66
14	153.94	35.56	993.15
16	201.06	40.64	1297.17
18	254.47	45.72	1641.73
20	314.16	50.80	2026.83
24	452.39	60.96	2918.63

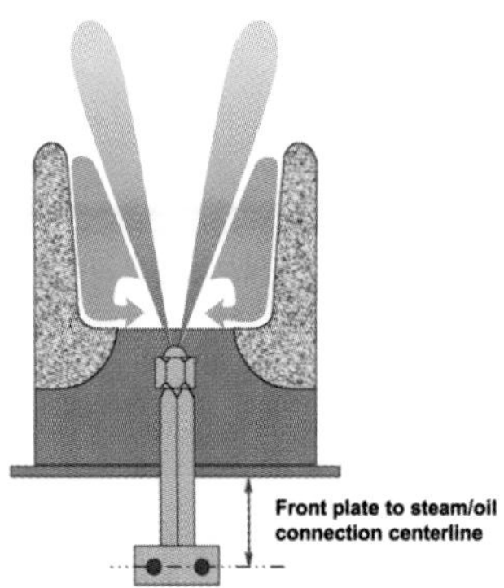

FIGURE 15.15 Oil tip in combination burner showing oil tip locations.

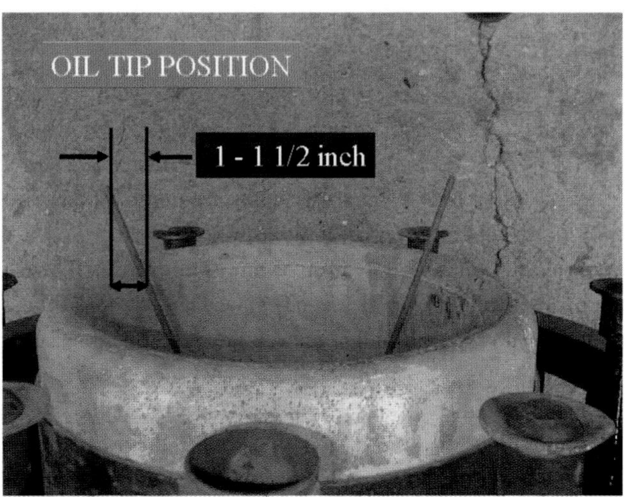

FIGURE 15.16 Welding rod in an oil tip.

On forced-draft, air-preheated fired heaters, the connection between the burner and the heater steel should be installed with a gasket to ensure that there is no leakage of hot air from the air plenum. Failure to seal the air may cause a safety hazard to operating personnel.

15.2.6 Burner Installation Inspection

Once the burners are bolted onto the heater or plenum and the surrounding refractory is placed, everything must be inspected to confirm that all dimensions and orientations are correct. In addition to checking the tile(s) as previously discussed, the positions and orientations of the gas tips, oil tip, and pilot tip must be checked. Any protective tape must be removed from the tips, and it is necessary to confirm that the ports are clear of foreign material.

15.2.6.1 Gas Tips

The position of the gas tips in relation to the flame holder must be correct in order to maintain stable burner operation. The flame holder may be a conical diffuser, a swirler, or a ledge in the secondary tile. The jets of fuel from the gas tip ports must be directed correctly, so the orientation of the tip ports is also important. The installer must position and orient the gas tips in accordance with the burner manufacturer's drawings.

15.2.6.2 Oil Tips

Figure 15.15 shows an oil tip located within an oil tile. Because there are different designs of oil tiles and tips used in burners, the drawings must be reviewed to determine the types used and the correct installation parameters. The dimension shown on the oil gun drawing from the burner's front plate to the centerline of the steam and oil connection is an estimate and must be checked after installation of the gun. The tip location with respect to tile features is also shown on the drawings. Another way to check the tip position within the tile is to insert welding rods into the oil tip exit ports (see Figure 15.16). These rods indicate the oil spray pattern that will be discharged from the exit ports. The rods should be about 1 to 1.5 in. (3 to 4 cm) from the edge of the oil tile when the oil gun is correctly positioned. This is a technique sometimes used by the operator to verify correct positioning.

15.2.6.3 Pilot Tip

The pilot tip is typically located close to the main gas tips so that the pilot flame will contact and ignite the gas discharged from the main tips. The burner drawing will show the correct pilot tip location. In Figure 15.17, the pilot is not properly located and should be repositioned correctly per the drawing in Figure 15.18. The pilot tip is usually at the plane of the diffuser flame holder. On spider gas tips, the pilot may be located 1 to 1.5 in. below the plane of the spider.

15.2.7 Air Control

The maximum air flow across the burner is primarily controlled by the pressure drop across the burner tile throat. The total air flow is adjusted with the burner air registers or dampers, the heater stack damper(s), or with dampers or speed control on fans. Air flow is usually adjusted to meet a target level of excess oxygen in the firebox.

15.2.7.1 Burner Air Registers

The common air register consists of two steel plates, each rolled and welded into a cylinder, with similar openings in each cylinder's sides. One cylinder is mounted so that it is stationary; the other cylinder is free to rotate (see Figure 15.19). When the free cylinder is rotated such that the openings match or are aligned, the maximum amount of air can flow through the register. As the free cylinder is rotated, the area for air flow through the openings is varied, as is the amount of air flowing to the burner. Air registers should be exercised after installation and periodically thereafter to verify functionality. If binding is noted, it should be corrected; try lubrication with graphite or removal of foreign material between the steel cylinders.

Older burners may have cast iron registers that operate as above. Other registers may consist of vanes whose positions are fixed or variable, depending on burner design. Changing the position of the vanes will vary the air flow rate to the burner.

After all interior work on the heater has been completed, the registers should be inspected to ensure operability and that loose debris has been cleared.

15.2.7.2 Burner Air Dampers

On some newer burners, the air registers just described have been replaced by a damper on each burner's air inlet box. The damper can be a single- or multiple-bladed design to control the incoming air flow. Seals can be specified and installed on the blade(s) in an attempt to minimize air flow when the damper is closed.

The dampers should be exercised after initial installation and periodically thereafter to verify functionality. Also check that any loose debris has been cleared from the air passages. The seals should be inspected after a period in service and adjusted so that they seal properly when the damper is in the closed position.

15.2.7.3 Fan Dampers

Fan dampers are used to control fan volumetric capacity and should be installed near the inlets of forced-draft or induced-draft fans. These are usually multiple blade units, with either parallel or opposed blade operation, that mount in the fan inlet box. An alternate design is the inlet vane assembly that mounts directly on the fan air inlet.

FIGURE 15.17 VYD burner gas tip in a diffuser with a pilot.

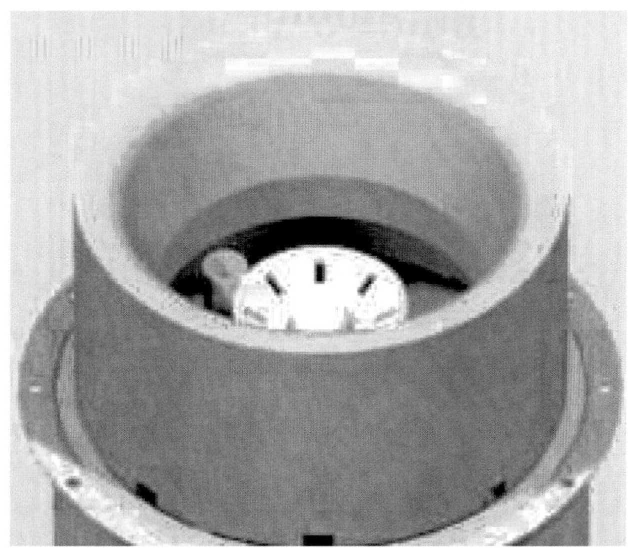

FIGURE 15.18 VYD sketch showing the diffuser cone and the pilot tip.

Exercise and check these dampers for proper function after installation and after any repair of the damper or actuator. If these dampers fail to operate correctly, the burners may operate with either far too much air or insufficient air for complete combustion.

15.2.7.4 Stack Dampers

Stack dampers are installed in heater stacks to control the pressure within the heater fireboxes. The dampers may be single- or multi-blade units, depending on the diameter of the stack. The negative pressure within the firebox at the burner

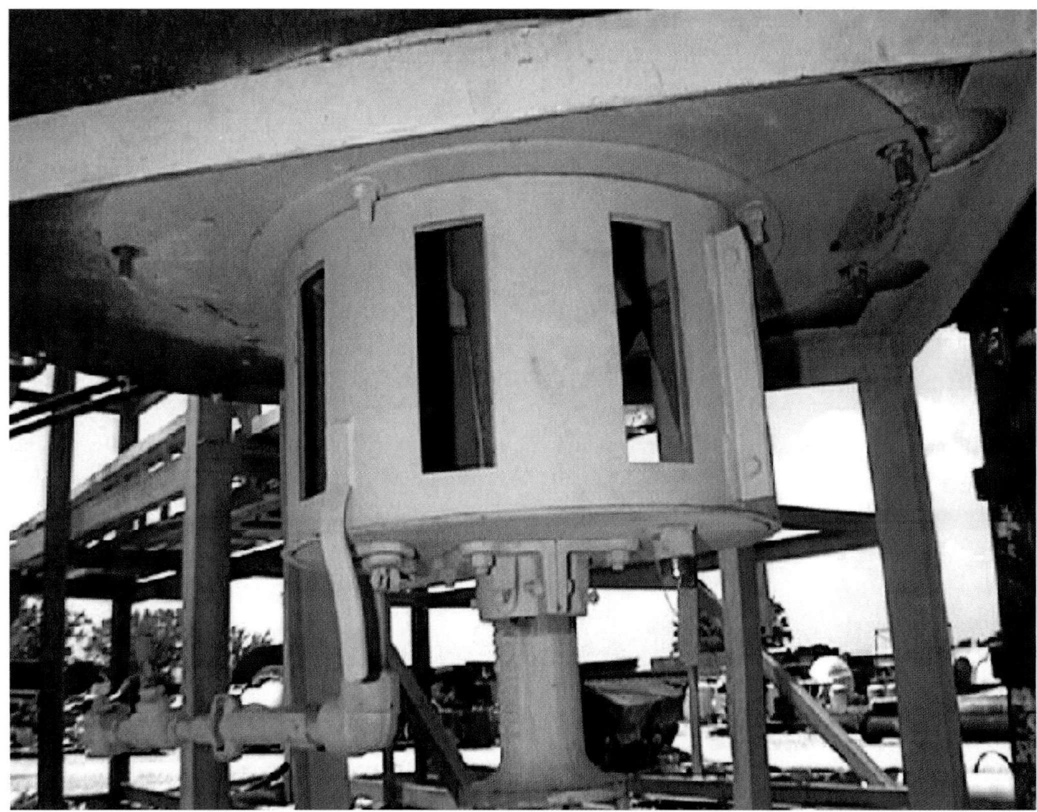

FIGURE 15.19 Example air register.

level(s) defines the driving force to push air through the burners and supply the air for combustion. These dampers should be exercised for proper function after installation, after any repair of the damper or actuator, and after any combustion upset event that could overheat the damper. Refractory has been known to fall from stacks, even during installation, and cause the stack damper to not function correctly. Inspect to ensure that no debris affects the operation of the damper or restricts the flow of flue gas from the heater.

15.2.8 Fuel Piping Design

The fuel piping from the header to the burner should leave the header vertically upward and then run to the burner (see Figure 15.20). This reduces the possibility that liquids in a gas fuel system or solids in a liquid fuel system will leave the header and enter burners. Liquids in a gas burner can cause fouling and plugging of the tip ports or orifice and can, in large amounts, extinguish the flame. Solids in a liquid fuel system will plug the passages in oil atomizers and tips.

The fuel oil piping for a system handling no. 6 or heavier fuel should be an insulated circulating system, as shown in Figure 15.21. This type of system allows the oil temperature and viscosity to be more reliably maintained at the levels required for adequate atomization. Lighter fuel oils with less concern for viscosity control may or may not need to be heated (see Figure 15.22).

15.2.8.1 Line Sizing

Good engineering design attempts to equalize the flows in the fuel lines leading from a common manifold to individual burners. Both momentum and friction effects on flow distribution from the manifold should be considered. Equalizing the burner flows promotes uniform flame patterns and even heat distribution within the firebox.

In heated fuel oil systems, it is common to circulate more oil than is burned to minimize the lowering of oil temperature at the burners due to heat losses from the piping. The proportion of oil returned unburned to the fuel preparation and heating facilities decreases as the amount of oil burned increases. This is due primarily to the reduction in the ratio of insulation surface area to oil volume circulated as the oil flow and piping size increase. The return flow will vary from about three-quarters of the total flow in small systems to one-third of total flow in very large systems or, for every barrel burned, 0.5 to 3 barrels will be returned, depending on the

FIGURE 15.20 Typical fuel gas piping system.

specific gravity of the oil. If time allows, a heat transferred/heat balance comparison calculation provides the required flows. The excess or return flow provides heat to ensure against solidification of heavy residual oils, possibly requiring piping disassembly and cleaning or total replacement. Lighter oil fuels with less concern for viscosity control and no need to heat the fuel do not require a return system.

15.3 MAINTENANCE

Burner performance typically deteriorates with operating time due to fouling, plugging, and wear on the components. The fouling, plugging, and wear reduce the effectiveness of the fuel and air mixing and can affect the flame and heat flux patterns in ways that reduce the heater efficiency and heating capability. The burner parts that usually require maintenance to avoid serious performance loss or safety issues include the fuel gas tips, the fuel gas orifice and mixer (premix designs), fuel oil tips and atomizers, flame stabilizer cone or tile ledge, pilot tip, and burner tile. The air registers and dampers should require less frequent maintenance. Operators and inspection personnel should inspect all of these when developing a maintenance worklist.

Many burner maintenance activities can be completed while the heater continues to operate. Some are simple and may occur fairly frequently. Among these are removal of oil guns, gas tips, and pilot burners for cleaning, the removal of center-fired raw gas guns, and maintenance on the registers/dampers of burners that are not in plenums. Complete removal of an inactive burner on a natural-draft heater can be done while the heater operates, but safety issues must be addressed. The heater must be kept operating steadily so that the firebox pressure stays negative. Procedures to minimize the exposure of personnel, proper protective clothing, tools to handle the burner (particularly if it is a heavy floor-mounted unit), and temporary sealing of the burner opening must all

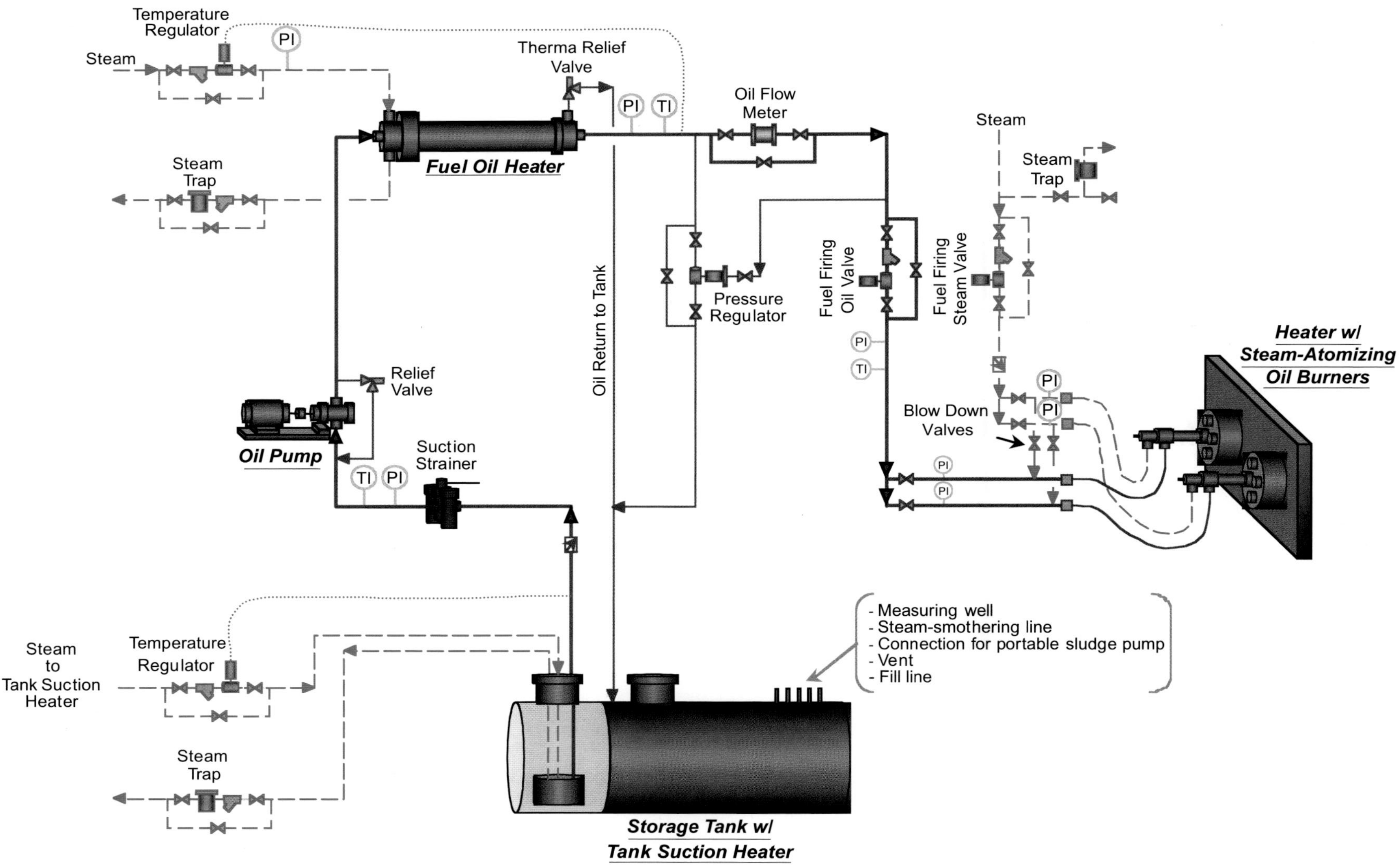

FIGURE 15.21 Typical heavy fuel oil piping system.

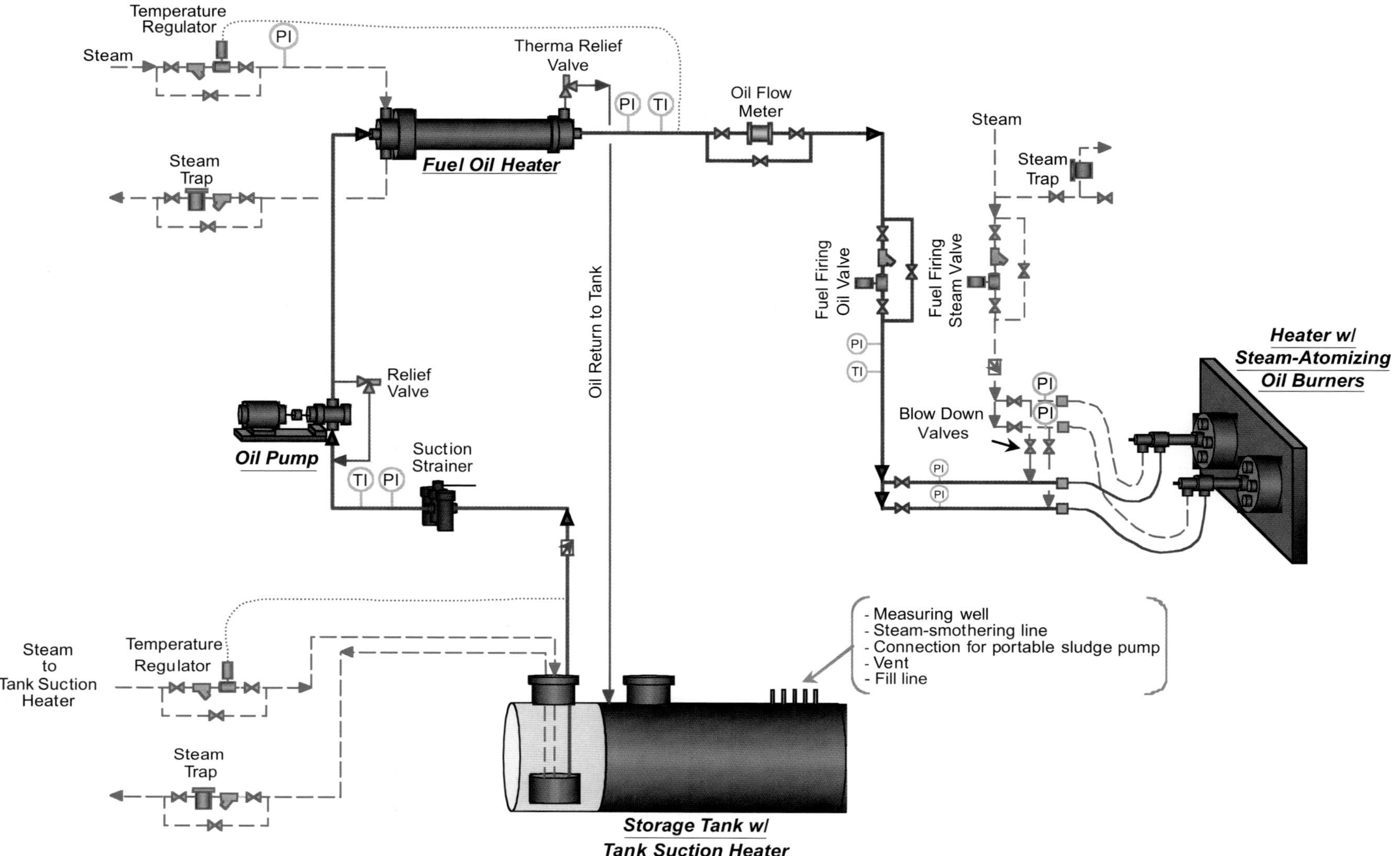

FIGURE 15.22 Typical light fuel oil piping system.

FIGURE 15.23 Typical fuel gas tips.

be addressed. Depending on burner construction, all parts can be removed and repaired, or all but the secondary and tertiary tiles are removable during service.

Burner maintenance for planned shutdowns can usually be identified by problems observed during operation or by conditions visually observed within the firebox. Many problems with burners develop because of broken or poorly maintained parts. See Chapter 17 for discussion of several such problems. Repairs of damaged tile are a common shutdown item that may be identified by observed damage or by flame problems. Other readily identified work items include inoperable registers and dampers, damaged tips and risers on combination or raw gas burners, damaged diffusers (flame holders), and whole burners and/or plenums damaged by fire or oil spill.

Replacement parts should be ordered from the original manufacturer so as to maintain the appropriate quality and durability. Copied parts normally do not have the same tolerances as the original parts and often cause inferior operation.

15.3.1 Gas Tip and Orifice Cleaning

The fuel gas tips (see Figure 15.23) and the fuel gas orifice have drilled ports that direct the flow of fuel into the air stream and combustion zone. These ports must be kept free of foreign material that would reduce the effective port size. If the ports become partially or completely plugged, the amount and distribution of fuel entering the combustion zone vary from the design intent, and combustion problems will likely occur. Sources of foreign materials that plug tips and orifices include:

- pipe scale and gums from the fuel gas piping
- amine compounds from the fuel gas hydrogen sulfide removal process
- coking of condensed heavy or unsaturated hydrocarbons in the fuel gas
- polymers that form inside the burner heater risers or tips
- hydrocarbon mists that vaporize or react in hot risers or fuel tips

Foreign materials must be carefully removed from the fuel orifices so as not to alter the orifice dimensions. Replace fuel tips if any orifices exceed the specified diameter by more than one to two twist drill sizes. The type of material plugging the ports determines how the burner part should be cleaned.

If the material is pipe scale or gums, a twist drill of the same size as the orifice or tip ports is used for cleaning. The drill bit is gently twisted manually and pushed through the plugged port to remove the deposit. If the material is not easily removed, soak the burner part in a solvent to loosen the deposit, and try again with the twist drill. Never use a power tool with the twist drill because it is likely to enlarge the port(s); this will lead to increased fuel gas flow and flame or combustion problems. An alternative that is often successful is to use a welder's file to gently remove the scale from the orifices. If the scale cannot be removed, the fuel tip or orifice must be replaced.

If the source of the foreign material is an amine compound, the orifices can be cleaned with wet steam or hot water, because amines are water soluble. Amine plugging is fairly common in the tips of premix-type burners (i.e., spiders or radiant wall tips). If the amine plugging occurs frequently, the ports can be cleaned without removing the burner by shutting off the fuel gas valve and injecting steam into the burner. The length of the steam injection period depends on the amount of deposit in the burner manifold, risers, gas tips or orifices, and associated piping. Note that injection cleaning may not clear all tip ports equally.

Coking of hydrocarbons in the fuel gas tips may be so severe that the tips must be replaced. Light coking can be removed with a twist drill as described earlier.

Plugging with polymers and as a result of heavy hydrocarbon vapors and high temperatures in the fuel line and tip can sometimes be removed by soaking the tips in a hydrocarbon solvent, followed by cleaning with a twist drill. Severe fouling may require replacement of the tip or orifice.

Coking, polymer, and solid deposits can be removed or converted to easily removed compounds by oxidization in a small, high-temperature furnace. Many tips can be cleaned at one time with this technique.

15.3.2 Oil Tip and Atomizer Cleaning

Fuel oil tips are more difficult to clean than gas tips and often require more frequent cleaning. Tips in light fuel oil service may have only a slight carbon-like deposit on the surface. This can be easily removed with a wire brush. Tips in heavy fuel oil service will usually have a tenacious hard deposit on the surface and in the ports. These tips should be removed from the burner when fouling is observed and placed in a naphtha or diesel oil bath to soften the deposits. Wire brushing, steam cleaning, or the use of a twist drill as described earlier, individually or in combination, may be effective in removing the deposits. Do not use a power drill to clean the ports or a power tool to clean the oil tip surface. Any nicks or indentations on the surface of the oil tip act as a site to collect oil and accelerate the tip coking problem.

Coking and deposits on the surface of the tip can be removed without taking the oil burner out of service in natural-draft and in some forced-draft applications, depending on the design of the burner. A spoon-like implement on a long handle can be used to reach through the air register and scrape off the deposit while oil and atomizing medium continue to flow. The deposit must be identified early, before the fouling becomes extreme, for this to be successful.

An oil gun atomizer is shown in Figure 15.24. There are two orifices in the atomizer: one is for steam and can supply steam to three or four ports, and the other is for oil supply to one to four ports. The steam orifice can become plugged with pipe scale that is easily removed. If heavy oil has entered the steam orifice, due to a (probably momentary) pressure imbalance on either the steam or oil, the atomizer should be soaked in a solvent followed by blowing steam through the orifice to remove remaining oil or solid deposit.

When removing the atomizer from the oil gun, the atomizer must be handled carefully to protect the labyrinth seal. This seal separates the oil from the steam by a series of rings on the atomizer with a tight tolerance between the atomizer and the oil gun body. Sometimes, the atomizer is galled and seizes and cannot be removed from the gun without breaking the atomizer. If this is a continuing problem, the atomizer metallurgy should be changed to a different metallurgy than that of the oil tip.

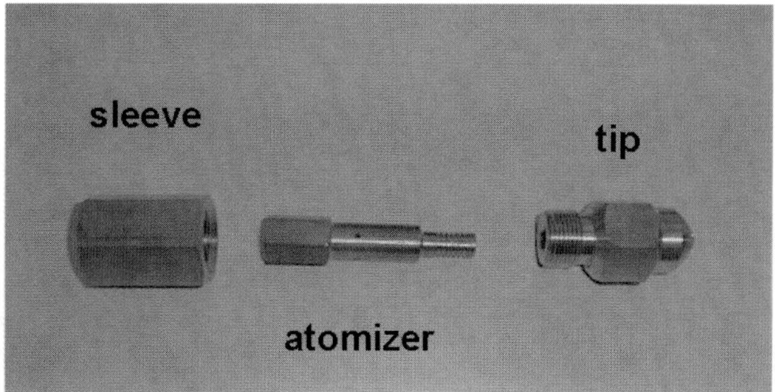

FIGURE 15.24 Example oil gun atomizer.

FIGURE 15.25 Catalyst deposit within an oil burner tile.

15.3.3 Tile

Burner tiles are difficult to repair and return to service successfully because the unique shape of the tile surface is difficult to reproduce. Also, the refractory of the tile has undergone phase transformation in service, and refractory repairs usually do not adhere to the surface for very long. Maintenance of burner tiles is generally limited to inspection of the tile condition and replacement when the tile is identified as damaged. Any major cracks, particularly in wall- or roof-mounted burner tile where the tile pieces are likely to fall to the floor, are reason to replace the tile. Multiple cracks in a section provide evidence of crushing due to restrained expansion and are reason for replacement.

The primary or oil tile should be checked. If it is badly pitted or cracked, the recirculation of gases within the tile is uneven and coking can occur on the tile. This coking can lead to oil dripping and spillage from the burner. Figure 15.25 depicts a catalyst deposit within an oil tile; the catalyst entered as a component of the blended fuel oil. If this is observed, the tile must be removed and cleaned.

15.3.4 Flame Stabilizer

Several flame stabilizer designs are used to hold a stable flame in the combustion zone in the burner. These designs include

FIGURE 15.26 Typical diffuser cone.

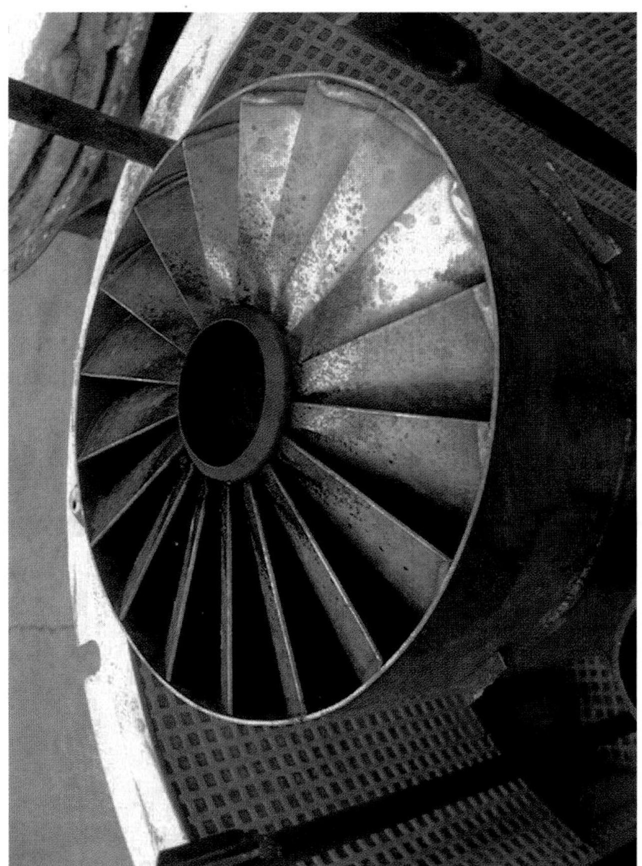

FIGURE 15.27 Typical spin diffuser.

FIGURE 15.28 Example of a damaged stabilizer.

the diffuser cones (see Figure 15.26), spin diffusers (see Figure 15.27), tile ledges, and tapered tiles in natural-draft burner designs. All of these must be inspected, condition and dimensions compared to the appropriate drawings, and the device replaced or repaired if it will no longer act as a reliable stabilizer. A stabilizer in poor condition, or missing, can result in unsatisfactory flame shape or an unsafe flame than can lift off the burner and leave the zone where combustion is initiated. Figure 15.28 depicts a damaged stabilizer that should be replaced. Operation with this stabilizer would result in a lopsided flame pattern with part of the flame lifting off the burner. Replacement of the diffuser cone can usually be done with the heater in operation.

15.3.5 Air Registers and Dampers

Air registers and dampers are used to vary and control the amount of air flowing through the burner. If the registers or dampers are not adjustable, the targeted level of excess oxygen at the burner cannot be maintained over the desired range of operation. This will result in either too much or too little oxygen, both resulting in inefficient combustion and the latter resulting in a possibly unsafe or damaging operation.

All registers and dampers should be exercised periodically. If dampers are inoperable, check the actuator and repair if necessary. Dampers are often inaccessible during operation and repairs must wait until the heater is shut down. If an air register is not operable, it can often be removed while the heater remains in service. Clean the register of foreign material such as refractory pieces wedged in the register, oil that has spilled, insulation that blocks the air flow, and rust or sand. If the register is deformed, possibly due to efforts to operate it, or the operating handle is broken, repair or replace the part.

15.3.6 Pilot Burners

There are several types of pilot burners used in the process industry, but the most common is the small heat release premix burner type. The basic parts include the pilot tip (possibly with a wind-resistant shield), the gas mixer and mixing tube, and the gas orifice. The gas orifice is commonly 1/16 in. (1.6 mm) in diameter and is easily plugged by pipe scale. This is cleaned by hand with a twist drill as previously described.

If inspection reveals a damaged pilot tip (see Figure 15.29), the tip should be replaced (see Figure 15.30).

If the pilot is equipped with electronic ignition (see Figure 15.31), the ignition rod should be inspected to ensure that the spark arc is properly located for ignition of the pilot gas. Check for unwanted electrical grounds that will inhibit proper arcing.

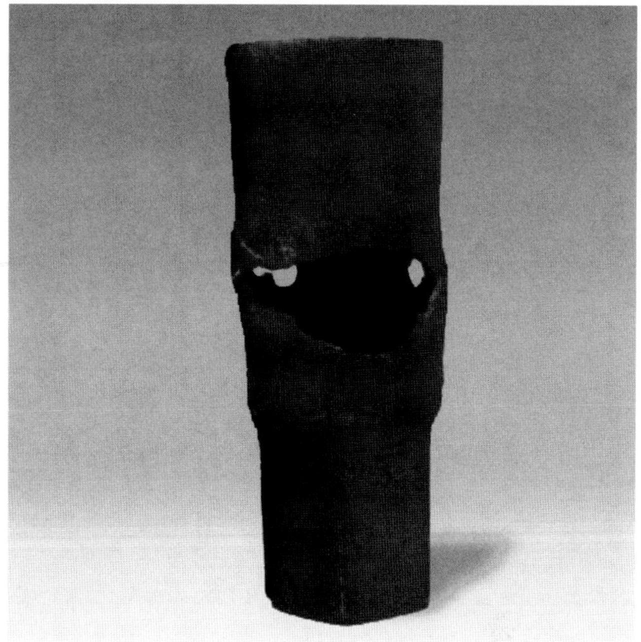

FIGURE 15.29 Example of a damaged pilot tip.

REFERENCES

1. API Standard 560, Fired Heaters for General Refinery Service, American Petroleum Institute, Washington, D.C., 1996.
2. John Zink Burner School Course Notes, October 2000, copyrighted 2000.
3. E.A. Barrington, *Fired Process Heaters*, Course Notes, copyright 1999.

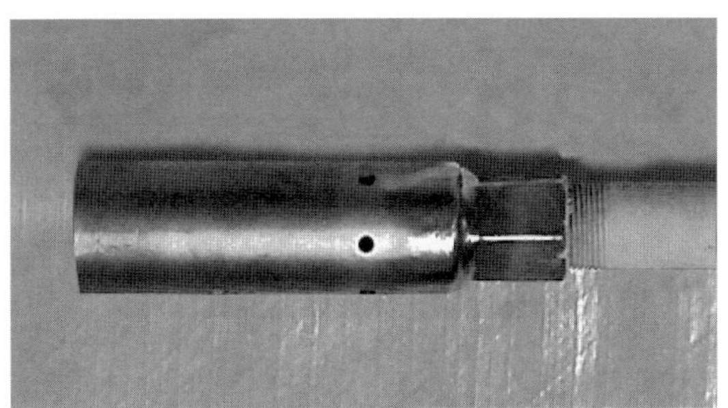

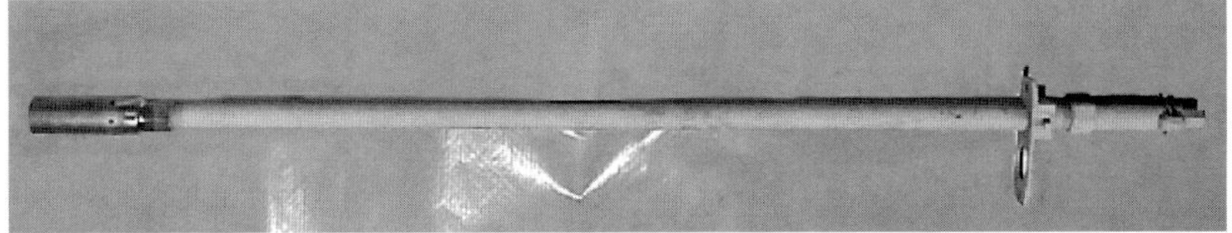

FIGURE 15.30 New ST-1S pilot tip without an electronic ignitor.

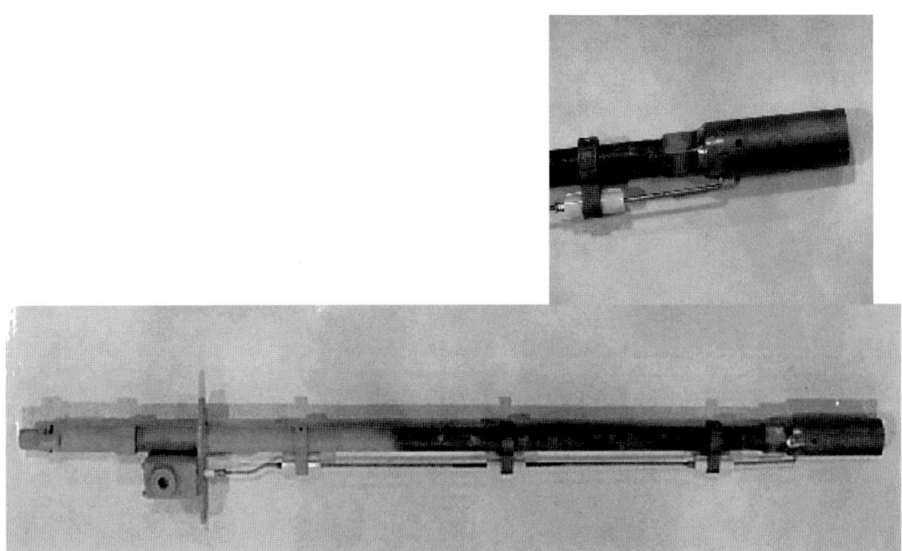

FIGURE 15.31 New ST-1SE pilot with electronic ignitor.

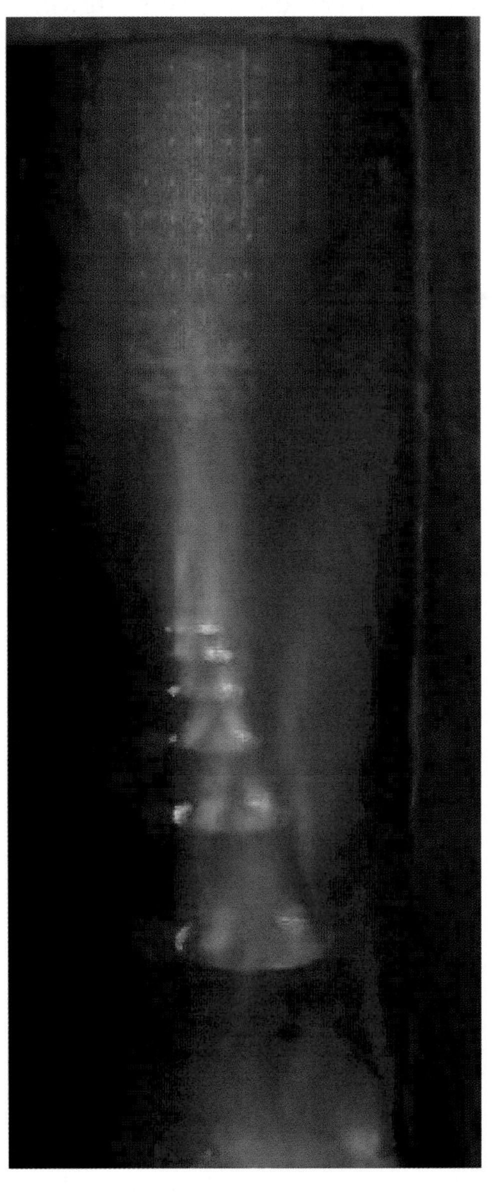

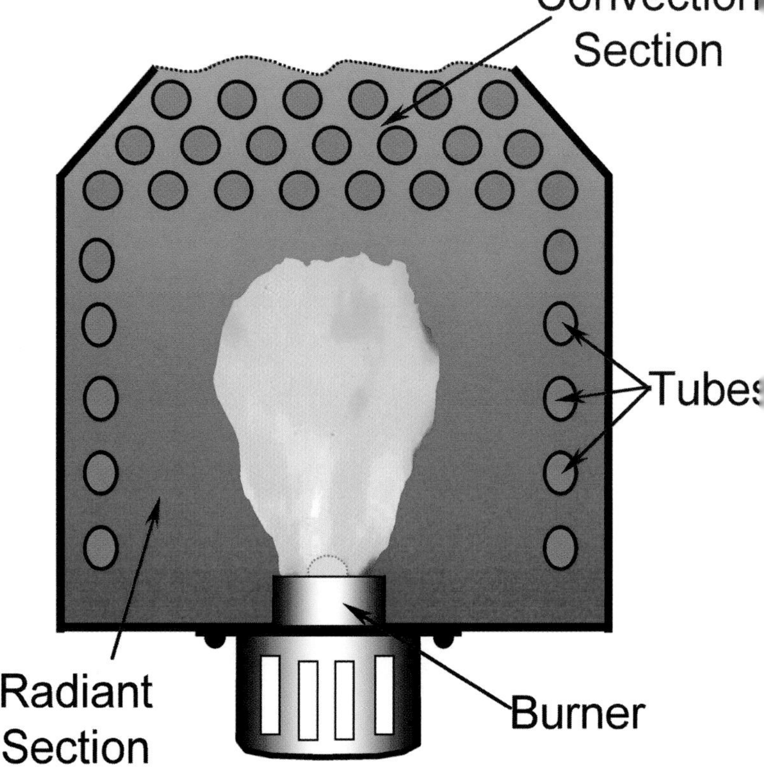
Convection
Section
Tubes
Radiant
Section
Burner

Chapter 16
Burner/Heater Operations

Roger H. Witte and Eugene A. Barrington

TABLE OF CONTENTS

16.1 Burner/Heater Operation 470

16.2 Measurements 470

16.2.1 Draft 470

16.2.2 Excess Air or Excess Oxygen 471

16.2.3 Fuel Flow 473

16.2.4 Fuel Pressure 473

16.2.5 Fuel Temperature 476

16.2.6 Combustion Air Temperature 476

16.2.7 Flue Gas Temperatures 476

16.2.8 Process Tube Temperature 477

16.2.9 Process Fluid Parameters 479

16.3 The Heater and Appurtenances 479

16.3.1 Burner 479

16.4 Further Operational Considerations 488

16.4.1 Target Draft Level 488

16.4.2 Target Excess Air Level 489

16.4.3 Heater Turndown Operation 492

16.4.4 Inspection and Observations Inside the Heater 493

16.4.5 Inspection and Observations Outside the Heater 495

16.4.6 Heater Performance Data 496

16.4.7 Developing Startup and Shutdown Procedures for Fired Heaters 497

16.4.8 Developing Emergency Procedures for Fired Heaters 499

References 499

16.1 BURNER/HEATER OPERATION

The governing principles for fired process heater and burner operation are:

1. operate safely
2. protect the environment
3. avoid damage to the fired equipment
4. satisfy the processing heat requirements
5. maximize heater efficiency

To accomplish these goals, procedures are established to guide heater startup, continuing operation, efficiency improvement, handling of emergencies, and operation of the fuel and air supply systems. In each case, the operator must refer to measurements to properly control the combustion reaction and monitor the performance of the heater.

The most important measurements for safely controlling combustion are draft (or pressure within the casing), excess air (measured as excess oxygen), fuel flow and pressure, and liquid fuel and atomization medium pressure and temperature. The operator should also be aware of the combustion air temperature. For proper heater operation, the operator must also monitor the temperature of the hot gases at the exit from the firebox (bridgewall temperature), the temperature of the flue gases entering the stack, the visual appearance of the flames and tubes, the temperature of the process tubes, the appearance and condition of the refractory, the process fluid flow rate in each pass, and the process fluid pressure drop and outlet temperature for each pass.

16.2 MEASUREMENTS

16.2.1 Draft

Draft is defined in API Standard 560 as the negative pressure of the flue gas measured at any point within the heater.[1] Draft can be expressed as inches (in.) of water, millimeters (mm) of water, or kiloPascals (kPa). Negative pressure or draft occurs because the hot flue gas within the confined volume of the heater and its appurtenances (e.g., ducts, stacks, air preheater, etc.) is less dense than the surrounding atmospheric air. All other factors being equal, the hotter the flue gas and/or the colder the surrounding air, the greater the difference in densities and the greater the draft or negative pressure within the heater. The difference in densities causes air to flow into the heater, through the burners or through other openings, and the hot flue gases to flow out of the heater.

Draft loss is the pressure drop of air or flue gas as it flows through ducts, burners, firebox volume, or air preheaters, and across tube banks. In burner terminology, the draft loss across the burner is the pressure drop of the combustion air as it flows through the throat of the burner tile. In a natural-draft heater, the burner draft loss is the difference between the pressure in the firebox at the burner elevation and the atmospheric pressure at that elevation. In a forced-draft heater, the burner draft loss is the difference between the pressure in the windbox or plenum (often positive) and the pressure in the firebox, both at the burner elevation.

Most of the process heaters within HPCI operate with a negative pressure in the firebox. Because the firebox is not completely sealed, if there is any air leakage it will be outside air leaking into the firebox rather than combustion gases leaking out of the firebox. Positive pressure inside the heater can cause flue gas leakage and damage to the furnace casing and structure. A positive pressure can even create a safety hazard to operating personnel.

Almost all fired process heaters should operate with a negative static pressure, or draft, throughout the flue gas path. This draft should be measured at specific points (see Figure 16.1). Most important is to measure and control the draft at the location of highest pressure within the heater; this typically occurs at the roof of the radiant section (or firebox). The draft is the lowest at this point, and maintaining a slight negative pressure at this point normally ensures a negative pressure throughout the heater. Another location for draft measurement is at the elevation(s) of the burners. This is checked to ensure that all burners have an adequate draft loss available to supply the necessary combustion air flow. The third important location for draft measurement is at the flue gas outlet from the convection section, often located below the stack damper. By combining this measurement with the draft value at the roof of the firebox, in many common heater designs, one can determine the draft loss across the convection tube bank. This can help identify the occurrence of damage or excessive fouling in the convection section.

Draft (static pressure) can be measured with an inclined manometer (see Figure 16.2) or with a dial gauge manometer. Draft transmitters can be mounted externally at the firebox roof level to provide remote draft indication or recording. Once the draft target at the firebox roof elevation has been properly determined and made known to the operator, this is the only draft that requires frequent monitoring. The static pressures within the firebox and convection section will always be less than this value in a properly designed heater. The draft at the firebox roof is controlled by adjusting the damper in the stack or, if an induced-draft fan is provided, by adjusting the fan damper or speed.

Erratic draft readings can be caused by pulsating flames or by sample lines that leak, are plugged, or contain water from the products of combustion.

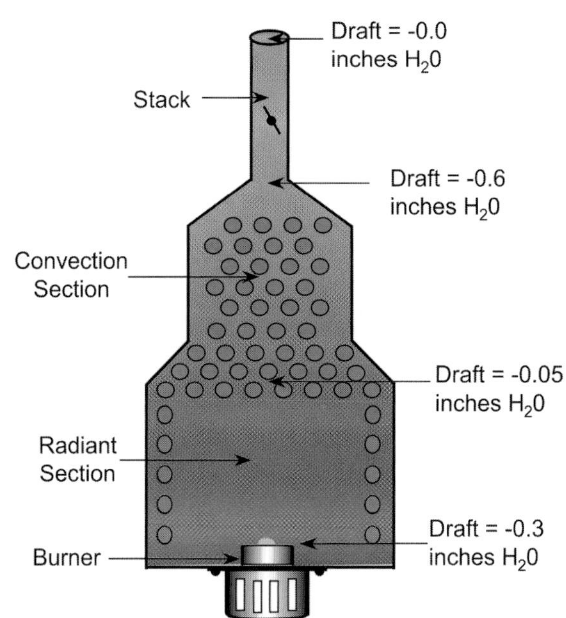

FIGURE 16.1 Typical draft measurement points.

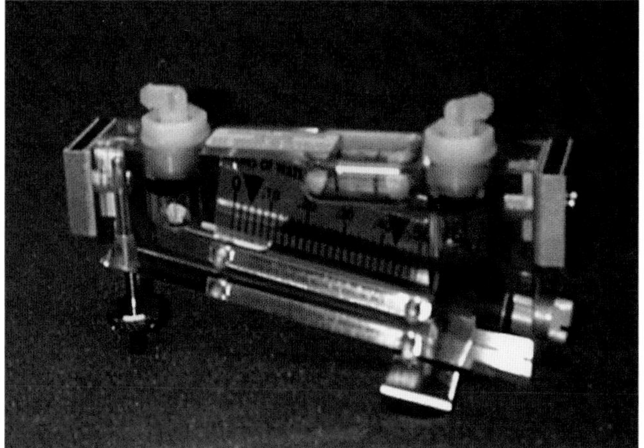

FIGURE 16.2 Inclined manometer.

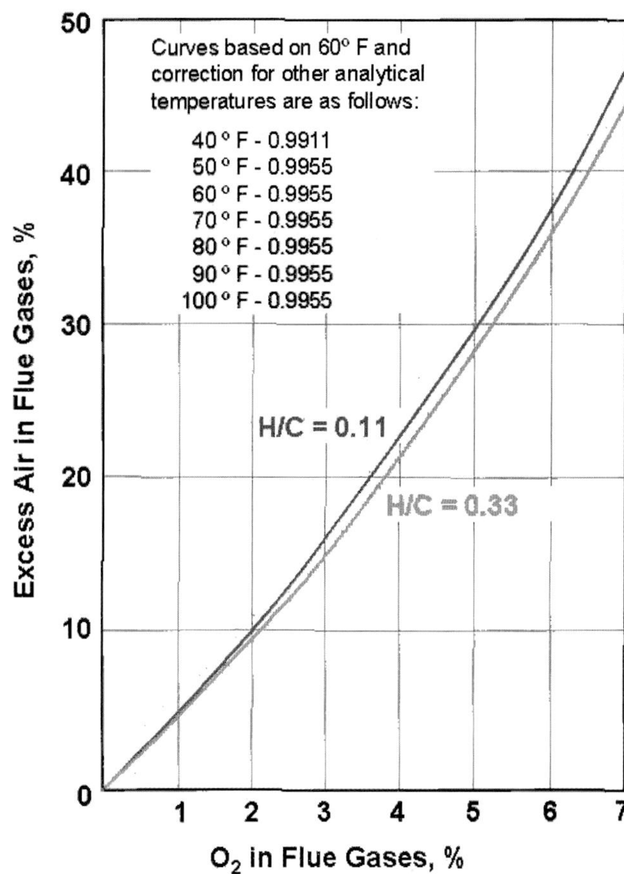

FIGURE 16.3 Excess air indication by oxygen content.

16.2.2 *Excess Air or Excess Oxygen*

Excess air is defined in API Standard 560 as the amount of air above the stoichiometric requirement for complete combustion, expressed as a percentage. The excess oxygen is the amount of oxygen in the incoming air not used during combustion and is related to percentage excess air as shown in Figure 16.3.[2] Excess oxygen is easy to measure and is used as a proxy for excess air. If there is an excess of oxygen in the flue gas, good fuel/air mixing at the burner, and a stable flame observed in the firebox, the operator can be reasonably assured that combustion is complete at the point in the heater where excess oxygen is measured.

Because excess oxygen is monitored to ensure complete combustion, it is best to sample the flue gas at a location that is most representative of the combustion at the burners. The sampling must also be at a location where the flue gas is representative of the combustion as a whole in a multi-burner heater (i.e., where the flue gas is well mixed). The location that best satisfies these criteria and the correct sample point for controlling the combustion reaction is at the flue gas outlet from the radiant section. The most common point is at the top of the radiant section (see Figure 16.4).

Because the heater operates under negative pressure, any openings will allow air to leak into the heater. The air leaking into the heater that does not pass through the burners cannot participate in the combustion process. The oxygen analysis used to determine the excess air cannot differentiate between air that enters via the burners and air leaking into the heater. The amount of air leaking into the heater is typically low in a well-maintained, well-operated radiant section, but convection sections usually have many more sources of air infiltration. Therefore, sampling for excess oxygen at the flue gas

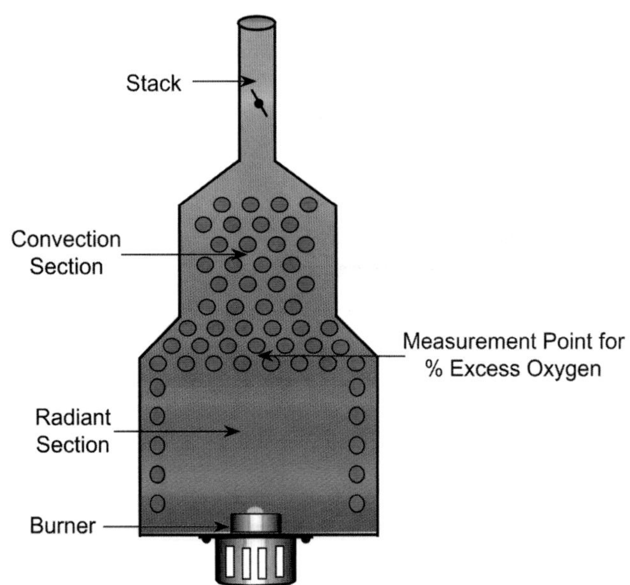

FIGURE 16.4 Location for measuring excess oxygen.

exit from the firebox, rather than at the stack inlet, gives values most representative of the combustion process. The combustion process is what the operator can affect by register and stack damper adjustments. If the operator depends on an excess oxygen reading taken at the stack, that reading may indicate an excess oxygen level that does not exist at the burner level. Thus, the adjustments to reduce excess oxygen that the operator would typically make, such as throttling air registers, could lead to insufficient combustion oxygen at the burners as well as unburned hydrocarbons and CO (carbon monoxide) in the flue gases.

The unburned hydrocarbons and CO may lead to a condition known as "afterburn." Afterburn is the term given to combustion that occurs near or within the convection tube bank. If hot, unburned combustibles leave the burner area due to inadequate air flow through the burners, they can burn wherever they come in contact with oxygen within the heater. Because the amount of air leaking into the heater is typically large at the convection section, this is where the afterburn is most likely to occur.

One additional consideration further improves the accuracy of oxygen analysis. It has been found that air that enters through casing openings tends to stay near the firebox walls as it flows to the exit. If the flue gas is sampled further into the flowing stream, the gas is more representative of the combustion at the burners. Therefore, it is recommended to sample through a probe that extends typically 18 in. (46 cm) or more from the wall into the flue gas.

The preceding paragraphs discuss the determination of the combustion efficiency, that is, the efficiency of converting the hydrocarbons into carbon dioxide. The operator can directly affect the combustion efficiency by managing the excess oxygen. At times, it is desired to calculate the overall heater efficiency, or the amount of heat transferred into the process fluid from the amount of heat released by the fuel. For this purpose, and for some emissions reporting, the oxygen and excess air in the flue gas are determined by sampling downstream of the last heat transfer surface or in the stack before the flue gases are released to the atmosphere. See API Standard 560 for the calculation procedure.

Today, oxygen analysis is done with electronic instruments. Portable analyzers (see Figure 16.5) that will measure oxygen, carbon monoxide, carbon dioxide, NOx, nitrogen dioxide, hydrocarbons, and smoke in the flue gas are available. When purchasing a portable analyzer, it is important to realize that sample lines will be long, from the sampling point to grade, and a sturdy built-in pump with high head capability is required to ensure rapid, accurate readings. Portable analyzers typically have a dessicant chamber at the flue gas inlet port. This removes water vapor and protects the analyzer cells, increasing cell life and reliability. The instrument therefore provides a "dry" analysis of the flue gas, without the moisture contributed by combustion and incoming air humidity. Portable analyzers are used for "spot" analyses and monitoring, but are not suitable for continuous flue gas analysis.

In situ extractive analyzers are mounted on the heater and continuously measure oxygen in the flue gas and, with added features, can also measure combustibles. These can be manufactured to withstand flue gas sampling temperatures up to 3000°F, although the typical standard construction is limited to about 1800°F. The oxygen measurement response time is a few seconds, a speed that allows the output signal to be used for automatic control of the combustion air supply. The sensitivity is such that fuel composition changes can be noticed almost immediately. The modern *in situ* oxygen analyzer is so reliable that it typically requires instrument technician attention and a calibration check about once a month. The combustibles analyzer feature requires more frequent attention and is not commonly added unless very low excess air operation is planned. The *in situ* continuous analyzer is mounted so as to sample the flue gas leaving the firebox, thus providing information on the quality of combustion. The oxygen and combustibles contents, in percent, are based on the actual flue gas composition and therefore the instrument provides a "wet" analysis.

The justification for controlling the excess oxygen in the heater is shown in Figure 16.6 for fuel gas and Figure 16.7 for liquid fuel oil. The information required to determine the

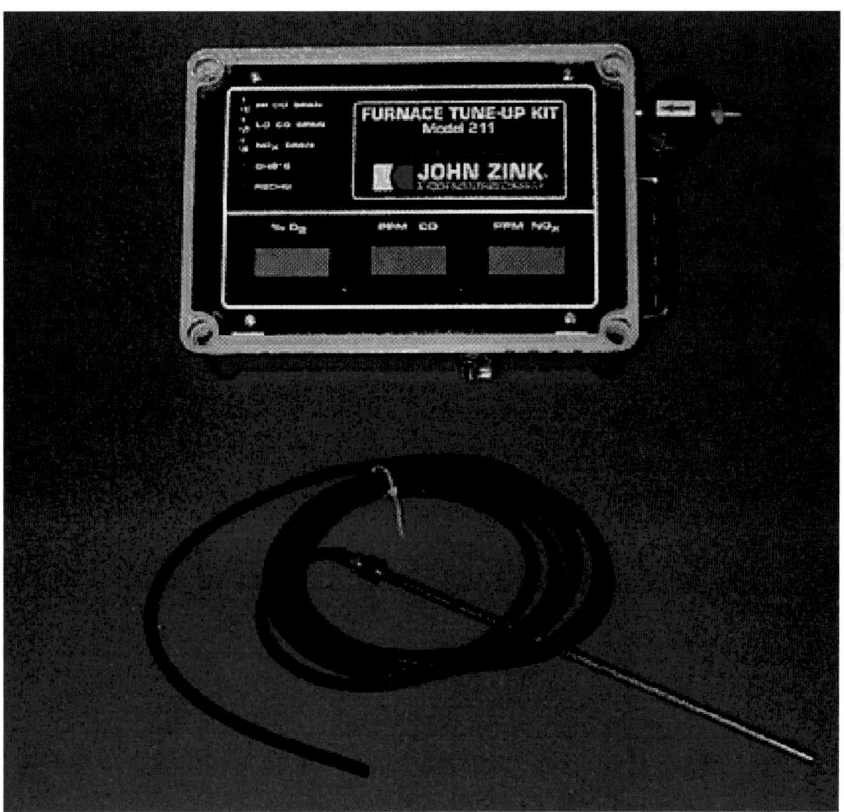

FIGURE 16.5 Oxygen analyzer.

annual savings achievable includes the heat release per burner or the total heat release for the heater; the operating excess oxygen; the stack temperature; and the cost of the fuel.

Oscillating or erroneous oxygen analyzer readings can be caused by leaking or plugged sample lines. Liquid water condensed from the products of combustion of gases entering the analyzer may result in instrument damage.

16.2.3 Fuel Flow

The fuel flow — or rate of heat release — is one of the most important controlled variables in a process heater. Each operator should be aware of the maximum design heat release or the maximum heat release that has been proven by successful operation. As operation approaches this maximum, it is important to intently monitor critical variables and be alert to the effects of small variations.

The fuel flow control valve most often acts to control the process bulk outlet temperature, modulating the rate of fuel input to maintain or change the desired temperature. Common valve control loops use cascade control techniques to minimize the effect of pressure and composition fluctuations in the fuel supply system. The control valve should be carefully sized so that it reacts quickly without imposing an excessive pressure drop on the fuel supply to the burners. A poorly sized valve may encourage operation with the valve bypass open, thereby reducing the effectiveness of any burner management system.

16.2.4 Fuel Pressure

The pressure of the fuel, whether gas or liquid, is the major energy source used within the burner to effect the required mixing of the fuel and the air. The design burner pressure for gas fuel will typically be 15 to 30 psig (1 to 2 barg) at the maximum design heat release with the design fuel composition as measured at the burner (see Figure 16.8). For liquid fuels, the design burner pressure may be 100 to 150 psig (6.8 to 10 barg) at design conditions. Higher fuel pressures allow a greater range of heat release, known as turndown. Turndown is the ratio of maximum heat release to the minimum heat release. The atomization medium pressure will depend on the type of oil gun used. Either a constant atomization medium pressure of about 70 to 250 psig (5 to 17 barg) or a pressure controlled to 15 to 30 psi (1 to 2 bar) above the fuel pressure is typical. The pressures of both the

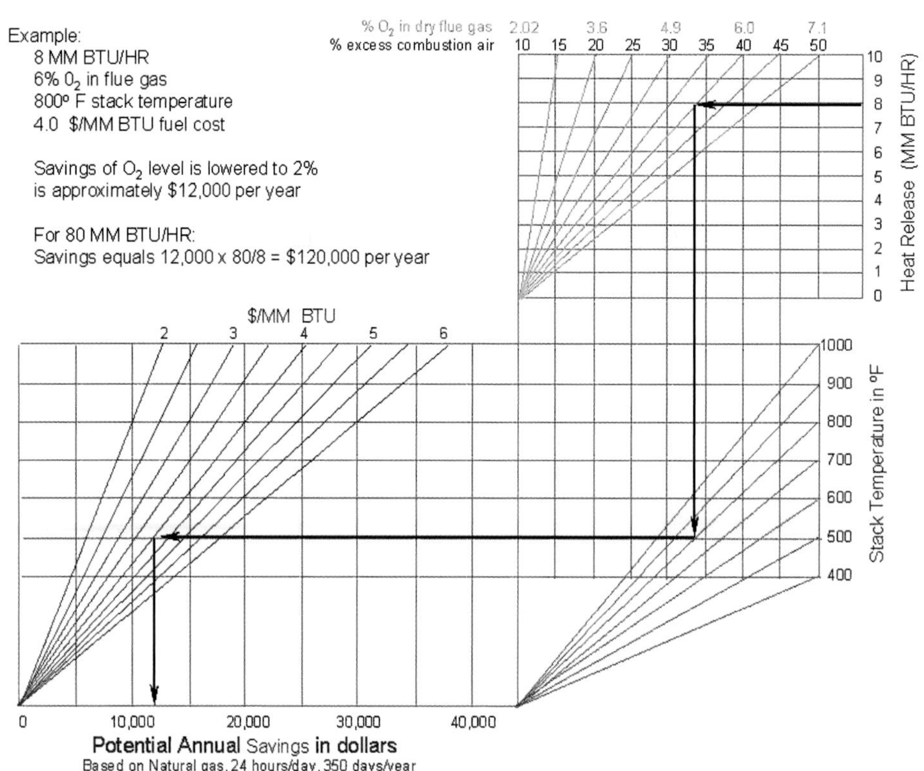

FIGURE 16.6 Cost of operating with higher excess oxygen levels (natural gas).

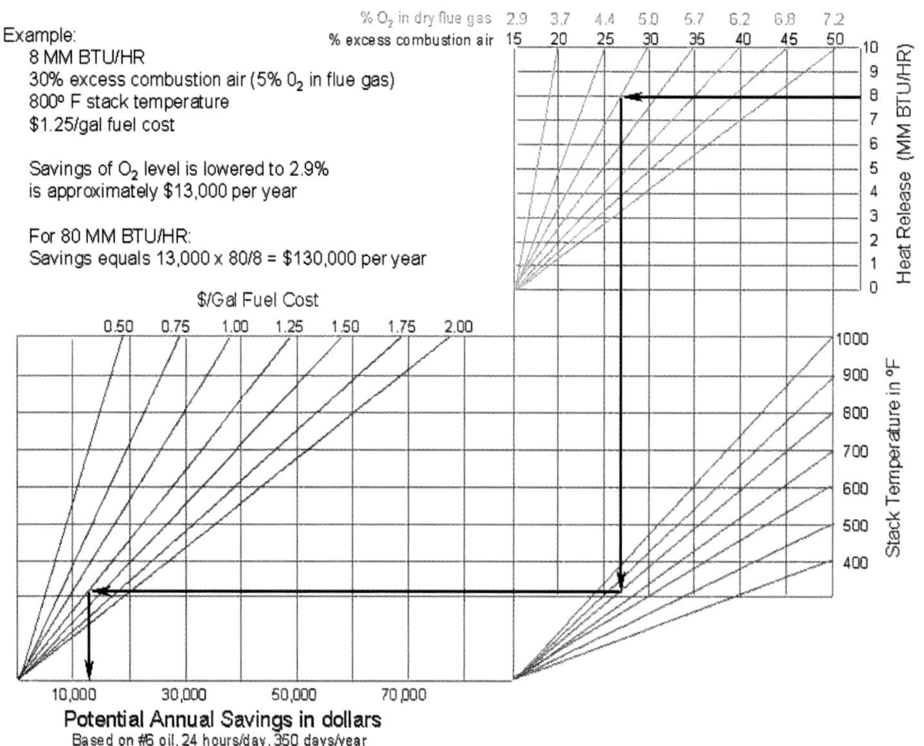

FIGURE 16.7 Cost of operating with higher excess oxygen levels (No. 6 oil).

FIGURE 16.8 Fuel gas pressure measurement.

fuel and any atomization medium should be monitored at a point downstream of any control valve and close to the burner block valves. The pressure of the fuel and atomization medium must be checked with the individual burner block valves fully opened. The fuel pressure at pilot burners will typically be regulated at a constant pressure, usually 5 to 15 psig (0.35 to 1.05 barg), depending on the pilot model employed.

The burner manufacturer normally provides a curve of the fuel pressure versus heat release (or fuel flow) for each burner and for each specified fuel composition. This curve (see Figure 16.9) will indicate the maximum and minimum recommended fuel pressures. These pressures should not be exceeded unless adequate burner performance has been proven beyond these extremes by field observations and testing.

Upper and lower fuel pressure limits are set at levels that prevent the heater from operating outside the range desired by the customer. The burner's upper and lower pressure limits at which instability of the combustion may occur can be determined during a performance test of the burner at the burner manufacturer's facility. After the burners are installed in the heater, the limits should be rechecked. Usually, it is only practical to check the low pressure limit, which may anyway be the most important. The pressure limit is used to

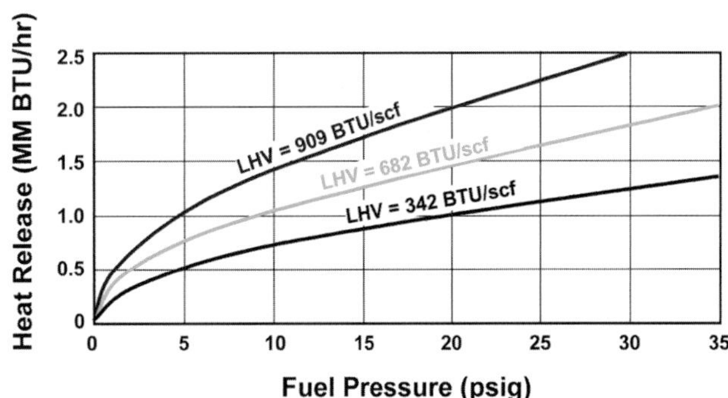

FIGURE 16.9 Graph of fuel pressure vs. heat release.

select the pressure at which low fuel pressure causes the burner management system to interrupt the fuel supply to the burners in order to maintain safe heater operations.

To check the low fuel pressure limit, install a pressure gage at one burner on the pipe between the burner and its block valve. With the heater firing steadily under normal conditions of draft and heat release, slowly throttle the valve while watching the flame. The pressure at which flame instability is initially observed is the low pressure limit. Adjust the low fuel pressure alarm 1 to 1.5 psig above the low pressure limit and the fuel trip 0.5 to 1.0 psig above the low pressure limit.

Fuel composition changes can affect the measured fuel pressure. With gas fuels, if the density and heating value decrease, the controls will call for a higher burner pressure to pass more fuel gas and provide the required heat release. If the gas fuel density and heating value increase, perhaps due to adding propane or butane, the controls will reduce the burner pressure. The energy for mixing in the burner will be decreased and the flame appearance will likely change. A good operator will visually check the flame quality after a significant fuel composition or pressure change and make the necessary adjustments that maintain adequate flame quality, good efficiency, and good combustion.

If fuel or atomization medium pressures change substantially, the controls and the supply source should be checked, along with a visual check of the flames. Atomization pressure will be controlled from the fuel pressure in those oil guns where a pressure difference is maintained.

16.2.5 Fuel Temperature

Fuel gas temperature can be at any value below the autoignition temperature. Typically, the fuel gas temperature is 80 to 180°F (25 to 82°C). The temperature is specified for burner orifice and tip sizing, and any change that varies the density of the gas will affect the pressure and volume of fuel flowing at a given pressure. See Chapter 4 for correcting the fuel flow for the fuel temperature changes. Normally, any changes will cause corrective action by the fuel control valve and a burner pressure increase or decrease (see Section 16.2.3).

If low ambient and fuel temperatures occur together with supplementing of higher molecular weight gas components, expect liquid condensation in the fuel line. Measurements have shown that as much as one-half of the heavy gases condense under extreme conditions. If proper knockout facilities are not included, the liquid can extinguish burners.

The variation of temperature of heavier (low API gravity) liquid fuels is used to control the oil viscosity. Most manufacturers require a viscosity of about 200 SSU (43 centistokes) at the atomizer. The fuel lines must be well insulated to maintain the oil temperature between the fuel heater and the burners. Damaged or missing insulation should be repaired or replaced. Figure 16.10 is a typical plot of fuel oil viscosity versus temperature for many types of liquid fuels. The viscosity measured at two temperatures can be plotted and the line extended to the desired viscosity to determine the required fuel oil temperature.

16.2.6 Combustion Air Temperature

The burner designer sets the tile air flow area in the burner on the basis of a design air density, determined by the air supply pressure and temperature. The higher the air temperature, the lower the weight (moles) of oxygen in a given volume of air. If the air temperature is significantly greater than design, the amount of oxygen flowing through the tile decreases, and the desired amount of fuel cannot be burned without experiencing incomplete combustion and the possibility of some operating problems. A significantly lower than design air temperature can lead to an increase in air flow and excess oxygen, leading to a reduction in efficiency. In either case, an adjustment of the stack or fan dampers or the burner air registers must be made.

Typically, the natural draft air temperature is the ambient air temperature that varies from –30°F to 150°F (–35°C to 66°C). This temperature range does not normally cause any problem in the operation of the burners. Today many heaters have air preheat systems installed to reduce the amount of fuel required. The air preheat system will normally require a forced-draft blower and an induced-draft blower to control the combustion air and flue gas flows and pressures within the heater. The preheated air temperatures range from 300 to 850°F (150 to 450°C). Some processes can provide combustion air temperatures as high as 1200°F (650°C).

16.2.7 Flue Gas Temperatures

The flue gas temperatures of primary interest are the bridgewall temperature (i.e., the temperature of the flue gas leaving the radiant section and entering the convection section) and the temperature of flue gases leaving the convection section and entering the stack. The bridgewall temperature is indicative of the radiant section performance and the degree of fouling of the radiant section tubes. The bridgewall temperature will rise as fouling occurs on the radiant section tubes or as excess oxygen in the firebox increases. The excess oxygen can be adjusted with the burner registers and dampers, but removing fouling deposits in the radiant section tubes usually requires heater shutdown. The stack gas temperature is a rough measure of the overall heater efficiency; as it rises, the heater efficiency decreases. A rising stack temperature is

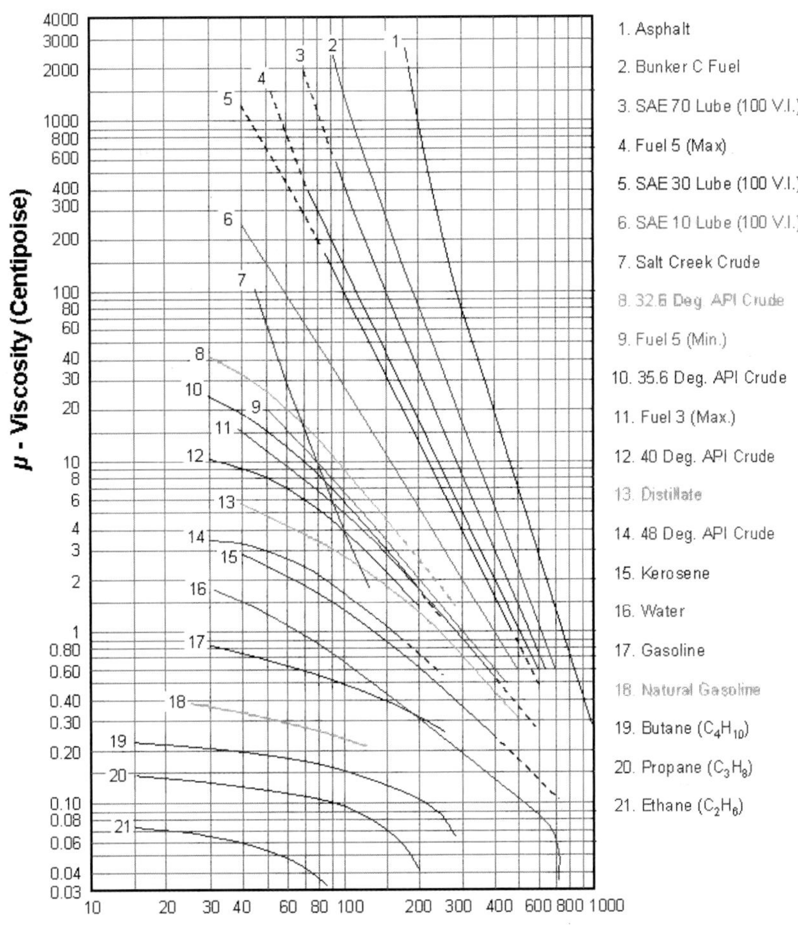

FIGURE 16.10 Viscosity vs. temperature for a range of hydrocarbons.

most commonly caused by high excess air, afterburning due to high levels of combustibles entering the convection section, or convection tube fouling, which lowers heat transfer from the stack flue gases.

These temperatures are commonly measured using thermocouples inserted in fixed thermowells. Great accuracy is not important, but observation of trends will give the best information. This is good because the radiant, convection, and conduction heat transfer to and from a thermowell in these services, combined with the typical arrangements in a heater, cause readings that are many degrees lower than the actual flue gas temperatures. If good accuracy is required for a test run, for debottlenecking studies or heater efficiency studies, these temperatures should be measured with a suction pyrometer, also known as a velocity thermocouple (see Reed,[2] p. 46; see also Figure 16.11).

16.2.8 Process Tube Temperature

The process heater tube materials become weaker and less able to withstand internal pressure as the temperature of the tubes increases. To provide safe operation and satisfactory tube life, avoiding process tube ruptures, the tube-wall or tube-skin temperatures must be limited. The basic reference for defining the allowable tube-skin temperatures is API Standard 530.[3]

Tube metal temperatures can be measured in several ways. Most general-service process heaters with alloy tubes will have tube-skin thermocouples installed at selected locations. These will measure and usually provide a record of temperatures of the tube external surface at the selected installation points. The shielded style of tube-skin thermocouple has been found to give the greatest accuracy with acceptable life.

Infrared thermography gives a photograph of the heater firebox interior where the temperature of the tubes and refractory

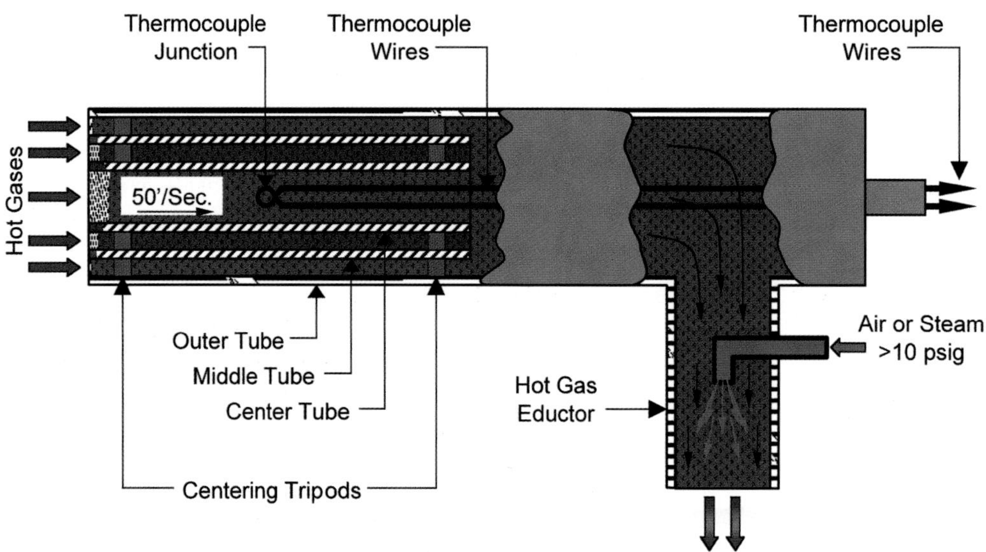

FIGURE 16.11 Velocity thermocouple.

can be accurately inferred from color or contrast. This technique requires special equipment, appropriate training, firing with gas fuel, and fairly generous sight-port sizing. This is not practical as a continuous, or even frequent, monitoring technique. It is limited to identifying hot tube areas caused by flame impingement, internal process tube fouling, or erratic hot flue gas flow patterns. It can be used for monitoring the progress of fouling, confirming the accuracy of tube-skin thermocouples, and identifying problem burners.

Tube temperatures in high-intensity or high-temperature heaters are generally monitored with infrared pyrometers. These hand-held optical instruments are aimed at the selected tube and measure the infrared energy entering the instrument. Using an assumed tube metal emissivity, the instrument circuits electronically convert the incident energy to an indicated metal temperature. The heaters commonly monitored in this way are those in high-temperature pyrolysis (e.g., ethylene production) and steam-hydrocarbon reforming (e.g., ammonia and hydrogen production) services.

The infrared pyrometer can accurately measure, within 50°F (28°C), high tube temperatures over the entire visible length of the tube. This is typically done twice a shift, or six times a day. The infrared pyrometer is not restricted to monitoring point locations as are tube-skin thermocouples, and measurements can be easily taken by a trained operator. The use of the infrared pyrometer is limited to high-temperature services because of instrument construction. The infrared energy entering the instrument comes not only from the tube, but also from the refractory, adjacent tubes, and hot gases. While most of the hot gas contribution can be removed by filters, the variety of contributions means that, mathematically, the errors become of the same magnitude as the indicated temperature in "cooler" fireboxes, and the results are unreliable.

High tube temperatures due to internal fouling can often be visually observed. A reddish or silvery spot on the tube is an indication of localized overheating.

When hot spots are identified, operation of the heater should be modified to keep the tube from becoming hotter. Also, tube cleanout should be considered. There are three generally applicable techniques for avoiding further increases in the tube temperature. One option is to increase the process flow in the pass containing the hot tube. The increased flow rate increases the convection heat transfer inside the tube and removes heat from the tube wall more rapidly. This should be undertaken with knowledge of the impact on the other passes, where flow is reduced. The temperatures of tubes in these passes can be expected to rise and must therefore be carefully observed.

The second option is to reduce the radiating effectiveness of the hot gases in the radiant zone by increasing the excess air. The radiating capability of the hot gases is inversely proportional to the concentration of symmetrical molecules, such as oxygen and nitrogen. As the concentration of these increases, the energy radiated from the gases to the radiant tubes decreases. Because hot spots usually occur in the radiant zone or firebox, this can be an effective action, particularly if firing gas fuel. The impact of this action on the remainder of the heater is an increase in the amount and temperature of the flue gases entering the convection section. The convection section temperatures and heat absorption will rise. Duty has been shifted from the radiant section to the convection section.

The second option will be of limited effectiveness if the heat transfer to the hot spot is largely radiation from a flame. In such a case, the most effective action is to reduce or eliminate the heat input from that flame by throttling or closing the burner valve. This action should be reported to management because it has the effect of possibly shifting the hot spot to another location. Section 16.3.1.7 discusses the problems and limitations involved in throttling burner valves.

16.2.9 Process Fluid Parameters

The process fluid flow rate, pressure drop, and outlet temperature are measured with conventional instruments. The flow rate through each individual pass is monitored in the majority of heaters. In some cases, the overall flow is measured and the flow to individual passes is governed by restriction orifices. In other designs, where the addition of measuring elements and control valves would impose an uneconomical pressure drop, careful distribution manifold design equalizes pass flows as much as possible.

The flow of the process fluid acts to cool the tubes and maintain an acceptable metal temperature. A reduction in flow will lead to a rise in the tube-skin temperature and, often, a phase change or chemical degradation in the fluid. An unexpected formation of vapor in a normally liquid-filled tube pass will increase the pressure drop and cause further reduction in flow and cooling capability with further increase in tube temperature.

Chemical degradation of hydrocarbons typically generates some solids and some vapor. This may devalue a product, as when the overheating discolors a lubricating oil component. More frequently, it will cause formation of a solid layer of material, referred to as coke, on the tube internal surface. This layer impedes the conduction of heat from the tube to the fluid, and the tube temperature must increase to maintain the amount of heat absorbed by the fluid. The pressure drop in the pass may increase measurably. Ultimately, the tube metal temperature rises to a level where corrective action is needed, usually a cleaning of some type, before the tube fails.

Both of the above effects can be largely avoided if the flow rate of the process fluid is kept near the design rate. The pass flows should be nearly equal and not vary by more than 10% when the heater is firing at greater than 75% of the rated heat release. At lower heat releases, the heat flux is unlikely to cause the problems mentioned above.

Thermocouples are often installed in the outlets of each pass. These can be used to check the pass flow controls and indicators. If the flow in a pass is reduced, without a change in the pass heat absorption, the outlet temperature will rise. Conversely, if the flow is increased, the outlet temperature will drop. If a change in pass outlet temperature occurs without a corresponding and appropriate change in the indicated pass flow, the operator should immediately investigate. An instrument problem is likely, and a flow interruption or reduction, with potential tube overheating, is possible.

16.3 THE HEATER AND APPURTENANCES

The typical heater consists of two major process sections: the radiant section and the convection section. The radiant section, also called the firebox, includes the burners and air plenum(s), tubes containing the process fluid, supports for the tubes, and refractory. The major mode of heat transfer to the tubes is radiation, although some convection heat transfer also occurs. The convection section contains the shock (or shield) tubes, convection tubes, refractory, and tube supports and hangars. The major mode of heat transfer is convection, although radiation also occurs. The radiant and convection sections are enclosed by steel casing and structural members that support the refractory, anchor the tube supports, and structurally support the tubes and stack(s). The casing, the structure steel supports, refractory, along with header boxes, ducts, burner tile, and stack(s), are collectively known as the heater setting. Other appurtenances may include an air preheater, a forced- and/or induced-draft fan, instruments, and necessary fuel and process piping and valves, along with safety features such as smothering or purge steam connections.

This section describes the components, reviews their proper operation or condition, discusses any visual clues, and suggests corrective actions to identified problems. Burner problems and corrective actions are discussed in detail in Chapter 17.

16.3.1 Burner

The burner is the mechanical device that mixes the fuel and air, initiates combustion, shapes the flame, and releases the heat required by the process. There are many different fuel gas burners installed in heaters. These burners are all of two basic types: the premix burner and the raw gas or diffusion burner, both described in Chapter 11. The common oil burner is a variant of the diffusion burner type, in which the combustion is heterogeneous (fuel and air phases differ) rather than homogeneous (fuel and air phases are the same). The Lo NOx burner can be either a premix or a raw gas (diffusion) burner. Each type of burner may differ in the way it adjusts combustion air flow to the burner and in the flame pattern

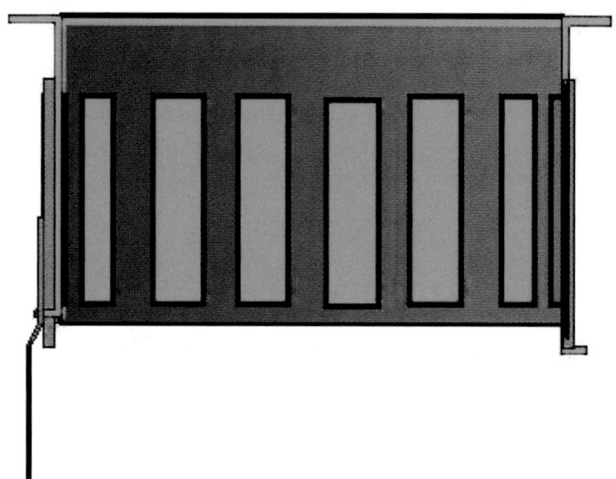

FIGURE 16.12 Air control device schematic.

FIGURE 16.13 Picture of air control device.

produced. An operator must know what to expect regarding flame patterns so that he or she can visually identify abnormal situations and make the appropriate adjustments.

Natural-draft burners are designed to provide the required combustion air flow across the burner with very low pressure drop, usually less than one in. of water column. The primary source for mixing the fuel and air is the kinetic energy in the fuel stream (fuel pressure is measured in psig, kg/nm^3, or kilopascals [kPa]). These burners can also be applied to forced-draft or balanced-draft systems with air blowers. The pressure drop in the forced-draft system can achieve 3 to 4 in. of water column.

Forced-draft burners utilize the high pressure drop of combustion air across the burner from forced-draft blowers. The air-side pressure drop may be up to 20 in. of water column. Mixing energy is significant in both the fuel gas pressure and the air stream pressure drop. Because of the high energy of mixing for the fuel and air, a forced-draft burner can release several times the heat release available from a natural-draft burner of the same footprint.

16.3.1.1 Air Flow Control

Most burners are equipped with some type of device to control the amount of combustion air flowing into the burner. This device is commonly known as an air register. In a simple configuration, it may consist of two metal cylinders with equal openings cut in each. As in Figure 16.12, one cylinder rotates and the other cylinder is stationary. As the free cylinder is rotated, the area for air to flow into the burner changes, causing variation in the air flow across the burner. The air control device can also be a damper located in the burner air inlet path (see Figure 16.13). Either the register or damper, whichever is applied, is also used to block air flow to the burner, for example, when the fuel flow is stopped and the operator wants to stop the air flow. Good operating practice calls for closing the air register or damper of any burner that is out of service. This forces all air entering the heater at the burner level to contact fuel in the active burners and improves the mixing energy, thereby improving the heater efficiency or fuel utilization.

Other types and combinations of registers are also applied. Some forced-draft burners will have a fixed air register whose function is to swirl the incoming air to the burner. In this application, the air flow will be varied by modulation of a fan damper or the fan speed.

16.3.1.2 Premix Burner

The combustion air flow on a natural-draft premix burner is controlled with a primary air door and a secondary air register. A 100% premix burner has no secondary air supply and uses only a primary air door. The primary air door is located on the mixer, or eductor, outside of the heater (see Figure 16.14). This door is adjusted to vary the air flow area so that the primary or premixed air flow is maximized and the flame on the burner tip is stable. The gas tip is in the heater, centered in the burner tile. The ideal flame will initiate and appear to stabilize in a position within 0.5 to 1.5 in. (1.3 to 3.8 cm) of the tip. If the flame appears to be less than 0.5 in. (1.3 cm) from the tip or even resting on the tip, the flame speed (or ignition velocity) is close to the flowing velocity of the air/gas mixture leaving the tip. If the flame speed exceeds the flowing velocity, a flashback to the orifice at the entrance of the mixer will occur, with a flame outside the heater and in the operating area. This may damage the burner and restrict its heat release (see Chapter 17.4 for corrective action). The

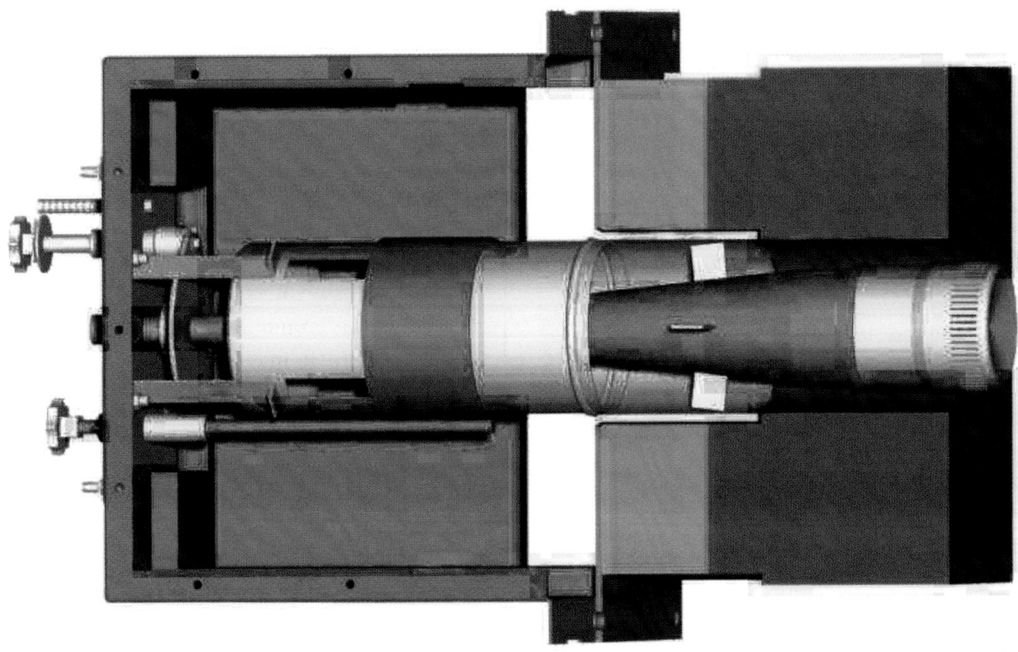

FIGURE 16.14 Premix burner.

appropriate primary air door adjustment will position the flame on the burner tip to the ideal position. If the flame is more than 1.5 in. (3.9 cm) from the burner tip, the gas/air mixture velocity leaving the tip holes is too high, and the flame is at risk of lifting off the burner and extinguishing (see Section 17.8 for corrective action). The flame liftoff from the gas tip and the flame extinguishing indicate flame instability. It is evident that the gas/air mixture volume and the flame speed are balanced in a properly operating burner, and that this balance is affected by the primary air door adjustment.

The secondary air register is positioned open or closed to adjust the excess air, as monitored by an oxygen analyzer, to meet the targeted excess oxygen. If the secondary air register is fully closed, any control of excess air occurs via the primary air door. Both the primary air door and secondary air registers must be in good working order to operate the heater efficiently.

A good natural-draft premix burner produces a very compact and short flame as compared to a raw gas or diffusion flame burner. Natural-draft premix burners operate best with a fuel gas of constant composition or specific gravity. These burners can be designed to handle gases of up to 90% (by volume) hydrogen content. The normal turndown of a premix burner is 3 or 4:1.

Forced-draft premix burners also mix the fuel and air before the mixture enters the combustion chamber. In this case, the premix chamber and the combustion chamber are both part of the burner, separated by a restriction orifice. The velocity through the orifice must be maintained above a minimum value or combustion can travel into the premix chamber, potentially causing damage. The operating parameters of a forced-draft premix burner are specific to the burner, and the burner manufacturer should provide complete operating guidance.

16.3.1.3 Diffusion or Raw Gas Burner

The conventional natural-draft diffusion or raw gas burner has one air register or damper controlling the air flow across the burner. Because there is no primary register or air door, air flow control on the diffusion gas burner is with a secondary air register or damper which is opened or closed to meet the excess oxygen target, depending on the oxygen analyzer reading.

Raw gas burners are preferred over premix burners for wide variations in fuel gas compositions and for high burner turndown requirements in forced/balanced-draft systems. Typically in these systems, the forced-draft fan damper or motor speed control is used to adjust the heater excess oxygen toward the target. The individual burner registers/dampers are adjusted to balance the air flow equally to all active burners. The goal is to equalize the air flow within 10 to 15% between active burners to ensure equal heat release from each burner and even heat distribution within the firebox and between tube passes. The forced-draft fan dampers and the burner registers/dampers must be in good working order and easily operable over their full range to operate the heater efficiently.

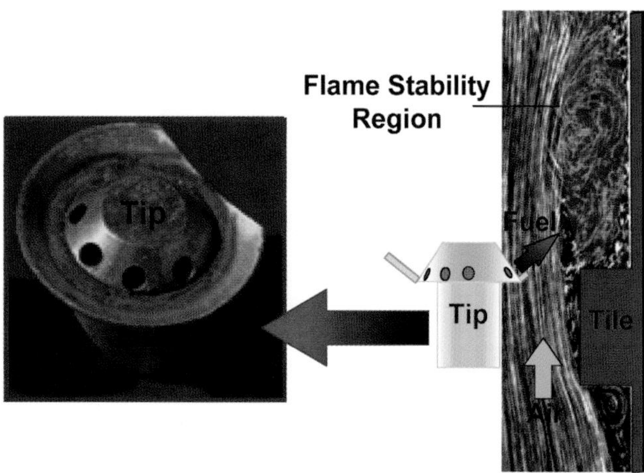

FIGURE 16.15 Burner ignition ledge.

A good raw gas burner flame will be attached to a diffuser or tile ledge to maintain a stable burning flame (see Figure 16.15). The flame will be larger than a premix flame of equal heat release because the mixing of the fuel and air is not as rapid as the fuel/air mixing in a premix burner. The tip drilling and the tile design will govern the flame shape. Narrow tip drilling promotes longer flames with a narrow shape; wider drilling favors bushier and wider flames.

Because the fuel ports in the tips of raw gas burners are small, as compared to premix burner tip ports, they are susceptible to plugging. Scale from the fuel piping, carryover from treating processes, and reactive components in the fuel can all contribute to plugging. Plugging is a common cause of irregular flames, flame liftoff, flame instabilities, high fuel pressure, and insufficient heat release. Refer to Chapter 17 for corrective actions.

The raw gas burner can satisfactorily handle a wide range of fuel gas compositions, including unlimited hydrogen concentrations and inert gas components (e.g., water vapor, carbon dioxide, nitrogen). A raw gas burner may have turndowns of up to 10:1, while premix burners are limited to 3 or 4:1. Any liquids should be removed from the fuel gas before the fuel gas reaches the burner. Fuel gases containing mists and significant concentrations of liquid hydrocarbons will likely cause tip fouling or coking problems and high gas pressures. The liquid hydrocarbons may cause coking within the gas tip, resulting in tip cracking.

16.3.1.4 Combination Burners

Combination burners are diffusion flame burners designed with the ability to fire either gas or liquid fuels, or a combination of the two. When firing gas and liquid fuel in combination, the total amount of fuel must be limited to match the combustion air supply to the burner. Hence, the flow measurement of each fuel being fired is very important to the operation of a combination burner. When gas and liquid fuels are fired in combination, it is recommended that the minor fuel flow be base loaded and that the outlet process temperature controller control the major fuel flow. The result is that each burner has the same flow of fuel or combination of fuels. The flame patterns should be the same on all the burners if the fuel flow and air are the same.

When firing in combination, the resulting flame length is normally 20% larger than if firing a single fuel. This is a result of the gas fuel mixing faster with the combustion air than the liquid fuel. The mixing of gas and air is rapid because both are in the gas phase. The oil flame, with atomized oil droplets spraying into the air flow, is a heterogeneous process. The fuel phase (liquid) is not the same as the air phase (gas); the combustion reaction is slower than when fuel gas is burned. The liquid fuel must receive heat from the combustion products to vaporize components that then burn homogeneously. The remaining material burns as a liquid on the droplet surface and as a solid when all liquids are consumed.

The majority of natural-draft combination burners will have a separate air register providing air that flows through the primary or oil tile. This is termed the primary register and is sized to provide 20 to 30% of the air required by the oil for complete combustion. The remainder of the air for liquid fuel combustion is provided through a conventional secondary air register. The secondary register also passes the air for fuel gas combustion. Inlet dampers can be substituted for registers in individual and common plenum designs.

The combination burner for low air pressure drop has a tile shape with a tile ledge (see Figure 16.15). The tile ledge is used to stabilize and hold the flame on the burner from the minimum to the maximum firing rate. On some high air pressure drop burners, the flame is stabilized with a spin diffuser or diffuser cone.

The flame envelope of an oil flame is normally the same as a gas flame of similar heat release. The oil flame will be bright yellow and opaque, a good radiator for heat transfer, and will cool relatively rapidly due to the radiation heat loss. The challenge with an oil flame is to mix the liquid droplets with the combustion air and complete combustion before the flame cools to below the ignition temperature. When the liquid droplets fall below the ignition temperature, soot and smoke are generated. If smoky flames are observed and excess oxygen is adequate, poor atomization or wet atomization steam may be causing the problem (see Chapter 17.11).

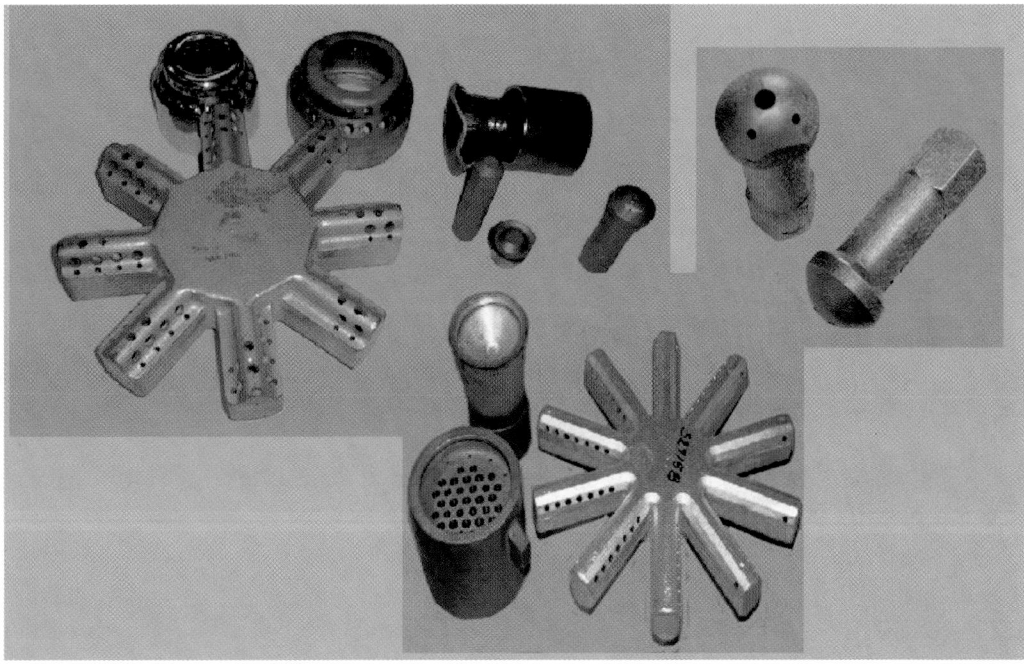

FIGURE 16.16 Gas tips.

16.3.1.5 Pilot Burners

Pilot burners are premix gas burners. Pilots may or may not be installed on a burner, depending on the requirements of the operator. When pilots are required, they are part of the burner assembly. The pilot may use the same air supply as the main burner, or it may be located outside the main burner air source to ensure stable operations.

The primary function of the pilot is to provide a small source of heat input for the ignition of the main burner fuel. The pilot may be shut off after lighting the main burner, or may continue to operate after the main burner is lit. The pilot must have a stable flame and be located in the correct position near the main burner fuel discharge for ignition of the main fuel.

16.3.1.6 Flame Patterns

The flame pattern from a burner is developed jointly by the heater designer and the burner manufacturer. The burner manufacturer selects the gas or oil tip drilling pattern, the type of diffuser (if applied), and the tile shape to achieve the desired flame. See Figures 16.16 and 16.17 for different gas and oil tip drillings. Figure 16.18 shows how variation in gas tip drilling pattern can create short and bushy or long and narrow flames. Oil tip drillings can be tailored to produce similar flames. There are many different types and shapes of fuel tips. Each fuel tip is drilled with a given pattern to meter the fuel and to inject the fuel into the combustion air. With the tile shape, diffuser, and the fuel tip drilling pattern, the burner manufacturer can obtain a predictable flame pattern from the burner. The fuel tip drill patterns can be varied based on testing and experience to obtain short, bushy flames or long, narrow flames depending on the requirements of the process heater (see Table 16.2 for typical flame patterns expected).

The burner tile shape and condition are critical to obtaining the desired flame shape. Round tile shapes provide round flame patterns; rectangular tile shapes provide flat flame patterns. Missing tile, or poorly maintained tile with holes and cracks, can cause a poor flame pattern. Substitution of tiles with ones of a different design can restrict air flow, result in poor fuel/air mixing, or cause flame instability because a tile ledge or other feature is missing or incorrectly located.

The flame patterns and dimensions and the impact of different tip drillings are determined by the burner manufacturer and can be confirmed by performance testing at the manufacturer's test facilities. Today, multi-burner testing is available at some burner manufacturers to determine the effect of the interaction between burners and the flue gas circulation within the firebox on the flame patterns. It should be noted that the flame dimensions observed in a single burner test, particularly a diffusion flame burner, will rarely be observed in a multi-burner firebox. Variation of the fuel and air flow between burners, flame interactions, flue gas circulation currents, and air leakage all act to vary the flame pattern between burners.

FIGURE 16.17 Oil tips.

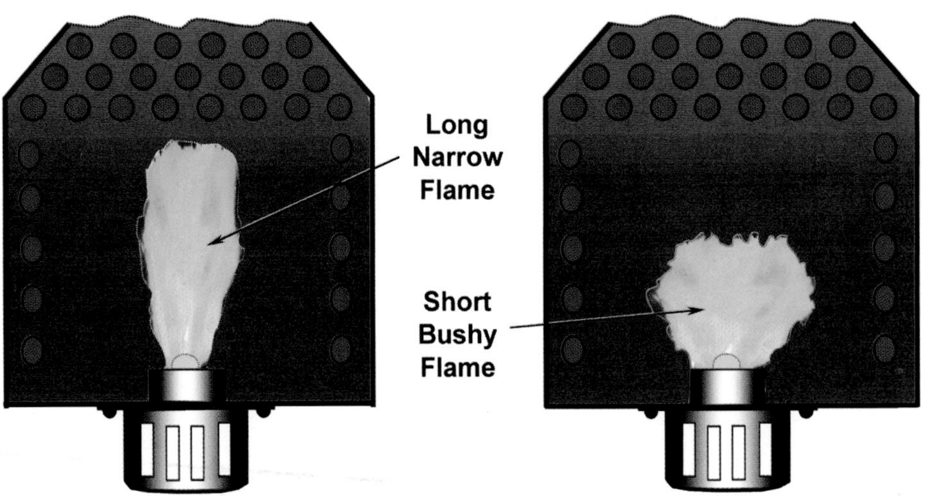

FIGURE 16.18 Long, narrow and short, bushy flame.

The flame patterns within the heater should be visually inspected as often as necessary, and any change in shape or dimensions of an individual flame should be noted. The flames from all active burners of the same size should be uniform because they all have the same heat release, the same air flow across the burner, and the same fuel flow at each burner. Table 16.1 can be used to estimate flame dimensions in order to evaluate observed flame patterns to determine if the flame patterns are typical and what is expected by the burner manufacturer.

Flame patterns are designed to stay within an envelope that is a safe distance from the process tubes and the refractory. Figure 16.19 shows a typical flame envelope, X-Y-Z, inside a cabin heater. API 560 cites some standards and recommendations as to burner-to-tube and burner-to-refractory. Some users apply greater clearances between burners and between burners and tubes than the API 560 recommendations. The burners installed on existing heaters many years ago may not comply with current API 560 recommendations. If installation of burners into an old heater are to comply with API 560, the

TABLE 16.1 Static Draft Effect per Foot of Height

Temperature (°F)	Inches
300°	0.0044
400°	0.0056
500°	0.0066
600°	0.0074
700°	0.0080
800°	0.0086
900°	0.0090
1000°	0.0094
1100°	0.0097
1200°	0.0100

TABLE 16.2 Typical Flame Dimensions for Different Burner Types

	Natural-Draft Burners	Forced-Draft Burners	LO NOx Burners
Height (ft/10^6 Btu/hr)	0.8–1.2	0.2–1.0	0.8–2.5
Diameter (ft)	Tile Size × 1.5	Tile Size × 1.5	Tile Size × 1.5

number of burners installed may need to be evaluated. More burners at a lower heat release will provide smaller flames.

Any flames that are visually different or unusual, either in dimension or stability, should be investigated and any problem corrected. Chapter 17 discusses several visual indications of flame problems, along with appropriate corrective actions.

In the combustion reaction, there are many different types of ions formed, depending on the fuel components being burned. The ions are excited by the high flame temperatures and radiate at wavelengths that are visible to the human eye. These visible wavelengths are the definition of the flame patterns on the burners. In Figure 16.20, sodium ions in the flame radiate in bright yellows and oranges. The color of the flame is affected by the flame temperature, the intermediate ions formed during the combustion reaction, and the amount of carbon particles in the flame. Regardless of the flame color, if combustion is completed, the amounts of heat released and heat available for heat transfer to the process fluid in the tubes are the same.

Figure 16.21 shows an example of a good flame within the firebox, while Figure 16.22 provides an example of a very bad flame pattern. Good flame patterns alone do not protect against localized tube overheating. Flue gas circulation flows within the firebox and uneven firing of burners will affect the distribution of heat to tubes.

FIGURE 16.19 Typical flame envelope with x-y-z axes.

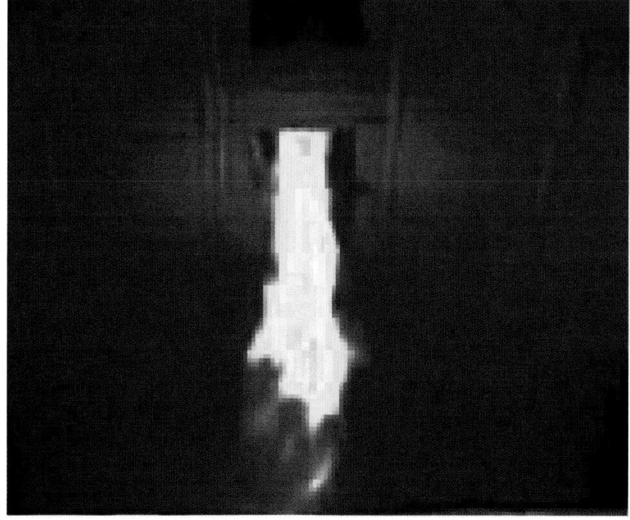

FIGURE 16.20 Sodium ions in the flame.

16.3.1.7 Burner Block Valves

All burners are installed with manual block valves on the individual fuel line (and atomization medium line, if provided) to each burner. Good operating practice on most

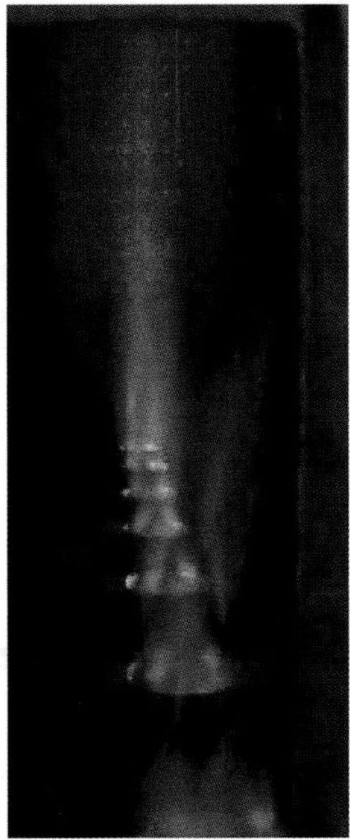

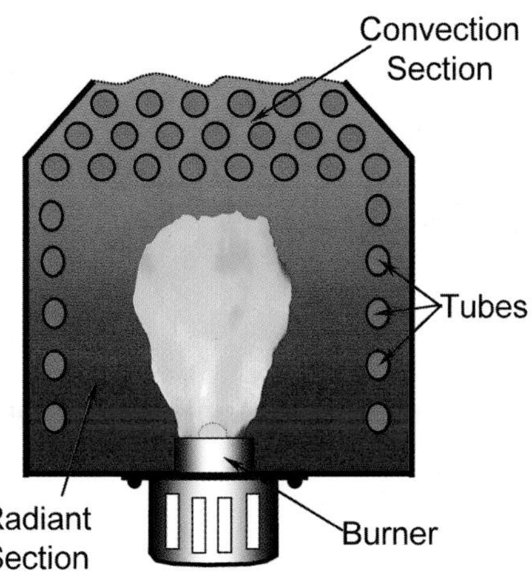

FIGURE 16.21 Example of a good flame within the firebox.

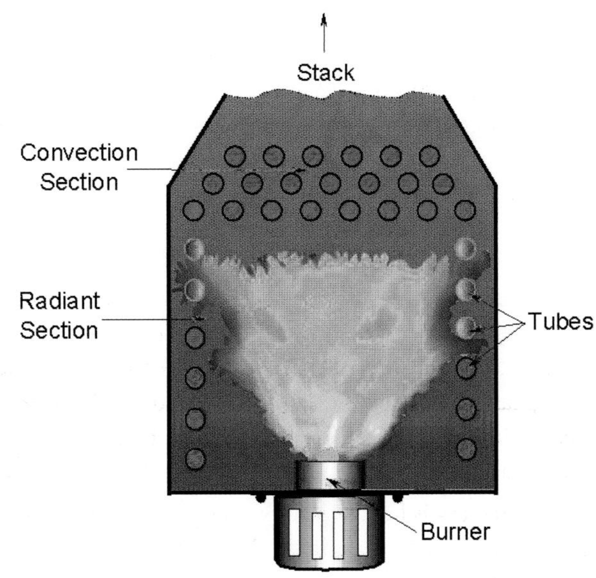

FIGURE 16.22 Example of a very bad flame pattern in a firebox.

heaters is to keep these valves fully open or fully closed and never to throttle the amount of fuel to the burner. The fuel to the burners should be regulated with the main fuel control valve. The fuel (and atomizing medium) block valve is used to stop fuel flow to a burner taken out of service in order to maintain adequate fuel pressure on active burners; to balance the heat release in the firebox; or to repair or maintain the burner taken out of service.

If the operator uses individual burner block valves to control flame patterns at selected burners, an uneven distribution of heat within the firebox may result. One possible consequence is overheating of tubes at some locations, causing coking of the tube contents and potential tube failure. It is also common for an operator to react to a high tube temperature alarm by throttling the block valve on the burner closest to the alarming tube-skin thermocouple. While this stops the alarm and reduces the temperature at the alarming thermocouple location, the total fuel input does not change. The fuel pressure and input on the other active burners increase and tube-wall temperatures rise in response. The new locations of high tube-wall temperature likely are not at one of the few thermocouples, and tube overheating goes undetected until failure or severe fouling occurs. The operator should adhere to the best practice of leaving the burner valves fully open or fully closed, and should react to the alarm in other ways to cool the tube.

If block valves on burners are partially closed, the fuel pressure on the burners varies, depending on valve positions. If the heater controls reduce the fuel flow, one or more burners may experience a fuel pressure below the stable limit and may extinguish. The low pressure alarm and trip instruments on the burner fuel manifold will not sense the same low pressure as at the burner(s) and will not protect the heater as expected with the burner management control system.

The burner block valves on high-temperature heaters, with 1800 to 2200°F fireboxes, such as in pyrolysis and reforming services, are often operated differently. These heaters may have many small wall-mounted burners heating vertical tubes located between the burner walls and operate with tube metal temperatures near the metal strength limit. When a high tube temperature is identified, it is common and accepted practice to reduce the fuel to the burner opposite the hot spot by throttling the burner valve. Tube temperatures are monitored over the full length with a pyrometer; thus, the effect of throttling a burner can be identified quickly and reliably. Also, the burners are numerous and small; throttling a few has little impact on the heat distribution. If the heater controls modulate the fuel such that fuel pressure at a throttled burner drops below the stability limit of a single burner, flameout may occur. The amount of unburned fuel entering the firebox is small and ignites and burns because of the elevated firebox temperature. Thus, the combination of small heat release, many burners, tube temperature monitoring over the full length of all radiant tubes, and high firebox temperature (well above the fuel gas autoignition temperature) leads to a safe practice and no equipment endangerment.

16.3.1.8 Casing and Refractory

Modern heaters have a steel plate casing supported by structural steel columns, often called buckstays. The steel, in turn, supports the internal refractory, the tubes, the access platforms, and in many designs the convection section and stack. Some heaters may have a separate structure for the convection section and stack.

The casing is normally not designed for airtight construction. There are typically open seams, sight doors, openings for tubes and manifolds, and doors for maintenance access in addition to the burner openings. Because the heater operates under a slight negative pressure, air will leak into the heater through all openings. It is desirable, for best operating efficiency and proper burner operation, to have all air enter the heater through the burners. Therefore, all doors should be kept tightly closed during operation and other openings should be minimized, or sealed, as much as possible.

Unwanted openings capable of leaking air into the heater can be located by smoke testing using smoke generators (or smoke bombs), usually placed inside the idle firebox. Cracks and seams thus identified are most successfully sealed with commonly available silicone caulk. One should repair openings due to warped or displaced plates.

If the fuel being burned in the heater contains sulfur, the casing is subject to acid corrosion. The sulfur partially converts to sulfur trioxide, which then combines with water vapor and condenses, at normal casing temperatures, to form dilute sulfuric acid. This sulfuric acid condensation and potential corrosion can best be reduced by coating the steel casing internal surfaces with an acid-resistant barrier suitable for the exposure temperature, by raising the casing temperature with external insulation, or by using a monolithic castable refractory lining. Operation to minimize excess oxygen will reduce sulfur trioxide formation and acid corrosion.

Sight doors, mounted on the casing, provide visual access to the firebox interior. The operator can monitor the burner flames, the temperature and condition of tubes, and the temperature and condition of the refractory. The sight port doors should ideally be sized and located so that all burners and the full length of all radiant section tubes can be observed. If the sight doors are not adequate to fully observe the inside of the firebox, additional sight doors should be added to ensure good visual observations inside the heater.

Before opening a sight door, the draft at the door elevation should be checked to ensure a negative pressure. A rag should be held at the door to confirm the flow of air into the heater. Hot gas exiting through a sight door, under positive firebox pressure, will blow the rag outward and can injure the unprotected observer. Proper protective gear should be worn when opening sight doors for inspecting heaters.

The refractory protects the steel casing from the heat of the combustion process and provides insulation to reduce the heat loss to the ambient air. The types of refractory used to line the casing steel include ceramic fiber, monolithic castable, brick, and plastic (a moldable form of brick). All have service temperature limits that must be observed in the selection of the proper lining for the heater. The refractory manufacturers provide recommended maximum exposure temperatures for each material; if this temperature is exceeded, the refractory undergoes a phase change and weakens. API Standard 560 provides guidance in refractory design and selection, including service temperature recommendations.

The refractory is held in place by (usually metallic) refractory anchors of various designs. The anchor design, spacing, and attachment are critical to obtaining a satisfactory refractory installation and service life. Most refractory failures can ultimately be traced to anchor failure. These failures are due to anchors overheating, corrosion, or inadequate anchor spacing (too far apart) to adequately hold the refractory.

Refractory failure may first be evident by debris on the heater floor. With severe refractory failure, hot areas and discolored paint on the outside of the casing may be observed. If this occurs, the casing can be cooled with a low-pressure steam spray or water flood until repairs can be completed.

When viewing the firebox refractory, color can be an indication of even heat distribution within the firebox. Refractory color can also be a very rough measure of the refractory temperature. Dark streaks on the refractory provide evidence of air leaking into the firebox and cooling the refractory.

16.4 FURTHER OPERATIONAL CONSIDERATIONS

16.4.1 Target Draft Level

A target draft is established at the point of highest flue gas pressure within the heater. The target value is selected to minimize air leaking into the heater and to provide adequate differential pressure or draft loss at the burner level for necessary air flow across all burners. The typical fired heater data sheet defines the draft at the arch, or radiant zone roof, of 0.1 in. (2.5 mm) of water. This is not a design limit, it is the reference value from which draft losses are calculated and flue gas passages sized by the heater designer. This is also the location at which the operator normally monitors and controls the heater draft performance. Each heater should have facilities to monitor the draft available at the arch or at the location where the draft (or negative pressure) within the heater is at a minimum.

Good efficient operation minimizes air leaking into the heater through paths other than the burners. Minimizing the arch level draft minimizes the differential pressure between the outside air and the flue gases within the heater. This minimizes the driving force pushing air through the openings that may allow air to enter the heater. Air leaking into the heater is an important consideration when setting a draft target. The higher the draft, the more air leaks into the heater. The lower the draft, the less air leaks into the heater. The cost for air leaking into the heater can be seen by the following example:

16.4.1.1 Example

What does air leakage cost at a draft of 0.5 in. (1.3 cm) of water and a stack temperature of 750°F (400°C)? Assume that one 4-in. (10 cm) by 6-in. (15 cm) peep door is left open. How much air would leak into the heater? Figure 16.16 indicates that at 0.5 in. (1.3 cm) of water, the airflow is 170,000 ft^3/hr-ft^2. If the fuel being burned is methane, the LHV is 910 Btu/scf. Using the following formula:

$$Q = wc_p\Delta t \qquad (16.1)$$

and a fuel cost of \$3.50/10^6 Btu, the cost of additional fuel is approximately \$12,000 per year. Hence, reducing the air leaking into a furnace significantly reduces the operating cost of a plant.

Remember that the burners require a draft loss or pressure drop to force combustion air through the tile and registers. The available draft decreases with height — that available at the floor of the firebox will be greater than that at the roof. This is shown in the typical static draft profile (Figure 16.23). Thus, burners at or near the floor will usually have ample draft available to exceed the required draft loss; but for burners located high on the walls or on the roof, there may not be enough draft. In such a case, the draft target should be set to satisfy the highest level burners at the design heat release, plus a small margin for ambient temperature, fuel composition, and operational changes.

The consequence of too much draft in the firebox is excessive air leaking into the heater and a lower heater efficiency. The additional air adds to the flue gases and increases the draft loss across the convection section and stack. Too little draft can restrict the air flow through burners, sometimes enough to cause flames to exceed the designed flame envelope (flame impingement), flame instability, and the formation of carbon monoxide. If the firebox pressure becomes positive at the arch, hot gases will flow out through openings, potentially damaging the casing, weakening refractory anchors, creating unsafe conditions, and restricting heat release.

The draft is controlled by modulating the flow of flue gas out of the firebox, usually via the stack and stack damper. The mechanism of control is stack damper positioning or variation of an induced-draft fan speed (or fan inlet damper position). The dampers are typically single- or multiple parallel blade designs. The draft is measured with a liquid inclined manometer (see Figure 16.2), a dial gauge, or a transmitter instrument.

The draft target selected will be a function of how the draft is controlled. If the draft is controlled automatically, using a draft transmitter and a pneumatic or electric operator attached to the damper shaft, a lower draft target (e.g., 0.05 in. [1.25 mm] or less of water column) is practical. If the draft is controlled manually using a manometer, a higher draft target (e.g., 0.1 in. [2.3 mm] of water column) may be required. Even further reduction in draft target can be achieved if additional information is provided to the automatic controller. Such information may include oxygen trim, setting a minimum value of excess air in the firebox, process inlet flow and temperature to anticipate changes in firing rate, and even ambient air temperature if large and relatively sudden swings are expected. Inputs such as these allow the damper to anticipate the movements that will be required to hold the target draft and will stabilize the burner operation.

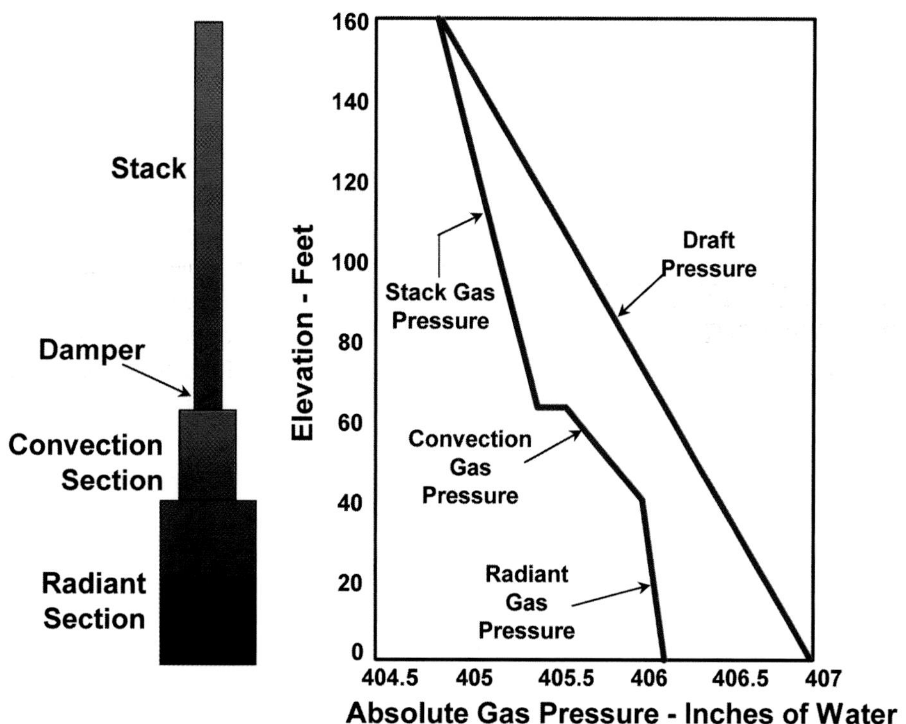

FIGURE 16.23 Typical draft profile in a natural draft heater.

16.4.2 Target Excess Air Level

It is not possible to recommend a single target excess air level for all fired heaters. The condition of the heater, the type and composition of the fuel fired, the level and variability of process operation, and the ambient conditions all affect the achievable target.

The target excess air level can be established by following a structured procedure. The recommended procedure for establishing excess air level targets for a natural-draft heater is given below. The procedures for forced- or balanced-draft heaters are similar. Essentially, the procedure is that which is used to adjust a heater to the maximum possible combustion efficiency, or to optimize the heater performance. The important instruments required to be in good working order and proper calibration include the fuel flow meter, the process outlet temperature indicator or recorder, the draft gauge or controller, the flue gas oxygen analyzer, and the carbon monoxide analyzer. The analyzers must be located so as to sample the flue gases leaving the radiant section.

Begin the procedure with all possible burners in service and firing equally on one fuel, satisfactory flame appearance, correct fuel temperature and pressure, steady process operation, and all potential limits (such as tube metal temperature, etc.) monitored and recorded. Close the air registers (or dampers) on all out-of-service burners. Equalize the air flow to all active burners by adjusting all registers to the same opening. Check the oxygen and carbon monoxide levels of the flue gas and ensure that the draft available at the firebox roof is on target.

Next, slightly close the air registers equally on the active burners or the common plenum air supply damper. The draft will probably increase above the target because the incoming air and the flue gas amounts are lowered, reducing the friction losses in flow through the heater. Measure the excess oxygen and carbon monoxide in the flue gas. Observe the flame condition and monitor the other instruments for satisfactory operation with no approaching limits. Adjust the draft to the target value with the stack damper.

Continue to close the burner air registers or plenum dampers in slight increments while holding the targeted draft level with stack damper adjustments. Measure the excess oxygen and carbon monoxide levels in the flue gases leaving the firebox after each burner air register or plenum damper adjustment. The minimum practical excess oxygen level for the heater at the current firing conditions and with the current fuel composition is reached when the draft reading is on target and any stack damper adjustment or burner air register adjustment (or plenum damper) closure causes the carbon monoxide in the firebox to exceed 100 ppm. At this point, there is insufficient air available to burn any additional input of the current fuel. Any further increase in fuel will increase the carbon monoxide level and cause the firebox to become "flooded," or "fuel rich." The minimum excess oxygen level is achieved.

If the operator or engineer has many years of experience in combustion and there is no carbon monoxide analyzer available, the minimum practical excess oxygen level can be determined by closing the burner air registers (or plenum dampers) and keeping the draft on target with the stack damper. When the firebox becomes flooded or fuel rich, the experienced engineer or operator will observe no increase and probably a decrease in process outlet temperature when the fuel flow increases.

The targeted excess oxygen level for continuous operations should be set at 1 to 2% greater than the minimum practical level determined above for natural-draft heaters, and 0.5 to 1% greater for forced-draft operations. The higher targeted excess oxygen level allows for variations in the fuel composition, variation in ambient air conditions, and variations in firing rates for the heater. Highly sophisticated heater instrumentation and control systems can safely allow lower excess oxygen levels than would be suitable for simple automatic or manual systems. The target excess oxygen level should provide safe and steady heater operations with good flame patterns (no flame impingement), good flame stability, and tube-skin and firebox temperatures within the limits set by the operator for the heater. If heat input requirements or fuel composition changes are significant, then the carbon monoxide should be monitored. If the carbon monoxide is observed to exceed the 100-ppm limit, then the minimum practical excess oxygen level should be reestablished for the current conditions. Tables 16.3 and 16.4 indicate typical excess air volumes that should be achieved with this procedure.

TABLE 16.3 Typical Excess Air Values for Gas Burners

Type of Furnace	Burner System
Natural draft	10–15%
Forced draft	5–10%

TABLE 16.4 Typical Excess Air Values for Liquid Fuel Firing

Operation	Fuel	Excess Air
Natural draft	Naphtha	10–15%
	Heavy fuel oil	15–20%
	Residual fuel oil	15–20%
Forced draft	Naphtha	10–15%
	Heavy fuel oil	10–15%
	Residual fuel oil	10–20%

With the target draft and the target excess oxygen established, the operator is now ready to make any adjustments needed to keep the heater operating as efficiently as possible. The flowcharts in Figures 16.24 (natural draft) and 16.25 (forced draft) will guide the operator through the necessary adjustments on the heater to achieve set targets on a continual operating basis. For example, the operator can control the preset targets as follows.

The target draft has been determined to be –0.05 in. of water from the previous discussion. The target excess oxygen has been determined to be 3.0%. In a natural-draft heater, the stack damper and the burner air register are adjusted to control the oxygen and the draft. In a forced-induced draft system, the forced-draft damper on the inlet to the forced-draft blower and the induced-draft damper on the induced-draft blower are used to adjust the excess oxygen and draft. See Figures 16.24 and 16.25 for the logic diagram for tuning natural-draft and a balanced-draft heaters, respectively.

For this example, the operator begins in the "START" box of Figure 16.24. The pressure (draft) is measured on the heater at –0.14 in. (–3.6 mm) of water. The pressure is below the target of –0.05 in. of water, and the logic box indicates "HIGH." The excess oxygen is measured on the heater at 5%. The chart indicates that if the excess oxygen is also above the target of 3%, then the excess oxygen is in the "HIGH" box. The flowchart indicates that the corrective action required is to close the stack damper on the heater. When the stack damper is closed, the pressure within the heater goes from –0.14 to –0.05 in. (–3.6 to –1.3 mm) of water. The logic chart indicates a return to "START." The draft (pressure) is now on target, so go to box "ON TARGET." The excess oxygen measured in the field is 3.9%. The excess oxygen is still above the target of 3%. The operator goes to the "HIGH" box.The corrective action indicated is to close the air register or damper on the burner and return to "START."

The draft is measured again and determined to be –0.07 in. (–1.8 mm) of water, above the target again. The logic chart indicates to check the excess oxygen. The new excess oxygen reading is 3.2%. The logic chart indicates to close the damper again and return to "START." Return to "START" and measure the draft again. The new draft reading is 0.5 in. (13 mm) of water. The draft is on target; thus, the logic chart indicates to measure the excess oxygen again. The excess oxygen reads 3.0% and is therefore on target. The logic chart indicates "Good Operations." The tuning of the natural-draft heater has been completed. How many of dollars were saved? The charts in Figures 16.6 or 16.7 can be used to determine the savings. The forced-draft heater

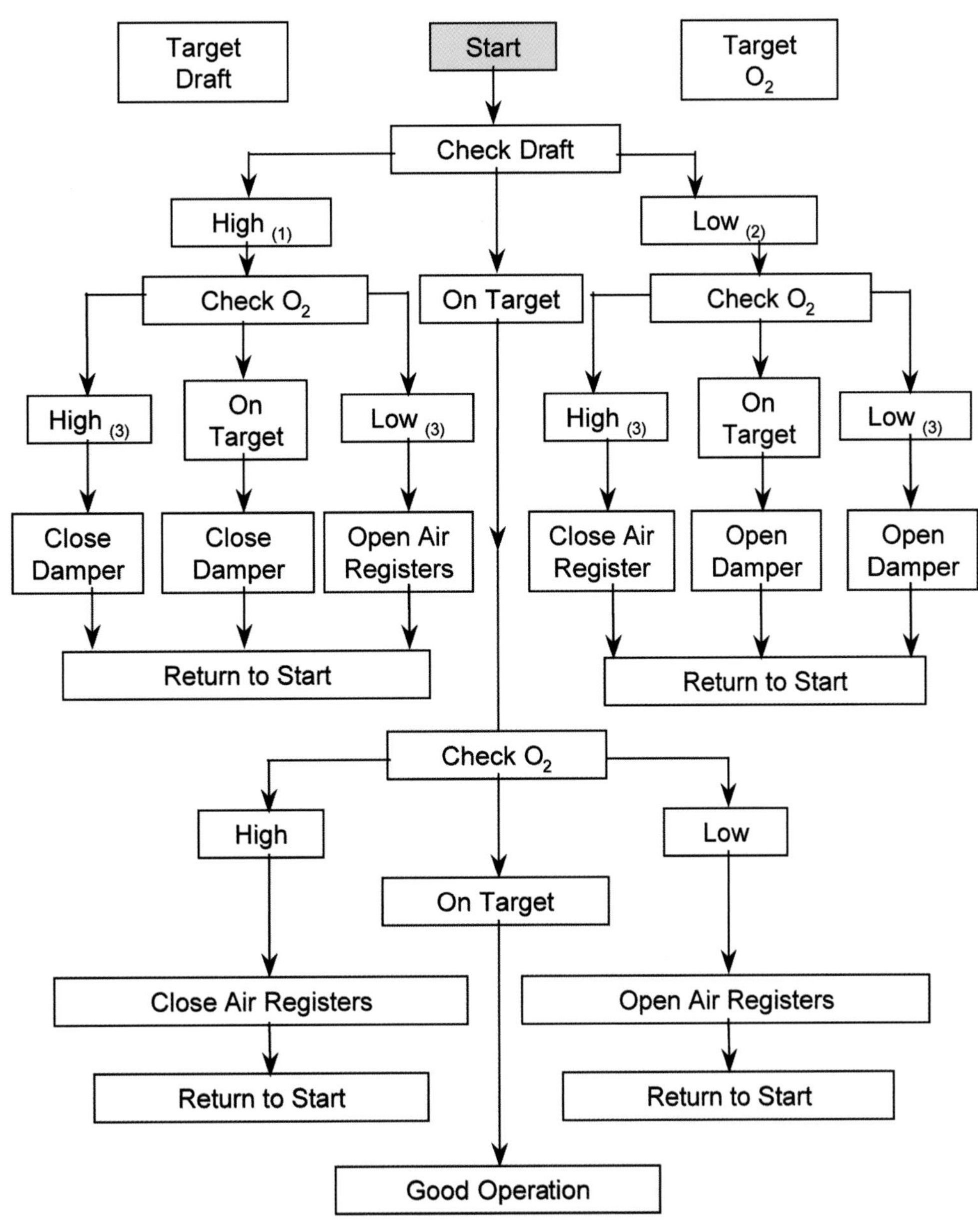

FIGURE 16.24 Logic diagram for tuning natural-draft heater.

diagram shown in Figure 16.25 is used to make adjustments on a balanced-draft heater.

While the operator should be encouraged to operate to the excess oxygen target as determined above, there may be conditions where it is desirable to operate with excess oxygen well above the target. One is mentioned in Chapter 16.2.8 in conjunction with mitigating high tube-skin temperature. Here, increasing the excess oxygen reduced the radiating effectiveness of the firebox gases, lowered the radiation to the radiant tubes, and lowered the heat flux and temperature of the tubes. The effect is to increase the amount of flue gas and the bridgewall temperature, increasing the amount and temperature of gas flowing to the convection section. Duty is shifted from the radiant to the convection section. Similarly, for heaters where waste heat steam generation occurs in the convection section, trouble with the plant steam boilers may make it desirable to generate more steam in the waste heat generation coils. Increasing the excess oxygen (excess air) above the target value will increase convection section duty and generate more steam.

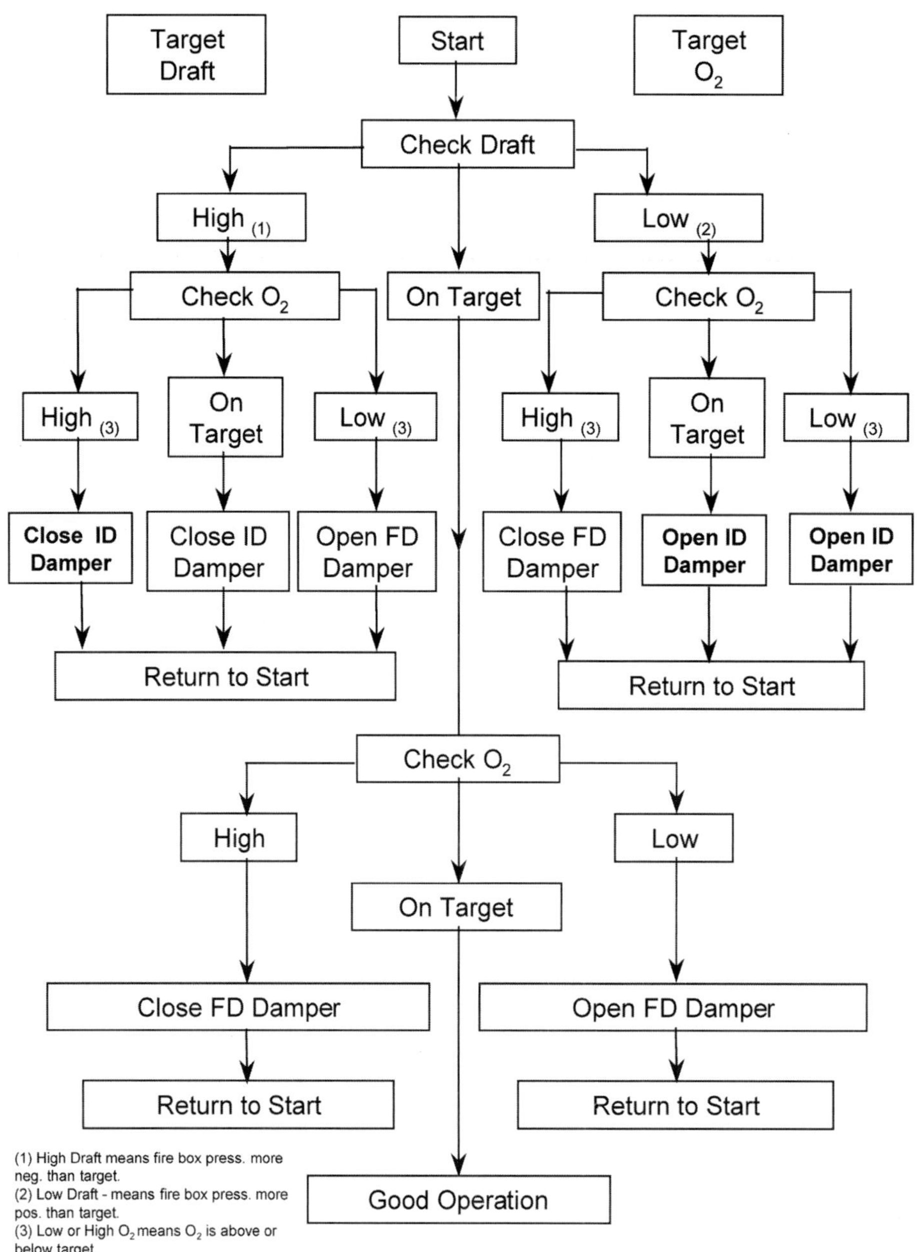

FIGURE 16.25 Logic diagram for tuning balanced-draft heater.

16.4.3 Heater Turndown Operation

During normal or design-level heater operation, the operator strives for even heat distribution in the firebox by equalizing the air and fuel to all burners while maintaining good, stable flames. The same goals apply to turndown operation. When operating at reduced heat-release levels, the operator will remove burners from service in a selected pattern, closing air registers and fuel valves, so as to maintain adequate fuel pressure well above any low-pressure trip limits. Flames will be kept stable and within the normal flame shape and envelope.

Burners are turned off in a pattern that maintains an even heat release throughout the firebox. The heater designer arranges burners so that each radiant tube pass "faces" the same number of burners. It is wise to maintain this practice when reducing the number of active burners. This helps to ensure that each pass receives the same amount of heat and avoids having to overly bias pass flow rates to equalize outlet temperatures. If the heater configuration allows (usually vertical tubes) and if tube metal temperatures are nearing upper limits, the best burners to deactivate are usually those opposite the hottest outlet tubes in each pass.

In natural-draft heaters with diffusion flame burners, the excess oxygen can be held nearly constant with air register or damper adjustments from 100% capacity to approximately 75% capacity. Then the excess oxygen level begins to increase as the heater (burner) heat release is further reduced. Forced-draft burners, with high kinetic energy in the air stream, are usually capable of maintaining the excess air level down to 50%, or even lower, of the design heat release.

In natural-draft heaters with premix burners, the excess oxygen will follow the fuel gas pressure as the heater duty is reduced. The excess oxygen will remain the same from 100% duty to 25% of duty on a 100% premix burner, and 100% to 50% of duty on a 50% premix and 50% secondary air burner. The mixing of the fuel and air is affected mainly by the primary air mixer efficiency.

16.4.4 Inspection and Observations Inside the Heater

Much of successful heater operation depends on frequent and knowledgeable visual checks of the equipment. In most applications, visual surveillance of the heater as often as necessary to ensure safe and optimum operations is considered good practice. Visual observations should also be made after significant load changes, atmospheric disturbances, fuel gas composition changes, and utility system upsets. A checklist of what to observe should be developed for each heater. The visual inspection of a heater should include the inside of the heater, the outside of the heater, and observation of the performance data of the heater.

16.4.4.1 Flame Pattern

A check into the firebox should show even flame patterns and good stability on all active burners. The size, color, and shape of the flames should be the same because all active burners have the same fuel pressure (heat release) and draft (combustion air flow). Any uneven flame patterns, any flames that are unstable, or any flames impinging on tubes indicate a problem that needs to be corrected. See Chapter 17 for a full discussion on troubleshooting that problem.

A hazy appearance in the firebox may be an indication of high carbon monoxide levels and may cause wildly swirling flames extending to the arch. If the carbon monoxide levels are high enough, afterburning may occur in the convection section, possibly resulting in tube support failure and nesting of the tubes in the convection section. The draft loss across the convection section should be increasing, depending on how much nesting of tubes is taking place. The afterburning may severely oxidize the extended surface (fins or studs) on the convection tubes such that heat transfer capability is lost, or may cause tube overheating leading to fouling or failure. The stack temperature with afterburning in the convection section should be increasing significantly above the design temperature. Check the excess oxygen, the carbon monoxide level, the process outlet temperature, and the fuel flow.

The operator should reduce the fuel input when high carbon monoxide levels are indicated or afterburning is known to occur in the convection section until an excess of oxygen is attained. Then, opening the stack damper or burner air registers (or dampers) should further increase the excess oxygen. Now increase the fuel flow rate to satisfy the process heat input requirements. **Do not increase excess air without first reducing the fuel flow to the point where an excess of oxygen is observed.**

16.4.4.2 Flame Stability

In observing the flame patterns inside the firebox, the flame stability should be noted. An unstable flame front will be oscillating on the burner tip as the fuel is mixed with the air. The flame front should always be very near the ignition ledge, diffuser, or fuel tip. If the flame front is detached from the ignition ledge, diffuser, or fuel tip, as in Figure 16.26, the flame is unstable and may be on the verge of being extinguished. When the flame is extinguished and fuel is injected into the heater and mixes with the air with no flame, an unsafe and potentially explosive condition exists within the heater.

16.4.4.3 Process Tubes

The process tubes should be periodically checked visually for evidence of localized hot spots, tube displacement, and process leaks. Tube hot spots may appear as red or silver spots and indicate temperatures approaching or exceeding the mechanical limit of the tube material. The immediate cause of the hot spot is usually a fouling deposit on the internal tube wall. This deposit may be the result of flame impingement, over-firing, uneven distribution of active burners, or concentrated heat input from flue gas circulation currents. Hot spots need to be continually monitored; if allowed to continue, they will ultimately cause tube failure. The process tube maximum temperature limits should be known to the operator and monitored to protect against tube overheating and failure.

Tubes that are out of position may bow due to overheating or because of loss of a tube support or guide. Overheating may be caused by over-firing, by concentration of active burners in the firebox, or by flame impingement. The bowing may be accompanied by internal tube fouling. Look for metal parts from a broken support or guide on the heater floor.

FIGURE 16.26 Unstable flame.

FIGURE 16.27 Broken burner tile.

A process tube leak may initially show as a small flame or wisp of smoke at the tube surface. Flames or smoke may be traveling through the tube-sheet into the firebox from a header box. The process tube leak acts as fuel and consumes oxygen. This may result in long flames from the burners, due to the reduction in excess oxygen, and afterburning in the convection section. The oxygen analyzer will often show a steep change to lower oxygen levels and the measured carbon monoxide levels will increase.

Black smoke from the stack indicates a major process tube failure or incomplete combustion of the fuel. The latter occurs most frequently when firing liquid fuels. The operator must check the flame appearance to determine whether the problem is a tube failure or poor combustion at one or more of the burners. Also check whether high molecular weight components have been added to the fuel gas system.

16.4.4.4 Refractory and Tube Support Color

The color of the refractory and tube supports can be an indication of high temperatures in the firebox. Tube supports should appear uniform in color if the heat distribution is uniform within the heater. If some tube supports appear red or visibly hotter than other tube supports, then there is uneven heat distribution within the heater. If the refractory is not uniform in color, then there may also be uneven heat distribution within the heater. Uneven firing on active burners or poor burner distribution, flame impingement, or flue gas circulation patterns are all suspects and should be checked.

16.4.4.5 Burner Refractory and Diffuser Condition

Broken burner tiles, burner tiles that have deteriorated from chemical attack, improper installation of the tile, and burner diffuser conditions should be noted (see Figure 16.27). Tile or diffuser in poor condition can lead to poor flames and possible impingement. The diffuser can normally be replaced while the heater is in service, but tile replacement may require heater shutdown.

16.4.4.6 Air Leaks

When observing inside the heater, look for air leaks around sight doors and other areas where air leakage into the heater is possible. In hot fireboxes, air leaks may emerge as a dark line or streak on the refractory surface. See Figure 16.28. The higher the air leakage into the heater, the higher the operating cost and likelihood of poor flame patterns. Idled burners should have air registers completely closed. Air leaks should be sealed when identified.

16.4.5 *Inspection and Observations Outside the Heater*

16.4.5.1 Stack Damper

The stack damper position should be noted to determine if there is sufficient control of the draft within the firebox. If the stack damper is a single-blade damper and is over 75% open, then there is probably very little capacity left to increase the draft (negative pressure) within the firebox. However, if this is a multi-blade damper, there is probably sufficient control remaining to control the targeted draft and get more capacity out of the heater.

Check the arrow indicating the stack damper position. Does it match the position shown in the control room? Does the damper move when the actuator is adjusted, or is it stuck? Most dampers control adequately over a range of positions between 20 to 80% open. Compare the present position to this range to get an idea of the amount of draft control available.

If a black plume is observed being emitted from the stack, look for a tube rupture, a burner with poor atomization, or insufficient excess air, possibly due to a fuel composition change. A blue plume from the stack indicates the burning of sulfur in the fuel. Operation changes will not affect this plume.

16.4.5.2 Burner Block Valves

Check the positions of the individual burner block valves. Observe those not fully open or closed. Most heaters should be operated with these valves either fully open if the burner is in operation, or fully closed if the burner is out of service. Some special heater designs allow throttling of individual valves to control the tube temperatures. However, in most heaters, the throttling of these valves can result in poor flames as well as compromised safety systems and tube temperature monitoring.

16.4.5.3 Pressure Gauges

The pressure gauges should be located on the fuel line to provide a pressure reading to indicate the amount of fuel flowing to each burner. The fuel pressure should be recorded and monitored in the control room and, if possible, it should be trended. The local pressure gauges should be in good working condition to give an accurate pressure reading.

16.4.5.4 Heater Shell or Casing Condition

The heater shell or casing should be inspected periodically to determine if there are any hot spots developing on the shell or casing. Hot spots on the heater shell are an indication of refractory failure within the firebox. If left uncorrected, the heater efficiency will decrease because of the higher shell heat loss, and casing failure can lead to the heater being shut down.

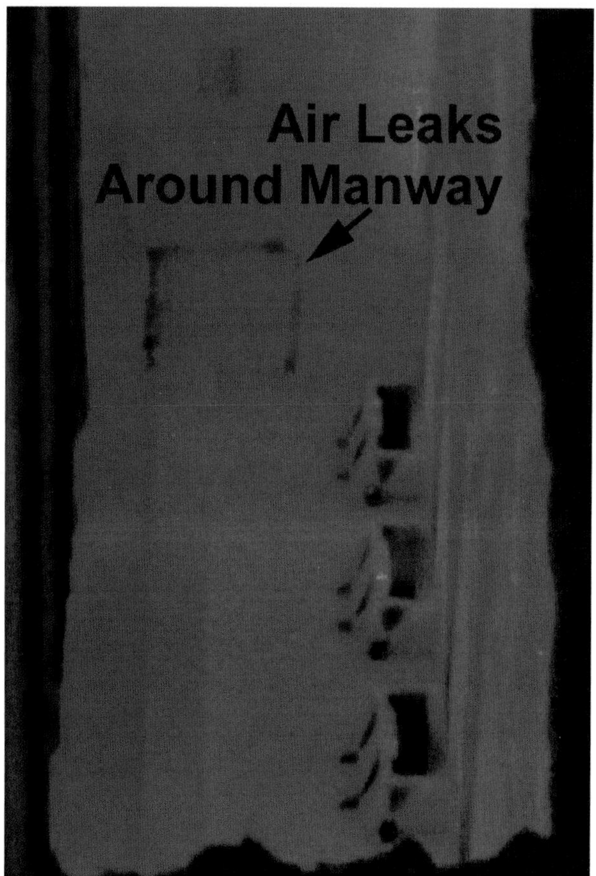

FIGURE 16.28 Dark line or streak on hot refractory surface indicating air leaks.

Hot spots on the casing are the result of internal refractory failure, possibly due to positive pressure in the firebox. Check the draft in the heater, and, if necessary, make damper adjustments to regain a positive draft (negative pressure) at the arch. The hot spot can be cooled with steam or water spray.

16.4.5.5 Burner Damper Position

The burner damper position or air register position should be at the same opening for all burners in operation. They should be equally open on the active burners and closed on all burners out of service. Check natural draft premix burners for backfiring and correct immediately by adjusting the air door.

16.4.5.6 Burner Condition

The general condition of the burner should be observed. A burner in good condition can be operated at very low excess oxygen levels if the heater is in good condition and sealed properly. Burners in poor condition will require higher excess oxygen levels to operate safely and should be removed and repaired.

16.4.5.7 Piping

Inspect the fuel and atomizing medium piping and the process piping for leaks. Also inspect piping for stresses due to expansion. Check the insulation and tracing condition on all fuel piping to ensure that it is adequate. Listen for any gas leaks. Gas leaks may also be detected by the odor of the gas. Check the air lines on all pneumatic instruments and control valves to ensure no leaking or loss of functionality.

16.4.5.8 Controllers

Check the control air and power air lines at all valves for leaks, and note the pressures and the positions of the actuator stems. This can provide an indication of the range of operation available on process flow, draft regulation, fuel input, and air supply.

16.4.5.9 Air Leaks

Look for open sight port doors, warped explosion doors, idle burners with open registers, and other locations of obvious air leakage into the heater. Air leaking into the heater increases the operating cost and may affect flame patterns.

16.4.6 Heater Operation Data

Every heater should have established control points on certain operating variables that are selected to ensure safe operation and maximum heater utilization. Operators must ensure that these control limits are not exceeded. Items that may be considered limits will include some of the following and should be monitored frequently.

16.4.6.1 Process Flow

The process flow rate should be monitored and trended. The process flow is an indication of the amount of work the heater is to perform. If the process flow is at design conditions, then the burners can be expected to be operating at design conditions. If the process flow is above design conditions, then the burners can be expected to be operating above design conditions.

16.4.6.2 Pass Flow

Check each pass flow. Pass minimum flow limits are established to avoid maldistribution of process flow through the tubes in each pass. Flows that are below the limit can cause coking of interior tube surfaces or overheating and tube failure.

16.2.4.3 Tube Metal Temperature

Check tube metal temperatures. Tube maximum temperature limits are established to protect against tube overheating and failure.

16.2.4.4 Stack Temperature

Check stack temperature. Stack gas temperature low limits are intended to protect against condensation and corrosion; high limits may help protect air preheater and stack materials from failure. This temperature can also be used to trend the heater to determine efficiency changes and identify convection section heat transfer problems.

16.2.4.5 Bridgewall Temperature

Bridgewall or radiant section inlet flue gas temperature limits may protect against overfiring and overheating of convection section tubes and supports.

16.2.4.6 Process Inlet and Outlet Temperature and Pressure

Check and monitor the process stream inlet and outlet temperature limits to help prevent overfiring, protect process stream quality and avoid unwanted reactions, and identify upstream problems, such as heat exchanger fouling. Check the process inlet and outlet pressure to maintain the equipment integrity and to help identify internal tube deposits.

16.2.4.7 Draft Targets

Draft targets are set to maintain negative pressures within the heater and avoid safety and structural problems while ensuring conditions that allow adequate air flow through the burners.

16.2.4.8 Excess Oxygen Targets

Excess oxygen, measured in the flue gas leaving the firebox, will provide guidance on adjusting burner registers (or dampers) to reach the excess oxygen target level. These adjustments must also consider the draft target. Trends in excess oxygen can indicate changes in fuel composition, tube leaks, and deterioration of heater operation.

16.2.4.9 Fuel Pressure, Temperature, and Flow Rate

Check the fuel pressure at the heater. Compare the fuel flow from the burner capacity curve to the fuel flow measured by the heater instruments. Fuel flow should be recorded and trended. When adjusting excess oxygen, the fuel flow should be observed to identify when the heater is "fuel rich" and the excess oxygen is low. An increase in fuel flow together with no increase or even a decrease in process outlet temperature identifies a "fuel rich" condition. Tube leaks may be identified with careful attention to trends of the fuel flow.

The fuel temperature should be monitored. The liquid fuel temperature at the burner is most critical to good combustion of the liquid fuel. If the temperatures are outside the design limits, then the temperature should be corrected to that designed by the burner manufacturer for the burner. The fuel gas temperatures are normally not monitored.

16.2.4.10 Atomization Medium, Pressure, and Temperature

The atomization medium pressure and temperature should be noted to ensure agreement with the liquid fuel gun being used. If the atomization medium is steam, the steam lines should be insulated and have adequate steam traps to ensure dry steam to the oil gun. The atomization pressure should be as shown on the capacity curve of the equipment. Any pressures outside the norm should be investigated and corrected.

16.2.411 Fuel Composition

Fuel composition (sometimes indicated by specific gravity) monitoring can be a "leading indicator" for adjustments of excess oxygen in the heater by adjusting burner registers or plenum damper.

16.2.4.12 Stack Emissions

Stack emissions, such as NOx and combustibles (including CO and VOC), are monitored to help ensure compliance with regulations and operating permits. A change may indicate an improperly adjusted burner, inadequate air flow to the burners, unstable flames, or a change in fuel composition.

16.4.7 Developing Startup and Shutdown Procedures for Fired Heaters

Each plant will need to develop startup and operating procedures for its particular heater and operations. These procedures should be followed for startup and shutdown to ensure a safe and efficient operation. The following procedures are only intended as a guide to help develop startup and operating procedures for a plant.

The startup and operating procedures are intended to ensure avoidance of an explosive mixture occurring in the firebox and to establish stable flames and process flows. Startup procedures will vary with the degree of automation and instrumentation on the heater. The procedures should address the issues listed below.

1. Fuel system and burner preparation:
   a. Confirm that the burners are properly installed (Chapter 15), with the tips properly positioned and oriented and no blockage in the fuel ports or the air flow passages. Isolation valves and line blinds in the fuel systems must be closed, including those at each burner.
   b. The fuel lines and valves may be pressure-tested to identify leaks or may be blown out to remove scale and debris. Fuel gas lines should then be confirmed as full of hydrocarbon fuel, not noncombustible gas. The fuel gas knockout vessel should be drained of all liquids. The circulation in the fuel oil system must be established and confirmation obtained that the heat tracing and atomizing medium systems are functional. Confirm adequate pressure in the fuel and atomizing medium supply systems.
   c. Adjust the burner registers to confirm operability over the full range. Leave them set at the 100% open position(s) for purging the heater.
2. Check the heater for readiness:
   a. Visually check that there is no debris from maintenance or construction left inside the heater. All doors (maintenance, header box, sight, explosion) are to be closed and protective refractory placed where applicable.
   b. Adjust the stack damper over the full range to confirm operability and leave it fully open.
3. Establish flow through the process tubes:
   a. Multi-pass heaters with vertical radiant tubes and liquid feed require a special procedure that involves filling one pass at a time. This is to ensure that full and stable flow is reached in each pass during operation. Heater manufacturer guidelines and instructions should be followed to ensure that each pass is flowing properly.
   b. During the filling process, check the operation of the pass flow indicators, valves, and controllers. If any pass does not have the correct flow indicated or if flow fluctuates, analyze and correct the problem before beginning the purge and burner lighting steps. It is often helpful to place the pass flow control valves on manual operation until the flow reaches about 75% of the normal flow value. This will overcome the reset windup problems that occur with some control systems and valves with minimum flow stops.
4. Purge any accumulated combustibles from the firebox:
   a. Use steam or fan-supplied air to purge the firebox for at least five volume changes. Normally, this takes approximately 15 to 20 minutes, or until a steam plume appears at the top of the stack. Avoid excessive steaming; a long exposure to steam and condensate can damage the refractory.
   b. Check the operation of the draft gauge during the purge period to ensure that the draft gauge is operable and reading the draft within the heater.
5. Light the pilot burners:
   a. First check that all pilot and main burner individual block valves are closed.
   b. Reduce the purge flow 75% and sample the firebox in several locations with an explosimeter. If safe, remove the pilot gas line blinds, and confirm that the pilot gas pressure regulator is set accurately and in accordance with the required pressure for the pilot to work properly. Close the burner air register or damper to a position of 15 to 25% open. Close the stack damper to 25% open. Light the pilots individually; if the first pilot does not light, briefly repurge and try to light another. Light all pilots and keep them burning steadily for 15 minutes to confirm their flame stability. Relight any that extinguish.

6. Light the main burners:
   a. Visually recheck that all main burner individual fuel valves are closed. Remove any blinds in the main fuel supply line. Check the fuel pressure controller setting for proper operation of the temperature control valve (TCV). Place the TCV in the "manual" mode on initial startup because operations will be below the automatic controller range. Set the TCV to 15% open, open the main fuel supply block valves, and open one individual burner block valve on a burner with a stable lighted pilot. If lightoff does not occur within 5 to 9 seconds, close the block valve and repurge the heater (see step 4). If lightoff does not occur within 5 to 9 seconds, check the positioning of the pilot with respect to the main burner tip.
   b. After the first burner is confirmed to be burning with a good stable flame, discontinue the firebox purge flow.
   c. Go to another burner with a stable operating pilot and repeat the lightoff procedure. A repurge is not necessary if the first burner is lit and this one fails to light. Wait 5 minutes and try a different burner. The heat from the first burner will cause enough draft to safely dilute and remove the fuel from the failed burner. Continue lighting main burners that have stable pilot flames, manually opening the temperature control valve after each to maintain adequate fuel pressure (minimum firing pressure shown on the burner capacity curve) at the active burners. Failure to do this will cause flame instability and likely failure, requiring a repurge of the heater if all the pilots and burners are extinguished. Light burners in a pattern that distributes active burners evenly throughout the firebox.
   d. If operating in the natural-draft mode, increase the fuel rate slowly to warm the heater and establish the draft that allows more air to enter the heater. Failure to establish adequate draft can cause a fuel-rich, potentially explosive mixture to develop in the firebox. Check the draft at the top of the firebox radiant section to ensure that air is flowing through the burners.
   e. Visually do a frequent check for stable flames on both main and pilot burners.
   f. When the process outlet temperature warms to within the range of the temperature controller, place the fuel control valve on automatic operation.
   g. Typical allowable warm-up rates for heaters vary from 100 to 200°F (55 to 110°C) per hour for heaters with plug headers, to up to 350°F per hour for heaters with fully welded coils. The lower rates for plug headers will avoid excessive thermal stresses that can cause leaks at plugs or header attachments. The above temperatures are flue gas temperatures measured at the bridgewall.
   h. Warm-up rates may be limited by the curing requirements for new refractory. If new refractory other than ceramic fiber has been installed since previous operation, the manufacturer or installer of the material will recommend a firing procedure to carefully dry and cure the material to achieve the optimum properties. If only Thermbond refractory patches are installed, the warm-up rate should not exceed 500°F per hour.

7. Periodic checks:
   a. Periodically during the startup, check the fuel and pass flows, the individual pass outlet temperatures, fuel pressure, draft at the highest pressure point in the firebox, flue gas oxygen content (to ensure correct instrument operation and good combustion) at the radiant section outlet, and firebox and tube-skin temperatures.
   b. Draft should be monitored frequently by the operator when heat input is increasing. As the heat input is increased, the air registers (or burner dampers) on a natural-draft heater and the stack damper will need to be opened.
   c. Place the pass flow controllers on automatic operation at about 75% of normal flow (see step 3). Watch the pass outlet temperatures closely. A high or uneven pass outlet temperature can be due to a low pass flow, flame impingement, or uneven firing of the burners. This must be analyzed and corrected quickly to avoid damage or curtailment of the run length due to internal tube fouling. The pass flow indicators may not be reliable; the outlet temperature may be the best information. If flow stoppage or reduction is suspected or a major pass outlet temperature discrepancy cannot be corrected, quickly extinguish the main burners until the problem is resolved. The pilots can usually be kept lit under these circumstances, thereby avoiding having to purge the heater again.
   d. Be aware that without process flow, the tube-wall temperature will rise to within about 200°F (110°C) of the firebox operating temperature, after the main burners are extinguished. Opening the stack damper(s) fully will increase the flow of cooling air. If the process flow is interrupted, do not reintroduce fluid into the tubes until they have cooled to below 900°F (485°C).
   e. Shutdown of the heater is far simpler. Gradually reduce the heat input, taking burners out of service in order to hold adequate fuel pressure on the active burners. Do not close the registers on burners taken out of service. The rate of cooling should be similar to the heating rate in step 6 above. Purge the contents from the tubes when they have cooled and fires are extinguished. Close all fuel and atomizing medium valves. Open the stack and air plenum dampers and the air registers to increase the flow of air and the rate of cooling of the heater. Install line blinds in the fuel and process lines as required by safe practice.

16.4.8 Developing Emergency Procedures for Fired Heaters

An emergency can be defined as an off-design condition that, if not properly handled, will result in major damage to or destruction of equipment and personnel injury or death. Emergency procedures most commonly address the problem of rupture of a tube containing a flammable material while the heater is operating. A more common and more easily handled event is an incident in which unburned combustibles collect in an active firebox due to errors in fuel and air handling. In the latter case, the appropriate procedure is to gradually reduce the firing rate, without increasing the air supplied to the firebox, until complete combustion is reestablished. (Adding air first to a combustible mixture at ignition temperature could cause a detonation and heater destruction.) Then, the fuel and air can be increased to satisfy the process requirements.

Some events that can result from a tube rupture and for which the proper mitigating actions must be developed include the following:

- melting or vaporization of the tubes and supports
- detonation in the firebox
- convection section collapse into the radiant section
- heater collapse due to support structure failure
- flaming oil pool spreading to other areas, putting additional equipment at risk
- explosive vapor cloud forming around the heater and possibly igniting
- damage to a stack used by several heaters
- rapid shutdown of heater and unit causing leaking flanges due to thermal shock
- rapid depressuring from high pressures causing upset of catalyst beds and distillation column trays

Some actions are almost always appropriate and should be considered when developing emergency procedures. These include the following:

1. Always leave the stack damper in position or try to open it fully, if possible from a remote location.
2. Attempt to minimize air entering the heater; close air plenum dampers if possible.
3. Turn on firebox smothering (snuffing) steam to cool the fire.
4. Activate firewater monitors and hoses to quench any spilled oil, to protect adjacent equipment with fogging sprays, and to cool the structure and stack to avoid possible collapse.

Aim at containing the fire inside the heater. Slowly, using steam or nitrogen, purge the contents of tubes in the failed pass into the firebox without losing containment. The tubes in the other passes should be purged of contents in the normal manner to avoid a rupture in another pass adding fuel to the conflagration. If the failed pass contained hydrogen, allow the contents to dissipate into the firebox without purging. A hydrogen fire may become so hot that tubes melt or vaporize, and it is important to give the smothering steam a chance to cool the firebox.

It is likely that there will be unburned hydrocarbon gases or vapors in the heater firebox. **Beware** of the instinctive reaction to cut off the fuel to the burners. The combustion air freed by such action could result in an explosive mixture in the heater, resulting in violent destruction. Avoid creating explosive mixtures in the affected heater and in any ducting or stack common with other heaters. These other heaters, if any, should be kept firing at low rates and with the absolute minimum excess air until they can be shut off and their air registers or plenum dampers closed. Then introduce smothering steam to them as well.

If valves are available, isolate the heater to minimize the amount of flammables that can flow from other equipment. Begin depressurizing the plant as soon as possible to minimize the amount of fuel that is available.

Keep the smothering steam flow on and the combustion air flow blocked until the heater cools to below 600°F. Test the firebox for combustibles with an explosimeter. If no combustibles are found, the plenum damper or air registers can be opened to increase the rate of cooling.

REFERENCES

1. API Standard 560, Fired Heaters for General Refinery Service, American Petroleum Institute, Washington, D.C., 1996.
2. R.D. Reed, *Furnace Operations*, 3rd ed., Gulf Publishing, Houston, TX, 1981.
3. API Standard 530, "Calculation of Heater-Tube Thickness in Petroleum Refineries," American Petroleum Institute, Washington, D.C., 1996
4. John Zink Burner School Course Notes, Sept. 2000, copyrighted 2000.
5. E.A. Barrington, *Fired Process Heaters*, Course Notes, copyrighted 1999.

Chapter 17
Troubleshooting

Roger H. Witte and Eugene A. Barrington

TABLE OF CONTENTS

17.1 Introduction ... 502
17.2 Pulsating Flame ... 502
17.3 Flame Impingement on Tubes ... 503
17.4 Flashback ... 505
17.5 Irregular Flame Patterns ... 505
17.6 Oil Spillage ... 507
17.7 Long Smoky Flames ... 507
17.8 Main Burner Fails to Light-Off or Extinguishes While in Service ... 508
17.9 Leaning Flames ... 509
17.10 High Fuel Pressure ... 510
17.11 High Stack Temperature ... 511
17.12 Overheating of Convection Section ... 511
17.13 "Motorboating" When Firing Oil ... 512
17.14 Flame Lift-Off ... 513
17.15 Pilot Burner Fails to Ignite or Extinguishes While in Service ... 514
17.16 Smoke Emission from the Stack ... 514
17.17 High NOx Emissions ... 515
17.18 Summary ... 519
References ... 519

17.1 INTRODUCTION

Diagnosing and solving problems with burners on heaters in the hydrocarbon and petrochemical industries often seem to be as much an art as a science. It is a basic scientific assumption that the principles of physics, chemistry, fluidics, hydraulics, and combustion do not change. Yet the myriad variables in a typical petrochemical operation sometimes make it appear that the equipment has a personality. The complexity of the sciences, multiplied by the many-staged processes in a typical plant, causes problems to occur that were not, and could not have been, anticipated by the design engineers. Moreover, although scientific principles remain the same, equipment changes with use. Parameters that may have been designed correctly change with time.

With all of the complexity of conditions that may occur, it is still the operator's job to "keep it running." When production suffers because of the inability of equipment to operate at required capacity, costs go up and profit or product margins decrease. Frequently, the operator of the heater must be trained and use knowledge of the equipment and process unit to make adjustments that bring operations back to the required capacity desired by plant management.

It is essential that troubleshooting be done in a systematic, well-organized fashion. Effective and safe troubleshooting involves four basic steps:

1. Recognizing the problem
2. Observing indications of the problem
3. Identifying the effects and the cost of the problem on the operation
4. Identifying solutions for the problem and taking corrective action

The initial indication that a problem exists may come from controls and instruments on the furnace, or from direct observation of conditions within or outside the furnace. Changes in process temperature, process pressure drop, excess oxygen, draft, fuel pressure, stack temperature, and emission levels are typically first observed with instruments. Changes in the noise being emitted from the heater or flame patterns, flame impingement, oil spillage, flame instability, and flashback are usually observed directly during inspection of the heater.

When a problem is noted, it is necessary to evaluate its likely effect on the process or product being produced. Some solutions may require the heater to be shut down for the problem to be resolved. Then, plant management must determine the economic value of meeting contracts for products vs. operating the equipment until it fails and has to be shut down for repairs at a higher cost. On the other hand, if the problem might result in large fines from controlling governmental agencies, significant damage to equipment, or danger to personnel, then a solution must be employed immediately — whatever the cost.

If a solution is going to be employed, a careful study of symptoms being exhibited by the furnace should be conducted and a probable cause for the symptoms should be identified. It is important that the possible causes be carefully identified lest an attempted solution escalate the seriousness of the problem rather than solve it.

Once a cause has been determined, standard procedures should be followed to solve the problem. All personnel involved should be aware of the problem, the planned corrective actions, the ways that safety is addressed, the expected results, and the proper action to take should the problem worsen or not be solved. During normal operations, care should be taken to make incremental changes and adjustments to parameters controlling the combustion process. However, under some conditions, a change must be made quickly and with confidence to prevent additional operating problems.

The following chapter sections describe typical problems, their indicators, their likely effects on operations, and their causes and standard solutions. Additional references are available on troubleshooting.[1–5]

17.2 PULSATING FLAME

17.2.1 Indications of the Problem

The pulsating flame phenomenon is sometimes called "woofing" or "breathing." Rather than a steady, stable flame of constant volume, the flame is pulsing and changes in size with the woofing or breathing. During the change in the flame pattern, the operator can hear low-frequency noise being emitted from the furnace. The flame oscillates along its axis and may periodically extend through the burner to the outside of the heater.

A woofing noise may be heard in the area of the heater. The woofing or breathing noise is a very low-frequency noise that is different from normal operational noise around the heater. The operator who has been around combustion equipment can immediately identify the change in sound. The noise will continue to increase to such a violent condition that even the most inexperienced operator can identify the noise. Visual inspection of the flame pattern when the woofing first starts reveals an unsteady and erratic flame — varying in length and volume. Under extreme woofing conditions, the sides of the furnace may be observed flexing.

17.2.2 The Cause and Effect on Operation

The effect on operation during the woofing condition is normally just a passing condition if corrected immediately.

However, if the woofing condition is not resolved and is allowed to continue, the intensity of the pressure surges within the furnace may cause the refractory insulation to begin to break up and fall onto the heater floor. The burner tile may even begin to deteriorate and fall apart. The heater vibration can break piping, tubing, and instruments. The incomplete combustion that occurs causes a drop in heat release, and the heater cannot fulfill its required duty. Hence, the heater may have to be shut down for major repairs.

The usual cause of the pulsating flame is lack of oxygen in the combustion reaction. When the oxygen or air flow is inadequate, the flame will search for oxygen alternately inside the heater as the air flows across the burner and outside the heater as it becomes starved for air. As the flame moves into the heater, a pressure front is generated, again causing the air flow to cease, and the flame moves back to the burner for oxygen. The movement of the flame and pressure continue to increase in intensity and with such force as to cause damage to the heater. In extreme cases the flame may oscillate so far from the burner that combustion is extinguished and the flame is lost.

The condition of insufficient oxygen in the combustion zone of the furnace may exist even when the oxygen levels measured in the stack flue gas indicate that sufficient oxygen is available. Air can leak into the heater through the flanges between the convection section and the radiant section, openings in the heater shell, sight ports left open, manway flanges, and corrosion holes in the heater steel shell. Air leakage into the heater will add oxygen to the flue gases, causing inaccurate indications of oxygen available for combustion in the throat of the burner.

17.2.3 Corrective Action

As soon as a pulsating flame is observed, the firing rate should be immediately reduced to establish sufficient oxygen for the combustion reaction to go to completion. The operator must reduce the firing rate until the pulsating flame has stopped and no woofing noise is heard. When no woofing noise is heard and the flames are not pulsating, the operator should observe good stable combustion. There should be a measurable excess of oxygen in the firebox gases. Then the operator can open the air registers on the burner and increase the stack damper's opening to adjust the excess oxygen and draft to the correct levels required for the firing rate desired. The firing rate can then be increased to the burner capacity or to the required heat release requested by the heater control system.

The combustion air **should not** be increased before cutting back on the fuel and establishing a stable flame. Increasing the air before cutting back on the fuel may fill the furnace with a large combustible mixture of fuel and air, which might result in a detonation and damage to the equipment.

If the air registers or the stack damper are fully opened, then the heater has reached its maximum capacity. Any wind blowing under the heater or across the top of the stack may interrupt the air flow to the burners and start the pulsating flame problem. Sometimes, a windscreen or fence around the furnace is necessary to prevent the wind effects on heaters operating at maximum capacity in high-wind locations. Another solution would be to consider changing the raw gas burners to premix burners, the latter not being affected by the wind conditions but by the fuel gas pressure. If the flue gas analyzer is in the stack, consider relocating it to the firebox flue gas exit so that air leaking into the heater does not give a false indication of the oxygen available at the burner.

17.3 FLAME IMPINGEMENT ON TUBES

17.3.1 Indications of the Problem

The most direct indication of flame impingement is visual observation by the operator of the flames contacting the external tube surface inside the firebox. The operator may also observe tubes with a cherry-red color or bulges in the tube walls. Indirect indications that flame impingement has occurred include higher pressure drop on the process side because of coke deposition on the tube walls, higher firing rate due to the loss of heat transfer because of coke deposition on the tube walls, and an increase in the bridgewall and stack temperatures.

If flame impingement is suspected but cannot be directly observed, several infrared photographs of the tubes should be taken to determine if there are any high tube skin temperatures as a result of direct or indirect flame impingement. Also, components of the flame radiating at frequencies not visible to the naked eye may be contacting the tubes. To establish if the flame is contacting the tubes, the operator can inject some baking soda (Na^+) or activated carbon particles into the combustion air. The flame temperature causes the Na^+ ion to get excited and glow a bright yellow or the carbon to burn with a yellow flame. The glowing Na^+ ion or the burning carbon will show the flame pattern being emitted from the burner for a short time span and will also indicate if there is any flame impinging on the tubes.

17.3.2 The Cause and Effect on Operation

During normal operation, the process fluid flowing through the tubes will provide sufficient cooling of the tube surface to cause the tube color to be essentially black in contrast to the

FIGURE 17.1 Corroded gas tip.

tube hangers or brackets. The tube hangers or brackets that support the tubes will normally be black or slightly red. Flame impingement on the tubes can create hot spots, causing the tubes to appear red or orange in color. The tube color indicates excessive tube wall temperatures that may result in localized coke formation. The layer of coke insulates the tube wall from the cooling effects of the process fluid and allows the metal temperature to rise. The insulating effect creates two undesirable conditions: (1) heat transfer to the process fluid is impeded, thereby reducing efficiency; and (2) the tube is inadequately cooled by the process fluid, resulting in hot spots, more coke deposition, and eventually tube rupture within the heater.

Some process liquids do not coke when overheated, but form vapor. If not considered in the heater design, the vapor may significantly increase the resistance to flow and the pressure drop. The lowered flow rate combined with film boiling at the location of impingement will reduce the heat transfer coefficient and raise the local tube temperature.

A possible cause of flame impingement on the tubes may be a deficiency of combustion air in the combustion reaction, causing the flame to search for additional combustion air within the firebox. The deficiency of combustion air for the combustion reaction may be a result of overfiring or of air leaking into the firebox and not flowing across the burner air orifice. Air leakage through other openings on the heater does not mix well with the combustion air moving across the throat of the burner. Hence, the flame is searching for additional oxygen from the leaking air. Overfiring may result from instrument failure, installation of wrong burner tips, or a change in fuel gas composition, affecting the flame pattern. Another possible cause may be that the burner firing ports are eroded (see Figure 17.1) such that the fuel is being injected through the combustion air flow directly toward the tube. Another possible cause may be that flue gas recirculation within the firebox is preventing the flame from forming an acceptable pattern in its allotted space. Flue gas circulation may be pushing some burner flames onto the tube surface.

17.3.3 Corrective Action

The heater operator must check the excess oxygen level within the firebox at the top of the radiant section to be sure that the instrument is operating correctly, that air is not leaking into sample lines, and that all air leaks on the radiant section are sealed. Then the operator can increase the amount of combustion air by opening the air registers or stack dampers to allow more air to flow across the burner. If fuel oil is being fired, the operator may need to check the atomizing steam pressure to ensure complete atomization and efficient combustion of the oil. The furnace should be checked regularly for air leaks by inspecting the heater shell, the areas around the tubes entering and leaving the firebox, the convection section flanges, the sight ports, and flanges around manways.

If the fuel is oil, the oil tips should be cleaned regularly to prevent coke buildup and clogging. The exit ports on the oil tip should be checked for erosion. Both oil and gas-fired burners should be checked for proper alignment and position of the fuel tips within the throat.

If the flue gas circulation patterns within the firebox are pushing the flames into the tubes, then a Reed wall[6] or division wall[7] may need to be installed in the heater firebox. The Reed wall redirects the flue gas circulation pattern within the heater while the division wall interrupts the circulation. Each allows the burner flame pattern to develop in the space designed for the flame pattern.

Damaged or missing burner tile sections can cause unequal distribution of air within the burner. This will lead to fuel-rich zones, locally longer flame segments, and the potential to lean toward and into the tubes. Check the tile condition and repair if necessary.

Impingement may be overcome by changing the flame shape. For example, if the firebox dimensions allow a longer flame, the burner tip port included angle can be reduced to obtain a more slender flame and move the flame envelope further from the tubes.

17.4 FLASHBACK

17.4.1 Indications of the Problem

Flashback is the phenomenon that occurs in premix burners when the flame velocity is greater than the velocity of the flowing mixture and the flame front propagates back through the mixer or venturi to the orifice or area where the fuel and primary air are being mixed. Then the fuel and primary air continue to burn in the venturi or mixer. Flashback is most likely to occur in burners using fuels having a high ratio between the upper and lower explosive limits of the fuel. Table 17.1 reveals that the gases most susceptible to flashback include carbon disulfide, acetylene, ethylene oxide, hydrogen, hydrogen sulfide, and ethylene. Another reason for flashback may be that the gas tip design is not optimized for the fuel that is being burned. If the velocity of the gas and air exiting the gas tip is very low, because of a large-diameter firing port, then the velocity of the flame front may be greater than the velocity of the fuel/air mixture exiting the tip and flashback will occur.

When flashback occurs within the mixer or venturi, a flame will be observed burning in the venturi or mixer. If flashback has occurred sometime in the past, the mixer or venturi will show signs of oxidation on the outside of the cast iron venturi. When flashback occurs in a premix burner, there is little doubt in the operator's mind that flashback is occurring within the burner. A sharp barking noise in the mixer is continually emitted until corrected.

17.4.2 Effect on Operation

When flashback occurs and remains uncorrected, the burner mixer or venturi is damaged from the high temperatures generated within the mixer from the combustion reaction that is occurring. The damage to the burner parts will result in higher maintenance costs. When flashback occurs, the capacity of the burner is restricted; if flashback occurs on many burners within the heater, the outlet temperature of the process cannot be obtained. The burning inside the venturi tube and intermittent open flame constitute a safety hazard.

17.4.3 The Cause and Corrective Action

The solution to the problem will vary depending on what is available to the operator. The operator should immediately check the gas tip discharge port and the main gas orifice to ensure that both are clean. If the gas orifice is dirty, the fuel flow may be reduced to the point of creating a flashback condition. If the gas tip discharge port is dirty, the flow of the fuel/air is reduced on the exit discharge port to the point of creating the flashback condition. Hence, the operator must clean both the gas metering orifice and the gas tip discharge port to allow the fuel/air velocity to remain higher than the flame velocity so that no flashback will occur. If the fuel composition or process heat requirement changes, resulting in lower operating pressures, burners must be shut off to raise the fuel gas pressure or new orifices may be required in the mixers or venturi to keep the fuel gas pressure at a level sufficient to prevent flashback.

TABLE 17.1 Ratio of the Upper and Lower Explosive Limits and Flashback Probability in Premix Burners for Various Fuels

Fuel	Ratio[1]	Probability[2]
Acetone	5.01	1.67
Acetylene	32.00	Infinite
Acrylonitrile	5.57	1.85
Ammonia	1.71	0.57
Aromatics (Mean)	5.00	1.66
Butadiene	5.75	1.92
Butane	4.52	1.51
Butylene	4.88	1.63
Carbon Disulfide	40.00	Infinite
Carbon Monoxide	5.93	1.97
Cyanogen	6.45	2.15
Ethane	4.02	1.34
Ethyl Alcohol	5.77	1.92
Ethyl Chloride	3.70	1.23
Ethylene	10.04	3.33
Ethylene Oxide	26.66	Infinite
Gasoline (Mean)	5.06	1.68
Hydrocyanic Acid	7.14	2.38
Hydrogen	18.55	Infinite
Hydrogen Sulphide	10.60	3.52
Methane	3.00	1.00
Methyl Alcohol	5.43	1.81
Methyl Chloride	2.26	0.75
Naphtha	5.45	1.81
Oil Gas	6.84	2.28
Propane	5.25	1.75
Propylene	5.55	1.85
Vinyl Chloride	5.42	1.80

[1] Ratio of the upper and lower explosive limits.

[2] Probability of flash-back as compared to methane.

Source: R.D. Reed, A new approach to design for radiant heat transfer in process work, *Petroleum Engineer*, August, C7-C10, 1950.

If raising the fuel pressure does not resolve the flashback problem, one can look to the flame velocity of the fuel/air mixture for another solution. The flame velocity is related to the percentage of air in the fuel gas. A change in this percentage, by adjusting the burner air door, can raise or lower the flame velocity. Try reducing the primary air flow to lower the mixture flame velocity. Adjust the secondary air register to maintain the target excess oxygen level.

17.5 IRREGULAR FLAME PATTERNS

17.5.1 Indications of the Problem

On a single burner, the flame pattern is nonsymmetrical. An irregular flame pattern implies that the flame varies in length

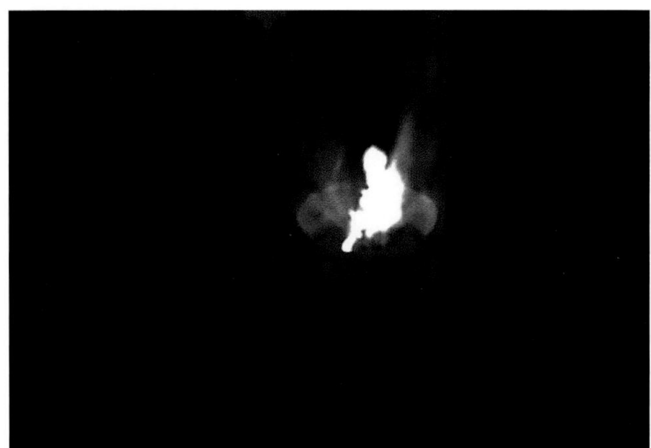

FIGURE 17.2 Irregular flame pattern.

or shape across the width of a burner. For multiple burners in a heater, the flame pattern from one burner to another burner is not of the same shape or volume for the same conditions.

Irregular flame patterns can appear in two ways. A single burner may have one side of the flame pattern longer than the other side (see Figure 17.2), or part of the flame may be emerging at a different angle from the main flame. In a multi-burner furnace, some burners may have a longer flame pattern than other burners in the furnace when the same fuel pressure and same air register opening is being provided at each of the burners.

17.5.2 Effect on Operation

In either the single burner or multi-burner case, the irregular pattern may cause the furnace to have hot spots on the tubes. In the case of multi-burner furnaces, the burners with the longer flame lengths may create flame impingement on the process tubes while the other burners have short, compact flames and cause no hot spots on the tubes. Irregular flame patterns can result in higher operating cost or afterburn due to incomplete combustion of the fuel being burned.

17.5.3 The Cause and Corrective Action

In the case of an irregular flame from a single burner (see Figure 17.2), the problem may be a dirty burner tip, an eroded tip, or a tip improperly oriented with respect to the refractory. In a dirty burner gas tip, some orifices may be partially plugged while others are operating normally. If part of the flame is emerging at a different angle from the main flame, tip erosion may have occurred. Improperly placed tips may be too close, too far, or turned at the wrong orientation with respect to the refractory tile ignition ledge. The irregular flame may also be caused by a foreign material, such as refractory or plenum insulation, in the burner throat or by a damaged tile. Either will affect the local mixing of fuel and air, delaying mixing in part of the throat and lengthening the flame on that burner.

In the case of several burners in the same firebox, the problem may be more complicated. Some of the burners may be plugged while others are operating normally. The air registers may be at different settings, causing uneven air flow across each of the burners. A poorly designed plenum may be supplying more combustion air to some of the burners and less to others or more to one side of the burners. Maintenance or installation personnel may have placed the burner tips in an incorrect position with respect to the refractory ledge. Burner tips designed to be oriented in a specific direction with respect to other tips or the refractory ledge may be turned to the wrong position, causing flame-to-flame interference. The refractory may be damaged, causing poor or uneven air circulation around the burner tips.

Obviously, it is good operational practice to maintain the burner tips and ensure that they do not become dirty or plugged. Many irregular flame patterns are corrected by simply keeping the burner tips in good operating condition. While cleaning the tips, the orifices should be inspected to ensure that no erosion has occurred that would change the burning characteristics of the tip. If the fuel is oil, some of the orifices on the tip may become completely plugged with coke. Manually inserting a twist drill bit into the orifice and blowing steam through the orifice should clean the orifice. Do not use powered drills to clean orifices. The use of a power drill may change the size or shape of the orifice. A larger or different-sized orifice then flows fuel at a different rate thereby causing the tip to produce an irregular-shaped flame.

In multi-burner furnaces, all of the burner tips should be inspected, cleaned, and of the same orifice size. All air registers should be set to the same approximate opening. All fuel pressures to the burners should be at the same level. If the flames are still irregular, check the manufacturer's drawings for the burner fuel tip position and orientation. Then check each tip on each burner and ensure that the tips are at the correct height and orientation with respect to the ignition ledge. Inspect the burner refractory tile for damage, especially in the immediate area of the tips. Some tips are designed to operate in close proximity to the refractory, and a difference of ±0.25 in. (6.4 mm) can cause irregular flames.

Air plenum distribution problems are difficult to diagnose. Such problems usually occur when the plenum is not correctly sized. As a result, some burners get more combustion air and other burners are starved for combustion air. The operator needs to ensure that all burner air registers have the same size opening. The designer should ensure that the air plenums are properly sized for good air distribution to all burners. In some cases, it may be beneficial to have the air plenums modeled so that they are correctly sized. In large plenums, benefit may be obtained by adding air inlets to improve the distribution of air.

17.6 OIL SPILLAGE

17.6.1 Indications of the Problem

On natural draft heaters, the operator observes fuel oil dripping and forming pools of oil under the heater. Spilling in an air plenum can cause a plenum fire, damaging burners and the heater structure. The steel air plenum cannot be touched by the operator's hand because of the excessively high skin temperatures. The paint on the air plenum is observed to be blistering and peeling. In extreme cases, the operator may observe oil burning under the heater where the oil has dripped and formed pools of oil at grade.

17.6.2 The Cause and Effect on Operation

An open pool of oil beneath an operating furnace is clearly a fire hazard that must be immediately corrected to ensure operator and equipment safety. If the pool of oil under the heater is allowed to ignite and burn, damage may occur to the heater. Second, the oil is not all being consumed in the combustion reaction, causing higher fuel operating costs.

The main cause of oil spillage or dripping from the burner is normally the poor atomization of the fuel oil when the temperature of the oil is lower than specified or the atomizer is damaged or plugged. At low fuel oil temperatures with higher oil viscosity, the oil droplets leaving the oil gun are larger. The large oil droplets may contact the burner tile and begin flowing down the burner onto the grade.

The temperature of the fuel oil must be at a viscosity of 200 SSU (45 cs) or less at the burner, not at the fuel oil heater outlet. If the fuel oil lines from the heated fuel oil supply tank to the burner oil gun connection are not insulated properly, the fuel oil temperature can be too low at the burner, making the viscosity too high. Even with higher steam atomization pressure, the cold fuel oil may not be broken into small enough droplets to burn effectively.

The burner fuel oil tip and atomizer may have suffered erosion from particles present in the fuel oil. The eroded tip will allow the oil to be injected into the tile at a larger angle than design. The oil then contacts the sides of the refractory tile and begins leaking fuel oil down onto the burner and dripping to the grade under the burner.

Not all spills are due to atomization or fuel tip problems. If the oil gun is not inserted far enough into the burner, the spray will contact the tile and cause coking on the tile as well as oil drips or spills. This sometimes happens when the oil gun from one heater is mistakenly installed in another. The operator should have manufacturer drawings available to check the oil tip location and should have the oil guns for each heater readily identified.

17.6.3 Corrective Action

Operators should be aware of indications of poor atomization so that action can be taken before oil spills occur. Showers of "sparklers" or "fireflies" leaving the flame indicate poor atomization or water droplets in the oil spray. Smoky flames, fouling of the burner tip, and uncontrollable flames reaching the convection section may also be indicative of atomization problems.

The operator should check the fuel temperature at the burner and control the temperature to a viscosity of 200 SSU (45 Centistokes) or less. The operator should inspect all parts of the fuel oil system, including the fuel oil gun and atomizer. The fuel oil tip and atomizer should be cleaned, and no foreign material should be on the tip or inside the fuel oil tip or atomizer. The fuel oil tip should be inspected for erosion and replaced if necessary. If the fuel oil tip orifice is +2 drill sizes larger than the design, then the oil tip should be replaced with a new case-hardened oil tip to provide longer life.

The operator should check the burner manufacturer's drawings to ensure the oil tip is located in accordance with manufacturer instructions. The steam-to-oil differential pressure should also be checked and set in accordance with the burner manufacturer's instructions. The steam traps on the atomizing steam line should be operating correctly to ensure that no condensate or water is present in the atomization steam; steam line insulation should be in good repair.

17.7 LONG SMOKY FLAMES

17.7.1 Indications of the Problem

Visual observation inside the firebox reveals long, dirty, smoky flames, possibly reaching into the convection section of the heater. Smoke may be observed exiting the stack. The flame, rather than being confined in the flame pattern space within the firebox, may be impinging on the process tubes in the convection section. The combustion zone may appear hazy, rather than bright and clear.

The operator of the heater may notice that the process outlet temperature cannot be achieved. The stack temperature may be above the design specifications.

17.7.2 Effect on Operation

Long, smoky flames are the result of incomplete combustion and can consume far more fuel than necessary to achieve the desired heat transfer and outlet temperature. Soot (unburned carbon) will be deposited on tube surfaces in the convection section, reducing heat transfer. Coke formation may be observed in the burner tile. The required process outlet temperatures may not be achieved.

17.7.3 Corrective Action

The corrective actions are aimed at ensuring that adequate air is provided to each burner and that this air is mixed quickly and completely with the fuel so as to achieve rapid combustion and smaller flame volume. Air that is not mixed with fuel at the burner passes into the less turbulent firebox where mixing of the fuel and air is less likely and ignition temperatures may be too low for combustion.

The operator should check the draft and excess oxygen level at the top of the radiant section of the heater to determine if there is sufficient oxygen to burn the fuel. If the excess oxygen is lower than the target, the draft and excess oxygen should be adjusted to the correct levels.

The next item to be checked by the operator is the temperature of the oil at the burner. The higher the specific gravity of the fuel oil, the more critical the fuel oil temperature is to atomization in the oil burner. If the viscosity is above 200 SSU (45 cs), atomization will suffer.

The operator should determine that the oil gun and atomizer are clean and no foreign material is plugging either the oil or steam orifices within the oil gun. The fuel oil and steam must be delivered to the oil gun at the burner manufacturer's design pressure and temperature. There are many different types of oil guns used in the burning of fuel oil, and each has different pressures for fuel and steam to make it work successfully. For no. 6 fuel oil and heavier, all fuel oil and all steam lines must be insulated to reduce loss of heat and temperature. The steam traps on the steam line should be checked to ensure that the condensate is being removed from the steam and the steam remains dry.

The oil tip should be positioned in the oil tile or diffuser cone in accordance with the burner manufacturer's drawings and instructions. Improper location may cause the oil–steam spray discharging from the oil gun to collapse and produce a dense spray. This spray cannot mix with the combustion air quickly so as to burn with a short flame. The flame becomes long and smoky.

If the burners still show long flames and most of the burners have the same flame length, the firing rate needs to be checked. Firing at above the rated heat release, particularly with closely sized burners or with some burners out of service, will cause a shortage of oxygen at the burners and long flames that smoke. Place more burners in service or reduce the fuel input. Contact the burner manufacturer if larger burners are required to increase the heat release.

Also check the size of the ports in the burner tips. The ports may be enlarged or the wrong tips may have been installed during maintenance. In the latter case, the tips may even vary from burner-to-burner. Enlarged ports allow uneven fuel input for the available air. Replace any damaged or improper tips.

If changing the operating variables and cleaning the tips and atomizers, together with replacing damaged parts, do not improve the flame, contact the burner manufacturer. He may recommend a different design for the atomizer and tip, particularly if fuel characteristics are not close to the original design.

When firing gas at low excess air, if flames become long and smoky, suspect a change in the fuel gas composition. The substitution of higher heating value, heavier components along with a lowered burner pressure will increase the oxygen requirement. The air registers and possibly the stack damper must be opened to provide more air.

17.8 MAIN BURNER FAILS TO LIGHT-OFF OR EXTINGUISHES WHILE IN SERVICE

17.8.1 Indications of the Problem

The operator follows the usual purge and ignition procedures for lighting a burner, but there is no indication of main burner ignition. After the heater is in operation, one or more burners flame out.

17.8.2 Effect on Operation

If a burner fails to light, the process unit outlet temperature may not be achieved or the heater startup may be delayed. Additionally, if the burner fails to ignite or flames out, the furnace may fill with a dangerous mixture of gas and air that can result in an explosion in the firebox.

17.8.3 Corrective Action

The most common cause of ignition failure is improper positioning of the pilot burner in relation to the main burner. If the pilot is not located so that its flame is directed into the fuel/air mixture leaving the main burner, the ignition temperature is not achieved and the main flame is not initiated. The operator should check all components of the pilot and burner to ensure that they are positioned according to the specifications in the manufacturer's design drawings.

Check that fuel gas is being supplied to the burner. Before startup, the fuel supply line may be purged or pressure tested with an inert gas. If this gas is not completely displaced with fuel, the operator will be attempting ignition of an inert material. Procedures should be in place for removing the inert gas and checking for the presence of a flammable material.

Check for closed or blocked valves, fuel gas line blinds, plugged strainers or filters, and plugged burner ports. Plugging is a frequent problem during startup if the fuel piping is new or just revised due to foreign debris left in the lines. Old piping may have internal scale that dislodges and enters the burners, thereby plugging the ports. The burner ports and fuel lines may need to be cleaned. A strainer should be installed to minimize plugging if this problem occurs frequently.

When burners flame out in service, the problem is almost always an interruption of either fuel or air being supplied to the burner. Fuel interruptions can be caused by instrument failures such as the closing and reopening of a fuel control valve. If the burner pilot is lit, the fuel should re-ignite with only a "puff" or minor detonation. If there is no pilot, the fuel may not ignite and fuel gas will continue to flow; a flammable mixture may accumulate and explosion may result.

Gas burners are likely to be extinguished if a significant amount of liquid enters the burner as anything other than a mist. Gas burners have no ability to atomize liquid into finely divided droplets that will burn. Hence, the flame is lost if the liquid is discharged through the burner. Liquid may be amine carryover from treating facilities that remove hydrogen sulfide; from (during cold weather) condensed heavy gases such as propane or butane added to supplement the normal fuel supply; or condensed water vapor. Facilities to remove liquids, such as a knockout drum, should be considered.

All burners may flame out if the air supply is interrupted either by dampers opening and closing or by pressure surges within the firebox. Ensure that conditions allow an appropriate air supply to flow across the burners at all times.

Too much air flow can also cause loss of ignition or burner flameout problems. Too much air may result in a nonflammable mixture at the burner. The large excess of air may also quench the fuel–gas mixture below the ignition temperature. If this is suspected, close the air registers and/or the stack damper to reduce the excess air in the firebox to the targeted levels or to a minimum level if during startup.

When firing oil, the atomizing medium can extinguish the flame if the fuel pressure is too low, if fuel temperature is too low, or if atomizing medium operating conditions are not correct. Check the atomizing medium pressure, temperature, and quality, as well as the fuel oil temperature, pressure, and flow, and correct if necessary. Monitor these conditions regularly.

17.9 LEANING FLAMES

17.9.1 Indications of the Problem

In floor or ceiling-fired furnaces, the flames may lean to one side rather than burning in a vertical line. In wall-fired furnaces, the flames may lean to the side rather than firing horizontally and curling upward.

Observation of the flame pattern inside the firebox reveals that the centerline of the flame does not follow the designed path as specified by the burner manufacturer. The flame is commonly expected to propagate in the general direction of the centerline of the air orifice or refractory tile. However, in some burners, the flame may be designed to propagate in the general direction of the gas orifice.

17.9.2 Effect on Operation

Flames that do not have the designed pattern and direction can create problems such as impingement on the process tubes which results in hot spots and may eventually cause tube rupture.

17.9.3 Causes and Corrective Action

Refer to Sections 17.3, 17.5, and 17.7 for some thoughts on corrective action.

Incorrect positioning and orientation of the burner tip(s) with respect to the refractory walls or floor of the firebox can cause leaning flames. This may be due to incorrect burner installation or to deformation of the heater steel. Both should be checked and corrected. The wrong burner tip, an improperly drilled tip, or a tip turned in the wrong direction can also cause flames to lean. All should be checked and the manufacturer's drawings referenced to ensure that orientation and locations are correct.

Improperly installed or damaged refractory can affect the direction of flame propagation. When observed, such sections should be repaired. Damaged diffuser cones can cause leaning and uneven flame patterns (see Figure 17.3) and should be replaced. Damaged burner tile should be replaced if it is contributing to flame leaning.

Circulating currents of the flue gas within the furnace can be a cause of leaning flames. The circulation can be observed by injecting baking soda or activated carbon into the burner. The glowing particles trace the flue gas circulation currents within the heater. The most effective correction is the installation of division or Reed walls, as discussed earlier.

FIGURE 17.3 Damaged diffuser cone.

Air infiltration when operating with minimum excess oxygen can cause flames to lean toward the source of the leaking air. Air leakage points should be identified with smoke tests and sealed.

17.10 HIGH FUEL PRESSURE

17.10.1 Indications of the Problem

The pressure gage on the fuel line at the burner measures a pressure higher than the burner capacity curve indicates for the required heat release. Typical process burners are designed for maximum heat release at 15 to 25 psig (1.0 to 1.7 barg). Hence, if the pressure observed is 30 to 40 psig (2.0 to 2.7 barg), the fuel pressure may be too high for the heat release required if the fuel composition is per design. Typical fuel oil pressure for maximum heat release is around 100 psig (6.8 barg). Hence, if the operating fuel oil pressure is 150 psig (10.2 barg), the operator is experiencing high fuel oil pressures and needs to determine the problem.

17.10.2 Effect on Operation

Higher than normal fuel gas or fuel oil pressure can cause irregular flame patterns and hot spots on the process tubes within the furnace. Again, the flame impingement and hot spots may lead to process tube rupture and unit shutdown. With higher than designed fuel pressures observed, the furnace might not achieve the desired outlet temperature.

The fuel gas pressure is normally designed at 15 to 25 psig (1.0 to 1.7 barg) for most process heaters. These pressures ensure reasonable orifice diameters to reduce fouling problems during operation with typical plant fuels and provide reasonable pressures for fuel/air mixing at turndown.

Under normal circumstances, increasing the fuel pressure above these levels will increase the fuel flow to the combustion process without problems as long as there is an excess of oxygen in the flue gases. However, if the oxygen analyzer does not indicate there is a sufficient excess of oxygen for additional fuel, then the fuel pressure should not be increased.

17.10.3 Solution and Corrective Action

First, look to the pressure gage at the burner fuel manifold. Experience tells us that pressure gage accuracy deteriorates with time. Replace the pressure gage with a new or reconditioned and calibrated gage. Also confirm the pressure gage location. It is not unusual to be using a pressure reading from a gage near or even upstream of the fuel control valve as representative of the pressure at the burner. This can lead to serious error in the "measured" pressure value. A gage at this location is measuring the pressure at the burner plus the pressure drop in the piping plus the pressure drop in any valves, which may be throttled — not the pressure at the burner. Make sure that the gage is properly located and, if in heavy oil service, properly sealed, insulated, and steam-traced.

Next, look to the flow measuring instrument from which the heat release is calculated. If the current fuel gas composition and specific gravity do not match the properties used to determine the meter factor, the flow indicated may be inaccurate, and the calculated heat release is wrong. If this is the case, the heating value of the fuel has also probably changed. A new burner curve using the new fuel properties should be obtained from the burner manufacturer.

If a problem is still evident, some plugging in the fuel delivery system, strainers, the burner orifice, or tip ports, the wrong-sized fuel lines, and lack of adequate insulation and tracing on heavy oil piping are all possible causes of the restriction of fuel flow. All should be checked and cleaned or corrected. The burner parts should also be checked for damage or incorrect drilling.

The operator may be attempting to fire too much fuel on too few burners. Place more burners in service to reduce the fuel pressure. Fully open the fuel valves on all active burners.

In some circumstances, the higher than designed fuel pressure is acceptable as long as there is sufficient excess oxygen to burn the fuel being injected into the heater, the flame patterns are acceptable, there is no flame impingement, and the CO level is below 100 ppm.

17.11 HIGH STACK TEMPERATURE

17.11.1 Indications of the Problem

Under normal furnace operation, the flue gas temperature measured in the stack will be near the operating temperature predicted by the furnace designer at the design duty. High stack temperatures indicate decreased heater efficiency, higher fuel consumption, and increased operating cost. High stack temperatures can indicate excessive heat in the convective section of the heater. Long-term operation at higher than normal stack temperatures can damage the furnace, especially in the roof area of the convective section.

17.11.2 Effect on Operation

High stack temperatures indicate reduced heater efficiency. The higher the stack temperature, the lower the efficiency of the heat recovery to the process and the higher the operating cost to produce a product. Long-term operation at higher than normal stack temperatures may result in damage to the furnace stack and convection section.

High stack temperatures may indicate afterburning in the convection section. The afterburning may be caused by lack of excess oxygen within the burner throat and firebox. Air leaks into the convection section and the temperature of the flue gases is sufficient to cause the completion of the combustion reaction to occur (afterburning) in the convection section of the heater. The afterburning will result in the destruction of the extended surface on the tubes in the convection section of the heater. The loss of the extended surfaces results in a loss of heat transfer in the convection section. The afterburning within the convection section may ultimately result in tube failures in the convection section of the heater.

17.11.3 Corrective Action

The stack temperature is controlled by the amount of excess oxygen in the flue gases. Hence, the operator can reduce the stack temperature and lower the excess oxygen in the flue gases by controlling the stack draft and burner air registers or dampers. This should be the first action by the operator.

The operator must then determine if there are any air leaks into the furnace and plug all possible cracks and openings leaking air. The excess oxygen at the top of the radiant section must then be checked to ensure that there is sufficient excess oxygen at the burner to burn all the hydrocarbon fuel being injected. If there is insufficient excess oxygen, then the afterburning may take place in the convection section, resulting in higher than normal stack temperatures.

If there is sufficient excess oxygen within the heater to ensure all hydrocarbons are burning completely in the radiant section of the heater, and if the air leaks are minimal, then the operator must check for foreign deposits on the convection section tubes. If fuel oil is being fired, the operator may need to blow the soot from the convection section tubes. At times, a rather simple technique will clean up an externally fouled convection over time, depending on what has fouled the tubes and the flue gas temperature. Changing from oil firing to gas firing, together with a modest increase in excess air, has been observed to remove fouling deposits within a few weeks. Apparently, the deposits oxidize slowly and disappear up the stack as gases. The operator must also determine if there is foreign material deposited between the fins on the convection section tubes or fin damage from previous afterburns. The operator will also need to check the fuel oil temperature to ensure proper atomization of the liquid fuel within the burner flame envelope, as well as check to ensure there is no water in the atomizing steam. Poor atomization will result in large fuel droplets that will carry into the convection section, deposit on the convection section tubes, and cause high stack temperatures to occur.

Sometimes the high stack gas temperature is caused by the heater construction. Hot flue gas will follow the path of least resistance to the stack entrance. This is usually at the ends of the convection section tubes, where the surface to absorb heat is minimal. Unless steps are taken to minimize this bypassing, which may occur in header boxes or at tube ends in the main flue gas passage, a significant amount of hot gas reaches the stack without transferring much heat to the convection tubes.

Also, if enough stacks or flue gas offtakes from the convection are not provided, the profile of flow across the tubes is not even.[7] Flue gas in the zones of high velocity flow escapes from the convection section at greater than design temperatures.

17.12 OVERHEATING OF THE CONVECTION SECTION

17.12.1 Indications of the Problem

Upon visually inspecting the inside of the firebox, there is a lot of refractory lying on the floor and in the burners. The heater shell and structure on the convection section show signs of overheating. The shock tubes and shock tube hangers are failing. The draft at the top of the radiant section is at a positive pressure. When the sight ports are opened, hot flue gases are forced out, thereby causing a safety hazard to the operator.

17.12.2 *Effect on Operation*

The convection section is designed to remove heat from the hot flue gases exiting the radiant section at the bridgewall temperature of 1200 to 2000°F (650 to 1100°C) and transfer the heat from the flue gases to the process liquid by convection. If the hot flue gases leak out of the cracks in the convection section, then the structural steel and heater convection section shell are overheated, resulting in damage to the heater.

The hot flue gases leaking out of the cracks in the heater shell result in overheating the carbon steel heater shell and structural steel supporting the convection section of the heater above the radiant section. With the overheating of the heater shell, the refractory anchors are damaged and leave nothing to support the refractory in the arch section of the heater, and the refractory falls to the heater floor. As a result, more heat reaches the structural steel, and finally the convection section falls into the radiant section of the heater and the unit is shut down for repair.

17.12.3 *Corrective Action*

The operator must first obtain the draft reading at the top of the radiant section and compare this to the targeted draft reading. If the draft reading indicates positive pressure, then the hot flue gases exiting all cracks must be stopped before the carbon steel shell is overheated and the anchors holding the refractory are damaged. When the anchors holding the refractory are damaged and can no longer hold the refractory, the refractory falls onto the floor of the heater. Upon observing the positive pressure in the heater, the operator must immediately open the stack damper on a natural-draft heater, or the induced-draft fan damper on a balanced-draft system if there is a measured excess of oxygen at the firebox exit. If there is no excess of oxygen, reduce the firing rate until an excess is attained, then increase fuel input while maintaining the draft and excess oxygen at the targets. A negative pressure must be established to eliminate the overheating of the convection section.

If the stack damper or induced-draft fan damper is completely opened, then the operator must reduce the firing rate on the heater such that there is a slightly negative pressure at the top of the radiant section of the heater. Also check the excess oxygen and reduce it to the target value. High excess air can lead to positive firebox pressure.

17.13 "MOTORBOATING" WHEN FIRING OIL

17.13.1 *Indications of the Problem*

When firing light fuel oil such as naphtha or propane, a motorboating sound may be emitted from the heater. The motorboating sound is caused by the alternate flowing of vapor and liquid from the fuel oil gun in the burner. The light fuel oil flashes and causes the flow of liquid to cease for a moment until the flashed vapor is emitted, goes back to liquid flow for a moment, and then back to vapor. The flames will have a rapidly pulsating appearance. The alternate flowing of vapor and liquid causes a fluttering sound similar to that of a motor boat. The motorboating sound is more pronounced by low fuel flow rate, low fuel pressure, and low initial boiling point oil.

The motorboating sound being emitted is normally experienced when light fuel oil with an IBP of 150 to 250°F (66 to 120°C) is being burned. The burning of heavy fuel oil does not normally produce the motorboating sound because the IBP of heavy fuel oil is usually greater than 400°F (200°C) and the heavy fuel oil will not vaporize at conditions within the oil gun.

17.13.2 *Effect on Operation*

The noise is more of a nuisance than a problem, possibly requiring the use of hearing protection by operators. The flame position is more of a concern since any flame instability has the potential to further deteriorate flame quality and possibly cause the flame to extinguish.

17.13.3 *Solution and Corrective Action*

The operator must first check the fuel oil pressure and atomization steam pressure to ensure they are within the operating parameters for the heat release being demanded by the furnace. If the pressures are within the operating parameters, then the operator will need to check the fuel oil bypass valve to ensure it is completely closed and not leaking any steam through the valve. If steam is leaking through the bypass valve, the steam will cause the fuel oil to be heated above the flashpoint by direct contact.

If there is no steam leaking across the bypass valve, then the operator needs to ensure that the fuel oil does not have any water being injected with the oil. Water will have a lower boiling point than the fuel oil and will vaporize and may cause the flame to be extinguished. The loss of flame may result in very unsafe operating conditions.

The operator can eliminate the motorboating noise from the burner by turning off a burner in a multi-burner installation.

The fuel oil pressure should increase to keep the same heat release in the furnace. The increase in pressure may eliminate the vaporization within the oil gun and stop the motorboating sound. The best solution for correcting the problem is to contact the burner manufacturer and change the oil gun design from a concentric tube design to a dual tube design. The dual tube design separates the oil and the steam until they reach the oil tip. Hence, the residence time for transferring heat from the steam to the light fuel oil is minimal, and vaporization of the light fuel oil is eliminated.

17.14 FLAME LIFT-OFF

17.14.1 *Indications of the Problem*

The first indication of a flame lift-off problem is when the operator observes that the flame is detached from the burner when inspecting the flame patterns inside the firebox. Normally, flame lift-off occurs at one or two burners in a multi-burner installation and not at all the burners at the same time. The process outlet temperature, the excess oxygen in the furnace, the draft in the furnace, the fuel pressures, and the noise from the operation of the burner will give no indication of any flame lift-off problem.

17.14.2 *Effect on Operation*

Flame lift-off from the burner is a very significant safety hazard. If the lift-off is extreme, there may be a total loss of flame at the burner and unburned fuel will be injected into the firebox. If the refractory remains at a sufficiently high temperature or if the pilot remains lit, then re-ignition may occur. Re-ignition may also be initiated by adjacent burner flames. The re-ignition may cause a minor or major explosion, depending on the amount of fuel injected into the firebox, with the extent of damage dependent on the heater design and configuration. If the explosion within the firebox is minor, the explosion doors will open and relieve the internal pressure built up within the firebox. If there are no explosion doors on the heater to relieve the internal pressure buildup, the heater may be damaged by an explosion. If the explosion within the firebox is a major explosion, the complete heater may be torn apart, resulting in the heater and process being shut down. The loss of the heater will cause a loss of product and hence a loss of profits being generated from the process unit. In the most severe explosions within the firebox, there may be loss of life or injury to operating personnel.

17.14.3 *Corrective Action*

Flame speed and air/fuel delivery speed must be balanced to ensure that the flame is attached to the tip rather than rising above it.

When flame lift-off from a burner is first observed, immediately reduce the fuel pressure on that burner by partially closing the specific burner block valve. If the flame continues to lift-off, the operator should completely shut off the burner. In either case, take the following corrective actions. Flame lift-off produces unsafe conditions and should be corrected immediately before a more serious condition occurs.

The alignment and positioning of the gas burner tip and the oil burner tip should be checked to ensure they are correctly installed and positioned in accordance with the burner manufacturer's drawings and specifications. The gas tips should be located in relation to the tile ledge or flame holder as shown on the drawings, and the oil tip should be located in relation to the inlet tile throat or diffuser as per the drawings. If the oil tip is too high in relation to the tile or flame holder, the oil flame may lift-off.

The fuel gas firing ports and ignition ports should be checked to ensure they are not plugged. If they are plugged, the ports should be cleaned by manually inserting a twist drill the same size as the port and twisting the drill to remove all foreign material in the port.

If the burner is a premix burner design, then the primary air door should be adjusted to the correct position. On a premix burner, the gas/air mixture exits the gas tip at a given velocity. The flame burning above the firing port has a flame velocity that is traveling in the opposite direction, that is, it is trying to get back to the source of the fuel and air mixture. If the gas/air velocity is too high and the flame speed is too low, then the flame begins lift-off from the burner gas tip. To correct the exit velocity from the firing port, the primary air door is closed to reduce the amount of primary combustion air entering with the gas, hence, a reduction in the gas/air velocity. With the reduced gas/air velocity, the flame reattaches to the burner gas tip or the flame holder.

If an oil-fired burner is experiencing flame lift-off, the operator will need to adjust the steam atomization pressure per the burner manufacturer's instructions. If the atomizing steam pressure is too high, the oil flame will tend to lift-off from the burner. Raw gas burners that utilize a stabilizing cone can experience lift-off if the cone is missing or damaged. Shut off any such burner where the cone is partially or completely missing or is improperly installed, and replace the diffuser cone or flame holder.

17.15 PILOT BURNER FAILS TO IGNITE OR EXTINGUISHES WHILE IN SERVICE

17.15.1 Indications of the Problem

Pilot burners are usually small premix gas burners designed to ignite the main burner while releasing only a small amount of heat. Because pilot burners have small ports, they are susceptible to plugging. Many users use a very clean, dedicated, and reliable pilot gas fuel to eliminate this problem. The pilot flame should remain stable and burn throughout the range of main burner operation. It is intended to re-ignite the main burner when the main fuel flow is reestablished, before a potentially damaging accumulation of combustible gases can occur.

In a problem situation, although the operator follows the appropriate purge and ignition procedures, the pilot burner may fail to light or may not continue to burn once lit. There may be the loss of the pilot burner flame during operation.

17.15.2 Effect on Operation

If the pilot does not light upon heater startup, the corresponding main burner is not placed in operation, or the main burner must be lit using a handheld torch. The latter is a less satisfactory method because a large gas supply valve must be opened rather than the small pilot gas supply valve. If a mistake is made, a relatively large amount of unburned gas enters the firebox, compared to the small amount from a pilot. This large amount of unburned gas is an unsafe condition and has the potential to form an explosive mixture within the firebox.

17.15.2 Corrective Action

Ensure that the fuel gas is flowing to the pilot. If the fuel piping has been pressure tested or purged with an inert gas, this gas must be completely displaced with the fuel before a successful light-off can occur. Also check for closed valves or blinds, plugging of the small pilot gas lines, or a plugged strainer or filter. Since the ports in the pilot are small, they are also susceptible to plugging. Check for a plugged orifice and clean it if it appears plugged.

Pilots are lit using a handheld torch or a spark ignitor. If the lighting torch flame is unstable or is not properly positioned, the torch flame might not contact the flammable mixture leaving the pilot tip. Adjust the torch flame to ensure stability, and take care in the positioning of the flame relative to the pilot tip. If spark ignition is used, clean the spark plug whenever possible and replace insulators on the ignition rod. Before startup, check that the ignitor sparks and that unwanted electrical grounding is avoided.

Sometimes the pilot gas–air mixture leaving the tip is not in the flammable region, usually too lean to support combustion. Close the primary air door to enrich the mixture. The pilot gas may be of the wrong composition or pressure to induce adequate primary air flow. Check both the gas composition and the pilot fuel pressure against the manufacturer's specifications and correct either if necessary.

A natural draft premix pilot will not stay lit if the firebox pressure at the pilot location is positive. Adjust dampers to obtain a negative pressure at the burners. With forced draft burners where the pressure at the pilot may be positive, it is good practice to provide a reliable pressurized source of combustion air or a special pilot design. Consult with the burner manufacturer for the correct pilot in this service.

The pilot may be wet, from condensed steam after firebox purging, for example, and refuse to light. Allow the pilot to dry or dry it with the handheld torch before introducing fuel. Wind may also make pilot lighting and continuous operation difficult. Consult with the burner manufacturer for a wind resistant pilot tip.

Manually light the pilot and observe the pilot flame for stability. If the pilot flame appears to be too high above the pilot tip, close the primary air door on the pilot mixer. If the flame appears to be burning with a yellow flame, open the primary air door on the mixer. If a thermocouple or a flame rectification rod is being used to monitor the pilot, the operator needs to ensure these devices are providing correct information and are functioning properly.

17.16 SMOKE EMISSION FROM THE STACK

17.16.1 Indication of the Problem

Smoke appears at the top of the stack.

17.16.2 Effect on Operation and Equipment

Smoke indicates either incomplete combustion or a process tube rupture if the process fluid is a hydrocarbon. As a regulated emission, continued smoking could lead to sanctions, fines, and termination of operations.

17.16.3 Corrective Action

If smoke appears while burning oil, suspect poor atomization. Check the fuel oil temperature, the atomizing medium conditions, oil and atomizing medium pressures, and the condition of the oil gun. Refer to Sections 17.6 and 17.10. Consider a switch to gas fuel until the problem is resolved.

Check the excess oxygen in the firebox to identify any deficiency. Determine if there has been any change in fuel composition. When operating at low excess air conditions, a change in fuel gas composition — adding some higher molecular weight gases — can initiate smoking because of a lack of increase in air flow.

A process tube leak, spilling hydrocarbon contents into the firebox, will create an incomplete combustion condition and smoke at the stack. This situation requires a heater shutdown and a steam purge into the firebox to cool any conflagration and to reduce the amount of air and the intensity of combustion.

17.17 HIGH NOx EMISSIONS

17.17.1 Indication of the Problem

The measured NOx concentration in the flue gas leaving the stack exceeds the permitted operating levels.

17.17.2 Effect on Operation

Continued operation above permitted emission levels can result in sanctions or loss of the operating permit.

17.17.3 Corrective Action

Thermal NOx formation can be significantly reduced by burner technology, so reduced emissions may be obtained by a change of burners. See Figure 17.8 for the effect of burner model on NOx emissions. Since high flame temperatures favor increased NOx formation, reducing the peak flame temperature will reduce the rate of NOx production. The rate of NOx production is also a function of fuel composition. If the fuel contains chemically-bound nitrogen, NOx emissions will increase in proportion to nitrogen concentration. Figure 17.7 shows the effect of bound nitrogen, usually found in liquid fuels, on NOx production. Figures 17.4 through 17.6 show the effect of other variables — excess oxygen, combustion air temperature, firebox temperature — on NOx formation.

To control the NOx properly the operator must first understand the type and model of the burner installed in the heater and the design features provided. Types of burners designed to reduce NOx emissions include staged air burners, staged fuel burners, Ultra LoNOX burners, and COOL technology burners. Since heater operation may depend on low NOx emissions, it is important that the burner manufacturer have extensively tested the selected burner, have a reputation for providing a durable product, and offer reliable field service.

The first thing to do if NOx emissions appear high is check the analyzer. Like all instruments of this type, it can

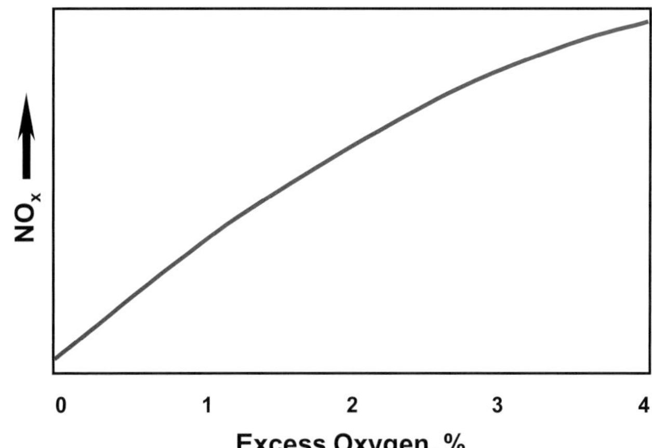

FIGURE 17.4 Effect of excess oxygen on NOx in raw gas burners.

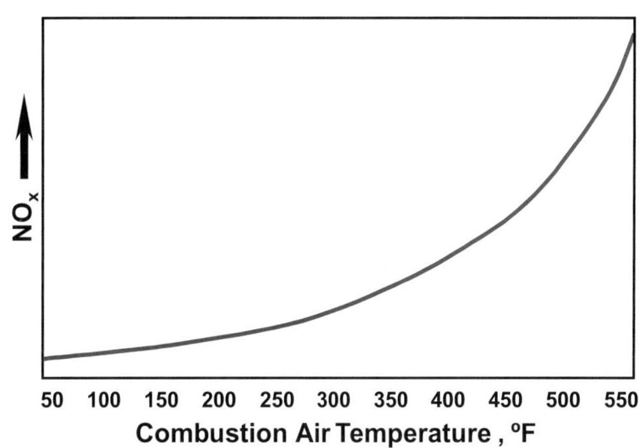

FIGURE 17.5 Effect of combustion air temperature on NOx.

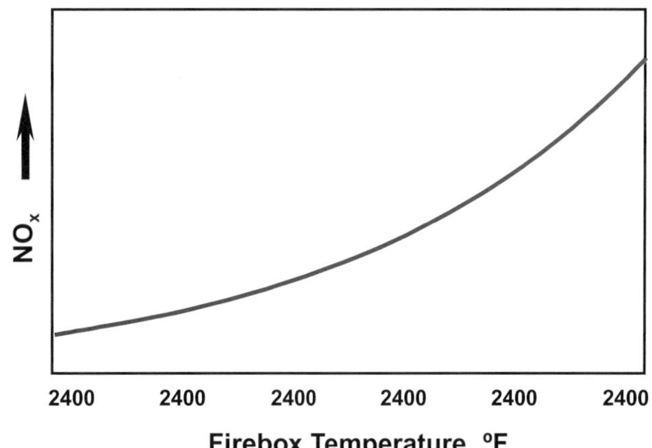

FIGURE 17.6 Effect of firebox temperature on NOx.

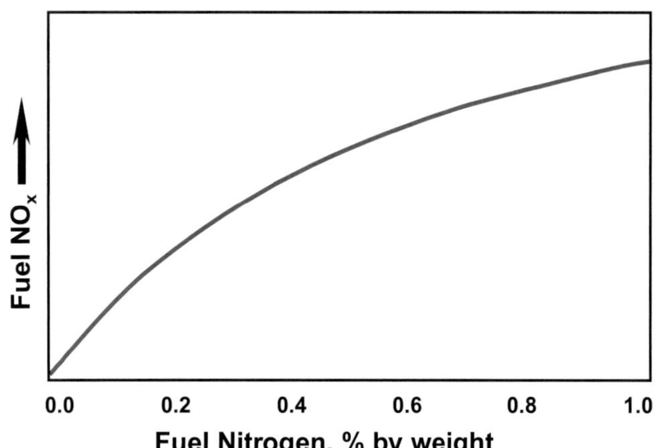

FIGURE 17.7 Effect of bound nitrogen in the liquid fuel on NOx.

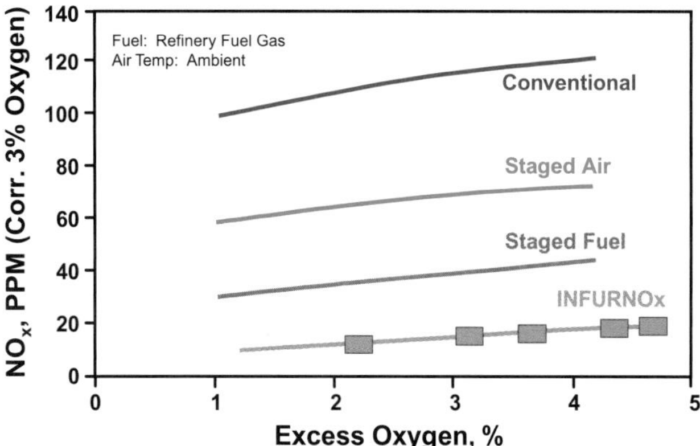

FIGURE 17.8 Effect of burner model on NOx.

be out of calibration or the sampling system may not be operating correctly.

Next check the level of excess oxygen in the flue gas at the firebox exit. If this level is above the target, follow the procedure to reduce excess oxygen. Seal all possible air leaks to make controlling the excess oxygen easier. Adjust the draft to the target value to minimize the amount of air leaking into the heater and the excess oxygen at the burners. Adjust registers in accordance with the burner manufacturer's recommendations for the specific burner model.

Check the fuel composition and the firing rate to identify changes that may raise flame temperature and NOx emission. Fuel gas components such as hydrogen burn with a high flame temperature, potentially forming more NOx than a natural gas flame. The concentration of chemically-bound nitrogen in fuel oil typically increases with the oil density, i.e., heavy oils will typically have more nitrogen than light oils.

As firing rates increase, larger flames may result. The larger flame may have a core that does not lose heat and cool as rapidly compared to smaller flames typical of lower firing rates. Some new technologies overcome this disadvantage of larger flames.

Burners designed for lowered NOx production typically require that the fuel, air, and any firebox gases or flue gases injected into the flame envelope be well-mixed and held close to the design proportions. Anything that will cause the burner to operate outside of the design parameters will tend to increase the production of NOx. Potential problems of this sort and respective solutions are listed below.

- Partially plugged or cracked fuel tip — clean or replace the tip.
- Fuel tip oriented or positioned improperly — refer to the manufacturer's drawing and correct the positioning.
- Improper tips installed (usually during maintenance) — refer to the manufacturer's drawing for the proper tip description and replace in accordance with the recommended positioning.
- Damaged primary flame holder or stabilizer — replace the damaged part.
- Damaged or improperly installed burner tile — replace or properly reinstall the tile.
- Unstable flame — see information in this chapter for corrective actions.
- Air passages partially plugged disrupting uniform air or recirculation flows — remove the material plugging the passages.
- Partially plugged or damaged flue gas orifices — clean or replace as appropriate.
- Temperature or pressure of the fuel or any injected material not in accordance with the design — adjust the operating parameters to the recommended values.

The pilot burner is not a low-NOx type of burner. If the pilot is too large or unstable, it may make a relatively large contribution to the measured NOx. Look at the possibility of safely reducing the pilot heat release.

The adjustment of the registers to control and proportion the combustion air varies with the type or model of low-NOx burner. Descriptions of register adjustment to reduce NOx follow. The fuel injection nozzle parameters — sizing, location, orientation — are determined by the burner manufacturer, and no adjustments should be made by the operator unless needed to conform to the manufacturer's drawings.

If the NOx emissions remain above the expected level after following the directions in this section, consult with the burner manufacturer for other possible solutions.

17.17.3.1 Staged Air Burner

The staged air burner is utilized for either gas or liquid fuel firing. This type of burner normally has three air registers to control the flow rate and distribution of combustion air through the burner, and only one fuel injection nozzle. However, some designs have only two air registers.

If the burner has three air registers or dampers they are termed the primary, secondary, and tertiary air registers (dampers); each must be correctly adjusted to successfully minimize NOx production. The primary air register is normally on burners that have liquid fuel firing capability, and it controls the air flow to the oil tile. The secondary air register controls the combustion air flow through the burner throat. The tertiary air register controls the flow of combustion air that bypasses the burner throat and is then mixed into the flame in the secondary combustion zone of the burner. See the sketch in Figure 17.9b for this type of burner.

To begin adjustment, the primary air register should be as open as possible and still provide a stable flame. The secondary and tertiary air registers should each be approximately 50% open. Further adjustments of the secondary air register position are made while referring to the oxygen analyzer output. The NOx analyzer is used to adjust the tertiary air register to the optimum opening. If the NOx target cannot be reduced with secondary and tertiary register adjustments, then the primary air register must be closed and the secondary and tertiary register adjustments repeated.

In staged air burner models with only two air registers, the registers are labeled the secondary and tertiary air registers. The oxygen analyzer will be used to guide adjustment of the secondary air register, and the NOx analyzer output will guide tertiary register positioning.

17.17.3.2 Staged Fuel Burner

The staged fuel burner is normally applied only with gas fuels to reduce NOx formation. However, it can be designed to stage gaseous fuels while having the independent capability of firing liquid fuels with standard, not low-NOx, burner technology. The staged gaseous fuel burner utilizes one combustion air register or damper to control the air flow across the burner and two sets of fuel injection nozzles to stage the fuel flow into the combustion reaction (see Figure 17.10a). The burner air register or damper position is set with reference to the output of the oxygen analyzer at the flue gas exit from the radiant section.

17.17.3.3 Ultra Low-NOx Burner

The Ultra Low-NOx burner utilizes the staged fuel concepts and also inspirates flue gases from the radiant section into the primary and secondary combustion reaction zones (see Figure 17.11).

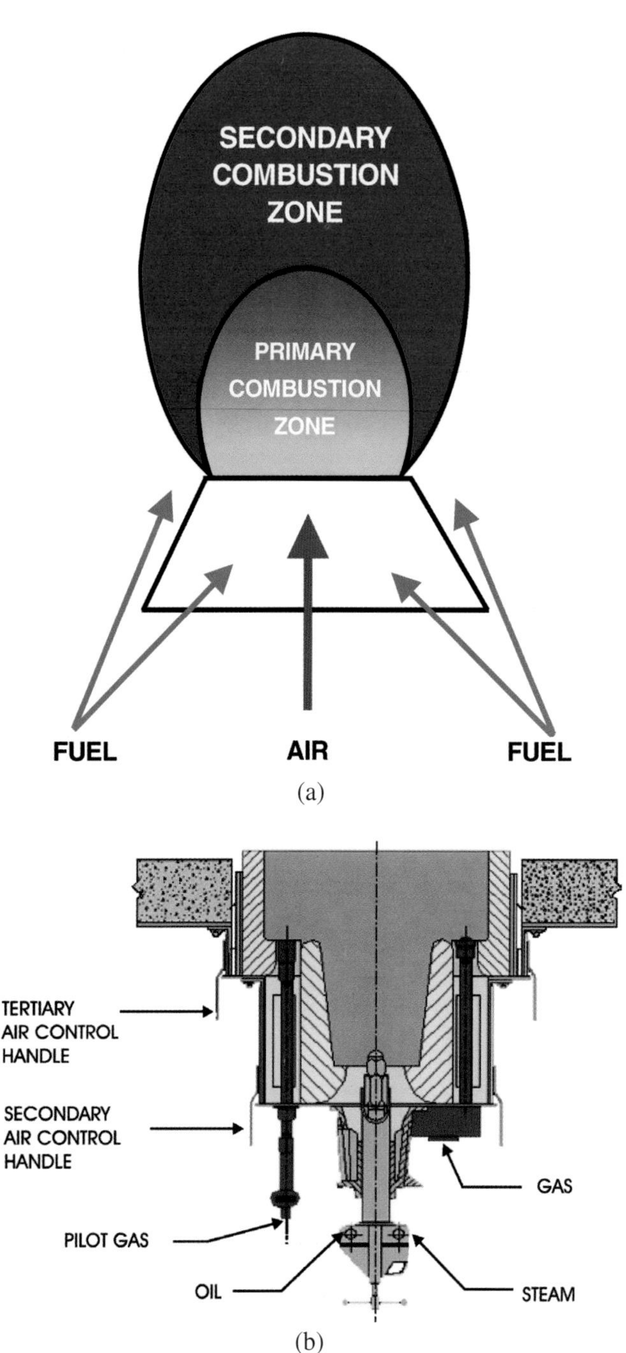

FIGURE 17.9 (a) Air staging. (b) Staged air burner.

The introduction of the largely inert flue gases into the combustion zones of the burner reduces the peak flame temperature and NOx formation. The combination of staging the gaseous fuel and the injection of flue gases results in a burner with the currently lowest possible NOx.

This burner type has one air register or damper to adjust the airflow across the burner. The burner has both primary

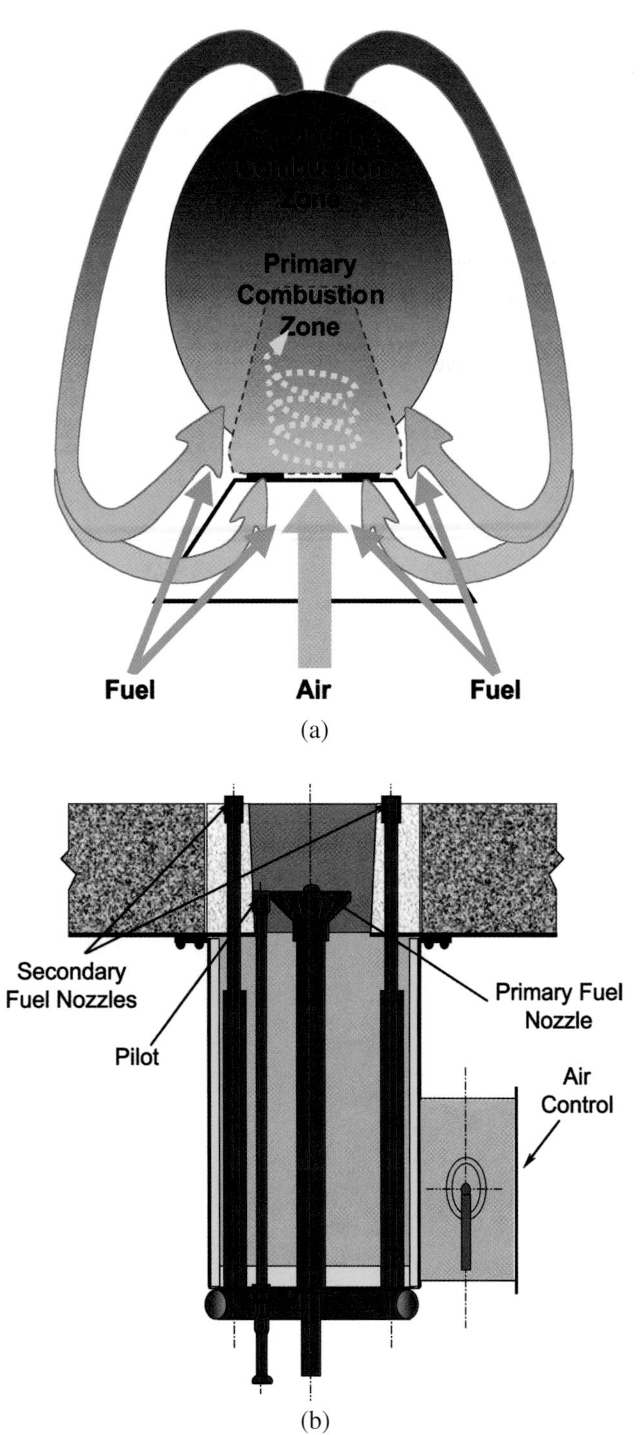

FIGURE 17.10 Staged fuel burner.

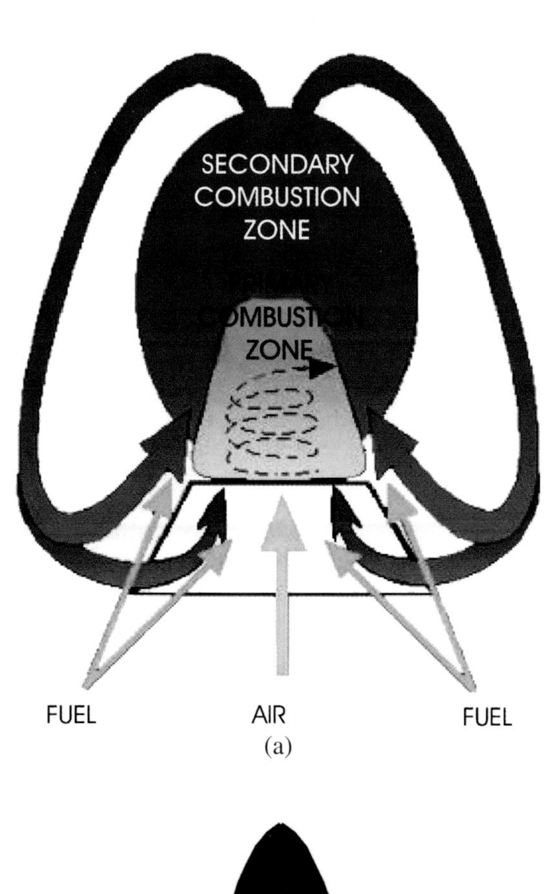

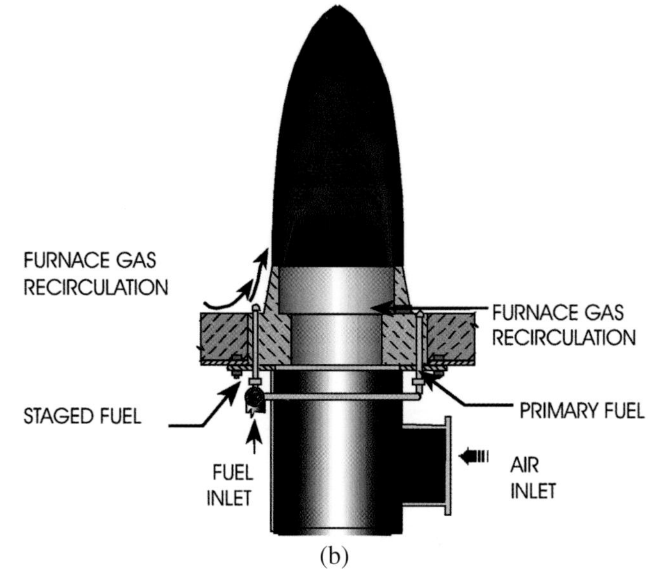

FIGURE 17.11 InfurNOx burner technology.

and secondary fuel injection nozzles that act to inject fuel gas into the combustion zone together with flue gas from the firebox. The burner air register or damper position is set with reference to the output of the oxygen analyzer located at the flue gas exit from the radiant section.

17.17.3.4 COOL TECHNOLOGY Burner

When inert components are mixed with the fuel gas, the flame temperature is lowered and the rate of NOx formation is reduced. COOL TECHNOLOGY burners take advantage of this and require that the amount of inert gas injected with the fuel gas be controlled within limits set by the burner manufacturer.

TABLE 17.2 Troubleshooting for Gas Burners

Trouble	Causes	Solutions
Burners go out	Gas/air mixture too lean, or too much draft	Reduce total air. Reduce primary air.
Flame flashback	Low gas pressure	Shut off burners to raise the fuel gas pressure to the operating burners; reducing the burner orifice size can also be helpful.
	High hydrogen concentration in fuel gas	Reduce primary air; tape the primary air shut if flashback continues; a new burner or tip drilling may be required.
Insufficient heat release	Low gas flow	Increase gas flow; increase burner tip orifice size; make sure that sufficient air will be available for the increased fuel rate.
	Desired heat release exceeds design capacity	Larger burner tips or new burners may be required.
Pulsating fire or breathing (flame alternately ignites and goes out)	Lack of oxygen/draft	Reduce firing rate immediately; establish complete combustion at lower rate; open stack damper and/or air registers to increase air and draft; reduce fuel before increasing air.
Erratic flame	Lack of combustion air	Adjust air register and/or stack damper.
	Incorrect position of burner tip	Locate tips per manufacturer's drawings.
	Damaged burner block	Repair burner block to manufacturer's tolerances.
Gas flame too long	Excessive firing	Reduce firing rates.
	Too little primary air	Increase primary air; decrease secondary air.
	Worn burner tip	Replace tip.
	Tip drilling angle too narrow	Change to wide drilling angle tip.
Gas flame too short	Too much primary air	Increase secondary air; decrease primary air.
	Tip drilling angle too wide	Change to narrow drilling angle tip.

Check the ports on the injection device to ensure that they are clear so that the inert gas is being added in the quantities designed by the burner manufacturer. Ensure that the damper on the inert gas supply plenum is open. If flue gas is the inert material being injected, be sure that the flue gas headers are properly insulated and that the fuel gas mixture is kept at the proper operating conditions.

COOL TECHNOLOGY burners have one air register or damper to adjust the airflow across the burner. The burner has both primary and secondary fuel injection nozzles that act to inject the fuel gas and inspirate firebox flue gases into the combustion zone. The burner air register is set with reference to the output of the oxygen analyzer located at the flue gas exit from the radiant section.

17.18 SUMMARY

In summary, troubleshooting burners in a furnace involves: (1) observing the problem; (2) identifying the problem; (3) determining the effect on the operation of the furnace; and (4) determining the solution and the corrective action that should be taken to correct the problem. If there is a problem that cannot be identified and resolved, one should consult with the burner manufacturer to obtain advice on how to proceed.

Tables 17.2 and 17.3 give summaries of the most frequently occurring troubleshooting operations for gas- and oil-fired burners, respectively.

REFERENCES

1. Burners for Fired Heaters in General Refinery Services, API Publication 535, 1st ed., July 1995.
2. N.P. Lieberman, *Troubleshooting Process Operations*, 3rd ed., Penn Well Publishing, Tulsa, OK, 1991.
3. R.A. Meyers, *Handbook of Petroleum Refining Processes*, 2nd ed., McGraw-Hill, New York, 1997.
4. W. Bartok, and A.F. Sarofim, Eds., *Fossil Fuel Combustion: A Source Book*, John Wiley & Sons, 1991.
5. John Zink Co. LLC, *John Zink Burner School Notes*, Tulsa, OK, Sept. 16-18, 1998.
6. R.D. Reed, A New Approach to Design for Radiant Heat Transfer in Process Work, *Petroleum Engineer*, August, C-7–C-10, 1950.
7. Fired Heaters for General Refinery Services, API Standard 560, second ed., Sept., 1995.
8. E.A. Barrington, *Fired Process Heaters*, course notes, 1999.

TABLE 17.3 Troubleshooting for Oil Burners

Trouble	Causes	Solutions
Burners dripping; coke deposits on burner blocks; coking of burner tip when firing fuel oil only	Improper atomization due to high oil viscosity, clogging of burner tip, insufficient atomizing steam, improper location of burner tip	Check fuel oil type; increase fuel temperature to lower viscosity to proper level; check composition for heavier fractions; clean or replace burner tip; confirm burner tip is in proper location; increase atomizing steam; place tip in location as per burner drawings.
Failure to maintain ignition	Too much atomizing steam	Reduce atomizing steam until ignition is stabilized; during start-up, have atomizing steam on low side until ignition is well established.
	Too much primary air at firing rates	Reduce primary air.
	Too much moisture in atomizing steam	Ensure appropriate insulation is on steam lines; confirm steam traps are functioning; adjust quality of atomizing steam to appropriate levels.
Coking of oil tip when firing oil in combination	High rate of gas with a low rate of oil; high heat radiation to the fuel oil tip	Increase atomizing steam to produce sufficient cooling effect to avoid coking; reduce gas firing rate; dedicate individual burners to either fuel.
	Incorrect oil gun position	Place tip in locations as per burner drawings. If this fails, readjust burner tip ±0.5 in. until coking ceases.
Erratic flame	Lack of combustion air	Adjust air damper or register.
	Plugged burner gun	Clean burner gun.
	Worn burner gun	Replace burner gun.
	High rate of gas firing while firing a low rate of oil	Reduce gas rate; dedicate burners to either fuel.
	Damaged burner block	Repair burner block.
Excess smoke at stack	Insufficient atomizing steam	Increase atomizing steam.
	Moisture in atomizing steam	Requires knockout drum or increase in superheat, check steam line insulation.
	Low excess air	Increase excess air.

Chapter 18
Duct Burners

Peter F. Barry and Stephen L. Somers

TABLE OF CONTENTS

18.1 Introduction 524

18.2 Applications 524

- 18.2.1 Cogeneration 524
- 18.2.2 Air Heating 524
- 18.2.3 Fume Incineration 526
- 18.2.4 Stack Gas Reheat 526

18.3 Burner Technology 526

- 18.3.1 In-duct or Inline Configuration 526
- 18.3.2 Grid Configuration (Gas Firing) 527
- 18.3.3 Grid Configuration (Liquid Firing) 527
- 18.3.4 Design Considerations 528
- 18.3.5 Maintenance 539
- 18.3.6 Accessories 540
- 18.3.7 Design Guidelines and Codes 540

18.1 INTRODUCTION

Linear and in-duct burners were used for many years to heat air in drying operations before their general use in cogeneration systems. Some of the earliest systems premixed fuel and air in an often complicated configuration that fired into a recirculating process air stream. The first uses in high temperature, depleted oxygen streams downstream of gas turbines in the early 1960s provided additional steam for process use in industrial applications and for electrical peaking plants operating steam turbines. As gas turbines have become larger and more efficient, duct burner supplemental heat input has increased correspondingly.

Linear burners are applied where it is desired to spread heat uniformly across a duct, whether in ambient air or oxygen-depleted streams. In-duct designs are more commonly used in fluidized bed boilers and small cogeneration systems.

18.2 APPLICATIONS

18.2.1 Cogeneration

18.2.1.1 Introduction

Cogeneration implies simultaneous production of two or more forms of energy, most commonly electrical (electric power), thermal (steam, heat transfer fluid, or hot water), and pressure (compressor). The basic process involves combustion of fossil fuel in an engine (reciprocating or turbine) that drives an electric generator, coupled with a recovery device that converts heat from the engine exhaust into a usable energy form. Production of recovered energy can be increased independently of the engine through supplementary firing provided by a special burner type known as a duct burner. Most modern systems will also include flue gas emission control devices. A typical plant schematic is shown in Figure 18.1. Aerial views of typical combined cycle electric power plants are shown in Figure 18.2 and Figure 18.3.

Reciprocating engines (typically diesel cycle) are used in smaller systems (10 MW and lower) and offer the advantage of lower capital and maintenance costs but produce relatively high levels of pollutants. Turbine engines are used in both small and large systems (3MW and above) and, although more expensive, generally emit lower levels of air pollutants.

Fossil fuels used in cogeneration systems can consist of almost any liquid or gaseous hydrocarbon, although natural gas and various commercial-grade fuel oils are most commonly used. Mixtures of hydrocarbon gases and hydrogen found in plant fuel systems are often used in refining and petrochemical applications. Duct burners are capable of firing all fuels suitable for the engine/turbine, as well as many that are not, including heavy oils and waste gases.

Heat recovery for large systems is usually accomplished by convective heat transfer in a boiler (commonly referred to as a heat recovery steam generator, also known by the acronym HRSG). Smaller systems utilize either a steam or hot water boiler, or, alternatively, some type of air-to-air heat exchanger or direct transfer to a process.

Supplementary firing is often incorporated into the boiler/HRSG design as it allows increased production of steam as demanded by the process. The device that provides the supplementary firing is a duct burner, so called because it is installed in the duct connecting the engine/turbine exhaust to the heat recovery device, or just downstream of a section of the HRSG superheater (see Figures 18.4 and 18.5). Oxygen required for the combustion process is provided by the turbine exhaust gas (TEG).

18.2.1.2 Combined Cycle

Combined cycle systems incorporate all components of the simple cycle configuration with the addition of a steam turbine/generator set powered by the HRSG. This arrangement is attractive when the plant cannot be located near an economically viable steam user. Also, when used in conjunction with a duct burner, the steam turbine/generator can provide additional power during periods of high or "peak" demand.

18.2.2 Air Heating

Duct burners are suitable for a wide variety of direct-fired air heating applications where the physical arrangement requires mounting inside a duct, and particularly for processes where the combustion air is at an elevated temperature and/or contains less than 21% oxygen. Examples include:

- *Fluidized bed boilers* (see Figure 18.6): where burners are installed in combustion air ducts and used only to provide heat to the bed during startup. At cold conditions, the burner is fired at maximum capacity with fresh ambient air; but as combustion develops in the bed, cross-exchange with hot stack gas increases the air temperature and velocity. Burners are shut off when the desired air preheat is reached and the bed can sustain combustion unaided.
- *Combustion air blower inlet preheat:* where burners are mounted upstream of a blower inlet to protect against thermal shock caused by ambient air in extremely cold climates (–40°F/°C and below). This arrangement is only suitable when the air will be used in a combustion process as it will contain combustion products from the duct burner.

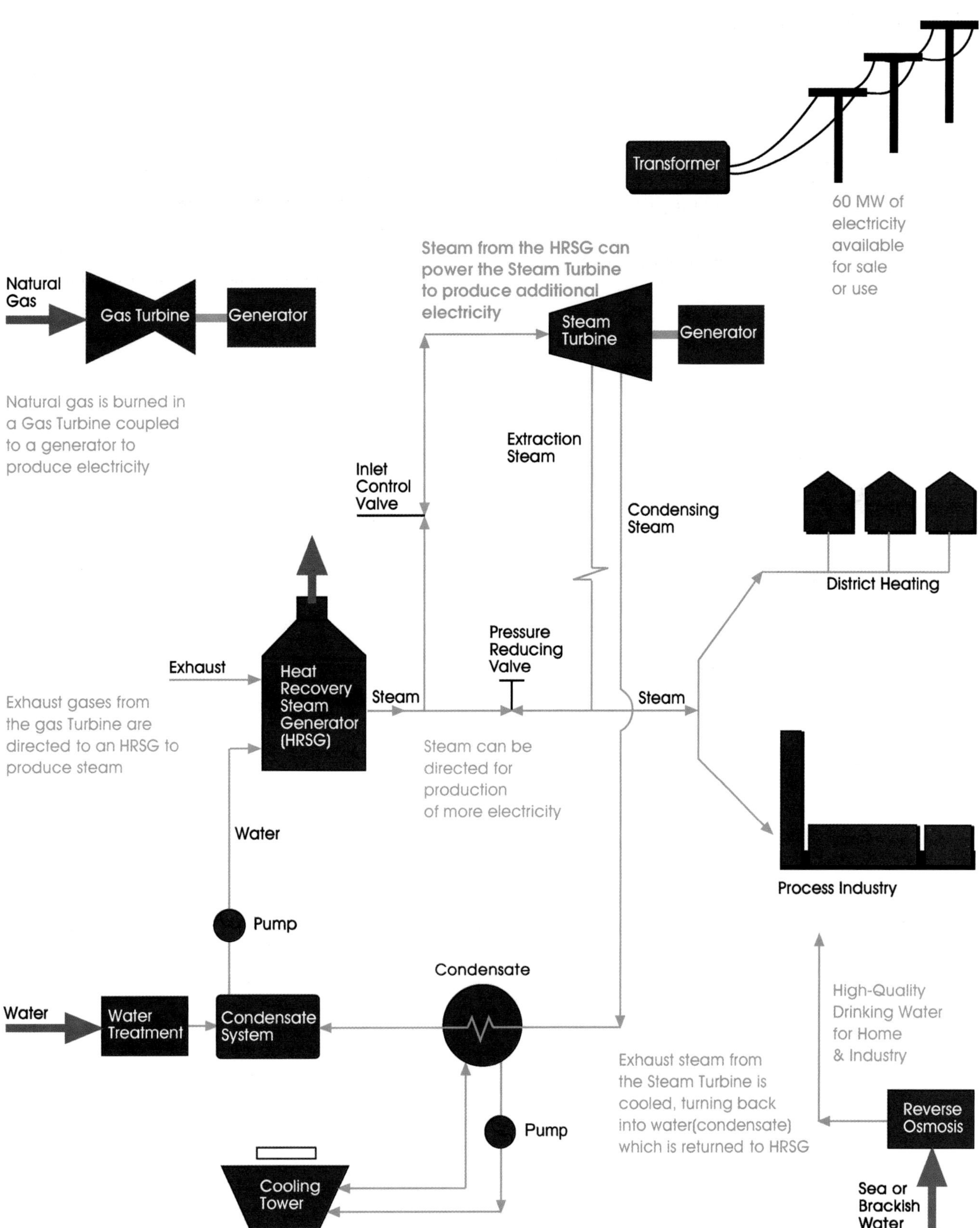

FIGURE 18.1 Typical plant schematic.

FIGURE 18.2 Cogeneration at Teesside, England. (Courtesy of Nooter/Eriksen.)

- *Drying applications:* where isolation of combustion products from the work material is not required, such as certain paper and wallboard manufacturing operations.

18.2.3 Fume Incineration

Burners are mounted inside ducts or stacks carrying exhaust streams primarily composed of air with varying concentrations of organic contaminants. Undesirable components are destroyed, both by an increase in the gas stream bulk temperature and through contact with localized high temperatures created in the flame envelope. Particular advantages of the duct burner include higher thermal efficiency as no outside air is used, lower operating cost as no blower is required, and improved destruction efficiency resulting from distribution of the flame across the duct section with grid-type design.

18.2.4 Stack Gas Reheat

Mounted at or near the base of a stack, heat added by a duct burner will increase natural draft, possibly eliminating a need for induced draft or eductor fans. In streams containing a large concentration of water vapor, the additional heat can also reduce or eliminate potentially corrosive condensation inside the stack. A source of ambient augmenting combustion air is often added if the stack gas oxygen concentration is low. This arrangement may also provide a corollary emissions reduction benefit (see Section 18.2.3).

18.3 BURNER TECHNOLOGY

18.3.1 In-duct or Inline Configuration

Register or axial flow burner designs are adapted for installation inside a duct. The burner head is oriented such that the flame will be parallel to and co-flow with the air or TEG stream, and the fuel supply piping is fed through the duct side wall, turning 90° as it enters the burner (see Figure 18.7). Depending on the total firing rate and duct size, one burner may be sufficient, or several may be arrayed across the duct cross-section. Inline burners typically require more air/TEG pressure drop, produce longer flames, and offer a less uniform heat distribution than grid-type. On the other hand, they are more flexible in burning liquid fuels, can be more easily

FIGURE 18.3 Combination (oil and gas) fired duct burners at Dahbol, India. (Courtesy of Enron.)

modified to incorporate augmenting air, and sometimes represent a less expensive option for high firing rates in small ducts without sufficient room for grid elements.

18.3.2 Grid Configuration (Gas Firing)

A series of linear burner elements that span the duct width are spaced at vertical intervals to form a grid. Each element is comprised of a fuel manifold pipe fitted with a series of flame holders (or wings) along its length. Fuel is fed into one end of the manifold pipe and discharged through discrete multi-port tips attached at intervals along its length, or through holes drilled directly into the pipe. Gas ports are positioned such that fuel is injected in co-flow with the TEG. The wings meter the TEG or air flow into the flame zone, thus developing eddy currents that anchor ignition. They also shield the flame in order to maintain suitably high flame temperatures, thereby preventing excessive flame cooling that might cause high emissions. Parts exposed to TEG and the flame zone are typically of high-temperature alloy construction (see Figures 18.8 and 18.9).

18.3.3 Grid Configuration (Liquid Firing)

As with the gas-fired arrangement, a series of linear burner elements comprised of a pipe and flame holders (wings) span the duct width. However, instead of multiple discharge points along the pipe length, liquid fuel is injected downstream of the element through the duct sidewall, and directed parallel to the flame holders (cross-flow to the TEG). This configuration utilizes the duct cross-section for containment of the flame length, thus allowing a shorter distance between the burner and downstream boiler tubes (see Figure 18.10). The injection device, referred to as a side-fired oil gun, utilizes a mechanical nozzle supplemented by low-pressure air (2 to 8 psi) to break the liquid fuel into small droplets (atomization) that will vaporize and readily burn. Although most commonly used for light fuels, this arrangement is also suitable for some heavier fuels where the viscosity can be lowered by heating. In some cases, high pressure steam may be required, instead of low-pressure air, for adequate atomization of heavy fuels.

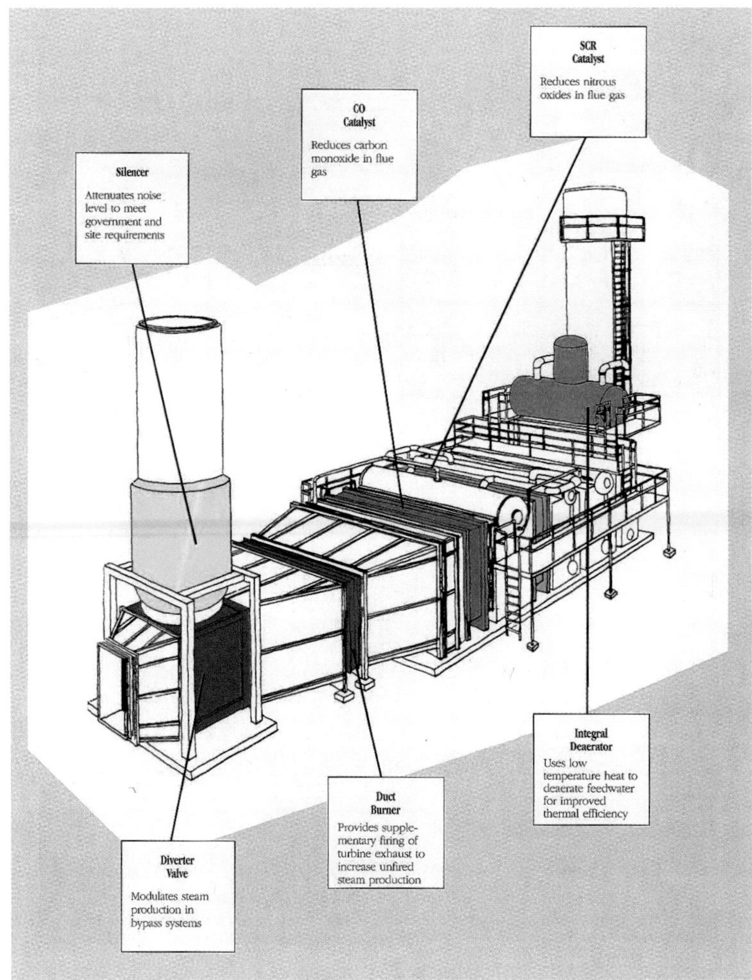

FIGURE 18.4 Typical location of duct burners in an HRSG. (Courtesy of Deltak.)

18.3.4 Design Considerations

18.3.4.1 Fuels

18.3.4.1.1 Natural Gas

Natural gas is by far the most commonly used fuel because it is readily available in large volumes throughout much of the industrialized world. Because of its ubiquity, its combustion characteristics are well-understood, and most burner designs are developed for this fuel.

18.3.4.1.2 Refinery/Chemical Plant Fuels

Refineries and chemical plants are large consumers of both electrical and steam power, which makes them ideal candidates for cogeneration. In addition, these plants maintain extensive fuel systems to supply the various direct- and indirect-fired processes, as well as to make the most economical use of residual products. This latter purpose presents special challenges for duct burners because the available fuels often contain high concentrations of unsaturated hydrocarbons with a tendency to condense and/or decompose inside burner piping. The location of burner elements inside the TEG duct, surrounded by high-temperature gases, exacerbates the problem. Plugging and failure of injection nozzles can occur, with a corresponding decrease in online availability and an increase in maintenance costs.

With appropriate modifications, however, duct burners can function reliably with most hydrocarbon-based gaseous fuels. Design techniques include insulation of burner element manifolds, insulation and heat tracing of external headers and pipe trains, and fuel/steam blending. Steam can also be used to periodically purge the burner elements of solid deposits before plugging occurs.

18.3.4.1.3 Low Heating Value

By-product gases produced in various industrial processes such as blast furnaces, coke ovens, and flexicokers, or from

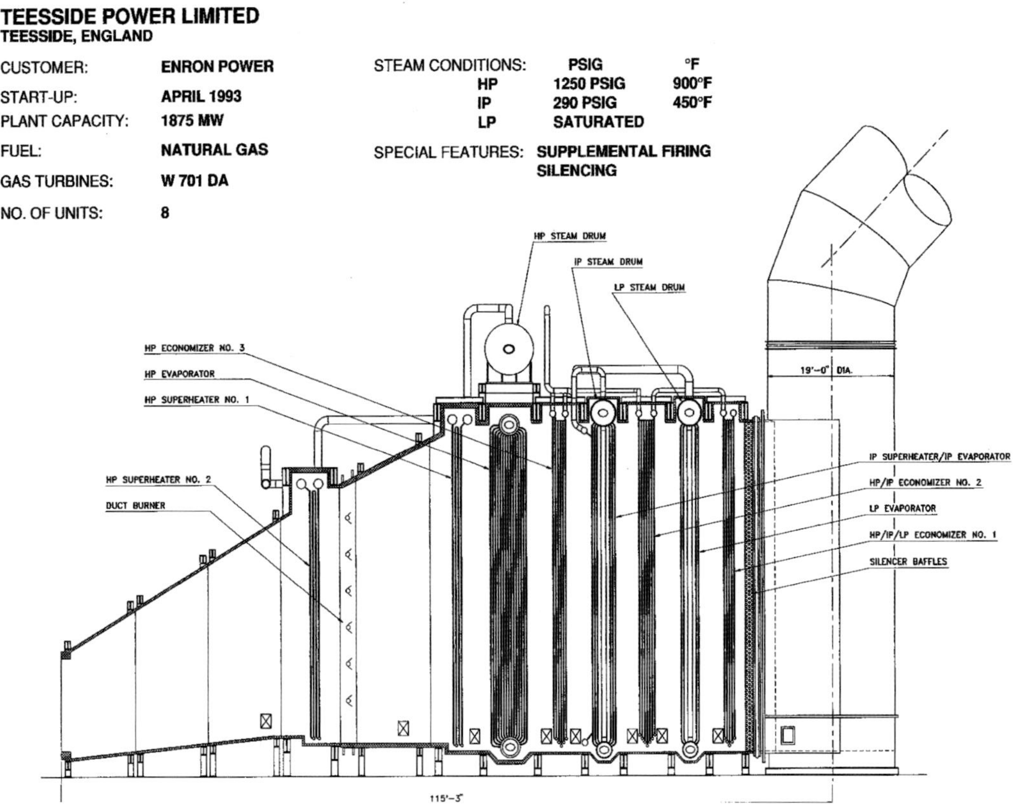

FIGURE 18.5 Schematic of HRSG at Teesside, England. (Courtesy of Nooter/Eriksen.)

mature landfills, contain combustible compounds along with significant concentrations of inert components, thus resulting in relatively low heating values (range of 50 to 500 Btu/scf). These fuels burn more slowly and at lower temperatures than conventional fuels, and thus require special design considerations. Fuel pressure is reduced to match its velocity to flame speed, and some form of shield or "canister" is employed to provide a protected flame zone with sufficient residence time to promote complete combustion before the flame is exposed to the quenching effects of TEG.

Other considerations that must be taken into account are moisture content and particulate loading. High moisture concentration results in condensation within the fuel supply system, which in turn produces corrosion and plugging. Pilots and ignitors are particularly susceptible to the effects of moisture because of small fuel port sizes, small igniter gap tolerance, and the insulation integrity required to prevent "shorting" of electrical components. A well-designed system might include a knock-out drum to remove liquids and solids, insulation and heat-tracing of piping to prevent or minimize condensation, and low-point drains to remove condensed liquids. Problems are usually most evident after a prolonged period of shutdown.

FIGURE 18.6 Fluidized bed startup duct burner.

Solid particulates can cause plugging in gas tip ports or other fuel system components and should therefore be removed to the maximum practical extent. In general, particle size should be no greater than 25% of the smallest port, and overall loading should be no greater than 5 ppm by volume (weight).

FIGURE 18.7 An inline burner.

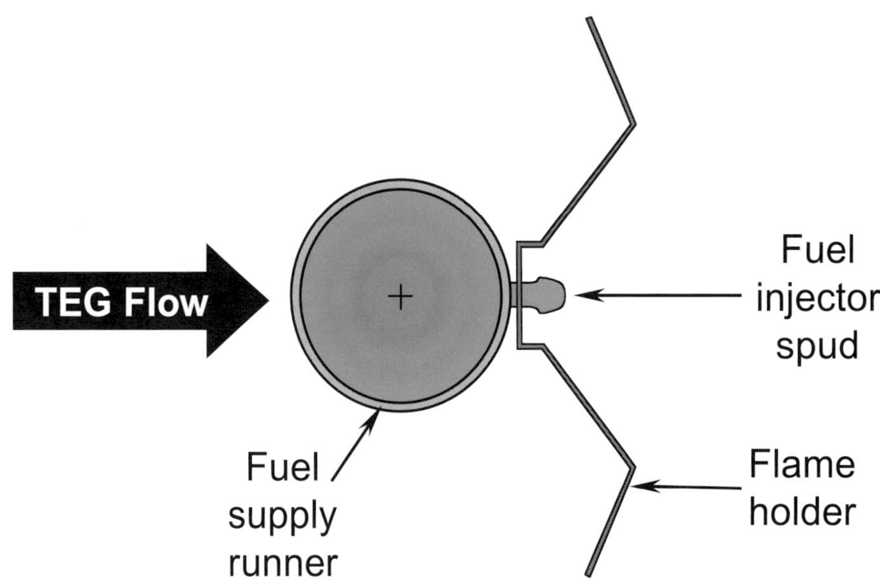

FIGURE 18.8 Linear burner elements.

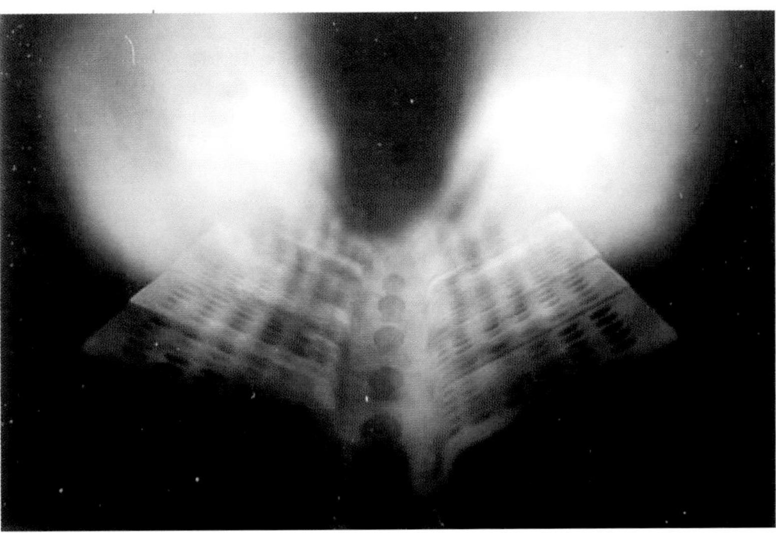

FIGURE 18.9 Gas flame from a grid burner.

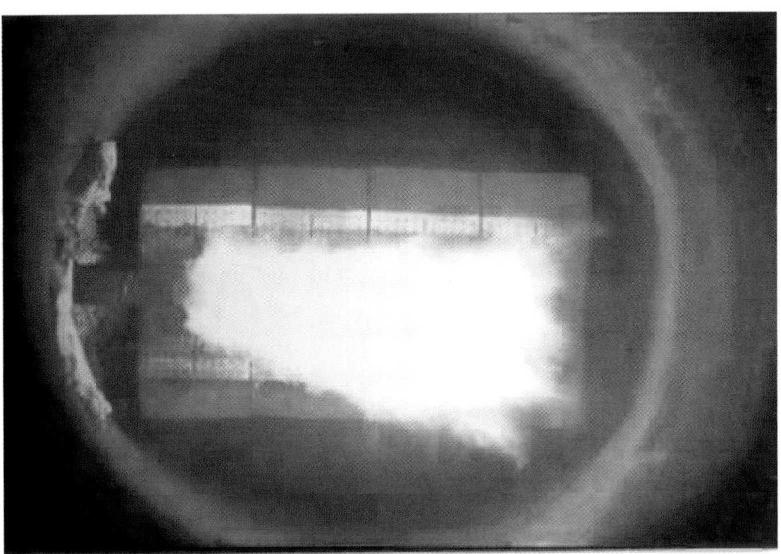

FIGURE 18.10 Oil flame from a side-fired oil gun.

18.3.4.1.4 Liquid Fuels

In cogeneration applications, duct burners are commonly fired with the same fuel as the turbine, which is typically limited to light oils such as No. 2 or naphtha. For other applications, specially modified side-fired guns or an inline design can be employed to burn heavier oils such as No. 6 and some waste fuels.

18.3.4.2 Combustion Air and Turbine Exhaust Gas

18.3.4.2.1 Temperature and Composition

When used for supplementary firing in HRSG cogeneration applications, the oxygen required for the combustion reaction is provided by the residual in the turbine exhaust gas instead of from a new, external source of air. Because this gas is already at an elevated temperature, duct burner thermal efficiency can exceed 90% as very little heat is required to raise the combustion products temperature to the final fired temperature. TEG contains less oxygen than fresh air, typically between 11 and 16% by volume, which, in conjunction with the TEG temperature, will have a significant effect on the combustion process. As the oxygen concentration and TEG temperature become lower, emissions of CO and unburned hydrocarbons occur more readily, eventually progressing to combustion instability. The effect of low oxygen concentration can be

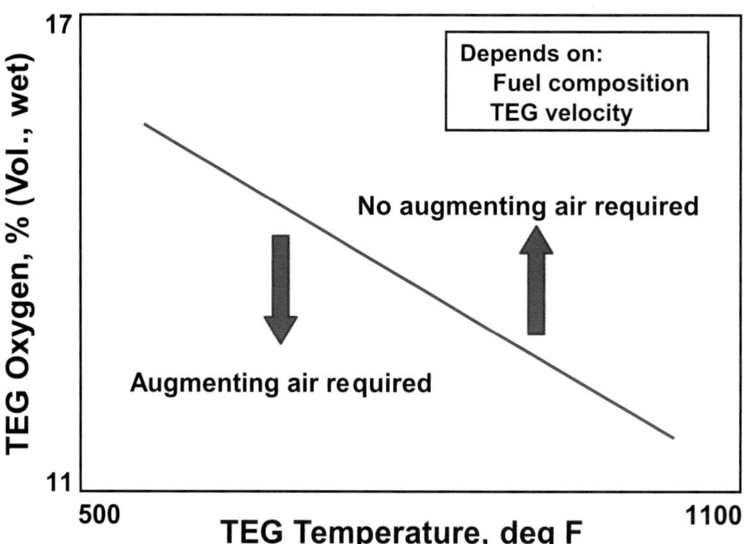

FIGURE 18.11 Approximate requirement for augmenting air.

partially offset by higher temperatures; conversely, higher oxygen concentrations will partially offset the detrimental effects of low TEG temperatures. This relationship is depicted graphically in Figure 18.11. Duct burner emissions are discussed in more detail elsewhere in this chapter.

18.3.4.2.2 Turbine Power Augmentation

During periods of high electrical demand, various techniques are employed to increase power output, and most will increase the concentration of water vapor in TEG. The corresponding effect is a reduction in TEG oxygen concentration and temperature with consequent effects on duct burner combustion. Depending on the amount of water vapor used, CO emissions may simply rise, or in extreme cases the flame may become unstable. The former effect can be addressed with an allowance in the facility operating permit or by increasing the amount of CO catalyst in systems so equipped. The latter requires air augmentation, a process whereby fresh air is injected at a rate sufficient to raise the TEG oxygen concentration to a suitable level.

18.3.4.2.3 Velocity and Distribution

Regardless of whether TEG or fresh air is used, velocity across flame stabilizers must be sufficient to promote mixing of the fuel and oxygen, but not so great as to prevent the flame from anchoring to the burner. Grid-type configurations can generally operate at velocities ranging from 20 to 90 fps (feet per second) and pressure drops of less than 0.5 in. water column. Inline or register burners typically require velocities of 100 to 150 fps with a pressure drop of 2 to 6 in. water column.

Grid burners are designed to distribute heat uniformly across the HRSG or boiler tube bank, and thus require a reasonably uniform distribution of TEG or air to supply the fuel with oxygen. Inadequate distribution causes localized areas of low velocity, resulting in poor flame definition along with high emissions of CO and unburned hydrocarbons. Turbine exhaust flow patterns, combined with rapidly diverging downstream duct geometry, will almost always produce an unsatisfactory result that must be corrected by means of a straightening device. Likewise, the manner in which ambient air is introduced into a duct can also result in flow maldistribution, requiring some level of correction. Selection and design of flow-straightening devices are discussed elsewhere in this chapter (see Figure 18.12).

In instances where bulk TEG or air velocity is lower than required for proper burner operation, flow straightening alone is not sufficient and it becomes necessary to restrict a portion of the duct cross-section at or near the plane of the burner elements, thereby increasing the "local" velocity across flame holders. This restriction, also referred to as blockage, commonly consists of unfired runners or similar shapes uniformly distributed between the firing runners to reduce the open flow area.

Inline or register burners inject fuel in only a few (or possibly only one) positions inside the duct, and can therefore be positioned in an area of favorable flow conditions, assuming the flow profile is known. On the other hand, downstream heat distribution is less uniform than with grid designs, and flames may be longer. As with grid-type burners, in some cases it may be necessary to block portions of the duct at or just upstream of the burners to force a sufficient quantity of TEG or air through the burner.

18.3.4.2.4 Ambient Air Firing (Air-only Systems and HRSG Backup)

Velocity and distribution requirements for air-systems are similar to those for TEG, although inlet temperature is not a concern because of the relatively higher oxygen concentration. As with TEG applications, the burner elements are exposed to the products of combustion, so material selection must take into account the maximum expected fired temperature.

Ambient (or fresh) air backup for HRSGs presents special design challenges. Because of the temperature difference between ambient air and TEG, designing for the same mass flow and fired temperature will result in velocity across the burner approximately one third that of the TEG case. If the cold condition velocity is outside the acceptable range, it will be necessary to add blockage, as described elsewhere in this chapter. Fuel input capacity must also be increased to provide heat required to raise the air from ambient to the design firing temperature. By far the most difficult challenge is related to flow distribution. Regardless of the manner in which backup air is fed into the duct, a flow profile different from that produced by the TEG is virtually certain. Flow-straightening devices can therefore not be optimized for either case, but instead require a compromise design that provides acceptable results for both. If the two flow patterns are radically different, it may ultimately be necessary to alter the air injection arrangement independently of the TEG duct-straightening device.

18.3.4.3 Augmenting Air

As turbines have become more efficient and more work is extracted in the form of, for example, electricity, the oxygen level available in the TEG continues to get lower. To some extent, a correspondingly higher TEG temperature provides some relief for duct burner operation.

In some applications, however, an additional oxygen source may be required to augment that available in the TEG when the oxygen content in the TEG is not sufficient for combustion at the available TEG temperature. If the mixture adiabatic flame temperature is not high enough to sustain a robust flame in the highly turbulent stream, the flame may become unstable.

The problem can be exacerbated when the turbine manufacturer adds large quantities of steam or water for NOx control and power augmentation. A corresponding drop in the TEG temperature and oxygen concentration occurs because of dilution. The TEG temperature is also reduced in installations where the HRSG manufacturer splits the steam superheater and places tubes upstream of the duct burner.

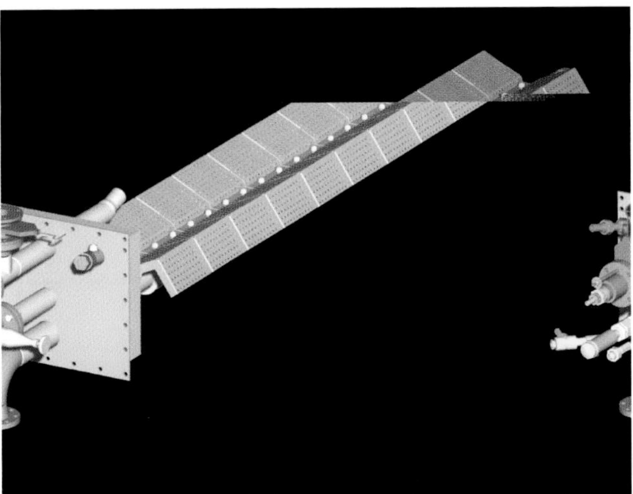

FIGURE 18.12 Duct burner arrangement.

With their research and development facilities, manufacturers have defined the oxygen requirement with respect to TEG temperature and fuel composition, and are able to quantify the amount of augmenting air required under most conditions likely to be encountered. It is usually not practical to add enough air to the turbine exhaust to increase the oxygen content to an adequate level. Specially designed runners are therefore used to increase the local oxygen concentration. In cases where augmenting air is required, the flow may be substantial: from 30 to 100% of the theoretical air required for the supplemental fuel.

The augmenting air runner of one manufacturer consists of a graduated air delivery tube designed to ensure a constant velocity across the length of the tube. Equal distribution of augmenting air across the face of the tube is imperative. The augmenting air is discharged from the tube into a plenum and passes through a second distribution grid to further equalize flow. The air passes through perforations in the flame holder, where it is intimately mixed with the fuel in the primary combustion zone. This intimate mixing ensures corresponding low CO and UHC emissions under most conditions likely to be encountered. Once the decision has been made to supply augmenting air to a burner, it is an inevitable result of the design that the augmenting air will be part of the normal operating regime of the combustion runner.

18.3.4.4 Equipment Configuration and TEG/Combustion Air Flow Straightening

The turbine exhaust gas/combustion air velocity profile at the duct burner plane must be within certain limits to ensure good combustion efficiency; in cogeneration applications, this is rarely achieved without flow-straightening devices. Even in

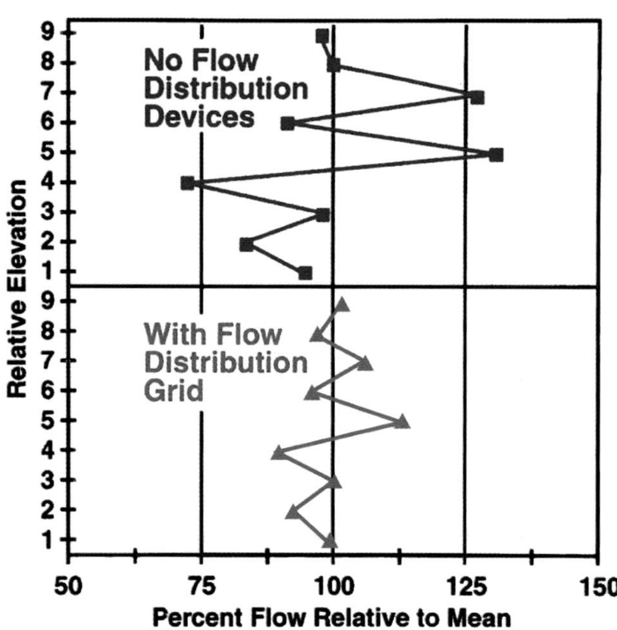

FIGURE 18.13 Comparison of flow variation with and without straightening device.

FIGURE 18.14 Physical model of burner.

non-fired configurations, it may be necessary to alter the velocity distribution to make efficient use of boiler heat transfer surface. Figure 18.13 shows a comparison of flow variation with and without flow straightening.

Duct burners are commonly mounted in the TEG duct upstream of the first bank of heat transfer tubes, or they may be nested in the boiler superheater between banks of tubes. In the former case, a straightening device would be mounted just upstream of the burner, while in the latter it is mounted either upstream of the first tube bank or between the first tube bank and (upstream of) the burner. Although not very common, some HRSG design configurations utilize two stages of duct burners with heat transfer tube banks in between, and a flow-straightening device upstream of the first burner. Such an arrangement is, however, problematic because the TEG downstream of the first-stage burner may not have the required combination of oxygen and temperature properties required for proper operation of the second-stage burner.

Perforated plates that extend across the entire duct cross-section are most commonly used for flow straightening because experience has shown they are less prone to mechanical failure than vane-type devices, even though they require a relatively high pressure drop. The pattern and size of perforations can be varied to achieve the desired distribution. Vanes can produce comparable results with significantly less pressure loss but require substantial structural reinforcement to withstand the flow-induced vibration inherent in HRSG systems. Regardless of the method used, flow pattern complexity — particularly in TEG applications — usually dictates the use of either physical or computational fluid dynamic (CFD) modeling for design optimization.

18.3.4.4.1 Physical Modeling

TEG/air flow patterns are determined by inlet flow characteristics and duct geometry, and are subject to both position and time variation. Design of an efficient (low pressure loss) flow-straightening device is therefore not a trivial exercise, and manual computational methods are impractical. For this reason, physical models, commonly 1:6 or 1:10 scale, are constructed, and flow characteristics are analyzed by flowing air with smoke tracers or water with polymer beads through the model (see Figure 18.14). Although this method produces reliable results, tests conducted at ambient conditions (known as "cold flow") are not capable of simulating the buoyant effects that may occur at elevated temperatures.

18.3.4.4.2 Computational Fluid Dynamic (CFD) Modeling

Flow modeling with CFD, using a computer-generated drawing of the inlet duct geometry, is capable of predicting flow pattern and pressure drop in the turbine exhaust flow path. The model can account for swirl flow in three dimensions, accurately predict pressure drop, and subsequently help design a suitable device to provide uniform flow. The CFD model must be quite detailed to calculate flow patterns incident and through a perforated grid or tube bank while also keeping the overall model solution within reasonable computation time. Combustion effects can be included in the calculations at the cost of increased computation time. Figure 18.15 shows a sample result of CFD modeling performed on a HRSG inlet duct.

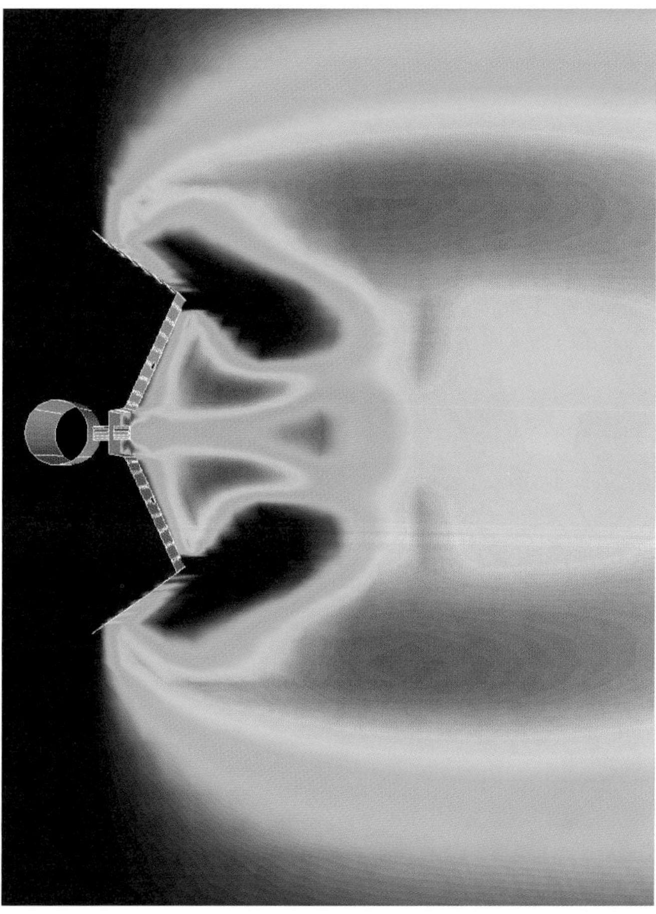

FIGURE 18.15 Sample result of CFD modeling performed on an HRSG inlet duct.

18.3.4.5 Wing Geometry: Variations

18.3.4.5.1 Flameholders

Design of the flame stabilizer, or flameholder, is critical to the success of supplementary firing. Effective emission control requires that the TEG be metered into the flame zone in the required ratio to create a combustible mixture and ensure that the combustion products do not escape before the reactions are completed. In response to new turbine and HRSG design requirements, each duct burner manufacturer has proprietary designs developed to provide the desired results.

18.3.4.5.2 Basic Flameholder

In its basic form, a fuel injection system and a zone for mixing with oxidant are all that is required for combustion. For application to supplemental firing, the simple design shown in Figure 18.16 consists of an internal manifold or "runner," usually an alloy pipe with fuel injection orifices spaced along the length. A bluff body plate, with or without perforations, is attached to the pipe to protect the flame zone from the turbulence in the exhaust gas duct. The low-pressure zone pulls the flame back onto the manifold. This low-cost runner may overheat the manifold, causing distortion of the metallic parts. Emissions are unpredictable with changing geometry and CO is usually much higher than the current typically permitted levels of under 0.1 lb/MMBtu.

18.3.4.5.3 Low Emissions Design

Modifications to the design for lower emission performance generally have a larger cross-section in the plane normal to the exhaust flow. The increased blocked area protects the fuel injection zone and increases residence time. The NOx is reduced by the oxygen-depleted TEG and the CO/UHC is reduced by the delayed quenching. The correct flow rate of TEG is metered through the orifices in the flameholder, and the fuel injection velocity and direction are designed to enhance combustion efficiency. The flame zone is pushed away from the internal manifold ("runner" pipe), creating space for cooling TEG to bathe the runner and flameholder and enhance equipment life.

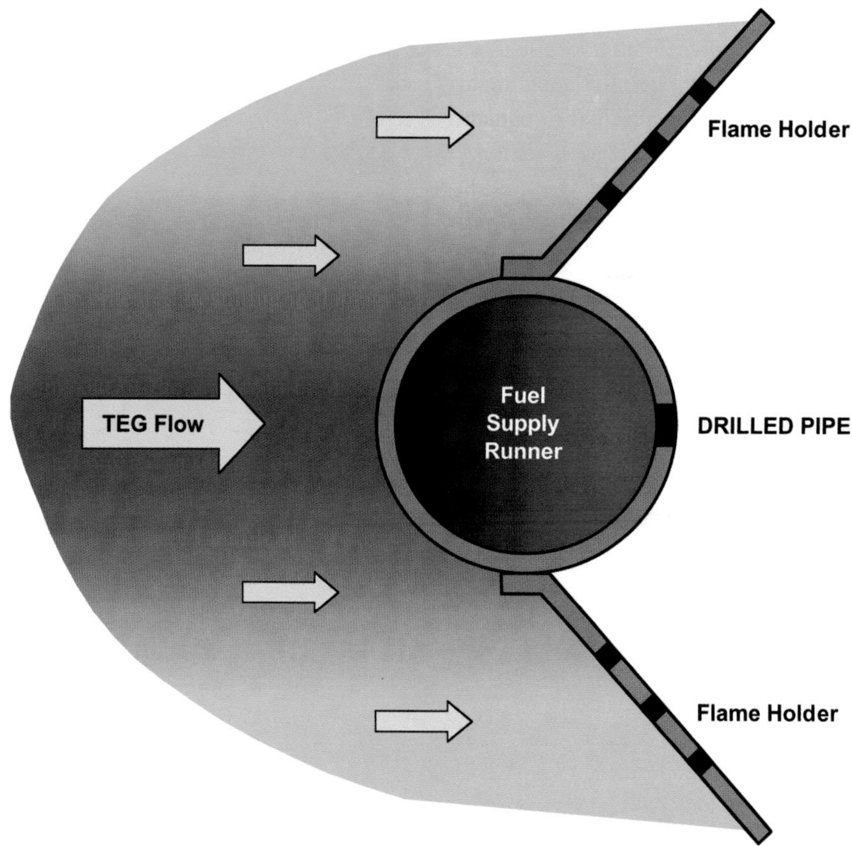

FIGURE 18.16 Drilled pipe duct burner.

FIGURE 18.17 Low emission duct burner.

Each manufacturer approaches the geometry somewhat differently. One manufacturer uses cast alloy pieces welded together to provide the required blockage. These standard pieces often add significant weight and are difficult to customize to specific applications. Hot burning fuels, such as hydrogen, may not receive the cooling needed to protect the metal from oxidation. Alternately, fuels subject to cracking, such as propylene, may not have the oxygen needed to minimize coke buildup.

Another manufacturer supplies custom designs to accommodate velocity extremes, while maintaining low emissions. In the design shown in Figure 18.17, the flameholder is optimized with CFD and research experimentation to enhance mixing and recirculation rate. Special construction materials are easily accommodated. This supplier also uses removable fuel tips with multiple orifices, which can be customized to counteract any unexpected TEG flow distribution discovered after commercial operation. Figure 18.18 depicts the flow patterns of air/TEG and fuel in relation to the duct burner flameholder.

18.3.4.6 Emissions

18.3.4.6.1 NOx and NO vs. NO_2

Formation of NO and NO_2 is the subject of ongoing research to understand the complex reactions (see Chapter 6). Potentially, several oxides of nitrogen (NOx) can be formed

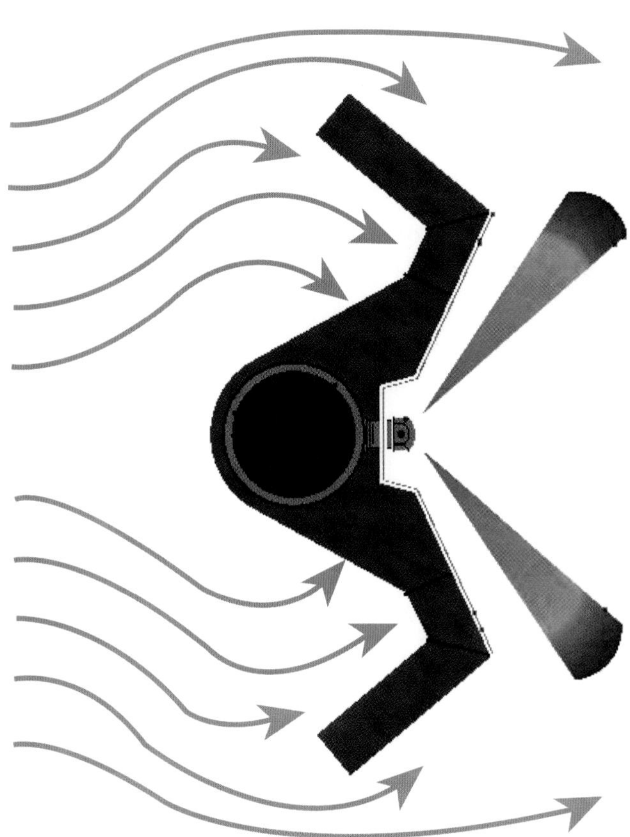

FIGURE 18.18 Flow patterns around flame stabilizer.

during the combustion process, but only nitric oxide (NO) and nitrogen dioxide (NO_2) occur in significant quantities. NO is colorless and NO_2 has a reddish-brown color.

In the elevated temperatures found in the flame zone in a typical HRSG turbine exhaust duct, NO formation is favored almost exclusively over NO_2 formation. Turbine exhaust NOx is typically 95% NO and 5% NO_2. In the high-temperature zone, NO_2 dissociates to NO by the mechanism of:

$$NO_2 + O + Heat \rightarrow NO + O_2$$

However, after the TEG exits the hot zone and enters the cooling zone at the boiler tubes, reaction slows and the NO_2 is essentially fixed. At the stack outlet, the entrained NO is slowly oxidized to NO_2 through a complex photochemical reaction with atmospheric oxygen. The plume will be colorless unless the NO_2 increases to about 15 ppm, at which time a yellowish tint is visible. Care must be taken in duct burner design because NO can also be oxidized to NO_2 in the immediate post-flame region by reactions with hydroperoxyl radicals:

$$NO + HO_2 \rightarrow NO_2 + OH$$

if the flame is rapidly quenched. This quenching can occur because of the large quantity of excess TEG commonly present in duct burner applications. Conversion to NO_2 may be even higher at fuel turndown conditions where the flame is smaller and colder. NO_2 formed in this manner can contribute to "brown plume" problems and may even convert some of the turbine exhaust NO to NO_2.

There are two principle mechanisms in which nitrogen oxides are formed.

1. *Thermal NOx:* The primary method is thermal oxidation of atmospheric nitrogen in the TEG. NOx formed in this way is called thermal NOx. As the temperature increases in the combustion zone and surrounding environment, increased amounts of N_2 from the TEG are converted to NO. Thermal NOx formation is most predominant in the peak temperature zones of the flame.
1. *Fuel-bound nitrogen NOx:* The secondary method utilized to form NOx is the reaction of oxygen with chemically bound nitrogen compounds contained in the fuel. NOx formed in this manner is called fuel NOx. Large amounts of NOx can be formed by fuels that contain molecularly bound nitrogen (e.g., amines and mercaptans). If a gaseous fuel such as natural gas contains diluent N_2, it simply behaves as atmospheric nitrogen and will form NOx only if it disassociates in the high-temperature areas. However, if the gaseous fuel contains, for example, ammonia (NH_3), this nitrogen is considered bound. In the low concentrations typically found in gaseous fuels, the conversion to NOx is close to 100% and can have a major impact on NOx emissions.

Bound nitrogen in liquid fuel is contained in the long carbon chain molecules. Distillate oil is the most common oil fired in duct burners as a liquid fuel. The fuel-bound nitrogen content is usually low, in the range of 0.05 weight percent. Conversion to NOx is believed to be 80 to 90%. For No. 6 oil, containing 0.30 weight percent nitrogen, the conversion rate to NOx would be about 50%. Other heavy waste oils or waste gases with high concentrations of various nitrogen compounds may add relatively high emissions. Consequently, fuel NOx can be a major source of nitrogen oxides and may predominate over thermal NOx.

The impact of temperature on NOx production in duct burners is not as pronounced as in, for example, fired heaters or package boilers. One reason is that both the bulk fired temperature and the adiabatic flame temperature are lower than in fired process equipment.

When used to provide supplementary firing of turbine exhaust, duct burners are generally considered to be "low NOx" burners. Because the turbine exhaust contains reduced oxygen,

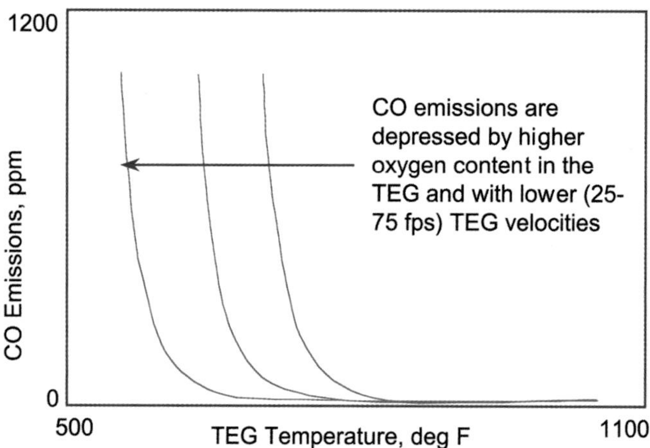

FIGURE 18.19 Effect of conditions on CO formation.

the peak flame temperature is reduced and the reaction speed for O_2 and N+ to form NOx is thus lowered. The burners also fire into much lower average bulk temperatures — usually less than 1600°F (870°C) — than process burners or fired boilers. The high-temperature zones in the duct burner flames are smaller due to large amounts of flame quenching by the excess TEG. Finally, mixing is rapid and therefore retention time in the high-temperature zone is very brief.

The same duct burner, when used to heat atmospheric air, is no longer considered "low NOx," because the peak flame temperature approaches the adiabatic flame temperature in air.

Clearly, operating conditions have a major impact on NO formation during combustion. To properly assess NOx production levels, the overall operating regime must be considered, including TEG composition, fuel composition, duct firing temperature, and TEG flow distribution.

18.3.4.6.2 Visible Plumes

Stack plumes are caused by moisture and impurities in the exhaust. Emitted NO is colorless and odorless, and NO_2 is brownish in color. If the NO_2 level in the flue gas exceeds about 15 to 20 ppm, the plume will take on a brownish haze. NOx also reacts with water vapor to form nitrous and nitric acids. Sulfur in the fuel may oxidize to SO_3 and condense in the stack effluent, causing a more persistent white plume

18.3.4.6.3 CO, VOC, SOx, and Particulates

Carbon monoxide: Carbon monoxide (CO), a product of incomplete combustion, has become a major permitting concern in gas turbine-based cogeneration plants. Generally, CO emissions from modern industrial and aero-derivative gas turbines are very low, in the range of a few parts per million (ppm). There are occasional situations in which CO emissions from the turbine increase due to high rates of water injection for NOx control or operation at partial load, but the primary concern is the sometimes large CO contribution from supplementary firing. The same low-temperature combustion environment that suppresses NOx formation is obviously unfavorable for complete oxidation of CO to CO_2. Increased CO is produced when fuels are combusted under fuel-rich conditions or when a flame is quenched before complete burnout. These conditions (see Figure 18.19) can occur if there is poor distribution of TEG to the duct burner, which causes some burner elements to fire fuel-rich and others to fire fuel-lean, depending on the efficiency of the TEG distribution device. The factors affecting CO emissions include:

- turbine exhaust gas distribution
- low TEG approach temperature
- low TEG oxygen content
- flame quench on "cold" screen tubes
- improperly designed flame holders that allow flame quench by relatively cold TEG
- steam or water injection

Unburned hydrocarbons (UHCs): In the same fashion as carbon monoxide generation, unburned hydrocarbons (UHCs) are formed in the exhaust gas when fuel is burned without sufficient oxygen, or if the flame is quenched before combustion is complete. UHCs can consist of hydrocarbons (defined as any carbon-hydrogen molecule) of one carbon or multiple carbon atoms. The multiple carbon molecules are often referred to as long-chain hydrocarbons. Unburned hydrocarbons are generally classified in two groups:

1. unburned hydrocarbons as methane
2. non-methane hydrocarbons or volatile organic compounds (VOCs)

The reason for the distinction and greater concern for VOCs is that longer chain hydrocarbons play a greater role in the formation of photochemical smog. VOCs are usually defined as molecules of two carbons or greater, and are sometimes considered to be three carbons or greater. These definitions are set by local air quality control boards and vary across the United States.

UHCs can only be eliminated by correct combustion of the fuel. However, hydrocarbon compounds will always be present in trace quantities, regardless of how the HRSG system is operated.

Sulfur dioxide: Sulfur dioxide (SO_2) is a colorless gas that has a characteristic smell in concentrations as low as 1 ppm. SO_2 is formed when sulfur (S) in the fuel combines with oxygen (O_2) in the TEG. If oxygen is present (from excess of combustion) and the temperature is correct, the sulfur will further combine and be converted to sulfur trioxide (SO_3). These oxides of sulfur are collectively known as SOx.

Except for sulfur compounds present in the incoming particulate matter, all of the sulfur contained in the fuel is converted to SO_2 or SO_3. Sulfur dioxide will pass through the boiler system to eventually form the familiar "acid rain" unless a gas-side scrubbing plant is installed. Sulfur trioxide can, in the cooler stages of the gas path, combine with moisture in the exhaust gas to form sulfuric acid (H_2SO_4), which is highly corrosive and will be deposited in ducts and the economizer if the exhaust gas is below condensing temperatures. Natural gas fuels are fortunately very low in sulfur and do not usually cause a problem. However, some oil fuels and plant gases can be troublesome in this respect.

Particulate matter (PM): Particulate emissions are formed from three main sources: ash contained in liquid fuels, unburned carbon in gas or oil, and SO_3. The total amount of particulate is often called TSP (total suspended particulate). There is concern for the smaller sized portion of the TSP, as this stays suspended in air for a longer period of time. The PM-10 is the portion of the total particulate matter that is less than 10 microns (1×10^{-6} m) in size. Particles smaller than PM-10 are on the order of smoke.

Typical NOx and CO emissions for various fuels are shown in Table 18.1.

TABLE 18.1 Typical NOx and CO Emissions From Duct Burners

Gas	NOx ($lb/10^6$ Btu fired)	CO ($lb/10^6$ Btu fired)
Natural gas	0.1	0.08
Hydrogen gas	0.15	0.00
Refinery gas	0.1–0.15	0.03–0.08
Plant gas	0.11	0.04–0.01
Flexicoker gas	0.08	0.01
Blast furnace gas	0.03–0.05	0.12
Producer gas	0.05–0.1	0.08
Syn fuels	0.08–0.12	0.08
Propane	0.14	0.14
Butane	0.14	0.14

Note: NOx emissions from butane and propane can be modified by direct steam injection into a gas or burner flame. CO emissions are highly dependent on TEG approach temperature and HRSG fired temperature.

18.3.5 Maintenance

1. *Normal wear and tear:* If nothing has been replaced in the past 5 years and the burner (or turbine/HRSG set) is operated fairly continuously, it is likely that some tips and wings may require replacement.
2. *Damage due to misuse, system upsets or poor maintenance practices:* Older systems designed without sufficient safety interlocks (TEG trip, high temperature) sometimes expose parts to excessively high temperatures, which results in wing warpage and oxidation failure.
3. *Fuel quality/composition:* Some refinery fuels or waste fuels contain unsaturated components and/or liquid carryover. Eventually, these compounds will form solids in the runner pipes or directly in tips, which results in plugging.

The following are some items to look for when operational problems are encountered:

- *Plugged gas ports,* which are evidenced by gaps in the flame or high fuel pressure: Gas ports may simply consist of holes drilled into the element manifold pipe, or they may be located in individual removable tips. Designs of the former type may be re-drilled or else the entire manifold pipe must be replaced. Discrete tips can be replaced individually as required.
- *Warped flame holders (wings):* Some warping is normal and will not affect flame quality, but excessive deformation such as "curling" around the gas ports will degrade the combustion and emissions performance. Most grid-type burner designs permit replacement of individual flameholder segments.
- *Oxidation of flame holders (wings) or portions of flame holders:* If more than one-third of the flameholder is missing, it is a good candidate for replacement. Fabricated and cast designs are equally prone to oxidation over time. Most grid-type burner designs permit replacement of individual flameholder segments.
- *Severe sagging of runner pipes (grid design only):* If the manifold pipe is no longer supported at both ends, it should be replaced. Beyond that relatively extreme condition, sagging at midspan in excess of approximately 2 to 3 in. (5 to 7 cm) should be corrected by runner replacement and/or installation of an auxiliary support.

18.3.6 Accessories

18.3.6.1 Burner Management System

All fuel-burning systems should incorporate controls that provide for safe manual light-off and shutdown, as well as automatic emergency shutdown upon detection of critical failures. Control logic may reside in a packaged flame safeguard module, a series of electromechanical relays, a programmable logic controller (PLC), or a distributed control system (DCS). At a minimum, the duct burner management system should include the following:

- flame supervision for each burner element
- proof of completed purge and TEG/combustion air flow before ignition can be initiated
- proof of pilot flame before main fuel can be activated
- automatic fuel cutoff upon detection of flame failure, loss of TEG/combustion air, and high or low fuel pressure

Other interlocks designed to protect downstream equipment can also be included, such as high boiler tube temperature or loss of feedwater.

18.3.6.2 Fuel Train

Fuel flow to the burners is controlled by a series of valves, safety devices, and interconnecting piping mounted on a structural steel rack or skid. A properly designed fuel train will include, at a minimum, the following:

- at least one manual block valve
- two automatic block valves in series
- one vent valve between the automatic block valves (gas firing only)
- flow control valve
- high and low fuel pressure switches
- two pressure gages, one each at the fuel inlet and outlet

Depending on the custom and operating requirements at a particular plant, pressure regulation, flow measurement devices, and pressure transmitters can also be incorporated. See Figures 18.20 through 18.27 for typical duct burner fuel system piping arrangements.

18.3.7 Design Guidelines and Codes

18.3.7.1 NFPA 8506 (National Fire Protection Association)

First issued in 1995, this standard has become the *de facto* guideline for heat recovery steam generators in the United States and many other countries that have not developed their own national standards. Specific requirements for burner safety systems are included, but as stated in the foreword, NFPA 8506 does not encompass specific hardware applications, nor should it be considered a "cookbook" for the design of a safe system. Prior to the issuance of NFPA 8506, designers often adapted NFPA boiler standards to HRSGs, which resulted in design inconsistencies.

18.3.7.2 FM (Factory Mutual)

An insurance underwriter that publishes guidelines on combustion system design, Factory Mutual also "approves" specific components such as valves, pressure switches, and flame safeguard equipment that meet specific design and performance standards. Manufacturers are given permission to display the FM symbol on approved devices. Although FM approval may be required for an entire combustion control system, it is more common for designers to simply specify the use of FM-approved components.

18.3.7.3 UL (Underwriters Laboratories)

Well-known in the United States for its certification of a broad range of consumer and industrial electrical devices, UL authorizes manufacturers to display their label on specific items that have demonstrated compliance with UL standards. Combustion system designers will frequently require the use of UL-approved components in burner management systems and fuel trains. Approval can also be obtained for custom-designed control systems, although this requirement generally applies only to a few large cities and a few regions in the United States.

18.3.7.4 ANSI B31.1 and B31.3 (American National Standards Institute)

These codes address piping design and construction. B31.1 is incorporated in the NFPA 8506 guideline, while B31.3 is generally used only for refining/petrochemical applications.

18.3.7.5 Others

The following may also apply to duct burner system designs, depending on the country where equipment will be operated.

- National Electrical Code (NEC)
- Canadian Standards Association (CSA)
- International Electrotechnical Commission (IEC)
- European Committee for Electrotechnical Standardization (CENELEC)

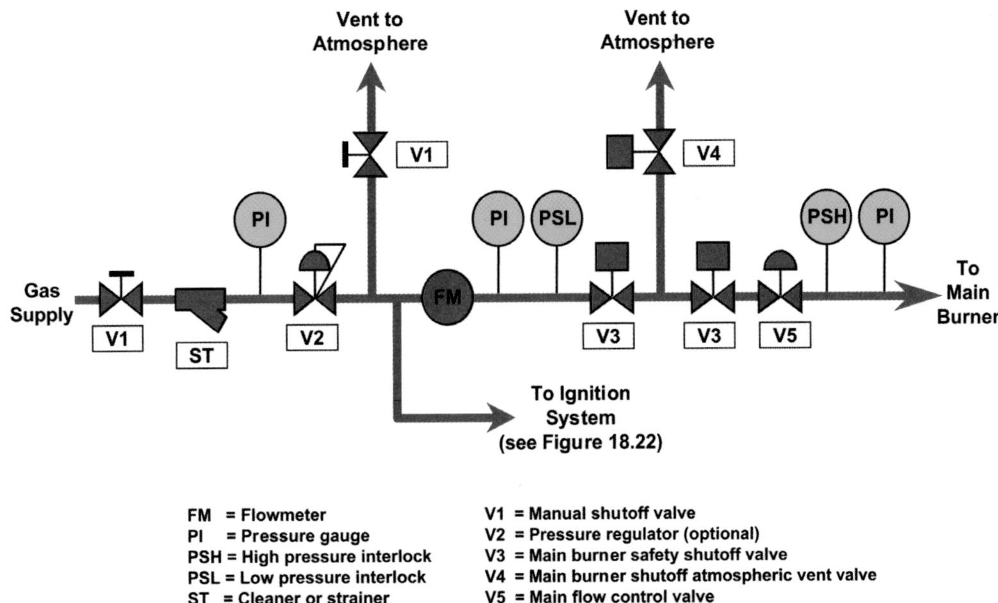

FIGURE 18.20 Typical main gas fuel train: single element or multiple elements firing simultaneously.

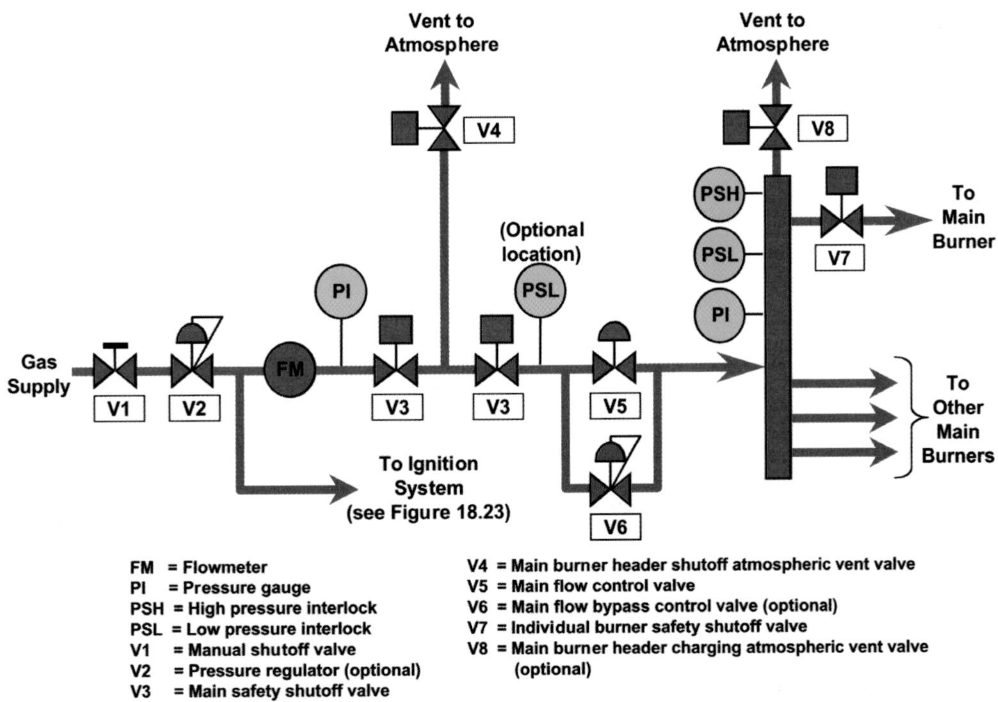

FIGURE 18.21 Typical main gas fuel train: multiple elements with individual firing capability.

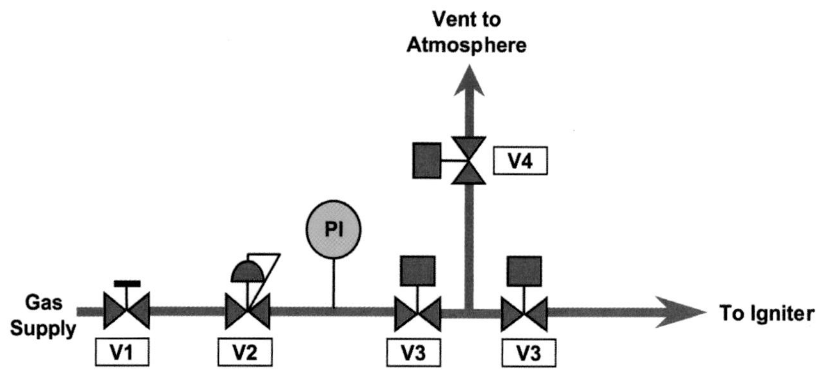

FIGURE 18.22 Typical pilot gas train: single element or multiple elements firing simultaneously.

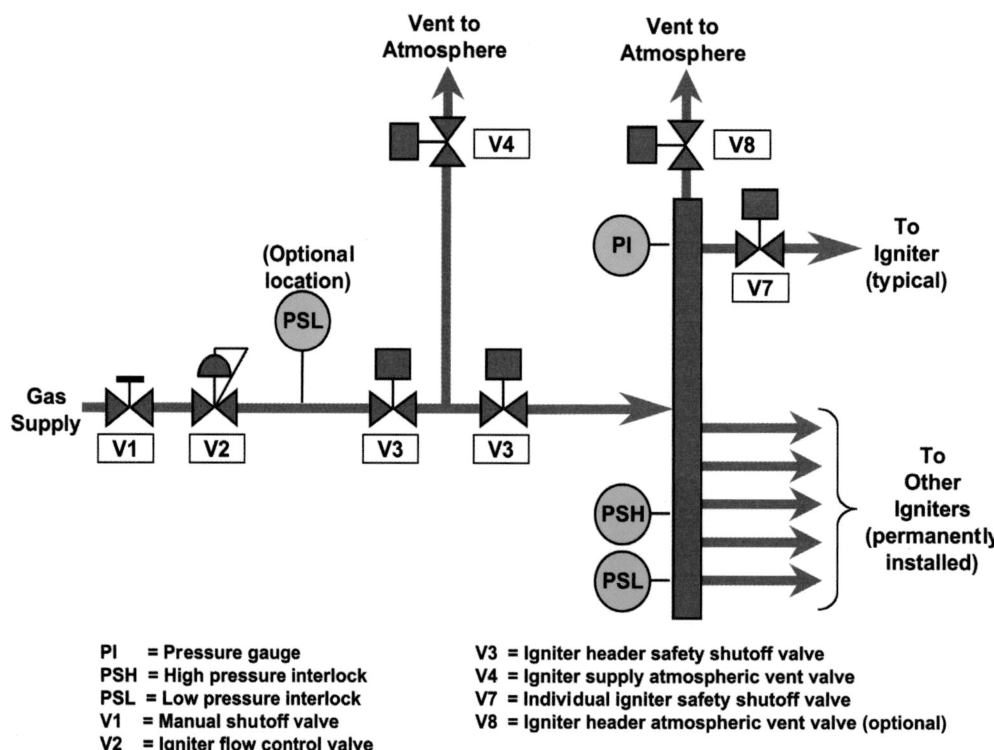

FIGURE 18.23 Typical pilot gas train: multiple elements with individual firing capability.

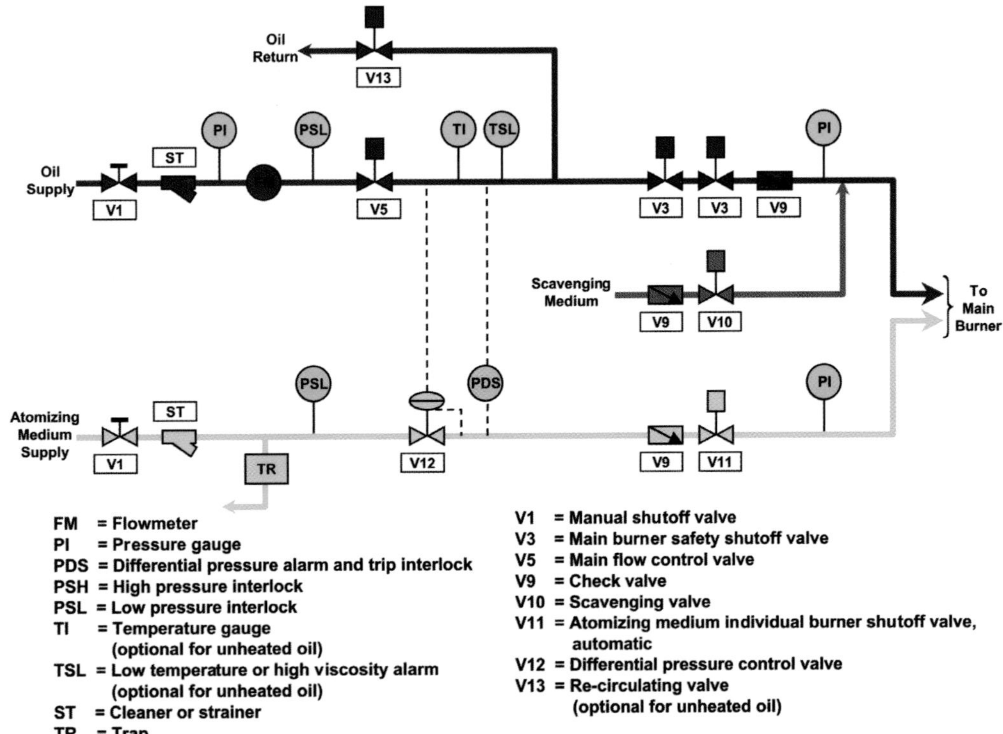

FIGURE 18.24 Typical main oil fuel train: single element.

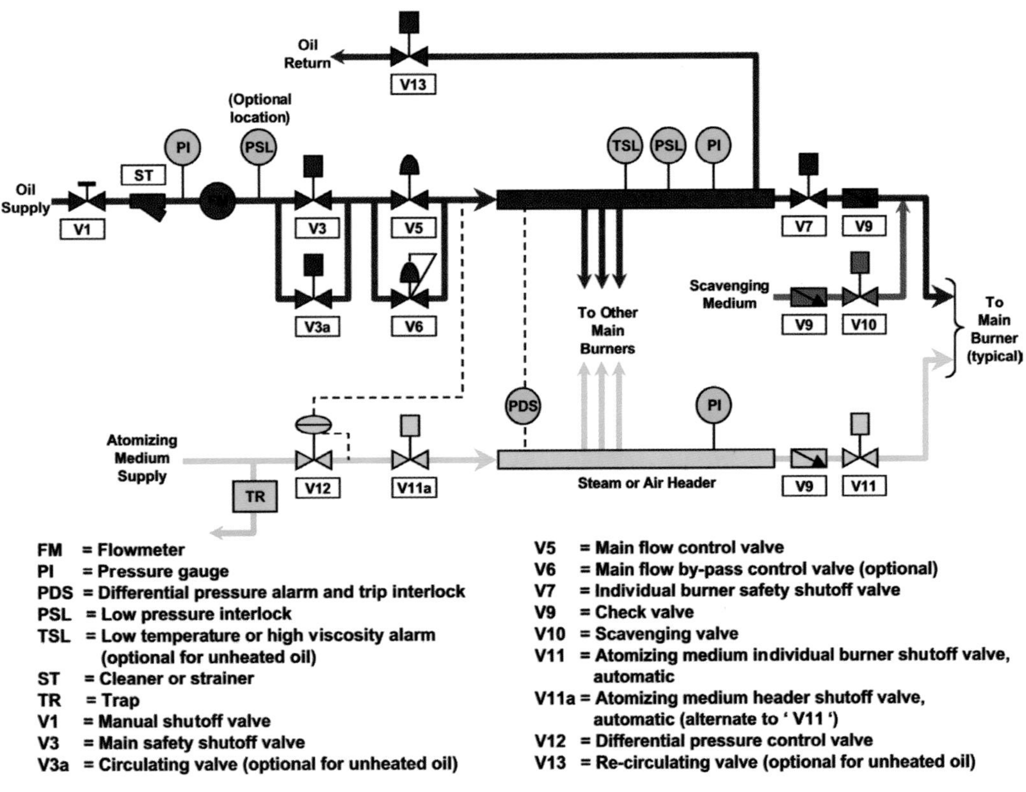

FIGURE 18.25 Typical main oil fuel train: multiple elements.

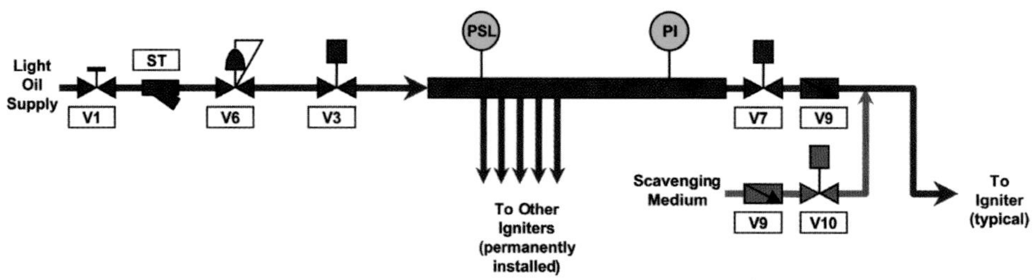

PI = Pressure gauge
PSL = Low pressure interlock
ST = Cleaner or strainer
V1 = Manual shutoff valve
V3 = Igniter safety shutoff valve
V6 = Igniter flow control valve
V7 = Individual igniter safety shutoff valve
V9 = Check valve
V10 = Scavenging valve

FIGURE 18.26 Typical pilot oil train: single element.

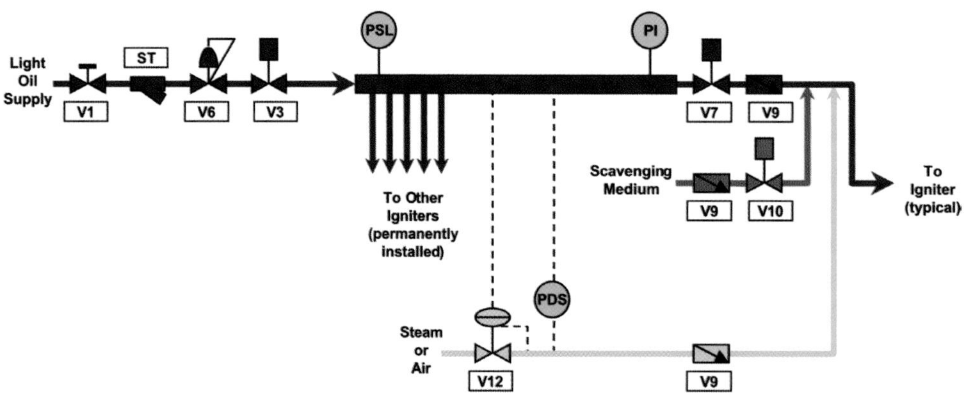

PI = Pressure gauge
PDS = Differential pressure alarm and trip interlock
PSL = Low pressure interlock
ST = Cleaner or strainer

V1 = Manual shutoff valve
V3 = Igniter safety shutoff valve
V6 = Igniter flow control valve
V7 = Individual igniter safety shutoff valve
V9 = Check valve
V10 = Scavenging valve
V12 = Differential pressure control valve

FIGURE 18.27 Typical pilot oil train: multiple elements.

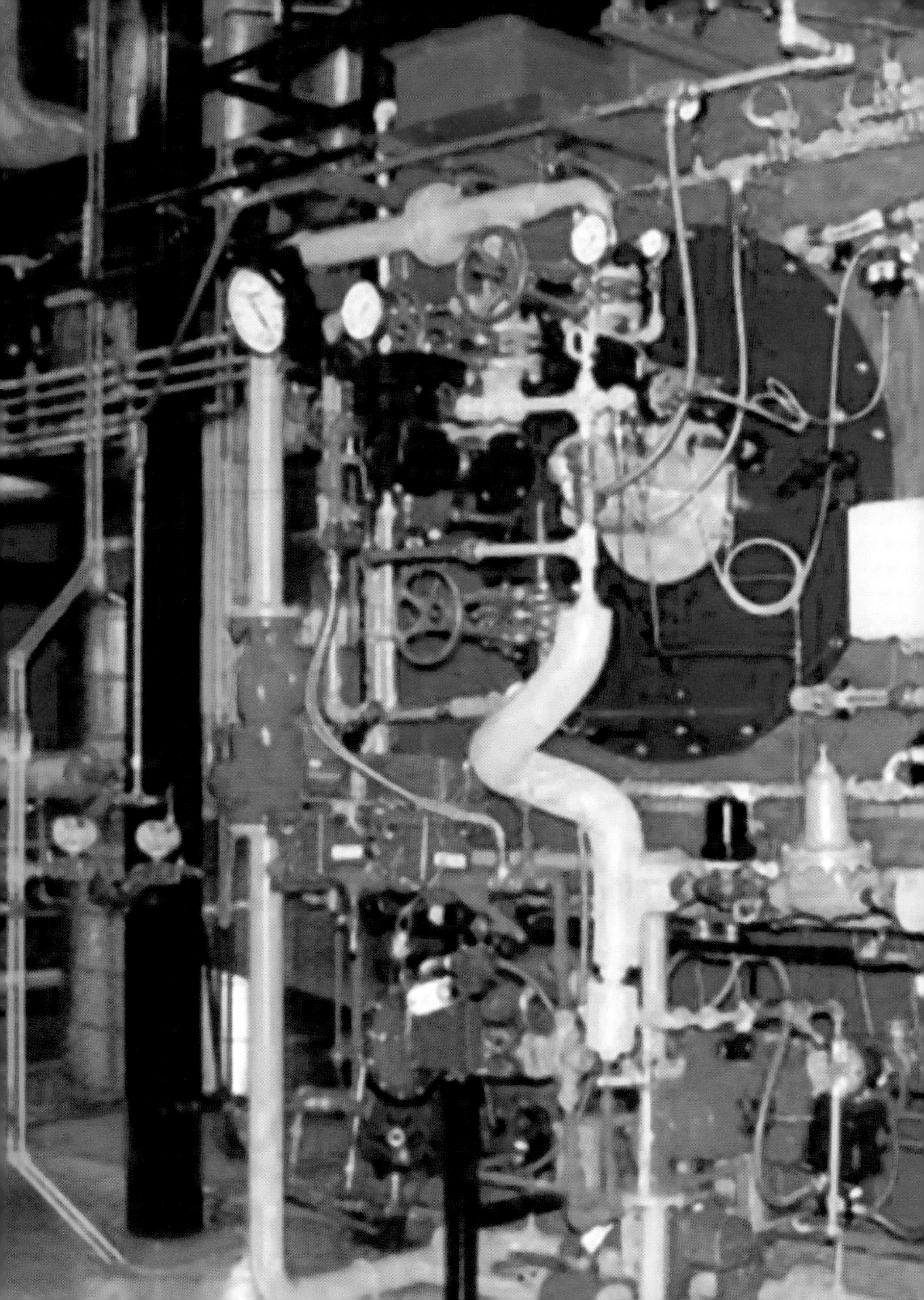

Chapter 19
Boiler Burners

Lev Tsirulnikov, John Guarco, and Timothy Webster

TABLE OF CONTENTS

19.1 Boiler-Specific Burner Requirements 548

- 19.1.1 Conventional Burner Technology for Boilers 548
- 19.1.2 Low-NOx Burner Technology for Boilers 550
- 19.1.3 Staged Burner Design Philosophy 551
- 19.1.4 Design Features of Low NOx Burners 552
- 19.1.5 Effects of Burner Retrofits on Boiler Performance 553

19.2 Boiler Design Impacts on NOx Emissions Correlations 553

- 19.2.1 Boiler Design 553
- 19.2.2 Excess Air 555
- 19.2.3 Boiler Load Influence on NOx 558
- 19.2.4 Boiler/System Condition Impacts on Combustion and NOx Formation 561

19.3 Current State-of-the-Art Concepts for Multi-Burner Boilers 563

- 19.3.1 Combustion Optimization 563
- 19.3.2 Methods to Reduce NOx Emissions 568

19.4 Low-NOx burners for Packaged Industrial Boilers 575

19.5 Ultra-low Emission Gas Burners 580

- 19.5.1 Background 580
- 19.5.2 Implementation of the NOx Formation Theory for Ultra-low-NOx Emissions 580
- 19.5.3 Ultra-low Emissions Burner Design 581
- 19.5.4 Ultra-low Emissions Burner Operation 582

19.6 Atomizers for Boiler Burners 584

19.1 BOILER-SPECIFIC BURNER REQUIREMENTS

The requirements for furnaces and burners of gas/oil-fired utilities and industrial boilers are based on long-term operational experience and comprehensive testing of different new designs and retrofits of boiler equipment. These requirements are changing with the times. Figure 19.1 shows typical utility boilers, and a typical single-burner industrial boiler is shown in Figure 19.2.

Prior to the establishment of NOx emission rate control requirements in the United States and other countries, performance data for furnace and burner design of utility and industrial boilers usually had to match the following common accepted requirements:

1. high reliability during long-term operation
2. simplicity and reliability of gas and oil fuel ignition
3. high flame stability while firing either gas or oil fuels, even at full turndown
4. provision for designed superheated and reheated steam temperatures while firing either gas or oil fuels at full turndown during long-term operation
5. high thermal efficiency and low concentrations of incomplete combustion products (ICPs) while firing either gas or oil fuels using comparatively low excess air, even at full turndown
6. low combustion air system pressure drop, especially the burner register draft loss
7. simplicity of burner/windbox maintenance and adjustment
8. ease of automatic mode operation and fuel changeover between gas and oil
9. provision of restricted flame dimensions to match the dimensions of an existing or newly designed furnace
10. allowing operation with no flame impingement on either the target or side walls of the furnace that promoted reliability of all high-temperature heat exchange surfaces — especially water-wall, superheater, and reheater tubes

To match this requirement for any boiler application, it was necessary to carefully review how a retrofit burner design should be implemented for each particular retrofit furnace. Prior to the establishment of NOx emission rate control requirements, burner performance was typically evaluated as a function of ICP concentration levels at low excess air conditions. While firing either oil or gas fuels in preheated air, optimal low excess air burner designs provided carbon monoxide (CO), hydrogen (H_2), and unburned hydrocarbon (UHC) concentrations of less than 200 ppm at excess oxygen (O_2) levels ranging between 0.2 and 0.6% at full load, as well as O_2 < 1% over a 3:1 turndown range. In many European countries (e.g., Germany, France, Italy, and Belgium), the low excess air gas/oil burners acceptable condition was considered to be CO < 1000 ppm and opacity < 6%. The opacity numbers were measured using the Ringelman or Baharac methods.

In some countries, the burner performance was evaluated specific to each boiler design. For example, in Russia, all industrial and relatively small utility boilers with steam flow capacities of up to 220,000 lb/hr (100,000 kg/hr) with preheated air were designed with up to six burners, each with rated heat inputs of up to 170×10^6 Btu/hr (50 MW_t). Larger utility boilers having capacities from 440,000 to 1,800,000 lb/hr (200,000 to 820,000 kg/hr) superheated steam flow were designed with up to eight burners, each with rated heat inputs of up to 400×10^6 Btu/hr (120 MW_t). All ICP concentrations in the flue gas, including both CO and H_2, were required to be less than 1000 ppm at O_2 < 1%. To obtain lower ICP values, burner air velocities were increased to at least 160 ft/s (50 m/s), with 300 ft/s (90 m/s) preferred.

Until recently, there were two operational parameters used to regulate the combustion processes in boilers to achieve the required operational characteristics: the variation of fuel and air input, and the regulation of reheated/superheated steam temperature. The many years of operating this way resulted in significant difficulties, including:

1. reduction in reliability due to frequent failures of high-temperature heat exchange surfaces
2. loss of reheated/superheated steam temperature during full turndown operation
3. overfiring of the unit

Experience has shown that controlling the combustion processes to allow maintenance of key operational parameters within a given range, including superheated and reheated steam temperatures, best solved these problems. Combustion control yielded the most efficient and reliable boiler operation, independent of load, fuel type, or other conditions.

19.1.1 Conventional Burner Technology for Boilers

Older generations of burners, such as the register, or swirl burner shown in Figure 19.3, were considered very reliable and were the staple of the industry for many years. They consisted of several main parts, including a diffuser, air doors, a throat ring, and a fuel supply system. They operated on a principle of precisely controlling the swirl of the burner by adjusting the air doors, either open or closed. In an ideal situation, each burner in a multi-burner windbox should receive an equal amount of air mass flow. However, in the real world, the burner-to-burner mass flow distribution varied up to ±30% of the average mass flow. The typical compensation for deviations in mass flow was to close down on the air doors of the burners receiving too much air. This in turn affected the

FIGURE 19.1 Typical utility boilers. (Courtesy of Florida Power & Light)

FIGURE 19.2 Typical single-burner industrial boiler. (Courtesy of North Carolina Baptist Hospital)

FIGURE 19.3 Swirl burner.

swirl of the burner, taking it away from the optimal single-burner performance setting.

Flame length, which depended largely on heat input, was the critical burner design parameter during this time period. Flame impingement on either the target or side walls was not allowed. Figure 19.4 shows the average flame length for natural gas and No. 6 oil as a function of burner heat input in an experimental furnace. Other parameters include preheated air of ~570°F (300°C) and $O_2 < 1\%$. The oil was mechanically atomized at ~280 to 300 psig (19 to 20 barg) and at 248 to 266°F (120 to 130°C).

Conventional gas and gas/oil burners usually produce flames with fixed parameters. However, the combustion processes can by modified by varying the aerodynamic and chemical structure of flows fed into the furnace via burners. In this case, in addition to the typical requirements listed above, the burner design should provide the following possibilities.

1. the ability to change the direction of or reverse the fuel/air mixture swirl to cause interaction of the flame vortices; this simplifies the problem of increasing the furnace wall heat absorption rate when switching from oil to gas while retaining the same O_2 in the furnace
2. the ability to change the swirl intensity, flame length, flow density, and static pressure across the furnace cross-section, as well as along the length of the combustion zone
3. the ability to vary the flue gas recirculation (FGR) rates to control the primary firing zone to enable both stable combustion and low emission formation

Regulating not only the aerodynamic properties of the flame, but also the chemical structure and emission properties of the gas and oil combustion products, can further increase the effect of burners on combustion processes. The chemical structure and emission properties of a flame can be varied using controlled fuel input.

19.1.2 Low-NOx Burner Technology for Boilers

The objective of the modern typical boiler burner retrofit is to reduce NOx. Two NOx reduction techniques for boilers are combustion modification and back-end (post-treatment) cleanup (see Chapter 6). Combustion modifications include low-NOx burners, low excess air, over-fired air, flue gas recirculation (FGR), fuel-induced recirculation, reburn, and water tempering. Back-end cleanup techniques include selective catalytic reduction (SCR) and selective non-catalytic reduction (SNCR). This chapter focuses on the combustion modification techniques of low-NOx burners, low excess air, over-fired air, flue gas recirculation, and fuel-induced recirculation.

When the United States and other countries mandated NOx control, in addition to the requirements listed in the previous section, this new requirement became the dictating feature that determined the quality of burner/furnace performance. As a result, many low-NOx burners, designed for a plethora of different boiler design and heat input capacities, firing gas and/or oil with preheated and ambient air, were developed in the United States and many other countries.

The situation is complicated by numerous contradictions between the operational baseline conditions for the existing boiler on the one hand, and the new NOx requirements on the other hand. For example, all NOx reduction combustion methods will increase flame length and redistribute the heat flux between radiant and convective heating surfaces. However, the flame length is restricted by the current furnace dimensions. Also, increasing heat flux in the superheater/reheater surfaces is not allowed because it will degrade the lifetime of these surfaces. Decreasing the total area of these surfaces increases their reliability, but reduces the boiler thermal efficiency.

Another contradiction is related to the desire to get simultaneous NOx reduction and reduction of incomplete combustion products (ICPs). Lowering the O_2 will decrease NOx but increase ICPs. To reduce ICP concentrations, O_2 must be increased and fuel/air mixing must be improved. However, these actions will tend to increase NOx. Solving these problems simultaneously is very difficult for both existing and new design furnaces, especially on utility boilers.

There are other complications on dual-fuel boilers. It is necessary to provide the designed reheated and superheated steam temperatures over long-term operation, while firing fuels generating different flame properties over the full turndown load range. The luminescence of an oil flame is significantly higher

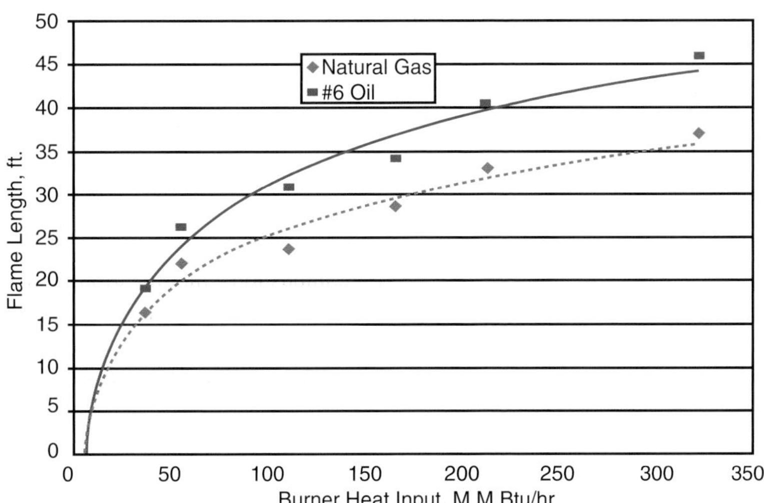

FIGURE 19.4 Average flame length as a function of burner heat input.

than that of a gas flame (see Chapter 3). Assuming all other conditions are the same, furnace heat absorption is less and the furnace exit gas temperature is higher while firing gas. Higher exit gas temperatures create conditions for more intense thermal NOx generation in all industrial boilers and most utility applications for 50 to 150 MW_e power units. On larger capacity utility boilers, the NOx is higher while firing gas than while firing oil. A more detailed explanation will be given in Section 19.3.2.3. Typically, the NOx while firing oil is higher due to fuel-bound nitrogen (N_f). Fuel-bound nitrogen usually ranges from 0.2 to 0.45% wt, but can be as high as 0.7 to 1% in No. 6 oil. Because the effectiveness of various NOx reduction methods is different with gas and oil firing, it makes sense — in principle — to implement different NOx reduction methods for each fuel. However, doing so would complicate boiler operations so much that such designs would not be acceptable to the customer. Attempts to implement NOx reduction methods for both fuels create situations where different methods are selected for the same boiler design installed at different power plants. The method chosen depends on many operational parameters, including annual consumption and seasonal distribution of each fuel, local climate, annual average load, levels and frequency of load peaks, and stack height.

Today's low-NOx burner relies on control of the combustion air in several component streams, as well as the controlled injection of fuel into the air streams at selected points for maintaining stable, attached flames with low NOx generation. Typical venturi-style, low-NOx burners are shown in Figures 19.5 and 19.6. Primary and secondary air enters the burner radially through the venturi and exits the burner axially with a primary air swirl defined by the fixed blade axial swirler. The swirler determines the size and strength of the recirculation zone. A tertiary air stream flows between the venturi base and the burner throat quarl. Tertiary air separates some of the combustion air from the main flame, effectively staging combustion and reducing NOx. For natural gas firing, fuel can be introduced through internal or external pokers or gas rings, and can also be injected through a central gas pipe with multiple orifices at the furnace end. A single conventional atomized burner (oil gun), located along the burner centerline, typically supplies the oil. The oil gun may use dual fluid, mechanical, or rotary cup atomization.

19.1.3 Staged Burner Design Philosophy

Most of today's low-NOx burner designs implement staged combustion principles as an effective way to reduce NOx. The staged burner should also be designed to provide the maximum degree of flexibility in achieving high burner turndown, low NOx, and improved flame shaping capability. The design basis of a staged burner is to develop a stratified flame structure with specific sections of the flame operating fuel-rich and other sections operating fuel-lean. The burner design thus provides for the internal staging of the flame to achieve NOx reductions while maintaining a stable flame.

Controlling combustion stoichiometry to fuel-rich conditions inhibits NOx production, especially in the burner's flame front. Operating the flame fuel-rich also reduces the burner NOx dependence on the burner zone heat release (BZHR) rate, which is discussed in Section 19.2.1. This is especially important in applications where very high BZHR near full load can result in an exponential increase in NOx.

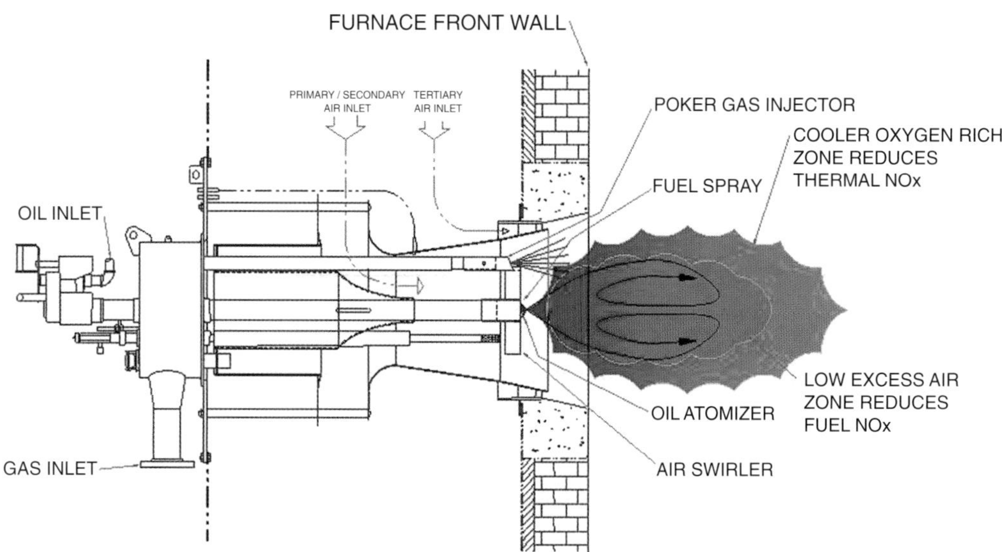

FIGURE 19.5 A typical low-NOx burner, venturi-style.

FIGURE 19.6 A typical-low NOx burner, venturi-style (second example).

In addition to controlling NOx formation, operating under fuel-rich conditions aids in the production of combustion intermediates that can result in the destruction of previously formed NOx. In a reducing environment, NO can act as an oxidizer to react with these combustion intermediates, resulting in the reduction of NO to N_2. Therefore, the NO necessarily formed to satisfy the requirements of establishing a strong flame front can be scavenged by the process and reduced to N_2.

To achieve complete fuel burnout at minimum O_2, the burner design must provide direct interaction of fuel-lean zones with the center fuel-rich sections. This ensures that the "rich" products of combustion from the center flame pass through the oxidizing zone for complete fuel burnout. The burner design allows control of the stoichiometry of the oxidizing zone to range from being pure air to having varying degrees of excess oxygen. This controls NOx formation by causing the fuel burnout to occur in the form of a premixed flame rather than a diffusion flame. The operational flexibility is provided in the burner design to control stoichiometry to achieve minimum NOx.

Flame stability should also be a design focus for a staged burner. The stability criterion is to maintain a minimum turndown capability on a gas-firing ratio of 8:1 over a broad range of excess O_2. To achieve this objective, it is necessary to carefully select the burner throat velocity and the velocity profile.

The burner flame front should be a fuel-rich premixed flame designed to increase flame speed and, in turn, flame stability. Combustion stoichiometry, approaching a perfect premixed flame, is established at the burner flame front. This condition broadens the flammability range of the fuel and ensures the maintenance of a flammable mixture over a broad range of burner throat velocities and fuel injection rates.

Controlling air distribution within the burner throat improves flame-shaping capability. Provisions are made in the burner design to allow the burner to operate as a low excess air burner as well as a low-NOx burner with an extended flame envelope.

19.1.4 Design Features of Low-NOx Burners

Some of the standard design features of low-NOx burners are:

1. flame stability at low excess air rates for reliable, energy efficient boiler operation
2. high turndown ratios for a wide range of boiler operations

3. well-distributed air flow to control the flame envelope and provide even heat flux
4. known flame length and diameter to suit furnace firing lane without impinging on boiler tubes or furnace walls
5. elimination of combustion-induced vibrations of the boiler, due to the aerodynamics of the register design and turbine blade diffuser
6. a strong flame front established within a maximum of 0.5 diffuser diameters of the face of the diffuser, as depicted in Figure 19.7 (the flame front should not move during changes in the firing rate, thus providing a stable flame for scanning, and resulting in reliable operation)
7. for gaseous fuel firing, a multiple-poker gas burner assembly with "poker shoes" (gas nozzles) oriented to provide internal fuel staging

19.1.5 Effects of Burner Retrofits on Boiler Performance

Low-NOx burners have different flame characteristics from their predecessors, the low excess air burners. Therefore, there exists the probability that the differences will affect boiler performance. Performance changes would only occur by an alteration of the heat absorption pattern in the furnace, thus affecting the amount of heat going to the boiler backpass. The overall furnace heat absorption could either increase or decrease, due to factors such as flame distance to walls, flame temperature, flame emissivity, and characteristics of ash deposits on furnace walls. The impact on boiler performance will vary by unit and by fuel. In some cases, there will be no impact; but in most cases, there will be either a positive or negative impact experienced. Historically, however, there has been no net effect on boiler performance that would be considered extreme. Atomizer tip design can drastically impact boiler performance, but total boiler performance is always a balancing act between key variables such as NOx and water wall temperatures.

To assess the impact on boiler performance, certain information is typically monitored during the testing of low-NOx burners. Important data includes superheated and reheated steam temperatures, superheater and reheater tube surface temperatures, and any operational parameters that affect or control steam temperatures. The boiler performance variation resulting from a change in the overall O_2 level is not considered in this analysis. The actual impacts on boiler efficiencies are relatively small, and there are many complicating factors such as furnace and heat recovery area cleanliness, fuel composition, sootblower availability, operational variability, etc.

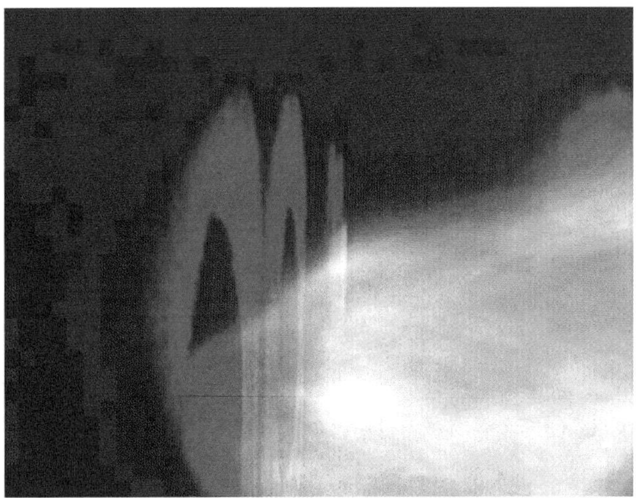

FIGURE 19.7 A strong flame front established within a maximum of 0.5 diffuser diameters of the face of the diffuser.

19.2 BOILER DESIGN IMPACTS ON NOx EMISSIONS CORRELATIONS

The two major mechanisms for the formation of NOx are: (1) NOx formed by the oxidation of N_f (fuel NOx) and (2) the thermal fixation of atmospheric nitrogen (thermal NOx). Separate NOx correlations have been developed for each. These correlations can be combined to predict the total NOx emissions for a selected burner design based on the fuel nitrogen content and boiler design.

The conversion of N_f to NOx is dependent on oxygen availability and N_f content. Thus, for a given burner system at constant excess air, the key variable controlling fuel NOx formation is N_f content. Previous studies have shown that N_f conversion efficiency is inversely proportional to the nitrogen content of the fuel. High conversion efficiencies are observed with low N_f, and low efficiencies are seen with high N_f.

19.2.1 Boiler Design

Based on the strong dependency of thermal NOx on flame zone temperature, thermal NOx formation for wall-fired boilers has been correlated with the ratio of heat input to furnace size and the number of firing walls. For this correlation, the ratio of heat input to the furnace burner zone area is defined as the burner zone heat release (BZHR) rate. The BZHR represents the boiler heat release rate divided by the water-cooled surface area in the burner zone (10^6 Btu/hr-ft^2), and is a measure of the "bulk furnace temperature." The NOx created within this zone is dependent on this bulk temperature by an exponential relationship, as discussed in previous chapters.

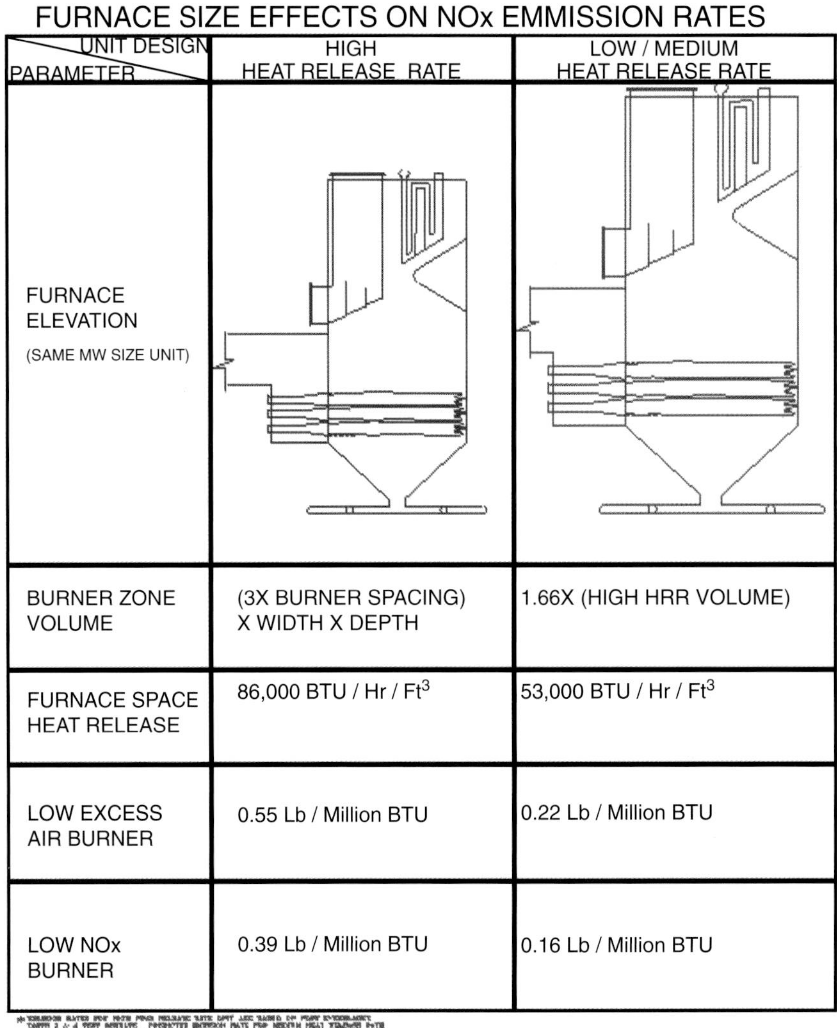

FURNACE SIZE EFFECTS ON NOx EMMISSION RATES

UNIT DESIGN / PARAMETER	HIGH HEAT RELEASE RATE	LOW / MEDIUM HEAT RELEASE RATE
FURNACE ELEVATION (SAME MW SIZE UNIT)		
BURNER ZONE VOLUME	(3X BURNER SPACING) X WIDTH X DEPTH	1.66X (HIGH HRR VOLUME)
FURNACE SPACE HEAT RELEASE	86,000 BTU / Hr / Ft^3	53,000 BTU / Hr / Ft^3
LOW EXCESS AIR BURNER	0.55 Lb / Million BTU	0.22 Lb / Million BTU
LOW NOx BURNER	0.39 Lb / Million BTU	0.16 Lb / Million BTU

FIGURE 19.8 The effects of boiler design on NOx.

The burner zone is defined as the six-sided surface bounded by the furnace walls and imaginary horizontal planes located one burner row spacing above the top row and below the bottom row of burners. The area of any division walls within this volume is also included in the calculation of the BZHR. A correction is made for re-radiation from any refractory on the floor, walls, or the burner throats. An example of the effects of boiler design on NOx is shown in Figure 19.8, where the same heat input is placed in two drastically differently sized boilers.

The correlation of thermal NOx with BZHR rate has been developed using an extensive database of gas and oil fired utility boilers. Figure 19.9 shows the NOx of various boilers included in the database on oil and gas, respectively. NOx of industrial boilers with comparable degrees of NOx control were found to be consistent with the BZHR rate correlations.

The correlations of N_f conversion and thermal NOx can be used to extrapolate burner NOx from one boiler to another. The increase in NOx observed with fuel oil compared to natural gas, which contains no nitrogen, is used to estimate fuel NOx for the selected burner. N_f conversion is then calculated based on the nitrogen content. The correlation of N_f conversion can then be used to project fuel NOx for fuels containing differing amounts of nitrogen. Thermal NOx is determined from NOx with natural gas. The correlation with BZHR is used to project the measured thermal NOx for the selected burner from one boiler to another. Total projected NOx emissions for the burner can then be determined by adding the fuel NOx and thermal NOx contributions for the fuel and boiler of interest. The amount of air preheat can also have a dramatic effect on NOx, CO, and particulate emission rates, as well as on flame stability.

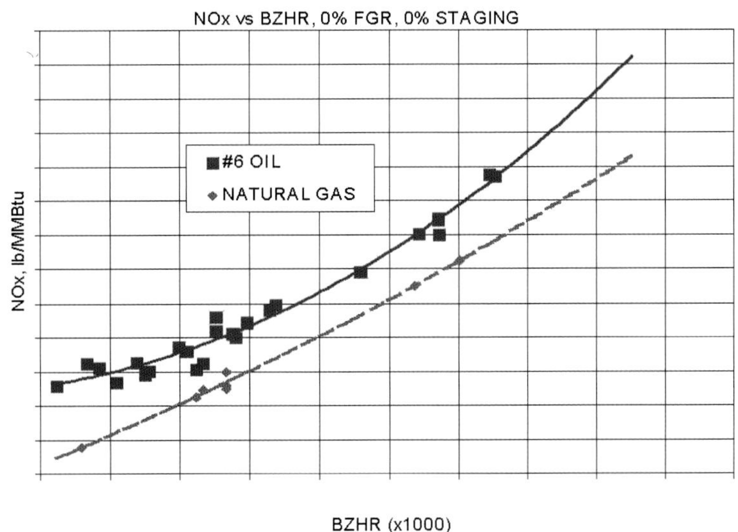

FIGURE 19.9 The NOx vs. BZHR of various boilers included in the database on oil and gas, respectively.

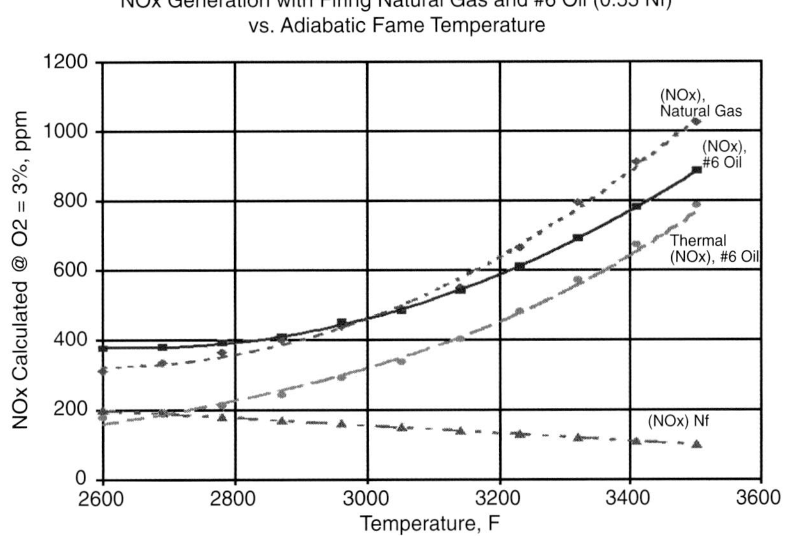

FIGURE 19.10 NOx generation with firing natural gas and No. 6 oil (0.5% N_f) vs. adiabatic flame temperature.

19.2.2 Excess Air

19.2.2.1 Theoretical Effect of Excess Air on NOx

It is well-known that thermal NOx formation primarily depends on the flame temperature, excess air in the flame, and residence time. However, the flame temperature is also dependent on excess air. As shown in Figure 19.11, demonstrating the experimental data for natural gas combustion under regular boiler conditions, with an excess air factor α value close to unity, increased excess air causes NOx to rise considerably; then, as excess air is further increased, the increase in NOx slows down, reaches a maximum, and then NOx is reduced as excess air is further increased.

According to some experimental studies, no maximum is attained and the NOx = f(α) dependence approaches the stabilized section of the exponential curve. A difference between the two mentioned contradicting results brings different approaches to optimized low-NOx combustion evaluations. That is why it is important to discuss the existence of the NOx maximum and to consider it in some detail. For simplicity, one can consider a natural gas (containing no bound

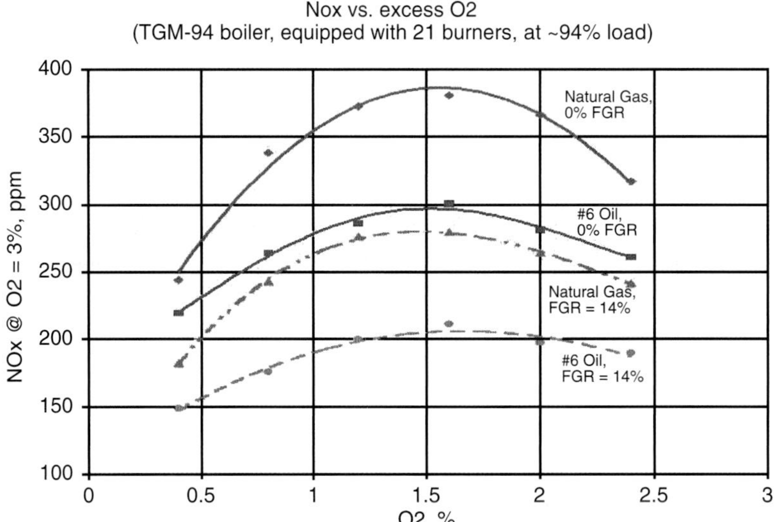

FIGURE 19.11 Excess O_2 influence on NOx formation.

nitrogen) combustion process, as a result of which NOx is formed only from nitrogen in the combustion air. It can be seen in Figures 19.10 and 19.11 that with a deficiency of oxidant, NOx depends more on α than on temperature; and when $\alpha > 1$ (mostly when $\alpha > 1.2$), when the rate of combustion reactions increases, the effect of temperature proves to be the dominant one. Therefore, despite the fact that the theoretical combustion temperature in the region of $\alpha > 1$ systematically decreases with increased α, the NOx increases as long as the thermal NOx formation is not made more difficult due to a considerable decrease in the temperature level. In this case, there is also a decrease in the NOx in the flue gas volume due to dilution by the excess air. When firing a fuel oil that contains N_f, it is unlikely that the nature of the NOx vs. O_2 curve considered above will change; but in all probability, the downward branch of the curve will be flatter.

When comparing the well-known empirical dependency of the ICP component concentrations on α with the empirical NOx data as a function of α, confirmed with the theoretically calculated dependencies on α shown in Figure 19.10, it was found that the excess air factor α value at which furnace ICP losses virtually disappear almost coincides with the values of α corresponding to the NOx maximum. It follows that the NOx dependence on excess air has the form of an extreme function with a maximum NOx value corresponding to that α value at which virtually complete fuel combustion is attained under the given conditions. Hence, if NOx and α are determined by reliable and sufficiently accurate methods, the absence of an experimentally established maximum on the test curve NOx vs. O_2 (or α) leads to an assumption that the fuel is not completely burned, indicating the need to not only improve the combustion process but also to check the validity of the methods used to measure ICP and NOx in the given α range.

19.2.2.2 Empirical Evidence of the Effect of Excess Air on NOx

Detailed empirical data sets were obtained under identical operational conditions at approximately full (~94%) load while firing natural gas and No. 6 oil (containing 0.22 to 0.32% N_f) with preheated air on the two neighboring 165-MW_e utility boilers of the TGM-94 model (~1,100,000 lb/hr or 500,000 kg/hr superheated steam flow), installed at the same power plant. The furnaces were balanced draft and had air in-leakage of 8 to 10%. The two boilers were almost identical except for one difference: the first boiler was equipped with 21 burners, rated at ~100×10^6 Btu/hr (30 MW_t) heat input each, having a single swirled air flow channel, and the second boiler was equipped with nine burners, rated at ~230×10^6 Btu/hr (67 MW_t) heat input each, having two air flow channels, one of which is a swirled portion consisting of ~15% of the total air flow. On both boilers, the burners were installed in three rows on the front wall (3×7 and 3×3, respectively). The boilers were equipped with an FGR system designed to supply up to 14% FGR flow (isolated from the air flow), supplied directly to the furnace through slots located on the target wall, opposite the lower burner row. The test data while firing either or both fuels, both with and without FGR implementation and shown in Figures 19.11 and 19.12 for the 21- and 9-burner

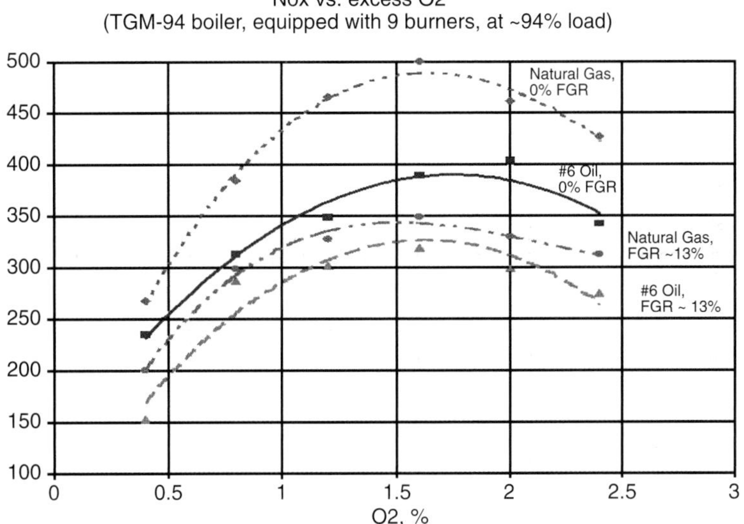

FIGURE 19.12 NOx vs. excess O_2 (the TGM-94 boiler equipped with 9 burners, at ~94% load).

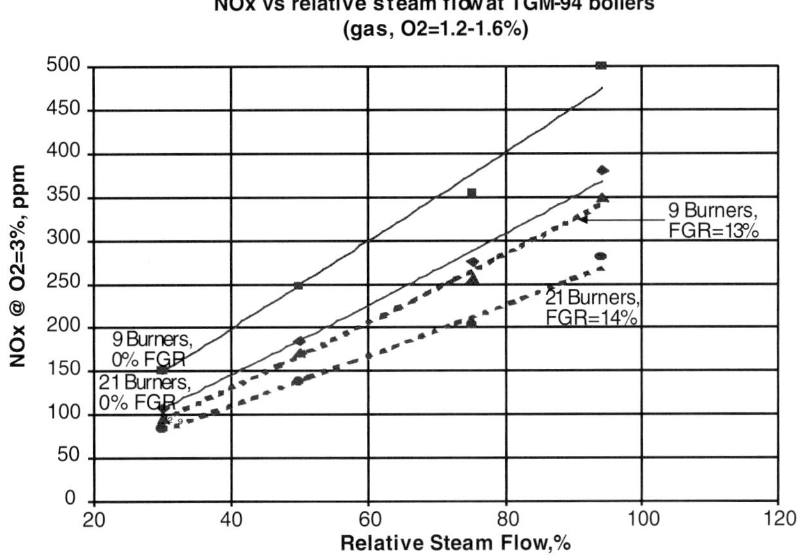

FIGURE 19.13 NOx vs. relative steam flow at the TGM-94 boilers (natural gas, O_2 = 1.2–1.6%).

boilers confirm the well-known dependencies of NOx concentrations on O_2 and load, respectively.

Differences can be seen between the maximum NOx numbers, and, accordingly, between shapes of the curves, obtained on these two boilers. These discrepancies are associated with distinctions in the combustion processes related to the differences between the mentioned burner designs, heat inputs, and numbers of burners. The measured flame temperatures while firing gas under full load with all other conditions being equal show that on the boiler equipped with nine burners, the maximum flame temperature is ~160°F (70°C) higher, resulting in a significant NOx difference of 100 to 120 ppm (NOx is ~20 to 24% higher than for the boiler equipped with nine burners). The increased temperature resulted in a stronger dependency of NOx on O_2 on the boiler equipped with nine burners, both with and without FGR, as demonstrated in Figure 19.12.

The above data, as well as other test data obtained from 150 to 800 MW_e utility boilers, under similar operational conditions, illustrate a significantly lower NOx level on No. 6 oil (even while containing N_f of up to ~0.7%) in comparison with NOx numbers on gas. An explanation of this fact contradicting the empirical data obtained on industrial and

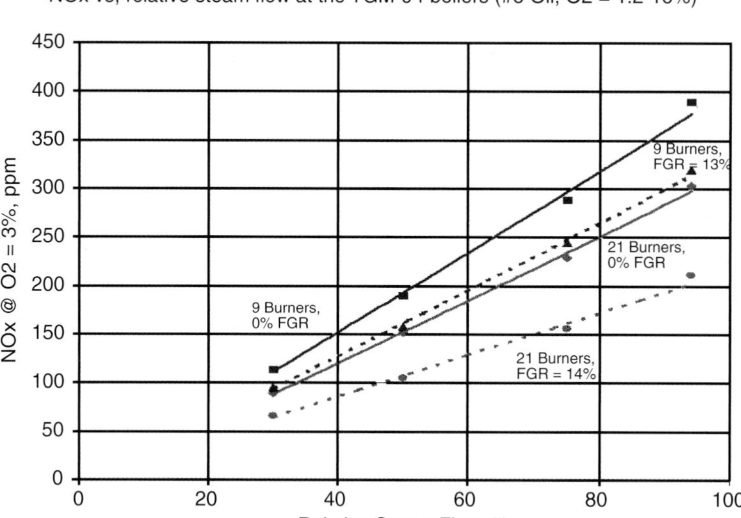

FIGURE 19.14 NOx vs. relative steam flow at the TGM-94 boilers (No. 6 oil, O_2 = 1.2–1.6%).

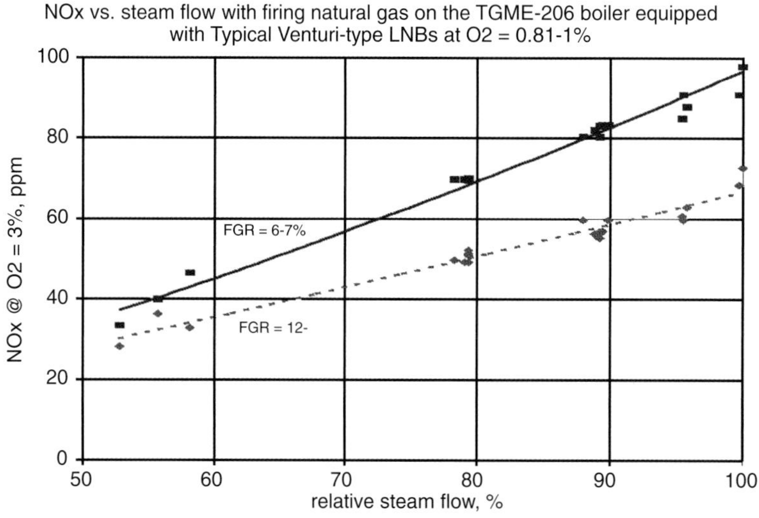

FIGURE 19.15 NOx vs. relative steam flow with firing natural gas on the TGME-206 boiler equipped with Todd Combustion low-NOx Dynaswirl burners at O_2 = 0.8–1%.

comparatively smaller utility boilers (where NOx numbers are higher while firing oil) is presented in Section 19.3.2.3.

19.2.3 Boiler Load Influence on NOx

The relationship between NOx and load was investigated on many utility boilers over their full load ranges, while firing either or both fuels. A detailed investigation was performed on the two TGM-94 boilers described above (see Section 19.2.2). The test results obtained at ~94, 75, 50, and ~30% loads, with and without FGR, are shown in Figures 19.13 (gas) and 19.14 (No. 6 oil). With O_2 = 1.2–1.6%, over the entire load range, CO concentrations were in the 50- to 150-ppm range while firing gas. When firing No. 6 oil on both boilers, over the entire load range and under the same O_2, CO and opacity did not exceed 100 ppm and 10%, respectively.

Figure 19.15 shows the data of combined (load and FGR) influence on NOx on a 200-MW_e boiler of the TGME-206 model, having no furnace air in-leakage, equipped with 12 low-NOx venturi-style burners installed in two rows on the rear wall. The boiler was tested at loads ranging from 100 to 53% while firing natural gas at comparatively low O_2 (0.8 to 1%) and with 6 to 7% and 12 to 14% FGR flows. The

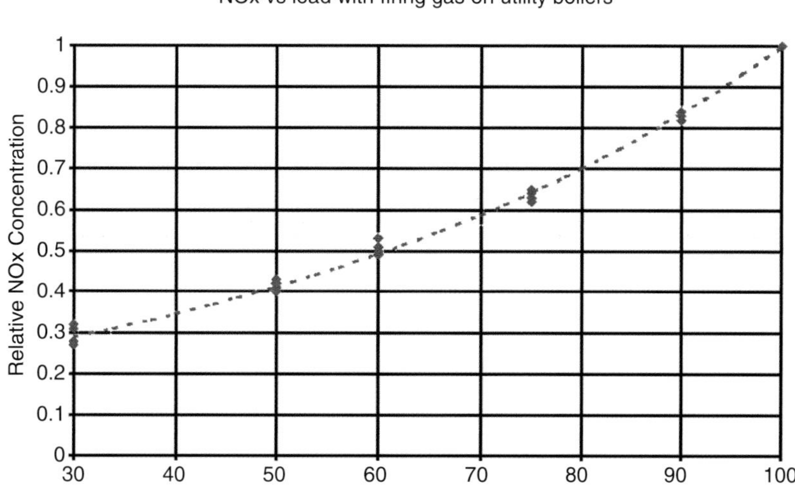

FIGURE 19.16 NOx vs. load with firing natural gas on utility burners.

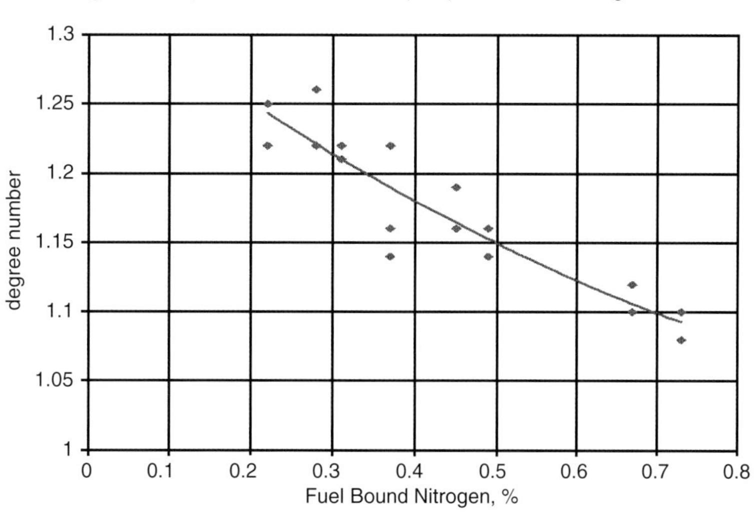

FIGURE 19.17 Degree of the power function NOx = f (load) vs. bounded nitrogen in No. 6 oil.

FGR was mixed with the combustion air upstream of the windbox. Under all above conditions, CO concentrations were less than 20 ppm.

Confirmation of these empirical dependencies on many utility boilers, equipped with various numbers, designs, and arrangements of burners, while firing gas and oil with preheated air (500 to 700°F, or 260 to 370°C), is shown in Figures 19.13 to 19.15. All available data obtained on 150- to 800-MW_e boilers, firing natural gas in the load range of 100 to 30%, have been generalized in Figure 19.16 as a dependence of relative NOx (defined as a ratio of NOx numbers at current loads to the NOx number at full load) on relative load. It is seen that relative NOx is a power function of relative load to the ~1.25 degree. While firing natural gas, significant changes in the O_2 range, FGR flow level, preheated air and FGR temperatures, and other operational conditions provide comparatively small derivations from the mentioned average degree number.

With firing No. 6 oil under similar operational conditions on utility boilers, in general, the power function degree number depends on N_f concentration in the oil: the higher the N_f, the lower the degree number. Corresponding test data, obtained at 17 boilers of 150 (three), 165 (eight), 210 (one),

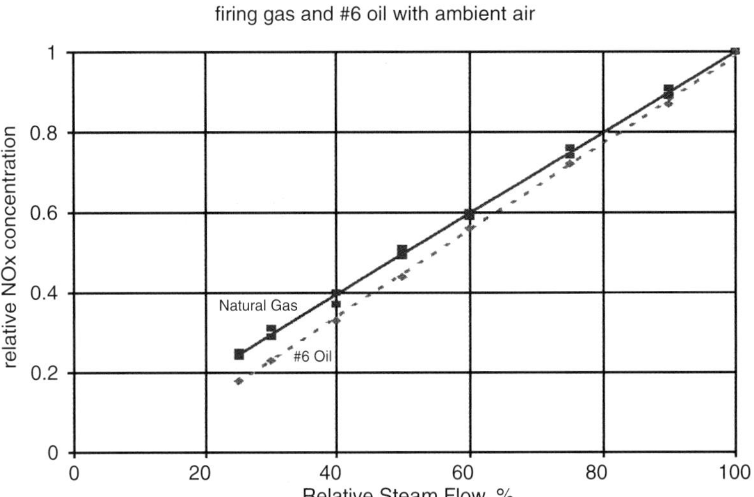

FIGURE 19.18 Relative NOx vs. relative load on industrial boilers firing natural gas and No. 6 oil with ambient air.

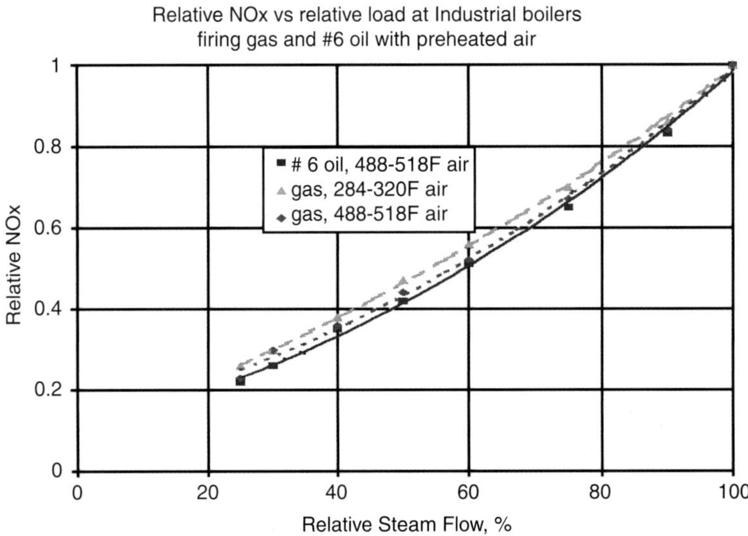

FIGURE 19.19 Relative NOx vs. relative load on industrial boilers firing natural gas and No. 6 oil with preheated air.

and 300 (five) MW_e utility boilers at full load with an O_2 range of 0.6 to 1.2%, are shown in Figure 19.17. With the exception of one 150-MW_e boiler, all boilers were equipped with FGR systems of various design. Increasing FGR flow from 0 to 14% (on the 165- and 210-MW_e boilers) and to 18% (on the other boilers) influences deviations from the average test data curve but does not change the established dependence.

Similar dependencies have also been established on industrial single-burner and multi-burner boilers, while firing either gas and/or oil, with ambient and preheated air, and without staged combustion. With ambient air, the shapes of the experimental curves, established on the 22,000 to 385,000 lb/hr (10,000 to 175,000 kg/hr) steam flow boilers, with firing both gas and No. 6 oil (0.22 to 0.49% N_f) in the 100 to 25% load range, are similar to a linear dependence as indicated in Figure 19.18. On preheated air (300 to 500°F [150 to 260°C]), all curves are power functions with the ~1.25 and (1.1–1.2) degree numbers on gas and No. 6 oil, respectively, as in Figure 19.19.

A clear conclusion was made based on the above empirical test data: as load decreases (i.e., as the flame temperature is reduced due to BZHR), the NOx level also decreases. A reduction in the original NOx level (full load) limits the potential opportunity to achieve a required NOx reduction

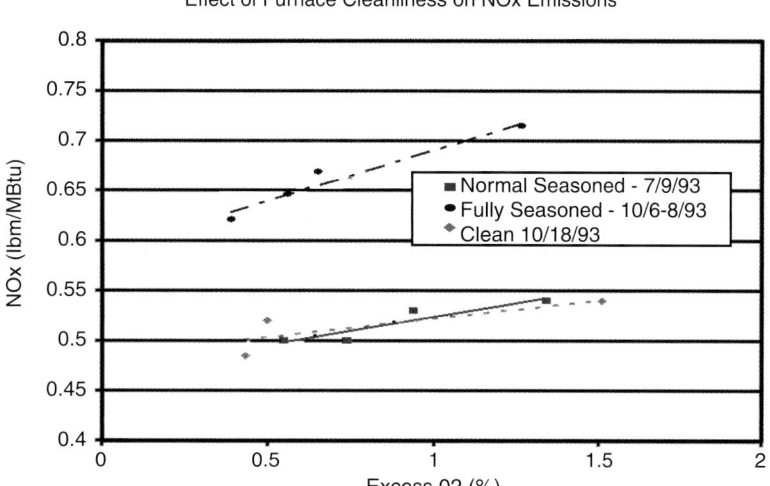

FIGURE 19.20 Effect of furnace cleanliness on NOx emissions.

at lower loads, resulting in a reduction in the effectiveness of all NOx reduction methods. This conclusion was made based on evaluations of data obtained on utility boilers, while using all known NOx reduction methods. For example, on the TGME-206 boiler with 12 venturi burners, ~14% FGR implementation provides ~70 and ~56% NOx reduction at 100 and 80% loads, respectively. For industrial boilers with ambient air, this effectiveness reduction is a little more pronounced, due to the significantly lower flame temperature level in industrial boiler furnaces.

19.2.4 Boiler/System Condition Impacts on Combustion and NOx Formation

A new philosophy for NOx reduction is emerging among multi-burner power boiler users. This philosophy views the entire boiler as a combustion system where NOx reduction can be accomplished and maintained in a three-step process:

1. installation of new, low-NOx burners to deliver the required NOx reduction
2. improved maintenance protocols to sustain the achieved NOx reductions
3. optimized operational protocols to achieve the best overall combustion and boiler performance

Boiler design parameters such as air preheater (APH) design, superheater (SH) and reheater (RH) configurations, and the amount of refractory in the furnace also have major impacts on the amount of NOx reduction achievable on a given boiler.

The cornerstone of this NOx reduction philosophy is still — as it has been for years — the low-NOx burner. Proven low-NOx burners, along with a physical windbox model, provide the required NOx reduction; but sustaining this NOx reduction requires the improved maintenance and optimized operational protocols. With a balanced air flow distribution, balanced fuel flow distribution, and a set of proven low-NOx burners, the combustion system's foundation is established for the desired NOx reduction.

However, the NOx reduction provided by the low-NOx burner is only the first part of the process. The boiler must be properly maintained to sustain the NOx reduction. Proper maintenance must address boiler cleanliness, the percentage of air in-leakage, and oil heater maintenance. The maintenance protocols affecting boiler cleanliness include proper soot blowing practices on a daily basis and boiler washes during outages.

19.2.4.1 Boiler Cleanliness

The impact of boiler cleanliness on NOx became very apparent when the results of NOx testing while firing residual fuel oil, on the same unit, showed a dramatic 23% increase in NOx as compared to results of NOx testing performed just months earlier, with the same atomizer and at the same unit conditions. A visual inspection of the boiler indicated that it was well-seasoned (or dirty). A waterwall wash was performed during the next available outage, about a week later. After the outage, another NOx test was performed, and the results of this testing showed a 25% decrease in NOx as compared to the pre-outage results. The effect of furnace cleanliness on NOx is shown in Figure 19.20.

The second indication of the impact of boiler cleanliness on NOx came when a sister unit started up after the low-NOx burner retrofit. The startup testing measured residual fuel oil

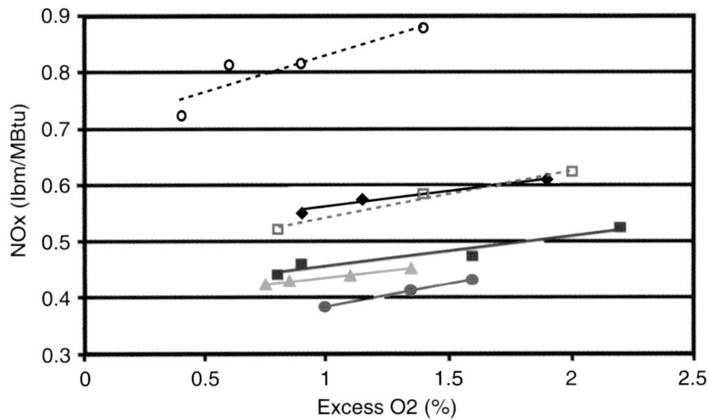

FIGURE 19.21 Effect of HRA cleanliness on NOx emissions.

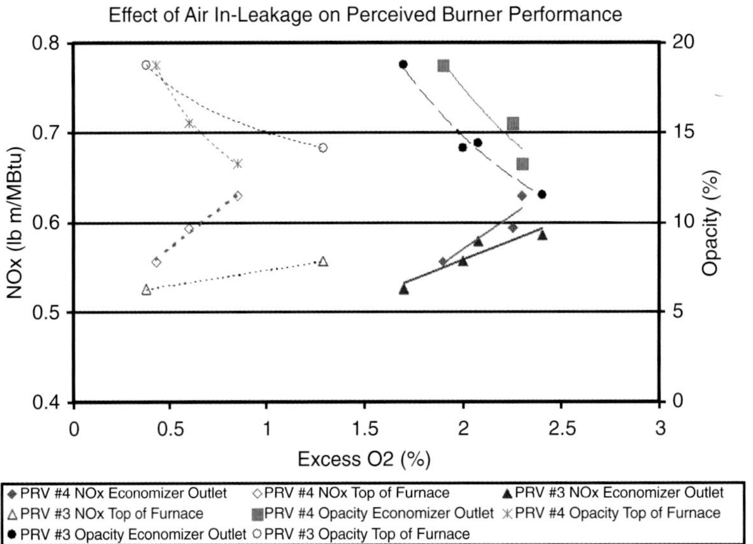

FIGURE 19.22 Effect of air in-leakage on the burner performance.

(RFO) NOx higher than any that had been previously measured with identical low-NOx burners and atomizers at the first unit.

As the data sets from the units were analyzed, it became apparent that the relative level of furnace cleanliness could be correlated to the flue gas temperature entering the air preheater (APHGIT). It was also found that the second unit had an extremely high APHGIT. The unit's soot-blowing practices were analyzed to determine the level of heat recovery area (HRA) cleanliness, and it was found that many of the soot blowers were out of service. Upon minimizing the number of soot blowers out of service, the RFO NOx on the sister unit fell to a level equal to those measured with identical low-NOx burners and atomizers at the first unit. Figure 19.21 exemplifies the impact of heat recovery area cleanliness on NOx.

19.2.4.2 Furnace Air In-leakage Influence

The impact of air in-leakage on burner performance is a phenomenon of balanced draft boilers. There is a significant difference between NOx generation conditions existing in forced-draft (pressurized) and balanced-draft furnaces. In pressurized furnaces under typical single-stage (unbiased) conditions, all

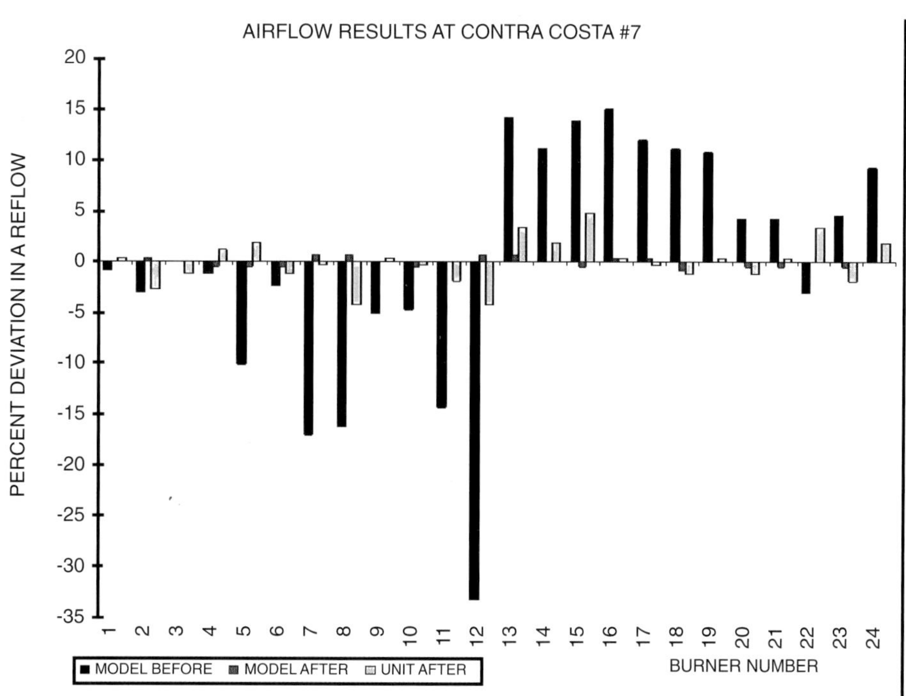

FIGURE 19.23 Improvement of mass flow distribution to burners (differences within ±2%).

the combustion air flow enters the furnace through the burners, and the O_2 required for complete combustion can be minimized. In balanced-draft furnaces, there is usually an air in-leakage of at least 3 to 5% (it sometimes exceeds 10%), sometimes leading to sub-stoichiometric combustion air flow coming through the burners. In-leakage air participates in the combustion process as well, providing complete combustion at slightly higher O_2 levels for many applications.

If the in-leakage location is within the furnace, the air deficit in the burners provides a self-stage combustion that complicates ICP burning out on one hand, but can reduce NOx on the other hand (again depending on the in-leakage location). With other conditions remaining constant in pressurized and balanced furnaces, an NOx reduction of up to 15 to 18% is available. This was established by testing three sets of identical utility boilers: (1) 210-MW_e boilers with 12 burners on the rear wall, (2) 300-MW_e opposed fired boilers with 16 burners, and (3) 150-MW_e opposed boilers with six burners. A comparison was made of NOx data measured in these similar utility boiler sets, one operating with a balanced draft and one operating with a forced draft, while firing gas at full load and with ~0% FGR flow.

However, if the in-leakage location is downstream of the furnace exit, in-leakage can have detrimental effects on the perceived performance of the burners. O_2 measurements taken at the top of the furnace can be approximately 0.8% lower than the control room O_2 measurements taken at the economizer outlet. In many cases, the mentioned difference between O_2 measurements is higher. This air in-leakage between the furnace exit and the economizer outlet can shift the NOx vs. O_2 and opacity vs. O_2 curves upward along the O_2 axis, as shown in Figure 19.22.

19.2.4.3 Fuel Oil Temperature

Problems with the main fuel oil heaters can reduce the maximum fuel oil temperature, therefore raising the minimum viscosity attainable. The increased viscosity will increase the Sauter mean diameter (see Chapter 8) of the atomized droplets, thus causing higher opacity. The increased droplet size will increase the excess O_2 required for complete combustion. NOx will increase due to the increased excess oxygen levels. The viscosity required by most burner vendors is 80 to 100 SSU.

19.3 CURRENT STATE-OF-THE-ART CONCEPTS FOR MULTI-BURNER BOILERS

19.3.1 Combustion Optimization

19.3.1.1 Windbox Air Flow Modeling

Extensive experience in the application of oil and gas firing equipment to a wide range of boiler designs has led to the conclusion that, especially on multiple burner installations, it

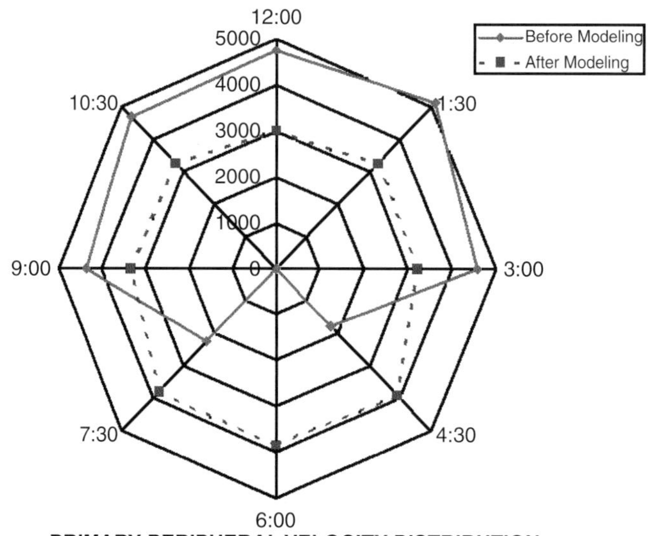

FIGURE 19.24 Improvement of peripheral air flow distribution to burners (deviations ±10%).

is imperative to achieve proper air distribution to each burner in order to control flame shape, flame length, excess air level, and overall combustion efficiency.

Proper air flow distribution consists of an even combustion air mass flow distribution, even burner entrance peripheral flow distribution, the elimination of tangential velocities within each burner, as well as an even combustion air O_2 content by balancing the FGR distribution to each burner. Considering that air in the combustion process accounts for approximately 94% of the mass flow, numerous observations on boiler combustion systems have shown that correct air distribution and peripheral entry condition are key factors in the achievement of high performance (low NOx, low O_2, and low CO). The concept of equal stoichiometry at each burner results in the minimal O_2, NOx, and CO. The most direct way to achieve this is to ensure equal distribution of air and fuel to each burner. Equal air distribution is difficult because it requires a reliable and repeatable flow measuring system in each burner, as well as a means to correct the air flow without disrupting the peripheral inlet distribution or adding swirl to the air flow.

The purpose of each objective relates to a specific burner performance parameter as described below.

To achieve the lowest emissions of NOx, CO, opacity, and particulates, at the minimum excess O_2, equalization of the mass flow of air to each burner is required. Mass flow deviations should be minimized to enable lower post-combustion O_2, CO, and NOx concentrations. The lowest post-combustion O_2 concentration possible is constrained by the burner most starved for air. This starved burner will generate a high CO concentration and, consequently, the total O_2 must be raised to minimize the formation of CO in that burner. By equalizing the air flow to each burner and ensuring that the fuel flow is equal, the O_2 can be lowered until the CO starts to increase equally for all burners. Lower O_2 has additional benefits of lower NOx formation and higher thermal efficiency. The goal is to bring the mass flow differences for each burner (in the model) to within ±2% of the mean, as shown in Figure 19.23.

Flame stability is probably the most important aspect of the model that appeals to the boiler owner. Flame stability is enhanced in the model by controlling two parameters: perimeter air inlet distribution and inlet swirl number (flame stability is primarily controlled in the burner design but must be supported by proper inlet conditions). The equalization of the peripheral air velocity at the burner inlet will result in equal mass flow of air around and through the periphery of the swirler. The flame stabilizer will tend to equalize any remaining flow deviations because of the high velocity developed in this region of the burner throat. The result of this equal air mass flow distribution through and around the swirler will be a fully developed and balanced air vortex at the center of the outlet of the swirler. Flame stability and turndown of the burner depend on the condition of this vortex and its attachment to the swirler. Unequal peripheral inlet velocity distributions result in an asymmetrical vortex, leading to a flame that has poor combustion performance and is more sensitive to operating conditions, turndown may be limited, combustion induced vibrations may be experienced, FGR may cause

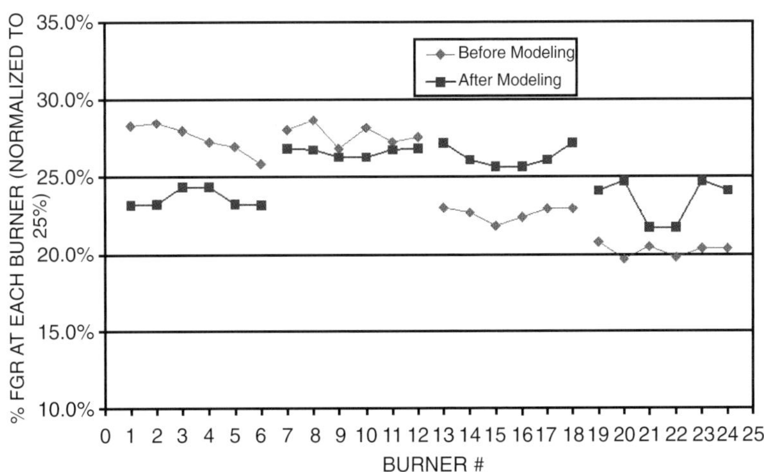

FIGURE 19.25 Improvement of FGR flow distribution to burners.

flame instability at lower loads, light-off by the ignitor may be more difficult, and flame scanning may exhibit increased sensitivity. The goal is to reduce peripheral air flow deviations to ±10%, as shown in Figure 19.24.

The swirl number is an indication of the rotational flow entering the burner. The creation of swirling air is a fundamental requirement of all burners. Louvered burners create this swirl by rotating the entire air mass. Unfortunately, this creates a problem at high turndown rates. At low loads (e.g., 10%), excess O_2 is typically 11 to 13%. By swirling the entire air mass, the fuel is diluted to the point where flame stability becomes marginal. Swirling air entering louvered burners (not created by the burner louvers) can cause differing burner-to-burner register settings to match swirl intensity at each burner. The differing register positions consequently affect the air mass flow at each burner.

An axial flow burner operates on the principle of providing axial air flow through the burner and developing a controlled limited vortex (swirl) of primary air at the face of the smaller centrally located swirler. This concept maintains a stable flame at the core of the burner by limiting dilution at high turndown rates. The secondary air that passes outside the swirler, however, is most effective if it is not swirling (which is the concept behind "axial flow" burners). Swirling secondary air increases the dilution of the fuel and limits turndown. The goal of both the louvered burner and the axial flow burner is to eliminate any tangential velocities entering the burner.

The thermal NOx from a burner increases exponentially with an increase in flame temperature. The introduction of FGR into the combustion air increases the overall mass of the reactants — and hence the products — in the combustion process. The increased mass, as well as the increased reactant diffusion time requirement, reduces the overall flame temperature. The burner with the least amount of FGR will theoretically have the highest flame temperature, and will therefore have the highest NOx. Likewise, the burner with the highest amount of FGR will theoretically have the lowest NOx. However, due to the exponential nature of the NOx/temperature relationship, given an equal deviation (e.g., ±5%), the higher NOx values from the low FGR burners will outweigh the lower NOx values from the high FGR burners. Minimizing the FGR deviations, as shown in Figure 19.25, will even out the flame temperatures and therefore minimize the NOx formation from each burner.

Equal inlet velocities and elimination of swirl through each burner are crucial to burner performance. Because low-NOx burners rely on injection of fuel at precise locations within burner air flow, it is imperative that the proper air flow be present at these locations. Likewise, for optimum performance, the only swirl present must be that created by the burner itself.

No burner can be expected to simultaneously correct imbalances in the draft system *and* precisely control fuel/air mixing to minimize NOx formation. Air flow modeling prepares the air flow for the burner, allowing the burner to precisely control fuel/air mixing for maximum NOx reduction. This approach has freed burner designers to focus solely on NOx control, thereby increasing the effectiveness of known control techniques to their maximum extent.

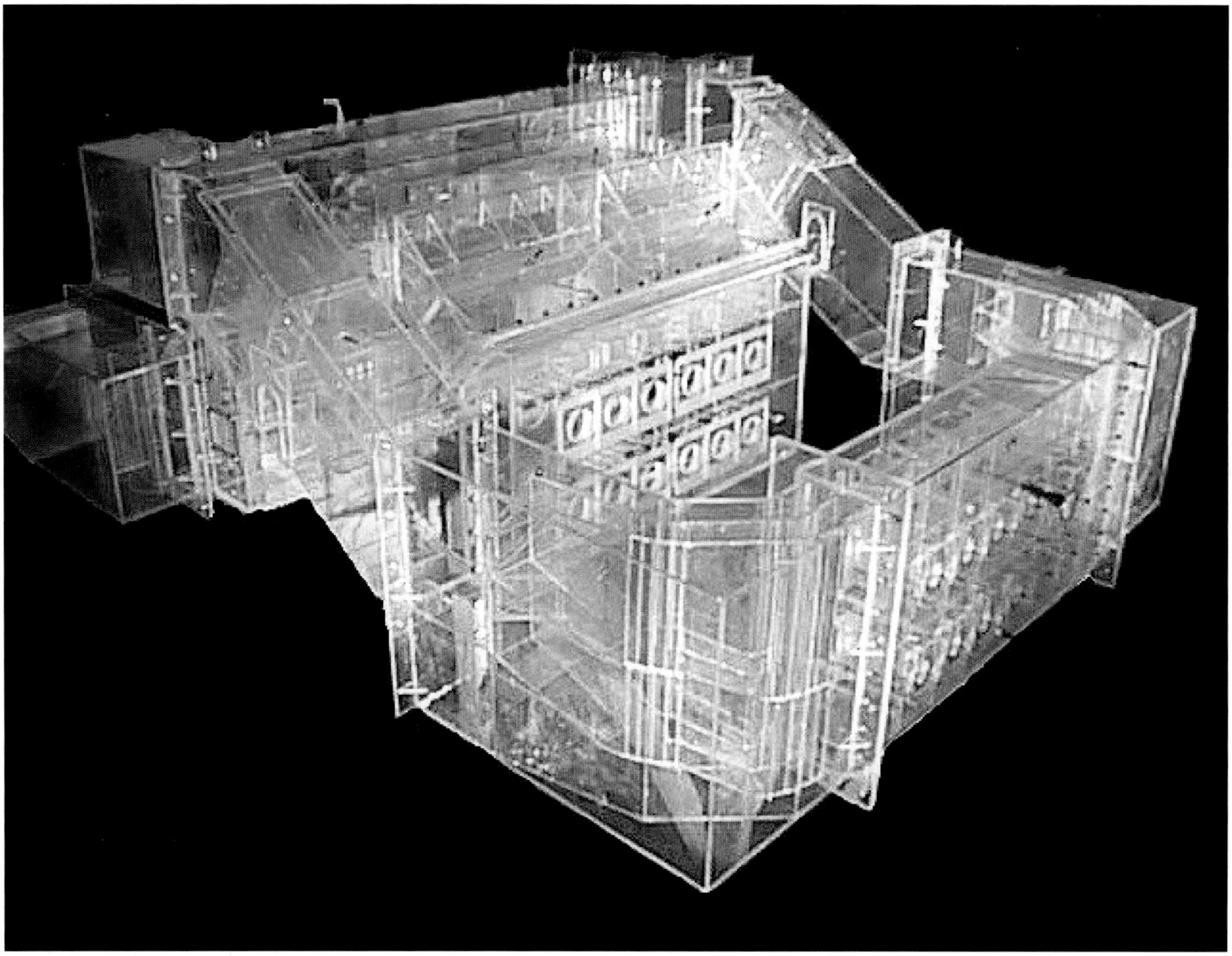

FIGURE 19.26 A scaled, physical, aerodynamic simulation model.

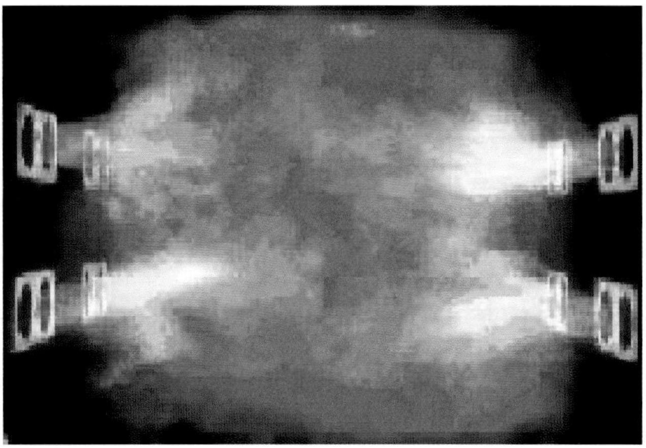

FIGURE 19.27 Flame-to-flame similarity of appearance.

To achieve these goals, a scaled, physical, aerodynamic simulation model, similar to the one shown in Figure 19.26, can be performed based on the physical dimensions and flow rates within the field unit. The goals of the model are to enhance flame stability; increase turndown; improve flame appearance; reduce NOx, CO, opacity, and particulate at the minimum O_2 level; and minimize mal-distributions of air flow that can occur during both normal and unusual operating conditions. A scale model allows for full observation and photographic recording of the air flow within the scale model version of the new windbox/burner configuration and existing ductwork.

The goals of the model are primarily accomplished by installing secondary air duct and windbox baffles. The modeler determines the location of baffles and turning vanes

FIGURE 19.28 Premixing the FGR flow with the combustion air upstream of the windbox.

within the combustion air/FGR supply system. As an added criteria, the windbox modifications must provide the minimal amount of impact on overall combustion air/FGR supply system pressure drop. This is to minimize the effects on existing fan performance. The major constraint in achieving these objectives is the air distribution arrangement and windbox internal dimensions. The result of a windbox model can be seen in the flame-to-flame similarity of appearance, as shown in Figure 19.27.

19.3.1.2 Fuel Flow Balancing Techniques

Fuel flow balance is just as important as air flow in reducing O_2. A rough balance of fuel flow distribution is relatively easy to achieve by balancing pressure drops in the fuel header, or in other words, equalizing the fuel pressure at each burner, thus ensuring that each burner is receiving the same amount of fuel.

Once the air flow has been balanced by a windbox model and the fuel pressures at each burner have been equalized, the unit is in the proper starting condition for the concept of fuel balancing to be taken a step further to reduce O_2 and maximize boiler efficiency. This goes back to the concept that there is probably one burner that is limiting the O_2 reduction. Even when the field data indicate that the air mass flow is within ±5% of average, and all the burners indicate equal fuel pressures (assuming that equal fuel pressures means equal fuel flow), the burner that is –5% in air flow will be running at an O_2 level that is approximately 1% lower than average.

For further balancing, the unit should first be brought to maximum continuous rating (MCR) conditions, and O_2 should be lowered to a point where the CO is in the 200 to 400 ppm range on gaseous fuel, or opacity is in the 12 to 14% by EPA Method 9 for oil. The burner with the highest CO/opacity should be found and the fuel to that burner reduced slightly, resulting in a reduction in the overall CO/opacity level. This is an iterative process and should be repeated until any further fuel adjustments either show no effect or increase the CO/opacity level.

The trick to this method of burner optimization is to find the burner with the highest CO/opacity. Ideally, on a multi-burner boiler, there should be a measurement grid at the economizer outlet that has been mapped out for burner stratification and measurement probes placed along the burner centerlines. CO measurements taken from this grid, at the conditions specified above, typically give a clear indication of the burner with the lowest amount of stoichiometric air. Lacking a mapped-out grid, when the boiler is brought to the conditions outlined above, a visual inspection of the flames will typically indicate the "problem burner" as the one that has the "dirtiest" or "sootiest" flame.

19.3.2 Methods to Reduce NOx Emissions

Combustion modification techniques that have been developed to reduce NOx emissions include low excess air operation, fuel and air staged burner design, staged combustion (off-stoichiometric or biased firing), reduced air preheat, flue gas recirculation, fuel-induced recirculation, reburn, and water tempering. As discussed in Section 19.2.2, low excess air operation limits the oxygen availability in the combustion zone and is highly effective in controlling fuel NOx formation and, to a lesser extent, thermal NOx.

The NOx problem forced the development of various completely different furnace designs, usually implementing low-NOx burners and FGR systems, sometimes containing over-fired air ports for supply of air and/or air/FGR mixtures to the furnace space located above the burners. There are single-wall burner utility boilers where air ports are located on the target wall. Also, there are low-NOx industrial and utility boilers containing special furnace devices for steam/water injection in the combustion zone and for fuel injection in the post-combustion zone (a kind of reburning).

Actually, any of the mentioned methods is capable of reducing NOx by up to 35 to 40%, but it cannot satisfy the strictest of today's NOx requirements. These applications require implementation of at least two NOx reduction methods. It is important to note that, as with the example of boiler load and FGR in Section 19.2.3, the efficiency of the implemented methods depends on their sequence. The first NOx reduction method is much more efficient than the second one. This conclusion has been confirmed on many utility boiler retrofits where up to four NOx reduction methods were used. That is why a combination using more than three methods on the same boiler is typically inefficient.

19.3.2.1 NOx Reduction by FGR Implementation

It is well-known that thermal NOx can be effectively controlled by reducing the flame temperature. The most effective technique is the use of FGR, which when mixed with the combustion air acts as a diluent to decrease the flame temperature, and sometimes (at high FGR rates) even increases the flame luminescence. FGR with lower temperatures will be more effective in flame temperature reduction.

Usually, FGR is taken after an economizer where temperatures range between 500 and 800°F (260 to 430°C), the O_2 range is typically between 1 and 3%, and actual ICP concentrations are negligible. Usually, an FGR fan is required to supply FGR flow to the combustion air flow or to the furnace.

There are also applications where a comparatively cold exit flue gas (240 to 320°F [120 to 160°C]) is taken from the ID fan outlet and supplied, for example, to the FD fan inlet or to the windbox. Usually, this flue gas contains much more excess O_2, especially on the boilers equipped with air preheaters having comparatively large air leakage characteristics. Empirical data show that if this air leakage can be minimized, this "cool" FGR can provide 25 to 30% greater relative NOx reduction as compared to the "hot" FGR, but it has a more severe impact on boiler thermal efficiency reduction. FGR flow can also be induced into the combustion air forced-draft fan inlet, allowing the use of FGR without requiring a separate FGR fan.

FGR can be taken from the furnace zone adjacent to the burner exit, and can be induced back into the combustion air flow, also allowing the use of FGR without requiring a separate FGR fan. Unlike regular, comparatively cold FGR flows, this flue gas has temperatures of at least 2000 and 2200°F (1100 and 1200°C) with gas and oil firing, respectively; it consists primarily of unreacted air (O_2 ~ 16–19%) and ICP. This method is completely different from the ones described above because its effectiveness cannot be associated with significant changes in the flame temperature conditions or lowering O_2 concentration in the combustion air. The effectiveness of this method relies on interactions between reagents participating in NO formation and both radicals and ICP that are present in high concentrations, which slow the NO formation reactions rates, thereby reducing NOx output.

Flue gas can be recirculated in a number of ways: directly to the furnace through slots located under, above, around, or between the burners; through over-fired air ports or ports located on the target wall; premixing the FGR flow with the combustion air upstream of the windbox (called premix FGR), typically accomplished via a sparger section as depicted in Figure 19.28; mixing the FGR at the burner exit (called plenum FGR); and premixing the FGR with the gaseous fuel (on oil, the flue gas would pass through the fuel gas piping). This technique of flue gas entrainment is called fuel dilution and uses no additional fan power because it is typically induced into the fuel stream by the fuel pressure itself, or by pressure energy supplied via an intermediate media such as steam if the available fuel pressure is not high enough. Also, FGR can be supplied to the boiler hopper; this method is typically used only for steam temperature control.

Various levels of NOx reductions can be achieved at the same FGR rates using these different techniques. Figure 19.29 (illustrating test data from 150, 165, 200, and 300 MW_e boilers) and Figure 19.30 (800 MW_e boiler) show full load gas firing test data received on utility boilers equipped with different burner designs, numbers, and arrangements, with O_2 = 0.8–1.4%.

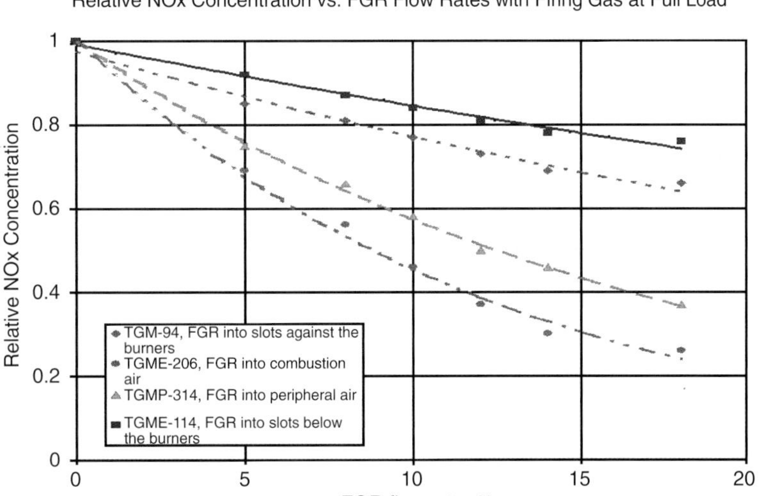

FIGURE 19.29 Relative NOx concentration vs. FGR flow rate with firing natural gas in the utility boilers at full load.

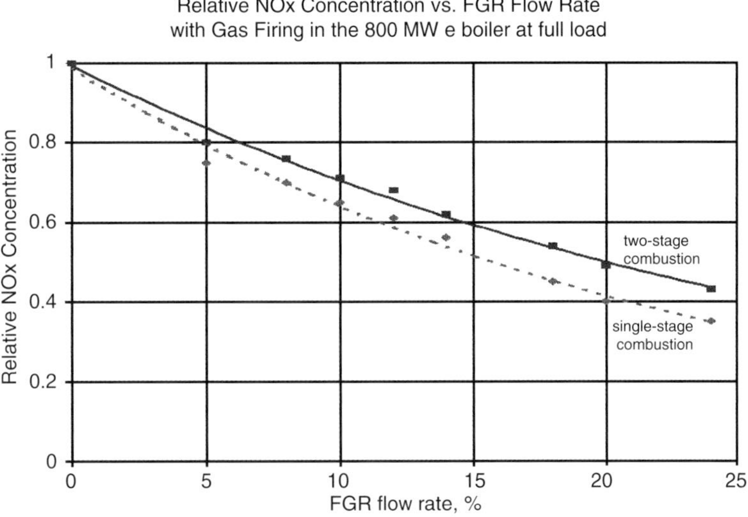

FIGURE 19.30 Relative NOx concentration vs. FGR flow rate with firing natural gas in the 800 MW_e boiler at full load.

It should be noted that all curves shown in these two figures are exponential, meaning that it would make sense to determine an optimal operational FGR rate. This optimal limit will be different for each boiler application because it depends on the boiler design (and accordingly on FGR impact on the boiler operational parameters). The data presented show that the typical optimal FGR rate range for NOx reduction is 14 to 18%. The NOx reductions experienced through FGR implementation, as described above, explain why it is the most frequently implemented NOx reduction technique on newly designed and low-NOx retrofit boilers.

Figure 19.29 shows that the most effective conventional FGR supply system for low NOx is to provide a high-quality mixture of the FGR and combustion air, as with the venturi-style, low-NOx burners installed at the 200 MW_e TGME-206 boiler (single-wall burner installation), where 10, 12, and 18% FGR flows provided 54, 63, and 74% NOx reduction, respectively. A comparatively high NOx reduction efficiency is also reached on the 300 MW_e TGME-314 boiler model (16 burners installed in two rows on opposed firing walls), where straight FGR flow is mixed with the burner air flow. Methods used to supply FGR directly into the furnace, implemented on the

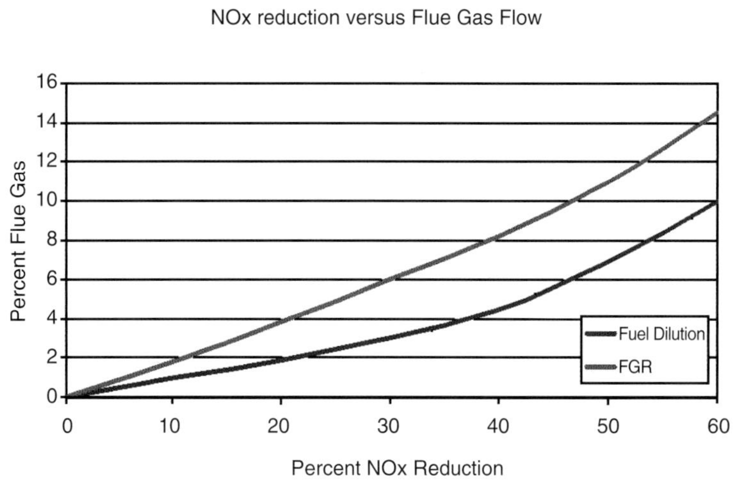

NOx reduction versus Flue Gas Flow

Percent Flue Gas: 0, 2, 4, 6, 8, 10, 12, 14, 16
Percent NOx Reduction: 0, 10, 20, 30, 40, 50, 60
Fuel Dilution
FGR

FIGURE 19.31 NOx reduction vs. flue gas flow.

TGM-114 (opposite wall burner installation, 150 MW_e) and the TGM-94 (single wall burner installation, 165 MW_e), are dramatically less effective.

Figure 19.30 demonstrates the test data obtained on the 800 MW_e boiler of model TGME-204 (36 low-NOx burners installed in three rows on two opposed firing walls) with full load single- and two-stage gas combustion (with 33% air supplied through over-fired air ports located above the top burners) at O_2 = 0.8–1.2%. As previously discussed, it is only natural that the NOx reduction provided by premixing FGR with combustion air flow is less effective with two-stage combustion when the flame temperature level is significantly lower. However, the 46 and 57% NOx reductions achieved with 18 and 24% FGR flow rates, respectively, during two-stage gas combustion can still be considered a comparatively high efficiency because it is the second NOx reduction method implemented (see Section 19.3.2). At full load on this boiler, the total NOx reduction exceeds 80% (from 600–620 to 90–95ppm), and CO < 200 ppm at O_2 = 0.8–1.2%. This NOx reduction was achieved utilizing a combination of low NOx burner retrofit, two-stage gas combustion, premixed FGR, and water injection in the primary and secondary combustion air flows. The same scope of work was repeated on three neighboring 800 MW_e boilers of the same design installed at this power plant. On all four units, total NOx reductions of 81 to 85% were achieved.

Similar full load NOx reduction data has been obtained on packaged industrial boilers equipped with a single venturi-style, low-NOx burner, firing gas with ambient air. The data are presented in Figure 19.31.

Recent developments have shown that, when firing gaseous fuels, using fuel dilution (mentioned above) to induce flue gas into the gaseous fuel is truly the most effective method of FGR supply (~25% more effective, as shown in Figure 19.31). As per the test data, while natural gas firing at 2.5 to 4.5% O_2, NOx can be reduced by 66% with 12 to 13% fuel dilution flow.

The reason why fuel dilution provides more efficient NOx reduction as compared to conventional FGR supply methods is as follows.

When natural gas (conditionally CH_4) is fired with combustion air, without the use of FGR or fuel dilution, every CH_4 molecule encounters a certain quantity of oxygen molecules that is dependant on the excess O_2 level, the mixing quality of the fuel with the air, and the diffusion rates of the CH_4 and O_2 at the flame front region, as shown in Figure 19.32(a). The combustion reaction rates (including reactions for incomplete combustion product burnout, prompt NOx, and thermal NOx generation) are dependent on the actual excess O_2, the combustion air temperature, and the fuel/air mixture quality. These variables also determine, along with the given burner/furnace design, the heat release and heat distribution within the furnace, which in turn determines the shape of the temperature profile curve, as well as the maximum temperature value and its location within the given furnace. Concurrently, these specific features of the temperature conditions have a major impact on the determination of the above combustion reaction rates.

When CH_4 is fired with combustion air containing typical FGR, with all other combustion conditions being the same as in the previous example, every CH_4 molecule will encounter

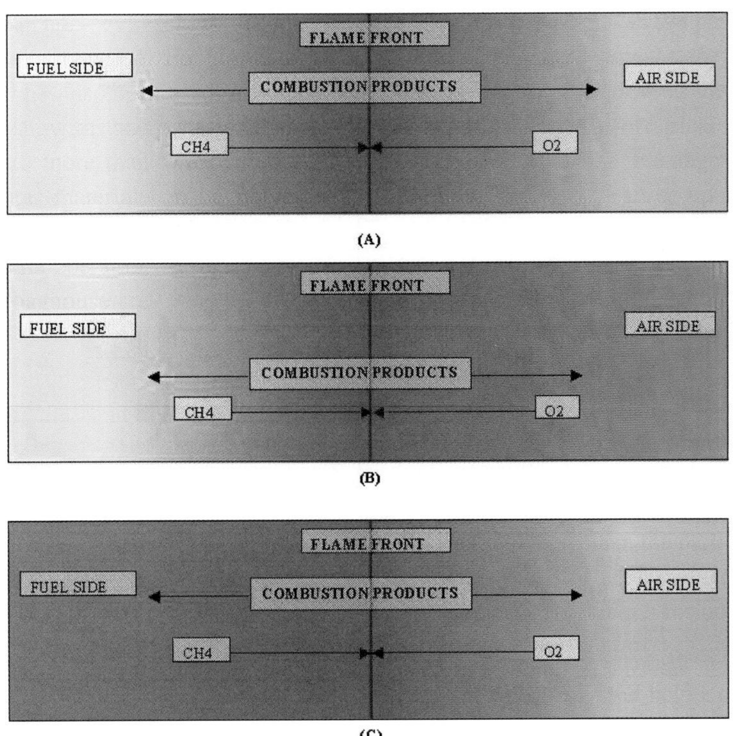

FIGURE 19.32 Full load NOx reduction data has been obtained on packaged industrial boilers equipped with a single venturi-style, low-NOx burner, firing gas with ambient air. (a) When natural gas (conditionally CH_4) is fired with combustion air, without the use of FGR or fuel dilution, (b) when CH_4 is fired with combustion air containing typical FGR, with all other combustion conditions being the same as in the previous example; (c) when CH_4 is diluted with flue gas at the 1:1 ratio, with all other combustion conditions being the same as in the previous examples.

a lower quantity of oxygen molecules than it would without FGR. The FGR impedes the diffusion of O_2 toward the flame front, as depicted by the increased combustion product content on the air side (slight purple coloring on the air side) in Figure 19.32(b). The difference between the quantity of oxygen values in these two cases depends on both the FGR composition and the FGR flow rate. For example, for a boiler design operating under a positive furnace pressure (i.e., no air in-leakage in either the furnace or the convective path) and providing an equal distribution of the combustion products in both the furnace exit cross-section as well as in the convective path cross-sections, the FGR composition will be similar to the combustion product composition leaving the furnace. Because the combustion air makes up ~94% of the mass flow within the combustion process, by supplying an FGR flow rate of ~11%, the oxygen concentration in the combustion air/FGR mixture is reduced by the same 11% (being only slightly diluted; this is why there is only a slight purple coloring on the air side in Figure 19.32(b)). Also, the flame temperature level in the furnace is reduced primarily due to the addition of the FGR mass. The magnitude of the temperature reduction caused by this phenomenon is dependent on the FGR temperature; relatively high temperature FGR flows will provide less flame temperature reduction than relatively low temperature FGR flows. The combination of decreased O_2 concentration and lowered flame temperature level results in reduced reaction rates of all the combustion reactions mentioned above, including those reactions responsible for NOx generation in the given furnace.

When CH_4 containing fuel dilution is fired with combustion air in the same furnace, combustion conditions change significantly from the examples previously discussed; because the CH_4 is only ~6% of the mass flow within the combustion process, the fuel dilution has a much greater dilution effect on the CH_4. If the fuel dilution rate is the same as the FGR rate previously discussed (11%), the dilution is equivalent to a 2 fuel dilution:1 fuel molecular ratio, as evidenced by the heavy orange coloring on the fuel side in Figure 19.32(c). Due to this phenomenon, when a diluted fuel fires, the flame temperature becomes significantly less in comparison with the second case, related to natural gas firing in the combustion air containing the same FGR flow. The reduced flame temperature influences all of the above combustion reactions rates, including reactions that are responsible for NOx generation in the given furnace. This is the explanation for why the NOx reduction efficiency of fuel dilution is so much greater than that of FGR. On the other hand, the mentioned flame temperature reduction (associated with the substitution of FGR with fuel dilution of the same percentage and temperature) reduces heat release in the furnace, increases the furnace exit flue gas temperature, and increases heat release in the boiler convective path, especially in the superheater and reheater sections. Also, the flue gas temperature will be increased throughout the convective path, including the stack flue gas temperature, which reduces the boiler thermal efficiency.

Fuel dilution can also be implemented in combination with FGR premixed in the combustion air and/or steam injection. A combination of fuel dilution and steam injection can provide a total NOx reduction of up to ~82%.

Low flue gas recirculation rates (<20%) typically have a relatively minor impact on boiler performance (in many cases, on utility boilers it can be compensated for by increasing the turbine's efficiency; on industrial boilers, this loss is unrecoverable). However, high FGR rates (>20%) can significantly impact boiler performance, especially the redistribution of heat flux between radiant surfaces and high-temperature convective heat exchange surfaces of utility boilers and, accordingly, on superheated/reheated steam temperature. Excessive use of flue gas recirculation can result in significantly increased superheated and reheated steam temperatures, or significantly increased superheat and reheat attemperation spray flows, the latter affecting boiler efficiency. In addition, power expenses recalculated for thermal efficiency losses are estimated at 0.03 to 0.05% per 1% FGR. These facts explain why FGR was (before NOx requirements were established) typically only used to control and regulate reheated and superheated steam temperatures at minimum loads. A typical older boiler would be designed to provide the required steam temperature and flows without FGR at full load; but at partial loads (usually $\lesssim$70%), FGR was switched on to maintain steam temperatures. FGR flow rates increased as loads continued to decrease. Modern boiler designs incorporate FGR for both NOx reduction and steam temperature control. At higher loads (100 to 80%), FGR is typically used for NOx reduction; however, as above, at partial loads, FGR is typically used for steam temperature control.

Very large amounts of premixed FGR, above 45%, can affect flame stability with conventional burner designs, while plenum FGR has been tested on some burners up to 70% with no noticeable effect on flame stability. Depending on the FGR supply method, the FGR/air mixing quality will be different, and, accordingly, flame stability, vibration, and other negative issues of combustion (especially in the case of firing natural gas in ambient air) can be met — even at low FGR amounts. However, these problems are not typically experienced when FGR flow rates are below 14 to 18%.

Alternative techniques to reduce flame temperature involve reduced air preheat and water or steam injection. Water or steam injection will result in efficiency losses that are unrecoverable, while reduced air preheat can be achieved with no loss in efficiency if steam surface changes are made.

19.3.2.2 Multi-stage Combustion on Utility Boilers

Staged combustion involves delaying the mixing of fuel and air, and is effective for both thermal and fuel NOx control. Typically, staged combustion creates an initial fuel-rich combustion zone with air added downstream to complete combustion. Staged combustion can be achieved using low-NOx burners (internal staging), over-fired air ports, operating existing burners with biased fuel firing, or with burners-out-of-service (BOOS). Biased fuel firing for a large, wall-fired unit is implemented by operating the lower burners with reduced combustion air and/or increased fuel to produce a fuel-rich zone in the lower portion of the furnace. The upper burners are operated with a corresponding increase in air or decrease in fuel to complete fuel burnout. An extreme form of staging involves operation with an upper row of burners out of service or its equivalent over-fired air ports. In this case, fuel is shut off to these burners and the burners are operated with air only.

The principle of stagewise fuel combustion involves the arbitrary division of the flame into two or more stages. In most cases, in the first high-temperature stage, combustion takes place with an excess air factor of less than unity; in subsequent stages, which have a comparatively low temperature level, secondary combustion of the previous ICPs occurs with relatively high excess air. Thus, NOx formation is retarded in the first stage because there is insufficient oxygen in the reaction zone, and in the subsequent stages due to the relatively low temperature of the flame. This has established the advantages of stagewise combustion as an NOx reduction method over other methods.

There are various methods of performing staged combustion in furnaces. The best known and most studied methods are based on the following three options:

1. gas/oil combustion with a significant air deficit in all burners and supplying a certain part of the total air flow directly to the furnace through ports or slots located above the burners
2. redistribution of air flows between the burners located at different rows
3. redistribution of fuel flows between the burners located at different rows

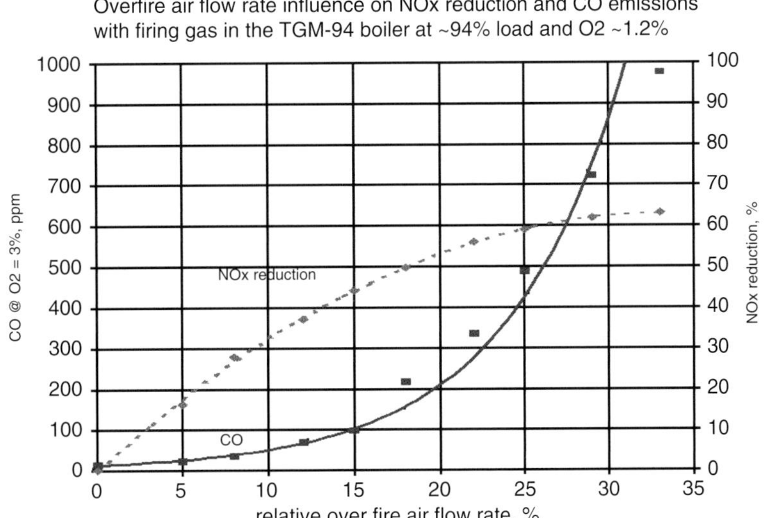

FIGURE 19.33 Overfire air flow rate influence on NOx reduction and CO emission with firing natural gas in the TGM-94 boiler at ~94% load and O_2 ~1.2%.

The mentioned options are directed to get a "vertical" fuel/air unbalance in the furnace: excess fuel and air deficit in the lower part and, on the contrary, excess air and fuel deficit in the top part of the combustion zone. However, when a boiler changes over to stagewise fuel combustion with a "vertical" unbalance in the fuel/air ratio, concentrations of all ICPs, including benz(*a*)pyrene and other carcinogenic substances, increase. Much test data exist confirming the above conclusion. An example is shown in Figure 19.33, where test data obtained on a model TGM-94 boiler are presented. It can be seen that a 50% NOx reduction while firing gas is reached at ~18% over fire air flow, but CO has increased to ~200 ppm. A further increase in the overfire air flow to 25% results in a ~60% NOx reduction; however, the CO has increased to ~400 ppm. Increases in CO concentrations and other ICPs are typical of "vertical" staged combustion. The data for this boiler also indicated that operation while firing gas at ~18% overfire air flow reduced the boiler efficiency by ~0.85%.

Another system of stagewise combustion, with a "horizontal" unbalance of the fuel/air ratio, gives rise to substantial reductions of ICP emissions. However, it results in increased NOx. This method of stage combustion utilizes the interaction of fresh combustion air jets directed from ports or slots located opposite the burners, with fuel radicals and already formed ICP in flames generated by burners receiving sub-stoichiometric air. It intensifies the combustion process and especially burn out while firing oil, as illustrated in Figure 19.34 where test data were obtained on a TGM-84 (~930,000 lb/hr or 420,000 kg/hr steam flow) cogenerating boiler firing No. 6 oil. The data show that this method provides reductions in opacity, carcinogenic substances, and a small CO reduction from ~38 to ~24 ppm. However, NOx increases very quickly; on average, 1% of the secondary air flow brings a relative NOx increase of ~1.5%. The boiler data from this testing indicated that this method can be applied to applications requiring increased reheated/superheated steam temperatures. Moreover, this method increases the boiler thermal efficiency and reduces operational power expenses due to a decrease in the air resistance of the air path. On the above boiler, ~15% secondary air flow increased the total efficiency by 1.1%. The data indicate that an optimally designed three-stage fuel combustion system could be developed with simultaneous "vertical" and "horizontal" imbalances of the fuel/air ratio.

Using calculations explaining the above data as a basis for preliminary design work, a new furnace was developed. The furnace had three burner rows and slots above the upper row located on the furnace front wall. FGR could either be premixed with the combustion air going through the burners and slots; supplied directly into the furnace through slots located on the rear wall, opposite the lower burners; or in both locations. The primary air flow was supplied through the burners, and secondary and tertiary air flows were redirected to the slots located on the rear and front walls, respectively.

Tests were performed over a boiler load range of 100 to 40%, $\alpha = 1.05$–1.11 firing gas, and at full load, $\alpha = 1.03$–1.1, firing No. 6 oil. To obtain reliable data on the NOx reduction effectiveness of stagewise combustion, tests were carried out in alternate modes, with measurements being made during single-stage combustion under the same load, excess air

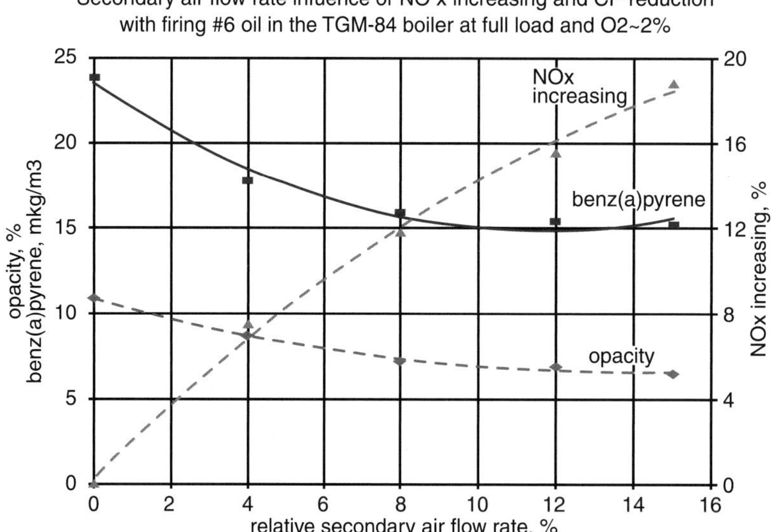

FIGURE 19.34 "Horizontal" imbalance intensifies the combustion process and especially burnout while firing oil.

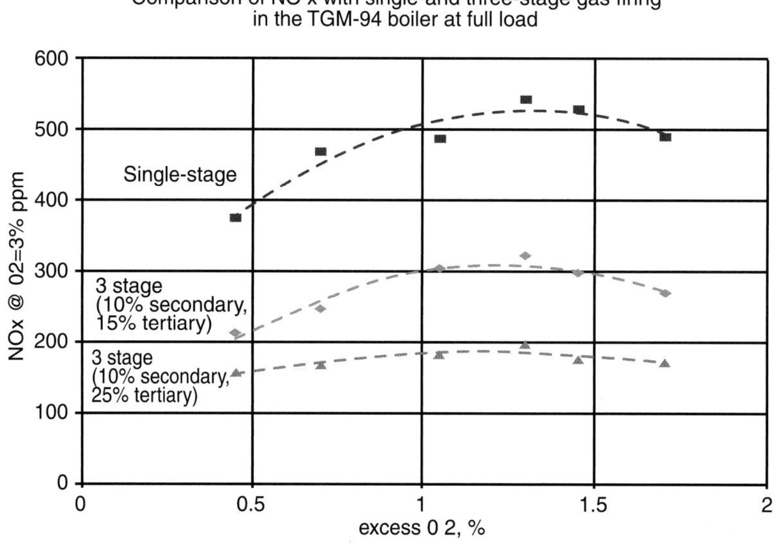

FIGURE 19.35 Data comparing single- and three-stage gas combustion.

factor, FGR, and other combustion conditions. Comparative tests when firing oil were carried out with oil supplied from the same tank. All results were compared with the baseline conditions of single-stage combustion, in which the secondary and the tertiary air flow dampers were fully throttled, it being conditionally assumed that their relative flows were ~0.

The test data comparing single- and three-stage gas combustion are shown in Figure 19.35. It is seen that transferring from single- to three-stage combustion at ~10% secondary air flow yields NOx reductions of ~40 and ~63%, with tertiary air flows of 15 and 25%, respectively. This NOx reduction is accompanied by a corresponding CO increase from ~135 to 165 ppm during single-stage combustion to 260–290 and 350–400 ppm, accordingly. Under the above conditions, operation with FGR was not required at any point over the entire load range. The decrease in thermal efficiency was 0.3 to 0.4%, much less dramatic than either two-stage combustion (with a "vertical" imbalance) or FGR fan operation. The same retrofit was performed on six neighboring TGM-94 boilers installed at this power plant.

19.3.2.3 Special Features of Low-NOx Gas/Oil Combustion on Utility Boilers

It was shown in Sections 19.1.2 and 19.2.2.2 that there are some utility boilers where NOx is higher while firing natural gas than that experienced while firing No. 6 oil. For example, data obtained on the TGM-94 boilers presented in Figures 19.11 to 19.14 demonstrates that the NOx while gas firing is ~24 to 27% higher than NOx data obtained while firing No. 6 oil under the same load (~94%), O_2 (1.4 ± 0.2%), and other operational conditions.

This phenomenon (i.e., NOx while firing gas is higher than NOx data obtained while firing No. 6 oil) has been experienced during testing of many former U.S.S.R. utility boilers in the size range of 165 to 800 MW_e. In some cases, NOx while firing gas exceeded the NOx while firing No. 6 oil (even containing 0.67 to 0.73% N_f) by ~50%. On the contrary, on utility boilers in the size range of 25 to 80 MW_e, the NOx while firing No. 6 oil exceeded the NOx while firing gas by ~25%. Moreover, as N_f values are increased (from 0.22 to 0.73%), the greater is the margin between the NOx while firing No. 6 oil and the NOx while firing gas. Likewise, on industrial boilers, the NOx was almost always higher while firing No. 6 oil, and differences of 24 to 36% and 12 to 18% were typically experienced with ambient and preheated air, respectively. However, intermediate results were obtained while testing utility boilers in the size range of 100 to 150 MW_e, where differences between the NOx numbers for the two fuels are less than ±10%. The controversial data can be explained in terms of the well-known influence of the flame temperature on NOx formation.

While firing natural gas containing no N_f, NOx can only be generated by the oxidation of nitrogen contained in the combustion air. This NOx generation can only occur by one of two mechanisms: (1) thermal NOx formation due to flame temperatures of at least 3000°F (1650°C), mostly in the burner zone heat release (BZHR) space — its concentration rises with temperature according to an exponential dependence; and (2) prompt NOx formation in the beginning of flame where the temperature level is significantly lower; under usual combustion conditions, its concentration is much less than the thermal NOx concentration. Prompt NOx has a slight dependency on temperature. Under certain furnace conditions required for low-NOx combustion, the thermal NOx concentration can be reduced to the prompt NOx concentration level — or even less. However, these conditions have little influence on prompt NOx.

While firing natural gas (conditionally consisting of only methane), the maximum combustion temperature level depends only on excess air and the combustion air temperature, as shown in Figure 19.10. The corresponding NOx concentration curves, calculated for excess air factor $\alpha = 1$, are shown in Figure 19.10.

When firing oil containing N_f, there are two sources of NOx formation: air nitrogen and fuel nitrogen. In addition to the two mechanisms available for gas firing, a third mechanism is available as well: oxidation of the partially destructed nitrogen-containing compositions that occur in the beginning of the flame, whose concentrations decrease slightly with increasing temperature.

Figure 19.10 shows calculated NOx curves for both nitrogen sources, along with their summed values. It shows that oxidation of 0.5% N_f can generate NOx of up to ~200 ppm. In considering the processes in the comparatively cold furnaces designed for industrial and small utility boilers (50 to 100 MW_e), the above 200 ppm will far outweigh the thermal NOx formation. Because these processes are occurring at relatively low temperatures, they indicate relatively low NOx levels from these boilers and an oil firing NOx level much higher than the gas firing NOx level. In the 120 to 150 MW_e boilers, temperatures can reach 2900 to 3100°F (1600 to 1700°C). Figure 19.10 indicates that the curves calculated for gas and oil firing intersect within this temperature interval, indicating relatively equal NOx levels while firing either gas or oil. For units greater than 165 MW_e, the flame temperature can reach between 3200 and 3500°F (1800 to 1900°C). This temperature increase results in a gas firing NOx level much higher than the oil firing NOx level.

The above explanation can be confirmed with the data presented in Figure 19.36, where the ratio between NOx numbers obtained with gas and oil firing are considered as a function of the furnace space heat release. Furnace designs for increasingly higher capacity utility boilers are usually (with very few exceptions) accompanied by a significantly increasing heat release as related to the furnace volume. Thus, the design of these boilers allows the analysis of the ratio between NOx numbers and this design parameter. Despite experimental data deviations from the average empirically established curve, the strong character of this dependence cannot be doubted.

19.4 LOW-NOx BURNERS FOR PACKAGED INDUSTRIAL BOILERS

There are a few state-of-the-art low-NOx burner (SLNB) designs developed for typical industrial packaged boiler applications, including those of several burner manufacturers (e.g., Todd Combustion, Coen, Forney, Alzeta, Peabody, Pillard, Babcock Hitachi, and others) and consulting companies (e.g., IGT, J.Lang, RJM, EPT, and others) in both the United States and other countries. A schematic for a typical

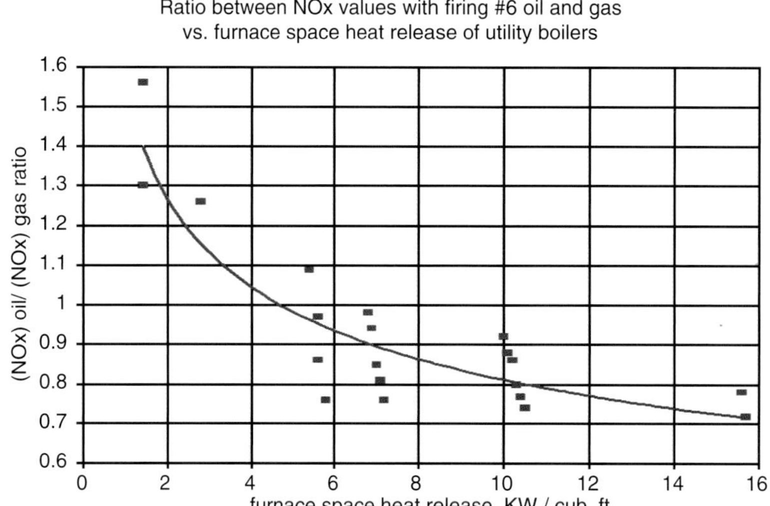

FIGURE 19.36 The ratio between NOx numbers obtained with gas and oil firing are considered as a function of the furnace space heat release.

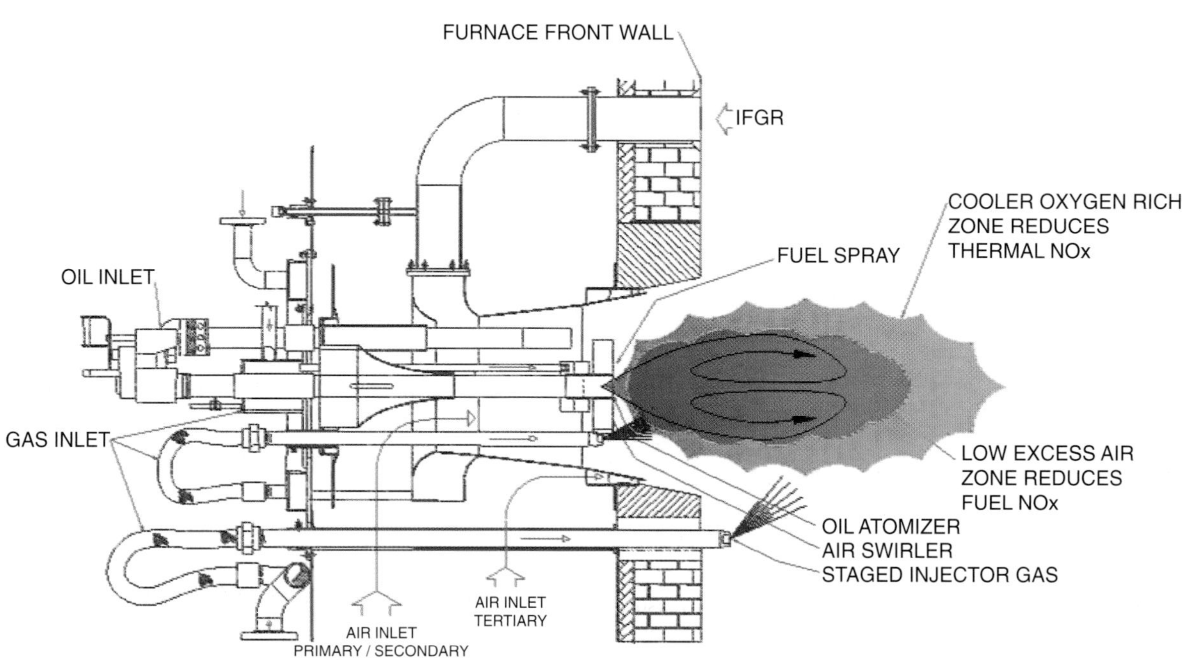

FIGURE 19.37 The Todd Combustion low-NOx gas/oil burner.

packaged SLNB is depicted in Figure 19.37. This burner was developed using a standard venturi-type low-NOx burner as a prototype. By comparing the schematic for this burner with the schematic for a typical standard venturi-type low-NOx burner shown in Figure 19.5, it can be seen that whereas the air channels are the same as in the standard low-NOx burner, the SLNB has two completely independent gas supply units: the center fire gas (CFG) and a set of outer gas injectors, replacing the inner gas poker assembly implemented in the prototype. Also, the SLNB utilizes internal furnace flue gas recirculation (IFGR) instead of the external flue gas recirculation that was used by the prototype and other standard low-NOx burners.

The SLNB shown in Figure 19.37 has been installed and tested at four industrial retrofit boilers where all existing auxiliary equipment, including fans, dampers, and combustion control systems, was kept. Before installation, the burner

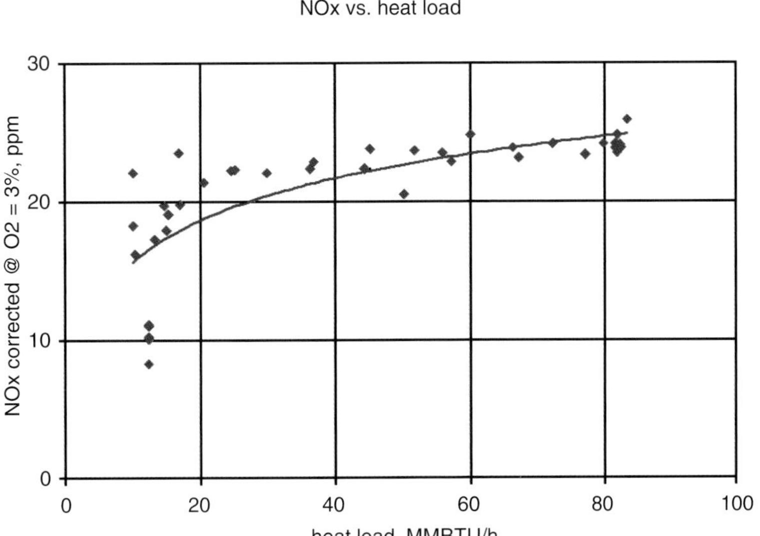

FIGURE 19.38 NOx vs. heat load with firing natural gas with ambient air in the packaged industrial boiler equipped with the Todd Combustion low NOx burner.

and windbox were modeled in order to provide equal air distribution. The SLNBs have designed heat inputs ranging from 28 to 134 × 10^6 Btu/hr (8.2 to 39.3 MW_t). At full load, the burners usually operate at moderate gas pressures (< 6 psig) and at reasonable register draft loss (RDL < 7 in. or 180 mm w.c.). Test data obtained on two industrial boilers (one with ambient air and one with preheated air) are described and compared below.

The first application was a 70,000 lb/hr (32,000 kg/hr) B&W FM 103-70 boiler generating 250 psig (17 barg) saturated steam, firing natural gas with ambient air. The burner was designed for 84.8 × 10^6 Btu/hr (25 MW_t) at full load. Over a 100 to 25% load range, the installed burner has demonstrated excellent flame stability, high reliability, and highly efficient operation while firing both natural gas and No. 6 oil as fuels. Measurement has shown that the noise level is less than 85 dBA. Vibration of boiler/burner parts is at regularly accepted levels.

NOx emission test data obtained on the above boiler is illustrated in Figure 19.38. These data sets indicate that under optimal combustion conditions, the SLNB typically generates NOx in the 23 to 28 ppm range and CO < 200 ppm (typically < 100 ppm). Due to the advanced gas staging provided by this burner, O_2 concentration changes in the range of 2.6 to 5.1% had practically no impact on NOx and a relatively low impact on CO (Figure 19.39). At 25% load, the emission data were close to the data obtained at full load, as shown in Figures 19.38 and 19.39. These results were numerously repeatable over a 7-month testing period.

The emission test data sets obtained on this boiler, while firing No. 6 oil (0.22% N_f) at full load, are shown in Figure 19.40. At O_2 = 3.5%, the following data are recorded: NOx ~ 200 ppm, CO ~ 100 ppm, and opacity < 10%. At partial loads, NOx is sufficiently less at the same O_2, CO, and opacity.

A significant NOx reduction (~15 to 20%) is realized through the use of 4 to 5% IFGR while firing gas with ambient air. The IFGR contains relatively high ICP concentrations and has a temperature of about 1800 to 2000°F (980 to 1100°C). The same relative NOx reduction due to IFGR was experienced while firing No. 6 oil, as shown in Figure 19.41. This figure indicates an NOx reduction of ~20%. Although the IFGR temperature is higher as compared with gas firing (~200°F against 2000°F), relatively higher ICP concentrations contained in this flow provide the mentioned NOx reduction effect. For both fuels, over the entire tested load range, there was no significant influence of IFGR on CO. There was also no influence on opacity while firing No. 6 oil.

The second installation of the SLNB shown in Figure 19.37 was on a B&W FM boiler designed to generate 100,000 lb/hr (45,000 kg/hr) superheated steam at 415 psig (28 barg) and 600°F (320°C), while firing natural gas with ~415°F (213°C) preheated air. The specific features of this application as compared to the prior application are the following:

- 60% higher heat input (134 × 10^6 Btu/hr vs. 84.8 × 10^6 Btu/hr, or 39 MW_t vs. 25 MW_t)
- 45% hotter furnace (BZHR: 245 × 10^3 Btu/ft^2-hr vs. 168 × 10^3 Btu/ft^2-hr)

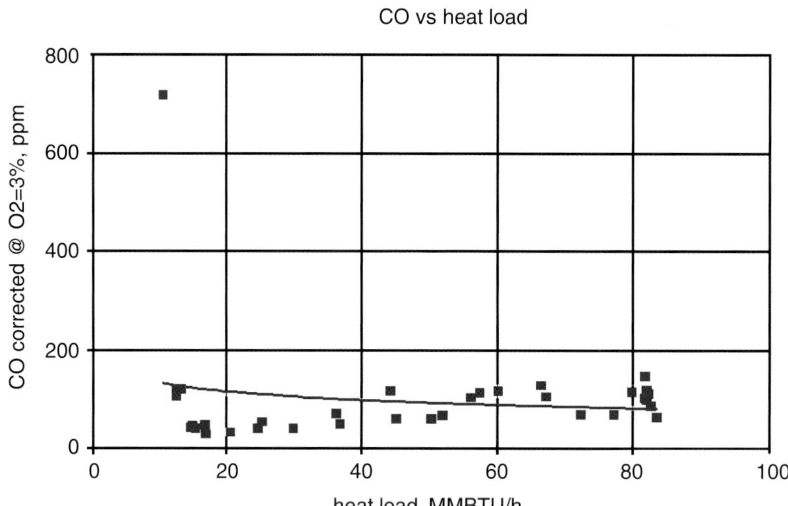

FIGURE 19.39 CO vs. heat load with firing natural gas with ambient air in the packaged industrial boiler equipped with the Todd Combustion low NOx burner.

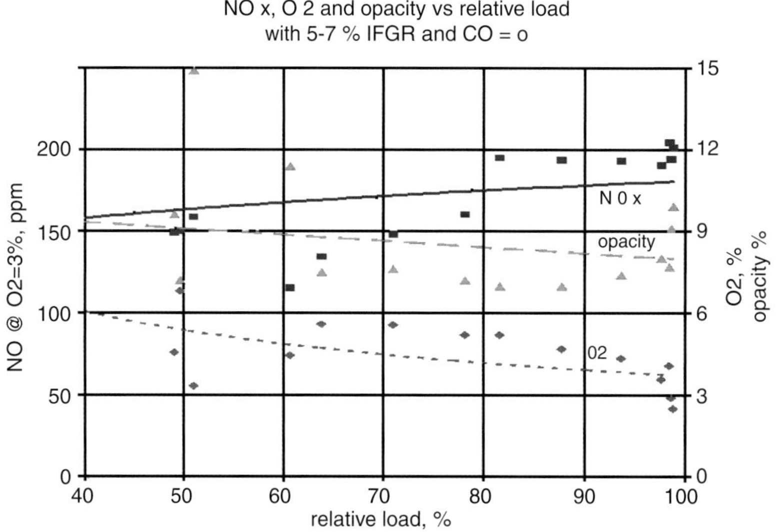

FIGURE 19.40 NOx, O_2, and opacity vs. relative heat load with firing No. 6 oil with ambient air in the packaged industrial boiler equipped with the Todd Combustion low NOx burner.

- pre-heated air (415 vs. 80°F, or 213 vs. 27°C)
- excess air 15% (vs. 23%)
- significantly lower CO level required (100 vs. 200 ppm at O_2 = 3%)
- required NOx = 0.22 lb/10^6 Btu (182 ppm at O_2 = 3%)
- available gas pressure < 10 psig (vs. ~15 psig, or 1 barg)
- available register draft loss = 7.25 in. (18 cm) w.c.

For this application, the IFGR was used to reduce NOx from 160–180 ppm to 120–150 ppm (14 to 18% reduction), while CO was kept in the range of 30 to 90 ppm. This reduction effect is very similar to the ambient air data where IFGR provided a 15 to 20% NOx reduction. These data confirm that the effectiveness of IFGR, established with ambient air, is similar to preheated air.

The full load test data shown in Figure 19.42 have confirmed a previous conclusion based on the data obtained on ambient air, that the ratio between injector and total gas flow rates is the major parameter dictating NOx reduction under

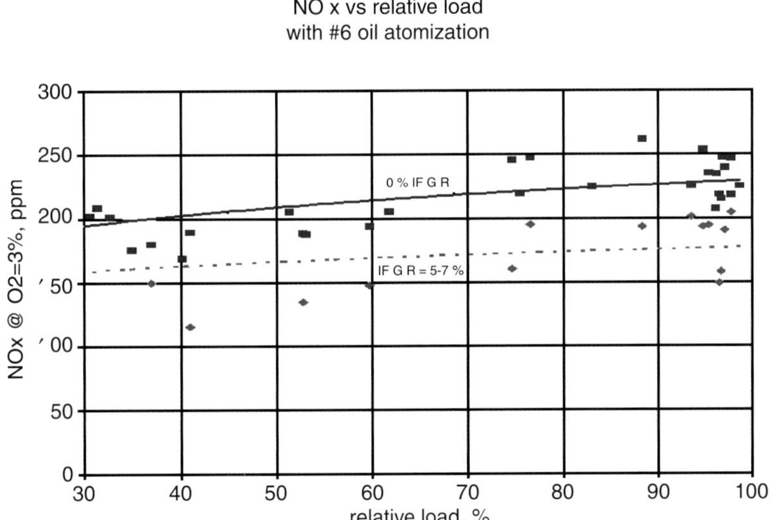

FIGURE 19.41 Internal FGR impact on NOx with firing No. 6 oil with ambient air in the 30–100% load range.

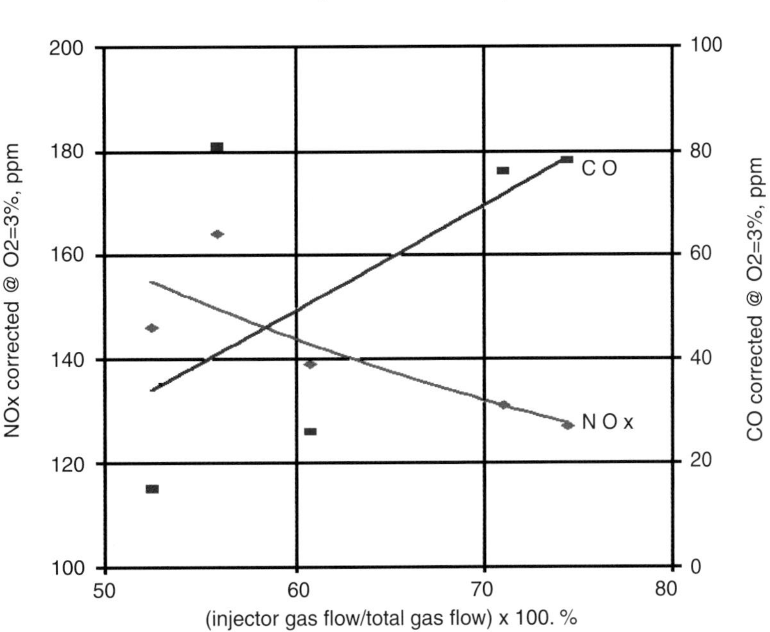

FIGURE 19.42 NOx and CO emissions vs. injector gas flow rate/total gas flow rate ratio, at the boiler equipped with the Todd Combustion low-NOx burner firing natural gas with preheated air at full load.

the natural gas staged combustion conditions formed with SLNB implementation on packaged boilers — not only on ambient air but also on preheated air.

There was an unsuccessful attempt to find an O_2 influence on NOx in the data, meaning that a regular dependence of NOx on O_2 does not work for the gas staged combustion conditions formed by an SLNB. This confirms the conclusion made based on the test data for ambient air.

Unlike the typical well-documented test data obtained with conventional low-NOx burners, indicating an NOx level that

FIGURE 19.43 An ultra-low emissions burner.

increases with load, this was not experienced during any of the SLNB testing. For the preheated air application, the full load NOx was actually slightly lower as compared to the data at partial loads. In any case, the test data described above have shown that in the 4:1 regular emission turndown, burner characteristics have matched today's requirements for both ambient and preheated air.

19.5 ULTRA-LOW EMISSION GAS BURNERS

19.5.1 Background

NOx formation fundamentals were used to design an ultra-low-NOx burner with a gas injection and mixing system radically different from all other commercially available low-NOx burners. The goal was to produce a burner capable of reliable operation with single-digit NOx generation while firing nitrogen-free gaseous fuels, such as natural gas. A new gas mixing approach was incorporated into an established burner geometry that had been optimized over the years at the International Flame Research Foundation (IJmuiden, The Netherlands) to provide an extremely stable flame.

Previous low-NOx burner designs focused primarily on the reduction of NOx through techniques to reduce the formation of thermal NOx. Techniques such as fuel-air staging, flue gas recirculation, and steam injection serve to lower NOx emissions into the 20 to 30 ppm range by lowering the peak flame temperatures. These methods do little to address the other mechanism for NOx formation — prompt NOx — and thus are not able to reliably reach single-digit NOx levels. Because some of these techniques also rely on delaying combustion and lowering reaction temperatures, incomplete combustion leads to increased CO and VOC emissions.

In developing a new burner technology, the control of CO and VOC emissions in addition to NOx was targeted, and an ultra-low emissions burner, shown in Figure 19.43, was the result. In addition to reliably controlling NOx emissions below 9 ppm, CO emissions are kept below 25 ppm, and VOC emissions are kept to less than 3 ppm.

19.5.2 Implementation of the NOx Formation Theory for Ultra-low-NOx Emissions

The relationship between temperature, stoichiometry, and NOx formation was used as the basis for design of an ultra-low emissions burner for natural gas and other nitrogen-free fuels. Thermal NOx emissions are the major source of NOx from these fuels, with NOx being created through the following high-temperature reactions of atmospheric nitrogen and oxygen from the combustion air (see Chapter 6):

$$N + O_2 = NO + O$$

$$N + OH = NO + H$$

$$N_2 + O = NO + N$$

Thermal NOx formation can be reduced by controlling the peak flame temperature. While thermal NOx is dependent to some extent on oxygen availability, if the temperature can be lowered sufficiently, thermal NOx from a natural gas flame can be reduced to less than 1 ppm. Figure 19.44 shows the relationship between the adiabatic flame temperature and thermal NOx formation.

In attaining ultra-low-NOx levels, prompt NOx also becomes a significant emissions source. Under fuel-rich conditions, particularly when stoichiometry is under 0.6, both HCN and NH_3 can be formed through the extremely rapid reaction of CH with N_2 to form HCN and N. The following reactions are considered:

$$CH + N_2 = HCN + N$$

$$N + H_2 = NH + H$$

$$NH + H_2 = NH_3$$

$$HCN + O_2 = NO + HCO$$

Below a stoichiometry of 0.5, almost all NOx formed is attributable to prompt NOx. The rate of formation of prompt NOx is very rapid, being complete in under 1 ms. Although prompt NOx is temperature sensitive, the temperature sensitivity is not as great as with thermal NOx. Unlike thermal NOx, simply lowering the peak flame temperatures will not reduce the prompt NOx into the single-digit

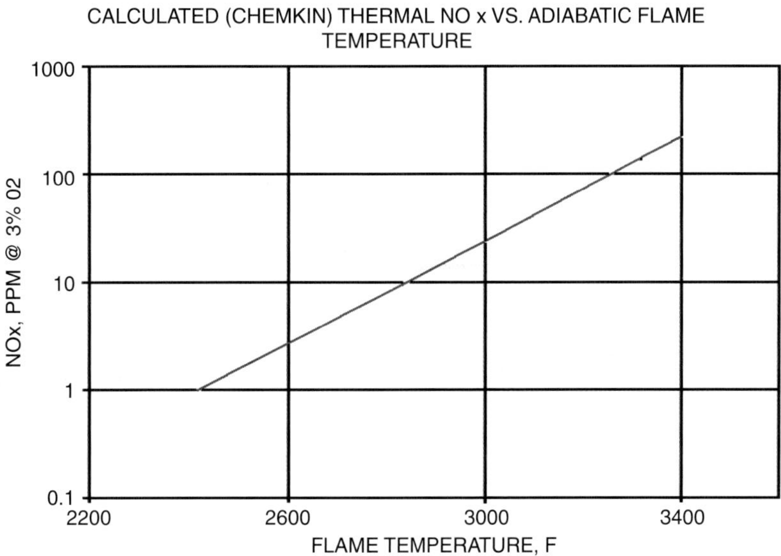

FIGURE 19.44 Relationship between the adiabatic flame temperature and thermal NOx formation.

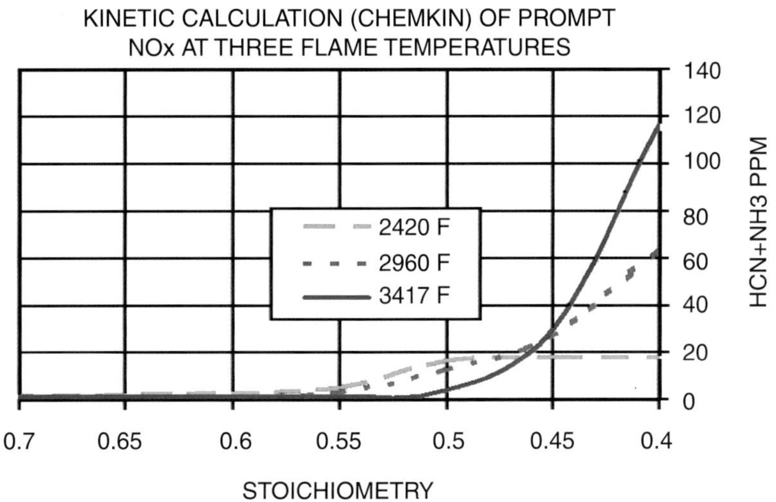

FIGURE 19.45 HCN and NH_3 formation at three flame temperatures.

range. As indicated by the curves in Figure 19.45, depicting HCN and NH_3 formation at three flame temperatures, under fuel-rich conditions and a temperature of 2400°F (1300°C), 20 ppm of prompt NOx still remain. To further control the formation of prompt NOx, it is necessary to take steps in the burner design to minimize the formation of sub-stoichiometric regions within the flame.

19.5.3 Ultra-low Emissions Burner Design

Most conventional low-NOx burners utilize staged combustion to delay the mixing of fuel and air. By creating an initial fuel-rich combustion zone and adding air downstream to complete combustion, oxygen availability is limited, peak flame temperature is lowered, and thermal NOx formation is reduced. It can be further reduced through other techniques, such as the addition of FGR or steam injection. However, the 20 ppm of prompt NOx created in the initial fuel-rich zone remains. It is this prompt NOx formation that has prevented conventional low-NOx burners from achieving sub-10 ppm NOx levels.

By "starting over" with NOx formation fundamentals, it was determined that the most direct method of achieving very low NOx emissions from a natural gas flame is to: (1) avoid fuel-rich regions with their corresponding potential for

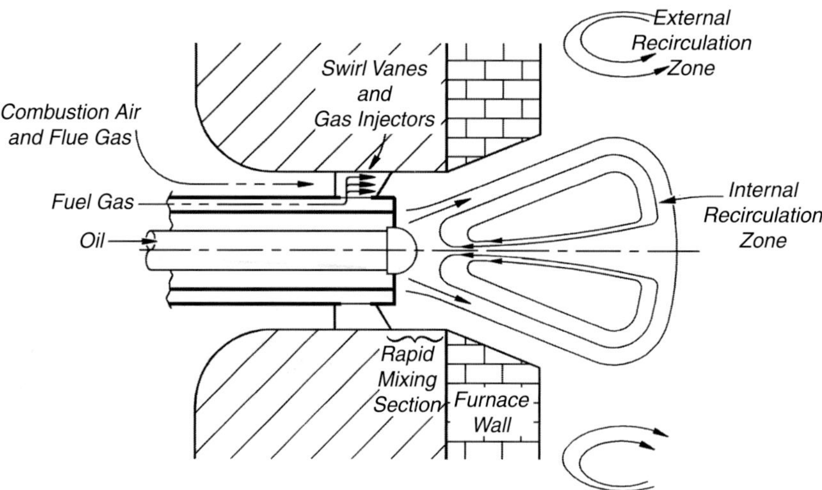

FIGURE 19.46 A nearly uniform fuel/air mixture at the ignition point.

prompt NOx, and (2) lower the flame temperature to reduce thermal NOx to the desired level. To accomplish this, a burner design that avoids fuel-rich regions by rapidly mixing gaseous fuel and air near the burner exit was developed. The rapid mixing results in a nearly uniform fuel/air mixture at the ignition point (Figure 19.46), which virtually eliminates prompt NOx formation. This rapid and complete combustion is also what results in the virtual elimination of both CO and VOC formation by the burner. Thermal NOx is then minimized using FGR, which is mixed with combustion air upstream of the burner, to control flame temperature. In effect, the burner performs like a pre-mix burner with one important distinction: because the fuel is added inside the burner, just upstream of the refractory throat, the extremely small premixed volume eliminates the possibility of flashback inherent in pre-mix burner designs.

Contrary to conventional LNB theory, increasing excess air *reduces* NOx formation in the ultra-low emissions burner. Because the burner employs near-perfect mixing, the fuel already has access to all of the oxygen required at the ignition point, and increasing excess air just serves to reduce the peak flame temperatures. Therefore, excess air has the same cooling effect as FGR, which provides major advantages in high excess-air combustion applications where FGR is impractical or unavailable. In boiler applications where FGR is available, it is preferred due to its lower impact on the boiler efficiency than high excess air. The rapid mix design is also what allows the burner to operate with preheated combustion for increased efficiency and still retain its single-digit emissions performance, by simply increasing the FGR or excess air rates to compensate for the higher air temperature.

19.5.4 Ultra-low Emissions Burner Operation

The basic ultra-low emissions burner consists of a parallel-flow air register with no moving parts. Combustion air premixed with FGR enters the register, and the entire mixture then passes through a set of axial swirl vanes. These vanes, which are attached to a central gas reservoir, have hollow bases that are perforated for gas injection. Thus, the swirl vanes also serve as the gas injectors and provide the burner's near-perfect fuel/air mixing (Figure 19.47).

For burner heat inputs greater than about 40×10^6 Btu/hr (12 MW_t), a second parallel-flow air sleeve surrounds the basic burner register. The outer sleeve contains a second set of gas injector vanes attached to an outer gas reservoir (Figure 19.48). These vanes, however, do not impart any swirl to the air flow. Air flow through the inner and outer burners is designated as "primary" and "secondary" air flow, respectively. Both the primary/swirled and secondary/axial zones operate with the same, near-perfect mixing.

In addition to sub-9 ppm NOx, another benefit of the rapid mix design is an extremely stable flame. Swirler geometry, burner internal geometry, and quarl expansion are matched to promote internal recirculation of a large amount of hot combustion gases. This enables the burner to operate at lower flame temperatures and NOx levels than other burners, with a "blow-off" point of about 3 ppm NOx. The ultra-low emissions burner flame remains stable at 60% FGR; therefore, the 25 to 30% FGR rate typically necessary for sub-9 ppm NOx does not begin to approach the burner's performance limits. The rapid combustion also results in a very short flame length. It is approximately half that of a staged combustion burner, which eliminates the potential for flame impingement, one of the most common problems experienced with conventional LNB retrofits.

For oil firing, the ultra-low emissions burner uses a conventional center-mounted atomizer gun assembly. The burner operates as a conventional, staged combustion LNB when firing oil. While there is no rapid mixing, fuel staging is provided by advanced atomizer designs, and air staging is provided by the burners' primary and secondary air flows. This allows emissions performance on oil firing equivalent to any other low-NOx burner available.

NOx emission data from the ultra-low emissions burner firing into a 4×10^6 Btu/hr (1.2 MW_t) firetube boiler for ambient, 300°F (150°C) preheat, and 500°F (260°C) preheat as a function of FGR rate are shown in Figure 19.49. NOx without FGR was a function of air preheat level and ranged from approximately 80 ppm with ambient air to 200 ppm with 500°F (260°C) preheat. The FGR rate required to produce a given NOx level varied with air preheat level; but, independent of the preheat level, NO_x could be reduced to approximately 3 ppm.

Similar results were obtained in a 25,000 lb/hr (11,000 kg/hr) watertube boiler (30×10^6 Btu/hr or 9 MW_t). The NOx emissions without FGR were higher for the larger burner but when enough FGR was added to reduce NOx below 20 ppm, both size burners exhibited very similar performance. Similar characteristics were also observed with air preheat on a 100×10^6 Btu/hr (30 MW_t) test furnace. Again, the larger burner produced higher NOx without FGR; but once FGR was added, the NOx from the 4×10^6 Btu/hr and 100×10^6 Btu/hr (1.2 and 30 MW_t) burners were almost identical for a given FGR rate.

Based on this test work, the ultra-low emissions burner has been applied to boilers ranging from 25,000 to 230,000 lb/hr (11,000 to 100,000 kg/hr) or 30 to 275×10^6 Btu/hr (9 to 80 MW_t). Repeatable performance has been observed across the various applications and size ranges, whether using FGR or excess air for NOx reduction, and single-digit emissions performance has been consistently achieved. Figure 19.50 shows the actual performance data taken from the application of an ultra-low emissions burner on a new 230,000 lb/hr (100,000 kg/hr) "A" type Nebraska Boiler. The ultra-low emissions burner can also be used on two-burner applications where NFPA 8501 guidelines are being followed. An example of this is shown in Figure 19.51.

Repeatable performance of this burner design has been demonstrated across a wide range of applications and sizes, and single-digit emissions performance has been consistently achieved. The operational performance of the burner has shown that ultra-low emissions technology has been proven as a reliable alternative to the control of combustion emissions through the use of flue gas cleaning technologies, such as

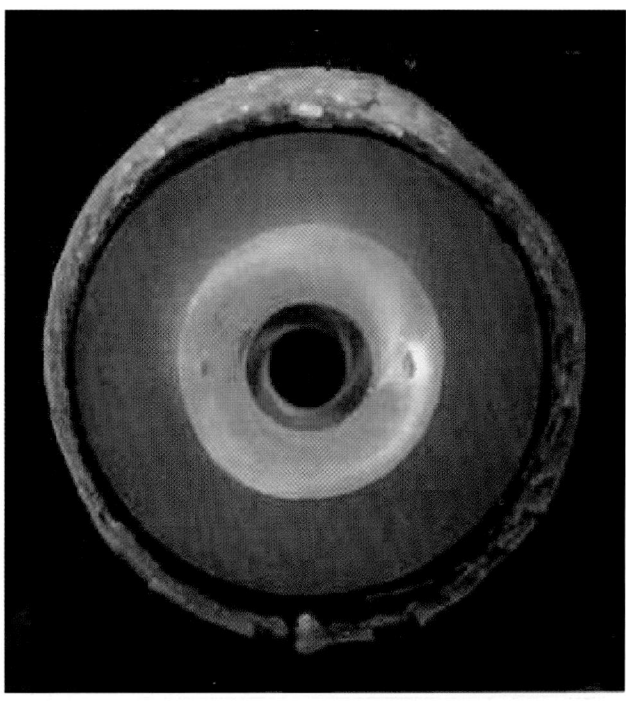

FIGURE 19.47 Swirl vanes also serve as the gas injectors, and provide the burner's near-perfect fuel/air mixing.

FIGURE 19.48 Outer sleeve contains a second set of gas injector vanes attached to an outer gas reservoir.

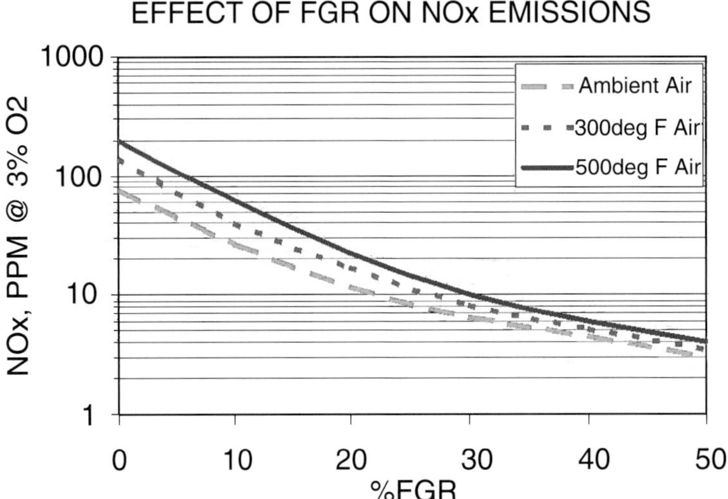

FIGURE 19.49 NOx emission from the ultra-low emissions burner firing into the firetube boiler for ambient, 300°F preheat, and 500°F preheat as a function of FGR rate.

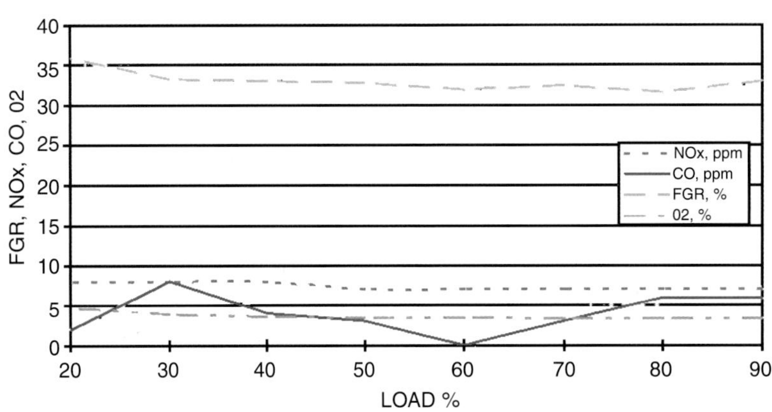

FIGURE 19.50 Data taken from the application of an ultra-low emissions burner on a new 230,000 lb/hr "A" type Nebraska boiler.

SCR or SNCR. In addition, the use of an ultra-low emissions burner can be accomplished at much lower capital installation costs, lower annual operating costs, and without the added complication of ammonia handling and emissions.

19.6 ATOMIZERS FOR BOILER BURNERS

Boiler burners are designed to utilize the many types of mechanical, dual fluid, and occasionally rotary cup oil atomizers presently used in most power plants. Most of these atomizers are fitted onto oil "guns" similar to the one shown in Figure 19.52. For dual fluid, these range from steam assist, air blast, or other "external mix" atomizers (depicted in Figure 19.53), the most common of which is the "Y-Jet," to "internal mix" atomizers. The atomizing medium is typically either steam or pressurized air.

Internal mix atomizers typically require a relatively large amount of atomizing medium consumption, on the order of 0.2 to 0.3 lb medium per 1 lb fuel, but provide excellent atomization quality, with a typical SMD of 75 μm at a fuel viscosity of 100 SSU. The supply pressure for the atomizing medium must be kept above the oil pressure to prevent oil from entering the atomizing medium supply lines. Operating pressures for internal mix atomizers range between 85 and 300 psig (5.8 and 20 barg) for the oil, with the atomizing medium pressure typically regulated to a differential pressure 15 to 30 psig (1 to 2 barg) higher than the oil. The turndown

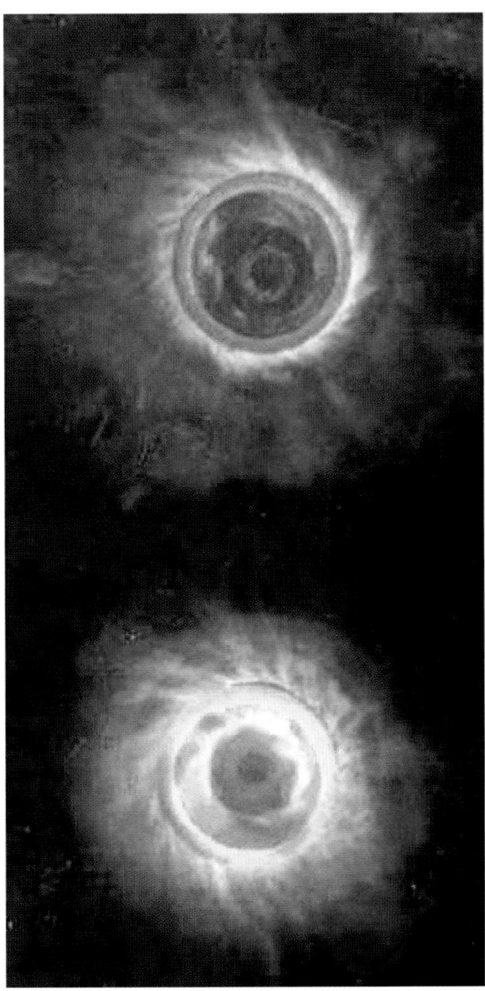

FIGURE 19.51 The ultra-low emissions burner can also be used on two-burner applications where NFPA 8501 guidelines are being followed.

ratio for these atomizers, depending on the oil supply pressure, can be as high as 10:1.

External mix atomizers such as the Y-Jet require much less steam consumption, on the order of 0.05 to 0.07 lb medium per 1 lb fuel. However, the atomization quality is slightly reduced with a typical SMD of 125 μm at a fuel viscosity of 100 SSU. Due to the external mixing of these atomizers, there is very little chance for oil to enter the atomizing medium supply lines. Therefore, these atomizers do not require that the supply pressure for the atomizing medium be kept above the oil pressure. Operating pressures for Y-Jet type atomizers range between 85 and 300 psig (5.8 and 20 barg) for the oil, with the atomizing medium pressure typically regulated to a constant pressure of 85 to 150 psig (5.8 to 10 barg). The turndown ratio for these atomizers, depending on the oil supply pressure, can be as high as 10:1.

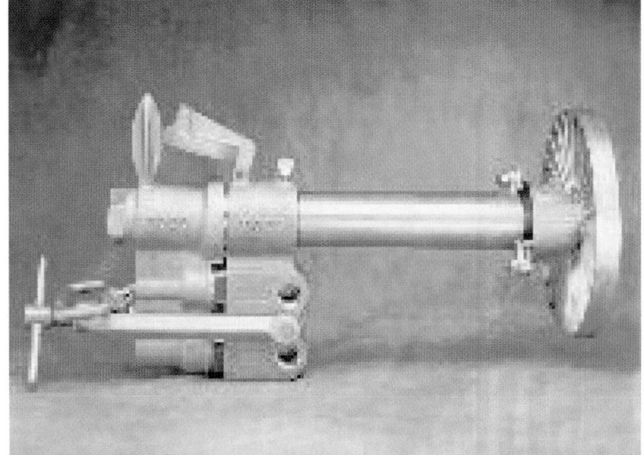

FIGURE 19.52 Atomizers are fitted onto oil "guns."

The two main types of mechanical atomizers are simplex atomizers and spill-return atomizers. Oil enters a mechanical atomizer through slots that expel the oil tangentially into a circular "whirling" chamber. The vortex created by the tangential entry into the whirling chamber is essential for good atomization quality. For a spill-return flow atomizer, a portion of the supply flow will return back into the oil gun through a series of holes located at the back of the atomizer. The remaining oil exits the atomizer through either a single hole or a series of holes located at the front end (furnace side) of the atomizer. The "return" oil will be recirculated through the oil supply system. Simplex atomizers are simply a spill-return atomizer without the return flow. Burner performance has been improved using specially designed return flow atomizers to maximize combustion efficiency and minimize NOx.

These atomizers produce good atomization quality at full load, with a typical SMD of 125 μm at a fuel viscosity of 100 SSU. However, atomization degrades rapidly with these atomizer types and limiting turndown. Operating oil pressures for simplex-type atomizers range between 300 and 600 psig (20 and 40 barg). The turndown ratio for these atomizers, depending on the oil supply pressure, is very low — typically 2:1.

The delivered flow supplied by a spill-return atomizer is dependent on the return "back" pressure, which is essentially the same pressure the atomizer would require if it were operating in simplex mode. The differential between supply and return oil pressures, which can range between 100 and 350 psig (7 and 24 barg) allows these atomizers a greater turndown range — up to 4:1. Operating supply pressures for spill-return type atomizers typically range between 400 and 10,000 psig (30 and 700 barg).

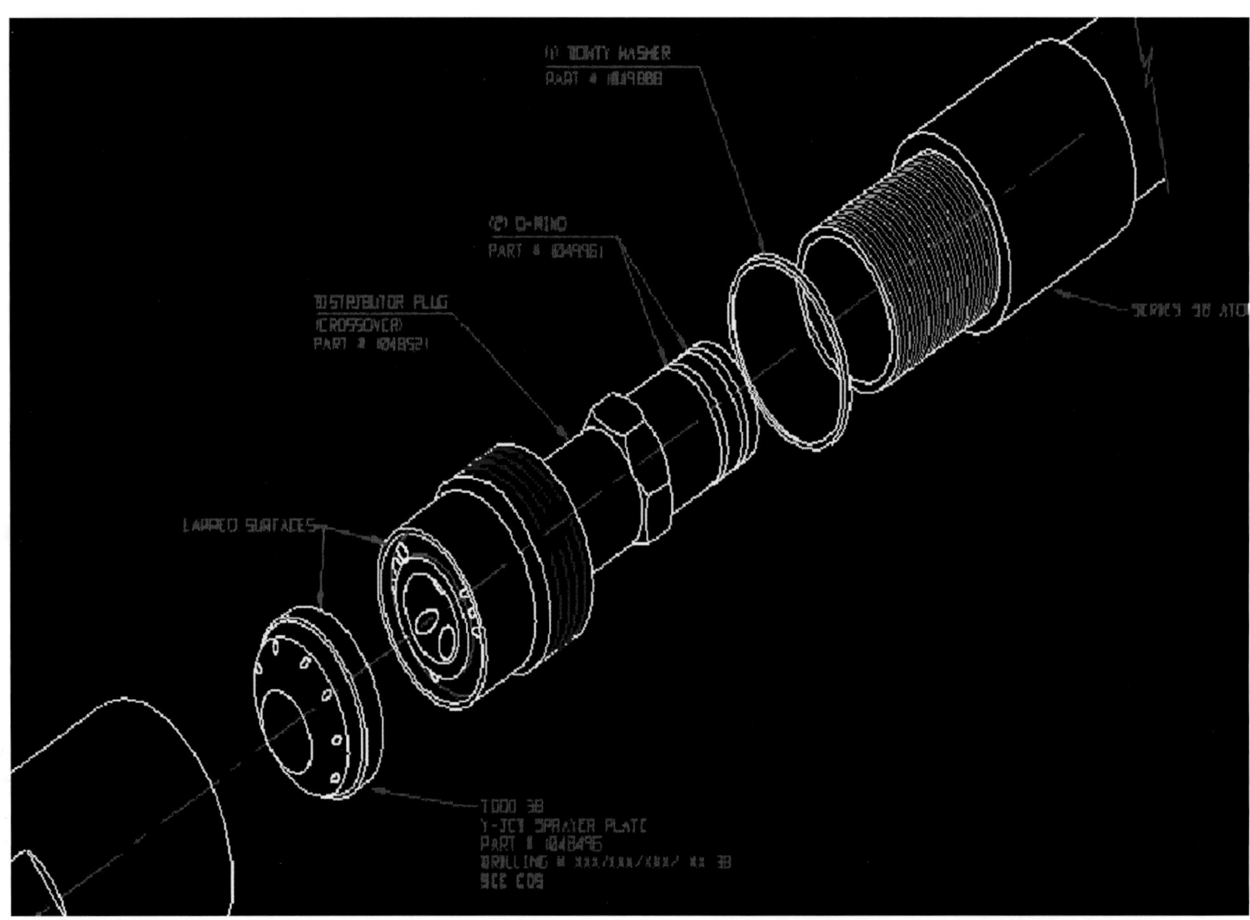

FIGURE 19.53 Steam assist, air blast, or other "external mix" atomizers.

Chapter 20
Flares

Robert Schwartz, Jeff White, and Wes Bussman

TABLE OF CONTENTS

20.1 Flare Systems ... 590
- 20.1.1 Purpose ... 590
- 20.1.2 Objective of Flaring ... 590
- 20.1.3 Applications ... 591
- 20.1.4 Flare System Types ... 591
- 20.1.5 Major System Components ... 593

20.2 Factors Influencing Flare Design ... 594
- 20.2.1 Flow Rate ... 594
- 20.2.2 Gas Composition ... 594
- 20.2.3 Gas Temperature ... 595
- 20.2.4 Gas Pressure Available ... 596
- 20.2.5 Utility Costs and Availability ... 596
- 20.2.6 Safety Requirements ... 597
- 20.2.7 Environmental Requirements ... 597
- 20.2.8 Social Requirements ... 598

20.3 Flare Design Considerations ... 598
- 20.3.1 Reliable Burning ... 599
- 20.3.2 Hydraulics ... 599
- 20.3.3 Liquid Removal ... 600
- 20.3.4 Air Infiltration ... 601
- 20.3.5 Flame Radiation ... 601
- 20.3.6 Smoke Suppression ... 603
- 20.3.7 Noise/Visible Flame ... 604
- 20.3.8 Air/Gas Mixtures ... 605

20.4 Flare Equipment 605

20.4.1 Flare Burners 605

20.4.2 Pilots, Ignitors, and Monitors 614

20.4.3 Knockout Drums 618

20.4.4 Liquid Seals 619

20.4.5 Purge Reduction Seals 620

20.4.6 Enclosed Flares 622

20.4.7 Flare Support Structures 623

20.4.8 Flare Controls 624

20.4.9 Arrestors 629

20.5 Flare Combustion Products 629

20.5.1 Reaction Efficiency 630

20.5.2 Emissions 631

20.5.3 Dispersion 632

References 633

20.1 FLARE SYSTEMS

During the operation of many hydrocarbon industry plants, there is the need to control process conditions by venting gases and/or liquids. In emergency circumstances, relief valves act automatically to limit equipment overpressure. For many decades of the last century, process vents and pressure relief flows were directed, individually or collectively, to the atmosphere unburned. Gases separated from produced oil were also vented to the atmosphere unburned. The custom of unburned venting began to change in the late 1940s when increased environmental awareness and safety concerns created the desire to convert vents to continuously burning flares.

Burning brought about the need for pilots and pilot ignitors and the need for awareness of the design factors and considerations imposed on a system by a flame at the exit. In many cases, the desirable flaring of the gases was accompanied by objectionable dense black smoke as shown in Figure 20.1. In addition to their development of flare pilots and ignition systems, industry pioneers John Steele Zink and Robert Reed[1] invented the first successful smokeless flare burner (Figure 20.2) in the early 1950s. This invention was an important point in the transition from unburned vents to flaring and from vent pipes to burners specifically designed for flare applications.

While the combustion fundamentals discussed earlier in this book continue to apply, flare burners differ from process and boiler burners in several respects, including:

1. The composition of the gases handled by a flare often vary over a much wider range.
2. Flares are required to operate over a very large turndown ratio (maximum emergency flow down to the purge flow rate).
3. A flare burner must operate over long periods of time without maintenance.
4. Flare burners operate at high levels of excess air as compared to other burners.
5. Many flare burners have an emergency relief flow rate that produces a flame hundreds of feet long with a heat release of billions of Btu per hour (Figure 20.3).

20.1.1 Purpose

The wide range of applications for flares throughout the hydrocarbon and petrochemical industries challenges plant owners and designers as well as the flare equipment designer. The purpose of this chapter is to provide an understanding of the design considerations and factors influencing flare system and equipment design. The most frequently used flaring techniques and associated equipment are also discussed.

20.1.2 Objective of Flaring

Regardless of the application, flare systems have a common prime objective: safe, effective disposal of gases and liquids…at an affordable cost. There should be constant awareness that flare system design and operation must never compromise the prime objective.

FIGURE 20.1 Typical early 1950s flare performance.

FIGURE 20.2 An early model smokeless flare.

FIGURE 20.3 Major flaring event. (Note that the stack height is over 300 ft (90 m).)

20.1.3 Applications

Within the hydrocarbon and petrochemical industries, from the drilling site to the downstream petrochemical plant and at many facilities in between, flares are utilized to achieve the prime objective. Individual flare design capacity can range from less than 100 to more than 10 million lb/hr. Material released into a flare system is often a mixture of several constituents that can vary from hydrogen to heavy hydrocarbons and may at times include inert gases. Some of the heavy hydrocarbons may be in the gaseous state when released into the system, but will be condensed as they cool.

While this chapter focuses on flares and flaring in the hydrocarbon industry, many of the subjects discussed also relate to flare applications in other industries.

The design requirements for a given facility are seldom identical to those of any other facility. This variation, plus the wide range of flare applications and site conditions, often requires that the flare system be custom designed.

20.1.4 Flare System Types

Flares for service in the hydrocarbon and petrochemical industries are generally of the following types or combinations thereof:

- single point
- multi-point
- enclosed

20.1.4.1 Single-Point Flares

Single-point flares can be designed with or without smoke suppression equipment and are generally oriented to fire

FIGURE 20.4 Typical elevated single-point flare.

FIGURE 20.5 Typical pit flare installation.

upward, with the discharge point at an elevated position relative to the surrounding grade and/or nearby equipment (Figure 20.4). Occasionally, a single-point flare is positioned to fire more or less horizontally, usually over a pit or excavation (Figure 20.5). Horizontal flares are generally limited to drilling and production applications where there is a high probability of nonrecoverable liquids.

20.1.4.2 Multi-point Flares

Multi-point flares are used to achieve improved burning by routing the gas stream to a number of burning points. For refinery or petrochemical plant applications, multi-point flares are usually designed to achieve smokeless burning of all flows. Such flares often divide the multiple burning points into stages to facilitate better burning. The multiple burning points can be arranged in arrays located at or near grade (Figure 20.6) or at an elevated position (Figure 20.7).

FIGURE 20.6 A grade-mounted, multi-point LRGO flare system.

FIGURE 20.7 Elevated multi-point LRGO flare system.

20.1.4.3 Enclosed Flares

Enclosed flares are constructed so as to conceal the flame from direct view. Additional benefits can be the reduction in noise level in the surrounding community and minimization of radiation. Capacity can be the system maximum, but is often limited to a flow rate that will allow the connected facility to start up, shutdown, and operate on a day-to-day basis without exposed flame flaring. Multiple enclosed flares are sometimes used to achieve the desired hidden flame capacity. Each of the two units shown in Figure 20.8 is designed for 100 metric tons/hr of waste gas from ethylene furnaces during startups.

20.1.4.4 Combination Systems

A common combination is an enclosed flare of limited capacity paired with an elevated flare (Figure 20.9) that is sized for the maximum anticipated flow to the system. Such a pairing results in a flare system that only has an exposed flame during major upset or failure events. Other pairings, such as an elevated flare with a multi-point flare, have also been used (Figure 20.11).

20.1.5 Major System Components

Each flare system type has its own set of required components. In addition, systems may include components that are noted below as being optional. These optional components address special needs, such as smoke suppression or liquid removal. Optional equipment can also help reduce cost or aid system operation. It should be noted that there is a difference between the flare burner, or burners, required for each type of flare system.

20.1.5.1 Single-Point Flares

For a single-point flare, the major required and optional components are:

1. flare burner, with or without smoke suppression capability:
   a. one or more pilots
   b. pilot ignitor(s)
   c. pilot flame detector(s)
2. support structure, piping, and ancillary equipment
3. purge reduction device (optional)
4. knockout drum (optional)
5. liquid seal (optional)
6. auxiliary equipment:
   a. smoke suppression control (optional)
   b. blower(s) (optional)
   c. flow, composition, heating value, or video monitor (optional)

20.1.5.2 Multi-point Flares

For a multi-point flare, the major required and optional components are:

1. two or more multi-point flare burners
2. pilot(s), pilot ignitor(s), and pilot flame detector(s)
3. if elevated, support structure and ancillary equipment
4. a fence to limit access and reduce flame radiation and visibility (optional)
5. knockout drum (optional)
6. liquid seal (optional)
7. piping
8. auxiliary equipment:
   a. staging equipment and instrumentation (optional)
   b. smoke suppression means where very large turndown is required
   c. flow, composition, heating value, or video monitoring (optional)

FIGURE 20.8 Multiple ZTOF installation in an ethylene plant.

FIGURE 20.9 Combination ZTOF and elevated flare system.

20.1.5.3 Enclosed Flares

For an enclosed flare, the major required and optional components are:

1. flare burners, with or without smoke suppression capability
2. pilot(s), pilot ignitor(s), and pilot flame detector(s)
3. enclosure/structure with protective refractory lining
4. a fence to limit access
5. knockout drum (optional)
6. liquid seal (optional)
7. piping and optional heat shielding
8. auxiliary equipment:
   a. staging equipment and instrumentation (optional)
   b. flow, composition, heating value, or video monitoring (optional)

20.2 FACTORS INFLUENCING FLARE DESIGN

As one approaches the specification of a flare system, there must be an awareness of certain factors that influence size, safety, environmental compliance, and cost. Major factors influencing flare system design[2] include:

- flow rate
- gas composition
- gas temperature
- gas pressure available
- utility costs and availability
- safety requirements
- environmental requirements
- social requirements

Information regarding each of these factors is normally available to the plant designer and/or the plant owner. These factors define the requirements of the flare system and should be made available to the flare designer as early in the design process as possible.

In reviewing the list of factors, it can be seen that the first four factors are all determined by the source(s) of the gas being vented into the flare header. The next factor is related to the design of the facility itself and its location. Safety, environmental, and social requirements all relate to regulatory mandates, the owner's basic practices, and the relationship between the facility and its neighbors. A discussion of each factor is provided below.

20.2.1 Flow Rate

The flare system designer relies heavily on the flow data provided. Therefore, the data must realistically reflect the various flow scenarios. Overstatement of the flows will lead to oversized equipment, which increases both capital and operating costs, and can lead to shorter service life. Understatement can result in an ineffective or unsafe system.

Flow rate obviously affects such things as the mechanical size of flare equipment. Its influence, however, is much broader. For example, increased flow generally results in an increase in thermal radiation from an elevated flare flame, which in turn will have a direct impact on the height and location of a flare stack.

The maximum emergency flow rate can occur during a major plant upset such as the total loss of electrical power or cooling water. However, some processes have their maximum flow rate under less obvious emergency conditions such as partial loss of electrical power whereby, for example, pumps continue to supply feedstock to a disabled section of the plant.

The duration of the maximum flow rate can affect flare system design in a number of ways. For example, the length of time a worker is exposed to heat from the flare flame can affect the choice of allowable heat flux. Usually, a very short duration relief into a flare system can result in a relatively high allowable radiation. In contrast, a very long duration, high flow relief may require a lower design allowable radiation level.

In the past, the maximum flow rate was sometimes determined by summing the flow rates of each of the connected relief devices. This approach resulted in an unrealistically large maximum flow rate because the assumption that all the connected devices relieve simultaneously is often false. Modern plant design and analysis tools such as dynamic simulation allow the process designer to define more appropriately the maximum flow rate to the flare. Careful attention to the design of control and electrical power systems can significantly reduce flare loads as well.[3]

In addition to the maximum flow conditions, it is also important to explicitly define any flow conditions under which the flare is expected to burn without smoke. These flow conditions can come from process upsets, from incidents such as a compressor trip-out, or from various operations of the plant, including startup, shutdown, and blowdown of certain equipment. Attempts to shortcut the establishment of factually based smokeless burning scenarios by setting the smokeless flow rate as a percentage of the maximum emergency flow rate can lead to disappointment or needless expense.

20.2.2 Gas Composition

Gas composition can influence flare design in a number of ways. The designer should be given the gas composition for each of the flow conditions identified above and for any special gases that may be in use, such as pilot fuel or purge gas. By studying the gas composition, its combustion characteristics and the identity of potential flue gas components can be determined.

For example, the composition reveals the relative presence of hydrogen and carbon. The weight ratio of hydrogen to carbon in a gas is one of the parameters that can indicate the smoking tendency of the gas.[4] The influence of the weight ratio of hydrogen to carbon, often referred to as the H/C ratio, on the smoking tendency is illustrated in Figure 20.10. Figure 20.10 shows the flame produced by burning three different gases using the same flare equipment and operating conditions. The flame produced by burning a 25 MW well head natural gas (H/C = 0.27) is clean, as shown in Figure 20.10(a). The flame in Figure 20.10(b) reflects the burning of propane (H/C = 0.22). Note that the propane flame has some trailing smoke, has a yellow color much closer to the flare burner, and is more opaque when compared to the natural gas flame. The dense black smoke and dark flame shown in Figure 20.10(c) was produced by burning propylene (H/C = 0.17). Note that a portion of the flame is being cloaked or shrouded by the smoke. The fact that the smoke hides part of the flame must be accounted for when calculating the radiation from the flare flame.

(a)

(b)

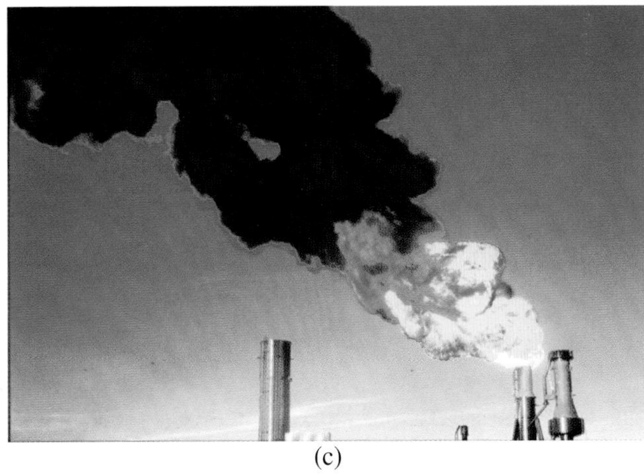

(c)

FIGURE 20.10 Comparison of the flame produced by burning (a) 25 MW well head natural gas, (b) propane, and (c) propylene.

Composition analysis also will reveal the presence of nonhydrocarbons such as hydrogen sulfide or inerts. Such gases might require special metallurgies or design considerations such as ground level concentration analysis. Composition combined with the flow rate allows determination of the volume flow or mass flow of gases to be handled by the flare system.

The practice of defining a stream by its bulk properties alone (molecular weight [MW], lower heating value [LHV], upper/lower explosive limits [UEL/LEL], etc.) can disguise safety hazards or prevent equipment and operating cost reductions that would otherwise be recognized by the flare designer. For example, a 28 to 30 MW gas could be ethane, ethylene, nitrogen, carbon monoxide, air, or even a mixture of hydrogen and xylene. A flare system to handle some fixed amount of 28 MW gas would have very different design and operating characteristics, depending on the actual gas composition involved. Ethylene may tend to smoke, but remains lit and stable at high exit velocities. Carbon monoxide, on the other hand, will not smoke, but can be difficult to keep lit even at moderate to low exit velocities. Radiation, possible relief gas enrichment, purge requirements, and the potential for condensation at ambient temperatures are other examples of the impact that gas composition can have on flare design.

20.2.3 Gas Temperature

In addition to the impact of relief gas temperature on thermal expansion, gas volume, and metallurgical requirements, there is the more subtle effect of gas temperature on the potential for some components of the gas to condense. Possible condensation or two-phase flow necessitates liquid removal equipment to avoid a higher smoking tendency and/or the possibility of a burning liquid rain. Condensation at low or no flow conditions will result in formation of a vacuum condition in the

FIGURE 20.11 Combination elevated LRGO and Utility flare system.

flare header and the resulting potential to draw air in through the flare tip or through piping leaks. Liquid seals are sometimes used to address this hazard. However, gas temperature can affect liquid seal design and operation. Hot gases will tend to boil off the seal fluid, sometimes very suddenly. On the other hand, extremely cold gases present a freezing scenario that could completely block waste gas flow.

While a flare stack may appear to be unrestrained and therefore free to expand, there can be mechanical design challenges as a result of large gas temperature variations. Header piping growth, relative movement of utility piping, and stack guy wire tensioning are just three areas where problems can arise. Both high and low temperatures have the potential to create issues that affect the design of the stack. In cases where the relief gas source pressure is extremely high, the plant designer should account for cooling by expansion across the relief or vent valve. Where the gas temperature at the source is significantly different from ambient, it is advisable to estimate the heat loss or gain through the flare header walls from the source to the flare stack and to determine the resulting gas temperature at the flare. Attention to such details can result in a reduced cost for the stack.

20.2.4 Gas Pressure Available

The gas pressure available for the flare is determined by hydraulic analysis of the complete pressure relief system from the vent or pressure-relieving devices to the flare burner. Each major flowing condition should be analyzed to determine the pressure at each relief or vent in each branch of the flare header. This pressure is usually limited by the lowest allowable back-pressure on any relief valve in the system. The limit applies to all flowing conditions, regardless of whether or not the limiting relief valve contributes to the flowing condition under study.

In most flare systems, much of the system pressure drop is due to flare header piping losses with little pressure drop remaining for the flare burner. Such system designs may not maximize the value of the gas pressure in promoting smokeless burning. Smokeless burning can be enhanced by converting as much of the gas pressure available as possible into gas momentum. In addition, redistributing the system pressure drop to provide more pressure at the flare tip can reduce the overall system cost.

Another benefit of taking a greater pressure drop across the flare burner is the increase in gas density in the flare header, which can lead to a smaller flare header size and reduced piping cost. More pressure at the flare tip generally means a smaller flare burner and, consequently, lower purge flows. The enhancement of smokeless burning and the decrease in purge gas requirements both reduce the day-to-day operating cost. Both capital and operating costs can be reduced in this manner.

Available gas pressure at the flare can be defined as total pressure at the flare inlet, or as static pressure in a specific size inlet pipe. Static pressure is the pressure applied by the gas to the walls of the pipe. That is, this is the pressure sensed by a pressure gage mounted on a simple nozzle in the side of the pipe. This is also the pressure that determines gas density. Total pressure is the sum of the static pressure and the velocity pressure at a given point in the piping (e.g., the stack inlet). When static pressure is used to define available gas pressure, the plant designer should also specify the anticipated inlet size.

20.2.5 Utility Costs and Availability

In many cases, the momentum of the gas stream alone is not sufficient to provide smokeless burning. In such cases, it is necessary to add an assist medium to increase the overall momentum to the smokeless burning level. The most common medium is steam, which is injected through one or more groups of nozzles. An alternative to steam is the use of a large volume of low-pressure air furnished by a blower. Local energy costs, availability, reliability, and weather conditions must be taken into account in selecting the smoke-suppression medium.

Purge and pilot gas must be supplied to the flare at all times. The amount of each gas required is related to the size of the flare system. The composition of the purge gas and/or the composition of the waste gas can also influence the purge gas requirement. Pilot gas consumption can be affected by the combustion characteristics of the waste gases. The gases used for purge gas and to fuel the flare pilots should come from the most reliable source available.

Purge gas can, in principle, be any noncorrosive gas that does not contain oxygen and does not go to dew point at any expected conditions. An attractive option for purge gas may be a mixture of nitrogen and a non-hydrogen-containing fuel gas such as natural gas or propane. For example, a 300 Btu/scf mixture of nitrogen and propane can be effective as a purge medium. Such a mixture presents a number of benefits when compared to fuel gas alone, including:

- reduced CO_2 emissions
- potential cost savings if nitrogen is less expensive
- higher reliability because either supply alone can function as purge gas
- reduced wear-and-tear on the flare burner

20.2.6 Safety Requirements

Almost every aspect of flare design involves some safety concerns. Safety concerns include thermal radiation from the flare flame, reliable ignition, hydraulic capacity, air infiltration, and flue gas dispersion. Certain aspects of safety are dictated by the basic practices of the owner. For example, the owner's safety practices usually set the allowable radiation from the flare flame to people or equipment. Therefore, it is not surprising that the allowable radiation level will vary from owner to owner.

A common point of variation involves the treatment of solar radiation relative to the allowable level. Experience has shown that solar radiation need not be considered in the majority of designs. In practice, consideration of solar radiation is a complex issue that does not lend itself to a simple solution. The solar radiation question involves a number of variables and is site specific. By way of example, it would be appropriate to include solar radiation in the design basis if there is substantial likelihood that a worker can become exposed to the maximum flare radiation and the sun's radiation in an additive manner.

There are several sources for guidance on the allowable radiation level. The most widely referenced is American Petroleum Institute (API) Recommended Practice (RP) 521.[5] Most specifications call for a maximum radiation level of 1500 Btu/hr/ft^2 (4.73 kW/m^2) for emergency flaring conditions. Some specifications define an additional radiation level limit of 500 Btu/hr/ft^2 (1.58 kW/m^2) for unprotected individuals during long-duration flaring events. Special consideration should be given to radiation limits for flares located close to potential public access areas along the plant boundary where public exposure could occur. See Section 20.3.5 for more detail on flame radiation.

Reliable ignition at the flare tip is one of the most fundamental safety requirements, ensuring that gases released to the flare are burned in a defined location. Dependable burning also ensures destruction of potentially toxic releases. The prime objective (Section 20.1.2) demands reliable burning of the flare. The subject is covered in more detail in Sections 20.3.1 and 20.4.2.

Hydraulics of the flare system determine back-pressure on relief valves. Improper initial system sizing or subsequent additions to the flare relief loads can prevent a unit from achieving its maximum relief rate when necessary and create an over-pressure risk in the plant. Section 20.3.2 provides further discussion on the effect of relief valve selection on the design of the flare system.

Prevention of air infiltration should be a consideration when developing operations and maintenance plans for the flare system and connected equipment. Air sources include the flare tip exit, loop seals on vessels, low point drains, high point vents, and flanges. These issues are discussed in Section 20.3.4.

20.2.7 Environmental Requirements

Flares can affect their environment by generating smoke, noise, or combustion products. Regulatory agencies sometimes define limits in some or all of these areas. In many cases, it is necessary to inject an assist medium such as steam in order to achieve smokeless burning and to meet smoke emission regulations. The injection of the steam and the turbulence created by the mixing of steam, air, and gas cause the emission of sound. The sound level at various points inside and outside the plant boundary is often subject to regulation.

Other environmental concerns are the reaction efficiency and flue gas emissions. Pioneering tests conducted by the John Zink Company established that a properly designed and operated flare burner will have a combustion efficiency of more than 98%.[6] Emissions of NOx, CO, and unburned hydrocarbons (UHCs) were also determined during these tests. NOx, CO, and UHC emission factors for flares are available in AP-42.[7] These emission factors are widely accepted by regulatory authorities as a basis for flare permit emissions estimates. For

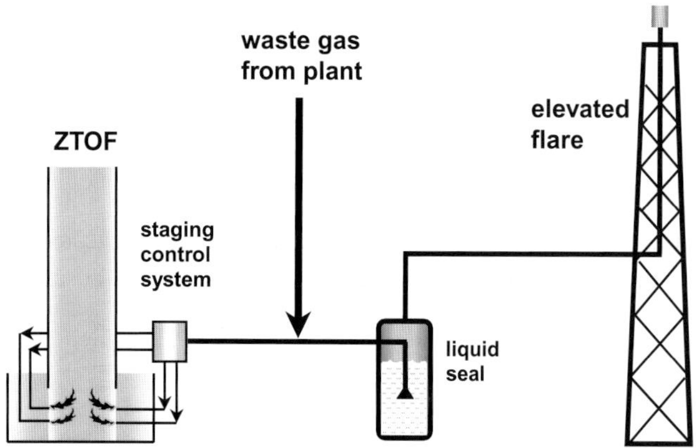

FIGURE 20.12 General arrangement of a staged flare system, including a ZTOF and an elevated flare.

FIGURE 20.13 John Zink Co. test facility in Tulsa, Oklahoma.

emissions estimates of SOx, it is often assumed that 100% of the available sulfur is converted to SO_2.

20.2.8 Social Requirements

Most emergency flare systems include a flare stack that is the tallest, or one of the tallest, structures in the plant. As a result, the flare flame is visible for great distances. Although the plant owner has complied with all environmental regulations, the flare system may not meet the expectations of the plant's neighbors. Public perception of the purpose and performance of the flare can place more stringent requirements on the flare design. For example, a smokeless flame may meet the regulatory requirements, but might be objectionable to the neighbors due to light or noise.

The John Zink Company recognized the need to reduce elevated flaring more than 30 years ago and invented the world's first successful enclosed ground flare for the elimination of day-to-day visible flaring. Since that time, many plants have included a Zink Thermal Oxidizer Flare (ZTOF) in their flare system. The general arrangement of a flare system that incorporates a ZTOF for day-to-day flaring rates and an elevated flare for emergency rates is shown in Figure 20.12. The liquid seal in the system acts to divert flow to the ZTOF until it reaches its maximum capacity. Any additional flow will pass through the liquid seal and be burned at the elevated flare tip. An installation of such a system is shown in Figure 20.9.

20.3 FLARE DESIGN CONSIDERATIONS

Having received information on the system defining factors set forth above, the flare designer must now apply his/her expertise to the following design considerations. A given project may require inclusion of all or only a few of these points, depending on the nature of the system information disclosed and the scope of the project. The flare designer must consider how decisions relating to each factor will affect the entire flare system as well as all the other factors. The prime objective of safe, effective disposal of gases and liquids should be used as a guiding principle as each appropriate consideration is incorporated into the overall flare design. The design considerations are:

1. reliable burning
2. hydraulics
3. liquid removal
4. air infiltration
5. flame radiation
6. smoke suppression
7. noise/visible flame
8. air/gas mixtures

Successful selection and operation of flare equipment require a clear understanding of these design considerations. The success and cost-effectiveness of a flare design are dependent on the skill and experience of the flare expert and his/her access to the latest state-of-the-art design tools and equipment. A key development tool is the ability to conduct flare tests at the high flow rates experienced in real plant operations. Facilities capable of conducting tests of this magnitude (Figure 20.13) represent a substantial capital investment and take on the characteristics of a complex process plant flare system. Insight into each consideration is set forth below.

20.3.1 *Reliable Burning*

Venting of waste gases can happen anytime during plant operation. Therefore, an integrated ignition system is required that can immediately initiate *and* maintain stable burning throughout the period of waste gas flow. Stable burning must be ensured at all flow conditions. An integrated ignition system includes one or more pilot(s), a pilot ignitor, pilot monitor(s), and a means to stabilize the flame.

In principle, all flares should have a continuous pilot flame to ensure reliable burning. This is especially true of refinery, petrochemical, and production field flares because they cannot be shut down unless and until the entire plant shuts down. In addition, such flares may be online for weeks or months before an unpredictable event that creates an immediate need for reliable ignition. Notable exceptions are landfill flares or biogas flares that operate continuously at substantial rates and include flame monitoring systems that automatically shut off waste gas flow in case of flame failure. Discontinuous pilots should only be considered in cases where *all* of the following conditions apply:

- The main flame remains lit and stable without a pilot at all design conditions.
- The main flame is monitored.
- The flare is shut down automatically on main flame failure.
- The flare shutdown does not create a safety hazard in the plant.

The number of pilots required can vary, depending on the size and type of the flare burner and its intended use. Flare pilots are usually premixed burners designed such that pilot gas and air are mixed together at a point remote from the flare burner exit and delivered through a pipe to the pilot tip for combustion. This ensures that the pilot flame is not affected by conditions at the flare burner exit (e.g., the presence of flue gas, inert gas, or steam). Pilot gas consumption varies according to the specific flaring requirements. However, there is a practical lower limit to the pilot gas consumption.

A pilot monitor is often required to verify the pilot flame. As a safety consideration, pilot ignition is usually initiated from a position remote from the flare stack. Either a flame front generator or direct spark pilot ignition can be used, depending on the system requirements. Further discussion of this important safety aspect is provided in Section 20.4.2.

There is a complex relationship between flare tip exit velocity, gas composition, tip design, and the maintenance of stable burning. There are a number of advantages in using the highest exit velocity possible, including minimum equipment size and optimal flame shape. In addition, because high discharge velocity tends to improve air mixing with a resultant reduction in soot formation, one can see that maximizing discharge velocity can help improve smokeless performance. It is important to note that discharge velocity can be constrained by the gas pressure available or concerns about flame stability. In some circumstances, such as VOC control, the discharge velocity may be limited by regulation. Early research on flare system design suggested limiting discharge velocity to 0.2 Mach due to stability concerns. It was later suggested that a discharge velocity of 0.5 Mach or higher could be used if proper flame stabilization techniques were employed. Flame stabilization techniques have been successfully employed for exit velocities of Mach 1 or greater.

Waste gas composition can significantly affect the allowable exit velocity. For example, a properly designed flare burner can maintain stable burning of propane at Mach 1 or greater. On the other hand, if the propane is mixed with a large quantity of inert gas, the maximum exit velocity must be limited to a much lower Mach number in order to ensure stable burning.

20.3.2 *Hydraulics*

Most flare systems consist of multiple relief valves discharging into a common flare manifold or header system. A key item influencing the flare system design is the allowable relief valve back-pressure. The system pressure drop from each relief valve discharge through the flare tip must not exceed the allowable relief valve back-pressure for all system flow conditions. The allowable back-pressure is typically limited to about 10% of the minimum relief valve upstream set pressure for conventional relief valves. The allowable relief valve back-pressure can be increased by the use of balanced pressure relief valves. Balanced valves can accept a back-pressure of about 30% of upstream set pressure in most cases. Where there is a wide variation in the allowable relief valve back-pressures, it may be economical to use separate high- and low-pressure flare headers.

Increasing the allowable relief valve back-pressure can have several effects on the flare system components, including:

- smaller manifold and header piping
- smaller knockout and liquid seal drums
- smaller flare size, giving lower purge rates and enhanced operating life
- significant reduction or elimination of utilities required for smokeless burning through the utilization of increased pressure energy at the flare tip

As mentioned in Section 20.2.4, each major flowing condition should be analyzed to verify that no relief source is over-pressured. In some applications, a large number of different flowing conditions can occur. To simplify the process

(a)

(b)

(c)

FIGURE 20.14 Liquid carryover from an elevated flare. (a) Start of flaring event. (b) Liquid fallout and flaming rain from flare flame. (c) Flaming liquid engulfs flare stack.

of identifying those cases that are likely to govern the hydraulics, a comparative measure of flow rates is useful. The volumetric equivalent, or V_{eq}, is one measure used to identify the hydraulically controlling case:

$$V_{eq} = Q\sqrt{\frac{\text{MW}}{29} \times \frac{T}{520}} \tag{20.1}$$

where V_{eq} = Volumetric equivalent, SCFH
Q = Waste gas flow, SCFH
MW = Waste gas molecular weight
T = Waste gas temperature, R

V_{eq} is the volumetric flow of air that would produce the same velocity pressure as the waste gas flow in the same size line. While this method gives general guidance, it should not replace a more thorough hydraulic analysis.

Properly utilized, a higher allowable pressure drop for the flare system provides an opportunity for capital cost savings, operating cost savings, and reduced downtime due to longer equipment life. While the capital cost savings are most apparent on entirely new flare systems, all of these savings can be realized on existing systems as well.

20.3.3 Liquid Removal

Inherent in many flare systems is the potential for either liquid introduction into, or the formation of hydrocarbon or water vapor condensate in, the flare header. Allowing this liquid-phase material to reach the combustion zone may make operation more difficult. For example, hydrocarbon droplets small enough to be entrained by waste gas and carried into the flame usually burn incompletely, forming soot, and, as a result, reduce the smokeless capacity of the flare. If the droplets become larger, they may be able to fall out of the main flame envelope. In addition, events have been reported where a mostly liquid stream has been discharged from the flare.

Figure 20.14 shows an offshore flare that received liquids in the manner described above. The last two photos, (b) and (c), were taken only a few minutes apart and illustrate how rapidly this situation can deteriorate.

Incorporation of a properly designed and operated knockout drum into the flare system can minimize these problems. There are three basic types of knockout drums that can be incorporated into a flare system: a horizontal settling drum, a vertical settling drum, and a cyclone separator. For more information on each of these types of knockout drums, refer to Section 20.4.3.

Regardless of the knockout drum concept, the holding capacity of the drum should be carefully considered. An overfilled knockout drum can obstruct gas flow to the flare, resulting in over-pressure to upstream systems. In the extreme case, an overfilled knockout drum can result in blowing large volumes of liquids up the flare stack. Liquid draw off capacity must be adequate to prevent overfilling of the drum. In addition, a backup pump and drive means should be considered. Liquid recovered from the knockout drum must be carefully disposed of or stored. Flare header piping must be sloped properly to prevent low point pockets where liquids can accumulate.

20.3.4 Air Infiltration

Infiltration of air into a flare system can lead to flame burnback, which in turn could initiate a destructive detonation in the system. Often, burnback can only be observed at night. Air can enter the flare system by one or more of the following scenarios:

- through stack exit by buoyant exchange, wind action, or contraction
- through leaks in piping connections
- as a component of the waste gas

Prevention measures are available to address each of the air infiltration mechanisms.

Purge gas is often injected into the flare system to prevent air ingression through the stack exit. The quantity of purge gas required is dependent on the size and design of the flare, the composition of the purge gas, and the composition of any waste gas that could be present in the system following a vent or relief event. In general, the lower the density of the gas in the flare stack, the greater the quantity of purge gas necessary for the safety of the system. The purge gas requirement can be reduced using a conservation device such as a John Zink Airrestor or Molecular Seal. The cost and availability of the purge gas will guide the choice of such a device.

Contraction of gas in the flare system occurs due to the cool-down following the flaring of hot gases. The rate of contraction will accelerate dramatically if the cooling leads to condensation of components of the contained gas. Contraction risk can be minimized by use of the Tempurge system.[8] Tempurge senses conditions in the flare header and initiates the introduction of extra purge gas to offset contraction.

An elevated flare stack filled with lighter than air gas will have a negative pressure at the base created by the difference in density between the stack gas and the ambient air. The gas density in the stack is related to the molecular weight of the gas and its temperature. Equation (20.2) defines the pressure at the base of a flare stack at very low flow conditions such as purge.

$$P_{\text{base}} = \frac{27.7H\left(\rho_{\text{gas}} - \rho_{\text{amb}}\right)}{144} \tag{20.2}$$

where P_{base} = Static pressure at base of stack, in. w.c.
H = Height of stack above inlet, ft
ρ_{gas} = Gas density in stack, lb-m/ft^3
ρ_{amb} = Density of atmospheric air, lb-m/ft^3

If a negative static pressure exists at the base of the stack, then at low flows the entire flare header system will be under negative pressure. Operation of the flare system under negative pressure greatly increases the potential of air infiltration into the header system through leaks, open valves, or flanges, or through the tip exit by decanting in the stack. Such leakage is known to have occurred during the servicing of relief valves.

Installation of a liquid seal in the system can produce positive flare header pressure although the pressure downstream of the seal is negative. This greatly reduces the potential of air leakage into the system. Because a liquid seal can also be a barrier to air entering the header from the flare stack, locating the liquid seal in the base of the stack offers maximum protection of the header system. In this position, the liquid seal can also be designed to isolate the flare ignition source from the flare header and the process units.

Oxygen-containing gases should be segregated from the main flare system. Waste gases that contain oxygen present a special design challenge. The risk of flashback in systems handling such gases can be minimized through the use of flame/detonation arrestors, special liquid seals, and/or the use of specialized flare burners. The presence of more than a trace amount of oxygen (more than 1% by volume) in a waste gas stream creates a separate design consideration discussed in Section 20.3.8.

20.3.5 Flame Radiation

As the waste gases are burned, a certain portion of the heat produced is transferred to the surroundings by thermal radiation. Safe design of a flare requires careful consideration of the thermal radiation. The radiation limits discussed in Section 20.2.6 can become the basis for determining the height of the flare stack and its location. For a given set of flare flow conditions, the radiation limits can usually be met by adjustment of the flare stack overall height and/or the use of a limited access area around the flare. The flare height and/or size of the limited access area can affect the economics of the plant. For plants with limited plot area (or for ships), an enclosed flare can be employed to meet radiation restrictions. Water spray curtains have also been used to control radiation on offshore platforms.

In Chapter 3, radiative heat transfer was described in theoretical terms. Radiation from a flame to another object is determined by:

- flame temperature
- concentrations of radiant emitters in the flame (e.g., CO_2, H_2O, and soot)
- size, shape, and position of the flame
- location and orientation of the target object relative to the flame
- characteristics of the intervening space between the flame and the object

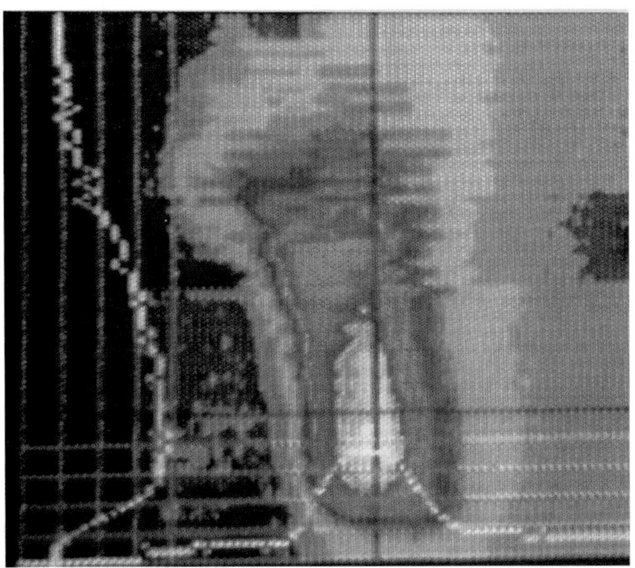

FIGURE 20.15 Thermogram of a flare flame.

Calculations based on theory may be feasible within the well-defined confines of a furnace operating at a steady condition. Unfortunately, most of these factors cannot be accurately defined for a flare flame in the open air.

Because temperature appears in the radiation equation to the fourth power, it is clearly a dominant factor. Despite its importance, the temperature of a flare flame is extremely difficult to measure or estimate. An error of only 10% in absolute temperature affects the calculated radiant heat transfer by over 40%. Observers have noted variations in local flame temperature as high as 1000°C (1800°F) between the core and the cooler outer surface of an open burning flame. Figure 20.15 shows a thermogram of a flare flame. In the thermogram, white represents the highest temperature and dark blue the lowest. Thus, only the small, bright yellow zone is at a high temperature. The temperature falls rapidly as one approaches the outer edge of the flame envelope.

The temperature of a flame is influenced by its interaction with its surroundings. The availability of ambient air causes the outer portions of the flame envelope to cool. In addition, the flame will radiate both to cold outer space and to relatively warmer objects on Earth. Therefore, it is not surprising that observations indicate peak flare flame temperatures far less than the calculated adiabatic flame temperature. To approach flare radiation from a theoretical basis, local flame temperature, which varies substantially throughout the flare flame, would need to be predicted with greater accuracy than present tools allow.

The other factors listed are also very difficult to determine. The concentrations of substances that are radiant emitters vary a great deal from point to point within a flare flame, which creates problems in predicting beam lengths and emissivity. The detailed shape of a flare flame is much more convoluted and chaotic than any simple geometric approximation can represent. Wind fluctuations cause the flame to move constantly, so concentrations, temperature, and relative positions are always changing. Atmospheric absorption and scattering depend on transient and unpredictable weather conditions such as ambient temperature, humidity, fog, rain, etc.

To overcome these difficulties, engineers have historically estimated radiation by treating flare flames as point sources, using heat release as a basis for emissive power and empirical radiant fractions in lieu of true radiant emissivity. Figure 20.16 shows the general geometry assumptions that affect the point source approach. The classic API radiation equation represents this approach in its simplest form:

$$K = \frac{\tau F Q}{4\pi D^2} \tag{20.3}$$

where K = Radiation, Btu/hr-ft^2
τ = Atmospheric transmissivity
F = Radiant fraction
Q = Heat release, Btu/hr
D = Distance from the heat epicenter to the object, ft

Many of the complexities of the full theoretical treatment are lumped into the empirically determined radiant fraction. This factor includes flame temperature effects, gas and soot emissivity, mean beam length, and other flame shape issues. The distance factor disguises a number of subtleties that arise as a result of flame shape prediction, including flame length, flame trajectory, and position of the heat epicenter. Nevertheless, this type of simplified approach has been used in one form or another to estimate radiation from flare flames for many years.

Several published methods are available for preliminary estimation of flare radiation and stack heights. An article by Schwartz and White[9] presents a detailed discussion of flare radiation prediction and a critical review of published methods. Based on Example 2 in the referenced paper, Figure 20.17 provides a visual comparison of the stack heights determined by each of several radiation methods and the relative equipment cost associated with each stack height. Plant designers and users alike must be cognizant that traditional methods of calculating radiant heat intensities are neither consistently conservative nor consistently optimistic. Long ago, the John Zink Company recognized the limitations and risks associated with the traditional methods and undertook the development of proprietary methods for radiation prediction. The latest

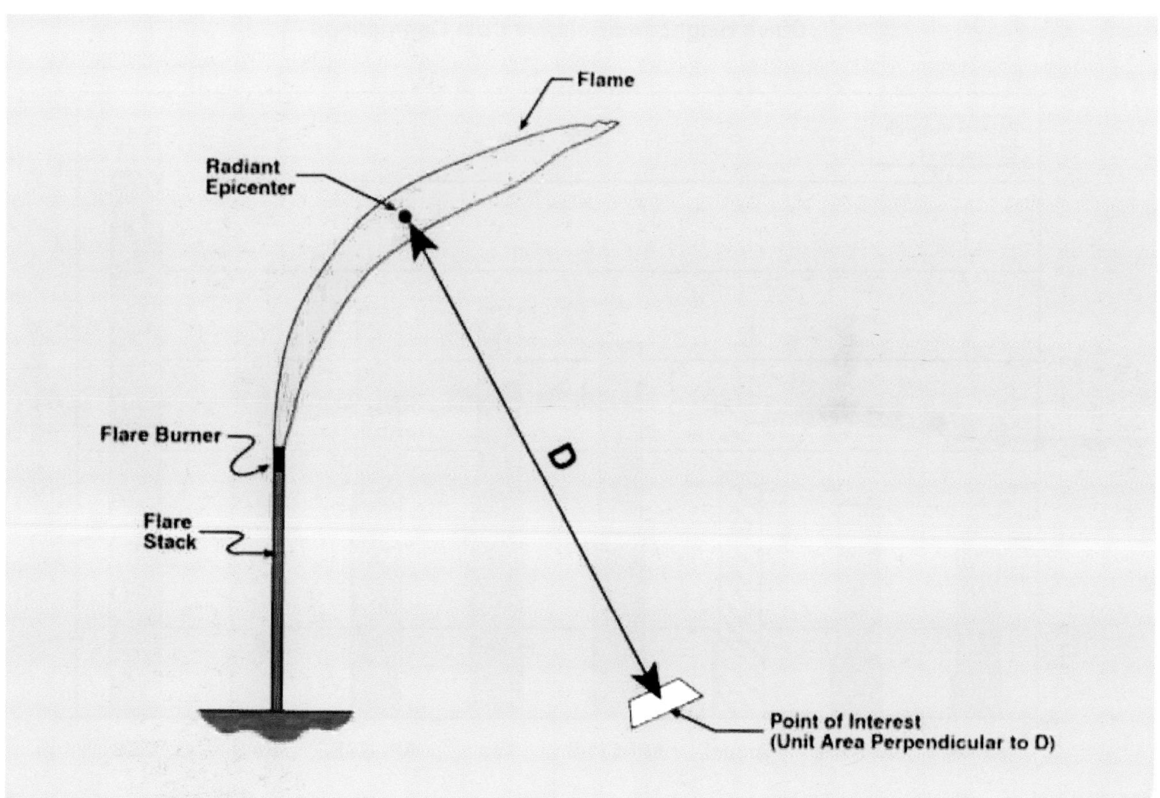

FIGURE 20.16 API radiation geometry.

prediction methods capture the effect of flare burner design, gas quantity and composition, various momenta, smokeless burning rate and smoke formation on the flame shape, and radiant characteristics.

20.3.6 Smoke Suppression

Smokeless burning is a complex issue that involves many of the system defining factors discussed in Section 20.2. In choosing the best smoke suppression method, the flare designer is guided by his/her experience in interpreting the job-specific information received relative to each of these factors.

Smokeless burning, in general, occurs when the momentum produced by all of the employed energy sources educts and mixes sufficient air with the waste gas. For smokeless burning, a key issue is the momentum of the waste gas as it exits the flare burner. In some cases, the waste gas stream is available at a pressure that, if properly transformed, can provide the required momentum. If the waste gas pressure (momentum) is not adequate for smokeless burning, the flare designer must enlist assistance from another energy source (e.g., steam or low-pressure air). In some cases, a combination of energy sources can be effective.

Briefly, energy transformation entails conversion of the internal energy (pressure) of the waste gas to kinetic energy (velocity). Designs for high-pressure flares (5 to 10 psig or more) exist that require no supplemental assist medium. Systems employing this technique have been very successful and enjoy low operating cost and an excellent service life.

Steam injection is the most common technique for adding momentum to low-pressure gases. In addition to adding momentum, steam also provides the smoke suppression benefits of gas dilution and participation in the chemistry of the combustion process. The effectiveness of steam is demonstrated in the series of photographs shown in Figure 20.18. In frame (a), there is no steam injection; in frame (b), steam injection has just begun; and in frame (c), steam injection has achieved smokeless burning.

Some plants have steam available at several different pressure levels. There is often an operating cost advantage to using low-pressure steam (30 to 50 psig). The plant designer must balance this operating cost advantage against the increased piping costs associated with low-pressure steam. Also, while the flare may achieve the design smokeless rate at the maximum steam pressure, steam consumption at turndown firing rates below the maximum may be higher than expected.

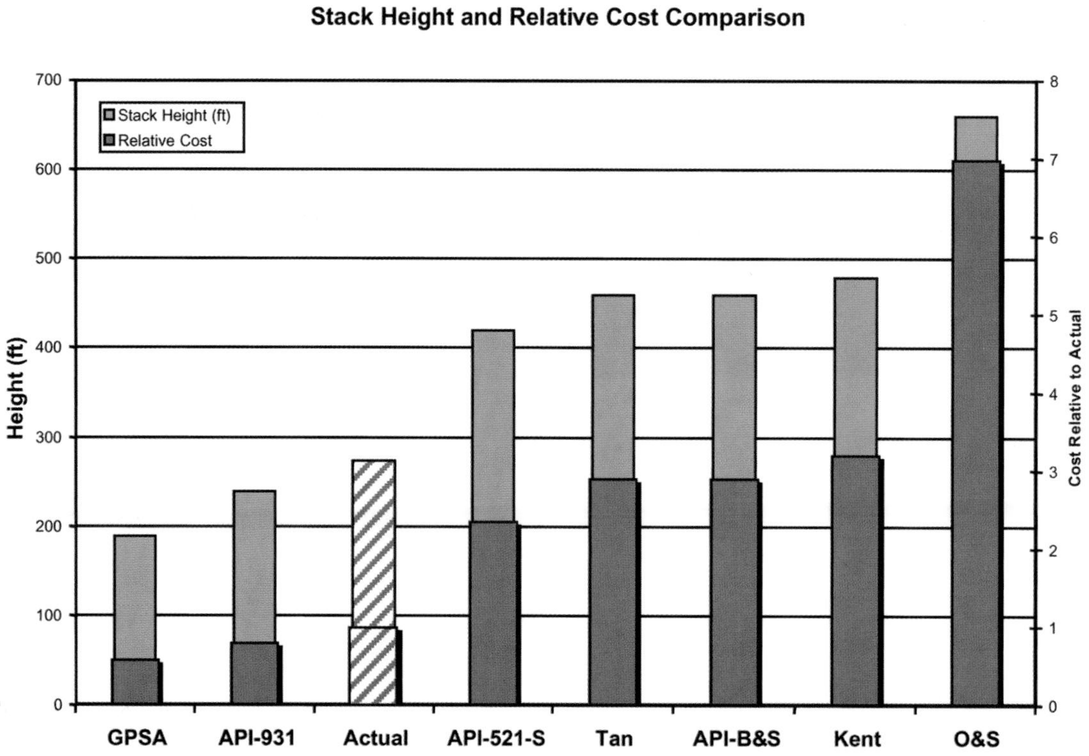

Stack Height and Relative Cost Comparison

FIGURE 20.17 Comparison of stack height and relative cost for various radiation calculation methods.

Because most flaring events involve relatively low flow rates, performance under these turndown conditions must be carefully considered.

Low-pressure (0.25 to 1.0 psig) air is utilized in cases where gas pressure is low and steam is not available. The supplied air adds momentum and is a portion of the combustion air required. Figure 20.19 shows another series of photographs that illustrates the effectiveness of air assistance. Frame (a) shows the flare with no assist air. The blower is turned on in frame (b), but because the blower requires some time to reach full speed, the complete effect of air injection is not seen until frame (d). The gas being flared in this case is propylene.

Generally, the blower supplies only a fraction of the combustion air required by the smokeless flow condition. For most designs, 15 to 50% of the stoichiometric air requirement is blown into the flame. The remainder of the required air is entrained along the length of the flare flame.

20.3.7 Noise/Visible Flame

The energy released in flare combustion produces heat, ionized gas, light, and sound. Most plants are equipped with elevated flares that by their nature broadcast flaring sound to the plant and to the surrounding neighborhood. In some cases, the sound level becomes objectionable and is considered to constitute noise. Flaring sound is generated by at least three mechanisms:

1. by the gas jet as it exits the flare burner and mixes with surrounding air
2. by any smoke suppressant injection and associated mixing
3. by combustion

Upstream piping and valves associated with the source of the relief gas may also create substantial noise levels that are carried along the flare header and exit through the flare tip.

At the maximum smokeless flaring rate of a steam- or air-assisted flare burner, gas jet noise is usually a minor contributor. The noise generated by the second mechanism can be mitigated by use of low noise injectors, mufflers, and careful distribution of suppressant. The Steamizer™ flare burner shown in Figure 20.20 is of low noise design with additional noise reduction coming from a muffler concept first developed for use on enclosed ground flares. Careful design can reduce flaring noise levels by a factor of 75% or more (6 dB or more).

Where the light from a flare flame is objectionable, an enclosed flare is a good selection. A properly sized enclosed flare can eliminate visible flame for all cases except emergencies. An equal benefit of an enclosed flare is the reduction of flaring noise.

20.3.8 Air/Gas Mixtures

Waste gas streams containing air/gas mixtures can generally be divided into two types. The first type is comprised of systems that are expected to contain air/gas mixtures. Examples include landfills, gasoline loading terminals, and medical equipment sterilization facilities. The second type is potentially more dangerous in that air is not expected in the composition of vents and reliefs. An example is venting air from a vessel or tank at the beginning of a pre-startup purging cycle.

Flare systems that handle air/gas mixtures usually involve a number of special safety considerations. The special considerations, which relate primarily to the increased risk of flashback, include:

1. safety interlocks to prove purge gas and pilots on startup
2. automatic shutdown on loss of purge gas or pilots
3. higher than normal purge rates to maintain burner exit velocity and prevent burnback
4. limited turndown range to maintain burner exit velocity and prevent burnback
5. use of detonation and/or flame arrestors
6. special operational practices

20.4 FLARE EQUIPMENT

While evaluating the general design considerations set forth above, the flare designer must also begin equipment selection. Overall design considerations and specific equipment selections are interrelated aspects of the system design process. The various major system components listed in Section 20.1.5 are discussed here in more detail.

20.4.1 Flare Burners

Although they are installed at the end of the flare system, flare burners are among the first items considered in the system design. At this point it should be clear that substantial benefits are attached to the use of most of the available pressure at the flare burner. The exit of the flare burner is the point where the flow determinate pressure drop usually occurs. Designs range from simple utility flares to enclosed multipoint staged systems, and from non-assisted to multiple steam injectors to multiple blower air-assisted designs. Regardless of the tip design, adequate ignition means must be available to ensure that the prime objective is achieved.

20.4.1.1 Non-assisted or Utility

The simplest types are the non-assisted, or utility, flare burners. These burners consist essentially of a cylindrical barrel with attachments for enhanced flame stability (flame retention means) and pilots to initiate and maintain ignition of the relief gases. A typical non-assisted flare is shown in Figure 20.21. Utility flares are usually flanged for ease of replacement. The horizontally mounted versions are usually referred to as burn pit flare tips (Figure 20.5). In both cases, a turbu-

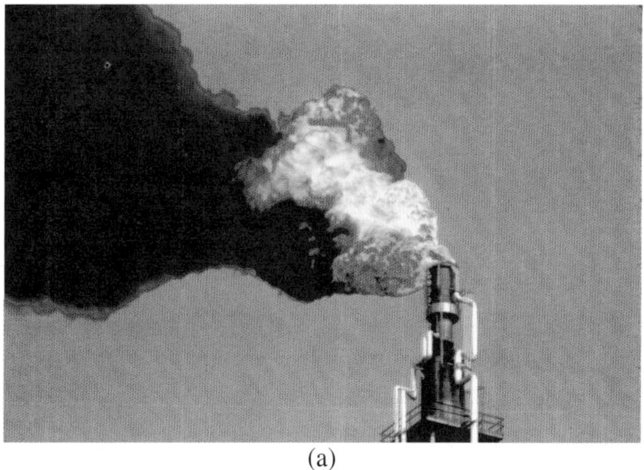

(a)

(b)

(c)

FIGURE 20.18 Effectiveness of steam in smoke suppression: (a) no steam, (b) starting steam, and (c) smokeless.

FIGURE 20.19 Effectiveness of air in smoke suppression: (a) no blower air, (b) start blower, (c) air flow increasing, and (d) smokeless.

lent diffusion flame is produced. The flame may be an attached or detached stable flame. The exit velocity of a flare burner is dependent on the waste gas composition, the specific design of the flare burner, and the allowable pressure drop. In some cases, the exit velocity can safely reach Mach 1. It should be noted that some flares or flare relief cases are subject to regulations that limit the exit velocity.

Optional features that can extend equipment service life include windshields and refractory lining. The Zink double refractory (ZDR) severe service flare tips (Figure 20.22) use refractory linings internally and externally to protect the tip against both internal burning and flame pulldown outside the tip. Alternatively, center steam is sometimes used to help avoid internal burning instead of an internal refractory lining to extend tip life. This approach is most effective in climates where freezing is not an issue. Center steam is a relatively inefficient means to control smoke because it does not entrain air, which is normally an essential part of any smoke suppression strategy.

20.4.1.2 Simple Steam Assisted

The first smokeless flares were adaptations of simple utility flares. This basic design has been improved over the years with multi-port nozzles to reduce steam injection noise, optimized injection patterns to improve steam efficiency, and optional center steam injection to reduce damaging internal burnback. Figure 20.23 shows a modern example of this design. A steam manifold, often referred to as the upper steam manifold or ring, is mounted near the exit of the flare tip. The steam ring can be designed for steam supply pressures normally ranging from 30 to 150 psig. Several steam injectors extend from the manifold and direct jets of steam into the waste gas as it exits the flare tip. The steam jets inspirate air from the surrounding atmosphere and inject it into the gas with high levels of turbulence. These jets also act to gather, contain, and guide the gases exiting the flare tip. This prevents wind from causing flame pulldown around the flare tip. Injected steam, educted air, and relief gas combine to form a

FIGURE 20.20 Steamizer™ steam-assisted smokeless flare.

mixture that burns relief gas without smoke. The maximum smokeless rate achievable with a given flare burner depends on a number of factors, including the gas composition, the amount of steam available, and the gas and steam pressures.

However, there are inherent design limitations in this type of flare tip. The steam injectors are located close to the exit of the flare tip, so it becomes very difficult to muffle the steam noise produced by the high-pressure jets. Any muffler for the upper steam noise would need to be able to withstand direct flame impingement in adverse winds and it could tend to interfere with air being drawn in by the steam jets. Furthermore, as the tip size increases, it becomes more difficult for the steam/air mixtures to reach the center of the flame. Finally, the perimeter of the flare tip only increases linearly with tip size, while the flow area of the flare tip increases with the square of tip size, as shown in Figure 20.24. Therefore, as flare tip size increases, the need for air (a function of the flow area) quickly outstrips the ability to educt air (a function of the perimeter.) This fundamental characteristic of a simple steam-assisted flare limits the maximum effective size of such a flare burner.

20.4.1.3 Advanced Steam Assisted

To overcome the limitations and other shortcomings of the simple steam-injected design, an advanced steam-assisted flare was invented that uses multiple steam injection points.[10] In addition to the upper steam manifold, a set of external-internal tubes is utilized to deliver a steam/air mixture to the

FIGURE 20.21 Typical non-assisted flare.

FIGURE 20.22 Zink double refractory (ZDR) severe service flare tip.

core of the waste gas exit (Figure 20.25). The presence of these tubes, properly distributed across the tip exit, increases the effective perimeter available for air access to the waste gas (Figure 20.26).

The external-internal tubes start outside the wall or barrel of the flare tip, pass through openings in the wall, and turn upward, terminating at the tip exit. Welding seals the point where the tube penetrates the wall. Steam jets inject steam into the inlet of these tubes, inspirating large amounts of air. The steam/air mixture exits the tubes at high velocity, delivering momentum, dilution steam, combustion air, and turbulence into the base of the flame. Recent innovations have increased the effectiveness of the tubes allowing greater smokeless capacities. New, enhanced steam injectors have increased the air eduction efficiency. Upper steam injection

FIGURE 20.23 Simple steam-assisted flare.

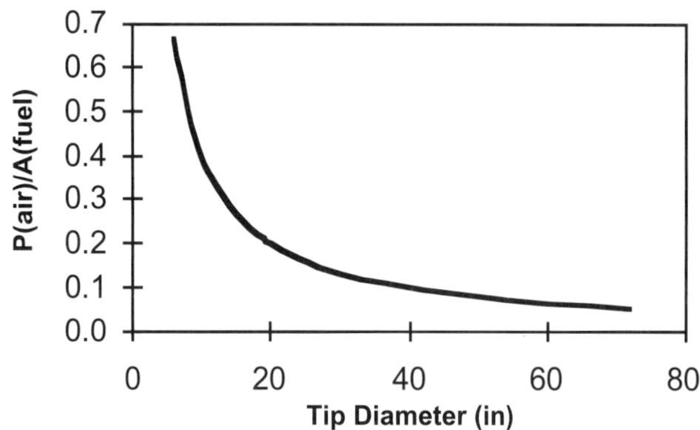

FIGURE 20.24 Perimeter:area ratio as a function of tip size for a simple steam-assisted flare.

further enhances smokeless combustion through increased turbulence and mixing and by mitigating adverse wind effects.

The advanced steam-assisted flare design incorporates several smoke suppression strategies: increased perimeter, higher momentum, more combustion air, greater turbulence for mixing, dilution and chemical interaction by steam, and molding of the flame to resist wind effects. Each of these strategies helps to reduce smoke; in combination, they produce some of the highest smokeless rates available in single-point flares.

New flare systems can achieve smokeless rates of more than 500,000 lb/hr (230,000 kg/hr) of gases that are generally considered difficult to burn cleanly. An example of a state-of-the-art flare burner design is shown in Figure 20.27. Improved muffler designs and redistribution of steam can give noise levels much lower than earlier models. In some cases, the steam jet noise can be totally neutralized through injector design and the use of new muffling techniques.

As the capacity, size, and smokeless capability of flare burners have increased, more emphasis has been placed on sophisticated design tools that can predict noise and smokeless performance. Design tools of this type can be developed and validated using large-scale tests in a facility such as shown in Figure 20.13.

20.4.1.4 Low-Pressure Air-assisted

Not all plants have large amounts of steam available for use by the flare. Some plants prefer not to use steam to avoid freezing problems; others cannot commit water to make steam for smoke control; and still others choose not to install a boiler. To meet this need, a series of air-assisted flare designs were invented.[11]

Generally, the air-assisted flare burner consists of a gas burner mounted in an air plenum at the top of the flare stack (Figure 20.28). Relief gas is delivered to the burner by a gas riser pipe running coaxially up the center of the flare stack. Low-pressure air is delivered to the burner from one or more blowers located near the base of the flare stack. The air flows upward through the annular space between the flare stack and the gas riser.

The first air-assisted flare applications were associated with operations some distance from the main plant or totally remote from plant utilities support. Early air flares were often designed to flare small to moderate flow rates. The success of these flares led to the use of air assist on flares of greater capacity. More recently, air-assisted flares have come into use as the flare for large process facilities. The Flame Similarity Method and the related near field mixing region models discussed in Chapter 8 are examples of the design tools necessary for cost-effective application of air-assisted flares. Today, air flare designs are available with demonstrated tip life spans of 5 to 10 years. Smokeless rates above 150×10^6 standard cubic feet per day (SCFD) are available for saturated hydrocarbons such as production facility reliefs.

Figure 20.29 shows an example of the latest air flare design. Waste gas exits the burner in one or more narrow annular jets, each surrounded by assist air. This design makes good use of the perimeter:area ratio concept discussed above in the context of steam-assisted flares.

20.4.1.5 Energy Conversion

In the smokeless flaring discussions above, the focus centered on adding energy from an outside source to boost the overall energy level high enough to achieve smokeless burning. An advantage is gained if an outside source is not required. This is the case with energy conversion flare burners. Such burners are also referred to as high-pressure flare burners or multi-point flare burners. Where they can be

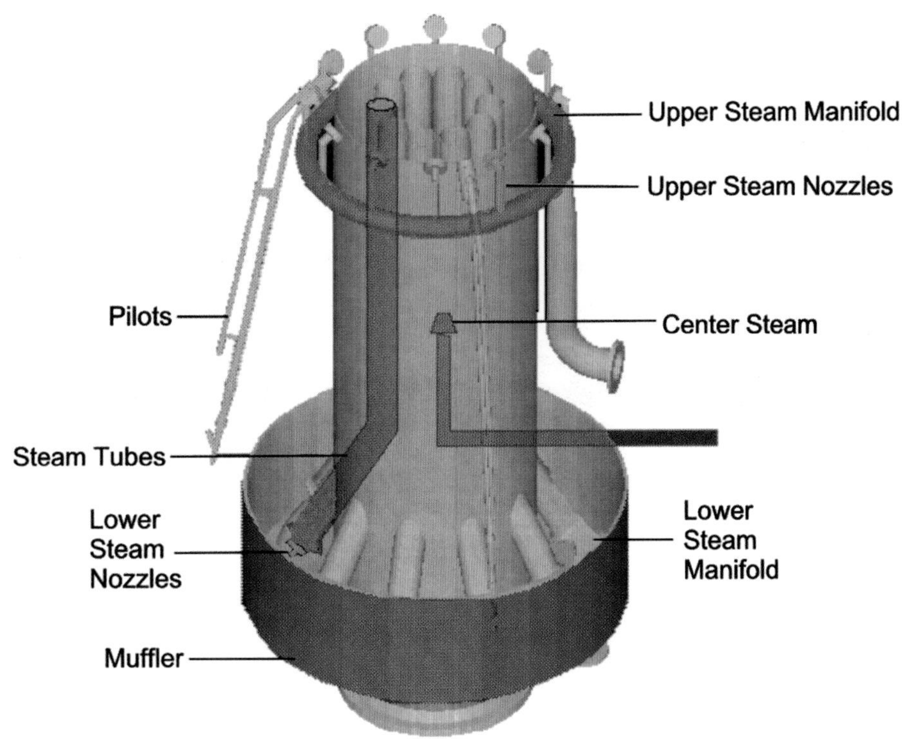

FIGURE 20.25 Schematic of an advanced steam-assisted flare.

employed, the use of energy conversion flare burners can provide a significant reduction in flare operating cost.

There are two distinct groups of energy conversion burners. The first group is distinguished by having a single inlet and relatively close grouping of the waste gas discharge points. The other group employs a means beyond energy conversion to achieve smokeless flaring. In both groups, an underlying principle is the conversion of the static pressure of the waste gas, at the burner, to jet velocity and ultimately into momentum.

Another concept employed by both groups is the division of the incoming gas stream into multiple burning points. This concept is sometimes referred to as the "firewood principle" because of an illustrative analogy. The "firewood principle" considers the situation where a tree has been cut down and is to be used as fireplace fuel. If a large section of the tree trunk is placed in the fireplace, it will be difficult to ignite and burn because the ratio of fuel to surface (air exposure) is very large. This situation can be improved by splitting the trunk section into smaller pieces, thus gaining a better fuel:surface ratio. Obviously, a balance must be reached between the effort expended to split the wood (think of it as cost) and the improved ability to burn (think of this as smokeless burning ability).

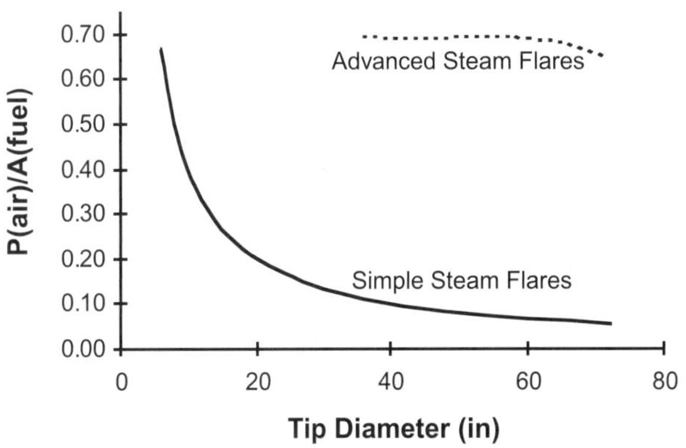

FIGURE 20.26 A comparison of the perimeter:area ratio for simple and advanced steam-assisted flares.

Applications of flare burners that employ energy conversion to achieve smokeless burning require special attention to flow turndown cases. Some flares, usually those associated with oil production activities, operate at or near their maximum design flow most of the time. Energy conversion flares for production applications, such as the commercially available Hydra flare shown in Figure 20.30, offer a controlled

FIGURE 20.27 State-of-the-art Steamizer™ flare burner and muffler.

flame shape and reduced flame radiation. These features are particularly attractive for platform-mounted flares.

The opposite is normally true of flare burners serving refineries or petrochemical plants. These flares usually have a very small load (purge gas plus any leakage) with an occasional short-duration intermediate load. In energy conversion flares, the gas pressure vs. flow characteristic follows the same rules as an orifice. Above critical pressure drop, the flow/pressure characteristic varies with the ratio of the absolute pressures. At or below the critical pressure drop, the flow/pressure characteristic follows a square relationship. For example, a flow reduction of 50% will reduce the gas pressure drop to 25% of the full flow value.

The turndown ratios experienced by refinery and petrochemical flares would reduce the gas pressure to such a degree that energy for smokeless burning would not be present. This problem has been overcome using a staging concept.[12] The staging concept divides the burners into stages or groups of burners, with the first stages having a smaller number of burners than the last stages. The flow to each stage

FIGURE 20.28 Air-assisted smokeless flare with two blowers in a refinery.

or group of burners is controlled by a valve that operates in an on/off manner as directed by a control system. The valve to the first stage is usually open all the time. The control system principle is to proportion the number of burners in service to the gas flow. In effect, this allows the burners to operate within a certain pressure range (see Figure 20.31) so that at least the minimum energy level for smokeless burning is always present.

The operation of a multi-point flare can be spectacular, as shown in Figure 20.31. This series of photographs illustrates the addition of burner stages as the gas flow to the flare increases. In the final frame, the flare achieves a smokeless burning rate of more than 550,000 lb/hr (250,000 kg/hr). The commercially available LRGO flare system shown is surrounded by a fence to exclude personnel and animals from the flare area. In cases where land space is limited, an enclosing fence, such as shown in Figure 20.32, can be employed to reduce radiation to the surroundings and to reduce visibility and noise. The largest flare system in the world employs an LRGO system to handle more than 10 million pounds of waste gas per hour, smokelessly.

Design of an energy conversion flare system, either single point or multi-point, involves issues that are not a consideration for other flare burner types. Questions such as burner or gas jet spacing must be resolved. For example, the ability of a given burner to be lit and to light its neighbors is of paramount importance. Figure 20.31(f) captures this cross-lighting feature in progress. However, spacing burners solely on the basis of cross-lighting may restrict air flow and hinder smokeless burning. Gas properties are even more critical when the design depends on energy conversion alone to achieve smokeless burning.

FIGURE 20.29 Annular air flare.

20.4.1.6 Endothermic

Some flare applications involve gases with a high inert gas content. When the inert content is high enough, the combustion reaction becomes endothermic, meaning that some external source of heat is required to sustain the reaction. Crude oil recovery by CO_2 injection, incinerator bypasses, coke ovens, and steel mills are examples of activities that generate gases that require additional fuel to maintain the main burner flame. Such gases often contain significant amounts of toxic materials such as H_2S, NH_3, CO, or various gases normally sent to an incinerator. Flaring has been recognized for many years as an adequate method of disposing of such gases. Substantially complete destruction of such gases protects the community and the environment.

The earliest endothermic flares consisted of simple non-assisted flare tips with fuel gas enrichment of the waste gas upstream of the flare to ensure that the mixture arriving at the tip was burnable. This system was simple but imposed a high fuel cost on the facility. An alternative design supplied a premixed supplemental fuel/air mixture to an annulus around the flare tip. Combustion of this mixture supplied heat and ignition to the waste gas as it exited the flare tip. This design had a limited supplemental fuel gas turndown before burnback occurred in the annulus, thus requiring a full on or off operation of the supplemental gas.

Today's high energy costs provide an incentive to reduce such fuel usage. The RIMFIRE® flare burner (Figure 20.33)

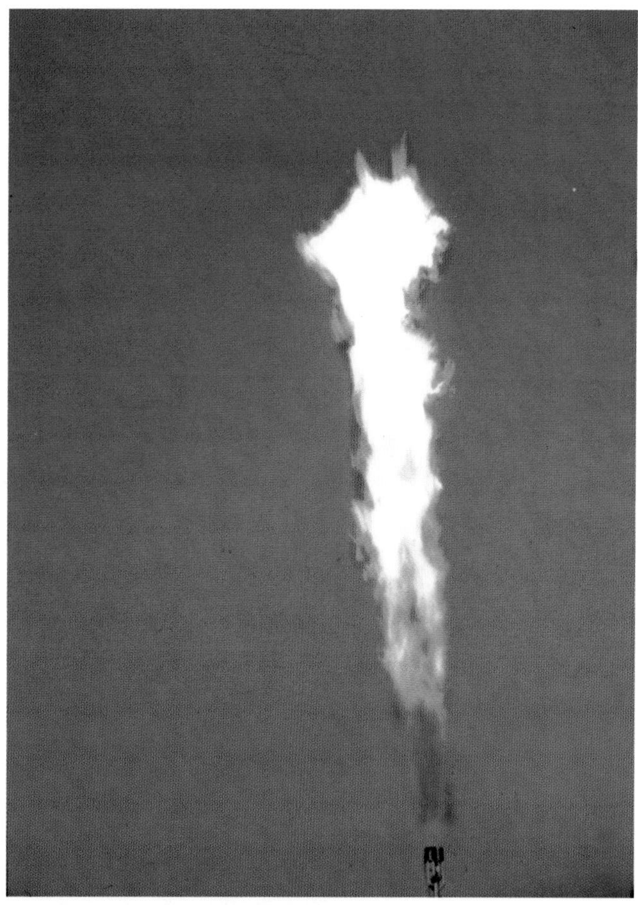

FIGURE 20.30 Hydra flare burner in an offshore location.

(a)

(b)

(c)

(d)

(e)

(f)

(g)

FIGURE 20.31 LRGO staging sequence during a flaring event from inception (a) to full load (g).

is a modern endothermic flare with an air-assisted supplemental gas burner surrounding the waste gas exit.[13] Using supplemental fuel to build a strong "forced-draft" ignition flame, the amount of enrichment gas required to sustain ignition is reduced. The amount of supplemental fuel required depends on the flare burner size and service.

The RIMFIRE® flare was originally developed for the CO_2 injection fields in the western United States. Vents and reliefs

FIGURE 20.32 Multi-point LRGO system with a radiation fence. (Courtesy of Maria Celia, Setal Engineering, Sao Paulo, Brazil).

FIGURE 20.33 A RIMFIRE® endothermic flare.

from the oil recovery system sent highly inert materials to the flare. In this service, a non-assisted tip would require enrichment of the relief gases to an LHV of about 300 Btu/scf. Using the RIMFIRE® flare burner, the LHV requirement was reduced to approximately 180 Btu/scf.

The total fuel requirement, enrichment plus supplemental fuel, for the RIMFIRE® is substantially lower than the fuel requirement for other designs. Today, the RIMFIRE® burner is used in a broad range of applications where enrichment of the waste gas is required.

FIGURE 20.34 OWB liquid flare test firing 150 gpm.

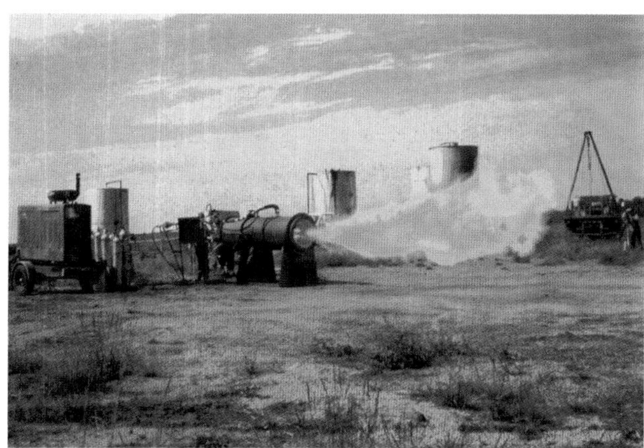

FIGURE 20.35 Forced-draft Dragon liquid flare.

20.4.1.7 Special Types

Sections 20.4.1.1 through 20.4.1.6 have described the wide variety of flare burners available to the flare designer today. In general, these burners are used to burn gaseous waste material and utilize certain smoke suppression techniques. There are, however, other flares that are designed to burn liquid hydrocarbons or to use a liquid as the smoke suppressant. Some examples are given below.

Figures 20.34 and 20.35 show two examples of liquid flares having quite different designs. The OWB flare (Figure 20.34) has multiple burners combined into a single unit. At the time of the photograph, the flare was burning oil with a heat release of about 1 billion Btu/hr. The OWB tips allow a modular approach to design for a specific capacity. In addition, a very

FIGURE 20.36 Poseidon flare: water-assisted Hydra.

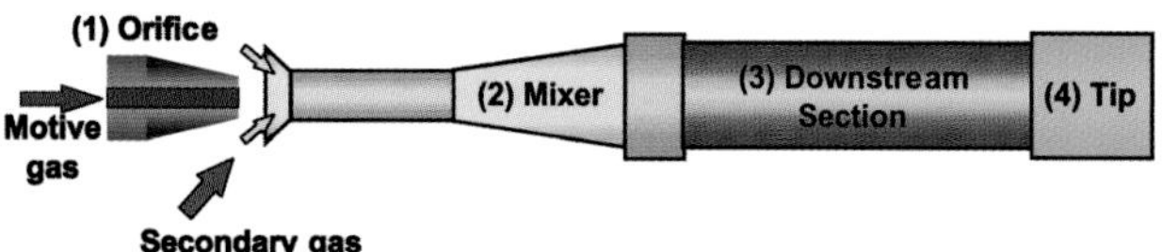

FIGURE 20.37 Fundamental pilot parts.

wide turndown can be obtained by simply turning off flow to some of the tips. Typical applications are oil well testing and spill cleanup.

The Dragon flare (Figure 20.35) uses one or more burners and is equipped with a blower to improve mixing for smokeless burning. This flare is employed in destruction of surplus or off-spec product or waste oil.

The success of steam as a smoke suppressant has prompted the use of water to prevent smoke. Over the years, a number of designs using water for smoke suppression have been invented. In certain situations, water can be used with great benefit. A case in point is the offshore installation of a Poseidon flare shown in Figure 20.36. This flare utilizes water to enhance smokeless burning, reduce thermal radiation, and decrease noise. The installation shown in the photograph achieved a 13-dBA reduction in noise and a 50% drop in radiation compared to the previous conventional flare.[14]

20.4.2 Pilots, Ignitors, and Monitors

Prior to 1947, venting of unburned hydrocarbons to the atmosphere was an industry practice. After 1947, regulations required hydrocarbons to be burned or "flared." Early methods to light a flare included hoisting a burning, oily rag to the top of the flare, shooting over the top of the flare an arrow with a burning rag attached, or shooting a signal flare over the top. These methods, however, were not reliable or safe. In 1949, the John Zink Company developed the first pilot to light and continually burn the vented gases from a flare. Since then, there have been a number of improvements in pilot design, ignition, and monitoring. This section discusses pilots and methods used for pilot ignition and monitoring.

20.4.2.1 Pilots

A flare pilot is a premixed burner system designed to operate over a narrow heat release range. As a burner, the pilot must (1) meter the fuel and air, (2) mix the fuel with the air, (3) mold the desired flame shape, and (4) maintain flame stability. Typically, pilots consist of four fundamental parts: a mixer or venturi, a gas orifice, a downstream section that connects the mixer and the tip, and a tip as illustrated in Figure 20.37. All components of a pilot are carefully designed to work together as a system to achieve proper performance. A change in any component will affect the balance of the system and hence the operation of the pilot.

In operation, the pressure energy of the pilot fuel is used to aspirate ambient air into the mixer inlet, mix the gas and air, and propel the mixture through the downstream section and out the pilot tip. The key goals for a properly designed pilot itself are to:

1. be capable of reliable ignition
2. provide pilot flame stability
3. prevent the pilot flame from being extinguished
4. provide a long service life

An application goal of the pilot is to ignite the waste gas exiting the flare burner.

If the volume of air aspirated into the pilot falls outside the flammability limits of the pilot fuel gas, the pilot will not operate properly. For example, methane requires 5.7 to 19 volumes of air per volume of fuel in order to burn. If a pilot is operating below this volumetric air:fuel ratio limit, air external to the pilot must be available in order to burn the fuel. If the pilot tip is engulfed with inert gases from the flare, due to purge gas or flue gas from the flare flame, then the pilot cannot be lit, nor will it burn. Conversely, if a pilot is operating with methane and the volumetric air:fuel ratio is above or near 19, the pilot will be difficult to light and may be unstable in windy conditions.

The operational environment of a flare pilot requires that the pilot be able to withstand rain, wind, heat from the flare flame, and direct flame contact. Common pilot problems are failure to light and burn with a stable flame and flashback.

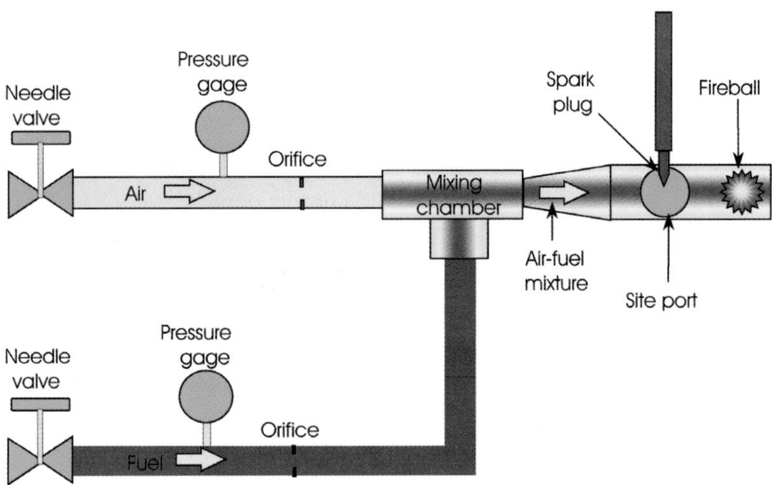

FIGURE 20.38 Conventional flame-front generator.

20.4.2.2 Pilot Ignition

The most common method of lighting elevated flare pilots is by the use of a flame-front generator (FFG). An FFG is a device designed to produce a fireball, which travels inside a pipe from the point of ignition to the pilot, thereby lighting the pilot. There are three fundamental types of FFGs: conventional, slipstream, and self-inspirating.

20.4.2.2.1 Conventional FFG

A conventional FFG is illustrated in Figure 20.38. A combination venturi mixer/ignition chamber is connected by a 1-in. pipe to the pilot. The pipe length can be 5000 ft (1500 m) or more, but the pipe must be 1 in. (2.5 cm) and no heavier than sch. 80. The ignition sequence starts by flowing air and gas to the venturi mixer. In this system, the flow rates of air and gas are each controlled and monitored using a needle valve and pressure gage. The ignition chamber is located immediately downstream of the mixing zone. A spark plug in the ignition chamber can ignite the air/fuel mixture. The resulting fireball travels through the pipe until it reaches the pilot. The fireball then ignites the air/fuel mixture generated by the pilot as it exits the pilot tip.

It is important to let the FFG line fill completely with the flammable air/fuel mixture before the spark is generated. If the FFG line is not completely filled with the air/fuel mixture, the fireball will extinguish before it reaches the pilot and will not light the pilot. Long FFG lines may take minutes to completely fill with a flammable air/fuel mixture, depending on the air/fuel mixture flow rate and FFG line size. For example, if the velocity of the air/fuel mixture in the FFG line is 50 ft/s (15 m/s) and the FFG line is 1000 ft (300 m) long, the air/fuel mixture should flow for about 20 to 30 seconds before attempting to light the pilot. Each attempt to light the pilot in this example should allow 20 to 30 seconds for the line to refill with the air/fuel mixture.

20.4.2.2.2 Slipstream FFG

The slipstream FFG directs a portion of the air/fuel mixture generated by the pilot venturi to a tube located adjacent to the pilot, as shown in Figure 20.39. The slipstream travels through the tube and exits near the pilot tip. A high-energy discharge ignitor probe is used to ignite the mixture, generating a fireball within the slipstream line that in turn ignites the pilot. The main advantages that this system has over the conventional FFG system are quick pilot relight, no flame-front lines, and no compressed air required. The main disadvantages are that critical components are located at the flare tip and, therefore, inaccessible without a flare shutdown, and that the electrical wire leading from the transformer to the spark probe is limited to approximately 750 ft (230 m). Many of the pilots equipped with a slipstream FFG are also equipped with a conventional FFG and the associated piping. In this case, the conventional FFG is used as an installed backup ignition system.

20.4.2.2.3 Self-inspirating FFG

The self-inspirating FFG is a system in which an air/fuel mixture is generated at grade using an eductor system, as shown in Figure 20.40. This eductor is separate from the main pilot venturi mixer. A spark, generated just downstream of the ignitor eductor, creates a fireball inside the ignition line that leads to the pilot tip. The main advantage

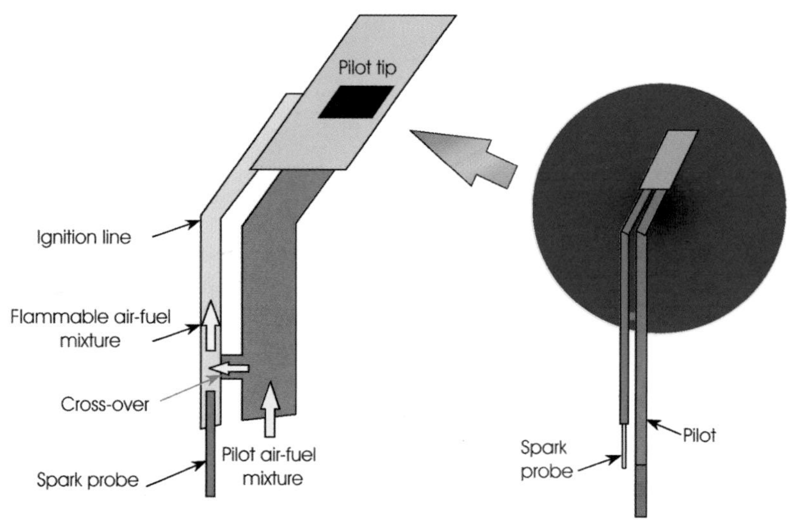

FIGURE 20.39 Slipstream flame-front generator.

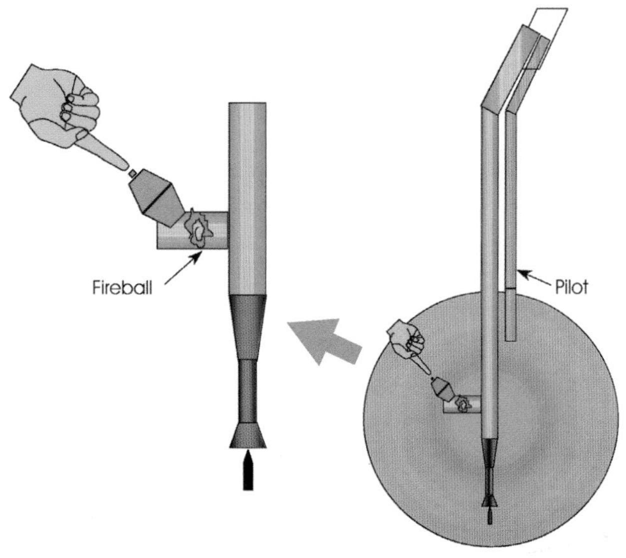

FIGURE 20.40 Self-inspirating flame-front generator.

that this system has over the conventional FFG system is that compressed air is not required. The advantage over the slipstream FFG is that the critical parts are accessible during flare operation. The disadvantage of this system, however, is that the maximum distance of the ignition line is limited to approximately 200 ft (60 m). The exact distance, however, can vary, depending on the fuel pressure available, composition of the fuel, diameter and wall roughness of the ignition line, and ambient air density.

20.4.2.3 Monitors

Verification that a flare pilot is burning is an important and in some cases mandatory requirement. The pilot's remote location and inaccessibility during flare operation make flame verification difficult. A brief review of flare pilot monitoring methods illustrates the difficulties.

Most pilot fuels produce a low luminosity flame because the gas mixture at the pilot tip contains close to 100% of the air required by the fuel. It can be very difficult to see a pilot flame during the day. Viewing at night is generally more successful. If the pilot is ignited using a conventional FFG, opening the fuel valve of the FFG can enhance visual sighting, day or night. The added fuel will produce a larger and more luminous flame at the pilot. After the pilot flame has been sighted, the extra fuel should be shut off.

By nature, one immediately associates flame with heat. In fact, a flame produces heat, ionized gas, light, and sound. The technique for verifying a pilot flame by sensing the flame-generated heat with a thermocouple has been used for many years. In the thermocouple technique, the thermocouple junction is placed in a position to sense the heat generated by the pilot flame. A balance must be struck between a high exposure to heat with possible rapid thermocouple burnout and a lower exposure with a slower response time. The thermocouple is connected to a temperature switch or a computer that indicates pilot failure if the temperature drops below a set point. In most cases, a shutdown is required to replace a burned out thermocouple.

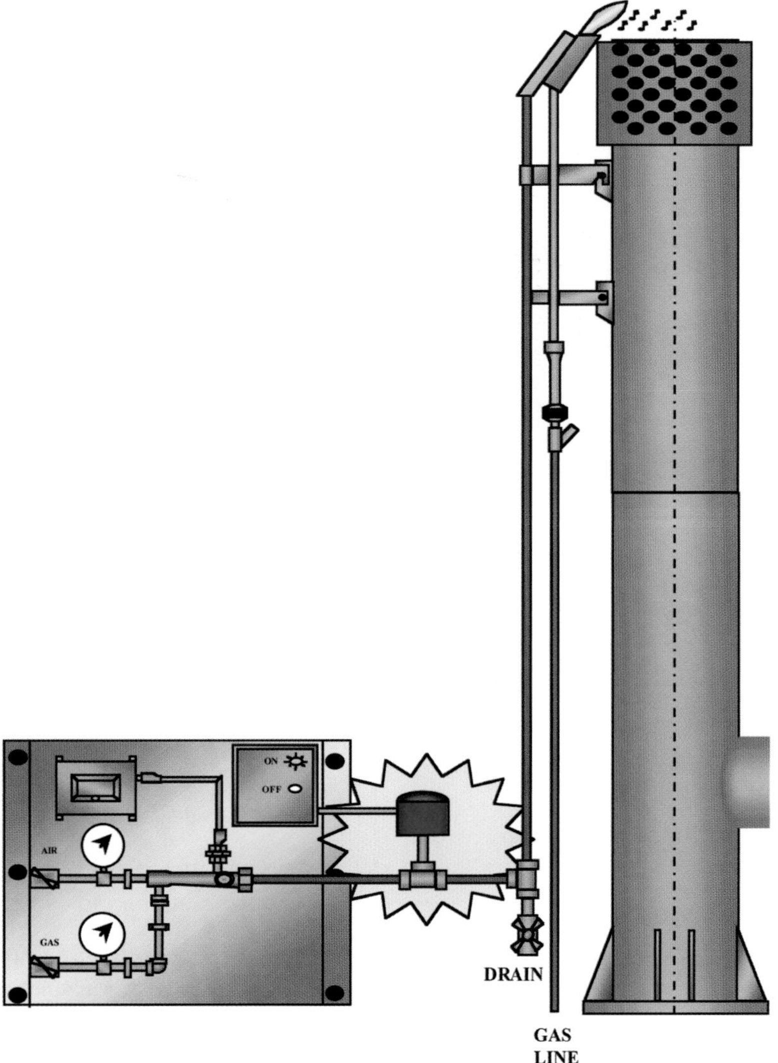

FIGURE 20.41 SoundProof acoustic pilot monitor.

Other techniques have sought to verify a flame using flame ionization or optical scanning. The flame ionization method requires two elements located in the pilot flame. The presence or absence of a flame is detected by a change in resistance between the elements. Like a thermocouple, these elements cannot be maintained during flare operation. The use of flame ionization to monitor flare pilots is limited.

Optical sensing benefits from an accessible grade-level location. Most optical sensors employed on flare pilots use one or more infrared wave bands to sense the presence of a flame. Ultraviolet sensors, which are frequently used on process and boiler burners, are generally limited to use on enclosed ground flares. Optical methods may be unable to distinguish pilots from the main flame or one pilot from another. In addition, the optical path can be obscured by heavy rain, fog, or snow, or the movement of the top of the flare stack may move the flame out of the sensor's field of view.

The newest flame monitoring technique detects the pilot flame remotely using the overlooked flame characteristic of sound generation.[15] This system consists of an acoustic sensor and a signal processor. The sensor listens to the pilot sounds through the flame-front generator line much as a doctor uses a stethoscope to listen to a heart. Acoustic data is conveyed from the sensor to the signal processor via a cable. The signal processor analyzes the acoustic data and signals the pilot flame status. An acoustic monitor system is shown in Figure 20.41. An acoustic pilot monitor can distinguish its connected pilot from nearby sound sources such as other pilots, steam injection, and combustion of the flare. Weather conditions do not adversely affect the monitor.

FIGURE 20.42 Horizontal settling drum at the base of an air assisted flare.

FIGURE 20.43 Cyclone separator. (Note the frost indicating the flow path of the low-temperature liquid as it is removed.)

20.4.3 Knockout Drums

There are three basic types of knockout drums that can be incorporated into a flare system: a horizontal settling drum, a vertical settling drum, and a cyclone separator.

Horizontal settling drums are large drums in which droplets are allowed sufficient residence time to separate from the gas by gravity. API RP-521[5] provides detailed design guidelines for this type of drum. Figure 20.42 shows a typical example of this design. The pressure drop across these drums is relatively low. Drums of this type are particularly useful for removing liquids within or near the process units that may send liquids to the flare header. It is common for the maximum liquid level to be at the drum centerline, thus allowing 50% of the total vessel volume to be used for temporary liquid storage during a relief.

Vertical settling drums work in a similar fashion. In designing vertical settling drums, careful attention must be focused on droplet terminal velocity because this velocity determines the drum diameter. Also, the volume available for storage of liquid during a relief is limited by the elevation of the flare header piping.

Any small droplets that pass through the knockout drum can agglomerate to form larger droplets in the flare system downstream of the knockout drum. Locating the knockout drum very near the base of the flare stack, or incorporating it into the stack base, can minimize this problem. Although the pressure drop required for the settling drums is generally low, the required drum diameter can become impractical to shop-fabricate if the flow rate is high.

Elimination of very small liquid droplets cannot be accomplished through a simple reduction in gas stream velocity. Cyclone separation is best for small droplet removal. Mist eliminators, utilizing centrifugal force, can be very effective when incorporated into the base of the flare stack. They are smaller in diameter than horizontal or vertical settling drums and usually provide high liquid removal efficiency at the expense of a greater pressure drop. The frost on the outside of the drum in Figure 20.43 vividly illustrates the liquid flow pattern.

Agglomeration of droplets downstream of a cyclone separator is generally less of a problem than it is for the settling drum designs. The typical settling drum is designed to remove droplets larger than 300 to 600 microns at the smokeless flow rate. Droplet sizes at the maximum flow rate can be over 1000 microns in some cases. By comparison, the droplets exiting the cyclone are much smaller, typically 20 to 40 microns, and the droplet size remains low throughout the operating range.

However, available volume for liquid storage in a cyclone separator is generally small compared to the vertical settling drum because a substantial length is required for the vapor space in this design. When substantial liquid loads are anticipated, horizontal settling drums are usually provided at the upstream end of the flare header to catch most of the liquid volume. The mist eliminator in the base of the flare stack removes the remaining liquid and minimizes problems with agglomeration.

20.4.4 Liquid Seals

A liquid seal is a device that uses a liquid, such as water or glycol/water mix, as a means of providing separation of a gas (or vapor) conduit into an upstream section and a downstream section. The physical arrangement of a typical liquid seal is shown in Figure 20.44. Gas flows through the seal when the gas pressure on the upstream side of the seal is equal to or greater than the pressure represented by the seal leg submergence plus any downstream back-pressure. The submergence depth used depends on the purpose of the seal. Present practice involves submergence depths ranging from about 2 in. (5 cm) to over 120 in. (300 cm).

Liquid seals are found in many types of combustion systems, including flares, due to the fact that liquid seals can be used to accomplish any of a variety of goals:

- prevent downstream fluids from contaminating the upstream section
- pressurize the upstream section
- divert gas flow
- provide a safe relief bypass around a control valve
- arrest a flame front or detonation

20.4.4.1 Prevent Upstream Contamination

One of the most frequent uses of liquid seals is to prevent air infiltration into the downstream section from propagating to the upstream section. Properly operated liquid seals provide a safeguard against the formation of an explosive mixture in the flare header by acting as a barrier to backflow.

20.4.4.2 Pressurize Upstream Section

As discussed in Section 20.3.4, a negative pressure can exist at the base of the flare stack at low flow conditions. By injecting purge gas upstream of the liquid seal, the upstream section is pressurized to a level related to the submergence depth. As a result, any leaks in the upstream section will flow gas out of the flare header rather than air into the header.

20.4.4.3 Diverting Gas Flow

Liquid seals are often used to divert gas flow in a preferred direction. An example is a staged flare system involving an enclosed flare and an elevated flare such as the one shown in

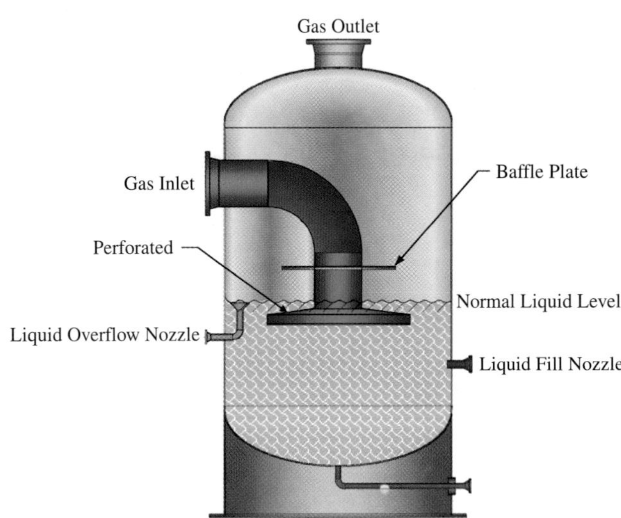

FIGURE 20.44 Schematic of a vertical liquid seal.

Figure 20.12. The liquid seal in the line to the elevated flare diverts all flows to the enclosed flare until the pressure drop caused by gas flow through the enclosed flare system exceeds the submergence depth.

20.4.4.4 Control Valve Bypass

A control valve in flare header service can represent a safety hazard should it fail to open when required. A liquid seal bypass around the control valve protects the plant against possible failure of the control valve to open during a relief. When the upstream pressure reaches the submergence depth plus any back-pressure from the elevated flare, waste gas will begin to flow through the liquid seal to the elevated flare — whether the control valve is open or not.

20.4.4.5 Liquid Seals as Arrestors

Liquid seals that are used as flame arrestors generally fall into three categories as follows:

1. seals designed to handle incoming combustible gases that do not contain air (or oxygen); example: a refinery flare
2. seals designed to handle incoming vapor streams that contain a mixture of combustible gases and air (or oxygen); example: the fuel vapor/air stream produced during tank truck or barge loading operations
3. seals designed to handle either ethylene oxide (ETO) or acetylene; example: the vent seal for the ETO gases from a sterilizer used for treating medical supplies

FIGURE 20.45 "Smoke signals" from a surging liquid seal.

20.4.4.6 Design Factors

Once the purpose of the liquid seal has been established, the designer can produce a suitable liquid seal design. Several factors influence general liquid seal design, including:

- seal vessel orientation (horizontal or vertical)
- seal vessel diameter
- seal leg submergence depth
- configuration of the end portion of the seal leg (seal tip or seal head)
- space above the liquid level
- type and size of outlet
- seal fluid selection

A complete discussion of all these factors is beyond the scope of this text.

Liquid seals have been known to cause pulsating flows to flares. Pulsating flow, in turn, makes smoke control difficult and creates a fluctuating noise and light source that can become a nuisance to neighbors. Figure 20.45 shows the "smoke signal" effect that can occur in extreme cases of such flow patterns. The fundamental cause of this pulsating flow is the bi-directional flow that occurs when the displaced liquid in the vessel moves to replace gas bubbles released at the gas exit of the seal head or inlet pipe. As the submergence depth increases, the buoyant forces acting on the liquid increase and the potential for violent movement by the seal fluid grows. Larger vessel diameters also increase the potential for liquid sloshing, which is another driver of pulsation.

Properly designed internals can reduce such pulsations by controlling the bi-directional gas flow and movement of the liquid. Robert Reed produced some of the earliest internals for liquid seals in the late 1950s. Since that time, a number of alternative designs have been developed[16] that improve on the basic idea. Several of the designs that have been used over the years are shown in Figure 20.46. Today, designs are available that can convert an existing horizontal knockout drum into a combination liquid seal/knockout drum with enough submergence depth to enable flare gas recovery.

The two fluids normally used for a liquid seal are water and glycol/water mix. Whenever possible, water is preferred. The use of hydrocarbon liquids (other than glycol) is strongly discouraged.

20.4.5 Purge Reduction Seals

Air infiltration into a flare system through the flare burner was discussed in Section 20.3.4. Most systems are designed to combat air infiltration into the tip and riser by purging. The amount of purge gas required to prevent air from entering into the system can be quite large, especially in the case where light gases are present. A high purge rate may pose several disadvantages. First is the cost of the purge gas; second, the heat from the combustion of the purge gas can be damaging to the flare burner; and third, burning more gas than is absolutely necessary increases the emissions level of the plant. Adding a purge reduction device to the flare system can mitigate these disadvantages. Normally, such devices are installed just above or immediately below the flare burner-mounting flange so as to maximize the air exclusion zone.

Purge reduction devices are intended to improve the effectiveness of the purge gas so that the amount required to protect the system can be reduced. Purge reduction devices, often referred to as seals, are based on the use of either of two basic strategies: (1) density difference (sometimes called a density seal), or (2) trap and accelerate (sometimes called a velocity seal). The discussion below addresses the principle of each seal type and its advantages and disadvantages.

A velocity seal is shown in Figure 20.47. The principle of the velocity seal is to trap air as it enters the flare tip, reverse its direction, and carry it out of the tip with accelerated purge gas. Tests on flare stacks, large and small, have demonstrated that air enters a flare tip along the inner wall of the tip. In the velocity seal, a shaped trap is placed on the inner wall of the tip. The trap intercepts the incoming air and turns it back toward the tip exit. At the same time, the shape of the trap acts to accelerate the purge gas. The accelerated purge gas and outflowing air meet at the exit of the seal device and flow out the tip. Without the accelerated purge gas, the trap will only delay air entry, not reduce it.

Compared to a density seal, a velocity seal is relatively small and has low capital cost. The velocity seal will reduce the purge gas requirement but the reduction is tempered by

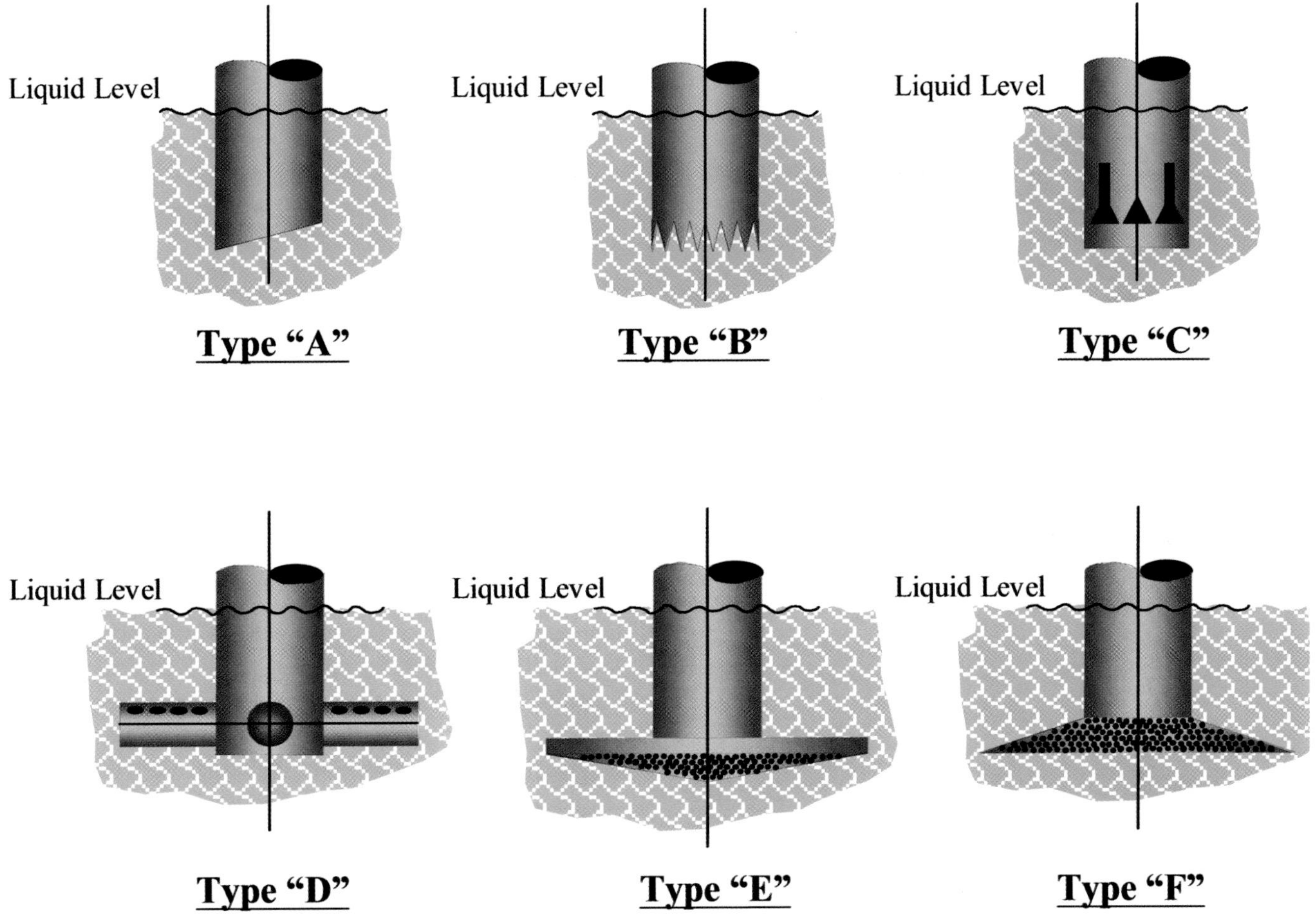

FIGURE 20.46 Various liquid seal head types: "A" Beveled end, "B" Sawtooth, "C" Slot and triangle (after API RP-521), "D" Arms with ports on upper surfaces, "E" Downward facing perforated cone, "F" Upward facing perforated cone.

the amount of oxygen allowed below the seal. A velocity seal requires more purge gas than a density seal. An additional disadvantage of a velocity seal occurs when purge gas flow is interrupted. In this event, the oxygen level in the riser begins to increase almost immediately.

The arrangement of a density-type purge reduction device is illustrated in Figure 20.48. As the gas flows upward through the riser, it is directed through two annular 180° turns, thus forming spaces where lighter- or heavier-than-air gases are trapped. The density difference between the trapped purge gas and air forms a barrier to air movement. Only diffusion will allow the air to work its way through the barrier. Thus, a purge rate sufficient to constantly refresh the gas at the gas/air interface is all that is required. This purge rate is much lower than the rate required for a velocity seal (which will have some level of oxygen below it). A density difference as small as nitrogen to air is sufficient for the seal to function. The lighter (or heavier) the purge gas, the more effective the seal becomes. Tests have shown that the oxygen level below a properly purged density-type seal will be zero. If the purge gas flow to a density seal is interrupted, air will begin to penetrate the gas by diffusion. However, the diffusion process is slow and a significant time will pass before air enters the riser.

The density seal requires the smaller purge gas rate and enjoys the lower operating cost. A lower purge gas requirement also means less heat around the flare tip and lower emissions.

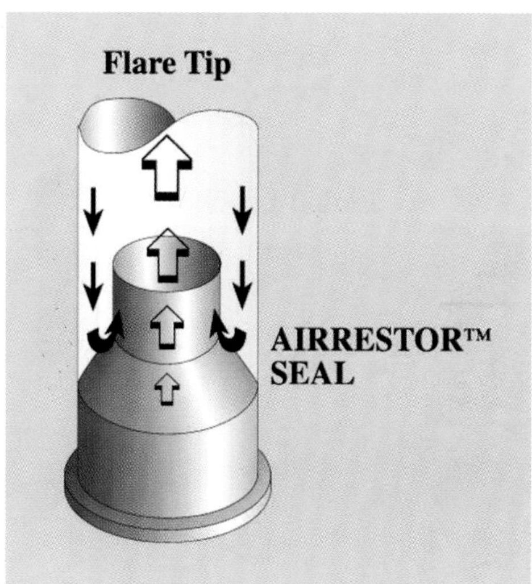

FIGURE 20.47 Airrestor velocity-type purge reduction seal.

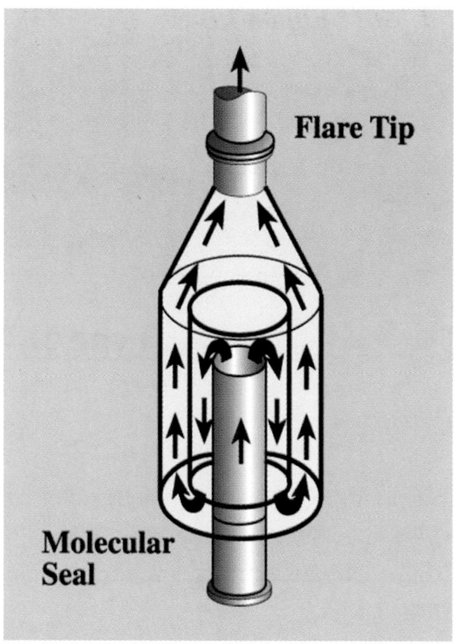

FIGURE 20.48 Molecular Seal density-type purge reduction seal.

However, the physical size of the density seal makes its capital cost larger.

20.4.6 Enclosed Flares

The desire to hide flaring activities dates back to the 1950s. Flare vendors and users tried for several years to design enclosed ground flares and failed, sometimes spectacularly. In one case, the smoke generated by a ground flare was so dense that it forced plant shutdown. In another case, smoking caused the shutdown of a major highway and noise broke windows at a great distance.

A different design concept yielded the first successful enclosed ground flares in 1968 when two units designated as ZTOFs (Zink thermal oxidizer flares) were placed in service by Caltex.[17] These enclosed ground flares and dozens of additional units constructed over the ensuing years led to the development of the modern enclosed ground flare.

Enclosed ground flares use a refractory-lined combustion chamber to contain the entire flame, rendering it invisible to neighbors. A schematic diagram of an enclosed ground flare is shown in Figure 20.49. The combustion chamber is generally cylindrical, but can be rectangular, hexagonal, or other shapes formed from flat panels. Cylindrical sections are generally favored. Flat panels are used in some cases to reduce shipping costs or to optimize field assembly.

A ZTOF is essentially a giant, direct-fired air heater. Air required for combustion and for temperature control enters the combustion chamber by natural draft after passing through a burner opening. Elevated temperatures in the combustion chamber reduce the density of the flue gases inside and produce draft according to Eq. (20.2). This draft is the motive force that drives combustion products out the top of the stack and draws air in through the burner openings. Optimizing the use of the available energy is an essential part of the proper design of a ZTOF.

Most ZTOFs are designed to handle substantially more than the stoichiometric air requirement. The excess air is used to quench the flame temperature. This reduces the required temperature rating for the refractory lining, which is a significant part of the overall cost of the system. Although the quenched flue gas temperature may be 1600°F (900°C) or lower, the refractory lining — at least in the lower section of the stack — should be selected for a higher service temperature because local temperatures may be higher than the final flue gas temperature.

When a ZTOF is used as the first stage of a flare system, there is the potential to deliver more waste gas to the ZTOF than it is designed to handle. Overfiring the ZTOF can result in flame and/or smoke out the top of the stack. There are usually two safeguards to prevent this from happening. First, when the pressure drop created by the gas flow through the ZTOF system exceeds the setpoint of the diversion device (whether liquid seal or valve), excess gas automatically flows to the elevated flare. Second, most ZTOFs are equipped with thermocouples to monitor the stack temperature. When the stack exit temperature exceeds the design level, a temperature switch initiates an automatic shutdown, either partial or total, of the ZTOF burner system. Gas flow is sent to the other parts

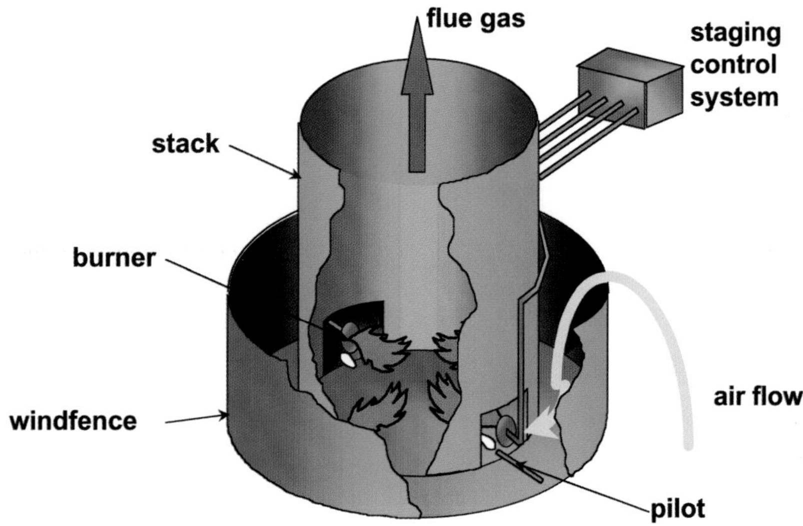

FIGURE 20.49 Schematic of a ZTOF.

of the plant flare system until the cause of the overfiring condition can be identified and corrected.

ZTOFs have been designed with capacities ranging from less than 100 lb/hr (45 kg/hr) to more than 100 metric tons/hr. Combustion chambers vary from 3 ft (1 m) to more than 50 ft (15 m) across and may be over 100 ft (30 m) tall.

To maximize the benefit of the available combustion volume, ZTOFs are usually equipped with multi-point burner systems. As discussed in Section 20.4.1.5, breaking up the gas flow into many small flames improves burner performance. ZTOF systems frequently operate at pressure levels consistent with liquid seal depths. As a result, the available energy from the waste gas is reduced. Thus, staging the ZTOF burner systems is even more important to maintain good performance at turndown conditions. When steam-assisted burners are used in ZTOFs, steam efficiencies are substantially higher than open air flares, resulting in lower day-to-day steam consumption.

On small units, adjusting the air openings feeding air into the combustion chamber can control temperature in the combustion chamber. Temperature control is common in landfill flares, biogas flares, and vapor combustors in gasoline loading terminals. Proper temperature control minimizes emissions from these units, which in some cases run continuously.

The windfence used to manage air flow into the combustion chamber can be designed to muffle combustion and steam noise generated by the burners. The refractory-lined combustion chamber may also absorb high-frequency noise and serves to block the direct line-of-sight path for noise transmission from the flames in the chamber. By providing clean, quiet, invisible disposal of day-to-day reliefs, use of a ZTOF allows plant operation in harmony with its neighbors.

20.4.7 Flare Support Structures

The combination of the heat released at maximum design flow and the owner's instructions on allowable incident radiation poses a design challenge that is often solved by elevating the flare burner. (See Sections 20.2.6 and 20.3.5 for discussion of the factors and design considerations that influence the determination of the required height.) Once the height has been calculated, the design focus turns to the selection of the type of flare structure to use.

In principle, there are three basic support structure concepts plus a variant that can be very useful in certain circumstances. The concepts and the variant are:

- self-supported
- guy wire supported
- derrick supported
- derrick with provision for lowering the riser and flare burner

A self-supported structure (Figure 20.50) requires the least land space and can easily accommodate a liquid seal or knockout drum, or both, in the base section. Varying the diameter and thickness of the structure at various elevations absorbs wind loads. Potential undamped vibration is avoided by varying the length and diameter of sections of the structure. Generally, self-supported structures are not cost-effective at heights above about 250 ft (76 m).

Perhaps the most common means of supporting an elevated flare burner is a riser that is held in line by guy wires (Figure 20.51). Usually, there are three sets of guy wires spaced 120° apart. The number of guy wires arranged vertically at a given location is dependent on the height of the

FIGURE 20.50 Self-supported flare.

structure, wind loads, and the diameter of the riser. Guy wire supported-structures require the greatest land space commitment. Overall heights can reach 600 ft or more.

Where land area is of high value or limited availability, a derrick structure can be employed. The derrick itself (Figure 20.52) acts as a guide to keep the riser in line. In general, derricks are designed with three or four sides and have been utilized at heights greater than 650 ft (200 m).

Flare burners on very tall support structures and flares located in remote areas are difficult to maintain or replace due to the limited number of cranes that can service the required elevation. In such cases, a derrick variation often referred to as a demountable derrick is employed. The design of the demountable derrick (Figure 20.53) allows the riser and attached flare burner to be lowered to the ground, either as a single piece or in multiple sections. An additional advantage of a demountable derrick is its ability to support more than one full-size riser.

With the exception of demountable derricks, any flare burner support structure with a height of more than about 50 ft (15 m) above grade typically includes a 360° platform for use during maintenance. If a 360° platform is used, it is often located just below the flare burner-mounting flange. The structure will also provide support for ladders, required step-off platforms, and utilities piping. Some flare structures, due to their height and location, may require aircraft warning markings such as paint or lights.

A number of factors enter into the selection of the structure: physical loads, process conditions, land space available, cost of land, availability of cranes, and the number of risers to be considered. The selection process can be simplified by using the guide shown in Figure 20.54. The guide asks a series of questions that can be answered "yes" or "no," with the answer influencing the next question. While the yes/no answers appear to lead to an absolute answer, there are subtleties that can promote an alternative. For example, the desire to locate a liquid seal or knockout drum, or both, in the base of the stack may make a self-supported design attractive.

The guide refers to situations in which there will be one waste gas riser (R1) or two waste gas risers (R1 and R2). If there are two risers, the guide questions the size of the second riser as compared to R1/3. Typically, a second riser with a size of R1/3 or less will be a small-capacity flare serving a vent system or incinerator bypass. Such a small flare, often referred to as a "piggyback flare," will be supported by the main flare or its support structure. If the second riser is greater than R1/3, it will be treated, for structural design purposes, as a second major flare. Cases involving more than two risers are good candidates for a demountable derrick structure.

20.4.8 Flare Controls

Flare systems are often associated with flare headers that collect gases discharged from relief valves and other sources. A flare is called upon to operate properly during upset and malfunction conditions that impact control systems throughout the plant, including power failure and instrument air failure. Therefore, controls on flare systems must be used with discretion to ensure that the flare will continue to operate safely even if its controls fail. Flare controls can help provide effective smokeless performance, low noise operation, and other desirable characteristics during normal day-to-day operation.

Many of the controls used in flare systems are associated with pilots, ignition, and pilot monitoring and have already been discussed in Section 20.4.2. This section discusses steam control, burner staging, level controls, and purge control.

FIGURE 20.51 Guy wire-supported flare.

FIGURE 20.52 Derrick-supported flare.

FIGURE 20.53 Demountable derrick.

20.4.8.1 Typical Steam Control Valve

Reliable steam control is an important part of the smoke suppression strategy for steam-assisted flares. The simplest steam control system consists of a manual valve that an operator uses to adjust steam flow to the flare tip. Most plants prefer not to dedicate an operator to manage the steam use of their flares. Instead, steam control valves are equipped with remote positioning equipment that allows an operator in the control room to adjust steam flow while performing other, more profitable duties. Figure 20.55 depicts a typical steam control valve station.

The steam control valve on a flare can operate almost completely closed for extended periods of time. As a result, wear on the valve seat becomes a maintenance issue. To allow for removal and maintenance while the flare is in operation, block valves are recommended both upstream and downstream of the control valve. To operate the flare smokelessly during control valve maintenance, a bypass line with a manual valve is installed around the control valve and its block valves. A pressure gage should be installed downstream of the control valve station to provide the operator with a tool for diagnosing control issues and a guide for manual control, when needed.

Most steam-assisted flares require a minimum steam flow for two reasons. First, a minimum steam flow keeps the steam line from the control valve to the flare burner warm and ready for use. It also minimizes problems with condensate in that line. Second, a minimum steam flow keeps the steam manifold on the flare burner cool ("cooling steam") in case a low flow flame attaches to the steam equipment. To maintain the minimum steam flow, a second bypass line is installed with a metering orifice sized for the minimum flow and a pair of block valves for maintenance of the orifice.

Steam traps are mandatory wherever condensate can accumulate in the steam piping. Many steam injector designs use relatively small orifices, at least in part to reduce audible

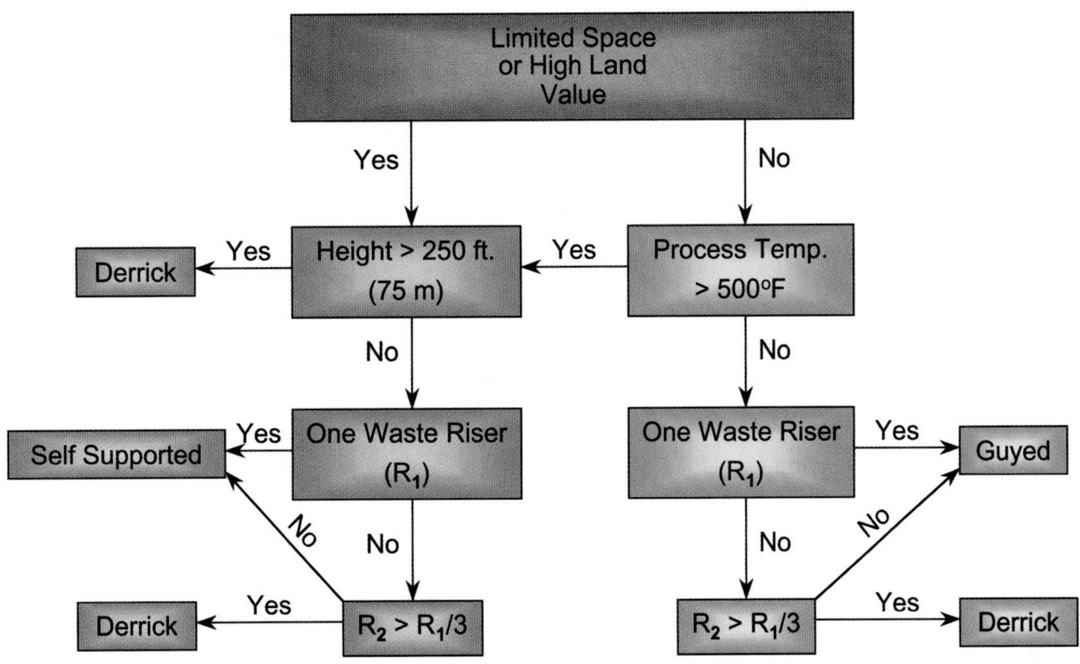

FIGURE 20.54 Flare support structure selection guide.

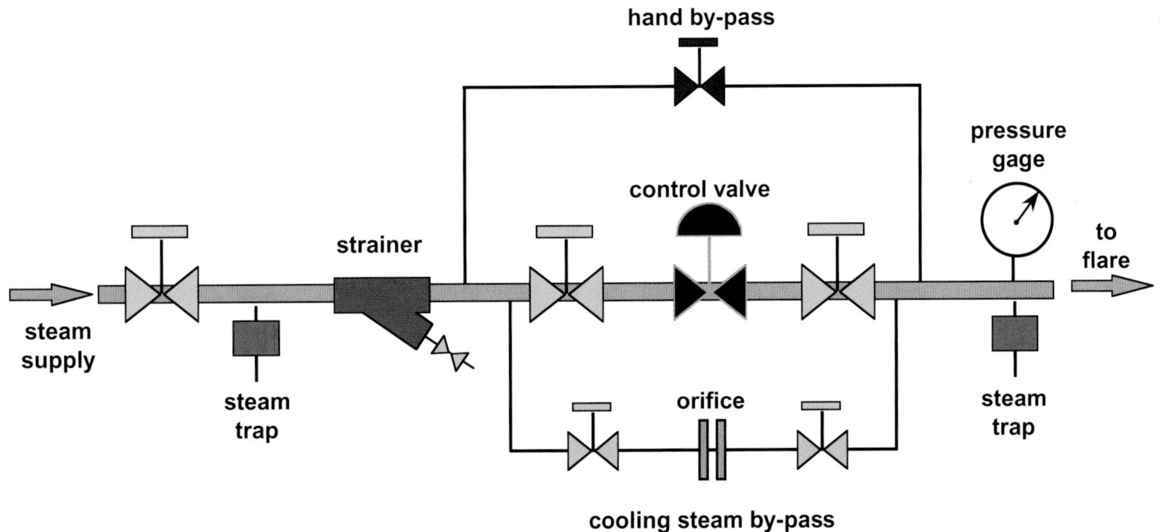

FIGURE 20.55 Steam control valve station.

noise. Therefore, a steam line strainer is recommended. If the orifices are very small, all stainless steel steam piping may be appropriate.

20.4.8.2 Automatic Steam Control

As the flow or composition of waste gas sent to the flare varies, the amount of steam required for smoke suppression changes. Many plants adjust the steam requirement based on periodic observations by an operator in the control room looking at a video image from a camera aimed at the flare. Any smoking condition will be quickly corrected by an increase in steam flow to the flare. However, when the gas flow begins to subside, the flare flame continues to look "clean" to the operator. Therefore, some time may pass before the operator reduces the steam flow. As a result, this method of smoke control tends to result in oversteaming

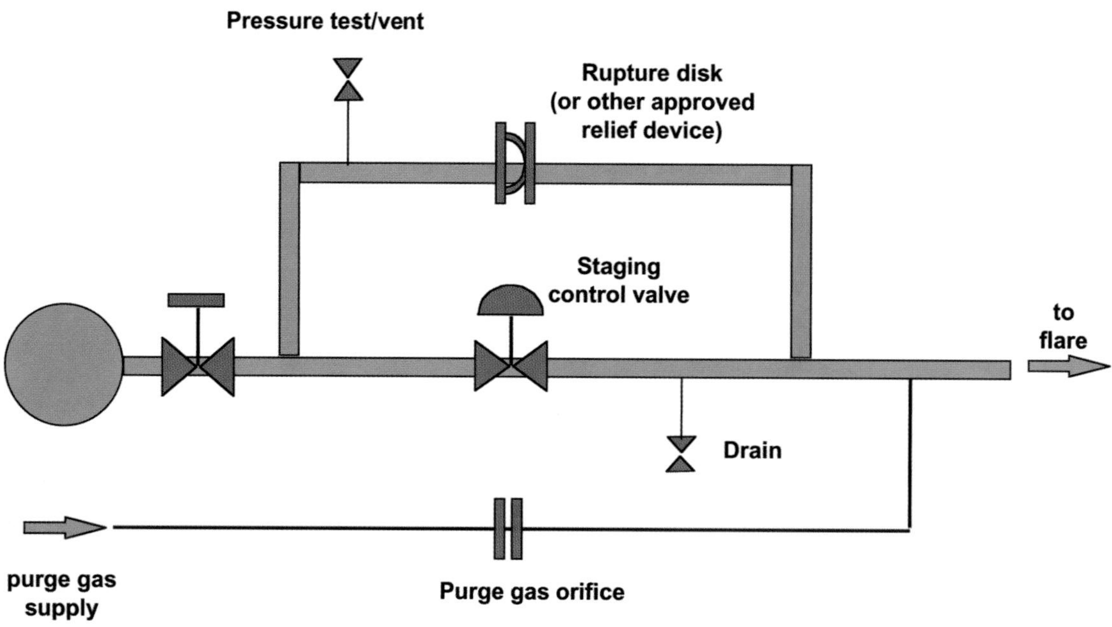

FIGURE 20.56 Staging control valve assembly.

of the flare, which in turn produces excessive noise and unnecessary steam consumption.

Optical sensing systems are available to monitor the condition of the flare flame and adjust the steam flow continuously. Automatic optical sensing equipment can effectively control steam flow to maintain a consistent flame appearance with minimum steam usage and minimum noise.

20.4.8.3 Typical Staging Control Valve

Energy conversion flare types such as the LRGO discussed in Section 20.4.1.5 are designed to operate smokelessly when the gas pressure is above a certain level. Two key operational goals for such systems using energy conversion burners are to: (1) maintain the gas pressure above the minimum level required for smokeless performance, and (2) prevent back-pressure from exceeding the allowable design level. To achieve these goals, a staging control system is used that starts and stops flow to various groups of burners based on the incoming gas flow. Depending on the application, the staging valve used to accomplish this can be installed in any of a number of possible configurations, ranging from a single valve to a complex system of bypasses and block valves. Figure 20.56 shows a typical staging valve assembly.

The main staging valve can be either fail closed or fail open, depending on the safety considerations governing the system as a whole. Generally, the staging valves for an enclosed ground flare are fail closed when another flare is available to handle emergency relief loads. Such staging valve assemblies usually do not include a bypass device.

Staging valves, especially on the last stages, of a multipoint staged plant flare system are usually designed to fail open when the bypass device is a rupture disk. If the bypass device is easily reclosed or resets automatically, the staging valve can be designed to fail closed. The bypass device, shown in the figure as a rupture disk, can also be a relief valve or a liquid seal.

20.4.8.4 Level Controls

Flare systems often include vessels such as knockout drums or liquid seals that can contain liquid levels that must be monitored and/or controlled for safe operation. Liquid level is controlled in knockout drums to prevent overfilling, as discussed in Section 20.3.3. In some cases, it is also important to prevent too low a level. When all the liquid in a knockout drum is removed, it becomes possible for waste gases in the flare header to migrate into the drain system, creating a possibly explosive mixture and a serious safety hazard. Instrumentation generally consists of one or more level switches or transmitters often mounted together with gage glasses to simplify setpoint adjustments and to allow visual monitoring or manual control.

Liquid seal level control presents a number of challenges not found in other level control applications. In normal operation, when gas is flowing through the liquid seal, the surface of the liquid is violently agitated. Small-scale wave action, spraying, and foam generation also create special requirements for liquid seal level control systems.

Liquid seals in flare service can accumulate a certain amount of hydrocarbon condensate. Such condensates are generally lighter than water and affect level control and safety. The presence of such condensates in the liquid seal creates the potential to generate hydrocarbon droplets in the waste gas flowing to the flare tip. As discussed in Section 20.3.3, this can become a safety hazard. To protect against this hazard, flare liquid seals are often equipped with hydrocarbon skimming systems that remove accumulated condensate from the liquid surface, in some cases automatically.

Loop seals are used to prevent gases from escaping a vessel while allowing liquids to be removed automatically. Hydrocarbon skimming systems on flare liquid seals often utilize loop seals, such as shown in Figure 20.57, to provide constant removal of liquids. Loop seal design guidance is provided by API RP-521.[5] Some additional concerns include:

- Loss of loop seal fluid by evaporation or overpressure can result in waste gas entering the sewer system and/or escaping to atmosphere through the anti-siphon break at the top of the outboard loop.
- Freezing of loop seal fluid can result in overfilling the vessel and/or accumulation of hydrocarbon condensate in the vessel.
- Elevation of the top of the outboard loop must be at or below the controlled liquid level.
- If liquids of differing densities are anticipated, the elevation of the outboard loop must be low enough to allow the lightest liquid to push out the heaviest liquid.

20.4.8.5 Purge Controls

Purge gas injection is one of the most important safety features in a flare system. A common method for controlling the flow of purge gas is the use of a metering orifice and a supply of purge gas with regulated pressure. A typical arrangement is shown in Figure 20.58. Safety features should include an effective strainer to prevent pluggage of the metering orifice, a flow transmitter with an alarm for low flow condition, and a supply pressure substantially higher than any anticipated flare header pressure.

20.4.9 Arrestors

Dry arrestors such as flame arrestors or detonation arrestors have limited application in plant or production flare systems. This is due to the concern that the small passages of an arrestor could become plugged, leading to increased back-pressure on the relieving or venting source. In the worst case, the source could become over-pressured. A few of the sources of concern regarding plugging are:

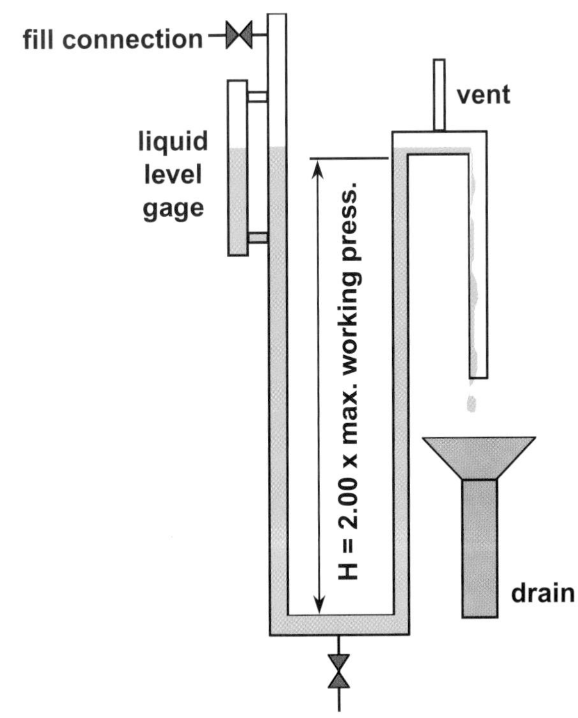

FIGURE 20.57 Loop seal.

- scale or debris carried by high waste gas velocities in the flare header
- gas compositions that include compounds prone to polymerize
- two-phase flow carrying liquids or condensate into the arrestor

There are some limited circumstances where the use of a dry arrestor may be acceptable. For example, an arrestor may be used on systems that handle relatively clean, dry material or systems that can be easily shut down for maintenance should an over-pressure be detected. As with all flare system issues, careful attention to the safety aspect of the prime objective is required.

20.5 FLARE COMBUSTION PRODUCTS

An industrial flare is the most suitable and widely used technology for disposing of large quantities of organic vapor releases. In the 1980s, the U.S. Environmental Protection Agency (U.S. EPA) conducted several tests to determine the destruction and combustion efficiency of an industrial flare operating under normal conditions. Based in part on these tests, the U.S. EPA made several rulings concerning the design of a flare. The purpose of this section is to discuss the results of these tests and U.S. EPA rulings on flare design.

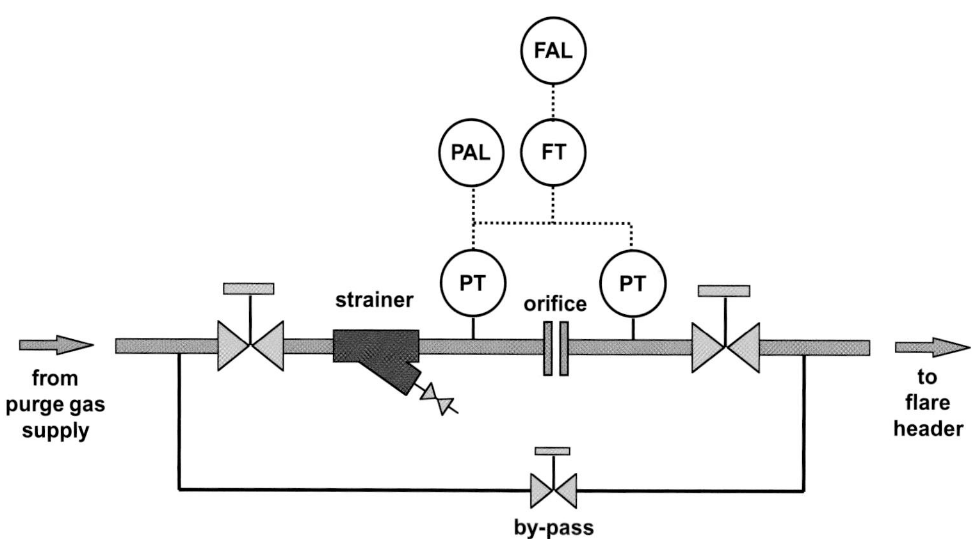

FIGURE 20.58 Purge control station.

20.5.1 *Reaction Efficiency*

The terms *combustion efficiency* and *destruction efficiency* have frequently and mistakenly been considered synonymous. In fact, these two concepts are quite different. A flare operating with a combustion efficiency of 98% can achieve a destruction efficiency in excess of 99.5%.

20.5.1.1 Definition of Destruction and Combustion Efficiency

Destruction efficiency is a measure of how much of the original hydrocarbon is destroyed, that is, broken down into non-hydrocarbon forms, specifically carbon monoxide (CO), carbon dioxide (CO_2), and water vapor (H_2O). The destruction efficiency can be calculated by a carbon balance as follows:

$$\%DE = \frac{CO_2 + CO}{CO_2 + CO + UHC} \times 100 \tag{20.4}$$

The term %DE is the percent destruction efficiency, and CO_2, CO, and UHC are the volume concentrations of carbon dioxide, carbon monoxide, and unburned hydrocarbons (as methane) in the plume at the end of the flare flame, respectively. Notice that if no unburned hydrocarbons escaped the flame, the destruction efficiency would be 100%.

Combustion efficiency is a measure of how much of the original hydrocarbon burns completely to carbon dioxide and water vapor. Using the carbon balance approach, combustion efficiency can be calculated as:

$$\%CE = \frac{CO_2}{CO_2 + CO + UHC} \times 100 \tag{20.5}$$

where %CE is the percent combustion efficiency. Notice that even if no unburned hydrocarbons escape the flame, the combustion efficiency can be less than 100% because CO represents incomplete combustion. It is evident from Eqs. (20.4) and (20.5) that the combustion efficiency will always be less than or equal to the destruction efficiency.

20.5.1.2 Technical Review of Industrial Flare Combustion Efficiency

In 1983, the John Zink Company and the Chemical Manufacturers Association (CMA) jointly funded a research project aimed at determining the emissions of flares operating under normal, real-world conditions.[18] Various mixtures of crude propylene and nitrogen were used as the primary fuel, with waste gas lower heating values (LHVs) varying from approximately 80 to 2200 Btu/scf and flow rates up to 3000 lb/hr. Tests were conducted using steam-assisted, air-assisted, and non-assisted flares. These tests concluded that flares operating with a normal, stable flame achieve combustion efficiencies greater than or equal to other available control technologies.

The U.S. EPA ruled that a flare can achieve a combustion efficiency of 98% or greater if the exit velocity of the organic waste stream, at the flare tip, is within the following limits:[19]

- Non-assisted and steam-assisted flares:
  - If 200 Btu/scf < LHV < 300 Btu/scf,

$$V_{max}(ft/s) = 60 \times \left(\frac{T(°F)+460}{520}\right) \quad (20.6)$$

  - If 300 Btu/scf < LHV < 1000 Btu/scf,

$$V_{max}(ft/s) = anti\log_{10}\left(\frac{LHV+1209.6}{849.1}\right) \times \left(\frac{T(°F)+460}{520}\right) \quad (20.7)$$

  - If LHV > 1000 Btu/scf,

$$V_{max}(ft/s) = 400 \times \left(\frac{T(°F)+460}{520}\right) \quad (20.8)$$

- Air-assisted flare:
  - If LHV > 300 Btu/scf,

$$V_{max}(ft/s) = \frac{(329.5+LHV)}{11.53} \times \left(\frac{T(°F)+460}{520}\right) \quad (20.9)$$

20.5.1.2.1 Hydrogen Enrichment

Because hydrogen has a lower volumetric LHV than organic gases commonly combusted in flares, the U.S. EPA amended 40 CFR 60 to include an allowance for hydrogen content. The EPA believes that hydrogen-fueled flares, meeting the maximum velocity limitation as shown below, will achieve a combustion efficiency of 98% or greater.

$$V_{max}(m/s) = \left(X_{H_2} - K_1\right) \times K_2 \times \left(\frac{T(°C)+273}{293}\right) \quad (20.10)$$

where

V_{max} = Maximum permitted velocity, m/s
K_1 = Constant, 6.0 vol% hydrogen
K_2 = Constant, 3.9 (m/s)/vol% hydrogen
X_{H2} = Vol% hydrogen, on a wet basis

Equation (20.10) should only be used for flares that have a diameter of 3 in. or greater, are non-assisted, and have a hydrogen content of 8.0% (by volume) or greater.

Example 20.1

Given: A steam-assisted flare is burning a gas with an LHV of 450 Btu/scf and 160°F (71°C).

Find: The maximum exit velocity of the gas at the flare tip to achieve a combustion efficiency of 98% or greater according to the U.S. EPA ruling.

Solution: Because the LHV of the fuel is 450 Btu/scf, Eq. (20.7) is used to determine the maximum velocity:

$$V_{max}(ft/s) = \left[anti\log_{10}\left(\frac{450+1209.6}{849.1}\right)\right] \times \left(\frac{160+460}{520}\right) = 107.4\ ft/s \quad (20.11)$$

20.5.2 Emissions

Industrial flares have been endorsed by the Clean Air Act Amendments to be one of the acceptable control technologies that can effectively destroy organic vapors. The U.S. EPA AP-42 guidance document[7] suggests that a properly operated flare, with a combustion efficiency of 98% or greater, will emit UHC, CO, and NOx at the following rates:

- UHC = 0.14 lb/MMBtu fired
- CO = 0.37 lb/MMBtu fired
- NOx = 0.068 lb/MMBtu fired

In 40 CFR 60.8c, the regulations indicate that during periods of startup, shutdown, or malfunction, emissions above the regulated limit may not be considered a violation. This includes UHC, CO, and NOx. This could also be interpreted to mean that exit velocity limits do not apply under these conditions. However, according to 40 CFR 60.10(c), states can make their own rules regarding flare operations as long as they are more stringent than the U.S. EPA ruling. As regulations are in a constant state of flux, the reader should determine the current regulations for the plant site in question.

Example 20.2

Given: Following the conditions of Example 20.1, assume that the flare is properly designed and operated according to the U.S. EPA ruling.

Find: The pounds of UHC, CO, and NOx emitted in 1 year if the flare is burning the waste gas at a rate of 10,000 lb/hr (4500 kg/hr), 50 times per year, for 1 hour during each event (neglect the emissions contribution from the pilots and purge gas). The density of the gas is 0.05 lb/scf.

Solution: First determine how many Btus are released in 1 year (yr):

$$\frac{HR}{year} = 450\frac{BTU}{scf} \times \frac{1\ scf}{0.05\ lb} \times \frac{10000\ lb}{hr} \times \frac{1\ hr}{Event} \times \frac{50\ Events}{year} = 4500 \times 10^6 \frac{Btu}{yr} \tag{20.12}$$

The pounds of UHC, CO, and NOx emitted in 1 year are then calculated as follows:

$$UHC = 4500 \times 10^6 \frac{Btu}{yr} \times \frac{0.14\ lb}{1 \times 10^6\ Btu} = 630 \frac{lb}{yr} \tag{20.13}$$

$$CO = 4500 \times 10^6 \frac{Btu}{yr} \times \frac{0.37\ lb}{1 \times 10^6\ Btu} = 1670 \frac{lb}{yr} \tag{20.14}$$

$$NO_x = 4500 \times 10^6 \frac{Btu}{yr} \times \frac{0.068\ lb}{1 \times 10^6\ Btu} = 310 \frac{lb}{yr} \tag{20.15}$$

20.5.3 Dispersion

If a flare fails to properly dispose of toxic, corrosive, or flammable vapors, it could pose a serious health hazard to personnel in the vicinity and the community downwind of the release. Sax[20] provides extensive information on many compounds that are sometimes sent to a flare. It is important for owners to have a dispersion analysis of their flare performed. A dispersion analysis is a statistical method used to estimate a downwind concentration of a gas vented to the atmosphere or emitted from a flare flame. Dispersion models are widely used in the industry and have been used in the past to size flare stack heights, estimate worst-case scenarios from emergency releases, and determine potential odor problems.

Mathematical modeling of stack gas dispersion began in the 1930s. At that time, these models were somewhat simplified. Today, however, through the advent of computers, these models have become more sophisticated and able to capture much more detail of the dispersion problem. The purpose of this section is to discuss the general concepts used for estimating the ground-level concentration (GLC) of a pollutant emitted from a flare.

When a pollutant is emitted from a flare, it is dispersed as it moves downwind by atmospheric turbulence and, to a lesser extent, by molecular diffusion, as illustrated in Figure 20.59. The GLC of a pollutant downwind of the flare depends on how fast the pollutant is spreading perpendicular to the direction of the wind and on the height of the plume above the ground.[21] The rate at which a pollutant is dispersing, in turn, depends on such factors as the wind speed, time of day, cloudiness, and type of terrain. The Gaussian dispersion model was one of the first models developed to estimate GLC. This model assumes that the concentration of the pollutants, in both the crosswind and vertical directions, takes the form of a Gaussian distribution about the centerline of the plume and is written as follows:

$$C = \left(\frac{Q}{U\sigma_z\sigma_y\pi}\right)\exp\left(\frac{-H^2}{2\sigma_z^2}\right)\exp\left(\frac{-y^2}{2\sigma_y^2}\right) \tag{20.16}$$

where

C	=	Predicted GLC concentration, g/m^3
Q	=	Source emission rate, g/s
U	=	Horizontal wind speed at the plume centerline height, m/s
H	=	Plume centerline height above ground, m
σ_y and σ_z	=	Standard deviations of the concentration distributions in the crosswind and vertical directions, respectively, m
y	=	crosswind distance, m (see Figure 20.59)

This Gaussian dispersion model was derived assuming a continuous buoyant plume, single-point source, and flat terrain. Beychok[22] discusses the shortcomings of Gaussian dispersion models. Beychok suggests that it is realistic to expect Gaussian dispersion models to consistently predict real-world dispersion plume concentrations within a factor that may be as high as 10. Gaussian dispersion models, however, are useful in that they can give a rough and fairly quick estimation and comparison of pollutant levels from elevated point sources.

The accuracy of a Gaussian dispersion model depends on how well one can determine the plume rise, H, at any given downwind distance and dispersion coefficients, σ_y and σ_z. A standard atmospheric stability classification method, known as the Pasquill-Gifford-Turner classification, is widely used in GLC models. This method categorizes the stability of the atmosphere into six classes that vary from very unstable (class A) to very stable (class F). An atmosphere that is stable has low levels of turbulence and will disperse a pollutant more slowly than an unstable atmosphere. The dispersion coefficients, σ_y and σ_z, are dependent on the amount of turbulence in the atmosphere and are, therefore, related to the atmospheric stability class. For more information on the equations describing the dispersion coefficients, see Turner.[23]

The plume height is defined as the vertical distance from the plume centerline to grade, as illustrated in Figure 20.59. There are several variables that can affect the plume height. These variables are divided into two categories: emission factors and meteorological factors. The emission factors

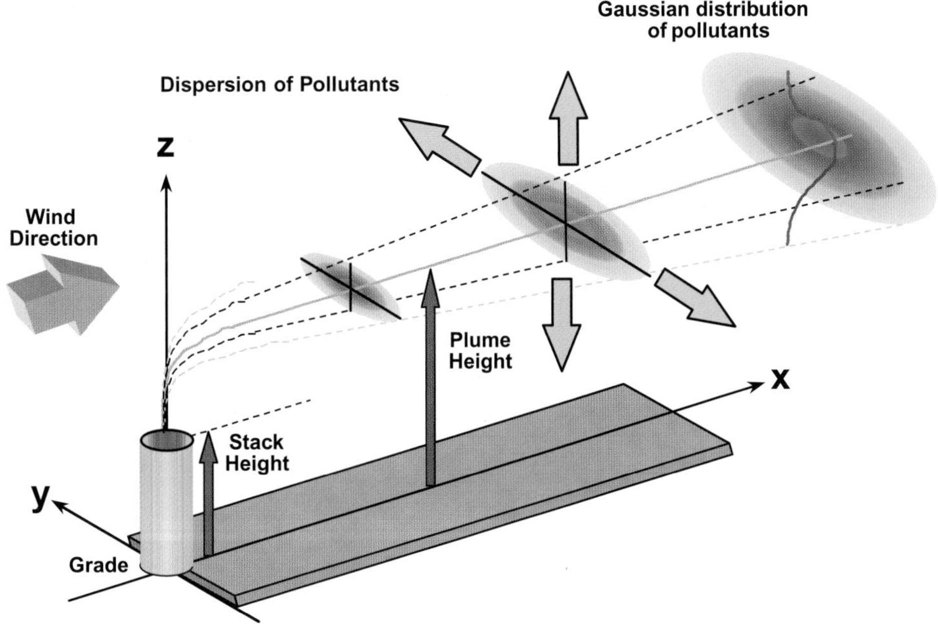

FIGURE 20.59 Geometry for dispersion calculations.

include the (1) stack gas exit velocity, (2) stack exit diameter, (3) stack height, and (4) temperature of the emitted gas. The meteorological factors include the (1) wind speed, (2) air temperature with height, (3) shear of the wind with height, (4) atmospheric stability, and (5) terrain. None of the equations reported in the literature for estimating plume heights, however, take into account all the emission and meteorological factors. For a review of these equations, see Moses, Strom, and Carson.[24]

GLC analysis is very complex because the results can depend on so many variables, as briefly discussed above. In the past, engineers and scientists have described GLC modeling as an art rather than a science. However, this paradigm is shifting due to more sophisticated computer models.[25] Due to the complexity of these models, one should consult an expert when requiring GLC analysis.

REFERENCES

1. J.S. Zink and R.D. Reed, Flare Stack Gas Burner, U.S. Pat. 2,779,399, January 29, 1957.
2. R.E. Schwartz and S.G. Kang, Effective design of emergency flaring systems, *Hydrocarbon Engineering*, February 1998.
3. S.H. Kwon et al., Improve flare management, *Hydrocarbon Processing*, July 1997.
4. R.D. Reed, *Furnace Operations*, 3rd ed., Gulf Publishing, Houston, TX, 1981.
5. API RP-521, 4th ed., American Petroleum Institute, Washington, D.C., March 1997.
6. M.R. Keller and R.K. Noble, RACT for VOC — A Burning Issue, *Pollution Engineering*, July 1983.
7. U.S. EPA, Compilation of Air Pollutant Emissions Factors, Vol. 1, Stationary Point and Area Sources, AP-42, 4th ed., 1985, Supplement F: Section 11.5, 9/91, Industrial Flares.
8. J.S. Zink, R.D. Reed, and R.E. Schwartz, Temperature-Pressure Activated Purge Gas Flow System for Flares, U.S. Pat. 3,901,643, August 26, 1975.
9. R.E. Schwartz and J.W. White, Flare Radiation Prediction: A Critical Review, John Zink Company, Tulsa, OK, 1996.
10. R.D. Reed, Flare Stack Burner, U.S. Pat. 3,429,645, February 25, 1969.
11. R.D. Reed, R.K. Noble, and R.E. Schwartz, Air Powered Smokeless Flare, U.S. Pat. 3,954,385, May 4, 1976.
12. R.D. Reed, J.S. Zink, and H.E. Goodnight, Smokeless Flare Pit Burner and Method, U.S. Pat. 3,749,546, July 31, 1973.

13. R.E. Schwartz and R.K. Noble, Method and Apparatus for Flaring Inert Vitiated Waste Gases, U.S. Pat. 4,664,617, May 12, 1987.

14. W.R. Bussman and D. Knott, Unique concept for noise and radiation reduction in high-pressure flaring, Offshore Technology Conference, Houston, TX, May 2000.

15. R.E. Schwartz, L.D. Berg, and W. Bussman, Flame Detection Apparatus and Methods, U.S. Pat. 5,813,849, September 29, 1998.

16. J.S. Zink, R.D. Reed, and R.E. Schwartz, Apparatus for Controlling the Flow of Gases, U.S. Pat. 3,802,455, April 9, 1974.

17. H. Glomm, Anordnung und Betrieb von Notabblasesystemen (blow down systems), *Rohrleitungstechnik in der Chemishen Industrie*, 199, 18-28, 1967.

18. Chemical Manufacturers Association, A Report on a Flare Efficiency Study, March 1983.

19. U.S. EPA, Code of Federal Regulations, Title 40, Part 60, Standards of Performance for New Stationary Sources.

20. N.I. Sax and R.J. Lewis Sr., *Dangerous Properties of Industrial Materials*, 7th ed., Van Nostrand Reinhold, New York, 1989.

21. M. Miller and R. Liles, Air modeling, *Environmental Protection*, September 1995.

22. M. Beychok, Error propagation in stack gas dispersion models, *The National Environmental Journal*, January/February 1996.

23. B. Turner, *Workbook of Atmospheric Dispersion Estimates*, U.S. Environmental Protection Agency, 1970.

24. H. Moses, G.H. Strom, and J.E. Carson, Effects of meteorological and engineering factors on stack plume rise, *Nuclear Safety*, 6(1), 1-19, 1964.

25. C. Seigneur, Understanding the basics of air quality modeling, *Chemical Engineering Progress*, 68, 1992.

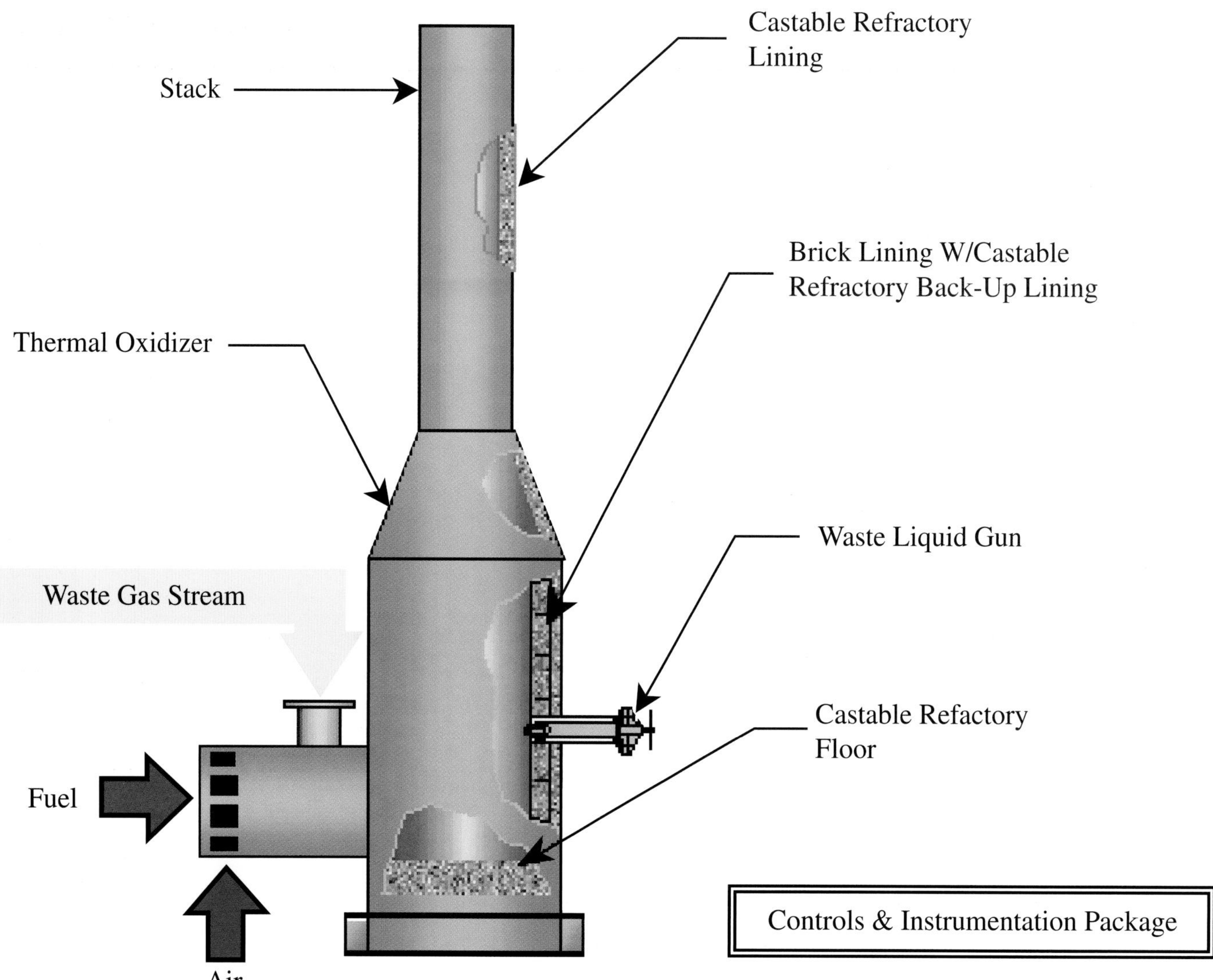
Castable Refractory
Lining
Stack
Brick Lining W/Castable
Refractory Back-Up Lining
Thermal Oxidizer
Waste Liquid Gun
Waste Gas Stream
Castable Refactory
Floor
Fuel
Air
Controls & Instrumentation Package

Chapter 21
Thermal Oxidizers

Paul Melton and Karl Graham

TABLE OF CONTENTS

21.1 Introduction 638
21.2 Combustion Basics 639
- 21.2.1 Material and Energy Balance 639
- 21.2.2 Oxidizing/Reducing Combustion Processes 639
- 21.2.3 NOx Formation 639
- 21.2.4 Carbon Monoxide 639
- 21.2.5 Acid Gases 639
- 21.2.6 Particulate 640

21.3 Basic System Building Blocks 640
- 21.3.1 Burners 640
- 21.3.2 Furnace/Thermal Oxidizer/Incinerator/Combustion Chamber 644
- 21.3.3 Refractory 645
- 21.3.4 Catalytic Systems 646
- 21.3.5 Flue Gas Processing Methods 647

21.4 Blowers 670
21.5 Control Systems and Instrumentation 675
21.6 System Configurations 677
- 21.6.1 Non-acid Gas Endothermic Waste Gas/Waste Liquid System 677
- 21.6.2 Non-acid Gas Exothermic Waste Gas/Waste Liquid System 681
- 21.6.3 Acid Gas Systems 682
- 21.6.4 Salts/Solids systems 686
- 21.6.5 NOx Minimization or Reduction Systems 688

21.7 Conclusion 689
21.8 Nomenclature 689
References 689

21.1 INTRODUCTION

Although improvements are continually made to the efficiency of the many chemical and mechanical manufacturing methods used to produce an ever-increasing number of compounds and products, unwanted by-products result from virtually all methods. Such by-products can exist in the vapor, liquid, or solid phase. Many are hydrocarbons, although some non-hydrocarbon materials also exist. Regardless, the many different by-products must be safely contained or destroyed to prevent potential environmental damage.

The by-products come from many different industrial sectors. Petroleum and natural gas production and refining, petrochemical manufacturing, pulp and paper production, agricultural chemicals, pharmaceuticals, distilling, automobiles, plastics molding, and carbon fiber and fiber optics production are just a few of the diverse areas that produce wastes. The by-products can be the remains of the normally less than 100% efficient chemical processes used to create hydrocarbon-based products. The wastes can be impurities and catalysts in feedstocks that are not consumed during the process. Manufacturing processes in widely varying fields often require significant ventilation resulting in contaminated air that must be treated. By-products can vary from a few parts per million (ppm) in air or water to nearly 100% concentration of a hydrocarbon.

Many methods of elimination are available. Smaller amounts of nonsoluble solids and liquids can be put into sealed drums and isolated in secure landfills. Larger amounts of liquids, primarily contaminated water, have been injected into deep wells. For all practical purposes, vapors cannot be stored and must be treated as generated. Activated carbon, for example, can be used to adsorb organic materials from vapor. Stripping and absorption can also remove contaminants from liquid and vapor streams. Filtration methods are used to remove solid materials from vapors and liquids. None of these methods, however, actually destroys the waste material. Chemical and biological treatments are used to destroy organic waste but may not be the most cost-effective alternative for rapidly and efficiently treating large amounts of material. The most effective method of rapidly eliminating a high percentage of hydrocarbon contaminants is to oxidize the organic materials at an elevated temperature (at or above 1500°F/800°C). Such high-temperature oxidation is known as combustion or thermal oxidation. For some contaminated air streams, effective oxidation can also be achieved at lower temperatures using a catalyst to increase the oxidation reaction rate.

High-temperature thermal oxidation quickly and efficiently destroys hydrocarbon-based waste materials, converting the carbon and hydrogen to carbon dioxide and water. However, during this process, some elements in wastes (e.g., sulfur and chlorine) form compounds (e.g., sulfur dioxide and hydrogen chloride) that, if present in sufficient quantities, must be recovered or removed from the combustion products by post-oxidation methods to meet various federal, state, and local air quality guidelines. Other elements or compounds in wastes (e.g., sodium, sodium chloride, catalysts, or other inert solids) will produce particulates that, if present in sufficient quantity, will also have to be removed by post-oxidation treatment. If nitrogen is organically bound to any waste compound, special staged oxidation methods may have to be used to prevent formation of excessive amounts of NOx. The same methods can be used to break down existing NOx that is part of a waste.

A number of factors determine the design of thermal oxidation systems. Process variables such as the waste composition and flow rate affect the size, materials of construction, and stability of the system. Economic considerations often impact decisions of capital expenditure vs. operating costs, as is the case with determining the feasibility of heat recovery systems. Regulations set the required destruction efficiencies, emission rates, and acceptable ground-level concentrations. Very few applications have identical specifications. Thus, many systems are custom designed to satisfy the process, economic, and regulatory requirements of a particular application.

Regardless of the specific design, most thermal oxidation systems consist of some or all of the following components:

1. a device and method to supply oxygen to the process to initiate and sustain oxidation (burner or catalyst)
2. a vessel (combustion chamber/thermal oxidizer) to contain the waste hydrocarbons during oxidation
3. a heat recovery (heat exchanger or boiler) and/or flue gas conditioning system
4. emission control equipment (filters, scrubbers, etc.) to treat the flue gas prior to discharge to the atmosphere
5. an elevated exhaust point (stack) through which the flue gas can be dispersed into the atmosphere
6. control hardware and logic to automatically maintain and monitor the various process parameters to ensure safe operation

The purpose of this chapter is to provide a better understanding of the use of thermal oxidation to destroy fume and liquid wastes. To accomplish this, the following are needed.

- An explanation of the basic practical thermal oxidation and post-oxidation processes and components that can be combined into complete systems to destroy hydrocarbon wastes and treat the combustion products to achieve required emission limits.
- Examples of practical complete system configurations that can be applied to treat different waste compositions and combinations.

21.2 COMBUSTION BASICS

The prerequisites for, the actions during, and the results of combustion must be known to design a thermal oxidation system equipment that will achieve the destruction and removal efficiencies needed to protect the environment.

21.2.1 Material and Energy Balance

To correctly choose and design the components of a thermal oxidation system, a certain amount of process information must be developed. For the burner, the designer must know the amount and properties of fuel/waste and the amount of air required to provide stable, effective combustion. For the combustion chamber, the operating temperature, volume, and properties of the combustion products must be determined. For post-combustion treatment of the combustion products such as heat recovery or flue gas conditioning, the mass flow and composition of the combustion products must be known. For emission control applications such as acid gas removal or particulate removal, the amount of the pollutant must be known to predict the level of removal required. By completing material and energy balances at different equilibrium points throughout the overall process, all of this information can be generated. This information is then used to configure systems and to design and dimension the individual components.

21.2.2 Oxidizing/Reducing Combustion Processes

Oxidation reactions occur during both excess oxygen (excess air) and sub-stoichiometric (starved air) processes. These processes are used separately and in combination to destroy hydrocarbon wastes. The high-temperature, excess O_2 process is by far the most commonly used. The excess O_2 process converts virtually all of the carbon and hydrogen in hydrocarbon wastes to harmless CO_2 and H_2O. To be certain the conversion is maximized, combustion air supplied to this process is greater than the theoretical amount needed to oxidize the compounds. If sulfur is present, the excess air also provides oxygen to convert it to SO_2/SO_3 (SOx). Additionally, some nitrogen is converted to NO/NO_2 (NOx). These compounds have a detrimental effect on the environment and are regulated with respect to stack emission levels.

The sub-stoichiometric (reducing) process operates with less than 100% of the theoretical amount of O_2 needed to oxidize the elements of the hydrocarbon compounds. Some of the oxygen is combined with the carbon and hydrogen to form CO_2 and H_2O. However, partial reactions also occur, forming CO and H_2 as well as other products of incomplete combustion (PICs). SOx and NOx production are significantly reduced, but eventually the incomplete combustion products must be completely oxidized. The final oxidation is completed at temperatures significantly lower than flame temperature, thereby minimizing NOx formation, although SOx is formed.

21.2.3 NOx Formation

In Chapter 6, NOx formation is discussed at length. Basically, three mechanisms for NOx formation are present. **Thermal NOx** is formed by the high-temperature reaction of nitrogen with oxygen. Formation increases exponentially with increasing operating temperature. At greater than 2000°F (1100°C), it is usually the primary source of NOx if the waste does not contain nitrogen compounds. **Prompt NOx** is formed by hundreds of rapid reactions between nitrogen, oxygen, and hydrocarbon radicals, intermediate species formed during the combustion process. Prompt NOx can be a large contributor in lower temperature combustion processes. **Fuel NOx** is formed by the excess-oxygen combustion of organic compounds containing nitrogen. However, as noted in the preceding section, NOx formation can be sufficiently reduced by initially utilizing the substoichiometric combustion process to destroy the waste.

21.2.4 Carbon Monoxide

Carbon monoxide (CO) is produced by all combustion reactions in relatively small amounts, especially if proper burner design is followed. As the products of combustion travel through the high temperature combustion chamber, part of that CO will be oxidized by the excess O_2 in the combustion products. By the time the combustion products exit the chamber, the resulting CO is at levels well below 100 ppmv. If large amounts of CO are produced because of poor combustion at the burner, only part of it will be oxidized as the gases pass through the combustion chamber, and the outlet concentration will be much greater than 100 ppmv. A high level of CO in the combustion products exiting a combustion chamber is a direct indication of poor combustion.

Some wastes contain CO as a component. Because the recognized auto-ignition temperature of CO is greater than 1200°F (650°C), very little is oxidized in lower temperature systems (less than 1400°F or 800°C) using lower residence times (smaller vessels) often used to minimize equipment costs.

21.2.5 Acid Gases

Many waste gases and liquids contain sulfur or chlorine compounds. When these compounds are oxidized, the sulfur is converted to SO_2/SO_3 and, with sufficient moisture, the chlorine converts mostly to HCl and a small amount of free Cl_2. To meet emission limits, these compounds must be removed from the flue gas. HCl can be recovered or removed, while SO_2 is normally just removed. Although sulfur and chlorine are the most common acid-producing components, phosphorus, fluorine, and bromine are occasionally encountered. Phosphorus can cause corrosion problems in heat recovery equipment because of high dew points. HF is very reactive and will attack virtually every part of a system until neutralized. Bromine in organic compounds is very difficult to convert to HBr for easy removal. Much of it goes to Br_2 in normal oxidizing conditions. A special process must be used to achieve high conversion to HBr to allow high removal efficiency.

21.2.6 Particulate

Particulate can exist in waste streams as inert solids that remain after the waste material is oxidized. Common examples would be NaCl in wastewater and catalyst material carried over in the off-gas stream from a catalytic cracking unit. Organic materials can contain elements that remain after the organic waste is burned. Examples include elements such as sodium or silicon in compounds catalyst materials such as cobalt, manganese, and nickel. Depending on the point of introduction into the system, the particulates formed will vary in size from several microns to sub-micron in diameter.

21.3 BASIC SYSTEM BUILDING BLOCKS

A simple thermal oxidation system may consist of only combustion components, that is, a burner mounted on an integral vertical combustion chamber and stack. A complicated system may include the combustion components and all possible heat recovery and flue gas treatment components such as boilers, hot oil heaters, waste preheaters, flue gas conditioning equipment, acid gas and particulate removal equipment, and a stack. A catalytic system with a preheater, burner, catalytic oxidation chamber, and stack lies somewhere between the simple and complicated systems. Each of the components is a stand-alone process block. When necessary, the blocks are combined to build a complete thermal oxidation system. To properly utilize the building blocks, an explanation of the components as well as the advantages and disadvantages is necessary.

21.3.1 Burners

The burner is the component required to mix and ignite the fuel (and waste, if capable of sustained combustion) and air, and to provide a stable flame with appropriate shape and combustion characteristics throughout the design operating range of the system. The basic parts of a burner include a pilot to provide the initial source of ignition; assemblies to introduce the fuel, waste, and air; and the means to ensure flame stability once lit. Burners are used over a wide range of heat releases and can burn gas and/or liquid fuels and combustible waste streams. The mixing of fuel/waste with combustion air is accomplished by the combination of air velocity through the burner (usually referred to as "pressure drop") and the velocity and distribution of the combustible material as it is introduced into the burner.

21.3.1.1 Pilots

A pilot is essentially a very small burner that provides the ignition source for the main burner fuel. Pilots normally utilize natural gas or propane gas that is mixed with air in the pilot assembly. Two methods of mixing the air and fuel are used. An inspirated air pilot utilizes pilot gas pressure drop through an orifice at the entrance of a venturi assembly to educt air into the venturi throat where it mixes with the pilot gas. A pressurized air pilot must have air available at a pressure greater than the operating pressure of the burner. Pressurized air is metered by use of an orifice or valve into a small mixing chamber where it mixes with the pilot gas, also metered into the chamber through an orifice. The premixed air and fuel then travel through a tube to a special high-temperature pilot tip where it is ignited.

There are two common methods for pilot ignition. The most simple ignition method utilizes a high-voltage (> 6000 V) electric spark to ignite the air/fuel mixture at the pilot tip. This arrangement is inexpensive, but heat exposure over time can damage the spark delivery hardware and lead to ignition difficulties. A common alternative is the flame-front ignition system, in which a spark, located outside the burner, ignites an air/fuel mixture flowing to the pilot tip through a steel pipe of about 1 in. (2.5 cm) diameter. The spark initiates a flame front that travels through the flowing mixture in the pipe finally emerging near the area of the pilot tip, igniting the pilot flame. Only the open pipe from which the flame front emerges to ignite the pilot air/fuel mixture is exposed to heat. The end of that flame-front pipe is high-temperature stainless steel and is far less likely to be damaged and result in ignition difficulties.

21.3.1.2 Fuel Introduction

To be considered a fuel, the material must have sufficient heating value to sustain stable combustion once ignited. The material can be a gas or a liquid. To quickly and efficiently burn, a gas need only be mixed with the appropriate amount of air, but a liquid must first be atomized into fine droplets and then mixed with air. Unlike the pilot, main burners used in thermal oxidizers almost never use premixed air and fuel. Therefore, a method must be employed to quickly mix the main fuel and air. The most effective method is to separate the main fuel flow, whether gas or atomized liquid, into smaller "jets" of flow using a tip or multiple tips with orifices (ports) drilled at the proper size and orientation. This serves two purposes: (1) more individual jets provide more fuel surface area exposed to the combustion air, regardless of the jet velocity; and (2) as the jets of fuel exit the tip at significant velocity, combustion air is drawn into the rapidly dispersing fuel jet. Thus, air is quickly mixed with the fuel. Once the fuel/air mixture is ignited, a mechanism is required to provide flame stability, that is, continuous ignition of the fuel/air mixture near the point of fuel introduction. This is usually accomplished by establishing an airflow disturbance adjacent to some of the fuel exit ports. The flow disturbance creates localized recirculation of a portion of the reacting flame constituents, thus continuously igniting the main fuel as the fuel mixes with air.

21.3.1.2.1 Gas Tips

For fuel gas or higher-pressure combustible waste gas introduction, specially designed tips made of heat-resistant alloys are utilized. They are mounted on the end of pipes, which are often removable through the front of the burner, so the tips can be externally accessed for maintenance and replacement. Based on the type and amount of gas to be introduced and the amount of gas pressure available at the tips, a specific number of firing ports are drilled into each tip. Smaller ports, known as ignition ports, are also drilled into each tip. A very important purpose of the firing and ignition ports is to direct and shape the gas discharge from the tips, thereby directing and shaping the flame. For lower heat release burners, a single gas tip, located at the center of the burner, is often used. For higher heat release burners, multiple gas tips, arranged symmetrically around the circumference of the burner, are used, just as multiple ports on a single tip are used, to increase the rate of mixing of the fuel and air thereby increasing the oxidation reaction rate. Rapid oxidation of fuel is important because it must be burned before nonflammable wastes, such as contaminated water, can be introduced into the system. The gas pressure drop through the tips is usually in the range of 10 to 25 psig (0.7 to 1.7 barg).

21.3.1.2.2 Liquid Tips

Liquid fuel and liquid waste tips serve the same purposes as gas tips but are more complex. Mechanical atomization, requiring a 200 to 300 psig (14 to 20 barg) liquid pressure drop at the tip, can be used. However, the turndown for mechanical atomization is only about 3:1. At that point, the atomized liquid droplets become larger than preferred for optimum burning. To maintain the small droplet size over a larger operating range and reduce the amount of liquid pressure needed, an assist medium such as steam or air is utilized. Droplet sizes similar to that achieved in a mechanical tip at 200 to 300 psig (14 to 20 barg) can be achieved in an assist medium gun at only 60 to 100 psig (4 to 7 barg) pressure drop for both fluids. Lower pressure operation also has the important advantage that it is accomplished with larger liquid passages through the tip, making it less susceptible to plugging.

21.3.1.3 Waste Introduction

Waste gases and waste liquids that are not capable of burning as a stable fuel are usually introduced downstream of the burner. However, although a waste does not have sufficient heating value to sustain stable combustion, it is still classified as endothermic or exothermic for a specific operating temperature. Waste is considered to be endothermic if the hydrocarbon content (heating value) is small and much more than a minimum amount of auxiliary fuel must be burned to maintain the required operating temperature in the combustion chamber. If the heating value is high enough that a cooling medium must be added to control the maximum operating temperature, the waste is considered exothermic. A waste can be exothermic at a lower operating temperature but endothermic for a higher operating temperature. Waste liquids are normally available at higher pressure and can be atomized into the system using hardware (tips, etc.) similar to that used for liquid fuels. Waste gases, on the other hand, are normally available only at lower pressures, so injection hardware must be designed for the lower pressure drop. Waste gas injection "tips" are often simply open pipes. As with fuel gas, more pipes will better distribute the waste gas and mix it more rapidly with available oxygen. Endothermic wastes may not support stable combustion but a significant amount of organic material (heating value) may still be present. If so, liquid and gas waste injection hardware may have annular spaces around them for local introduction of air to react with the organic material. The air entering through the annulus also serves to cool the hardware as well as providing some or all of the air required for oxidation. Positioning of the waste injection hardware is critical. Low organic content wastes, whether liquid or gas, should be introduced downstream of the burner flame zone so as not to impair oxidation of the fuels and result in formation of carbon monoxide, unburned hydrocarbons, or worse, soot. If the waste is air that is only slightly contaminated with organic material, it may be used as the combustion air source, and introduced directly through the burner. This is the method of waste introduction for a catalytic system.

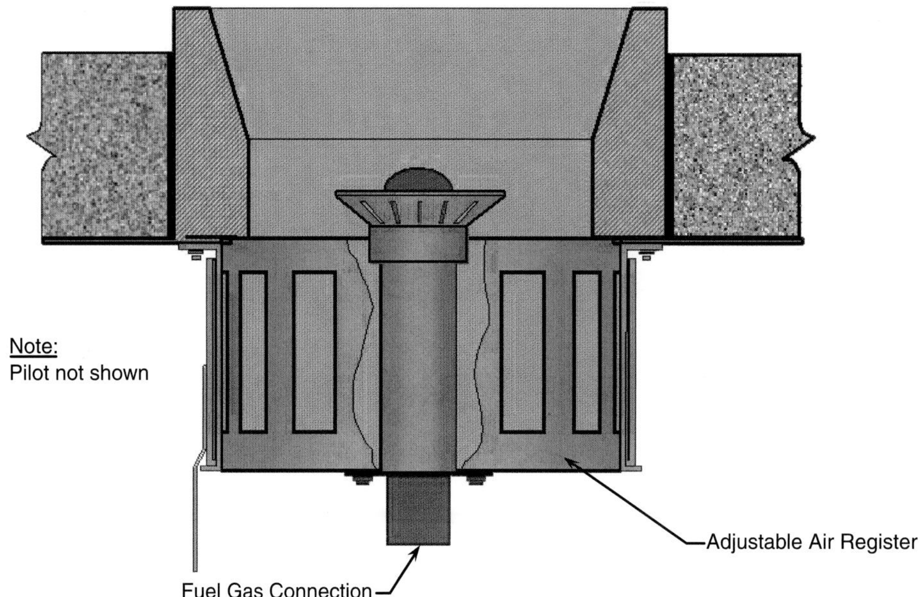

FIGURE 21.1 Typical natural-draft burner.

21.3.1.4 Low Pressure Drop Burners

Low pressure drop burners are designed to operate with a very low pressure drop/air velocity through the combustion zone. Generally, the motive force that "pulls" the air through the burner is created by the buoyancy of hot flue gas in the stack relative to the cooler atmospheric air. This "pull," or "draft," created depends on stack height, flue gas temperature, flue gas flow rate, and stack diameter. Because the draft is a natural result of system configuration, it is designated "natural" draft, and burners that utilize this motive force are natural-draft burners. Natural-draft systems typically generate 0.15 to 1.5 in. (3.8 to 38 mm) w.c. (water column) negative pressure at the stack base. The majority (75 to 80%) of the drop occurs as the air passes through the throat of the burner into the combustion zone. At such low pressure drop, the velocity of the air at the burner throat can be no more than 15 to 50 ft/s (4.6 to 15 m/s). As a result, the burner flow-area-vs.-air-flow ratio is relatively large, and air flow control and distribution can be more difficult. With little energy available in the form of combustion air velocity, and because a short flame is usually desired, multiple gas tips are normally utilized with a fuel gas fired natural-draft burner. For liquid fueled natural-draft burners, an increased number of ports and wider-angle port orientations for the liquid tip are methods used to shorten the flame.

Natural-draft systems are impractical if any equipment (e.g., boiler or scrubber) with substantial pressure drop is required downstream of the furnace. Hence, most incinerator systems require at least one blower to move the gas through the system. Figure 21.1 illustrates a typical natural-draft burner configuration.

21.3.1.5 Medium to High Pressure Drop (Forced-Draft) Burners

Medium to high pressure drop burners operate at much higher pressure drops than natural-draft burners because the combustion air supplied to the burner is pushed, or "forced," into the system by a blower or compressor. Thus, such burners are known as forced-draft burners. Generally, the medium pressure drop burner would be designed for 1.0 to 8.0 in. (2.5 to 20 cm) W.C. pressure drop. A high pressure drop burner may be designed for 30 in. (76 cm) W.C. or more pressure drop. With much more energy available to provide fuel and air mixing, the heat-release-to-flame-length ratio can be much greater for both gas and liquid fuels. Low pressure combustible waste gases can be more easily introduced because the greater energy of the air provides a large portion of the fuel-to-air mixing. Combustible waste liquids that are more difficult to burn are more easily burned in a higher pressure drop burner. With more pressure available, the flow-area-vs.-air-flow ratio is smaller, so the burner is smaller. Also, proper combustion air flow distribution is more easily achieved. Some of the available pressure can be used to impart an angular velocity component to the combustion air to further aid in mixing. Figures 21.2 and 21.3

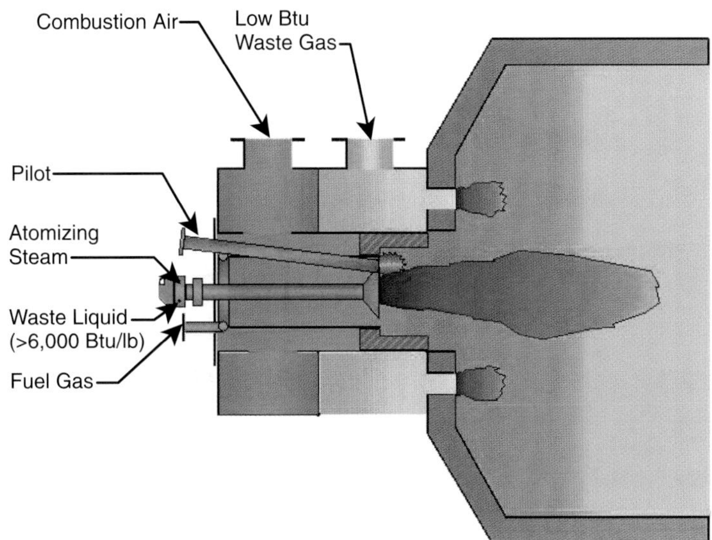

FIGURE 21.2 Typical medium pressure drop burner.

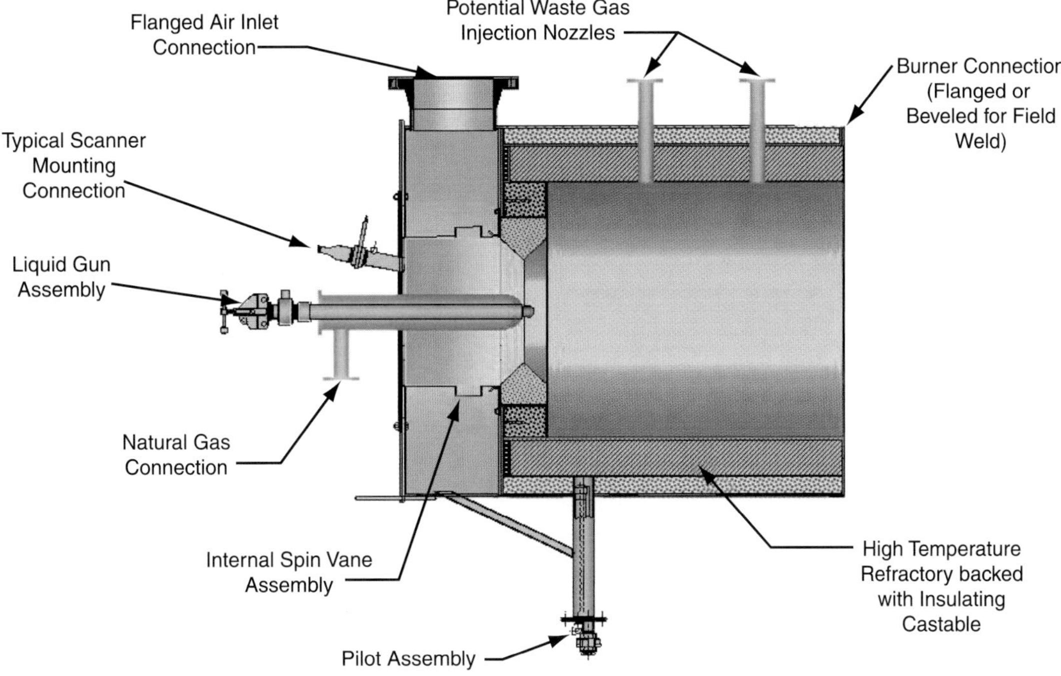

FIGURE 21.3 Typical high pressure drop burner.

are typical configurations for medium and high pressure drop burners, respectively.

21.3.1.6 Combination Gas and Liquid Fuel Burners

Combination fuel burners are virtually the same as dedicated gas or liquid burners. The difference is the special care required when locating the liquid and gas tips. A typical configuration would be for the liquid injection point to be at the center of the burner, through a stabilizer cone, with gas tips located at the circumference of the circular throat of the burner. This configuration is shown in Figure 21.2. By separating the fuels, better air mixing and faster burning occurs. If the gas tips are located near the liquid tip, the gas will consume the air more readily, delaying mixing of air with the atomized liquid. This would delay the oxidation reaction of

TABLE 21.1 Typical Thermal Oxidizer Operating Conditions

Waste Type	Operating Temp. (°F)	Retention Time (seconds)	Typical DRE
Lean gases containing hydrocarbons or sulfur compounds	1200–1400	0.3–0.6	>99
Lean gases containing common chlorinated solvents	1600–1800	1	>99.9
Liquid streams	1600–2000	1.0–2.0[a]	>99.99
Halogenated hydrocarbon liquids	1800–2000	1.5–2.0[a]	>99.99
Dioxins and polychlorinated biphenyls (PCBs)	2200	2	>99.9999

[a] Extra time for liquids to allow for droplet evaporation.

the liquid resulting in a longer flame zone. In general, the faster the gas and liquid fuels are burned, the sooner noncombustible wastes can be introduced to the system.

21.3.2 Furnace/Thermal Oxidizer/Incinerator/Combustion Chamber

Although combustible fuels and wastes are introduced through the burner in high-temperature oxidation systems, noncombustible wastes must be introduced downstream of the flame zone. Once all the fuel, waste, and air are combined, several intimately connected and simultaneous conditions must exist to achieve the required destruction removal efficiency (DRE). The mixture must be (1) exposed to a sufficiently high Temperature (2) for an adequate period of Time (3) in a relatively Turbulent environment to enable the oxidation reactions to reach the degree of completion needed to achieve the waste destruction efficiency required. These conditions are known as the "Three Ts" in the combustion industry. The vessel that provides the environment for all these conditions is known by various designations as a furnace, thermal oxidizer, combustion chamber, or incinerator. For consistency, this vessel is referred to as a thermal oxidizer (T.O.). To provide the environment needed for the three conditions to be optimized, the T.O. must have the correct volume and geometry.

21.3.2.1 Size

The optimal residence Time, or working volume, needed to meet the required DRE in the T.O. is a function of many factors, the most important of which are waste composition, waste characteristics, degree of Turbulence, DRE required, and T.O. operating Temperature. The working volume is generally considered to be from the point at which the final amount of waste is introduced, to the point nearest the incinerator outlet where the operating temperature is measured. Size can also be influenced by the capital and/or operating costs, which are not directly related to the destruction efficiency of hydrocarbons. Additionally, modifying design factors to reduce size can increase undesirable emissions, such as NOx.

Some complex hydrocarbons in waste require longer residence Times and/or higher Temperatures than a simple sulfur compound such as H_2S. Longer Time means a larger vessel, which increases capital cost. Higher Temperature means more fuel usage, unless the waste is exothermic, which increases operating cost.

For the same operating temperature, a waste stream containing primarily water with a hydrocarbon contaminant will require more time for the mixture to vaporize and raise the hydrocarbon to its oxidation point than would a vapor waste stream containing the same hydrocarbon contaminant. Again, a longer time means a larger, more expensive T.O.

If the flow in the T.O. is not relatively Turbulent or contains areas of stagnant flow, a longer residence Time will be required for the available oxygen to come into contact with the organic material so it can be oxidized and destroyed. This poor mixing of waste with air delays the onset of oxidation. A larger, more costly vessel would be needed to achieve the desired DRE. T.O. designers are careful to avoid such areas of stagnant flow to minimize wasted volume whenever possible.

A high DRE requirement may result in the need for longer residence time and/or higher operating temperature, assuming the mixing is adequate. The greater the T.O. temperature, the greater the oxidation reaction rate. Again, a higher temperature may require additional fuel, adding to the operating cost. A higher temperature can also result in the need for more expensive refractory material. On the other hand, the increased reaction rate could reduce the residence time needed and result in a smaller volume T.O., thus decreasing capital cost. For wastes with high heating values, intentionally increasing the operating temperature of a T.O. in order to reduce T.O. size can be economically attractive.

Increasing the temperature has the above-mentioned pros and cons, but an overall drawback is that higher temperatures may lead to higher nitrogen oxide (NOx) emissions.

It is obvious that for T.O. design, it ultimately becomes a trade-off between the capital cost of increased residence time and the costs and problems associated with an increase in operating temperature.

Table 21.1 shows typical T.O. operating conditions for a variety of cases. Although ranges of operating conditions have been developed by testing, experience with actual operating systems allows specific conditions to be set on a case-by-case basis, depending on past experience with similar waste streams.

21.3.2.2 Flow Configuration

The orientation and flow configuration of the T.O. must accommodate the user's space restrictions, the characteristics of the waste being burned, the downstream flue gas treatment requirements, and again, provide the most cost-effective solution. The T.O. can be arranged for vertical up-flow, horizontal flow, or vertical down-flow of the combustion products. Vertical up-flow is preferred for the simplest situations when the combustion products are vented directly to atmosphere with no downstream treatment needed. This is also the arrangement of choice if the equipment located immediately downstream of the T.O. requires an elevated entry. Horizontal flow is the most utilized configuration when a heat recovery boiler or other equipment with side entry near grade is located immediately downstream of the T.O. Vertical down-flow is required in many systems when a waste is burned that contains ash-forming materials or salts to prevent the accumulation of these materials in the T.O. On-line solids collection and removal equipment must be installed at the base of the T.O. for this case.

Most T.O. vessels are cylindrical in design with length-to-inside diameter of refractory (L:D) ratios ranging from 2:1 to 4:1. For a cylinder of a specific volume, the surface area is the least when the L:D is to 1:1. As L:D increases for the same volume, the surface area of the cylinder increases. From an L:D of 1:1 to 2:1, the increase is almost 26%; from 1:1 to 3:1, the increase is more than 44%; and from 1:1 to 4:1, the increase is almost 59%. The result is that the cost of the cylinder increases as L:D increases. Also, the greater the L:D, the longer the T.O., thus requiring a larger plot area for a horizontal configuration. A positive result of greater L:D is a smaller cross-sectional flow area, increasing flue gas velocity. As discussed in the previous section, the greater velocity helps minimize dead zones in the T.O., improving mixing and the DRE, which enables use of a smaller working volume. A smaller volume reduces equipment costs. Another important consideration that affects L:D is that flame or liquid impingement on the T.O. will cause incomplete combustion and refractory deterioration. A larger diameter or length may be required to maximize DRE or equipment life, regardless of the oxidation reaction rate.

21.3.3 *Refractory*

Virtually all T.O. vessels are internally lined with heat-resistant refractory material. Installation of a refractory lining provides three important consequences.

1. The steel shell is protected from the high-temperature environment inside the T.O.
2. An extremely hot external surface is avoided.
3. The oxidation process is insulated against heat loss so that the vessel is a reasonably adiabatic chamber.

Refractories used in T.O. vessels are primarily ceramic materials made from combinations of high-melting oxides such as aluminum oxide or alumina (Al_2O_3), silicon dioxide or silica (SiO_2), or magnesium oxide or magnesia (MgO). Refractories containing primarily alumina and silica are "acid" refractories and are by far the most common type used for T.O. linings. Refractories containing large amounts of MgO are "basic" refractories and are used for their good resistance to specific reactive ash components, particularly alkali metal compounds, which result from burning some inorganic salt-laden wastes. Because MgO refractories are significantly more expensive than the alumina and silica materials, the T.O. configuration is often optimized to allow use of and to maximize the life of the less-alkali-resistant alumina and silica refractories.

Refractories can be further divided into "hard" and "soft" categories, which applies to their state when ready for service. Hard refractories can be further categorized as bricks, plastics, or castables. Brick refractory is available in a wide variety of compositions ranging from high-alumina-content aluminosilicates to magnesites. The binding material in brick refractory can be calcium cement based or phosphoric acid based. A brick lining is held in place by gravity and/or the compressive forces resulting from proper placement (as in the construction of an arch). The linings must be installed in a vessel by skilled craftsmen and require more time to install, especially if special shapes have to be assembled by cutting bricks. Because of its high density and low porosity (good penetration resistance to molten or refractory-attacking materials), brick typically offers the best abrasion and corrosion resistance of any refractory. However, the high density results in the brick usually being heavy (120 lb/ft^3 or more) and the insulating value being lower, resulting in greater lining thickness to achieve the same thermal resistance. An additional consequence of a thicker lining is that a larger, more expensive T.O. shell is required to maintain the needed inside diameter.

Plastic refractories have similar alumina and silica content as brick and are so-called because the binder in plastic refractories — usually a water-wetted clay but also available with a phosphoric acid base — is not set and the material is very malleable or "plastic" in the ready-to-install condition. Once in place, however, the binder is set by exposure to air or to heat. Plastic refractories are shipped ready-to-install in sealed containers to provide shelf life. Once opened, the material must be used immediately. Because the refractory is in a "plastic" state, it can be forced or "rammed" into place and can be formed into almost any shape needed. It is held in position by an anchor system that consists of a metal piece welded to the shell and a prefired anchor (a special ceramic refractory shape) held by the metal portion attached to the shell. The anchor extends through the lining to the surface of the refractory. Also, because plastic refractory also is dense (140 lb/ft^3 or more), has low porosity, and is relatively easy to install, it is often used in difficult-to-brick places when a high-temperature, corrosive environment is expected. The total cost of plastic refractory, based on material and installation, insulating capabilities, and erosion and corrosion resistance is usually less than that of comparable composition brick, especially if the final shape needed is unusual.

Castable refractories consist of fireclay or high-alumina-content aggregates that are held together in a matrix of hydraulic calcium aluminate cement. Castable refractories are the least expensive to install among the hard refractories. Castable refractory is shipped in bags like dry cement, mixed with water prior to installation (a variable that can affect its final properties), and either poured or gunned (slightly dampened and blown through a nozzle) into place. The castable refractory is held in place by alloy steel anchors that are welded to the furnace shell. The castables used for incinerator applications usually weigh between 50 lb/ft^3 and 120 lb/ft^3. Compared to the other hard refractories, castables generally have the best insulating properties and the poorest corrosion and erosion resistance. To minimize lining thickness, a layer of insulating castable refractory is often installed as a backup to the brick layer, which is the internal surface of a T.O. The brick provides the resistance to high temperature, corrosion, and abrasion, while the castable provides the insulation qualities needed to reduce overall refractory thickness.

Soft refractories are composed of ceramic fibers formed into a blanket, a soft block module, or stiff board. They remain soft when in service. The blanket and board are usually held in place with stainless or other high temperature alloy anchors (or pins) welded to the inside T.O. wall. They are easily installed by pressing onto the steel shell with the pins projecting through. Self-locking washers are then placed on the pins to keep the material from coming loose. The block modules have an internal frame that is attached to an anchor welded to the shell. Soft refractories are much lighter (usually less than 12 lb/ft^3), are much better insulators, and can be heated rapidly without fear of damage because of thermal shock. Thermal shock is the rapid thermal expansion of the surface of hard refractory. That layer then separates and falls off, reducing the refractory thickness. Soft refractories are limited to 2300°F (1260°C), are susceptible to erosion, and do have poor resistance to alkali liquids and vapors. Ceramic fiber refractory is very cost effective in certain applications.

Common problems related to all refractories include the following:

1. Improper operation or design can lead to thermal shock or erosion damage.
2. Normal acid gases (SO_2 or HCl) in the combustion products can raise the gas dew-point temperature to as high as 400°F (200°C). Excessive refractory thickness, rain, or extremely cold weather can result in a furnace shell temperature below the dew point, which can result in acid condensation and its associated corrosion. To minimize weather-related effects, a ventilated, sheet-metal rain shield is often used to prevent rain contact and limit external convective heat transfer.
3. Flame impingement on refractory surfaces can result in higher-than-expected temperatures, frequent temperature fluctuations, and locally reducing conditions, all of which can shorten refractory life.
4. Liquid impingement on hot refractory will cause spalling and erosion, which decrease refractory life.
5. At higher temperatures, salts (those containing Na, Ca, K, etc.) and alkaline-earth oxides (e.g., K_2O, Na_2O, CaO, and MgO) will react with most acid refractories. The result of these reactions can be a loss of mechanical strength, crumbling, or even a "fluxing" (i.e., liquefaction) of the exposed surface. In any case, refractory life is shortened.

21.3.4 Catalytic Systems

In a typical thermal incinerator, waste destruction occurs in the flame or T.O. because of high-temperature, gas-phase oxidation reactions. In a catalytic unit, waste destruction occurs within a catalyst bed at much lower temperatures via surface oxidation reactions. The lower operating temperature required in a catalytic unit is its advantage because the lower temperature reduces the need for auxiliary fuel to maintain furnace temperature, thus lowering the operating cost. Another advantage to lower operating temperature catalytic oxidation is that NOx formation during oxidation is reduced. The lower temperature also eliminates the need for internal refractory lining, reducing the shell diameter. However, use of a stainless steel shell and external insulation is necessary and cancels any reduced operating cost. A major limitation of catalytic oxidation is that the catalyst is susceptible to damage from certain compounds in the waste or from overheating. Most waste streams for which catalytic oxidation is considered are contaminated air streams that have more than enough O_2 to complete combustion. It is possible to treat an inert gas stream, such as nitrogen contaminated by a small amount of hydrocarbon, but enough air must be blended with the gas stream prior to entering the catalyst bed to give approximately 2% O_2 in the flue gas after oxidation has occurred. This greatly increases the volume and overall capital and operating costs. Another limitation is that a DRE greater than 99% requires a significant amount of catalyst, which increases the capital cost. Recuperative heat exchangers are often used downstream of the catalytic unit to preheat the incoming waste gas to further reduce the fuel requirement.

Catalysts used in catalytic oxidation systems actually consist of a ceramic or stainless steel base material (carrier or support structure) covered with a thin coating of catalyst material. The more surface available for contact with the waste gas, the greater the amount of oxidation reaction. The catalyst material is generally one of two types: noble metal or transition metal oxide. The noble metals are generally preferred. Catalyst type is also based on the ability of the catalyst to resist chemical deactivation (poisoning) from compounds present in waste streams. Typical compounds responsible for poisoning are HCl, HBr, HF, and SO_2 (reversible poisons) and elements such as Pb, Bi, Hg, As, Sb, and P (irreversible poisons). While catalysts have been formulated that will retain their activity in the presence of many of these poisons, there is no single catalyst that is best for all applications. An additional reversible situation is fouling by fine particulate, which could be fine rust particles, refractory dust, or particulate in the waste stream. For this reason, refractory is not used upstream of the catalyst bed, and the vessel material upstream of the catalyst is often made of stainless steel. Fine particulate quite simply covers the surface of the catalyst, reducing the amount of surface area available for reaction. When the DRE has degraded too much, the particulate can often be removed by removing and washing the catalyst blocks or by washing in place.

There are generally two types of catalyst carrier media: ceramic beads and honeycomb monoliths. Virtually all new applications utilize catalysts that are applied onto honeycomb monoliths because they require less pressure drop and allow more flexibility in furnace design and orientation. If a bead catalyst is used, the flow must usually be in a vertical (up or down) direction. The honeycomb monoliths can be installed in any orientation and are usually found in horizontal flow catalytic oxidizers, which are easier to maintain due to access.

Below some minimum threshold temperature, all oxidation catalysts become ineffective. At higher temperatures, the oxidation rate increases rapidly until the rate becomes limited only by the catalyst surface available for interaction with the waste gas. The temperature at which this rapid increase occurs varies, depending on the hydrocarbon, but is typically between 400°F (200°C) and 700°F (400°C) and is usually referred to as the "ignition" temperature. Catalytic units are designed such that the inlet temperature to the catalyst bed is maintained above the ignition temperature. As the waste moves through the catalyst bed, oxidation occurs and the gas and catalyst temperatures rise. The temperature rise in the catalyst bed depends on the heating value of the waste stream. If subjected to temperatures between 1200°F (650°C) and 1350°F (730°C) for very long, many catalysts will begin to suffer significant damage as a result of sintering. Sintering is the melting and coalescence of the active catalyst material, which results in a loss of available catalyst surface area and, consequently, a loss of catalytic activity. The rate of sintering increases rapidly with increasing temperature. A catalyst that shows the first signs of damage at 1200°F (650°C) will likely be severely damaged in a matter of hours at 1500°F (820°C). Therefore, for long-term operation and best DRE, the catalyst bed needs to be maintained above the temperature at which high-rate reactions occur but below the temperature at which significant sintering occurs. Typical catalyst outlet temperatures are in the range of 600°F (320°C) to 1000°F (540°C).

Destruction efficiency in a catalytic incinerator depends on the waste gas composition, catalyst type and configuration, waste gas temperature at the entrance to the catalyst bed, and the amount of time the contaminant is exposed to the catalyst (catalyst surface area). Changes in destruction efficiency are achieved by changing the amount of catalyst, or by changing the waste flow rate for a given amount of catalyst, either of which changes the effective exposure (or residence) time. Residence time in a catalyst bed is often expressed as its inverse and is called space velocity (volumetric flow rate of waste, SCFH/catalyst bed volume, ft^3). Typically, catalytic units are designed with space velocities of less than 30,000, inlet temperatures less than 700°F (370°C), and outlet temperatures less than 1200°F (650°C). Practical catalytic systems typically achieve destruction efficiencies of 90 to 99%.

Catalytic oxidizers must be configured to provide well-mixed, uniform waste gas flow at the catalyst bed entrance and to avoid flame impingement on the catalyst bed by the heat-up burner. As noted previously, units constructed from catalyst-coated monoliths can be oriented in any flow direction but are usually mounted in horizontal flowing units to facilitate catalyst loading and maintenance. Systems using either a fixed or a fluidized bed of beads are mounted vertically. Figure 21.4 shows a typical horizontal system with a preheat exchanger.

21.3.5 Flue Gas Processing Methods

In some incineration systems, the flue gas does not require treatment to reduce emissions of acid gases or particulate, and heat recovery is not economical. In such cases, the hot flue gas leaving the furnace is vented directly to the atmosphere.

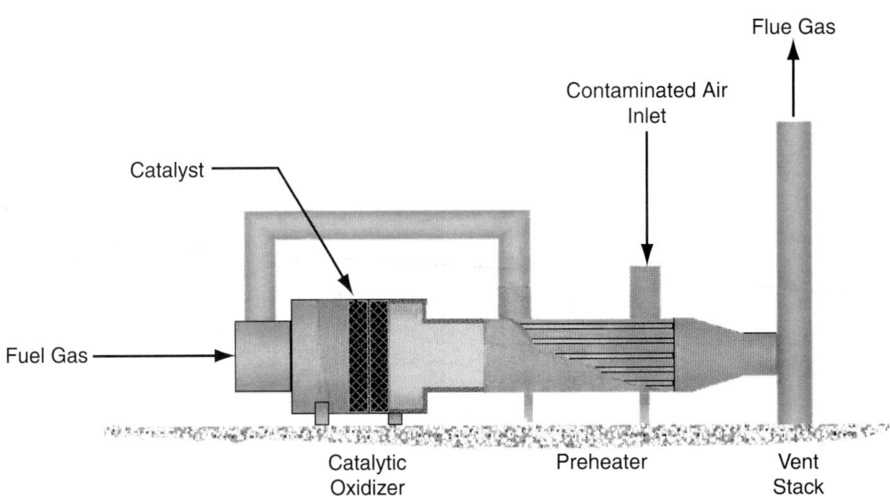

FIGURE 21.4 Typical horizontal system with a preheat exchanger.

For many systems, however, some form of flue gas cooling is utilized, either by heat recovery and/or conditioning of the flue gas before it enters downstream equipment. In general, flue gas cooling is accomplished indirectly or directly. Indirect cooling is achieved by heat transfer from a higher temperature mass to a lower temperature mass through the use of heat recovery devices such as boilers, recuperative preheat exchangers, heat-transfer fluid exchangers, or regenerative preheat exchangers. Heat recovery devices remove heat from the flue gas to lower the temperature but do not change the mass flow rate. Direct cooling is accomplished by adding a cooler material directly to the flue gas to complete the necessary heat transfer. Adding cooling material to the flue gas increases the total mass flow rate as well as reduces the overall total gas temperature. The added material can be water, air, or recycle flue gas, depending on the downstream equipment. Emission control procedures include wet and dry particulate removal, wet and dry acid gas removal, and NOx removal.

21.3.5.1 Cooling by Heat Recovery

Heat recovery can be in the form of steam with either firetube or watertube boilers, recuperative heat exchangers, process oil heaters, or regenerative preheat systems. Such heat transfer devices act indirectly on the flue gas so that as its temperature is reduced, only the heat content is changed — not the composition.

21.3.5.1.1 Boilers

There are two basic types of boilers: firetube and watertube. In a firetube boiler, the combustion products pass through the inside of the boiler tubes while water is evaporated on the outside. Conversely, in a watertube boiler, the hot combustion products flow over the outside of the tubes while the steam is generated on the inside of the tubes. Cooling flue gas from 1800 to 500°F (1000 to 260°C) with a boiler can result in substantial steam production. Adding an economizer downstream of the boiler will recover even more heat by reducing the flue gas temperature to about 350°F (180°C). (An economizer is a lower temperature heat exchanger used to heat the boiler feedwater from its normal supply temperature of about 220°F (100°C) before it is injected into the boiler.)

21.3.5.1.1.1 Watertube boiler There are several important differences between the watertube and firetube boilers. A watertube boiler (Figure 21.5) is generally less expensive to build for applications that require high steam pressure (i.e., > 700 psig or 48 barg) and/or large steam flows (i.e., > 50,000 lb/hr or 23,000 kg/hr). Extended surfaces (finned tubes) and superheaters are more easily incorporated into watertube boilers, often resulting in smaller space requirements. Most importantly, the heat transfer surfaces of a watertube boiler are accessible to soot blowers (high-pressure steam or air lances) used for periodic cleaning of the flue gas side of the tubes to prevent loss of efficiency due to fouling by nonmolten particulate resulting from waste combustion. Thus, for an application that requires high-pressure steam production from a large flow of combustion products containing significant amounts of particulate, the boiler design of choice is the watertube. The typical flue gas pressure drop through a watertube boiler is 2 to 6 in. (5 to 15 cm) w.c. Thus, the watertube boiler is also used when pressure drop must be minimized.

21.3.5.1.1.2 Firetube boiler Firetube boilers have the important advantage that virtually all the surfaces in the boiler

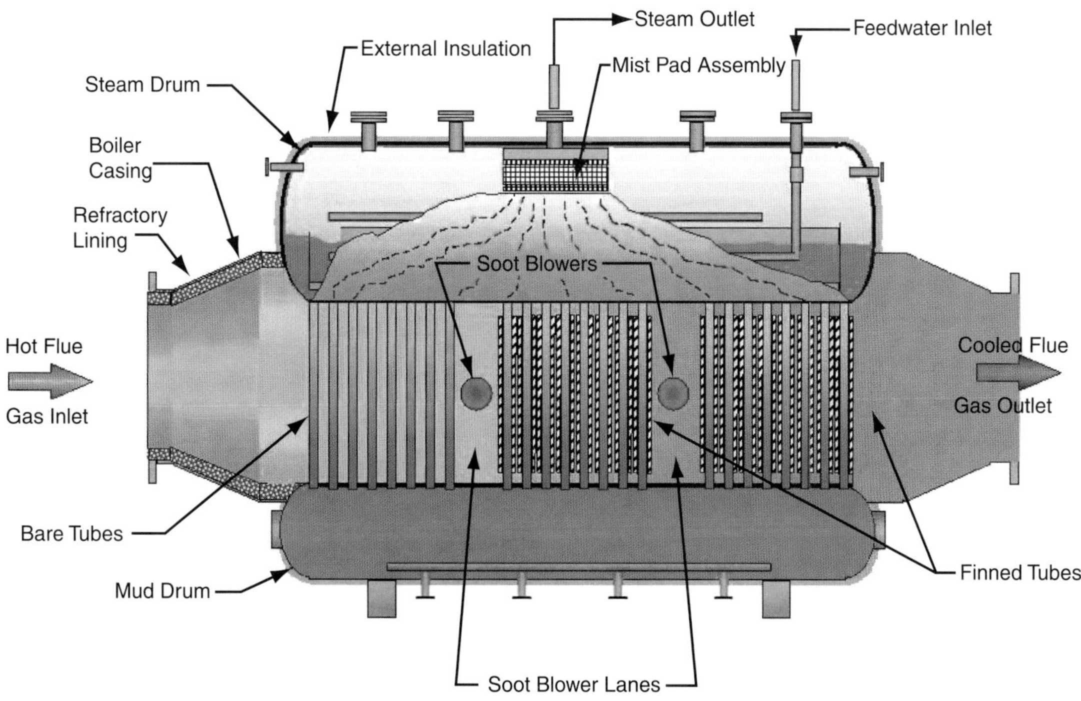

FIGURE 21.5 Watertube boiler.

are maintained at the steam temperature. Thus, there are no cold spots on which acid can condense. For this reason, firetube boilers are well-suited to those applications where low to medium pressure steam is to be produced from an acidic flue gas. The typical flue gas pressure drop through a firetube boiler is 8 to 12 in. (20 to 30 cm) w.c. Figure 21.6 shows a typical firetube boiler.

There are times when the flue gas contains molten particulate but it is cost-effective to cool the flue gas to "freeze" molten particulate so the remaining heat can be recovered in a boiler. Even if the flue gas is cooled to 1200°F (650°C), for example, significant heat recovery is still available. It is important to understand that the choice of a quenching medium used upstream of a boiler will affect both the size of the equipment and the heat recovery. If clean, recycled flue gas from the outlet of the system is used for quenching, a relatively large mass is required because of its high initial temperature (350 to 500°F/180 to 260°C). The flow through the rest of the system could be twice the mass of the flue gas from the T.O. If water is used to quench, a much smaller mass is required, resulting in a much smaller mass flow through the rest of the system. However, using recycled flue gas for the quenching medium results in much greater heat recovery efficiency than if water quench is used because the added water leaves the system with its latent heat of vaporization, as well as the sensible heat at the exit temperature.

21.3.5.1.2 Recuperative Preheat Exchanger

If no steam is needed, waste heat boilers are not a viable heat recovery option. However, if a low heating value waste gas is being treated and a large amount of auxiliary fuel is needed to maintain the operating temperature for the required DRE, a preheat exchanger can be used to minimize the auxiliary fuel requirement by transferring heat from the flue gas to the incoming waste gas or combustion air. Figure 21.7 shows a typical all-welded shell-and-tube heat exchanger. The furnace exhaust flows through the tube in the exchanger while the waste gas or combustion air flows around the tubes inside the shell. Up to 70% of the energy released in the furnace can be recovered economically by this method. Normal recovery efficiencies are in the 55 to 60% range. Structural limitations (thermal expansion) typically constrain the hot flue gas temperature to no more than 1600°F (870°C). However, more expensive U-tube type heat exchangers exist that can tolerate higher temperatures.

Recuperative exchangers can also be of plate-and-frame type construction. This type can withstand higher temperature expansion differences because of its non-welded construction. However, some leakage of waste gas into the clean combustion products will occur, increasing the unburned waste

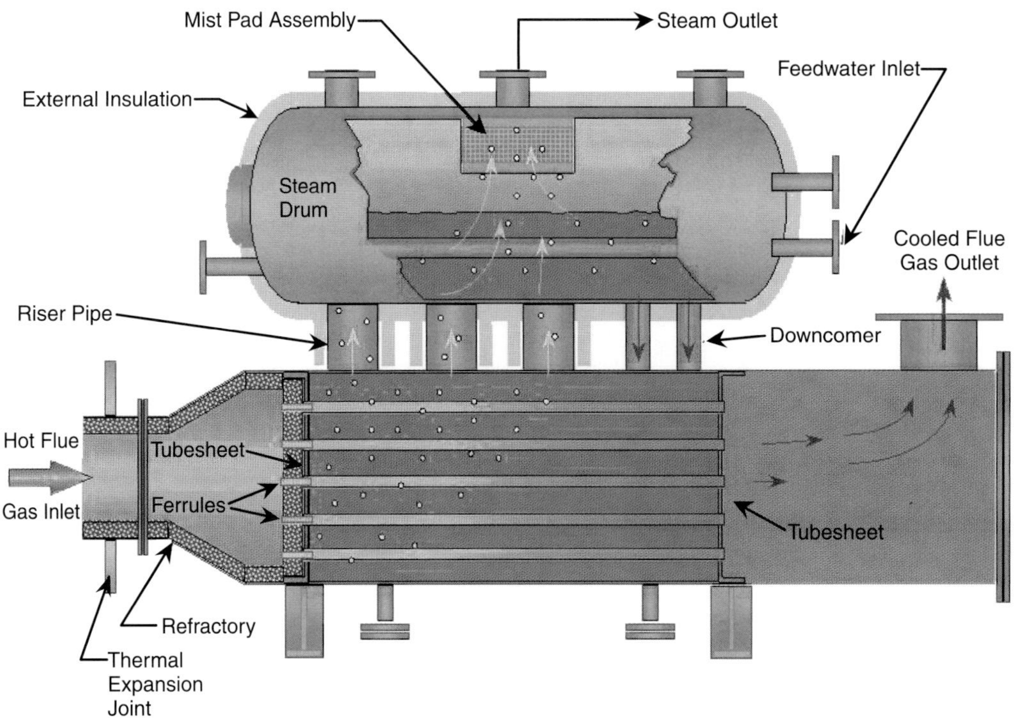

FIGURE 21.6 Firetube boiler.

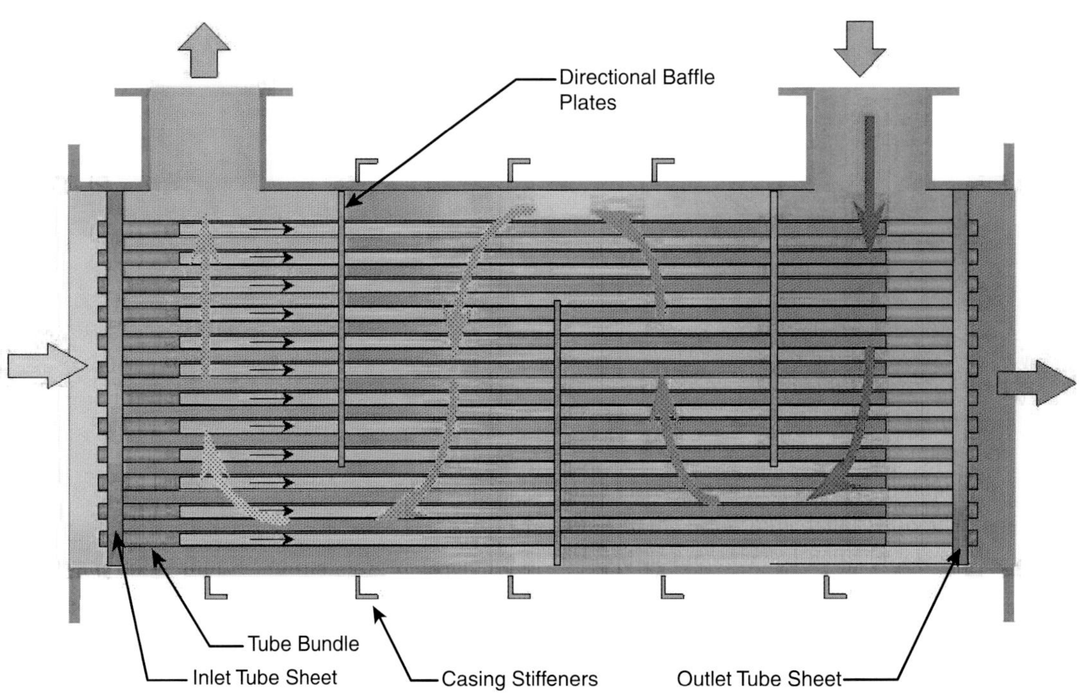

FIGURE 21.7 Typical all-welded shell-and-tube heat exchanger.

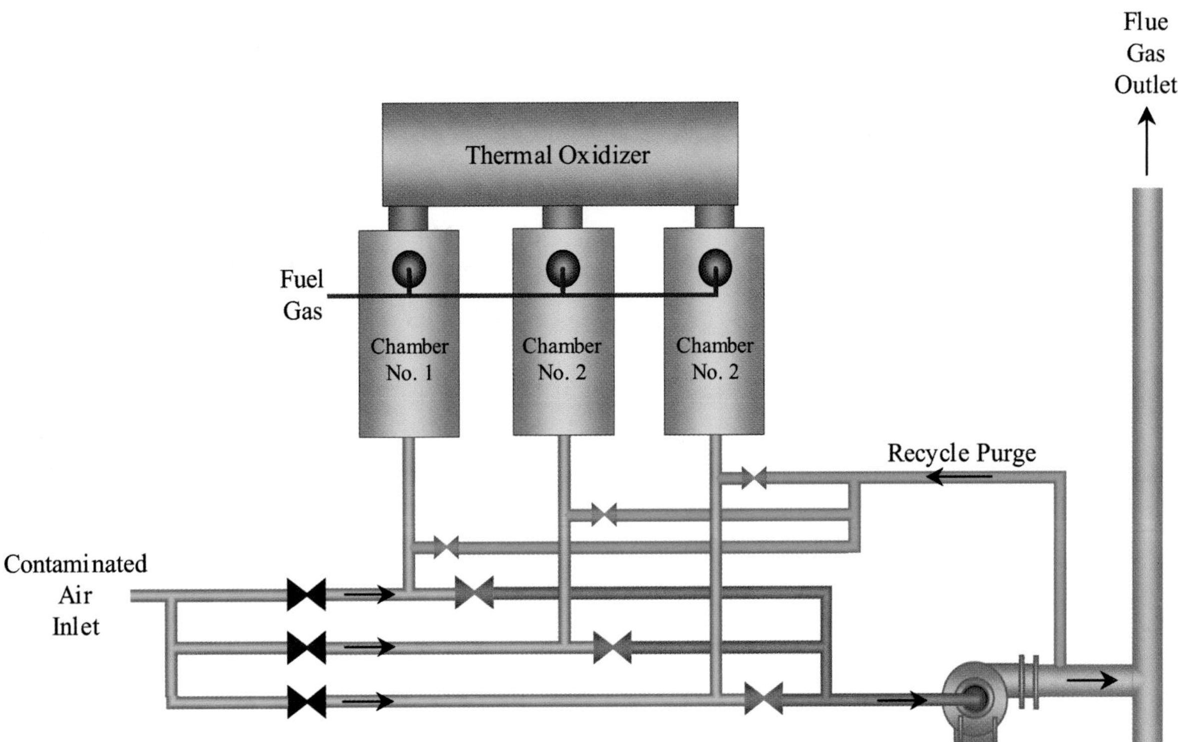

FIGURE 21.8 Regenerative preheat exchanger.

hydrocarbon emission to the point that the apparent DRE is not as required. Also, although the initial leakage may be low enough when the exchanger is first put into service, it will likely increase with time, particularly in systems that operate intermittently (shut down daily or weekly). To avoid leakage problems, the more expensive shell-and-tube type heat exchangers with all-welded construction should be used when high DRE is required.

21.3.5.1.3 Regenerative Preheat Exchanger

A T.O. system with regenerative preheat exchange consists of a refractory-lined T.O. connected to three or more vessels containing a ceramic packing (often ceramic scrubber bed packing) that alternately functions to preheat the waste and to cool the flue gas exiting the T.O. section. Figure 21.8 shows the configuration of such a system. This system is primarily for contaminated air stream. The temperature of the gases is measured at the inlet and outlet of each ceramic-packed vessel. Numerous valves must be used to control the direction of flow at all times during operation. The system is somewhat larger and more expensive than a recuperative system, but it is much more efficient. Up to 95% heat recovery is possible if the incoming hydrocarbon content is very low. However, the normal rates are in the 85–90% range. The potential fuel savings make these systems attractive for large waste flows that have little heating value.

In operation, the waste gas flows into the system through a hot bed of packing (Chamber 1) before it enters the T.O. The incoming waste gas temperature is monitored at the hot end (nearest the T.O.) of Chamber 1. The flue gas exiting the T.O. flows through an identical but cooler bed of packing (Chamber 2) before it is vented to the atmosphere. When the packing in Chamber 2 has absorbed heat to the point that the exit gas temperature rises above a preset maximum, typically 300 to 350°F (150 to 180°C), the hot gas is redirected to the third bed of packing (Chamber 3), which was out of service and is cool. At the same time, the incoming waste gas is switched from Chamber 1 to Chamber 2, which is now the hot bed, to pick up stored heat before flowing into the T.O. Chamber 1, which is now temporarily out of service, was in preheat service when the flows were switched. Consequently, Chamber 1 is filled with untreated waste gas. The waste gas in Chamber 1 is purged into the T.O. with "cool" recycle flue gas while it is out of heat exchange service. If the most recently used incoming bed was not purged, or if only two beds were used, a bed full of waste gas would be vented at each flow reversal. The result would be similar to the plate

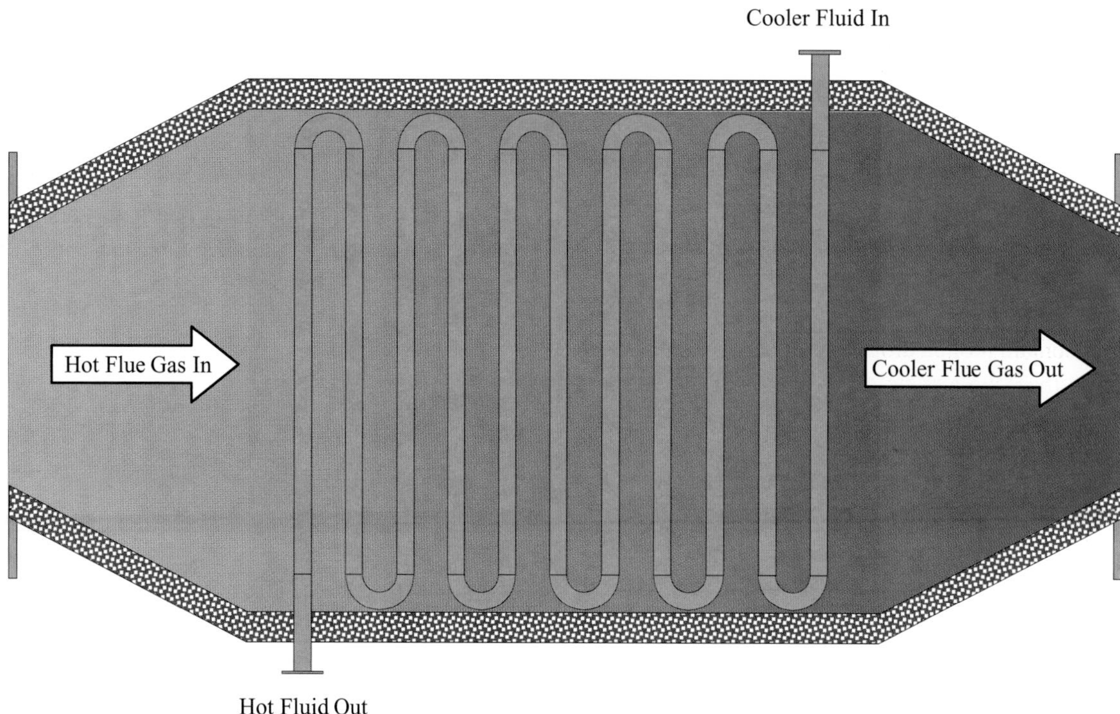

FIGURE 21.9 Organic fluid transfer system configuration.

exchanger problem above in that the apparent DRE would not be high enough to meet the emission regulations. Thus, at least three beds are required to allow the cool inlet bed to be purged before it becomes an outlet bed. In practice, some large systems are constructed of five or more beds to overcome shipping restrictions and to allow prefabrication.

21.3.5.1.4 Organic Heat-Transfer Fluid Heat Exchangers

Many plants use organic heat-transfer fluids (hot oil) to provide a controlled heat source for plant processes. If the T.O. system can be operated with or without waste so that a continuous heat source is available, the flue gas can be used to heat this fluid in a shell-and-tube design exchanger. However, there may be periods when the system is generating more flue gas than is needed because the heat demand of the hot oil is reduced. Also, during normal operation, the temperature of the flue gas entering the exchanger may be limited to some maximum because of the properties of the fluid. For such cases, the system may have to be designed to bypass some or all of the hot flue gas directly to the atmosphere and to cool the flue gas by mixing in some ambient air to avoid coking the organic fluid. Figure 21.9 shows the necessary configuration.

Briefly, the waste is fired into a horizontal T.O. that is connected directly into the base of a hot vent stack. Another connection at the base of the stack allows flue gas to be drawn out of the stack into a duct connected to the hot oil heat exchanger. An induced-draft blower is located downstream of the hot oil heater to "pull" the flue gas from the stack and through the exchanger. The cool flue gas exiting the exchanger is then pushed through a cool duct and injected into the hot stack at least two stack diameters about the base-connection centerlines. A hot oil temperature controller monitors the fluid temperature and modulates the flue gas flow to maintain the desired temperature.

To cool the incoming hot flue gas temperature, a flue gas temperature controller monitors the hot oil heater inlet temperature and operates a valve that regulates the amount of ambient air drawn into the stack-to-oil heater duct.

This particular configuration can be used for any similar heat transfer device when more flue gas is available than the exchanger can process.

21.3.5.2 Cooling Without Heat Recovery

Flue gas cooling by means other than heat removal is often necessary. When the flue gas has to be processed to remove emissions, it must be cooled to a temperature that will not harm the downstream equipment nor reduce the efficiency of the downstream equipment. If particulate is to be removed by a dry process such as an electrostatic precipitator, the flue gas must usually be cooled (conditioned) to below 650°F

(340°C). If it is to pass through a baghouse, it will have to be conditioned to 400°F (200°C) or less. For wet particulate removal or wet acid gas removal, the flue gas will likely have to be cooled (quenched) to its saturation temperature for treatment.

21.3.5.2.1 Conditioning Section

The flue gas exiting a T.O. is at such a high temperature that often it must be cooled before entering downstream equipment to prevent damage to that equipment. In some cases, flue gas may contain molten droplets of material that must be cooled below their melting point (frozen) so they will not adhere to downstream boiler tubes or other cooler surfaces. In other words, the "condition" of the flue gas must sometimes be altered before it can be further treated. Removing heat with heat recovery devices (indirect conditioning) has already been discussed. As noted, removing heat reduces the temperature but does not change the mass flow rate or composition. This section reviews direct conditioning heat transfer methods that reduce the temperature and also change the mass flow rate as a result of adding a cooling material to the flue gas to which heat is transferred.

The cooling medium that adds the least mass to the flue gas is water. Each pound of water sprayed into the flue gas absorbs almost 1000 Btu as it vaporizes (heat of vaporization), as well as sensible heat. For example, assuming no other heat losses occur, cooling 10,000 lb/hr (4500 kg/hr) of flue gas from 1800 to 600°F (980 to 320°C) requires about 2655 lb/hr (1200 kg/hr) of water at 70°F (21°C). Cooling the same amount of flue gas with 70°F (21°C) air requires about 26,190 lb/hr (11,900 kg/hr) of air — almost 10 times the cooling mass. If the flue gas was only cooled to 1200°F (650°C), the water required is about 1110 lb/hr (500 kg/hr) and the air needed is about 6150 lb/hr (2800 kg/hr) — only about 5.5 times more mass. Minimizing the flue gas flow to downstream equipment is normally desired so the smallest size, lowest cost equipment can be used.

Because 100% of the water injected should vaporize, the conditioning section can be oriented in any direction. However, the vertical up or down flow configuration is usually utilized. This configuration also allows features to be used (e.g., a hopper at the base of the unit, regardless of flow direction) to collect and remove water online if any of the removable spray tips fail to properly atomize the water. The hopper will also provide storage volume for particulate that may not exit the conditioning section.

If the purpose of the conditioning section is to cool flue gas to only 1200°F (650°C) to freeze molten particulate before the flue gas goes into a waste heat recovery device, more heat will be available for recovery if air or cool recycle flue gas is used to provide the cooling instead of water. As noted in an earlier heat recovery section, although the flue gas flow when using air or cooled recycled flue gas will be greater than with water, and the downstream equipment size will be larger, increasing equipment cost, the value of the additional heat recovered will likely exceed the additional cost of larger equipment in only a few years. Also, because of the higher flue gas temperature, the entire conditioning section would be refractory lined.

If the flue gas is being quenched to low temperature (400 to 600°F, or 200 to 320°C), the vessel is usually made of carbon steel with internal refractory lining for part of its length at the hot inlet end, and external insulation but no refractory for the rest of its length. External insulation is used at the cooler end to prevent condensation in the cool unlined portion.

The hot flue gas is usually passed through a reduced diameter, refractory-lined section to increase the velocity just before the atomized water is injected. This is done to improve mixing and heat transfer, which increases the evaporation rate of the atomized water droplets traveling through the conditioning section. The conditioning section outlet temperature is continuously monitored and the water flow adjusted to maintain the desired temperature. Although the temperature of the conditioned flue gas may be well above the saturation temperature, if the flue gas has to be cooled to less than 400°F (200°C), so much cooling water may cause the particulate to become damp and stick to the outlet duct between the conditioning section and the dry particulate removal device. Also, the chances of condensation on the walls of the outlet duct and dry particulate removal device increase as the water content of the flue gas increases. To reduce the possibility of such problems, the flue gas is sometimes cooled the last 100 to 150°F (38 to 66°C) by the addition of ambient air.

Figure 21.10 is a general representation of a vertical, downflow conditioning section that can use water, air, etc. as the cooling medium.

21.3.5.2.2 Saturation Quench Section

If hot flue gas is to be treated by a wet particulate or acid gas removal device, it is usually best to complete the heat transfer portion by quenching to full saturation before the subsequent mass transfer process is initiated. Although this is not always done, most mass transfer equipment is designed for fully saturated flue gas. The saturation temperature could typically be as high as 210°F (100°C) and usually is no less than about 135°F (57°C), depending on the composition and temperature of the flue gas when it enters the quench section.

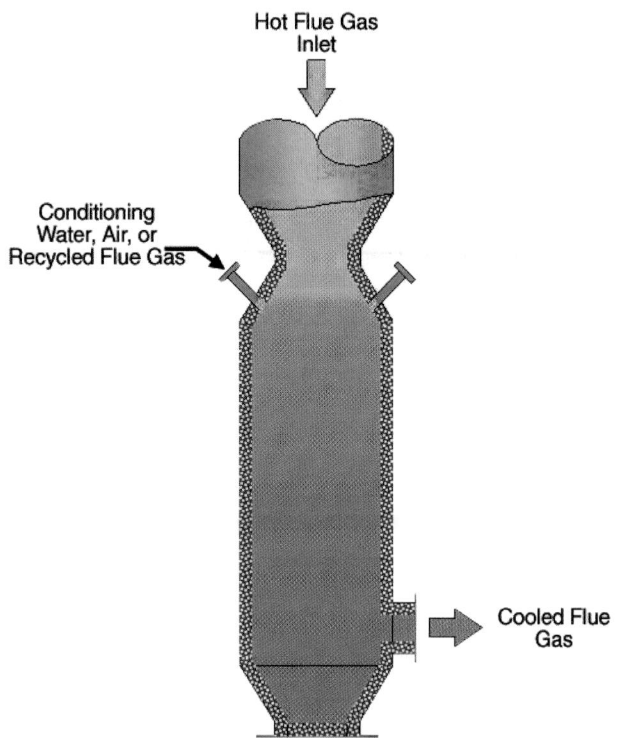

FIGURE 21.10 Vertical, down-flow conditioning section.

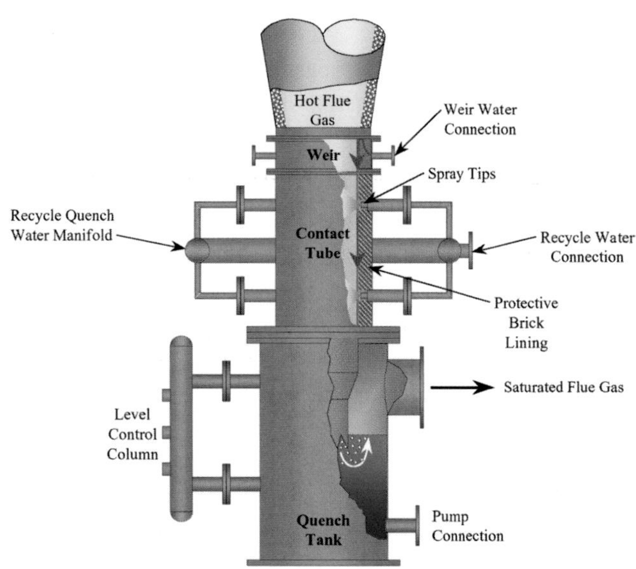

FIGURE 21.11 Direct spray contact quench.

Three basic saturation quench configuration options exist:

1. direct spray contact quench section
2. submerged quench section
3. combination adjustable plug-type quench and venturi scrubber section with an integral droplet separator

A description of the three options follows, along with advantages and disadvantages of each. In each configuration a steep temperature gradient will exist between the furnace exit gas and the quenched gas. As waste rates to the incinerator change, the location of this gradient will shift slightly, presenting a challenge to the hardware designer because radical temperature variations may eventually result in refractory damage. For this reason, a hot-to-cool water-cooled metal interface is often used in this area.

21.3.5.2.2.1 Direct spray contact quench As shown in Figure 21.11, the hot flue gas flows downward through an annular overflow assembly (weir), then through a brick-lined water spray contact duct (contactor tube) fitted with recirculating water spray guns, then through a downcomer tube (downcomer), and into the water collection and droplet separator vessel (quench tank).

The weir is the hot-to-cool junction in the system. It is used to feed fresh (make-up) water into the quench system and to form a wetted-wall in the contactor tube and downcomer. The wetted-wall provides a cooling film in the contactor tube and downcomer. Material of construction for the weir is stainless steel unless more severe service (chlorinated waste) requires use of a more corrosion-resistant metal such as hastelloy, zirconium, or inconel.

The contactor tube, located directly under the weir, is a short, brick-lined spray tower into which a large quantity of recycled water (approximately 4 times the calculated amount needed for saturation) is injected. The spray guns must provide efficient atomization and penetration across the full cross-section of the contactor tube if quenching is to be effective and virtually complete saturation of flue gases achieved. The high-temperature flue gas is cooled, primarily through vaporization of atomized recycle water, to near the adiabatic saturation temperature. If the flue gas contains particulate, some will drop out in the quench tank. The recycle water may absorb some acid gas as long as the equilibrium concentration of acid in the water is less than in the flue gas. The contactor tube shell can be constructed of carbon steel with a corrosion barrier, special corrosion-resistant alloys, or fiberglass-reinforced plastic (FRP).

After the nearly saturated flue gas and extra water exit the contactor tube, they pass through the downcomer, which extends to the water level in the quench tank. The downcomer

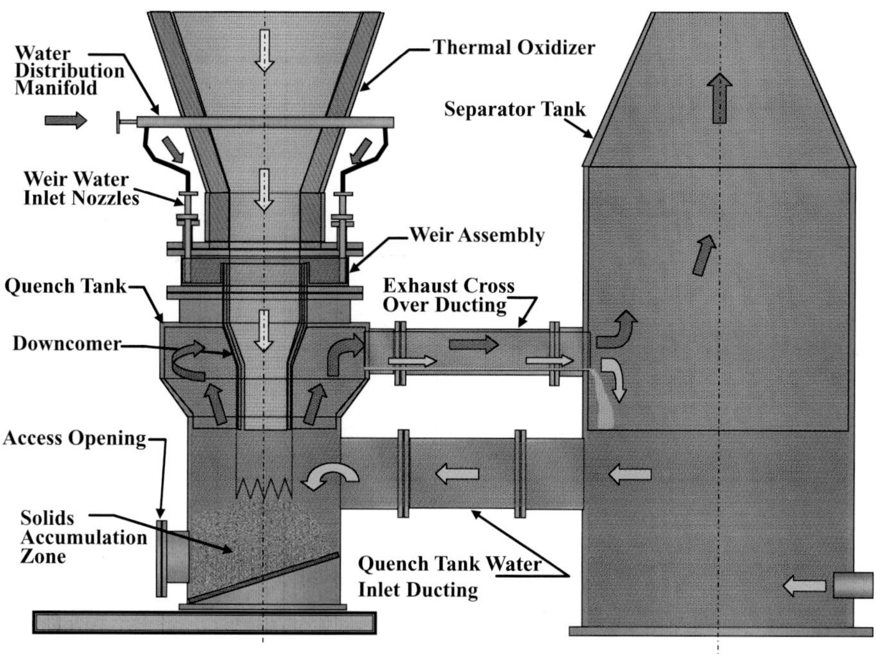

FIGURE 21.12 Submerged quench.

is smaller in diameter than the contactor tube, thus increasing the velocity and mixing energy of the flue gas and atomized recycle water to complete the cooling and saturation of the flue gas. The flue gas makes a 180° direction change at the downcomer outlet as it turns to travel upward in the annular space between the downcomer and the quench tank wall to the quench tank outlet. The higher velocity through the downcomer and subsequent 180° direction change, along with the low velocity in the annulus, allow water droplets to be separated from the saturated flue gas. The downcomer should be constructed of a corrosion-resistant metal such as stainless steel or other more exotic alloys as required for the service. The quench tank can be made of carbon steel with a corrosion barrier, special corrosion-resistant alloys, or FRP.

Although the primary purpose of the quench tank is as the collection sump for recycling water, it also captures some of the particulate coming from the T.O. Thus, the configuration of the quench tank must allow for continuous, if necessary, removal of solids and continuous withdrawal of recycle water. To accomplish this, the base of the quench tank should be conical, with a solids blowdown nozzle at the base and the suction point for the recycle pump(s) in the conical section. Furthermore, the recycle line should be equipped with a dual-basket strainer upstream of the pump(s) and a cyclone separator (hydroclone) downstream of the strainer. The strainer openings are sized to remove only large pieces from the recycle water that might damage the pumps, while the hydroclone removes the smaller pieces that might plug the atomizing spray tips. A continuous blowdown from the hydroclone carries out the collected solids, but the strainer must be manually cleaned periodically. Blowdown from the base of the conical section can be automatic or manual.

The advantages of the direct spray contact quench are:

1. faster, more efficient cooling of the flue gas as a result of the large heat transfer surface area of the atomized water
2. very low pressure drop across the entire section, usually less than 2 in. (5 cm) w.c.
3. efficient water droplet separation from the flue gas
4. large, open flow area unlikely to be affected by any obstruction

Disadvantages include loss of cooling flow as a result of pump failure and potential plugging of the spray tips. However, a hydroclone separator is used, plugging is minimized. Also, utilizing two recycle pumps with auto-start on the spare when flow drops below a minimum greatly reduces the potential loss of flow.

21.3.5.2.2.2 Submerged quench This configuration, Figure 21.12, is mechanically similar to the direct-spray design, but instead of atomizing water to provide a large contact surface area between the water and hot flue gas, this method divides the hot gas flow and "bubbles" it through the water. To accomplish this, the hot flue gas enters the quench section, traveling

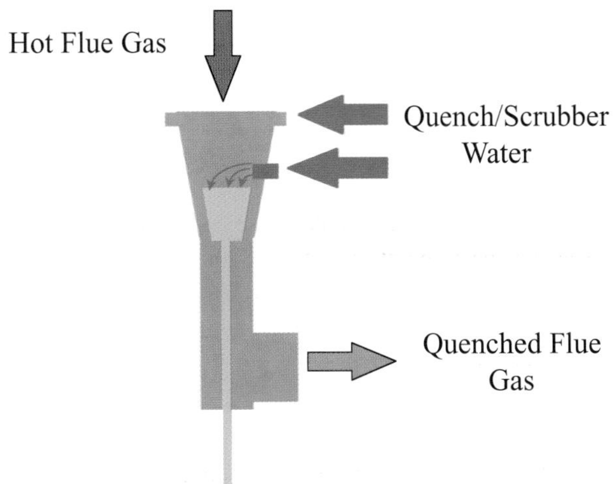

FIGURE 21.13 Adjustable-plug venturi quench.

downward through an annular overflow section (weir), directly into a downcomer tube (downcomer) that extends several feet below the liquid surface in the separator vessel (quench tank). The lower, cylindrical portion of the downcomer contains a number of smaller holes through which the hot gas exits the downcomer. As the flue gas bubbles upward through the water, cooling occurs. Often, an additional droplet separator vessel is used. The total pressure drop across this section is usually between 24 and 30 in. (61 and 76 cm) w.c.

The weir is again the hot-to-cool interface in the system. It is used to feed fresh (make-up) water into the quench system and, unlike the spray quench, is also where the large volume of recirculated water from the quench tank is introduced. All the water flows down the wall of the downcomer, flooding and, therefore, cooling the downcomer wall. The material of construction for the weir is usually stainless steel for non-acid gas service and hastelloy, zirconium, or inconel for acid gas, such as chlorinated hydrocarbon service.

The downcomer is usually an extension of the weir for this process. It is open-ended and extends several feet below the water level in the quench tank. Although the downcomer is open at the bottom, the flue gas actually passes through holes cut in the circumference of the lower portion of the downcomer, well below the water surface but above the open end. As the flue gas passes through the openings, it is distributed evenly and separated into smaller volumes that further divide into smaller bubbles, providing the contact surface area needed to quickly evaporate water and quench the flue gas to saturation temperature. The openings must be properly sized and located at the proper depth below the surface for the downcomer to function properly. The downcomer is fabricated from the same material as the weir.

The quench tank serves as the source of the water for quenching. Although a significant amount of water is entering the vessel through the downcomer, the internal design of the quench tank ensures proper internal recirculation of the water and prevention of excessive gasification of the water in any location. The bubbling flue gas creates significant turbulence and splashing at the surface, making droplet separation in the quench tank more difficult, and often a second vessel is used as a separator to remove droplets before the cooled flue gas travels to the next flue gas treatment section. The separated water usually flows back into the quench tank by gravity from the separator. As with the direct spray quench, if particulate enters the quench section, some will remain in the quench tank, and some of the acid gas will be absorbed by the recycle water. The configuration of the base of the quench tank and recycle pumping system components (strainer and hydroclone) should be similar to that of the direct spray quench to allow for continuous removal of solids from the bottom of the vessel as well as continuous withdrawal of particulate from the recycle water to prevent filling the overflow weir with solids. The quench tank (and separator, if used) can be made of carbon steel with a corrosion barrier, special corrosion-resistant alloys, or FRP, depending on the service.

The primary advantages of the submerged quench are:

1. No spray tips are required.
2. There is a large flow area that is unlikely to be affected by obstructions.
3. Even with pump failure and loss of recirculation water, little or no downstream equipment damage is likely to occur because the high liquid level in the quench tank will prevent hot gas from traveling downstream.

Disadvantages include high pressure drop across the section and, often, another vessel must be used for effective droplet separation.

21.3.5.2.2.3 Adjustable-plug venturi quench This downward flow-oriented device, Figure 21.13, combines both heat and mass transfer. The upper portion of this one-piece section utilizes the familiar hot-to-cool interface, an annular overflow weir followed by a short converging inlet tube, then the adjustable plug/throat section and an outlet tube. After passing through the adjustable-plug venturi, the quenched flue gas and water then travel through a separator/quench tank and on to the next section.

Once again, a weir is the hot-to-cool interface in the system. It is used to feed fresh (make-up) water into the quench system and recirculate some of the needed large volume of recycle water.

A converging duct, wetted by the weir overflow, provides an entrance for the flue gas to the throat section. A tapered, vertically adjustable plug varies the throat area, changing the gas velocity and pressure drop across the section. A large volume of recycle water is added to the center of the throat through a pipe that directs the water onto the center of the adjustable plug. The water and flue gas pass through the annular space, between the plug and the throat wall. The high velocity (up to 500 ft/s, or 150 m/s) at the throat provides the energy needed to atomize and mix the water with the flue gas, cooling the flue gas to saturation. Downstream of the throat, a 90° elbow directs the quenched flue gas and water to a separator/quench tank. A shaft extends from the plug, through the bottom of the elbow, and is attached to an actuator, which is used to automatically adjust the position of the plug to maintain a set pressure drop regardless of flow variation. The entire venturi assembly can be fabricated using stainless steel, acid-resistant metals, or, in some cases, a combination of FRP and metals.

The separator/quench tank usually has a tangential inlet to help separate the droplets and water more effectively from the saturated flue gas. Recycle water and solids removed from the flue gas collect in the separator/quench tank, so vessel configuration and recycle pumping configuration should also be very similar to that of the two previously described quench processes. The separator/quench tank can be built of the same materials noted above.

Because particulate in the flue gas is more effectively wetted (resulting in more being captured) in this quench section than the two previously described, this method of quench is sometimes used when particle removal is required or desired, especially if the particulate is relatively large and easy to remove. Greater acid gas absorption is also more likely with this method of quench because of the extra mixing energy at the throat.

The total pressure drop across this section could vary between 20 and 70 in. (51 and 180 cm) w.c., depending on whether the device is used primarily for quench or if particulate removal is needed. Advantages of the adjustable-plug venturi quench are:

1. Quenching and scrubbing in one section reduces plot space.
2. No spray tips are required.

Disadvantages include (1) higher pressure drop than other quenching methods, (2) loss of cooling water flow as a result of pump failure, (3) more complexity of design, including the moving plug and shaft, with a shaft seal that must be maintained, and (4) the solids removal efficiency of the combination quench/venturi may not be sufficient to meet the particulate emission limit and a second particulate removal device may be needed, adding to the system pressure drop. Another potential disadvantage is that pieces of refractory material falling from the refractory-lined duct or T.O. above the venturi could block the annular space in the venturi section. This is not as likely if the adjustable plug is operated in the automatic mode to maintain a preset pressure drop. In the case of material falling from above, blocked flow area would cause an increase in pressure drop above the setpoint, resulting in the plug being moved upward to open the annular space, allowing most pieces to flush through the throat. Another potential disadvantage is that very large pieces will not pass through the plug /throat annulus as they would with the direct spray or submerged quench designs.

21.3.5.3 Particulate/Acid Gas Removal

Equipment used to remove pollutants from T.O. system flue gas streams are known as air pollution control (APC) devices. The most common pollutants that result from burning liquid and gaseous wastes, which require removal, are particulate matter and acid gases such as SOx and HCl. The most commonly used equipment for the removal of these pollutants are dry removal devices such as baghouses and electrostatic precipitators, and wet removal devices such as venturi scrubbers and packed columns. This section provides a brief summary of the design considerations associated with these devices. More detailed information can be found in References 1–3.

21.3.5.3.1 Particulate Removal: Dry

The most common methods of dry particulate removal are filtration and electrostatic attraction collection. Reasonable, cost-effective efficiencies are achieved even when submicron particles must be removed. High pressure drop cyclonic separation is also available but is rarely used with combustion systems.

Dry particulate recovery is often preferred if the material recovered is to be re-utilized. If a wet process is used to recover catalyst particles, the particles have to be separated from the water and dried before any purification or refining process can begin. Dry removal minimizes the volume and weight of material that must be handled after recovery. Much less water is used in the dry removal process making dry removal more attractive in locations where water is scarce or expensive. For the same removal efficiency, dry removal requires much less pressure drop than wet removal of the same removal efficiency.

21.3.5.3.1.1 Filtering device (baghouse) A common method of dry particle recovery is by collection on the surface of fabric bags (baghouses). The principal design parameters for a baghouse, assuming a particle size distribution for the particulate is known, are fabric type, air-to-cloth ratio, and cleaning

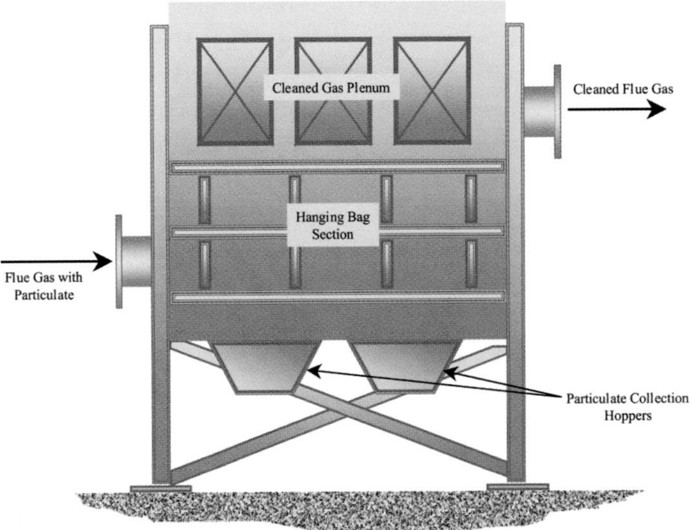

FIGURE 21.14 Baghouse.

method used. Gas to be cleaned enters the baghouse, flows through the bags from the outside (depositing the particles on the outside surfaces of the bags), flows inside the bags up to the clean gas plenum, and out to a stack or to another treatment device. Many bag fabrics are available. The fabric weave is tight enough that some of the particles are initially captured in or on the surface. Once a base coat of particulate (filter cake) has been collected, an even finer filter medium than the original fabric now exists, allowing high efficiency capture of even small particles (up to 99.5% for 0.1-micron particles and up to 99.99% for 1- to 10-micron particles) with a relatively low pressure drop of about 6 in. (15 cm) w.c. As the particulate accumulates, the pressure drop increases and eventually the bags must be cleaned. The commonly used cleaning methods are pulse-jet, shaker, or flow reversal. The dislodged dust falls to the bottom of the baghouse and is removed during operation through special valves.

None of the baghouse design parameters is independent, and all are based on testing and previous experience. Fabrics may be woven or felted and can be made from a variety of materials. Fabrics differ in their particle capture efficiencies, corrosion resistance, erosion resistance, temperature range, pressure drop, strength, durability, and the ease with which they can be cleaned. The fabric is chosen based on exhaust gas conditions, type and size distribution of the particles to be filtered, particle loading, and cleaning method.

A typical bag used in a T.O. system would be a 6-in. (15-cm) diameter by 10-ft (3-m) long cylinder closed at the bottom, supported by an internal wire cage and suspended from a tube sheet forming the top of the dirty gas chamber. The number of bags is set by the desired gas:cloth ratio, which usually falls between 2 and 5 ACFM/ft^2.

The choice of a cleaning method (pulse-jet, shaker, or flow reversal) can be determined by the fabric strength and durability, but is often a compromise between capital and operating expenses. The pulse-jet technology has relatively high energy requirements but can usually operate at higher filtration rates, thus requiring a smaller filter area. Conversely, the shaker and flow reversal technologies typically have lower energy requirements and lower filtration rates, leading to larger filter areas. Pulse-jet is the cleaning method normally used for T.O. systems. To minimize air usage, bags are not all cleaned simultaneously. Also, care must be taken to avoid excessive cleaning, which removes the filter cake, reducing collection effectiveness until it builds up again.

Baghouses are sometimes compartmentalized so that valves can isolate the compartment or section of the baghouse being cleaned. While this action requires increasing the number of bags by 50% and adding inlet and outlet manifolding and valves, it does reduce the amount of material being drawn back to the filter immediately after cleaning and increases the period of time needed between cleaning each section.

In general, baghouses are used only in applications where the gas is dry and the temperature is below 450°F (230°C). A typical T.O. system application would utilize a pulse-jet baghouse operated at 400°F (200°C), with a Nomex fabric designed for an air-to-cloth ratio of between 2 and 5 ACFM/ft^2 and a pressure drop of 5 to 10 in. (13 to 25 cm) w.c. Figure 21.14 is a general representation of a baghouse.

Baghouses are usually used in non-acid gas service and are therefore usually constructed of carbon steel. For non-halogen acid gas service, stainless steel has also been used. Any material can be used as long as it can withstand occasional short-term temperature excursions to more than twice the design flue gas temperature. Bag filter material can be polyester, polyaramid, cellulose, fiberglass, Nomex, Gortex, or any other proven fiber that meets the operational criteria.

The advantages of using a baghouse are:

1. The particle removal efficiency is high.
2. The pressure drop to collect the material is low.
3. The material is collected "dry" and does not have to be separated from water.
4. Although it contains some moisture, the recovered material is basically at its final volume and weight.

The disadvantages are:

1. The baghouse is relatively expensive and occupies a lot of plot space.
2. Field construction may be required for large flue gas flows.
3. The large baghouse surface area requires extensive insulation to minimize acid gas or moisture condensation.
4. The maximum treatable flue gas temperature is about 450°F (230°C).

21.3.5.3.1.2 Electrostatic precipitator An electrostatic precipitator (ESP) is a device that removes particles from a gas stream by means of electrostatic attraction. A high voltage potential, usually applied to weighted vertical hanging wires (emitters), causes the particles to be charged. Once charged, the particles are exposed to grounded collecting electrodes (plates) to which the particles are attracted, separating (precipitating) them from the flue gas. The particles must then be separated from the collecting plate, while minimizing re-entrainment, and removed from the ESP collection hopper.

An ESP is a large, often rectangular-shaped, chamber containing numerous flat parallel collecting plates with emitter wires located midway between the plates (see Figure 21.15). The flue gas entering the ESP must be uniformly distributed across the chamber for effective treatment. As the flue gas passes between the plates, the high voltage potential (40 to 50 kV) carried by the emitter wires creates a corona discharge, making a large number of both positive and negative gas ions. The positive ions are attracted to the negatively charged emitter wires, leaving the space between the plates rich in negative ions. Particles passing through the negative ion-rich space quickly acquire a negative charge. Smaller particles are, however, more difficult to charge. (*Note:* Negative discharge electrodes are normally used for industrial ESPs because of the higher potentials available and more predictable performance.)

The force that moves the particles to the collection plates results from the charge on the particles and the strength of the electrical field between the emitters and collectors. For smaller particles, the electrical field strength must be greater to remove the same percentage of smaller particles as bigger particles. The force can be up to several thousand times the acceleration of gravity so that the particles move rapidly to the collection surfaces. Because most of the particles retain a portion of their negative charge even after contacting the collecting plate, some remain on the plates until a physical action dislodges them.

The weight of the particulate on the plates causes some of the particulate to fall into the collection hoppers, but the remainder must be dislodged by vibration (or rapping) of the plates and emitters. Once collected in the hoppers at the bottom of the chamber, the particulate is removed by the same means as particulate is removed from baghouses.

Design considerations include:

1. electrical characteristics of the particulate (i.e., how well it will accept and hold a charge [the particles must have a resistivity in the range of 10^4 to 10^{10} ohm-cm for efficient removal by electrostatic means])
2. gas and particle velocity (very important with sub-micron particulate), including gas velocity in the unit, drift velocity of the particulate induced by the electric field, and particle settling velocity
3. gas distribution
4. electrical sectionalization (i.e., the increase in power input in sequential zones or cells through the length of the ESP to achieve the desired removal efficiency)
5. particle re-entrainment

The flue gas must also contain readily ionizable species such as O_2, CO_2, and SO_2. Particle resistivity can be a strong function of the flue gas temperature, composition, and moisture content. Thus, ESP performance can be quite sensitive to changes in upstream process conditions.

As with the baghouse, most of the design parameters have been developed empirically and then fit to equations to help the engineer develop the physical equipment design. The result is a chamber containing the correct number and size (length and width) of collector plates spaced appropriately to allow locating emitters between the plates. The capital cost of the ESP is directly related to that physical information. Both sides of each collector plate functions as collecting surface area (CSA), which is also referred to as specific collector area (SCA) or specific collector surface (SCS). This area is often expressed in terms of surface area per 1000 ACFM of flue gas through the ESP. Depending on the particle-size distribution and other particle-related parameters, removal efficiencies of more than 99% can be achieved at less than 300 ft^2/1000 ACFM of

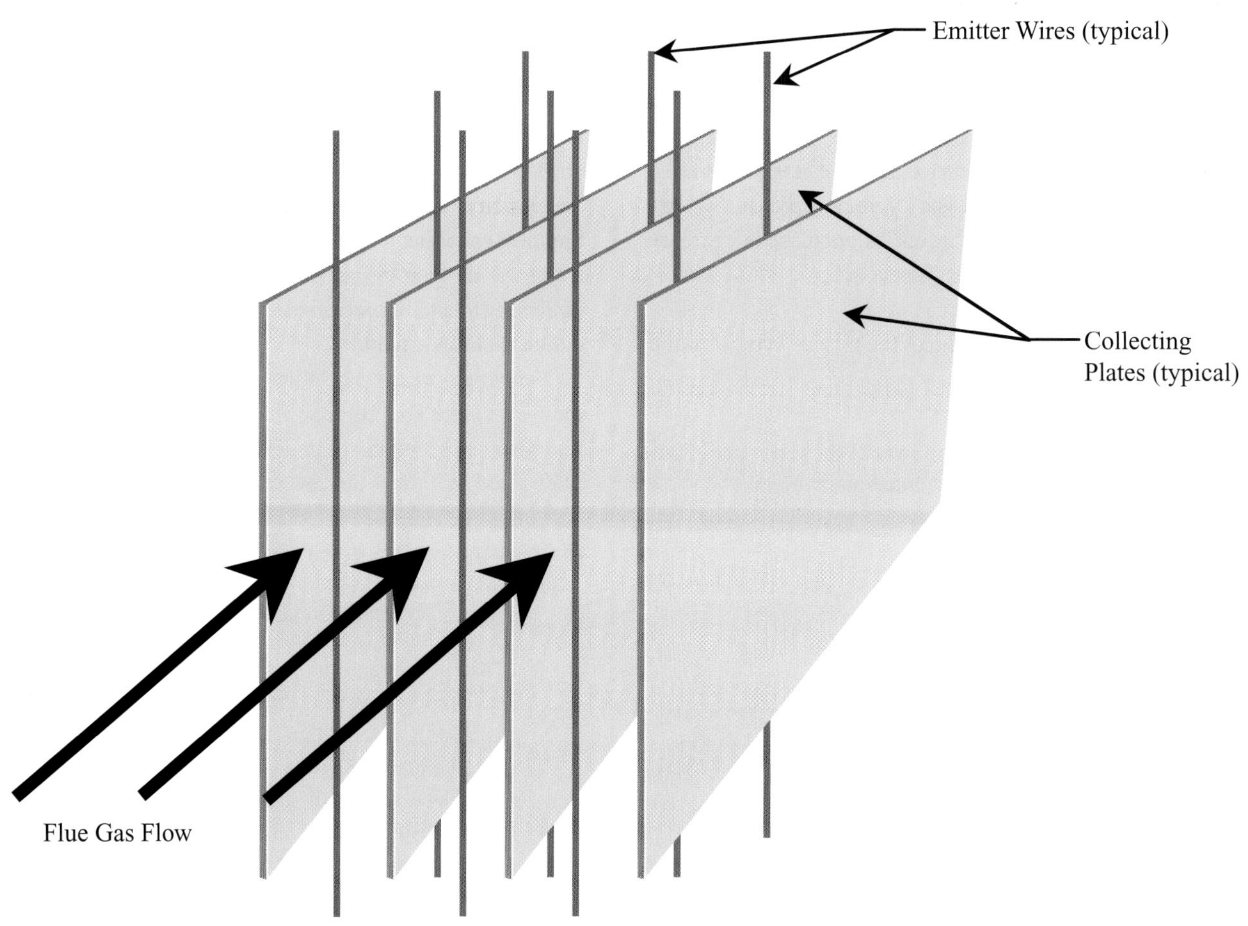

FIGURE 21.15 Dry electrostatic precipitator.

collector area, but can be upward of 900 ft²/1000 ACFM for very high particulate loading.

Because the temperature of the flue gas entering the ESP is normally greater than 650°F (340°C), corrosion is not usually a problem if the exterior is well-insulated and sealed from rainwater intrusion. Therefore, materials of construction would be carbon steel or stainless steel for the emitters, and carbon steel for the plates and casing.

Advantages of using an ESP include:

1. the desire for dry particle collection
2. relatively low energy usage
3. very low pressure drop
4. small particle removal

Disadvantages include:

1. relatively high capital cost
2. relatively large space requirement
3. field construction may be required for units with large flows
4. the large surface area encourages heat loss, leading to potential acid gas dew-point problems with certain wastes
5. multiple stages may have to be used to achieve the high removal efficiency.

21.3.5.3.2 Particulate Removal: Wet

The most common methods of wet particulate removal utilized in T.O. systems are venturi-type devices and wet electrostatic precipitators (WESPs) for very small particles. Reasonable removal can be achieved with the simple venturi-type device if particles are larger than 1 micron. For smaller particles, multi-stage venturi-type devices with subcooling can be effective at high pressure drop. To treat a flue gas containing a high percentage of very small particles, a WESP is the better choice. The pressure drop is also much less than with a venturi-type device.

21.3.5.3.2.1 Venturi-type scrubber The common feature of all venturi-type devices is a constricted passage or "throat" that increases gas velocity to achieve a desired pressure drop. Flue gas pressure drops range from 20 to 100 in. (51 to 250 cm) w.c. depending on various factors. If the flow rate varies significantly, the cross-sectional area of the throat must be adjusted to maintain the necessary velocity (pressure drop). Some configurations use a fixed throat but increase or decrease recycle flue gas flow to maintain the constant flow rate necessary to sustain the design pressure drop. Liquid is injected either in the throat or just upstream of the throat. Typical liquid injection rates are in the range of 10 gal/1000 ft^3 of gas. Because of the high scrubbing water flow rate, venturi scrubbers require a liquid recirculation system. The recirculation system will include equipment to control the blowdown rate to maintain the total (suspended and dissolved) solids content of the recycle water at about 5% by weight. Additional equipment can also be utilized to cool the recirculated water and control pH.

Typical particle removal efficiencies (when operated at saturation with a relatively high pressure drop) are > 90% for particles with aerodynamic diameters of ≥ 1 micron and ±50% for particles with aerodynamic diameters of 0.5 to 1.0 micron. Overall particle collection efficiency obviously depends on the particle size distribution and will range from 80% to above 99%. For a properly designed, sub-cooled system with efficient droplet separation, the overall particle collection efficiency should be ≥ 99%.

The basic principle of operation for the venturi-type particulate-scrubbing device is to provide small water droplets that will capture (wet and surround with water by inertial impaction) the particulate matter suspended in saturated flue gas. Given time, the small water-encapsulated particle droplets will then agglomerate (i.e., droplets contact other droplets and combine to form larger droplets). The larger droplets can be separated from the flue gas downstream of the venturi by a cyclonic separator, a mist eliminator, a settling chamber, or by a combination of two or all three separation methods. Assuming the flue gas is fully quenched to saturation, the overall removal efficiency depends how effectively the particles are wetted, how much droplet agglomeration time is provided, and how effectively the larger droplets are separated from the flue gas. As with other types of particulate removal devices, many of the design parameters have been developed from empirical data.

Inertial impaction of the particle into the droplet is the dominant mechanism for removal of larger particles with an aerodynamic diameter greater than 1 micron. Primarily, two things determine the effectiveness of capture: the relative velocity difference between the particle and the water droplet, and the droplet diameter. A greater relative velocity difference improves impaction effectiveness. Similarly, a smaller water droplet also improves impaction effectiveness.

The relative velocity difference between the particles and the liquid droplets increases with higher flue gas pressure drop (i.e., increased energy consumption), which increases the velocity of the flue gas and the particles suspended in it. Smaller scrubbing water droplets can be produced by mechanically atomizing the water or by using an atomizing medium. The shearing effect of the high-velocity flue gas also atomizes the water more finely.

If the flue gas is not fully saturated with water vapor, scrubbing fluid will be vaporized until saturation is achieved and removal performance will be poorer. Vaporization of the scrubbing fluid causes problems such as reducing the amount of scrubbing fluid available for particle removal, forming new particles from previously captured particulate matter as the droplet evaporates, and reducing the diffusive capture because of a net flow of gas away from the evaporating particle.

The vertical, downflow adjustable plug venturi described previously utilizes mostly pressure drop to shear/atomize the water and to provide the relative velocity difference. Other types use a combination of pressure drop and water atomization to achieve the velocity difference and smaller droplet size. Figure 21.16 shows a horizontal venturi scrubber with atomized water injection upstream of the throat.

Although particles larger than 1 micron are more easily removed, smaller, sub-micron size, particles are much more difficult to capture. The capture effectiveness can be enhanced by (1) sub-cooling saturated flue gas to below the saturation temperature, and (2) using colder water for scrubbing. The purpose is to take advantage of thermophoresis and diffusiophoresis effects to produce a directional preference in the Brownian motion *toward* the target droplet by these submicron particles.

Basically, thermophoresis is the migration of a particle away from a higher temperature zone and toward a lower temperature zone.

Diffusiophoresis is a more complicated phenomenon that occurs when a mixture of particles of varying weight exist in a gas stream and a concentration gradient within the gas stream occurs for the heavier particles. Diffusion of the heavier particles from the higher concentration zone to the lower concentration zone occurs in accordance with Ficke's law. The net motion of the lighter particles is also altered toward the low concentration zone due to the momentum imparted during collisions with the heavier particles traveling in that direction. By injecting colder water and sub-cooling the flue gas below saturation temperature, a temperature gradient and water vapor concentration gradient are created,

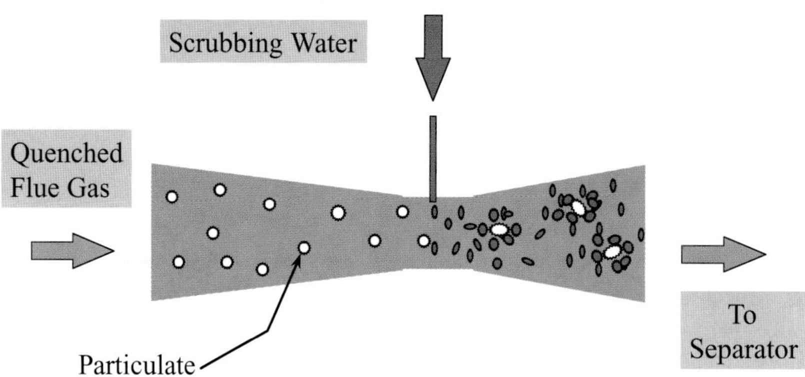

FIGURE 21.16 Horizontal venturi scrubber.

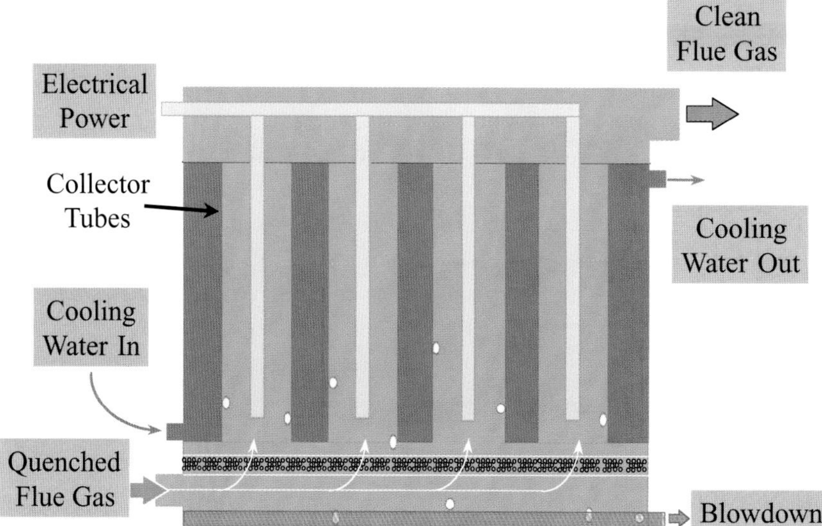

FIGURE 21.17 Wet electrostatic precipitator.

resulting in a net increase in particle motion toward the target water droplets.[1,2]

Advantages of the venturi scrubber include:

1. reasonably good particulate removal performance
2. relatively simple design
3. usually lower capital cost
4. relatively low plot space requirement

Disadvantages are (1) high energy consumption (i.e., high flue gas pressure drop), (2) high removal efficiency for small particles requires additional cost of sub-cooling, (3) continuous blowdown to maintain a low total solids content would be difficult if water was scarce, and (4) the removed solids are in a large volume of water which has to be treated.

21.3.5.3.2.2 Wet electrostatic precipitator A wet electrostatic precipitator (WESP) functions very similar to the previously described dry ESP. It removes particles from a flue gas stream by means of electrostatic attraction. As before, particles must first be charged, and, as in the dry ESP, vertically oriented emitter wires are used to generate the corona. Once charged, the particles are drawn to the oppositely charged, grounded collecting electrodes, which for the WESP are tubes through which the emitter wires hang. The particulate collects on the inside surface of the tubes. The particles are then removed from the collection surface while minimizing re-entrainment. The primary process difference is that the flue gas must be saturated when it enters the emitter/collector section.

Flue gas coming into the WESP first enters a chamber under the vertically oriented collector tube section (see Figure 21.17). The saturated flue gas is uniformly distributed before reaching the tubes. The collector tube section is similar in construction to a shell-and-tube heat exchanger in that it has an inlet and outlet tubesheet and a sealed casing (shell) around the tubes. The emitter wires are positioned at the centerline of each tube. Ambient-temperature water is circulated through the shell side of the collector tube section, ensuring that the collector tube temperature is less than the temperature of the saturated flue gas. This causes water vapor in the flue gas to condense on the tubes and flow downward into the chamber under the collector tube section. As the wetted particles are charged and move to the tube wall, they are washed down the tube by the condensing water. Occasionally, the power is shut off to part of the collector tube section and water is sprayed downward through that section of tubes to further clean them. The distribution section in the inlet chamber can also be designed to absorb acid gases similar to a packed column.

The design criteria for emitting and collecting electrode areas in a wet ESP are similar to those for the dry ESP. However, the potential for sparking/arcing is greater in the WESP, so proper spacing between emitters and tubes, or anything else, is very important.

Because the WESP is wet, materials of construction would be stainless steel or Hastelloy (or equivalent) for emitters and tubes, and stainless steel or FRP for the housing and water collection section.

Advantages of the WESP are:

1. high removal efficiency for sub-micron particles
2. low gas-side pressure drop, usually less that 6 in. (15 cm) w.c. (normally about 4 in. or 10 cm w.c.) if the distribution section is designed to absorb acid gases as well as distribute flow
3. can absorb acid gases
4. "cool" service
5. less problems with re-entrainment because the wet particles stick to the tube wall

Disadvantages include (1) greater capital cost than most other wet removal devices, (2) larger plot space required, (3) more complicated operation than with other equipment, (4) multiple stages may have to be used to achieve extremely low particulate emission levels if the particulate loading is high, and (5) more maintenance may be required due to corrosion and complexity.

21.3.5.3.3 Acid Gas Removal: Dry

Acid gases can be removed from flue gas by reaction with or adsorption by dry alkaline materials such as limestone/lime. For most waste incineration systems utilized by waste by-product generators, dry removal is not cost-effective and does not achieve the removal efficiency needed for the amount of combustion products generated. Dry removal is used primarily on large power boiler applications that have a high flue gas volume. Some of the drawbacks are:

1. The reactant/adsorbent must be disposed of after use, so by removing the acid gas, a solid waste is created.
2. The removal efficiency is not high.
3. The capital cost of equipment needed for dry inject is quite high.

For these reasons, no detailed description of dry systems will be covered.

21.3.5.3.4 Acid Gas Removal: Wet

The most common methods of wet acid gas removal used in T.O. systems are packed columns and the previously described venturi-type devices. Packed columns contain packing material that distributes the water over a large surface area for contact with the flue gas. The venturi-type devices utilize many small water droplets to provide the large amount of liquid surface area for contact by flue gas. Each has its advantages and disadvantages.

21.3.5.3.4.1 Packed column A packed column is the device of choice most often used to recover or remove acid gases from a flue gas stream. The device consists of a vertical, usually cylindrical, vessel containing a section filled with packing material supported by internal hardware. Recirculated water is pumped to the top of the packed section and flows downward through the packing and collects in the base of the vessel. Flue gas, quenched to its saturation temperature, enters the vessel under the packed section and flows upward through the packing and out to the atmosphere or to another treatment section. Figure 21.18 shows a simple packed column.

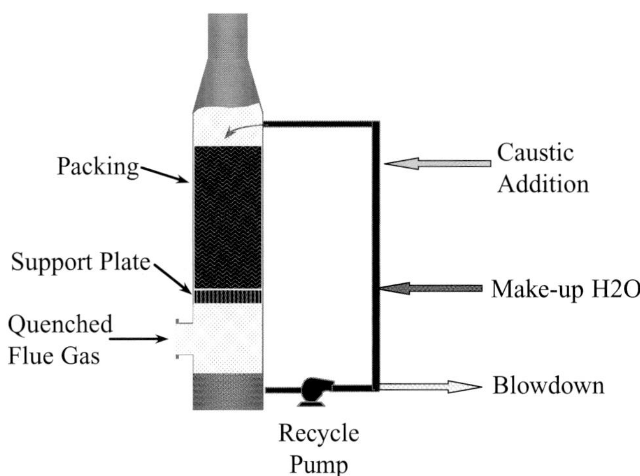

FIGURE 21.18 Simple packed column.

The purpose of the packed column is to transfer compounds from the flue gas to an absorbing liquid. For combustion systems, the compounds are acid gases such as HCl or SO_2. The absorbing liquid is typically water or a weak caustic solution. As the liquid flows downward through the randomly placed packing, it is distributed over the large amount of packing surface. The counter-current flow of flue gas traveling in the open spaces between packing comes into contact with a large amount of liquid for a relatively long period of time. As the flue gas flows through the absorbing liquid, soluble gases are dissolved into it. Just as in wet particulate removal cases, the flue gas must be saturated prior to coming into contact with the absorbing liquid to eliminate evaporation during contact.

The rate of acid gas absorption at any point within the packed section is limited by the mass transfer rate across the gas-phase boundary layer. Thus, a large mass transfer coefficient, a high concentration of the pollutant in the flue gas, a low concentration of the dissolved pollutant in the absorbing liquid (i.e., low pollutant vapor pressure over the absorbing liquid), and a large amount of interfacing contact area all increase the rate of absorption. Counter-current flow of flue gas to the absorbing liquid, which puts the cleanest absorbing liquid in contact with the cleanest flue gas, is the primary reason absorption can be so effective in a packed column. The packing type, the flue gas velocity, the type of substance (i.e., acid gas) to be dissolved into the absorbing liquid, and the type of absorbing liquid determine the mass transfer coefficient. The amount of interfacial area is determined by the type of packing in the column and by how well the packing is "wetted." Operational testing has been used to develop virtually all of the packing performance parameters.

The upstream incineration process determines the concentration of acid gas in the flue gas entering the column, the acceptable outlet gas concentrations are set by environmental regulations, and the liquid purge stream concentration is either specified by the customer or determined by design to achieve adequate acid gas removal. The diameter of a packed column is typically designed to give a superficial gas velocity of 7 to 10 ft/s (2 to 3 m/s). This range of velocities is high enough to create enough pressure drop (about 0.3 to 0.5 in., or 8 to 13 mm w.c. per foot of bed depth) to prevent poor flue gas flow distribution (channeling), yet low enough to avoid flooding. A recirculation rate of 7 to 10 gpm of liquid per square foot of bed cross-sectional area is usually required to ensure adequate "wetting" of the packing. The remaining variables, which are determined by the designer, are column height, packing type, and the type and temperature of absorbing liquid.

Packed columns using water to recover acid are commonly referred to as "absorbers," and those operated with a caustic solution to remove acid are commonly referred to as "scrubbers."

Absorbers yield an acid solution and are typically used in applications in which the acid solution can be used in a process or when the waste water treatment plant can make use of the acid solution to neutralize a caustic waste. Packed bed absorbers can be designed to produce acid purge streams of up to 22% HCl by weight (the azeotropic maximum concentration), assuming the HCl vapor concentration in the flue gas leaving the absorber is sufficiently high. The concentration of HCl in the flue gas determines the maximum strength of the acid blowdown stream. A typical 10-ft (3-m) deep absorber section could remove about 99% of the HCl present in the flue gas while producing a 2% HCl blowdown stream.

Scrubbers not only remove acid gas from the flue gas, but also neutralize the dissolved acid. By adding NaOH to maintain the pH between 6.8 and 8.0 in the absorbing liquid, the vapor pressure of the acid gas in the outlet flue gas is greatly reduced, thus increasing the absorption rate. The advantages of the scrubber over the absorber are a less corrosive blowdown stream, and either a greater acid gas removal for the same blowdown rate or a much lower blowdown rate for the same percentage of acid gas removal. A typical 10-ft (3-m)-deep scrubber could remove 99.9% of the HCl present in the flue gas and produce a blowdown stream containing up to 5% by weight total solids, most of it being NaCl.

The neutralized effluent from a scrubber also contains some sodium hypochlorite, which results from NaOH reacting with the free chlorine generated during the combustion process. The sodium hypochlorite is a strong oxidant and may require treatment to meet effluent requirements. A reducing agent such as SO_2 or a solution of sodium bisulfate can be added to the recirculating stream in the scrubber to reduce the sodium hypochlorite:sodium chloride ratio. It is important to remember that CO_2 in the flue gas also reacts with NaOH. The reaction rate, however, is relatively low until the pH of the recirculating water becomes greater than 8.0. Above that concentration, the CO_2/NaOH reaction rate increases significantly, greatly increasing the consumption of NaOH. Proper pH control of the recycle water is an economic necessity.

In some applications, both acid production and a very high level of acid gas removal are required. This can be accomplished with a two-stage system. Figure 21.19 illustrates this application. The first stage is an absorber that removes 80 to 95% of the acid gas and produces an acid blowdown stream. The second stage is a scrubber, which achieves the desired level of acid gas removal and, with a significant portion of the acid gas removed in the absorber section, produces a relatively low volume blowdown stream. Such a system is

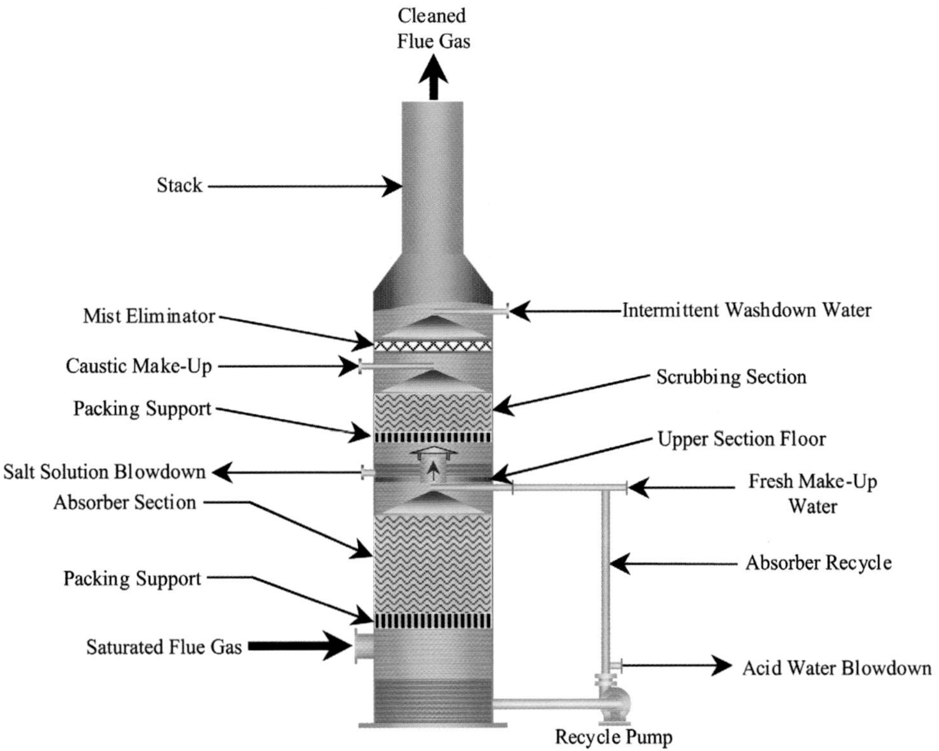

FIGURE 21.19 Two-stage acid gas removal system.

advantageous when (1) there is a need to achieve very high levels of acid gas removal, and/or (2) there is a use for the acidic blowdown stream or, because of the amount of acid gas to neutralize, it is more economical to neutralize it externally with a less expensive reagent such as lime (CaO) or slaked lime ($Ca(OH)_2$).

In some applications, the flue gas quench can be incorporated in the packed column. For this condition, the lower section of the vessel and the packing material must be able to withstand both high temperatures and acidic conditions. A high-temperature and acid gas-resistant lining must be used in the lower portion of the vessel, and the packing material and internal support hardware must be made of ceramic or graphite. Figure 21.20 represents this application.

While the packed column is primarily designed for acid gas removal, it will also remove some particulate matter. However, the basic mechanism of entrained particle removal is inertial separation from the gas stream, followed by entrapment in the absorbing liquid. Because the gas velocities through the packed section are far too low for effective inertial separation and entrainment of particles with an aerodynamic diameter of less than 10 microns, a packed column is not an effective particulate removal device.

Common materials of construction for a packed column assembly are FRP for the vessel, and ceramic, FRP, plastics (such as polypropylene, PVC, CPVC, Teflon, Kynar, etc.), or some combination of these for the packing, internal support hardware, and water circulation equipment. If the recirculating liquid is sub-cooled to enhance recovery/removal, the heat exchanger would have to be built of similar corrosion-resistant materials.

The advantage of using a packed column include:

1. high removal efficiency capability
2. low pressure drop/energy cost (< 10 in. or 25 cm w.c. drop for a 10-ft- or 3-m-deep packed section)
3. no moving parts in the column itself

Disadvantages are (1) poor small particulate removal capability, (2) dependent on recirculation pumps, caustic feed pumps, etc. to operate, and (3) momentary loss of quench can cause "meltdown."

21.3.5.3.4.2 Venturi scrubber The venturi scrubber described earlier as a particulate removal device can also function as an acid gas removal device. One of the important factors for effective mass transfer is intimate contact between the flue gas and the liquid used to absorb the acid gases; the venturi scrubber can provide relatively good gas-to-liquid contact. However,

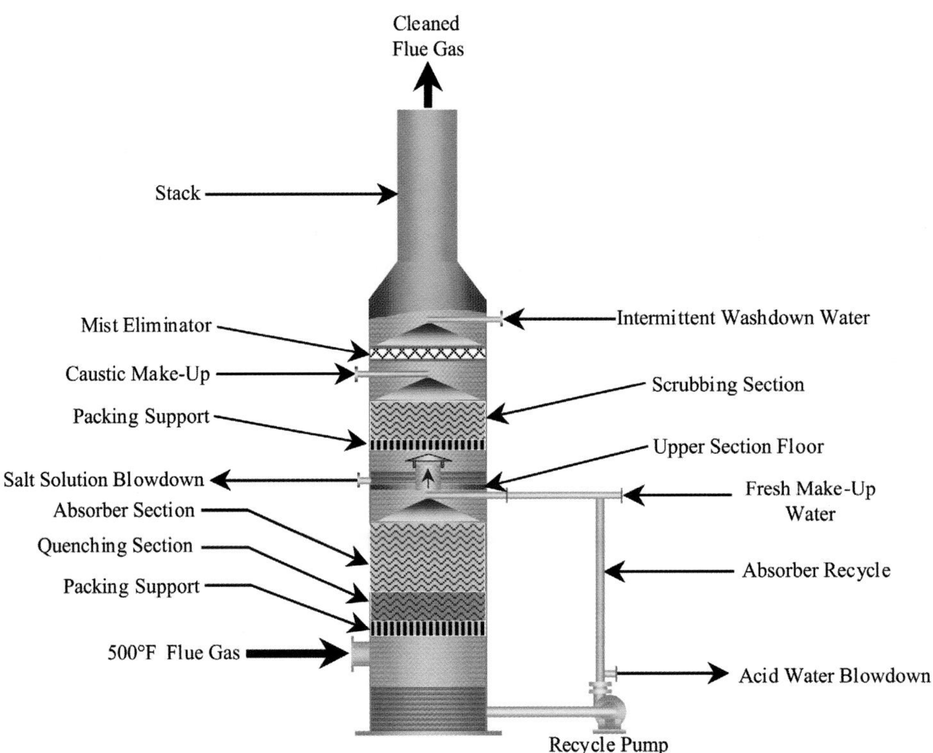

FIGURE 21.20 Combination quench/two-stage acid removal system.

because of other factors such as acid gas solubility and solubility rates (which are affected by gas film and liquid film resistance), and the difference in concentration of the pollutant in the flue gas compared to the concentration of the dissolved pollutant in the absorbing liquid (i.e., low pollutant vapor pressure over the absorbing liquid), the venturi scrubber, although good, is not the best overall choice for acid gas removal.

HCl is a good example of a highly soluble acid gas. It easily absorbs into the water, creating hydrochloric acid. If necessary, the HCl can then be reacted with sodium hydroxide (NaOH) added to the recycle water to form NaCl.

SO_2 is an example of an acid gas that is only moderately soluble in water. Therefore, it must be in contact with the recycle water for a longer period of time for high removal efficiency. Adding NaOH to the recycle water will improve the solubility by decreasing liquid film resistance, but a longer period of contact is still needed to achieve high removal efficiency. The concentration of NaOH must be as low as possible to avoid excessive reaction of NaOH with CO_2.

Because of design, the venturi utilizes a relatively short gas/liquid contact time. The short period of time reduces the amount of absorption, especially for moderately soluble acid gases. Also, because the flue gas and the absorbing fluid have to travel in the same direction (co-current flow), at the venturi outlet the cleanest gas is exposed to recycle water that has the highest concentration of acid gas, which further reduces the effectiveness of the scrubber. Despite these drawbacks, a properly designed venturi system, including the separation equipment, can still effectively remove much more than 90% of the acid gas from a flue gas stream.

Advantages of the venturi as an acid gas removal device include:

1. relatively effective for highly soluble acid gases
2. low capital cost
3. low plot space requirement

Disadvantages of the venturi as an acid gas removal device consist of (1) only moderately effective for lower solubility acid gases, (2) high operating cost (pressure drop), (3) short gas-to-liquid contact period, and (4) co-current flow, which minimizes the difference in concentration of the pollutant in the flue gas compared to the concentration of the dissolved pollutant in the absorbing liquid.

21.3.5.4 NOx Control Methods

NOx can be controlled during or after the combustion process. The most effective method will be determined by the NOx emission allowed and by the capital and operational costs of various methods.

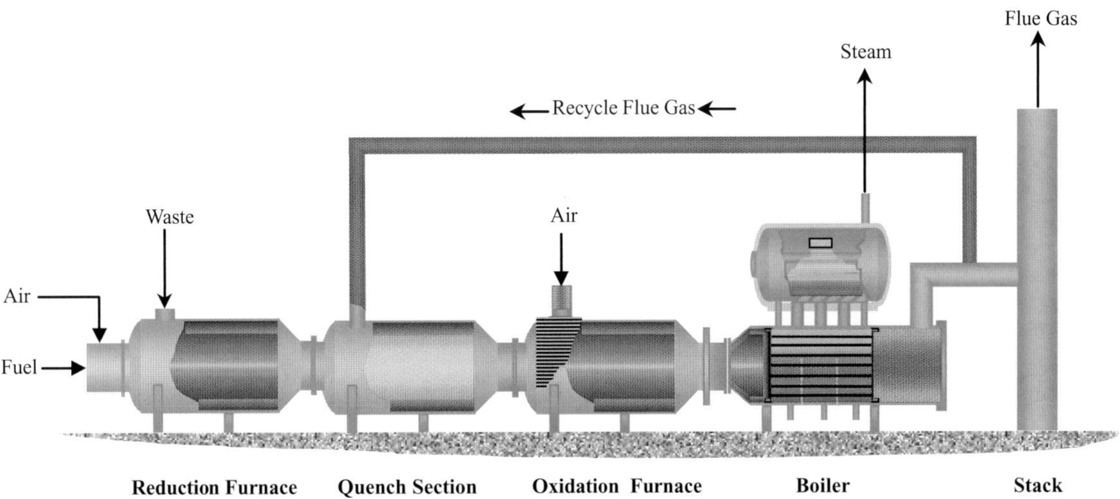

FIGURE 21.21 Three-stage NOx reduction process.

Two primary mechanisms for the formation of NOx in combustion systems exist. They are (1) high-temperature dissociation of molecular nitrogen (N_2) and molecular oxygen (O_2) and subsequent reaction of the independent radicals to NOx (thermal NOx), and (2) any oxidation reaction of bound nitrogen in organic compounds (i.e., acrylonitrile, ammonia, nitrobenzene, hydrogen cyanide, amines, etc.) to NOx. Thermal NOx occurs just downstream of burners in the flame zone, where peak temperatures occur. NOx from organic-bound nitrogen is formed at any location during the oxidation reaction.

Limiting, or preventing, the high-temperature formation from occurring during combustion reduces NOx production by the first mechanism. This is basically accomplished by reducing peak temperatures with different burner design variations. Thermal NOx control methods using burners are discussed in previous chapters and will not be covered here. It is important to understand that many of the burner design methods can also be used to minimize thermal NOx formed in the burners of T.O. systems.

The second mechanism is much more difficult to prevent because combustion of organics with bound nitrogen in the presence of excess oxygen at any temperature will result in NOx production. In addition, the NOx produced will usually be much greater than the burner-generated thermal NOx. Also, some waste streams contain NOx that is not destroyed in an excess oxygen environment. For these conditions, reducing burner-generated NOx is virtually inconsequential. Either the combustion process has to be changed to reduce the NOx exiting the combustion section, or the flue gas must be treated after the NOx is formed (post-combustion).

If the waste streams do not contain large quantities of noncombustible materials such as air, water, or other inerts, a modified combustion process, utilizing sub-stoichiometric oxidizing conditions, can be used very effectively. Although many different post-combustion treatment methods are available — including some wet techniques that are very efficient — the most commonly used methods chosen for typical thermal oxidation systems are selective noncatalytic reduction (SNCR) and selective catalytic reduction (SCR).

21.3.5.4.1 Combustion Process Modification

For wastes and/or fuels with a high concentration of bound nitrogen, or for wastes that contain NOx, the single most effective practical process modification is a form of staged-air combustion. The most common implementation of staged-air combustion is accomplished in a three-stage combustion process (Reed, 1981),[3] as shown in Figure 21.21. As noted above, this method may not be the most cost-effective for NOx reduction if the waste stream(s) containing the NOx-producing compounds are mostly air, water, or inerts because of the large amount of auxiliary fuel it would take to operate such a unit. If steam is needed in the plant, it is possible that the high fuel use could be justified based on the amount of steam generated. NOx reduction by this method is achieved in the first stage by combusting the nitrogen-bound compound or NOx itself in a high-temperature atmosphere that is deficient in oxygen (i.e., sub-stoichiometric or reducing). This results in dissociation of the organic compound and reaction of the released nitrogen atoms to N_2. The flue gas is cooled in the second stage, resulting in a lower peak temperature in the third stage. The lowered peak temperature minimizes reformation/formation of thermal NOx.

Fuel, waste(s), and less-than-stoichiometric combustion air are introduced into the first stage (reduction furnace) to produce a high-temperature (2000 to 2800°F, 1100 to 1500°C) reducing atmosphere. The excess combustible material in the high-temperature reducing zone provides the driving force for the reduction of the oxides of nitrogen and conversion of bound nitrogen to N_2 instead of NOx. Some inert (low or no oxygen content) such as recycle flue gas, steam, or water can also be introduced into the reducing zone to allow more consistent control of the operating temperature so as not to exceed the limits of the refractory. The optimum level of oxygen deficiency in this stage depends on the waste composition. Although the oxygen is usually supplied by combustion air, the oxygen in NOx and in waste gas streams will also be utilized. The residence time in this stage is usually in the range of 0.5 to 1.0 s. The primary components of the gas leaving this stage are CO, H_2, CO_2, N_2, and H_2O. Most of the fuel-bound nitrogen is converted to N_2, with the remaining present primarily as very low levels of HCN, NH_3, and NO.

The hot flue gas then enters the second process stage (the quench chamber) by passing through a venturi mixing section. An inert cooling medium, as described above, is injected through multiple openings in the venturi throat to quickly mix with the flue gas and decrease the temperature to 1300 to 1750°F (700 to 950°C). The temperature must be high enough so that rapid ignition of the combustibles occurs by simply adding air, but low enough to limit the temperature achieved in the final oxidation stage to less than 2000°F (1100°C). Although cooling is rapid, sufficient time must be allowed in this section to ensure that the bulk gas temperature is uniform.

As the cooled flue gas exits the quench section, air in excess of stoichiometric is introduced at the entrance into the third stage (the oxidation zone), again using a venturi mixing section. In this final stage, the carbon monoxide, hydrogen, and any remaining hydrocarbons produced in the first stage are oxidized. The flue gas cooling step performed prior to introduction of the oxidation air controls the peak oxidizing operating temperature, thereby limiting formation of thermal NOx. These process steps often result in levels of NOx less than 150 ppmv (parts per million by volume) at excess oxygen conditions of less than 1 to 2% (dry). The residence time in this stage is usually in the range of 0.5 to 1.0 s.

The flue gas can then be treated by any of the previously discussed methods or exhausted directly to the atmosphere for dispersion.

A wide variety of bound-nitrogen-containing gas and liquid wastes can be incinerated using this treatment method, especially if the waste contains significant heating value. Nitrogen in the chemical waste can be in the organic or inorganic form. Examples of the organic form are HCN amines, nitriles, and nitroaromatics; examples of the inorganic form are NH_3 and NOx. A small process stream containing some quantity of NOx can be treated to produce a cleaner stream that contains comparably less NOx than the original waste. This method is not suitable for NOx reduction in flue gas from large combustion processes, such as utilities. It also is not appropriate for waste streams containing large quantities of air, water, or inerts. Either case would require an excessive amount of auxiliary fuel to operate the system.

Three-stage systems using the process described above have been operating in a variety of industrial applications for more than 30 years. Destruction efficiencies of incoming components of more than 99.99% are achieved. Carbon monoxide is usually in the 50 to 100 ppmv (dry, corrected to 3% O_2) range. NOx in the flue gases can vary from 50 to 200 ppmv, dry, corrected to 3% O_2, depending on the composition of the waste stream being treated.

A two-stage process modification can also be used if the NOx level does not have to be as low. The same first-stage reduction furnace is still used. The difference starts at the reduction furnace outlet. Instead of adding an inert cooling medium and then adding just enough combustion air to oxidize combustibles and maintain 1.5% excess O_2, a large amount of air is introduced to cause the oxidation reactions to occur *and* to limit the oxidation section outlet temperature to 1800°F (1000°C) or less. This process is shown in Figure 21.22.

The high excess O_2 (up to 10%) causes more equilibrium NOx to form in the oxidation section than would form in the lower O_2 three-stage system. Also, the method of control is slightly different in that the two-stage system requires the use of a sometimes maintenance-intensive combustibles analyzer to measure and control the combustibles level in the reduction furnace. For the three-stage system, the combustibles level is controlled by measuring the differential temperature between the quench section and the outlet of the oxidizing section, thus avoiding the instrument maintenance. Another minor drawback is that recycle flue gas cannot be used to control temperature in the reduction furnace, which reduces heat recovery efficiency in the event that heat recovery is used. Despite of these less positive items, the two-stage system is still far better than a single-stage, oxidizing-only combustion process for wastes containing bound nitrogen.

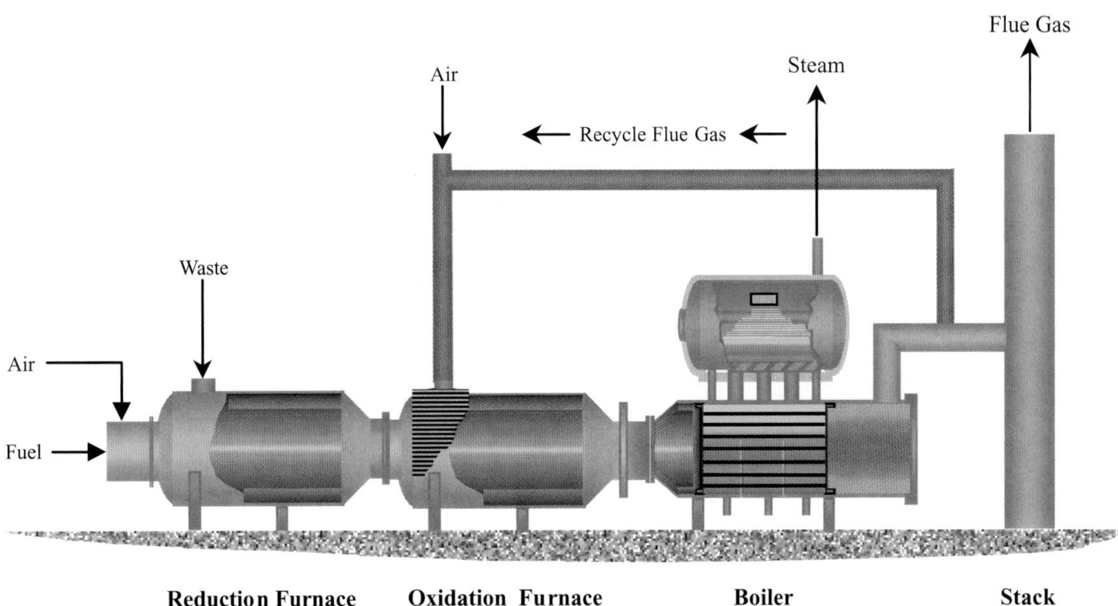

FIGURE 21.22 Two-stage NOx reduction process.

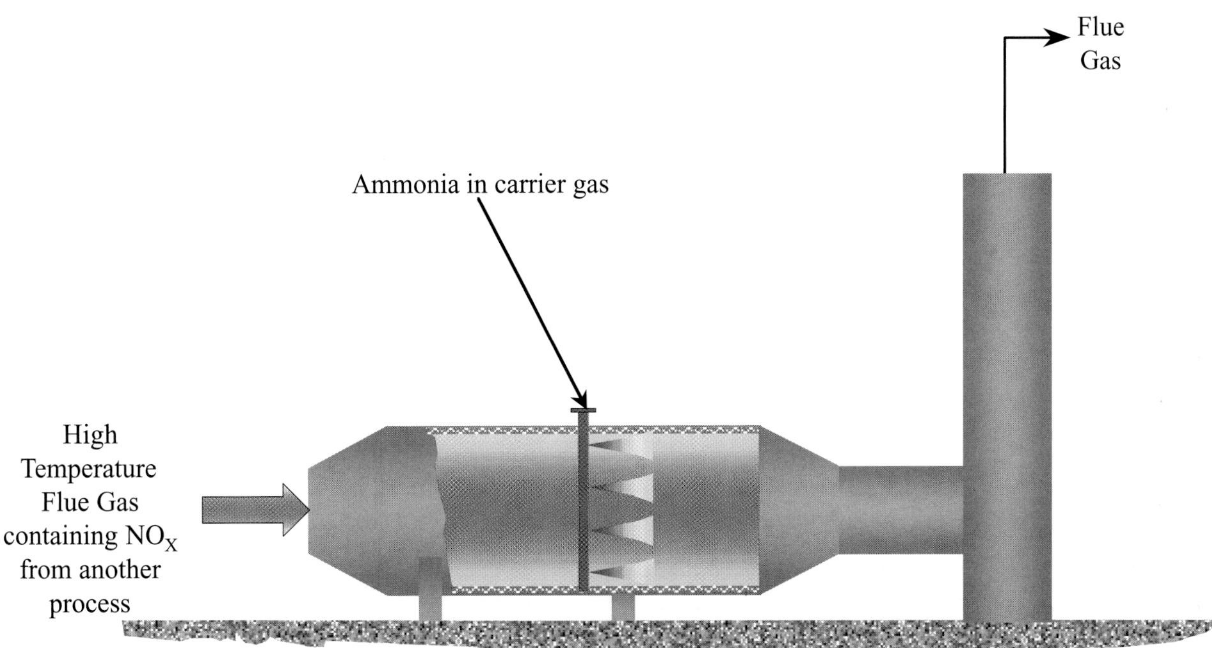

FIGURE 21.23 Selective noncatalytic reduction system.

21.3.5.4.2 Selective Noncatalytic Reduction

The selective noncatalytic reduction (SNCR) process for NOx reduction is one in which a compound, added at the end of the combustion process zone (post combustion), selectively reacts with NOx without the aid of a catalyst. The most commonly used process utilizes ammonia (NH_3) as the additive. Figure 21.23 shows such a system. It must be mixed uniformly into the hot flue gas in the presence of excess O_2. The overall net reaction, which occurs by way of a complex free radical chain reaction, is:

$$NO + NH_3 + \frac{1}{4}O_2 \rightarrow N_2 + \frac{3}{2}H_2O$$

Concurrently with the NO reduction reaction, NH_3 is oxidized to form NO following the overall reaction:

$$NH_3 + \frac{5}{4}O_2 \rightarrow NO + \frac{3}{2}H_2O$$

Although the reduction and oxidation reactions are in competition and are both very temperature sensitive, there is a relatively narrow but maintainable range of NH_3-to-NO and temperature in which the balance between reduction and oxidation is favorable and both reactions proceed toward completion. The temperature range for best reaction is 1600°F (900°C) to 1900°F (1000°C), which is the basic operating temperature range of a 99.99% DRE system. The amount of ammonia used is about 0.5 lb (0.2 kg) per pound of NOx, as NO_2, so the overall NH_3 volume is quite low relative to the total flue gas volume. To achieve the mixing required, the NH_3 must be combined with a carrier gas such as steam or air for injection. The total volume of the carrier gas should be about 1.5% of the total flue gas. For optimum results, it should be distributed uniformly by use of an injection grid to be rapidly mixed. Because most (90 to 95%) of the NOx formed is NO, effective reaction with NH_3 can reduce overall NOx significantly.

For waste combustion systems for which process modification, by burner modification or by staged combustion, is not a viable alternative to reduce NOx, NH_3 injection can be reasonably effective. The process is relatively predictable, although some operational trial runs are usually needed to establish the optimum operating conditions. High operational temperature is necessary, but that fits in with the high DRE required by most waste combustion systems. SNCR can be used in systems containing particulates with little or no process degradation. The NH_3 injection process can also be utilized at the end of a modified combustion process, such as a staged combustion system, to further reduce NOx.

NOx reduction of 50 to 70% is possible, but if that is still greater than the allowable emission, NH_3 injection is not the complete solution to the situation. Even with optimized operation, some small amount of NH_3 slip (nonreacted NH_3 in the flue gas) will occur. Also, 1 to 2% of the reduced NO will produce N_2O. Urea is also used as a reactant. It works in a similar manner with similar reduction capabilities.

21.3.5.4.3 Selective Catalytic Reduction

The selective catalytic reduction (SCR) process is very similar to SNCR except the use of a catalyst increases the reduction of NOx. The increased reduction is primarily the result of increased reaction between NH_3 and NOx (both NO and NO_2). In addition, the increased reactions occur at much lower temperatures (400 to 850°F or 200 to 450°C) than the noncatalyzed SNCR reactions. The specific reactions cited for SCR include:

$$4NO + 4NH_3 + O_2 \rightarrow 4N_2 + 6H_2O$$

$$6NO + 4NH_3 \rightarrow 5N_2 + 6H_2O$$

$$2NO_2 + 4NH_3 + O_2 \rightarrow 3N_2 + 6H_2O$$

$$6NO_2 + 8NH_3 \rightarrow 7N_2 + 12H_2O$$

$$NO + NO_2 + 2NH_3 \rightarrow 2N_2 + 3H_2O$$

The NH_3 must still be uniformly mixed with the flue gas upstream of the catalyst section using a multi-point grid for best results. Catalyst materials are primarily vanadium and wolfram derivatives, but others are being used and developed such as alumina/titania, rhodium, etc. See Figure 21.24 for a common configuration.

SCR can achieve as high as 95% reduction of NOx because it enables reaction with NO_2 as well as NO. The high reduction is achieved at lower temperature operation, such as might be found downstream of a heat recovery section. A heat recovery or cooling section protects the catalyst from over-temperature damage. The NH_3 slip is normally less than in the SNCR method.

The catalyst surface can be fouled by particulate or possibly poisoned by some materials, but the catalyst material is relatively resistant to those as well as erosion damage. Also, in the unlikely event that a high temperature upset condition occurs, the catalyst could be damaged. Catalysts are usually guaranteed for between two and four years, at which time they must be replaced to maintain maximum NOx reduction. However, if multiple layers/sections of catalyst are used, instead of replacing all the catalyst, replacement of sections can often be alternated and still maintain adequate removal efficiency.

21.4 BLOWERS

Blowers (also referred to as fans) are used to overcome the pressure drop required to move air/flue gas through a T.O. system consisting of multiple flue gas treating processes. Although blowers are also used to move waste gases, this section will discuss only combustion air and flue gas blowers. Location of the blower is a major factor in blower selection. If a blower is located at the front end of a system, "pushing" flow through, the process is described as forced draft and the blower will be handling clean air. The forced-draft blower will draw in air at ambient pressure and raise it to the pressure needed to push combustion products through the system.

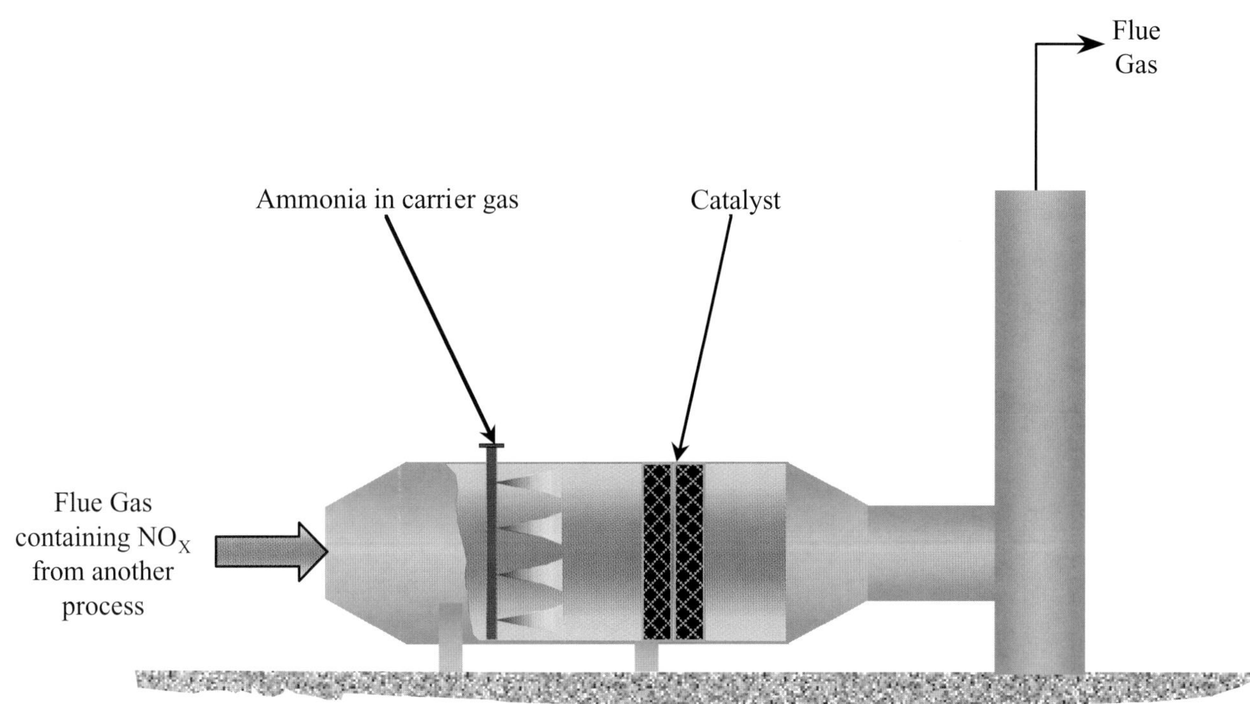

FIGURE 21.24 Common catalyst configuration.

If a blower is located at the back end of a system, "pulling" flow through, the process is described as induced draft. At that point, the gas handled will be combustion products, cooled sufficiently for the fan material of construction (by heat recovery or quenching). For this case, the blower draws in conditioned flue gas at a pressure below ambient (vacuum) and raises it to just greater than ambient to exhaust it to the atmosphere.

Occasionally, a blower will be located at the front end, "pushing" through part of a system, while a blower at the outlet "pulls" through the rest of the components. In that case, the process is called balanced draft. Obviously, the same criteria for blower selection applies as noted above.

When the blower location is determined, the final selection is based on inlet/outlet composition and volumetric flow rate, inlet/outlet pressure requirements, inlet/outlet temperature, and flow, pressure, and temperature ranges during operation.

Two general classes of blowers exist for moving gas volumes: axial and centrifugal. Axial blowers use propellers to move the gas parallel to the axis of rotation of the blower. Although axial blowers are inexpensive to buy and install and occupy little space, they are very limited in pressure capability (less than 20 in. or 51 cm w.c.) and very noisy at higher pressures. For these reasons, axial-type blowers are seldom used for multi-component T.O. systems and are not examined further in this chapter.

Centrifugal blowers draw gas (air) into the center of the blower wheel at the axis of rotation where it enters the spaces between the paddle-like blades that impart radial-outward, as well as angular, velocity to the gas, generating pressure. Centrifugal blowers can generate well over 100 in. (250 cm) w.c. pressure, more than sufficient for almost any multi-component system. As the gas is centrifuged to the periphery of the housing, more gas is drawn into the blade space.

There are three distinct types of centrifugal blowers. The difference is basically the blade configuration used. The three basic blade types are straight (or radial), forward curved, and backward curved. Figure 21.25a illustrates the types of fan wheel designs available. Each has its advantages and disadvantages relative to the other two. Table 21.2 summarizes those relative characteristics.

Proper selection of a centrifugal blower is not accomplished by any single straightforward formula. Experience, usage, and careful evaluation of each application are necessary to ensure proper fan selection. It is always best to work with blower manufacturers to get the most cost-effective recommendation.

Another aspect of proper blower selection is utilizing the best-suited method of flow control for the blower to match the combustion system. Three primary methods are used: discharge damper control, inlet vane control, and fan speed

Fan Wheel Designs

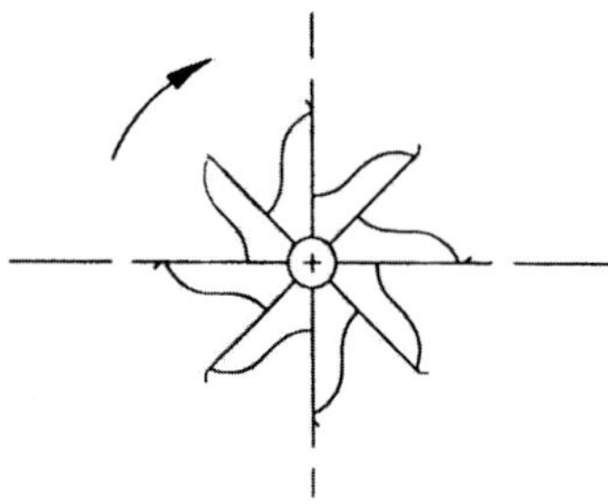

Paddlewheel
- Open design/no shroud
- 60-65% static efficiency
- Inexpensive design
- Good for high temperature or highly erosive applications
- Medium to high pressure

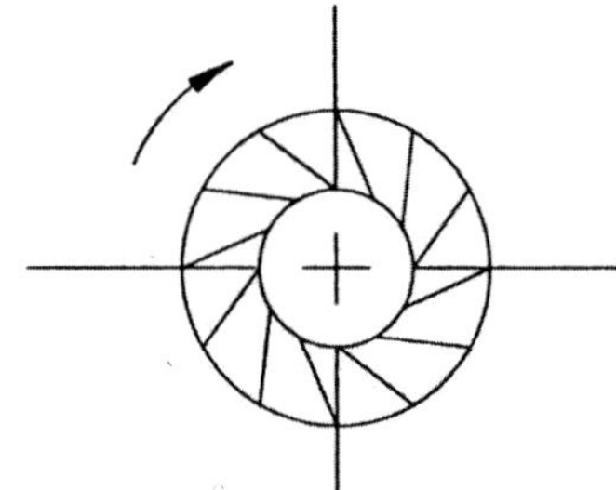

Backward Inclined
- Static efficiency to 80%
- Low to medium tip speed capabilities

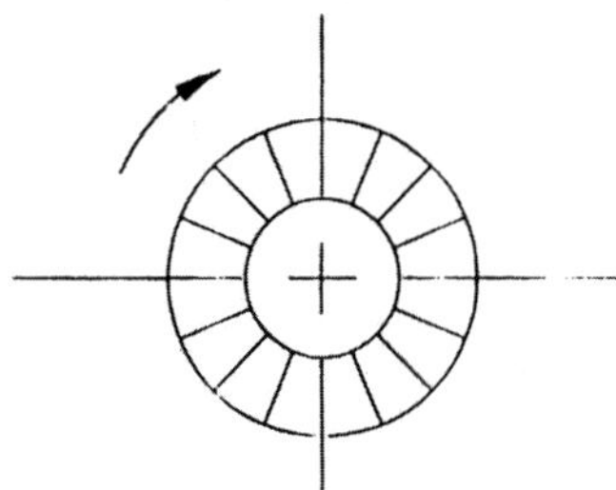

Radial Blade
- Static efficiency to 75%
- High tip speed capabilities
- Reasonable running clearance
- Best for erosive or sticky particulate

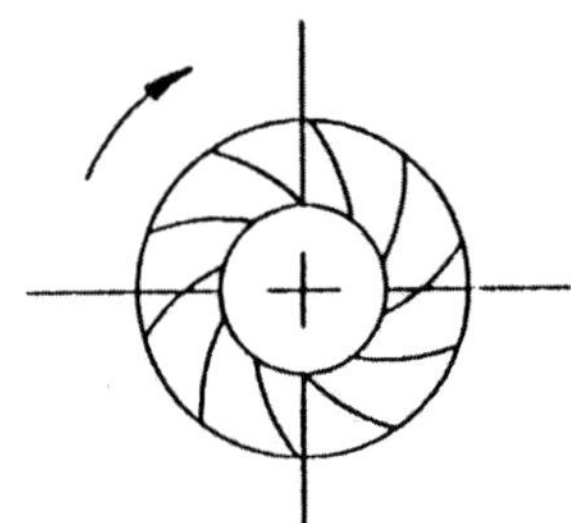

Backward Curved
- Medium to high tip speed capabilities
- High efficiency to 83%
- Clean or dirty airstreams
- Solid one-piece blade design

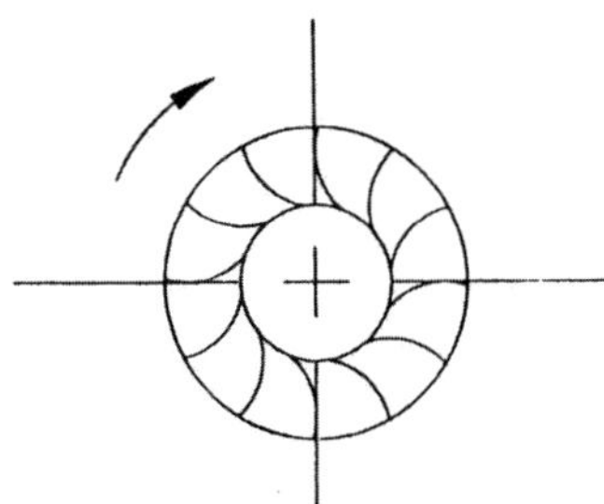

Radial Tip
- Static efficiency to 75%
- Medium to high tip speed capabilities
- Running clearance tighter than radial blade but not as critical as backward inclined and airfoil
- Good for high particulate airstream

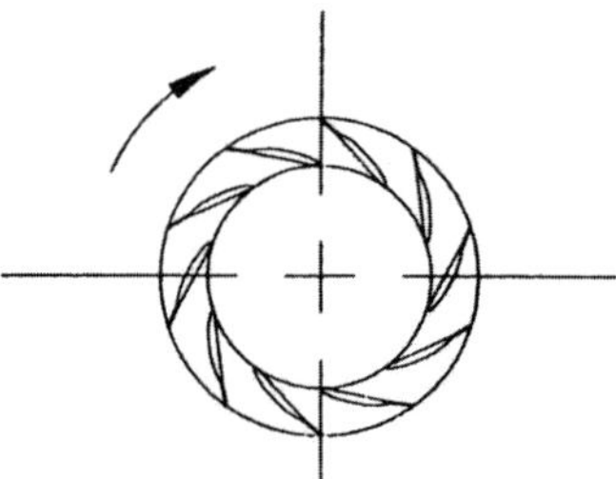

Airfoil
- Static efficiency to 87%
- Medium to high tip speed capabilities
- Relatively tight running clearances

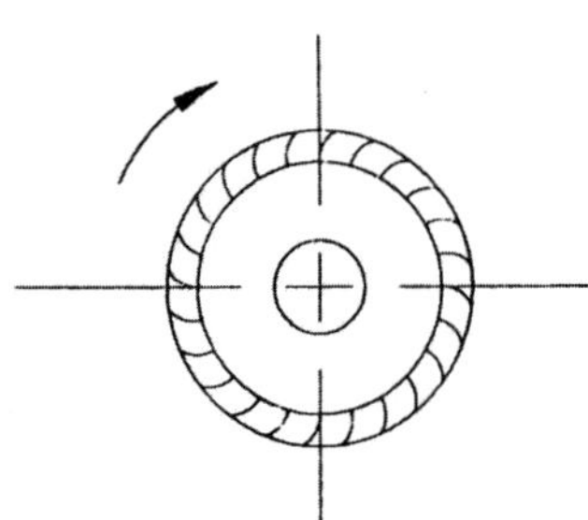

Forward Curved (Sirrocco)
- Smallest diameter wheel for a given pressure requirement
- High volume capability
- 55-65% static efficiency
- Often used for high temperatures

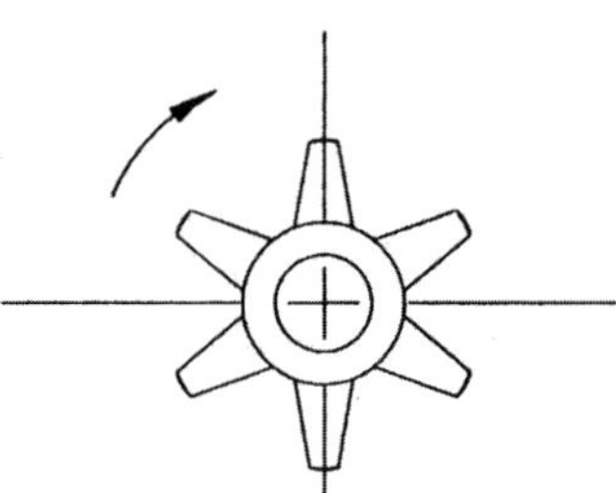

Axial Flow
- High volume, low pressure
- 35-50% static efficiency
- High temperature furnace recirc. applications
- Reversing flow capability
- Airflow parallel to shaft axis

FIGURE 21.25a Fan wheel designs. (Robinson Industries, Zelienople, PA. With permission.)

TABLE 21.2 Relative Characteristics of Centrifugal Blowers

	Radial	Forward Curved	Backward Curved
Efficiency	Medium	Medium	High
Tip speed	High	Medium	Medium
Size[a]	Small	Medium	Large
Initial cost[b]	Small	Medium	Large
HP curve	Medium rise	Medium rise	Power limiting
Accept corrosion coating	Excellent	Fair to poor	Good (thin coat)
Abrasion resistance	Good	Medium	Medium
Sticky material handling	Good	Poor	Medium
High temperature capability	Excellent	Good	Good
Running clearance	Liberal	Medium	Minimum req'd
Operation without diffuser	Not as efficient	Must use	Good efficiency
Noise level	High	Medium	Low
Stability/non-surge range[c]	Medium 20%–100%	Poor 40%–100%	Medium 20%–100%

(a) Size is based on fans at the same speed, volume, and pressure.
(b) Cost is based on fans at the same speed, volume, and pressure.
(c) More a function of operating point along a curve than fan type.

control. The pressure-flow-horsepower curves are shown in Figure 21.25b.

A discharge damper consists of one or more sliding or pivoting blades (such as a butterfly damper) that reduce flow area in a duct. Closing the damper increases resistance to flow and reduces flow. However, when flow is reduced, the operating condition of the blower (and the point on the fan curve) is shifted to lower flow but also to a corresponding higher pressure. This pressure is greater than needed, so the pressure drop taken across the damper wastes it. The blower performance curve is not changed by using a discharge damper. The horsepower usage ratio is reduced, but by less than the flow ratio change (see Figure 21.25c).

Inlet vane control is accomplished with a special damper that consists of multiple adjustable vanes oriented radially from the centerline of the damper, which is located at the

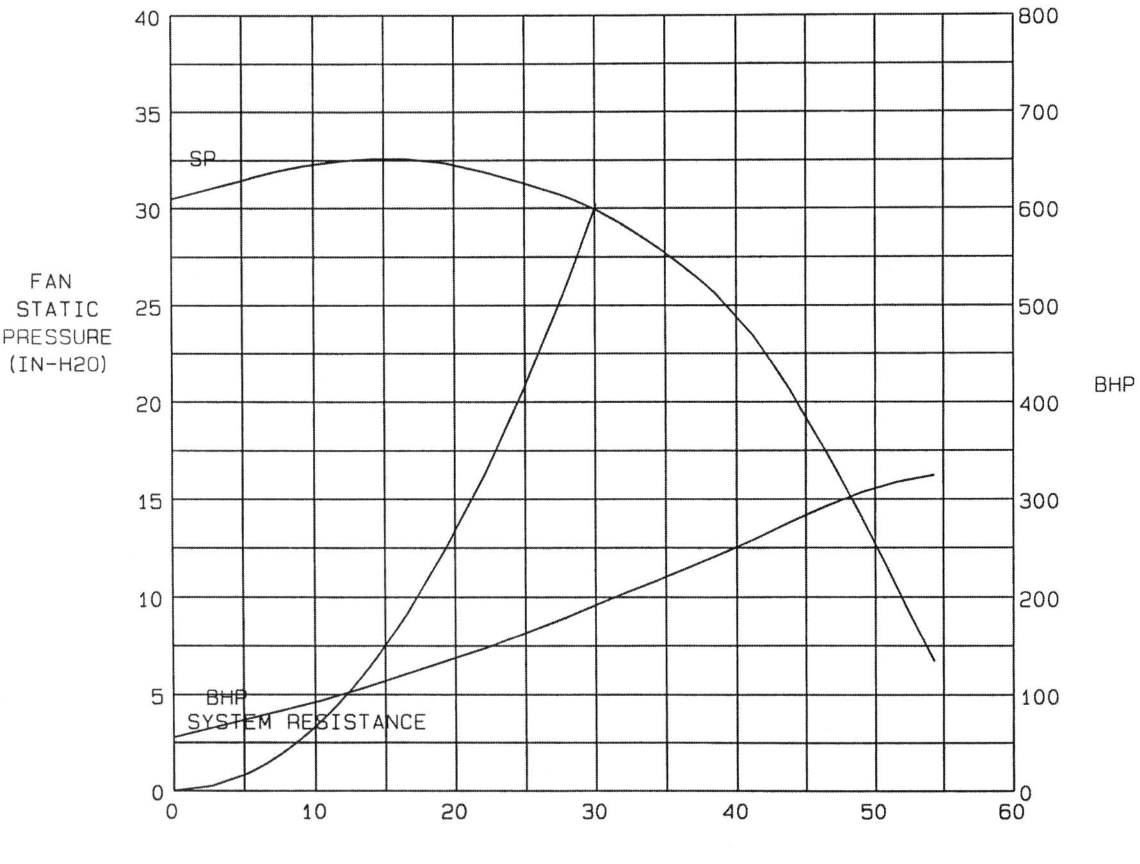

FIGURE 21.25b Radial blade operating curve for 1780 RPM, 70°F, and 0.075 lb/ft^3 density. (Robinson Industries, Zelienople, PA. With permission.)

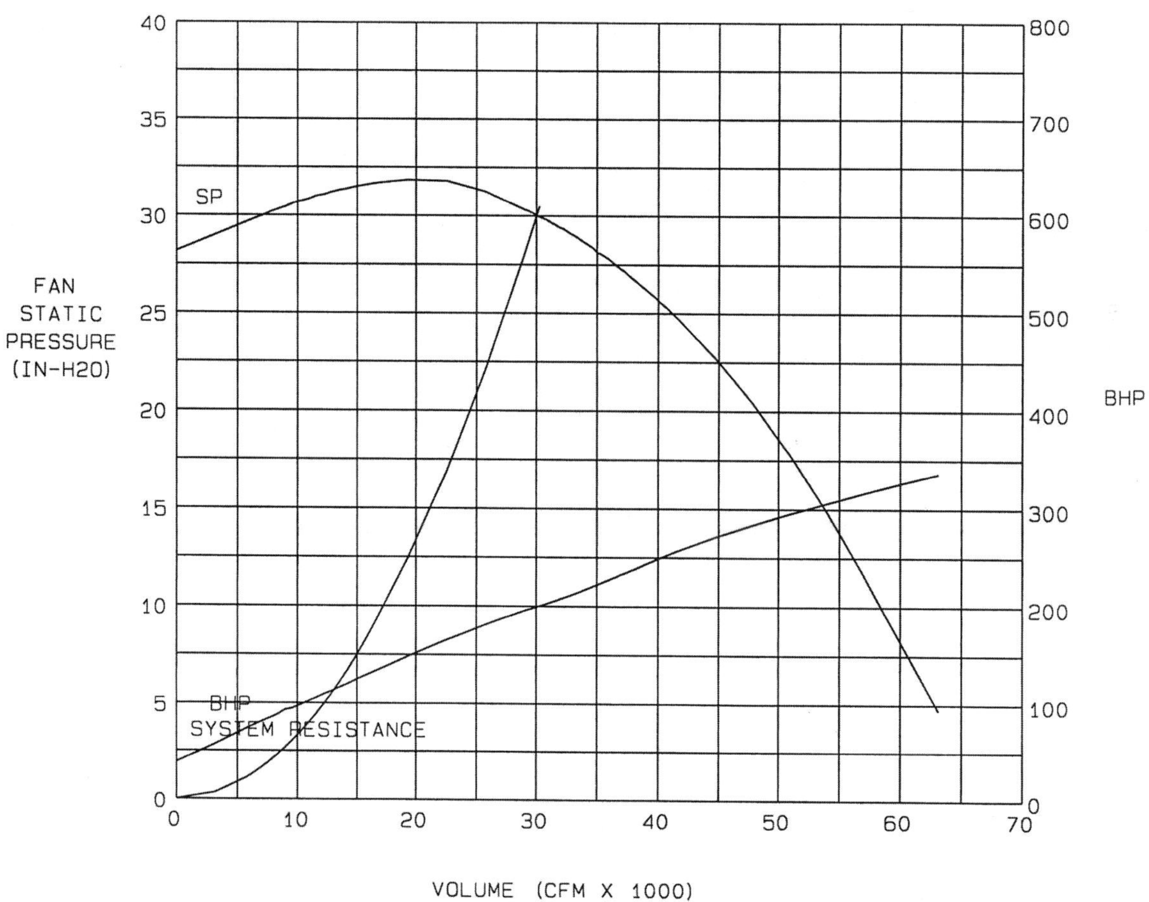

FIGURE 21.25c Forward tip blade operating curve for 1780 RPM, 70°F, and 0.075 lb/ft³ density. (Robinson Industries, Zelienople, PA. With permission.)

blower inlet. As the vanes are rotated (adjusted) closed, reducing the amount of air allowed into the blower inlet, the entering air is given an angular velocity vector (spin) in the direction of rotation of the blower wheel. This spin modifies the basic characteristics of pressure output and power input, resulting in new and reduced pressure and horsepower characteristics. As the vanes are further closed, the flow of air is further reduced but the spin is increased. This further reduces the pressure and horsepower characteristics. Effectively, the inlet vanes change the blower performance curve so that the horsepower reduction ratio is actually greater than the flow ratio change. A specially designed inlet box with a parallel blade damper directing the flow to one side of the box, effectively providing rotation at the blower inlet, provides very similar results (see Figure 21.25d).

Fan speed control for most combustion system blowers is accomplished utilizing variable-speed drivers (motors or turbines). Ideally, if a blower is controlled by varying its speed, there is little wasted energy. The theory behind variable speed control is that the volume of air flowing is proportional to the blower speed, the pressure developed is proportional to the square of the speed, and the horsepower required is proportional to the cube of the speed. Thus, unlike inlet vane control, for which the blower curves start at the same low flow point for each vane setting and change the end point for the high flow, variable speed control results in a completely separate performance curve for each blower speed. The net effect is that the horsepower reduction ratio is even greater than with inlet vane control (see Figure 21.25e).

As with blowers, the method of flow control is not necessarily based simply on cost or efficiency. Often, a combination of two methods is needed, such as discharge dampers controlling flow to different parts of a combustion system while the pressure upstream of the dampers is maintained at a constant point by use of an inlet vane damper.

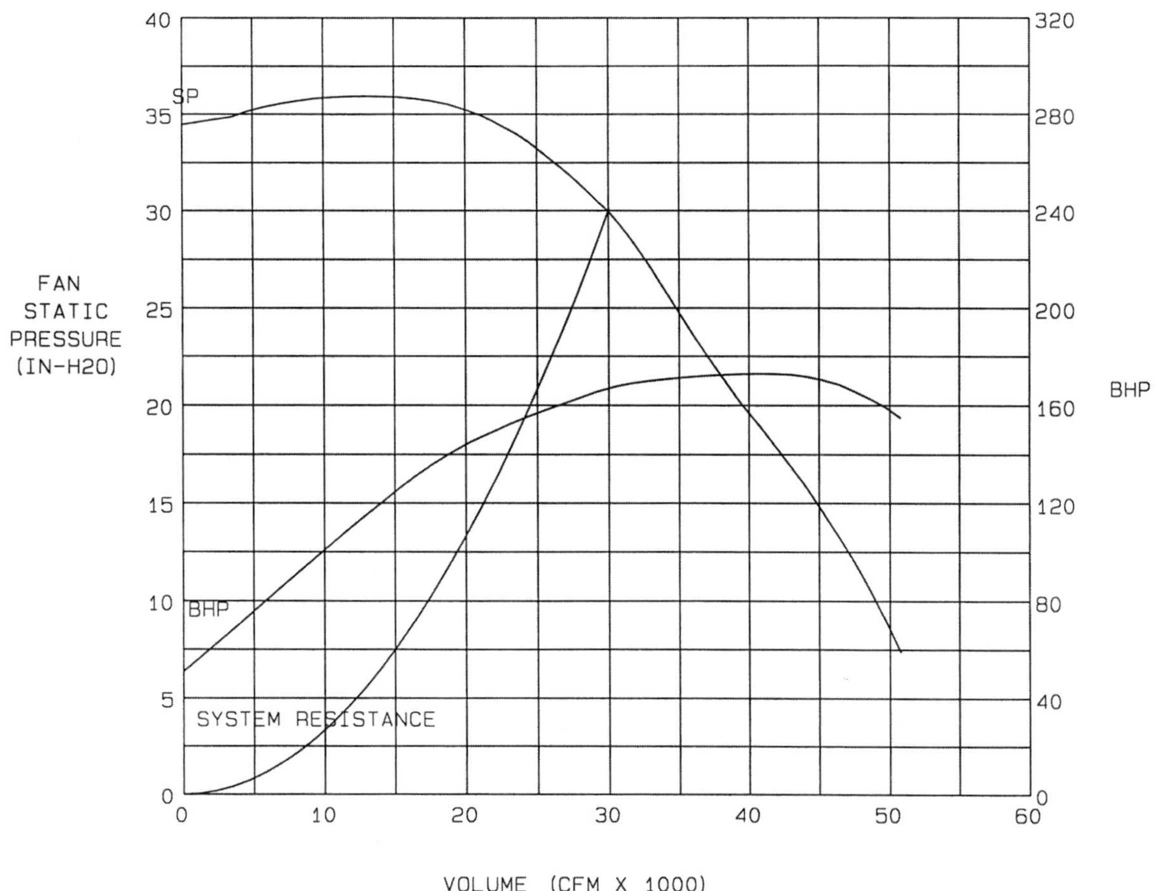

FIGURE 21.25d Backward curved blade operating curve for 1780 RPM, 70°F, and 0.075 lb/ft^3 density. (Robinson Industries, Zelienople, PA. With permission.)

21.5 CONTROL SYSTEMS AND INSTRUMENTATION

All thermal oxidizer systems require some sort of control, if only a flame failure switch to ensure waste and fuel shutoff when needed. Controls can be classified into flame safeguard and process control functions.

Flame safeguard requirements evolved from insurance and general safety regulations for fuel-fired burners in general. This part of a control system is usually designed to satisfy detailed rules published by the National Fire Protection Association (NFPA), Industrial Risk Insurers (IRI), and Factory Mutual (FM). They are meant to ensure that the fuel and waste flows to an incinerator are stopped if the flame is lost and that the furnace is fully purged of combustibles prior to ignition, so that the potential for explosions is eliminated. Generally, the major components of these systems include:

1. flame sensor (flame scanner, flame rod, or other method): to ensure that either a stable flame is maintained (normal operation) or that no flame is present (during purge)
2. fuel supply pressure switches: to ensure that the fuel supply is within the design range
3. air supply flow or pressure switches: to be certain an adequate supply of combustion air is available
4. automatic shut-off block valves: for the fuel flow and waste flow

Process controls are provided to keep the system operating within boundaries to meet legal emission requirements and to protect the equipment from operational damage. Some form of automatic temperature control is normally used to adjust fuel and quench (air, water, or steam) flow to the unit so that the waste is burned properly without exceeding the refractory lining temperature limits. In many cases, the outlet O_2 concentration is monitored and used to control the combustion air flow. Downstream flue gas treating equipment will be

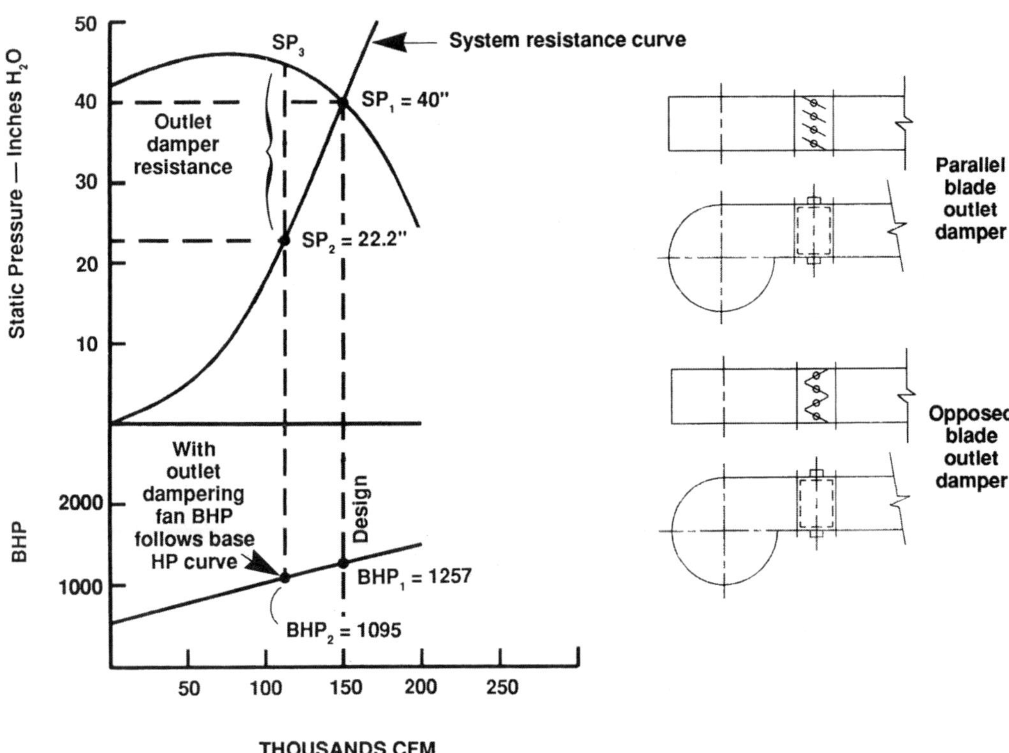

FIGURE 21.25e Outlet damper flow control. (Robinson Industries, Zelienople, PA. With permission.)

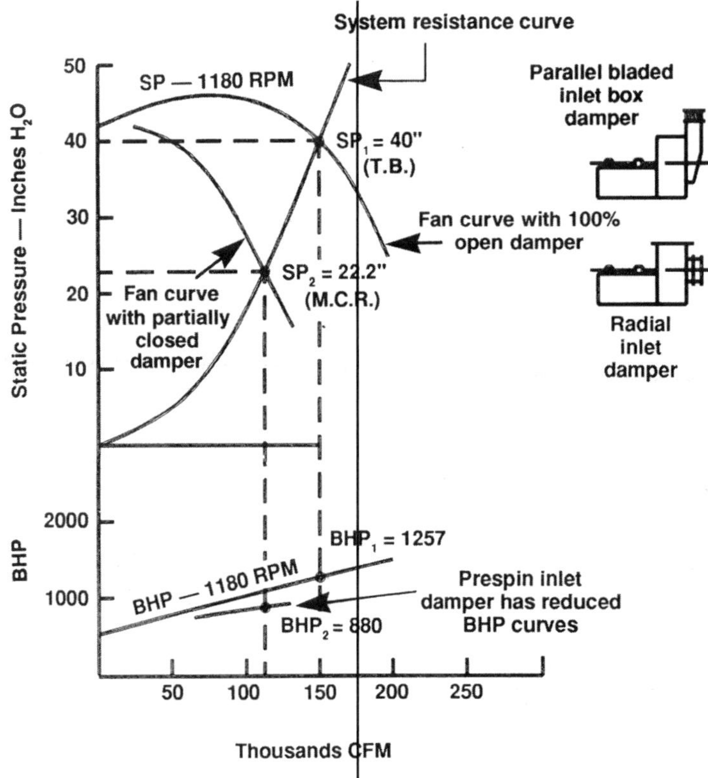

FIGURE 21.25f Radial inlet damper/inlet box damper flow control. (Robinson Industries, Zelienople, PA. With permission.)

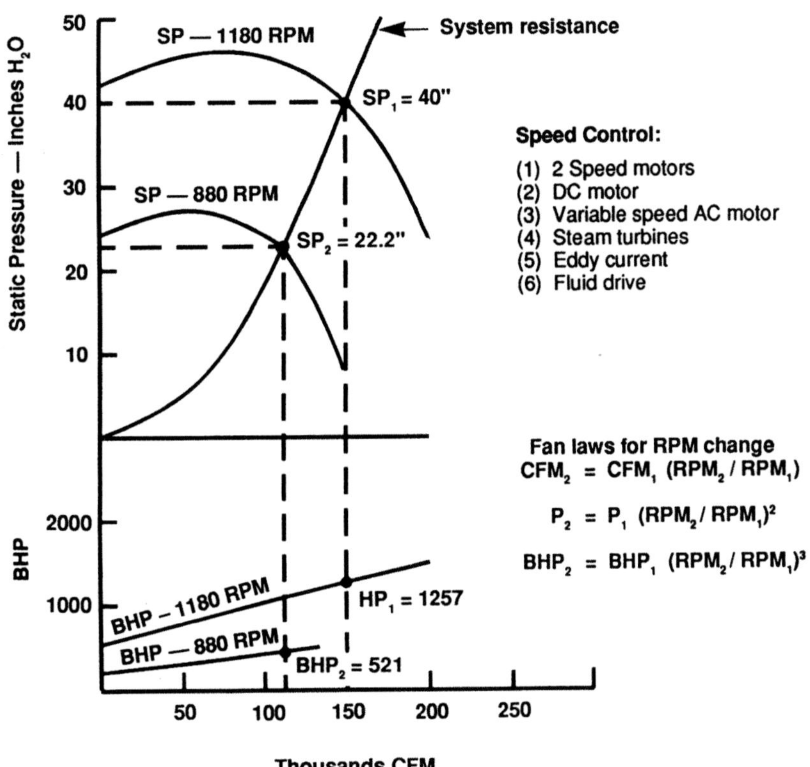

FIGURE 21.25g Blower speed control. (Robinson Industries, Zelienople, PA. With permission.)

built for much lower temperature than the T.O., so flue gas cooling process parameters (boiler water level, water spray flow rate, etc.) and the resulting flue gas temperature are always monitored to avoid expensive thermal damage. Automatic steps (system shutdown, hot gas diversion, etc.) are in place to deal with any failure. For scrubbers, the flow rate and pH of the circulating liquids are controlled automatically to ensure proper removal of acid gases. Less obvious are controls applied to deal with specific variations in incinerator waste feed streams. In some applications, the waste flow and composition are expected to change abruptly. When waste flow and composition are expected to change abruptly, the control method required to maintain effective system performance (or even flame stability) could become very complex. The speed of analyzer or thermocouple response often plays a major part in control system design. In these cases, control system (and burner) design experience is absolutely critical.

21.6 SYSTEM CONFIGURATIONS

The previous sections of this chapter covered the basics of the combustion process, the individual components that make up the combustion section, and the post-combustion flue gas treatment sections. The choice of these components for any total system depends primarily on the nature of the waste stream to be destroyed and the emission requirements for the flue gas ultimately exhausted to the atmosphere. This section is a discussion of suggested system configurations designed for seven of the most common types of waste streams. The types are:

1. non-acid gas endothermic waste gas/waste liquid
2. non-acid gas exothermic waste gas/waste liquid
3. sulfur-bearing acid gas (includes pulp and paper)
4. chlorine-bearing acid gas
5. down-fired "salts" (i.e., solids that melt)
6. vertical/horizontal combustible solids
7. NOx minimization or reduction

21.6.1 Non-acid Gas Endothermic Waste Gas/Waste Liquid System

Endothermic waste gas and waste liquid are often handled similarly. Examples are waste gas containing primarily inert materials such as N_2 or CO_2 with some hydrocarbon contamination, and waste liquid that is water with some small amount of organic material. For these cases, a system would be required to introduce significant amounts of fuel and air to bring the waste components to the temperature needed to oxidize them. The simplest system option is a vertical thermal oxidizer with either a natural- or forced-draft burner. The flue gas exhausts directly to the atmosphere.

Often, it is economically desirable to recover heat from the flue gas of such systems by making steam from a waste liquid or waste gas T.O. system or by preheating the waste gas stream to reduce fuel usage. Preheating a waste liquid is not usually cost-effective.

Also, if the waste gas is air contaminated with a small amount of hydrocarbon and is well below the lower flammability (or explosive) limit (LEL), standard thermal oxidation is still used, but other methods such as catalytic oxidation and regenerative oxidation can be more cost effective.

In all cases, fuel firing capability is required to maintain the correct outlet temperature except in the case of the catalytic system, where the fuel firing is often used only to heat up the unit prior to introducing the contaminated air.

Figure 21.26 shows a simple thermal oxidizer. Fuel and combustion air enters the burner while the inert waste gas or inert waste liquid is effectively introduced past the fuel/air combustion zone. Because no heat recovery or flue gas treatment is needed, the most cost-effective design is a refractory-lined, vertical up-flow unit. The refractory-lined vessel functions as both the combustion chamber, for residence time, and the stack, to disperse the flue gas to the atmosphere. The operating temperature is maintained at the minimum possible to achieve DRE so the least amount of fuel will be used. For discussion purposes, a waste gas stream of 100,000 lb/hr (45,000 kg/hr) of N_2 at 80°F (27°C) with no hydrocarbon content is considered. The operating temperature chosen is 1500°F (820°C). To heat the waste to the chosen operating temperature, the calculated fuel requirement, ignoring heat losses, is slightly greater than 65×10^6 Btu/hr (19 MW).

Recovering some of the fuel heating value put into a T.O. system can be more cost-effective when compared to the simple system, especially if a large amount of fuel is needed. If a T.O. is to be located in a plant that needs steam for process use, a system configuration represented by Figure 21.27 is better to use than the simple vertical T.O. system. The refractory-lined T.O. itself is horizontal so flue gas can exit directly into a boiler (flue gas cooler), which can be either a firetube or a watertube configuration. As noted previously in this chapter, firetube boilers are normally recommended for smaller units with lower pressure steam needs, while watertube boilers are the preferred type for larger systems with higher steaming rates or higher steam pressures. For the same N_2 waste case

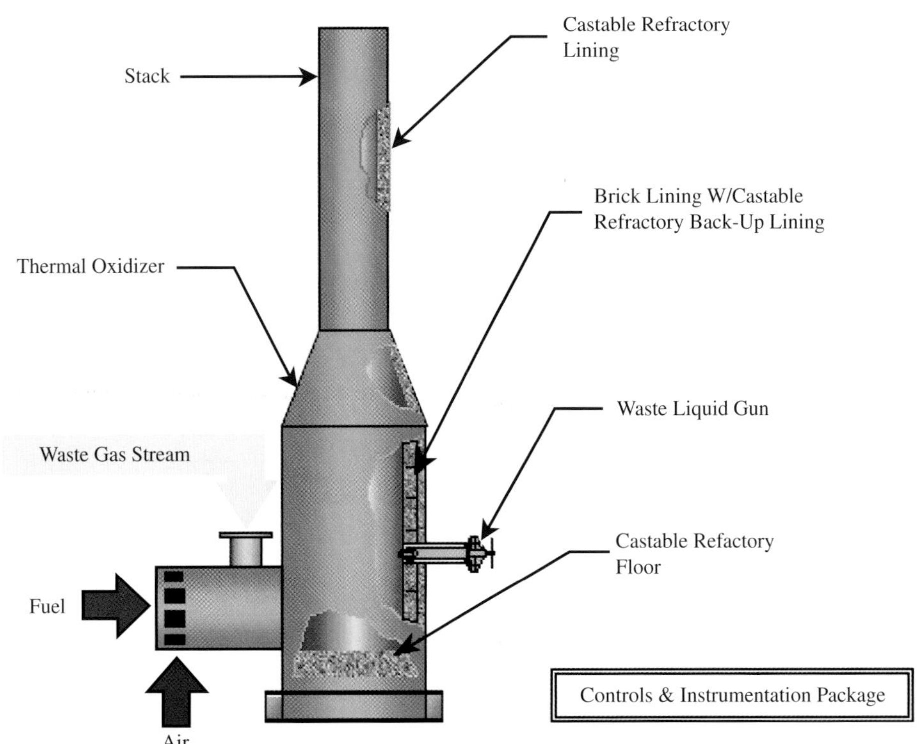

FIGURE 21.26 Simple thermal oxidizer.

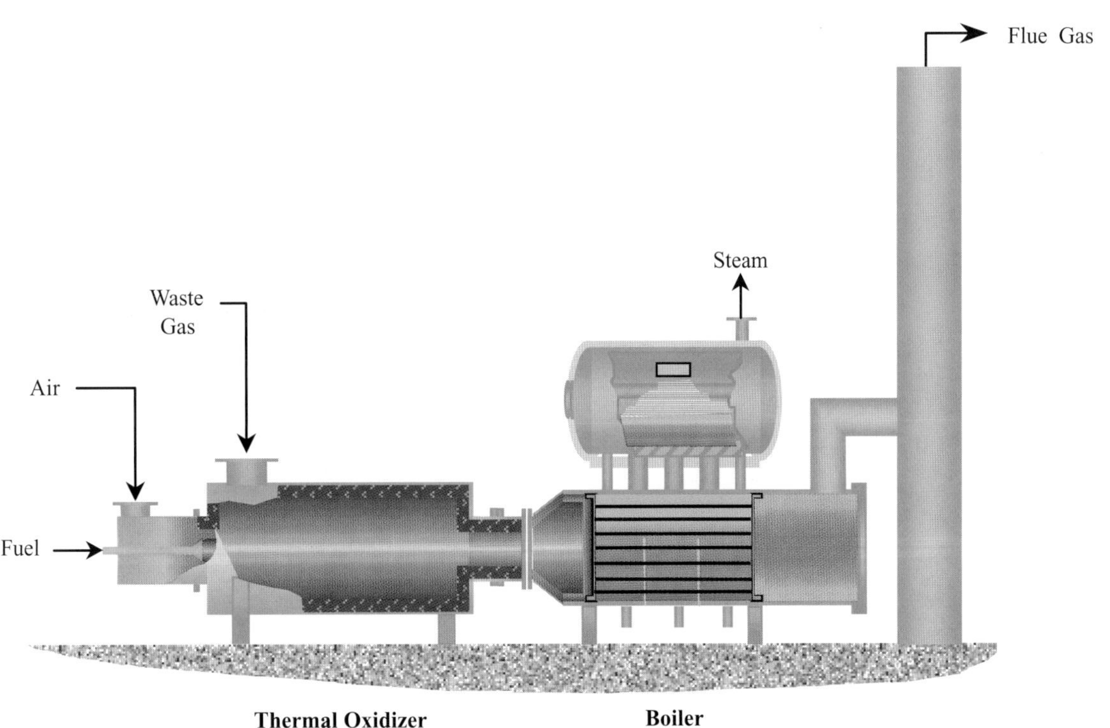

FIGURE 21.27 Thermal oxidizer system generating steam.

noted above, a boiler could produce as much as 44,500 lb/hr (20,000 kg/hr) of 150 psig (10 barg) steam by cooling the flue gas from 1500 to 500°F (820 to 260°C) (a recovery efficiency of about 70%). By utilizing an economizer to further cool the flue gas to 350°F (180°C), about 50,000 lb/hr (23,000 kg/hr) of steam could be produced (a recovery efficiency of more than 80%). Because the flue gas temperature is 500°F (260°C) or less, an economical, unlined carbon steel vent stack is used.

If no steam is needed, the best system is one that reduces the fuel required by transferring heat from the outlet hot flue gas from the T.O. to the waste gas. Figure 21.28 illustrates such a system. The process is usually referred to as a recuperative process because energy is being removed from the flue gas and put back into the system by heating the waste gas before it enters the T.O. Usually, the refractory-lined T.O. is horizontal for the same reasons as the boiler system. Some smaller systems are built utilizing a vertical up-flow T.O. with the preheat exchanger mounted on the top end of the T.O. and the stack on top of the exchanger. For the same waste case described above (100,000 lb/hr or 45,000 kg/hr of N_2), preheating from 80 to 900°F (27 to 480°C) by transferring heat from the flue gas would reduce the fuel needed to maintain 1500°F (820°C) in the T.O. from more than 65×10^6 Btu/hr (19 MW) to less than 29×10^6 Btu/hr (8.5 MW). That is a fuel reduction of about 56%. At only \$3.50 per 10^6 Btu/hr of fuel fired, the savings in fuel cost would be about \$1,000,000 per year compared to the simple vertical T.O. The amount of heat transferred from the flue gas would drop the temperature from 1500°F to about 900°F (820°C to about 480°C), which corresponds to a heat recovery of about 42%. Because less fuel is required, less combustion air (operating cost) is required, and the amount of flue gas is reduced. This reduces the size of the T.O. (capital cost). The vent stack would have to have refractory for this example unless it was fabricated using a heat-resistant alloy that could withstand the greater than 900°F (480°C) flue gas temperature.

Overall, the value of the steam generated and the reduction in fuel used during the first year will typically pay for the additional equipment needed to make steam or preheat the waste gas.

When the waste gas is contaminated air, the configuration of the simple system and the boiler and preheat systems is nearly the same as when the waste gas is a non-oxygen-bearing inert. The only real difference in the simple system is that it is forced draft. Both the boiler and recuperative systems are already forced (or induced) draft to overcome the pressure drop of the heat transfer component. Because the waste contains the O_2 needed to burn the fuel, no additional combustion air is needed for any of the systems, so the fuel requirement for these cases is reduced.

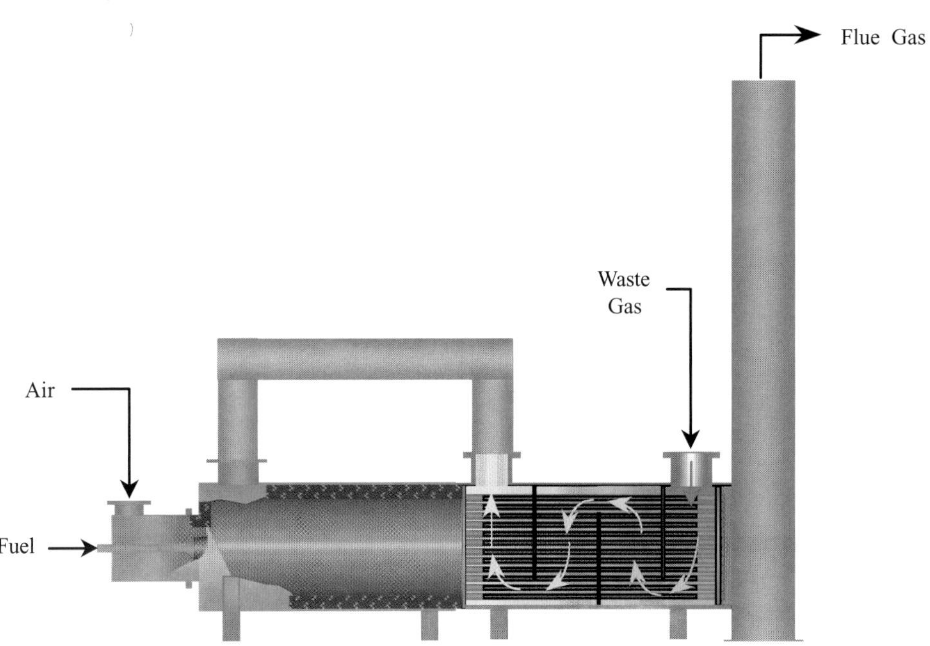

FIGURE 21.28 Heat recovery thermal oxidation system.

For example, 100,000 lb/hr (45,000 kg/hr) of 80°F (27°C) contaminated air with no hydrocarbon content can be heated to 1500°F (820°C) with only 39 × 10^6 Btu/hr (11 MW) of fuel. Cooling the flue gas to 350°F (180°C) can generate about 30,000 lb/hr (14,000 kg/hr) of 150 psig (10 barg) steam. For the recuperative case, by preheating the contaminated air from 80 to 900°F (27 to 480°C), the fuel required would be a little more than 17 × 10^6 Btu/hr (5 MW). The amount of heat transferred from the flue gas would drop its temperature from 1500°F (820°C) to about 735°F (391°C), which corresponds to a heat recovery of about 54%.

However, if the hydrocarbon concentration in the air is too great, care must be taken to preheat the waste air to a safe margin below the lowest accepted auto-ignition point. If some of the hydrocarbon oxidizes in the preheat exchanger, severe over-temperature damage can occur if the exchanger metallurgy is not capable of the higher temperature. For a simple example, at a concentration of 1% methane (CH_4) in air, the calculated temperature rise is greater than 450°F (230°C). If preheat to 900°F (480°C) is intended, an additional temperature rise of 450°F (230°C) yields 1350°F (730°C). If the T.O. outlet temperature is 1500°F (820°C), the average tube temperature at the exchanger inlet is about 1450°F (790°C), 250°F (120°C) greater than the design. In general, the maximum heating value of contaminated air for a T.O. system using recuperative heat recovery is about 20 Btu/ft^3.

When the hydrocarbon content of the contaminated air is low enough, two other systems become more economically attractive when comparing the sum of the capital and operating costs. Those are catalytic oxidation with recuperative heat recovery and thermal oxidation with regenerative heat recovery. The catalytic process (Figure 21.4), is usually used for lower volumes of particulate-free contaminated air than the regenerative process. For the catalytic process, the maximum heating value of the contaminated air must be limited, not just to remain below LEL, but also to prevent overheating the catalyst, which would be damaged rapidly at greater than 1300°F (700°C). The unlined but externally insulated chamber upstream of the catalyst receives and evenly distributes the approximately 600°F (320°C) preheated contaminated air to the catalyst section. Downstream of the catalyst, the chamber is sometimes lined because the temperature after reaction may be 1000 to 1300°F (540 to 700°C) before passing into the heat exchanger.

For the regenerative process, a short-term over-temperature problem does not seem to be as great because the refractory in the T.O. portion and the ceramic media in the heat recovery chambers can withstand operating temperatures greater than 2000°F (1100°C) to produce higher DRE. However, rapid cycling of the recovery chambers will result in excessive wear and tear on the valves and ceramic media. Furthermore, the temperature downstream of the chambers in the carbon steel ductwork can exceed the maximum allowable.

In general, the maximum heating value of contaminated air for a catalytic system with recuperative heat recovery is about 13 Btu/ft^3. For a regenerative system, which has even

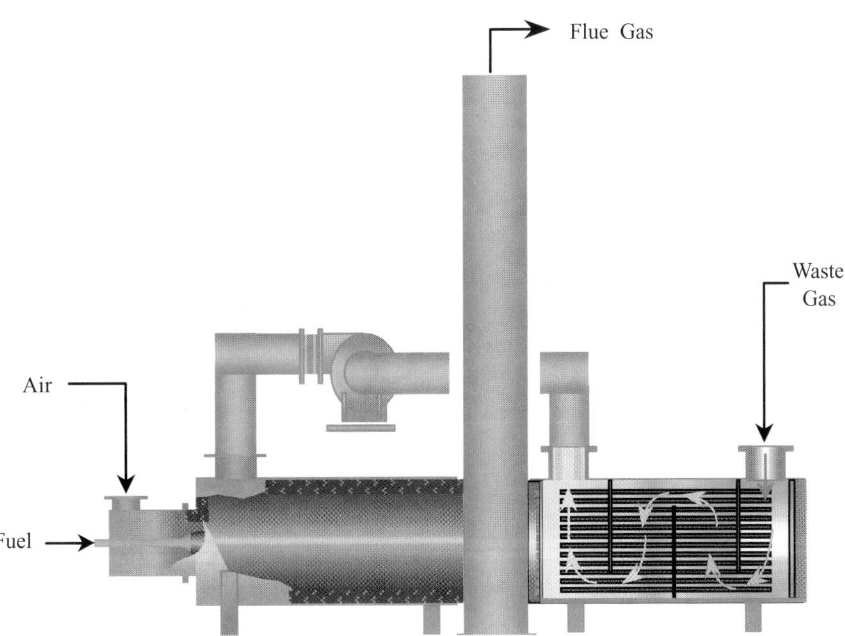

FIGURE 21.29 Bypass recuperative system.

greater heat recovery efficiency, the maximum heat content is about 7 Btu/ft^3. For greater organic content, more air can be added to dilute the overall heat content but that increases the volumetric flow rate, which increases the capital cost of either type system. Also, some of the flue gas can be bypassed in a recuperative system so that less heat is available for heat transfer. Such a bypass system is shown by Figure 21.29.

Also, depending on the DRE required, strong consideration should be given to using a fired boiler that consists of a radiant and a convective section. It is essentially a "cold-wall" T.O. with a boiler. Heat recovery is greater and the NOx emission will be reduced.

At times, the contaminated air may contain enough hydrocarbon to be greater than 25% of LEL. In such cases, use of explosion safeguards such as flame arrestors, detonation arrestors, or liquid seals is recommended.

21.6.2 *Non-acid Gas Exothermic Waste Gas/Waste Liquid System*

Incineration systems for the disposal of high-heating-value hydrocarbon wastes (greater than 200 Btu/ft^3 for gases and greater than 5000 Btu/lb for liquids) that do not have a substantial halogen, ash, nitrogen, or sulfur content are relatively simple. The combustion of these wastes provides more than enough heat to maintain the furnace above the desired operating temperature so that no auxiliary fuel is required during normal operation. If the flue gas from these wastes does not require treatment before discharge, such systems generally consist of a simple natural- or forced-draft burner mounted on a vertical refractory-lined T.O. similar to the non-acid endothermic system (Figure 21.26). If heat recovery in the form of steam or hot oil is desirable, a boiler or oil heater can be used downstream of a horizontal refractory-lined T.O. configured similar to the system shown in Figure 21.27. In any case, the maximum temperature in the T.O. section will have to be held below a set maximum by using a direct cooling medium.

High-heating-value hydrocarbon wastes, whether gaseous or liquid, have characteristics very similar to those of fuels. They are generally as easy to burn and are typically injected directly into the burner. The burner can be of low, medium, or high intensity, depending on the waste being burned, the destruction efficiency desired, and what, if any, post-combustion treatment is utilized. Liquid wastes are atomized with medium-pressure steam or compressed air and are nearly always fired through the throat of the burner. Waste gases with higher heating values can be fired through a single gun or through multiple tips. Fuel is used only to heat up the system in most cases.

The high heating value of these wastes produces high flame temperatures ($\geq$ 2800°F or 1500°C). To achieve the desired destruction efficiency, the combustion chambers for these wastes are generally maintained at relatively high temperatures ($\geq$ 1800°F or 1000°C) and have residence times of greater than or equal to 1 second. Destruction efficiency of organic compounds is 99.99% or greater when using 20 to 25% excess combustion air. To avoid damage to the unit, excessive temperature is controlled by cooling the products of combustion with additional air, water, recycled flue gas, or steam injected into the T.O.

Additionally, the high operating temperatures result in excessive NOx formation. Therefore, being able to limit the maximum local temperature at any point in the T.O. must also be a consideration. In some cases, the cooling medium can be injected into multiple locations, both in the burner portion and in the T.O. to limit thermal NOx formation. Low NOx burner techniques can also apply with these wastes.

The refractory used in the T.O. partially depends on the waste type and the operating temperature. For gases, a castable refractory of sufficient thickness to protect the T.O. shell is adequate. A liquid-burning incinerator, however, can be lined with a firebrick backed with an insulating castable to withstand potential impingement of flame or liquid. The difficult-to-brick areas are usually still castable-lined but with a higher density material. Also, because excess heat is available, less concern is shown for minimizing heat loss through the T.O. Maintaining a low stack temperature for personnel protection may require thicker refractory or a personnel protection shield.

Because the exothermic T.O. systems are usually forced-draft, stack height is not required for them to provide draft in the simple system, and the exit gas velocity is often greater than 50 ft/sec. In addition, the high temperature of the flue gas exiting the simple unit has sufficient buoyancy to carry the flue gas to high altitude. Consequently, a tall stack is not necessary for flue gas dispersion, and stack height becomes a matter of customer choice based on surrounding structures and/or dispersion modeling.

As with the endothermic wastes, heat recovery can be performed with either a boiler or a hot oil heater. With heat recovery, the flue gas is typically vented at less than 600°F (320°C) through an unlined steel stack. Although the buoyancy is not as great because of lower flue gas temperature, exhaust dispersion is rarely an issue. Depending on the flue gas temperature, external insulation of the stack may be desirable to prevent water condensation and possible corrosion on the inside of the stack.

21.6.3 Acid Gas Systems

Acid gas systems are so called because the wastes treated contain components that, as a result of the oxidation reaction to destroy the waste, produce acid compounds such as SO_2/SO_3 and HCl. Emission of these acid compounds is limited by national and local air quality permitting agencies and must be removed before the flue gas is exhausted into the atmosphere. The systems that complete the destruction and removal are acid gas systems.

21.6.3.1 Sulfur-Bearing Hydrocarbon Systems

During the early stages of the oil and gas refining process, sulfur compounds, primarily in the form of hydrogen sulfide (H_2S), are removed. The H_2S is then converted to elemental sulfur by the Claus process. The final "clean" by-product of that process is known as tail gas. Tail gas is mostly N_2, CO_2, and water vapor. However, although the efficiency of the Claus process has improved over the years, some of sulfur compounds still remain. Because sulfur compounds have a very strong odor, often likened to "rotten eggs," even a small amount is detectable by the human nose. By destroying a high percentage of the sulfur compounds, the concentration in the flue gas is reduced to less than the detectable limit. A majority of the thermal oxidation systems supplied for sulfur-bearing waste streams have been simple units for the treatment of tail gas. The sulfur compounds in tail gas include H_2S, sulfur dioxide (SO_2), carbonyl sulfide (COS), carbon disulfide (CS_2), and elemental sulfur vapors. A small amount of CO and hydrocarbon is also usually present. Thermal oxidizers are very effective for odor control of wastes containing mercaptans and other odoriferous sulfur compounds.

Sulfur plant tail gas incinerators are generally designed to operate with natural draft. A stack tall enough to create the necessary amount of draft is used to provide air flow to the burner. The burners are designed for pressure drops from 0.25 in. (6.4 mm) w.c. to more than 1.0 in. (2.5 cm) w.c. Waste heat recovery boilers are also utilized occasionally. For those cases, medium pressure drop, forced-draft burners are used.

The simple incineration process is nearly the same as the simple endothermic configuration described earlier. Castable type refractory is usually adequate for the temperature and environment expected. As before, the lower portion of the stack is the residence time section. One difference from the simple endothermic system is that an internal coating may be applied to the vessel shell before the refractory is installed to protect the steel shell from weak sulfuric acid attack. Another alternative is to add an external rainshield around the vessel to keep the shell temperature above the dew point to prevent condensation.

Because of the low auto-ignition temperatures (generally less than 700°F or 370°C), destroying sulfur-bearing compounds is very easy. Traditionally, a T.O. temperature of 1000 to 1200°F (540 to 650°C) and a residence time of 0.6 to 1.0 seconds was used for all sulfur plant tail gases. In the 1980s, the need to reduce the H_2S content to less than 5 ppmv required an increase in operating temperature to 1400°F

(760°C) in some cases. The higher temperature ensures a higher degree of destruction of the sulfur compounds. It also ensures that the fuel use is greater.

During the oxidation of sulfur compounds, a small amount (typically 1 to 5%) of the sulfur dioxide is further oxidized to SO_3. The extent of conversion depends on a number of conditions such as the temperature profile the flue gas experiences, the amount of SO_2 in the flue gas, the potential catalytic action of alumina refractory material, etc. Once formed, SO_3 reacts at a temperature between 450 and 650°F (230 and 340°C), with water in the gas stream, to form sulfuric acid. Sulfuric acid can cause several problems. First, it raises the dew point substantially, sometimes to temperatures above 400°F (200°C). Sulfuric acid condensation can lead to rapid corrosion of carbon steel surfaces. Second, when quenched rapidly, sulfuric acid can form an extremely fine aerosol that is difficult to remove in a packed bed scrubber. Third, the fine aerosol can create a visible plume that has a bright blue-white hue.

The flue gas treatment depends on the sulfur content of the waste and the regulatory requirements. Typically, tail gas incinerators have such a small SO_2/SO_3 emission that no flue gas treatment is used. The maximum allowable ground level concentration (GLC) of SO_2 can be achieved by utilizing a tall stack. Stack heights of 100 to 300 ft (30 to 90 m) are common. However, if the requirement is to meet a maximum stack emission instead of a GLC, something else must be done.

For applications in which emission limits are very stringent or that produce higher SO_2 concentration, a scrubber is used to remove a good portion of the SO_2 before the flue gas is dispersed into the atmosphere. To accomplish this, the flue gas from the T.O. must first be cooled by a boiler, then quenched to its adiabatic saturation temperature by one of the quenching methods described earlier, and then passed through a packed bed scrubber. If steam were not needed, the flue gas would be quenched directly to saturation. SO_2 is less soluble than HCl in water, and therefore more difficult to remove. Adding sodium hydroxide (NaOH), often referred to as "caustic," is normally used to enhance the removal efficiency and to convert the SO_2 sulfates and sulfites for further treatment. The pH of the recirculated scrubbing solution must be no greater than 8.0 or a significant amount of caustic will react with the CO_2 in the flue gas, wasting caustic. Also, because SO_2 is more difficult to remove, taller packed beds are required when compared to similar applications (same inlet concentration and removal efficiency) scrubbing HCl. The presence of a significant quantity of SO_3 in the flue gases may also necessitate the use of a mist eliminator downstream of the scrubber.

Because of the low temperatures and lower oxygen content, NOx formation is fairly limited and is not normally a consideration. However, when operating at a T.O. temperature of only 1200°F (650°C), any carbon monoxide coming in with the tail gas will be only partially (40% or less) destroyed. At 1400°F (760°C), substantially more is oxidized (more than 80%), but CO destruction is still not high. At lower temperatures, increased residence time (large T.O.) can provide greater destruction, but at greater cost.

Pulp and paper plants also generate waste gas containing sulfur compounds. Those wastes are handled in the same manner. One difference in the equipment is in the material of construction. While the refineries find carbon steel an acceptable material, the pulp and paper industry often prefers stainless steel for many of their installations.

21.6.3.2 Halogenated Hydrocarbon Systems

While sulfur is probably the most common acid gas waste constituent found in petroleum refining, chlorine (Cl_2) is probably the most common halogen encountered in petrochemical plant wastes. Because it is commonly found, this chapter section discusses chlorinated hydrocarbon treatment only. Of the other halogens, fluorine is also relatively common. It converts even more readily to hydrogen fluoride (HF), while bromine and iodine have much lower reaction rates (conversion to HBr and HI) and must be handled differently.

Chlorine is added to many hydrocarbon feedstocks to formulate numerous useful compounds. Wastes containing chlorinated hydrocarbons can be in the form of gas or liquid. Waste gases can be air based, inert based, or organic based; waste liquids can be organic or water based. In each of these cases, the waste can contain chlorinated hydrocarbons such as vinyl chloride, methyl chloride, chlorobenzene, polychlorinated biphenyls (PCB), etc. A typical incineration system for a relatively high-heating-value chlorinated waste consists of a horizontal T.O. to destroy the wastes, a firetube boiler to cool the flue gas for further treatment (and for heat recovery), a direct spray contact quench to cool the flue gas to saturation, a packed column to remove HCl, and a stack to vent the cleaned flue gas to the atmosphere. Figure 21.30 shows this process. The boiler may not be used if no steam is needed. In that case, the T.O. could be vertical up-flow with a 180° turn into the quench section. Figure 21.31 would apply.

Depending on the composition/heating value, wastes can be fired through the burner as a fuel or added peripherally into the T.O. The type of chlorinated hydrocarbon waste and the destruction efficiency required dictates the incinerator operating temperature, typically between 1600 and 2200°F (870 and 1200°C). Residence time varies from 1.0 to

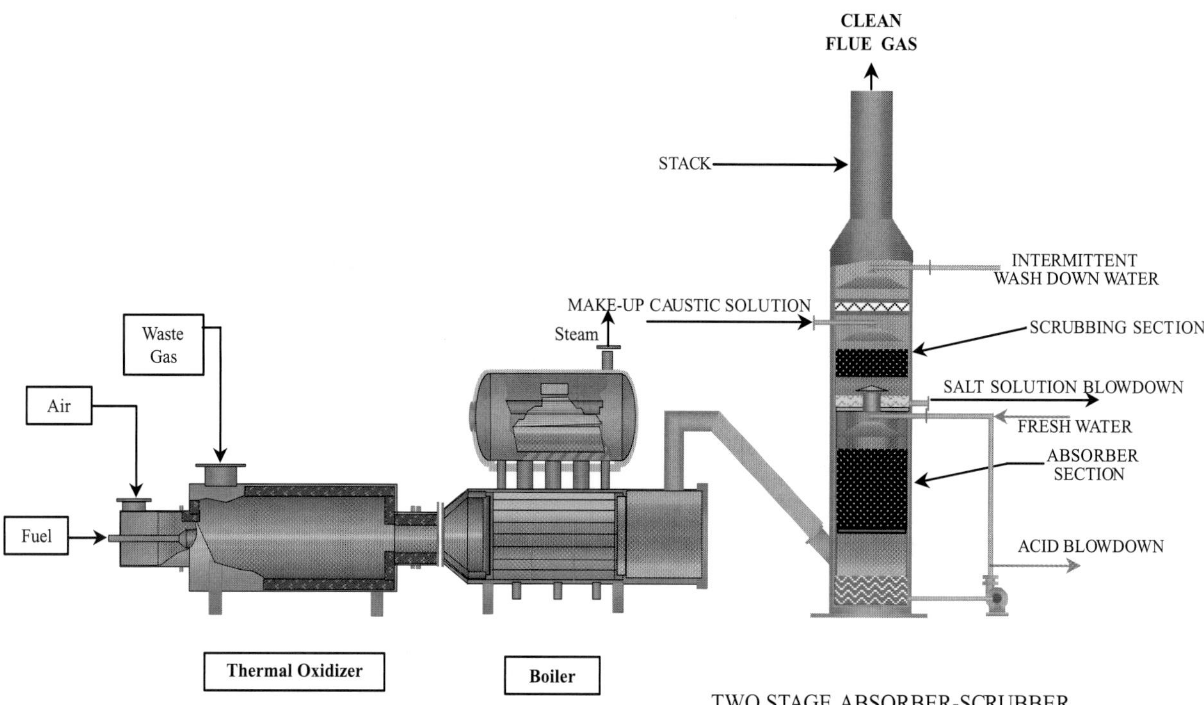

FIGURE 21.30 Horizontal thermal oxidizer with firetube boiler and HCl removal system.

2.0 seconds. Generally, air-based waste streams have lower organic content and require lower destruction efficiency, which can be accomplished beginning at 1600°F (870°C). On the other hand, wastes containing PCB require a destruction efficiency of 99.9999%. This level of destruction efficiency is usually accomplished at a temperature of 2200°F (1200°C) with a residence time of up to 2.0 seconds. The majority of waste streams containing chlorinated hydrocarbons require a destruction efficiency of 99.99%, which is usually obtained at temperatures of 1800 to 2000°F (1000 to 1100°C) and at residence times of 1.0 to 1.5 seconds.

The T.O. refractory can be as simple as a ceramic fiber blanket for air-based fume streams with low operating temperature requirements, or as elaborate as a high alumina firebrick with an insulating firebrick backup for waste liquid streams with high operating temperature requirements. Plastic refractory is used in places where brick is not easily installed. Castable refractory with a calcium oxide binder is generally avoided because the HCl in the flue gases can react with the binder and cause the refractory to degrade.

The oxidation process produces HCl and some free chlorine gas, along with the normal combustion products. It is important to limit the quantity of Cl_2 produced because it is very corrosive at higher temperatures and also because it forms sodium hypochlorite, a strong oxidant, when the combustion products are scrubbed with sodium hydroxide. The presence of hypochlorite in the blowdown stream from the caustic scrubber may require special treatment, as noted in the section discussing packed column scrubbers. Cl_2 formation can be minimized by shifting the reaction equilibrium away from Cl_2 formation and toward HCl formation. As can be seen from the reaction equilibrium vs. operating temperature curve (Figure 21.32), the equilibrium can be shifted by increasing the T.O. temperature, by increasing the water vapor concentration, or by decreasing the excess oxygen concentration. In practice, some excess O_2 is required to maintain highly efficient oxidation reactions so only changes to the temperature and water concentration are used.

A flue gas cooler/waste heat boiler is often used in chlorinated hydrocarbon systems. As noted previously (Section 21.3.5.1.1), a firetube boiler is preferred to a watertube boiler because all of the heat transfer surfaces in a firetube boiler can be maintained at temperatures above the dew point (< 250°F or < 120°C). Of course, to do this, the steam pressure in the boiler must be high enough to have the saturation temperature sufficiently above the dew point of flue gas. The material of construction for the boiler tubes is carbon steel, as it is for the rest of the boiler.

The flue gas must be quenched to saturation before the absorption/scrubbing step to remove HCl. When quenching

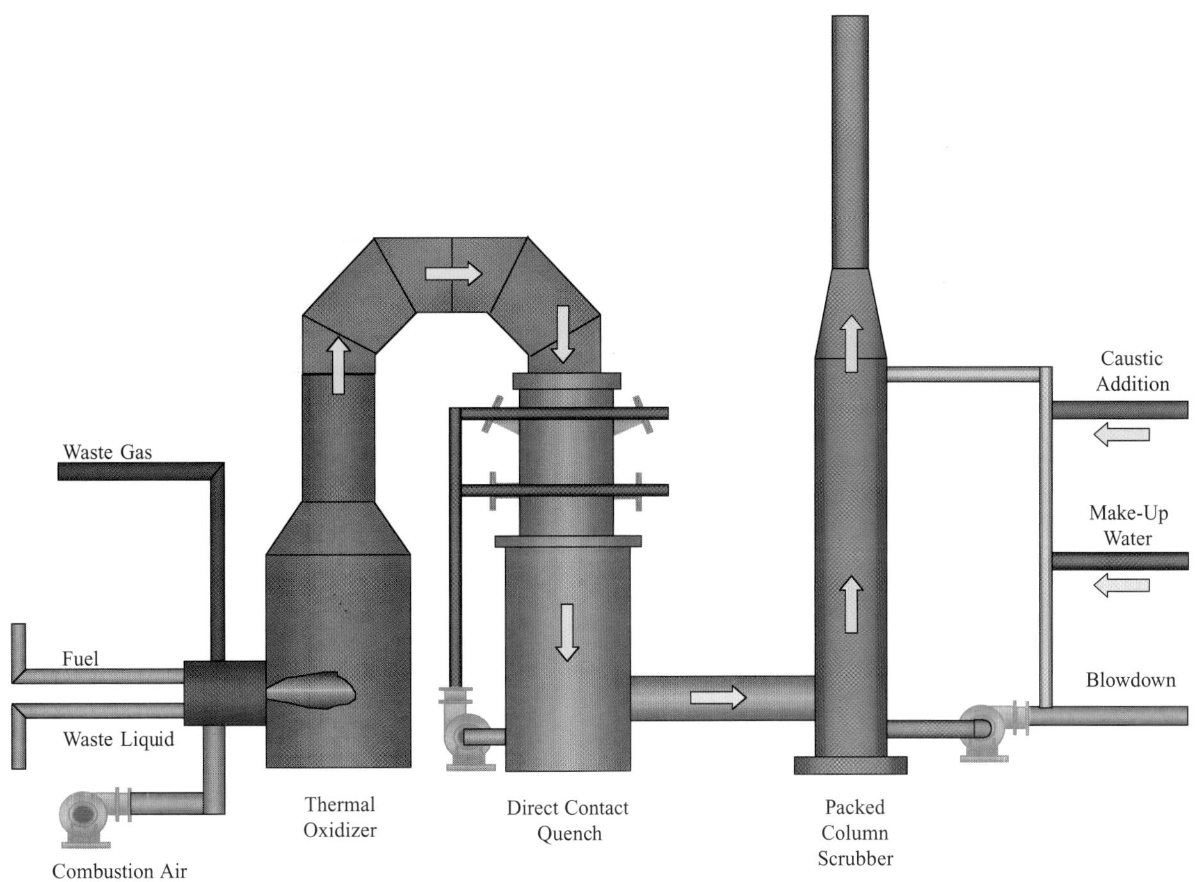

FIGURE 21.31 Vertical thermal oxidizer with 180° turn quench section.

follows a boiler, it can be carried out in the bottom section of the packed column. A two-stage column should be used for this case if HCl is being removed using caustic. With no waste heat boiler, the flue gas is quenched in a direct contact quench. Quench systems are often fabricated using reinforced plastic (FRP), with protection against hot flue gases provided by dry-laid brick lining. Some users prefer metal fabrication such as Hastelloy.

The quantity of HCl in the flue gas dictates whether a single-stage absorber or a two-stage absorber/scrubber is used. For smaller systems where a relatively small quantity of HCl is present in the flue gas, a single-stage water absorber or a caustic scrubber can be used, with sufficient blowdown to maintain the concentration of HCl in the recycle water low enough to allow it to absorb the incoming HCl. If caustic scrubbing is used, the blowdown should maintain the dissolved and suspended solids (mostly NaCl) at no more than 5%, but preferably at 3%. Again, the pH should be kept below 8.0. The single-stage absorber produces a dilute acid stream and the single-stage scrubber produces a stream containing NaCl. For larger systems and those that contain larger quantities of HCl, it is not cost-effective to use large quantities of water or caustic to remove the HCl. In such systems, a two-stage removal system is used. The first stage is an absorber from which the majority of the acid is discharged as a concentrated solution of HCl (up to about 20% by weight). The HCl and Cl_2 remaining in the flue gas after it passes through the absorber are removed in the downstream scrubber, where caustic is used as the scrubbing reagent. HCl removal efficiency of as high as 99.9% can be achieved, although most applications require only about a 99% removal.

While oxidation of chlorinated hydrocarbons produces HCl with a little free Cl_2, fluorinated hydrocarbons produce even less and are easier to remove than HCl. However, burning brominated compounds results in as little as half of the bromine being converted to HBr. Iodine compounds are worse. Free bromine and iodine are much more difficult to remove than Cl_2. Brominated and iodine compounds must be treated completely differently. Sub-stoichiometric oxidation is necessary to drive the reaction toward HBr and HI.

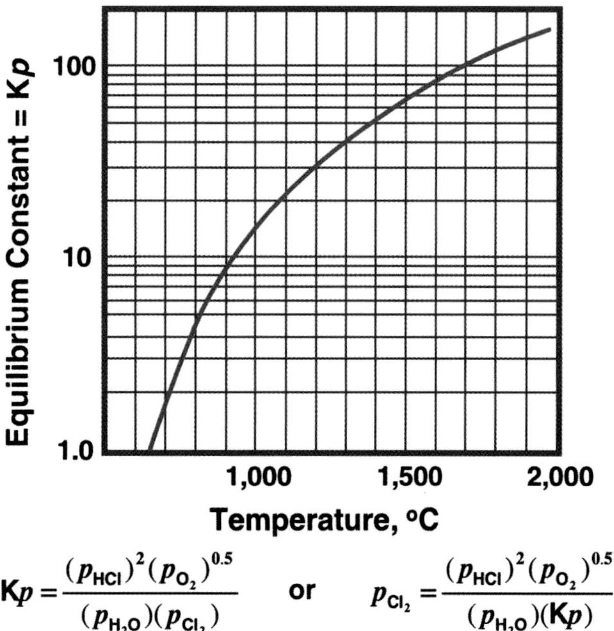

FIGURE 21.32 Chlorine reaction equilibrium vs. operating temperature.

21.6.4 Salts/Solids Systems

Solids system configurations are determined by the condition of the material coming into and out of the combustion section. Either a liquid or a gas can carry the solid. If the material has a high melting point (> 2400°F or 1300°C) and comes in with water, it will likely pass through the combustion section virtually unaffected. If the solid is part of an organic liquid, it may melt in the flame zone, but will return to solid phase very quickly after leaving the flame zone. If flow patterns and vessel orientation are wrong, the material will often collect in the equipment. Regardless, if the particulate emission excluded the allowable amount, a particulate removal device must be employed. Examples are catalyst fines in organic or aqueous liquids, metal-machining dust in air, titanium dioxide (an opaque, white additive to many products) in water or organic liquid, and some organic salts such as sodium acetate (which oxidizes to form organic combustion products and sodium oxides, which have high melting points).

If the material comes into the vessel as an organic solid, it may very well be completely oxidized before it exits. However, organic materials often contain non-organic compounds that may be high-melting-point inerts or lower-melting-point materials such as silica (~2000°F or 1100°C), or alkali metals inorganic salts such as NaCl (~1500°F or 820°C). If the solid is organic, certain configurations allow quicker, more complete oxidation than others. If lower-melting-point materials are present, the system must be designed to handle molten material. Some examples of organic solids containing non-organic material are sawdust and rice hulls. Both can be conveyed into a combustion system with air as the carrier, and both have good heating value. However, both contain a large amount of silica. When both are burned, the silica forms very small molten particles of silicon dioxide in the flame zone. When cooled, these fine particles of "fume silica" have a tendency to collect inside equipment and are difficult to remove from the flue gas.

If the solid material is an alkali metal inorganic salt such as NaCl, Na_2SO_4, CaCl, or KCl, in water, for example, the total system design must be based on molten material in the flue gas. Because virtually all cases involve molten material, the inorganic salt case is reviewed here.

Most of the wastes in this category are salt-contaminated liquids. The waste streams are both organic and water based. Because the waste is a hydrocarbon liquid or because the water-based waste liquid contains a hydrocarbon, thermal oxidation is often the best method for disposal.

Unfortunately, because most salt-containing wastes are liquid, they usually have to be oxidized at higher temperatures to rapidly achieve high destruction efficiency. This presents a problem with the refractory in the T.O. Alkali material attacks the binder in refractory materials. Higher temperatures increase the rate of attack. The design must balance the need for higher temperature (i.e., higher destruction efficiency) with the need to prolong refractory life. Temperatures ranging from 1600 to 2000°F (870 to 1100°C) and residence times of 1.0 to 1.5 seconds are commonly used.

The system best-suited for molten salts is illustrated in Figure 21.33. The incinerator is usually vertical, with the burner mounted on the top and firing downward. The vertical design is highly desirable because it does not allow the molten salts to accumulate on the refractory, as it would in a horizontal oxidizer. This is extremely important, as any accumulation of salts in the oxidizer can drastically reduce refractory life.

Basic (MgO) brick has very good resistance to salt attack. However, it is very expensive, highly susceptible to thermal shock and hydration, and will quickly erode if hit by a stream of water. High alumina brick has been used with varying results. In some cases, a 90% alumina brick, with its low porosity, has been found to be more effective against salt attack than a 60% alumina brick. Unfortunately, while the higher alumina brick is two to three times the cost of the lower alumina brick, it does not last two or three times longer. In general, using the lowest porosity, 60 to 70% alumina brick and the proper burner/T.O. configuration provides the most cost-effective service.

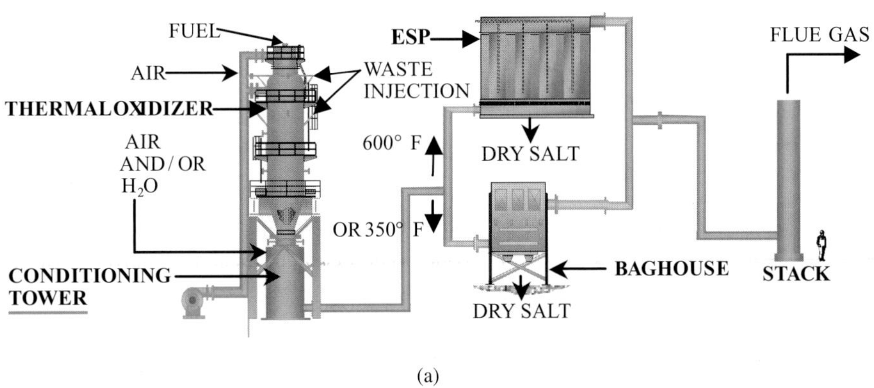

(a)

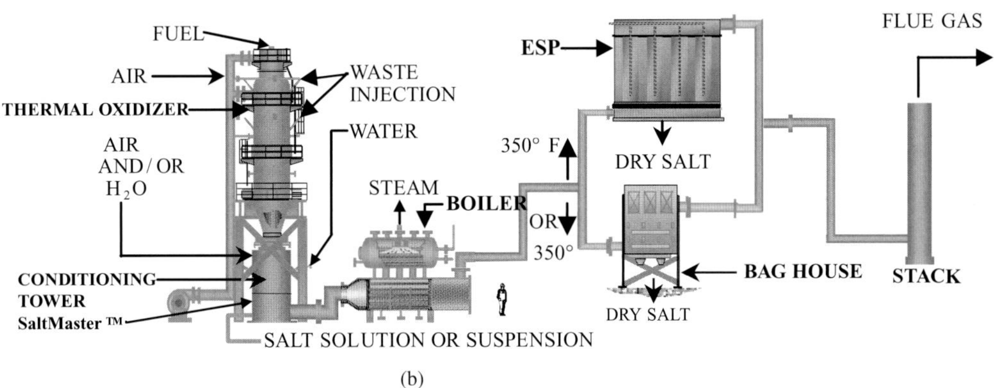

(b)

FIGURE 21.33 Molten salt system.

For organic wastes, moderate to low-pressure-drop burners are preferred, because a high-pressure-drop burner, with its attendant turbulence, tends to centrifuge the salts in the waste toward the refractory, causing it to deteriorate more rapidly. A moderate-pressure-drop burner with no cyclonic action tends to keep the salts in suspension and away from the incinerator walls. The less salt that contacts the refractory, the slower the refractory degrades. The hydrocarbon liquid is fired through the burner.

Water-based wastes are injected into the T.O. downstream of the burner. Because the water in these wastes must be evaporated and the bulk mass raised to near the oxidation point before combustion of organic materials can begin, it is necessary to burn some amount of auxiliary fuel. The amount of auxiliary fuel burned depends on the operating temperature and the waste composition. A similar moderate-pressure-drop burner is also used for this condition. The auxiliary fuel must be allowed to burn to near completion before injecting the aqueous waste stream into the burner combustion products. Incomplete combustion of the auxiliary fuel can cause soot formation and incomplete combustion of the waste.

Although rare, waste heat recovery boilers have been used in some systems. To utilize heat recovery, the flue gas must be conditioned to a temperature below the melting point of the salt, freezing it before it contacts the cooler heat transfer surface. This makes the salt friable so that the accumulation is easier to remove from the heat transfer surface. A water-tube-type boiler is used to allow online cleaning by soot blowers. Figure 21.34 shows this configuration.

Before the flue gas can be vented to the atmosphere, particulate matter must be reduced to the concentration allowable by emission standards. This is accomplished with a baghouse, an ESP a WESP, or a venturi scrubber. Each of these particle removal devices has been discussed in previ-

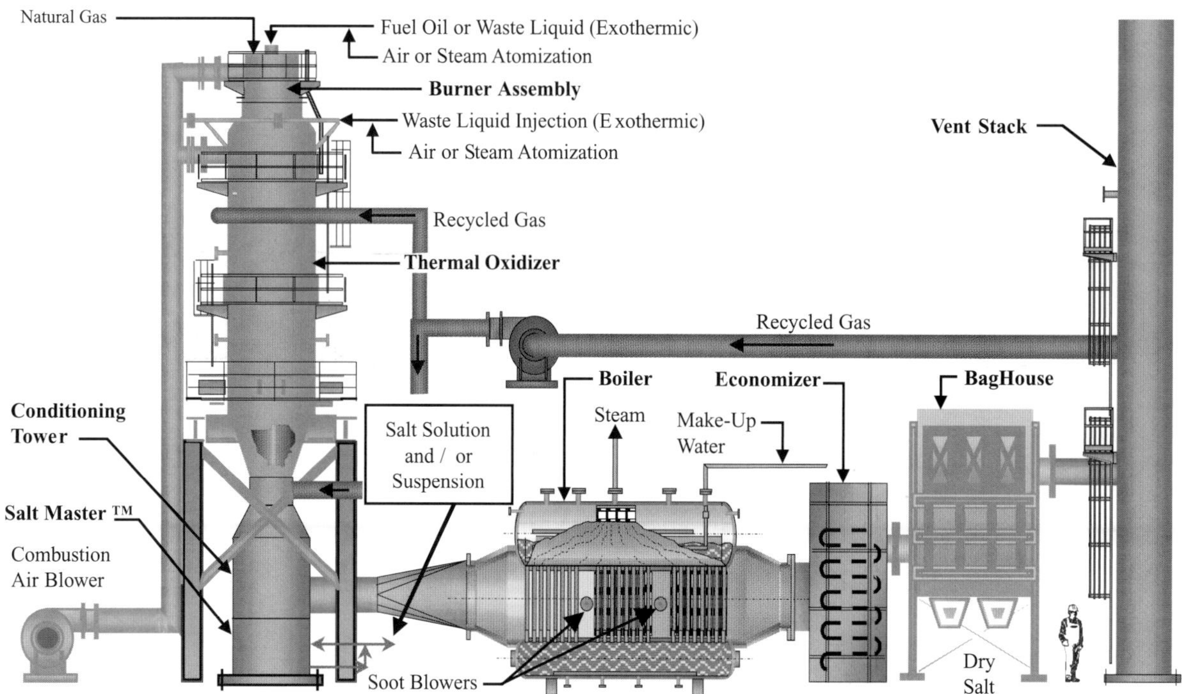

FIGURE 21.34 Online cleaning with soot blowers.

ous chapter sections. As noted, proper flue gas conditioning or quenching, for the baghouse or ESP, or for the WESP or venturi scrubber, must be accomplished upstream of any air pollution control equipment.

Excessive salt in the flue gas from a salts unit will form a nondissipating visible plume. Consequently, the flue gas from the saturated system often receives intense public scrutiny. For this reason, some operators choose to install equipment to eliminate the visible water vapor plume. This is accomplished either by cooling and removing virtually all the water vapor in the saturated flue gas, or by removing a substantial portion of it and then heating the flue gas stream so that the plume becomes less visible. De-pluming is costly, both in terms of capital and operating costs, and the results are not always satisfactory.

For both water- and organic-based liquid streams, 99.99% DRE is normally required. A particulate matter concentration of 0.015 grains per dry standard cubic foot (DSCF) of flue gas, corrected to a specific excess O_2 level, is also normally required.

21.6.5 NOx Minimization or Reduction Systems

The theory behind the combustion-process-modification method of minimizing or reducing NOx was covered in previous sections. This section reviews only the three-stage system configuration.

Consider a waste stream consisting of chlorinated hydrocarbon, amine, and other hydrocarbons. Oxidizing this waste mixture in a standard excess-air T.O. would produce high NOx and HCl emissions. A three-stage NOx system would provide the NOx reduction needed, but the HCl generated would be an issue. By simply adding a packed column scrubber, which is capable of quenching the flue gas in the base of the scrubber, to the end of the NOx system, the problem is solved. Figure 21.35 illustrates this configuration.

The combustion air blower, burner, reduction furnace, quench section, oxidation air blower, and oxidation section are very similar to those supplied for a "normal" three-stage NOx system. Carbon steel construction is still acceptable. It is necessary to add a full-length rainshield or internal anticorrosion lining to the three vessels to protect against HCl corrosion. Also, the carbon steel duct between the boiler and the packed column and the entire recycle duct to the reduction furnace and quench section must be externally insulated and sealed from weather so the unlined ductwork steel temperature will stay well above the dew point. The recycle blower should be carbon steel with sealed external insulation. The packed column should be FRP.

The refractory is very similar to that in the "normal" NOx system with the exception that the presence of HCl must be

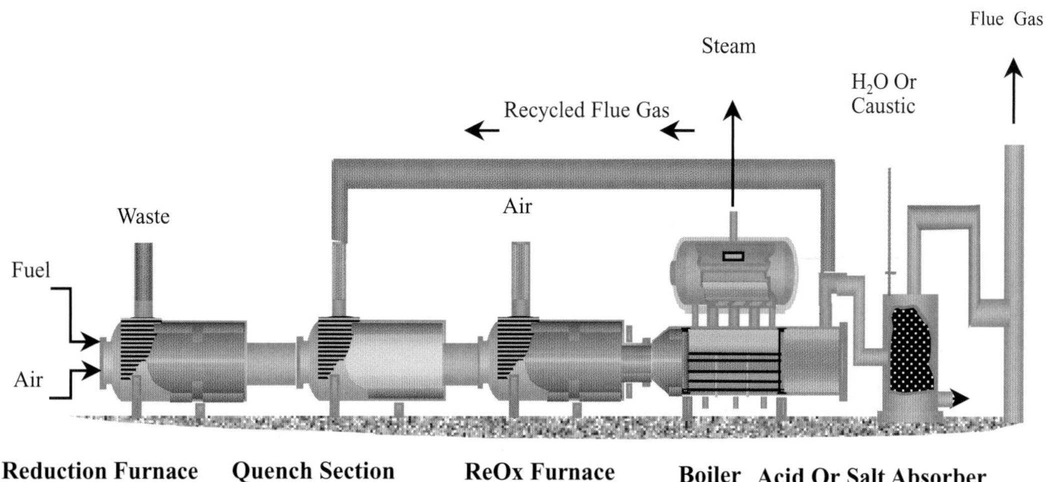

FIGURE 21.35 Three-stage NOx system with packed column scrubber.

considered. The castable refractory should be reviewed to ensure that the material with the lowest CaO content is utilized without compromising the refractory strength or operating capability.

Special consideration should be given to the instrumentation used in the NOx reduction described here as it must be capable of operation in the anticipated HCl/Cl_2 environment.

21.7 CLOSING

The purpose of presenting this chapter was to attempt to provide a better understanding of the use of thermal oxidation processes to destroy vapor and liquid wastes. Although specific design details were not included, written explanations and diagrams provided an overview of the multi-faceted subject.

21.9 NOMENCLATURE

APC Air pollution control
CSA Collecting surface area
DRE Destruction and removal efficiency
ESP Electrostatic precipitator
FRP Fiberglass-reinforced plastic
LEL Lower explosive (flammability) limit
PCB Polychlorinated biphenyl
SCR Selective catalytic reduction
SCS Specific collector surface
SNCR Selective noncatalytic reduction
T.O. Thermal oxidizer
WESP Wet electrostatic precipitator

REFERENCES

1. C.D. Cooper and F.L. Alley, *Air Pollution Control, A Design Approach*, 2nd ed., Waveland Press, Prospect Heights, IL, 1994.
2. R.C. Flagan and J.H. Seinfeld, *Fundamentals of Air Pollution Engineering*, Prentice-Hall, Englewood Cliffs, NJ, 1988.
3. W. Licht, *Air Pollution Control Engineering, Basic Calculations for Particulate Collection*, Marcel Dekker, Inc., New York, 1980.
4. R.D. Reed, *Furnace Operations*, 3rd ed., Gulf Publishing, Houston, TX, 1981.
5. H.L. Gutzwiller, *Fan Performance and Design*, Robinson Industries, Inc., January 2000.

Appendices

Appendices

Appendix A — Physical Properties of Materials

Table A.1 Areas and Circumferences of Circles and Drill Sizes 695

Table A.2 Physical Properties of Pipe 704

Table A.3 Physical Properties of Tubing 709

Table A.4 SAE Grades for Steel Bolts 711

Table A.5 ASTM Grades for Steel Bolts 712

Table A.6 Properties for Metric Steel Bolts, Screws, and Studs 713

Appendix B — Properties of Gases and Liquids

Table B.1 Combustion Data for Hydrocarbons 715

Table B.2 Thermodynamic Data for Common Substances 716

Table B.3 Properties of Dry Air at Atmospheric Pressure 717

Table B.4 Properties of Gases and Vapors in English and Metric Units 719

Appendix C — Common Conversions 725

Appendix A
Physical Properties of Materials

TABLE A.1 Areas and Circumferences of Circles and Drill Sizes

Drill Size	Diameter (in.)	Circumference (in.)	Area (in.)	Area (ft)
80	0.0135	0.042 41	0.000 143	0.000 000 9
79	0.0145	0.045 55	0.000 165	0.000 001 1
1/64"	0.0156	0.049 09	0.000 191	0.000 001 3
78	0.0160	0.050 27	0.000 201	0.000 001 4
77	0.0180	0.056 55	0.000 254	0.000 001 8
76	0.0200	0.062 83	0.000 314	0.000 002 2
75	0.0210	0.065 97	0.000 346	0.000 002 4
74	0.0225	0.070 69	0.000 398	0.000 002 8
73	0.0240	0.075 40	0.000 452	0.000 003 1
72	0.0250	0.078 54	0.000 491	0.000 003 4
71	0.0260	0.081 68	0.000 531	0.000 003 7
70	0.0280	0.087 96	0.000 616	0.000 004 3
69	0.0292	0.091 73	0.000 670	0.000 004 7
68	0.0310	0.097 39	0.000 755	0.000 005 2
1/12"	0.0313	0.098 18	0.000 765	0.000 005 3
67	0.0320	0.100 53	0.000 804	0.000 005 6
66	0.0330	0.103 67	0.000 855	0.000 005 9
65	0.0350	0.109 96	0.000 962	0.000 006 7
64	0.0360	0.113 10	0.001 018	0.000 007 1
63	0.0370	0.116 24	0.001 075	0.000 007 5
62	0.0380	0.119 38	0.001 134	0.000 007 9
61	0.0390	0.122 52	0.001 195	0.000 008 3
60	0.0400	0.125 66	0.001 257	0.000 008 7
59	0.0410	0.128 81	0.001 320	0.000 009 2
58	0.0420	0.131 95	0.001 385	0.000 009 6
57	0.0430	0.135 09	0.001 452	0.000 010 1
56	0.0465	0.146 08	0.001 698	0.000 011 8
3/64"	0.0469	0.147 26	0.001 73	0.000 012 0
55	0.0520	0.163 36	0.002 12	0.000 014 7
54	0.0550	0.172 79	0.002 38	0.000 016 5
53	0.0595	0.186 93	0.002 78	0.000 019 3
1/16"	0.0625	0.196 35	0.003 07	0.000 021 3
52	0.0635	0.199 49	0.003 17	0.000 022 0
51	0.0670	0.210 49	0.003 53	0.000 024 5
50	0.0700	0.219 91	0.003 85	0.000 026 7
49	0.0730	0.229 34	0.004 19	0.000 029 1
48	0.0760	0.238 76	0.004 54	0.000 031 5
5/64"	0.0781	0.245 44	0.004 79	0.000 033 3
47	0.0785	0.246 62	0.004 84	0.000 033 6

TABLE A.1 (continued) Areas and Circumferences of Circles and Drill Sizes

Drill Size	Diameter (in.)	Circumference (in.)	Area (in.)	Area (ft)
46	0.0810	0.254 47	0.005 15	0.000 035 8
45	0.0820	0.257 61	0.005 28	0.000 036 7
44	0.0860	0.270 18	0.005 81	0.000 040 3
43	0.0890	0.279 60	0.006 22	0.000 043 2
42	0.0935	0.293 74	0.006 87	0.000 047 7
3/32"	0.0937	0.294 52	0.006 90	0.000 047 9
41	0.0960	0.301 59	0.007 24	0.000 050 3
40	0.0980	0.307 88	0.007 54	0.000 052 4
39	0.0995	0.312 59	0.007 78	0.000 054 0
38	0.1015	0.318 87	0.008 09	0.000 056 2
37	0.1040	0.326 73	0.008 49	0.000 059 0
36	0.1065	0.334 58	0.008 91	0.000 061 9
7/64"	0.1094	0.343 61	0.009 40	0.000 065 2
35	0.1100	0.345 58	0.009 50	0.000 066 0
34	0.1110	0.348 72	0.009 68	0.000 067 2
33	0.1130	0.355 00	0.010 03	0.000 069 6
32	0.1160	0.364 43	0.010 57	0.000 073 4
31	0.1200	0.376 99	0.011 31	0.000 078 5
1/8"	0.1250	0.392 70	0.012 27	0.000 085 2
30	0.1285	0.403 70	0.012 96	0.000 090 1
29	0.1360	0.427 26	0.014 53	0.000 100 9
28	0.1405	0.441 39	0.015 49	0.000 107 7
9/64"	0.1406	0.441 79	0.015 53	0.000 107 9
27	0.1440	0.442 39	0.016 29	0.000 113 1
26	0.1470	0.461 82	0.016 97	0.000 117 9
25	0.1495	0.469 67	0.017 55	0.000 121 9
24	0.1520	0.477 52	0.018 15	0.000 126 0
23	0.1540	0.483 81	0.018 63	0.000 129 4
5/32"	0.1562	0.490 87	0.019 17	0.000 133 1
22	0.1570	0.493 23	0.019 36	0.000 134 4
21	0.1590	0.499 51	0.019 86	0.000 137 9
20	0.1610	0.505 80	0.020 36	0.000 141 4
19	0.1660	0.521 51	0.021 64	0.000 150 3
18	0.1695	0.532 50	0.022 56	0.000 156 7
11/64"	0.1719	0.539 96	0.023 20	0.000 161 1
17	0.1730	0.543 50	0.023 51	0.000 163 2
16	0.1770	0.556 06	0.024 61	0.000 170 9
15	0.1800	0.565 49	0.025 45	0.000 176 7
14	0.1820	0.571 77	0.026 02	0.000 180 7
13	0.1850	0.581 20	0.026 88	0.000 186 7
3/16"	0.1875	0.589 05	0.027 61	0.000 191 7
12	0.1890	0.593 76	0.028 06	0.000 194 8
11	0.1910	0.600 05	0.028 65	0.000 199 0
10	0.1930	0.606 33	0.029 40	0.000 203 2
9	0.1960	0.615 75	0.030 17	0.000 209 5

TABLE A.1 (continued) Areas and Circumferences of Circles and Drill Sizes

Drill Size	Diameter (in.)	Circumference (in.)	Area (in.)	Area (ft)
8	0.1990	0.625 18	0.031 10	0.000 216 0
7	0.2010	0.631 46	0.031 73	0.000 220 4
13/64"	0.2031	0.638 14	0.032 41	0.000 224 8
6	0.2040	0.640 89	0.032 69	0.000 227 0
5	0.2055	0.645 60	0.033 17	0.000 230 3
4	0.2090	0.656 59	0.034 31	0.000 238 2
3	0.2130	0.669 16	0.035 63	0.000 247 5
7/32"	0.2187	0.687 22	0.037 58	0.000 261 0
2	0.2210	0.694 29	0.038 36	0.000 266 4
1	0.2280	0.716 28	0.040 83	0.000 283 5
A	0.2340	0.735 13	0.043 01	0.000 298 7
15/64"	0.2344	0.736 31	0.043 14	0.000 299 6
B	0.2380	0.747 70	0.044 49	0.000 308 9
C	0.2420	0.760 27	0.046 00	0.000 319 4
D	0.2460	0.772 83	0.047 53	0.000 330 1
E = 1/4"	0.2500	0.785 40	0.049 09	0.000 340 9
F	0.2570	0.807 39	0.051 87	0.000 360 2
G	0.2610	0.819 96	0.053 50	0.000 371 5
17/64"	0.2656	0.834 41	0.055 42	0.000 384 9
H	0.2660	0.835 67	0.055 57	0.000 385 9
I	0.2720	0.854 52	0.058 11	0.000 403 5
J	0.2770	0.870 22	0.060 26	0.000 418 5
K	0.2810	0.882 79	0.062 02	0.000 430 7
9/32"	0.2812	0.883 57	0.062 13	0.000 431 5
L	0.2900	0.911 06	0.066 05	0.000 458 7
M	0.2950	0.926 77	0.068 35	0.000 474 7
19/64"	0.2969	0.932 66	0.069 22	0.000 480 7
N	0.3030	0.951 90	0.071 63	0.000 500 7
5/16"	0.3125	0.981 75	0.076 70	0.000 532 6
O	0.3160	0.992 75	0.078 43	0.000 544 6
P	0.3230	1.014 74	0.081 94	0.000 569 0
21/64"	0.3281	1.030 8	0.084 56	0.000 587 2
Q	0.3320	1.043 0	0.086 57	0.000 601 2
R	0.3390	1.065 0	0.090 26	0.000 626 8
11/32"	0.3437	1.079 8	0.092 81	0.000 644 5
S	0.3480	1.093 3	0.095 11	0.000 660 5
T	0.3580	1.124 7	0.100 6	0.000 699 0
23/64"	0.3594	1.129 0	0.101 4	0.000 704 4
U	0.3680	1.156 1	0.106 4	0.000 738 6
3/8"	0.3750	1.178 1	0.110 5	0.000 767 0
V	0.3770	1.184 4	0.111 6	0.000 775 2
W	0.3860	1.212 7	0.117 0	0.000 812 7
25/64"	0.3906	1.227 2	0.119 8	0.000 832 2
X	0.3970	1.247 2	0.123 8	0.000 859 6

TABLE A.1 (continued) Areas and Circumferences of Circles and Drill Sizes

Drill Size	Diameter (in.)	Circumference (in.)	Area (in.)	Area (ft)
Y	0.4040	1.269 2	0.128 2	0.000 890 2
13/32"	0.4062	1.276 3	0.129 6	0.000 900 1
Z	0.4130	1.297 5	0.134 0	0.000 930 3
27/64"	0.4219	1.325 4	0.139 8	0.000 970 8
7/16"	0.4375	1.3745	0.1503	0.001 044
29/64"	0.4531	1.4235	0.1613	0.001 120
15/32"	0.4687	1.4726	0.1726	0.001 198
31/64"	0.4844	1.5217	0.1843	0.001 280
1/2"	0.5000	1.5708	0.1964	0.001 364
33/64"	0.5156	1.6199	0.2088	0.001 450
17/32"	0.5313	1.6690	0.2217	0.001 539
35/64"	0.5469	1.7181	0.2349	0.001 631
9/16"	0.5625	1.7672	0.2485	0.001 726
37/64"	0.5781	1.8162	0.2625	0.001 823
19/32"	0.5938	1.8653	0.2769	0.001 923
39/64"	0.6094	1.9144	0.2917	0.002 025
5/8"	0.6250	1.9635	0.3068	0.002 131
41/64"	0.6406	2.0126	0.3223	0.002 238
21/32"	0.6562	2.0617	0.3382	0.002 350
43/64"	0.6719	2.1108	0.3545	0.002 462
11/16"	0.6875	2.1598	0.3712	0.002 578
23/32"	0.7188	2.2580	0.4057	0.002 818
3/4"	0.7500	2.3562	0.4418	0.003 068
25/32"	0.7812	2.4544	0.4794	0.003 329
13/16"	0.8125	2.5525	0.5185	0.003 601
27/32"	0.8438	2.6507	0.5591	0.003 883
7/8"	0.8750	2.7489	0.6013	0.004 176
29/32"	0.9062	2.8471	0.6450	0.004 479
15/16"	0.9375	2.9452	0.6903	0.004 794
31/32"	0.9688	3.0434	0.7371	0.005 119
1"	1.0000	3.1416	0.7854	0.005 454
1 1/16"	1.0625	3.3379	0.8866	0.006 157
1 1/8"	1.1250	3.5343	0.9940	0.006 903
1 3/16"	1.1875	3.7306	1.1075	0.007 691
1 1/4"	1.2500	3.9270	1.2272	0.008 522
1 5/16"	1.3125	4.1233	1.3530	0.009 396
1 3/8"	1.3750	4.3170	1.4849	0.010 31
1 7/16"	1.4375	4.5160	1.6230	0.011 27
1 1/2"	1.5000	4.7124	1.7671	0.012 27
1 9/16"	1.5625	4.9087	1.9175	0.013 32
1 5/8"	1.6250	5.1051	2.0739	0.014 40
1 11/16"	1.6875	5.3014	2.2365	0.015 53
1 3/4"	1.7500	5.4978	2.4053	0.016 70
1 13/16"	1.8125	5.6941	2.5802	0.017 92

TABLE A.1 (continued) Areas and Circumferences of Circles and Drill Sizes

Drill Size	Diameter (in.)	Circumference (in.)	Area (in.)	Area (ft)
1 7/8"	1.8750	5.8905	2.7612	0.019 18
1 15/16"	1.9375	6.0868	2.9483	0.020 47
2"	2.0000	6.2832	3.1416	0.021 82
2 1/16"	2.0625	6.4795	3.3410	0.023 20
2 1/8"	2.1250	6.6759	3.5466	0.024 63
2 3/16"	2.1875	6.8722	3.7583	0.026 10
2 1/4"	2.2500	7.0686	3.9761	0.027 61
2 5/16"	2.3125	7.2649	4.2000	0.029 17
2 3/8"	2.3750	7.4613	4.4301	0.030 76
2 7/16"	2.4375	7.6576	4.6664	0.032 41
2 1/2"	2.5000	7.8540	4.9087	0.034 09
2 9/16"	2.5625	8.0503	5.1572	0.035 81
2 5/8"	2.6250	8.2467	5.4119	0.037 58
2 11/16"	2.6875	8.4430	5.6727	0.039 39
2 3/4"	2.7500	8.6394	5.9396	0.041 25
2 13/16"	2.8125	8.8357	6.2126	0.043 14
2 7/8"	2.8750	9.0323	6.4918	0.045 08
2 15/16"	2.9375	9.2284	6.7771	0.047 06
3"	3.0000	9.4248	7.0686	0.049 09
3 1/16"	3.0625	9.6211	7.3662	0.051 15
3 1/8"	3.1250	9.8175	7.6699	0.053 26
3 3/16"	3.1875	10.014	7.9798	0.055 42
3 1/4"	3.2500	10.210	8.2958	0.057 36
3 5/16"	3.3125	10.407	8.6179	0.059 85
3 3/8"	3.3750	10.603	8.9462	0.062 13
3 7/16"	3.4375	10.799	9.2806	0.064 45
3 1/2"	3.5000	10.996	9.6211	0.066 81
3 9/16"	3.5625	11.192	9.9678	0.069 22
3 5/8"	3.6250	11.388	10.321	0.071 67
3 11/16"	3.6875	11.585	10.680	0.074 17
3 3/4"	3.7500	11.781	11.045	0.076 70
3 13/16"	3.8125	11.977	11.416	0.079 28
3 7/8"	3.8750	12.174	11.793	0.081 90
3 15/16"	3.9375	12.370	12.177	0.084 56
4"	4.0000	12.566	12.566	0.087 26
4 1/16"	4.0625	12.763	12.962	0.090 02
4 1/8"	4.1250	12.959	13.364	0.092 81
4 3/16"	4.1875	13.155	13.772	0.095 64
4 1/4"	4.2500	13.352	14.186	0.098 52
4 5/16"	4.3125	13.548	14.607	0.101 4
4 3/8"	4.3750	13.745	15.033	0.104 3
4 7/16"	4.4375	13.941	15.466	0.107 4
4 1/2"	4.5000	14.137	15.904	0.110 4
4 9/16"	4.5625	14.334	16.349	0.113 5

TABLE A.1 (continued) Areas and Circumferences of Circles and Drill Sizes

Drill Size	Diameter (in.)	Circumference (in.)	Area (in.)	Area (ft)
4 5/8"	4.6250	14.530	16.800	0.1167
4 11/16"	4.6875	14.726	17.257	0.1198
4 3/4"	4.7500	14.923	17.721	0.1231
4 13/16"	4.8125	15.119	18.190	0.1263
4 7/8"	4.8750	15.315	18.665	0.1296
4 15/16"	4.9375	15.512	19.147	0.1330
5"	5.0000	15.708	19.635	0.1364
5 1/16"	5.0625	15.904	20.129	0.1398
5 1/8"	5.1250	16.101	20.629	0.1433
5 3/16"	5.1875	16.297	21.135	0.1468
5 1/4"	5.2500	16.493	21.648	0.1503
5 5/16"	5.3125	16.690	22.166	0.1539
5 3/8"	5.3750	16.886	22.691	0.1576
5 7/16"	5.4375	17.082	23.221	0.1613
5 1/2"	5.5000	17.279	23.758	0.1650
5 9/16"	5.5625	17.475	24.301	0.1688
5 5/8"	5.6250	17.671	24.851	0.1726
5 11/16"	5.6875	17.868	25.406	0.1764
5 3/4"	5.7500	18.064	25.967	0.1803
5 13/16"	5.8125	18.261	26.535	0.1843
5 7/8"	5.8750	18.457	27.109	0.1883
5 15/16"	5.9375	18.653	27.688	0.1923
6"	6.0000	18.850	28.274	0.1963
6 1/8"	6.1250	19.242	29.465	0.2046
6 1/4"	6.2500	19.649	30.680	0.2131
6 3/8"	6.3750	20.028	31.919	0.2217
6 1/2"	6.5000	20.420	33.183	0.2304
6 5/8"	6.6250	20.813	34.472	0.2394
6 3/4"	6.7500	21.206	35.785	0.2485
6 7/8"	6.8750	21.598	37.122	0.2578
7"	7.0000	21.991	38.485	0.2673
7 1/8"	7.1250	22.384	39.871	0.2769
7 1/4"	7.2500	22.777	41.283	0.2867
7 3/8"	7.3750	23.169	42.718	0.2967
7 1/2"	7.5000	23.562	44.179	0.3068
7 5/8"	7.6250	23.955	45.664	0.3171
7 3/4"	7.7500	24.347	47.173	0.3276
7 7/8"	7.8750	24.740	48.707	0.3382
8"	8.0000	25.133	50.266	0.3491
8 1/8"	8.1250	25.525	51.849	0.3601
8 1/4"	8.2500	25.918	53.456	0.3712
8 3/8"	8.3750	26.301	55.088	0.3826
8 1/2"	8.5000	26.704	56.745	0.3941
8 5/8"	8.6250	27.096	58.426	0.4057

TABLE A.1 (continued) Areas and Circumferences of Circles and Drill Sizes

Drill Size	Diameter (in.)	Circumference (in.)	Area (in.)	Area (ft)
8 3/4"	8.7500	27.489	60.132	0.4176
8 7/8"	8.8750	27.882	61.862	0.4296
9"	9.0000	28.274	63.617	0.4418
9 1/8"	9.1250	28.667	65.397	0.4541
9 1/4"	9.2500	29.060	67.201	0.4667
9 3/8"	9.3750	29.452	69.029	0.4794
9 1/2"	9.5000	29.845	70.882	0.4922
9 5/8"	9.6250	30.238	72.760	0.5053
9 3/4"	9.7500	30.631	74.662	0.5185
9 7/8"	9.8750	31.023	76.589	0.5319
10"	10.0000	31.416	78.540	0.5454
10 1/8"	10.1250	31.809	80.516	0.5591
10 1/4"	10.2500	32.201	82.516	0.5730
10 3/8"	10.3750	32.594	84.541	0.5871
10 1/2"	10.5000	32.987	86.590	0.6013
10 5/8"	10.6250	33.379	88.664	0.6157
10 3/4"	10.7500	33.772	90.763	0.6303
10 7/8"	10.8750	34.165	92.886	0.6450
11"	11.0000	34.558	95.033	0.6600
11 1/8"	11.1250	34.950	97.205	0.6750
11 1/4"	11.2500	35.343	99.402	0.6903
11 3/8"	11.3750	35.736	101.6	0.7056
11 1/2"	11.5000	36.128	103.9	0.7213
11 5/8"	11.6250	36.521	106.1	0.7371
11 3/4"	11.7500	36.914	108.4	0.7530
11 7/8"	11.8750	37.306	110.8	0.7691
12"	12.0000	37.699	113.1	0.7854
12 1/4"	12.2500	38.485	117.9	0.819
12 1/2"	12.5000	39.269	122.7	0.851
12 3/4"	12.7500	40.055	127.7	0.886
13"	13.0000	40.841	132.7	0.921
13 1/4"	13.2500	41.626	137.9	0.957
13 1/2"	13.5000	42.412	143.1	0.995
13 3/4"	13.7500	43.197	148.5	1.031
14"	14.0000	43.982	153.9	1.069
14 1/4"	14.2500	44.768	159.5	1.109
14 1/2"	14.5000	45.553	165.1	1.149
14 3/4"	14.7500	46.339	170.9	1.185
15"	15.0000	47.124	176.7	1.228
15 1/4"	15.2500	47.909	182.7	1.269
15 1/2"	15.5000	48.695	188.7	1.309
15 3/4"	15.7500	49.480	194.8	1.352
16"	16.0000	50.266	201.1	1.398
16 1/4"	16.2500	51.051	207.4	1.440

TABLE A.1 (continued) Areas and Circumferences of Circles and Drill Sizes

Drill Size	Diameter (in.)	Circumference (in.)	Area (in.)	Area (ft)
16 1/2"	16.5000	51.836	213.8	1.485
16 3/4"	16.7500	52.622	220.4	1.531
17"	17.0000	53.407	227.0	1.578
17 1/4"	17.2500	54.193	233.7	1.619
17 1/2"	17.5000	54.978	240.5	1.673
17 3/4"	17.7500	55.763	247.5	1.719
18"	18.0000	56.548	254.5	1.769
18 1/4"	18.2500	57.334	261.6	1.816
18 1/2"	18.5000	58.120	268.8	1.869
18 3/4"	18.7500	58.905	276.1	1.920
19"	19.0000	59.690	283.5	1.969
19 1/4"	19.2500	60.476	291.0	2.022
19 1/2"	19.5000	61.261	298.7	2.075
19 3/4"	19.7500	62.047	306.4	2.125
20"	20.0000	62.832	314.2	2.182
20 1/4"	20.2500	63.617	322.1	2.237
20 1/2"	20.5000	64.403	330.1	2.292
20 3/4"	20.7500	65.188	338.2	2.348
21"	21.0000	65.974	346.4	2.405
21 1/4"	21.2500	66.759	354.7	2.463
21 1/2"	21.5000	67.544	363.1	2.521
21 3/4"	21.7500	68.330	371.5	2.580
22"	22.0000	69.115	380.1	2.640
22 1/4"	22.2500	69.901	388.8	2.700
22 1/2"	22.5000	70.686	397.6	2.761
22 3/4"	22.7500	71.471	406.5	2.823
23"	23.0000	72.257	415.5	2.885
23 1/4"	23.2500	73.042	424.6	2.948
23 1/2"	23.5000	73.828	433.7	3.012
23 3/4"	23.7500	74.613	443.0	3.076
24"	24.0000	75.398	452.4	3.142
24 1/4"	24.2500	76.184	461.9	3.207
24 1/2"	24.5000	76.969	471.4	3.274
24 3/4"	24.7500	77.755	481.1	3.341
25"	25.0000	78.540	490.9	3.409
25 1/4"	25.2500	79.325	500.7	3.477
25 1/2"	25.5000	80.111	510.7	3.547
25 3/4"	25.7500	80.896	520.8	3.616
26"	26.0000	81.682	530.9	3.687
26 1/4"	26.2500	82.467	541.2	3.758
26 1/2"	26.5000	83.252	551.6	3.830
26 3/4"	26.7500	84.038	562.0	3.903
27"	27.0000	84.823	572.6	3.976
27 1/4"	27.2500	85.609	583.2	4.050

TABLE A.1 (continued) Areas and Circumferences of Circles and Drill Sizes

Drill Size	Diameter (in.)	Circumference (in.)	Area (in.)	Area (ft)
27 1/2"	27.5000	86.394	594.0	4.125
27 3/4"	27.7500	87.179	604.8	4.200
28"	28.0000	87.965	615.8	4.276
28 1/4"	28.2500	88.750	626.8	4.353
28 1/2"	28.5000	89.536	637.9	4.430
28 3/4"	28.7500	90.321	649.2	4.508
29"	29.0000	91.106	660.5	4.587
29 1/4"	29.2500	91.892	672.0	4.666
29 1/2"	29.5000	92.677	683.5	4.746
29 3/4"	29.7500	93.463	695.1	4.827
30"	30.0000	94.248	706.9	4.909
31"	31.0000	97.390	754.8	5.241
32"	32.0000	100.53	804.3	5.585
33"	33.0000	103.67	855.3	5.940
34"	34.0000	106.81	907.9	6.305
35"	35.0000	109.96	962.1	6.681
36"	36.0000	113.10	1017.9	7.069
37"	37.0000	116.24	1075.2	7.467
38"	38.0000	119.38	1134.1	7.876
39"	39.0000	122.52	1194.6	8.296
40"	40.0000	125.66	1256.6	8.727
41"	41.0000	128.81	1320.3	9.168
42"	42.0000	131.95	1385.4	9.621
43"	43.0000	135.09	1452.2	10.08
44"	44.0000	138.23	1520.5	10.56
45"	45.0000	141.37	1590.4	11.04
46"	46.0000	144.51	1661.9	11.54
47"	47.0000	147.66	1734.9	12.04
48"	48.0000	150.80	1809.6	12.57
49"	49.0000	153.94	1885.7	13.10
50"	50.0000	157.08	1963.5	13.64

TABLE A.2 Physical Properties of Pipe

Nominal pipe size, OD, in.	Schedule Number a	Schedule Number b	Schedule Number c	Wall Thick-ness, in.	I.D., in.	Inside Area, sq. in.	Metal Area, sq. in.	Sq. Ft. outside surface, per ft	Sq. Ft. inside surface, per ft	Weight per ft, lb	Weight of water per ft, lb	Moment of Inertia, in.⁴	Section modulus, in.³	Radius gyration, in.
	…	…..	10S	0.049	0.307	0.0740	0.0548	0.106	0.0804	0.186	0.0321	0.00088	0.00437	0.1271
1/8	40	Std	40S	0.068	0.269	0.0568	0.0720	0.106	0.0705	0.245	0.0246	0.00106	0.00525	0.1215
0.405	80	XS	80S	0.095	0.215	0.0364	0.0925	0.106	0.0563	0.315	0.0157	0.00122	0.00600	0.1146
	…	…..	10S	0.065	0.410	0.1320	0.0970	0.141	0.1073	0.330	0.0572	0.00279	0.01032	0.1694
1/4	40	Std	40S	0.088	0.364	0.1041	0.1250	0.141	0.0955	0.425	0.0451	0.00331	0.01230	0.1628
0.540	80	XS	80S	0.119	0.302	0.0716	0.1574	0.141	0.0794	0.535	0.0310	0.00378	0.01395	0.1547
	…	…..	10S	0.065	0.545	0.2333	0.1246	0.177	0.1427	0.423	0.1011	0.00586	0.01737	0.2169
3/8	40	Std	40S	0.091	0.493	0.1910	0.1670	0.177	0.1295	0.568	0.0827	0.00730	0.02160	0.2090
0.675	80	XS	80S	0.126	0.423	0.1405	0.2173	0.177	0.1106	0.739	0.0609	0.00862	0.02554	0.1991
	…	…..	10S	0.083	0.674	0.3570	0.1974	0.220	0.1765	0.671	0.1547	0.01431	0.0341	0.2692
	40	Std	40S	0.109	0.622	0.3040	0.2503	0.220	0.1628	0.851	0.1316	0.01710	0.0407	0.2613
1/2	80	XS	80S	0.147	0.546	0.2340	0.3200	0.220	0.1433	1.088	0.1013	0.02010	0.0478	0.2505
0.840	160	…..	…	0.187	0.466	0.1706	0.3830	0.220	0.1220	1.304	0.0740	0.02213	0.0527	0.2402
	…	XXS	…	0.294	0.252	0.0499	0.5040	0.220	0.0660	1.714	0.0216	0.02425	0.0577	0.2192
	…	…..	5S	0.065	0.920	0.6650	0.2011	0.275	0.2409	0.684	0.2882	0.02451	0.0467	0.349
	…	…..	10S	0.083	0.884	0.6140	0.2521	0.275	0.2314	0.857	0.2661	0.02970	0.0566	0.343
3/4	40	Std	40S	0.113	0.824	0.5330	0.3330	0.275	0.2157	1.131	0.2301	0.0370	0.0706	0.334
1.050	80	XS	80S	0.154	0.742	0.4320	0.4350	0.275	0.1943	1.474	0.1875	0.0448	0.0853	0.321
	160	…..	…	0.218	0.614	0.2961	0.5700	0.275	0.1607	1.937	0.1284	0.0527	0.1004	0.304
	…	XXS	…	0.308	0.434	0.1479	0.7180	0.275	0.1137	2.441	0.0641	0.0579	0.1104	0.284
	…	…..	5S	0.065	1.185	1.1030	0.2553	0.344	0.3100	0.868	0.478	0.0500	0.0760	0.443
	…	…..	10S	0.109	1.097	0.9450	0.4130	0.344	0.2872	1.404	0.409	0.0757	0.1151	0.428
1	40	Std	40S	0.133	1.049	0.8640	0.4940	0.344	0.2746	1.679	0.374	0.0874	0.1329	0.421
1.315	80	XS	80S	0.179	0.957	0.7190	0.6390	0.344	0.2520	2.172	0.311	0.1056	0.1606	0.407
	160	…..	…	0.250	0.815	0.5220	0.8360	0.344	0.2134	2.844	0.2261	0.1252	0.1903	0.387
	…	XXS	…	0.358	0.599	0.2818	1.0760	0.344	0.1570	3.659	0.1221	0.1405	0.2137	0.361
	…	…..	5S	0.065	1.530	1.839	0.326	0.434	0.401	1.107	0.797	0.1038	0.1250	0.564
	…	…..	10S	0.109	1.442	1.633	0.531	0.434	0.378	1.805	0.707	0.1605	0.1934	0.550
1-1/4	40	Std	40S	0.140	1.380	1.496	0.669	0.434	0.361	2.273	0.648	0.1948	0.2346	0.540
1.660	80	XS	80S	0.191	1.278	1.283	0.881	0.434	0.335	2.997	0.555	0.2418	0.2913	0.524
	160	…..	…	0.250	1.160	1.057	1.107	0.434	0.304	3.765	0.458	0.2839	0.342	0.506
	…	XXS	…	0.382	0.896	0.631	1.534	0.434	0.2346	5.214	0.2732	0.341	0.411	0.472
	…	…..	5S	0.065	1.770	2.461	0.375	0.497	0.463	1.274	1.067	0.1580	0.1663	0.649
	…	…..	10S	0.109	1.682	2.222	0.613	0.497	0.440	2.085	0.962	0.2469	0.2599	0.634
1-1/2	40	Std	40S	0.145	1.610	2.036	0.799	0.497	0.421	2.718	0.882	0.310	0.326	0.623
1.900	80	XS	80S	0.200	1.500	1.767	1.068	0.497	0.393	3.631	0.765	0.391	0.412	0.605
	160	…..	…	0.281	1.338	1.406	1.429	0.497	0.350	4.859	0.608	0.483	0.508	0.581
	…	XXS	…	0.400	1.100	0.950	1.885	0.497	0.288	6.408	0.412	0.568	0.598	0.549
	…	…..	5S	0.065	2.245	3.960	0.472	0.622	0.588	1.604	1.716	0.315	0.2652	0.817
	…	…..	10S	0.109	2.157	3.650	0.776	0.622	0.565	2.638	1.582	0.499	0.420	0.802
2	40	Std	40S	0.154	2.067	3.360	1.075	0.622	0.541	3.653	1.455	0.666	0.561	0.787
2.375	80	XS	80S	0.218	1.939	2.953	1.477	0.622	0.508	5.022	1.280	0.868	0.731	0.766
	160	…..	…	0.343	1.689	2.240	2.190	0.622	0.442	7.444	0.971	1.163	0.979	0.729
	…	XXS	…	0.436	1.503	1.774	2.656	0.622	0.393	9.029	0.769	1.312	1.104	0.703

TABLE A.2 (continued) Physical Properties of Pipe

Nominal pipe size, OD, in.	Schedule Number			Wall Thick-ness, in.	I.D., in.	Inside Area, sq. in.	Metal Area, sq. in.	Sq. Ft. outside surface, per ft	Sq. Ft. inside surface, per ft	Weight per ft, lb	Weight of water per ft, lb	Moment of Inertia, in.^{4}	Section modulus, in.^{3}	Radius gyration, in.
	a	b	c											
	...	.....	5S	0.083	2.709	5.76	0.728	0.753	0.709	2.475	2.499	0.710	0.494	0.988
	...	.....	10S	0.120	2.635	5.45	1.039	0.753	0.690	3.531	2.361	0.988	0.687	0.975
2-1/2	40	Std	40S	0.203	2.469	4.79	1.704	0.753	0.646	5.793	2.076	1.530	1.064	0.947
2.875	80	XS	80S	0.276	2.323	4.24	2.254	0.753	0.608	7.661	1.837	0.193	1.339	0.924
	160	.....	...	0.375	2.125	3.55	2.945	0.753	0.556	10.01	1.535	2.353	1.637	0.894
	...	XXS	...	0.552	1.771	2.46	4.030	0.753	0.464	13.70	1.067	2.872	1.998	0.844
	...	.....	5S	0.083	3.334	8.73	0.891	0.916	0.873	3.03	3.78	1.301	0.744	1.208
	...	.....	10S	0.120	3.260	8.35	1.274	0.916	0.853	4.33	3.61	1.822	1.041	1.196
3	40	Std	40S	0.216	3.068	7.39	2.228	0.916	0.803	7.58	3.20	3.02	1.724	1.164
3.500	80	XS	80S	0.300	2.900	6.61	3.020	0.916	0.759	10.25	2.864	3.90	2.226	1.136
	160	.....	...	0.437	2.626	5.42	4.210	0.916	0.687	14.32	2.348	5.03	2.876	1.094
	...	XXS	...	0.600	2.300	4.15	5.470	0.916	0.602	18.58	1.801	5.99	3.43	1.047
	...	.....	5S	0.083	3.834	11.55	1.021	1.047	1.004	3.47	5.01	1.960	0.980	1.385
3-1/2	...	.....	10S	0.120	3.760	11.10	1.463	1.047	0.984	4.97	4.81	2.756	1.378	1.372
4.000	40	Std	40S	0.226	3.548	9.89	2.68	1.047	0.929	9.11	4.28	4.79	2.394	1.337
	80	XS	80S	0.318	3.364	8.89	3.68	1.047	0.881	12.51	3.85	6.28	3.14	1.307
	...	.....	5S	0.083	4.334	14.75	1.152	1.178	1.135	3.92	6.40	2.811	1.249	1.562
	...	.....	10S	0.120	4.260	14.25	1.651	1.178	1.115	5.61	6.17	3.96	1.762	1.549
4	40	Std	40S	0.237	4.026	12.73	3.17	1.178	1.054	10.79	5.51	7.23	3.21	1.510
4.500	80	XS	80S	0.337	3.826	11.50	4.41	1.178	1.002	14.98	4.98	9.61	4.27	1.477
	120	.....	...	0.437	3.626	10.33	5.58	1.178	0.949	18.96	4.48	11.65	5.18	1.445
	160	.....	...	0.531	3.438	9.28	6.62	1.178	0.900	22.51	4.02	13.27	5.90	1.416
	...	XXS	...	0.674	3.152	7.80	8.10	1.178	0.825	27.54	3.38	15.29	6.79	1.374
	...	.....	5S	0.109	5.345	22.44	1.868	1.456	1.399	6.35	9.73	6.95	2.498	1.929
	...	.....	10S	0.134	5.295	22.02	2.285	1.456	1.386	7.77	9.53	8.43	3.03	1.920
5	40	Std	40S	0.258	5.047	20.01	4.30	1.456	1.321	14.62	8.66	15.17	5.45	1.878
5.563	80	XS	80S	0.375	4.813	18.19	6.11	1.456	1.260	20.78	7.89	20.68	7.43	1.839
	120	.....	...	0.500	4.563	16.35	7.95	1.456	1.195	27.04	7.09	25.74	9.25	1.799
	160	.....	...	0.625	4.313	14.61	9.70	1.456	1.129	32.96	6.33	30	10.8	1.760
	...	XXS	...	0.750	4.063	12.97	11.34	1.456	1.064	38.55	5.62	33.6	12.1	1.722
	...	.....	5S	0.109	6.407	32.20	2.231	1.734	1.677	5.37	13.98	11.85	3.58	2.304
	...	.....	10S	0.134	6.357	31.70	2.733	1.734	1.664	9.29	13.74	14.4	4.35	2.295
6	40	Std	40S	0.280	6.065	28.89	5.58	1.734	1.588	18.97	12.51	28.14	8.5	2.245
6.625	80	XS	80S	0.432	5.761	26.07	8.40	1.734	1.508	28.57	11.29	40.5	12.23	2.195
	120	.....	...	0.562	5.501	23.77	10.70	1.734	1.440	36.39	10.30	49.6	14.98	2.153
	160	.....	...	0.718	5.189	21.15	13.33	1.734	1.358	45.30	9.16	59	17.81	2.104
	...	XXS	...	0.864	4.897	18.83	15.64	1.734	1.282	53.16	8.17	66.3	20.03	2.060
	...	.....	5S	0.109	8.407	55.5	2.916	2.258	2.201	9.91	24.07	26.45	6.13	3.01
	...	.....	10S	0.148	8.329	54.5	3.94	2.258	2.180	13.40	23.59	35.4	8.21	3.00
	20	.....	...	0.250	8.125	51.8	6.58	2.258	2.127	22.36	22.48	57.7	13.39	2.962
	30	.....	...	0.277	8.071	51.2	7.26	2.258	2.113	24.70	22.18	63.4	14.69	2.953
	40	Std	40S	0.322	7.981	50.0	8.40	2.258	2.089	28.55	21.69	72.5	16.81	2.938
8	60	.....	...	0.406	7.813	47.9	10.48	2.258	2.045	35.64	20.79	88.8	20.58	2.909
8.625	80	XS	80S	0.500	7.625	45.7	12.76	2.258	1.996	43.39	19.80	105.7	24.52	2.878
	100	.....	...	0.593	7.439	43.5	14.96	2.258	1.948	50.87	18.84	121.4	28.14	2.847
	120	.....	...	0.718	7.189	40.6	17.84	2.258	1.882	60.63	17.60	140.6	32.6	2.807
	140	.....	...	0.812	7.001	38.5	19.93	2.258	1.833	67.76	16.69	153.8	35.7	2.777

TABLE A.2 (continued) Physical Properties of Pipe

Nominal pipe size, OD, in.	Schedule Number a	b	c	Wall Thick-ness, in.	I.D., in.	Inside Area, sq. in.	Metal Area, sq. in.	Sq. Ft. outside surface, per ft	Sq. Ft. inside surface, per ft	Weight per ft, lb	Weight of water per ft, lb	Moment of Inertia, in.^{4}	Section modulus, in.^{3}	Radius gyration, in.
	...	XXS	...	0.875	6.875	37.1	21.30	2.258	1.800	72.42	16.09	162	37.6	2.757
	160	....	...	0.906	6.813	36.5	21.97	2.258	1.784	74.69	15.80	165.9	38.5	2.748
	...	.....	5S	0.134	10.482	86.3	4.52	2.815	2.744	15.15	37.4	63.7	11.85	3.75
	...	.....	10S	0.165	10.420	85.3	5.49	2.815	2.728	18.70	36.9	76.9	14.3	3.74
	20	.....	...	0.250	10.250	82.5	8.26	2.815	2.683	28.04	35.8	113.7	21.16	3.71
	...	.....	...	0.279	10.192	81.6	9.18	2.815	2.668	31.20	35.3	125.9	23.42	3.70
	30	.....	...	0.307	10.136	80.7	10.07	2.815	2.654	34.24	35.0	137.5	25.57	3.69
10	40	Std	40S	0.365	10.020	78.9	11.91	2.815	2.623	40.48	34.1	160.8	29.9	3.67
10.750	60	XS	80S	0.500	9.750	74.7	16.10	2.815	2.553	54.74	32.3	212	39.4	3.63
	80	.....	...	0.593	9.564	71.8	18.92	2.815	2.504	64.33	31.1	244.9	45.6	3.60
	100	.....	...	0.718	9.314	68.1	22.63	2.815	2.438	76.93	29.5	286.2	53.2	3.56
	120	.....	...	0.843	9.064	64.5	26.24	2.815	2.373	89.20	28.0	324	60.3	3.52
	140	.....	...	1.000	8.750	60.1	30.6	2.815	2.291	104.13	26.1	368	68.4	3.47
	160	.....	...	1.125	8.500	56.7	34.0	2.815	2.225	115.65	24.6	399	74.3	3.43
	...	.....	5S	0.165	12.420	121.2	6.52	3.34	3.25	19.56	52.5	129.2	20.27	4.45
	...	.....	10S	0.180	12.390	120.6	7.11	3.34	3.24	24.20	52.2	140.5	22.03	4.44
	20	.....	...	0.250	12.250	117.9	9.84	3.34	3.21	33.38	51.1	191.9	30.1	4.42
	30	.....	...	0.330	12.090	114.8	12.88	3.34	3.17	43.77	49.7	248.5	39.0	4.39
	...	Std	40S	0.375	12.000	113.1	14.58	3.34	3.14	49.56	49.0	279.3	43.8	4.38
12	40	.....	...	0.406	11.938	111.9	15.74	3.34	3.13	53.53	48.5	300	47.1	4.37
12.750	...	XS	80S	0.500	11.750	108.4	19.24	3.34	3.08	65.42	47.0	362	56.7	4.33
	60	.....	...	0.562	11.626	106.2	21.52	3.34	3.04	73.16	46.0	401	62.8	4.31
	80	.....	...	0.687	11.376	101.6	26.04	3.34	2.978	88.51	44.0	475	74.5	4.27
	100	.....	...	0.843	11.064	96.1	31.5	3.34	2.897	107.20	41.6	562	88.1	4.22
	120	.....	...	1.000	10.750	90.8	36.9	3.34	2.814	125.49	39.3	642	100.7	4.17
	140	.....	...	1.125	10.500	86.6	41.1	3.34	2.749	139.68	37.5	701	109.9	4.13
	160	.....	...	1.312	10.126	80.5	47.1	3.34	2.651	160.27	34.9	781	122.6	4.07
	10	.....	...	0.250	13.500	143.1	10.80	3.67	3.53	36.71	62.1	255.4	36.5	4.86
	20	.....	...	0.312	13.376	140.5	13.42	3.67	3.5	45.68	60.9	314	44.9	4.84
	30	Std	...	0.375	13.250	137.9	16.05	3.67	3.47	54.57	59.7	373	53.3	4.82
	40	.....	...	0.437	13.126	135.3	18.62	3.67	3.44	63.37	58.7	429	61.2	4.80
	...	XS	...	0.500	13.000	132.7	21.21	3.67	3.4	72.09	57.5	484	69.1	4.78
	...	.....	...	0.562	12.876	130.2	23.73	3.67	3.37	80.66	56.5	537	76.7	4.76
14	60	.....	...	0.593	12.814	129.0	24.98	3.67	3.35	84.91	55.9	562	80.3	4.74
14.000	...	.....	...	0.625	12.750	127.7	26.26	3.67	3.34	89.28	55.3	589	84.1	4.73
	...	.....	...	0.687	12.626	125.2	28.73	3.67	3.31	97.68	54.3	638	91.2	4.71
	80	.....	...	0.750	12.500	122.7	31.2	3.67	3.27	106.13	53.2	687	98.2	4.69
	...	.....	...	0.875	12.250	117.9	36.1	3.67	3.21	122.66	51.1	781	111.5	4.65
	100	.....	...	0.937	12.126	115.5	38.5	3.67	3.17	130.73	50.0	825	117.8	4.63
	120	.....	...	1.093	11.814	109.6	44.3	3.67	3.09	150.67	47.5	930	132.8	4.58
	140	.....	...	1.250	11.500	103.9	50.1	3.67	3.01	170.22	45.0	1127	146.8	4.53
	160	.....	...	1.406	11.188	98.3	55.6	3.67	2.929	189.12	42.6	1017	159.6	4.48
	10	.....	...	0.250	15.500	188.7	12.37	4.19	4.06	42.05	81.8	384	48	5.57
	20	.....	...	0.312	15.376	185.7	15.38	4.19	4.03	52.36	80.5	473	59.2	5.55
	30	Std	...	0.375	15.250	182.6	18.41	4.19	3.99	62.58	79.1	562	70.3	5.53
	...	.....	...	0.437	15.126	179.7	21.37	4.19	3.96	72.64	77.9	648	80.9	5.50
	40	XS	...	0.500	15.000	176.7	24.35	4.19	3.93	82.77	76.5	732	91.5	5.48
	...	.....	...	0.562	14.876	173.8	27.26	4.19	3.89	92.66	75.4	813	106.6	5.46
	...	.....	...	0.625	14.750	170.9	30.2	4.19	3.86	102.63	74.1	894	112.2	5.44

TABLE A.2 (continued) Physical Properties of Pipe

Nominal pipe size, OD, in.	Schedule Number a	b	c	Wall Thickness, in.	I.D., in.	Inside Area, sq. in.	Metal Area, sq. in.	Sq. Ft. outside surface, per ft	Sq. Ft. inside surface, per ft	Weight per ft, lb	Weight of water per ft, lb	Moment of Inertia, in.⁴	Section modulus, in.³	Radius gyration, in.
16	60	.....	...	0.656	14.688	169.4	31.6	4.19	3.85	107.50	73.4	933	116.6	5.43
16.000	...	.....	...	0.687	14.626	168.0	33.0	4.19	3.83	112.36	72.7	971	121.4	5.42
	...		...	0.750	14.500	165.1	35.9	4.19	3.8	122.15	71.5	1047	130.9	5.40
	80	.....	...	0.842	14.314	160.9	40.1	4.19	3.75	136.46	69.7	1157	144.6	5.37
	...	.....	...	0.875	14.250	159.5	41.6	4.19	3.73	141.35	69.1	1193	154.1	5.36
	100	.....	...	1.031	13.938	152.6	48.5	4.19	3.65	164.83	66.1	1365	170.6	5.30
	120	.....	...	1.218	13.564	144.5	56.6	4.19	3.55	192.29	62.6	1556	194.5	5.24
	140	.....	...	1.437	13.126	135.3	65.7	4.19	3.44	223.50	58.6	1760	220.0	5.17
	160	.....	...	1.593	12.814	129.0	72.1	4.19	3.35	245.11	55.9	1894	236.7	5.12
	10	.....	...	0.250	17.500	240.5	13.94	4.71	4.58	47.39	104.3	549	61.0	6.28
	20	.....	...	0.312	17.376	237.1	17.34	4.71	4.55	59.03	102.8	678	75.5	6.25
	...	Std	...	0.375	17.250	233.7	20.76	4.71	4.52	70.59	101.2	807	89.6	6.23
	30	.....	...	0.437	17.126	230.4	24.11	4.71	4.48	82.06	99.9	931	103.4	6.21
	...	XS	...	0.500	17.000	227.0	27.49	4.71	4.45	93.45	98.4	1053	117.0	6.19
	40	.....	...	0.562	16.876	223.7	30.8	4.71	4.42	104.75	97.0	1172	130.2	6.17
	...	.....	...	0.625	16.750	220.5	34.1	4.71	4.39	115.98	95.5	1289	143.3	6.15
18	...	.....	...	0.687	16.626	217.1	37.4	4.71	4.35	127.03	94.1	1403	156.3	6.13
18.000	60	.....	...	0.750	16.500	213.8	40.6	4.71	4.32	138.17	92.7	1515	168.3	6.10
	...	.....	...	0.875	16.250	207.4	47.1	4.71	4.25	160.04	89.9	1731	192.8	6.06
	80	.....	...	0.937	16.126	204.2	50.2	4.71	4.22	170.75	88.5	1834	203.8	6.04
	100	.....	...	1.156	15.688	193.3	61.2	4.71	4.11	207.96	83.7	2180	242.2	5.97
	120	.....	...	1.375	15.250	182.6	71.8	4.71	3.99	244.14	79.2	2499	277.6	5.90
	140	.....	...	1.562	14.876	173.8	80.7	4.71	3.89	274.23	75.3	2750	306	5.84
	160	.....	...	1.781	14.438	163.7	90.7	4.71	3.78	308.51	71.0	3020	336	5.77
	10	.....	...	0.250	19.500	298.6	15.51	5.24	5.11	52.73	129.5	757	75.7	6.98
	...	.....	...	0.312	19.376	294.9	19.30	5.24	5.07	65.40	128.1	935	93.5	6.96
	20	Std	...	0.375	19.250	291.0	23.12	5.24	5.04	78.60	126.0	1114	111.4	6.94
	...	.....	...	0.437	19.126	287.3	26.86	5.24	5.01	91.31	124.6	1286	128.6	6.92
	30	XS	...	0.500	19.000	283.5	30.6	5.24	4.97	104.13	122.8	1457	145.7	6.90
	...	.....	...	0.562	18.876	279.8	34.3	5.24	4.94	116.67	121.3	1624	162.4	6.88
20	40	.....	...	0.593	18.814	278.0	36.2	5.24	4.93	122.91	120.4	1704	170.4	6.86
20.000	...	.....	...	0.625	18.750	276.1	38.0	5.24	4.91	129.33	119.7	1787	178.7	6.85
	...	.....	...	0.687	18.626	272.5	41.7	5.24	4.88	141.71	118.1	1946	194.6	6.83
	...	.....	...	0.750	18.500	268.8	45.4	5.24	4.84	154.20	116.5	2105	210.5	6.81
	60	.....	...	0.812	18.376	265.2	48.9	5.24	4.81	166.40	115.0	2257	225.7	6.79
	...	.....	...	0.875	18.250	261.6	52.6	5.24	4.78	178.73	113.4	2409	240.9	6.77
	80	.....	...	1.031	17.938	252.7	61.4	5.24	4.70	208.87	109.4	2772	277.2	6.72
	100	.....	...	1.281	17.438	238.8	75.3	5.24	4.57	256.10	103.4	3320	332	6.63
	120	.....	...	1.500	17.000	227.0	87.2	5.24	4.45	296.37	98.3	3760	376	6.56
	140	.....	...	1.750	16.500	213.8	100.3	5.24	4.32	341.10	92.6	4220	422	6.48
	160	.....	...	1.968	16.064	202.7	111.5	5.24	4.21	379.01	87.9	4590	459	6.41
	10	.....	...	0.250	23.500	434	18.65	6.28	6.15	63.41	188.0	1316	109.6	8.40
	...	.....	...	0.312	23.376	430	23.20	6.28	6.12	78.93	186.1	1629	135.8	8.38
	20	Std	...	0.375	23.250	425	27.83	6.28	6.09	94.62	183.8	1943	161.9	8.35
	...	.....	...	0.437	23.126	420	32.4	6.28	6.05	109.97	182.1	2246	187.4	8.33
	...	XS	...	0.500	23.000	415	36.9	6.28	6.02	125.49	180.1	2550	212.5	8.31
24	30	.....	...	0.562	22.876	411	41.4	6.28	5.99	140.80	178.1	2840	237.0	8.29
24.000	...	.....	...	0.625	22.750	406	45.9	6.28	5.96	156.03	176.2	3140	261.4	8.27
	40	.....	...	0.687	22.626	402	50.3	6.28	5.92	171.17	174.3	3420	285.2	8.25
	...	.....	...	0.750	22.500	398	54.8	6.28	5.89	186.24	172.4	3710	309	8.22

TABLE A.2 (continued) Physical Properties of Pipe

Nominal pipe size, OD, in.	Schedule Number a	b	c	Wall Thick-ness, in.	I.D., in.	Inside Area, sq. in.	Metal Area, sq. in.	Sq. Ft. outside surface, per ft	Sq. Ft. inside surface, per ft	Weight per ft, lb	Weight of water per ft, lb	Moment of Inertia, in.^{4}	Section modulus, in.^{3}	Radius gyration, in.
	60	.....	...	0.968	22.064	382	70.0	6.28	5.78	238.11	165.8	4650	388	8.15
	80	.....	...	1.218	21.564	365	87.2	6.28	5.65	296.36	158.3	5670	473	8.07
	100	.....	...	1.531	20.938	344	108.1	6.28	5.48	367.40	149.3	6850	571	7.96
	120	.....	...	1.812	20.376	326	126.3	6.28	5.33	429.39	141.4	7830	652	7.87
	140	.....	...	2.062	19.876	310	142.1	6.28	5.20	483.13	134.5	8630	719	7.79
	160	.....	...	2.343	19.314	293	159.4	6.28	5.06	541.94	127.0	9460	788	7.70
	10	.....	...	0.312	29.376	678	29.1	7.85	7.69	98.93	293.8	3210	214	10.50
30	20	.....	...	0.500	29.000	661	46.3	7.85	7.59	157.53	286.3	5040	336	10.43
30.000	30	.....	...	0.625	28.750	649	57.6	7.85	7.53	196.08	281.5	6220	415	10.39

a = ASA B36.10 Steel-pipe schedule numbers
b = ASA B36.10 Steel-pipe nominal wall-thickness designations
c = ASA B36.19 Stainless-steel-pipe schedule numbers

TABLE A.3 Commercial Copper Tubing*

The following table gives dimensional data and weights of copper tubing used for automotive, plumbing, refrigeration, and heat exchanger services. For additional data see the standards handbooks of the Copper Development Association, Inc., the ASTM standards, and the "SAE Handbook."

Dimensions in this table are actual specified measurements, subject to accepted tolerances. Trade size designations are usually by actual OD, except for water and drainage tube (plumbing), which measures 1/8-in. larger OD. A 1/2-in. plumbing tube, for example, measures 5/8-in. OD, and a 2-in. plumbing tube measures 2 1/8-in. OD.

KEY TO GAGE SIZES

Standard-gage wall thicknesses are listed by numerical designation (14 to 21), BWG or Stubs gage. These gage sizes are standard for tubular heat exchangers. The letter *A* designates SAE tubing sizes for automotive service. Letter designations *K* and *L* are the common sizes for plumbing services, soft or hard temper.

OTHER MATERIALS

These same dimensional sizes are also common for much of the commercial tubing available in aluminum, mild steel, brass, bronze, and other alloys. Tube weights in this table are based on copper at 0.323 lb/in^3. For other materials the weights should be multiplied by the following approximate factors:

aluminum	0.30	monel	0.96
mild steel	0.87	stainless steel	0.89
brass	0.95		

Size, OD		Wall Thickness			Flow Area		Metal Area,	Surface Area		Weight,
in.	mm	in.	mm	gage	in.^{2}	mm^2	in.^{2}	Inside, ft^2/ft	Outside, ft^2/ft	lb/ft
1/8	3.2	.030	0.76	A	0.003	1.9	0.012	0.017	0.033	0.035
3/16	4.76	.030	0.76	A	0.013	8.4	0.017	0.034	0.049	0.058
1/4	6.4	.030	0.76	A	0.028	18.1	0.021	0.050	0.066	0.080
1/4	6.4	.049	1.24	18	0.018	11.6	0.031	0.038	0.066	0.120
5/16	7.94	.032	0.81	21A	0.048	31.0	0.028	0.065	0.082	0.109
3/8	9.53	.032	0.81	21A	0.076	49.0	0.033	0.081	0.098	0.134
3/8	9.53	.049	1.24	18	0.060	38.7	0.050	0.072	0.098	0.195
1/2	12.7	.032	0.81	21A	0.149	96.1	0.047	0.114	0.131	0.182
1/2	12.7	.035	0.89	20L	0.145	93.6	0.051	0.113	0.131	0.198
1/2	12.7	.049	1.24	18K	0.127	81.9	0.069	0.105	0.131	0.269
1/2	12.7	.065	1.65	16	0.108	69.7	0.089	0.97	0.131	0.344
5/8	15.9	.035	0.89	20A	0.242	156	0.065	0.145	0.164	0.251
5/8	15.9	.040	1.02	L	0.233	150	0.074	0.143	0.164	0.285
5/8	15.9	.049	1.24	18K	0.215	139	0.089	0.138	0.164	0.344
3/4	19.1	.035	0.89	20A	0.363	234	0.079	0.178	0.196	0.305
3/4	19.1	.042	1.07	L	0.348	224	0.103	0.174	0.196	0.362
3/4	19.1	.049	1.24	18K	0.334	215	0.108	0.171	0.196	0.418
3/4	19.1	.065	1.65	16	0.302	195	0.140	0.162	0.196	0.542
3/4	19.1	.083	2.11	14	0.268	173	0.174	0.151	0.196	0.674
7/8	22.2	.045	1.14	L	0.484	312	0.117	0.206	0.229	0.455
7/8	22.2	.065	1.65	16K	0.436	281	0.165	0.195	0.229	0.641
7/8	22.2	.083	2.11	14	0.395	255	0.206	0.186	0.229	0.800
1	25.4	.065	1.65	16	0.594	383	0.181	0.228	0.262	0.740
1	25.4	.083	2.11	14	0.546	352	0.239	0.218	0.262	0.927
1 1/8	28.6	.050	1.27	L	0.825	532	0.176	0.268	0.294	0.655

* Compiled and computed.

From: The CRC Handbook of Mechanical Engineering, CRC Press, Boca Raton, FL, 1998.

TABLE A.3 (continued) Commercial Copper Tubing*

Size, OD		Wall Thickness			Flow Area		Metal Area, in²	Surface Area		Weight, lb/ft
in.	mm	in.	mm	gage	in.²	mm²		Inside, ft²/ft	Outside, ft²/ft	
1 1/8	28.6	.065	1.65	16K	0.778	502	0.216	0.261	0.294	0.839
1 1/4	31.8	.065	1.65	16	0.985	636	0.242	0.293	0.327	0.938
1 1/4	31.8	.083	2.11	14	0.923	596	0.304	0.284	0.327	1.18
1 3/8	34.9	.055	1.40	L	1.257	811	0.228	0.331	0.360	0.884
1 3/8	34.9	.065	1.65	16K	1.217	785	0.267	0.326	0.360	1.04
1 1/2	38.1	.065	1.65	16	1.474	951	0.294	0.359	0.393	1.14
1 1/2	38.7	.083	2.11	14	1.398	902	0.370	0.349	0.393	1.43
1 5/8	41.3	.060	1.52	L	1.779	1148	0.295	0.394	0.425	1.14
1 5/8	41.3	.072	1.83	K	1.722	1111	0.351	0.388	0.425	1.36
2	50.8	.083	2.11	14	2.642	1705	0.500	0.480	0.628	1.94
2	50.8	.109	2.76	12	2.494	1609	0.620	0.466	0.628	2.51
2 1/8	54.0	.070	1.78	L	3.095	1997	0.449	0.520	0.556	1.75
2 1/8	54.0	.083	2.11	14K	3.016	1946	0.529	0.513	0.556	2.06
2 5/8	66.7	.080	2.03	L	4.77	3078	0.645	0.645	0.687	2.48
2 5/8	66.7	.095	2.41	13K	4.66	3007	0.760	0.637	0.687	2.93
3 1/8	79.4	.090	2.29	L	6.81	4394	0.950	0.771	0.818	3.33
3 1/8	79.4	.109	2.77	12K	6.64	4284	1.034	0.761	0.818	4.00
3 5/8	92.1	.100	2.54	L	9.21	5942	1.154	0.897	0.949	4.29
3 5/8	92.1	.120	3.05	11K	9.00	5807	1.341	0.886	0.949	5.12
4 1/8	104.8	.110	2.79	L	11.92	7691	1.387	1.022	1.080	5.38
4 1/8	104.8	.134	3.40	10K	11.61	7491	1.682	1.009	1.080	6.51

TABLE A.4 Standard Grades of Bolts

Part a: SAE Grades for Steel Bolts

SAE grade no.	Size range incl.	Proof strength,† kpsi	Tensile strength,† kpsi	Material	Head marking
1	$\frac{1}{4}$–$1\frac{1}{2}$			Low- or medium-carbon steel	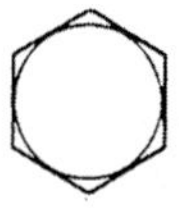
2	$\frac{1}{4}$–$\frac{3}{4}$	55	74		
	$\frac{7}{8}$–$1\frac{1}{2}$	33	60		
5	$\frac{1}{4}$–1	85	120	Medium-carbon steel, Q & T	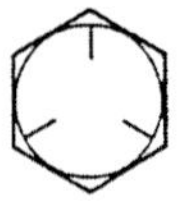
	$1\frac{1}{8}$–$1\frac{1}{2}$	74	105		
5.2	$\frac{1}{4}$–1	85	120	Low-carbon martensite steel, Q & T	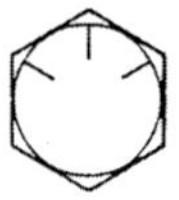
7	$\frac{1}{4}$–$1\frac{1}{2}$	105	133	Medium-carbon alloy steel, Q & T‡	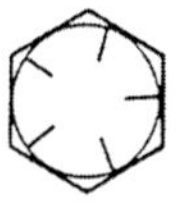
8	$\frac{1}{4}$–$1\frac{1}{2}$	120	150	Medium-carbon alloy steel, Q & T	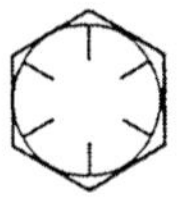
8.2	$\frac{1}{4}$–1	120	150	Low-carbon martensite steel, Q & T	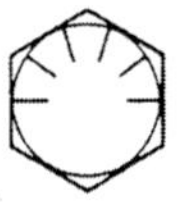

†Minimum values.
‡Roll threaded after heat treatment.
SOURCES: See "Helpful Hints," by Russell, Burdsall & Ward Corp., Mentor, Ohio 44060; and Chap. 23.

From: The CRC Press Handbook of Mechanical Engineering, CRC Press, Boca Raton, FL, 1998.

TABLE A.5 Standard Grades of Bolts

Part b: ASTM Grades for Steel Bolts

ASTM designation	Size range incl.	Proof strength,† kpsi	Tensile strength,† kpsi	Material	Head marking
A307	$\frac{1}{4}$ to 4			Low-carbon steel	
A325 type 1	$\frac{1}{2}$ to 1 $1\frac{1}{8}$ to $1\frac{1}{2}$	85 74	120 105	Medium-carbon steel, Q & T	A325
A325 type 2	$\frac{1}{2}$ to 1 $1\frac{1}{8}$ to $1\frac{1}{2}$	85 74	120 105	Low-carbon martensite steel, Q & T	A325
A325 type 3	$\frac{1}{2}$ to 1 $1\frac{1}{8}$ to $1\frac{1}{2}$	85 74	120 105	Weathering steel, Q & T	A325
A354 grade BC				Alloy steel, Q & T	BC
A354 grade BD	$\frac{1}{4}$ to 4	120	150	Alloy steel, Q & T	
A449	$\frac{1}{4}$ to 1 $1\frac{1}{8}$ to $1\frac{1}{2}$ $1\frac{3}{4}$ to 3	85 74 55	120 105 90	Medium-carbon steel, Q & T	
A490type	$\frac{1}{2}$ to $1\frac{1}{2}$	120	150	Alloy steel, Q & T	A490
A490type 3				Weathering steel, Q & T	A490

† Minimum value.

Sources: See "Helpful Hints," by Russell,Burdsall&Ward Corp.,Mentor,Ohio 44060; and Chapter 23.

From: The CRC Press Handbook of Mechanical Engineering, CRC Press, Boca Raton, FL, 1998.

TABLE A.6 Standard Grades of Bolts

Part c: Metric Mechanical Property Classes for Steel Bolts, Screws, and Studs

Property class	Size range incl.	Proof strength, MPa	Tensile strength, MPa	Material	Head marking
4.6	M5–M36	225	400	Low- or medium-carbon steel	4.6
4.8	M1.6–M16	310	420	Low- or medium-carbon steel	4.8
5.8	M5–M24	380	520	Low- or medium-carbon steel	5.8
8.8	M16–M36	600	830	Medium-carbon steel, Q & T	8.8
9.8	M1.6–M16	650	900	Medium-carbon steel, Q & T	9.8
10.9	M5–M36	830	1040	Low-carbon martensite steel, Q & T	10.9
12.9	M1.6–M36	970	1220	Alloy steel, Q & T	12.9

sources: "Helpful Hints," by Russell, Burdsall & Waard Corp,. Mentor, Ohio 44060; see also Chapter 23 and SAEstandard J1199, and ASTM standard F568.

From: The CRC Press Handbook of Mechanical Engineering, CRC Press, Boca Raton, FL, 1998.

Appendix B
Properties of Gases and Liquids

TABLE B.1 Combustion Data for Hydrocarbons*

Hydrocarbon	*Formula*	*Higher heating value (vapor), Btu/lb$_m$*	*Theor. air/fuel ratio, by mass*	*Max flame speed, ft/sec*	*Adiabatic flame temp (in air), °F*	*Ignition temp (in air), °F*	*Flash point, °F*	*Flammability limits (in air), % by volume*	
				PARAFFINS OR ALKANES					
Methane	CH_4	23875	17.195	1.1	3484	1301	gas	5.0	15.0
Ethane	C_2H_6	22323	15.899	1.3	3540	968–1166	gas	3.0	12.5
Propane	C_3H_8	21669	15.246	1.3	3573	871	gas	2.1	10.1
n-Butane	C_4H_{10}	21321	14.984	1.2	3583	761	−76	1.86	8.41
iso-Butane	C_4H_{10}	21271	14.984	1.2	3583	864	−117	1.80	8.44
n-Pentane	C_5H_{12}	21095	15.323	1.3	4050	588	< −40	1.40	7.80
iso-Pentane	C_5H_{12}	21047	15.323	1.2	4055	788	< −60	1.32	9.16
Neopentane	C_5H_{12}	20978	15.323	1.1	4060	842	gas	1.38	7.22
n-Hexane	C_6H_{14}	20966	15.238	1.3	4030	478	−7	1.25	7.0
Neohexane	C_6H_{14}	20931	15.238	1.2	4055	797	−54	1.19	7.58
n-Heptane	C_7H_{16}	20854	15.141	1.3	3985	433	25	1.00	6.00
Triptane	C_7H_{16}	20824	15.141	1.2	4035	849	—	1.08	6.69
n-Octane	C_8H_{18}	20796	15.093	—	—	428	56	0.95	3.20
iso-Octane	C_8H_{18}	20770	15.093	1.1	—	837	10	0.79	5.94
				OLEFINS OR ALKENES					
Ethylene	C_2H_4	21636	14.807	2.2	4250	914	gas	2.75	28.6
Propylene	C_3H_6	21048	14.807	1.4	4090	856	gas	2.00	11.1
Butylene	C_4H_8	20854	14.807	1.4	4030	829	gas	1.98	9.65
iso-Butene	C_4H_8	20737	14.807	1.2	—	869	gas	1.8	9.0
n-Pentene	C_5H_{10}	20720	14.807	1.4	4165	569	—	1.65	7.70
				AROMATICS					
Benzene	C_6H_6	18184	13.297	1.3	4110	1044	12	1.35	6.65
Toluene	C_7H_8	18501	13.503	1.2	4050	997	40	1.27	6.75
p-Xylene	C_8H_{10}	18663	13.663	—	4010	867	63	1.00	6.00
				OTHER HYDROCARBONS					
Acetylene	C_2H_2	21502	13.297	4.6	4770	763–824	gas	2.50	81.0
Naphthalene	$C_{10}H_8$	17303	12.932	—	4100	959	174	0.90	5.9

* Based largely on: *Gas Engineers' Handbook*, American Gas Association, Inc., Industrial Park, 1967.

REFERENCES

American Institute of Physics Handbook, 2nd ed., D.E. Gray, Ed., McGraw-Hill Book Company, NY, 1963.

Chemical Engineer's Handbook, 4th ed., R.H. Perry, C.H. Chilton, and S.D. Kirkpatrick, Eds., McGraw-Hill Book Company, NY, 1963.

Handbook of Chemistry and Physics, 53rd ed., R.C. Weast, Ed., The Chemical Rubber Company, Cleveland, OH, 1972; gives the heat of combustion of 500 organic compounds.

Handbook of Laboratory Safety, 2nd ed., N.V. Steere, Ed., The Chemical Rubber Company, Cleveland, OH, 1971.

Physical Measurements in Gas Dynamics and Combustion, Princeton University Press, 1954.

Note: For heating value in J/kg, multiply the value in Btu/lb$_m$ by 2324. For flame speed in m/s, multiply the value in ft/s by 0.3048.
From: The CRC Press Handbook of Mechanical Engineering, CRC Press, Boca Raton, FL, 1998.

TABLE B.2 Enthalpy of Formation, Gibbs Function of Formation, and Absolute Entropy of Various Substances at 298 K and 1 atm

$\bar{h}_f^\circ$ **and** $\bar{g}_f^\circ$ **(kJ/kmol),** $\bar{s}^\circ$ **(kJ/kmol•K)**

Substance	Formula	$\bar{h}_f^\circ$	$\bar{g}_f^\circ$	$\bar{s}^\circ$
Carbon	C(s)	0	0	5.74
Hydrogen	H_2(g)	0	0	130.57
Nitrogen	N_2(g)	0	0	191.50
Oxygen	O_2(g)	0	0	205.03
Carbon monoxide	CO(g)	–110,530	–137,150	197.54
Carbon dioxide	CO_2(g)	–393,520	–394,380	213.69
Water	H_2O(g)	–241,820	–228,590	188.72
	H_2O(l)	–285,830	–237,180	69.95
Hydrogen peroxide	H_2O_2(g)	–136,310	–105,600	232.63
Ammonia	NH_3(g)	–46,190	–16,590	192.33
Oxygen	O(g)	249,170	231,770	160.95
Hydrogen	H(g)	218,000	203,290	114.61
Nitrogen	N(g)	472,680	455,510	153.19
Hydroxyl	OH(g)	39,460	34,280	183.75
Methane	CH_4(g)	–74,850	–50,790	186.16
Acetylene	C_2H_2(g)	226,730	209,170	200.85
Ethylene	C_2H_4(g)	52,280	68,120	219.83
Ethane	C_2H_6(g)	–84,680	–32,890	229.49
Propylene	C_3H_6(g)	20,410	62,720	266.94
Propane	C_3H_8(g)	–103,850	–23,490	269.91
Butane	C_4H_{10}(g)	–126,150	–15,710	310.03
Pentane	C_5H_{12}(g)	–146,440	–8200	348.40
Octane	C_8H_{18}(g)	–208,450	17,320	463.67
	C_8H_{18}(l)	–249,910	6610	360.79
Benzene	C_6H_6(g)	82,930	129,660	269.20
Methyl alcohol	CH_3OH(g)	–200,890	–162,140	239.70
	CH_3OH(l)	–238,810	–166,290	126.80
Ethyl alcohol	C_2H_5OH(g)	–235,310	–168,570	282.59
	C_2H_5OH(l)	–277,690	174,890	160.70

Source: Adapted from Wark, K. 1983. *Thermodynamics*, 4th ed. McGraw-Hill, New York, as based on JANAF Thermochemical Tables, NSRDS-NBS-37, 1971; *Selected Values of Chemical Thermodynamic Properties*, NBS Tech. Note 270-3, 1968; and *API Research Project 44*, Carnegie Press, 1953.

From: The CRC Handbook of Thermal Engineering CRC Press, Boca Raton, FL, 2000.

TABLE B.3 Properties of Dry Air at Atmospheric Pressure

Symbols and Units:

K = absolute temperature, degrees Kelvin
deg C = temperature, degrees Celsius
deg F = temperature, degrees Fahrenheit
ρ = density, kg/m^3
c_p = specific heat capacity, kJ/kg·K
c_p/c_v = specific heat capacity ratio, dimensionless
μ = viscosity, N·s/m^2 × 10^6 (For N·s/m^2 (= kg/m·s) multiply tabulated values by 10^{-6})
k = thermal conductivity, W/m·k × 10^3 (For W/m·K multiply tabulated values by 10^{-3})
Pr = Prandtl number, dimensionless
h = enthalpy, kJ/kg
V_s = sound velocity, m/s

Temperature			Properties							
K	deg C	deg F	ρ	c_p	c_p/c_v	μ	k	Pr	h	V_s
100	−173.15	−280	3.598	1.028		6.929	9.248	.770	98.42	198.4
110	−163.15	−262	3.256	1.022	1.420 2	7.633	10.15	.768	108.7	208.7
120	−153.15	−244	2.975	1.017	1.416 6	8.319	11.05	.766	118.8	218.4
130	−143.15	−226	2.740	1.014	1.413 9	8.990	11.94	.763	129.0	227.6
140	−133.15	−208	2.540	1.012	1.411 9	9.646	12.84	.761	139.1	236.4
150	−123.15	−190	2.367	1.010	1.410 2	10.28	13.73	.758	149.2	245.0
160	−113.15	−172	2.217	1.009	1.408 9	10.91	14.61	.754	159.4	253.2
170	−103.15	−154	2.085	1.008	1.407 9	11.52	15.49	.750	169.4	261.0
180	−93.15	−136	1.968	1.007	1.407 1	12.12	16.37	.746	179.5	268.7
190	−83.15	−118	1.863	1.007	1.406 4	12.71	17.23	.743	189.6	276.2
200	−73.15	−100	1.769	1.006	1.405 7	13.28	18.09	.739	199.7	283.4
205	−68.15	−91	1.726	1.006	1.405 5	13.56	18.52	.738	204.7	286.9
210	−63.15	−82	1.684	1.006	1.405 3	13.85	18.94	.736	209.7	290.5
215	−58.15	−73	1.646	1.006	1.405 0	14.12	19.36	.734	214.8	293.9
220	−53.15	−64	1.607	1.006	1.404 8	14.40	19.78	.732	219.8	297.4
225	−48.15	−55	1.572	1.006	1.404 6	14.67	20.20	.731	224.8	300.8
230	−43.15	−46	1.537	1.006	1.404 4	14.94	20.62	.729	229.8	304.1
235	−38.15	−37	1.505	1.006	1.404 2	15.20	21.04	.727	234.9	307.4
240	−33.15	−28	1.473	1.005	1.404 0	15.47	21.45	.725	239.9	310.6
245	−28.15	−19	1.443	1.005	1.403 8	15.73	21.86	.724	244.9	313.8
250	−23.15	−10	1.413	1.005	1.403 6	15.99	22.27	.722	250.0	317.1
255	−18.15	−1	1.386	1.005	1.403 4	16.25	22.68	.721	255.0	320.2
260	−13.15	8	1.359	1.005	1.403 2	16.50	23.08	.719	260.0	323.4
265	−8.15	17	1.333	1.005	1.403 0	16.75	23.48	.717	265.0	326.5
270	−3.15	26	1.308	1.006	1.402 9	17.00	23.88	.716	270.1	329.6
275	+1.85	35	1.285	1.006	1.402 6	17.26	24.28	.715	275.1	332.6
280	6.85	44	1.261	1.006	1.402 4	17.50	24.67	.713	280.1	335.6
285	11.85	53	1.240	1.006	1.402 2	17.74	25.06	.711	285.1	338.5
290	16.85	62	1.218	1.006	1.402 0	17.98	25.47	.710	290.2	341.5
295	21.85	71	1.197	1.006	1.401 8	18.22	25.85	.709	295.2	344.4
300	26.85	80	1.177	1.006	1.401 7	18.46	26.24	.708	300.2	347.3
305	31.85	89	1.158	1.006	1.401 5	18.70	26.63	.707	305.3	350.2
310	36.85	98	1.139	1.007	1.401 3	18.93	27.01	.705	310.3	353.1
315	41.85	107	1.121	1.007	1.401 0	19.15	27.40	.704	315.3	355.8
320	46.85	116	1.103	1.007	1.400 8	19.39	27.78	.703	320.4	358.7

* Condensed and computed from: Tables of Thermal Properties of Gases, National Bureau of Standards Circular 564, U.S. Government Printing Office, November 1955.

From: The CRC Press Handbook of Thermal Engineering, CRC Press, Boca Raton, FL, 2000.

TABLE B.3 (continued) Properties of Dry Air at Atmospheric Pressure

Temperature			*Properties*							
K	*deg C*	*deg F*	ρ	c_p	c_p/c_v	μ	k	*Pr*	*h*	V_s
325	51.85	125	1.086	1.008	1.400 6	19.63	28.15	.702	325.4	361.4
330	56.85	134	1.070	1.008	1.400 4	19.85	28.53	.701	330.4	364.2
335	61.85	143	1.054	1.008	1.400 1	20.08	28.90	.700	335.5	366.9
340	66.85	152	1.038	1.008	1.399 9	20.30	29.28	.699	340.5	369.6
345	71.85	161	1.023	1.009	1.399 6	20.52	29.64	.698	345.6	372.3
350	76.85	170	1.008	1.009	1.399 3	20.75	30.03	.697	350.6	375.0
355	81.85	179	0.994 5	1.010	1.399 0	20.97	30.39	.696	355.7	377.6
360	86.85	188	0.980 5	1.010	1.398 7	21.18	30.78	.695	360.7	380.2
365	91.85	197	0.967 2	1.010	1.398 4	21.38	31.14	.694	365.8	382.8
370	96.85	206	0.953 9	1.011	1.398 1	21.60	31.50	.693	370.8	385.4
375	101.85	215	0.941 3	1.011	1.397 8	21.81	31.86	.692	375.9	388.0
380	106.85	224	0.928 8	1.012	1.397 5	22.02	32.23	.691	380.9	390.5
385	111.85	233	0.916 9	1.012	1.397 1	22.24	32.59	.690	386.0	393.0
390	116.85	242	0.905 0	1.013	1.396 8	22.44	32.95	.690	391.0	395.5
395	121.85	251	0.893 6	1.014	1.396 4	22.65	33.31	.689	396.1	398.0
400	126.85	260	0.882 2	1.014	1.396 1	22.86	33.65	.689	401.2	400.4
410	136.85	278	0.860 8	1.015	1.395 3	23.27	34.35	.688	411.3	405.3
420	146.85	296	0.840 2	1.017	1.394 6	23.66	35.05	.687	421.5	410.2
430	156.85	314	0.820 7	1.018	1.393 8	24.06	35.75	.686	431.7	414.9
440	166.85	332	0.802 1	1.020	1.392 9	24.45	36.43	.684	441.9	419.6
450	176.85	350	0.784 2	1.021	1.392 0	24.85	37.10	.684	452.1	424.2
460	186.85	368	0.767 7	1.023	1.391 1	25.22	37.78	.683	462.3	428.7
470	196.85	386	0.750 9	1.024	1.390 1	25.58	38.46	.682	472.5	433.2
480	206.85	404	0.735 1	1.026	1.389 2	25.96	39.11	.681	482.8	437.6
490	216.85	422	0.720 1	1.028	1.388 1	26.32	39.76	.680	493.0	442.0
500	226.85	440	0.705 7	1.030	1.387 1	26.70	40.41	.680	503.3	446.4
510	236.85	458	0.691 9	1.032	1.386 1	27.06	41.06	.680	513.6	450.6
520	246.85	476	0.678 6	1.034	1.385 1	27.42	41.69	.680	524.0	454.9
530	256.85	494	0.665 8	1.036	1.384 0	27.78	42.32	.680	534.3	459.0
540	266.85	512	0.653 5	1.038	1.382 9	28.14	42.94	.680	544.7	463.2
550	276.85	530	0.641 6	1.040	1.381 8	28.48	43.57	.680	555.1	467.3
560	286.85	548	0.630 1	1.042	1.380 6	28.83	44.20	.680	565.5	471.3
570	296.85	566	0.619 0	1.044	1.379 5	29.17	44.80	.680	575.9	475.3
580	306.85	584	0.608 4	1.047	1.378 3	29.52	45.41	.680	586.4	479.2
590	316.85	602	0.598 0	1.049	1.377 2	29.84	46.01	.680	596.9	483.2
600	326.85	620	0.588 1	1.051	1.376 0	30.17	46.61	.680	607.4	486.9
620	346.85	656	0.569 1	1.056	1.373 7	30.82	47.80	.681	628.4	494.5
640	366.85	692	0.551 4	1.061	1.371 4	31.47	48.96	.682	649.6	502.1
660	386.85	728	0.534 7	1.065	1.369 1	32.09	50.12	.682	670.9	509.4
680	406.85	764	0.518 9	1.070	1.366 8	32.71	51.25	.683	692.2	516.7
700	426.85	800	0.504 0	1.075	1.364 6	33.32	52.36	.684	713.7	523.7
720	446.85	836	0.490 1	1.080	1.362 3	33.92	53.45	.685	735.2	531.0
740	466.85	872	0.476 9	1.085	1.360 1	34.52	54.53	.686	756.9	537.6
760	486.85	908	0.464 3	1.089	1.358 0	35.11	55.62	.687	778.6	544.6
780	506.85	944	0.452 4	1.094	1.355 9	35.69	56.68	.688	800.5	551.2
800	526.85	980	0.441 0	1.099	1.354	36.24	57.74	.689	822.4	557.8
850	576.85	1 070	0.415 2	1.110	1.349	37.63	60.30	.693	877.5	574.1
900	626.85	1 160	0.392 0	1.121	1.345	38.97	62.76	.696	933.4	589.6
950	676.85	1 250	0.371 4	1.132	1.340	40.26	65.20	.699	989.7	604.9
1 000	726.85	1 340	0.352 9	1.142	1.336	41.53	67.54	.702	1 046	619.5
1 100	826.85	1 520	0.320 8	1.161	1.329	43.96			1 162	648.0
1 200	926.85	1 700	0.294 1	1.179	1.322	46.26			1 279	675.2
1 300	1 026.85	1 880	0.271 4	1.197	1.316	48.46			1 398	701.0
1 400	1 126.85	2 060	0.252 1	1.214	1.310	50.57			1 518	725.9
1 500	1 220.85	2 240	0.235 3	1.231	1.304	52.61			1 640	749.4
1 600	1 326.85	2 420	0.220 6	1.249	1.299	54.57			1 764	772.6
1 800	1 526.85	2 780	0.196 0	1.288	1.288	58.29			2 018	815.7
2 000	1 726.85	3 140	0.176 4	1.338	1.274				2 280	855.5
2 400	2 126.85	3 860	0.146 7	1.574	1.238				2 853	924.4
2 800	2 526.85	4 580	0.124 5	2.259	1.196				3 599	983.1

TABLE B.4 Chemical, Physical, and Thermal Properties of Gases: Gases and Vapors, Including Fuels and Refrigerants, English and Metric Units

Common name(s)	*Acetylene (Ethyne)*	*Butadiene*	*n-Butane*	*Isobutane (2-Methyl propane)*
Chemical formula	C_2H_2	C_4H_6	C_4H_{10}	C_4H_{10}
Refrigerant number	—	—	*600*	*600a*
CHEMICAL AND PHYSICAL PROPERTIES				
Molecular weight	26.04	54.09	58.12	58.12
Specific gravity, air = 1	0.90	1.87	2.07	2.07
Specific volume, ft³/lb	14.9	7.1	6.5	6.5
Specific volume, m³/kg	0.93	0.44	0.405	0.418
Density of liquid (at atm bp), lb/ft³	43.0		37.5	37.2
Density of liquid (at atm bp), kg/m³	693.		604.	599.
Vapor pressure at 25 deg C, psia			35.4	50.4
Vapor pressure at 25 deg C, MN/m²			0.024 4	0.347
Viscosity (abs), lbm/ft·sec	6.72×10^{-6}		4.8×10^{-6}	
Viscosity (abs), centipoises[a]	0.01		0.007	
Sound velocity in gas, m/sec	343	226	216	216
THERMAL AND THERMODYNAMIC PROPERTIES				
Specific heat, c_p, Btu/lb·deg F or cal/g·deg C	0.40	0.341	0.39	0.39
Specific heat, c_p, J/kg·K	1 674.	1 427.	1 675.	1 630.
Specific heat ratio, c_p/c_v	1.25	1.12	1.096	1.10
Gas constant R, ft-lb/lb·deg R	59.3	28.55	26.56	26.56
Gas constant R, J/kg·deg C	319	154.	143.	143.
Thermal conductivity, Btu/hr·ft·deg F	0.014		0.01	0.01
Thermal conductivity, W/m·deg C	0.024		0.017	0.017
Boiling point (sat 14.7 psia), deg F	−103	24.1	31.2	10.8
Boiling point (sat 760 mm), deg C	−75	−4.5	−0.4	−11.8
Latent heat of evap (at bp), Btu/lb	264		165.6	157.5
Latent heat of evap (at bp), J/kg	614 000		386 000	366 000
Freezing (melting) point, deg F (1 atm)	−116	−164.	−217.	−229
Freezing (melting) point, deg C (1 atm)	−82.2	−109.	−138	−145
Latent heat of fusion, Btu/lb	23.		19.2	
Latent heat of fusion, J/kg	53 500		44 700	
Critical temperature, deg F	97.1		306	273.
Critical temperature, deg C	36.2	171.	152.	134.
Critical pressure, psia	907.	652.	550.	537.
Critical pressure, MN/m²	6.25		3.8	3.7
Critical volume, ft³/lb			0.070	
Critical volume, m³/kg			0.004 3	
Flammable (yes or no)	Yes	Yes	Yes	Yes
Heat of combustion, Btu/ft³	1 450	2 950	3 300	3 300
Heat of combustion, Btu/lb	21 600	20 900	21 400	21 400
Heat of combustion, kJ/kg	50 200	48 600	49 700	49 700

[a]For N·sec/m² divide by 1 000.

Note: The properties of pure gases are given at 25°C (77°F, 298 K) and atmospheric pressure (except as stated).
From: The CRC Press Handbook of Thermal Engineering, CRC Press, Boca Raton, FL, 2000.

TABLE B.4 (continued) Chemical, Physical, and Thermal Properties of Gases: Gases and Vapors, Including Fuels and Refrigerants, English and Metric Units

Common name(s)	*1-Butene (Butylene)*	*cis-2-Butene*	*trans-2-Butene*	*Isobutene*
Chemical formula	C_4H_8	C_4H_8	C_4H_8	C_4H_8
Refrigerant number	—	—	—	—
CHEMICAL AND PHYSICAL PROPERTIES				
Molecular weight	56.108	56.108	56.108	56.108
Specific gravity, air = 1	1.94	1.94	1.94	1.94
Specific volume, ft³/lb	6.7	6.7	6.7	6.7
Specific volume, m³/kg	0.42	0.42	0.42	0.42
Density of liquid (at atm bp), lb/ft³				
Density of liquid (at atm bp), kg/m³				
Vapor pressure at 25 deg C, psia				
Vapor pressure at 25 deg C, MN/m²				
Viscosity (abs), lbm/ft·sec				
Viscosity (abs), centipoises[a]				
Sound velocity in gas, m/sec	222	223.	221.	221.
THERMAL AND THERMODYNAMIC PROPERTIES				
Specific heat, c_p, Btu/lb·deg F or cal/g·deg C	0.36	0.327	0.365	0.37
Specific heat, c_p, J/kg·K	1 505.	1 368.	1 527.	1 548.
Specific heat ratio, c_p/c_v	1.112	1.121	1.107	1.10
Gas constant R, ft-lb/lb·deg F	27.52			
Gas constant R, J/kg·deg C	148.			
Thermal conductivity, Btu/hr·ft·deg F				
Thermal conductivity, W/m·deg C				
Boiling point (sat 14.7 psia), deg F	20.6	38.6	33.6	19.2
Boiling point (sat 760 mm), deg C	−6.3	3.7	0.9	−7.1
Latent heat of evap (at bp), Btu/lb	167.9	178.9	174.4	169.
Latent heat of evap (at bp), J/kg	391 000	416 000.	406 000.	393 000.
Freezing (melting) point, deg F (1 atm)	−301.6	−218.	−158.	
Freezing (melting) point, deg C (1 atm)	−185.3	−138.9	−105.5	
Latent heat of fusion, Btu/lb	16.4	31.2	41.6	25.3
Latent heat of fusion, J/kg	38 100	72 600.	96 800.	58 800.
Critical temperature, deg F	291.			
Critical temperature, deg C	144.	160.	155.	
Critical pressure, psia	621.	595.	610.	
Critical pressure, MN/m²	4.28	4.10	4.20	
Critical volume, ft³/lb	0.068			
Critical volume, m³/kg	0.004 2			
Flammable (yes or no)	Yes	Yes	Yes	Yes
Heat of combustion, Btu/ft³	3 150	3 150.	3 150.	3 150.
Heat of combustion, Btu/lb	21 000	21 000.	21 000.	21 000.
Heat of combustion, kJ/kg	48 800	48 800.	48 800.	48 800.

[a]For N·sec/m² divide by 1 000.

TABLE B.4 (continued) Chemical, Physical, and Thermal Properties of Gases: Gases and Vapors, Including Fuels and Refrigerants, English and Metric Units

Common name(s)	*Carbon dioxide*	*Carbon monoxide*	*Ethane*	*Ethylene (Ethene)*
Chemical formula	CO_2	CO	C_2H_6	C_2H_4
Refrigerant number	744	—	170	1 150
CHEMICAL AND PHYSICAL PROPERTIES				
Molecular weight	44.01	28.011	30.070	28.054
Specific gravity, air = 1	1.52	0.967	1.04	0.969
Specific volume, ft^3/lb	8.8	14.0	13.025	13.9
Specific volume, m^3/kg	0.55	0.874	0.815	0.87
Density of liquid (at atm bp), lb/ft^3	—		28.	35.5
Density of liquid (at atm bp), kg/m^3	—		449.	569.
Vapor pressure at 25 deg C, psia	931.			
Vapor pressure at 25 deg C, MN/m^2	6.42			
Viscosity (abs), lbm/ft·sec	9.4×10^{-6}	12.1×10^{-6}	$64. \times 10^{-6}$	6.72×10^{-6}
Viscosity (abs), centipoises[a]	0.014	0.018	0.095	0.010
Sound velocity in gas, m/sec	270.	352.	316.	331.
THERMAL AND THERMODYNAMIC PROPERTIES				
Specific heat, c_p, Btu/lb·deg F or cal/g·deg C	0.205	0.25	0.41	0.37
Specific heat, c_p, J/kg·K	876.	1 046.	1 715.	1 548.
Specific heat ratio, c_p/c_v	1.30	1.40	1.20	1.24
Gas constant R, ft-lb/lb·deg F	35.1	55.2	51.4	55.1
Gas constant R, J/kg·deg C	189.	297.	276.	296.
Thermal conductivity, Btu/hr·ft·deg F	0.01	0.014	0.010	0.010
Thermal conductivity, W/m·deg C	0.017	0.024	0.017	0.017
Boiling point (sat 14.7 psia), deg F	−109.4[b]	−312.7	−127.	−155.
Boiling point (sat 760 mm), deg C	−78.5	−191.5	−88.3	−103.8
Latent heat of evap (at bp), Btu/lb	246.	92.8	210.	208.
Latent heat of evap (at bp), J/kg	572 000.	216 000.	488 000.	484 000.
Freezing (melting) point, deg F (1 atm)		−337.	−278.	−272.
Freezing (melting) point, deg C (1 atm)		−205.	−172.2	−169.
Latent heat of fusion, Btu/lb	—	12.8	41.	51.5
Latent heat of fusion, J/kg	—		95 300.	120 000.
Critical temperature, deg F	88.	−220.	90.1	49.
Critical temperature, deg C	31.	−140.	32.2	9.5
Critical pressure, psia	1 072.	507.	709.	741.
Critical pressure, MN/m^2	7.4	3.49	4.89	5.11
Critical volume, ft^3/lb		0.053	0.076	0.073
Critical volume, m^3/kg		0.003 3	0.004 7	0.004 6
Flammable (yes or no)	No	Yes	Yes	Yes
Heat of combustion, Btu/ft^3	—	310.		1 480.
Heat of combustion, Btu/lb	—	4 340.	22 300.	20 600.
Heat of combustion, kJ/kg	—	10 100.	51 800.	47 800.

[a]For N·sec/m^2 divide by 1 000.

TABLE B.4 (continued) Chemical, Physical, and Thermal Properties of Gases: Gases and Vapors, Including Fuels and Refrigerants, English and Metric Units

Common name(s)	*Hydrogen*	*Methane*	*Nitric oxide*	*Nitrogen*
Chemical formula	H_2	CH_4	NO	N_2
Refrigerant number	702	50	—	728
CHEMICAL AND PHYSICAL PROPERTIES				
Molecular weight	2.016	16.044	30.006	28.013 4
Specific gravity, air = 1	0.070	0.554	1.04	0.967
Specific volume, ft^3/lb	194.	24.2	13.05	13.98
Specific volume, m^3/kg	12.1	1.51	0.814	0.872
Density of liquid (at atm bp), lb/ft^3	4.43	26.3		50.46
Density of liquid (at atm bp), kg/m^3	71.0	421.		808.4
Vapor pressure at 25 deg C, psia				
Vapor pressure at 25 deg C, MN/m^2				
Viscosity (abs), lbm/ft·sec	6.05×10^{-6}	7.39×10^{-6}	12.8×10^{-6}	12.1×10^{-6}
Viscosity (abs), centipoises[a]	0.009	0.011	0.019	0.018
Sound velocity in gas, m/sec	1 315.	446.	341.	353.
THERMAL AND THERMO-DYNAMIC PROPERTIES				
Specific heat, c_p, Btu/lb·deg F or cal/g·deg C	3.42	0.54	0.235	0.249
Specific heat, c_p, J/kg·K	14 310.	2 260.	983.	1 040.
Specific heat ratio, c_p/c_v	1.405	1.31	1.40	1.40
Gas constant R, ft-lb/lb·deg F	767.	96.	51.5	55.2
Gas constant R, J/kg·deg C	4 126.	518.	277.	297.
Thermal conductivity, Btu/hr·ft·deg F	0.105	0.02	0.015	0.015
Thermal conductivity, W/m·deg C	0.018 2	0.035	0.026	0.026
Boiling point (sat 14.7 psia), deg F	−423.	−259.	−240.	−320.4
Boiling point (sat 760 mm), deg C	20.4 K	−434.2	−151.5	−195.8
Latent heat of evap (at bp), Btu/lb	192.	219.2		85.5
Latent heat of evap (at bp), J/kg	447 000.	510 000.		199 000.
Freezing (melting) point, deg F (1 atm)	−434.6	−296.6	−258.	−346.
Freezing (melting) point, deg C (1 atm)	−259.1	−182.6	−161.	−210.
Latent heat of fusion, Btu/lb	25.0	14.	32.9	11.1
Latent heat of fusion, J/kg	58 000.	32 600.	76 500.	25 800.
Critical temperature, deg F	−399.8	−116.	−136.	−232.6
Critical temperature, deg C	−240.0	−82.3	−93.3	−147.
Critical pressure, psia	189.	673.	945.	493.
Critical pressure, MN/m^2	1.30	4.64	6.52	3.40
Critical volume, ft^3/lb	0.53	0.099	0.033 2	0.051
Critical volume, m^3/kg	0.033	0.006 2	0.002 07	0.003 18
Flammable (yes or no)	Yes	Yes	No	No
Heat of combustion, Btu/ft^3	320.	985.	—	—
Heat of combustion, Btu/lb	62 050.	2 2900.	—	—
Heat of combustion, kJ/kg	144 000.		—	—

[a]For N·sec/m^2 divide by 1 000.

TABLE B.4 (continued) Chemical, Physical, and Thermal Properties of Gases: Gases and Vapors, Including Fuels and Refrigerants, English and Metric Units

Common name(s)	*Nitrous oxide*	*Oxygen*	*Propane*	*Propylene (Propene)*
Chemical formula	N_2O	O_2	C_3H_8	C_3H_6
Refrigerant number	*744A*	*732*	*290*	*1 270*
CHEMICAL AND PHYSICAL PROPERTIES				
Molecular weight	44.012	31.998 8	44.097	42.08
Specific gravity, air = 1	1.52	1.105	1.52	1.45
Specific volume, ft^3/lb	8.90	12.24	8.84	9.3
Specific volume, m^3/kg	0.555	0.764	0.552	0.58
Density of liquid (at atm bp), lb/ft^3	76.6	71.27	36.2	37.5
Density of liquid (at atm bp), kg/m^3	1 227.	1 142.	580.	601.
Vapor pressure at 25 deg C, psia			135.7	166.4
Vapor pressure at 25 deg C, MN/m^2			0.936	1.147
Viscosity (abs), lbm/ft·sec	10.1×10^{-6}	13.4×10^{-6}	53.8×10^{-6}	57.1×10^{-6}
Viscosity (abs), centipoises[a]	0.015	0.020	0.080	0.085
Sound velocity in gas, m/sec	268.	329.	253.	261.
THERMAL AND THERMODYNAMIC PROPERTIES				
Specific heat, c_p, Btu/lb·deg F or cal/g·deg C	0.21	0.220	0.39	0.36
Specific heat, c_p, J/kg·K	879.	920.	1 630.	1 506.
Specific heat ratio, c_p/c_v	1.31	1.40	1.2	1.16
Gas constant R, ft-lb/lb·deg F	35.1	48.3	35.0	36.7
Gas constant R, J/kg·deg C	189.	260.	188.	197.
Thermal conductivity, Btu/hr·ft·deg F	0.010	0.015	0.010	0.010
Thermal conductivity, W/m·deg C	0.017	0.026	0.017	0.017
Boiling point (sat 14.7 psia), deg F	−127.3	−297.3	−44.	−54.
Boiling point (sat 760 mm), deg C	−88.5	−182.97	−42.2	−48.3
Latent heat of evap (at bp), Btu/lb	161.8	91.7	184.	188.2
Latent heat of evap (at bp), J/kg	376 000.	213 000.	428 000.	438 000.
Freezing (melting) point, deg F (1 atm)	−131.5	−361.1	−309.8	−301.
Freezing (melting) point, deg C (1 atm)	−90.8	−218.4	−189.9	−185.
Latent heat of fusion, Btu/lb	63.9	5.9	19.1	
Latent heat of fusion, J/kg	149 000.	13 700.	44 400.	
Critical temperature, deg F	97.7	−181.5	205.	197.
Critical temperature, deg C	36.5	−118.6	96.	91.7
Critical pressure, psia	1 052.	726.	618.	668.
Critical pressure, MN/m^2	7.25	5.01	4.26	4.61
Critical volume, ft^3/lb	0.036	0.040	0.073	0.069
Critical volume, m^3/kg	0.002 2	0.002 5	0.004 5	0.004 3
Flammable (yes or no)	No	No	Yes	Yes
Heat of combustion, Btu/ft^3	—	—	2 450.	2 310.
Heat of combustion, Btu/lb	—	—	21 660.	21 500.
Heat of combustion, kJ/kg	—	—	50 340.	50 000.

[a] For $N{\cdot}sec/m^2$ divide by 1 000.

TABLE B.4 (continued) Chemical, Physical, and Thermal Properties of Gases: Gases and Vapors, Including Fuels and Refrigerants, English and Metric Units

Common name(s)	*Sulfur dioxide*
Chemical formula	*SO_2*
Refrigerant number	*764*
CHEMICAL AND PHYSICAL PROPERTIES	
Molecular weight	64.06
Specific gravity, air = 1	2.21
Specific volume, ft^3/lb	6.11
Specific volume, m^3/kg	
Density of liquid (at atm bp), lb/ft^3	42.8
Density of liquid (at atm bp), kg/m^3	585.
Vapor pressure at 25 deg C, psia	56.6
Vapor pressure at 25 deg C, MN/m^2	0.390
Viscosity (abs), lbm/ft·sec	8.74×10^{-6}
Viscosity (abs), centipoises[a]	0.013
Sound velocity in gas, m/sec	220.
THERMAL AND THERMODYNAMIC PROPERTIES	
Specific heat, c_p, Btu/lb·deg F or cal/g·deg C	0.11
Specific heat, c_p, J/kg·K	460.
Specific heat ratio, c_p/c_v	1.29
Gas constant R, ft-lb/lb·deg F	24.1
Gas constant R, J/kg·deg C	130.
Thermal conductivity, Btu/hr·ft·deg F	0.006
Thermal conductivity, W/m·deg C	0.010
Boiling point (sat 14.7 psia), deg F	14.0
Boiling point (sat 760 mm), deg C	−10.
Latent heat of evap (at bp), Btu/lb	155.5
Latent heat of evap (at bp), J/kg	362 000.
Freezing (melting) point, deg F (1 atm)	−104.
Freezing (melting) point, deg C (1 atm)	−75.5
Latent heat of fusion, Btu/lb	58.0
Latent heat of fusion, J/kg	135 000.
Critical temperature, deg F	315.5
Critical temperature, deg C	157.6
Critical pressure, psia	1 141.
Critical pressure, MN/m^2	7.87
Critical volume, ft^3/lb	0.03
Critical volume, m^3/kg	0.001 9
Flammable (yes or no)	No
Heat of combustion, Btu/ft^3	—
Heat of combustion, Btu/lb	—
Heat of combustion, kJ/kg	—

[a]For N·sec/m^2 divide by 1 000.

Appendix C Common Conversions

1 Btu =	252.0 cal
	1055 J
1 Btu/ft^3 =	0.00890 cal/cm^3
	0.0373 MJ/m^3
1 Btu/hr =	0.0003931 hp
	0.2520 kcal/hr
	0.2931 W
1,000,000 Btu/hr =	0.293 MW
1 Btu/hr-ft^2 =	0.003153 kW/m^2
1 Btu/hr-ft-°F =	1.730 W/m-K
1 Btu/hr-ft^2-°F =	5.67 W/m^2-K
1 Btu/lb =	0.5556 cal/g
	2326 J/kg
1 Btu/lb-°F =	1 cal/g-°C
	4187 J/kg-K
1 cal =	0.003968 Btu
	4.187 J
1 cal/cm^2-sec =	3.687 Btu/ft^2-sec
	41.87 kW/m^2
1 cal/cm-sec-°C =	241.9 Btu/ft-hr-°F
	418.7 W/m-K
1 cal/g =	1.80 Btu/lb
	4187 J/kg
1 cal/g-°C =	1 Btu/lb-°F
	4187 J/kg-K
1 centipoise =	2.421 lbm/hr-ft
1 cm^2/sec =	100 centistokes
	3.874 ft^2/hr
1 ft =	0.3048 m
1 ft^2/sec =	0.0929 m^2/sec
1 g/cm^3 =	1000 kg/m^3
	62.43 lb/ft^3
	0.03613 lb/in.^{3}
1 hp =	33,000 ft-lb/min
	550 ft-lb/sec
	641.4 kcal/hr
	745.7 W
1 in. =	2.540 cm
	25.40 mm
1 J =	0.000948 Btu
	0.239 cal
	1 W/sec
1 kcal =	3.968 Btu
	1000 cal
	4187 J
1 kcal/hr =	3.968 Btu/hr
	1.162 J/sec
1 kcal/m^3 =	0.1124 Btu/ft^3
	4187 J/m^3
1 kg =	2.205 lb
1 kg/hr-m =	0.00278 g/sec-cm
	0.672 lb/hr-ft
1 kg/m^3 =	0.06243 lb/ft^3
1 kW =	3413 Btu/hr
	1.341 hp
	660.6 kcal/hr
1 kW/m^2 =	317.2 Btu/hr-ft^2
1 kW/m^2-°C =	176.2 Btu/hr-ft^2-°F
1 lb =	0.4536 kg
1 lb/ft^3 =	0.0160 g/cm^3
	16.02 kg/m^3
1 lbm/hr-ft =	0.413 centipoise
1 m =	3.281 ft
1 mm =	0.03937 in.
1 m^2/sec =	10.76 ft^2/sec
1 mton =	1000 kg
	2205 lb
1 MW =	3,413,000 Btu/hr
	1000 kW
1 therm =	100,000 Btu
1 W =	1 J/sec
1 W/m-K =	0.5778 Btu/ft-hr-°F

TEMPERATURE CONVERSIONS

°C = 5/9 (°F – 32) °F = 9/5°C + 32

K = °C + 273.15 °R = °F + 459.67

From: Baukal, C.E., *Heat Transfer in Industrial Combustion*, CRC Press, Boca Raton, FL, 2000.

Glossary

3 "T"'s: Time, Temperature, Turbulence

5 "M"'s: Meter, Mix, Maintain, Mold, Minimize

absolute pressure: the pressure measured relative to a perfect vacuum. Absolute pressures are always positive. The British units for absolute pressure are psia.

atmospheric pressure: the force exerted per unit area by the atmospheric gases at the earth's surface. Atmospheric pressure varies with altitude. At sea level it is about 14.7 pounds per square inch. At one mile above sea level, it is about 12.2 pounds per square inch.

atomization: the process whereby a volume of liquid is converted into a multiplicity of small drops. The principal goal is to produce a high surface area to mass ratio so that the liquid will vaporize quickly and thus be susceptible to combustion.

atomizer: part of an oil gun which breaks up the fuel oil flow into tiny particles by both mechanical means and the use of an atomizing medium. The oil and atomizing medium mix together in the atomizer and then flow to the oil tip to be discharged into the furnace.

audible sound: vibrations in a gas, liquid, or solid with components falling in the frequency range of 16 Hz to 20 kHz.

beta ratio (β): for a single orifice the beta ratio is the ratio of the orifice bore diameter to that of the upstream pipe diameter. However, since in burner designs typically there is more than one orifice at a riser pipe exit, the beta ratio is equal to the square root of the ratio between the total area of the fuel ports to that of the upstream pipe area.

Btu (British Thermal Unit): standard measure of *energy* in the British unit system. 1 Btu is the amount of *heat* required to raise the temperature of liquid water by 1°F.

burner: a device which combines fuel and air in proper proportions for combustion and which enables the fuel–air mixture to burn stably to give a specified flame size and shape.

burner block: also called "burner tile," "muffler block," or "quarl." The specially formed refractory pieces which mount around the burner opening inside the furnace. The burner block forms the burner's airflow opening and helps stabilize the flame.

burner capacity: amount of heat release a burner can deliver (i.e., amount of fuel which can be completely burned through a burner) at a given set of operating conditions.

calorie: the amount of *energy* required to raise the temperature of 1 gram of water by 1°C. The kilocalorie (kcal) is a typical unit of measure in the process industry, 1 kcal = 1000 calories.

combustion: the rapid reaction of fuel and oxidant (usually oxygen in air) to produce light, heat, and noise. Major products of combustion for hydrocarbon fuels (e.g., natural gas, refinery gas, fuel oils) are carbon dioxide (CO_2) and water vapor (H_2O). Trace products include carbon monoxide (CO) and nitrogen oxides (NO and NO_2), which are pollutants.

combustion efficiency: the fraction of carbon in the fuel that is converted to CO_2 in the flue gas, customarily expressed as a percent.

conduction: the transfer of heat by molecular collision. This process is more efficient in metals and other thermal "conductors" and poorer in fluids and insulators such as refractories.

convection: the transfer of heat or mass by large scale fluid movements. When the process occurs due to density and temperature differences, it is termed natural convection. When the process occurs due to external devices (such as fans), it is termed forced convection.

convection section: the part of a furnace between the radiant section and the stack. The area is filled with tubes or pipes which carry a process stream and which absorb heat via convection heat transfer from the hot gases passing through the area on their way out the stack. The convection section forms an obstacle to the combustion gas flow and can greatly affect furnace draft in the radiant section of the furnace.

dB(A): "A" weighted average of the sound pressure levels over the entire frequency band. Intended to be a more accurate representation of how a human hears the sound.

Decibel: unit of sound pressure or power. Abbreviation is 'dB'. 1W of sound power is equal to 120 dB. A Log10 scale relates the unit of Watts and dB. Consequently, an increase or decrease of 10 W equates to a 10 dB difference, while a change of 100W equates to a 20 dB difference.

decibel (dB): unit of measure for sound pressure level. Developed by Bell Laboratories.

diffusion (raw gas) flame: combustion state controlled by mixing phenomena. Fuel and air diffuse into one another until a flammable mixture ratio is achieved.

emissivity: the efficiency with which a material radiates thermal energy, expressed as a fraction between 0 and 1.

excess air: the amount of air needed by a burner which is in excess of the amount required for perfect or stoichiometric combustion. Some amount of excess air, depending on the available fuel/air mixing energy, is required to assure thorough mixing of the fuel and air for complete combustion.

flame speed: the rate at which a flame can propagate in a combustible mixture. If the flame speed is lower than the speed of the reacting flow, the flame may *lift off* the burner. If the flame speed is higher than the speed of the reacting flow the flame may *flash back* into the burner.

flammability limits (upper and lower): the upper and lower bounds of the fuel/air mixture which will support combustion. The upper flammability limit indicates the maximum fuel concentration in air that will support combustion. The lower flammability limit indicates the minimum fuel concentration in air that will support combustion. Outside these bounds the mixture does not burn.

flashback: phenomenon occurring only in pre-mix gas burners when the flame speed overcomes the gas–air mixture flow velocity exiting the gas tip. The flame rushes back to the gas orifice and can make an explosive sound when flashback occurs. Flashback is most common when hydrogen is present in fuel gas.

flashing: the process whereby a drop in pressure or increase in temperature causes vaporization.

fuel NOx: NOx that is formed from nitrogen that is organically bound to the fuel molecule. Fuel NOx is most often a problem with liquid fuel or coal firing. Once the nitrogen has been cracked from the fuel molecule, the mechanism follows basically the same path as the prompt NOx mechanism.

furnace arch: uppermost part of a radiant furnace (also called the "Bridgewall," a term which came from the original furnace designs and has remained in use). The last area in an upflow furnace before the convection section.

furnace draft: the negative air pressure generated by buoyancy of hot gases inside a furnace. The temperature difference between gases within the furnace and in the atmosphere along with furnace and stack height basically determine the amount of draft generated by a furnace. Draft is generally measured in negative inches of water column ("-w.c."; 27.7 inches w.c. = 1 psig).

furnace or fired heater: a piece of process equipment which is used to heat any of the various process streams in refineries and chemical plants. Furnaces most commonly utilize direct combustion of fuels to generate the required heat.

gage pressure: the pressure measured relative to the local atmospheric pressure. Gage pressure may be negative. A negative gage pressure is known as a *suction* or *vacuum*.

gas tip: the part of a burner which discharges the gas fuel via one or more openings into the furnace. The size, arrangement, and angular disposition of the openings in the tip have a major effect on the size and shape of the flame.

heat liberation: amount of heat released during combustion of fuels. One of the criteria for determining what burner to use in an application. (Also called heat release.)

Higher Heating Value (HHV): the theoretical heat the combustion process can release if the fuel and oxidant are converted with 100% efficiency to CO_2 and liquid H_2O.

ignition temperature: the temperature required to initiate combustion.

laminar flow: very smooth flow in which all the molecules are traveling in generally the same direction. For internal flows, it occurs at Reynolds numbers less than 2000.

lift-off: this condition occurs when the fuel or fuel/air mixture velocity is too high, thus allowing the fuel to exit the stabilizing zone before it has achieved its ignition temperature.

Lower Heating Value (LHV): the theoretical heat the combustion process can release if the fuel and oxidant are converted with 100% efficiency to CO_2 and H_2O vapor. The process industries generally prefer to use LHV.

mixer: the part of a pre-mix burner (also "Gas–Air Mixer") which uses the kinetic energy of the high velocity fuel gas stream to draw in part or all of the air required by the burner for combustion.

noise: any undesirable sound.

normal cubic meter (Nm^3): the quantity of a gas that is present in 1 m^3 at the thermodynamic conditions of 1 atm and 0°C. For an ideal gas there are 22.41 Nm^3 in 1 kmol.

NOx: any combination of nitrogen and oxygen in a compound form. The most common in terms of environmental considerations is NO, which constitutes 90% of combustion NOx emissions, and NO_2. All NO is eventually converted to NO_2 in the atmosphere. Hence, most regulations are written to assume that the NOx which is emitted is in the form of NO_2. NOx emissions are influenced by many factors, including furnace temperature, flame temperature, burner design, combustion air temperature, nitrogen content of liquid fuels, ammonia content of gas fuels, and other factors.

oil block: usually a monolithic block located at the center of a burner assembly. The oil block acts to stabilize the oil flame. (Also call the "Oil Tile")

oil gun: the assembly of parts in a burner which provides atomized fuel oil mixture to the furnace for burning.

oil tip: part of the oil gun which discharges the atomized fuel oil mixture into the furnace through multiple openings. The hole pattern in the tip has a great effect on the flame size and shape.

orifice discharge coefficient (C_d): the ratio of the actual flow through an orifice to that of the theoretical or isentropic flow through the orifice. Basically, this parameter is a measure of the orifice efficiency. Values are dimensionless and range from 0.61 for a thin-plate orifice to 0.85 for thick-plate square-edged orifices, and up to 0.90–0.95 for tapered orifices.

PAH (polycyclic aromatic hydrocarbons): the carcinogenic byproducts of some very sub-stoichiometric combustion processes. Usually absent in process burner flames.

particulates: the residue left over from coal and fuel oil combustion.

Pascals: a unit of pressure. One Pascal (Pa) is equal to a force of one Newton per square meter.

pre-mixed flame: before ignition, the fuel and air are intimately mixed. The combustion process is controlled by heat conduction and diffusion of radicals.

pressure, gas: the force exerted per unit area on a surface created by the collision of gas molecules with that surface.

pressure, static: the pressure of a gas measured at a point where the gas velocity is zero.

pressure, total: the sum of the static pressure and the velocity pressure of a gas.

pressure, velocity or dynamic: the pressure of a flowing gas attributed to the impact of the gas molecules resulting from the velocity of the gas flow. $P_V = \rho V^2/2g_c$, where ρ is the density of the flowing gas, V is the velocity, and g_c is the gravitational constant.

prompt NOx: NOx formed at the initial stages of combustion that can not be explained by either the thermal mechanism or the fuel NOx mechanism. The prompt NOx mechanism requires the CH radical as an intermediate, so the fuel must have carbon present to create prompt NOx (see Chapter 6).

radiant section: the part of a process heater into which the burners fire. Tubes mounted in this area of the furnace receive heat principally via direct radiation from both burner flames and furnace refractory. Physical volume arrangement of the radiant section has a great effect on burner choice and required flame patterns.

radiation: all warm bodies emit light (electromagnetic radiation – mostly infrared). When this radiation is absorbed or emitted by a body, heat is transferred and termed "heat transfer by radiation." Such heat transfer requires a line of sight (view factor) and is proportional to the fourth power of the absolute temperature difference between bodies and the emissivity of the bodies (see Chapter 3).

ratio of specific heats (k): also known as the *isentropic coefficient*. Is equal to the quotient of the heat capacity at constant pressure and the heat capacity at constant volume (C_p/C_v). This parameter is tabulated for many pure components at standard conditions, but is technically dependent on the gas composition and temperature. The values are dimensionless and range from 1.0 to 1.6.

sonic flow: when the flow velocity is equal to the speed of sound. The point at which the flow turns sonic is called the *critical pressure*. This transition occurs at about 12.2 psig for natural gas at 60°F.

sound: an alteration of density in an elastic medium that propagates through the medium (see Chapter 7).

sound frequency: the number of pressure waves that pass by in a given time. Hertz is the unit of frequency. One wave per second is one Hertz.

sound power level: the intensity of the sound given off in all the directions at the source.

sound pressure level: a measure of the acoustical "disturbance" produced at a point removed from the source.

sound wave: a wave moving at the speed of sound in a given medium.

SSU (seconds, Saybolt Universal): units of kinematic viscosity.

stack: the "chimney" or "flue" of a furnace. The stack contains the damper which controls furnace draft.

stack loss: the fraction of total heat which exits with the flue gas through the stack. The quantity is customarily expressed as a percent of the total heat input. The stack loss is directly proportional to the stack exit temperature; the higher the temperature, the greater the stack loss.

staged air: NOx reduction technique predominantly used for fuel oil firing. The fuel is injected into a fuel-rich primary zone. This stoichiometry helps to control the fuel NOx mechanism. When firing gas, staged air produces higher NOx emissions than staged fuel.

staged fuel: NOx reduction technique whereby a small portion of the fuel is injected in a lean primary combustion zone. The flue products from this region flow to the secondary combustion zone where the remainder of the fuel is burned out. The lengthening of the flame creates cooler flame temperatures, thus lowering the thermal NOx.

standard cubic foot (SCF): the quantity of a gas that is present in 1 ft^3 at the thermodynamic conditions of 14.696 psia and 60°F. For an ideal gas there are 379.7 SCF in 1 lb-mol.

steam quality: the fraction of saturated steam that is in the vapor state.

theoretical flame temperature (adiabatic flame temperature): the temperature the flame can achieve if it transfers no heat to its surroundings.

thermal conductivity: the ability of a material to conduct heat, expressed as thermal power conducted per unit temperature and thickness. Metals and other thermal "conductors" have a large thermal conductivity. Refractories and other thermal "insulators" have a low thermal conductivity.

thermal efficiency: the fraction of total heat input absorbed by the material being heated. The quantity is customarily expressed as a percent.

thermal NOx: NOx formed via the Zeldovich mechanism. The rate limiting step in this mechanism is the formation of the O radical. This occurs only at high temperatures (above about 2400°F). Hence the term thermal NOx, since it is NOx produced in the highest temperature regions of the flame (see Chapter 6).

thermo acoustic efficiency: equal to the sound power level/heat release. A value used to characterize the amount of combustion noise emitted from a flame. Defined as the ratio of the acoustical power emitted from the flame to the total heat release of the flame. Approximately equal to 1×10^{-6} for premixed and turbulent flames and equal to 1×10^{-9} for diffusion and laminar flames.

tramp air: any air which enters (infiltrates) the furnace through leaks. This air may be measured by the O_2 analyzer and often contributes to the burning of the fuel.

turbulent flow: characteristically random flow patterns that form eddies from large to small scales. For internal flows, it occurs at Reynolds numbers greater than 4000. Turbulence is integral to the mixing process between the fuel and air for combustion.

UHC: any unburned hydrocarbon that is emitted in a combustion process. Also termed VOC (volatile organic compound).

velocity thermocouple (suction pyrometer): a device for measuring furnace gas temperature. It is comprised of a thermocouple which has been recessed into an insulating shroud, and a suction device such as an eductor which aspirates large volumes of furnace gas through the shroud and past the thermocouple. The high velocity of gas ensures good convective heat transfer to the thermocouple. The shroud blocks radiant exchange between the thermocouple and the surrounding furnace. The velocity thermocouple represents the most accurate means to measure flue gas temperature. Bare thermocouples are unacceptable for this purpose, being in error often by more than 100°F due to radiation losses.

view factor: the fraction of one surface that is visible to another (see Chapter 3).

Watt: unit of measure for *power*, equal to 1 Joule of energy per second.

Index

Index

A

Absolute viscosity, 119, 122
Accident, factors contributing to, 328
ACERC, see Advanced Combustion Engineering Research Center
Acetaldehyde, 216
Acetone, 216
Achimedes' principle, 129
Acid gas(es), 639
  composition of used in CFD study, 309
  removal, 657, 663, 665
  systems, 682
Acrylonitrile, 667
Actuator, 385
Additives, 199
Adiabatic equilibrium reaction process, 54
Adiabatic flame temperature (AFT), 46, 184, 185, 213
Adjustable-plug venturi quench, 656
Adsorption, with solid desiccants, 158
Advanced Combustion Engineering Research Center (ACERC), 291
Aeration rate, 274
Aerospace applications, industrial combustion and, 8
AFT, see Adiabatic flame temperature
Afterburn, 472
Agency approvals and safety, 379
Air
  ambient atmospheric, 352
  -assisted flare
    burner, 608
    horizontal settling drum at base of, 618
  augmenting, 533
  blast, 586
  control, 358, 456
  control device
    picture of, 480
    schematic, 480
  dampers, 464
  density, combustion, 264
  excess, 471, 555, 556
  flare(s), 272
    annular, 611
    comparison of to prediction, 276
  flow
    balancing of by windbox model, 567
    control, 480
    distribution, to burners, 564
    rate, air valve position vs., 392
    rate, O_2 trim of, 395
  /fuel mixture, bigger explosions of, 235
  -to-fuel ratio, 38, 257
  /gas mixtures, 605
  heating, 524
  infiltration, 601
  leaks, 494
  low excess, 200
  metering, 356
  mixer assembly, 355
  -only systems, 533
  pollution, environmental regulations limiting, 362
  preheat, 62
    effects, 55
    reduction, 200
    temperature, 195
  preheater (APH), 438, 561, 562
  pressurized, 584
  properties of dry, 715
  -quality regulations, 198
  registers, 458, 464
  staging, 517
  temperature, combustion, 476
  valve
    butterfly-type combustion, 391
    characterizer, 392, 393
  VOC-laden, 280
Alcohols, 216
Alkanes
  combustion data for, 45
  non-methane, 404
  properties of, 713
Alkenes
  combustion data for, 45
  properties of, 713
Alkylation, 5, 13
Ambient air firing, 533
Ambient atmospheric air, 352
American National Standards Institute (ANSI), 540
American Petroleum Institute (API), 190, 339, 358, 597
Ammonia injection, use of to control NOx, 413
Analog controllers, 394
Analog control systems, 376
Analog devices, 385
Analog loop, 378
Analysis of variance (ANOVA), 409, 411
  for factorial design, 414
  with separate model effects, 409
  table, testing separately for bias in, 413
Analytical instruments, types of, 389
Analyzer(s)
  in situ extractive, 472
  oxygen, 473
  portable, 472
Annular air flare, 611
Annunciators, 383
ANOVA, see Analysis of variance
ANSI, see American National Standards Institute
APH, see Air preheater
API, see American Petroleum Institute
Aromatic hydrocarbons, 215
Aromatics, 177
  combustion data for, 45
  properties of, 713
Arrestors, 629
Asphalt, 5
Asphyxiation, 214
Atmospheric attenuation, 231, 246
Atmospheric distillation, 5
Atmospheric pressure, 148, 715–716
Atomic weights, list of for common elements, 36–37
Atomizer(s)
  eternal mix, 586
  fitted onto oil guns, 585
  internal mix, 584
  spill-return, 585
Atomizing medium block valve, 486
Autoignition
  ethylene oxide plant explosion caused by, 343
  temperature, of methane, 56
Automatic steam control, 627
Autooxidation, 343
Available heat, 64, 65
Averaged velocity, 140
Aviation gasolines, 175
Avogadro's number, 35
Axial flow burners, 565

B

Back surfaces, radiant exchanges between, 90
Baghouse, 657, 658
Balanced-draft heater, logic diagram for tuning, 492
Benzene, 216
Bernoulli equation, 130, 131
Bessel functions, 80
Beta ratio, 153
Bias, 412
Blackbody(ies)
  emissive power graph, 103
  radiation, 88, 91, 95
  source, 90
Blockage, 532
Blocking generator, 418
Blood-borne pathogens, 346

Blower(s), 670
- centrifugal, relative characteristics of, 673
- inlet, 674
- location, 671
- speed control, 677

Bluff body, 67
Boiler(s), 648
- cleanliness, 561
- design, impacts of on NOx emissions correlations, 553
- firetube, 648, 650
- fluidized bed, 524
- load, influence of on NOx, 558
- low-NOx retrofit, 569
- multi-burner, 559
- municipal solid waste, 413
- Nebraska, 584
- performance, effects of burner retrofits on, 553
- single-burner industrial, 549
- -specific burner requirements, 546
- tube, 218, 532
- utility, 549, 572
- watertube, 649

Boiler burners, 547–586
- atomizers for boiler burners, 584–586
- boiler design impacts on NOx emissions correlations, 553–563
  - boiler design, 553–554
  - boiler load influence on NOx, 558–561
  - boiler/system condition impacts on combustion and NOx formation, 561–563
  - excess air, 555–558
- boiler-specific burner requirements, 548–553
  - conventional burner technology for boilers, 548–550
  - design features of low NOx burners, 552–553
  - effects of burner retrofits on boiler performance, 553
  - low-NOx burner technology for boilers, 550–551
  - staged burner design philosophy, 551–552
- current state-of-the-art concepts for multi-burner boilers, 563–575
  - combustion optimization, 563–568
  - methods to reduce NOx emissions, 568–575
- low-NOx burners for packaged industrial boilers, 575–580
- ultra-low emission gas burners, 580–584
  - background, 580
  - implementation of NOx formation theory for ultra-low-NOx emissions, 580–581
  - ultra-low emissions burner design, 581–582
  - ultra-low emissions burner operation, 582–584

Boiling points, general fraction, 177
Bolts, standard grades of, 709, 710, 711
Boltzmann constant, 88
BOOS, see Burner out-of-service
Boussinesq hypothesis, 292
Box heater, 17
Breathing, 502
Bridgewall temperature, 496
Bulirsch-Stoer technique, 282
Buoyancy, 129
Burnback, 67
Burner(s), 15–28
- air-assisted flare, 608
- air dampers, 457
- air registers, 457
- arrangements, 20
- axial flow, 565
- block(s)
  - for floor-mounted service, 367
  - material, 370
  - valves, 485, 495
- boiler, see Boiler burners
- bolt circle, on new heater, 451
- capacity curves burner manufacturers use for sizing, 262
- CFD simulation of, 306
- combination, 418, 482, 643
- combustion
  - instability noise, 237
  - noise, 236
  - roar, 237
- competing priorities, 15–16
- condition, 495
- configuration, by heater type, 12
- connection of to heater, 455
- controller, 374
- conventional, round flame, 367
- COOL TECHNOLOGY, 518, 519
- crude unit, 19
- damper position, 495
- design factors, 16–21
- development of for petrochemical applications, 291
- diffusion, 25, 67, 481, 482
- downfired, 369
- drawing, 450
- duct, 524, 534
  - arrangement, 533
  - drilled pipe, 536
  - fluidized bed startup, 529
  - low emission, 536
- eduction processes in premixed, 257
- elevation, 470
- firing heavy oil, 180
- flame lift-off from, 513
- flare, 605
- flat flame, 368
- forced draft, 358, 437, 439, 480
- free-standing flat flame, 368
- fuel flows of, 67
- fuel-staged, 405, 407
- furnace combination, 111
- gas, 23
  - troubleshooting for, 519
  - typical excess air values for, 490
- general burner types, 21–27
- geometry, 310, 405
- grid
  - gas flame from 531
  - heat distribution by, 532
- heat input, average flame length as function of, 551
- high pressure drop, 643
- horizontal floor-fired, 14
- ignition
  - attempt, 336
  - ledge, 482
- improper installation of, 451
- improvement of mass flow distribution to, 563
- in-duct, 524
- inline, 530
- installation, 450, 456
- light-off sequence, simplified flow diagram of standard, 377
- low-NOx, 552, 575
- low pressure drop, 642
- maintenance activities, 459
- management system, 540
- manufacturers, 111, 262
- materials, 369
- medium pressure drop, 643
- mounted on top of heater, 452
- mounting, 451
- natural draft, 4, 357, 358, 480
- noise
  - abatement techniques, 242
  - curve, 231
  - example, 244
- NOx, 202
  - example, 402
  - response from, 404
- oil
  - common, 479
  - -fired, experiencing flame lift-off, 513
  - needing service, 28
  - troubleshooting for, 520
- operation, ultra-low emissions, 582
- out-of-service (BOOS), 202, 573
- partially premixed, 25
- performance, 459
- physical model of, 534
- pilot, 465, 483, 514
- piping, 453
- plenum, viewing oil flame through, 174
- ports, 509
- potential problems, 27–28
- pre-installation work, 450
- premix, 21, 22, 480, 481
- pressure drop, 252, 260
- problem, 568
- radiant wall, 103, 260, 369
- raw gas, 206, 357, 364, 366, 481, 482, 515
- refractory, 494, 506
- requirements, boiler-specific, 546
- retrofits, effects of on boiler performance, 553
- role of
  - in furnace, 112
  - in heat transfer, 110
- sound pressure level, 237
- staged

air, 25, 517
design philosophy, 551
fuel, 26, 517, 518
systems, ZTOF, 623
technology, 526
boiler, 548
NOx, 518
test setup, 433
throat, 354, 455
tile
broken, 494
ledge in, 361
picture of showing multiple tile pieces, 454
support for, 367
tip, plots of contours of streamfunction with increasing backpressure at, 322
Todd Combustion low NOx, 578, 579
turndown, 425
types, 21, 362
ultra-low emissions, 580, 583
ultra low-NOx, 517
unmodified, 308
utility, NOx vs. load with firing natural gas on, 559
venturi-style, 552, 558
wall-fired, 15, 368
zone heat release (BZHR), 551, 554, 575
Burner design, 263, 351–370, 432
air control, 358–359
air metering, 356–358
forced draft, 358
natural draft, 357
burner types, 362–367
oil or liquid firing, 364–367
premix and partial premix gas, 363–364
raw gas or nozzle mix, 364
combustion, 352–353
final, 312
fuel metering, 353–356
gas fuel, 353–354
liquid fuel, 354–356
ignition, 360–361
initial, 312
materials selection, 369–370
mixing fuel/air, 359–360
co-flow, 360
cross-flow, 360
entrainment, 360
flow stream disruption, 360
mounting and direction of firing, 367–369
conventional burner, round flame, 367–368
downfired, 369
flat flame burner, 368
radiant wall, 369
patterned and controlled flame shape, 361–362
problems, 27
pollutants, 362
ultra-low emissions, 581
Burner/heater operations, 469–499
heater and appurtenances, 479–488
air flow control, 480
burner block valves, 485–487
casing and refractory, 487–488
combination burners, 482
diffusion or raw gas burner, 481–482
flame patterns, 483–485
pilot burners, 483
premix burner, 480–481
measurements, 470–479
combustion air temperature, 476
draft, 470
excess air or excess oxygen, 471–473
flue gas temperatures, 476–477
fuel flow, 473
fuel pressure, 473–476
fuel temperature, 476
process fluid parameters, 479
process tube temperature, 477–479
operational considerations, 488–499
developing startup and shutdown procedures for fired heaters, 497–498
developing emergency procedures for fired heaters, 499
heater operation data, 496–497
heater turndown operation, 492–493
inspection and observations inside heater, 493–494
inspection and observations outside heater, 495–496
target draft level, 488
target excess air level, 489–491
Burner testing, 431–447
burner test setup, 433–437
application, 434
test furnace selection criteria, 434–437
instrumentation and measurements, 437–442
emissions analysis, 438–441
flame dimensions, 441
fuel flow rate metering, 441
furnace gas temperature measurement, 438
heat flux, 442
measuring air-side pressure and temperature, 437–438
test matrix, 442–444
definition of data to be collected, 444
heater operation specifications, 442
performance guarantee specifications, 442–443
Burning
points, division of incoming gas stream into, 609
smokeless, 603
velocity, 296
Butane lighter, shadow photograph of burning, 236
Butylene dehydrogenation, 14
BZHR, see Burner zone heat release

C

Cabin heater, 16, 18, 28
Calcium chloride, dehydration with, 158
California Air Resources Board (CARB), 197
Canadian Standards Association (CSA), 540
Capacity curves, 262, 263, 265
CARB, see California Air Resources Board
Carbon dioxide, 99, 218
Carbon monoxide (CO), 439, 538
analyzer, 389
formation, 60
/opacity level, 567
oxidation reaction, 314
Carrier gas, ammonia in, 669
Cartesian coordinates, 76, 134
Cartesian differential equation set, 294
Castable refractory, 646, 684, 689
Catalyst configuration, 671
Catalytic cracking, 5
Catalytic hydrocracking, 5
Catalytic hydrotreating/hydroprocessing, 5
Catalytic oxidizers, 647
Catalytic reforming, 5, 13
Categorical factor, 418
CEMs, see Continuous emissions monitors
CEN, see European Committee for Standardization
CENELEC, see European Committee for Electrotechnical Standardization
Center fire gas (CFG), 576
Central-composite designs, 419
Central tube wall process heater, 16
Centrifugal blowers, relative characteristics of, 673
Ceramic temperature, 281
CFD, see Computational fluid dynamics
CFG, see Center fire gas
Chemical
heat release, 137
industry, major fired heater applications in, 14
manufacturing process, by-products of, 63
process industry (CPI), 288, 361
reaction(s)
global, 204
modeling of, 252
reactors, idealized, 258
Chemical Manufacturers Association (CMA), 630
Chemiluminescent analyzer, 197
CHEMKIN, 125, 258
Chlorinated hydrocarbons, 309
Chloroform, 216
Choked flow, 147, 275
mass flux, 275
test rig, 147
Circles, areas and circumferences of, 693–701
Clean Air Act Amendments, 631
CMA, see Chemical Manufacturers Association
CMC, see Conditional moment closure
CO, see Carbon monoxide
Coal, 72
Coking, 11, 13, 462
Cold flow furnace, 264
Colebrook formula, 151
Collecting surface area, 659
Combination burners, 418, 482, 643
Combustible waste gas streams, 160

Combustion
air, 19
blower inlet preheat, 524
density, 264
preheating of, 63
temperature, 26, 476
basics, 639
chemistry
of hydrocarbons, 34
models, 299
as controlled process, 352
control strategies, 303, see also Combustion controls
data, for hydrocarbons, 45
efficiency, 630
equipment, see also Combustion equipment, experimental design for
designers, 224
noise, 243
gases, polynomial expression for, 125
-generated particles, 217
heat of, 38
instability noise, 234
kinetics, 60
modeling, 258
modification, 194, 199, 201
non-premixed, 299
optimization, 563
overall sound pressure level from, 235
premixed, 301
process(es), 58
approximations involved in modeling, 296
modification, 667
product(s)
flare, 629
temperature, 59
reactions, molar ratios for, 37
roar, flare, 234
safety, see Combustion safety
staged, 200, 573
submodels, CFD-based, 296
substoichiometric, 46, 47
sustainable, for methane, 353
systems, scaling of from laboratory scale to industrial scale, 291
tetrahedron, 330
turbulent, regimes of, 299
waste, 670
Combustion controls, 373–399
controllers, 394–398
control schemes, 389–393
fully metered cross limiting, 392–393
parallel positioning, 389–392
fundamentals, 374–383
agency approvals and safety, 379–380
analog control systems, 376–378
control platforms, 374–376
discrete control systems, 376
failure modes, 378–379
pipe racks and control panels, 381–383
primary measurement, 383–389
analog devices, 385–389
discrete devices, 383–385
tuning, 398–399
Combustion equipment, experimental design for, 401–429
accounting for fuel mixtures, 420–424
building $2n$-level factorials from two-level factorials, 424
combining mixture and factorial designs, 423–424
experimental designs for mixtures, 421–423
orthogonal mixture designs, 423
combining domain knowledge with SED, 424–427
practical considerations, 425
semi-empirical models, 425–426
sequential experimental strategies, 426–427
important statistics, 408–412
analysis of variance, 409
ANOVA with separate model effects, 409–411
F-distribution, 409
pooling insignificant effects, 411–412
standard errors of effects, 412
linear algebra primer, 427–429
identity, inverse, and transpose, 428–429
matrix addition, 429
matrix multiplication, 428
Taylor and MacLaurin series approximations, 427–428
linear transformations, 407–408
matrix solution, 406–407
method of least squares, 405–406
power of SED, 402–404
second-order designs, 419–420
central-composite designs, 419–420
practical considerations, 420
statistical experimental design principles, 404–405
two-level factorial designs, 412–419
foldover, 416–417
including categorical factors, 418–419
interactions, 412
method of steepest ascent, 416
orthogonal blocking, 417–418
pure error and bias, 412–414
screening designs, 415–416
serial correlation and lurking factors, 416
two-level fractional factorials, 414–415
Combustion safety, 327–349
design engineering, 339–346
fire extinguishment, 344
flammability characteristics, 339–342
ignition control, 342–344
safety documentation and operator training, 344–346
overview, 329–339
codes and standards, 338–339
combustion tetrahedron, 330–331
definitions, 329
explosion hazards, 334–337
fire hazards, 331–334
process hazard analysis, 337–338
sources of information, 346–347
Composite wall, 74
Compressibility factor, 127
Compressible flow, 144, 145
Computational fluid dynamics (CFD), 82, 110, 142, 265, 288
-based combustion submodels, 296
burner simulation, 306
code, 293
engineer, 289
model, 302, 534
background, 291
for design optimization, 534
elements in, 290
ethylene pyrolysis furnace, 305
rendered view of, 303
specialist, 289
study, acid gas used in, 309
vendors, 303
Computational fluid dynamics based combustion modeling, 287–325
case studies, 305–319
ethylene pyrolysis furnace, 305–306
incineration of chlorinated hydrocarbons, 309–319
sulfur recovery reaction furnace, 307–309
venturi eductor optimization, 319
xylene reboiler, 306
combustion submodels, 296–302
combustion chemistry models, 299–301
pollutant chemistry models, 301
radiation models, 297–299
solution algorithms, 297
turbulence models, 301–302
model background, 291
simulation model, 291–296
convergence criteria, 295
modeling basis, 296
model validation, 295–296
solution technique, 295
transport equations, 292
turbulence equations, 292–294
solution methodology, 302–305
analysis of results, 303–305
problem setup, 302–303
solution convergence, 303
Conditional moment closure (CMC), 300, 301
Conduction
heat transfer, 71, 137
one-dimensional steady-state, 73
in radial coordinate systems, 76
steady-state, 77
transient, 77, 80
unsteady-state, 78
Conductivity analyzer, 389
Configuration factor, 92
Conservation
energy, 134, 137, 275, 289
mass, 35, 133, 134, 274, 289
momentum, 133, 134, 136, 253, 275
species, 134, 138
Constant surface heat flux, 80
Constant velocity, 134
Contact resistance, definition of, 75
Continuous emissions monitors (CEMs), 207

Controlled variable (CV), 394, 395
Controller(s), 394
 analog, 394
 modes of operation of, 397
 output mode, 393
 pass flow, 498
Control panel(s), 381
 inside of, 382
 large, 382
 small, 382
Control platforms, 374
Control schemes, 389
Control signal, fuel flow rate versus, 392
Control system(s)
 analog, 376
 discrete, 375
 distributed, 375
 instrumentation and, 675
Control valve, 397
 body, 385
 bypass, 619
 characteristics, 385
 typical staging, 628
Control volumes, 132, 133
Convection, 82, 511
 coefficient, 79, 84
 correlation, for flow in circular tube, 86
 forced, 104
 heat transfer
  in banks of tubes, 87
  coefficient, 78
 laminar flow, 83
 mixed, 82
Convective heat transfer, 137
Conventional burner, round flame, 367
Converging–diverging nozzle, 255
Conversions, common, 723
Cooler fireboxes, 478
Cooling, Newton's law of, 82, 281
COOL TECHNOLOGY burner, 518, 519
Copper tubing, commercial, 707–708
Coriolis flow meter, 388
Correction factor, 95
CPI, see Chemical process industry
Cracking furnaces, 34
Crocco similarity, 296
Cross fluctuating velocity terms, 142
Crude oil, 173, 175, 181
Crude unit burners, 19
CSA, see Canadian Standards Association
Current-to-pressure transducer, 385
CV, see Controlled variable
Cycloalkanes, 177
Cyclone separator, 618, 655
Cylinder, temperature distribution in, 76
Cylindrical coordinate system, 77
Cylindrical differential equation set, 294
Cylindrical furnace, radiation heat transfer in, 107
Cylindrical heater, 17

D

Damköhler number, 291
Damper(s), 465, 488
 air, 464
 burner air, 457
 fan, 457
 plenum, 489, 490
 radial inlet, 676
 stack, 457, 495
Data, definition of to be collected, 444
DCS, see Distributed control system
Decibel, 226
Defining contrast, 415
Degrees of freedom (DF), 408
Dehydration, with calcium chloride, 158
Delayed coking, 5
Demountable derrick, 626
Density, definition of, 118
Derivative gain, 396
Derrick
 demountable, 626
 -supported flare, 626
Design engineering, 339
Destruction efficiency, 630, 681–682
Destruction and removal efficiency (DRE), 280, 644
DF, see Degrees of freedom
Diesel engine exhaust, 352
Differential formulation, 134
Diffuser
 condition, 494
 cone, 464, 510
Diffuse surfaces, view factors for, 92
Diffusion
 burner, 25, 67, 481
 flame(s), 299
  burners, 482
  painting of, 24
 -mixed flames, 21
Digital signal processing (DSP), 228
Dioxins, 219
Direct numerical simulations (DNS), 297
Direct spray contact quench, 645
Discharge coefficient, 148, 152, 153
Discrete control systems, 375
Discrete devices, 383
Dispersion, 632, 633
Distillation, 180
Distributed control system (DCS), 375, 540
DNS, see Direct numerical simulations
Domain knowledge, combining of with SED, 424
Double-block-and-bleed, for fuel supply, 379
Downcomer, 656
Downfired burner, 369
Down-fired tests, test furnace for simulation of, 435
Draft, 130
 expression of, 470
 level, 488
 loss, 470
 targets, 496
 type, 26
DRE, see Destruction and removal efficiency
Drier off-gas, 352
Drilled pipe duct burner, 536
Drilling and exploration techniques, 167
Drill sizes, areas and circumferences of, 693–701
Droplet size
 comparison, between standard and newer oil gun, 279
 distribution, 278
Dry air, properties of at atmospheric pressure, 715–716
DSP, see Digital signal processing
Duct burner(s), 523–544
 applications, 524–526
  air heating, 524–526
  cogeneration, 524
  fume incineration, 526
  stack gas reheat, 526
 arrangement, 533
 drilled pipe, 536
 flame, 7
 fluidized bed startup, 529
 in large duct, 7
 low emission, 536
 technology, 526–544
  accessories, 540
  design considerations, 528–539
  design guidelines and codes, 540–544
  grid configuration (gas firing), 527
  grid configuration (liquid firing), 527
  in-duct or inline configuration, 526–527
  maintenance, 539
Dynamic pressure, 131

E

Early flame zone, 317
Eddy
 -breakup model, 299
 diffusivity, definition of, 293
 formation, 360
Eduction processes, 252
 in pilots, 256
 in premixed burners, 257
Eductor(s)
 modeling of, 252
 optimization, venturi, 319
 throat, re-circulation region in, 320
EF, see Exposure factor
Electrical heating, 202
Electromagnetic radiation, spectrum of, 89
Electromechanical relays, 540
Electronic ignitor
 ST-1S pilot tip without, 466
 ST-1SE pilot tip with, 466
Electronic linkage, 390
Electrostatic precipitator (ESP), 659, 660
Ellipsoidal radiometer schematic, 443
Emergency flow rate, maximum, 594
Emergency shutdown (ESD), 338
Emission(s), 631
 analysis, 438
 guarantees, 443

noise, 443
NOx, high, 515
particulate, 442
pollutant, 432
regulations, 111
Emissivity(ies)
of carbon dioxide, 99
for various surfaces, 100
of water vapor, 99
Enclosed flares, 241, 622
Endwall fired burner arrangement, 20
Energy
conservation of, 134, 137, 275, 289
conversion, 608
flux, 73
kinetic, 276, 293
minimum ignition, 342
thermal
turbulent, 294
Energy Information Administration, 168
Enthalpy of formation, 714
Environmental requirements, 597
Equation(s)
Bernoulli, 130, 131
Favre-averaged, 292, 301
ideal gas, 127
Le Chatelier, 339
momentum, 142
Navier–Stokes, 145, 292
partial differential, 295
radiation transport, 297
radiative transfer, 97
Redlich–Kwong, 128
Reynolds-averaged conservation, 301
of state, 126
steady-state conduction, 77
transport, 292
turbulence, 292
wall to wall, 288
Equilibrium
chemistry assumption, 300
reaction process, adiabatic, 54
thermodynamics and, 54
ESD, see Emergency shutdown
ESP, see Electrostatic precipitator
Esters, 216
Eternal mix atomizers, 586
Ethers manufacture, 5
Ethylbenzene dehydrogenation, 14
Ethylene
cracking furnace, 111
furnaces, test furnace for simulation of, 434
hydration, 14
oxide (ETO), 343, 619
plant, multiple ZTOF installation in, 593
/polyethylene gases, 186
pyrolysis furnace
CFD model of, 305
rendered view inside, 304
ETO, see Ethylene oxide
European Committee for Electrotechnical Standardization (CENELEC), 540
European Committee for Standardization (CEN), 339
Event trees, 337
Excess air, 471, 555
empirical evidence of effect of on NOx, 556
level, 489
species concentration vs., 39
stoichiometric ratio and, 38
Excess oxygen, 471
Expansion joints, 455
Experimental fluid dynamics, 288
Explosion(s)
ethylene oxide plant, 343
furnace, 336
hazards, 334
piping, 335
potential dangers caused by, 328
source of in furnaces, 337
stack, 335
tank, 335
unburned fuel accumulations causing, 393
Exposure factor (EF), 232

F

Factorial(s)
design(s)
ANOVA for, 414
combining mixture and, 423
two-level, 412
saturated, 415
two-level, 424
Factor space, 416
Factory Mutual (FM), 540, 675
Failure modes, 378
and effect analysis (FMEA), 337
safety concerns and, 390
Fan(s), 670
dampers, 457
noise, 237, 243
wheel designs, 672
Fanno flow, 148
Fault trees, 337
Favre-averaged conservation equations, 301
Favre-averaged transport equations, 292
F-distribution, 409, 410–411
Feedback, 378
Feedforward, 378, 379
FFG, see Flame-front generator
FFT, see Fourier transform analysis
FGR, see Flue gas recirculation
Fiberglass-reinforced plastic (FRP), 654
Fieldbus, 375
Finite-rate chemistry, 296
Fire(s)
common ignition source of, 343
diluted fuel, 571
extinguishment, 344
hazards, 331
ignition sources of major, 342
plenum, 507
potential dangers caused by, 328
tetrahedron, 330
Firebox(es)
consequence of too much draft in, 488
cooler, 478
excess oxygen level within, 504
good flame within, 486
temperature, effect of on NOx, 515
very bad flame pattern in, 486
Fired heater(s), 9–15
applications, in chemical industry, 14
design, 70
developing emergency procedures for, 499
developing startup and shutdown procedures for, 497
major refinery processes requiring, 13
process heaters, 11–15
reformers, 10–11
tube rupture in, 332
Firetube boiler, 648, 650
Firewood principle, 609
Firing
combination, 482
mounting and direction of, 367
Fixed effects, 417
Fixed heater size distribution, 12
Flame
burst, 105
color observed in, 267
diffusion, 21, 24, 299
dimensions, 432, 441, 442, 443
duct burner, 7
envelope
of oil flame, 485
with x-y-x axes, 485
equivalence ratio, 193
fireflies leaving, 507
-to-flame similarity of appearance, 566
flat, 363, 368
-front generator (FFG), 615
self-inspirating, 615, 616
slipstream, 615, 616
gas, 104, 105, 531
gaseous fuel, 165
holder, 67, 360, 361
oxidation of, 539
warped, 539
ignition, forced-draft, 612
impingement, 9, 333, 493
possible cause of on tubes, 504
on tubes, 28, 503
infrared thermal image of in furnace, 98
inside thermal oxidizer, 317
leaning, 509
lift-off, 513
long, narrow, 484
long smoky, 507
luminous, photographic view of, 101
nonluminous, photographic view of from John Zink gas burner, 102
oil, flame envelope of, 482
pattern, 483, 493
in firebox, 486
irregular, 505, 506
pilot, stable ignition of, 336
presence of soot in, 298
process heater oil, 276
properties, 61
pulled toward wall, 28
pulsating, 502

quality, 476
radiation, 101, 105, 107, 598
round, 362, 367
scanners, 384
short, bushy, 484
simulated reforming gas, 169
smoking tendencies, predicting, 266
sparklers leaving, 507
speeds, 67
stability, 493, 564
stabilizer, 360, 361, 463
cones, 369
design of, 534
flow patterns around, 537
temperature, 61, 101
adiabatic, 46
reduction, 559
transfer of heat from, 58
unstable, 494
visible, 604
Flameholder, 534
Flamelet, 296, 299, 300
Flamesheet, 296
Flammability
characteristics, 339, 340
limits, for gas mixtures, 66
Flare(s), 5, 589–634
air-assisted smokeless, 610
annular air, 275, 611
applications for throughout hydrocarbon industry, 590
available gas pressure at, 596
burner(s), 605
RIMFIRE®, 611, 613
state-of-the-art Steamizer™, 610
capacity, smokeless, 266
combustion
instability noise, 236
roar, 234
combustion products, 629–633
dispersion, 632–633
emissions, 631–632
reaction efficiency, 630–631
controls, 624
derrick-supported, 626
design considerations, 598–605
air/gas mixtures, 605
air infiltration, 601
flame radiation, 601–603
hydraulics, 599–600
liquid removal, 600–601
noise/visible flame, 604
reliable burning, 599
smoke suppression, 603–604
early model smokeless, 591
enclosed, 241, 622
equipment, 605–629
enclosed flares, 622–623
flare burners, 605–614
flare controls, 624–629
flare support structures, 623–624
knockout drums, 618–619
liquid seals, 619–620
pilots, ignitors, and monitors, 614–618
purge reduction seals, 620–622
factors influencing flare design, 594–598
environmental requirements, 597–598
flow rate, 594
gas composition, 594–595
gas pressure available, 596
gas temperature, 595–596
safety requirements, 597
social requirements, 598
utility costs and availability, 596–597
flame, thermogram of, 602
forced-draft Dragon liquid, 613
guy wire-supported, 625
high-pressure, 244, 245
horizontal, 592
modeling of air-assisted, 270
multi-point, 592
noise
abatement techniques, 239
effect of distance on, 246
level, engineer measuring, 235
non-smoking, 267
offshore oil rig, 7
pilot, 614, 616
pressure relief vessel venting to, 131
self-supported, 624
single-point, 591, 593
smokeless
operation, 266
rates, modeling of, 252
steam, 272
smoking, 267
stack, flaming liquid engulfing, 600
steam, 237, 254, 268
control valve on, 626
with steam eductor, 254
support structures, 623
systems, 590–594
applications, 591
flare system types, 591–593
major system components, 593–594
objective of flaring, 590
purpose, 590
tip(s)
exit velocity, 599
Zink double refractory severe service, 606
vendors, 273, 622
Flarestack explosion, due to improper purging, 336
Flaring
event, 591
objective of, 590
Flashback, 67, 505
Flash point, of liquid, 179
Flat flame burner, 368
Flat-shaped flame, 363
Flexicoking waste gas, 162
Flow(s)
air, balancing of by windbox model, 567
choked, 275
compressible, 144
configuration, 644
control, method of, 674
controller gains, 396
Fanno, 148
fluid, kinetic energy in, 258
fuel, 473
incompressible, 136
inviscid, 145, 290
laminar, 138, 141
meter(s), 387
Coriolis, 388
magnetic, 388
orifice, 388
positive displacement, 388
turbine, 388
vortex shedder, 387
pass, 496
patterns, around flame stabilizer, 537
pulsating, 620
quasi-one-dimensional isentropic, 145
rate, FGR, 571
reversal, 658
stream disruption, 360
switches, 384
turbulent, 138
types of, 138
viscous, 290
volume equivalent of, 185
Flowmeter, ultrasonic, 388
Fluctuating velocity, 140
Flue gas(es), 72, 471
condition of, 653
cooled, 652, 668
counter-current flow of, 664
drawn through U.S. EPA sampling train, 440
flow
minimizing, 653
NOx reduction vs., 570
hot, 511, 512
measured pollutant concentration, 190
oxygen concentration in, 425
processing methods, 647
recirculation (FGR), 20, 22, 550
flow distribution, improvement of to burners, 565
flow level, 559
flow rate, 571
hot, 568
implementation, NOx reduction by, 568
recycle, 668
shearing effect of high-velocity, 661
ST, 130
temperature, 476
Fluid(s)
absolute viscosity vs. temperature for, 122
coking/flexicoking, 5
density, 118, 139
dynamics, theoretical, 288
flow
conservation of mass for, 135
kinetic energy of, 258
Newtonian, 136, 292
packet, velocity of, 130
properties, 118
velocity profile of, flowing along solid surface, 119
viscosity, 139

Fluid dynamics, fundamentals of, 117–154
flow types, 138–148
compressible flow, 144–148
turbulent and laminar flow, 138–144
fluid properties, 118–128
density, 118
equations of state, 126–128
specific heat, 125–126
viscosity, 119–125
fundamental concepts, 128–138
Bernoulli equation, 130–132
control volumes, 132–134
differential formulation, 134–138
hydrostatics, 128–130
pressure drop fundamentals, 148–153
basic pressure concepts, 148–149
discharge coefficient, 152–153
loss coefficient, 151–152
roughness, 149–151
Fluidized bed
boilers, 524
startup duct burner, 529
Fluidized catalytic cracking, 13
FM, see Factory Mutual
FMEA, see Failure modes and effect analysis
Foldover, 416
Forced convection, 104
Forced draft
air preheater, 438
burners, 358, 437, 439, 480
ignition flame, 612
Formaldehyde, 216
FORTRAN, 282
calculation and standardized storage of chemical kinetic data, 258
conversion of thermodynamic information into NASA polynomials using, 125
Forward tip blade operating curve, 674
Fourier number, 78
Fourier transform analysis (FFT), 228
Fourier transform infrared (FTIR), 440
gas analyzer, 440
system, 440
Fractional factorials, 412, 414
Free jet, 142
entrainment, 143
interaction with surrounding fluids, 143
structure, 143
Freestream velocity, 87
Friction
coefficient, 86
factor, 84, 85, 150
FRP, see Fiberglass-reinforced plastic
FTIR, see Fourier transform infrared
Fuel(s), 157–187
/air
mixing of, 359
ratio, horizontal imbalances of, 573
unbalance, vertical, 573
at ambient temperature, 61
blend
composition, 205
effects, 57
-bound NOx mechanism, 61
chemical properties of, 717–722
CO and unburned, 214
combustion, stagewise, 572
composition, 195, 205, 437, 496
dilution, 571
distribution, liquid, 279
fires, diluted, 571
flow, 473
balancing techniques, 567
rate, control signal vs., 392
rate metering, 441
free jet of, 363
gas, 353
firing ports, 513
piping system, 459
pressure measurement, 475
tips, 209, 462
gaseous, 18, 158–165
combustible waste gas streams, 160–165
liquefied petroleum gas, 159
molecular weights and stoichiometric coefficients for common, 37
natural gas, 158–159
photographs of gaseous fuel flames, 165
physical properties of, 165
refinery gases, 159–160
gas property calculations, 183–185
derived quantities, 185
flammability limits, 184–185
lower and higher heating values, 184
molecular weight, 183–184
specific heat capacity, 184
viscosity, 185
high inert composition, 359
hot burning, 536
hydrocarbon-based gaseous, 528
injection nozzle parameters, 516
injector spud, 530
introduction, 640
leaks, 344
liquid, 165–183
high viscosity, 365
history, 165–172
liquid naphtha, 179
oil recovery, 173–175
oils, 178–179
physical properties of liquid fuels, 179–183
production, refining, and chemistry, 175–178
vaporization of, 354
metering, 353, 354
mixtures, accounting for, 420
nonluminous, 18
nozzles, 266
oils, 72
piping system, heavy, 460
piping system, light, 461
temperature, 563
viscosity of, 183
orifice-to-eductor throat diameter, 319
piping design, 458
preheat temperature, 56
pressure, 213, 473, 496
drop, 443
graph of heat release vs., 475
high, 510
pretreatment, 199
rapid oxidation of, 641
refinery/chemical plant, 528
selection of test, 434
source, 344
spray, 277
-staged burner, 405, 407
staging strategies, 291
storage tanks, test, 436
supply, double-block-and-bleed for, 379
switching, 198
system preparation, 497
temperature, 476
train, 540
oil, 543
typical main gas, 541
typical flared gas compositions, 185–186
ethylene/polyethylene gases, 186
oil field/production plant gases, 186
other special cases, 186
refinery gases, 186
valves, 380
Fully metered cross limiting scheme, 392
Fume incineration, 526
Fundamentals, 33–67
combustion kinetics, 60–61
fuel-bound NOx mechanism, 61
prompt-bound NOx mechanism, 61
reaction rate, 60–61
thermal NOx formation, 60
conservation of mass, 35
equilibrium and thermodynamics, 47
flame properties, 61–67
available heat, 64
flame speeds, 67
flame temperature, 61–64
flammability limits for gas mixtures, 66–67
minimum ignition energy, 64–66
general discussion, 54–59
air preheat effects, 55–57
fuel blend effects, 57–59
ideal gas law, 35–38
net combustion chemistry of hydrocarbons, 34–35
overview of combustion equipment and heat transfer, 34
stoichiometric ratio and excess air, 38–46
adiabatic flame temperature, 46
heat of combustion, 38–45
substoichiometric combustion, 46, 47–53
uses for combustion, 34
Furans, 219
Furnace(s)
air in-leakage influence, 562
burner combination, 111
cleanliness, effect of on NOx emissions, 561
cold flow, 264
cracking, 34
cylindrical, radiation heat transfer in, 107
design, terrace wall, 111
ethylene
cracking, 111
pyrolysis, 304, 305

test furnace for simulation of, 434
explosions in, 336
gas, 290, 296
flow patterns, 109
radiation, 105
recirculation, schematic, 201
temperature measurement, 438
heat transfer, 104
hot oil, 14
hydrogen reforming, 111
infrared thermal image of flame in, 98
low temperature, 109
manufacturers, 111
process, heat transfer in, 102
reforming, 34
role of burner in, 112
source of explosions in, 337
sulfur recovery reaction, 307
temperature, 191
test
selection criteria, 434
for simulation of down-fired tests, 435
for simulation of up-fired tests, 435
/thermal oxidizer/incinerator/combustion chamber, 644
tube, 8
vertical tube, 104
wall
cross section of, 108
heat transfer through, 108

G

Gage
pressure, 149
sizes, 707
Gas, see also Gases and liquids, properties of; Natural gas
acid, 639, 657, 663
analyzer, FTIR, 440
behavior of, 127
burner(s), 23
troubleshooting for, 519
typical excess air values for, 490
typical premixed, 23
center fire, 576
chemical properties of, 717–722
combustion
data comparing single- and three-stage, 574
polynomial expression for, 125
composition, 54, 55, 594
emissivity, 96
ethylene/polyethylene, 186
-fired furnaces, 290, 296
firing, grid configuration, 527
flame(s), 104, 105, 531
flammability characteristics of, 340
flexicoking waste, 162
flow, diverting, 619
flue, 72, 471
hot, 511, 512
recirculation, 550
shearing effect of high-velocity, 661
temperature, 476
fuel, 353
capacity curve, 355
train, typical main, 541
geometries, mean beam lengths for, 100
ignition characteristics of, 340
injectors, 583
jet
Mach number of, 238
mixing noise, 238
noise, 238, 241, 242
kinetic theory of, 122, 126
lighter-than-air, 130
lines, pilot, 514
mass flow rate of secondary, 253
mixtures
flammability limits for, 66
multi-component, 339
viscosity of, 120
oil field/production plant, 186
oxygen-containing, 601
partial premix, 363
premix, 363
pressure available, 596
property(ies)
calculations, 183
of selected, 72
purge, 597, 601
radiation
absorption and emission in, 93
properties, 298
recirculation, 20, 200
refinery, 159, 186
comparison of to test blend, 437
example, 437
simulated refinery, 39, 48
specification sheet, test procedure, 445–446
stack, reheat, 526
temperature, 438, 595
tip(s), 456, 462, 641
cleaning, 462
corroded, 504
trapped purge, 621
treatment, 5
turbine exhaust, 352, 531
valve data, 392
waste, 599
wet fuel, 160
Gaseous fuel(s), 158
flames, photographs of, 165
mixtures
physical constants of typical, 163
volumetric analysis of, 163
molecular weights and stoichiometric coefficients for common, 37
physical properties of, 165
Gases and liquids, properties of, 713–722
chemical, physical, and thermal properties of gases, 717–722
combustion data for hydrocarbons, 713
enthalpy of formation, Gibbs function of formation, and absolute entropy of various substances, 714
properties of dry air at atmospheric pressure, 715–716
Gasolines, aviation, 175
Gauss constant, 143
Gibbs function of formation, 714
GLC, see Ground-level concentration
Glossary, 727–728
Gravitational body force, 136, 137
Gravitational constant, 130
Gray/diffuse surfaces, radiant exchange between, 91
Grease, 5
Grid burner(s)
gas flame from, 531
heat distribution by, 532
Ground-level concentration (GLC), 632, 683
Guy wire-supported flare, 625

H

Halogenated hydrocarbon systems, 683
Hard refractory, 645
Hastelloy, 685
Hazardous waste incinerators, 291
Hearing
threshold of, 226
typical range of human, 225
Heat, see also Heat transfer
capacity, 125, 182
of combustion, 38
conduction, 71
damage, 331
exchangers, 70, 652
flux, 104, 442
recovery
area (HRA), 562
cooling by, 648
cooling without, 652
efficiencies (HREs), 282
thermal oxidation system, 680
regenerator performance, modeling of, 252
release, 185, 602
chemical, 137
graph of fuel pressure vs., 475
graphical representation of, 66
Heater(s)
balanced-draft, 492
box, 17
burner mounted on top of, 452
cabin, 16, 18, 28, 34
collapse, 499
connection of burner to, 455
cutout, 451
cylindrical, 17
existing, 450
fired, 9
developing emergency procedures for, 499
developing startup and shutdown procedures for, 497
major refinery processes requiring, 13
tube rupture in, 332
natural-draft, logic diagram for tuning, 491

new, 450
operation
data, 496
specifications, 442
preparation, 450
process, 11, 12
refinery, 11
refractory, 453
shell, 495
shutdown of, 498
turndown operation, 492
warped steel on shell of, 451
Heat transfer, 69–114
conductive, 71, 137
convection, 82–87
laminar flow convection, 83–84
Newton's law of cooling, 82
turbulent external flow, 85–87
turbulent internal flow, 84–85
heat transfer in process furnaces, 102–112
analysis of radiation heat transfer, 106–107
flame radiation, 105
furnace gas flow patterns, 109–110
furnace gas radiation, 105–106
heat transfer in process tube, 109
heat transfer through wall of furnace, 108–109
refractory surface radiation, 106
role of burner in heat transfer, 110–112
mechanisms of, 70
in packed bed, 281
in process furnaces, 102
in process tube, 109
radiation, 87–102
blackbody radiation/Planck distribution, 88–90
equation of radiative transfer, 97–101
infrared temperature measurement, 92–93
mean-beam-length method, 95–97
radiant exchange between black surfaces, 90–91
radiant exchange between gray/diffuse surfaces, 91
radiation in absorbing/emitting/scattering media, 93–95
radiation emitted by flame, 101–102
view factors for diffuse surfaces, 92
role of burner in, 110
thermal conductivity, 71–82
one-dimensional steady-state conduction, 73–77
transient conduction, 77–82
Heavy fuel oil piping system, 460
Heavy oils, 178, 180
Helium balloon, attached to ground, 129
HHV, see Higher heating value
HIGH box, 490
Higher heating value (HHV), 184
High pressure drop burner, 643
High-pressure flare, 244, 245
High viscosity liquid fuels, 365
Horizontal flares, 592
Horizontal floor-fired burners, 14
Hot flue gas, 130
Hot oil furnace, 14
Hottel's assumption, 95
HPI, see Hydrocarbon processing industry
HRA, see Heat recovery area
HREs, see Heat recovery efficiencies
Human ear, cross section of, 225
Human error, 338
Hybrid systems, 375
Hydraulics, 139, 599
Hydrocarbon(s)
-based gaseous fuels, 528
chemical degradation of, 479
chlorinated, 309
coking of, 462
combustion data for, 45
industry, 4, 590
long-chain, 538
net combustion chemistry of, 34
paraffinic, 269
processing industry (HPI), 361
properties of, 713
reforming, 14
systems
halogenated, 683
sulfur-bearing, 682
temperature vs. viscosity for, 123
unburned, 433, 439, 538, 548, 597
vapor pressures for light, 341
viscosity vs. temperature for range of, 477
Hydrodesulfurization, 13
Hydrogen
bubble technique, 150
-to-carbon weight ratio, 165
cyanide, 667
enrichment, 631
production, 8
purification, 161
reforming furnace, 111
Hydrostatics, 128
Hydrotreating, 13

I

ICPs, see Incomplete combustion products
Ideal gas
equation, 127
law, 35, 126, 263
Idealized chemical reactors, 258
Identity matrix, 428
IEC, see International Electrochemical Commission
IFGR, see Internal flue gas recirculation
Ignition, 360
characteristics, of liquids and gases, 340
control, 342
energy, minimum, 64
flame, forced-draft, 612
fuel pressure, unsuccessful, 336
graphical representation of, 66
ledge, burner, 482
pilot, 615, 640
ports, 513
sources, of major fires, 342
spontaneous, 342
transformers, 384
zone, in natural draft burners, 360
Ignitors, 614
Incendiarism, 342
Incineration, upstream, 664
Inclined manometer, 129
Incomplete combustion products (ICPs), 548, 550
Incompressible flow, 136
In-duct burners, 524
Industrial equipment, noise-producing, 224
Industrial insurance carriers, 339
Industrial noise pollution, 231
Industrial Risk Insurers (IRI), 339, 675
Infrared pyrometer, 478
Infrared temperature measurement, 92
Inline burner, 530
In-scattering, 97
Insignificant effects, pooling of, 411
In situ extractive analyzers, 472
Installation and maintenance, 449–466
installation, 450–459
air control, 456–458
burner installation inspection, 456
burner mounting, 451–453
burner pre-installation work, 450–451
connection of burner to heater, 455–456
fuel piping design, 458–459
preparation of heater, 450
tile installation, 453–455
maintenance, 459–465
air registers and dampers, 464–465
flame stabilizer, 463–464
gas tip and orifice cleaning, 462
oil tip and atomizer cleaning, 463
pilot burners, 465
tile, 463
Internal flue gas recirculation (IFGR), 576, 578
Internal mix atomizers, 584
Internal mix twin fluid atomizer, 364, 365
Internal tube fouling, 493
International Electrochemical Commission (IEC), 540
Inviscid flows, 145, 290
IRI, see Industrial Risk Insurers
Isomerization, 5
Isotropic turbulence, 142

J

Jet(s), 142
engine, flow around air intake of, 144
high-pressure, 607

K

Karlovitz number, 299
KE, see Kinetic energy
Ketones, 216

K factors, 132
Kiln off-gas, 352
Kinematic viscosity, 120
Kinetic energy (KE), 276
 definition of turbulent, 293
 in fluid flow, 258
Kinetic theory of gases, 126
Knockout drums, types of, 618

L

Laminar flow, 138
 convection, 83
 of smoke, 141
Laminar regime, friction factor for flow in, 150
Laplacian operator, 73
Large control panel, 382
Large eddy simulations (LES), 297
Laser Doppler velocimetry (LDV), 295
LDV, see Laser Doppler velocimetry
Le Chatelier equation, 339
LES, see Large eddy simulations
Level controls, 628
LFL, see Lower flammability limit
LHV, see Lower heating value
Liftoff, 67
Light fuel oil piping system, 461
Lightning, 342
Light oils, 178
Linear algebra, 427
Linear burner elements, 530
Linear transformations, 407
Line sizing, 458
Liquefied petroleum gas (LPG), 338
Liquid(s), see also Gases and liquids, properties of
 autoignition temperature, 343
 firing, 364, 527
 flammability characteristics of, 340
 flash point of, 179
 fuel(s), 165
  atomization system, 354
  atomizer/spray tip configurations, 356
  capacity curve, 357
  distribution, 279
  firing, typical excess air values for, 490
  high viscosity, 365
  physical properties of, 179
  vaporization of, 354
 ignition characteristics of, 340
 low-viscosity, 366
 naphtha, 179
 oxidizers, 330
 pour point of, 179
 removal, 600
 seals, 619, 621
 tips, 641
 viscosity for multiple constituent, 120
Liquified petroleum gas (LPG), 159
Literature review, 7–9
 combustion, 8
 combustion in process industries, 9
 process industries, 9
Long-chain hydrocarbons, 538
Loop seals, 629
Loss coefficient(s), 151
 definition of, 152
 for various fittings, 152
Lower flammability limit (LFL), 184, 339
Lower heating value (LHV), 184, 436, 437, 630
Low pressure drop burners, 642
Low-viscosity liquids, 366
LPG, see Liquefied petroleum gas
Lube oil, 5
Luminous flame, photographic view of, 101
Lumped capacitance, 77, 78, 79
Lurking variables, 416

M

Mach number, 144, 146, 276
 flows, 292
 gas jet, 238
 maximum, 255
 relation, 147
MacLaurin series approximation, 427, 428
Magnetic flow meter, 388
Maintenance procedures, refinery damaged due to improper, 346
Manipulated variable (MV), 394
Manometer
 inclined, 129
 U-tube, 128, 129
Manometry, 128
Manufacturing tolerances, 153
Mass
 conservation, 133, 135, 274, 289
 flow distribution, improvement of to burners, 563
 selective detector (MSD), 440
Mathematical modeling, of combustion systems, 251–284
 burner pressure drop, 260–266
 education processes, 252–258
  education processes in pilots, 256–257
  education processes in premixed burners, 257–258
  steam flare education modeling, 254–256
 flare smokeless operation, 266–273
  application to steam flares, 268–270
  modeling air-assisted flares, 270–273
  predicting flame smoking tendencies, 266–268
 idealized chemical reactors and combustion modeling, 258–260
  perfectly stirred reactor, 259
  plug flow reactor, 258–259
  systems of reactors, 259–260
 oil gun capacities, 273–277
  results, 277
  two-phase flow analytical development, 274–277
 oil gun development, 277–280
 overview, 252
 regenerative thermal oxidizer performance, 280–282
Matrix
 addition, 429
 arithmetic, rules of, 406
 multiplication, 428
Maximum continuous rating (MCR) conditions, 567
Maxwell relation, 144, 145, 146
MCR conditions, see Maximum continuous rating conditions
Mean beam length, 95, 96, 100
Mean droplet size, 277, 278
Mean square residual (MSR), 408
Mean squares (MS), 409
Mechanical linkage, 389
Mechanical stops, 386
Medium pressure drop burner, 643
Metering
 air, 356
 fuel, 353
Methane
 auto-ignition temperature of, 56
 concentration, predicted centerline profiles for excess air, 316
 consumption, 314, 317
 graph of sustainable combustion for, 353
 oxygen–nitrogen system, 258
 stoichiometric O_2:CH_4 ratio for, 55
Methanol, 216
Method of least squares, 405
Method of steepest ascent, 416, 417
Middle ear, bones in, 224
MIE, see Minimum ignition energy
Minimum ignition energy (MIE), 64, 342
Minutes per repeat (MPR), 396
Mixed convection, 82
Mixed-is-burnt assumption, 296
Mixtures, experimental designs for, 421
Model
 aerodynamic simulation, 566
 burner, 516, 534
 CFD, 302, 305
 combustion chemistry, 299
 development, regenerative thermal oxidizer, 280
 eddy breakup, 299
 flamelet, 300
 plug flow reactor, 259
 pollutant chemistry, 301
 radiation, 297
 second-order, 428
 semi-empirical, 425
 smokeless nozzle scaling, 270
 turbulence, 292, 301
 validation, 295
 weighted-sum-of-gray-gases, 298
 windbox, air flow balanced by, 567
Modeling
 air-assisted flares, 270
 burner pressure drop, 252
 chemical reactions, 252
 computational fluid dynamic, 534, see also Computational fluid dynamics based combustion modeling

eductors, 252
flare smokeless rates, 252
heat regenerator performance, 252
oil gun performance and improvement, 252
physical, 534
sample results of simplified, 261
steam flare eduction, 254
Molar ratios, for combustion reactions, 37
Molecular weight, 183
Molten salt system, 687
Molten substances, 342
Momentum
conservation, 133, 253, 275
equation, 142
flux normal to control surface, 255
Monitors, 614
Monte Carlo method, 297
Moody diagram, 151
Motorboating, when firing oil, 512
MPR, see Minutes per repeat
MS, see Mean squares
MSD, see Mass selective detector
MSR, see Mean square residual
Muffler elbow, 265
Multi-burner boilers, 559
Multiple burner interaction, 243
Multi-point flares, 592
Municipal solid waste boiler, 413
MV, see Manipulated variable

N

Naphtha
distillation curve, 181
liquid, 179
NASA
equilibrium code, 126
nozzle performance data, 253
polynomials, 125
National Electrical Code (NEC), 338, 379, 540
National Fire Protection Association (NFPA), 338, 379, 675
Natural draft
burner, 4, 357, 358, 4, 480
gas burner, 27
heater, logic diagram for tuning, 491
Natural gas, 18, 72, 158, 528
components, commercial, 159
example pipeline quality, 158
flame produced by burning, 595
NOx generation with firing, 555
NOx vs. heat load with firing, 577
production, 638
reforming, 14
species concentration
vs. excess air for, 39
vs. stoichiometric ratio for, 48
Tulsa, 170, 171, 172
Navier–Stokes equations, 145, 292
Nebraska boiler, 584
NEC, see National Electrical Code
Newtonian fluid, 136, 292
Newton iteration, 259
Newton's law of cooling, 82, 281
Newton's second law, 137, 289
NFPA, see National Fire Protection Association
Nitric oxide (NO), 438
Nitrobenzene, 667
Nitrogen oxides (NOx), 38, 191
analyzer, 389
boiler load influence on, 558
burner(s)
low, 202
technology, 518
control
methods, 666
technique, reduction efficiencies for, 198
technologies, in process heaters, 199
degree of power functional, 559
effect of boiler design on, 554
effect of bound nitrogen in liquid fuel on, 516
effect of burner model on, 516
effect of firebox temperature on, 515
emission(s), 61
correlations, boiler design impacts on, 553
effect of furnace cleanliness of, 561
factors, for typical process heaters, 193
fuel composition effects on, 211
high, 515
methods to reduce, 568
test data, 577
theory for ultra-low-, 580
from ultra-low emissions burner, 584
empirical evidence of effect of excess air on, 556
formation, 60
excess O_2 influence on, 556
fundamentals, 581
prompt-, 61
thermal, 60
fuel-bound nitrogen, 537
as function of burner geometry, 405
generation, with firing natural gas, 555
important factors affecting, 192
influence of oxygen on, 408
mechanism, fuel-bound, 61
minimization, 688
municipal solid waste boiler using ammonia injection to control, 413
predictions, improving, 319
reduction, 565
data, 570
by FGR implementation, 568
philosophy, 561
regulations for, 196
relative steam flow vs., 557
response from burner, potential factors affecting, 404
strategies for reducing, 198
thermal, 537
NO, see Nitric oxide, 438
Noise, 223–249
abatement techniques, 239–243
burner noise abatement techniques, 242–243
fan noise abatement techniques, 243
flare noise abatement techniques, 239–242
valve and piping noise abatement techniques, 243
analysis of combustion equipment noise, 243–246
high-pressure flare, 244–246
multiple burner interaction, 243–244
burner combustion, 236, 237
combustion
equipment, 243
instability, 234
contributions, based on mathematical model, 245
emissions, 443
exposures, OSHA permissible, 234
fan, 237
flare combustion instability, 236
fundamentals of sound, 224–231
basics of sound, 224–228
measurements, 228–231
gas jet, 238, 242
glossary, 246–248
industrial noise pollution, 231–234
international requirements, 232–233
noise sources and environment interaction, 234
OSHA requirements, 232
mechanisms of industrial combustion equipment noise, 234–239
combustion roar and combustion instability noise, 234–237
fan noise, 237–238
gas jet noise, 238–239
valve and piping noise, 239
meter, schematic of, 228
pollution, 231, 234
radiating from valve, 240
reduction plenum, 451
screech, 245
shock-associated, 238
/visible flame, 604
Nonluminous flame, photographic view of from John Zink gas burner, 102
Non-smoking flares, 267
Normal forces, 136, 137
NOx, see Nitrogen oxides
Nozzle mix burner, 364
Null hypothesis, 409
Nusselt correlation, 282
Nusselt number, 84, 281

O

OASPL, see Overall sound pressure level
Occupational Safety and Health Act (OSHA), 232
HAZWOPER, 346
permissible noise exposures, 234
Octave bands, 230
Off-gas, 352
Offshore oil rig flare, 7

Oil(s), 178
  atomizer cleaning, 463
  burner(s)
    common, 479
    needing service, 28
    tile, 455, 463
    troubleshooting for, 520
  crude, 173, 175
  deposits, found in United States, 166
  derrick, 174
  discovery, successes of American, 166
  exploration, 169
  field/production plant gases, 186
  -fired burner, experiencing flame lift-off, 513
  firing, 364
  flame
    flame envelope of, 482
    viewing of through burner plenum, 174
  fuel, 72
    train, 543
    viscosity of, 183
  gun(s), 551
    atomizers fitted onto, 585
    capacities, 273
    development, 277
    droplet size comparison between standard and newer, 279
    modeling, 283
    oil–steam spray discharging from, 508
    performance and improvement, modeling of, 252
  heavy, 178, 180
  light, 178
  motorboating when firing, 512
  recovery, 173
  reserves, United States, 168
  residual, 178
  sour crude, 176
  spillage, main cause of, 507
  steam spray, discharge of from oil gun, 508
  sweet crude, 176
  tip(s), 456, 484
    cleaning, 463
    drillings, 483
  train, typical pilot, 544
  viscosity of mid-continent, 124
  well, capping of burning, 175
Olefins, 404
  combustion data for, 45
  properties of, 713
OPEC, see Organization of Petroleum Exporting Countries
Operator training and documentation, 345
Optical sensing, 617
Organic heat-transfer fluid heat exchangers, 652
Organization of Petroleum Exporting Countries (OPEC), 166
Orifice flow meter, 388
Orthogonal blocking, 416, 417
Orthogonal mixture designs, 423
Orthogonal subspace, 423
Osborn Reynolds' experimental apparatus, 140
Oscillation, 398
OSHA, see Occupational Safety and Health Act
Outlet damper flow control, 676
Out-scattering, 97
Overall heat transfer coefficient, 76
Overall sound pressure level (OASPL), 246
Oxidation reactions, 639
Oxidizer, 19
  catalytic, 647
  liquid, 330
  switching, 199
  thermal, see Thermal oxidizers
Oxygen
  analyzer, 389, 473
  concentration, in flue gas, 425
  condition
    excess, 309
    of insufficient, 503
    stoichiometric, 315
  -containing gases, 601
  content
    excess air indication by, 471
    TEG, 538
  -enriches streams, 352
  excess, 471
  fuel chemical compound rate of reaction with, 364
  -to-fuel ratio, 38
  levels, cost of operating wit higher excess, 474
  location for measuring excess, 472
  mass fractions, 311
  sensors, 336
  sources, 393
  trim, 393
Ozone, 438

P

Packed column, 663, 665
Paraffinic hydrocarbons, 269
Paraffins
  combustion data for, 45
  properties of, 713
Parallel positioning, 389
  electronically linked, 391
  mechanically linked, 390
  variation of, 391
Parallel stress, 136
Partial differential equations (PDEs), 295
Partial premix gas, 363
Particle
  carryover, 218
  entrainment, 217
  removal efficiencies, 661
Particulate(s), 217, 440, 640
  /acid gas removal, 657
  emissions, 442
  matter (PM), 539
  removal, 660
Pasquill–Gifford–Turner classification, 632
Pass flow, 496, 498
PCB, see Polychlorinated biphenyls
PCDD, see Polychlorinated dibenzo-*p*-dioxin
PCDF, see Polychlorinated dibenzofuran Polychlorinated dibenzo-*p*-dioxin (PCDD), 219
PDEs, see Partial differential equations
PDPA, see Phase Doppler Particle Analyzer
Perfectly stirred reactor (PSR), 259
Performance guarantee specifications, 442
Petrochemical
  industry, 4, 590
  manufacturing, 638
  process heaters, 291
Petroleum
  industry, quantitative listing of products made by U.S., 175
  refinery, typical, 4
  refining processes, 5
PFDs, see Process flow diagrams
PFR, see Plug flow reactor
PGI, see Power generation industry
PHA, see Process hazard analysis
pH analyzer, 389
Phase Doppler Particle Analyzer (PDPA), 279
Physical modeling, 534
Physical properties of materials, 693–711
  areas and circumferences of circles and drill sizes, 693–701
  commercial copper tubing, 707–708
  physical properties of pipe, 702–706
  standard grades of bolts, 709, 710, 711
PICs, see Products of incomplete combustion
P&IDs, see Piping and instrumentation diagrams
Pilot(s), 614
  burners, 465, 483, 514
  eduction processes in, 256
  flames, stable ignition of, 336
  gas
    air mixture, 514
    lines, 514
    train, 542
  ignition, 615, 640
  monitor, SoundProof acoustic, 617
  oil train, 544
  parts, 614
  tip, 456, 465
  VYD burner gas tip in diffuser with, 457
Pipe
  physical properties of, 702–706
  racks, 381
Pipeline transmission network, 158
Piping
  burner, 453
  explosions in, 335
  noise abatement techniques, 243
  upstream, 604
Piping and instrumentation diagrams (P&IDs), 345
Planck distribution, 88, 89
Plane wall, 73, 74
Plant schematic, typical, 525
Plastic
  fiberglass-reinforced, 654
  refractory, 645, 684
Platinum, use of in resistance temperature detectors, 387

PLC, see Programmable logic controller
Plenum
  burner mounted in common, 452
  damper, 489, 490
  fire, 507
  noise reduction, 451
Plug flow reactor (PFR), 258
Plugged gas ports, 539
Plugging, with polymers, 462
Plume heights, estimating, 633
PM, see Particulate matter
Pneumatic control valve, 385
Pollutant(s)
  accurate measurements of, 197
  chemistry models, 301
Pollutant emissions, 189–220, 432
  carbon dioxide, 218
  combustibles, 214–217
    CO and unburned fuel, 214–215
    volatile organic compounds, 215–217
  conversions, 190–191
  dioxins and furans, 219
  emission in hydrocarbon and petrochemical industries, 190
  nitrogen oxides, 191–214
    abatement strategies, 198–204
    field results, 204–214
    measurement techniques, 197–198
    regulations, 196–197
    theory, 192–196
  particulates, 217–218
    sources, 217–218
    treatment techniques, 218
  SOx, 219
Polychlorinated biphenyls (PCB), 683
Polychlorinated dibenzofuran (PCDF), 219
Polymerization, 5
Polymers, plugging with, 462
Polynomial expression, for combustion gases, 125
Portable analyzers, 472
Port mix twin fluid atomizers, 365
Positioner, 386
Position switches, 383
Positive displacement flow meters, 388
Potential energy, change in, 130
Pour point, of liquid, 179
Power generation industry (PGI), 5, 361
Prandtl–Meyer expansion waves, 147
Premix
  burner, 21, 22, 480, 481
  gas, 363
  metering orifice spud, 355
Pressure
  atmospheric, 148
  constant, 258
  definition of, 148
  drop, 148, 152
  dynamic, 131
  energy, change in, 130
  forces, 136, 137
  fuel, 213
  gage, 149, 495
  relief vessel, venting to flare, 131
  stagnation, 148
  static, 130, 131
  swing adsorption (PSA), 161
  switches, 383
  total, 130, 131
  transmitters, 387
  units of, 149
  upstream, 147
  velocity, 132
Problem burner, 568
Process
  flow diagrams (PFDs), 345
  fluid parameters, 479
  furnaces, heat transfer in, 102
  hazard analysis (PHA), 337, 344, 345
  heater(s), 11
    examples of, 12, 16
    heat balance, 21
    NOx control technologies in, 199
    oil flame, 276
  heat transfer tube, cross section of, 109
  industries, 4–7, 9
    hydrocarbon and petrochemical industries, 4–5
    power generation industry, 5–7
  thermal oxidation, 7
  tubes, 477, 493
  variable (PV), 395, 396, 397–398
Products of incomplete combustion (PICs), 309
Profibus, 375
Programmable logic controller (PLC), 374, 375, 380, 540
Prompt-NOx formation, 61
Propane, 23, 170, 595
Propylene, 23, 595
PSA, see Pressure swing adsorption
PSR, see Perfectly stirred reactor
Pulsating flame, 502
Pulse-jet, 658
Pulverized coal combustor, 291
Pure error, 412
Pure tone, 224
Purge
  controls, 629
  gas, 597, 601
  reduction seals, 620
PV, see Process variable
Pyrometer, infrared, 478

Q

Quasi-one-dimensional isentropic flow, 145
Quench/two-stage acid removal system, 666

R

Radial blade operating curve, 673
Radial coordinate systems, 76, 79
Radial inlet damper, 676
Radiant exchanges
  between black surfaces, 90
  between gray/diffuse surfaces, 91
Radiant intensity, 94
Radiant wall burner, 369
  photographic view of, 103
  picture of, 260
Radiation, 87
  in absorbing/emitting/scattering media, 93
  absorption, in gases, 93
  calculation methods, 604
  electromagnetic, 89
  exchange rate, net, 96
  flame, 105
  furnace gas, 105
  heat transfer, 70
    analysis of, 106
    correction factor, for mixtures of water vapor and carbon dioxide, 100
    in cylindrical furnace, 107
  intensity of, 94
  models, 297
  refractory surface, 106
  thermal, 88
  transport equation (RTE), 297
  wavelength of, 88
Radiative exchange, network representation of between surfaces, 92
Radiative heat transfer, 137
Radiometer schematic, ellipsoidal, 443
Radiosity, 89, 91
Random effects, 417
Random error, 408
Raw gas burner(s), 206, 364, 366, 481, 482
  effect of excess oxygen on NOx in, 515
  typical throat of, 357
Reaction
  efficiency, 630
  rate, 60
Reactor(s)
  plug flow, 258
  systems of, 259
Reboiler, xylene, 306, 307
Reburning, 201
Recuperative preheat exchanger, 649
Recycle flue gas, 668
Redlich–Kwong equation, 128
Reed wall, 504
Refinery
  /chemical plant fuels, 528
  damage of due to improper maintenance procedures, 346
  flow diagram, 6, 176
  fuel composition, 446
  gas(es), 159, 186
    comparison of to test blend, 437
    composition of typical, 160
    example, 437
  heaters, 11
Reformate extraction, 14
Reformers
  terrace wall, 10
  top-fired, 10
Reforming furnaces, 34
Refractory(ies)
  castable, 646, 684, 689
  failure, 487
  hard, 645
  plastic, 645, 684

soft, 645, 646
surface radiation, 106
Refrigerants, chemical properties of, 717–722
Regenerative thermal oxidizer (RTO), 280
bed retrofit applications, 282
model development, 280
burner pressure drops, 283
Regen tile, 366
Reheater (RH), 561
Relative NOx, 559
Relative roughness, 150
Relay system, 374
Renormalization Group (RNG), 302
Repeats per minute (RPM), 396
Residual fuel oil (RFO), 561–562
Residual oils, 178
Resistance temperature detectors (RTDs), 387
Reynolds averaging, 139, 292, 301
Reynolds number, 83, 138, 145, 180
Reynolds stress model (RSM), 302
RFO, see Residual fuel oil
RH, see Reheater
RIMFIRE® flare burner, 611, 613
RNG, see Renormalization Group
Rockets, industrial combustion and, 8
Rosin–Rammler relationship, 278
Rotatability, 420
Roughness, 149, 150
Round-shaped flame, 362
RPM, see Repeats per minute
RSM, see Reynolds stress model
RTD, see Resistance temperature detectors
RTE, see Radiation transport equation
RTO, see Regenerative thermal oxidizer
Runge–Kutta solver, shooting, 277
Run indicators, 384
Runner pipes, severe sagging of, 539
Rust preventatives, 175

S

Saddle data, 281
Safety
documentation, operator training and, 344
requirements, 597
shutoff valves (SSOVs), 379, 384
Salts/solids systems, 686
Sankey diagram, 64
Saturated factorial, 415
Saturation quench section, 653
Sauter mean diameter (SMD), 277–278
SBCR, see Selective noncatalytic reduction
SCA, see Specific collector area
Scaling functional, 272
Scanning electron microscopy (SEM), 441
SCAQMD, see South Coast Air Quality Management District
SCR, see Selective catalytic reduction
Screech noise, 239, 240, 245
Screening designs, 415
Scrubbers, 664, 665, 687
SCS, see Specific collector surface
Seal(s)
liquid, 619
loop, 629
purge reduction, 620
smoke signals from surging liquid, 620
Second law of thermodynamics, 71
Second-order designs, 419
Second-order regression model, 428
Seconds Saybolt Furol (SSF), 182
Seconds Saybolt Universal (SSU), 182
SED, see Statistical experimental design
Selective catalytic reduction (SCR), 203, 550, 667
Selective noncatalytic reduction (SNCR), 203, 204, 550, 667
Self-supported flare, 624
SEM, see Scanning electron microscopy
Semi-empirical models, 425
Semi-Implicit Method for Pressure Linked Equations (SIMPLE), 295
Semi-infinite solids, transient conduction in, 80
Sequential experimental strategy, 426
Serial correlation, 416
SG, see Specific gravity
SH, see Superheater
Shadow photograph, of burning butane lighter, 236
Shape factor, 92
Shear stress, 119, 136, 137
Shock-associated noise, 238
Shock waves, 245
Shooting Runge–Kutta solver, 277
Shutdown string, 381
Sidewall fired burner arrangement, 20
SIMPLE, see Semi-Implicit Method for Pressure Linked Equations
Simplex-centroid design, 421
Simplex designs, 415
Simulated refinery gas
species concentration vs. excess air for, 39
species concentration vs. stoichiometric ratio for, 48
Simulated reforming gas flame, 169
Simulation model, aerodynamic, 566
Single-point flares, 591, 593
Slipstream flame-front generator, 615, 616
Small control panel, 382
SMD, see Sauter mean diameter
Smoke
emission, from stack, 514
generation, 334
from incense, 138
laminar flow of, 141
suppression, 598, 603
Smokeless burning, 603
Smokeless flare
capacity, 266
early model, 591
of paraffinic hydrocarbons, 269
Smokeless nozzle scaling model, calibration of, 270
Smokeless steam flare, 272
Smoking flares, 266
SNCR, 669
Soft refractory, 645, 646
Soil fuel carryover, 218
Solenoid valve(s)
most common types of, 384
three-way, 386
Solid desiccants, adsorption with, 158
Solvent deasphalting, 5
Soot
blowers, online cleaning with, 688
formation
favorable condition for, 279
precursors to, 298
rates, 267
presence of in flame, 298
SOPs, see Standard operating procedures
Sound(s)
fundamentals of, 224
level(s)
meter, block diagram of, 229
overall, 229
of various sources, 234
low-frequency, 228
power level, 226
pressure level (SPL), 225, 226, 235
aircraft-to-ground propagation in, 231
burner, 237
spectrum, of high-pressure flare, 245
spectrum, 233
speed, definition of, 146
SoundProof acoustic pilot monitor, 617
Sour crude oil, 176
South Coast Air Quality Management District (SCAQMD), 196
Species, conservation of, 134, 138
Specific collector area (SCA), 659
Specific collector surface (SCS), 659
Specific gravity (SG), 182
Specific heat, 125, 182, 184
Spectral blackbody emissive power, 90
Spill-return flow atomizer, 585
Spin diffuser, 464
SPL, see Sound pressure level
Spontaneous ignition, 342
Square matrix, 429
SSF, see Seconds Saybolt Furol
SSOVs, see Safety shutoff valves
SSU, see Seconds Saybolt Universal
Stabilizer, damaged, 464
Stack(s)
damper, 495, 457
emissions, 496
explosions in, 335
gas reheat, 526
smoke emission from, 514
temperature, high, 511
Staged air burner, 25, 517
Staged burner design philosophy, 551
Staged combustion, 200, 573
Staged flare system, general arrangement of, 598
Staged fuel burner, 26, 517, 518
Stagnation pressure, 148
Standard deviation, 409
Standard errors of effects, 412
Standard operating procedures (SOPs), 345
Stanton number, 84, 85

START box, 490
Static electricity, 343
Static mixers, 160
Static pressure, 130, 131, 596
Static sparks, 342
Statistical experimental design (SED), 402
  combining domain knowledge with, 424
  contrast of classical experimentation and, 403
  principles, 404
Steady-state conduction equation, 77
Steam
  -assisted flare, 237, 241
  control, automatic, 627
  flare(s), 268
    eduction modeling, 254
    with steam eductor, 254
    tube layout, third-generation, 254
  flaring smoking tendencies, predicting, 269
  -to-hydrocarbon ratios, 269, 271
  injection, 201, 607
  manifold, 606
  reforming, composition of, 161
  superheater, 14
  trapped, 333
Steamizer™, 610
Stefan–Boltzmann law, 79, 90
Stoichiometric oxygen condition, 315
Stoichiometric ratio, excess air and, 38
Submerged quench, 655
Substoichiometric combustion, 46, 47
Suction pyrometer, 207, 386
Sulfur
  -bearing hydrocarbon systems, 682
  dioxide, 539
  oxides, 219
  recovery reaction furnace, 307
Superheater (SH), 34, 561
Supersonic eductor performance, analysis of, 254
Surface
  convection condition, 81
  exchange, 93
Sweet crude oil, 176
Sweetening/sulfur removal, 5
Swirler, 361
Swirl vanes, 583
Switches
  flow, 384
  position, 383
  pressure, 383
  temperature, 383
System shutdown
  local request required after, 380
  unsatisfactory parameter, 380

T

TAE, see Thermoacoustic efficiency
Tanks, explosions in, 335
Taylor series approximation, 427
TCV, see Temperature control valve
TDMA algorithm, 295
TE, see Thermal energy
TEG, see Turbine exhaust gas
Temperature
  bridgewall, 496
  control valve (TCV), 498
  distribution, in cylinder, 76
  drop, due to thermal contact resistance, 75
  firebox, effect of on NOx, 515
  flame, reduced, 559
  flue gas, 476
  fuel, 476, 563
  gas, 595
  indicating controller (TIC), 389
  process tube, 477
  stack, high, 511
  switches, 383
  tube, 487, 496
Terrace wall
  furnace design, 111
  reformers, 10, 436
Test
  fuel storage tanks, 436
  furnace
    selection criteria, 434
    for simulation of up-fired tests, 435
  matrix, 442, 444
  procedure
    gas specification sheet, 445–446
    example, 447
Tetracontane, 177
Theoretical fluid dynamics, 288
Thermal conductivity, 71, 81
Thermal contact resistance, temperature drop due to, 75
Thermal cracking, 13, 14
Thermal diffusivity, 78
Thermal energy (TE), 276
Thermal NOx, 60, 537
Thermal oxidizer(s), 8, 637–689
  basic system building blocks, 640–670
    burners, 640–644
    catalytic systems, 646–647
    flue gas processing methods, 647–670
    furnace/thermal oxidizer/incinerator/combustion chamber, 644–645
    refractory, 645–646
  blowers, 670–674
  combustion basics, 639–640
    carbon monoxide, acid gases, 639–640
    material and energy balance, 639
    NOx formation, 639
    oxidizing/reducing combustion processes, 639
    particulate, 640
  control systems and instrumentation, 675–677
  flame inside, 317
  geometric information describing, 314
  picture of, 260
  sample results of simplified modeling for, 262
  system configurations, 677–689
    acid gas systems, 682–685
    non-acid gas endothermic waste gas/waste liquid system, 677–681
    non-acid gas exothermic waste gas/waste liquid system, 681–682
    NOx minimization or reduction systems, 688–689
    salts/solids systems, 686–688
  system generating steam, 679
Thermal radiation, 88
Thermoacoustic efficiency (TAE), 235
Thermocouples, 386, 387, 478
Thermodynamics
  equilibrium and, 47
  Maxwell relation of classical, 144
  relations, 144
  second law of, 71
Thermophoresis, 661
Thermowell, 386
Three-component system, 422
Three-way solenoid valve, 386
Threshold of pain, 227
Throat velocity, 147
TIC, see Temperature indicating controller
Tile
  burner refractory, 506
  height, 455
  installation, 453, 454
  oil burner, 455, 463
  tolerances, 454
Time-averaged emission, 298
TNG, see Tulsa Natural Gas
Todd Combustion low NOx burner, 578, 579
Toluene, 216
Top-fired reformers, 10
Total dry products, 38
Total pressure, 130, 131
Total thermal resistance, 74
Total wet products, 38
Touchscreen, 376
Transducer, current-to-pressure, 385
Transformers, ignition, 384
Transient conduction, 77
Transmitters, pressure, 387
Transport
  characteristics of turbulent, 292
  equations, 292
  properties, 122
Troubleshooting, 501–520
  flame impingement on tubes, 503–504
    cause and effect on operation, 503–504
    corrective action, 504
    indications of problem, 503
  flame lift-off, 513
    corrective action, 513
    effect on operation, 513
    indications of problem, 513
  flashback, 505
    cause and corrective action, 505
    effect on operation, 505
    indications of problem, 505
  high fuel pressure, 510
    effect on operation, 510
    indications of problem, 510
    solution and corrective action, 510

high NOx emissions, 515–519
corrective action, 515–519
effect on operation, 515
indication of problem, 515
high stack temperature, 511
corrective action, 511
effect on operation, 511
indications of problem, 511
irregular flame patterns, 505–507
cause and corrective action, 506–507
effect on operation, 506
indications of problem, 505–506
leaning flames, 509–510
causes and corrective action, 509–510
effect on operation, 509
indications of problem, 509
long smoky flames, 507–508
corrective action, 508
effect on operation, 508
indications of problem, 507
main burner fails to light-off or extinguishes while in service, 508–509
corrective action, 508–509
effect on operation, 508
indications of problem, 508
motorboating when firing oil, 512–513
effect on operation, 512
indications of problem, 512
solution and corrective action, 512–513
oil spillage, 507
cause and effect on operation, 507
corrective action, 507
indications of problem, 507
overheating of convection section, 511–512
corrective action, 512
effect on operation, 512
indications of problem, 511
pilot burner fails to ignite or extinguishes while in service, 514
corrective action, 514
effect on operation, 514
indications of problem, 514
pulsating flame, 502–503
cause and effect on operation, 502–503
corrective action, 503
indications of problem, 502
smoke emission from stack, 514–515
corrective action, 514–515
effect on operation and equipment, 514
indications of problem, 514
Tube(s)
bank, constants of equation for, 87
failures, 331, 333
flame impingement on, 503
furnaces, used in hydrogen production, 8
metal temperature, 496
rupture, in fired heater, 332
support color, 494
temperatures, 487
-to-wall spacing, 110
venturi, 505
Tulsa Natural Gas (TNG), 170, 207, 208, 256, 437
Tuning, 398
Turbine
exhaust gas (TEG), 352, 524, 531
oxygen content, 538
quenching effects of, 529
flow meters, 388
power augmentation, 532
Turbulence
equations, 292
Leonardo da Vinci's view of, 140
model(s), 292, 301
/radiation interaction, 298, 299
Turbulent combustion regimes of, 299
Turbulent energy, 294
Turbulent external flow, 85
Turbulent flow, 138
Turbulent internal flow, 84
Turbulent kinetic energy, definition of, 293
Turndown operation, heater, 492
Two-level factorials, 424
design, 412
fractional, 414
Two-phase flow analytical development, 274

U

UFL, see Upper flammability limit
UHCs, see Unburned hydrocarbons
UL, see Underwriters Laboratories
Ultra-low emissions burner, 580, 581, 582, 583
Ultra low-NOx burner, 517
Ultrasonic flow meter, 388
Unburned fuel, 214
Unburned hydrocarbons (UHCs), 433, 434, 439, 538, 548, 597
Underwriters Laboratories (UL), 540
Universal gas constant, 126
Unsatisfactory parameter system shutdown, 380
Unstable flame, 494
Upfired burner arrangement, 20
Up-fired tests, test furnace for simulation of, 435
Upper flammability limit (UFL), 184, 339
Upstream pressure, 147
U.S. EPA, 197
sampling system schematic as recommended by, 198
sampling train, flue gas drawn through, 440
Utility
boilers, 549, 572
burners, NOx vs. load with firing natural gas on, 559
costs, 596
U-tube manometer, 128, 129

V

Vacuum distillation, 5
Valve(s)
atomizing medium block, 486
burner block, 485, 495
butterfly-type combustion air, 391
control, 385, 397
gas, data, 392
noise radiating from, 240
pneumatic control, 385
remote isolation, 333
safety shutoff, 379, 384
solenoid, 384, 386
steam control, 626
Van Driest hypothesis, 293
Vapor(s)
chemical properties of, 717–722
pressures, for light hydrocarbons, 341
Vaporization process, 337
Variable loop gain, 398
Variance, 408
VC process heater, see Vertical cylindrical process heater
Velocity
boundary layer thickness, 85
constant, 134
energy, change in, 130
gradient, mathematical expression of, 119
pressure, 132
thermocouple, 386, 387, 478
Venturi eductor optimization, 319
Venturi quench, adjustable-plug, 656
Venturi scrubber, 665, 687
Venturi-style burners, 558
Venturi-style low NOx burner, 522
Venturi tube, burning inside, 505
Vertical cylindrical (VC) process heater, 16, 34
Vertical tube furnace, oil and gas firing in, 104
View factor(s)
aligned parallel rectangles, 97
coaxial parallel disks, 97
computation of, 106
diffuse surfaces, 92
perpendicular rectangles with common edge, 97
two-dimensional geometries, 94
Visbreaking, 5, 13
Viscosity, 180, 185
absolute, 119
conversion table, 121
definition of, 119
fluid, 139
fuel oils, 183
kinematic, 120
mid-continent oils, 124
temperature vs. for hydrocarbons, 123
Viscous flow, 290
Visual Basic, 277
VOCs, see Volatile organic compounds
Volatile organic compounds (VOCs), 214, 215
Vortex shedder flow meter, 387

W

Wake area showing mixing vortices, 141
Wall
composite, 74
-fired burner, 15

-fired flat flame burner, 368
plane, 73
Reed, 504
to wall equations, 288
Warped flame holders, 539
Waste(s)
combustion systems, 670
gas
composition, 599
streams, combustible, 160
introduction, 641
salt-containing, 686
water-based, 687
Water
-based wastes, 687
exiting faucet at low velocity, 139
from faucet showing transition, 141
gas shift reaction, 47
injection, 201, 241
vapor, emissivity of, 99
Watertube boiler, 649
Wave patterns, development of orderly, 238
Weighted-sum-of-gray-gases (WSGG)
method, 306
model, 298
Welding, 335, 607
WESPs, see Wet electrostatic precipitators
Western States Petroleum Association (WSPA), 190
Wet electrostatic precipitators (WESPs), 660, 662
advantages of, 663
flue gas coming into, 662
Wien's displacement law, 90
Windbox
air flow modeling, 563
model, air flow balanced by, 567
Wing geometry, 534
Wobbe index, 436
Woofing, 502
Word length, 415
WSGG, see Weighted-sum-of-gray-gases
WSPA, see Western States Petroleum Association

X

Xylene
isomerization, 14
reboiler, 306, 307

Z

ZDR severe service flare tips, see Zink double refractory severe service flare tips
Zeldovich mechanism, 192
Zink double refractory (ZDR) severe service flare tips, 606, 607
Zink Thermal Oxidizer Flare (ZTOF), 598, 622
burner systems, 623
schematic of, 623
ZTOF, see Zink Thermal Oxidizer Flare